THE CRC HANDBOOK OF COMBINATORIAL DESIGNS

The CRC Press Series on

DISCRETE MATHEMATICS AND ITS APPLICATIONS

Series Editor

Kenneth H. Rosen, Ph.D.

AT&T Bell Laboratories

Published Titles

Cryptography: Theory and Practice, *Douglas R. Stinson*

Network Reliability, *Daryl D. Harms, Miroslav Kraetzl, Charles J. Colbourn, and Stanley J. Devitt*

Quadratics, *Richard A. Mollin*

Forthcoming Handbooks

Handbook of Discrete Mathematics, *Kenneth H. Rosen*

Standard Reference of Discrete Mathematics, *Kenneth H. Rosen*

Handbook of Graph Theory, *Jonathan Gross*

Handbook of Cryptography, *Alfred Menezes, Paul van Oorschot, and Scott Vanstone*

Handbook of Discrete and Computational Geometry, *Jacob E. Goodman and Joseph O'Rourke*

Forthcoming Textbooks and Monographs

Graph Theory with Computer Science Applications, *Jonathan Gross and Jay Yellen*

Error-Correcting Codes and Algebraic Curves, *Serguei Stepanov*

Frames and Resolvable Designs, *Steven Furino*

Introduction to Information Theory and Data Compression, *Darrel R. Hankerson, Greg A. Harris, and Peter D. Johnson*

THE CRC HANDBOOK OF COMBINATORIAL DESIGNS

EDITED BY

CHARLES J. COLBOURN
DEPARTMENT OF COMBINATORICS AND OPTIMIZATION
UNIVERSITY OF WATERLOO

JEFFREY H. DINITZ
DEPARTMENT OF MATHEMATICS AND STATISTICS
UNIVERSITY OF VERMONT

CRC Press
Boca Raton New York London Tokyo

Library of Congress Cataloging-in-Publication Data

The CRC handbook of combinatorial designs / edited by Charles J. Colbourn, Jeffrey H. Dinitz.
p. cm. -- (Discrete mathematics and its applications)
Includes bibliographical references and index.
ISBN 0-8493-8948-8
1. Combinatorial designs and configurations. I. Colbourn, C. J. (Charles J.), 1953- . II. Dinitz, Jeffrey H., 1952- . III. Series.
QA166.25.C73 1996 95-46197
511'.6--dc20 CIP

International Standard Book Number 0-8493-8948-8
Library of Congress Card Number 95-46197
Printed in the United States of America 1 2 3 4 5 6 7 8 9 0
Printed on acid-free paper

Preface

The Purpose of this Book

The *CRC Handbook of Combinatorial Designs* is a reference book of combinatorial designs, including constructions of designs, existence results, and properties of designs. The handbook is not a textbook introduction to combinatorial design theory, but rather a comprehensive, easy-to-access reference for facts about designs.

Organization of the Handbook

The presentation is organized into six main parts, each divided into chapters. The first three parts center around three main classes of combinatorial designs: balanced incomplete block designs, orthogonal arrays and latin squares, and pairwise balanced designs. Part IV then contains a large number of shorter chapters devoted to other classes of designs — naturally, many of these classes are closely connected with designs in the first three parts.

Part V concerns the design-theoretic aspects of a number of application areas. Part VI gives reference information for mathematical and computational background of particular use in design theory.

How to Use this Book

This handbook is not meant to be used in a sequential way, but rather for direct access to information about relevant topics. Each chapter makes references to other chapters for related information; in particular, the **See Also** section at the end of each chapter provides useful links.

There is an extensive index; it and the Table of Contents provide a first entry point for most material. This handbook can be used on its own as a reference book, or in conjunction with numerous excellent available texts in the subject. Although material at the research frontier is often included, the handbook emphasizes basic, useful information.

Combinatorial design theory has extensive interactions with numerous areas of mathematics: group theory, graph theory, number theory and finite fields, finite geometry, and linear algebra all have close connections with design theory. It also has applications ranging from routine to complex in numerous disciplines: experimental design, coding theory, cryptography, and computer science provide the main examples. The handbook treats all of these topics, providing information for design theorists to understand the applications, and for users of design theory to obtain entry points to the rest of the handbook and hence to the mathematics of combinatorial designs.

Notes for the User

A home page for the CRC Handbook of Combinatorial Designs *is being maintained.* Its URL is `http://www.emba.uvm.edu/~dinitz/hcd.html`. Notes, corrections, and updates for certain chapters will be maintained on this home page.

Fixed numbering conventions are used. This book has six main parts numbered using roman numerals I–VI. Each part is divided into chapters, which in turn have sections; all are numbered. So, for example, the third section in fourth chapter of the first part is §I.4.3. This is the manner to which it is referred when referenced from any *other* chapter. When referenced from within chapter I.4 it is referenced simply as §4.3. Occasionally, sections contain subsections. However, to avoid four level numbers, such subsections are *not* numbered; rather, the beginning of a subsection is given by the symbol '▷' preceding the title of the subsection.

Numbering of Theorems, Remarks, Corollaries, Tables, Algorithms, and the like is consecutive within each chapter. So, for example, in §I.4.2, we have Remark I.4.5 followed by Theorem I.4.6. As before the roman numeral is omitted if both the referenced Theorem and the reference to it are in the same chapter.

Definitions are given in shaded boxes and are not numbered.

Original references are not, in general, given. Instead, each chapter references surveys and textbooks in which the information is readily accessible in more detail. These sources should be consulted for more complete lists of references. Each chapter covers the *basic* material, but readers interested in material at the research frontier are typically given sufficient pointers to the literature to be able to find more specialized information.

The state of the subject is continually advancing. Each chapter presents, as far as possible, information that is current at the time of writing. Inevitably, improvements will occur! Readers aware of improvements on, or corrections to, the material presented here are invited to contact the author(s) of the chapter(s) involved and also the editors-in-chief.

Updates are being maintained. A library of pages on the World Wide Web is being maintained to report corrections and updates and to provide online versions of some of the material in the handbook.

Acknowledgments are due to many individuals. We must begin by acknowledging the excellent work done by the contributors. Their job was made difficult by the style and length restrictions placed upon them. Many had written hundreds of pages on their topic, yet were asked to summarize in a few pages. They also were asked to do this within the fairly rigid stylistic framework that is consistent now throughout the book. The editors were merciless in their attempt to uphold this style. Many chapters were edited extensively. We thank the authors for their cooperation.

Most chapters have profited from comments from a wide variety of readers, in addition to the authors and editors. Acknowledgments for assistance on individual chapters are lengthy, and are available on the *CRC Handbook of Combinatorial Designs* home page.

A compilation of this size can only be accomplished with the help of a small army of people. We were fortunate to have a helpful group of advisory editors, and excellent technical support from CRC Press, and the Universities of Vermont and Waterloo. We give special thanks to Debbie Street for her help in the area of statistical designs and to Vladimir Tonchev for expert advice on coding theory. Special thanks also go to Dan Archdeacon and to Norm Finizio who proofread much of the manuscript. In addition, for their support in compiling this handbook, we thank Julian Abel, Frank Bennett, Dave Dummit, Nora Konopka, Alan Ling, Ron Mullin, Bob Stern, Doug Stinson, Brett Tagendal, John van Rees, Wayne Yuhasz, and of course Hal and Daisy.

We thank our wives, Sue Dinitz and Karen Colbourn, and our children, Mike, Amy, and Tommy Dinitz and Susie Colbourn, for their support and love despite the many long hours that it took to assemble this handbook. At times we tried their patience, but they understood.

Charles J. Colbourn
Jeffrey H. Dinitz
December, 1995

To our wives and children:

Karen Colbourn
Susie Colbourn

Sue Dinitz
Mike Dinitz, Amy Dinitz, and Tommy Dinitz

Editors-in-Chief

Advisory Editors

Contributors

R. Julian R. Abel
University of New South Wales
Australia

Brian Alspach
Simon Fraser University
Canada
alspach@cs.sfu.ca

Lars D. Andersen
University of Aalborg
Denmark
lda@iesd.auc.dk

Ian Anderson
University of Glasgow
U.K.
ia@maths.gla.ac.uk

Lucien Beneteau
INSA
France
beneteau@insatlse.insa-tlse.fr

Frank E. Bennett
Mount St. Vincent University
Canada
bennett@ash.msvu.ca

Albrecht Beutelspacher
Universität Giessen
Germany
albrecht.beutelspacher
@math.uni-giessen.de

Jürgen Bierbrauer
Michigan Technological University
U.S.A.
jbierbra@math.mtu.edu

Andries E. Brouwer
Technical University of Eindhoven
Netherlands
aeb@cwi.nl

Yeow Meng Chee
University of Waterloo
Canada
ymchee@barrow.uwaterloo.ca

Leo G. Chouinard II
University of Nebraska–Lincoln
U.S.A.

Charles J. Colbourn
University of Waterloo
Canada
cjcolbou@math.uwaterloo.ca

R. Craigen
Fresno Pacific College
U.S.A.

Peter Danziger
Ryerson University
Canada
danziger@acs.ryerson.ca

Warwick de Launey
Victoria Barracks
Australia
warwick@defcen.gov.au

Jeffrey H. Dinitz
University of Vermont
U.S.A.
jeff.dinitz@uvm.edu

Anthony B. Evans
Wright State University
U.S.A.
tevans@desire.wright.edu

Norman J. Finizio
University of Rhode Island
U.S.A.
finizio@uriacc.uri.edu

STEVEN C. FURINO
University of Waterloo
Canada
scfurino@math.uwaterloo.ca

PETER B. GIBBONS
University of Auckland
New Zealand
p_gibbons@cs.auckland.ac.nz

CHRISTOPHER D. GODSIL
University of Waterloo
Canada
chris@bilby.uwaterloo.ca

K. GOPALAKRISHNAN
Wichita State University
U.S.A.
gopal@cs.twsu.edu

M. GREIG
Greig Consulting
Canada
greig@unixg.ubc.ca

HANS-DIETRICH O.F. GRONAU
Universität Rostock
Germany
gronau@zeus.
math.uni-rostock.de

HARALD GROPP
Universität Heidelberg
Germany
d12@ix.urz.uni-heidelberg.de

KATHERINE HEINRICH
Simon Fraser University
Canada
heinrich@cs.sfu.ca

STUART G. HOGGAR
University of Glasgow
U.K.
sgh@maths.gla.ac.uk

ROBERT JAJCAY
Indiana State University
U.S.A.
jajcay@laurel.indstate.edu

DIETER JUNGNICKEL
Universität Augsburg
Germany
dieter.jungnickel@math.uni-augsburg.de

A. DONALD KEEDWELL
University of Surrey
U.K.

HADI KHARAGHANI
University of Lethbridge
Canada
kharaghani@hg.uleth.ca

CHRISTOS KOUKOUVINOS
Tech. University of Athens
Greece
ckoukouv@atlas.uoa.ariadne-t.gr

EARL S. KRAMER
University of Nebraska–Lincoln
U.S.A.
kramer@tdesun.unl.edu

DONALD L. KREHER
Michigan Technological University
U.S.A.
kreher@math.mtu.edu

THOMAS J. KUNKLE
University of Charleston
U.S.A.
kunkle@math.cofc.edu

ESTHER R. LAMKEN
California Institute of Technology
U.S.A.
lamken@cco.caltech.edu

CHARLES F. LAYWINE
Brock University
Canada
claywine@spartan.ac.brocku.ca

C.C. LINDNER
Auburn University
U.S.A.
lindncc@mail.auburn.edu

ALAN C.H. LING
University of Waterloo
Canada
chaling@barrow.uwaterloo.ca

SPYROS S. MAGLIVERAS
University of Nebraska–Lincoln
U.S.A.
spyros@helios.unl.edu

ALIREZA MAHMOODI
University of Toronto
Canada
mahmoodi@cs.toronto.edu

RUDOLF MATHON
University of Toronto
Canada
combin@csri.utoronto.ca

ERIC MENDELSOHN
University of Toronto
Canada
mendelso@math.toronto.edu

RONALD C. MULLIN
University of Waterloo
Canada
rcmullin@math.uwaterloo.ca

STANLEY E. PAYNE
University of Colorado–Denver
U.S.A.
spayne@carbon.denver.colorado.edu

CHRISTIAN PIETSCH
Universität Rostock
Germany

ALEXANDER POTT
Universität Augsburg
Germany
alexander.pott@math.uni-augsburg.de

DONALD A. PREECE
University of Kent at Canterbury
U.K.
d.a.preece@ukc.ac.uk

DIJEN K. RAY-CHAUDHURI
Ohio State University
U.S.A.
dijen@math.ohio-state.edu

MARIALUISA J. DE RESMINI
Università di Roma "La Sapienza"
Italy
resmini@mat.uniroma1.it

C. A. RODGER
Auburn University
U.S.A.
rodgec1@mail.auburn.edu

PETER RODNEY
Carleton University
Canada
prodney@math.carleton.ca

ALEXANDER ROSA
McMaster University
Canada
rosa@mcmail.cis.mcmaster.ca

GORDON F. ROYLE
University of Western Australia
Australia
gordon@cs.uwa.edu.au

DINESH G. SARVATE
University of Charleston
U.S.A.
sarvated@ashley.cofc.edu

PAUL J. SCHELLENBERG
University of Waterloo
Canada
pjschell@math.uwaterloo.ca

JENNIFER SEBERRY
University of Wollongong
Australia
j.seberry@uow.edu.au

NABIL SHALABY
Memorial University of Newfoundland
Canada
nshalaby@beothuk.swgc.mun.ca

MOHAN S. SHRIKHANDE
Central Michigan University
U.S.A.
mohan.shrikhande@cmich.edu

KEN W. SMITH
Central Michigan University
U.S.A
ken.w.smith@cmich.edu

HONG-YEOP SONG
Yonsei University
Korea
hysong@bubble.yonsei.ac.kr

EDWARD SPENCE
University of Glasgow
U.K.
ted@maths.gla.ac.uk

DOUGLAS R. STINSON
University of Nebraska–Lincoln
U.S.A.
stinson@bibd.unl.edu

ANNE PENFOLD STREET
University of Queensland
Australia
aps@maths.uq.oz.au

DEBORAH J. STREET
University of Technology, Sydney
Australia
deborah@maths.uts.edu.au

HERBERT TAYLOR
University of Southern California
U.S.A.
herbt@ccrwest.org

JOSEPH A. THAS
University of Ghent
Belgium
jat@cage.rug.ac.be

VLADIMIR D. TONCHEV
Michigan Technological University
U.S.A.
tonchev@math.mtu.edu

TRAN VAN TRUNG
Universität Essen
Germany
trung@exp-math.uni-essen.de

G. H. JOHN VAN REES
University of Manitoba
Canada
vanrees@cs.umanitoba.ca

WALTER D. WALLIS
Southern Illinois University
U.S.A.
wdwallis@math.siu.edu

HUGH WILLIAMS
University of Manitoba
Canada
hugh_williams
@macmail.cs.umanitoba.ca

L. ZHU
Suzhou University
China

Contents

V Applications

PART I

BALANCED INCOMPLETE BLOCK DESIGNS AND t-DESIGNS

1 2-(v, k, λ) Designs of Small Order

RUDOLF MATHON
ALEXANDER ROSA

1.1 Definition and Basics

A *balanced incomplete block design* (BIBD) is a pair $(V, \mathcal{B})$ where V is a v-set and $\mathcal{B}$ is a collection of b k-subsets of V (*blocks*) such that each element of V is contained in exactly r blocks and any 2-subset of V is contained in exactly λ blocks. The numbers v, b, r, k, λ are *parameters* of the BIBD.

1.1 **Proposition** Trivial necessary conditions for the existence of a BIBD(v, b, r, k, λ) are (1) $vr = bk$, and (2) $r(k-1) = \lambda(v-1)$. Parameter sets that satisfy (1) and (2) are *admissible*.

A BIBD $(X, \mathcal{D})$ is a *subdesign* of a BIBD $(V, \mathcal{B})$ if $X \subseteq V$ and $\mathcal{D} \subseteq \mathcal{B}$. The subdesign is *proper* if $X \subset V$.

1.2 **Proposition** If a (v, k, λ) design has a proper (w, k, λ) subdesign, then $w \leq (v-1)/(k-1)$.

1.3 **Remark** The three parameters v, k, and λ determine the remaining two as $r = \frac{\lambda(v-1)}{k-1}$ and $b = \frac{vr}{k}$. Hence one often writes (v, k, λ) *design* to denote a BIBD(v, b, r, k, λ). The notation 2-(v, k, λ) *design* is also used, since BIBDs are t-designs with $t = 2$. See §I.3. When $\lambda = 1$, the notation $S(2, k, v)$ is also employed in the literature, since such BIBDs are Steiner systems. See §I.4. The notations $S_\lambda(2, k, v)$ or (v, k, λ) BIBD for a (v, k, λ) design are also in common use.

1.4 **Theorem** (*Fisher's inequality*) If a BIBD(v, b, r, k, λ) exists with $2 \leq k < v$, then $b \geq v$.

The *incidence matrix* of a BIBD $(V, \mathcal{B})$ with parameters v, b, r, k, λ is a $v \times b$ matrix $A = (a_{ij})$, in which $a_{ij} = 0$ when the ith element of V occurs in the jth block of $\mathcal{B}$, and $a_{ij} = 0$ otherwise.

1.5 **Theorem** If A is the incidence matrix of a (v, k, λ)-design, then $AA^T = (r - \lambda)I + \lambda J$ and $JA = k\hat{J}$, where I is a $v \times v$ identity matrix, J is a $v \times v$ all ones matrix, and $\hat{J}$ is a $v \times b$ all ones matrix. Moreover, any matrix A satisfying these conditions also satisfies $\lambda(v-1) = r(k-1)$ and $bk = vr$; when $k < v$, it is the incidence matrix of a (v, k, λ) design.

A BIBD $(V, \mathcal{B})$ with parameters v, b, r, k, λ is

complete	or *full* if it is simple and contains $\binom{v}{k}$ blocks.
decomposable	if $\mathcal{B}$ can be partitioned into two nonempty collections $\mathcal{B}_1$ and $\mathcal{B}_2$ so that $(V, \mathcal{B}_i)$ is a (v, k, λ_i) design for $i = 1, 2$.
derived	(from a symmetric design) $(X, \mathcal{D})$ if for some $D \in \mathcal{D}$, the collection of blocks $\{D' \cap D : D' \in \mathcal{D}, D \neq D'\} = \mathcal{B}$.
Hadamard	if $v = 4n - 1$, $k = 2n - 1$, and $\lambda = n - 1$ for some integer $n \geq 2$. See §IV.24.

m-multiple	if $v, \frac{b}{m}, \frac{r}{m}, k, \frac{\lambda}{m}$ are the parameters of a BIBD.
nontrivial	if $3 \leq k < v$.
quasi-symmetric	if every two distinct blocks intersect in either μ_1 or μ_2 elements; the block intersection graph is strongly regular. See §IV.37 and §IV.5.
residual	(of a symmetric design) $(X, \mathcal{D})$ if for some $D \in \mathcal{D}$, the collection of blocks $\{D' \setminus D : D' \in \mathcal{D}, D \neq D'\} = \mathcal{B}$.
resolvable	if there exists a partition R of its set of blocks B into *parallel classes*, each of which in turn partitions the set V; R is a *resolution*. See §I.6
simple	if it has no repeated blocks.
symmetric	if $v = b$, or equivalently $k = r$. See §I.5.

1.6 Proposition An additional trivial necessary condition for the existence of a resolvable BIBD is (3) $k|v$. A nontrivial condition is that $b \geq v + r - 1$, a result that extends the inequality in Theorem 1.4. See §I.6.5.

Two BIBDs (V_1, B_1), (V_2, B_2) are *isomorphic* if there exists a bijection $\alpha : V_1 \to V_2$ such that $B_1\alpha = B_2$. Isomorphism of resolutions of BIBDs is defined similarly. An *automorphism* is an isomorphism of a design with itself. The set of all automorphisms of a design forms a group, the *(full) automorphism group.* An *automorphism group* of the design is any subgroup of the full automorphism group.

1.7 Remark If $(V, \mathcal{B})$ is a BIBD(v, b, r, k, λ) with automorphism group G, the action of G partitions $\mathcal{B}$ into classes (*orbits*). A set of orbit representatives is a set of *starter blocks* or *base blocks*. Together with the group G, a set of base blocks can be used to define a design.

A BIBD $(V, \mathcal{B})$ with parameters v, b, r, k, λ and automorphism group G is

cyclic	if G contains a cycle of length v.
regular	if G contains a subgroup G' of order v which acts transitively on the elements.
k-rotational	if some automorphism has one fixed point and k cycles each of length $(v-1)/k$.
transitive	if for any two elements x and y, there is an automorphism mapping x to y.

The *complement* of a design $(V, \mathcal{B})$ is $(V, \overline{\mathcal{B}})$, where $\overline{\mathcal{B}} = \{V \backslash B : B \in \mathcal{B}\}$.

1.8 Proposition The complement of a design with parameters (v, b, r, k, λ) is a design with parameters $(v, b, b-r, v-k, b-2r+\lambda)$.

1.9 Remarks

1. In view of Proposition 1.8, one usually considers designs with $v \geq 2k$ and obtains the rest by taking the complement.
2. *Complement* has also been used, when $(V, \mathcal{B})$ is a simple 2-(v, b, r, k, λ) design, to denote the 2-$(v, \binom{v}{k} - b, \binom{v-1}{k-1} - r, k, \binom{v-2}{k-2} - \lambda)$ design obtained by taking all k-subsets *not* in $\mathcal{B}$ as blocks. The term *supplement* is used here for this second kind of complementation.

1.2 Small Examples

1.10 Remark To conserve space, designs are displayed in a $k \times b$ array in which each column contains the elements (taken from the decimal digits and roman letters) forming a block. For a discussion of how to find the automorphism group of each of these designs, see Example VI.9.43.

1.11 Example The unique $(6, 3, 2)$ design.

```
0000011122
1123423433
2345554545
```

1.12 Table The four nonisomorphic $(7, 3, 2)$ designs.

1:	00000011112222 11335533443344 22446655666655	2:	00000011112222 11335533443344 22446656565656
3:	00000011112222 11334533453344 22456646565656	4:	00000011112223 11234523453344 23456664565656

1.13 Table The 10 nonisomorphic $(7, 3, 3)$ designs.

1:	000000000111111222222 111333555333444333444 222444666555666666555	2:	000000000111111222222 111333555333444333444 222444666556566566556
3:	000000000111111222222 111333455333445333444 222445666456566566556	4:	000000000111111222222 111333455333445333444 222445666466556556566
5:	000000000111111222222 111333445333445333445 222456566456566456566	6:	000000000111111222223 111233455233445333444 223445666546566566565
7:	000000000111111222223 111233455233445333444 223445666645566566556	8:	000000000111111222223 111233445233455333444 223455666644566566556
9:	000000000111111222223 111233445233445333454 223456566546566456665	10:	000000000111111222233 111223445223345334544 234356566465656456656

1.14 Table The four nonisomorphic $(8, 4, 3)$ designs.

1:	00000001111222 11123342334334 22554666455455 34675777677766	2:	00000001111222 11123342334334 22554666455455 34675777767676
3:	00000001111222 11123342334334 22554566465455 34677677577676	4:	00000001111224 11122332233335 24645454545466 35767767667577

1.15 Example The unique $(9, 3, 1)$ design.

```
000011122236
134534534547
268787676858
```

1.16 Table The 36 nonisomorphic $(9,3,2)$ designs.

1:	000000001111112222223344 113355773344663344556655 224466885577888866777788	2:	000000001111112222223344 113355773344663344556655 224466885578787866787878
3:	000000001111112222223344 113355673344673344556655 224467885568787867687878	4:	000000001111112222223344 113355773344663344555656 224466885758786768788877
5:	000000001111112222223344 113355773344663344555656 224466885758786867787887	6:	000000001111112222223344 113355673344673344555656 224467885657887868678787
7:	000000001111112222223344 113355673344673344555656 224467885658787867687887	8:	000000001111112222223344 113355673344673344555656 224467885658787867688778
9:	000000001111112222223344 113355673344663344555756 224467885858776767886878	10:	000000001111112222223344 113355773344563344565656 224466885768786857877887
11:	000000001111112222223344 113355673344573344565656 224467885678687856788787	12:	000000001111112222223344 113355673344573344565656 224467885678687856877887
13:	000000001111112222223344 113355673344573344565656 224467885867686758788787	14:	000000001111112222223344 113355673344573344565656 224467885867686758877887
15:	000000001111112222223344 113346673345573344555656 224557884676887868677887	16:	000000001111112222223344 113345673345673344556655 224567884586787867687878
17:	000000001111112222223344 113345673345673344555656 224567884856786778687887	18:	000000001111112222223344 113345673345663344555756 224567884857786778686887
19:	000000001111112222223344 113345673345663344555657 224567884857787867686788	20:	000000001111112222223344 113345673345563344566755 224567884876785678877868
21:	000000001111112222223344 113345673345563344576655 224567884876785867687788	22:	000000001111112222223344 113345673345563344575656 224567884786876857687887
23:	000000001111112222223344 113345673345563344565756 224567884876786758876887	24:	000000001111112222233333 112445672445674455644556 233567883567887868778687
25:	000000001111112222233333 112445672445674455644556 233567883568787867878687	26:	000000001111112222233334 112445672345673455644565 233567884567888768778678
27:	000000001111112222233334 112445672345673455644565 233567884568787868778687	28:	000000001111112222233334 112445672345673455644565 233567884568788767878678
29:	000000001111112222233334 112445672345673455644565 233567884576886878778876	30:	000000001111112222233334 112445672345673455644565 233567884576888678778678
31:	000000001111112222233334 112445672345673455644565 233567884586787768868877	32:	000000001111112222233334 112445672345673455644565 233567884586788768767788
33:	000000001111112222233344 112355672344673345645655 234467885568787868776878	34:	000000001111112222233344 112355672344673345645556 234467885658786778887867
35:	000000001111112222233344 112345672345673345645655 234567885468787876886778	36:	000000001111112222233344 112345672345673345645655 234567885468787886776878

1.17 Table The 11 nonisomorphic $(9,4,3)$ designs.

1:	000000001111122223	2:	000000001111122223
	111233562334433444		111233562334433444
	225544777556666555		225544777565656565
	346678888787878786		346678888788787786
3:	000000001111122223	4:	000000001111122223
	111233462334533444		111233462334533444
	225545777456666555		225545777456666555
	346768888788778687		346768888877878687
5:	000000001111122223	6:	000000001111122223
	111233462334533444		111233462334533444
	225545677457666555		225545677457666555
	346778888688778687		346788788678878687
7:	000000001111122223	8:	000000001111122223
	111233462334533444		111233462334533444
	225545777465656565		225545676465757565
	346768888788787687		346778888877868876
9:	000000001111122223	10:	000000001111122223
	111233462334533444		111233452334533444
	225545676465757565		225546677467656565
	346788788778868786		346787888578888677
11	000000001111122234		
	111223352233533446		
	246454674745656557		
	357867886887887768		

1.18 Table The three nonisomorphic $(10,4,2)$ designs.

1:	000000111122233	2:	000000111122233	3:	000000111122233
	112356234534544		112356234534544		112345234534545
	244778767658656		244778747658656		246867867647667
	356899899899787		356899998889797		357989979858998

1.19 Table The unique $(11,5,2)$ design and the unique $(13,4,1)$ design.

00000111223	0000111223345
11234236354	1246257364789
24567457465	385a46b57689a
35889898677	9c7ba8cb9cabc
769aaaa99a8	

1.20 Table The two nonisomorphic $(13,3,1)$ designs.

1:	00000011111222223334445556	2:	00000011111222223334445556
	13579b3469a3467867868a7897		13579b3469a3467867868a7897
	2468ac578bc95acbbacc9bbac9		2468ac578bc95abcbcac9babc9

1.21 Table The 80 nonisomorphic $(15,3,1)$ designs.

1:	00000001111112222223333444455556666	2:	00000001111112222223333444455556666
	13579bd3478bc3478bc789a789a789a789a		13579bd3478bc3478bc789a789a789a789a
	2468ace569ade65a9edbcdecbeddebcedcb		2468ace569ade65a9edbcdecbededcbdebc
3:	00000001111112222223333444455556666	4:	00000001111112222223333444455556666
	13579bd3478bc3478bc789a789a789a789a		13579bd3478bc3478bc789a789a789a789a
	2468ace569ade65a9edbcdedebcedcbcbed		2468ace569ade65a9edbcdedbecedcbcebd

```
 5: 00000001111112222223333444455556666
    13579bd3478bc3478bc789a789a789a789a
    2468ace569ade65a9edbceddecbcbdeedbc
 6: 00000001111112222223333444455556666
    13579bd3478bc3478bc789a789a789a789a
    2468ace569ade65a9edbceddecbcdbeebdc
 7: 00000001111112222223333444455556666
    13579bd3478bc3478bc789a789a789a789a
    2468ace569ade65a9edbdececbdcedbdbce
 8: 00000001111112222223333444455556666
    13579bd3478bc34789c78ab789a789a789a
    2468ace569ade65abedc9deedcbdebcbcde
 9: 00000001111112222223333444455556666
    13579bd3478bc34789c78ab789a789a789a
    2468ace569ade65abedc9deedbcdecbbcde
10: 00000001111112222223333444455556666
    13579bd3478bc34789c78ab789a789a789a
    2468ace569ade65abedc9deedcbdcbebedc
11: 00000001111112222223333444455556666
    13579bd3478bc34789c789a789a78ab789a
    2468ace569ade65abedcdbeecdbd9cebecd
12: 00000001111112222223333444455556666
    13579bd3478bc34789c789a789a789a78ab
    2468ace569ade65abedcdbeecdbbecdd9ce
13: 00000001111112222223333444455556666
    13579bd3478bc34789c789a789a78ab789a
    2468ace569ade65abedcebdedcbd9cebcde
14: 00000001111112222223333444455556666
    13579bd3478bc34789c789a78ab789a789a
    2468ace569ade65abedcebdd9ceedcbbcde
15: 00000001111112222223333444455556666
    13579bd3478bc34789a78ac789a789b789a
    2468ace569ade65bcdee9bddbecadcecebd
16: 00000001111112222223333444455556666
    13579bd3478bc34789a789a78bc789a789a
    2468ace569ade65bcdeedcba9eddebccbed
17: 00000001111112222223333444455556666
    13579bd3478bc34789a789a789c78ab789a
    2468ace569ade65bcdeedcbabedd9cecebd
18: 00000001111112222223333444455556666
    13579bd3478bc34789a78ac789b789a789a
    2468ace569ade65bcede9bdadcedebccbde
19: 00000001111112222223333444455556666
    13579bd3478bc34789a789a78ab789c789a
    2468ace569ade65bdceebdcc9deaebddceb
20: 00000001111112222223333444455556666
    13579bd3478bc34789a78ac789b789a789a
    2468ace569ade65bdece9bdacdedbcecebd
21: 00000001111112222223333444455556666
    13579bd3478ac34789b789a789a78ab789c
    2468ace569bde65acedbdcecedbd9ceeabd
22: 00000001111112222223333444455556666
    13579bd3478ac34789b789a789c789a78ab
    2468ace569bde65acedbdceeabdcedbd9ce
23: 00000001111112222223333444455566667
    13579bd3478bc34589c789a58ab789789aa
    2468ace569ade67abedcdbed9cebececdbd
24: 00000001111112222223333444455566667
    13579bd3478bc34589c789a589a78978aba
    2468ace569ade67abedcdbebecdecdd9ceb
25: 00000001111112222223333444455566667
    13579bd3478bc34589c789a589a78978aba
    2468ace569ade67abedcedbecbdbdcd9cee
26: 00000001111112222223333444455566667
    13579bd3478bc3458ac789a589a789789ab
    2468ace569ade67b9edcbededbcacddecbe
27: 00000001111112222223333444455566667
    13579bd3478bc34589a78ab589c789789aa
    2468ace569ade67bcded9ceaebdcdeebcdb
28: 00000001111112222223333444455566667
    13579bd3478bc34589a78ab589c789789aa
    2468ace569ade67bcedd9ceaebdedccbdeb
29: 00000001111112222223333444455566667
    13579bd3478bc34589a789a58bc789789aa
    2468ace569ade67bcedebdca9eddeccdbeb
30: 00000001111112222223333444455566667
    13579bd3478bc34589a78ab589c789789aa
    2468ace569ade67bdced9ceabedcedecbdb
31: 00000001111112222223333444455566667
    13579bd3478bc34589a789a58bc789789aa
    2468ace569ade67bdcedbeca9edcedecbdb
32: 00000001111112222223333444455566667
    13579bd3478bc34589a78ab589c789789aa
    2468ace569ade67bdcec9deaebddceebdcb
33: 00000001111112222223333444455566667
    13579bd3478bc34589a78ab589c789789aa
    2468ace569ade67bdecc9deaebdecddbceb
34: 00000001111112222223333444455566667
    13579bd3478bc34589a789a589c78978baa
    2468ace569ade67bdecdbceaebdecdc9edb
35: 00000001111112222223333444455566678
    13579bd3478bc34569a689a579a78a789bc
    2468ace569ade78bcdedbecedcbc9daebed
36: 00000001111112222223333444455566678
    13579bd3478bc34569a689a579a78a789bc
    2468ace569ade78bceddbceecdbd9caebed
37: 00000001111112222223333444455566678
    13579bd3478bc34569a68ab579c78978aa9
    2468ace569ade78bced9dceaebdceddbebc
38: 00000001111112222223333444455566678
    13579bd3478bc34569a68ab579a78978ac9
    2468ace569ade78bdce9cdecedbadebecdb
39: 00000001111112222223333444455566678
    13579bd3478bc34569a68ab579c78978aa9
    2468ace569ade78bdce9dceabedecdcebdb
40: 00000001111112222223333444455566678
    13579bd3478bc34569a689a579a78a789bc
    2468ace569ade78bdcecbedecdbd9caebed
41: 00000001111112222223333444455566678
    13579bd3478bc34569a689c57ab78a789a9
    2468ace569ade78bdceabed9dceecdcebbd
42: 00000001111112222223333444455566678
    13579bd3478bc34569a68ab579c78978aa9
    2468ace569ade78bdec9cdeaebddeccbebd
```

```
43: 0000000111111222222333344444555666678
    13579bd3478bc34569a68ac579b78978aa9
    2468ace569ade78bdec9ebdacdeedcbcedb
44: 0000000111111222222333344444555666678
    13579bd3478bc34569a689a579a78a789bc
    2468ace569ade78bdeccbdeedcbc9daebed
45: 0000000111111222222333344444555666678
    13579bd3478bc34569a68ac579a78978ab9
    2468ace569ade78becd9bededbcacdcdbee
46: 0000000111111222222333344444555666678
    13579bd3478bc34569a68ab579c78978aa9
    2468ace569ade78becd9dceabedceddcbeb
47: 0000000111111222222333344444555666678
    13579bd3478bc34569a68ab579c78978aa9
    2468ace569ade78bedc9cdeabeddeccdbeb
48: 0000000111111222222333344444555666678
    13579bd3478bc34568a69ab578c78979aa9
    2468ace569ade79becd8dceabedcdedcbeb
49: 0000000111111222222333344444555666679
    13579bd3478bc34568a689a578a78978abc
    2468ace569ade79becddebccdbeadec9bed
50: 0000000111111222222333344444555666678
    13579bd3478bc34568a69ac578a78979ab9
    2468ace569ade79becd8bededbcadccdbee
51: 0000000111111222222333344444555666678
    13579bd3478bc34568968ab579c79a78aa9
    2468ace569ade7abecd9dce8ebddcecbdbe
52: 0000000111111222222333344444555666678
    13579bd3478bc345689689a579c79a78aab
    2468ace569ade7abecdcdeb8ebddceb9dce
53: 0000000111111222222333344444555666678
    13579bd3478bc345689689a579c79a78aab
    2468ace569ade7abecdbdec8ebddcec9dbe
54: 0000000111111222222333344444555666679
    13579bd3478bc345689689a578c78a78aab
    2468ace569ade7abecdbecd9ebdcded9cbe
55: 0000000111111222222333344444555666678
    13579bd3478bc34568969ab578c78a79aa9
    2468ace569ade7abecd8cde9ebdcdedbcbe
56: 0000000111111222222333344444555666678
    13579bd3478bc345689689a579c79a78aab
    2468ace569ade7abedcdceb8ebdcdeb9cde
57: 0000000111111222222333344444555666678
    13579bd3478bc34568968ab579c79a78aa9
    2468ace569ade7abedc9cde8ebdcdedbcbe
58: 0000000111111222222333344444555666678
    13579bd3478bc345689689a579c79a78aab
    2468ace569ade7abedcbced8ebdcded9cbe
59: 000000011111122222233334444455556667a
    13579bd3478bc345689689a578a789789cb
    2468ace569ade7ebacdcbeddb9caecedbde
60: 000000011111122222233334444455556667a
    13579bd3478bc34568968a9578a789789cb
    2468ace569ade7ebacddecbcb9dadeebcde
61: 0000000111111222222333344444555556666
    13579bd3478ac34789b789c789a78ab789a
    2468ace569bde65aecdbaedecdbd9cecdbe
62: 0000000111111222222333344444555566667
    13579bd3478ac34589b78ab589a789789ca
    2468ace569bde67adcec9edbedcdceeabdb
63: 0000000111111222222333344444555566667
    13579bd3478ac34589b78ab589a789789ca
    2468ace569bde67aecdd9cebcdecdeeabdb
64: 0000000111111222222333344444555566667
    13579bd3478ac34589a789a589b78b789ca
    2468ace569bde67cdbeecdbaecdd9ebaedc
65: 0000000111111222222333344444555666678
    13579bd3478ac34569b68ac579a789789ba
    2468ace569bde78dacee9bdbedcacecdbde
66: 0000000111111222222333344444555666678
    13579bd3478ac34568968ac579a79b789ba
    2468ace569bde7bdacee9bd8edcacecdbde
67: 0000000111111222222333344444555666679
    13579bd3478ac34568a68ac578b789789ab
    2468ace569bde79cbdee9bdaecdbeddacce
68: 0000000111111222222333344444555666678
    13579bd3478ac34569b68ac579a789789ba
    2468ace569bde78adcee9bdbedccdeacbde
69: 0000000111111222222333344444555666679
    13579bd3478ac3456a868ac578b789789ab
    2468ace569bde79bcede9bdaecddecbadce
70: 000000011111122222233334444455556667a
    13579bd3478ac34568968ab5789789789cb
    2468ace569bde7dbaece9cdceabadebcdde
71: 000000011111122222233334444455556667a
    13579bd3478ac34568968ab5789789789cb
    2468ace569bde7cabdee9cddeabbecacdde
72: 000000011111122222233334444455556667a
    13579bd3478ac34568968ab5789789789cb
    2468ace569bde7dcbaee9cdaecbbedadcde
73: 0000000111111222222333344444555666679
    13579bd3478ac34568a689a578b78c789ab
    2468ace569bde7c9bdeecdbae9dbeddacce
74: 000000011111122222233334444455556667a
    13579bd3478ac345698689a578b789789cb
    2468ace569bde7abdecedbcce9daedbacde
75: 000000011111122222233334444455556667a
    13579bd3478ac345689689a578b789789cb
    2468ace569bde7cdbaeedbcae9dbecacdde
76: 0000000111111222222333344444555666678
    13579bd3478ac34569b68ab578978979aac
    2468ace569bde7da8cee9cdcbaedebcdbed
77: 0000000111111222222333344444555666678
    13579bd3478ac34569a68ab578979b789ac
    2468ace569bde7d8cebe9cdacebdcebaded
78: 0000000111111222222333344444555666678
    13579bd3478ac34569b689c578b78a79aa9
    2468ace569bde7ac8edeabd9cdedebbdcec
79: 0000000111111222222333344444555666678
    13579bd3478ac34568b689c57ab79a789a9
    2468ace569bde79ecadaebd8dcecdbbdeec
80: 0000000111111222222333344444555666678
    13579bd3469ac34578b678a58ab78979c9a
    2468ace578bde96aecdbcded9cebecaeddb
```

1.22 Table Properties of the 80 STS(15)s ((15,3,1) BIBDs). $|G|$ is the order of the automorphism group. CI is the *chromatic index*, the minimum number of colors with which the blocks can be colored so that no two intersecting blocks receive the same color; when CI = 7, the design is resolvable. PC is the number of parallel classes. Sub is the number of (7,3,1)-subdesigns, and Pa is the number of Pasch configurations (four triples on six points).

#	$\|G\|$	CI	PC	Sub	Pa	#	$\|G\|$	CI	PC	Sub	Pa
1	20160	7	56	15	105	2	192	8	24	7	73
3	96	9	8	3	57	4	8	9	8	3	49
5	32	8	16	3	49	6	24	8	12	3	37
7	288	7	32	3	33	8	4	9	4	1	37
9	2	9	2	1	31	10	2	9	6	1	31
11	2	9	6	1	23	12	3	9	1	1	32
13	8	9	4	1	33	14	12	9	0	1	37
15	4	8	8	1	25	16	168	9	0	1	49
17	24	8	12	1	25	18	4	9	4	1	25
19	12	7	16	1	17	20	3	9	1	1	20
21	3	9	1	1	20	22	3	8	4	1	17
23	1	9	1	0	18	24	1	9	0	0	19
25	1	9	1	0	20	26	1	9	0	0	23
27	1	9	3	0	14	28	1	9	2	0	15
29	3	9	0	0	19	30	2	9	3	0	14
31	4	9	5	0	18	32	1	9	2	0	13
33	1	9	1	0	12	34	1	9	1	0	12
35	3	9	0	0	13	36	4	9	1	0	10
37	12	9	5	0	6	38	1	9	4	0	9
39	1	9	1	0	12	40	1	9	0	0	13
41	1	9	1	0	12	42	2	9	5	0	8
43	6	9	3	0	10	44	2	9	1	0	8
45	1	9	2	0	9	46	1	9	2	0	7
47	1	9	1	0	10	48	1	9	1	0	8
49	1	9	2	0	7	50	1	8	7	0	6
51	1	9	2	0	9	52	1	9	0	0	9
53	1	9	1	0	10	54	1	9	2	0	11
55	1	9	2	0	9	56	1	9	1	0	8
57	1	8	4	0	5	58	1	9	3	0	8
59	3	9	0	0	13	60	1	9	6	0	7
61	21	7	7	1	14	62	3	9	0	0	7
63	3	8	6	0	7	64	3	9	3	0	10
65	1	9	2	0	7	66	1	9	3	0	6
67	1	8	4	0	5	68	1	9	1	0	6
69	1	8	4	0	5	70	1	9	2	0	9
71	1	9	2	0	5	72	1	9	4	0	5
73	4	9	9	0	6	74	4	9	3	0	8
75	3	8	6	0	7	76	5	9	1	0	10
77	3	9	1	0	2	78	4	9	9	0	6
79	36	8	17	0	6	80	60	9	11	0	0

1.23 Table The five nonisomorphic $(15,7,3)$ designs.

```
1:
000000011112222
111335533443344
222446655666655
37b797978787878
48c8a8a9a9aa9a9
59dbddbbccbbccb
6aeceecdeededde
```

```
2:
000000011112222
111335533443344
222446655666655
37b797978787878
48c8a8a9a9aa9a9
59dbddbbccbcbbc
6aeceecdeeddeed
```

```
3:
000000011112222
111335533443344
222446655666655
37b797978787878
48c8a8a9aa99aa9
59dbddbbccbcbbc
6aeceecdedeeded
```

```
4:
000000011112222
111335533443344
222446655666655
37b797879787878
48c8a9a8aa99aa9
59dbdbcdbcbcbbc
6aecedeecdeeded
```

```
5:
000000011112222
111335533443344
222446656565656
37b797977888877
48c8a8a9aa99aa9
59dbddbbccbcbbc
6aeceecdedeeded
```

1.24 Example The unique $(16,4,1)$ design.

```
0000011112222333456
147ad456945684567897
258be7b8cc79a98abbac
369cfadefefbddcfefed
```

1.25 Table The three nonisomorphic $(16,6,2)$ designs.

```
1:
0000001111222334
1123452345345455
2667896789877666
37aabcdbaa998987
48bddeeccbabccba
59cefffedfeddef
```

```
2:
0000001111222334
1123452345345455
2667896789877666
37aabcdbaa998987
48bddeeccbacbbca
59cefffedfdeedf
```

```
3:
0000001111222334
1123452345345455
2667896789877666
37aabcdbaa998987
48bddeeccbcabbac
59cefffeddfeefd
```

1.26 Table The six nonisomorphic $(19,9,4)$ designs.

```
1:
0000000001111122223
1111233562334433444
2225544777556666555
3346678888787878786
499acab99a9b9aa9b99
5aebdcdabddcbcbccaa
6bfegefcdefecdfedbd
7cgfhghfehgghfgfege
8dhiiiihgiihighifih
```

```
2:
0000000001111122223
1111233562334433444
2225544777556666555
3346678888787878786
499acab99a9ba99ab99
5aebdcdabcdcbcbdcaa
6bfegefdcffeddeedbc
7cgfhghefhgggehfegf
8dhiiiihgiihhiigfih
```

```
3:
0000000001111122223
1111233562334433444
2225544777565656565
3346678888788787786
499acab99a99abba999
5aebdcdabbcdcccdbaa
6bfegefdcgfeddeedcb
7cgfhghfehhgefffefg
8dhiiiighiiihgghiih
```

```
4:
0000000001111122223
1111233562334433444
2225544777565656565
3346678888788787786
499acab99a9ab99ba99
5aebdcdabdbccccdbaa
6bfegefdcegfddeecbd
7cgfhghfehhgeffffge
8dhiiiighiihgiighih
```

```
5:  0000000001111122223
    1111233462334533444
    2225545777456666555
    3346768888788778687
    499aca9a9abb999ab99
    5aebdcdbbcdccabdcaa
    6bfeegfdcffeddeedbc
    7cgfghhefhggeghffge
    8dhiiiihgiihihighif
```

```
6:  0000000001111122223
    1111233452334533444
    2225546677457656565
    3346787888568888677
    499aaab99bc9a99ab99
    5aebcddbcddcbabccaa
    6bfegefcdefecdfeddb
    7cgfhgghehhgfggfefe
    8dhiiihifiiighihgih
```

1.27 Table The 16 nonisomorphic $(25, 4, 1)$ designs with nontrivial automorphism group.

```
1:  00000000111111222222333334444555566677889 9aabbil
```

```
1:  000000001111112222223333344445555666778899aabbil
    134567ce34578cd34568de468bh679f78ag79b9aabcddecejm
    298dfbhkea6g9kf7c9afkg5cgfihdgifchi8ejjcjdfhgfghkn
    iaolgmjnmbohnljonblhmjjdlknmeklnekmkinlimimonooloo
2:  00000000111111222222333334444555566667778889abil
    13457bce34589cd3456ade489eh6acf7bdg79ab9ab9abdecjm
    2689gdfka76fekg798fckh5cfgidghichfi8chjjdfgjehfgkn
    iloahnjmbmohlnjobngmljjdknmeklnekmlkinmnilmlioooo
3:  00000000111111222222333334444555566667778889abil
    13457bde34589ce3456acd489eh6acf7bdg79ab9ab9abcdejm
    2689gcfka76fdkg798fehk5cgfidhgicfhi8hcjjfdejgfghkn
    iloahmjnbmohnljobngljmjdkmneknleklmkminlniimlooooo
4:  00000000111111222222333334444555566667778889abil
    13457aef3458bcf34569dg489dh69ef7acg79bc9acabdcdejm
    2689bcjha769djg7b8aejh5cbfkdagkebhk8gieihdifefghkn
    ilodgknmemohklnocnfkmljgminhnilflimkjnmljnmjlooooo
5:  000000001111112222223333344445555666778899abcfil
    134567cd34578de34568ce468ag67bh789f79aab9bacdedgjm
    298abhgkba69fhk79bgakf5cdfiedgicehi8djejjcbfghehkn
    ieolfmjnmcognjlondlhmjjhlknmfklngkmkinilmiolmnoooo
6:  0000000011111122222233333444455555666677778899abcde
    123468cj23479af3458bg456ah57bi78bd89ce9adabacghiff
    5ad97fek6be8gdl7ch9em89fcn6gdkfickgjdlhmeekblhijjg
    oignbhmlkjhcinmlfjdonmeikoajlonlgmomhnkoijnfolmnok
7:  00000000111111222222333333444445555556666677 7889al
    124789bh2589ace39abde47abcf8bcde79cdf78adg89b9cabm
    365egifj46fhjgi5gikhf6ihjegjikfggkjehfhekiadcbdcdn
    dcaolmnk7bolmnk8olmnj9nolmknolmhmnolilmnojkjheifgo
8:  00000000111111222222333333444455566667777889aacee
    12459bdf24569cg3458bh4568bi59bh9ac78ab89cgaddbedfl
    3786cihjd78haek96kfgja7fgclgkdiimfi9djebkjcjffhmgn
    oanegklmifbmljnecolmnjdkhnmlmeojnhnoglmhloiknokoio
9:  000000001111112222223333344445555677889abcdefgh
    13456789345678a34567894679c67ad68be9aab9bdecijkijk
    2fbcdgiadg9jcfbaehficb5b8eg89ch7adfchdfgefghmlllmn
    onklehjmmlikehnjnmgkdloilhkmjfinkgjolomnokijnnmooo
10: 000000001111112222223333344445556657 7889abccdfgh
    13456789345678a345678b467ad679e69c9b89aabfghdeeijk
    2jadebchekbdc9f9ciaedg5f8bg8gbf7afcihjdiemllikjlmn
    onmkgfliilnmhgjljmhnfkokmchnidhlegojjkokonnmnmlooo
```

```
11: 000000001111112222222333334444555566778899acdefgh
    1345678b3456789345678a467bc679d68ae9daebcabbkijijk
    2iace9fgdjbgcah9cibhdf5a8fg8bfh79gfcfdgehedclmllmn
    ojmdhnklekniflmlekmjgnokmhnniglljhmojokoiikjmnnooo
12: 000000001111112222222333334444555566778899accdfgh
    1345678a345678b34567894679e67ac68bd9fagbfabbdeeijk
    2j9ebcmickal9dibdielaj5a8df8beg79chcgdhehgfhjiklmn
    olhgdfnkhmfnegjgfnhmckomkininjljlkmokoiojnmlnmlooo
13: 000000001111112222222333334444555566677899accdfgh
    134578ae34568bc345679d467ad78be689c7898ababbdeeijk
    2f6b9cjg9g7dakh8ahbeif5lbigl9ifamjffhcgdegfhikjlmn
    omkdhlnienimfljjclgnmkonckhmdjhenkgjiokoolnmnmlooo
14: 000000001111112222222333334444555566677899accdfgh
    1345789c34568ad34567be467ad78be689c7898ababbdeeijk
    2h6beajg9f7bckh8agd9if5lbifl9jgamiffhcgdegfhjiklmn
    omidnflkenjglmikclmhnjoncjgmdkhenkhkjoioonmlnmlooo
15: 000000001111112222222333334444555566677889abcdeil
    1345689b345679a34578ab467ad789e689c7bc9daeghffghjm
    2cg7afdi8dhfbjcf6eg9kc59ebhacbfdbag8ehcfdgijkijkkn
    lnkjohemknigomejinhomdmligoljhoklfonjmkmimnnnllloo
16: 000000001111112222222333334444555566677889abcdeil
    134568ab345679b345789a467ae789c689d7be9cadfghfghjm
    2cf7g9dk8dgafich6ebfcj59dbgaebhcbaf8dfegchkijijkkn
    lnjihoemjnkohmeiknogdmmlkfoligojlhonimjmkmnnnllloo
```

1.3 Parameter Tables

1.28 **Table** Admissible parameter sets of nontrivial BIBDs with $r \leq 41$ and $k \leq v/2$, updated from [94]. Earlier listings of BIBDs by Hall [51], Takeuchi [127], DiPaola-Wallis-Wallis [37], and Kageyama [73] and papers by Hanani [52] and Wilson [139] are frequently referenced. Multiples of known designs are included; although their existence is trivially implied, information concerning their number and resolvability usually is not.

The admissible parameter sets of nontrivial BIBDs satisfying $r \leq 41$, $3 \leq k \leq v/2$ and conditions (1), (2) of Proposition 1.1 are ordered lexicographically by r, k, and λ (in this order). The column "Nd" contains the number $Nd(v, b, r, k, \lambda)$ of pairwise nonisomorphic BIBD(v, b, r, k, λ) or the best known lower bound for this number. The column "Nr" contains a dash (-) if condition (3) of Proposition 1.6 is not satisfied. Otherwise it contains the number Nr of pairwise nonisomorphic resolutions of BIBD(v, b, r, k, λ)'s or the best known lower bound for this number. The number of nonisomorphic resolvable BIBDs is not necessarily Nr, as two nonisomorphic resolutions can have isomorphic underlying (resolvable) designs. To illustrate the difference, there are seven nonisomorphic resolutions of BIBD(15,35,7,3,1)s but only four nonisomorphic resolvable BIBD(15,35,7,3,1)s (see Example I.6.6). The symbol ? indicates that the existence of the corresponding BIBD (resolvable BIBD, respectively) is in doubt. The meaning of symbols that occur in the column "Comments" is as follows:

m#x	m-multiple of an existing BIBD #x
m#x*	m-multiple of #x which does not exist or whose existence is undecided
R#x (D#x)	residual (derived) design of #x which exists

R#x* (D#x*)	residual (derived) design of #x which does not exist or whose existence is undecided
#x+#y	union of two designs on the same set of elements
#x↓#y	design #x is a design $(V, \mathcal{B})$ with parameters (v, b, r, k, λ); design #y is a design $(V, \mathcal{D})$ with parameters $(v, b', b-r, k+1, r-\lambda)$; add a new point ∞ to each block of $\mathcal{B}$ to obtain $\widehat{\mathcal{B}}$; then $(V, \widehat{\mathcal{B}} \cup \mathcal{D})$ is a design with parameters $(v+1, b+b', b, k+1, r)$.
PG (AG)	projective (affine) geometry (see §VI.7)
×1	BIBD does not exist by Bruck-Ryser-Chowla (BRC) Theorem (see §I.5.2)
×2	BIBD is a residual of a BIBD that does not exist by the BRC theorem, and $\lambda = 1$ or 2
×3	resolvable BIBD does not exist by Bose's condition (see Theorem I.6.14).
HD	resolvable BIBD$(4t, 8t-2, 4t-1, 2t, 2t-1)$ exists since there exists a symmetric (Hadamard) BIBD$(4t-1, 4t-1, 2t-1, 2t-1, t-1)$; see §IV.24.

In the column "Ref", typically no references are given for designs that are multiples of known BIBDs, or derived or residual designs. Often used is a trivial formula giving $Nd(v, mb, mr, k, m\lambda) \geq n+1$ provided $Nd(v, b, r, k, \lambda) \geq n$, $m \geq 2, n \geq 1$ (and similarly for Nr).

The column headed "Where?" gives a pointer to the theorem or proposition number where an explicit construction can be found in this handbook for a design with this parameter set. Pointers are generally not given for multiple, derived, or residual designs. Only one pointer is given; there may be additional constructions elsewhere.

No	v	b	r	k	λ	Nd	Nr	Comments	Ref	Where?
1	7	7	3	3	1	1	-	PG(2,2)		I.5.3
2	9	12	4	3	1	1	1	R#3,AG(2,3)		1.15
3	13	13	4	4	1	1	-	PG(2,3)		1.19
4	6	10	5	3	2	1	0	R#7,×3		1.11
5	16	20	5	4	1	1	1	R#6,AG(2,4)		1.24
6	21	21	5	5	1	1	-	PG(2,4)		IV.12.22
7	11	11	5	5	2	1	-			1.19
8	13	26	6	3	1	2	-		[89]	1.20
9	7	14	6	3	2	4	-	2#1,D#20	[106]	1.12
10	10	15	6	4	2	3	-	R#13	[106]	1.18
11	25	30	6	5	1	1	1	R#12,AG(2,5)		
12	31	31	6	6	1	1	-	PG(2,5)		IV.12.22
13	16	16	6	6	2	3	-		[41]	1.25
14	15	35	7	3	1	80	7	PG(3,2)	[73, 89]	1.21
15	8	14	7	4	3	4	1	R#20,$AG_2(3,2)$	[73, 41]	1.14
16	15	21	7	5	2	0	0	R#19*,×2		
17	36	42	7	6	1	0	0	R#18*,×2,AG(2,6)		
18	43	43	7	7	1	0	-	×1,PG(2,6)		
19	22	22	7	7	2	0	-	×1		
20	15	15	7	7	3	5	-	$PG_2(3,2)$	[41]	1.23
21	9	24	8	3	2	36	9	2#2,D#40	[91]	1.16
22	25	50	8	4	1	18	-		[78, 124]	1.27
23	13	26	8	4	2	2461	-	2#3	[111]	
24	9	18	8	4	3	11	-	D#41	[41]	1.17
25	21	28	8	6	2	0	-	R#28*,×2		
26	49	56	8	7	1	1	1	R#27,AG(2,7)		
27	57	57	8	8	1	1	-	PG(2,7)		IV.12.22

No	v	b	r	k	λ	Nd	Nr	Comments	Ref	Where?
28	29	29	8	8	2	0	-	×1		
29	19	57	9	3	1	$\geq 1.1 \times 10^9$	-		[98]	IV.10.4
30	10	30	9	3	2	960	-	D#54	[26, 40, 60]	IV.10.65
31	7	21	9	3	3	10	-	3#1	[90]	1.13
32	28	63	9	4	1	$\geq$145	$\geq$7		[73, 17, 92]	III.1.4
33	10	18	9	5	4	21	0	R#41,×3	[41]	IV.10.69
34	46	69	9	6	1	?	-			
35	16	24	9	6	3	18920	-	R#40	[121]	I.5.21
36	28	36	9	7	2	7	0	R#39,×3	[6]	
37	64	72	9	8	1	1	1	R#38,AG(2,8)		
38	73	73	9	9	1	1	-	PG(2,8)		IV.12.22
39	37	37	9	9	2	4	-		[6]	I.5.25
40	25	25	9	9	3	78	-		[36]	I.5.25
41	19	19	9	9	4	6	-		[41]	1.26
42	21	70	10	3	1	$\geq 2 \times 10^6$	$\geq$192		[73, 31, 140]	IV.10.4
43	6	20	10	3	4	4	1	2#4	[73, 47]	
44	16	40	10	4	2	22859	$\geq$10	2#5	[111]	
45	41	82	10	5	1	$\geq$5	-		[86]	IV.10.8
46	21	42	10	5	2	$\geq$35	-	2#6	[134]	IV.10.21
47	11	22	10	5	4	4393	-	2#7,D#63	[15]	
48	51	85	10	6	1	?	-			
49	21	30	10	7	3	3809	0	R#54,×3	[51, 73, 119]	
50	36	45	10	8	2	0	-	R#53*,×2		
51	81	90	10	9	1	7	7	R#52,AG(2,9)	[32, 79]	
52	91	91	10	10	1	4	-	PG(2,9)	[32, 79]	IV.12.22
53	46	46	10	10	2	0	-	×1		
54	31	31	10	10	3	151	-		[119]	I.5.25
55	12	44	11	3	2	$\geq 10^6$	$\geq$2	D#84	[73, 30, 113]	IV.10.65
56	12	33	11	4	3	$\geq$17172470	$\geq$1	D#85*	[51, 73, 121]	IV.10.67
57	45	99	11	5	1	$\geq$16	?		[92]	IV.10.22
58	12	22	11	6	5	11603	1	R#63,HD	[51, 73, 111]	
59	45	55	11	9	2	$\geq$16	0	R#62,×3	[73, 34]	
60	100	110	11	10	1	0	0	R#61*,AG(2,10)	[80]	
61	111	111	11	11	1	0	-	PG(2,10)	[80]	
62	56	56	11	11	2	$\geq$5	-		[63]	I.5.25
63	23	23	11	11	5	1102	-		[51, 59, 123]	IV.12.22
64	25	100	12	3	1	$\geq 10^{14}$	-		[89]	IV.10.4
65	13	52	12	3	2	$\geq$92714	-	2#8,D#96	[92]	
66	9	36	12	3	3	22521	$\geq$395	3#2	[88]	
67	7	28	12	3	4	35	-	4#1	[47]	
68	37	111	12	4	1	$\geq$3	-		[28]	IV.10.6
69	19	57	12	4	2	$\geq$423	-		[86]	IV.10.7
70	13	39	12	4	3	$\geq$3702	-	3#3,D#97	[92, 120]	
71	10	30	12	4	4	$\geq$1759613	-	2#10	[90]	
72	25	60	12	5	2	$\geq$28	$\geq$28	2#11	[68]	
73	61	122	12	6	1	?	-			
74	31	62	12	6	2	$\geq$72	-	2#12	[70]	
75	21	42	12	6	3	$\geq$1	-			IV.10.10
76	16	32	12	6	4	$\geq$111	-	2#13	[67]	
77	13	26	12	6	5	$\geq$2572156	-	D#98	[51, 111]	
78	22	33	12	8	4	?	-	R#85*		
79	33	44	12	9	3	$\geq$3375	-	R#84	[127, 95]	
80	55	66	12	10	2	0	-	R#83*,×2		
81	121	132	12	11	1	$\geq$1	$\geq$1	R#82,AG(2,11)		
82	133	133	12	12	1	$\geq$1	-	PG(2,11)		IV.12.22
83	67	67	12	12	2	0	-	×1		

No	v	b	r	k	λ	Nd	Nr	Comments	Ref	Where?
84	45	45	12	12	3	$\geq$3752	-		[95]	IV.12.22
85	34	34	12	12	4	0	-	×1		
86	27	117	13	3	1	$\geq 10^{11}$	$\geq$909	AG(3,3)	[73, 27, 89]	IV.10.4
87	40	130	13	4	1	$\geq 10^{6}$	$\geq$1	PG(3,3)	[73]	IV.10.6
88	66	143	13	6	1	$\geq$1	?		[35]	I.2.34
89	14	26	13	7	6	$\geq$17896	0	R#98,×3	[127, 111]	
90	27	39	13	9	4	$\geq$8071	68	R#97,$AG_2(3,3)$	[73, 82, 120]	
91	40	52	13	10	3	?	0	R#96*,×3		
92	66	78	13	11	2	$\geq$2	0	R#95,×3	[73, 5]	
93	144	156	13	12	1	?	?	R#94*,AG(2,12)		
94	157	157	13	13	1	?	-	PG(2,12)		
95	79	79	13	13	2	$\geq$2	-		[5]	I.5.25
96	53	53	13	13	3	0	-	×1		
97	40	40	13	13	4	$\geq$389	-	$PG_2(3,3)$	[120]	IV.12.22
98	27	27	13	13	6	208310	-		[123]	IV.12.22
99	15	70	14	3	2	$\geq$685521	$\geq$21	2#14,D#140	[92]	
100	22	77	14	4	2	$\geq$7921	-		[38]	IV.10.7
101	8	28	14	4	6	2310	4	2#15	[111]	
102	15	42	14	5	4	$\geq$207	?	2#16*,D#141*	[57, 77]	IV.10.69
103	36	84	14	6	2	$\geq$5	$\geq$2	2#17*	[69, 138]	IV.10.70
104	15	35	14	6	5	$\geq$117	-	D#142	[51, 86]	IV.10.70
105	85	170	14	7	1	?	-			
106	43	86	14	7	2	$\geq$4	-	2#18*	[1]	IV.10.21
107	29	58	14	7	3	$\geq$1	-			IV.10.21
108	22	44	14	7	4	$\geq$34	-	2#19*	[109]	IV.10.21
109	15	30	14	7	6	$\geq$57810	-	2#20,D#143	[90]	
110	78	91	14	12	2	0	-	R#113*,×2		
111	169	182	14	13	1	$\geq$1	$\geq$1	R#112,AG(2,13)		
112	183	183	14	14	1	$\geq$1	-	PG(2,13)		IV.12.22
113	92	92	14	14	2	0	-	×1		
114	31	155	15	3	1	$\geq 6 \times 10^{16}$	-		[92]	IV.10.4
115	16	80	15	3	2	$\geq 10^{13}$	-	D#169	[92]	IV.10.5
116	11	55	15	3	3	$\geq$436800	-		[92]	IV.10.14
117	7	35	15	3	5	109	-	5#1	[121]	
118	6	30	15	3	6	6	0	3#4	[47, 92]	
119	16	60	15	4	3	$\geq 6 \times 10^{5}$	$\geq 6 \times 10^{5}$	3#5,D#170	[92]	
120	61	183	15	5	1	$\geq$10	-		[28]	IV.10.8
121	31	93	15	5	2	$\geq$1	-			IV.10.9
122	21	63	15	5	3	$\geq 10^{9}$	-	3#6	[18]	
123	16	48	15	5	4	$\geq$11	-	D#171	[51, 20]	
124	13	39	15	5	5	$\geq$30	-		[51, 131]	IV.10.9
125	11	33	15	5	6	$\geq$127	-	3#7	[14, 67]	
126	76	190	15	6	1	$\geq$1	-		[105]	I.2.35
127	26	65	15	6	3	$\geq$1	-		[51]	IV.10.73
128	16	40	15	6	5	$\geq$15	-	D#172	[51, 86]	
129	91	195	15	7	1	$\geq$2	?		[11]	IV.10.57
130	16	30	15	8	7	$\geq 9 \times 10^{7}$	$\geq$5	R#143,$AG_3(4,2)$,HD	[73, 75]	
131	21	35	15	9	6	$\geq 10^{4}$	-	R#142	[51, 21]	
132	136	204	15	10	1	?	-			
133	46	69	15	10	3	?	-			
134	28	42	15	10	5	$\geq$3	-	R#141*	[84]	
135	56	70	15	12	3	$\geq$4	-	R#140	[49]	
136	91	105	15	13	2	0	0	R#139*,×2		
137	196	210	15	14	1	0	0	R#138*,×2,AG(2,14)		
138	211	211	15	15	1	0	-	×1,PG(2,14)		
139	106	106	15	15	2	0	-	×1		

No	v	b	r	k	λ	Nd	Nr	Comments	Ref	Where?
140	71	71	15	15	3	≥8	-		[49]	I.5.25
141	43	43	15	15	5	0	-	×1		
142	36	36	15	15	6	≥25634	-		[122]	IV.12.22
143	31	31	15	15	7	≥1266891	-	$PG_3(4,2)$	[107]	IV.12.11
144	33	176	16	3	1	$\geq 10^{13}$	≥28		[92, 133]	IV.10.4
145	9	48	16	3	4	16585031	≥9	4#2	[121]	
146	49	196	16	4	1	≥236	-		[28, 48]	IV.10.6
147	25	100	16	4	2	≥17	-	2#22		
148	17	68	16	4	3	≥1	-	D#185	[127]	
149	13	52	16	4	4	≥2462	-	4#3		
150	9	36	16	4	6	$\geq 1.25 \times 10^8$	-	2#24	[121]	
151	65	208	16	5	1	≥2	≥1		[23, 28]	IV.10.8
152	81	216	16	6	1	?	-			
153	21	56	16	6	4	≥1	-	2#25*	[52]	
154	49	112	16	7	2	≥1	≥1	2#26		
155	113	226	16	8	1	?	-			
156	57	114	16	8	2	≥1362	-	2#27	[70]	
157	29	58	16	8	4	≥2	-	2#28*	[127]	IV.10.21
158	17	34	16	8	7	≥11	-	D#186	[127, 132]	
159	145	232	16	10	1	?	-			
160	25	40	16	10	6	≥43	-	R#172	[71, 121]	
161	33	48	16	11	5	≥19	0	R#171,×3	[73, 20]	
162	177	236	16	12	1	?	-			
163	45	60	16	12	4	≥1	-	R#170	[9]	
164	65	80	16	13	3	?	0	R#169*,×3		
165	105	120	16	14	2	?	-	R#168*		
166	225	240	16	15	1	?	?	R#167*,AG(2,15)		
167	241	241	16	16	1	?	-	PG(2,15)		
168	121	121	16	16	2	?	-			
169	81	81	16	16	3	?	-			
170	61	61	16	16	4	≥1	-		[9]	I.5.25
171	49	49	16	16	5	≥12146	-		[76]	I.5.25
172	41	41	16	16	6	≥115307	-		[121]	I.5.25
173	18	102	17	3	2	$\geq 4 \times 10^{14}$	≥1	D#217*	[127, 53, 92]	IV.10.65
174	52	221	17	4	1	≥206	≥1		[28]	IV.10.6
175	35	119	17	5	2	≥1	≥1		[127, 1]	IV.10.9
176	18	51	17	6	5	≥3	≥2	D#218*	[127, 72]	IV.10.70
177	35	85	17	7	3	≥2	?		[1, 55, 92]	
178	120	255	17	8	1	≥94	≥1		[110, 116]	
179	18	34	17	9	8	$\geq 10^3$	0	R#186,×3	[73, 127, 21]	
180	52	68	17	13	4	≥1	0	R#185,×3	[73, 37]	
181	120	136	17	15	2	0	0	R#184*,×2		
182	256	272	17	16	1	≥189	≥189	R#183,AG(2,16)	[64, 65]	
183	273	273	17	17	1	≥22	-	PG(2,16)	[64, 65, 87]	IV.12.22
184	137	137	17	17	2	0	-	×1		
185	69	69	17	17	4	≥1	-			I.5.25
186	35	35	17	17	8	≥108131	-		[122]	IV.12.22
187	37	222	18	3	1	$\geq 10^{10}$	-		[127, 83]	IV.10.4
188	19	114	18	3	2	$\geq 2 \times 10^9$	-	2#29,D#231*	[92]	
189	13	78	18	3	3	$\geq 3 \times 10^9$	-	3#8	[92]	
190	10	60	18	3	4	≥961	-	2#30		
191	7	42	18	3	6	418	-	6#1	[111, 121]	
192	28	126	18	4	2	≥139	≥8	2#32		
193	10	45	18	4	6	≥14819	-	3#10	[92]	
194	25	90	18	5	3	$\geq 10^{17}$	$\geq 10^{17}$	3#11	[92]	
195	10	36	18	5	8	≥135922	7	2#33	[73, 90]	

No	v	b	r	k	λ	Nd	Nr	Comments	Ref	Where?
196	91	273	18	6	1	≥ 4	-		[25, 102]	IV.10.10
197	46	138	18	6	2	≥ 1	-	2#34*		IV.10.10
198	31	93	18	6	3	$\geq 10^{22}$	-	3#12	[92]	
199	19	57	18	6	5	≥ 1	-	D#232*		I.6.28
200	16	48	18	6	6	$\geq 10^{8}$	-	3#13,2#35	[92]	
201	28	72	18	7	4	≥ 392	?	2#36	[67]	
202	64	144	18	8	2	≥ 121	≥ 121	2#37	[68]	
203	145	290	18	9	1	?	-			
204	73	146	18	9	2	≥ 3500	-	2#38	[70]	
205	49	98	18	9	3	?	-			
206	37	74	18	9	4	≥ 852	-	2#39	[67]	
207	25	50	18	9	6	≥ 79	-	2#40		
208	19	38	18	9	8	≥ 7	-	2#41,D#233		
209	55	99	18	10	3	?	-			
210	100	150	18	12	2	?	-			
211	34	51	18	12	6	≥ 2	-	R#218*	[45]	
212	85	102	18	15	3	?	-	R#217*		
213	136	153	18	16	2	?	-	R#216*		
214	289	306	18	17	1	≥ 1	≥ 1	R#215,AG(2,17)		
215	307	307	18	18	1	≥ 1	-	PG(2,17)		IV.12.22
216	154	154	18	18	2	?	-			
217	103	103	18	18	3	0	-	$\times 1$		
218	52	52	18	18	6	0	-	$\times 1$		
219	39	247	19	3	1	$\geq 10^{44}$	≥ 88		[83, 126]	IV.10.4
220	20	95	19	4	3	≥ 1	≥ 3	D#270	[127, 30]	IV.10.67
221	20	76	19	5	4	≥ 1	≥ 2	D#271*	[52, 127, 30]	IV.10.69
222	96	304	19	6	1	≥ 1	?		[103]	I.2.36
223	153	323	19	9	1	?	?			I.2.16
224	20	38	19	10	9	$\geq 10^{16}$	3	R#233,HD	[75]	
225	39	57	19	13	6	?	0	R#232*,$\times 3$		
226	96	114	19	16	3	?	0	R#231*,$\times 3$		
227	153	171	19	17	2	0	0	R#230*,$\times 2$		
228	324	342	19	18	1	?	?	R#229*,AG(2,18)		
229	343	343	19	19	1	?	-	PG(2,18)		
230	172	172	19	19	2	0	-	$\times 1$		
231	115	115	19	19	3	?	-			
232	58	58	19	19	6	0	-	$\times 1$		
233	39	39	19	19	9	≥ 38	-		[10]	IV.24.18
234	21	140	20	3	2	$\geq 5 \times 10^{14}$	≥ 79	2#42,D#307	[92]	
235	9	60	20	3	5	≥ 330	≥ 9	5#2		
236	6	40	20	3	8	13	1	4#4	[60]	
237	61	305	20	4	1	≥ 18132	-		[127, 28]	IV.10.48
238	31	155	20	4	2	≥ 1	-		[127]	IV.10.7
239	21	105	20	4	3	≥ 1	-	D#308*	[127]	IV.10.7
240	16	80	20	4	4	$\geq 6 \times 10^{5}$	$\geq 6 \times 10^{5}$	4#5	[92]	
241	13	65	20	4	5	$\geq 10^{3}$	-	5#3	[92]	
242	11	55	20	4	6	≥ 1	-		[127]	IV.10.7
243	81	324	20	5	1	≥ 1	-		[127]	IV.10.22
244	41	164	20	5	2	≥ 6	-	2#45		
245	21	84	20	5	4	$\geq 10^{9}$	-	4#6,D#309	[92]	I.6.28
246	17	68	20	5	5	≥ 1	-		[23]	IV.10.9
247	11	44	20	5	8	≥ 4394	-	4#7		
248	51	170	20	6	2	≥ 1	-	2#48*	[52]	IV.10.76
249	21	70	20	6	5	≥ 1	-	D#310	[52]	IV.10.10
250	21	60	20	7	6	≥ 3810	≥ 1	2#49,D#311*	[1]	
251	36	90	20	8	4	≥ 2	-	2#50*	[1, 136]	

No	v	b	r	k	λ	Nd	Nr	Comments	Ref	Where?
252	81	180	20	9	2	$\geq$1169	$\geq$1169	2#51	[68]	
253	181	362	20	10	1	?	-			
254	91	182	20	10	2	$\geq$46790	-	2#52	[70]	
255	61	122	20	10	3	?	-			
256	46	92	20	10	4	?	-	2#53*		
257	37	74	20	10	5	$\geq$1	-		[127]	IV.10.21
258	31	62	20	10	6	$\geq$152	-	2#54		
259	21	42	20	10	9	$\geq$4	-	D#312	[127, 12]	
260	111	185	20	12	2	?	-			
261	45	75	20	12	5	?	-			
262	141	188	20	15	2	?	-			
263	57	76	20	15	5	?	-	R#271*		
264	36	48	20	15	8	$\geq$1	-		[118]	
265	76	95	20	16	4	$\geq$1	-	R#270	[127, 23]	
266	171	190	20	18	2	?	-	R#269*		
267	361	380	20	19	1	$\geq$1	$\geq$1	R#268,AG(2,19)		
268	381	381	20	20	1	$\geq$1	-	PG(2,19)		IV.12.22
269	191	191	20	20	2	?	-			
270	96	96	20	20	4	$\geq$2	-		[115]	IV.12.22
271	77	77	20	20	5	0	-	$\times$1		
272	43	301	21	3	1	$\geq 5 \times 10^{64}$	-		[52, 83]	IV.10.4
273	22	154	21	3	2	$\geq 3 \times 10^{9}$	-	D#336*	[92]	IV.10.65
274	15	105	21	3	3	$\geq 10^{15}$	$\geq 10^{11}$	3#14	[92]	
275	8	56	21	3	6	3077244	-		[121]	IV.10.66
276	7	49	21	3	7	1508	-	7#1	[111, 121]	
277	64	336	21	4	1	$\geq$12048	$\geq$1		[28, 56]	IV.10.6
278	8	42	21	4	9	5201971	10	3#15	[121]	
279	85	357	21	5	1	$\geq$10	$\geq$1	PG(3,4)	[28]	IV.10.57
280	15	63	21	5	6	$\geq$2211	$\geq$149	3#16*	[93]	IV.10.9
281	106	371	21	6	1	$\geq$1	-		[101]	I.2.37
282	36	126	21	6	3	$\geq$1	$\geq$1	3#17*	[52, 1]	IV.10.70
283	22	77	21	6	5	$\geq$1	-	D#337,#6$\downarrow$#153	[52]	IV.10.10
284	16	56	21	6	7	$\geq$1	-	#13+#128	[117]	
285	127	381	21	7	1	?	-			
286	64	192	21	7	2	$\geq$1	-		[2]	I.2.42
287	43	129	21	7	3	$\geq$1	-	3#18*	[139]	IV.10.21
288	22	66	21	7	6	$\geq$1	-	3#19*,D#338*	[52]	I.6.29
289	19	57	21	7	7	$\geq$1	-		[139]	I.6.28
290	15	45	21	7	9	$\geq 10^{8}$	-	3#20	[92]	
291	57	133	21	9	3	?	-			
292	190	399	21	10	1	?	?			
293	22	42	21	11	10	$\geq$2	0	R#312,$\times$3	[73, 127, 50]	
294	232	406	21	12	1	?	-			
295	274	411	21	14	1	?	-			
296	92	138	21	14	3	?	-			
297	40	60	21	14	7	?	-	R#311*		
298	295	413	21	15	1	?	-			
299	50	70	21	15	6	$\geq$1	-	R#310	[62]	
300	64	84	21	16	5	$\geq$157	$\geq$157	R#309,$AG_2(3,4)$	[127, 66]	
301	85	105	21	17	4	?	0	R#308*,$\times$3		
302	120	140	21	18	3	?	-	R#307*		
303	190	210	21	19	2	?	0	R#306*,$\times$3		
304	400	420	21	20	1	?	?	R#305*,AG(2,20)		
305	421	421	21	21	1	?	-	PG(2,20)		
306	211	211	21	21	2	?	-			
307	141	141	21	21	3	0	-	$\times$1		

No	v	b	r	k	λ	Nd	Nr	Comments	Ref	Where?
308	106	106	21	21	4	0	-	$\times 1$		
309	85	85	21	21	5	≥ 213964	-	$PG_2(3,4)$	[66]	IV.12.22
310	71	71	21	21	6	≥ 2	-		[62]	I.5.25
311	61	61	21	21	7	0	-	$\times 1$		
312	43	43	21	21	10	≥ 2	-		[50]	IV.12.22
313	45	330	22	3	1	$\geq 6 \times 10^{76}$	≥ 84		[52, 83, 92]	IV.10.4
314	12	88	22	3	4	≥ 20476	≥ 3	2#55		
315	34	187	22	4	2	≥ 1	-		[52]	IV.10.7
316	12	66	22	4	6	$\geq 1.7 \times 10^6$	≥ 1	2#56		
317	45	198	22	5	2	≥ 17	?	2#57		
318	111	407	22	6	1	≥ 1	-		[103]	I.2.38
319	12	44	22	6	10	≥ 11604	≥ 400	2#58	[14]	
320	133	418	22	7	1	?	?			
321	45	110	22	9	4	≥ 1353	?	2#59	[67]	
322	100	220	22	10	2	≥ 1	?	2#60*	[24]	III.5.12
323	221	442	22	11	1	?	-			
324	111	222	22	11	2	?	-	2#61*		
325	56	112	22	11	4	≥ 2696	-	2#62	[67]	
326	45	90	22	11	5	?	-			
327	23	46	22	11	10	≥ 1103	-	2#63,D#351		
328	287	451	22	14	1	?	-			
329	45	66	22	15	7	?	0	R#338*	[32]	
330	56	77	22	16	6	≥ 3	-	R#337	[130]	
331	133	154	22	19	3	?	0	R#336*,$\times 3$		
332	210	231	22	20	2	0	-	R#335*,$\times 2$		
333	441	462	22	21	1	0	0	R#334*,$\times 2$,AG(2,21)		
334	463	463	22	22	1	0	-	$\times 1$,PG(2,21)		
335	232	232	22	22	2	0	-	$\times 1$		
336	155	155	22	22	3	?	-			
337	78	78	22	22	6	≥ 3	-		[130]	I.5.25
338	67	67	22	22	7	0	-	$\times 1$		
339	24	184	23	3	2	$\geq 3 \times 10^9$	≥ 1	D#404*	[53, 92]	IV.10.65
340	24	138	23	4	3	≥ 1	≥ 1	D#405*	[52, 8]	IV.10.67
341	24	92	23	6	5	≥ 1	≥ 1	D#406*	[52, 39]	IV.10.70
342	70	230	23	7	2	≥ 1	?		[2]	IV.10.72
343	24	69	23	8	7	≥ 1	?	D#407	[52]	IV.10.72
344	70	161	23	10	3	?	?			
345	231	483	23	11	1	?	?			
346	24	46	23	12	11	$\geq 4 \times 10^{17}$	≥ 129	R#351,HD	[127, 59, 75]	
347	231	253	23	21	2	0	0	R#350*,$\times 2$		
348	484	506	23	22	1	0	0	R#349*,$\times 2$,AG(2,22)		
349	507	507	23	23	1	0	-	$\times 1$,PG(2,22)		
350	254	254	23	23	2	0	-	$\times 1$		
351	47	47	23	23	11	≥ 1	-			IV.12.22
352	49	392	24	3	1	$\geq 6 \times 10^{14}$	-		[52, 83]	IV.10.4
353	25	200	24	3	2	$\geq 10^{14}$	-	2#64,D#438*		
354	17	136	24	3	3	≥ 4968	-		[92]	IV.10.14
355	13	104	24	3	4	$\geq 10^8$	-	4#8	[92]	
356	9	72	24	3	6	$\geq 10^7$	$\geq 10^5$	6#2	[92]	
357	7	56	24	3	8	5413	-	8#1	[111, 121]	
358	73	438	24	4	1	$\geq 10^7$	-		[16]	IV.10.48
359	37	222	24	4	2	≥ 4	-	2#68		
360	25	150	24	4	3	$\geq 10^{22}$	-	3#22,D#439*	[92]	
361	19	114	24	4	4	≥ 424	-	2#69		
362	13	78	24	4	6	$\geq 10^8$	-	6#3	[92]	
363	10	60	24	4	8	≥ 1759614	-	4#10	[90]	

No	v	b	r	k	λ	Nd	Nr	Comments	Ref	Where?
364	9	54	24	4	9	$\geq 10^6$	-	3#24	[92]	
365	25	120	24	5	4	$\geq 10^{17}$	$\geq 10^{17}$	4#11,D#440	[92]	
366	121	484	24	6	1	≥ 1	-		[52, 102]	IV.10.22
367	61	244	24	6	2	≥ 1	-	2#73*		IV.10.10
368	41	164	24	6	3	≥ 1	-			IV.10.10
369	31	124	24	6	4	$\geq 10^{22}$	-	4#12	[92]	
370	25	100	24	6	5	≥ 1	-	D#441	[52]	
371	21	84	24	6	6	≥ 1	-	3#25*,2#75		
372	16	64	24	6	8	$\geq 10^8$	-	4#13	[92]	
373	13	52	24	6	10	≥ 2572157	-	2#77		
374	49	168	24	7	3	$\geq 10^{52}$	$\geq 10^{52}$	3#26	[92]	
375	169	507	24	8	1	?	-			
376	85	255	24	8	2	?	-			
377	57	171	24	8	3	$\geq 10^{63}$	-	3#27	[92]	
378	43	129	24	8	4	≥ 1	-		[139]	IV.10.21
379	29	87	24	8	6	≥ 1	-	3#28*	[52]	IV.10.21
380	25	75	24	8	7	≥ 1	-	D#442*	[139]	IV.10.22
381	22	66	24	8	8	≥ 1	-	2#78*		IV.10.21
382	33	88	24	9	6	≥ 3376	-	2#79		
383	55	132	24	10	4	?	-	2#80*		
384	25	60	24	10	9	≥ 3	-	D#443	[139, 29]	
385	121	264	24	11	2	≥ 365	≥ 365	2#81	[68]	
386	265	530	24	12	1	?	-			
387	133	266	24	12	2	≥ 9979200	-	2#82	[70]	
388	89	178	24	12	3	?	-			
389	67	134	24	12	4	≥ 1	-	2#83*	[1]	IV.10.21
390	45	90	24	12	6	≥ 3753	-	2#84		
391	34	68	24	12	8	≥ 1	-	2#85*	[1]	IV.10.21
392	25	50	24	12	11	≥ 10	-	D#444	[139, 29]	
393	105	180	24	14	3	?	-			
394	85	136	24	15	4	?	-			
395	46	69	24	16	8	≥ 1	-	R#407	[61]	
396	69	92	24	18	6	?	-	R#406*		
397	115	138	24	20	4	?	-	R#405*		
398	161	184	24	21	3	?	-	R#404*		
399	49	56	24	21	10	≥ 1	-		[139]	IV.10.23
400	253	276	24	22	2	0	-	R#403*,×2		
401	529	552	24	23	1	≥ 1	≥ 1	R#402,AG(2,23)		
402	553	553	24	24	1	≥ 1	-	PG(2,23)		IV.12.22
403	277	277	24	24	2	0	-	×1		
404	185	185	24	24	3	0	-	×1		
405	139	139	24	24	4	?	-			
406	93	93	24	24	6	0	-	×1		
407	70	70	24	24	8	≥ 7	-		[42, 61]	I.5.25
408	51	425	25	3	1	$\geq 6 \times 10^{53}$	≥ 9419		[83, 126]	IV.10.4
409	6	50	25	3	10	19	0	5#4	[60, 92]	
410	76	475	25	4	1	≥ 32	≥ 1		[28, 56]	IV.10.6
411	16	100	25	4	5	$\geq 10^6$	$\geq 10^6$	5#5	[92]	
412	101	505	25	5	1	≥ 1	-			IV.10.49
413	51	255	25	5	2	≥ 1	-		[52]	IV.10.9
414	26	130	25	5	4	≥ 1	-	D#471*	[52]	I.6.28
415	21	105	25	5	5	$\geq 10^9$	-	5#6	[92]	
416	11	55	25	5	10	≥ 3337	-	5#7		
417	126	525	25	6	1	≥ 2	≥ 1		[52, 32, 99]	IV.10.77
418	176	550	25	8	1	?	?			
419	226	565	25	10	1	?	-			

No	v	b	r	k	λ	Nd	Nr	Comments	Ref	Where?
420	76	190	25	10	3	?	-			
421	46	115	25	10	5	≥1	-		[136]	
422	26	65	25	10	9	≥19	-	D#472	[52, 96]	
423	276	575	25	12	1	?	?			
424	26	50	25	13	12	≥1	0	R#444,×3	[73, 127]	
425	351	585	25	15	1	?	-			
426	51	85	25	15	7	?	-			
427	36	60	25	15	10	≥1	-	R#443	[52]	
428	51	75	25	17	8	?	0	R#442*,×3		
429	76	100	25	19	6	≥1	0	R#441,×3		
430	476	595	25	20	1	?	-			
431	96	120	25	20	5	≥1	-	R#440		
432	126	150	25	21	4	?	0	R#439*,×3		
433	176	200	25	22	3	?	0	R#438*,×3		
434	276	300	25	23	2	?	0	R#437*,×3		
435	576	600	25	24	1	?	?	R#436*,AG(2,24)		
436	601	601	25	25	1	?	-	PG(2,24)		
437	301	301	25	25	2	?	-			
438	201	201	25	25	3	?	-			
439	151	151	25	25	4	0	-	×1		
440	121	121	25	25	5	≥1	-		[112]	I.5.25
441	101	101	25	25	6	≥1	-			IV.12.22
442	76	76	25	25	8	0	-	×1		
443	61	61	25	25	10	≥1	-		[19]	I.5.25
444	51	51	25	25	12	≥1	-		[127]	IV.24.24
445	27	234	26	3	2	$\geq 10^{11}$	≥910	2#86,D#511*		
446	40	260	26	4	2	$\geq 10^{6}$	≥1	2#87		
447	14	91	26	4	6	≥4	-		[52, 100]	I.4.29
448	105	546	26	5	1	≥1	?		[52]	III.2.2
449	66	286	26	6	2	≥1	≥1	2#88	[43]	
450	27	117	26	6	5	≥1	-	D#512*	[52]	IV.10.70
451	14	52	26	7	12	≥17897	≥1	2#89	[51]	
452	92	299	26	8	2	?	-			
453	27	78	26	9	8	≥8072	≥13	2#90,D#513		
454	235	611	26	10	1	?	-			
455	40	104	26	10	6	≥1	?	2#91*	[1]	
456	66	156	26	11	4	≥494	?	2#92	[67]	
457	144	312	26	12	2	≥1	?	2#93*	[24]	III.5.12
458	313	626	26	13	1	?	-			
459	157	314	26	13	2	?	-	2#94*		
460	105	210	26	13	3	?	-			
461	79	158	26	13	4	≥940	-	2#95	[67]	
462	53	106	26	13	6	≥1	-	2#96*	[139]	IV.10.21
463	40	80	26	13	8	≥390	-	2#97		
464	27	54	26	13	12	≥208311	-	2#98,D#514		
465	40	65	26	16	10	≥1	-	R#472	[135]	
466	105	130	26	21	5	?	0	R#471*,×3		
467	300	325	26	24	2	0	-	R#470*,×2		
468	625	650	26	25	1	≥33	≥33	R#469,AG(2,25)	[74]	
469	651	651	26	26	1	≥17	-	PG(2,25)	[74]	IV.12.22
470	326	326	26	26	2	0	-	×1		
471	131	131	26	26	5	?	-			
472	66	66	26	26	10	≥1	-		[135]	I.5.25
473	55	495	27	3	1	$\geq 6 \times 10^{76}$	-		[83]	IV.10.4
474	28	252	27	3	2	$\geq 10^{55}$	-	D#564*	[92]	IV.10.5
475	19	171	27	3	3	$\geq 10^{17}$	-	3#29	[92]	

No	v	b	r	k	λ	Nd	Nr	Comments	Ref	Where?
476	10	90	27	3	6	$\geq 10^{12}$	-	3#30	[92]	
477	7	63	27	3	9	17785	-	9#1	[111, 121]	
478	28	189	27	4	3	$\geq 10^{32}$	$\geq 10^{26}$	3#32,D#565*	[92]	
479	55	297	27	5	2	≥ 1	≥ 1		[52, 39]	III.5.12
480	10	54	27	5	12	$\geq 10^{8}$	0	3#33	[92]	
481	136	612	27	6	1	≥ 1	-		[104]	I.2.39
482	46	207	27	6	3	$\geq 10^{28}$	-	3#34*	[52, 22]	IV.10.74
483	28	126	27	6	5	≥ 2	-	D#566*	[138]	IV.10.10
484	16	72	27	6	9	≥ 18921	-	3#35		
485	28	108	27	7	6	≥ 5047	≥ 1	3#36,D#567	[39, 67]	IV.10.71
486	64	216	27	8	3	$\geq 10^{77}$	$\geq 10^{77}$	3#37	[92]	
487	217	651	27	9	1	?	-			
488	109	327	27	9	2	≥ 1	-		[92]	IV.10.45
489	73	219	27	9	3	$\geq 10^{90}$	-	3#38	[92]	
490	55	165	27	9	4	≥ 1	-		[1]	IV.10.21
491	37	111	27	9	6	$\geq 10^{37}$	-	3#39	[92]	
492	28	84	27	9	8	≥ 3	-	D#568*	[52, 92, 138]	I.6.30
493	25	75	27	9	9	$\geq 10^{28}$	-	3#40	[92]	
494	19	57	27	9	12	$\geq 10^{16}$	-	3#41	[92]	
495	55	135	27	11	5	?	?			
496	100	225	27	12	3	?	-			
497	28	63	27	12	11	≥ 22	-	D#569	[71, 81]	
498	325	675	27	13	1	?	?			
499	28	54	27	14	13	$\geq 9 \times 10^{21}$	≥ 4	R#514,HD	[75, 128]	
500	190	342	27	15	2	?	-			
501	55	99	27	15	7	?	-			
502	460	690	27	18	1	?	-			
503	154	231	27	18	3	?	-			
504	52	78	27	18	9	≥ 2	-	R#513	[43]	
505	91	117	27	21	6	?	-	R#512*		
506	208	234	27	24	3	?	-	R#511*		
507	325	351	27	25	2	?	0	R#510*,×3		
508	676	702	27	26	1	?	?	R#509*,AG(2,26)		
509	703	703	27	27	1	?	-	PG(2,26)		
510	352	352	27	27	2	?	-			
511	235	235	27	27	3	0	-	×1		
512	118	118	27	27	6	0	-	×1		
513	79	79	27	27	9	≥ 1	-		[35]	I.5.25
514	55	55	27	27	13	≥ 1	-		[127, 23]	IV.24.18
515	57	532	28	3	1	$\geq 10^{90}$	≥ 1		[52]	IV.10.4
516	15	140	28	3	4	$\geq 10^{15}$	$\geq 10^{11}$	4#14	[92]	
517	9	84	28	3	7	≥ 330	≥ 9	7#2		
518	85	595	28	4	1	$\geq 10^{15}$	-		[52, 16]	IV.10.6
519	43	301	28	4	2	≥ 1	-			IV.10.7
520	29	203	28	4	3	≥ 1	-	D#586*		I.6.28
521	22	154	28	4	4	≥ 7922	-	2#100		
522	15	105	28	4	6	≥ 31300	-		[7]	IV.10.7
523	13	91	28	4	7	$\geq 10^{8}$	-	7#3	[92]	
524	8	56	28	4	12	≥ 2310	31	4#15	[60, 92, 111]	
525	15	84	28	5	8	≥ 104	≥ 1	4#16*,2#102	[1]	
526	141	658	28	6	1	?	-			
527	36	168	28	6	4	≥ 5	≥ 3	4#17*,2#103		
528	21	98	28	6	7	≥ 1	-	#75+#153		
529	16	70	28	6	10	≥ 118	-	2#104		
530	169	676	28	7	1	≥ 1	-		[55, 92]	IV.10.52
531	85	340	28	7	2	≥ 2	-	2#105*	[1, 46]	IV.10.21

No	v	b	r	k	λ	Nd	Nr	Comments	Ref	Where?
532	57	228	28	7	3	≥1	-		[52]	IV.10.21
533	43	172	28	7	4	≥4	-	4#18*,2#106		
534	29	116	28	7	6	≥1	-	2#107,D#587*		
535	25	100	28	7	7	≥1	-		[139]	IV.10.22
536	22	88	28	7	8	≥35	-	4#19*,2#108		
537	15	60	28	7	12	$\geq 10^8$	-	4#20	[92]	
538	50	175	28	8	4	≥1	-		[1]	IV.10.75
539	225	700	28	9	1	?	?			
540	85	238	28	10	3	?	-			
541	309	721	28	12	1	?	-			
542	78	182	28	12	4	?	-	2#110*		
543	45	105	28	12	7	≥1	-	#84+#163		
544	169	364	28	13	2	≥765	≥765	2#111	[68]	
545	365	730	28	14	1	?	-			
546	183	366	28	14	2	$\geq 10^9$	-	2#112	[70]	
547	92	184	28	14	4	?	-	2#113*		
548	53	106	28	14	7	≥1	-		[139]	IV.10.21
549	29	58	28	14	13	≥1	-	D#588	[139]	
550	36	63	28	16	12	≥32	-	R#569	[71, 81]	
551	477	742	28	18	1	?	-			
552	57	84	28	19	9	?	0	R#568*,×3		
553	561	748	28	21	1	?	-			
554	141	188	28	21	4	?	-			
555	81	108	28	21	7	≥1	-	R#567		
556	57	76	28	21	10	?	-			
557	99	126	28	22	6	?	-	R#566*		
558	162	189	28	24	4	?	-	R#565*		
559	225	252	28	25	3	?	0	R#564*,×3		
560	351	378	28	26	2	0	-	R#563*,×2		
561	729	756	28	27	1	≥7	≥7	R#562,AG(2,27)	[74]	
562	757	757	28	28	1	≥3	-	PG(2,27)	[32]	IV.12.22
563	379	379	28	28	2	0	-	×1		
564	253	253	28	28	3	?	-			
565	190	190	28	28	4	0	-	×1		
566	127	127	28	28	6	?	-			
567	109	109	28	28	7	≥1	-			IV.12.22
568	85	85	28	28	9	?	-			
569	64	64	28	28	12	≥4	-		[71]	IV.12.22
570	30	290	29	3	2	$\geq 2 \times 10^{51}$	≥1	D#655*	[53, 92]	IV.10.65
571	88	638	29	4	1	≥2	≥1		[52, 16, 56]	III.2.2
572	30	174	29	5	4	≥1	≥4	D#656	[52, 30]	
573	30	145	29	6	5	≥1	≥1	D#657*	[52, 3]	IV.10.70
574	175	725	29	7	1	?	?			
575	117	377	29	9	2	≥1	?		[54]	IV.10.57
576	30	87	29	10	9	≥1	?	D#658*	[1]	IV.10.72
577	117	261	29	13	3	?	?			
578	378	783	29	14	1	?	?			
579	30	58	29	15	14	≥1	0	R#588,×3	[73, 139]	
580	88	116	29	22	7	?	0	R#587*,×3		
581	175	203	29	25	4	?	0	R#586*,×3		
582	378	406	29	27	2	?	0	R#585*,×3		
583	784	812	29	28	1	?	?	R#584*,AG(2,28)		
584	813	813	29	29	1	?	-	PG(2,28)		
585	407	407	29	29	2	?	-			
586	204	204	29	29	4	?	-			
587	117	117	29	29	7	0	-	×1		

No	v	b	r	k	λ	Nd	Nr	Comments	Ref	Where?
588	59	59	29	29	14	≥ 1	-			IV.12.22
589	61	610	30	3	1	$\geq 2 \times 10^{24}$	-		[52, 83]	IV.10.4
590	31	310	30	3	2	$\geq 6 \times 10^{16}$	-	2#114,D#677*		
591	21	210	30	3	3	$\geq 10^{24}$	$\geq 10^{21}$	3#42	[92]	
592	16	160	30	3	4	$\geq 10^{13}$	-	2#115		
593	13	130	30	3	5	$\geq 10^{8}$	-	5#8	[92]	
594	11	110	30	3	6	≥ 436801	-	2#116		
595	7	70	30	3	10	54613	-	10#1	[90]	
596	6	60	30	3	12	34	1	6#4,3#43	[60, 92]	
597	46	345	30	4	2	≥ 1	-		[52]	IV.10.7
598	16	120	30	4	6	$\geq 10^{15}$	$\geq 10^{15}$	6#5	[92]	
599	10	75	30	4	10	≥ 29638	-	5#10	[92]	
600	121	726	30	5	1	≥ 1	-		[52]	IV.10.49
601	61	366	30	5	2	≥ 11	-	2#120		
602	41	246	30	5	3	$\geq 10^{46}$	-	3#45		
603	31	186	30	5	4	≥ 1	-	2#121,D#678*		
604	25	150	30	5	5	$\geq 10^{17}$	$\geq 10^{17}$	5#11	[92]	
605	21	126	30	5	6	$\geq 10^{24}$	-	6#6	[92]	
606	16	96	30	5	8	≥ 12	-	2#123		
607	13	78	30	5	10	≥ 31	-	2#124		
608	11	66	30	5	12	$\geq 10^{6}$	-	6#7	[92]	
609	151	755	30	6	1	≥ 1	-		[52, 139]	IV.10.42
610	76	380	30	6	2	≥ 1	-	2#126		
611	51	255	30	6	3	≥ 1	-	3#48*	[52]	IV.10.10
612	31	155	30	6	5	$\geq 10^{23}$	-	5#12,D#679	[92]	
613	26	130	30	6	6	≥ 1	-	2#127		
614	16	80	30	6	10	$\geq 10^{8}$	-	5#13	[92]	
615	91	390	30	7	2	≥ 3	?	2#129		
616	21	90	30	7	9	$\geq 10^{18}$	≥ 1	3#49	[1]	
617	36	135	30	8	6	≥ 3	-	3#50*	[1, 45, 138]	
618	16	60	30	8	14	$\geq 9 \times 10^{7}$	≥ 6	2#130		
619	81	270	30	9	3	$\geq 10^{108}$	$\geq 10^{108}$	3#51	[92]	
620	21	70	30	9	12	$\geq 10^{4}$	-	2#131		
621	271	813	30	10	1	?	-			
622	136	408	30	10	2	?	-	2#132*		
623	91	273	30	10	3	$\geq 10^{125}$	-	3#52	[92]	
624	55	165	30	10	5	≥ 1	-		[54]	IV.10.21
625	46	138	30	10	6	≥ 1	-	2#133*,3#53*	[1]	
626	31	93	30	10	9	≥ 152	-	3#54,D#680*		
627	28	84	30	10	10	≥ 5	-	2#134*	[52, 92, 138]	I.6.30
628	166	415	30	12	2	?	-			
629	56	140	30	12	6	≥ 5	-	2#135		
630	34	85	30	12	10	?	-			
631	91	210	30	13	4	?	?	2#136*		
632	196	420	30	14	2	?	?	2#137*		
633	421	842	30	15	1	?	-			
634	211	422	30	15	2	?	-	2#138*		
635	141	282	30	15	3	?	-			
636	106	212	30	15	4	?	-	2#139*		
637	85	170	30	15	5	?	-			
638	71	142	30	15	6	≥ 9	-	2#140		
639	61	122	30	15	7	≥ 1	-		[139]	IV.10.21
640	43	86	30	15	10	≥ 1	-	2#141*	[52]	IV.10.21
641	36	72	30	15	12	≥ 25635	-	2#142		
642	31	62	30	15	14	$\geq 10^{6}$	-	2#143,D#681		
643	171	285	30	18	3	?	-			

No	v	b	r	k	λ	Nd	Nr	Comments	Ref	Where?
644	286	429	30	20	2	?	-			
645	96	144	30	20	6	?	-			
646	58	87	30	20	10	?	-	R#658*		
647	301	430	30	21	2	?	-			
648	116	145	30	24	6	?	-	R#657*		
649	145	174	30	25	5	≥ 1	-	R#656	[115]	
650	261	290	30	27	3	?	-	R#655*		
651	406	435	30	28	2	0	-	R#654*,×2		
652	841	870	30	29	1	≥ 1	≥ 1	R#653,AG(2,29)		
653	871	871	30	30	1	≥ 1	-	PG(2,29)		IV.12.22
654	436	436	30	30	2	0	-	×1		
655	291	291	30	30	3	?	-			
656	175	175	30	30	5	≥ 2	-		[18, 115]	IV.12.22
657	146	146	30	30	6	0	-	×1		
658	88	88	30	30	10	0	-	×1		
659	63	651	31	3	1	$\geq 10^{42}$	≥ 82160	PG(5,2)	[83, 92]	IV.10.4
660	32	248	31	4	3	≥ 1	≥ 1	D#729*	[52, 8]	IV.10.67
661	125	775	31	5	1	≥ 1	≥ 1	AG(3,5)		III.2.2
662	156	806	31	6	1	≥ 1	≥ 1	PG(3,5)		III.2.2
663	63	279	31	7	3	≥ 1	≥ 1		[52, 1]	IV.10.71
664	32	124	31	8	7	≥ 1	≥ 1	D#730*	[52, 108]	IV.10.71
665	63	217	31	9	4	≥ 1	≥ 1		[1]	IV.10.71
666	280	868	31	10	1	?	?			
667	435	899	31	15	1	?	?			
668	32	62	31	16	15	$\geq 10^{28}$	≥ 1	R#681,$AG_4(5,2)$,HD	[75]	
669	63	93	31	21	10	?	0	R#680*	[32]	
670	125	155	31	25	6	$\geq 10^{12}$	$\geq 10^{12}$	R#679,$AG_2(3,5)$	[66]	
671	156	186	31	26	5	?	0	R#678*,×3		
672	280	310	31	28	3	?	0	R#677*,×3		
673	435	465	31	29	2	0	0	R#676*,×2		
674	900	930	31	30	1	0	0	R#675*,×2,AG(2,30)		
675	931	931	31	31	1	0	-	×1,PG(2,30)		
676	466	466	31	31	2	0	-	×1		
677	311	311	31	31	3	?	-			
678	187	187	31	31	5	0	-	×1		
679	156	156	31	31	6	$\geq 10^{17}$	-	$PG_2(3,5)$	[66]	IV.12.22
680	94	94	31	31	10	0	-	×1		
681	63	63	31	31	15	$\geq 10^{17}$	-	$PG_4(5,2)$	[66]	IV.12.22
682	33	352	32	3	2	$\geq 10^{13}$	≥ 1	2#144,D#775*		
683	9	96	32	3	8	$\geq 10^{7}$	$\geq 10^{5}$	8#2	[92]	
684	97	776	32	4	1	≥ 2	-		[16]	IV.10.48
685	49	392	32	4	2	≥ 237	-	2#146		
686	33	264	32	4	3	≥ 1	-	D#776*	[52]	I.6.28
687	25	200	32	4	4	$\geq 10^{22}$	-	4#22	[92]	
688	17	136	32	4	6	≥ 1	-	2#148		
689	13	104	32	4	8	$\geq 10^{8}$	-	8#3	[92]	
690	9	72	32	4	12	$\geq 10^{8}$	-	4#24	[92]	
691	65	416	32	5	2	≥ 3	≥ 1	2#151		
692	81	432	32	6	2	≥ 1	-	2#152*	[52]	III.5.9
693	33	176	32	6	5	≥ 1	-	D#777*	[52]	IV.10.10
694	21	112	32	6	8	≥ 1	-	4#25*,2#153		
695	49	224	32	7	4	$\geq 10^{52}$	$\geq 10^{52}$	4#26	[92]	
696	225	900	32	8	1	?	-			
697	113	452	32	8	2	≥ 1	-	2#155*	[43]	IV.10.51
698	57	228	32	8	4	$\geq 10^{63}$	-	4#27	[92]	
699	33	132	32	8	7	≥ 1	-	D#778	[52]	I.6.28

No	v	b	r	k	λ	Nd	Nr	Comments	Ref	Where?
700	29	116	32	8	8	≥ 3	-	4#28*,2#157		
701	17	68	32	8	14	≥ 12	-	2#158		
702	145	464	32	10	2	?	-	2#159*		
703	25	80	32	10	12	≥ 44	-	2#160		
704	33	96	32	11	10	≥ 20	?	2#161,D#779*		
705	177	472	32	12	2	?	-	2#162*		
706	45	120	32	12	8	≥ 1	-	2#163		
707	33	88	32	12	11	≥ 1	-	D#780*	[52]	
708	65	160	32	13	6	?	?	2#164*		
709	105	240	32	14	4	?	-	2#165*		
710	225	480	32	15	2	?	?	2#166*		
711	481	962	32	16	1	?	-			
712	241	482	32	16	2	?	-	2#167*		
713	161	322	32	16	3	?	-			
714	121	242	32	16	4	≥ 1	-	2#168*	[1]	IV.10.22
715	97	194	32	16	5	?	-			
716	81	162	32	16	6	?	-	2#169*		
717	61	122	32	16	8	≥ 1	-	2#170		
718	49	98	32	16	10	≥ 45	-	2#171		
719	41	82	32	16	12	≥ 115308	-	2#172		
720	33	66	32	16	15	≥ 1	-	D#781	[137]	
721	305	488	32	20	2	?	-			
722	369	492	32	24	2	?	-			
723	93	124	32	24	8	?	-	R#730*		
724	217	248	32	28	4	?	-	R#729*		
725	465	496	32	30	2	0	-	R#728*,$\times 2$		
726	961	992	32	31	1	≥ 1	≥ 1	R#727,AG(2,31)		
727	993	993	32	32	1	≥ 1	-	PG(2,31)		IV.12.24
728	497	497	32	32	2	0	-	$\times 1$		
729	249	249	32	32	4	?	-			
730	125	125	32	32	8	0	-	$\times 1$		
731	67	737	33	3	1	$\geq 10^{35}$	-		[52, 83]	IV.10.4
732	34	374	33	3	2	$\geq 10^{41}$	-	D#811*	[92]	IV.10.65
733	23	253	33	3	3	$\geq 2 \times 10^{14}$	-		[92]	IV.10.14
734	12	132	33	3	6	$\geq 10^{8}$	≥ 1	3#55	[92]	
735	7	77	33	3	11	155118	-	11#1	[90]	
736	100	825	33	4	1	≥ 2	≥ 1		[52, 16, 56]	III.2.2
737	12	99	33	4	9	$\geq 10^{10}$	≥ 1	3#56		
738	45	297	33	5	3	$\geq 10^{54}$	≥ 1	3#57	[2, 92]	
739	166	913	33	6	1	?	-			
740	56	308	33	6	3	≥ 1	-		[52]	III.5.9
741	34	187	33	6	5	≥ 1	-	D#812*	[52]	IV.10.70
742	16	88	33	6	11	≥ 1	-	#13+#484		
743	12	66	33	6	15	≥ 11604	≥ 12	3#58		
744	232	957	33	8	1	≥ 1	≥ 1		[33]	
745	45	165	33	9	6	≥ 35805	?	3#59	[67]	
746	100	330	33	10	3	≥ 1	?	3#60*	[24]	III.5.12
747	331	993	33	11	1	?	-			
748	166	498	33	11	2	?	-			
749	111	333	33	11	3	≥ 1	-	3#61*	[24]	III.5.12
750	67	201	33	11	5	≥ 1	-		[139]	IV.10.21
751	56	168	33	11	6	$\geq 10^{71}$	-	3#62	[92]	
752	34	102	33	11	10	≥ 1	-	D#813*	[52]	IV.10.72
753	31	93	33	11	11	≥ 1	-		[52]	I.6.28
754	23	69	33	11	15	≥ 1103	-	3#63		
755	364	1001	33	12	1	?	-			

No	v	b	r	k	λ	Nd	Nr	Comments	Ref	Where?
756	155	341	33	15	3	?	-			
757	496	1023	33	16	1	≥1	≥1		[116]	
758	34	66	33	17	16	≥1	0	R#781,×3	[73, 137]	
759	133	209	33	21	5	?	-			
760	56	88	33	21	12	?	-	R#780*		
761	694	1041	33	22	1	?	-			
762	232	348	33	22	3	?	-			
763	100	150	33	22	7	?	-			
764	78	117	33	22	9	?	-			
765	64	96	33	22	11	?	-	R#779*		
766	760	1045	33	24	1	?	-			
767	100	132	33	25	8	≥1	0	R#778,×3	[73, 50]	
768	144	176	33	27	6	?	-	R#777*		
769	232	264	33	29	4	?	0	R#776*,×3		
770	320	352	33	30	3	?	-	R#775*		
771	496	528	33	31	2	?	0	R#774*,×3		
772	1024	1056	33	32	1	≥11	≥11	R#773,AG(2,32)	[32]	
773	1057	1057	33	33	1	≥6	-	PG(2,32)	[32]	IV.12.24
774	529	529	33	33	2	?	-			
775	353	353	33	33	3	0	-	×1		
776	265	265	33	33	4	?	-			
777	177	177	33	33	6	?	-			
778	133	133	33	33	8	≥1	-		[50]	IV.12.22
779	97	97	33	33	11	?	-			
780	89	89	33	33	12	0	-	×1		
781	67	67	33	33	16	≥1	-		[137]	IV.12.22
782	69	782	34	3	1	$\geq 4\times 10^{41}$	≥1		[52, 56, 83]	IV.10.4
783	18	204	34	3	4	$\geq 4\times 10^{14}$	≥1	2#173		
784	52	442	34	4	2	≥207	≥1	2#174		
785	18	153	34	4	6	≥1	-		[52]	IV.10.68
786	35	238	34	5	4	≥2	≥2	2#175,D#869*		
787	171	969	34	6	1	≥1	-		[4]	I.2.40
788	18	102	34	6	10	≥4	≥3	2#176		
789	35	170	34	7	6	≥3	?	2#177,D#870*		
790	120	510	34	8	2	≥1	≥1	2#178		
791	18	68	34	9	16	$\geq 10^3$	≥1	2#179		
792	35	119	34	10	9	≥1	-	D#871*	[52]	
793	341	1054	34	11	1	?	?			
794	52	136	34	13	8	≥1	?	2#180		
795	35	85	34	14	13	≥1	-	D#872*	[1]	
796	120	272	34	15	4	?	?	2#181*		
797	256	544	34	16	2	≥477603	≥477603	2#182	[68]	
798	545	1090	34	17	1	?	-			
799	273	546	34	17	2	$\geq 10^{12}$	-	2#183	[70]	
800	137	274	34	17	4	?	-	2#184*		
801	69	138	34	17	8	≥1	-	2#185		
802	35	70	34	17	16	≥1854	-	2#186,D#873		
803	715	1105	34	22	1	?	-			
804	69	102	34	23	11	?	0	R#813*,×3		
805	154	187	34	28	6	?	-	R#812*		
806	341	374	34	31	3	?	0	R#811*,×3		
807	528	561	34	32	2	0	-	R#810*,×2		
808	1089	1122	34	33	1	0	0	R#809*,×2,AG(2,33)		
809	1123	1123	34	34	1	0	-	×1		
810	562	562	34	34	2	0	-	×1		
811	375	375	34	34	3	?	-			

No	v	b	r	k	λ	Nd	Nr	Comments	Ref	Where?
812	188	188	34	34	6	0	-	×1		
813	103	103	34	34	11	?	-			
814	36	420	35	3	2	$\geq 2\times10^{50}$	≥ 1	D#961*	[53, 92]	IV.10.65
815	15	175	35	3	5	$\geq 10^{15}$	$\geq 10^{11}$	5#14	[92]	
816	6	70	35	3	14	48	0	7#4	[60, 92]	
817	36	315	35	4	3	≥ 1	≥ 1	D#962*	[52, 8]	III.5.12
818	16	140	35	4	7	$\geq 10^{15}$	$\geq 10^{15}$	7#5	[92]	
819	8	70	35	4	15	≥ 2224	≥ 5	5#15		
820	141	987	35	5	1	≥ 1	-		[52]	I.2.33
821	71	497	35	5	2	≥ 1	-		[52]	IV.10.9
822	36	252	35	5	4	≥ 2	-	D#963*	[52, 138]	IV.10.69
823	29	203	35	5	5	≥ 1	-		[52]	I.6.28
824	21	147	35	5	7	$\geq 10^{24}$	-	7#6	[92]	
825	15	105	35	5	10	≥ 1	≥ 1	5#16*,#102+#280	[1]	
826	11	77	35	5	14	$\geq 10^{6}$	-	7#7	[92]	
827	36	210	35	6	5	≥ 1	≥ 1	#103+#282,D#964*	[52, 8]	
828	211	1055	35	7	1	?	-			
829	106	530	35	7	2	?	-			
830	71	355	35	7	3	≥ 1	-		[139]	IV.10.21
831	43	215	35	7	5	≥ 1	-	5#18*,#106+#287		
832	36	180	35	7	6	≥ 1	-	D#965*	[52]	I.6.28
833	31	155	35	7	7	≥ 5	-	$PG_2(4,2)$	[129]	I.6.28
834	22	110	35	7	10	≥ 1	-	5#19*,#108+#288		
835	16	80	35	7	14	≥ 1	-	#131↓#259		
836	15	75	35	7	15	$\geq 10^{9}$	-	5#20	[92]	
837	36	140	35	9	8	≥ 2	?	D#966*	[52, 138]	IV.10.72
838	316	1106	35	10	1	?	-			
839	106	371	35	10	3	?	-			
840	64	224	35	10	5	?	-			
841	46	161	35	10	7	?	-			
842	36	126	35	10	9	≥ 2	-	D#967*	[52, 138]	
843	22	77	35	10	15	≥ 1	-	#104↓#290		
844	176	560	35	11	2	?	?			
845	36	105	35	12	11	≥ 1	≥ 1	D#968*	[52, 97]	IV.10.72
846	456	1140	35	14	1	?	-			
847	92	230	35	14	5	?	-			
848	66	165	35	14	7	?	-			
849	36	90	35	14	13	≥ 1	-	D#969*	[52]	
850	246	574	35	15	2	?	-			
851	99	231	35	15	5	?	-			
852	36	84	35	15	14	≥ 1	-	D#970*, #142+#264		
853	176	385	35	16	3	?	?			
854	561	1155	35	17	1	?	?			
855	36	70	35	18	17	≥ 91	≥ 91	R#873,HD	[21, 137]	
856	96	168	35	20	7	?	-			
857	351	585	35	21	2	?	-			
858	141	235	35	21	5	?	-			
859	51	85	35	21	14	?	-	R#872*		
860	85	119	35	25	10	?	-	R#871*		
861	316	395	35	28	3	?	-			
862	136	170	35	28	7	?	-	R#870*		
863	64	80	35	28	15	?	-			
864	204	238	35	30	5	?	-	R#869*		
865	561	595	35	33	2	0	0	R#868*,×2		
866	1156	1190	35	34	1	?	?	R#867*,AG(2,34)		
867	1191	1191	35	35	1	?	-	PG(2,34)		

No	v	b	r	k	λ	Nd	Nr	Comments	Ref	Where?
868	596	596	35	35	2	0	-	×1		
869	239	239	35	35	5	?	-			
870	171	171	35	35	7	?	-			
871	120	120	35	35	10	?	-			
872	86	86	35	35	14	0	-	×1		
873	71	71	35	35	17	≥ 1	-		[137]	IV.12.22
874	73	876	36	3	1	$\geq 10^{34}$	-		[52, 83]	IV.10.4
875	37	444	36	3	2	$\geq 10^{10}$	-	2#187,D#991*		
876	25	300	36	3	3	$\geq 10^{25}$	-	3#64	[92]	
877	19	228	36	3	4	$\geq 10^{17}$	-	4#29	[92]	
878	13	156	36	3	6	$\geq 10^{17}$	-	6#8	[92]	
879	10	120	36	3	8	$\geq 10^{12}$	-	4#30	[92]	
880	9	108	36	3	9	≥ 330	≥ 10	9#2		
881	7	84	36	3	12	412991	-	12#1	[90]	
882	109	981	36	4	1	≥ 2	-		[16]	IV.10.48
883	55	495	36	4	2	≥ 1	-		[52]	III.5.9
884	37	333	36	4	3	$\geq 10^{40}$	-	3#68,D#992*	[92]	
885	28	252	36	4	4	$\geq 10^{32}$	$\geq 10^{26}$	4#32	[92]	
886	19	171	36	4	6	$\geq 10^{20}$	-	3#69		
887	13	117	36	4	9	$\geq 10^{8}$	-	9#3	[92]	
888	10	90	36	4	12	$\geq 10^{9}$	-	6#10	[92]	
889	145	1044	36	5	1	≥ 1	?		[52]	IV.10.57
890	25	180	36	5	6	$\geq 10^{38}$	$\geq 10^{38}$	6#11	[92]	
891	10	72	36	5	16	$\geq 10^{8}$	≥ 1	4#33,2#195	[92]	
892	181	1086	36	6	1	≥ 1	-		[52, 139]	IV.10.42
893	91	546	36	6	2	≥ 5	-	2#196		
894	61	366	36	6	3	≥ 1	-	3#73*	[52]	IV.10.10
895	46	276	36	6	4	≥ 1	-	4#34*,2#197		
896	37	222	36	6	5	≥ 1	-	D#993	[52]	
897	31	186	36	6	6	$\geq 10^{51}$	-	6#12	[92]	
898	21	126	36	6	9	≥ 1	-	3#75		
899	19	114	36	6	10	≥ 1	-	2#199		
900	16	96	36	6	12	$\geq 10^{19}$	-	6#13	[92]	
901	13	78	36	6	15	$\geq 10^{11}$	-	3#77	[92]	
902	217	1116	36	7	1	≥ 1	?		[85]	IV.10.21
903	28	144	36	7	8	≥ 5432	≥ 1	4#36	[1, 67]	
904	64	288	36	8	4	$\geq 10^{77}$	$\geq 10^{77}$	4#37	[92]	
905	22	99	36	8	12	≥ 1	-	3#78*	[52]	IV.10.21
906	289	1156	36	9	1	?	-			
907	145	580	36	9	2	≥ 1	-	2#203*	[46]	
908	97	388	36	9	3	≥ 1	-		[1]	IV.10.21
909	73	292	36	9	4	$\geq 10^{90}$	-	4#38	[92]	
910	49	196	36	9	6	≥ 5	-	2#205*	[58]	IV.10.22
911	37	148	36	9	8	$\geq 10^{37}$	-	4#39,D#994*	[92]	
912	33	132	36	9	9	$\geq 10^{39}$	-	3#79		I.6.28
913	25	100	36	9	12	$\geq 10^{28}$	-	4#40	[92]	
914	19	76	36	9	16	$\geq 10^{16}$	-	4#41	[92]	
915	325	1170	36	10	1	?	-			
916	55	198	36	10	6	?	-	3#80*,2#209*		
917	121	396	36	11	3	$\geq 10^{188}$	$\geq 10^{188}$	3#81	[92]	
918	397	1191	36	12	1	?	-			
919	199	597	36	12	2	?	-			
920	133	399	36	12	3	$\geq 10^{208}$	-	3#82	[92]	
921	100	300	36	12	4	?	-	2#210*		
922	67	201	36	12	6	≥ 1	-	3#83*	[139]	IV.10.21
923	45	135	36	12	9	$\geq 10^{57}$	-	3#84		

No	v	b	r	k	λ	Nd	Nr	Comments	Ref	Where?
924	37	111	36	12	11	$\geq$1	-	D#995*	[52]	I.6.28
925	34	102	36	12	12	$\geq$2	-	3#85*,2#211		IV.10.72
926	469	1206	36	14	1	?	-			
927	505	1212	36	15	1	?	-			
928	85	204	36	15	6	?	-	2#212*		
929	136	306	36	16	4	?	-	2#213*		
930	289	612	36	17	2	$\geq$3481	$\geq$3481	2#214	[68]	
931	613	1226	36	18	1	?	-			
932	307	614	36	18	2	$\geq 8 \times 10^{13}$	-	2#215	[70]	
933	205	410	36	18	3	?	-			
934	154	308	36	18	4	?	-	2#216*		
935	103	206	36	18	6	?	-	2#217*		
936	69	138	36	18	9	?	-			
937	52	104	36	18	12	?	-	2#218*		
938	37	74	36	18	17	$\geq$1	-	D#996	[52]	
939	685	1233	36	20	1	?	-			
940	115	207	36	20	6	?	-			
941	721	1236	36	21	1	?	-			
942	91	156	36	21	8	?	-			
943	49	84	36	21	15	?	-	R#970*		
944	253	414	36	22	3	?	-			
945	55	90	36	22	14	?	-	R#969*		
946	208	312	36	24	4	?	-			
947	70	105	36	24	12	?	-	R#968*		
948	91	126	36	26	10	?	-	R#967*		
949	105	140	36	27	9	?	-	R#966*		
950	973	1251	36	28	1	?	-			
951	145	180	36	29	7	?	0	R#965*,×3		
952	1045	1254	36	30	1	?	-			
953	175	210	36	30	6	?	-	R#964*		
954	217	252	36	31	5	?	0	R#963*,×3		
955	280	315	36	32	4	?	-	R#962*		
956	385	420	36	33	3	?	-	R#961*		
957	595	630	36	34	2	?	-	R#960*		
958	1225	1260	36	35	1	?	?	R#959*,AG(2,35)		
959	1261	1261	36	36	1	?	-	PG(2,35)		
960	631	631	36	36	2	?	-			
961	421	421	36	36	3	?	-			
962	316	316	36	36	4	0	-	×1		
963	253	253	36	36	5	?	-			
964	211	211	36	36	6	0	-	×1		
965	181	181	36	36	7	?	-			
966	141	141	36	36	9	?	-			
967	127	127	36	36	10	?	-			
968	106	106	36	36	12	0	-	×1		
969	91	91	36	36	14	0	-	×1		
970	85	85	36	36	15	?	-			
971	75	925	37	3	1	$\geq 10^{196}$	$\geq$1		[52, 50, 60]	IV.10.4
972	112	1036	37	4	1	$\geq$2	$\geq$1		[52, 50]	III.2.2
973	75	555	37	5	2	$\geq$1	?		[52]	IV.10.22
974	186	1147	37	6	1	$\geq$1	$\geq$1		[52, 43]	III.2.2
975	112	592	37	7	2	$\geq$1	?		[52]	
976	297	1221	37	9	1	?	?			
977	408	1258	37	12	1	?	?			
978	75	185	37	15	7	$\geq$1	$\geq$1		[108]	
979	112	259	37	16	5	?	?			

No	v	b	r	k	λ	Nd	Nr	Comments	Ref	Where?
980	630	1295	37	18	1	?	?			
981	38	74	37	19	18	≥ 1	0	R#996,×3	[73, 137]	
982	75	111	37	25	12	?	0	R#995*,×3		
983	112	148	37	28	9	?	?	R#994*		
984	186	222	37	31	6	≥ 1	0	R#993,×3	[73, 112]	
985	297	333	37	33	4	?	0	R#992*,×3		
986	408	444	37	34	3	?	0	R#991*,×3		
987	630	666	37	35	2	0	0	R#990*,×2		
988	1296	1332	37	36	1	?	?	R#989*,AG(2,36)		
989	1333	1333	37	37	1	?	-	PG(2,36)		
990	667	667	37	37	2	0	-	×1		
991	445	445	37	37	3	0	-	×1		
992	334	334	37	37	4	0	-	×1		
993	223	223	37	37	6	≥ 1	-		[112]	
994	149	149	37	37	9	?	-			
995	112	112	37	37	12	?	-			
996	75	75	37	37	18	≥ 1	-		[115]	IV.24.24
997	39	494	38	3	2	$\geq 10^{44}$	≥ 89	2#219,D#1071*		
998	58	551	38	4	2	≥ 1	-		[52]	III.5.9
999	20	190	38	4	6	≥ 1	≥ 4	2#220		
1000	20	152	38	5	8	≥ 1	≥ 3	2#221		
1001	96	608	38	6	2	≥ 1	≥ 1	2#222	[43]	
1002	39	247	38	6	5	≥ 1	-	D#1072*	[52]	IV.10.10
1003	77	418	38	7	3	≥ 2	?		[52, 1]	IV.10.57
1004	20	95	38	8	14	≥ 1	-		[52]	IV.10.72
1005	153	646	38	9	2	≥ 1	?	2#223*	[44, 92]	IV.10.22
1006	115	437	38	10	3	?	-			
1007	20	76	38	10	18	$\geq 10^{16}$	≥ 4	2#224		
1008	77	266	38	11	5	?	?			
1009	210	665	38	12	2	?	-			
1010	39	114	38	13	12	≥ 1	?	2#225*,D#1073*	[52]	IV.10.72
1011	96	228	38	16	6	?	?	2#226*		
1012	153	342	38	17	4	?	?	2#227*		
1013	324	684	38	18	2	≥ 1	?	2#228*	[24]	III.5.12
1014	685	1370	38	19	1	?	-			
1015	343	686	38	19	2	?	-	2#229*		
1016	229	458	38	19	3	?	-			
1017	172	344	38	19	4	?	-	2#230*		
1018	115	230	38	19	6	?	-	2#231*		
1019	77	154	38	19	9	?	-			
1020	58	116	38	19	12	?	-	2#232*		
1021	39	78	38	19	18	≥ 39	-	2#233,D#1074		
1022	666	703	38	36	2	?	-	R#1025*		
1023	1369	1406	38	37	1	≥ 1	≥ 1	R#1024,AG(2,37)		
1024	1407	1407	38	38	1	≥ 1	-	PG(2,37)		IV.12.24
1025	704	704	38	38	2	?	-			
1026	79	1027	39	3	1	$\geq 10^{56}$	-		[83]	IV.10.4
1027	40	520	39	3	2	$\geq 6 \times 10^{24}$	-	D#1163*	[92]	IV.10.5
1028	27	351	39	3	3	$\geq 10^{28}$	$\geq 10^{30}$	3#86	[92]	
1029	14	182	39	3	6	$\geq 2 \times 10^{34}$	-		[92]	IV.10.66
1030	7	91	39	3	13	≥ 417	-	13#1		
1031	40	390	39	4	3	$\geq 10^{33}$	$\geq 10^{33}$	3#87,D#1164*	[92]	
1032	40	312	39	5	4	≥ 1	≥ 1	D#1165*	[52, 1]	IV.10.69
1033	196	1274	39	6	1	?	-			
1034	66	429	39	6	3	≥ 1	≥ 1	3#88	[114]	
1035	40	260	39	6	5	≥ 1	-	D#1166*	[52]	IV.10.70

No	v	b	r	k	λ	Nd	Nr	Comments	Ref	Where?
1036	16	104	39	6	13	≥ 1	-	#13+#742		
1037	14	91	39	6	15	≥ 1	-		[52]	IV.10.70
1038	14	78	39	7	18	$\geq 10^{11}$	0	3#89	[92]	
1039	40	195	39	8	7	≥ 1	≥ 1	D#1167*	[52, 1]	IV.10.71
1040	105	455	39	9	3	?	-			
1041	27	117	39	9	12	$\geq 10^{17}$	$\geq 10^{17}$	3#90	[92]	
1042	40	156	39	10	9	≥ 1	?	3#91*,D#1168*	[52]	IV.10.72
1043	66	234	39	11	6	≥ 21584	?	3#92	[67]	
1044	144	468	39	12	3	≥ 1	?	3#93*	[24]	III.5.12
1045	40	130	39	12	11	≥ 1	-	D#1169*	[52]	IV.10.72
1046	469	1407	39	13	1	?	-			
1047	235	705	39	13	2	?	-			
1048	157	471	39	13	3	≥ 1	-	3#94*	[24]	III.5.12
1049	118	354	39	13	4	?	-			
1050	79	237	39	13	6	$\geq 10^{125}$	-	3#95	[92]	
1051	53	159	39	13	9	≥ 1	-	3#96*	[1]	
1052	40	120	39	13	12	$\geq 10^{33}$	-	3#97,D#1170	[92]	
1053	37	111	39	13	13	≥ 1	-		[52]	I.6.28
1054	27	81	39	13	18	≥ 208311	-	3#98		
1055	40	104	39	15	14	≥ 1	-	D#1171*	[52]	
1056	222	481	39	18	3	?	-			
1057	703	1443	39	19	1	?	?			
1058	40	78	39	20	19	≥ 1	≥ 1	R#1074,HD	[137]	
1059	196	364	39	21	4	?	-			
1060	976	1464	39	26	1	?	-			
1061	326	489	39	26	3	?	-			
1062	196	294	39	26	5	?	-			
1063	76	114	39	26	13	?	-	R#1073*		
1064	66	99	39	26	15	?	-			
1065	209	247	39	33	6	?	-	R#1072*		
1066	456	494	39	36	3	?	-	R#1071*		
1067	703	741	39	37	2	0	0	R#1070*,×2		
1068	1444	1482	39	38	1	0	0	R#1069*,×2,AG(2,38)		
1069	1483	1483	39	39	1	0	-	×1,PG(2,38)		
1070	742	742	39	39	2	0	-	×1		
1071	495	495	39	39	3	?	-			
1072	248	248	39	39	6	0	-	×1		
1073	115	115	39	39	13	0	-	×1		
1074	79	79	39	39	19	≥ 1	-		[115]	IV.12.22
1075	81	1080	40	3	1	$\geq 10^{48}$	$\geq 10^{7}$	AG(4,3)	[83, 92]	IV.10.4
1076	21	280	40	3	4	$\geq 10^{24}$	$\geq 10^{21}$	4#42	[92]	
1077	9	120	40	3	10	$\geq 10^{8}$	$\geq 10^{5}$	10#2	[92]	
1078	6	80	40	3	16	76	1	8#4,4#43	[92]	
1079	121	1210	40	4	1	$\geq 10^{6}$	-		[16, 18]	IV.10.48
1080	61	610	40	4	2	$\geq 10^{4}$	-	2#237		
1081	41	410	40	4	3	≥ 1	-	D#1192*		I.6.28
1082	31	310	40	4	4	≥ 1	-	2#238		
1083	25	250	40	4	5	$\geq 10^{23}$	-	5#22	[92]	
1084	21	210	40	4	6	≥ 1	-	2#239		
1085	16	160	40	4	8	$\geq 10^{15}$	$\geq 10^{15}$	8#5	[92]	
1086	13	130	40	4	10	$\geq 10^{14}$	-	10#3	[92]	
1087	11	110	40	4	12	≥ 1	-	2#242		
1088	9	90	40	4	15	$\geq 10^{6}$	-	5#24	[92]	
1089	161	1288	40	5	1	≥ 1	-		[52]	IV.10.22
1090	81	648	40	5	2	≥ 1	-	2#243		
1091	41	328	40	5	4	$\geq 10^{46}$	-	4#45,D#1193*		

No	v	b	r	k	λ	Nd	Nr	Comments	Ref	Where?
1092	33	264	40	5	5	≥ 1	-		[52]	IV.10.9
1093	21	168	40	5	8	$\geq 10^{24}$	-	8#6	[92]	
1094	17	136	40	5	10	≥ 1	-	2#246		
1095	11	88	40	5	16	$\geq 10^7$	-	8#7	[92]	
1096	201	1340	40	6	1	?	-			
1097	51	340	40	6	4	≥ 1	-	4#48*,2#248		
1098	21	140	40	6	10	≥ 1	-	5#25*,2#249		
1099	49	280	40	7	5	$\geq 10^{53}$	$\geq 10^{53}$	5#26	[92]	
1100	21	120	40	7	12	$\geq 10^{18}$	≥ 1	4#49,2#250		
1101	281	1405	40	8	1	?	-			
1102	141	705	40	8	2	?	-			
1103	71	355	40	8	4	≥ 1	-		[139]	IV.10.21
1104	57	285	40	8	5	$\geq 10^{63}$	-	5#27	[92]	
1105	41	205	40	8	7	≥ 1	-	D#1194*	[52]	I.6.28
1106	36	180	40	8	8	≥ 3	-	4#50*,2#251		
1107	29	145	40	8	10	≥ 1	-	5#28*,#157+#379		
1108	21	105	40	8	14	≥ 1	-		[52]	IV.10.21
1109	81	360	40	9	4	$\geq 10^{108}$	$\geq 10^{108}$	4#51	[92]	
1110	361	1444	40	10	1	?	-			
1111	181	724	40	10	2	≥ 2	-	2#253*	[1, 46]	IV.10.21
1112	121	484	40	10	3	≥ 1	-		[1]	IV.10.22
1113	91	364	40	10	4	$\geq 10^{125}$	-	4#52	[92]	
1114	73	292	40	10	5	≥ 1	-		[139]	IV.10.21
1115	61	244	40	10	6	≥ 10	-	2#255*	[13]	IV.10.21
1116	46	184	40	10	8	≥ 1	-	4#53*,2#256*	[1]	IV.10.21
1117	41	164	40	10	9	≥ 1	-	D#1195*	[52]	I.6.28
1118	37	148	40	10	10	≥ 1	-	2#257		
1119	31	124	40	10	12	≥ 152	-	4#54		
1120	25	100	40	10	15	≥ 1	-	#160+#384		
1121	21	84	40	10	18	≥ 5	-	2#259		
1122	441	1470	40	12	1	?	-			
1123	111	370	40	12	4	?	-	2#260*		
1124	45	150	40	12	10	≥ 1	-	2#261*,#84+#543		
1125	481	1480	40	13	1	?	?			
1126	105	300	40	14	5	?	-			
1127	561	1496	40	15	1	?	-			
1128	141	376	40	15	4	?	-	2#262*		
1129	81	216	40	15	7	≥ 1	-		[139]	
1130	57	152	40	15	10	?	-	2#263*		
1131	36	96	40	15	16	≥ 1	-	2#264		
1132	76	190	40	16	8	≥ 1	-	2#265		
1133	171	380	40	18	4	?	-	2#266*		
1134	361	760	40	19	2	≥ 7417	≥ 7417	2#267	[68]	
1135	761	1522	40	20	1	?	-			
1136	381	762	40	20	2	$\geq 3 \times 10^{16}$	-	2#268	[70]	
1137	191	382	40	20	4	?	-	2#269*		
1138	153	306	40	20	5	?	-			
1139	96	192	40	20	8	≥ 3	-	2#270		
1140	77	154	40	20	10	?	-	2#271*		
1141	41	82	40	20	19	≥ 1	-	D#1196	[137]	
1142	121	220	40	22	7	?	-			
1143	921	1535	40	24	1	?	-			
1144	231	385	40	24	4	?	-			
1145	93	155	40	24	10	?	-			
1146	65	104	40	25	15	?	-	R#1171*		
1147	1001	1540	40	26	1	?	-			

No	v	b	r	k	λ	Nd	Nr	Comments	Ref	Where?
1148	81	120	40	27	13	$\geq 10^{13}$	$\geq 10^{13}$	R#1170,AG$_3$(4,3)	[66]	
1149	217	310	40	28	5	?	-			
1150	91	130	40	28	12	?	-	R#1169*		
1151	1161	1548	40	30	1	?	-			
1152	291	388	40	30	4	?	-			
1153	117	156	40	30	10	?	-	R#1168*		
1154	156	195	40	32	8	?	-	R#1167*		
1155	221	260	40	34	6	?	-	R#1166*		
1156	273	312	40	35	5	?	-	R#1165*		
1157	351	390	40	36	4	?	-	R#1164*		
1158	481	520	40	37	3	?	0	R#1163*,×3		
1159	741	780	40	38	2	0	-	R#1162*,×2		
1160	1521	1560	40	39	1	?	?	R#1161*,AG(2,39)		
1161	1561	1561	40	40	1	?	-	PG(2,39)		
1162	781	781	40	40	2	0	-	×1		
1163	521	521	40	40	3	?	-			
1164	391	391	40	40	4	?	-			
1165	313	313	40	40	5	0	-	×1		
1166	261	261	40	40	6	0	-	×1		
1167	196	196	40	40	8	0	-	×1		
1168	157	157	40	40	10	0	-	×1		
1169	131	131	40	40	12	?	-			
1170	121	121	40	40	13	$\geq 10^{29}$	-	PG$_3$(4,3)	[66]	IV.12.22
1171	105	105	40	40	15	?	-			
1172	42	574	41	3	2	$\geq 6 \times 10^{24}$	≥ 1	D	[53, 92]	IV.10.65
1173	124	1271	41	4	1	≥ 2	≥ 1		[52, 16, 56]	III.2.7
1174	165	1353	41	5	1	≥ 1	≥ 1		[52, 3]	III.2.7
1175	42	287	41	6	5	≥ 2	≥ 2	D	[52, 1, 2]	IV.10.70
1176	42	246	41	7	6	≥ 1	≥ 1	D	[52, 1]	IV.10.71
1177	288	1476	41	8	1	≥ 1	≥ 1		[44]	I.6.31
1178	370	1517	41	10	1	?	?			
1179	247	779	41	13	2	?	?			
1180	42	123	41	14	13	≥ 1	?	D	[1]	
1181	247	533	41	19	3	?	?			
1182	780	1599	41	20	1	?	?			
1183	42	82	41	21	20	≥ 1	0	R#1196,×3	[137]	
1184	124	164	41	31	10	?	0	R#1195*,×3		
1185	165	205	41	33	8	?	0	R#1194*,×3		
1186	288	328	41	36	5	?	0	R#1193*,×3		
1187	370	410	41	37	4	?	0	R#1192*,×3		
1188	780	820	41	39	2	?	0	R#1191*,×3		
1189	1600	1640	41	40	1	?	?	R#1190*,AG(2,40)		
1190	1641	1641	41	41	1	?	-	PG(2,40)		
1191	821	821	41	41	2	?	-			
1192	411	411	41	41	4	?	-			
1193	329	329	41	41	5	?	-			
1194	206	206	41	41	8	0	-	×1		
1195	165	165	41	41	10	?	-			
1196	83	83	41	41	20	≥ 1	-		[137]	IV.12.22

1.4 See Also

§I.2	Existence results on designs with "small" block size.
§I.6	Resolvable designs.
§I.5	Symmetric designs.
§I.4	Steiner systems.
§III.2	PBD constructions make BIBDs.
[51]	A general introduction to combinatorics and in particular to design theory.

References

[1] R.J.R. Abel, Forty-three balanced incomplete block designs, *J. Combin. Theory A* 65 (1994), 252-267.

[2] R.J.R. Abel, On the Existence of Balanced Incomplete Block Designs and Transversal Designs, Ph.D. thesis, University of New South Wales, 1995.

[3] R.J.R. Abel, private communication.

[4] R.J.R. Abel and W.H. Mills, Some New BIBDs with $k = 6$ and $\lambda = 1$, *J. Combin. Des.* 3 (1995), 381-391.

[5] M. Aschbacher, On collineation groups of symmetric block designs, *J. Combin. Theory A* 11 (1971), 272-281.

[6] E.F. Assmus Jr., J.A. Mezzaroba and C.J. Salwach, Planes and biplanes, *Higher Combinatorics* (M. Aigner, ed.), Reidel, Dordrecht, (1977), 205-212.

[7] A. Baartmans, I.F. Blake and V.D. Tonchev, On the extendability of Steiner t-designs, *J. Combin. Des.* 1 (1993), 239-247.

[8] R.D. Baker, Resolvable BIBDs and SOLS, *Discrete Math.* 44 (1983), 13-29.

[9] H. Beker and F. Piper, Some designs which admit strong tactical decompositions, *J. Combin. Theory A* 22 (1977), 38-42.

[10] V.N. Bhat, Nonisomorphic solutions of some BIBDs, III, *J. Combin. Theory A* 12 (1972), 225-252.

[11] V.N. Bhat-Nayak, V.D. Kane, W.L. Kocay and R.G. Stanton, Settling some BIBD conjectures, *Ars Combin.* 16 (1983), 229-234.

[12] V.N. Bhat-Nayak and A. Wirmani-Prasad, Three new dicyclic solutions of (21,42,20, 10,9) designs, *J. Combin. Theory A* 40 (1985), 427-428.

[13] E. Billington, New cyclic (61,244,40,10,6) BIBDs, *Discrete Math.* 77 (1989), 51-53.

[14] D. Breach and A.R. Thompson, Reducible 2-(11,5,4) and 3-(12,6,4) designs, Mathematics Research Report 31/1983, University of Canterbury, Christchurch, N.Z.

[15] D. Breach, private communication.

[16] A.E. Brouwer, Some nonisomorphic BIBDs (4,1;v), Math. Centrum Amsterdam, ZW 102 (1977).

[17] A.E. Brouwer, Some unitals on 28 points and their embeddings in projective planes of order 9, *Geometries and Groups*, Proc. Colloq. Berlin 1981, Lecture Notes Math. 893, Springer-Verlag, New York, 1981, 183-188.

[18] A.E. Brouwer, private communication.

[19] A.E. Brouwer, An infinite series of symmetric designs, Math. Centrum Amsterdam Report, ZW 202/83.

[20] A.E. Brouwer and H.A. Wilbrink, A symmetric design with parameters 2-(49,16,5), *J. Combin. Theory A* 37 (1984), 193-194.

[21] F.C. Bussemaker, R. Mathon and J.J. Seidel, Tables of two-graphs, *Combinatorics and Graph Theory*, Proc. Conf. Calcutta 1980, Lecture Notes Math. 885, Springer-Verlag, New York 1981, 70-112.

[22] P.J. Cameron, Several 2-(46,6,3) designs, *Discrete Math.* 87 (1991), 89-90.

[23] W.H. Clatworthy and R.J. Lewyckyj, Comments on Takeuchi's table of difference sets generating balanced incomplete block designs, *Rev. Inst. Internat. Stat.* 36 (1968), 12-18.

[24] C.J. Colbourn, Transversal designs with block size eight and nine, *Europ. J. Combin.*, to appear.

[25] C.J. Colbourn and M.J. Colbourn, On cyclic Steiner systems S(2,6,91), *Abstracts Amer. Math. Soc.* 2 (1981), 463.

[26] C.J. Colbourn, M.J. Colbourn, J.J. Harms and A. Rosa, A complete census of (10,3,2) block designs and of Mendelsohn triple systems of order ten, III. (10,3,2) block designs without repeated blocks, *Congr. Numer.* 39 (1983), 211-234.

[27] C.J. Colbourn, S.S. Magliveras and R. Mathon, Transitive Steiner and Kirkman triple systems of order 27, *Math. Comput.* 58 (1992), 441-449+S23-S27.

[28] M.J. Colbourn, Cyclic block designs: Computational aspects of their construction and analysis, Computer Science, Univ. of Toronto, Technical Report 146/80.

[29] D.G. Corneil and R.A. Mathon, Algorithmic techniques for the generation and analysis of strongly regular graphs and other combinatorial configurations, *Ann. Discrete Math.* 2 (1978), 1-32.

[30] D.J. Curran and S.A. Vanstone, Doubly resolvable designs from generalized Bhaskar Rao designs, *Discrete Math.* 73 (1989), 49-63.

[31] I.J. Dejter, F. Franek and A. Rosa, A completion conjecture for Kirkman triple systems, *Utilitas Math.*, to appear.

[32] P. Dembowski, *Finite Geometries*, Springer-Verlag, New York, 1968.

[33] R.H.F. Denniston, Some maximal arcs in finite projective planes, *J. Combin. Theory* 6 (1969), 317-319.

[34] R.H.F. Denniston, On biplanes with 56 points, *Ars Combin.* 9 (1980), 167-179.

[35] R.H.F. Denniston, A Steiner system with a maximal arc, *Ars Combin.* 9 (1980), 247-248.

[36] R.H. Denniston, Enumeration of symmetric designs (25,9,3), *Ann. Discrete Math.* 15 (1982), 111-127.

[37] J.W. DiPaola, J.S. Wallis and W.D. Wallis, A list of (v, b, r, k, λ) designs for $r \leq 30$, *Congr. Numer.* 8 (1973), 249-258.

[38] F. Franek, R. Mathon, R.C. Mullin and A. Rosa, A census of dicyclic (v,4,2) designs for $v \leq 22$, *J. Combin. Math. Combin. Comput.* 8 (1990), 89-96.

[39] S. Furino, Y. Miao and J. Yin, *Frames and Resolvable Designs*, CRC Press, Boca Raton, to appear.

[40] B. Ganter, A. Gülzow, R. Mathon and A. Rosa, A complete census of (10,3,2)-block designs and of Mendelsohn triple systems of order ten, IV. (10,3,2) designs with repeated blocks, Math. Schriften Kassel 5/78.

[41] P. Gibbons, Computing techniques for the construction and analysis of block designs, Computer Science, University of Toronto, Technical Report 92/76.

[42] A. Golemac, Construction of new symmetric designs with parameters (70,24,8), *Discrete Math.* 120 (1993), 51-58.

[43] M. Greig, Some balanced incomplete block design constructions, *Congr. Numer.* 77 (1990), 121-134.

[44] M. Greig, Some group divisible design constructions, unpublished.

[45] M. Greig, Cut and paste constructions of BIBDs, unpublished.

[46] M. Greig, Designs from ovals and other configurations in projective planes, unpublished.

[47] H.-D.O.F. Gronau and J. Prestin, Some results on designs with repeated blocks, *Rostock. Math. Kolloq.* 20 (1982), 15-38.

[48] H. Gropp, The elementary abelian Steiner systems S(2,4,49), *Discrete Math.* 64 (1987), 87-90.
[49] W. Haemers, Eigenvalue Techniques in Design and Graph Theory (thesis), Tract 121, Math. Centrum Amsterdam 1980.
[50] M. Hall Jr., A survey of difference sets, *Proc. Amer. Math. Soc.* 7 (1956), 975-986.
[51] M. Hall Jr., *Combinatorial Theory*, Blaisdell, Waltham, Mass. 1967.
[52] H. Hanani, Balanced incomplete block designs and related designs, *Discrete Math.* 11 (1975), 255-369.
[53] H. Hanani, On resolvable balanced incomplete block designs, *J. Combin. Theory A* 17 (1974), 275-289.
[54] H. Hanani, private communication.
[55] H. Hanani, BIBDs with block-size seven, *Discrete Math.* 77 (1989), 89-96.
[56] H. Hanani, D.K. Ray-Chaudhuri and R.M. Wilson, On resolvable designs, *Discrete Math.* 3 (1972), 343-357.
[57] S.A. Hobart and W.G. Bridges, Remarks on (15,5,4) designs, in: *Coding Theory and Design Theory*, Part II, IMA Vol. Math. Appl. 21, Springer-Verlag, New York, 1990, 132-143.
[58] W.C. Huffman and V.D. Tonchev, The existence of extremal self-dual [50,25,10] codes and quasi-symmetric 2-(49,9,6) designs, preprint.
[59] N. Ito, J.S. Leon and J.Q. Longyear, Classification of 3-(24,12,5) designs and 24-dimensional Hadamard matrices, *J. Combin. Theory A* 31 (1981), 66-93.
[60] A.V. Ivanov, Constructive enumeration of incidence systems, *Ann. Discrete Math.* 26 (1985), 227-246.
[61] Z. Janko and Tran van Trung, The existence of a symmetric block design for (70,24,8), *Mitt. Math. Sem. Giessen* 165 (1984), 17-18.
[62] Z. Janko and Tran van Trung, Construction of two symmetric block designs for (71,21,6), *Discrete Math.* 55 (1985), 327-328.
[63] Z. Janko and Tran van Trung, A new biplane of order 9 with a small automorphism group, *J. Combin. Theory A* 42 (1986), 305-309.
[64] N.L. Johnson, The translation planes of order 16 that admit nonsolvable collineation groups, *Math. Zeitschr.* 185 (1984), 355-372.
[65] N.L. Johnson, private communication.
[66] D. Jungnickel, The number of designs with classical parameters grows exponentially, *Geom. Ded.* 16 (1984) 167-178.
[67] D. Jungnickel, Quasimultiples of biplanes and residual biplanes, *Ars Combin.* 19 (1985), 179-186.
[68] D. Jungnickel, Quasimultiples of projective and affine planes, *J. Geometry* 26 (1986), 172-181.
[69] D. Jungnickel, On the existence of small quasimultiples of affine and projective planes of arbitrary order, *Discrete Math.* 85 (1990), 177-189.
[70] D. Jungnickel and K. Vedder, Simple quasidoubles of projective planes, *Aequat. Math.* 34 (1987), 96-100.
[71] D. Jungnickel and V.D. Tonchev, On symmetric and quasi-symmetric designs with the symmetric difference property and their codes, *J. Combin. Theory A* 59 (1992), 40-50.
[72] S. Kageyama, A resolvable solution of BIBD(18,51,17,6,5), *Ars Combin.* 15 (1983), 315-316.
[73] S. Kageyama, A survey of resolvable solutions of balanced incomplete block designs, *Rev. Inst. Internat. Stat.* 40 (1972), 269-273.
[74] W. Kantor, private communication.
[75] W. Kocay and G.H.J. van Rees, Some non-isomorphic $(4t+4, 8t+6, 4t+3, 2t+2, 2t+1)$-BIBD's, *Discrete Math.* 92 (1991), 159-172.

[76] T. Kölmel, Symmetric (49,16,5) designs, *Ars Combin.* 38 (1994), 47-55.

[77] G.B. Khosrovshahi, A.Nowzari-Dalini and R.Torabi, Some simple and automorphism-free 2-(15,5,4) designs, *J. Combin. Des.* 3 (1995), 285-292.

[78] E.S. Kramer, S.S. Magliveras and R. Mathon, The Steiner systems S(2,4,25) with nontrivial automorphism group, *Discrete Math.* 77 (1989), 137-157.

[79] C.W.H. Lam, G. Kolesova and L. Thiel, A computer search for finite projective planes of order 9, *Discrete Math.* 92 (1991), 187-195.

[80] C.W.H. Lam, L. Thiel and S. Swiercz, The non-existence of finite projective planes of order 10, *Canad. J. Math.* 41 (1989), 1117-1123.

[81] C.W.H. Lam, L. Thiel and V.D. Tonchev, On quasi-symmetric 2-(28,12,11) and 2-(36,16,12) designs, *Des. Codes Crypt.* 5 (1995), 43-56.

[82] C.W.H. Lam and V.D. Tonchev, Classification of affine resolvable 2-(27,9,4) designs, preprint.

[83] C.C. Lindner and A. Rosa, On the existence of automorphism free Steiner triple systems, *J. Algebra* 34 (1975), 430-443.

[84] J.H. van Lint, V.D. Tonchev and I.N. Landgev, A new design, *Coding Theory and Design Theory, Part II, Design Theory*, Springer-Verlag, New York, 1990, 251-256.

[85] R. Mathon, Constructions for cyclic Steiner 2-designs, *Ann. Discrete Math.* 34 (1987), 353-362.

[86] R. Mathon, Nonisomorphic designs by simulated annealing, preprint.

[87] R. Mathon, On a new projective plane of order 16, preprint.

[88] R. Mathon and D. Lomas, A census of 2-(9,3,3) designs, *Australas. J. Combin.* 5 (1992), 145-158.

[89] R. Mathon, K.T. Phelps and A. Rosa, Small Steiner triple systems and their properties, *Ars Combin.* 15 (1983), 3-110.

[90] R. Mathon and C. Pietsch, On the family of 2-(7,3,λ) designs, preprint.

[91] R. Mathon and A. Rosa, A census of Mendelsohn triple systems of order nine, *Ars Combin.* 4 (1977), 309-315.

[92] R. Mathon and A. Rosa, Some results on the existence and enumeration of BIBDs, Math. Report 125-December-1985, McMaster University, 33 pp.

[93] R. Mathon and A. Rosa, On the (15,5,λ)-family of BIBDs, *Discrete Math.* 77 (1989), 205-216.

[94] R. Mathon and A. Rosa, Tables of parameters of BIBDs with $r \leq 41$ including existence, enumeration and resolvability results: An update, *Ars Combin.* 30 (1990), 65-96.

[95] R. Mathon and E. Spence, On symmetric 2-(45,12,3) designs, preprint.

[96] I. Matulić-Bedenić, K. Horvatić-Baldasar and E. Kramer, Construction of new symmetric designs with parameters (66,26,10), *J. Combin. Des.* 3 (1995), 405-410.

[97] R.B. McFeat and A. Neumaier, Some tuple systems constructions with applications to resolvable designs, *J. Combin. Theory A* 36 (1984), 92-100.

[98] B.D. McKay, private communication.

[99] B.D. McKay and R.G. Stanton, Isomorphism of two large designs, *Ars Combin.* 6 (1978), 87-90.

[100] N.S. Mendelsohn and S.H.Y. Hung, On the Steiner systems S(3,4,14) and S(4,5,15), *Utilitas Math.* 1 (1972), 5-95.

[101] W.H. Mills, A new block design, *Congr. Numer.* 14 (1975), 461-465.

[102] W.H. Mills, Two new block designs, *Utilitas Math.* 7 (1975), 73-75.

[103] W.H. Mills, The construction of balanced incomplete block designs with $\lambda = 1$, *Congr. Numer.* 20 (1977), 131-148.

[104] W.H. Mills, The construction of BIBDs using nonabelian groups, *Congr. Numer.* 21 (1978), 519-526.

[105] W.H. Mills, The construction of balanced incomplete block designs, *Congr. Numer.* 23 (1979), 73-86.

[106] H.K. Nandi, Enumeration of nonisomorphic solutions of balanced incomplete block designs, *Sankhya* 7 (1946), 305-312.

[107] C.W. Norman, Nonisomorphic Hadamard designs, *J. Combin. Theory A* 21 (1976), 336-344.

[108] S.C. Panandikar, On a family of resolvable BIB designs, *Calcutta Statist. Assoc. Bull.* 31 (1981), 123-128.

[109] E.T. Parker and Y.-C. Chen, Some designs, unpublished.

[110] T. Penttila, G.F. Royle and M.K. Simpson, Hyperovals in the known projective planes of order 16, *J. Combin. Des.*, to appear.

[111] C. Pietsch, Numbers of nonisomorphic block designs found with DESY, preprint.

[112] D.P. Rajkundlia, Some techniques for constructing infinite families of BIBDs, *Discrete Math.* 44 (1983), 64-96.

[113] G. Royle, 1000000 2-(12,3,2) designs, *Notices AMS* 10 (1989), 482.

[114] B.T. Rumov, The existence of some resolvable block designs with divisibility into groups, *Math. Notes* 19 (1976), 376-382.

[115] S.S. Sane, On a class of symmetric designs, in: *Proc. Sem. Combinatorics and Applications in Honour of Prof. S.S. Shrikhande on his 65th Birthday*, Indian Statistical Institute, December 1982, 292-302.

[116] E. Seiden, A method of construction of resolvable BIBD, *Sankhya (A)* 25 (1963), 393-394.

[117] S.S. Shrikhande, On a two-parameter family of balanced incomplete block designs, *Sankhya (A)* 24 (1962), 38-40.

[118] S. Sillito, An extension property of a class of balanced incomplete block designs, *Biometrika* 44 (1957), 278-279.

[119] E. Spence, A complete classification of symmetric (31,10,3) designs, *Des. Codes Crypt.* 2 (1992), 127-136.

[120] E. Spence, (40,13,4) designs derived from strongly regular graphs, in: *Advances in Finite Geometries and Designs* (J.W.P. Hirschfeld, D.R. Hughes, J.A. Thas; eds.), Oxford, 1991, 359-368.

[121] E. Spence, private communication.

[122] E. Spence, Regular 2-graphs on 36 vertices, preprint.

[123] E. Spence, Classification of Hadamard matrices of order 24 and 28, *Discrete Math.*, to appear.

[124] E. Spence, The complete classification of Steiner systems $S(2, 4, 25)$, preprint.

[125] R.G. Stanton and R.C. Mullin, Uniqueness theorems in balanced incomplete block designs, *J. Combin. Theory* 7 (1969), 37-48.

[126] D.R. Stinson and S.A. Vanstone, Some non-isomorphic Kirkman triple systems of order 39 and 51, *Utilitas Math.* 27 (1985), 199-205.

[127] K. Takeuchi, A table of difference sets generating balanced incomplete block designs, *Rev. Inst. Internat. Stat.* 30 (1962), 361-366.

[128] V.D. Tonchev, Hadamard matrices of order 28 with automorphism of order 13, *J. Combin. Theory A* 35 (1983), 43-57.

[129] V.D. Tonchev, Quasi-symmetric 2-(31,7,7) designs and a revision of Hamada's conjecture, *J. Combin. Theory A* 42 (1986), 104-110.

[130] V.D. Tonchev, Embedding of the Witt-Mathieu system S(3,6,22) in a symmetric 2-(78,22,6) design, *Geom. Ded.* 22 (1987), 49-75.

[131] V.D. Tonchev and R.V. Raev, Cyclic 2-(13,5,5) designs, *C.R. Acad. Bulg. Sci.* 35 (1982), 1205-1208.

[132] V.D. Tonchev and R.V. Raev, Cyclic 2-(17,8,7) designs and related doubly-even codes, *C.R. Acad. Bulg. Sci.* 35 (1982), 1367-1370.

[133] V.D. Tonchev and S.A. Vanstone, On Kirkman triple systems of order 33, *Discrete Math.* 106/107 (1992), 493-496.

[134] S. Topalova, Enumeration of 2-(21,5,2) designs with an automorphism of order 7, preprint.

[135] Tran van Trung, The existence of symmetric block designs with parameters (41,16,6) and (66,26,10), *J. Combin. Theory A* 33 (1982), 201-204.

[136] Tran van Trung, Maximal arcs and related designs, *J. Combin. Theory A* 57 (1991), 294-301.

[137] W.D. Wallis, A.P. Street and J.S. Wallis, *Combinatorics: Room Squares, Sum-Free Sets, Hadamard Matrices*, Lecture Notes Math. 292, Springer-Verlag, Berlin, 1972.

[138] M.A. Wertheimer, Designs in Quadrics, Ph.D. thesis, University of Pennsylvania, 1986.

[139] R.M. Wilson, Cyclotomy and difference families in elementary abelian groups, *J. Number Theory* 4 (1972), 17-47.

[140] R.M. Wilson, Nonisomorphic Steiner triple systems, *Math. Zeitschr.* 135 (1974), 303-313.

2 BIBDs with Small Block Size

R. Julian R. Abel
Malcolm Greig

2.1 Existence Results for BIBDs for Specific k and λ

2.1 **Remarks** The existence question for BIBDs with small k is: Given k and λ, for what values of v are Proposition I.1.1 (integrality conditions) and Theorem I.1.4 (Fisher's inequality) sufficient? This problem has been addressed for all λ when $k \leq 7$, and for a few values of λ when $k = 8$ or 9, but for no (k, λ) with $k \geq 10$. The notation $\mathrm{B}(k, \lambda)$ is used for the set of v values for which a (v, k, λ) BIBD is known. Theorem III.5.4 establishes that, for every fixed k and λ, there is a constant $C(k, \lambda)$, so that if $v > C(k, \lambda)$ and v satisfies $\lambda(v-1) \equiv 0 \pmod{k-1}$ and $\lambda v(v-1) \equiv 0 \pmod{k(k-1)}$, then there exists a BIBD(v, k, λ). Table 2.3 summarizes known results for all (k, λ) values where an attempt has been made to obtain a specific bound on $C(k, \lambda)$.

2.2 **Proposition** For any v, k, if $\lambda_{\min}$ is the minimum λ for which the integrality conditions are satisfied, then $\lambda_{\min}$ divides $k(k-1)$ (if k is odd) or $k(k-1)/2$ (if k is even).

2.3 **Table** A summary of known results.

k	λ	Possible Values of v	Minimum Possible $v > k$	Exceptions	Possible Exceptions	Largest Possible Exception
3	1	1,3 mod 6	7	none	none	
3	2	0,1 mod 3	4	none	none	
3	3	1 mod 2	5	none	none	
3	6	all	4	none	none	
4	1	1,4 mod 12	13	none	none	
4	2	1 mod 3	7	none	none	

k	λ	Possible Values of v	Minimum Possible $v > k$	Exceptions	Possible Exceptions	Largest Possible Exception
4	3	0,1 mod 4	5	none	none	
4	6	all	5	none	none	
5	1	1,5 mod 20	21	none	none	
5	2	1,5 mod 10	11	15	none	
5	4	0,1 mod 5	6	none	none	
5	5	1 mod 4	9	none	none	
5	10	1 mod 2	7	none	none	
5	20	all	6	none	none	
6	1	1,6 mod 15	31	36	Table 2.5	2031
6	2	1,6 mod 15	16	21	none	
6	3	1 mod 5	11	none	none	
6	5	0,1 mod 3	7	none	none	
6	15	all	7	none	none	
7	1	1,7 mod 42	43	43	Tables 2.6, 2.7	4387
7	2	1,7 mod 21	22	22	Table 2.8	994
8	1	1,8 mod 56	57	none	Tables 2.9, 2.10	3753
8	4	1 mod 7	15	none	Table 2.14	365
8	7	0,1 mod 8	9	none	none	
9	1	1,9 mod 72	73	none	Tables 2.11, 2.12	16497
9	8	0,1 mod 9	10	none	Table 2.15	1278

BIBDs with parameters $(15, 5, \lambda)$ and $(21, 6, \lambda)$ are known for all even $\lambda \geq 4$.

2.4 Theorem If $k = 7$ and $\lambda > 2$, then a $(v, 7, \lambda)$ BIBD exists for all v satisfying the necessary conditions except possibly when $v = 253$ and $\lambda = 4$ or 5.

2.5 Table Values of v for which existence of a $(v, 6, 1)$ BIBD remains undecided.

46	51	61	81	141	166	196	201	226	231	256	261
276	286	291	316	321	346	351	376	406	411	436	441
471	496	501	526	561	591	616	646	651	676	706	741
766	771	796	801	916	946	1221	1246	1251	1396	1456	1486
1521	1611	1671	1851	2031							

2.6 Table Values of t for which existence of a $(42t + 1, 7, 1)$ BIBD remains undecided.

2	3	5	6	12	14	17	19	22	27	29	33	37	39
42	43	47	59	61	62	66	67	83					

2.7 Table Values of t for which existence of a $(42t + 7, 7, 1)$ BIBD remains undecided.

3 4 6 19 29 34 39 44

2.8 Table Values of v for which existence of a $(v, 7, 2)$ BIBD remains undecided.

106 253 274 358 574 694 988 994

2.9 Table Values of t for which existence of a $(56t + 1, 8, 1)$ BIBD remains undecided.

2	3	4	5	6	7	14	19	20	21	22	24	25	26
27	28	31	32	34	35	39	40	46	52	59	61	62	67

2.10 Table Values of t for which existence of a $(56t + 8, 8, 1)$ BIBD remains undecided.

3 11 13 20 22 23 25 26 27 28

2.11 Table Values of t for which existence of a $(72t+1, 9, 1)$ BIBD remains undecided.

2	3	4	5	7	8	11	12	15	20	21	22	24	27
31	32	34	37	38	40	42	43	45	47	50	52	53	56
60	61	62	67	68	72	75	76	84	92	93	94	96	97
102	107	128	132	174	191	194	196	197	201	204	209	212	219

2.12 Table Values of t for which existence of a $(72t+9, 9, 1)$ BIBD remains undecided.

2	3	4	5	12	13	14	18	20	22	23	25	26	27
28	31	33	34	38	40	41	43	46	47	48	52	59	60
61	62	67	68	76	85	93	94	102	103	139	148	174	183
192	202	203	209	229									

2.13 Remark Greig (unpublished) has established that the necessary conditions for $(v,8,4)$ and $(v,9,8)$ BIBDs are sufficient for all but a small number of v values.

2.14 Table Values of v for which existence of a $(v, 8, 4)$ BIBD remains undecided.

22	85	99	141	148	155	274	323	358	365

2.15 Table Values of v for which existence of a $(v, 9, 8)$ BIBD remains undecided.

54	108	126	135	136	198	216	315	378	1278

2.16 Remark The construction for a (153,9,1) BIBD in [9] contains an error which apparently cannot be corrected. This invalidates recursive constructions which make use of this design, for instance those in [6, 7].

2.17 Remark There exists a k-GDD of type $k^{t(k-1)+1}$ (or equivalently, a $(k(k-1)t+k, k, 1)$ BIBD containing a parallel class) if any one of the following conditions is satisfied:

1. $k \leq 5$.
2. $k = 6$ and $t \neq 1, 2, 7, 9, 11, 39, 50$, or 51.
3. $k = 7$, t is not in Table 2.7, and $t \neq 24$.
4. $k = 8$ and t is not in Table 2.10.
5. $k = 9$, t is not in Table 2.12, and $t \neq 16, 92, 104, 147, 166, 187, 191$, or 205.

2.18 Remark k-GDDs of type $(k(k-1))^t$ can sometimes be useful for BIBD constructions. For $k = 3, 4$, and 5, k-GDDs of type $(k(k-1))^t$ exist for all $t \geq k$. For $k = 6$, not much is known, as most recursive constructions require the existence of k-GDDs of types $(k(k-1))^x$ and $(k(k-1))^{x+1}$ for some small value of x; at present, none is known for $k = 6$. Tables 2.19 – 2.22 summarize known results for $k = 6, 7, 8, 9$ and $t < 100$.

2.19 Table Values of $t < 100$ for which existence of a 6-GDD of type 30^t is known.

6	16	21	26	31	36	41	51	61	66	71	76	78	81	86	90	91	96

2.20 Table Values of $t < 100$ for which existence of a 7-GDD of type 42^t is known.

8	9	15	17	28	33	36	41	49	50	56	57	63	64	65	70	71	72
73	77	78	80	81	85	88	91	92	96	97	99						

2.21 Table Values of $t < 100$ for which existence of a 8-GDD of type 56^t is known.

8	9	10	15	16	17	29	30	33	36	37	41	43	44	49	50	51	53
54	55	57	58	63	64	65	71	72	73	74	79	80	81	82	83	84	85
86	87	88	89	90	91	92	93	94	95	96	97	98	99				

2.22 Table Values of $t < 100$ for which existence of a 9-GDD of type 72^t is known.

9	10	17	19	49	54	55	57	58	64	65	66	73	74	80	81	89	90	91

2.2 Recursive Constructions

2.23 Theorem A (v, k, λ) BIBD exists if any one of the following conditions holds:

1. $v = w + r/\lambda$, where there exists a $(r/\lambda, k, \lambda)$ BIBD and a resolvable $(w, k-1, \lambda)$ BIBD whose blocks form r parallel classes.
2. $v = (w-1)m + 1$, $\lambda = \lambda_1\lambda_2$, where a (k, λ_1)-GDD of type $(k-1)^{(w-1)/(k-1)}$, a $((k-1)m+1, k, \lambda_1\lambda_2)$ BIBD, and $\mathrm{TD}_{\lambda_2}(k, m)$ exist.
3. $v = v_1(v_2 - \alpha) + \alpha$, $\lambda = \lambda_1\lambda_2$ where α is 0 or 1, a $\mathrm{TD}_{\lambda_2}(k, v_2 - \alpha)$ exists, and (v_1, k, λ_1) and (v_2, k, λ) BIBDs both exist.

2.24 Remark Most recursive BIBD constructions start with a GDD (usually a truncated transversal design), use Wilson's Fundamental GDD construction (Theorem III.5.7) to obtain a larger GDD, and finally fill in the groups of the larger GDD. Some examples follow.

2.25 Example Suppose TD$(10, t)$ exists and $0 \le u \le t$. Then a $\{9, 10\}$-GDD of type t^9u^1 exists, and since 9-GDD of type 8^x exists for $x = 9, 10$, a 9-GDD of type $(8t)^9(8u)^1$ also exists. Therefore, if $\alpha = 1$ or 9 and $\{8t+\alpha, 8u+\alpha\} \subset \mathrm{B}(9, 1)$, then $8(9t+u)+\alpha \in \mathrm{B}(9, 1)$.

2.26 Example Suppose TD$(9, t)$ exists and $0 \le u \le t$. Then a $\{8, 9\}$-GDD of type t^8u^1 exists, and since a 7-GDD of type 42^x exists for $x = 8, 9$, a 7-GDD of type $(42t)^8(42u)^1$ also exists. Therefore, if $\alpha = 1$ or 7 and $\{42t + \alpha, 42u + \alpha\} \subset \mathrm{B}(7, 1)$, then $42(8t + u) + \alpha \in \mathrm{B}(7, 1)$.

2.27 Example Suppose TD$(8, t)$ exists and $0 \le u \le t$. Then a $\{7, 8\}$-GDD of type t^7u^1 exists, and since a $(7, 3)$-GDD of type 2^x exists for $x = 7, 8$, a $(7, 3)$-GDD of type $(2t)^7(2u)^1$ also exists. Therefore, if $\{2t + 1, 2u + 1\} \subset \mathrm{B}(7, 3)$, then $2(7t + u) + 1 \in \mathrm{B}(7, 3)$.

2.28 Example Suppose TD$(9, t)$ exists and $0 \le u, w \le t$. Then a $\{7, 8, 9\}$-GDD of type $t^7u^1w^1$ exists, and since $(7, 21)$-GDDs of type 2^x exist for $x = 7, 8, 9$, a $(7, 21)$-GDD of type $(2t)^7(2u)^1(2w)^1$ also exists. Therefore, if $\{2t + 1, 2u + 1, 2w + 1\} \subset \mathrm{B}(7, 21)$, then $2(7t + u + w) + 1 \in \mathrm{B}(7, 21)$.

2.29 Example Suppose there exists a resolvable $(w, k, 1)$ BIBD whose blocks form r parallel classes. Then if $x \le r$, adding each of x infinite points to a different parallel class gives a $\{k, k+1\}$-GDD of type $1^w x^1$. If $k, k-1$ are prime powers, then k-GDDs of type $(k-1)^t$ exist for $t = k, k+1$; hence a k-GDD of type $(k-1)^w((k-1)x)^1$ exists. Therefore, if $(k-1)x + 1 \in \mathrm{B}(k, 1)$, then $(k-1)(w + x) + 1 \in \mathrm{B}(k, 1)$.

2.30 Remark Theorem 2.31, the singular indirect product or incomplete TD construction, has been useful for BIBDs with $k \ge 6$ and $\lambda = 1$.

2.31 Theorem Suppose $v = k(w-f) + (f + (k-1)\cdot a)$ where a TD$(k, w-f+a)$–TD(k, a), a $(w, \{k\} \cup f^*, 1)$ PBD and a $(f + (k-1)a, k, 1)$ BIBD all exist. Then $v \in \mathrm{B}(k, 1)$. If, in addition, a k-GDD of type $k^{(f+(k-1)a)/k}$ and TD$(k, w - f + a)$–TD(k, a) – $(w - f)\cdot$TD$(k, 1)$ exist, then so does a k-GDD of type $k^{v/k}$.

2.32 Examples

1. Since a resolvable (65,5,1) BIBD exists, there is a PBD$(81, \{6\} \cup 16^*)$. A $(76, 6, 1)$ BIBD and a TD$(6, 77)$–TD$(6, 12)$ also exist; hence there exists a $(v, 6, 1)$ BIBD for $v = 466 = 6(81 - 16) + (76 = 16 + 5 \cdot 12)$.

2. Since a resolvable (28, 4, 1) BIBD exists, a PBD(37, $\{5\} \cup 9^*$) also exists. A 5-GDD of type 5^5 and a TD(5, 32)−TD(5,4)−28·TD(5, 1) also exist; this gives a 5-GDD of type $5^{v/5}$ for $v = 165 = 5(37 - 9) + (25 = 9 + 4 \cdot 4)$. Together with results in [4], this implies that 5-GDDs of type 5^x exist for all $x \equiv 1 \bmod 4$.

2.3 Miscellaneous Constructions

2.33 Example (Hanani) A 5-GDD of type 20^7, and a (141,5,1) design. Take $V = \mathbb{Z}_4 \times \mathbb{Z}_5 \times \mathbb{Z}_7$. Base blocks are $\{(A, 0, 0), (B, 2, y), (C, 2, 6y), (D, 3, 3y), (E, 3, 4y)\}$ all taken mod $(-, 5, 7)$ for $y = 1, 2, 4$ and eight values of (A, B, C, D, E): $(1, 0, 1, 0, 0)$, $(0, 2, 3, 0, 2)$, $(2, 1, 2, 1, 1)$, $(1, 3, 0, 1, 3)$, $(3, 1, 0, 2, 2)$, $(0, 3, 2, 2, 0)$, $(2, 2, 1, 3, 3)$ and $(3, 0, 3, 3, 1)$. This is a 5-GDD of type 20^7 on V. To obtain a (141,5,1) design on $V \cup \{\infty\}$, construct a (21,5,1) BIBD on $\mathbb{Z}_4 \times \mathbb{Z}_5 \times \{x\} \cup \{\infty\}$ for all x in $\mathbb{Z}_7$.

2.34 Example (Denniston) A (66,6,1) design. Let $V = (\mathbb{Z}_{13} \times \{0, 1, 2, 3, 4\}) \cup \{\infty\}$, and write $(a, x) \in V$ as a_x. Then develop the following eleven base blocks under $\mathbb{Z}_{13}$:

$$\begin{array}{cccccc}
2_0 & 5_0 & 4_1 & 9_1 & 0_3 & 6_3 \\
6_1 & 2_1 & 12_2 & 1_2 & 0_3 & 5_3 \\
5_2 & 6_2 & 10_0 & 3_0 & 0_3 & 2_3 \\
1_0 & 2_0 & 6_0 & 12_2 & 5_4 & 8_4 \\
3_1 & 6_1 & 5_1 & 10_0 & 2_4 & 11_4 \\
9_2 & 5_2 & 2_2 & 4_1 & 6_4 & 7_4
\end{array}
\qquad
\begin{array}{cccccc}
7_0 & 9_0 & 10_1 & 1_2 & 3_3 & 4_4 \\
8_1 & 1_1 & 4_2 & 3_0 & 9_3 & 12_4 \\
11_2 & 3_2 & 12_0 & 9_1 & 1_3 & 10_4 \\
2_3 & 6_3 & 5_3 & 4_4 & 12_4 & 10_4 \\
\infty & 0_0 & 0_1 & 0_2 & 0_3 & 0_4
\end{array}$$

2.35 Example (Mills) A (76,6,1) design. Let $V = \mathbb{Z}_{19} \times \{0, 1, 2, 3\}$, and write $(a, x) \in V$ as a_x. Then develop the following base blocks under $\mathbb{Z}_{19}$:

$$\begin{array}{cccccc}
0_0 & 0_1 & 1_1 & 3_1 & 14_1 & 10_3 \\
0_0 & 0_2 & 7_2 & 2_2 & 3_2 & 13_1 \\
0_0 & 0_3 & 11_3 & 14_3 & 2_3 & 15_2 \\
1_0 & 3_0 & 0_1 & 7_1 & 0_2 & 2_3 \\
7_0 & 2_0 & 0_2 & 11_2 & 0_3 & 14_1
\end{array}
\qquad
\begin{array}{cccccc}
11_0 & 14_0 & 0_3 & 1_3 & 0_1 & 3_2 \\
1_0 & 7_0 & 11_0 & 3_1 & 2_2 & 14_3 \\
3_0 & 2_0 & 14_0 & 12_1 & 8_2 & 18_3 \\
7_1 & 16_1 & 11_2 & 17_2 & 1_3 & 5_3 \\
13_1 & 17_1 & 15_2 & 5_2 & 10_3 & 16_3
\end{array}$$

2.36 Example (Mills) A (96,6,1) design over $\mathbb{Z}_{48} \times \{0, 1\}$. Develop the following base blocks using $\mathbb{Z}_{48}$:

$$\begin{array}{cccccc}
0_0 & 8_0 & 16_0 & 24_0 & 32_0 & 40_0 \\
0_1 & 8_1 & 16_1 & 24_1 & 32_1 & 40_1 \\
0_0 & 1_0 & 3_0 & 13_0 & 28_0 & 0_1 \\
0_0 & 4_0 & 11_0 & 17_1 & 36_1 & 38_1
\end{array}
\qquad
\begin{array}{cccccc}
0_0 & 5_0 & 19_0 & 1_1 & 24_1 & 42_1 \\
0_0 & 9_0 & 26_0 & 4_1 & 7_1 & 40_1 \\
0_0 & 6_0 & 8_1 & 9_1 & 18_1 & 22_1 \\
0_0 & 18_0 & 11_1 & 28_1 & 33_1 & 39_1
\end{array}$$

2.37 Example (Mills) A (106,6,1) design over $\mathbb{Z}_{53} \times \{0, 1\}$. Develop the following base blocks using $\mathbb{Z}_{53}$:

$$\begin{array}{cccccc}
0_0 & 1_0 & 3_0 & 11_0 & 38_0 & 0_1 \\
0_0 & 5_0 & 19_0 & 25_0 & 36_1 & 39_1 \\
0_0 & 7_0 & 29_0 & 8_1 & 16_1 & 48_1 \\
0_0 & 9_0 & 21_0 & 12_1 & 13_1 & 27_1
\end{array}
\qquad
\begin{array}{cccccc}
0_0 & 13_0 & 30_0 & 23_1 & 35_1 & 51_1 \\
0_0 & 4_0 & 28_1 & 30_1 & 37_1 & 47_1 \\
0_0 & 2_1 & 7_1 & 25_1 & 29_1 & 49_1
\end{array}$$

2.38 Example (Mills) A (111,6,1) design over the elements of the nonabelian group Γ of order 111 generated by a (order 37) and b (order 3), for which $bab^{-1} = a^{10}$. Multiply the blocks $\{1, a, a^3, a^7, a^{17}, b\}$ and $\{1, a^5, a^{19}b, a^{28}b, a^{10}b^2, a^{30}b^2\}$ by every element of Γ to produce 111 blocks. Add blocks $\{a^ix, a^iy, a^ibx, a^iby, a^ib^2x, a^ib^2y\}$ for $0 \le i < 37$, and $\{x, y\} \in \{(a^5, a^{33}), (a^9, a^{27}), (a^{10}, a^{23}), (a^{13}, a^{24}), (a^{26}, a^{34}\})$.

2.39 Example (Mills) A (136,6,1) design over ∞ plus the elements of a nonabelian group Γ of order 135. This group is generated by an element a of order 45 and an element b of order 3, for which $bab^{-1} = a^{16}$. Multiply the blocks

$$\{1, a^3, a^{15}, a^{35}, a^6b^2, a^{10}b^2\}, \{1, a^{22}, a^{11}b, a^{30}b, ab^2, a^{18}b^2\},$$
$$\{1, a^5, a^{18}b, a^{41}b, a^{13}b^2, a^{42}b^2\}, \{1, a^{11}, a^{17}, a^4b^2, a^5b^2, a^{28}b^2\}$$

by every element of Γ to produce 135 blocks. For $0 \le i \le 8$, $0 \le j \le 2$, add blocks $\{a^ib^j, a^{i+9}b^j, a^{i+18}b^j, a^{i+27}b^j, a^{i+36}b^j, \infty\}$; and $\{a^i, a^{i+1}, a^ib, a^{i+16}b, a^ib^2, a^{i+31}b^2\}$ for $0 \le i < 45$.

2.40 Example (Mills) A (171,6,1) design over the elements of the nonabelian group Γ of order 171 generated by an element a of order 57 and an element b of order 3, for which $bab^{-1} = a^7$. Multiply the blocks

$$\{1, a^{19}, a^{39}, a^{41}, a^{14}b, a^{38}b^2\}, \{1, a^{21}, a^{44}, a^{48}, a^{26}b, a^{11}b^2\}, \{1, a, a^{43}, a^8b^2, a^{15}b^2, a^{44}b^2\},$$
$$\{1, a^3, a^{31}, a^{23}b, a^{43}b, a^{36}b^2\}, \{1, a^{40}, a^{50}, a^{11}b, a^{25}b^2, a^{34}b^2\}$$

by every element of Γ to produce 171 blocks. Finally, for $\{x, y\} \in \{(1, a^{12}), (a^{37}, a^{42})\}$, add blocks $\{a^ix, a^iy, a^ibx, a^iby, a^ib^2x, a^ib^2y\}$ for $0 \le i < 57$.

2.41 Example (Abel) A (21,6,4) design over the elements of the nonabelian group Γ of order 21 generated by an element a of order 7 and an element b of order 3, for which $bab^{-1} = a^2$. Multiply $\{1, a^5, a^3b, a^3b^2, a^5b^2, a^6b^2\}$ and $\{1, a, a^2, a^4, a^3b, a^6b^2\}$ by every element of Γ to produce 21 blocks. Then, for each of $(x, y) \in \{(1, a), (1, a^3)\}$, add blocks $\{a^ix, a^iy, a^ibx, a^iby, a^ib^2x, a^ib^2y\}$ for $0 \le i < 7$.

2.42 Example (Abel) A (64,7,2) design over seven infinite points and the elements of a nonabelian group Γ of order 57, generated by a (order 19) and b (order 3), with $bab^{-1} = a^7$. Multiply $\{1, a^{10}, a^{12}, a^{14}, a^{18}, b, a^2b\}$ by Γ to produce 57 blocks. Next, for $\{x, y\} \in \{\{a, a^7\}, \{a, a^8\}, \{a^9, a^{10}\}, \{a^9, a^{14}\}, \{a^{12}, a^{15}\}, \{a^2, a^{13}\}, \{a^4, a^{14}\}\}$, add one of the infinite points to blocks $\{a^ix, a^iy, a^ibx, a^iby, a^ib^2x, a^ib^2y\}$ for $0 \le i < 19$. Finally form two (identical) blocks on the infinite points.

2.4 See Also

§I.5	See §I.5.2 for known nonexistence results for symmetric and quasi-residual BIBDs.
§IV.10	Difference family constructions provide most of the small ingredients needed for recursive constructions.
§IV.6	Resolvable designs are used in some recursive constructions for non-resolvable designs.
§VI.7	Several BIBDs come from special structures in affine and projective geometries.
[5]	An extensive textbook presentation of direct and recursive construction for BIBDs.
[10]	A major paper, proving existence results for $k \le 5$, for $k = 6$ with $\lambda \ne 1$, and for $k = 7$ with $\lambda \in \{6, 7, 42\}$.

References

[1] R.J.R. Abel, On the existence of BIBDs with $\lambda = 1$ and $6 \le k \le 10$, *J. Combin. Des.*, to appear.

[2] R.J.R. Abel and M. Greig, Balanced incomplete block designs with a block size of 7, preprint.

[3] R.J.R. Abel and W.H. Mills, On the existence of BIBDs with $k = 6$, and $\lambda = 1$, *J. Combin. Des.* 3 (1995), 381-391.
[4] A.M. Assaf, Modified group divisible designs, *J. Combin. Theory A* 68 (1994), 152-168.
[5] T. Beth, D. Jungnickel and H. Lenz, *Design Theory*, Cambridge University Press, London, 1986.
[6] B. Du, A few more BIBDs with $k = 8$ or 9, *Australas. J. Combin.* 10 (1994), 245-249.
[7] B. Du and L. Zhu, On the existence of $(v, 8, 1)$ BIBD, in *Combinatorial Designs and Applications*, Marcel Dekker, New York, 1990, 33-57.
[8] M. Greig, Recursive constructions of balanced incomplete block designs with block size of 7, 8 or 9, unpublished.
[9] M. Hall Jr., *Combinatorial Theory*, John Wiley and Sons, New York, 1986.
[10] H. Hanani, Balanced incomplete block designs and related designs, *Discrete Math.* 11 (1975), 255-369.
[11] H. Hanani, BIBDs with block size 7, *Discrete Math.* 77 (1989), 89-96.
[12] R. M. Wilson, An existence theory for pairwise balanced designs, III. Proof of the existence conjectures, *J. Combin. Theory A* 16 (1975), 71-79.
[13] S. Wu and J. Yin, Four new BIBDs with block size seven, *Discrete Math.* 113 (1993), 281-284.
[14] S. Wu and L. Zhu, Existence of BIBDs with block size 7, *Utilitas Math.* 38 (1990), 3-13.
[15] J. Yin, Existence of BIBDs with block size 8 and $\lambda = 7$, *J. Combin. Math. Combin. Comput.* 16 (1994), 193-198.

3 t-Designs, $t \geq 3$

Donald L. Kreher

3.1 Definitions

A t-(v, k, λ) *design* is a pair $(X, \mathcal{B})$ where X is a v-element set of *points* and $\mathcal{B}$ is a collection of k-element subsets of X (*blocks*) with the property that every t-element subset of X is contained in exactly λ blocks.

A design is *simple* if no two blocks are identical.

A t-(v, k, λ) design is also denoted by $S_\lambda(t, k, v)$. If $\lambda = 1$, then it can be omitted.

3.1 **Example** Let $X = \{0, 1, 2, 3, 4, 5, 6, 7, 8, 9, a, b\}$ and choose the 30 base blocks, shown with braces and commas omitted:

045678 023456 012346 023567 014567 014568 012568 023578 024569 014569
024579 02346a 01245a 02347a 02457a 03468b 02345b 01245b 01246b 02356b
01458b 01248b 04567b 02348b 02458b 04569b 02459b 03469b 0456ab 0235ab

Applying the automorphism $(0\ 1\ 2\ 3\ 4\ 5\ 6\ 7\ 8\ 9\ a)(b)$ yields 330 blocks which form a 4-$(12, 6, 10)$ design.

3.2 **Remark** A 2-(v, k, λ) design is a balanced incomplete block design. See §I.1.

3.2 Properties and Theorems

3.3 **Theorem** If $(X, \mathcal{B})$ is a t-(v, k, λ) design and S is any s-element subset of X, with $0 \leq s \leq t$, then the number of blocks containing S is

$$b_s = |\{B \in \mathcal{B} : S \subseteq B\}| = \lambda\binom{v-s}{t-s} \Big/ \binom{k-s}{t-s} .$$

In particular b_s is an integer and $(X, \mathcal{B})$ is a s-(v, k, b_s) design, for all $0 \leq s \leq t$.

Given integers $0 < t < k < v$, the smallest positive integer λ such that the value b_s in Theorem 3.3 is an integer for all $0 \leq s \leq t$ is the *minimal* λ and is denoted by λ_{min}. It follows that if a t-(v, k, λ) design exists, then λ_{min} divides λ.

3.4 **Remark** When $k = t + 1$, $\lambda_{min} = \gcd(v - t, \mathrm{lcm}(1, 2, \ldots, t + 1))$.

Given integers $0 < t < k < v$, if X is a v-element set and $\mathcal{B}$ is the collection of all k-element subsets of X, then $\mathcal{B}$ is a t-(v, k, λ_{max}) design, where $\lambda_{max} = \binom{v-t}{k-t}$. The value λ_{max} is the *maximum* λ and $(X, \mathcal{B})$ is the *complete design* or the *full design.*

Given a collection $\mathcal{B}$ of k-element subsets of a v-element set X and a point $x \in X$, five operations are defined:

1. $\textsc{comp}(\mathcal{B}) = \binom{X}{k} \setminus \mathcal{B}$, the *complement* of $\mathcal{B}$;
2. $\textsc{supp}(\mathcal{B}) = \{X \setminus B : B \in \mathcal{B}\}$, the *supplement* of $\mathcal{B}$, and
3. $\textsc{app}_\infty(\mathcal{B}) = \{B \cup \{\infty\} : B \in \mathcal{B}\}$, where ∞ is a new point not in X.
4. $\textsc{der}_x(\mathcal{B}) = \{B \setminus \{x\} : x \in B \in \mathcal{B}\}$, the *derivation of $\mathcal{B}$ with respect to* x, and
5. $\textsc{res}_x(\mathcal{B}) = \{B \in \mathcal{B} : x \notin B\}$, the *residual of $\mathcal{B}$ with respect to* x.

3.5 **Theorem** If $(X, \mathcal{B})$ is a simple t-(v, k, λ) design, then $(X, \textsc{comp}(\mathcal{B}))$ is a simple t-$(v, k, \lambda_{max} - \lambda)$ design.

3.6 **Theorem** If $(X, \mathcal{B})$ is a t-(v, k, λ) design and S is any s-element subset of X with $0 \leq s \leq t$, then the number of blocks that do not contain any point of S is

$$b(s) = |\{B \in \mathcal{B} : B \cap S = \emptyset\}| = \lambda\binom{v-s}{k} \Big/ \binom{v-t}{k-t} .$$

In particular $(X, \textsc{supp}(\mathcal{B}))$ is a t-$(v, v - k, b(t))$ design.

3.7 **Remark** In view of Theorems 3.5 and 3.6, one usually only considers t-(v, k, λ) designs with $t < k \leq v/2$ and simple designs only for $0 < \lambda \leq \lambda_{max}/2$.

3.8 **Theorem** If there exists a t-(v, k, λ) design $(X, \mathcal{B})$ and $x \in X$ is any point, then $(X \setminus \{x\}, \textsc{der}_x(\mathcal{B}))$ is a $(t-1)$-$(v-1, k-1, \lambda)$ design.

3.9 **Theorem** If there exists a t-(v, k, λ) design $(X, \mathcal{B})$ and $x \in X$ is any point, then $(X \setminus \{x\}, \textsc{res}_x(\mathcal{B}))$ is a $(t-1)$-$(v-1, k, \frac{\lambda(v-k)}{(k-t+1)})$ design.

3.10 **Remark** If $(X \cup \{\infty\}, \mathcal{B})$ is a t-(v, k, λ) design, then $\mathcal{B} = \textsc{app}_\infty(\textsc{der}_\infty(\mathcal{B})) \cup \textsc{res}_\infty(\mathcal{B})$. Thus t-designs may be broken down into their derived and residual designs.

A $(t+1)$-$(v+1, k+1, \lambda)$ design $(Y, \mathcal{A})$ is an *extension* of a t-(v, k, λ) design $(X, \mathcal{B})$ if $Y = X \cup \{\infty\}$ and $\mathcal{B} = \textsc{der}_\infty(\mathcal{A})$ where ∞ is a new point not in X.

3.11 **Theorem** (Alltop [3]) Let $(X, \mathcal{B})$ be a t-$(2k+1, k, \lambda)$ design.

1. If t is even, then $(X \cup \{\infty\}, \textsc{app}_\infty(\mathcal{B}) \cup \textsc{supp}(\mathcal{B}))$ is a $(t+1)$-$(2k+2, k+1, \lambda)$ design.
2. If t is odd and $\lambda = \lambda_{max}/2$, then $(X \cup \{\infty\}, \textsc{app}_\infty(\mathcal{B}) \cup \textsc{comp}(\textsc{supp}(\mathcal{B})))$ is a $(t+1)$-$(2k+2, k+1, \lambda)$ design.

3.12 Theorem If there exists a t-(v,k,λ_1) design $(X,\mathcal{B}_1)$ and a t-$(v,k+1,\lambda_2)$ design $(X,\mathcal{B}_2)$ such that $\lambda_1(v-t+1)/(k-t+1)=\lambda_1+\lambda_2$, then $(X\cup\{\infty\}, \text{APP}_\infty(\mathcal{B}_1)\cup\mathcal{B}_2)$ is a t-$(v+1,k+1,\lambda_1+\lambda_2)$ design.

3.13 Theorem (Tran van Trung [67]) If there exists a t-$(2k+1,k,\lambda)$ design, then a t-$(2k+2,k+1,\frac{\lambda(2k+2-t)}{(k+1-t)})$ design exists.

3.14 Theorem (Ganter, Pelikan and Teirlinck [25]) If there exists a simple t-(v,k,λ_1) design $(X,\mathcal{A})$ and a simple t-(v,k,λ_2) design $(X,\mathcal{B})$ such that $|\mathcal{B}|<\lambda_{max}/\lambda_1$, then $\mathcal{B}$ can be chosen to be disjoint from $\mathcal{A}$, and thus a simple t-$(v,k,\lambda_1+\lambda_2)$ design exists.

3.15 Example Let $X=\{0,1,2,3,4,5,6,7,8,9,a,b,c\}$ and choose the 66 base blocks:

013459 014567 012457 012346 012345 012367 023567 034567 024567 012458 023468
012468 012368 012568 012378 014568 034568 013478 023578 024578 012678 024569
012469 023459 014569 014579 012579 012479 023579 013679 012589 024589 034689
013469 035678 035679 01247a 035789 034789 036789 02357a 01467a 014789 01347a
02368a 02358a 02567a 01458a 01368a 024678 023478 023479 023469 025679 024679
025689 023689 023589 012356 013467 013458 013579 015678 013578 013569 013489

Applying the automorphism $\alpha=(0\ 1\ 2\ 3\ 4\ 5\ 6\ 7\ 8\ 9\ a\ b\ c)$ yields 858 blocks which form a 5-(13, 6, 4) design. Theorem 3.11 extends this design to a 6-(14, 7, 4) design. This is the smallest possible 6-design [43].

3.16 Remark A t-$(v,k,\lambda_{max}/2)$ design is equivalent to a halving of the complete design (see §3.4) so applying Theorem 3.41, Theorem 3.17 is obtained.

3.17 Theorem If there exists a t-$(v,t+1,\lambda_{max}/2)$ design, then a t-$(t+m(v-t),t+1,m\lambda_{max}/2)$ design exists for all integers $m\geq 0$.

3.18 Remark The 5-design and the 6-design in Example 3.15 are both halvings of the complete design, and so Theorem 3.17 applies to them; see Table 3.31.

3.19 Theorem (Driessen [24]) If there is a t-(v,k,λ) design $(X,\mathcal{B})$ such that any two blocks intersect in at most $k-(\ell+m+1)$ points, for fixed integers $\ell,m\geq 0$, then $(X,\mathcal{A})$ is a

$$t\text{-}(v,k+\ell-m,\lambda\tbinom{v-k}{\ell}\tbinom{k+\ell-m}{t}\tbinom{k}{m}/\tbinom{k}{t})$$

design, where $\mathcal{A}=\{(B\cup L)\setminus M : B\in\mathcal{B}, L\subseteq(X\setminus B), |L|=\ell, M\subseteq B, |M|=m\}$.

3.20 Remark A *Hadamard* 3-design is a 3-$(4n,2n,n-1)$ design. One exists whenever there is a $4n$ by $4n$ Hadamard matrix. The blocks of such a design intersect in 0 or n points.

3.21 Theorem If there is a Hadamard matrix of order $4n$, then for all integers $\ell,m\geq 0$ with $m+\ell<n$, there is a 3-$(4n,2n+\ell-m,(n-1)\binom{2n}{\ell}\binom{2n+\ell-m}{3}\binom{2n}{m}/\binom{2n}{3})$ design.

3.22 Remark If G is a group acting t-homogeneously or t-transitively on a set X and S is a k-element subset of X, then the orbit S^G of S under the action of G is a t-(v,k,λ) design with $\lambda=(|G|\binom{k}{t})/(|G_S|\binom{v}{t})$ and $|G_S|$ the number of elements of G that fix S.

3.23 Example Let q be a prime power and $n>0$ be an integer. Then $G=\text{PGL}_2(q^n)$ acts sharply 3-transitive on $X=\mathbb{F}_{q^n}\cup\{\infty\}$. If $S\subseteq X$ is the natural inclusion of $\mathbb{F}_q\cup\{\infty\}$, then the orbit of S under G is a 3-$(q^n+1,q+1,1)$ design. These designs are *spherical geometries*; when $n=2$ they are *inversive planes*, *Möbius planes*, or *circle geometries*.

3.24 Theorem (Hughes [31]) Let B be a subgroup of the multiplicative group of nonzero elements of $\mathbb{F}_q$. Then the orbit of $S=B\cup\{0,\infty\}$ under the action of $\text{PGL}_2(q)$ is one of the following:

1. 3-$(q^n+1, q+1, 1)$ where q is a prime power and $n \geq 2$, i.e. a spherical geometry;
2. 3-$(q+1, k+1, k(k+1)/2)$ if $(k-1)|(q-1)$, $k \nmid q$ and $k \neq 3$ or 5;
3. 3-$(q+1, 4, 3)$ for $q \equiv 1, 5 \pmod 6$; and
4. 3-$(q+1, 6, 5)$ for $q \equiv 1, 9, 13, 17 \pmod{20}$.

3.25 Remark The distribution of the 3-designs among the orbits of 4- and 5-element subsets under $\mathrm{PSL}_2(q)$ for $q \equiv 3 \pmod 4$ can be found in [20]; see also Theorem 3.43.

3.26 Theorem (Assmus and Mattson [4]) Let A be a $[v, x, d]$ code over $\mathbb{F}_q$ and let $B = A^{\perp}$ be the $[v, v-x, e]$ dual code. Let t be an integer less than d. Let k_0 and ℓ_0 be the largest integers satisfying

$$k_0 - \left\lceil \tfrac{k_0}{q-1} \right\rceil < d \text{ and } \ell_0 - \left\lceil \tfrac{\ell_0}{q-1} \right\rceil < f$$

where, if $q = 2$, then $k_0 = \ell_0 = v$. Suppose that the number of nonzero weights of B which are less than or equal to $(v-t)$ is itself less than or equal to $(d-t)$. Then for each weight k with $d \leq k \leq k_0$ the support vectors of the code words of weight k in A form a t-design and for each weight ℓ with $f \leq \ell \leq \min\{\ell_0, v-t\}$, the support of the codewords of weight ℓ in B form a t-design. See §V.1.4.

3.27 Example Take $v = 24$, $x = 12$, $q = 2$ and let $A = A^{\perp}$ be the extended binary Golay code. The only weights that occur in this code are 0, 8, 12, 16, and 24. Consequently, a 5-$(24, 8, 1)$ design is obtained from Theorem 3.26.

3.28 Remark A more thorough discussion of the relationship between codes and designs is given in §V.1.

3.29 Remark Although a better result is given in Theorem 3.40, Theorem 3.30 is the most celebrated theorem in t-design theory.

3.30 Theorem (Teirlinck [65]) Given integers t and v with $v \equiv t \pmod{(t+1)!^{2t+1}}$ and $v \geq t+1 > 0$, then a simple t-$(v, t+1, (t+1)!^{2t+1})$ design exist.

3.31 Table The known infinite families of simple t designs with $t \geq 3$ (derived and residual families are omitted).

t-(v,k,λ)	Conditions	Ref
3-$(n, 4, 1)$	$n \equiv 2$ or $4 \pmod 6$	[28]
3-$(q+1, 4, 2)$	$q \equiv 1 \pmod 3$	[31]
3-$(q+1, 4, 3)$	odd $q \geq 5$	[31]
3-$(3^n+1, 4, \lambda)$	$\lambda = 6k$ or $6k+1$, $0 \leq k \leq (3^{n-1}-1)/2$	[24]
3-$(q+1, 4, \lambda)$	odd $q \not\equiv 1 \pmod 6$, $\lambda = 6k$ or $6k+3$, $0 \leq k \leq (q-5)/6$	[24]
3-$(2^n+1, 4, \lambda)$	$\lambda = 6k$, $1 \leq k \leq (2^{n-1}-1)/3$	[24]
3-$(4n, 4, 2n-3)$	$n \geq 2$	[24]
3-$(n, 5, 5(n-4)/2)$	$n \equiv 2, 4 \pmod 6$	[24]
3-$(6n, 6, (3n-2)(3n-4)(3n-5)/4)$	$n \equiv 0, 3 \pmod 8$	[24]
3-$(2^n, 6+\ell-m, \lambda\binom{2^n-6}{\ell}\binom{6+\ell-m}{3}\binom{6}{m}/20)$	$n \geq 4$, $\ell+m \leq 2$, and $\lambda = \begin{cases}(2^n-4)/3 \text{ for } n \text{ even,}\\ (2^{n-1}-4)/3 \text{ for } n \text{ odd}\end{cases}$	[24]
3-$(q^f+1, q+1+\ell-m, \binom{q^f-q}{\ell}\binom{q+1+\ell-m}{3}\binom{q+1}{m}/\binom{q+1}{3})$	$f \geq 2$, and $\ell+m \leq q-2$	[24]
3-$(4n, 2n+\ell-m, (n-1)\binom{2n}{\ell}\binom{2n+\ell-m}{3}\binom{2n}{m}/\binom{2n}{3})$	$\ell+m \leq n-1$, and H$(4n)$ exists	[24]
3-$(2n, n, n-2)$	H$(4n)$ exists and $n \geq 5$ is odd; or $2n-1$ is a prime power	[24]
3-$(6k, 2k, (2k-1)\binom{3k-2}{k-2})$	$k \geq 2$	[24]

t-(v,k,λ)	Conditions	Ref
3-$(2n, k\binom{n}{s}\binom{n-s}{k-2s}2^{k-2s}\binom{k}{3}/\binom{2n}{3})$	$(2n-1)\mid\binom{k}{2}$, $2n-1>2k$, $s=\binom{k}{2}/(2n-1)$	[24]
4-$(2^f+1,5,\lambda)$	$\lambda\in\{5,20\}$, odd $f\geq 5$	[12]
4-$(2^f+1,6,10)$	for all odd $f\geq 5$	[10]
4-$(2^f+1,6,\lambda)$	$\lambda\in\{60,70,90,100,150,160\}$, $\gcd(f,6)=1$	[13]
4-$(2^f+1,7,70(2^f-5)/3)$	for $\gcd(f,6)=1$	[12]
4-$(2^f+1,8,35)$	$\gcd(f,6)=1$	[13]
4-$(2^f+1,9,84)$	for all $f\geq 5$	[11]
4-$(2^f+1,9,\lambda)$	$\lambda\in\{63,147\}$, $\gcd(f,6)=1$	[13]
4-$(2^n+1,2^m,(2^m-3)\Pi_{i=2}^{m-1}\frac{2^{n-i}-1}{2^{m-i}-1})$	$2<m<n$	[30]
4-$(2^n+1,2^{n-1}-1,(2^{n-1}-3)(2^{n-2}-1)(2^{n-1}-4))$	$n\geq 4$	[24]
4-$(2^n+1,2^m+1,(2^m+1)\Pi_{i=2}^{m-1}\frac{2^{n-i}-1}{2^{m-i}-1})$	$2<m<n-1$, $m\not\mid n$	[30]
4-$(2^n+1+s,2^{n-1}-1,\binom{2^n+s-3}{s}(2^{n-1}-1)(2^{n-2}-1)(2^{n-1}-4))$	$\frac{2^{n-1}-2}{n-1}>s+6$, $s\geq 2$, $n\geq 6$	[55]
4-$(2^n+1+s,2^m,\binom{2^n+s-3}{s}(2^m-3)\mu)$	for m sufficiently close to n, with m large enough so that $\binom{v}{k}/\binom{v+s}{s}>b_0(b_0-b_1)$ where $\mu=\Pi_{i=2}^{m-1}\frac{2^{n-i}-1}{2^{m-i}-1}$	[55]
4-$(2^n+1+s,2^m+1,\binom{2^n+s-3}{s}(2^m+1)\mu)$	for m sufficiently close to n, with m large enough so that $\binom{v}{k}/\binom{v+s}{s}>b_0(b_0-b_1)$ where $\mu=\Pi_{i=2}^{m-1}\frac{2^{n-i}-1}{2^{m-i}-1}$	[55]
4-$(9m+5,6,(27m^2+3m)/2)$	$m>0$,	[41]
5-$(2^n+2,2^{n-1}+1,(2^{n-1}-3)(2^{n-2}-1))$	$n\geq 4$,	[2]
5-$(2^n+3,2^{n-1}+1,(2^n-2)(2^{n-1}-3)(2^{n-2}-1))$	$n\geq 6$	[67]
5-$(2^n+4,2^{n-1}+2,(2^n-1)(2^n-2)(2^{n-2}-1))$	$n\geq 6$	[67]
5-$(2^n+5,2^{n-1}+2,2^n(2^n-1)(2^n-2)(2^{n-2}-1))$	$n\geq 6$	[67]
5-$(2^n+6,2^{n-1}+3,2^{n-1}(2^n+1)(2^n-1)(2^n-2))$	$n\geq 6$	[67]
5-$(2^n+2+s,2^n+1,\binom{2^n+s-3}{s}(2^{n-1}-3)(2^{n-2}-1))$	$N\geq 4$, $n\geq N$, $s>0$, $((2^N-N)/(N-1))>s+4$	[55]
6-$(6+8u,7,4u)$	$u>0$	[66]
6-$(23+16m,8,\frac{1}{2}\binom{16m+1}{2})$	$m>0$	[48]
t-$(t+\ell(t)\cdot N,t+1,m\ell(t))$	See Theorem 3.40.	[66]

3.32 Remark There is no known simple 4-$(4^m+1,5,2)$ design for any $m>0$. However, there exist such designs with repeated blocks for all $m>0$ [5].

3.33 Remark (Kramer [36]) A 6-$(17,7,2)$ design with repeated blocks can be constructed using orbits of 7-element subsets under the action of $G=\mathrm{PSL}_2(16)$ on the projective line $\mathbb{F}_{16}\cup\{\infty\}$. Only one orbit is repeated. More precisely take G to be the group generated by

(0 5 12 4 3 1 14 9 13 11 2 8 6 10 7);
(0 15)(1 2)(3 4)(5 6)(7 8)(9 10)(11 12)(13 14); and
(0)(1 2)(3 8)(4 6)(5 7)(9 13)(10 12)(11 14)(15 16).

Then the union of the orbits of the 3 base blocks $\{0,1,2,4,5,6,9\}$,$\{0,1,2,3,4,5,7\}$ and $\{0,2,3,4,6,8,9\}$ with the last one repeated twice form a 6-$(17,7,2)$ design.

3.34 Theorem (Wilson [73]) Given integers $0<t\leq k\leq v$, then there is an integer $n>0$ such that a t-(v,k,λ) design with repeated blocks exists whenever the conditions $\lambda\binom{v-i}{t-i}\equiv 0 \pmod{\binom{k-i}{t-i}}$ for $i=0,1,\ldots,t$, and $\lambda>n$ are satisfied.

3.35 Table Existence and nonexistence of 3-(v, k, λ) designs with **repeated blocks** for $k \leq v/2 \leq 15$. An entry of m in the set R for row 3-$(v, k, m\lambda_{min})$ indicates that a 3-$(v, k, m\lambda_{min})$ exists with repeated blocks, but no such design is known to exist as a simple design. Other 3-designs with repeated blocks may be constructed by taking the unions of these designs and designs in Table 3.37. An entry in U indicates that the design is not known to exist with or without repeated blocks. Explicit constructions for the designs may be found in [29].

3-$(14, 5, m5)$	$R = \{1\}$	3-$(16, 7, m5)$	$U = \{1\}$
3-$(17, 7, m7)$	$U = \{1\}$	3-$(20, 5, m2)$	$U = \{1,2\}$
3-$(21, 6, m4)$	$U = \{1\}$, $R = \{2,3\}$	3-$(21, 7, m15)$	$U = \{1,2,5\}$, $R = \{4\}$
3-$(21, 8, m84)$	$U = \{1\}$, $R = \{2,3\}$	3-$(22, 7, m)$	$U = \{3,7\}$
3-$(22, 8, m6)$	$U = \{1\}$	3-$(22, 10, m6)$	$U = \{1\ldots6,10,11,13\}$, $R = \{9,12\}$
3-$(23, 8, m8)$	$U = \{1\}$		
3-$(25, 5, m3)$	$R = \{1\}$	3-$(25, 6, m20)$	$R = \{1\}$
3-$(25, 7, m35)$	$R = \{1\}$	3-$(25, 8, m42)$	$R = \{1\}$
3-$(25, 9, m21)$	$U = \{1\}$, $R = \{3\}$	3-$(25, 10, m24)$	$U = \{1\}$, $R = \{2,3\}$
3-$(25, 11, m495)$	$R = \{1\}$	3-$(25, 12, m110)$	$R = \{1\}$
3-$(26, 8, m7)$	$U = \{1\}$, $R = \{2,3\}$	3-$(26, 9, m21)$	$U = \{1\}$, $R = \{2,3\}$
3-$(26, 10, m3)$	$U = \{2\ldots7,9,10,11,13,14\}$, $R = \{8,15\ldots22\}$	3-$(26, 11, m33)$	$U = \{1\}$, $R = \{2,3\}$
		3-$(26, 12, m55)$	$R = \{1\}$
3-$(26, 13, m11)$	$R = \{1\}$	3-$(27, 6, m4)$	$R = \{1\}$
3-$(27, 8, m168)$	$R = \{1\}$	3-$(27, 10, m72)$	$R = \{1\}$
3-$(27, 13, m66)$	$R = \{1\}$	3-$(28, 6, m10)$	$R = \{1\}$
3-$(28, 10, m20)$	$U = \{1\}$, $R = \{2,3\}$	3-$(28, 13, m22)$	$U = \{1,2,4,7\}$, $R = \{3,5\}$
3-$(29, 6, m20)$	$R = \{1\}$		
3-$(29, 7, m5)$	$U = \{1\}$, $R = \{2,3\}$	3-$(29, 8, m4)$	$U = \{1,3,5\}$, $R = \{4,6\}$
3-$(29, 9, m14)$	$U = \{1\}$	3-$(29, 13, m143)$	$R = \{1\}$
3-$(29, 14, m52)$	$U = \{1\}$, $R = \{2,3\}$	3-$(30, 6, m5)$	$R = \{1\}$
3-$(30, 7, m15)$	$R = \{1\}$	3-$(30, 8, m6)$	$U = \{1\ldots6\}$, $R = \{8\ldots13\}$
3-$(30, 9, m6)$	$U = \{1,2,3,4\}$, $R = \{5,6,8,9\}$	3-$(30, 11, m495)$	$R = \{1\}$
		3-$(30, 13, m429)$	$R = \{1\}$
3-$(30, 14, m39)$	$U = \{2,3\}$, $R = \{1\}$	3-$(30, 15, m13)$	$R = \{1\}$

3.36 Conjecture If there exists a t-(v, k, λ) design with $\lambda \leq \lambda_{max}/2$ having repeated blocks, then there exists such a design without repeated blocks.

3.3 Table of Simple t-Designs

3.37 Table Existence and nonexistence of **simple** t-(v, k, λ) designs for $3 \leq t < k \leq v/2 \leq 15$, and $\lambda < \lambda_{max}/2$. An entry of m in the set Y for row t-$(v, k, m\lambda_{min})$ indicates the existence of a t-$(v, k, m\lambda_{min})$ design. An entry in N indicates nonexistence. All other parameter situations are undecided. The upper limit $\lfloor \lambda_{max}/(2\lambda_{min}) \rfloor$ on m is given by LIM. The sets Y and N are followed by references and theorems among which the constructions or the nonexistence proofs can be found.

3-$(8, 4, m)$	$Y = \{1, 2\}$ LIM $= 2$, Theorem 3.11. (See Example I.4.27).
3-$(10, 4, m)$	$Y = \{1, 2, 3\}$ LIM $= 3$, Theorem 3.8.
3-$(10, 5, m3)$	$Y = \{1, 2, 3\}$ LIM $= 3$, Theorem 3.9.
3-$(11, 4, m4)$	$Y = \{1\}$ LIM $= 1$, [66].
3-$(11, 5, m2)$	$N = \{1\}$, $Y = \{2\ldots7\}$ LIM $= 7$, [58, 21, 15].
3-$(12, 4, m3)$	$Y = \{1\}$ LIM $= 1$, [64].
3-$(12, 5, m6)$	$Y = \{1, 2, 3\}$ LIM $= 3$, [15], Theorem 3.8.
3-$(12, 6, m2)$	$Y = \{1\ldots21\}$ LIM $= 21$, Theorem 3.11.
3-$(13, 4, m2)$	$Y = \{1, 2\}$ LIM $= 2$, [15].

3-$(13, 5, m15)$	$Y = \{1\}$ LIM $= 1$, [17].
3-$(13, 6, m20)$	$Y = \{1, 2, 3\}$ LIM $= 3$, [37].
3-$(14, 4, m)$	$Y = \{1 \ldots 5\}$ LIM $= 5$, [6, 15]. (See Example I.4.29).
3-$(14, 5, m5)$	$Y = \{2 \ldots 5\}$ LIM $= 5$, [15], Theorem 3.9.
3-$(14, 6, m5)$	$Y = \{1 \ldots 16\}$ LIM $= 16$, [15].
3-$(14, 7, m5)$	$Y = \{1 \ldots 33\}$ LIM $= 33$, Theorem 3.11.
3-$(15, 5, m6)$	$Y = \{1 \ldots 5\}$ LIM $= 5$, [15], Theorems 3.12, 3.3.
3-$(15, 6, m20)$	$Y = \{1 \ldots 5\}$ LIM $= 5$, Theorems 3.12, 3.8.
3-$(15, 7, m15)$	$Y = \{1 \ldots 16\}$ LIM $= 16$, [15], Theorem 3.12.
3-$(16, 4, m)$	$Y = \{1 \ldots 6\}$ LIM $= 6$, [50].
3-$(16, 5, m6)$	$Y = \{1 \ldots 6\}$ LIM $= 6$, Theorems 3.9, 3.8.
3-$(16, 6, m2)$	$N = \{1\}$, $Y = \{2 \ldots 71\}$ LIM $= 71$, [15], Theorems 3.11, 3.3, 3.8.
3-$(16, 7, m5)$	$Y = \{2 \ldots 71\}$ LIM $= 71$, [47, 15].
3-$(16, 8, m3)$	$Y = \{1 \ldots 214\}$ LIM $= 214$, Theorem 3.11.
3-$(17, 4, m2)$	$Y = \{1, 2, 3\}$ LIM $= 3$, Theorem 3.8.
3-$(17, 5, m)$	$Y = \{1 \ldots 45\}$ LIM $= 45$, [60, 15, 14].
3-$(17, 6, m4)$	$Y = \{1 \ldots 45\}$ LIM $= 45$, [15].
3-$(17, 7, m7)$	$Y = \{2 \ldots 71\}$ LIM $= 71$, [15].
3-$(17, 8, m14)$	$Y = \{1 \ldots 71\}$ LIM $= 71$, [15].
3-$(18, 4, m3)$	$Y = \{1, 2\}$ LIM $= 2$, [64].
3-$(18, 5, m15)$	$Y = \{1, 2, 3\}$ LIM $= 3$, Theorem 3.3.
3-$(18, 6, m5)$	$Y = \{1 \ldots 45\}$ LIM $= 45$, [15].
3-$(18, 7, m105)$	$Y = \{1 \ldots 6\}$ LIM $= 6$, Theorem 3.3.
3-$(18, 8, m21)$	$Y = \{1 \ldots 71\}$ LIM $= 71$, [15], Theorem 3.12.
3-$(18, 9, m7)$	$Y = \{1 \ldots 357\}$ LIM $= 357$, Theorem 3.11.
3-$(19, 4, m4)$	$Y = \{1, 2\}$ LIM $= 2$, Theorem 3.8.
3-$(19, 5, m30)$	$Y = \{1, 2\}$ LIM $= 2$, [15].
3-$(19, 6, m20)$	$Y = \{1 \ldots 14\}$ LIM $= 14$, [15].
3-$(19, 7, m35)$	$Y = \{1 \ldots 26\}$ LIM $= 26$, [47].
3-$(19, 8, m168)$	$Y = \{1 \ldots 13\}$ LIM $= 13$, Theorem 3.9.
3-$(19, 9, m28)$	$Y = \{1 \ldots 4, 6, 8, 9, 10, 11, 12, 14 \ldots 22, 24 \ldots 42, 44 \ldots 52, 54 \ldots 62, 64 \ldots 72, 74 \ldots 82, 84 \ldots 92, 94 \ldots 102, 104 \ldots 112, 114 \ldots 122, 124 \ldots 132, 134 \ldots 142\}$ LIM $= 143$, [47], Theorems 3.12, 3.9, 3.8.
3-$(20, 4, m)$	$Y = \{1 \ldots 8\}$ LIM $= 8$, [39].
3-$(20, 5, m2)$	$Y = \{3, 5, 6, 8, 9, 11, 12, 14, 15, 17, 18, 20, 21, 23 \ldots 30, 32, 33, 34\}$ LIM $= 34$, [39, 47].
3-$(20, 6, m10)$	$Y = \{1 \ldots 34\}$ LIM $= 34$, [39].
3-$(20, 7, m35)$	$Y = \{1 \ldots 34\}$ LIM $= 34$, [39].
3-$(20, 8, m14)$	$Y = \{1 \ldots 221\}$ LIM $= 221$, [39].
3-$(20, 9, m28)$	$Y = \{1, 2, 4, 7, 10, 12, 13, 16, 17, 19, 22, 24, 25, 28, 31, 34, 36, 37, 40, 43, 45, 46, 49, 51, 52, 54, 55, 58, 60, 61, 64, 66, 67, 68, 70, 72, 73, 76, 78, 79, 82, 84, 85, 88, 90, 91, 94, 96, 97, 100, 102, 103, 105, 106, 108, 109, 111, 112, 114, 115, 117 \ldots 121, 123, 124, 126, 127, 129, 130, 132, 133, 135, 136, 138, 139, 141, 142, 144, 145, 147, 148, 150, 151, 153, 154, 156, 157, 159, 160, 162, 163, 165, 166, 168 \ldots 172, 174, 175, 177, 178, 180, 181, 183, 184, 186, 187, 189, 190, 192, 193, 195, 196, 198, 199, 201, 202, 204, 205, 207, 208, 210, 211, 213, 214, 216, 217, 219, 220\}$ LIM $= 221$, [39], Theorems 3.14, 3.3, 3.8, 3.12, 3.21.
3-$(20, 10, m4)$	$Y = \{1 \ldots 2431\}$ LIM $= 2431$, Theorem 3.11.
3-$(21, 4, m6)$	$Y = \{1\}$ LIM $= 1$, [64].
3-$(21, 5, m3)$	$Y = \{1, 3 \ldots 6, 8 \ldots 13, 15 \ldots 25\}$ LIM $= 25$, [40, 47, 38], Theorem 3.12.
3-$(21, 6, m4)$	$Y = \{9, 10, 15, 17, 18, 24, 27, 30, 32, 33, 34, 36, 37, 40, 42, 44, 45, 51, 52, 54, 55, 59, 60, 62, 63, 67, 69, 70, 72, 74, 75, 78, 80, 81, 82, 84, 85, 87, 89, 90, 92, 94, 96, 99, 100, 102\}$ LIM $= 102$, [47, 38], Theorem 3.12.
3-$(21, 7, m15)$	$Y = \{3, 6, 9, 12, 15, 18, 21, 24, 27, 28, 30, 33, 36, 39, 40, 42, 45, 48, 51, 52, 54, 56, 57, 60, 63, 64, 66, 68, 69, 72, 75, 78, 80, 81, 84, 87, 90, 91, 93, 96, 99, 102\}$ LIM $= 102$, [38], Theorem 3.12.

3-$(21, 8, m84)$ $Y = \{6, 8, 14, 16, 22, 32, 38, 40, 46, 51\}$ LIM $= 51$, [38], Theorems 3.3, 3.12.

3-$(21, 9, m42)$ $Y = \{1, 2, 4, 7, 10, 12, 13, 16, 17, 19, 22, 24, 25, 28, 31, 34, 36, 37, 40, 43, 45, 46, 49, 51, 52, 54, 55, 58, 60, 61, 64, 66, 67, 68, 70, 72, 73, 76, 78, 79, 82, 84, 85, 88, 90, 91, 94, 96, 97, 100, 102, 103, 105, 106, 108, 109, 111, 112, 114, 115, 117...121, 123, 124, 126, 127, 129, 130, 132, 133, 135, 136, 138, 139, 141, 142, 144, 145, 147, 148, 150, 151, 153, 154, 156, 157, 159, 160, 162, 163, 165, 166, 168...172, 174, 175, 177, 178, 180, 181, 183, 184, 186, 187, 189, 190, 192, 193, 195, 196, 198, 199, 201, 202, 204, 205, 207, 208, 210, 211, 213, 214, 216, 217, 219, 220\}$ LIM $= 221$, Theorems 3.12, 3.3.

3-$(21, 10, m72)$ $Y = \{1, 2, 4, 7, 10, 12, 13, 16, 17, 19, 22, 24, 25, 28, 31, 34, 36, 37, 40, 43, 45, 46, 49, 51, 52, 54, 55, 58, 60, 61, 64, 66, 67, 68, 70, 72, 73, 76, 78, 79, 82, 84, 85, 88, 90, 91, 94, 96, 97, 100, 102, 103, 105, 106, 108, 109, 111, 112, 114, 115, 117...121, 123, 124, 126, 127, 129, 130, 132, 133, 135, 136, 138, 139, 141, 142, 144, 145, 147, 148, 150, 151, 153, 154, 156, 157, 159, 160, 162, 163, 165, 166, 168...172, 174, 175, 177, 178, 180, 181, 183, 184, 186, 187, 189, 190, 192, 193, 195, 196, 198, 199, 201, 202, 204, 205, 207, 208, 210, 211, 213, 214, 216, 217, 219, 220\}$ LIM $= 221$, Theorems 3.12, 3.3.

3-$(22, 4, m)$ $Y = \{1...9\}$ LIM $= 9$, Theorem 3.8.

3-$(22, 5, m3)$ $Y = \{1...28\}$ LIM $= 28$, Theorems 3.12, 3.9, 3.8.

3-$(22, 6, m)$ $Y = \{1...60, 68, 80, 85, 96, 97, 102, 112, 113, 114, 119, 120, 128, 129, 136, 152, 153, 156, 170, 171, 175, 180, 187, 190, 204, 209, 213, 221, 233, 238, 247, 249, 255, 272, 285, 289, 306, 323, 327, 340, 342, 357, 374, 380, 384, 391, 399, 403, 408, 422, 425, 437, 442, 456, 459, 475, 476\}$ LIM $= 484$, [35, 24, 14], Theorems 3.12, 3.8, 3.9.

3-$(22, 7, m)$ $N = \{1, 2\}$, $Y = \{4, 5, 6, 8, 10, 12, 16, 20, 24, 28, 32, 36, 40, 44, 48, 52, 56, 60, 64, 68, 72, 76, 80, 84, 88, 90, 92, 96, 136, 171, 204, 272, 285, 340, 342, 360, 408, 420, 424, 448, 452, 456, 476, 480, 484, 512, 513, 516, 544, 570, 612, 627, 630, 636, 680, 684, 720, 726, 748, 760, 798, 810, 816, 855, 876, 884, 936, 952, 969, 988, 1020, 1026, 1088, 1140, 1156, 1197, 1224, 1260, 1288, 1292, 1311, 1360, 1368, 1386, 1425, 1428, 1482, 1496, 1520, 1539, 1564, 1596, 1632, 1653, 1680, 1700, 1710, 1716, 1768, 1824, 1836, 1860, 1881, 1890, 1904, 1938\}$ LIM $= 1938$, [26, 24, 29, 14], Theorems 3.12, 3.9, 3.8, 3.14.

3-$(22, 8, m6)$ $Y = \{2, 3, 4, 6, 8, 10, 12, 14, 16, 18, 24, 27, 30, 42, 45, 48, 56, 72, 90, 114, 168, 180, 224, 226, 240, 242, 256, 258, 266, 336, 339, 360, 363, 378, 381, 405, 408, 608, 672, 760, 784, 800, 840, 858, 912, 930, 969\}$ LIM $= 969$, [24, 14], Theorems 3.12, 3.8, 3.9, 3.3.

3-$(22, 9, m42)$ $Y = \{1, 4, 6, 8, 9, 10, 24, 30, 36, 45, 48, 55, 56, 60, 80, 96, 110, 117, 118, 126, 127, 135, 136, 158, 166, 176, 206, 214, 224, 280, 286, 304, 310, 312, 318\}$ LIM $= 323$, [24, 14], Theorems 3.9, 3.8.

3-$(22, 10, m6)$ $Y = \{7, 8, 14, 16, 19, 38, 76, 96, 104, 130, 133, 190, 228, 247, 304, 323, 361, 390, 418, 456, 475, 528, 532, 585, 589, 624, 646, 672, 684, 703, 715, 728, 760, 817, 855, 874, 931, 960, 968, 969, 988, 1026, 1040, 1045, 1056, 1064, 1102, 1140, 1152, 1159, 1160, 1216, 1248, 1254, 1273, 1292, 1330, 1368, 1387, 1430, 1444, 1482, 1501, 1534, 1558, 1596, 1615, 1672, 1710, 1729, 1786, 1824, 1843, 1900, 1938, 1957, 1995, 2014, 2052, 2054, 2071, 2109, 2128, 2158, 2166, 2185, 2223, 2242, 2261, 2280, 2299, 2337, 2356, 2394, 2413, 2451, 2470, 2494, 2508, 2527, 2565, 2584, 2590, 2622, 2641, 2679, 2698, 2736, 2755, 2793, 2812, 2850, 2869, 2907, 2926, 2964, 2983, 3021, 3022, 3040, 3078, 3097, 3118, 3135, 3154, 3192, 3211, 3230, 3249, 3268, 3306, 3325, 3363, 3382, 3420, 3439, 3477, 3496, 3534, 3553, 3591, 3610, 3648, 3667, 3705, 3724, 3762, 3781, 3819, 3838, 3876, 3895, 3933, 3952, 3990, 4009, 4047, 4066, 4104, 4123, 4161, 4180\}$ LIM $= 4199$, [29, 14], Theorems 3.9, 3.12, 3.8, 3.14.

3-(22, 11, m9)	Y = {1, 2, 4, 8, 16, 19, 38, 76, 88, 96, 104, 133, 190, 228, 247, 304, 323, 361, 418, 456, 475, 532, 589, 646, 672, 684, 703, 715, 760, 817, 855, 874, 931, 968, 969, 976, 988, 1026, 1045, 1056, 1064, 1102, 1140, 1152, 1159, 1160, 1216, 1248, 1254, 1273, 1292, 1330, 1368, 1387, 1430, 1444, 1482, 1501, 1534, 1558, 1596, 1615, 1672, 1710, 1729, 1786, 1824, 1843, 1900, 1938, 1957, 1995, 2014, 2052, 2071, 2109, 2128, 2166, 2185, 2223, 2242, 2261, 2280, 2299, 2337, 2356, 2394, 2413, 2451, 2470, 2494, 2508, 2527, 2565, 2584, 2590, 2622, 2641, 2679, 2698, 2736, 2755, 2793, 2812, 2850, 2869, 2907, 2926, 2964, 2983, 3021, 3040, 3078, 3097, 3135, 3154, 3192, 3211, 3230, 3249, 3268, 3306, 3325, 3363, 3382, 3420, 3439, 3470, 3477, 3496, 3534, 3553, 3558, 3591, 3610, 3648, 3667, 3705, 3724, 3762, 3781, 3819, 3838, 3876, 3895, 3933, 3952, 3990, 4009, 4047, 4066, 4104, 4123, 4161, 4180} LIM = 4199, [24, 14], Theorems 3.14, 3.13, 3.9.
3-(23, 4, m4)	Y = {1, 2} LIM = 2, [17].
3-(23, 5, m10)	Y = {1...9} LIM = 9, Theorem 3.3.
3-(23, 6, m20)	Y = {1...28} LIM = 28, Theorems 3.12, 3.3.
3-(23, 7, m5)	Y = {1...24, 34, 42, 51, 68, 85, 102, 112, 113, 114, 119, 120, 128, 129, 136, 153, 156, 170, 171, 180, 187, 190, 204, 213, 221, 238, 247, 255, 272, 285, 289, 306, 323, 327, 340, 342, 357, 374, 380, 384, 391, 399, 408, 425, 442, 456, 459, 476} LIM = 484, Theorems 3.3, 3.12.
3-(23, 8, m8)	Y = {2...969} LIM = 969, [47].
3-(23,9,m12)	Y = {4, 5, 6,}$\dot{\cup}$\{m : $m \geq 8$ and $m \equiv 0 \pmod 2$}, [47], Theorem 3.3.
3-(23, 10, m120)	Y = {8, 10, 30, 45, 48, 55, 56, 80, 96, 110, 118, 158, 166, 304} LIM = 323, Theorems 3.3, 3.12.
3-(23, 11, m15)	Y = {8, 16, 19, 38, 76, 96, 104, 133, 190, 228, 247, 304, 323, 361, 418, 456, 475, 532, 589, 646, 672, 684, 703, 715, 760, 817, 855, 874, 931, 968, 969, 988, 1026, 1045, 1056, 1064, 1102, 1140, 1152, 1159, 1160, 1216, 1248, 1254, 1273, 1292, 1330, 1368, 1387, 1430, 1444, 1482, 1501, 1534, 1558, 1596, 1615, 1672, 1710, 1729, 1786, 1824, 1843, 1900, 1938, 1957, 1995, 2014, 2052, 2071, 2109, 2128, 2166, 2185, 2223, 2242, 2261, 2280, 2299, 2337, 2356, 2394, 2413, 2451, 2470, 2494, 2508, 2527, 2565, 2584, 2590, 2622, 2641, 2679, 2698, 2736, 2755, 2793, 2812, 2850, 2869, 2907, 2926, 2964, 2983, 3021, 3040, 3078, 3097, 3135, 3154, 3192, 3211, 3230, 3249, 3268, 3306, 3325, 3363, 3382, 3420, 3439, 3477, 3496, 3534, 3553, 3591, 3610, 3648, 3667, 3705, 3724, 3762, 3781, 3819, 3838, 3876, 3895, 3933, 3952, 3990, 4009, 4047, 4066, 4104, 4123, 4161, 4180} LIM = 4199, Theorems 3.3, 3.12.
3-(24, 4, m3)	Y = {1, 2, 3} LIM = 3, [64].
3-(24, 5, m30)	Y = {1, 2, 3} LIM = 3, [17].
3-(24, 6, m10)	Y = {1...66} LIM = 66, [47].
3-(24, 7, m105)	Y = {1...28} LIM = 28, [17].
3-(24, 8, m21)	Y = {1...484} LIM = 484, [47].
3-(24, 9, m84)	Y = {1, 2, 4, 5, 6, 8, 9, 10, 12, 14, 16, 18, 20, 22, 24, 26, 28, 30, 32, 34, 36, 38, 40, 42, 44, 46, 48, 50, 52, 54, 56, 58, 60, 62, 64, 66, 68, 70, 72, 74, 76, 78, 80, 82, 84, 86, 88, 90, 92, 94, 96, 98, 100, 102, 104, 106, 108, 110, 112, 114, 116, 118, 120, 122, 124, 126, 127, 128, 130, 132, 134, 135, 136, 138, 140, 142, 144, 146, 148, 150, 152, 154, 156, 158, 160, 162, 164, 166, 168, 170, 172, 174, 176, 178, 180, 182, 184, 186, 188, 190, 192, 194, 196, 198, 200, 202, 204, 206, 208, 210, 212, 214, 216, 218, 220, 222, 224, 226, 228, 230, 232, 234, 236, 238, 240, 242, 244, 246, 248, 250, 252, 254, 256, 258, 260, 262, 264, 266, 268, 270, 272, 274, 276, 278, 280, 282, 284, 286, 288, 290, 292, 294, 296, 298, 300, 302, 304, 306, 308, 310, 312, 314, 316, 318, 320, 322} LIM = 323, [24], Theorems 3.12, 3.3.
3-(24, 10, m180)	Y = {1, 8, 10, 30, 40, 41, 48, 56, 80, 96, 110, 118, 158, 166, 304} LIM = 323, [24], Theorems 3.12, 3.3.
3-(24, 11, m45)	Y = {1, 2, 4, 56, 66, 67, 385, 672, 770, 826, 1210, 1211, 1276, 1277, 2128} LIM = 2261, [24], Theorems 3.14, 3.3, 3.12.

3-$(24, 12, m5)$	$Y = \{1, 2, 4, 8, 16, 24, 32, 48, 56, 64, 112, 133, 144, 145, 266, 532, 672, 728, 931, 1330, 1596, 1729, 2128, 2261, 2527, 2926, 3192, 3325, 3724, 4123, 4356, 4357, 4500, 4501, 4522, 4704, 4788, 4921, 5005, 5320, 5719, 5985, 6118, 6517, 6776, 6783, 6916, 7182, 7315, 7392, 7448, 7714, 7980, 8064, 8113, 8120, 8512, 8736, 8778, 8911, 9044, 9310, 9576, 9709, 10010, 10108, 10374, 10507, 10738, 10906, 11172, 11305, 11704, 11970, 12103, 12502, 12768, 12901, 13300, 13566, 13699, 13965, 14098, 14364, 14497, 14763, 14896, 15162, 15295, 15561, 15694, 15827, 15960, 16093, 16359, 16492, 16758, 16891, 17157, 17290, 17458, 17556, 17689, 17955, 18088, 18130, 18354, 18487, 18753, 18886, 19152, 19285, 19551, 19684, 19950, 20083, 20349, 20482, 20748, 20881, 21147, 21280, 21546, 21679, 21945, 22078, 22344, 22477, 22610, 22743, 22876, 23142, 23275, 23541, 23674, 23940, 24073, 24339, 24472, 24738, 24871, 25137, 25270, 25536, 25669, 25935, 26068, 26334, 26467, 26733, 26866, 27132, 27265, 27531, 27664, 27930, 28063, 28329, 28462, 28728, 28861, 29127, 29260\}$ LIM $= 29393$, [24, 31], Theorems 3.14, 3.3, 3.13.
3-(25 [illegible]	$Y = \{1, 4, 5\}$ LIM $= 5$, [47].
3-(25, [illegible]	$Y = \{11, 14, 21 \ldots 35\}$ LIM $= 38$, Theorems 3.12, 3.8.
3-$(25, 6, m20)$	$Y = \{11, 14, 21 \ldots 35\}$ LIM $= 38$, Theorems 3.12, 3.9.
3-$(25, 7, m35)$	$Y = \{11, 22, 33, 38, 44, 55, 57, 66, 76, 77, 88, 95, 99\}$ LIM $= 104$, Theorems 3.3, 3.9.
3-$(25, 8, m42)$	$Y = \{11, 22, 33, 44, 55, 66, 77, 88, 99, 110, 114, 121, 132, 143, 154, 165, 171, 176, 187, 198, 209, 220, 228, 231, 242, 253, 264, 275, 285, 286, 297, 308\}$ LIM $= 313$, Theorems 3.12, 3.9.
3-$(25, 9, m21)$	$Y = \{2, 4, 11, 22, 33, 44, 55, 66, 77, 88, 99, 110, 121, 132, 143, 154, 165, 176, 187, 198, 209, 220, 231, 242, 253, 264, 275, 286, 297, 308, 319, 330, 341, 352, 363, 374, 385, 396, 407, 418, 429, 440, 451, 462, 473, 484, 495, 506, 517, 528, 539, 550, 561, 572, 583, 594, 605, 616, 627, 638, 649, 660, 671, 682, 693, 704, 715, 726, 737, 748, 759, 770, 781, 792, 803, 814, 825, 836, 847, 858, 869, 880, 891, 902, 913, 924, 935, 946, 957, 968, 979, 990, 1001, 1012, 1023, 1034, 1045, 1056, 1067, 1078, 1089, 1100, 1111, 1122, 1133, 1144, 1155, 1166, 1177, 1188, 1199, 1210, 1221, 1232, 1243, 1254, 1265, 1276, 1287, 1298, 1309, 1320, 1331, 1342, 1353, 1364, 1375, 1386, 1397, 1408, 1419, 1430, 1441, 1452, 1463, 1474, 1485, 1496, 1507, 1518, 1529, 1540, 1551, 1562, 1573, 1584, 1595, 1606, 1617, 1628, 1639, 1650, 1661, 1672, 1683, 1694, 1705, 1716, 1727, 1738, 1749, 1760, 1771\}$ LIM $= 1776$, [29], Theorems 3.12, 3.14.
3-$(25, 10, m24)$	$Y = \{11, 88, 110, 330, 440, 528, 616, 880, 1056, 1210, 1298, 1738, 1826, 3344\}$ LIM $= 3553$, Theorem 3.12.
3-$(25, 11, m495)$	$Y = \{8, 96, 110, 118, 304\}$ LIM $= 323$, Theorem 3.12.
3-$(25, 12, m110)$	$Y = \{56, 385, 672, 770, 826, 2128\}$ LIM $= 2261$, Theorem 3.12.
3-$(26, 4, m)$	$Y = \{1 \ldots 11\}$ LIM $= 11$, [28], Theorem 3.8.
3-$(26, 5, m)$	$Y = \{1, 3, 21, 22, 33, 43, 44, 55, 66, 77, 88, 92, 99, 100, 103, 105, 106, 110, 121, 126\}$ LIM $= 126$, [14], Theorems 3.8, 3.9, 3.12.
3-$(26, 6, m)$	$Y = \{1, 2, 4, 7, 42, 49, 56, 120, 121, 140, 147, 154, 231, 253, 308, 322, 385, 462, 483, 506, 529, 539, 552, 575, 598, 616, 621, 644, 667, 690, 693, 713, 735, 736, 742, 759, 770, 782, 784, 791, 805, 840, 847, 882\}$ LIM $= 885$, [14], Theorems 3.9, 3.12, 3.3, 3.14, 3.8.
3-$(26, 7, m35)$	$Y = \{1, 7, 8, 21, 28, 36, 46, 57, 58, 69, 92, 105, 106, 112, 113, 115, 119, 120\}$ LIM $= 126$, [14], Theorems 3.9, 3.3, 3.8.
3-$(26, 8, m7)$	$Y = \{76, 152, 253, 506, 532, 684, 759, 874, 1012, 1265, 1311, 1518, 1748, 1771, 2024, 2185, 2277\}$ LIM $= 2403$, [14], Theorems 3.12, 3.3, 3.9, 3.8.
3-$(26, 9, m21)$	$Y = \{228, 253, 506, 684, 759, 1012, 1265, 1518, 1771, 2024, 2277\}$ LIM $= 2403$, [14], Theorems 3.12, 3.9.
3-$(26, 10, m3)$	$N = \{1\}$, $Y = \{1012, 1265, 3795, 5060, 6072, 7084, 10120, 12144, 13915, 14927, 19987, 20999, 38456\}$ LIM $= 40859$, [16], Theorem 3.12.
3-$(26, 11, m33)$	$Y = \{184, 2208, 2530, 2714, 6992\}$ LIM $= 7429$, Theorem 3.12.

3-(26, 12, m55)	$Y = \{184, 2208, 2530, 2714, 6992\}$ LIM = 7429, Theorem 3.12.
3-(26, 13, m11)	$Y = \{1288, 8855, 15456, 17710, 18998, 48944\}$ LIM = 52003, Theorem 3.13.
3-(27, 4, m12)	$Y = \{1\}$ LIM = 1, [64].
3-(27, 5, m6)	$Y = \{1, 3, 4, 6, 7, 8, 10, 12, 14, 16, 18, 19, 20, 22, 23\}$ LIM = 23, [29], Theorems 3.3, 3.9, 3.12.
3-(27, 6, m4)	$Y = \{2, 42, 44, 66, 88, 110, 132, 154, 176, 177, 184, 198, 209, 210, 212, 220, 242, 252, 253\}$ LIM = 253, Theorems 3.3, 3.9, 3.12.
3-(27, 7, m21)	$Y = \{1, 2, 14, 16, 42, 92, 138, 184, 209, 210, 212, 224, 226, 230, 240, 253\}$ LIM = 253, [29], Theorems 3.3, 3.9, 3.12.
3-(27, 8, m168)	$Y = \{8, 28, 36, 46, 69, 92, 115\}$ LIM = 126, Theorem 3.3.
3-(27, 9, m28)	$Y = \{1, 2, 4, 253, 506, 684, 759, 1012, 1265, 1518, 1771, 2024, 2277\}$ LIM = 2403, [29], Theorems 3.12, 3.3, 3.14.
3-(27, 11, m99)	$Y = \{1, 2, 4, 92, 1104, 1265, 1357, 3496\}$ LIM = 3714, [29], Theorems 3.12, 3.14.
3-(27, 12, m44)	$Y = \{1, 2, 4, 8, 368, 4416, 5060, 5428, 13984\}$ LIM = 14858, [29], Theorems 3.12, 3.14.
3-(27, 13, m66)	$Y = \{368, 4416, 5060, 5428, 13984\}$ LIM = 14858, Theorem 3.12.
3-(28, 4, m)	$Y = \{1 \ldots 12\}$ LIM = 12, [50].
3-(28, 5, m30)	$Y = \{1 \ldots 5\}$ LIM = 5, [49], Theorems 3.9, 3.3.
3-(28, 6, m10)	$Y = \{20, 30, 40, 50, 60, 70, 80, 90, 95, 100, 110, 115\}$ LIM = 115, Theorems 3.3, 3.12.
3-(28, 7, m5)	$Y = \{2, 3, 4, 10, 210, 920, 1045, 1050, 1060, 1265\}$ LIM = 1265, [29], Theorems 3.3, 3.12, 3.14, 3.9.
3-(28, 8, m42)	$Y = \{1, 40, 66, 230, 345, 460, 575\}$ LIM = 632, [29], Theorems 3.3, 3.21.
3-(28, 9, m28)	$Y = \{1, 2, 4, 99, 900\}$ LIM = 3162, [29], Theorems 3.21, 3.3, 3.14.
3-(28, 10, m20)	$Y = \{11, 99, 2772\}$ LIM = 12017, Theorems 3.8, 3.21.
3-(28, 11, m495)	$Y = \{1, 2, 77\}$ LIM = 1092, Theorems 3.9, 3.21.
3-(28, 12, m55)	$Y = \{1, 2, 4, 5, 6, 8, 10, 12, 17, 98, 336, 460, 1025, 5520, 6006, 6325, 6785, 17480\}$ LIM = 18572, [29], Theorems 3.9, 3.8, 3.12, 3.14, 3.21.
3-(28, 13, m22)	$Y = \{3, 6, 12, 24, 48, 273, 392, 1840, 4100, 7098, 22080, 25300, 27140, 69920\}$ LIM = 74290, Theorems 3.9, 3.12, 3.14, 3.21.
3-(28, 14, m6)	$Y = \{1,2,4,8,16,32,64,128,196,1960, 8281, 9200, 20500, 110400, 126500, 132496, 135700, 349600\}$ LIM = 371450, Theorems 3.9, 3.13, 3.14, 3.21.
3-(29, 4, m2)	$Y = \{1 \ldots 6\}$ LIM = 6, [47].
3-(29, 5, m5)	$Y = \{1, 6, 13, 19, 25, 26, 32\}$ LIM = 32, [29], Theorems 3.3, 3.12, 3.9.
3-(29, 6, m20)	$Y = \{50, 65\}$ LIM = 65, Theorems 3.3, 3.9.
3-(29, 7, m5)	$Y = \{1150, 1235, 1495\}$ LIM = 1495, Theorems 3.3, 3.9.
3-(29, 8, m4)	$Y = \{2,4,7,8,13,14,5980,6325\}$ LIM = 8222, [29], Theorems 3.12,3.14,3.9.
3-(29, 9, m14)	$Y = \{2, 4, 8\}$ LIM = 8222, [29], Theorem 3.14.
3-(29, 10, m40)	$Y = \{1, 2, 4\}$ LIM = 8222, [29], Theorem 3.14.
3-(29, 11, m55)	$Y = \{1, 2, 4, 8, 13\}$ LIM = 14202, [29], Theorems 3.3, 3.14.
3-(29, 12, m110)	$Y = \{1, 2, 4, 13\}$ LIM = 14202, [29], Theorems 3.3, 3.14.
3-(29, 13, m143)	$Y = \{6, 12, 98, 460, 1025, 5520, 6325, 6785, 17480\}$ LIM = 18572, Theorems 3.3, 3.12.
3-(29, 14, m52)	$Y = \{392, 1840, 4100, 22080, 25300, 27140, 69920\}$ LIM = 74290, Theorems 3.3, 3.12.
3-(30, 4, m3)	$Y = \{1 \ldots 4\}$ LIM = 4, [64].
3-(30, 5, m3)	$Y = \{1, 10, 12, 45, 46, 57\}$ LIM = 58, [29], Theorems 3.12, 3.8, 3.3.
3-(30, 6, m5)	$Y = \{60, 225\}$ LIM = 292, Theorems 3.9, 3.3.
3-(30, 7, m15)	$Y = \{120, 450, 585\}$ LIM = 585, Theorems 3.3, 3.9.
3-(30, 8, m6)	$Y = \{7, 1380, 5175\}$ LIM = 6727, [29], Theorems 3.9, 3.3.
3-(30, 9, m6)	$Y = \{6, 7, 12, 14, 24\}$ LIM = 24667, [29], Theorems 3.12, 3.14.
3-(30, 10, m18)	$Y = \{1, 2, 4, 6, 8, 12, 16\}$ LIM = 24667, [29], Theorems 3.14, 3.12.
3-(30, 12, m55)	$Y = \{1 \ldots 4,6,8,12,16,39\}$ LIM = 42607, [29], Theorems 3.3,3.12,3.14.
3-(30, 14, m39)	$Y = \{882, 4140, 9225, 49680, 56925, 61065, 157320\}$ LIM = 167152, Theorems 3.3, 3.12.

3-$(30, 15, m13)$	Y = {3528, 16560, 36900, 198720, 227700, 244260, 629280} LIM = 668610, Theorem 3.13.
4-$(11, 5, m)$	Y = {1, 2, 3} LIM = 3, [68, 34, 15].
4-$(12, 5, m4)$	Y = {1} LIM = 1, [23].
4-$(12, 6, m2)$	N = {1, 3}, Y = {2, 4...7} LIM = 7, [45, 53], Theorems 3.8, 3.11, 3.13.
4-$(13, 5, m3)$	Y = {1} LIM = 1, Theorem 3.8.
4-$(13, 6, m6)$	Y = {2, 3} LIM = 3, [37], Theorem 3.12.
4-$(14, 6, m15)$	Y = {1} LIM = 1, [17].
4-$(14, 7, m20)$	Y = {1, 2, 3} LIM = 3, [15], Theorems 3.11, 3.13.
4-$(15, 5, m)$	N = {1}, Y = {3, 4, 5} LIM = 5, [57, 15, 46].
4-$(15, 6, m5)$	Y = {2, 3, 5} LIM = 5, [15], Theorem 3.9.
4-$(15, 7, m5)$	Y = {6...16} LIM = 16, [15].
4-$(16, 6, m6)$	Y = {2...5} LIM = 5, [15], Theorems 3.12, 3.8.
4-$(16, 7, m20)$	Y = {1...5} LIM = 5, [15], Theorem 3.9.
4-$(16, 8, m15)$	Y = {5...16} LIM = 16, [15], Theorem 3.13.
4-$(17, 5, m)$	Y = {4, 5} LIM = 6, [36].
4-$(17, 6, m6)$	Y = {1...6} LIM = 6, Theorems 3.9, 3.8.
4-$(17, 7, m2)$	N = {1}, Y = {3, 6...71} LIM = 71, [15], Theorem 3.8.
4-$(17, 8, m5)$	N = {1}, Y = {3, 6...33, 35...57, 59...71} LIM = 71, [26, 30, 15, 36], Theorems 3.12, 3.3, 3.9, 3.8.
4-$(18, 5, m2)$	Y = {1, 2, 3} LIM = 3, [15].
4-$(18, 6, m)$	N = {1}, Y = {5...33, 35, 36, 37, 39, 40, 42, 43} LIM = 45, [68, 15, 36].
4-$(18, 7, m28)$	Y = {1...6} LIM = 6, Theorem 3.3.
4-$(18, 8, m7)$	Y = {3, 5...71} LIM = 71, [36, 15], Theorems 3.12, 3.3.
4-$(18, 9, m14)$	Y = {3, 6...71} LIM = 71, [36], Theorem 3.13.
4-$(19, 6, m15)$	Y = {1, 2, 3} LIM = 3, [15], Theorem 3.12.
4-$(19, 7, m35)$	Y = {1...6} LIM = 6, Theorems 3.3, 3.8.
4-$(19, 8, m105)$	Y = {1...6} LIM = 6, Theorems 3.3, 3.9.
4-$(19, 9, m21)$	Y = {3, 6...71} LIM = 71, Theorems 3.12, 3.3, 3.8.
4-$(20, 5, m4)$	Y = {1, 2} LIM = 2, [47], Theorem 3.8.
4-$(20, 6, m30)$	Y = {1, 2} LIM = 2, [47, 39].
4-$(20, 7, m140)$	Y = {2} LIM = 2, [39].
4-$(20, 8, m70)$	Y = {1...13} LIM = 13, [39].
4-$(20, 9, m168)$	Y = {2...6, 8...12} LIM = 13, [39], Theorem 3.3.
4-$(20, 10, m28)$	Y = {6, 9, 11, 12, 14...22, 24...32, 34...42, 44...52, 54...62, 64...72, 74...82, 84...92, 94...102, 104...112, 114...122, 124...132, 134...142} LIM = 143, [39], Theorem 3.13.
4-$(21, 6, m2)$	Y = {6, 8, 17, 18, 20, 30, 34} LIM = 34, [47, 38], Theorems 3.12, 3.3.
4-$(21, 7, m10)$	Y = {12, 34} LIM = 34, [38], Theorem 3.12.
4-$(21, 8, m70)$	Y = {17} LIM = 17, Theorem 3.12.
4-$(21, 9, m14)$	Y = {34, 51, 68, 85, 102, 130, 136, 153, 154, 170, 187, 192, 204, 216} LIM = 221, [38], Theorems 3.12, 3.3.
4-$(21, 10, m28)$	Y = {12, 16, 24, 34, 36, 40, 46, 52, 54, 60, 64, 66, 68, 70, 72, 78, 82, 84, 85, 90, 94, 96, 102, 120, 132, 136, 144, 153, 170, 187, 202, 204} LIM = 221, [38], Theorems 3.12, 3.3.
4-$(22, 7, m4)$	Y = {102} LIM = 102, Theorem 3.3.
4-$(22, 8, m30)$	Y = {51} LIM = 51, Theorem 3.12.
4-$(22, 10, m42)$	Y = {34, 68, 85, 102, 136, 153, 170, 187, 204} LIM = 221, Theorem 3.12.
4-$(22, 11, m72)$	Y = {12, 16, 24, 34, 36, 40, 46, 52, 54, 60, 64, 66, 68, 70, 72, 78, 82, 84, 85, 90, 94, 96, 102, 120, 132, 136, 144, 153, 170, 187, 202, 204} LIM = 221, Theorem 3.13.
4-$(23, 5, m)$	Y = {1...9} LIM = 9, [47], Theorem 3.8.
4-$(23, 6, m3)$	Y = {1...28} LIM = 28, Theorems 3.9, 3.8.
4-$(23, 7, m)$	Y = {1...24, 34, 51, 68, 85, 102, 112, 113, 119, 120, 128, 129, 136, 153, 170, 180, 187, 204, 221, 238, 255, 272, 289, 306, 323, 340, 357, 374, 391, 408, 425, 442, 459, 476} LIM = 484, [68, 35, 24, 14], Theorems 3.9, 3.8.

4-$(23, 8, m2)$	$N = \{1\}$, $Y = \{2, 3, 4, 6, 8, 10, 12, 14, 16, 18, 45, 48, 180, 240, 242, 256, 258, 360, 363, 405, 408, 840, 858, 969\}$ LIM = 969, [59, 14], Theorems 3.9, 3.8, 3.12.
4-$(23, 9, m18)$	$Y = \{1, 6, 9, 10, 24, 30, 56, 60, 126, 127, 135, 136, 224, 280, 286, 304, 310\}$ LIM = 323, [24, 14], Theorems 3.9, 3.8.
4-$(23, 10, m42)$	$Y = \{8, 10, 30, 45, 48, 55, 56, 80, 96, 110, 118, 158, 166\}$ LIM = 323, [24, 14], Theorems 3.9, 3.8.
4-$(23, 11, m6)$	$N = \{1, 2\}$, $Y = \{8, 96, 104, 646, 672, 715, 1056, 1064, 1152, 1160, 1248, 1292, 1430, 1534, 1615, 1938, 2494, 2584, 2590, 2907, 3230, 3553, 3876\}$ LIM = 4199, [26, 14], Theorems 3.9, 3.8, 3.12.
4-$(24, 6, m10)$	$Y = \{1 \ldots 9\}$ LIM = 9, Theorem 3.3.
4-$(24, 7, m20)$	$Y = \{1 \ldots 28\}$ LIM = 28, Theorem 3.3.
4-$(24, 8, m5)$	$Y = \{1 \ldots 9, 22, 24, 120, 128, 129, 180, 204\}$ LIM = 484, Theorems 3.3, 3.12.
4-$(24, 9, m24)$	$Y = \{1, 6, 60, 135, 136, 280, 286\}$ LIM = 323, Theorem 3.3.
4-$(24, 10, m60)$	$Y = \{10, 30, 56\}$ LIM = 323, Theorem 3.3.
4-$(24, 11, m120)$	$Y = \{8, 55, 96, 110, 118\}$ LIM = 323, Theorem 3.3.
4-$(24, 12, m15)$	$N = \{1\}$, $Y = \{8, 96, 104, 646, 672, 715, 1056, 1064, 1152, 1160, 1248, 1292, 1430, 1534, 1615, 1938, 2494, 2584, 2590, 2907, 3230, 3553, 3876\}$ LIM = 4199, [26], Theorems 3.3, 3.13.
4-$(25, 6, m30)$	$Y = \{1, 2\}$ LIM = 7, [49].
4-$(25, 7, m70)$	$Y = \{1 \ldots 9\}$ LIM = 9, Theorem 3.12.
4-$(25, 8, m105)$	$Y = \{12\}$ LIM = 28, Theorem 3.12.
4-$(25, 9, m63)$	$Y = \{3, 68\}$ LIM = 161, Theorem 3.12.
4-$(25, 12, m45)$	$Y = \{56, 385, 672, 770, 826\}$ LIM = 2261, Theorem 3.12.
4-$(26, 6, m3)$	$Y = \{14, 21 \ldots 35\}$ LIM = 38, Theorem 3.8.
4-$(26, 7, m140)$	$Y = \{2 \ldots 5\}$ LIM = 5, Theorem 3.9.
4-$(26, 8, m35)$	$Y = \{38, 44, 57, 76, 95\}$ LIM = 104, Theorems 3.9, 3.12.
4-$(26, 9, m126)$	$Y = \{44\}$ LIM = 104, Theorem 3.12.
4-$(26, 13, m110)$	$Y = \{56, 385, 672, 770, 826\}$ LIM = 2261, Theorem 3.13.
4-$(27, 5, m)$	$Y = \{2 \ldots 11\}$ LIM = 11, Theorem 3.8.
4-$(27, 6, m)$	$Y = \{1, 21, 22, 33, 44, 55, 66, 77, 88, 99, 105, 106, 110, 121, 126\}$ LIM = 126, [22, 14], Theorems 3.9, 3.8.
4-$(27, 7, m7)$	$Y = \{1, 7, 8, 46, 69, 92, 105, 106, 112, 113, 115\}$ LIM = 126, [14], Theorems 3.9, 3.3, 3.8.
4-$(27, 8, m35)$	$Y = \{8, 28, 36, 46, 69, 92, 115\}$ LIM = 126, [14], Theorems 3.3, 3.8.
4-$(27, 9, m7)$	$Y = \{684, 1012\}$ LIM = 2403, [14], Theorem 3.12.
4-$(28, 5, m12)$	$Y = \{1\}$ LIM = 1, Theorem 3.8.
4-$(28, 6, m6)$	$Y = \{4, 6, 8, 10, 12, 14, 16, 18, 19, 20, 22, 23\}$ LIM = 23, Theorems 3.3, 3.9, 3.12.
4-$(28, 7, m4)$	$Y = \{2, 209, 210, 212, 253\}$ LIM = 253, Theorems 3.3, 3.9.
4-$(28, 8, m42)$	$Y = \{8, 46, 69, 92, 115\}$ LIM = 126, Theorem 3.3.
4-$(28, 9, m168)$	$Y = \{36\}$ LIM = 126, Theorem 3.3.
4-$(29, 5, m5)$	$Y = \{1\}$ LIM = 2, [47].
4-$(29, 6, m30)$	$Y = \{5\}$ LIM = 5, Theorem 3.3.
4-$(29, 7, m10)$	$Y = \{95, 115\}$ LIM = 115, Theorems 3.3, 3.12.
4-$(29, 11, m220)$	$Y = \{1\}$ LIM = 1092, Theorem 3.8.
4-$(29, 12, m495)$	$Y = \{1\}$ LIM = 1092, Theorem 3.9.
4-$(29, 13, m55)$	$Y = \{98, 1025\}$ LIM = 18572, Theorem 3.8.
4-$(29, 14, m22)$	$Y = \{392, 4100\}$ LIM = 74290, Theorem 3.9.
4-$(30, 5, m2)$	$Y = \{4, 5\}$ LIM = 6, [49], Theorem 3.8.
4-$(30, 6, m5)$	$Y = \{3, 25\}$ LIM = 32, [49], Theorem 3.9.
4-$(30, 7, m20)$	$Y = \{50, 65\}$ LIM = 65, Theorems 3.3, 3.9.
4-$(30, 8, m10)$	$Y = \{575\}$ LIM = 747, Theorem 3.9.
4-$(30, 12, m55)$	$Y = \{13\}$ LIM = 14202, Theorem 3.3.
4-$(30, 14, m143)$	$Y = \{98, 1025\}$ LIM = 18572, Theorem 3.3.
4-$(30, 15, m52)$	$Y = \{392, 4100\}$ LIM = 74290, Theorem 3.13.

5-$(12,6,m)$ $Y = \{1, 2, 3\}$ LIM $= 3$, Theorem 3.11.

5-$(13,6,m4)$ $Y = \{1\}$ LIM $= 1$, [43]. (See Example 3.15).

5-$(14,6,m3)$ $Y = \{1\}$ LIM $= 1$, [15].

5-$(14,7,m6)$ $Y = \{2, 3\}$ LIM $= 3$, Theorems 3.11, 3.13.

5-$(15,7,m15)$ $Y = \{1\}$ LIM $= 1$, [67].

5-$(16,6,m)$ $N = \{1\}$, $Y = \{3, 5\}$ LIM $= 5$, [15], Theorem 3.8.

5-$(16,7,m5)$ $Y = \{3\}$ LIM $= 5$, [15].

5-$(16,8,m5)$ $Y = \{6 \ldots 16\}$ LIM $= 16$, Theorem 3.11.

5-$(17,7,m6)$ $Y = \{2, 3, 4\}$ LIM $= 5$, [15, 67].

5-$(17,8,m20)$ $Y = \{3, 4\}$ LIM $= 5$, [67, 36].

5-$(18,6,m)$ $Y = \{4, 5\}$ LIM $= 6$, [36, 15].

5-$(18,7,m6)$ $Y = \{1 \ldots 6\}$ LIM $= 6$, [36].

5-$(18,8,m2)$ $N = \{1\}$, $Y = \{3, 7, 8, 15, 16, 20, 22, 23, 24, 30 \ldots 33, 38 \ldots 41, 46 \ldots 49, 52, 54 \ldots 57, 62 \ldots 65, 70, 71\}$ LIM $= 71$, [54, 36], Theorems 3.12, 3.8.

5-$(18,9,m5)$ $N = \{1\}$, $Y = \{3, 6 \ldots 33, 35 \ldots 57, 59 \ldots 71\}$ LIM $= 71$, [36, 15], Theorems 3.8, 3.11.

5-$(19,7,m7)$ $Y = \{4, 5, 6\}$ LIM $= 6$, [67, 15], Theorem 3.12.

5-$(19,8,m28)$ $Y = \{2 \ldots 5\}$ LIM $= 6$, [67], Theorem 3.8.

5-$(19,9,m7)$ $N = \{1\}$, $Y = \{3, 7, 8, 15, 16, 20, 22, 23, 24, 30 \ldots 33, 38 \ldots 41, 44, 46 \ldots 49, 52, 54 \ldots 57, 62 \ldots 65, 70, 71\}$ LIM $= 71$, [33, 15, 67], Theorem 3.9.

5-$(20,8,m35)$ $Y = \{1 \ldots 6\}$ LIM $= 6$, [39].

5-$(20,9,m105)$ $Y = \{1 \ldots 6\}$ LIM $= 6$, [39, 67].

5-$(20,10,m21)$ $Y = \{3, 6 \ldots 71\}$ LIM $= 71$, [39], Theorems 3.13, 3.11.

5-$(21,6,m4)$ $Y = \{2\}$ LIM $= 2$, Theorem 3.8.

5-$(21,7,m30)$ $Y = \{2\}$ LIM $= 2$, Theorem 3.9.

5-$(21,8,m280)$ $Y = \{1\}$ LIM $= 1$, Theorem 3.9.

5-$(21,9,m70)$ $Y = \{2, 4, 6, 8, 10, 12\}$ LIM $= 13$, [67].

5-$(21,10,m168)$ $Y = \{2, 4, 6, 8, 10, 12\}$ LIM $= 13$, [38, 67], Theorem 3.12.

5-$(22,7,m2)$ $Y = \{8, 30, 34\}$ LIM $= 34$, [49], Theorem 3.3.

5-$(22,8,m20)$ $Y = \{3, 4, 5, 17\}$ LIM $= 17$, [49], Theorem 3.3.

5-$(22,9,m70)$ $Y = \{3\}$ LIM $= 34$, [49].

5-$(22,10,m14)$ $Y = \{34, 48, 68, 102, 136, 170, 204\}$ LIM $= 221$, [49], Theorem 3.12.

5-$(22,11,m28)$ $Y = \{10, 12, 16, 24, 34, 36, 40, 46, 52, 54, 60, 64, 66, 68, 70, 72, 78, 82, 84, 85, 90, 94, 96, 102, 120, 132, 136, 144, 153, 170, 187, 202, 204\}$ LIM $= 221$, [49], Theorems 3.11, 3.13.

5-$(23,8,m8)$ $Y = \{51\}$ LIM $= 51$, Theorem 3.3.

5-$(23,11,m42)$ $Y = \{34, 68, 102, 136, 170, 204\}$ LIM $= 221$, Theorem 3.12.

5-$(24,6,m)$ $Y = \{1 \ldots 9\}$ LIM $= 9$, [22, 47, 24].

5-$(24,7,m3)$ $Y = \{1 \ldots 28\}$ LIM $= 28$, [24, 47].

5-$(24,8,m)$ $Y = \{1 \ldots 9, 24, 128, 129\}$ LIM $= 484$, [68, 35, 24, 14]. (See Example 3.27).

5-$(24,9,m6)$ $Y = \{1, 6, 60, 135, 136, 280, 286\}$ LIM $= 323$, [4, 24, 14].

5-$(24,10,m18)$ $Y = \{10, 30, 56\}$ LIM $= 323$, [24, 14].

5-$(24,11,m42)$ $Y = \{8, 55, 96, 110, 118\}$ LIM $= 323$, [24, 14].

5-$(24,12,m6)$ $N = \{1, 2\}$, $Y = \{8, 96, 104, 646, 672, 715, 1056, 1064, 1152, 1160, 1248, 1292, 1430, 1534, 1615, 1938, 2494, 2584, 2590, 2907, 3230, 3553, 3876\}$ LIM $= 4199$, [4, 14], Theorems 3.8, 3.11, 3.13.

5-$(25,7,m10)$ $Y = \{1 \ldots 9\}$ LIM $= 9$, Theorem 3.12.

5-$(25,9,m15)$ $Y = \{3\}$ LIM $= 161$, Theorem 3.12.

5-$(25,12,m120)$ $Y = \{8, 55, 96, 110, 118\}$ LIM $= 323$, Theorem 3.12.

5-$(26,13,m45)$ $Y = \{56, 385, 672, 770, 826\}$ LIM $= 2261$, Theorem 3.11.

5-$(27,7,m21)$ $Y = \{2 \ldots 5\}$ LIM $= 5$, Theorem 3.8.

5-$(27,8,m140)$ $Y = \{2 \ldots 5\}$ LIM $= 5$, Theorem 3.9.

5-$(28,6,m)$ $Y = \{2 \ldots 11\}$ LIM $= 11$, [44].

5-$(28,7,m)$ $Y = \{1, 105, 106\}$ LIM $= 126$, [22, 14].

5-$(28,8,m7)$ $Y = \{8, 46, 69, 92, 115\}$ LIM $= 126$, [14], Theorem 3.3.

5-(28, 9, m35) $Y = \{36\}$ LIM = 126, [14].
5-(29, 6, m12) $Y = \{1\}$ LIM = 1, Theorem 3.8.
5-(29, 7, m6) $Y = \{19, 23\}$ LIM = 23, Theorems 3.9, 3.12.
5-(30, 6, m5) $Y = \{1\}$ LIM = 5, [49].
5-(30, 7, m30) $Y = \{5\}$ LIM = 5, Theorem 3.3.
5-(30, 8, m20) $Y = \{5\}$ LIM = 260, [49].
5-(30, 12, m220) $Y = \{1\}$ LIM = 1092, [54].
5-(30, 14, m55) $Y = \{98, 1025\}$ LIM = 18572, [54].
5-(30, 15, m22) $Y = \{392, 4100\}$ LIM = 74290, Theorem 3.11.
6-(14, 7, m4) $Y = \{1\}$ LIM = 1, Theorem 3.11.
6-(17, 7, m) $N = \{1\}$ LIM = 5, Theorem 3.8.
6-(19, 9, m2) $N = \{1 \ldots 5\}$ LIM = 71, [26].
6-(20, 9, m28) $Y = \{4\}$ LIM = 6, [39].
6-(20, 10, m7) $N = \{1, 2\}$ LIM = 71, [26].
6-(22, 7, m4) $Y = \{2\}$ LIM = 2, [66].
6-(22, 8, m60) $Y = \{1\}$ LIM = 1, [48].
6-(23, 8, m4) $Y = \{17\}$ LIM = 17, Theorem 3.12.
6-(28, 8, m21) $Y = \{2 \ldots 5\}$ LIM = 5, [69].
6-(30, 7, m12) $Y = \{1\}$ LIM = 1, [66].
6-(30, 15, m44) $N = \{1\}$ LIM = 14858, [26].
7-(18, 8, m) $N = \{1\}$ LIM = 5, Theorem 3.8.
7-(20, 10, m2) $N = \{1 \ldots 5\}$ LIM = 71, Theorem 3.8.
7-(26, 11, m6) $N = \{1\}$ LIM = 323, [26].
8-(19, 9, m) $N = \{1\}$ LIM = 5, Theorem 3.8.
8-(27, 11, m3) $N = \{1\}$ LIM = 161, [26].
8-(27, 12, m6) $N = \{1, 2\}$ LIM = 323, [26].
8-(27, 13, m18) $N = \{1\}$ LIM = 323, [26].
8-(28, 13, m24) $N = \{1\}$ LIM = 323, [26].
9-(20, 10, m) $N = \{1\}$ LIM = 5, Theorem 3.8.
9-(28, 12, m3) $N = \{1\}$ LIM = 161, [26].
9-(28, 13, m6) $N = \{1, 2\}$ LIM = 323, [26].
9-(28, 14, m18) $N = \{1\}$ LIM = 323, [26].
9-(29, 14, m24) $N = \{1\}$ LIM = 323, [26].
10-(27, 12, m4) $N = \{1\}$ LIM = 17, [26].
10-(27, 13, m20) $N = \{1\}$ LIM = 17, [26].
10-(29, 13, m3) $N = \{1, 2, 3\}$ LIM = 161, [26].
10-(29, 14, m6) $N = \{1 \ldots 8\}$ LIM = 323, [26].
10-(30, 14, m15) $N = \{1\}$ LIM = 161, [26].
10-(30, 15, m24) $N = \{1 \ldots 4\}$ LIM = 323, [26].
11-(28, 13, m4) $N = \{1\}$ LIM = 17, [26].
11-(28, 14, m20) $N = \{1\}$ LIM = 17, [26].
11-(30, 14, m3) $N = \{1 \ldots 5\}$ LIM = 161, [26].
11-(30, 15, m6) $N = \{1 \ldots 8\}$ LIM = 323, [26].
12-(29, 14, m4) $N = \{1\}$ LIM = 17, [26].
12-(30, 15, m24) $N = \{1\}$ LIM = 17, [26].
13-(30, 15, m4) $N = \{1\}$ LIM = 17, Theorem 3.8.

3.38 Remarks There are other sporadic examples of t-designs known with $3 \leq t \leq 5$; see §V.1. The only known t-designs with $t \geq 6$ not already listed in Tables 3.31 and 3.37 were obtained by using $G = \mathrm{P\Gamma L}(2, 32)$ as an automorphism group. This group is generated by the two permutations:

$(1\ 2\ 4\ 8\ g)(3\ 6\ c\ o\ h)(5\ a\ k\ 9\ i)(7\ e\ t\ p\ j)(b\ m\ d\ q\ \ell)(f\ v\ u\ s\ n)$ and
$(1\ i\ v)(2\ \ell\ c)(3\ a\ t)(4\ w\ x)(5\ o\ e)(6\ 7\ h)(8\ p\ s)(9\ j\ k)(b\ f\ d)(g\ n\ u)(m\ 0\ q)$.

Six 6-(33, 8, 36) designs were obtained by Magliveras and Leavitt [56]. Base blocks for one of these designs are:

01234568 01235789 0123569a 1234678c 0123567a 013689ab
0124568a 0134678b 01345789 1234789a 0145678a

A 7-$(33, 8, 10)$ design was obtained by Betten, Kerber, Kohnert, Laue and Wasserman [8]. Base blocks for this design are:

01234589 01234569 01234568 0123456a 01235789 01356789 01345679 0123468a 0123458b
0125679a 012469ab 1245679b 01456789 013678ab 0123567a 01346789 01235689 0123579b
012345ab 01234689 0134579a 0134569a 0125689b 01234679 01345789 01234789 0124579b
012378ab 01245789 013478ab 0124678a 014679ab 134679ac 1234568a 0145678b 0145678a

Its derived design is a 6-$(32, 7, 10)$ design, and its residual is a 6-$(32, 8, 125)$ design. By Theorem 3.3, it is a 6-$(33, 8, 135)$ design. Several entries in Table 3.37 are obtained by applying the appropriate theorems.

Betten, Laue and Wassermann have very recently reported the existence of a 7-(26,8,6) design and a 7-(26,9,63) design with automorphism group PGL(2,25). The designs that arise from these have not been included in the tables.

3.4 Large Sets of t-Designs

A *large set* of t-(v, k, λ) designs, denoted $\mathrm{LS}[N](t, k, v)$, is a partition $[(X, \mathcal{B}_i)]_{i=1}^{N}$ of the complete design into N disjoint t-$(v, k, \lambda_{max}(t, k, v)/N)$ designs. Large sets are also denoted by $\mathrm{LS}_\lambda(t, k, v)$ to emphasize the value of λ. The λ can be omitted if it is one.

An $\mathrm{LS}[2](t, k, v)$ is a *halving* of the complete design.

The set of all v for which an $\mathrm{LS}[N](t, k, v)$ exists is denoted by $\mathrm{LS}[N](t, k)$.

If every t-(v, k, λ) design in the large set $[(X, \mathcal{B}_i)]_{i=1}^{N}$ has the group G as an automorphism group, the large set is a *uniformly-G* $\mathrm{LS}[N](t, k, v)$.

3.39 Example A uniformly-G LS[3](3, 5, 13) is given by the group

$$G = \langle (0\ 1\ 2\ 3\ 4\ 5\ 6\ 7\ 8\ 9\ a\ b\ c), (2\ 4\ 9)(3\ 7\ 6)(5\ 0\ b)(8\ 9\ c)\rangle$$

and the base blocks:

design 1: 12357 12359 1238a 1248a 1246b 1245c 12345 12468 12678 12459 1234b
design 2: 12456 12467 12349 1235c 12347 12378 12389 1235b 1239b 1246c 12350
design 3: 12457 12348 12478 12379 1237c 12356 12358 12368 12469 1234a 1234c

3.40 Theorem (Teirlinck [66]) For every natural number t let $\lambda(t) = \mathrm{lcm}(\binom{t}{m} | m = 1, 2, \ldots t))$, $\lambda^*(t) = \mathrm{lcm}(1, 2, \ldots t+1)$ and $\ell(t) = \prod_{i=1}^{t} \lambda(i) \cdot \lambda^*(i)$. Then for all $N > 0$, there is an $\mathrm{LS}[N](t, t+1, t + N \cdot \ell(t))$.

3.41 Theorem (Khosrovshahi and Ajoodani-Namini [32]) If there are $\mathrm{LS}[N](t, t+1, v)$ and $\mathrm{LS}[N](t, t+1, w)$, then there is also an $\mathrm{LS}[N](t, t+1, v+w-t)$.

3.42 Theorem (Qiu-rong Wu [74]) If there are large sets with parameters $\mathrm{LS}[N](t-1, t+1, v)$, $\mathrm{LS}[N](t-1, t+1, w)$, $\mathrm{LS}[N](t-1, t, v-1)$, $\mathrm{LS}[N](t-1, t, w-1)$, then there is also an $\mathrm{LS}[N](t-1, t+1, v+w-t)$.

3.43 Theorem (Cusack, Kreher and Graham [20]) Let $q > 3$ be a prime power with $q \equiv 3 \pmod 4$ and let $N > 1$ be a integer.

1. There is a uniformly-PSL$(2, q)$ $\mathrm{LS}[N](3, 4, q+1)$ if and only if $q \equiv 2 \pmod{3N}$, $q \equiv 11$ or $23 \pmod{24}$ and $N \leq \frac{q-2}{15}$ or $N = 5$ and $q = 47$.
2. There is a uniformly-PSL$(2, q)$ $\mathrm{LS}[N](3, 5, q+1)$ if and only if $q \equiv 3, 23, 27,$ or $47 \pmod{60}$ and N divides $(q-2)(q-3)/60$.

3.44 Table The known large sets of t-designs.

$\lambda = 1$:

- An LS$(1, k, v)$ exists if and only if k divides v, [7].
- An LS$(2, 3, v)$ exists if and only if $v \equiv 1$ or $3 \pmod 6$ and $v \neq 7$ [63].
- An LS(2,4,13) exists, [19]. This is the only known large set of $t-(v, k, 1)$ designs with $t \geq 2$ and $k \geq 4$.

$\lambda \geq 1$:

- For all integers $N, t > 0$, $\exists v$ such that an LS$[N](t, t+1, v)$ exists (Theorem 3.40).
- An LS$_{\lambda_{min}}(2, 3, v)$ exists if and only if $v \equiv 0, 2, 4, 5 \pmod 6$, [51, 52, 70, 71, 61].
- An LS$_{\lambda_{min}}(3, 4, v)$ exists if $v \equiv 0 \pmod 3$, [62].
- LS$_{\lambda_{min}}(4, 5, 20u + 4)$ exists if $\gcd(u, 30) = 1$, [66]. The LS$_4(3, 4, 23)$ in [17] is used.
- An LS$_{60}(4, 5, 60u + 4)$ exists if $\gcd(u, 60)$ is 1 or 2 [66].
- An LS$_4(4, 5, 12)$ exists [23].
- An LS$_{\lambda_{min}}(4, 5, v)$ exists for $v \in \{28, 44, 124\}$ [66].
- An LS$_{4u}(5, 6, 5 + 8u)$ exists for all positive integers u [66]. These are all halvings of the complete design and are minimal when $u \in \{1, 3, 5, 15\}$.
- An LS$_{4u}(6, 7, 6 + 8u)$ exists for all positive integers u [66]. These are all halvings of the complete design and have minimal λ when $u \in \{1, 3, 5, 7, 15, 21, 35, 105\}$.
- An LS$_{15}(4, 6, 14)$ exists [17].
- An LS$_{60}(6, 8, 22)$ exists [48].
- An LS$_{68}(6, 8, 23)$ exists [48].
- An LS$_{4(m+1)(16m+17)}(6, 8, 23 + 16m,)$ exists for all positive integers m [48].
- In [42] all parameter situations on less than or equal to 12 points are settled.

Nonexistence:

- If a simple t-(v, k, λ) design does not exist (see Table 3.37), then a LS$_\lambda(1, k, v)$ does not exist.
- There does not exist an LS$(2, 3, 7)$ nor an LS$(3, 4, 8)$ (folklore, see also [72]).
- There does not exist an LS$(3, 4, 10)$ [34, 57].
- There is *no known* LS$(3, 4, v)$ for $v > 4$, but Etzion and Hartman [27] have constructed $(v - 5)$ disjoint $3 - (v, 4, 1)$ designs for $v = 5 \cdot 2^n$. This leaves only two more to go for a large set!
- There are no known LS(t, k, v) designs with $t \geq 7$.

3.5 See Also

§I.1	2-(v, k, λ) designs of small order.
§I.4	Steiner systems.
§IV.24	Hadamard designs and matrices.
§IV.49	t-wise balanced designs.
§V.1	Codes.
§VI.1	Group theory.
§VI.9.2	Computer algorithms for finding t-designs.

References

[1] W.O. Alltop, An infinite class of 4-designs, *J. Combin. Theory A* 6 (1969), 320-322.
[2] W.O. Alltop, An infinite class of 5-designs, *J. Combin. Theory A* 12 (1972), 390-395.
[3] W.O. Alltop, Extending t-designs, *J. Combin. Theory A* 12 (1975), 177-186.

[4] E.F. Assmus Jr. and H.F. Mattson Jr., New 5-designs, *J. Combin. Theory* 6 (1969), 122-151.
[5] A.H. Baartmans, I. Bluskov and V.D. Tonchev, The Preparata codes and a class of 4-designs, *J. Combin. Des.* 2 (1994), 167-170.
[6] S. Bays and E. de Weck, Sur les systèmes de quadruples, *Comment. Math. Helvet.* 7 (1935), 222-241.
[7] Z. Baranyai, On the factorization of the complete uniform hypergraph, *Coll. Math. Soc. Janos Bolyai* 10 (1975), 91-108.
[8] A. Betten, A. Kerber, A. Kohnert, R. Laue and A. Wasserman, The discovery of simple 7-designs with automorphism group PΓL(2,32), preprint.
[9] K.N. Bhattacharya, A note on two-fold triple systems, *Sankhya* 6 (1943), 313-314.
[10] J. Bierbrauer, A new family of 4-designs, *Graphs Combin.*, to appear.
[11] J. Bierbrauer, A new family of 4-designs with block size 9, *Discrete Math.* 138 (1995), 113-117.
[12] J. Bierbrauer, Shadow and shade of designs 4-$(2^f+1,5,5)$, preprint.
[13] J. Bierbrauer, Some friends of Alltop's designs 4-$(2^f+1,5,5)$, preprint.
[14] I. Bluskov, private communication.
[15] A.E. Brouwer, Table of t-designs without repeated blocks, $2 \leq t \leq k \leq v/2$, $\lambda \leq \lambda^+/2$, Math. Centrum. Report ZN76, Amsterdam, 1977; unpublished update, 1986.
[16] P.J. Cameron, Extending symmetric designs, *J. Combin. Theory A* 14 (1973), 215-220.
[17] Y.M. Chee, C.J. Colbourn, S.C. Furino and D.L. Kreher, Large sets of disjoint t-designs, *Australas. J. Combin.* 2 (1990), 111-119.
[18] Y.M. Chee and D.L. Kreher, Simple 4-(21,5, λ) designs from the Frobenius group of order 171, *Ars Combin.* 36 (1993), 199-205.
[19] L.G. Chouinard, Partitions of the 4-subsets of a 13-set into disjoint projective planes, *Discrete Math.* 45 (1983), 296-300.
[20] C.A. Cusack, D.L. Kreher and S.W. Graham, Large sets of 3-designs from PSL(2,q), with block sizes 4 and 5, *J. Combin. Des.* 3 (1995), 147-160.
[21] M. Dehon, Non-existence d'un 3-design de parameters $\lambda = 2$, $k = 5$ et $v = 11$, *Discrete Math.* 15 (1976), 23-25.
[22] R.H.F. Denniston, Some new 5-designs, *Bull. London Math. Soc.* 8 (1976), 263-267.
[23] R.H.F. Denniston, A small 4-design, *Ann. Discrete Math.* 18 (1983), 291-294.
[24] L.H.M.E. Driessen, t-designs, $t \geq 3$, Technical Report, Department of Mathematics, Eindhoven University of Technology, Netherlands, 1978.
[25] B. Ganter, J. Pelikan and L. Teirlinck, Small sprawling systems of equicardinal sets, *Ars Combin.* 4 (1977), 133-142.
[26] W. Haemers, C. Weug and P. Delsarte, Linear programming bounds for codes and designs, Technical Note N96, M.B.L.E. Lab. de Recherches, Brussels, Belgium, 1974.
[27] T. Etzion and A. Hartman, Towards a large set of Steiner quadruple systems, *SIAM J. Disc. Math.* 4 (1991), 182-195.
[28] H. Hanani, On quadruple systems, *Canad. J. Math.* 12 (1960), 145-157.
[29] H. Hanani, A. Hartman and E.S. Kramer, On three-designs of small order, *Discrete Math.* 45 (1983), 75-97.
[30] X. Hubaut, Two new families of 4-designs, *Discrete Math.* 9 (1974), 247-249.
[31] D.R. Hughes, On t-designs and groups, *Am. J. Math.* 87 (1965), 761-778.
[32] G.B. Khosrovshahi and S. Ajoodani-Namini, Combining t-designs, *J. Combin. Theory A* 58 (1991), 26-34.
[33] E. Köhler, Über den allgemeinen Gray-Code und die Nichtexistenz einiger t-designs, *Graphen in Forschung und Unterricht, Festschrift K. Wagner* (1985), 102-111.
[34] E.S. Kramer and D.M. Mesner, Intersections among Steiner systems, *J. Combin. Theory A* 16 (1974), 273-285.

[35] E.S. Kramer and S.S. Magliveras, Some mutually disjoint Steiner systems, *J. Combin. Theory A* 17 (1974), 39-43.

[36] E.S. Kramer, Some t-designs for $t \geq 4$ and $v = 17, 18$, *Congr. Numer.* 14 (1975), 443-459.

[37] E.S. Kramer and D. Mesner, t-designs on hypergraphs, *Discrete Math.* 15 (1976), 263-296.

[38] E.S. Kramer, The t-designs using $M_{21} \simeq PSL_3(4)$ on 21 points, *Ars Combin.* 17 (1984), 191-208.

[39] E.S. Kramer, D.W. Leavitt and S.S. Magliveras, Construction procedures for t-designs and the existence of new simple 6-designs, *Ann. Discrete Math.* 26 (1985), 247-274.

[40] E.S. Kramer, An $S_3(3, 5, 21)$ using graphs, *Discrete Math.* 81 (1990), 223-224.

[41] E.S. Kramer, S.S. Magliveras and E.A. O'Brien, Some new large sets of t-designs, *Australas. J. Combin.* 7 (1993), 189-193.

[42] E.S. Kramer, S.S. Magliveras and D.R. Stinson, Some small large sets of t-designs, *Australas. J. Combin.* 3 (1991), 191-205.

[43] D.L. Kreher and S.P. Radziszowski, The existence of simple 6-(14,7,4) designs, *J. Combin. Theory A* 43 (1986), 237-243.

[44] D.L. Kreher and S.P. Radziszowski, Simple 5-(28,6,λ) designs from $PSL_2(27)$, *Ann. Discrete Math.* 34 (1987), 315-318.

[45] D.L. Kreher and S.P. Radziszowski, New t-designs found by basis reduction, *Congr. Numer.* 59 (1987), 155-164.

[46] D.L. Kreher, A 4-(15,5,5) design, in: *Coding Theory and Design Theory: Part II. Design Theory* (D.K. Ray-Chaudhuri; ed.), Inst. Math. Appl., Springer-Verlag, New York, 1990, 226.

[47] D.L. Kreher, Y.M. Chee, D. de Caen, C.J. Colbourn and E.S. Kramer, Some new simple t-designs, *J. Combin. Math. Combin. Comput.* 7 (1990), 53-90.

[48] D.L. Kreher, An infinite family of (simple) 6-designs, *J. Combin. Des.* 1 (1993), 277-280.

[49] D. Kreher, t-designs 1995, technical report, 1995.

[50] C.C. Lindner, A note on disjoint Steiner quadruple systems, *Ars Combin.* 3 (1977), 271-276.

[51] J.X. Lu, On large sets of disjoint Steiner triple systems I, II, and III, *J. Combin. Theory A* 34 (1983), 140-146, 147-155, and 156-182.

[52] J.X. Lu, On large sets of disjoint Steiner triple systems IV, V, and VI, *J. Combin. Theory A* 37 (1984), 136-163, 164-188, and 189-192.

[53] B. McKay and S.P. Radziszowski, The nonexistence of 4-(12,6,6) designs. in: *Computational and Constructive Design Theory* (W.D. Wallis; ed.), Kluwer, to appear.

[54] F.J. MacWilliams, A.M. Odlyzko, N.J.A. Sloane and H.N. Ward, Self-dual codes over GF(4), *J. Combin. Theory A* 25 (1978), 288-318.

[55] S.S. Magliveras and T.E. Plambeck, New infinite families of simple 5-designs, *J. Combin. Theory A* 44 (1987), 1-5.

[56] S.S. Magliveras and D.W. Leavitt, Simple 6-(33, 8, 36) designs from $P\Gamma L_2(32)$, in: *Computational Group Theory* (M.D. Atkinson, ed.), Academic Press, New York, 1984, 337-352.

[57] N.S. Mendelsohn and S.H.Y. Hung, On the Steiner systems S(3,4,14) and S(4,5,15), *Utilitas Math.* 1 (1972), 5-95.

[58] W. Oberschelp, Lotto-Garantiesysteme und Block-Pläne, *Mathematisch-Phys. Semesterberichte* XIX (1972), 55-67.

[59] D.K. Ray-Chaudhuri and R.M. Wilson, On t-designs, *Osaka J. Math.* 12 (1975), 737-744.

[60] T. Skolem, Note 16, *E. Netto, Kombinatorik. 2. Auflage, Teubner, Leipzig*, 1927.

[61] L. Teirlinck, On the maximum number of disjoint triple systems, *J. Geometry* 12 (1975), 93-96.
[62] L. Teirlinck, On large sets of disjoint quadruple systems, *Ars Combin.* 17 (1984), 173-176.
[63] L. Teirlinck, A completion of Lu's determination of the spectrum for large sets of disjoint Steiner triple systems, *J. Combin. Theory A* 57 (1991), 302-305.
[64] L. Teirlinck, On large sets of disjoint quadruple systems, *Ars Combin.* 12 (1984), 173-176.
[65] L. Teirlinck, Non-trivial t-designs exist for all t, *Discrete Math.* 65 (1987), 345-356.
[66] L. Teirlinck, Locally trivial t-designs and t-designs without repeated blocks, *Discrete Math.* 77 (1989), 345-356.
[67] Tran van Trung, On the construction of t-designs and the existence of some new infinite families of simple 5-designs, *Arch. Math.* 47 (1986), 187-192.
[68] E. Witt, Über Steinersche Systeme, *Abh. Math. Sem. Hamburg* 12 (1938), 265-275.
[69] B. Schmalz, The t-designs with prescribed automorphism group, new simple 6-designs, *J. Combin. Des.* 1 (1993), 125-170.
[70] S. Schreiber, Covering all triples on n marks by disjoint Steiner systems, *J. Combin. Theory A* 15 (1973), 347-350.
[71] S. Schreiber, Some balanced complete block designs, *Israel J. Math.* 18 (1974), 31-37.
[72] W.D. Wallis, Large sets and Room squares, *Bull. ICA* 12 (1994), 93-94.
[73] R.M. Wilson, The necessary conditions for t-designs are sufficient for something, *Utilitas Math.* 4 (1973), 207-215.
[74] Qiu-rong Wu, A note on extending t-designs, *Australas. J. Combin.* 4 (1991), 229-235.

4 Steiner Systems

Charles J. Colbourn
Rudolf Mathon

4.1 Basics

Given three integers t, k, v such that $2 \leq t < k < v$, a *Steiner system* $S(t,k,v)$ is a v-set V together with a family $\mathcal{B}$ of k-subsets of V (*blocks*) with the property that every t-subset of S is contained in exactly one block. Equivalently, an $S(t,k,v)$ is a t-$(v,k,1)$ design.

4.1 **Remark** The nineteenth century geometer Jakob Steiner posed the question of the existence of Steiner systems. Unaware of the earlier work of Kirkman, in 1853 Steiner first asked about $S(2,3,v)$:

Steiner's Combinatorische Aufgabe:
Welche Zahl, N, von Elementen hat die Eigenschaft, dass sich die Elementen so zu dreien ordnen lassen, dass je zwei in einer, aber nur in einer Verbindung vorkommen?

Steiner went on to ask about $S(t, t+1, v)$ systems.

4.2 **Remark** Suppose that an $S(t,k,v)$ exists. By choosing an element x, selecting all blocks containing x, and deleting x from each, one obtains an $S(t-1,k-1,v-1)$, a *derived design.*

4.3 **Lemma** (Numerical or Divisibility Conditions) An $S(t,k,v)$ exists only if $\binom{v-i}{t-i}/\binom{k-i}{t-i}$ is an integer for every $i=0,1,\ldots,t-1$.

4.4 **Theorem** An $S(t,k,v)$ exists only if $\binom{v}{t} \geq \binom{v}{\lfloor t/2 \rfloor}\binom{k}{t}$.

4.2 Infinite Families

An $S(2,3,v)$ is a *Steiner triple system* STS(v). See §4.5.

4.5 **Remark** See §I.2 and §I.1.2 for existence results and examples of small $S(2,k,v)$.

An $S(3,4,v)$ is a *Steiner quadruple system* SQS(v). See §4.6.

4.6 **Theorem** Known infinite families of $S(t,k,v)$ are:

1. $S(2,q,q^n)$, q a prime power, $n \geq 2$ (affine geometries);
2. $S(3,q+1,q^n+1)$, q a prime power, $n \geq 2$ (spherical geometries);
3. $S(2,q+1,q^n+\cdots+q+1)$, q a prime power, $n \geq 2$ (projective geometries);
4. $S(2,q+1,q^3+1)$, q a prime power (unitals);
5. $S(2,2^r,2^{r+s}+2^r-2^s)$, $2 \leq r < s$ (Denniston designs).

4.7 **Theorem** (Wilson [2]) Let q be a prime power. If an $S(3,q+1,v+1)$ and an $S(3,q+1,w+1)$ both exist, then there exists an $S(3,q+1,vw+1)$.

4.8 **Corollary** [10] If q is a prime power and an $S(3,q+1,v+1)$ exists, then an $S(3,q+1,qv+1)$ exists.

4.9 **Theorem** [10] If an $S(3,6,v)$ exists, then an $S(3,6,4v-2)$ exists.

4.10 **Theorem** [2] Let q be a prime power, and let u be a prime power satisfying the standard divisibility conditions. Then there is a constant n_0 depending only on q and u so that, for all $\ell \geq 0$ and $n \geq n_0$, there exists an $S(3,q+1,u^\ell q^n+1)$.

4.11 **Remark** Only finitely many $S(t,k,v)$ with $t > 3$ are known and none are known for $t \geq 6$.

4.3 Existence Table

4.12 **Table** Parameters for $S(t,k,t+n)$ with $n \leq 200$. The column headed "$t \leq$" gives the maximum value of t for which numerical necessary conditions are met and Theorem 4.4 holds. $\exists$ gives the largest $t \geq 2$ for which a system in the family is known. $\nexists$ gives the smallest t in the admissible range for which the system is known *not* to exist. For example, the second entry in the first column of the table ($n=13, t \leq 11, \exists = 3$) indicates that there exists an $S(3,4,16)$ (hence an $S(2,3,15)$) and that it is possible according to the necessary conditions that there exists an $S(t,t+1,t+13)$ for all $t \leq 11$. The table was constructed using the infinite families and recursive constructions of §4.2, along with the following remarks:

1. For results on designs with $t=2$, see also §I.2.
2. All designs with $t=5$ are displayed in §4.7, and all parameters of known Steiner 4-designs are derived from the known Steiner 5-designs.

3. The nonexistence of $S(4,5,15)$ and $S(4,6,18)$ (see [1]), $S(4,10,66)$ [9] and $S(3,13,145)$ [4] has been established.

n	$t \leq$	$\exists$	$\nexists$	n	$t \leq$	$\exists$	$\nexists$	n	$t \leq$	$\exists$	$\nexists$
				k = t + 1							
5	3	3		7	5	5		11	9	3	4
13	11	3		17	15	3		19	17	5	
23	21	3		25	3	3		29	27	3	
31	29	3		35	3	3		37	35	3	
41	39	3		43	41	5		47	45	3	
49	5	3		53	51	3		55	3	3	
59	57	3		61	59	3		65	3	3	
67	65	5		71	69	3		73	71	3	
77	5	3		79	77	3		83	81	3	
85	3	3		89	87	3		91	5	3	
95	3	3		97	95	3		101	99	3	
103	101	5		107	105	3		109	107	3	
113	111	3		115	3	3		119	5	3	
121	9	3		125	3	3		127	125	5	
131	129	3		133	5	3		137	135	3	
139	137	3		143	9	3		145	3	3	
149	147	3		151	149	3		155	3	3	
157	155	3		161	5	3		163	161	5	
167	165	3		169	11	3		173	171	3	
175	3	3		179	177	3		181	179	3	
185	3	3		187	9	3		191	189	3	
193	191	3		197	195	3					
				k = t + 2							
11	2	2		14	4	3	4	23	8	5	
26	2	2		35	2	2		38	16	2	
47	20	2		50	2	2		59	26	2	
62	28	3		71	2	2		74	34	2	
83	38	2		86	2	2		95	2	2	
98	4	3		107	50	2		110	2	2	
119	4	2		122	8	2		131	2	2	
134	4	2		143	8	2		146	2	2	
155	2	2		158	76	2		167	80	2	
170	2	2		179	86	2		182	4	2	
191	2	2		194	94	2					
				k = t + 3							
19	5	5		23	3	3		39	9	2	
43	3	2		59	15	2		63	3	2	
79	3	2		83	5	3		99	3	3	
103	13	3		119	3	2		123	7	3	
139	19	2		143	7	2		159	49	2	
163	3	2		179	55	2		183	3	2	
				k = t + 4							
29	2	2		34	3		2	44	2		
49	2			59	2			64	2	2	
74	3	2		79	2			89	3	2	
94	2	2		104	3	2		109	4	2	
119	2	2		124	6	2		134	2	2	
139	3			149	2	2		154	2	2	
164	2	2		169	2	2		179	3	2	
184	2	2		194	3						

n	$t \le$	$\exists$	$\nexists$	n	$t \le$	$\exists$	$\nexists$	n	$t \le$	$\exists$	$\nexists$
					k = t + 5						
41	2		2	47	3	3		83	2		
89	2	2		125	5			131	2		
167	5	2		173	3						
					k = t + 6						
55	2	2		62	12	3	4	111	2		
118	2	2		167	2			174	2		
					k = t + 7						
71	2	2		79	3	3		143	3		
151	2										
					k = t + 8						
89	2	2		98	4		2	134	2		
143	2			179	2			188	2		
					k = t + 9						
109	3		2	119	3	3					
					k = t + 10						
131	2	2		142	3		3	175	2		
					k = t + 11						
155	2			167	4	3					
					k = t + 12						
181	2	2		194	3		2				

4.4 Intersection Triangles

Let $(V, \mathcal{B})$ be an $S(t, k, v)$ and let $B \in \mathcal{B}$. Let i and j be nonnegative integers with $i+j \le k$, and let I and O be disjoint subsets of B of sizes i and j, respectively. Then the *i,j-intersection number* $\lambda_{i,j}$ is the number of blocks that contain all elements of I and no elements of O. The array $(\lambda_{i,j} : 0 \le i+j \le k)$ is the *intersection triangle* of the design.

4.13 Construction The definition of $\lambda_{i,j}$ does not depend on the choice of B, I, or O, since $\lambda_{i,j}$ can be computed as follows. By definition,

$$\lambda_{i,0} = \begin{cases} \binom{v-i}{t-i}/\binom{k-i}{t-i} & \text{when } 0 \le i \le t \\ 1 & \text{when } t < i \le k \end{cases}$$

and, since $\lambda_{i,j-1} = \lambda_{i+1,j-1} + \lambda_{i,j}$, for all $j \ge 1$ and $1 \le i+j \le k$, one can calculate $\lambda_{i,j} = \lambda_{i+1,j-1} - \lambda_{i,j-1}$.

4.14 Example The intersection triangle for $S(5, 8, 24)$. See Table 4.34.

759	506	330	210	130	78	46	30	30
253	176	120	80	52	32	16	0	
77	56	40	28	20	16	16		
21	16	12	8	4	0			
5	4	4	4	4				
1	0	0	0					
1	0	0						
1	0							
1								

4.15 Remark The number x_i of blocks that intersect a fixed block B in exactly i elements is $\binom{k}{i}\lambda_{i,k-i}$, and hence does not depend on B. Thus, for example, in an $S(5, 8, 24)$, for every block B, one finds 30 disjoint blocks, 448 blocks intersecting B in two elements, and 280 blocks intersecting B in four elements.

4.16 Theorem [8] For an $S(t,k,v)$, if $x_i = 0$ and $0 \le i < t$, then
1. $i = 0$ and the design is one of $S(4,7,23)$, $S(t,t+1,2t+3)$ when $t+3$ is prime, or a projective plane $S(2,q+1,q^2+q+1)$;
2. $i = 1$ and the design is one of $S(3,6,22)$, $S(5,8,24)$, or $S(t,t+1,2t+2)$ when $t+2$ is prime;
3. $i = 2$ and the design is $S(4,7,23)$; or
4. $i = 3$ and the design is $S(5,8,24)$.

4.17 Remark An analogous definition of intersection triangles can be made for t-designs with $\lambda > 1$; however, in that case, since the intersection numbers $\lambda_{i,0}$ for $i > t$ may depend on the block B and sets I and O chosen, one cannot guarantee that $\lambda_{i,j}$ for $t < i+j \le k$ is a constant. However, when $i+j \le t$, the intersection numbers for t-designs in general are again independent of the block and sets chosen.

4.5 Steiner Triple Systems

4.18 Theorem An STS(v) ((v, 3, 1) BIBD) exists if and only if $v \equiv 1, 3 \pmod 6$.

4.19 Remark There are many different proofs of Theorem 4.18. The Reverend Thomas P. Kirkman (see Example I.6.6) first proved it in 1847 (six years before Steiner posed the question of their existence!); Kirkman's solution is recursive. A proof by direct construction is known from Skolem sequences (Constructions IV.43.11 and IV.43.13). An elementary recursive proof employs a $v \to 2v+1$ construction and a $v \to 2v+7$ construction, along with an STS(3), an STS(9) and an STS(13).

4.20 Examples All nonisomorphic STS(v) for $v \in \{7, 9, 13, 15\}$ are given in §I.1.

4.21 Theorem An STS(v) whose automorphism group contains a (cyclic) automorphism of order v exists if and only if $v \equiv 1, 3 \pmod 6$ and $v \ne 9$. See §IV.10 and §IV.43.

A resolvable STS(v) is a *Kirkman triple system*. See §I.6.

4.22 Theorem A Kirkman triple system of order v exists for all positive integers $v \equiv 3 \pmod 6$.

If $(V, \mathcal{B})$ is an STS(v), $W \subseteq V$, $\mathcal{D} = \{B \in \mathcal{B} : B \subseteq W\}$, and $(W, \mathcal{D})$ is a Steiner triple system, then $(W, \mathcal{D})$ is a *subsystem* of $(V, \mathcal{B})$.

4.23 Theorem (Doyen-Wilson) If $v, w \equiv 1, 3 \pmod 6$ and $v \ge w \ge 1$, then there is a STS(v) containing a subsystem of order w if and only if $v = w$ or $v \ge 2w+1$.

A *large set* of STS(v), $v \equiv 1, 3 \pmod 6$, is a partition of the set of all $\binom{v}{3}$ 3-subsets of v elements into $v-2$ STS(v).

4.24 Theorem (Lu-Teirlinck) A large set of STS(v) exists if and only if $v \equiv 1, 3 \pmod 6$ and $v \ne 7$.

4.6 Steiner Quadruple Systems

4.25 Theorem An SQS(v) (3-(v, 4, 1) design) exists if and only if $v \equiv 2, 4 \pmod 6$.

4.26 **Remark** Theorem 4.25 can be proved using six recursive constructions: If there is an SQS(v), then SQS($2v$), SQS($3v-2$), SQS($4v-6$), and SQS($12v-10$) all exist. If, in addition, $v \equiv 2 \pmod{12}$, then an SQS($3v-8$) also exists. If instead $v \equiv 10 \pmod{12}$, then an SQS($3v-4$) exists. These constructions, together with an SQS(14) and a certain SQS(38), suffice to prove existence. See [11].

Let $V = \mathbb{Z}_2^n$, $n \geq 3$. Let $\mathcal{B}$ be the set of all sets of four distinct elements of V whose (vector) sum is zero. Then $(V, \mathcal{B})$ is an SQS(2^n), the *boolean quadruple system* of order 2^n.

4.27 **Example** The boolean quadruple system of order 8; the unique SQS(8).

000 001 010 011	000 001 100 101	000 001 110 111	000 010 100 110
000 010 101 111	000 011 100 111	000 011 101 110	100 101 110 111
010 011 110 111	010 011 100 101	001 011 101 111	001 011 100 110
001 010 101 110	001 010 100 111		

4.28 **Example** The unique SQS(10) is a graphical design. See §IV.23.4.

4.29 **Example** An SQS(14) (also a (14,4,6) BIBD). Let $V = \{x_i : x \in \mathbb{Z}_7,\ i \in \mathbb{Z}_2\}$. The blocks are obtained by applying the automorphisms $x_i \mapsto (x+1)_i$ and $x_i \mapsto (3x)_{i+1}$ to the five base blocks:

$$0_0, 1_0, 2_0, 4_0 \quad 0_0, 1_1, 2_1, 4_1 \quad 3_0, 4_0, 3_1, 4_1 \quad 4_0, 5_0, 2_1, 3_1 \quad 5_0, 6_0, 1_1, 2_1$$

4.30 **Theorem** For $v \leq 100$, an SQS(v) whose automorphism group contains a (cyclic) automorphism of order v exists when $v \equiv 2, 4 \pmod 6$, except when $v \in \{8, 14, 16\}$ and possibly when $v \in \{46, 56, 62, 70, 86, 94\}$.

4.31 **Theorem** Every Steiner triple system of order $v \leq 15$ is the derived system of some SQS($v+1$).

An SQS(v) is *resolvable* if its blocks can be partitioned into *parallel classes*, each of which is a set of blocks that partition the set of elements.

4.32 **Theorem** A resolvable SQS(v) exists for all $v \equiv 4, 8 \pmod{12}$ except possibly for $v \in \{220, 236, 292, 364, 460, 596, 676, 724, 1076, 1100, 1252, 1316, 1820, 2236, 2308, 2324, 2380, 2540, 2740, 2812, 3620, 3820, 6356\}$.

4.7 Steiner 5-Designs

4.33 **Remark** All known Steiner 5-designs are of order $q+1$ where $q \equiv 3 \pmod 4$ is a prime power. All known examples have automorphism group $\mathrm{PSL}_2(q)$; the $S(5,8,24)$ has the larger Mathieu group M_{24} as its automorphism group, and the unique $S(5,6,12)$ has the Mathieu group M_{12}. The $S(5,8,24)$ and its derived $S(4,7,23)$ are the *Witt designs*. Employ the representation of $\mathrm{PSL}_2(q)$ when $q \in \{11, 23, 47, 71, 83, 107, 131, 167\}$ acting on $\mathbb{Z}_q \cup \{\infty\}$ which contains the permutations generated by (1) $x \mapsto x+1 \pmod q$, (2) $x \mapsto \alpha x \pmod q$ where $\alpha = 2$ for $q \in \{23, 47, 71, 167\}$ and $\alpha = 3$ for $q \in \{11, 83, 107, 131\}$; and (3) $x \mapsto -\frac{1}{x}$. For $\mathrm{PSL}_2(27)$, an explicit set of five generators is:

(0 1 2)(3 4 5)(6 7 8)(9 10 11)(12 13 14)(15 16 17)(18 19 20)(21 22 23)(24 25 26)
(0 3 6)(1 4 7)(2 5 8)(9 12 15)(10 13 16)(11 14 17)(18 21 24)(19 22 25)(20 23 26)
(0 9 18)(1 10 19)(2 11 20)(3 12 21)(4 13 22)(5 14 23)(6 15 24)(7 16 25)(8 17 26)
(1 9 15 13 20 12 11 6 7 16 22 8 25)(2 18 21 26 10 24 19 3 5 23 17 4 14)
(0 ∞)(1 2)(3 11)(4 15)(5 12)(6 19)(7 24)(8 21)(9 14)(10 16)(13 17)(18 25)(20 23)(22 26)

4.34 Table Parameters of known $S(5,k,v)$, and an example of each. The $S(5,8,24)$ is unique; for $S(5,6,24)$ and $S(5,7,28)$, all examples having the PSL group as a group of automorphisms are listed. See [5, 6, 12, 13, 14] for further discussion of these parameters.

$S(5,6,12)$			
∞,0,1,2,3,5			
$S(5,8,24)$			
∞,0,1,2,3,5,14,17			
$S(5,6,24)$ (three nonisomorphic solutions)			
∞,0,1,2,3,12	∞,0,1,2,5,14	∞,0,1,3,7,10	
∞,0,1,2,3,12	∞,0,1,2,5,14	∞,0,1,3,7,21	
∞,0,1,2,3,13	∞,0,1,2,5,18	∞,0,1,3,4,16	
$S(5,7,28)$			
∞,0,1,2,3,4,5	∞,0,1,3,7,10,18		
$S(5,6,48)$			
∞,0,1,2,3,40	∞,0,1,2,5,8	∞,0,1,2,10,27	∞,0,1,2,13,26
∞,0,1,3,4,18	∞,0,1,3,8,14	∞,0,1,3,12,39	∞,0,1,4,20,41
$S(5,6,72)$			
∞,0,1,2,3,14	∞,0,1,2,5,17	∞,0,1,2,6,67	∞,0,1,2,7,46
∞,0,1,2,8,56	∞,0,1,2,9,68	∞,0,1,2,10,18	∞,0,1,2,27,31
∞,0,1,3,7,8	∞,0,1,3,10,51	∞,0,1,3,11,67	∞,0,1,3,12,46
∞,0,1,3,13,66	∞,0,1,3,17,38	∞,0,1,3,20,42	∞,0,1,3,31,65
∞,0,1,4,9,21	∞,0,1,6,10,49	∞,0,1,6,14,31	
$S(5,6,84)$			
∞,0,1,2,3,8	∞,0,1,2,5,46	∞,0,1,2,6,20	∞,0,1,2,9,26
∞,0,1,2,10,13	∞,0,1,2,11,59	∞,0,1,2,12,73	∞,0,1,2,15,22
∞,0,1,2,31,45	∞,0,1,2,41,57	∞,0,1,3,4,46	∞,0,1,3,11,49
∞,0,1,3,12,58	∞,0,1,3,14,15	∞,0,1,3,18,26	∞,0,1,3,22,35
∞,0,1,3,23,52	∞,0,1,3,25,48	∞,0,1,3,33,60	∞,0,1,3,44,69
∞,0,1,3,62,65	∞,0,1,4,9,11	∞,0,1,4,13,24	∞,0,1,4,22,50
∞,0,1,5,6,25	∞,0,1,5,8,40	∞,0,1,6,19,31	∞,0,1,6,48,52
$S(5,6,108)$			
∞,0,1,2,3,55	∞,0,1,2,5,28	∞,0,1,2,6,103	∞,0,1,2,7,19
∞,0,1,2,8,101	∞,0,1,2,9,68	∞,0,1,2,10,13	∞,0,1,2,11,98
∞,0,1,2,12,97	∞,0,1,2,14,95	∞,0,1,2,16,93	∞,0,1,2,22,87
∞,0,1,2,25,59	∞,0,1,3,4,9	∞,0,1,3,7,21	∞,0,1,3,8,24
∞,0,1,3,10,90	∞,0,1,3,12,86	∞,0,1,3,13,102	∞,0,1,3,14,42
∞,0,1,3,15,67	∞,0,1,3,16,94	∞,0,1,3,17,77	∞,0,1,3,19,64
∞,0,1,3,20,60	∞,0,1,3,26,91	∞,0,1,3,27,30	∞,0,1,3,28,45
∞,0,1,3,31,72	∞,0,1,3,33,62	∞,0,1,3,35,40	∞,0,1,3,44,71
∞,0,1,3,48,51	∞,0,1,3,56,98	∞,0,1,3,74,96	∞,0,1,3,88,100
∞,0,1,4,5,92	∞,0,1,4,11,69	∞,0,1,4,15,90	∞,0,1,4,17,31
∞,0,1,4,20,80	∞,0,1,4,21,32	∞,0,1,4,28,50	∞,0,1,4,35,44
∞,0,1,4,38,45	∞,0,1,4,46,60	∞,0,1,4,47,57	∞,0,1,4,51,63
∞,0,1,4,68,84	∞,0,1,4,76,83	∞,0,1,5,6,25	∞,0,1,5,8,80
∞,0,1,5,18,70	∞,0,1,5,26,62	∞,0,1,5,35,46	∞,0,1,6,26,30
∞,0,1,6,33,78	∞,0,1,6,39,66	∞,0,1,6,50,77	∞,0,1,7,23,85
∞,0,1,8,11,38	∞,0,1,9,15,73	∞,0,1,9,22,91	∞,0,1,9,30,99

$S(5,6,132)$			
∞,0,1,2,3,68	∞,0,1,2,5,99	∞,0,1,2,7,126	∞,0,1,2,8,33
∞,0,1,2,9,18	∞,0,1,2,10,105	∞,0,1,2,11,24	∞,0,1,2,12,22
∞,0,1,2,13,121	∞,0,1,2,15,118	∞,0,1,2,16,25	∞,0,1,2,17,74
∞,0,1,2,28,31	∞,0,1,2,32,94	∞,0,1,2,34,43	∞,0,1,2,36,51
∞,0,1,2,38,65	∞,0,1,2,40,48	∞,0,1,2,58,100	∞,0,1,2,60,112
∞,0,1,2,64,95	∞,0,1,3,4,9	∞,0,1,3,7,41	∞,0,1,3,8,63
∞,0,1,3,10,62	∞,0,1,3,11,52	∞,0,1,3,12,91	∞,0,1,3,13,55
∞,0,1,3,14,92	∞,0,1,3,15,69	∞,0,1,3,16,90	∞,0,1,3,17,124
∞,0,1,3,18,83	∞,0,1,3,19,27	∞,0,1,3,20,65	∞,0,1,3,21,100
∞,0,1,3,22,24	∞,0,1,3,23,51	∞,0,1,3,25,115	∞,0,1,3,28,106
∞,0,1,3,29,31	∞,0,1,3,33,43	∞,0,1,3,34,111	∞,0,1,3,40,57
∞,0,1,3,42,127	∞,0,1,3,61,84	∞,0,1,3,72,117	∞,0,1,3,98,110
∞,0,1,4,5,103	∞,0,1,4,11,102	∞,0,1,4,13,128	∞,0,1,4,14,84
∞,0,1,4,15,106	∞,0,1,4,19,48	∞,0,1,4,22,46	∞,0,1,4,23,114
∞,0,1,4,24,92	∞,0,1,4,26,112	∞,0,1,4,28,34	∞,0,1,4,29,81
∞,0,1,4,35,77	∞,0,1,4,39,63	∞,0,1,4,52,108	∞,0,1,4,61,62
∞,0,1,4,64,107	∞,0,1,4,93,125	∞,0,1,5,6,98	∞,0,1,5,11,83
∞,0,1,5,12,31	∞,0,1,5,14,70	∞,0,1,5,16,40	∞,0,1,5,19,30
∞,0,1,5,22,107	∞,0,1,5,24,35	∞,0,1,5,43,82	∞,0,1,5,52,60
∞,0,1,5,55,64	∞,0,1,5,79,95	∞,0,1,6,7,36	∞,0,1,6,10,53
∞,0,1,6,13,51	∞,0,1,6,20,49	∞,0,1,6,38,97	∞,0,1,7,8,83
∞,0,1,7,18,91	∞,0,1,7,24,97	∞,0,1,8,20,42	∞,0,1,8,35,94
∞,0,1,9,10,90	∞,0,1,9,19,97	∞,0,1,9,101,118	∞,0,1,9,107,114
∞,0,1,11,12,73	∞,0,1,11,53,117	∞,0,1,15,37,92	∞,0,1,16,54,68

$S(5,6,168)$			
∞,0,1,2,3,85	∞,0,1,2,5,113	∞,0,1,2,7,162	∞,0,1,2,8,25
∞,0,1,2,9,16	∞,0,1,2,10,131	∞,0,1,2,11,20	∞,0,1,2,12,92
∞,0,1,2,13,24	∞,0,1,2,14,78	∞,0,1,2,18,60	∞,0,1,2,19,150
∞,0,1,2,21,40	∞,0,1,2,22,94	∞,0,1,2,28,69	∞,0,1,2,31,75
∞,0,1,2,34,72	∞,0,1,2,36,107	∞,0,1,2,43,164	∞,0,1,2,45,106
∞,0,1,2,47,70	∞,0,1,2,56,86	∞,0,1,2,59,118	∞,0,1,3,4,83
∞,0,1,3,8,41	∞,0,1,3,10,30	∞,0,1,3,11,95	∞,0,1,3,12,68
∞,0,1,3,13,39	∞,0,1,3,14,135	∞,0,1,3,15,81	∞,0,1,3,17,147
∞,0,1,3,18,134	∞,0,1,3,19,132	∞,0,1,3,20,109	∞,0,1,3,21,51
∞,0,1,3,22,115	∞,0,1,3,23,123	∞,0,1,3,24,117	∞,0,1,3,25,156
∞,0,1,3,27,116	∞,0,1,3,28,131	∞,0,1,3,31,90	∞,0,1,3,33,69
∞,0,1,3,37,44	∞,0,1,3,40,45	∞,0,1,3,43,54	∞,0,1,3,48,110
∞,0,1,3,49,89	∞,0,1,3,52,73	∞,0,1,3,57,162	∞,0,1,3,61,153
∞,0,1,3,62,118	∞,0,1,3,63,99	∞,0,1,3,64,107	∞,0,1,3,65,100
∞,0,1,3,67,87	∞,0,1,3,70,76	∞,0,1,3,77,146	∞,0,1,3,79,165
∞,0,1,3,94,128	∞,0,1,3,98,119	∞,0,1,4,5,20	∞,0,1,4,13,45
∞,0,1,4,14,131	∞,0,1,4,15,59	∞,0,1,4,17,46	∞,0,1,4,18,98
∞,0,1,4,19,31	∞,0,1,4,21,95	∞,0,1,4,23,60	∞,0,1,4,24,130
∞,0,1,4,25,115	∞,0,1,4,26,104	∞,0,1,4,27,37	∞,0,1,4,29,148
∞,0,1,4,30,125	∞,0,1,4,32,97	∞,0,1,4,34,92	∞,0,1,4,39,73
∞,0,1,4,40,81	∞,0,1,4,41,48	∞,0,1,4,43,138	∞,0,1,4,47,124
∞,0,1,4,49,91	∞,0,1,4,50,63	∞,0,1,4,52,82	∞,0,1,4,55,152
∞,0,1,4,62,157	∞,0,1,4,64,99	∞,0,1,4,68,113	∞,0,1,4,71,143
∞,0,1,4,77,135	∞,0,1,4,79,149	∞,0,1,4,87,100	∞,0,1,4,136,142
∞,0,1,5,6,148	∞,0,1,5,11,34	∞,0,1,5,12,60	∞,0,1,5,16,108
∞,0,1,5,19,146	∞,0,1,5,21,48	∞,0,1,5,24,87	∞,0,1,5,37,141
∞,0,1,5,39,131	∞,0,1,5,41,49	∞,0,1,5,44,74	∞,0,1,5,66,163

∞,0,1,6,7,138	∞,0,1,6,13,75	∞,0,1,6,14,33	∞,0,1,6,15,65
∞,0,1,6,17,142	∞,0,1,6,20,64	∞,0,1,6,22,154	∞,0,1,6,23,104
∞,0,1,6,25,157	∞,0,1,6,35,131	∞,0,1,6,39,145	∞,0,1,6,43,139
∞,0,1,6,52,100	∞,0,1,6,66,78	∞,0,1,6,108,134	∞,0,1,6,147,159
∞,0,1,7,11,123	∞,0,1,7,12,128	∞,0,1,7,15,110	∞,0,1,7,20,71
∞,0,1,7,22,54	∞,0,1,7,23,107	∞,0,1,7,41,95	∞,0,1,7,50,78
∞,0,1,7,60,66	∞,0,1,7,73,159	∞,0,1,7,77,102	∞,0,1,7,82,119
∞,0,1,7,98,149	∞,0,1,7,115,133	∞,0,1,8,10,58	∞,0,1,8,13,124
∞,0,1,8,20,141	∞,0,1,8,34,44	∞,0,1,8,55,131	∞,0,1,8,63,77
∞,0,1,9,23,40	∞,0,1,9,39,159	∞,0,1,9,47,117	∞,0,1,9,53,155
∞,0,1,9,59,107	∞,0,1,9,89,157	∞,0,1,9,90,121	∞,0,1,9,96,131
∞,0,1,10,81,103	∞,0,1,11,15,149	∞,0,1,12,30,86	∞,0,1,13,40,142
∞,0,1,13,79,152	∞,0,1,15,31,119	∞,0,1,18,34,58	∞,0,1,33,34,81

4.35 Remark The number N of nonisomorphic $S(5,6,q+1)$ that are invariant under $\mathrm{PSL}_2(q)$:

$q+1$	48	72	84	108	132	168
N	459	≥ 4204	≥ 38717	≥ 71	≥ 55	≥ 4

For $q+1=72$ and 84, the numbers given are the total number of systems that can be constructed using only short orbits under the action of the group [7].

4.8 See Also

§I.2	Steiner 2-designs are BIBDs with index one.
§I.3	$S(t,k,v)$ are t-designs with index one.
§III.1	PBDs whose set of block sizes contains 3 are useful in constructing special classes of STS.
§IV.49	t-wise balanced designs are generalizations of Steiner systems.
§IV.22	An STS(v) is an edge-partition of the complete graph K_v into triangles.
§IV.25	Hall triple systems form a special class of Steiner triple systems.
§VI.7	Finite geometries furnish many examples of $S(t,k,v)$.
[1]	Textbook covering most of the material in this section.
[3]	A comprehensive monograph on triple systems.
[11]	A comprehensive survey of SQS.

References

[1] T. Beth, D. Jungnickel and H. Lenz, *Design Theory*, Cambridge University Press, London, 1986.

[2] J.L. Blanchard, A construction for Steiner 3-designs, *J. Combin. Theory A* 71 (1995), 60-67.

[3] C.J. Colbourn and A. Rosa, *Triple Systems*, Oxford University Press, to appear.

[4] P. Dembowski, Möbiusebenen gerader Ordnung, *Math. Ann.* 157 (1964), 179-205.

[5] R.H.F. Denniston, Some new 5-designs, *Bull. London Math. Soc.* 8 (1976), 263-267.

[6] M.J. Grannell, T.S. Griggs and R.A. Mathon, Some Steiner 5-designs with 108 and 132 points, *J. Combin. Des.* 1 (1993), 213-238.

[7] M.J. Grannell, T.S. Griggs and R.A. Mathon, Steiner systems $S(5,6,v)$ with $v=72$ and 84, preprint.

[8] B.H. Gross, Intersection triangles and block intersection numbers of Steiner systems, *Math. Zeitschr.* 139 (1974), 87-104.

[9] H.-R. Halder and W. Heise, *Einführung in die Kombinatorik*, Hanser Verlag, München, 1976.

[10] H. Hanani, A class of 3-designs, *J. Combin. Theory A* 26 (1979), 1-19.

[11] A. Hartman and K.T. Phelps, Steiner quadruple systems, in: *Contemporary Design Theory: A Collection of Surveys* (J.H. Dinitz, D.R. Stinson; eds.) John Wiley & Sons, New York, 1992, 205-240.

[12] R. Mathon, A tabu search for Steiner 5-designs, preprint.

[13] W.H. Mills, A new 5-design, *Ars Combin.* 6 (1978), 193-195.

[14] E. Witt, Die 5-fach transitiven Gruppen von Mathieu, *Abh. Math. Sem. Univ. Hamburg* 12 (1938), 256-264.

5 Symmetric Designs

TRAN VAN TRUNG

5.1 Preliminaries

5.1 **Theorem** (Fisher's inequality) If a (v, k, λ) BIBD exists, then $b \geq v$ (or equivalently, then $r \geq k$).

A BIBD with $b = v$ (or equivalently, $r = k$) is a *symmetric* (v, k, λ) design or simply a *symmetric design.* A symmetric design with $k = 1$ or $k = v - 1$ is *trivial.* This section deals only with nontrivial symmetric designs, i.e. $1 < k < v - 1$.

5.2 **Lemma** In a symmetric (v, k, λ) design each pair of blocks have λ points in common (the design is *linked*).

5.3 **Example** The paradigmatic example of a symmetric design is the $(7, 3, 1)$ design.

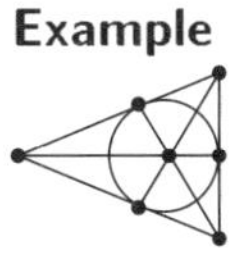

The point set is $X = \{0, \cdots, 6\}$ and the block set is $\mathcal{B} = \{013, 124, 235, 346, 450, 561, 602\}$. This design is the *Fano plane.* The seven blocks are presented by six lines and one circle.

The *complement* of a symmetric (v, k, λ) design $\mathcal{D}$ is a symmetric $(v, v-k, v-2k+\lambda)$ design, whose blocks are the complements of blocks of $\mathcal{D}$. The value $n = k - \lambda$ is the *order* of the design.

5.4 **Remarks**

1. In a symmetric design the parameter v satisfies $4n - 1 \leq v \leq n^2 + n + 1$.
2. The lower bound on v corresponds to a symmetric design $\mathcal{D}$ with parameters $(v = 4n - 1, k = 2n - 1, \lambda = n - 1)$ or its complement; and $\mathcal{D}$ is a *Hadamard design* of order n. See §IV.24.
3. The upper bound on v corresponds to a symmetric design $\mathcal{D}$ with parameters $(v = n^2 + n + 1, k = n + 1, \lambda = 1)$ or its complement; and $\mathcal{D}$ is a *finite projective plane* of order n. See §VI.7.2.

Let $\mathcal{D} = (V, \mathcal{B})$ be a BIBD, then the *dual* of $\mathcal{D}$ is $\mathcal{D}^* = (\mathcal{B}, V)$ where $b \in \mathcal{B}$ is contained in $v \in V$ if and only if v is contained in b in $\mathcal{D}$.

5.5 **Remark** The dual of a design is a design if and only if the design is symmetric. Also, the parameters of a symmetric design and its dual are the same, yet they are not necessarily isomorphic.

A *polarity* of a symmetric design $\mathcal{D}$ is an isomorphism π from $\mathcal{D}$ to $\mathcal{D}^*$ which has order 2 (i.e. π^2 is the identity mapping on $\mathcal{D}$).

5.2 The Bruck-Ryser-Chowla Theorem

5.6 **Remark** For a symmetric (v, k, λ) design to exist the arithmetic condition $\lambda(v-1) = k(k-1)$ must be satisfied. The most fundamental necessary condition on the existence of symmetric designs is due to Bruck, Ryser and Chowla [13, 16].

5.7 **Theorem** (*Bruck-Ryser-Chowla*) Let v, k, and λ be integers with $\lambda(v-1) = k(k-1)$ for which there exists a symmetric (v, k, λ) design.

1. If v is even, then $n = k - \lambda$ is a square.
2. If v is odd, then the equation $z^2 = nx^2 + (-1)^{(v-1)/2}\lambda y^2$ has a solution in integers x, y, z not all zero.

5.8 **Remarks**

1. The Bruck-Ryser-Chowla theorem implies that if a projective plane of order $n \equiv 1$ or 2 (mod 4) exists, then n is the sum of two squares of integers. The smallest orders of projective planes whose existence is still undecided are $n =$ 12, 15, 18, 20, 24, 26, 28.
2. At one time it was conjectured that the necessary conditions of the Bruck-Ryser-Chowla theorem are sufficient for the existence of symmetric designs. This conjecture is now known to be false. With an elaborate computer search, Lam, Thiel and Swiercz [31] showed that a symmetric (111,11,1) design cannot exist. See Remark VI.9.12.

5.9 **Conjecture** For every $\lambda > 1$, there exist only finitely many symmetric (v, k, λ) designs.

5.10 **Remarks**

1. If $\lambda = 2$, the only values of k for which symmetric $(v, k, 2)$ designs (*biplanes*) are known are $k =$ 4, 5, 6, 9, 11, 13. The values of k of the known symmetric designs with $\lambda = 3$ are $k =$ 6, 9, 10, 12, 15.
2. Arasu [1] shows the finiteness of the number of of symmetric designs for a given λ, when v is of a special form.

5.3 Automorphisms of Symmetric Designs

An *automorphism* of a design is a permutation of the point set that preserves the block set. A group obtained under composition of automorphisms is an *automorphism group*. The group of *all* automorphisms of a design $\mathcal{D}$ is the *full automorphism group* of $\mathcal{D}$, denoted by $Aut(\mathcal{D})$.

5.11 **Theorem** An automorphism of a symmetric design fixes the same number of points as blocks.

5.12 **Theorem** (Aschbacher [2]) Let $\mathcal{D}$ be a symmetric (v, k, λ) design and let p be a prime divisor of the order of $Aut(\mathcal{D})$. Then either p divides v or $p \leq k$.

5.13 Theorem (Multiplier theorem (Hall-Ryser) [22]) Suppose that there exists a symmetric (v, k, λ) design having an abelian group G of order v as a regular automorphism group. Suppose that p is a prime such that $p > \lambda$, p divides $n = k - \lambda$, and p is relatively prime to v. Then the mapping $g \to pg$, $g \in G$, is also an automorphism of the design. (See also §IV.12).

5.14 Theorem (Kantor [30]) Let $\mathcal{D}$ be a symmetric (v, k, λ) design with $k < v/2$ such that $Aut(\mathcal{D})$ acts doubly-transitive on points. Then $\mathcal{D}$ is one of the following:

1. a point-hyperplane design with $v = \frac{q^{m+1}-1}{q-1}$, $k = \frac{q^m-1}{q-1}$, and $\lambda = \frac{q^{m-1}-1}{q-1}$ (§VI.7),
2. the unique Hadamard design with $v = 11$, $k = 5$, and $\lambda = 2$ (§IV.24),
3. a unique design with $v = 176$, $k = 50$, and $\lambda = 14$ admitting the Higman-Sims group,
4. a unique design with $v = 2^{2m}$, $k = 2^{m-1}(2^m - 1)$, and $\lambda = 2^{m-1}(2^{m-1} - 1)$, for each $m \geq 2$.

5.15 Remark For further results about automorphisms of symmetric designs, see Hughes [24, 25] and Lander [32].

5.4 Quasi-Residual Designs and their Embedding

Let $\mathcal{D}$ be a symmetric (v, k, λ) design and let B be a block of $\mathcal{D}$. Remove B and all points in B from the other blocks. The result is a $(v - k, k - \lambda, \lambda)$ BIBD, a *residual design.*

5.16 Remark If a design with parameters (v, b, r, k, λ) is a residual design of a symmetric design, then $r = k + \lambda$.

A BIBD with $r = k + \lambda$ is a *quasi-residual design.* If such a design is a residual design, then it is *embeddable* or *residual*; otherwise, *nonembeddable.*

5.17 Remark A quasi-residual design with $\lambda = 1$ is an affine plane and is embeddable in a projective plane.

5.18 Theorem (Hall-Connor [20]) A quasi-residual design with $\lambda = 2$ is embeddable.

5.19 Theorem [8] For every $\lambda \geq 3$, there is a function $f(\lambda)$ such that any quasi-residual (w, k, λ) design with $k > f(\lambda)$ is embeddable. The function f is given by

$$f(\lambda) = \begin{cases} 76 & : \lambda = 3, \\ \frac{1}{2}(\lambda - 1)(\lambda^4 - 2\lambda^2 + \lambda + 2) & : 4 \leq \lambda \leq 9, \\ \frac{1}{2}(\lambda - 1)(\alpha + \sqrt{\alpha^2 + 4(\lambda - 1)}) - (\lambda - 1) & : 10 \leq \lambda, \end{cases}$$

where $\alpha = (\lambda - 1)(\lambda^2 - 3\lambda + 3)$.

5.20 Remarks

1. No example of a quasi-residual (w, k, λ) design with $k > f(\lambda)$ of Theorem 5.19 is known.
2. Example 5.21 gives a nonembeddable quasi-residual design with parameters $(16, 6, 3)$, discovered by Bhattacharya (1944) [7]. For more information on the existence of nonembeddable quasi-residual designs, see [47].
3. van Lint and Tonchev [34] prove a sufficient condition for nonembeddability of quasi-derived designs and also construct designs satisfying this condition.

5.21 Example A nonembeddable quasi-residual (16,6,3) design. (Look at the block intersection sizes to verify nonembeddability.)

1 2 3 4 5 6	1 2 3 4 7 8	1 2 9 10 12 13	1 3 10 11 12 15
1 4 9 13 14 16	1 5 7 10 14 15	1 5 7 11 13 16	1 6 8 9 11 14
1 6 8 12 15 16	2 3 9 10 11 16	2 4 12 13 14 15	2 5 6 9 15 16
2 5 8 11 13 15	2 6 7 11 12 14	2 7 8 10 14 16	3 4 11 14 15 16
3 5 6 10 13 14	3 5 8 9 12 14	3 6 7 12 13 16	3 7 8 9 13 15
4 5 7 9 11 12	4 5 8 10 12 16	4 6 7 9 10 15	4 6 8 10 11 13

5.5 Extensions of Symmetric Designs

If a t-design $\mathcal{D}^*$ is the derived design of some $(t+1)$-design $\mathcal{D}$, then $\mathcal{D}$ is an *extension* of $\mathcal{D}^*$, and $\mathcal{D}^*$ is *extendible.*

5.22 Example Every Hadamard $(4n-1, 2n-1, n-1)$ design $\mathcal{D}$ is extendible to a 3-$(4n, 2n, n-1)$ design $\mathcal{E}$, a *Hadamard 3-design.* The blocks of $\mathcal{E}$ consist of the complements of the blocks of $\mathcal{D}$ together with all of the blocks of $\mathcal{D}$ adjoined with a new point.

5.23 Theorem (Cameron [14]) If a symmetric (v, k, λ) design $\mathcal{D}$ is extendible, then one of the following holds:

1. $v = 4\lambda + 3$, $k = 2\lambda + 1$, i.e. $\mathcal{D}$ is a Hadamard design,
2. $v = (\lambda+2)(\lambda^2+4\lambda+2)$, $k = \lambda^2 + 3\lambda + 1$,
3. $v = 495$, $k = 39$, $\lambda = 3$.

5.24 Remarks

1. The known symmetric designs in the three cases of Theorem 5.23 are the known Hadamard designs in case 1; the projective plane of order 4, and biplanes with parameters (56,11,2) in case 2. The Hadamard designs and the projective plane of order 4 are extendible. It is claimed that no (56,11,2) biplane is extendible (Bagchi [3], [4]); however, Brouwer [10] found a fatal error in the proof.
2. In case 2, Sane, Shrikhande and Singhi [40] show a connection between extendability and existence of maximal arcs.

5.6 Known Symmetric Designs

Family 1 (Point-hyperplane designs)
$v = \frac{q^{m+1}-1}{q-1}$, $k = \frac{q^m-1}{q-1}$, $\lambda = \frac{q^{m-1}-1}{q-1}$, $n = q^{m-1}$, q a prime power and $m \geq 2$. Points and blocks of the designs are points and hyperplanes of the projective geometries PG(m, q). (See §VI.7.)

Family 2 (Hadamard designs)
$v = 4n-1$, $k = 2n-1$, $\lambda = n-1$. (See §IV.24.)

Family 3 (Chowla (1944)[15])
$v = 4t^2+1$, $k = t^2$, $\lambda = (t^2-1)/4$, $n = (3t^2+1)/4$, t an odd positive integer, v a prime. Designs are constructed via a difference set consisting of biquadratic residues modulo v. (See Example IV.12.15.)

Family 4 (Lehmer (1953)[33]) (See Example IV.12.15.)

a. $v = 4t^2+9$, $k = t^2+3$, $\lambda = (t^2+3)/4$, $n = 3(t^2+3)/4$, t an odd positive integer, v a prime. Designs are constructed via a difference set consisting of 0 and biquadratic residues modulo v.
b. $v = 8t^2+1 = 64u^2+9$, $k = t^2$, $\lambda = u^2$, $n = t^2-u^2$, t, u odd positive integers, v a prime. Designs are constructed via a difference set of octic residues modulo v.
c. $v = 8t^2+49 = 64u^2+441$, $k = t^2+6$, $\lambda = u^2+7$, $n = t^2-u^2-1$, t an odd positive integer, u an even integer, v a prime. Designs are constructed via a difference set consisting of 0 and octic residues modulo v.

Family 5 (Whiteman (1962)[49])
$v = pq$, $k = (pq-1)/4$, $\lambda = (pq-5)/16$, $n = (3pq+1)/16$, p and $q = 3p+2$ primes, k an odd square. Designs are constructed via a difference set modulo v. However, the only known examples are $p = 17$, $q = 53$, and $p = 46817$, $q = 140453$.

Family 6 (Menon (1962)[36])
$v = 4t^2$, $k = 2t^2-t$, $\lambda = t^2-t$, $n = t^2$. One of the constructions is based on a regular Hadamard matrix of size $4t^2$ with constant diagonal. The existence of such designs is known whenever there is a Hadamard design with parameters $(2t-1, t-1, t/2-1)$ or whenever $2t-1$ and $2t+1$ are both prime powers. Brouwer and van Lint (1983) construct such a design for $t = 7$. It is conjectured that designs exist for all values of t. (See Theorems IV.12.18 - 12.21 and §IV.24.)

Family 7 (Wallis (1971) [48])
$v = q^{m+1}(q^m+\cdots+q+2)$, $k = q^m(q^m+\cdots+q+1)$, $\lambda = q^m(q^{m-1}+\cdots+q+1)$, $n = q^{2m}$, q a prime power, m a positive integer. The construction is based on the existence of affine designs with parameters $(q^{m+1}, q^m, \frac{q^m-1}{q-1})$. A different construction due to McFarland (1973) [35] is based on noncyclic difference sets. (See §IV.12.3.)

Family 8 (Wilson (unpublished); Shrikhande and Singhi (1975)[44])
$v = m^3+m+1$, $k = m^2+1$, $\lambda = m$, $n = m^2-m+1$, where both $m-1$ and m^2-m+1 are prime powers. The construction is based on symmetric group-divisible designs and projective planes of order $m-1$.

Family 9 (Spence (1977)[41])
$v = 3^m(3^m+\varepsilon)/2$, $k = 3^{m-1}(3^m-\varepsilon)/2$, $\lambda = 3^{m-1}(3^{m-1}-\varepsilon)/2$, $n = 3^{2(m-1)}$, $\varepsilon = \pm 1$. Designs for $\varepsilon = -1$ are constructed via a noncyclic difference set. (See §IV.12.3.)

Family 10 (Rajkundlia (1978, 1983)[38, 39]; Mitchell (1979)[37])
$v = q^{m+1}-q+1$, $k = q^m$, $\lambda = q^{m-1}$, $n = q^m-q^{m-1}$, $q > 2$ a prime power, $m \geq 2$ an integer, whenever there exists an affine plane of order $q-1$. The construction is based on projective planes and group-divisible 1-designs.

Family 11 (Wilson (unpublished); Brouwer (1983)[11])
$v = 2(q^m+q^{m-1}+\cdots+q)+1$, $k = q^m$, $\lambda = q^{m-1}(q-1)/2$, $n = (q^m+q^{m-1})/2$, q an odd prime power, m a positive integer. The construction is based on symmetric group-divisible designs using projective geometries and Hadamard designs.

Family 12 (Spence (1993) [42] Jungnickel and Pott (1994) [29])
$v = p^s(q^{2m}-1)/(q-1)$, $k = q^{2m-1}p^{s-1}$, $\lambda = p^{s-1}q^{2m-2}(p^{s-1}-1)/(p-1)$, $n = p^{2s-2}q^{2m-2}$, p a prime, $q = (p^s-1)/(p-1)$ a prime power. The construction is based on projective geometries and ω-circulant matrices.

Special constructions

(133,33,8) (Hall (1956) [19]), via a difference set.

(176,50,14) (Higman (1969)[23]), from the Higman-Sims group. The latter acts on the design as a 2-transitive automorphism group.

(56,11,2) (Hall, Lane and Wales (1970)[21]), constructed in terms of the rank-3 group PSL(3, 4) and its associated strongly regular graph.

(79,13,2) (Aschbacher (1971)[2]), by using an automorphism group of order 110.
(71,15,3) (Haemers (1980) [17]; Beker and Haemers (1980) [5]), obtained from embedding a quasi-residual 2-(56,70,15,12,3) design. The latter is constructed using eigenvalue techniques.
(41,16,6) (Bridges, Hall and Hayden (1981) [9]), via coding theory and an automorphism group of order 15.
(66,26,10) (Tran van Trung (1982)[46]), using an automorphism group of order 55.
(49,16,5) (Brouwer and Wilbrink (1984) [12]), via an automorphism group of order 15.
(70,24,8) (Janko and Tran van Trung (1984) [26]), via automorphism group of order 42.
(78,22,6) (Janko and Tran van Trung (1985) [27]), using an automorphism group of order 39; (Tonchev(1987) [45]), via embedding of the Witt-Mathieu 3-(22,6,1) design.
(71,21,6) (Janko, Tran van Trung (1985) [28]), via automorphism group of order 21.
(160,54,18) (Spence, Tonchev and Tran van Trung (1993) [43]), using the incidence matrix of the point-hyperplane design of PG(3,3) and 4 × 4-circulant matrices.

5.7 Symmetric Designs with n ≤ 25

5.25 Table A listing of symmetric (v, k, λ) designs with $n = k - \lambda \le 25$, $k \le v/2$, and $4n - 1 < v < n^2 + n + 1$ satisfying $(v-1)\lambda = k(k-1)$ and the Bruck-Ryser-Chowla condition. If a design is known to exist, one solution is given. Parameters where existence is still undecided are marked by a question mark.

n	v	k	λ	Solution
4	16	6	2	Base block 0000 1000 0100 0010 0001 1111 (mod (2, 2, 2, 2)).
6	25	9	3	Points $A, B, C, D, E, F, G, 1_i, 2_i, 3_i, 4_i, 5_i, 6_i$, $i = 0, 1, 2$ (mod 3), Automorphisms $\rho = (A)(B)(C)(D)(E)(F)(G)(I_0\ I_1\ I_2)$, $I = 1, \ldots, 6$, $\tau_1 = (A\ D)(B\ C)(E)(F)(G)(1_i)(2_i)(3_i\ 6_{i+1})(4_i\ 5_{i+1})$, $\tau_2 = (A\ C)(B\ D)(E)(F)(G)(1_i)(2_i)(3_i\ 5_{i+1})(4_i\ 6_{i+1})$, Base blocks $1_0 1_1 1_2 3_0 3_1 3_2 5_0 5_1 5_2$, $2_0 2_1 2_2 3_0 3_1 3_2 6_0 6_1 6_2$, $EFG 1_0 1_1 1_2 2_0 2_1 2_2$, $EFG 3_0 3_1 3_2 4_0 4_1 4_2$, $ABE 1_0 2_0 3_0 4_0 5_0 6_0$, $ACF 1_0 2_2 3_1 4_0 5_2 6_1$, $ADG 1_2 2_0 3_1 4_0 5_1 6_2$.
7	37	9	2	Base block 1 7 9 10 12 16 26 33 34 (mod 37).
7	31	10	3	Points $A, B, C, 1_i, 2_i, 3_i, 4_i$, $i = 0, 1, 2, 3, 4, 5, 6$ (mod 7), Automorphisms $\rho = (A)(B)(C)(I_0\ I_1\ I_2\ I_3\ I_4\ I_5\ I_6)$, $I = 1, 2, 3, 4$, $\sigma = (A\ B\ C)(1_i\ 2_i\ 3_i)(4_i)$, Base blocks $A 1_1 1_6 2_2 2_5 3_3 3_4 4_1 4_2 4_4$, $1_3 1_6 1_5 2_3 2_6 2_5 3_3 3_6 3_5 4_0$, $ABC 1_0 1_1 1_2 1_3 1_4 1_5 1_6$.
9	56	11	2	Points $1_i, 2_i, 3_i, 4_i, 5_i, 6_i, 7_i, 8_i$, $i = 0, 1, \ldots, 6$ (mod 7), Automorphisms $\rho = (I_0\ I_1\ I_2\ I_3\ I_4\ I_5\ I_6)$, $I = 1, 2, 3, 4, 5, 6, 7, 8$, $\sigma = (1_i\ 2_{2i}\ 3_{4i})(4_i\ 5_{2i}\ 6_{4i})(7_i\ 7_{2i}\ 7_{4i})(8_i\ 8_{2i}\ 8_{4i})$,

n	v	k	λ	Solution
				$\tau = (1_i\ 4_{6i})(2_i\ 5_{6i})(3_i\ 6_{6i})(7_i\ 8_{6i})$, Base blocks $1_1 2_2 3_4 4_1 5_2 6_4 7_0 7_3 7_6 7_5 8_0$, $1_2 1_5 2_2 2_5 3_1 3_6 4_0 4_1 4_6 7_1 8_6$.
9	45	12	3	Base block 000 001 002 100 110 120 200 211 222 300 312 321 (mod (5, 3, 3)).
9	40	13	4	Base block 1 2 3 5 6 9 14 15 18 20 25 27 35 (mod 40).
9	36	15	6	Base block 11 22 33 44 55 01 02 03 04 05 10 20 30 40 50 (mod (6, 6)).
10	41	16	6	Points A, $1_i, 2_i, 3_i, 4_i, 5_i, 6_i, 7_i, 8_i$, $i = 0, 1, 2, 3, 4 \pmod 5$, Automorphisms $\rho = (A)(I_0\ I_1\ I_2\ I_3\ I_4)$, $I = 1, 2, 3, 4, 5, 6, 7, 8$, $\sigma = (A)(1_i\ 2_i\ 3_i)(4_i\ 5_i\ 8_i)(6_i)(7_i)$, Base blocks $A 1_0 1_1 1_2 1_3 1_4 2_0 2_1 2_2 2_3 2_4 3_0 3_1 3_2 3_3 3_4$, $A 1_1 1_4 2_2 2_3 3_0 4_1 4_2 4_3 4_4 6_1 6_4 7_2 7_3 5_0 8_0$, $1_2 1_3 2_2 2_3 3_2 3_3 4_0 4_2 4_3 5_0 5_2 5_3 8_0 8_2 8_3 6_0$, $1_1 1_4 2_1 2_4 3_1 3_4 7_1 7_2 7_3 7_4 4_2 4_3 5_2 5_3 8_2 8_3$, $3_0 3_2 3_3 2_0 2_1 2_4 6_0 6_1 6_4 4_2 4_3 8_1 8_4 7_1 7_4 5_0$.
11	79	13	2	Points $A, B, 1_i, 2_i, 3_i, 4_i, 5_i, 6_i, 7_i$, $i = 0, 1, \ldots, 10 \pmod{11}$, Automorphisms $\rho = (A)(B)(I_0\ I_1\ \cdots\ I_{10})$, $I = 1, 2, \ldots, 7$, $\sigma = (A)(B)(1_i\ 2_{4i}\ 3_{5i}\ 4_{9i}\ 5_{3i})\ (6_i\ 6_{4i}\ 6_{5i}\ 6_{9i}\ 6_{3i})(7_i\ 7_{4i}\ 7_{5i}\ 7_{9i}\ 7_{3i})$, $\tau = (A\ B)(1_i\ 1_{10i})(2_i\ 2_{10i})(3_i\ 3_{10i})$ $(4_i\ 4_{10i})(5_i\ 5_{10i})(6_i\ 6_{10i})(7_i\ 7_{10i})$, Base blocks $AB 6_0 6_1 6_2 6_3 6_4 6_5 6_6 6_7 6_8 6_9 6_{10}$, $AB 7_0 7_1 7_2 7_3 7_4 7_5 7_6 7_7 7_8 7_9 7_{10}$, $A 1_1 1_4 2_4 2_5 3_5 3_9 4_9 4_3 5_3 5_1 6_0 7_0$, $B 1_{10} 1_7 2_7 2_6 3_6 3_2 4_2 4_8 5_8 5_{10} 6_0 7_0$, $1_0 2_2 2_9 2_4 2_7 3_5 3_6 5_4 5_7 6_2 6_9 7_5 7_6$.
11	49	16	5	Points $A, B, C, D, 1_i, 2_i, 3_i, 4_i, 5_i, 6_i, 7_i, 8_i, 9_i$, $i = 0, 1, 2, 3, 4 \pmod 5$, Automorphisms $\rho = (A)(B)(C)(D)(I_0\ I_1 \cdots\ I_4)$, $I = 1, 2, \ldots, 9$, $\sigma = (A)(B\ C\ D)(1_i\ 2_i\ 3_i)(4_i\ 5_i\ 6_i)(7_i\ 8_i\ 9_i)$, Base blocks $A 7_0 7_1 7_2 7_3 7_4 8_0 8_1 8_2 8_3 8_4 9_0 9_1 9_2 9_3 9_4$, $B 4_0 4_1 4_2 4_3 4_4 5_0 5_1 5_2 5_3 5_4 9_0 9_1 9_2 9_3 9_4$, $AB 1_0 1_1 2_0 2_2 3_0 4_0 4_1 5_2 5_4 6_3 7_0 7_1 8_0 8_3$, $BC 1_0 1_2 2_3 3_0 3_1 4_1 4_3 6_2 6_3 7_3 7_4 8_2 9_2 9_4$, $1_0 1_1 1_2 1_3 2_0 3_4 4_4 5_0 5_4 6_2 6_4 7_2 8_2 8_3 9_1 9_4$.
12	71	15	3	Points $A, 1_i, 2_i, \ldots, 10_i$, $i = 0, 1, \ldots, 6 \pmod 7$, Automorphisms $\rho = (A)(I_0\ I_1\ \cdots\ I_6)$, $I = 1, 2, \ldots, 10$, $\tau = (A)(1_i)(2_i)(3_i\ 4_i)(5_i\ 6_i)(7_i\ 8_i)(9_i\ 10_i)$, $\sigma = (A)(1_i)(2_i)(3_i\ 5_i\ 7_i)(4_i\ 6_i\ 8_i)(9_i)(10_i)$, Base blocks $A 1_0 1_1 1_2 1_3 1_4 1_5 1_6 2_0 2_1 2_2 2_3 2_4 2_5 2_6$, $A 1_0 2_0 3_2 3_4 5_1 5_4 7_1 7_2 4_3 6_6 8_5 9_1 9_2 9_4$, $1_1 1_2 1_4 3_1 3_3 5_2 5_6 7_4 7_5 4_1 4_3 6_2 6_6 8_4 8_5$, $1_1 1_2 1_4 3_0 5_0 7_0 4_0 6_0 8_0 9_1 9_2 9_4 10_1 10_2 10_4$, $1_0 2_1 2_3 5_1 5_3 7_0 4_1 4_2 4_5 6_0 6_2 6_6 9_5 9_6 10_1$.

n	v	k	λ	Solution
12	61	16	4	Points $A, 1_i, 2_i, \ldots, 20_i$, $i = 0, 1, 2 \pmod 3$, Automorphisms $\rho = (A)(I_0\ I_1\ I_2)$, $I = 1, \ldots, 20$, $\sigma = (A)(1_i\ 2_i\ 3_i\ 4_i\ 5_i)(6_i\ 7_i\ 8_i\ 9_i\ 10_i)$ $(11_i\ 12_i\ 13_i\ 14_i\ 15_i)(16_i\ 17_i\ 18_i\ 19_i\ 20_i)$, Base blocks $A1_01_11_22_02_12_23_03_13_24_04_14_25_05_15_2$, $A1_01_11_26_07_18_19_011_012_113_114_016_017_118_119_0$, $1_02_13_14_010_110_28_18_26_07_011_111_212_112_220_018_0$, $1_02_13_14_015_115_213_113_211_012_016_116_217_117_210_08_0$, $1_02_13_14_020_120_218_118_216_017_06_16_27_17_215_013_0$.
13	81	16	3	?
13	69	17	4	Points $A, B, C, D, 1_i, 2_i, \ldots, 5_i$, $i = 0, \ldots, 12 \pmod{13}$, Automorphisms $\rho = (A)(B)(C)(D)(I_0\ I_1\ \cdots\ I_{12})$, $I = 1, 2, \ldots, 5$, $\tau = (A\ B\ C\ D)(1_i\ 2_i\ 3_i\ 4_i)(5_i)$, Base blocks $ABCD1_01_11_21_31_41_51_61_71_81_91_{10}1_{11}1_{12}$, $A1_11_31_92_22_62_53_43_{12}3_{10}4_84_{11}4_75_05_45_{12}5_{10}$, $1_01_11_31_92_02_12_32_93_03_13_33_94_04_14_34_95_0$.
14	121	16	2	?
15	71	21	6	Points $A, 1_i, \ldots, 10_i$, $i = 0, 1, \ldots, 6 \pmod 7$, Automorphisms $\rho = (A)(I_0\ I_1\ \cdots\ I_6)$, $I = 1, 2, \ldots, 10$, $\sigma = (A)(K_0)(K_1\ K_2\ K_4)(K_3\ K_6\ K_5)(4_i\ 5_{2i}\ 6_{4i})$ $(7_i\ 8_{2i}\ 9_{4i})$, $K = 1, 2, 3, 10$, Base blocks $1_01_11_21_31_41_51_62_02_12_22_32_42_52_63_03_13_23_33_43_53_6$, $1_02_03_03_13_23_44_05_06_07_07_27_38_08_48_69_09_19_510_110_210_4$, $2_12_22_43_13_23_44_34_44_55_65_15_36_56_26_67_37_68_68_59_59_3$, $1_01_31_61_52_03_04_14_55_25_36_46_67_37_47_58_68_18_39_59_29_6$, $1_31_61_53_13_23_44_14_34_65_25_65_56_46_56_37_08_09_010_310_610_5$, $A1_31_52_12_43_23_34_04_24_34_65_05_27_17_47_68_28_39_510_110_2$, $1_11_42_02_12_53_34_04_14_55_25_66_08_38_59_09_39_49_510_410_510_6$.
15	61	25	10	Points $A, 1_i, \ldots, 12_i$, $i = 0, 1, \ldots, 4 \pmod 5$, Automorphisms $\rho = (A)(I_0\ I_1\ \cdots\ I_4)$, $I = 1, 2, \ldots, 12$, $\tau = (A)(1_i\ 2_i\ 3_i\ 4_i\ 5_i)(6_i\ 7_i\ 8_i\ 9_i\ 10_i)(11_i)(12_i)$, Base blocks $1_01_11_21_31_42_02_12_22_32_43_03_13_23_33_44_04_14_24_34_45_05_15_25_35_4$, $A1_11_42_02_12_43_03_23_34_24_36_07_17_48_28_39_010_110_210_310_411_111_412_212_3$, $1_02_12_43_23_34_05_15_25_35_46_26_37_08_09_19_410_110_210_310_411_211_312_012_212_3$, $1_11_42_12_43_13_44_14_45_15_46_06_16_47_07_17_48_08_18_49_09_19_410_010_110_4$, $1_21_32_22_33_23_34_24_35_25_36_16_47_17_48_18_49_19_410_110_411_011_111_211_311_4$.
16	154	18	2	?
16	115	19	3	?
16	96	20	4	Base block 00000 01000 10000 11000 00001 00101 00011 00111 00002 01102 10012 11112 00003 01013 11103 10113 00004 10104 01114 11014 $(\mathrm{mod}\ (2, 2, 2, 2, 6))$.

n	v	k	λ	Solution
16	85	21	5	Base block 0 1 2 4 7 8 14 16 17 23 27 28 32 34 43 46 51 54 56 64 68 (mod 85).
16	78	22	6	Points $1_i, \ldots, 6_i$, $i = 0, 1, \ldots, 12$ (mod 13), Automorphisms $\rho = (I_0\ I_1\ \cdots\ I_{12})$, $I = 1, 2, \ldots, 6$, $\tau = (1_i)(6_i)(2_i\ 3_i)(4_i\ 5_i)$, Base blocks $1_0 1_1 1_3 1_9 1_4 1_{12} 1_{10} 2_2 2_6 2_5 3_2 3_6 3_5 4_4 4_{12} 4_{10} 5_4 5_{12} 5_{10} 6_4 6_{12} 6_{10}$, $2_0 2_2 2_6 2_5 2_7 2_8 2_{11} 1_1 1_3 1_9 3_4 3_{12} 3_{10} 4_2 4_6 4_5 5_7 5_8 5_{11} 6_7 6_8 6_{11}$, $4_0 4_2 4_6 4_5 4_7 4_8 4_{11} 1_7 1_8 1_{11} 2_2 2_6 2_5 3_7 3_8 3_{11} 5_1 5_3 5_9 6_4 6_{12} 6_{10}$, $6_0 6_1 6_3 6_9 6_4 6_{12} 6_{10} 1_1 1_3 1_9 2_1 2_3 2_9 3_1 3_3 3_9 4_2 4_6 4_5 5_2 5_6 5_5$.
16	70	24	8	Points $1_i, \ldots, 10_i$, $i = 0, 1, \ldots, 6$ (mod 7), Automorphisms $\rho = (I_0\ I_1\ \cdots\ I_6)$, $I = 1, 2, \ldots, 10$, $\sigma = (K_0)(K_1\ K_2\ K_4)(K_3\ K_6\ K_5)(4_i\ 5_{2i}\ 6_{4i})$ $(7_i\ 8_{2i}\ 9_{4i})$, $K = 1, 2, 3, 10$, $\tau = (3_i)(10_i)(1_i\ 2_i)(4_i\ 7_i)(5_i\ 8_i)(6_i\ 9_i)$, Base blocks $1_0 1_1 1_2 1_4 2_0 2_1 2_2 2_4 3_0 3_3 3_6 3_5 4_1 4_6 5_2 5_5 6_3 6_4 7_1 7_6 8_2 8_5 9_3 9_4$, $10_0 10_3 10_6 10_5 1_0 1_3 1_6 1_5 2_0 2_3 2_6 2_5 4_0 4_1 5_0 5_2 6_0 6_4 7_0 7_1 8_0 8_2 9_0 9_4$, $10_0 10_3 10_6 10_5 2_0 2_1 2_2 2_4 3_0 3_1 3_2 3_4 4_0 4_6 5_0 5_5 6_0 6_3 7_2 7_5 8_4 8_3 9_1 9_6$, $10_0 10_2 1_3 1_4 2_0 2_5 3_0 3_1 4_0 4_1 4_3 4_5 5_1 5_2 5_3 5_5 6_2 6_3 7_0 7_3 7_4 7_6 9_1 9_5$.
16	66	26	10	Points $1_i, \ldots, 6_i$, $i = 0, 1, \ldots, 10$ (mod 11), Automorphisms $\rho = (I_0\ I_1\ \cdots\ I_{10})$, $I = 1, 2, \ldots, 6$, $\sigma = (6_i)(1_i\ 2_i\ 3_i\ 4_i\ 5_i)$, Base blocks $1_1 1_3 1_4 1_5 1_9 2_2 2_6 2_7 2_8 2_{10} 3_2 3_6 3_7 3_8 3_{10} 4_1 4_3 4_4 4_5 4_9 6_1 6_3 6_4 6_5 6_9 5_0$, $1_2 1_6 1_7 1_8 1_{10} 2_2 2_6 2_7 2_8 2_{10} 3_2 3_6 3_7 3_8 3_{10} 4_2 4_6 4_7 4_8 4_{10} 5_2 5_6 5_7 5_8 5_{10} 6_0$.
16	64	28	12	Base block 0000 0001 0002 0003 0004 0005 0006 0010 0012 0013 0100 0101 0104 0110 0115 0116 1001 1002 1005 1012 1014 1016 1101 1103 1106 1113 1114 1115 (mod (2, 2, 2, 8)).
18	191	20	2	?
18	79	27	9	Points $A, 1_i, \ldots, 6_i$, $i = 0, 1, \ldots, 12$ (mod 13), Automorphisms $\rho = (A)(I_0\ I_1\ \cdots\ I_{12})$, $I = 1, 2, \ldots, 6$, $\tau_1 = (A)(1_i)(2_i)(3_i\ 4_i)(5_i\ 6_i)$, $\tau_2 = (A)(1_i)(2_i)(3_i\ 5_i)(4_i\ 6_i)$, Base blocks $A 1_0 1_1 1_2 1_3 1_4 1_5 1_6 1_7 1_8 1_9 1_{10} 1_{11} 1_{12} 2_0 2_1 2_2 2_3 2_4 2_5 2_6 2_7 2_8 2_9 2_{10} 2_{11} 2_{12}$, $A 1_0 1_1 1_3 1_9 2_0 2_1 2_3 2_9 3_1 3_3 3_9 3_4 3_{12} 3_{10} 4_2 4_6 4_5 5_1 5_3 5_9 5_4 5_{12} 5_{10} 6_2 6_6 6_5$, $1_1 1_3 1_9 2_4 2_{12} 2_{10} 2_8 2_{11} 2_7 3_1 3_3 3_9 3_8 3_{11} 3_7 4_2 4_6 4_5 5_2 5_6 5_5 6_1 6_3 6_9 6_8 6_{11} 6_7$, $1_4 1_{12} 1_{10} 1_8 1_{11} 1_7 2_1 2_3 2_9 3_1 3_3 3_9 3_8 3_{11} 3_7 4_1 4_3 4_9 4_8 4_{11} 4_7 5_2 5_6 5_5 6_2 6_6 6_5$.
19	211	21	2	?
19	155	22	3	?
19	101	25	6	Base block 1 5 16 19 24 25 31 36 37 52 54 56 58 68 71 78 79 80 81 84 87 88 92 95 97 (mod 101).
19	85	28	9	?
20	139	24	4	?

n	v	k	λ	Solution
20	121	25	5	Points $A, B, C, D, E, \infty_1, \ldots, \infty_{16}, 1_i, \ldots, 20_i$, $i = 0, \ldots, 4 \pmod 5$, Automorphisms $\rho = (A)(B)(C)(D)(E)(\infty_i)(I_0\ I_1\ \cdots\ I_4)$, $i = 1, \ldots, 16,\ I = 1, 2, \ldots, 20$, $\sigma = (\infty_1)(\infty_2\ \infty_5\ \infty_6\ \infty_7\ \infty_8)(\infty_3\ \infty_9\ \infty_{10}\infty_{11}\ \infty_{12})$ $(\infty_4\ \infty_{13}\ \infty_{14}\ \infty_{15}\ \infty_{16})(A\ 1_3\ 2_3\ 3_3\ 4_3)(B\ 1_4\ 2_4\ 3_4\ 4_0)$ $(C\ 1_2\ 2_2\ 3_0\ 4_4)(D\ 1_1\ 2_0\ 3_2\ 4_2)(E\ 1_0\ 2_1\ 3_1\ 4_1)$ $(5_i)(6_i\ 9_i\ 10_i\ 11_i\ 12_i)(7_i\ 13_i\ 14_i\ 15_i\ 16_i)(8_i\ 17_i\ 18_i\ 19_i\ 20_i)$, Base blocks $ABCDE1_01_11_21_31_42_02_12_22_32_43_03_13_23_33_44_04_14_24_34_4$, $ABCDE5_05_15_25_35_46_06_16_26_36_47_07_17_27_37_48_08_18_28_38_4$, $ABCDE9_09_19_29_39_412_012_112_212_312_414_014_114_214_314_4$ $15_015_115_215_315_4$, $ABCDE13_013_113_213_313_416_016_116_216_316_4$ $18_018_118_218_318_419_019_119_219_319_4$, $ABCDE17_017_117_217_317_420_020_120_220_320_4$ $10_010_110_210_310_411_011_111_211_311_4$, $A1_02_03_04_09_112_114_115_113_216_218_219_25_36_37_38_3$ $17_420_410_411_4\infty_1\infty_2\infty_3\infty_4$, $A1_02_03_04_013_116_118_119_117_220_210_211_29_312_314_315_3$ $5_46_47_48_4\infty_5\infty_8\infty_{10}\infty_{11}$, $A1_02_03_04_017_120_110_111_15_26_27_28_213_316_318_319_3$ $9_412_414_415_4\infty_9\infty_{12}\infty_{14}\infty_{15}$, $A1_02_03_04_05_16_17_18_19_212_214_215_217_320_310_311_3$ $13_416_418_419_4\infty_{13}\infty_{16}\infty_6\infty_7$.
21	131	26	5	?
21	109	28	7	Base block 0 1 3 5 7 9 15 16 21 22 25 26 27 35 38 45 48 49 63 66 73 75 78 80 81 89 97 105 (mod 109).
21	85	36	15	?
22	201	25	3	?
22	127	28	6	?
22	97	33	11	?
23	301	25	2	?
23	103	34	11	?
25	352	27	2	?
25	253	28	3	?
25	204	29	4	?
25	175	30	5	Base block 000 100 200 300 400 001 011 021 031 041 002 112 222 332 442 003 123 243 313 433 004 134 214 344 424 005 145 235 325 415 (mod $(5, 5, 7)$).
25	156	31	6	Base block 0 1 5 11 13 25 28 39 46 55 58 65 68 74 76 86 87 91 111 117 118 119 122 123 125 127 134 140 142 143 147 (mod 156).
25	133	33	8	Base block 1 4 5 14 16 19 20 21 25 38 54 56 57 64 66 70 76 80 83 84 91 93 95 98 100 101 105 106 114 123 125 126 131 (mod 133).
25	120	35	10	?
25	112	37	12	?

n	v	k	λ	Solution
25	105	40	15	?
25	100	45	20	Points $A, 1_i, \ldots, 9_i$, $i = 0, 1, \ldots, 10 \pmod{11}$, Automorphisms $\rho = (A)(I_0\ I_1\ \cdots\ I_{10})$, $I = 1, 2, \ldots, 9$, $\sigma = (A)(1_i\ 2_i\ 3_i\ 4_i)(5_i\ 6_i\ 7_i\ 8_i)(9_i)$, Base blocks $A1_01_11_21_31_41_51_61_71_81_91_{10}2_02_12_22_32_42_52_62_72_82_92_{10}$ $3_03_13_23_33_43_53_63_73_83_93_{10}4_04_14_24_34_44_54_64_74_84_94_{10}$, $A1_01_21_61_71_{10}1_82_02_12_32_92_52_43_03_23_63_73_{10}3_84_05_25_65_7$ $5_{10}5_86_16_36_96_56_47_17_37_97_57_48_28_68_78_{10}8_89_29_69_79_{10}9_8$, $1_11_31_91_51_42_12_32_92_52_43_23_63_73_{10}3_84_24_64_74_{10}4_85_05_15_3$ $5_95_55_46_06_26_66_76_{10}6_87_07_17_37_97_57_48_09_09_29_69_79_{10}9_8$, $1_21_61_71_{10}1_82_22_62_72_{10}2_83_23_63_73_{10}3_84_24_64_74_{10}4_85_05_2$ $5_65_75_{10}5_86_06_26_76_{10}6_87_07_27_67_77_{10}7_88_08_28_68_78_{10}8_89_0$.

5.8 See Also

§IV.12, IV.13	Symmetric designs are constructible from difference sets.
§IV.24	Hadamard designs are an important class of symmetric designs.
§IV.55	Youden squares are an arrangement of the blocks of a symmetric design.
§VI.8	Projective planes are another important class of symmetric designs.
[6]	A comprehensive book for combinatorial designs.
[18]	This second edition contains more results on symmetric designs.
[50]	*Semibiplanes* are connected incidence structures of points and lines, in which every two points lie on two common lines, and every two lines meet in two points. A semibiplane has the same number of points as lines; every line has k points, and every point lies on k lines.

References

[1] K.T. Arasu, Number theoretic consequences on the parameters of a symmetric design, *Discrete Math.* 65 (1987), 1-4.

[2] M. Aschbacher, On collineation groups of symmetric block designs, *J. Combin. Theory A* 11 (1971), 272-281.

[3] B. Bagchi, No extendable biplane of order 9, *J. Combin. Theory A* 49 (1988), 1-12.

[4] B. Bagchi, Corrigendum: No extendable biplane of order 9, *J. Combin. Theory A* 57 (1991), 162.

[5] H. Beker and W. Haemers, 2-designs having an intersection number $k-n$, *J. Combin. Theory A* 28 (1980), 64-81.

[6] T. Beth, D. Jungnickel and H. Lenz, *Design Theory*, Cambridge University Press, London, 1986.

[7] K.N. Bhattacharya, A new balanced incomplete block design, *Sci. Culture* 9 (1944), 508.

[8] R.C. Bose, S.S. Shrikhande and N.M. Singhi, Edge regular multigraphs and partial geometric designs with an application to the embedding of quasi-residual designs, *Atti del Convegni Lincei* 17 (1976), 49-81.

[9] W.G. Bridges, M. Hall Jr. and J.L. Hayden, Codes and designs, *J. Combin. Theory A* 31 (1981), 155-174.

[10] A.E. Brouwer, private communication.

[11] A.E. Brouwer, An infinite series of symmetric designs, Mathematisch Centrum, Amsterdam, Report, ZW 202/83.

[12] A.E. Brouwer and H.A. Wilbrink, A symmetric design with parameters 2-(49,16,5), *J. Combin. Theory A* 37 (1984), 193-194.

[13] R.H. Bruck and H.J. Ryser, The nonexistence of certain finite projective planes, *Canad. J. Math.* 1 (1949), 88-93.

[14] P.J. Cameron, Extending symmetric designs, *J. Combin. Theory A* 14 (1973), 215-220.

[15] S. Chowla, A property of biquadratic residues, *Proc. Nat. Acad. Sci. India A* 14 (1944), 45-46.

[16] S. Chowla and H.J. Ryser, Combinatorial problems, *Canad. J. Math.* 2 (1950), 93-99.

[17] W. Haemers, *Eigenvalue Techniques in Design and Graph Theory*, Mathematisch Centre Tracts 121, Mathematisch Centrum, Amsterdam, 1980.

[18] M. Hall Jr., *Combinatorial Theory*, 2nd Ed. John Wiley and Sons, New York, 1986.

[19] M. Hall Jr., A survey of difference sets, *Proc. Amer. Math. Soc.* 7 (1956), 975-986.

[20] M. Hall Jr. and W.S. Connor, An embedding theorem for balanced incomplete block designs, *Canad. J. Math.* 6 (1953), 35-41.

[21] M. Hall Jr., R. Lane and D. Wales, Designs derived from permutation groups, *J. Combin. Theory A* 8 (1970), 12-22.

[22] M. Hall Jr. and H.J. Ryser, Cyclic incidence matrices, *Canad. J. Math.* 3 (1951), 495-502.

[23] G. Higman, On the simple group of D.G. Higman and C.C. Sims, *Illinois J. Math.* 13 (1969), 74-80.

[24] D.R. Hughes, Regular collineation groups, *Proc. Amer. Math. Soc.* 8 (1957), 165-168.

[25] D.R. Hughes, Collineations and generalized incidence matrices, *Trans. Amer. Math. Soc.* 86 (1957), 284-296.

[26] Z. Janko and Tran van Trung, The existence of a symmetric block design for (70,24,8), *Mitt. Math. Sem. Gießen* 165 (1984), 17-18.

[27] Z. Janko and Tran van Trung, Construction of a new symmetric block design for (78,22,6) with the help of tactical decompositions, *J. Combin. Theory A* 40 (1985), 451-455.

[28] Z. Janko and Tran van Trung, Construction of two symmetric block designs for (71,21,6), *Discrete Math.* 55 (1985), 327-328.

[29] D. Jungnickel and A. Pott, A new class of symmetric (v, k, λ)-designs, *Des. Codes Crypt.* 4 (1994), 319-325.

[30] W.M. Kantor, Classification of 2-transitive symmetric designs, *Graphs Combin.* 1 (1985), 165-166.

[31] C.W.H. Lam, L. Thiel and S. Swiercz, The non-existence of finite projective planes of order 10, *Canad. J. Math.* 41 (1989), 1117-1123.

[32] E.S. Lander, *Symmetric Designs: An Algebraic Approach*, Cambridge University Press, London, 1983.

[33] E. Lehmer, On residue difference sets, *Canad. J. Math.* 5 (1953), 425-432.

[34] J.H. van Lint and V.D. Tonchev, A class of non-embeddable designs, *J. Combin. Theory A* 62 (1993), 252-260.

[35] R.L. McFarland, A family of difference sets in non-cyclic groups, *J. Combin. Theory A* 15 (1973), 1-10.

[36] P.K. Menon, On difference sets whose parameters satisfy a certain relation, *Proc. Amer. Math. Soc.* 13 (1962), 739-745.

[37] C.J. Mitchell, An infinite family of symmetric designs, *Discrete Math.* 26 (1979), 247-250.

[38] D.P. Rajkundlia, Some Techniques for Constructing New Infinite Families of Balanced Incomplete Block Designs, Ph.D. dissertation, Queen's University, Kingston, Canada, 1978.

[39] D.P. Rajkundlia, Some techniques for constructing infinite families of BIBD's, *Discrete Math.* 44 (1983), 61-96.

[40] S.S. Sane, S.S. Shrikhande and N.M. Singhi, Maximal arcs in designs, *Graphs Combin.* 1 (1985), 97-106.

[41] E. Spence, A family of difference sets in non-cyclic groups, *J. Combin. Theory A* 22 (1977), 103-106.

[42] E. Spence, A new family of symmetric 2-(v, k, λ) block designs, *Europ. J. Combin.* 14 (1993), 131-136.

[43] E. Spence, V.D. Tonchev and Tran van Trung, A symmetric 2-(160,54,18) design, *J. Combin. Des.* 1 (1993), 65-68.

[44] S.S. Shrikhande and N.M. Singhi, Construction of geometroids, *Utilitas Math.* 8 (1975), 187-192.

[45] V.D. Tonchev, Embedding of the Witt-Mathieu system S(3,6,22) in a symmetric 2-(78,22,6) design, *Geom. Ded.* 22 (1987), 49-75.

[46] Tran van Trung, The existence of symmetric block designs with parameters (41,16,6) and (66,26,10), *J. Combin. Theory A* 33 (1982), 201-204.

[47] Tran van Trung, Nonembeddable quasi-residual designs, *Contemp. Math.* 111 (1990), 237-278.

[48] W.D. Wallis, Construction of strongly regular graphs using affine designs, *Bull. Austr. Math. Soc.* 4 (1971), 41-49.

[49] A.L. Whiteman, A family of difference sets, *Illinois J. Math.* 6 (1962), 107-121.

[50] P. Wild, Generalized Hussain graphs and semibiplanes with $k \leq 6$, *Ars Combin.* 14 (1982), 147-167.

6 Resolvable and Near Resolvable Designs

R. Julian R. Abel
Steven C. Furino

6.1 Definitions and Examples

A *parallel class* or *resolution class* in a design is a set of blocks that partition the point set. A *partial parallel class* is a set of blocks that contain no point of the design more than once.

A *resolvable balanced incomplete block design* is a (v, k, λ) BIBD whose blocks can be partitioned into parallel classes. The notation (v, k, λ) RBIBD is commonly used.

6.1 **Example** A $(9, 3, 1)$ RBIBD. (Each column is a parallel class.)

$$\begin{array}{cccc} \{1,2,3\} & \{1,4,7\} & \{1,5,9\} & \{1,6,8\} \\ \{4,5,6\} & \{2,5,8\} & \{2,6,7\} & \{2,4,9\} \\ \{7,8,9\} & \{3,6,9\} & \{3,4,8\} & \{3,5,7\} \end{array}$$

An *affine design* (or *affine resolvable design*) is an RBIBD with the property that any two nonparallel lines intersect in a constant number of points.

A (K, λ)-*frame* is a (K, λ)-GDD $(X, \mathcal{G}, \mathcal{A})$ whose blocks can be partitioned into partial parallel classes each of which contains exactly the points of $X \backslash G$ for some $G \in \mathcal{G}$. The group type of the frame is the group type of the underlying GDD.

6.2 **Example** A $(\{3\}, 1)$-frame of type 2^4

groups	{2,3}	{4,7}	{5,9}	{6,8}
blocks	{4,5,6}	{2,5,8}	{2,6,7}	{2,4,9}
	{7,8,9}	{3,6,9}	{3,4,8}	{3,5,7}

A *near resolvable design*, $(v, k, k-1)$ NRB, is a $(v, k, k-1)$ BIBD with the property that the blocks can be partitioned into partial parallel classes missing a single point, and every point of the design is absent from exactly one class.

6.3 **Example** A $(7, 3, 2)$ NRB.

missing point	{ 0 }	{ 1 }	{ 2 }	{ 3 }	{ 4 }	{ 5 }	{ 6 }
blocks	{1,2,4}	{2,3,5}	{3,4,6}	{4,5,0}	{5,6,1}	{6,0,2}	{0,1,3}
	{3,5,6}	{4,6,0}	{5,0,1}	{6,1,2}	{0,2,3}	{1,3,4}	{2,4,5}

6.4 **Remark** The necessary conditions for the existence of a (v, k, λ) RBIBD are $\lambda(v-1) \equiv 0 \pmod{k-1}$ and $v \equiv 0 \pmod{k}$. The necessary condition for the existence of a $(v, k, k-1)$ NRB is $v \equiv 1 \pmod{k}$.

6.2 Equivalent Combinatorial Objects

6.5 **Theorem** The following equivalencies are known.

1. A $(v, k, 1)$ RBIBD is equivalent to a $(\{k\}, 1)$-frame of type $(k-1)^r$ where $r = \frac{v-1}{k-1}$.
2. A $(k^2, k, 1)$ RBIBD is equivalent to an affine plane. (See §VI.7.3.)
3. A $(4n, 2n, 2n-1)$ RBIBD is equivalent to a $(4n-1, 2n-1, n-1)$-Hadamard design. (See §IV.24).
4. A $(v, k, k-1)$ NRB is equivalent to a $(\{k\}, k-1)$-frame of type 1^v.

6.3 Kirkman Triple Systems

A *Kirkman triple system* of order v, $KTS(v)$, is a $(v, 3, 1)$ RBIBD, together with a resolution of its blocks into parallel classes.

6.6 **Example** In 1850, the Reverend Thomas P. Kirkman posed the question:

Kirkman's schoolgirl problem:
Fifteen young ladies in a school walk out three abreast for seven days in succession: it is required to arrange them daily, so that no two walk twice abreast.

This is equivalent to finding a KTS(15), i.e., a resolution of some Steiner triple system of order 15. Of the eighty nonisomorphic STS(15)s (Table I.1.21), exactly four are resolvable.

However, there are *seven* nonisomorphic *resolved* systems, i.e., seven KTS(15)s. These are given here, using the numbering of the underlying STSs, and displaying blocks as columns.

The Seven Solutions of the Kirkman Schoolgirl Problem							
#	Monday	Tuesday	Wednesday	Thursday	Friday	Saturday	Sunday
1a	adefg	abcdf	abcdg	abcfg	abcde	abcef	abceg
	bjhik	hemkj	jmehi	dlikh	fhlik	lidjh	ndhij
	cnmol	ignol	kofln	enjmo	gjomn	mkgon	ofklm
1b	adefg	abcdf	abcdg	abcfg	abcde	abceg	abcef
	bjhik	hemkj	jmehi	dilhj	flhij	ldikh	nhdik
	cnmol	ignol	kofln	ekonm	gnkmo	mfjno	ojglm
7a	adefg	abceg	abcdf	abcfg	abcde	abcef	abcdg
	bjikh	hdlkj	jemhi	dlihk	fmhij	lidhj	nheki
	comln	ifonm	kgnlo	enjmo	goknl	mkgon	ofjml
7b	adefg	abceg	abcdf	abcfg	abcde	abcdg	abcef
	bjikh	hdlkj	jemhi	dmhji	flikh	lheik	nidjn
	comln	ifonm	kgnlo	eoknl	gnjmo	mjfno	okglm
15a	abcfg	abcde	abceg	abcdf	adefg	abcdg	abcef
	dihmj	fhikl	ldkih	nejih	bhkji	hmejk	jldhi
	eklno	gjnmo	mfojn	ogmlk	conlm	iofnl	kngmo
15b	abcfg	abcde	abceg	abcdf	adefg	abcef	abcdg
	dihmj	fhikl	ldkih	nejih	bjhik	hmdkj	jlehi
	eklno	gjnmo	mfojn	ogmlk	cnmol	iognl	knfom
61	adefg	abcfg	abcde	abcdf	abcdg	abcef	abceg
	bijlh	dijhk	fhimk	heljk	jmehi	lkdhi	ndhij
	cknom	elmno	gjonl	ignom	kofln	mngoj	ofkml

6.7 Theorem A Kirkman triple system KTS(v) exists if and only if $v \equiv 3 \pmod 6$.

6.8 Remark The existence of Kirkman triple systems was a celebrated open problem throughout the period 1850-1970, until the first published solution was given by Ray-Chaudhuri and Wilson [9]. The first record of a solution appears to be that of Lu Jiaxi in Mongolia in 1965 [8]; unfortunately, the Great Cultural Revolution in China seems to have prevented publication of his work on the subject.

6.4 Standard Frame Constructions

6.9 Remark Almost all of the known asymptotic results for RBIBDs arise from frame type constructions.

6.10 Theorem Let $(X, \mathcal{G}, \mathcal{A})$ be an $(M,1)$-GDD and let $w : X \to N \cup \{0\}$. For each $A \in \mathcal{A}$ suppose there exists a (K,λ)-frame of type $\{w(x) : x \in A\}$. Then there exists a (K,λ)-frame of type $\left\{\sum_{x\in G} w(x) : G \in \mathcal{G}\right\}$.

6.11 Theorem Let $(X, \mathcal{G}, \mathcal{A})$ be a (K,λ)-frame. Let F be a new set of points, that is, $F \cap X = \emptyset$. Suppose that for each group $G \in \mathcal{G}$, there exists a (K,λ)-frame $(\mathcal{G} \cup F, \mathcal{H}_G \cup F, \mathcal{A}_G)$. Then there exists a (K,λ)-frame $(X', \mathcal{G}', \mathcal{A}')$ where

$$X' = X \cup F, \quad \mathcal{G}' = \left(\bigcup_{G\in\mathcal{G}} \mathcal{H}_G\right) \cup \{F\}, \quad \mathcal{A}' = \mathcal{A} \bigcup \left(\bigcup_{G\in\mathcal{G}} \mathcal{A}_G\right)$$

6.12 Theorem Let $(X, \mathcal{G}, \mathcal{A})$ be a $(\{k\},\lambda)$-frame. Suppose that for each group $G \in \mathcal{G}$, there exists a $(|G| + w, k, \lambda)$ RBIBD containing a (w, k, λ) RBIBD as a subdesign. Then there exists a $(|X| + w, k, \lambda)$ RBIBD containing a (w, k, λ) RBIBD as a subdesign.

6.5 Non-Existence and Asymptotic Existence Results

6.13 Remark All parameter sets for which an RBIBD is known not to exist are multiples of parameter sets of affine designs. The theorems in this section are useful for establishing nonexistence results.

6.14 Theorem If a (v, k, λ) RBIBD exists, then $b \geq v + r - 1$ where b is the number of blocks. When $b = v + r - 1$ (or equivalently, $r = k + \lambda$) the RBIBD is an affine design and any two nonparallel lines intersect in exactly $\frac{k^2}{v}$ points.

6.15 Remark The condition that $b \geq v + r - 1$ is known as *Bose's condition.*

6.16 Theorem Consider a (v, k, λ) affine design with replication number r whose nonparallel blocks intersect in $\mu = k^2/v$ points.

1. If $v = tk$ and $k \not\equiv 0 \pmod{t}$, then no $\left(tk, k, \frac{k-1}{t-1}\right)$ RBIBD exists.
2. If v and r are odd, then k is a perfect square.
3. If v is odd and r is even, then μ is a perfect square.
4. If $r \equiv 2 \pmod{4}$ or $r \equiv 3 \pmod{4}$, then the square free factor of $\frac{v}{k}$ has no prime divisor $p \equiv 3 \pmod{4}$.
5. If $\lambda \in \{1, 2\}$, then the affine design is a residual design. See §I.5.4.

6.17 Theorem Given any positive integers k and λ, there exists an integer $c(k, \lambda)$ such that a (v, k, λ) RBIBD exists whenever $v > c(k, \lambda)$ and the necessary conditions are satisfied.

6.18 Theorem Given any positive integer k, there exist $n(k) \in \mathbb{Z}^+$ such that a $(v, k, k-1)$ NRB exists whenever $v > n(k)$ and $v \equiv 1 \pmod{k}$.

6.6 Existence Results for RBIBDs for Specific k and λ

6.19 Table A summary of existence results.

k	λ	Exceptions	Possible Exceptions	Largest Possible Exception
3	1	none	none	
3	2	6	none	
4	1	none	none	
4	3	none	none	
5	1		see Table 6.21	4905
5	2	15	no explicit enumeration	50722390
5	4	10	see Table 6.22	195
6	5		see Table 6.23	594
6	10	none	none	
7	6	14	no explicit enumeration	33936
8	1		see Table 6.24	24480
8	7		see Table 6.25	1488
12	11		explicit enumeration in [6]	108768

6.20 Remark No specific bounds have been computed for any pairs of k and λ not specifically mentioned in Table 6.19.

6.21 Table Values of v for which the existence of $(v, 5, 1)$ RBIBDs remains open.

45	105	145	185	225	345	465	585	645	665	705
785	885	925	945	1045	1065	1145	1165	1185	1305	1385
1485	1545	1665	1905	2265	2385	2505	2745	2865	2985	3105
3225	3345	3585	3785	3945	4065	4185	4425	4665	4905	

6.22 Table Values of v for which the existence of $(v, 5, 4)$ RBIBDs remains open.

15 70 75 90 95 135 160 185 190 195

6.23 Table Values of v for which the existence of $(v, 6, 5)$ RBIBDs remains open.

48 60 138 174 240 318 330 336 414 594

6.24 Table Values of v for which the existence of $(v, 8, 1)$ RBIBDs remains open.

176	624	736	1128	1240	1296	1408	1464	1520	1576
1744	2136	2416	2640	2920	2976	3256	3312	3424	3760
3872	4264	4432	5216	5720	5776	6224	6280	6448	6896
6952	7008	7456	7512	7792	7848	8016	9752	10200	10704
10760	10928	11040	11152	11376	11656	11712	11824	11936	12216
12328	12496	12552	12720	12832	12888	13000	13280	13616	13840
13896	14008	14176	14232	21904	24480				

6.25 Table Values of v for which the existence of $(v, 8, 7)$ RBIBDs remains open.

24	48	56	160	168	192	224	240	312	336	384	408
440	448	504	552	560	616	744	760	984	1464	1488	

6.7 Existence Results for NRBs for Specific k

6.26 Table The spectrum of $(v, k, k-1)$ NRBs for $3 \leq k \leq 8$.

k	Possible Exceptions
3	none
4	none
5	none
6	55 145
7	85 99 141 148 183 246 267 274 295 365 386 589 596 603
8	129 161 177 305 321 385 465 481 497 545 553

6.27 Table Bounds on the existence of $(v, k, k-1)$ NRBs. N denotes the number of possible exceptions.

k	$n(k) \leq$	N
9	18145	142
10	9701	100
12	9061	53
15	126331	1116
16	10033	158
17	464373	2647

k	$n(k) \leq$	N
18	103609	463
22	716057	2643
24	58105	290
26	1238849	4530
28	121073	651
30	490141	1197

6.8 Some Explicit Constructions

6.28 Table Some $(v, k, k-1)$ NRBs. In each case, develop the blocks given over $\mathbb{Z}_v$ to obtain the v parallel classes. Adding the element 0 to each block prior to the development gives a $(v, k+1, k+1)$-design.

v, k	Base Near Parallel Classes
19,6	{1,7,8,11,12,18}, {2,3,5,14,16,17}, {4,6,9,10,13,15}
21,5	{1,2,3,5,10}, {4,8,11,15,16}, {6,12,14,17,20}, {7,9,13,18,19}
26,5	{1,4,8,18,22}, {2,9,11,19,24}, {3,6,16,17,23}, {5,7,10,21,25}, {12,13,14,15,20}
29,4	{1,12,17,28}, {2,5,24,27}, {4,10,19,25}, {8,9,20,21}, {11,13,16,18}, {3,7,22,26}, {6,14,15,23}
31,6	{1,5,6,25,26,30}, {3,13,15,16,18,28}, {8,9,14,17,22,23}, {4,7,11,20,24,27}, {2,10,12,19,21,29}
31,10	{1,2,4,8,15,16,23,27,29,30}, {3,6,7,12,14,17,19,24,25,28}, {5,9,10,11,13,18,20,21,22,26}
33,4	{1,8,18,19}, {2,3,24,31}, {4,13,21,23}, {5,11,17,20}, {6,10,26,29}, {7,12,15,16}, {9,22,28,30}, {14,25,27,32}
33,8	{1,11,12,15,20,22,26,28}, {2,5,9,14,17,18,19,29}, {3,4,16,21,23,24,27,31}, {6,7,8,10,13,25,30,32}
36,7	{1,13,18,24,28,32,34}, {2,12,14,25,30,31,33}, {3,6,9,10,15,23,35}, {4,5,8,16,17,19,26}, {7,11,20,21,22,27,29}
37,12	{1,6,8,10,11,14,23,26,27,29,31,36}, {2,9,12,15,16,17,20,21,22,25,28,35} {3,4,5,7,13,18,19,24,30,32,33,34}
41,4	{1,9,32,40}, {6,13,28,35}, {4,5,36,37}, {11,17,24,30}, {16,20,21,25}, {3,14,27,38}, {2,18,23,39} {12,15,26,29}, {8,10,31,33}, {7,19,22,34}
41,8	{1,3,9,14,27,32,38,40}, {2,6,13,18,23,28,35,39}, {4,5,12,15,26,29,36,37}, {8,10,11,17,24,30,31,33}, {7,16,19,20,21,22,25,34}
41,10	{1,4,10,16,18,23,25,31,37,40}, {6,14,15,17,19,22,24,26,27,35}, {2,5,8,9,20,21,32,33,36,39}, {3,7,11,12,13,28,29,30,34,38}

6.29 Example A (22,7,6) NRB. Elements are $\mathbb{Z}_{21} \cup \{\infty\}$, and 21 near parallel classes are obtained from $\{\infty$,1,5,6,7,15,18}, {2,3,4,8,10,13,17}, and {9,11,12,14,16,19,20} modulo 21. The last consists of the three distinct translates of {0,3,6,9,12,15,18}.

6.30 Example A (28,9,8) NRB. The elements are $\mathbb{Z}_{27} \cup \{\infty\}$, and 27 of the near parallel classes are obtained by developing $\{\infty$,1,2,5,7,14,16,23,24}, {9,12,13,15,17,19,20,25,26}, and {3,4,6,8,10,11,18,21,22} modulo 27. The last consists of the three distinct translates of {0,3,6,9,12,15,18,21,24}.

6.31 Example A resolvable (288,8,1) design. The element set is $(\mathbb{Z}_7 \times \mathbb{Z}_{41}) \cup \{\infty\}$. Obtain the first parallel class as

$$\{(0,9t),(0,32t),(1,3t),(1,38t),(2,t),(2,40t),(4,14t),(4,27t)\}$$

for $t = 1, 37, 16, 18, 10$, developing modulo $(7, -)$; then append the block

$$\{\infty,(0,0),(1,0),(2,0),(3,0),(4,0),(5,0),(6,0)\}$$

to complete the first parallel class. Develop this parallel class modulo $(-, 41)$ to obtain the resolution.

6.9 Orthogonal Resolutions

Let $(V, \mathcal{B})$ be a design. Let $\mathcal{P} = \{P_1, \ldots, P_s\}$ and $\mathcal{Q} = \{Q_1, \ldots, Q_t\}$ be partitions of $\mathcal{B}$ into partial parallel classes. Partitions $\mathcal{P}$ and $\mathcal{Q}$ are *orthogonal* if $|P_i \cap Q_j| \le 1$ for all $1 \le i \le s$ and $1 \le j \le t$.

6.32 Remark The situations most explored are when $\mathcal{P}$ and $\mathcal{Q}$ are both resolutions, or both near resolutions. When a design has two orthogonal resolutions, it is *doubly resolvable*; when it has two orthogonal near resolutions, it is *doubly near resolvable.*

6.33 Example A doubly near resolvable (10,3,2) design. One near resolution is the rows, the second is the columns.

		3,4,8	1,6,7				2,5,9		
			4,0,9	2,7,8				3,6,5	
3,8,9				0,1,5					4,7,6
1,2,6	4,9,5				0,8,7				
	2,3,7	0,5,6				1,9,8			
		1,7,9					4,6,8		0,2,3
			2,8,5		1,3,4			0,7,9	
				3,9,6		2,4,0			1,8,5
4,5,7					2,9,6		3,0,1		
	0,6,8					3,5,7		4,1,2	

6.34 Theorem For $v \equiv 3 \pmod 6$ and v sufficiently large, there exists a doubly resolvable Steiner triple system of order v.

6.35 Theorems (Lamken [5])

1. A doubly resolvable 2-(v,3,2) design exists if and only if $v \equiv 0 \pmod 3$ except when $v \in \{6, 9\}$ and possibly when $v \in \{72, 78, 90, 114, 117, 126\}$.
2. A doubly near resolvable 2-(v,3,2) design exists if and only if $v \equiv 1 \pmod 3$ except when $v = 7$, and possibly when $v \in \{34, 70, 85, 88, 115, 124, 133, 142\}$.

6.10 See Also

§I.5	Some affine designs exist only as residual designs of symmetric designs. See Theorem 6.16.
§IV.10	Block disjoint difference families can be used to construct RBIBDs. See [4] for examples and references. In §IV.10.6, several of the 1-rotational difference families are resolvable.
§IV.24	Hadamard designs are related via Theorem 6.5.
§IV.41	SOLSSOMs can be used to construct RBIBDs. See [2].
§VI.7	Affine geometries are the source of many RBIBDs.
[4]	Survey of techniques and results. Contains existence tables for RBIBDs with $10 \le v \le 200$, and an extensive bibliography.
[7]	English translation of Lu's asymptotic existence results together with a more modern treatment.
[10]	Contains first asymptotic proof covering all k. Deals with $\lambda = 1$.
[3]	A survey of orthogonal resolutions of triple systems and related cubical arrays.

References

[1] R.J.R. Abel and M. Greig, Resolvable balanced incomplete block designs with a block size of 8, preprint.

[2] R.D. Baker, Resolvable BIBDs and SOLS, *Discrete Math.* 44 (1983), 13-29.

[3] C.J. Colbourn and A. Rosa, Orthogonal resolutions of triple systems, *Australas. J. Combin.* 12 (1995), 259-269.

[4] S.C. Furino, Y. Miao, and J.X. Yin, *Frames and Resolvable Designs*, CRC Press, Boca Raton FL, to appear.

[5] E.R. Lamken, The existence of DR(v,3,2)-BIBDs, *J. Combin. Theory A* 72 (1995), 50-76.

[6] T.C.Y. Lee, Tools for Constructing RBIBDs, Frames, NRBs and SOLSSOMs, Ph.D. thesis, University of Waterloo, 1995.

[7] T.C.Y. Lee and S.C. Furino, A translation of J.X. Lu's "An existence theory for resolvable balanced incomplete block designs", *J. Combin. Des.* 3 (1995), 321-340.

[8] Lu Jiaxi, *Collected Works on Combinatorial Designs*, Inner Mongolia People's Press, Hunhot, Mongolia, 1990.

[9] D.K. Ray-Chaudhuri and R.M. Wilson, Solution of Kirkman's schoolgirl problem, *Proc. Symp. Pure Math. Amer. Math. Soc.* 19 (1971), 187-204.

[10] D.K. Ray-Chaudhuri and R.M. Wilson, The existence of resolvable block designs, in: *A Survey of Combinatorial Theory* (Srivastava et. al.; eds), North-Holland, Amsterdam, 1973, 361-375.

PART II

LATIN SQUARES, MOLS, AND ORTHOGONAL ARRAYS

1 Latin Squares

Charles J. Colbourn
Jeffrey H. Dinitz

1.1 Definitions and Examples

A *latin square* of *side* n is an $n \times n$ array in which each cell contains a single element from an n-set S, such that each element occurs exactly once in each row and exactly once in each column.

A latin square of side n (on the symbol set $1, 2, \ldots, n$ or on the symbol set $0, 1, \ldots, n-1$) is *reduced* or in *standard form* if in the first row and column the elements occur in natural order.

1.1 **Example** A reduced latin square of side 8 on the symbols $0, 1, \ldots, 7$.

0	1	2	3	4	5	6	7
1	0	3	4	5	6	7	2
2	3	5	0	6	7	4	1
3	4	0	7	1	2	5	6
4	5	6	1	7	0	2	3
5	6	7	2	0	3	1	4
6	7	4	5	2	1	3	0
7	2	1	6	3	4	0	5

Let L be an $n \times n$ latin square on symbol set E_3, with rows indexed by the elements of the n-set E_1 and columns indexed by the elements of the n-set E_2. Let $\mathcal{T} = \{(x_1, x_2, x_3) : L(x_1, x_2) = x_3\}$. Let $\{a, b, c\} = \{1, 2, 3\}$. The (a, b, c)-*conjugate* of L, $L_{(a,b,c)}$, has rows indexed by E_a, columns by E_b, and symbols by E_c, and is defined by $L_{(a,b,c)}(x_a, x_b) = x_c$ for each $(x_1, x_2, x_3) \in \mathcal{T}$.

1.2 **Example** A latin square of side 4 and its six conjugates.

1	4	2	3
2	3	1	4
4	1	3	2
3	2	4	1

(1,2,3)-conjugate

1	2	4	3
4	3	1	2
2	1	3	4
3	4	2	1

(2,1,3)-conjugate

1	3	2	4
2	4	1	3
4	2	3	1
3	1	4	2

(3,2,1)-conjugate

1	2	4	3
3	4	2	1
2	1	3	4
4	3	1	2

(2,3,1)-conjugate

1	3	4	2
3	1	2	4
2	4	3	1
4	2	1	3

(1,3,2)-conjugate

1	3	2	4
3	1	4	2
4	2	3	1
2	4	1	3

(3,1,2)-conjugate

1.3 **Remark** Any latin square has 1,2,3, or 6 distinct conjugates.

The *transpose* of a latin square L, denoted L^T, is the latin square which results from L when the role of rows and columns are exchanged (i.e., $L^T(i, j) = L(j, i)$).

A latin square L of side n is *symmetric* if $L(i,j) = L(j,i)$ for all $1 \le i,j \le n$.

1.4 **Remark** A latin square L is symmetric if and only if $L = L^T = L_{(2,1,3)}$. The square in Example 1.1 is symmetric.

1.2 Equivalent Objects

1.5 **Theorem** A latin square of side n is equivalent to

1. the multiplication table (*Cayley table*) of a quasigroup on n elements (§IV.36);
2. a transversal design of index one, $TD(3,n)$ (§II.2.2);
3. a $(3,n)$-net (§II.2.2);
4. an orthogonal array of strength two and index one, $OA(3,n)$ (§II.4);
5. a 1-factorization of the complete bipartite graph $K_{n,n}$ (§VI.4);
6. an edge-partition of the complete tripartite graph $K_{n,n,n}$ into triangles;
7. a set of n^2 mutually non-attacking rooks on a $n \times n \times n$ board;
8. a single error detecting code of word length 3, with n^2 words from an n–symbol alphabet ([5] p. 354).

1.3 Enumeration

Two latin squares L and L' of side n are *isotopic or equivalent* if there are three bijections from the rows, columns, and symbols of L to the rows, columns, and symbols, respectively, of L', that map L to L'.

Two latin squares L and L' of side n are *main class isotopic* if L is isotopic to any conjugate of L'.

Two latin squares L and L' of side n are *isomorphic* if there is a bijection $\phi : S \to S$ such that $\phi L(i,j) = L'(\phi(i), \phi(j))$ for every $i,j \in S$, where S is not only the symbol set of each square, but also the indexing set for the rows and columns of each square.

1.6 **Table** Enumeration of latin squares of side $n \le 8$ (see [5, 17] and Table 1.12).

$n =$	1	2	3	4	5	6	7	8
main classes	1	1	1	2	2	12	147	283,657
isotopy classes	1	1	1	2	2	22	563	1,676,267
isomorphism classes	1	1	1	2	6	109	?	?
reduced squares	1	1	1	4	56	9,408	16,942,080	535,281,401,856

1.7 **Table** Main classes of latin squares of side n, $4 \le n \le 7$. For each, an identifying number is given, followed by three integers giving the number of transversals, the number of 2×2 subsquares, and the number of 3×3 subsquares.

$n = 4$

0	1	2	3
1	0	3	2
2	3	0	1
3	2	1	0

4.1: 8,12,0

0	1	2	3
1	0	3	2
2	3	1	0
3	2	0	1

4.2: 0,4,0

$n = 5$

0	1	2	3	4
1	2	3	4	0
2	3	4	0	1
3	4	0	1	2
4	0	1	2	3

5.1: 15,0,0

0	1	2	3	4
1	0	3	4	2
2	3	4	0	1
3	4	1	2	0
4	2	0	1	3

5.2: 3,4,0

$n = 6$

0 1 2 3 4 5	0 1 2 3 4 5	0 1 2 3 4 5	0 1 2 3 4 5	0 1 2 3 4 5	0 1 2 3 4 5
1 0 3 2 5 4	1 0 3 2 5 4	1 0 3 4 5 2	1 0 3 4 5 2	1 0 3 4 5 2	1 0 3 4 5 2
2 3 4 5 0 1	2 3 4 5 0 1	2 3 0 5 1 4	2 3 0 5 1 4	2 3 1 5 0 4	2 3 4 5 0 1
3 2 5 4 1 0	3 2 5 4 1 0	3 4 5 0 2 1	3 4 5 0 2 1	3 4 5 1 2 0	3 4 5 2 1 0
4 5 0 1 2 3	4 5 0 1 3 2	4 5 1 2 0 3	4 5 1 2 3 0	4 5 0 2 3 1	4 5 0 1 2 3
5 4 1 0 3 2	5 4 1 0 2 3	5 2 4 1 3 0	5 2 4 1 0 3	5 2 4 0 1 3	5 2 1 0 3 4
6.1: 0,9,4	6.2: 32,9,0	6.3: 24,15,0	6.4: 8,7,0	6.5: 8,5,0	6.6: 0,9,4

0 1 2 3 4 5	0 1 2 3 4 5	0 1 2 3 4 5	0 1 2 3 4 5	0 1 2 3 4 5	0 1 2 3 4 5
1 0 3 2 5 4	1 0 3 2 5 4	1 0 3 2 5 4	1 0 3 2 5 4	1 0 3 4 5 2	1 2 0 4 5 3
2 4 0 5 1 3	2 4 0 5 1 3	2 4 0 5 1 3	2 4 0 5 3 1	2 3 1 5 0 4	2 0 1 5 3 4
3 5 1 4 0 2	3 5 1 4 0 2	3 5 1 4 2 0	3 5 4 0 1 2	3 5 4 1 2 0	3 5 4 1 0 2
4 2 5 0 3 1	4 2 5 1 3 0	4 3 5 1 0 2	4 2 5 1 0 3	4 2 5 0 1 3	4 3 5 2 1 0
5 3 4 1 2 0	5 3 4 0 2 1	5 2 4 0 3 1	5 3 1 4 2 0	5 4 0 2 3 1	5 4 3 0 2 1
6.7: 0,27,4	6.8: 0,19,0	6.9: 0,15,0	6.10: 8,11,0	6.11: 8,4,0	6.12: 0,0,4

$n = 7$

0 1 2 3 4 5 6	0 1 2 3 4 5 6	0 1 2 3 4 5 6	0 1 2 3 4 5 6	0 1 2 3 4 5 6
1 2 0 4 3 6 5	1 2 0 4 3 6 5	1 2 0 4 3 6 5	1 2 0 4 3 6 5	1 2 0 4 5 6 3
2 0 1 5 6 3 4	2 0 1 5 6 3 4	2 0 1 5 6 3 4	2 0 1 5 6 4 3	2 0 3 5 6 1 4
3 4 5 6 0 1 2	3 4 5 6 0 2 1	3 4 5 6 1 2 0	3 4 5 6 0 1 2	3 4 5 6 2 0 1
4 3 6 0 5 2 1	4 3 6 0 5 1 2	4 3 6 1 5 0 2	4 3 6 0 5 2 1	4 5 6 2 1 3 0
5 6 3 1 2 4 0	5 6 3 2 1 4 0	5 6 3 2 0 4 1	5 6 4 1 2 3 0	5 6 1 0 3 4 2
6 5 4 2 1 0 3	6 5 4 1 2 0 3	6 5 4 0 2 1 3	6 5 3 2 1 0 4	6 3 4 1 0 2 5
7.1: 3,18,1	7.2: 23,26,3	7.3: 63,42,7	7.4: 23,14,1	7.5: 19,6,0

0 1 2 3 4 5 6	0 1 2 3 4 5 6	0 1 2 3 4 5 6	0 1 2 3 4 5 6	0 1 2 3 4 5 6
1 2 3 0 5 6 4	1 2 3 4 5 6 0	1 0 3 2 5 6 4	1 0 3 2 5 6 4	1 0 3 2 5 6 4
2 3 4 5 6 0 1	2 3 4 5 6 0 1	2 3 0 1 6 4 5	2 3 0 1 6 4 5	2 3 0 1 6 4 5
3 0 5 6 1 4 2	3 4 5 6 0 1 2	3 4 5 6 0 1 2	3 4 5 6 0 1 2	3 4 5 6 0 2 1
4 5 6 1 3 2 0	4 5 6 0 1 2 3	4 2 6 5 1 0 3	4 2 6 5 1 3 0	4 2 6 5 1 3 0
5 6 0 4 2 1 3	5 6 0 1 2 3 4	5 6 1 4 3 2 0	5 6 4 0 3 2 1	5 6 1 4 3 0 2
6 4 1 2 0 3 5	6 0 1 2 3 4 5	6 5 4 0 2 3 1	6 5 1 4 2 0 3	6 5 4 0 2 1 3
7.6: 25,0,0	7.7: 133,0,0	7.8: 21,18,1	7.9: 30,16,1	7.10: 43,30,3

0 1 2 3 4 5 6	0 1 2 3 4 5 6	0 1 2 3 4 5 6	0 1 2 3 4 5 6	0 1 2 3 4 5 6
1 0 3 2 5 6 4	1 0 3 2 5 6 4	1 0 3 2 5 6 4	1 0 3 2 5 6 4	1 0 3 2 5 6 4
2 3 0 1 6 4 5	2 3 0 1 6 4 5	2 3 0 1 6 4 5	2 3 0 1 6 4 5	2 3 0 1 6 4 5
3 4 5 6 0 2 1	3 4 5 6 0 2 1	3 4 5 6 0 2 1	3 4 5 6 0 2 1	3 4 5 6 0 2 1
4 2 6 5 1 3 0	4 2 6 5 1 3 0	4 5 6 0 1 3 2	4 5 6 0 2 1 3	4 5 6 0 3 1 2
5 6 4 0 2 1 3	5 6 4 0 3 1 2	5 6 1 4 2 0 3	5 6 1 4 3 0 2	5 6 1 4 2 0 3
6 5 1 4 3 0 2	6 5 1 4 2 0 3	6 2 4 5 3 1 0	6 2 4 5 1 3 0	6 2 4 5 1 3 0
7.11: 43,18,3	7.12: 55,22,3	7.13: 13,18,1	7.14: 33,22,1	7.15: 15,22,1

0 1 2 3 4 5 6	0 1 2 3 4 5 6	0 1 2 3 4 5 6	0 1 2 3 4 5 6	0 1 2 3 4 5 6
1 0 3 2 5 6 4	1 0 3 2 5 6 4	1 0 3 2 5 6 4	1 0 3 2 5 6 4	1 0 3 2 5 6 4
2 3 0 4 6 1 5	2 3 0 4 6 1 5	2 3 0 4 6 1 5	2 3 0 4 6 1 5	2 3 0 4 6 1 5
3 2 5 6 0 4 1	3 2 5 6 0 4 1	3 2 5 6 0 4 1	3 2 5 6 0 4 1	3 2 5 6 0 4 1
4 5 6 0 1 2 3	4 5 6 0 1 2 3	4 5 6 0 1 3 2	4 5 6 1 2 3 0	4 5 6 1 3 0 2
5 6 4 1 2 3 0	5 6 4 1 3 0 2	5 6 4 1 3 2 0	5 6 4 0 1 2 3	5 6 4 0 1 2 3
6 4 1 5 3 0 2	6 4 1 5 2 3 0	6 4 1 5 2 0 3	6 4 1 5 3 0 2	6 4 1 5 2 3 0
7.16: 30,8,0	7.17: 14,15,0	7.18: 20,12,0	7.19: 22,11,0	7.20: 15,14,0

0 1 2 3 4 5 6	0 1 2 3 4 5 6	0 1 2 3 4 5 6	0 1 2 3 4 5 6	0 1 2 3 4 5 6
1 0 3 2 5 6 4	1 0 3 2 5 6 4	1 0 3 2 5 6 4	1 0 3 2 5 6 4	1 0 3 2 5 6 4
2 3 0 4 6 1 5	2 3 0 4 6 1 5	2 3 0 4 6 1 5	2 3 0 4 6 1 5	2 3 0 4 6 1 5
3 2 5 6 0 4 1	3 2 5 6 0 4 1	3 2 5 6 0 4 1	3 2 5 6 1 4 0	3 2 5 6 1 4 0
4 5 6 1 3 2 0	4 6 1 5 2 3 0	4 6 1 5 3 0 2	4 5 6 1 0 2 3	4 5 6 1 0 3 2
5 6 4 0 1 3 2	5 4 6 0 1 2 3	5 4 6 1 2 3 0	5 6 4 0 2 3 1	5 6 4 0 3 2 1
6 4 1 5 2 0 3	6 5 4 1 3 0 2	6 5 4 0 1 2 3	6 4 1 5 3 0 2	6 4 1 5 2 0 3
7.21: 24,11,0	7.22: 18,11,0	7.23: 22,13,0	7.24: 19,14,0	7.25: 33,18,0

0	1	2	3	4	5	6
1	0	3	2	5	6	4
2	3	0	4	6	1	5
3	4	5	6	0	2	1
4	2	6	5	1	0	3
5	6	1	0	3	4	2
6	5	4	1	2	3	0

7.26: 11,16,0

0	1	2	3	4	5	6
1	0	3	2	5	6	4
2	3	0	4	6	1	5
3	4	5	6	0	2	1
4	2	6	5	1	0	3
5	6	4	1	2	3	0
6	5	1	0	3	4	2

7.27: 13,10,0

0	1	2	3	4	5	6
1	0	3	2	5	6	4
2	3	0	4	6	1	5
3	4	5	6	0	2	1
4	5	6	0	1	3	2
5	6	4	1	2	0	3
6	2	1	5	3	4	0

7.28: 16,16,0

0	1	2	3	4	5	6
1	0	3	2	5	6	4
2	3	0	4	6	1	5
3	4	5	6	0	2	1
4	5	6	1	2	0	3
5	6	1	0	3	4	2
6	2	4	5	1	3	0

7.29: 21,16,0

0	1	2	3	4	5	6
1	0	3	2	5	6	4
2	3	0	4	6	1	5
3	4	5	6	0	2	1
4	5	6	1	2	0	3
5	6	4	0	1	3	2
6	2	1	5	3	4	0

7.30: 32,14,0

0	1	2	3	4	5	6
1	0	3	2	5	6	4
2	3	0	4	6	1	5
3	4	5	6	0	2	1
4	5	6	1	2	3	0
5	6	1	0	3	4	2
6	2	4	5	1	0	3

7.31: 15,12,0

0	1	2	3	4	5	6
1	0	3	2	5	6	4
2	3	0	4	6	1	5
3	4	5	6	0	2	1
4	6	1	5	2	0	3
5	2	6	1	3	4	0
6	5	4	0	1	3	2

7.32: 24,14,0

0	1	2	3	4	5	6
1	0	3	2	5	6	4
2	3	0	4	6	1	5
3	4	5	6	0	2	1
4	6	1	5	2	3	0
5	2	6	0	1	4	3
6	5	4	1	3	0	2

7.33: 15,12,0

0	1	2	3	4	5	6
1	0	3	2	5	6	4
2	3	0	4	6	1	5
3	4	5	6	1	0	2
4	2	6	5	0	3	1
5	6	4	1	3	2	0
6	5	1	0	2	4	3

7.34: 19,16,1

0	1	2	3	4	5	6
1	0	3	2	5	6	4
2	3	0	4	6	1	5
3	4	5	6	1	0	2
4	5	6	0	2	3	1
5	6	4	1	0	2	3
6	2	1	5	3	4	0

7.35: 24,12,0

0	1	2	3	4	5	6
1	0	3	2	5	6	4
2	3	0	4	6	1	5
3	4	5	6	1	0	2
4	5	6	0	2	3	1
5	6	4	1	3	2	0
6	2	1	5	0	4	3

7.36: 29,14,1

0	1	2	3	4	5	6
1	0	3	2	5	6	4
2	3	0	4	6	1	5
3	4	5	6	1	0	2
4	5	6	0	3	2	1
5	6	4	1	2	3	0
6	2	1	5	0	4	3

7.37: 14,10,0

0	1	2	3	4	5	6
1	0	3	2	5	6	4
2	3	0	4	6	1	5
3	4	5	6	1	0	2
4	5	6	1	0	2	3
5	6	4	0	2	3	1
6	2	1	5	3	4	0

7.38: 13,16,0

0	1	2	3	4	5	6
1	0	3	2	5	6	4
2	3	0	4	6	1	5
3	4	5	6	1	0	2
4	5	6	1	3	2	0
5	6	1	0	2	4	3
6	2	4	5	0	3	1

7.39: 9,8,0

0	1	2	3	4	5	6
1	0	3	2	5	6	4
2	3	0	4	6	1	5
3	4	5	6	1	0	2
4	5	6	1	3	2	0
5	6	4	0	2	3	1
6	2	1	5	0	4	3

7.40: 15,14,0

0	1	2	3	4	5	6
1	0	3	2	5	6	4
2	3	0	4	6	1	5
3	4	5	6	1	0	2
4	6	1	5	2	3	0
5	2	6	1	0	4	3
6	5	4	0	3	2	1

7.41: 20,12,0

0	1	2	3	4	5	6
1	0	3	2	5	6	4
2	3	0	4	6	1	5
3	4	5	6	1	2	0
4	6	1	5	3	0	2
5	2	6	1	0	4	3
6	5	4	0	2	3	1

7.42: 18,20,0

0	1	2	3	4	5	6
1	0	3	2	5	6	4
2	3	0	4	6	1	5
3	4	5	6	2	0	1
4	5	6	0	1	2	3
5	6	4	1	0	3	2
6	2	1	5	3	4	0

7.43: 16,10,0

0	1	2	3	4	5	6
1	0	3	2	5	6	4
2	3	0	4	6	1	5
3	4	5	6	2	0	1
4	5	6	1	0	2	3
5	6	1	0	3	4	2
6	2	4	5	1	3	0

7.44: 19,14,0

0	1	2	3	4	5	6
1	0	3	2	5	6	4
2	3	0	4	6	1	5
3	4	5	6	2	0	1
4	5	6	1	0	2	3
5	6	4	0	1	3	2
6	2	1	5	3	4	0

7.45: 22,18,0

0	1	2	3	4	5	6
1	0	3	2	5	6	4
2	3	0	4	6	1	5
3	4	5	6	2	0	1
4	6	1	5	0	2	3
5	2	6	1	3	4	0
6	5	4	0	1	3	2

7.46: 19,12,0

0	1	2	3	4	5	6
1	0	3	2	5	6	4
2	3	0	4	6	1	5
3	4	5	6	2	0	1
4	6	1	5	0	3	2
5	2	6	0	1	4	3
6	5	4	1	3	2	0

7.47: 14,10,0

0	1	2	3	4	5	6
1	0	3	2	5	6	4
2	3	0	4	6	1	5
3	4	5	6	2	0	1
4	6	1	5	0	3	2
5	2	6	1	3	4	0
6	5	4	0	1	2	3

7.48: 15,12,0

0	1	2	3	4	5	6
1	0	3	2	5	6	4
2	3	0	4	6	1	5
3	4	5	6	2	0	1
4	6	1	5	3	2	0
5	2	6	0	1	4	3
6	5	4	1	0	3	2

7.49: 23,10,0

0	1	2	3	4	5	6
1	0	3	2	5	6	4
2	3	0	4	6	1	5
3	4	6	5	0	2	1
4	2	5	6	1	0	3
5	6	4	1	2	3	0
6	5	1	0	3	4	2

7.50: 19,12,0

0	1	2	3	4	5	6
1	0	3	2	5	6	4
2	3	0	4	6	1	5
3	4	6	5	0	2	1
4	2	5	6	1	3	0
5	6	1	0	2	4	3
6	5	4	1	3	0	2

7.51: 15,10,1

0	1	2	3	4	5	6
1	0	3	2	5	6	4
2	3	0	4	6	1	5
3	4	6	5	0	2	1
4	2	5	6	1	3	0
5	6	4	1	2	0	3
6	5	1	0	3	4	2

7.52: 27,18,1

0	1	2	3	4	5	6
1	0	3	2	5	6	4
2	3	0	4	6	1	5
3	4	6	5	0	2	1
4	5	1	6	2	0	3
5	6	4	0	1	3	2
6	2	5	1	3	4	0

7.53: 34,18,0

0	1	2	3	4	5	6
1	0	3	2	5	6	4
2	3	0	4	6	1	5
3	4	6	5	0	2	1
4	5	1	6	3	0	2
5	6	4	1	2	3	0
6	2	5	0	1	4	3

7.54: 20,10,0

0	1	2	3	4	5	6
1	0	3	2	5	6	4
2	3	0	4	6	1	5
3	4	6	5	0	2	1
4	6	5	1	2	3	0
5	2	4	6	1	0	3
6	5	1	0	3	4	2

7.55: 25,12,0

0	1	2	3	4	5	6
1	0	3	2	5	6	4
2	3	0	4	6	1	5
3	4	6	5	1	0	2
4	5	1	6	2	3	0
5	6	4	0	3	2	1
6	2	5	1	0	4	3

7.56: 28,12,0

0	1	2	3	4	5	6
1	0	3	2	5	6	4
2	3	0	4	6	1	5
3	4	6	5	1	0	2
4	6	5	1	2	3	0
5	2	1	6	0	4	3
6	5	4	0	3	2	1

7.57: 21,12,0

0	1	2	3	4	5	6
1	0	3	2	5	6	4
2	3	0	4	6	1	5
3	4	6	5	2	0	1
4	5	1	6	0	3	2
5	6	4	0	1	2	3
6	2	5	1	3	4	0

7.58: 25,16,0

0	1	2	3	4	5	6
1	0	3	2	5	6	4
2	3	0	4	6	1	5
3	4	6	5	2	0	1
4	5	1	6	0	3	2
5	6	4	1	3	2	0
6	2	5	0	1	4	3

7.59: 21,10,0

0	1	2	3	4	5	6
1	0	3	2	5	6	4
2	3	0	4	6	1	5
3	4	6	5	2	0	1
4	5	1	6	3	2	0
5	6	4	0	1	3	2
6	2	5	1	0	4	3

7.60: 28,14,1

0	1	2	3	4	5	6
1	0	3	2	5	6	4
2	3	0	4	6	1	5
3	4	6	5	2	0	1
4	6	5	0	1	2	3
5	2	1	6	3	4	0
6	5	4	1	0	3	2

7.61: 26,10,0

0	1	2	3	4	5	6
1	0	3	2	5	6	4
2	3	0	4	6	1	5
3	4	6	5	2	0	1
4	6	5	0	1	3	2
5	2	1	6	0	4	3
6	5	4	1	3	2	0

7.62: 19,10,0

0	1	2	3	4	5	6
1	0	3	2	5	6	4
2	3	0	4	6	1	5
3	4	6	5	2	0	1
4	6	5	1	0	2	3
5	2	1	6	3	4	0
6	5	4	0	1	3	2

7.63: 24,14,0

0	1	2	3	4	5	6
1	0	3	2	5	6	4
2	3	0	4	6	1	5
3	4	6	5	2	0	1
4	6	5	1	0	2	3
5	2	4	6	1	3	0
6	5	1	0	3	4	2

7.64: 19,10,0

0	1	2	3	4	5	6
1	0	3	2	5	6	4
2	3	0	4	6	1	5
3	5	4	6	1	0	2
4	6	1	5	0	2	3
5	2	6	1	3	4	0
6	4	5	0	2	3	1

7.65: 17,14,0

0	1	2	3	4	5	6
1	0	3	2	5	6	4
2	3	0	4	6	1	5
3	5	4	6	2	0	1
4	6	1	5	0	3	2
5	4	6	0	1	2	3
6	2	5	1	3	4	0

7.66: 25,18,0

0	1	2	3	4	5	6
1	0	3	2	5	6	4
2	3	0	4	6	1	5
3	5	4	6	2	0	1
4	6	1	5	0	3	2
5	4	6	1	3	2	0
6	2	5	0	1	4	3

7.67: 16,12,0

0	1	2	3	4	5	6
1	0	3	2	5	6	4
2	3	0	4	6	1	5
3	5	4	6	2	0	1
4	6	1	5	3	2	0
5	2	6	0	1	4	3
6	4	5	1	0	3	2

7.68: 41,14,0

0	1	2	3	4	5	6
1	0	3	2	5	6	4
2	3	0	4	6	1	5
3	5	4	6	2	0	1
4	6	5	0	1	3	2
5	2	6	1	0	4	3
6	4	1	5	3	2	0

7.69: 27,14,0

0	1	2	3	4	5	6
1	0	3	2	5	6	4
2	3	0	4	6	1	5
3	5	4	6	2	0	1
4	6	5	1	0	3	2
5	2	6	0	1	4	3
6	4	1	5	3	2	0

7.70: 24,14,0

0	1	2	3	4	5	6
1	0	3	2	5	6	4
2	3	0	4	6	1	5
3	6	4	5	2	0	1
4	5	1	6	3	2	0
5	2	6	1	0	4	3
6	4	5	0	1	3	2

7.71: 33,18,0

0	1	2	3	4	5	6
1	0	3	2	5	6	4
2	3	1	0	6	4	5
3	4	5	6	0	1	2
4	2	6	5	3	0	1
5	6	4	1	2	3	0
6	5	0	4	1	2	3

7.72: 47,10,1

0	1	2	3	4	5	6
1	0	3	2	5	6	4
2	3	1	0	6	4	5
3	4	5	6	0	1	2
4	5	6	1	2	3	0
5	6	0	4	1	2	3
6	2	4	5	3	0	1

7.73: 24,9,1

0	1	2	3	4	5	6
1	0	3	2	5	6	4
2	3	1	0	6	4	5
3	4	5	6	0	1	2
4	6	0	5	3	2	1
5	2	6	4	1	3	0
6	5	4	1	2	0	3

7.74: 37,18,1

0	1	2	3	4	5	6
1	0	3	2	5	6	4
2	3	1	0	6	4	5
3	4	5	6	1	0	2
4	6	0	5	2	3	1
5	2	6	4	3	1	0
6	5	4	1	0	2	3

7.75: 24,15,1

0	1	2	3	4	5	6
1	0	3	2	5	6	4
2	3	1	0	6	4	5
3	4	6	5	0	2	1
4	2	5	6	1	3	0
5	6	0	4	2	1	3
6	5	4	1	3	0	2

7.76: 31,6,3

0	1	2	3	4	5	6
1	0	3	2	5	6	4
2	3	1	4	6	0	5
3	2	5	6	0	4	1
4	5	6	0	1	2	3
5	6	4	1	2	3	0
6	4	0	5	3	1	2

7.77: 21,10,0

0	1	2	3	4	5	6
1	0	3	2	5	6	4
2	3	1	4	6	0	5
3	2	5	6	0	4	1
4	5	6	0	1	3	2
5	6	4	1	3	2	0
6	4	0	5	2	1	3

7.78: 23,14,0

0	1	2	3	4	5	6
1	0	3	2	5	6	4
2	3	1	4	6	0	5
3	2	5	6	0	4	1
4	5	6	0	2	1	3
5	6	4	1	3	2	0
6	4	0	5	1	3	2

7.79: 23,10,0

0	1	2	3	4	5	6
1	0	3	2	5	6	4
2	3	1	4	6	0	5
3	2	5	6	0	4	1
4	5	6	1	2	3	0
5	6	4	0	3	1	2
6	4	0	5	1	2	3

7.80: 20,10,0

0	1	2	3	4	5	6
1	0	3	2	5	6	4
2	3	1	4	6	0	5
3	2	5	6	0	4	1
4	5	6	1	3	2	0
5	6	4	0	1	3	2
6	4	0	5	2	1	3

7.81: 25,10,0

0	1	2	3	4	5	6
1	0	3	2	5	6	4
2	3	1	4	6	0	5
3	2	5	6	0	4	1
4	5	6	1	3	2	0
5	6	4	0	2	1	3
6	4	0	5	1	3	2

7.82: 19,8,0

0	1	2	3	4	5	6
1	0	3	2	5	6	4
2	3	1	4	6	0	5
3	2	5	6	0	4	1
4	6	0	5	1	3	2
5	4	6	0	2	1	3
6	5	4	1	3	2	0

7.83: 19,8,0

0	1	2	3	4	5	6
1	0	3	2	5	6	4
2	3	1	4	6	0	5
3	2	5	6	0	4	1
4	6	0	5	1	3	2
5	4	6	1	3	2	0
6	5	4	0	2	1	3

7.84: 22,14,0

0	1	2	3	4	5	6
1	0	3	2	5	6	4
2	3	1	4	6	0	5
3	2	5	6	1	4	0
4	5	6	1	0	2	3
5	6	4	0	3	1	2
6	4	0	5	2	3	1

7.85: 12,11,0

0	1	2	3	4	5	6
1	0	3	2	5	6	4
2	3	1	4	6	0	5
3	4	5	6	0	1	2
4	5	6	0	1	2	3
5	6	4	1	2	3	0
6	2	0	5	3	4	1

7.86: 19,8,0

0	1	2	3	4	5	6
1	0	3	2	5	6	4
2	3	1	4	6	0	5
3	4	5	6	0	1	2
4	5	6	0	2	3	1
5	6	4	1	3	2	0
6	2	0	5	1	4	3

7.87: 19,8,0

0	1	2	3	4	5	6
1	0	3	2	5	6	4
2	3	1	4	6	0	5
3	4	5	6	0	1	2
4	5	6	0	3	2	1
5	6	0	1	2	4	3
6	2	4	5	1	3	0

7.88: 18,7,0

0	1	2	3	4	5	6
1	0	3	2	5	6	4
2	3	1	4	6	0	5
3	4	5	6	0	1	2
4	5	6	1	2	3	0
5	6	4	0	1	2	3
6	2	0	5	3	4	1

7.89: 17,7,0

0	1	2	3	4	5	6
1	0	3	2	5	6	4
2	3	1	4	6	0	5
3	4	5	6	0	1	2
4	5	6	1	3	2	0
5	6	4	0	2	3	1
6	2	0	5	1	4	3

7.90: 23,6,0

0	1	2	3	4	5	6
1	0	3	2	5	6	4
2	3	1	4	6	0	5
3	4	5	6	0	1	2
4	6	0	5	1	2	3
5	2	6	0	3	4	1
6	5	4	1	2	3	0

7.91: 22,12,0

0	1	2	3	4	5	6
1	0	3	2	5	6	4
2	3	1	4	6	0	5
3	4	5	6	0	1	2
4	6	0	5	1	2	3
5	2	6	1	3	4	0
6	5	4	0	2	3	1

7.92: 22,6,0

0	1	2	3	4	5	6
1	0	3	2	5	6	4
2	3	1	4	6	0	5
3	4	5	6	0	1	2
4	6	0	5	2	3	1
5	2	6	1	3	4	0
6	5	4	0	1	2	3

7.93: 14,9,0

0	1	2	3	4	5	6
1	0	3	2	5	6	4
2	3	1	4	6	0	5
3	4	5	6	0	2	1
4	6	0	5	2	1	3
5	2	6	1	3	4	0
6	5	4	0	1	3	2

7.94: 22,9,0

0	1	2	3	4	5	6
1	0	3	2	5	6	4
2	3	1	4	6	0	5
3	4	5	6	2	1	0
4	5	6	0	1	2	3
5	6	0	1	3	4	2
6	2	4	5	0	3	1

7.95: 12,10,0

0	1	2	3	4	5	6
1	0	3	2	5	6	4
2	3	1	4	6	0	5
3	4	5	6	2	1	0
4	6	0	5	1	3	2
5	2	6	1	0	4	3
6	5	4	0	3	2	1

7.96: 13,5,0

0	1	2	3	4	5	6
1	0	3	2	5	6	4
2	3	1	4	6	0	5
3	4	6	5	0	1	2
4	2	5	6	1	3	0
5	6	0	1	2	4	3
6	5	4	0	3	2	1

7.97: 7,6,0

0	1	2	3	4	5	6
1	0	3	2	5	6	4
2	3	1	4	6	0	5
3	4	6	5	0	1	2
4	5	0	6	2	3	1
5	6	4	0	1	2	3
6	2	5	1	3	4	0

7.98: 21,9,0

0	1	2	3	4	5	6
1	0	3	2	5	6	4
2	3	1	4	6	0	5
3	4	6	5	0	1	2
4	5	0	6	2	3	1
5	6	4	1	3	2	0
6	2	5	0	1	4	3

7.99: 25,10,0

0	1	2	3	4	5	6
1	0	3	2	5	6	4
2	3	1	4	6	0	5
3	4	6	5	0	2	1
4	5	0	6	1	3	2
5	6	4	0	2	1	3
6	2	5	1	3	4	0

7.100: 16,8,0

0	1	2	3	4	5	6
1	0	3	2	5	6	4
2	3	1	4	6	0	5
3	4	6	5	0	2	1
4	5	0	6	2	1	3
5	6	4	0	1	3	2
6	2	5	1	3	4	0

7.101: 25,12,0

0	1	2	3	4	5	6
1	0	3	2	5	6	4
2	3	1	4	6	0	5
3	4	6	5	0	2	1
4	5	0	6	3	1	2
5	6	4	1	2	3	0
6	2	5	0	1	4	3

7.102: 26,11,0

0	1	2	3	4	5	6
1	0	3	2	5	6	4
2	3	1	4	6	0	5
3	4	6	5	2	1	0
4	5	0	6	1	3	2
5	6	4	0	3	2	1
6	2	5	1	0	4	3

7.103: 22,7,0

0	1	2	3	4	5	6
1	0	3	2	5	6	4
2	3	1	4	6	0	5
3	4	6	5	2	1	0
4	5	0	6	3	2	1
5	6	4	0	1	3	2
6	2	5	1	0	4	3

7.104: 21,10,0

0	1	2	3	4	5	6
1	0	3	2	5	6	4
2	3	1	4	6	0	5
3	4	6	5	2	1	0
4	6	5	0	1	3	2
5	2	0	6	3	4	1
6	5	4	1	0	2	3

7.105: 30,10,0

0	1	2	3	4	5	6
1	0	3	2	5	6	4
2	3	1	4	6	0	5
3	4	6	5	2	1	0
4	6	5	1	0	3	2
5	2	0	6	1	4	3
6	5	4	0	3	2	1

7.106: 26,9,0

0	1	2	3	4	5	6
1	0	3	2	5	6	4
2	3	1	4	6	0	5
3	5	0	6	1	4	2
4	6	5	1	0	2	3
5	4	6	0	2	3	1
6	2	4	5	3	1	0

7.107: 16,10,0

0	1	2	3	4	5	6
1	0	3	2	5	6	4
2	3	1	4	6	0	5
3	5	0	6	1	4	2
4	6	5	1	2	3	0
5	4	6	0	3	2	1
6	2	4	5	0	1	3

7.108: 14,8,0

0	1	2	3	4	5	6
1	0	3	2	5	6	4
2	3	1	4	6	0	5
3	5	0	6	2	4	1
4	6	5	1	3	2	0
5	4	6	0	1	3	2
6	2	4	5	0	1	3

7.109: 25,16,0

0	1	2	3	4	5	6
1	0	3	2	5	6	4
2	3	1	4	6	0	5
3	5	4	6	0	1	2
4	6	0	5	1	2	3
5	4	6	0	2	3	1
6	2	5	1	3	4	0

7.110: 25,10,0

0	1	2	3	4	5	6
1	0	3	2	5	6	4
2	3	4	5	6	0	1
3	2	5	6	1	4	0
4	6	0	1	2	3	5
5	4	6	0	3	1	2
6	5	1	4	0	2	3

7.111: 14,8,0

0	1	2	3	4	5	6
1	0	3	2	5	6	4
2	3	4	5	6	0	1
3	2	5	6	1	4	0
4	6	1	0	2	3	5
5	4	6	1	0	2	3
6	5	0	4	3	1	2

7.112: 18,9,0

0	1	2	3	4	5	6
1	0	3	2	5	6	4
2	3	4	5	6	0	1
3	4	0	6	1	2	5
4	5	6	0	3	1	2
5	6	1	4	2	3	0
6	2	5	1	0	4	3

7.113: 28,9,0

0	1	2	3	4	5	6
1	0	3	2	5	6	4
2	3	4	5	6	0	1
3	4	0	6	1	2	5
4	5	6	1	2	3	0
5	6	1	0	3	4	2
6	2	5	4	0	1	3

7.114: 19,8,0

0	1	2	3	4	5	6
1	0	3	2	5	6	4
2	3	4	5	6	0	1
3	4	0	6	1	2	5
4	6	5	0	3	1	2
5	2	6	1	0	4	3
6	5	1	4	2	3	0

7.115: 19,10,0

0	1	2	3	4	5	6
1	0	3	2	5	6	4
2	3	4	5	6	0	1
3	4	0	6	1	2	5
4	6	5	1	2	3	0
5	2	6	4	0	1	3
6	5	1	0	3	4	2

7.116: 23,6,0

0	1	2	3	4	5	6
1	0	3	2	5	6	4
2	3	4	5	6	0	1
3	4	0	6	2	1	5
4	5	6	0	1	3	2
5	6	1	4	3	2	0
6	2	5	1	0	4	3

7.117: 21,10,0

0	1	2	3	4	5	6
1	0	3	2	5	6	4
2	3	4	5	6	0	1
3	4	0	6	2	1	5
4	5	6	1	3	2	0
5	6	1	4	0	3	2
6	2	5	0	1	4	3

7.118: 17,5,0

0	1	2	3	4	5	6
1	0	3	2	5	6	4
2	3	4	5	6	0	1
3	4	0	6	2	1	5
4	6	5	0	1	3	2
5	2	6	1	0	4	3
6	5	1	4	3	2	0

7.119: 17,7,0

0	1	2	3	4	5	6
1	0	3	2	5	6	4
2	3	4	5	6	0	1
3	4	0	6	2	1	5
4	6	5	0	1	3	2
5	2	6	1	3	4	0
6	5	1	4	0	2	3

7.120: 21,5,0

0	1	2	3	4	5	6
1	0	3	2	5	6	4
2	3	4	5	6	0	1
3	4	1	6	0	2	5
4	6	5	0	1	3	2
5	2	6	1	3	4	0
6	5	0	4	2	1	3

7.121: 18,10,0

0	1	2	3	4	5	6
1	0	3	2	5	6	4
2	3	4	5	6	0	1
3	4	1	6	0	2	5
4	6	5	0	2	1	3
5	2	6	1	3	4	0
6	5	0	4	1	3	2

7.122: 18,8,0

0	1	2	3	4	5	6
1	0	3	2	5	6	4
2	3	4	5	6	0	1
3	4	1	6	0	2	5
4	6	5	0	2	1	3
5	2	6	4	1	3	0
6	5	0	1	3	4	2

7.123: 17,8,0

0	1	2	3	4	5	6
1	0	3	2	5	6	4
2	3	4	5	6	0	1
3	4	1	6	0	2	5
4	6	5	1	2	3	0
5	2	6	0	1	4	3
6	5	0	4	3	1	2

7.124: 23,12,0

0	1	2	3	4	5	6
1	0	3	2	5	6	4
2	3	4	5	6	0	1
3	4	5	6	0	1	2
4	5	6	1	3	2	0
5	6	1	0	2	4	3
6	2	0	4	1	3	5

7.125: 13,2,0

0	1	2	3	4	5	6
1	0	3	2	5	6	4
2	3	4	5	6	0	1
3	4	5	6	1	2	0
4	6	1	0	2	3	5
5	2	6	1	0	4	3
6	5	0	4	3	1	2

7.126: 25,12,0

0	1	2	3	4	5	6
1	0	3	2	5	6	4
2	3	4	5	6	0	1
3	4	5	6	2	1	0
4	2	6	0	1	3	5
5	6	1	4	0	2	3
6	5	0	1	3	4	2

7.127: 21,10,0

0	1	2	3	4	5	6
1	0	3	2	5	6	4
2	3	4	5	6	0	1
3	4	5	6	2	1	0
4	6	1	0	3	2	5
5	2	6	1	0	4	3
6	5	0	4	1	3	2

7.128: 11,6,0

0	1	2	3	4	5	6
1	0	3	2	5	6	4
2	3	4	5	6	0	1
3	4	6	0	2	1	5
4	5	0	6	1	3	2
5	6	1	4	3	2	0
6	2	5	1	0	4	3

7.129: 21,7,0

0	1	2	3	4	5	6
1	0	3	2	5	6	4
2	3	4	5	6	0	1
3	6	0	4	2	1	5
4	2	5	6	1	3	0
5	4	6	1	0	2	3
6	5	1	0	3	4	2

7.130: 23,4,0

0 1 2 3 4 5 6	0 1 2 3 4 5 6	0 1 2 3 4 5 6	0 1 2 3 4 5 6	0 1 2 3 4 5 6
1 0 3 2 5 6 4	1 0 3 2 5 6 4	1 0 3 2 5 6 4	1 0 3 2 5 6 4	1 0 3 2 5 6 4
2 3 4 5 6 0 1	2 3 4 5 6 1 0	2 3 4 5 6 1 0	2 3 4 5 6 1 0	2 4 0 5 6 1 3
3 6 0 4 2 1 5	3 4 1 6 2 0 5	3 4 6 0 1 2 5	3 6 0 4 1 2 5	3 5 4 6 1 0 2
4 5 1 6 3 2 0	4 5 6 0 3 2 1	4 5 1 6 3 0 2	4 5 1 6 2 0 3	4 6 5 0 2 3 1
5 2 6 0 1 4 3	5 6 0 4 1 3 2	5 6 0 4 2 3 1	5 2 6 0 3 4 1	5 2 6 1 3 4 0
6 4 5 1 0 3 2	6 2 5 1 0 4 3	6 2 5 1 0 4 3	6 4 5 1 0 3 2	6 3 1 4 0 2 5
7.131: 32,8,0	7.132: 28,7,0	7.133: 26,8,0	7.134: 16,5,0	7.135: 31,11,0

0 1 2 3 4 5 6	0 1 2 3 4 5 6	0 1 2 3 4 5 6	0 1 2 3 4 5 6	0 1 2 3 4 5 6
1 0 3 2 5 6 4	1 0 3 2 5 6 4	1 0 3 2 5 6 4	1 0 3 2 5 6 4	1 0 3 2 5 6 4
2 4 0 5 6 1 3	2 4 0 5 6 1 3	2 4 0 5 6 1 3	2 4 0 5 6 1 3	2 4 0 5 6 3 1
3 5 4 6 1 0 2	3 5 4 6 1 2 0	3 5 6 4 2 0 1	3 6 1 4 0 2 5	3 5 4 6 1 0 2
4 6 5 1 2 3 0	4 3 6 1 2 0 5	4 3 1 6 0 2 5	4 5 6 0 1 3 2	4 3 6 0 2 1 5
5 2 6 0 3 4 1	5 6 1 4 0 3 2	5 6 4 0 1 3 2	5 3 4 6 2 0 1	5 6 1 4 0 2 3
6 3 1 4 0 2 5	6 2 5 0 3 4 1	6 2 5 1 3 4 0	6 2 5 1 3 4 0	6 2 5 1 3 4 0
7.136: 22,6,0	7.137: 23,10,0	7.138: 34,11,0	7.139: 36,13,0	7.140: 31,8,0

0 1 2 3 4 5 6	0 1 2 3 4 5 6	0 1 2 3 4 5 6	0 1 2 3 4 5 6	0 1 2 3 4 5 6
1 0 3 2 5 6 4	1 0 3 2 5 6 4	1 0 3 2 5 6 4	1 0 3 4 5 6 2	1 0 3 4 5 6 2
2 4 0 5 6 3 1	2 4 0 5 6 3 1	2 4 1 5 6 0 3	2 3 1 0 6 4 5	2 3 1 0 6 4 5
3 5 4 6 2 1 0	3 5 6 0 1 4 2	3 5 0 6 1 4 2	3 5 0 6 1 2 4	3 5 0 6 1 2 4
4 6 5 1 0 2 3	4 3 1 6 2 0 5	4 3 6 0 2 1 5	4 2 6 5 3 0 1	4 6 5 1 2 3 0
5 3 6 0 1 4 2	5 6 4 1 3 2 0	5 6 4 1 3 2 0	5 6 4 1 2 3 0	5 4 6 2 0 1 3
6 2 1 4 3 0 5	6 2 5 4 0 1 3	6 2 5 4 0 3 1	6 4 5 2 0 1 3	6 2 4 5 3 0 1
7.141: 45,16,0	7.142: 31,10,0	7.143: 20,4,0	7.144: 15,4,0	7.145: 11,2,0

0 1 2 3 4 5 6	0 1 2 3 4 5 6
1 0 3 4 5 6 2	1 0 3 4 5 6 2
2 3 4 0 6 1 5	2 3 4 6 0 1 5
3 5 6 2 1 4 0	3 4 6 5 2 0 1
4 2 5 6 3 0 1	4 5 1 2 6 3 0
5 6 0 1 2 3 4	5 6 0 1 3 2 4
6 4 1 5 0 2 3	6 2 5 0 1 4 3
7.146: 7,3,0	7.147: 15,1,0

1.8 **Theorem** ([22] p. 162) Let $L(n)$ denote the number of distinct latin squares of side n. Then $L(n) \sim (e^{-2}n)^{n^2}$ for $n \to \infty$.

> A latin rectangle is *normalized* (or *reduced*) if the first row and first column read $[1, 2, \ldots n]$ and $[1, 2, \ldots, k]$, respectively. $L(k, n)$ denotes the number of normalized $k \times n$ latin rectangles.

1.9 **Remark** It is easy to see that the total number of $k \times n$ latin rectangles is $n!\,(n-1)!\,L(k,n)/(n-k)!$.

1.10 **Theorem** ([11], p. 507) The number of $3 \times n$ latin rectangles is $(n!)^2$ times the coefficient of x^n in

$$e^{2x} \sum_{i=0}^{\infty} i! \frac{x^i}{(1+x)^{3i+3}}.$$

1.11 **Theorem** [10] The number of $k \times n$ latin rectangles is asymptotically

$$(n!)^k \left(\frac{n(n-1)\cdots(n-k+1)}{n^k} \right)^n \left(1 - \frac{k}{n}\right)^{-n/2} e^{-k/2}$$

as $n \to \infty$ with $k = o(n^{6/7})$.

1.12 Table (See [18]) $L(k,n)$ for $1 \le n \le 10$.

n	k	$L(k,n)$	n	k	$L(k,n)$	n	k	$L(k,n)$
1	1	1	7	1	1	9	1	1
2	1	1		2	309		2	16687
	2	1		3	35792		3	1034 43808
3	1	1		4	1293216		4	20 76245 60256
	2	1		5	11270400		5	11268 16430 83776
	3	1		6	16942080		6	12 95260 54043 81184
4	1	1		7	16942080		7	224 38296 79166 91456
	2	3	8	1	1		8	377 59757 09642 58816
	3	4		2	2119		9	377 59757 09642 58816
	4	4		3	1673792	10	1	1
5	1	1		4	4209 09504		2	1 48329
	2	11		5	27206 658048		3	81549 99232
	3	46		6	33 53901 89568		4	14717 45210 59584
	4	56		7	53 52814 01856		5	746 98838 30762 86464
	5	56		8	53 52814 01856		6	8 70735 40559 10037 09440
6	1	1					7	1771 44296 98305 41859 22560
	2	53					8	42920 39421 59185 42730 03520
	3	1064					9	75807 21483 16013 28114 89280
	4	6552					10	75807 21483 16013 28114 89280
	5	9408						
	6	9408						

1.13 Table [18] Estimates of $L(n,n)$ for $11 \le n \le 15$.

n	11	12	13	14	15
$L(n,n)$	5.36×10^{33}	1.62×10^{44}	2.51×10^{56}	2.33×10^{70}	1.5×10^{86}

1.4 Transversals

A *transversal* in a latin square of side n is a set of n cells, one from each row and column containing each of the n symbols exactly once.

A *partial transversal of length* k in a latin square of side n is a set of k cells, each from a different row and each from a different column and such that no two contain the same symbol.

1.14 Remark Some latin squares have no transversals at all. One such class of latin squares is the addition tables of $\mathbb{Z}_{2n}$ for every $n \geq 1$.

1.15 Remark See §IV.5.1 for some connections between complete mappings of groups (and quasigroups) and transversals in latin squares.

1.16 Proposition If a latin square L has a transversal, then any latin square isotopic to L also has a transversal.

1.17 Theorem [3] Every latin square of even order has an even number of transversals.

1.18 Conjecture (Ryser) Every latin square of odd order has a transversal.

1.19 Theorem [20] Every latin square of side n has a partial transversal of length k where $k \geq max\{n - \sqrt{n}, n - 5.53(\log n)^2\}$. See [6], pages 9-10.

1.20 Remark In order for a latin square of side n to have an orthogonal mate it must possess a set of n disjoint transversals.

5	1	7	3	4	0	6	2	8	9
1	2	3	4	5	6	7	8	9	0
7	3	4	5	6	2	8	9	0	1
3	4	5	6	7	8	9	0	1	2
4	5	6	7	8	9	0	1	2	3
0	6	2	8	9	5	1	7	3	4
6	7	8	9	0	1	2	3	4	5
2	8	9	0	1	7	3	4	5	6
8	9	0	1	2	3	4	5	6	7
9	0	1	2	3	4	5	6	7	8

The square on the left is a latin square of side 10 with 5504 different transversals [19]. This square has an estimated one million distinct orthogonal mates, but it has been proven not to be one a set of three mutually orthogonal latin squares of side 10.

A k-transversal in a latin square of side n is a set of nk cells, k from each row, k from each column in which every symbol occurs exactly k times.

1.21 Remark Little is known about k-transversals in latin squares. If a latin square possesses k disjoint transversals, then it possesses a k-transversal. The converse is false as it is easy to see that the addition table of $\mathbb{Z}_{2n}$ always has a 2-transversal, yet has no transversal.

1.22 Conjecture (Rodney) Every latin square has a 2-transversal.

1.5 Subsquares and Holes

If in a latin square L of side n the k^2 cells defined by k rows and k columns form a latin square of side k it is a latin *subsquare of L.*

1.23 Theorem A latin square of side n with a proper subsquare of side k exists if and only if $k \leq \left[\frac{n}{2}\right]$.

1.24 Theorem There exists a latin square of side n which contains latin subsquares of every side k, $k \leq \left[\frac{n}{2}\right]$, if and only if $1 \leq n \leq 7$ or $n = 9$ or 13.

An *incomplete latin square* ILS$(n; b_1, b_2, \ldots, b_k)$ is an $n \times n$ array A with entries from an $n-$set B, where $B_i \subseteq B$ for $1 \leq i \leq k$ with $|B_i| = b_i$, and $B_i \cap B_j = \emptyset$ for $1 \leq i, j \leq k$. Moreover,

1. each cell of A is empty or contains an element of B;
2. the subarrays indexed by $B_i \times B_i$ are empty (these subarrays are *holes*); and
3. the elements in row or column b are exactly those of $B \setminus B_i$ if $b \in B_i$, and of B otherwise.

A *partitioned incomplete latin square* PILS$(n; b_1, b_2, \ldots, b_k)$ is an incomplete latin square with $b_1 + b_2 + \cdots + b_k = n$.

1.25 Example A PILS$(7; 1, 1, 1, 2, 2)$.

	4	5	6	7	2	3
3		6	7	1	5	4
2	7		1	6	4	5
7	6	2			3	1
6	3	7			1	2
4	5	1	2	3		
5	1	4	3	2		

1.26 Remark Given an ILS$(n; b_1, b_2, \ldots, b_k)$ it is always possible to "fill in" the hole of size b_i, $1 \le i \le k$, with a latin square of side b_i (on the symbols B_i) to obtain a latin square of side n. The existence of a latin square of side n containing a latin subsquare of side k is equivalent to an ILS$(n; k)$.

1.27 Theorem A PILS$(n; b_1, b_2, b_3)$ exists if and only if $b_1 = b_2 = b_3$. Moreover, a PILS$(n; b_1, b_2, b_3, b_4)$ exists if and only if $b_1 = b_2 = b_3$ and $1 \le b_4 \le 2b_1$.

A subsquare of side 2 of a latin square is an *intercalate.* A square with no intercalates is an N_2 *square.*

1.28 Theorem There exists a latin square of side n with no intercalates if and only if $n \neq 2$ or 4.

1.29 Remark If n is odd, then the Cayley table of the cyclic group of order n (the square L defined by $L(i,j) = i + j \pmod{n}$) has no intercalates. Also, if A and B are N_2 squares, then so is their product.

1.30 Remark A discussion of a computer algorithm for finding N_2 latin squares can be found in Example VI.9.34.

1.31 Example A latin square of order 8 with no subsquares of side 2 and a latin square of order 12 with no proper subsquares.

1	2	3	4	5	6	7	8
2	3	1	5	6	7	8	4
3	1	4	6	7	8	2	5
4	6	8	2	1	3	5	7
5	8	2	7	3	4	6	1
6	5	7	1	8	2	4	3
7	4	5	8	2	1	3	6
8	7	6	3	4	5	1	2

1	2	3	4	5	6	7	8	9	*a*	*b*	*c*
2	3	4	5	6	1	8	9	*a*	*b*	*c*	7
3	1	5	2	7	8	4	*a*	6	*c*	9	*b*
4	5	6	7	1	9	*b*	*c*	8	3	2	*a*
5	6	2	8	*a*	7	9	*b*	*c*	4	1	3
6	*c*	8	1	3	*a*	2	7	*b*	9	4	5
7	8	1	*a*	*c*	*b*	5	4	2	6	3	9
8	9	*b*	3	4	*c*	*a*	6	5	1	7	2
9	*b*	7	*c*	2	5	1	3	4	8	*a*	6
a	7	*c*	*b*	9	4	6	1	3	2	5	8
b	4	*a*	9	8	3	*c*	2	7	5	6	1
c	*a*	9	6	*b*	2	3	5	1	7	8	4

1.32 Theorem There exists a latin square of side n with no proper subsquares if $n \neq 2^a 3^b$ or if $n = 3, 9, 12, 16, 18, 27, 81$ or 243. (See [13].)

1.33 Remarks If p is a prime, then the Cayley table of the cyclic group of order p has no proper subsquares. Also, if there exists a perfect 1-factorization of K_{n+1}, then there exists a latin square of side n which contains no subsquares. (See §VI.4.6.) The smallest n for which it is not known if there exists a subsquare-free latin square of side n is $n = 24$.

1.6 Completion and Embedding

An n by n array L with cells that are either empty or contain exactly one symbol is a *partial* latin square if no symbol occurs more than once in any row or column. A partial latin square is *symmetric* (or *commutative*) if cell (i,j) is occupied by x, then so is cell (j,i), for every $1 \le i, j \le n$. A partial latin square is *idempotent* if all of the cells on the main diagonal are occupied and furthermore cell (i,i) is occupied by i.

1.34 **Theorem** [21] A partial latin square of side n with at most $n-1$ filled cells can always be completed to a latin square of side n.

1.35 **Remark** Theorem 1.34 was conjectured by Evans in 1960 and is still referred to as the *Evans Conjecture.* A symmetric analogue is given in Theorem 1.36.

1.36 **Theorem** [2] A partial symmetric latin square of side n with admissible diagonal and with at most $n-1$ filled cells can always be completed to a symmetric latin square of side n.

1.37 **Remark** In Theorem 1.36, the diagonal is *admissible* if and only if the number of symbols occurring on the diagonal a number of times that is different from n modulo 2 is at most the number of empty diagonal cells. This is necessary, since the parity must be changed for every such symbol. Theorem 1.36 addresses the sufficiency.

If the first k rows of a partial latin square ($k \leq n$) are all filled and the remaining cells are all empty, then L is a k by n *latin rectangle.*

1.38 **Theorem** A k by n latin rectangle, $k < n$, can always be completed to a latin square of side n.

1.39 **Theorem** Let L be a partial latin square of order n in which cell (i,j) is filled if and only if $i \leq r$ and $j \leq s$. Then L can be completed to a latin square of order n if and only if $N(i) \geq r+s-n$ for $i = 1, 2, \ldots, n$, where $N(i)$ denotes the number of elements in L that are equal to i.

An $n \times n$ partial latin square P is *embedded* in a latin square L if the upper $n \times n$ left corner of L agrees with P.

1.40 **Example** The 4×4 partial latin square P is embedded in the 5×5 latin square L.

$$P = \begin{array}{|cccc|} \hline 1 & 4 & \cdot & \cdot \\ \cdot & \cdot & 3 & 4 \\ \cdot & 1 & 2 & \cdot \\ \cdot & \cdot & \cdot & 1 \\ \hline \end{array} \qquad L = \begin{array}{|ccccc|} \hline 1 & 4 & 5 & 2 & 3 \\ 2 & 5 & 3 & 4 & 1 \\ 3 & 1 & 2 & 5 & 4 \\ 5 & 3 & 4 & 1 & 2 \\ 4 & 2 & 1 & 3 & 5 \\ \hline \end{array}$$

1.41 **Theorem** (Evans) A partial $n \times n$ latin square can be embedded in a $t \times t$ latin square for every $t \geq 2n$.

1.42 **Remark** Theorem 1.41 is the best possible general embedding result, as it is possible to construct for every $n \geq 4$ a partial $n \times n$ latin square which can not be embedded in a latin square less than twice its size. See §IV.17 for related results on embedding partial latin squares.

1.7 Diagonal Latin Squares

Let L be a latin square of side n. Then the *jth right diagonal* of L is the set of n cells of L, $\{(i, j+i) : i = 0, 1, \ldots, n-1 \pmod{n}\}$. The *$j$th left diagonal* of L is the set of n cells of L, $\{(i, j-i) : i = 0, 1, \ldots, n-1 \pmod{n}\}$.

A latin square L of side n is a *diagonal latin square* if both its 0th right and $(n-1)$st left diagonals are transversals.

1.43 **Theorem** [9] A diagonal latin square of side n exists if and only if $n > 3$.

1.44 Example Diagonal latin squares of sides 4, 5, 6, and 8. A diagonal latin square of side 7 can be found in Example IV.5.45.

0	2	3	1
1	3	2	0
2	0	1	3
3	1	0	2

1	3	4	0	2
3	2	0	1	4
0	4	3	2	1
2	0	1	4	3
4	1	2	3	0

1	2	3	4	5	0
0	5	4	3	2	1
2	4	0	1	3	5
3	1	5	2	0	4
5	3	1	0	4	2
4	0	2	5	1	3

0	2	4	6	7	1	3	5
1	7	5	3	2	0	6	4
2	4	6	0	1	3	5	7
3	1	7	5	4	2	0	6
4	6	0	2	3	5	7	1
5	3	1	7	6	4	2	0
6	0	2	4	5	7	1	3
7	5	3	1	0	6	4	2

1.45 Remark A discussion of orthogonal diagonal latin squares can be found in §II.2.7.

A latin square L of side n is a *Knut Vik design* if every right and left diagonal of L is a transversal.

1.46 Theorem [12] A Knut Vik design of side n exists if and only if n is not divisible by either 2 or 3.

1.47 Remark The connection between Knut Vik designs and complete mappings is given in Theorem IV.5.44 and a Knut Vik design of side 7 is given in Example IV.5.45.

A latin square L of side n is a *crisscross latin square* if all the jth right diagonals for even j and all the left j diagonals for odd j are transversals.

1.48 Theorem [15] Crisscross latin squares of side n exist if and only if $n \equiv 0 \pmod 4$.

1.49 Example The squares of sides 4 and 8 in Example 1.44 are crisscross latin squares.

1.8 Even and Odd Latin Squares

A *row inversion* in a latin square $L = (a_{i,j})$ is a pair of numbers in a row which are out of order, i.e., $a_{i,j} > a_{i,k}$ with $j < k$. A latin square is *row even* (respectively *row odd*) if it contains an even (respectively odd) number of row inversions. The concepts of *column inversion*, *column even*, and *column odd* are defined similarly. A latin square is *even* (respectively *odd*) if the sum of the number of row inversions and the number of column inversions is even (respectively odd).

RELS(n), ROLS(n), CELS(n), COLS(n), ELS(n), and OLS(n) denote the numbers of row even, row odd, column even, column odd, even, and odd latin squares of side n.

1.50 Remark When n is an odd integer, by permuting a pair of rows or a pair of columns, one easily sees that RELS(n) = ROLS(n), CELS(n) = COLS(n), and ELS(n) = OLS(n).

1.51 Theorem [14, 16] Let n be an even integer. Then RELS(n) $\neq$ ROLS(n) if and only if ELS(n) $\neq$ OLS(n).

1.52 Table [16] The number of even and odd reduced latin squares of side $2n$ for $1 \leq n \leq 4$.

$n =$	2	4	6	8
even	1	4	5856	270746124288
odd	0	0	3552	264535277568

1.53 Conjecture (Alon and Tarsi 1986) If n is even, then ELS(n) $\neq$ OLS(n).

1.54 Remark A motivation for the study of even and odd latin squares came from the *Dinitz conjecture*: For each positive integer n and for any collection of n-element sets $S_{i,j}$, an $n \times n$ partial latin square can be found with the (i,j) entry taken from $S_{i,j}$. Alon and Tarsi [1] prove that if $\text{ELS}(n) \neq \text{OLS}(n)$ for all even integers n, then the Dinitz conjecture holds for all even n. In 1995, the Dinitz conjecture was proved by Galvin [8] using different methods. Also, the Alon-Tarsi conjecture has been proven for $n = p + 1$ with p a prime by Drisko.

1.9 Conflict-Free Latin Squares

1.55 Remarks In assigning array elements to memory devices in parallel computation, it is important that array elements that are to be retrieved at the same time be stored in different memory units, permitting parallel access. The attempt to retrieve two array elements from the same device simultaneously is a *memory conflict*. When storing an $n \times n$ array A, one forms an $n \times n$ array S, the *skewing scheme*; the (i,j) entry of S indicates in which memory unit the (i,j) entry of A is to be stored.

A *template* is a set of ordered pairs of the form $(\rho + i, \gamma + j)$, where ρ and γ are formal parameters and i and j are integers. An *instance* of the template is obtained by choosing integer values for ρ and γ.

If T is a template with parameters ρ and γ, and S is an $n \times n$ skewing scheme, then S is *conflict-free for T* if for every instance I of T, and any distinct pairs (a,b) and (c,d) in I that index elements of S, the (c,d) entry of S differs from the (a,b) entry of S.

1.56 Remarks Rows and columns form important classes of templates. Conflict-free for rows is precisely row latinicity, and for columns it is column latinicity. Thus a skewing scheme that is conflict-free for rows and columns is a partial latin square. In the application, the number of memory units is limited, and hence one prefers skewing schemes with a minimum number of symbols. When the number of symbols is n, and the skewing scheme is conflict-free for rows and columns, the scheme is a latin square. Diagonal latin squares arise when one requires diagonals to be conflict-free. The most extensively studied cases are for rectangular templates; see [4, 7].

1.10 See Also

§II.2	Mutually orthogonal latin squares.
§II.3	Gives the connection with incomplete transversal designs.
§IV.20	Frequency squares are squares with the latin property with more than one symbol in each entry.
§IV.36	The multiplication tables of quasigroups are latin squares.
§IV.40	Self-orthogonal latin squares.
§IV.48	Tuscan squares (and their relatives such as row complete latin squares) are squares where each row is a permutation and distance properties are studied.
[5]	An old but comprehensive reference for latin squares.
[6]	A newer reference for latin squares than [5].

References

[1] N. Alon and M. Tarsi, Colorings and orientations of graphs, *Combinatorica* 12 (1992), 125-134.

[2] L.D. Andersen and A.J.W. Hilton, Symmetric Latin square and complete graph analogues of the Evans conjecture, *J. Combin. Des.* 2 (1994), 197-252.

[3] K. Balasubramanian, On transversals in latin squares, *Linear Alg. Applic.* 131 (1990), 125-129.

[4] C.J. Colbourn and K.E. Heinrich, Conflict-free access to parallel memories, *J. Parallel Distributed Computing* 14 (1992) 193-200.

[5] J. Dénes and A.D. Keedwell, *Latin Squares and Their Applications*, English Universities Press, London, 1974.

[6] J. Dénes and A.D. Keedwell, *Latin Squares: New Developments in the Theory and Applications*, North-Holland, Amsterdam, 1991.

[7] D.L. Erickson and C.J. Colbourn, Conflict-free access to rectangular subarrays of constant perimeter, *Interconnection Networks and Mapping and Scheduling Parallel Computations*, ACM/DIMACS, to appear.

[8] F. Galvin, The list chromatic index of a bipartite multigraph, *J. Combin. Theory B* 63 (1995), 153-158.

[9] E. Gergely, A simple method for constructing doubly diagonalized latin squares, *J. Combin. Theory A* 16 (1974), 266-272.

[10] C.D. Godsil and B.D. McKay, Asymptotic enumeration of latin rectangles, *J. Combin. Theory B* 48 (1990), 19-44.

[11] I.P. Goulden and D.M. Jackson, *Combinatorial Enumeration*, John Wiley and Sons, 1983.

[12] A. Hedayat, A complete solution to the existence and nonexistence of Knut Vik designs, *J. Combin. Theory A* 22 (1977), 331-337.

[13] K. Heinrich, Latin squares with and without subsquares of prescribed type, in [6], 101-148.

[14] R. Huang and G-C. Rota, On the relations of various conjectures on Latin squares and straightening coefficients, *Discrete Math.* 128 (1994), 225-236.

[15] F.K. Hwang, Crisscross latin squares, *J. Combin. Theory A* 27 (1979), 371-375.

[16] J.C.M. Janssen, On even and odd latin squares, *J. Combin. Theory A* 69 (1995), 173-181.

[17] G. Kolesova, C.W.H. Lam and L.Thiel, On the number of 8×8 latin squares, *J. Combin. Theory A* 54 (1990), 143-148.

[18] B.D. McKay and E. Rogoyski, Latin squares of order 10, *Electron. J. Comb.* 2 (1995), #N3, (4 pages).

[19] E.T. Parker, Computer investigations of orthogonal latin squares of order 10. *Proc. Sympos. Appl. Math* 15 (1963), 73-81.

[20] P.W. Shor, A lower bound for the length of a partial transversal in a latin square, *J. Combin. Theory A* 33 (1982), 1-8.

[21] B. Smetaniuk, A new construction on Latin squares - I: A proof of the Evans conjecture, *Ars Combin.* 11 (1981), 155-172.

[22] J.H. van Lint and R.M. Wilson, *A Course in Combinatorics*, Cambridge University Press, Cambridge, 1992.

2 Mutually Orthogonal Latin Squares (MOLS)

R. Julian R. Abel
Andries E. Brouwer
Charles J. Colbourn
Jeffrey H. Dinitz

2.1 Definition and Example

Two latin squares L and L' of the same order are *orthogonal* if $L(a,b) = L(c,d)$ and $L'(a,b) = L'(c,d)$, implies $a = c$ and $b = d$.

2.1 **Remark** An equivalent definition for orthogonality: Two latin squares of side n $L = (a_{i,j})$ (on symbol set S) and $L' = (b_{i,j})$ (on symbol set S') are *orthogonal* if every element in $S \times S'$ occurs exactly once among the n^2 pairs $(a_{i,j}, b_{i,j})$, $1 \le i, j \le n$.

A set of latin squares $L_1, \ldots, L_m$ is *mutually orthogonal*, or a set of *MOLS*, if for every $1 \le i < j \le m$, L_i and L_j are orthogonal. These are also referred to as *POLS*, *pairwise orthogonal* latin squares.

2.2 **Example** Three mutually orthogonal latin squares of side 4.

1	2	3	4
4	3	2	1
2	1	4	3
3	4	1	2

1	2	3	4
3	4	1	2
4	3	2	1
2	1	4	3

1	2	3	4
2	1	4	3
3	4	1	2
4	3	2	1

2.2 Equivalent Objects

An *orthogonal array* $\mathrm{OA}(k, s)$ is a $k \times s^2$ array with entries from an s-set S having the property that in any two rows, each (ordered) pair of symbols from S occurs exactly once.

2.3 **Remark** A far more general definition of orthogonal arrays can be found in §II.4.1

2.4 **Remark** Let $\{L_i : 1 \le i \le k\}$ be a set of k MOLS on symbols $\{1, \ldots, n\}$. Form a $(k+2) \times n^2$ array $A = (a_{ij})$ whose columns are $(i, j, L_1(i,j), L_2(i,j), \ldots, L_k(i,j))^T$ for $1 \le i, j \le k$. Then A is an orthogonal array, $\mathrm{OA}(k+2, n)$. This process can be reversed to recover k MOLS of side n from an $\mathrm{OA}(k+2, n)$, by choosing any two rows of the OA to index the rows and columns of the k squares.

2.5 **Example** An OA(5, 4) derived from the 3 MOLS in Example 2.2.

$$\begin{pmatrix} 1111222233334444 \\ 1234123412341234 \\ 1234432121433412 \\ 1234341243212143 \\ 1234214334124321 \end{pmatrix}$$

A *transversal design* of *order* or *groupsize* n, *blocksize* k, and *index* λ, denoted $\mathrm{TD}_\lambda(k,n)$, is a triple $(V,\mathcal{G},\mathcal{B})$, where

1. V is a set of kn elements;
2. $\mathcal{G}$ is a partition of V into k classes (the *groups*), each of size n;
3. $\mathcal{B}$ is a collection of k-subsets of V (the *blocks*);
4. every unordered pair of elements from V is either contained in exactly one group or is contained in exactly λ blocks, but not both.

When $\lambda = 1$, one writes simply $\mathrm{TD}(k,n)$.

2.6 **Remark** Let A be an $\mathrm{OA}(k,n)$ on the n symbols in X. On $V = X \times \{1,\ldots,k\}$ (a set of size kn), form a set $\mathcal{B}$ of k-sets as follows. For $1 \le j \le n^2$, include $\{(a_{i,j},i) : 1 \le i \le k\}$ in $\mathcal{B}$. Then let $\mathcal{G}$ be the partition of V whose classes are $\{X \times \{i\} : 1 \le i \le k\}$. Then $(V,\mathcal{G},\mathcal{B})$ is a $\mathrm{TD}(k,n)$. This process can be reversed to recover an $\mathrm{OA}(k,n)$ from a $\mathrm{TD}(k,n)$.

2.7 **Example** A $\mathrm{TD}(5,4)$ derived from the $\mathrm{OA}(5,4)$ in Example 2.5.

On element set $\{1,2,3,4\} \times \{1,2,3,4,5\}$, the blocks are

{ 11,12,13,14,15 }	{ 11,22,23,24,25 }	{ 11,32,33,34,35 }	{ 11,42,43,44,45 }
{ 21,12,43,34,25 }	{ 21,22,33,44,15 }	{ 21,32,23,14,45 }	{ 21,42,13,24,35 }
{ 31,12,23,44,35 }	{ 31,22,13,34,45 }	{ 31,32,43,24,15 }	{ 31,42,33,14,5 }
{ 41,12,33,24,45 }	{ 41,22,43,14,35 }	{ 41,32,13,44,25 }	{ 41,42,23,34,15 }

A $\mathrm{TD}_\lambda(k,n)$ $(V,\mathcal{G},\mathcal{B})$ is *α-resolvable* if its blocks can be partitioned into sets $\mathcal{B}_1,\ldots,\mathcal{B}_s$, where every element of V occurs exactly α times in each $\mathcal{B}_i$. The classes $\mathcal{B}_1,\ldots,\mathcal{B}_s$ are *α-parallel classes*. When $\alpha = 1$ the design is *(completely) resolvable* and the classes are *parallel classes*.

An $\mathrm{RTD}_\lambda(k,n)$ is a λ-resolvable $\mathrm{TD}_\lambda(k,n)$.

2.8 **Remark** Let $(V,\mathcal{G},\mathcal{B})$ be a $\mathrm{TD}_\lambda(k+1,n)$. Let $G \in \mathcal{G}$ and $G = \{g_1,\ldots,g_n\}$. Form $\mathcal{B}_i = \{B\backslash\{g_i\} : g_i \in B \in \mathcal{B}\}$, and let $\widehat{\mathcal{B}} = \bigcup_{i=1}^n \mathcal{B}_i$. Then $(V \setminus G, \mathcal{G} \setminus \{G\}, \widehat{\mathcal{B}})$ is an $\mathrm{RTD}_\lambda(k,n)$. Its λ-parallel classes are $\mathcal{B}_1,\ldots,\mathcal{B}_n$. This process can be reversed to form a $\mathrm{TD}_\lambda(k+1,n)$ from an $\mathrm{RTD}_\lambda(k,n)$.

2.9 **Example** A resolvable $\mathrm{TD}(4,4)$ derived from the $\mathrm{TD}(5,4)$ in Example 2.7.

On element set $\{1,2,3,4\} \times \{2,3,4,5\}$, the blocks are

{ 12,13,14,15 }	{ 22,23,24,25 }	{ 32,33,34,35 }	{ 42,43,44,45 }
{ 12,43,34,25 }	{ 22,33,44,15 }	{ 32,23,14,45 }	{ 42,13,24,35 }
{ 12,23,44,35 }	{ 22,13,34,45 }	{ 32,43,24,15 }	{ 42,33,14,25 }
{ 12,33,24,45 }	{ 22,43,14,35 }	{ 32,13,44,25 }	{ 42,23,34,15 }

Each row is a parallel class.

A *(k,n)-net* is a pair $(X,\mathcal{C})$ where X is a set of n^2 elements, and $\mathcal{C}$ is a set of kn subsets (*blocks*) of X each of size n with the property that two distinct blocks intersect in at most one element. Moreover, the blocks of $\mathcal{C}$ can be partitioned into k parallel classes each containing n blocks.

2.10 **Remark** Let $(V,\mathcal{G},\mathcal{B})$ be a $\mathrm{TD}(k,n)$. Interchanging the roles of elements and blocks gives a (k,n)-net.

2.11 **Example** The (5,4)-net obtained from the $\mathrm{TD}(5,4)$ in Example 2.7 is an affine plane of order four.

2.12 Theorem The existence of k MOLS of side n is equivalent to the existence of:

1. a transversal design of index one, blocksize $k+2$, and groupsize n, namely a $\mathrm{TD}(k+2,n)$;
2. a resolvable transversal design of index one, blocksize $k+1$ and groupsize n, namely an $\mathrm{RTD}(k+1,n)$;
3. a $(k+2,n)$-net;
4. an orthogonal array of strength two, index one, degree $k+2$ and order n, namely an $\mathrm{OA}(k+2,n)$;
5. an edge-partition of the complete $(k+2)$-partite graph $K_{n,\ldots,n}$ into complete subgraphs of order $k+2$.

A set of $n-1$ MOLS of side n is a *complete set* of MOLS.

2.13 Theorem The existence of a complete set of MOLS of side n is equivalent to the existence of:

1. a projective plane of order n (§VI.7.2), and
2. an affine plane of order n (§VI.7.3).

2.14 Remark It is not known whether there exists a complete set of MOLS of side n for any n that is not a power of a prime. There is no complete set of MOLS of side 6 or 10. The smallest unknown case is $n = 12$.

2.3 Basic Facts about N(n)

$N(n)$ is the maximum number of latin squares in a set of MOLS of side n.

2.15 Remark

Euler's 36 Officers Problem:

A very curious question, which has exercised for some time the ingenuity of many people, has involved me in the following studies, which seem to open a new field of analysis, in particular the study of combinations. The question revolves around arranging 36 officers to be drawn from 6 different ranks and also from 6 different regiments so that they are ranged in a square so that in each line (both horizontal and vertical) there are 6 officers of different ranks and different regiments.

Euler went on to conjecture that such an $n \times n$ array does not exist for $n = 6$, nor does one exist whenever $n \equiv 2 \pmod 4$. This was known as the *Euler conjecture* until its disproof (see Theorem 2.34).

2.16 Remark By convention, $N(0) = N(1) = \infty$.

2.17 Theorem For every $n > 1$, $1 \le N(n) \le n-1$.

2.18 Theorem (MacNeish [19]) $N(n \times m) \ge \min\{N(n), N(m)\}$.

2.19 Theorem If $q = p^e$ is a prime power, then $N(q) = q-1$.

2.20 Construction Construct the complete set of MOLS of order q for Theorem 2.19 as follows: For each $\alpha \in \mathbb{F}_q \setminus \{0\}$, define the latin square $L_\alpha(i,j) = i + \alpha j$, where $i,j \in \mathbb{F}_q$ and the algebra is performed in $\mathbb{F}_q$. The set of latin squares $\{L_\alpha | \alpha \in \mathbb{F}_q \setminus \{0\}\}$ is a set of $q-1$ MOLS of side q.

2.21 Corollary If $n = p_1^{e_1} p_2^{e_2} \ldots p_k^{e_k}$ where each p_i is a prime, then $N(n) \geq min\{p_i^{e_i} - 1 | i = 1, 2, \ldots k\}$.

2.22 Theorem A latin square of side n has an orthogonal mate ($N(n) \geq 2$) if and only if it contains n disjoint transversals (equivalently, the corresponding TD$(3, n)$ is resolvable).

2.23 Theorem Let $s \equiv 1$ or $2 \pmod 4$ and assume that s is not the sum of two squares.

1. Let t be the largest positive integer for which $\frac{1}{2}t^4 - t^3 + t^2 + \frac{1}{2}t - 1 < s$. Then $N(s) \leq s - t - 2$.
2. [20] Let t be the largest positive integer for which $8t^3 + 18t^2 + 8t + 4 - 2\rho(t^2 - t - 1) + \frac{9}{2}\rho(\rho - 1)(t - 1) < 3s$, where $\rho \equiv t + 1 \pmod 3$ and $\rho \in \{0, 1, 2\}$. Then $N(s) \leq s - t - 2$.

2.24 Example $N(n) \leq n - 4$ for $n = 14, 21, 22$;
$N(n) \leq n - 5$ for $n = 30, 33, 38, 42, 46, 54, 57, 62, 66, 69, 70, 77, 78$;
$N(n) \leq n - 6$ for $n = 86, 93, 94$.

2.25 Theorem [12] $\lim_{n\to\infty} N(n) = \infty$. In fact, for n sufficiently large, $N(n) \geq n^{\frac{1}{14.8}}$ [8].

2.4 MOLS of Small Side

2.26 Remark Sets of k orthogonal latin squares of side n for $n \leq 10$ where k is the largest number such that $N(n) \geq k$ are displayed in this section. Also displayed are the constructions which yield the squares of sides $n \leq 56$ that attain the lower bound for $N(n)$ in the cases where n is not a prime power.

2.27 Theorem $N(3) = 2$.

```
1 2 3    1 2 3
2 3 1    3 1 2
3 1 2    2 3 1
```

2.28 Theorem $N(4) = 3$. (The set of 3 MOLS of order 4 is given in Example 2.2).

2.29 Theorem $N(5) = 4$.

```
0 1 2 3 4    0 1 2 3 4    0 1 2 3 4    0 1 2 3 4
1 2 3 4 0    2 3 4 0 1    3 4 0 1 2    4 0 1 2 3
2 3 4 0 1    4 0 1 2 3    1 2 3 4 0    3 4 0 1 2
3 4 0 1 2    1 2 3 4 0    4 0 1 2 3    2 3 4 0 1
4 0 1 2 3    3 4 0 1 2    2 3 4 0 1    1 2 3 4 0
```

2.30 Theorem $N(6) = 1$.

2.31 Theorem $N(7) = 6$.

```
0 1 2 3 4 5 6    0 1 2 3 4 5 6    0 1 2 3 4 5 6
1 2 3 4 5 6 0    2 3 4 5 6 0 1    3 4 5 6 0 1 2
2 3 4 5 6 0 1    4 5 6 0 1 2 3    6 0 1 2 3 4 5
3 4 5 6 0 1 2    6 0 1 2 3 4 5    2 3 4 5 6 0 1
4 5 6 0 1 2 3    1 2 3 4 5 6 0    5 6 0 1 2 3 4
5 6 0 1 2 3 4    3 4 5 6 0 1 2    1 2 3 4 5 6 0
6 0 1 2 3 4 5    5 6 0 1 2 3 4    4 5 6 0 1 2 3
```

0	1	2	3	4	5	6
4	5	6	0	1	2	3
1	2	3	4	5	6	0
5	6	0	1	2	3	4
2	3	4	5	6	0	1
6	0	1	2	3	4	5
3	4	5	6	0	1	2

0	1	2	3	4	5	6
5	6	0	1	2	3	4
3	4	5	6	0	1	2
1	2	3	4	5	6	0
6	0	1	2	3	4	5
4	5	6	0	1	2	3
2	3	4	5	6	0	1

0	1	2	3	4	5	6
6	0	1	2	3	4	5
5	6	0	1	2	3	4
4	5	6	0	1	2	3
3	4	5	6	0	1	2
2	3	4	5	6	0	1
1	2	3	4	5	6	0

2.32 Theorem $N(8) = 7$.

0	1	2	3	4	5	6	7
1	0	3	2	5	4	7	6
2	3	0	1	6	7	4	5
3	2	1	0	7	6	5	4
4	5	6	7	0	1	2	3
5	4	7	6	1	0	3	2
6	7	4	5	2	3	0	1
7	6	5	4	3	2	1	0

0	1	2	3	4	5	6	7
2	3	0	1	6	7	4	5
4	5	6	7	0	1	2	3
6	7	4	5	2	3	0	1
5	4	7	6	1	0	3	2
7	6	5	4	3	2	1	0
1	0	3	2	5	4	7	6
3	2	1	0	7	6	5	4

0	1	2	3	4	5	6	7
3	2	1	0	7	6	5	4
6	7	4	5	2	3	0	1
5	4	7	6	1	0	3	2
1	0	3	2	5	4	7	6
2	3	0	1	6	7	4	5
7	6	5	4	3	2	1	0
4	5	6	7	0	1	2	3

0	1	2	3	4	5	6	7
4	5	6	7	0	1	2	3
5	4	7	6	1	0	3	2
1	0	3	2	5	4	7	6
7	6	5	4	3	2	1	0
3	2	1	0	7	6	5	4
2	3	0	1	6	7	4	5
6	7	4	5	2	3	0	1

0	1	2	3	4	5	6	7
5	4	7	6	1	0	3	2
7	6	5	4	3	2	1	0
2	3	0	1	6	7	4	5
3	2	1	0	7	6	5	4
6	7	4	5	2	3	0	1
4	5	6	7	0	1	2	3
1	0	3	2	5	4	7	6

0	1	2	3	4	5	6	7
6	7	4	5	2	3	0	1
1	0	3	2	5	4	7	6
7	6	5	4	3	2	1	0
2	3	0	1	6	7	4	5
4	5	6	7	0	1	2	3
3	2	1	0	7	6	5	4
5	4	7	6	1	0	3	2

0	1	2	3	4	5	6	7
7	6	5	4	3	2	1	0
3	2	1	0	7	6	5	4
4	5	6	7	0	1	2	3
6	7	4	5	2	3	0	1
1	0	3	2	5	4	7	6
5	4	7	6	1	0	3	2
2	3	0	1	6	7	4	5

2.33 Theorem $N(9) = 8$.

0	1	2	3	4	5	6	7	8
1	2	0	4	5	3	7	8	6
2	0	1	5	3	4	8	6	7
3	4	5	6	7	8	0	1	2
4	5	3	7	8	6	1	2	0
5	3	4	8	6	7	2	0	1
6	7	8	0	1	2	3	4	5
7	8	6	1	2	0	4	5	3
8	6	7	2	0	1	5	3	4

0	1	2	3	4	5	6	7	8
2	0	1	5	3	4	8	6	7
1	2	0	4	5	3	7	8	6
6	7	8	0	1	2	3	4	5
8	6	7	2	0	1	5	3	4
7	8	6	1	2	0	4	5	3
3	4	5	6	7	8	0	1	2
5	3	4	8	6	7	2	0	1
4	5	3	7	8	6	1	2	0

0	1	2	3	4	5	6	7	8
3	4	5	6	7	8	0	1	2
6	7	8	0	1	2	3	4	5
4	5	3	7	8	6	1	2	0
7	8	6	1	2	0	4	5	3
1	2	0	4	5	3	7	8	6
8	6	7	2	0	1	5	3	4
2	0	1	5	3	4	8	6	7
5	3	4	8	6	7	2	0	1

0	1	2	3	4	5	6	7	8
4	5	3	7	8	6	1	2	0
8	6	7	2	0	1	5	3	4
7	8	6	1	2	0	4	5	3
2	0	1	5	3	4	8	6	7
3	4	5	6	7	8	0	1	2
5	3	4	8	6	7	2	0	1
6	7	8	0	1	2	3	4	5
1	2	0	4	5	3	7	8	6

0	1	2	3	4	5	6	7	8
5	3	4	8	6	7	2	0	1
7	8	6	1	2	0	4	5	3
1	2	0	4	5	3	7	8	6
3	4	5	6	7	8	0	1	2
8	6	7	2	0	1	5	3	4
2	0	1	5	3	4	8	6	7
4	5	3	7	8	6	1	2	0
6	7	8	0	1	2	3	4	5

0	1	2	3	4	5	6	7	8
6	7	8	0	1	2	3	4	5
3	4	5	6	7	8	0	1	2
8	6	7	2	0	1	5	3	4
5	3	4	8	6	7	2	0	1
2	0	1	5	3	4	8	6	7
4	5	3	7	8	6	1	2	0
1	2	0	4	5	3	7	8	6
7	8	6	1	2	0	4	5	3

0	1	2	3	4	5	6	7	8
7	8	6	1	2	0	4	5	3
5	3	4	8	6	7	2	0	1
2	0	1	5	3	4	8	6	7
6	7	8	0	1	2	3	4	5
4	5	3	7	8	6	1	2	0
1	2	0	4	5	3	7	8	6
8	6	7	2	0	1	5	3	4
3	4	5	6	7	8	0	1	2

0	1	2	3	4	5	6	7	8
8	6	7	2	0	1	5	3	4
4	5	3	7	8	6	1	2	0
5	3	4	8	6	7	2	0	1
1	2	0	4	5	3	7	8	6
6	7	8	0	1	2	3	4	5
7	8	6	1	2	0	4	5	3
3	4	5	6	7	8	0	1	2
2	0	1	5	3	4	8	6	7

2.34 Theorem (Parker; see [6]) $N(10) \geq 2$. See also Remark IV.28.11 for a pair with four disjoint common transversals, and Example IV.40.13 for a self-orthogonal example.

0	4	1	7	2	9	8	3	6	5
8	1	5	2	7	3	9	4	0	6
9	8	2	6	3	7	4	5	1	0
5	9	8	3	0	4	7	6	2	1
7	6	9	8	4	1	5	0	3	2
6	7	0	9	8	5	2	1	4	3
3	0	7	1	9	8	6	2	5	4
1	2	3	4	5	6	0	7	8	9
2	3	4	5	6	0	1	8	9	7
4	5	6	0	1	2	3	9	7	8

0	7	8	6	9	3	5	4	1	2
6	1	7	8	0	9	4	5	2	3
5	0	2	7	8	1	9	6	3	4
9	6	1	3	7	8	2	0	4	5
3	9	0	2	4	7	8	1	5	6
8	4	9	1	3	5	7	2	6	0
7	8	5	9	2	4	6	3	0	1
4	5	6	0	1	2	3	7	8	9
1	2	3	4	5	6	0	9	7	8
2	3	4	5	6	0	1	8	9	7

2.35 Theorem [18] $N(12) \geq 5$.

The first row of each of the five squares L_1, L_2, L_3, L_4, L_5 is given. Each square is obtained by developing the first row over $\mathbb{Z}_2 \times \mathbb{Z}_6$. The group element (x, y) is written as xy.

L_1:	00	01	02	03	04	05	10	11	12	13	14	15
L_2:	00	03	10	01	13	15	02	12	05	04	11	14
L_3:	00	12	01	15	05	13	03	14	02	11	10	04
L_4:	00	04	15	14	02	11	12	10	13	01	03	05
L_5:	00	10	12	02	11	01	13	15	04	14	05	03

2.36 Theorem [30] $N(14) \geq 3$.

1	2	3	4	5	6	7	8	9	a	b	c	d	e
2	1	6	a	4	d	8	c	e	5	7	9	b	3
3	4	1	7	b	5	e	9	d	2	6	8	a	c
4	d	5	1	8	c	6	2	a	e	3	7	9	b
5	c	e	6	1	9	d	7	3	b	2	4	8	a
6	b	d	2	7	1	a	e	8	4	c	3	5	9
7	a	c	e	3	8	1	b	2	9	5	d	4	6
8	7	b	d	2	4	9	1	c	3	a	6	e	5
9	6	8	c	e	3	5	a	1	d	4	b	7	2
a	3	7	9	d	2	4	6	b	1	e	5	c	8
b	9	4	8	a	e	3	5	7	c	1	2	6	d
c	e	a	5	9	b	2	4	6	8	d	1	3	7
d	8	2	b	6	a	c	3	5	7	9	e	1	4
e	5	9	3	c	7	b	d	4	6	8	a	2	1

1	2	3	4	5	6	7	8	9	a	b	c	d	e
4	a	7	1	3	8	c	2	6	b	e	5	9	d
5	e	b	8	1	4	9	d	3	7	c	2	6	a
6	b	2	c	9	1	5	a	e	4	8	d	3	7
7	8	c	3	d	a	1	6	b	2	5	9	e	4
8	5	9	d	4	e	b	1	7	c	3	6	a	2
9	3	6	a	e	5	2	c	1	8	d	4	7	b
a	c	4	7	b	2	6	3	d	1	9	e	5	8
b	9	d	5	8	c	3	7	4	e	1	a	2	6
c	7	a	e	6	9	d	4	8	5	7	1	b	3
d	4	8	b	2	7	a	e	5	9	6	3	1	c
e	d	5	9	c	3	8	b	2	6	a	7	4	1
2	1	e	6	a	d	4	9	c	3	7	b	8	5
3	6	1	2	7	b	e	5	a	d	4	8	c	9

1	2	3	4	5	6	7	8	9	a	b	c	d	e
e	3	b	d	8	a	6	5	2	c	9	4	7	1
2	1	4	c	e	9	b	7	6	3	d	a	5	8
3	9	1	5	d	2	a	c	8	7	4	e	b	6
4	7	a	1	6	e	3	b	d	9	8	5	2	c
5	d	8	b	1	7	2	4	c	e	a	9	6	3
6	4	e	9	c	1	8	3	5	d	2	b	a	7
7	8	5	2	a	d	1	9	4	6	e	3	c	b
8	c	9	6	3	b	e	1	a	5	7	2	4	d
9	e	d	a	7	4	c	2	1	b	6	8	3	5
a	6	2	e	b	8	5	d	3	1	c	7	9	4
b	5	7	3	2	c	9	6	e	4	1	d	8	a
c	b	6	8	4	3	d	a	7	2	5	1	e	9
d	a	c	7	9	5	4	e	b	8	3	6	1	2

2.37 Theorem [29] $N(15) \geq 4$.

Take the four cyclic squares L_1, L_2, L_3, L_4 whose first rows are given by

L_1:	1	0	2	14	3	13	4	12	5	11	6	10	7	8	9
L_2:	1	14	3	11	6	9	8	7	10	4	13	12	5	0	2
L_3:	1	10	7	13	4	2	0	6	11	9	8	3	14	12	5
L_4:	1	6	11	10	7	0	2	5	12	14	3	9	8	4	13

2.38 Remark Many of the constructions for MOLS in the remainder of this section use difference matrices, quasi-difference matrices or $V(m, t)$ vectors. To go from these objects to a set of MOLS see the constructions given in §IV.11.

2.39 Theorem [32] $N(18) \geq 3$. Consider the matrix

$$\left(\begin{array}{c|c|ccccccccc} 0 & 0 & 10 & 1 & 8 & 5 & 7 & 0 & 4 & 6 & 13 \\ 7 & 1 & 2 & 3 & 11 & 9 & 12 & - & - & - & - \\ \hline 1 & 7 & 12 & 0 & 6 & 2 & 3 & 8 & 9 & 10 & 5 \\ 8 & 4 & 11 & - & - & - & - & 13 & 7 & 4 & 1 \end{array}\right).$$

Form a new array by replacing every column $(a,b,c,d)^T$, except the first column, by the pair of columns $(a,b,c,d)^T$ and $(b,a,d,c)^T$. This is a quasi-difference matrix with $n=14$, $u=4$, $\lambda=1$, and $\mu=0$; form the corresponding $\mathrm{OA}(4,18)$. It is resolvable, as follows. Four parallel classes are given by extending the orbits of $(0,0,0,0)^T$, $(0,7,1,8)^T$, $(0,1,7,4)^T$, and $(1,0,4,7)^T$ modulo 14 with the four parallel classes of the resolvable $\mathrm{OA}(4,4)$ on the fixed points. Fourteen parallel classes are formed by developing the parallel class consisting of the remaining 18 columns modulo 14. Thus there is an $\mathrm{OA}(5,18)$.

2.40 Theorem [31, 4] $N(20) \geq 4$. Consider the matrix

$$\begin{pmatrix} - & 7 & 13 & 1 & 16 & 9 & 2 \\ 0 & 1 & 15 & 7 & 17 & 6 & 14 \\ 0 & 11 & 10 & 11 & 5 & 4 & 3 \\ 7 & - & 13 & 16 & 1 & 2 & 9 \\ 1 & 0 & 15 & 17 & 7 & 14 & 6 \\ 11 & 0 & 10 & 5 & 11 & 3 & 4 \end{pmatrix}.$$

Each column $(a,b,c,d,e,f)^T$ is replaced by columns $(a,b,c,d,e,f)^T$, $(b,c,a,f,d,e)^T$, and $(c,a,b,e,f,d)^T$ to obtain a quasi-difference matrix with $n=19$, $u=1$, $\lambda=1$, and $\mu=1$.

2.41 Theorem [23] $N(21) \geq 5$. Take the five cyclic squares whose first rows are given by

$$\begin{array}{lcccccccccccccccccccccc} L_1: & 0 & 8 & 9 & 11 & 14 & 18 & 16 & 5 & 1 & 19 & 17 & 20 & 4 & 10 & 3 & 12 & 2 & 15 & 6 & 13 & 7 \\ L_2: & 0 & 9 & 11 & 14 & 18 & 8 & 5 & 1 & 19 & 17 & 16 & 4 & 10 & 3 & 12 & 20 & 15 & 6 & 13 & 7 & 2 \\ L_3: & 0 & 11 & 14 & 18 & 8 & 9 & 1 & 19 & 17 & 16 & 5 & 10 & 3 & 12 & 20 & 4 & 6 & 13 & 7 & 2 & 15 \\ L_4: & 0 & 14 & 18 & 8 & 9 & 11 & 19 & 17 & 16 & 5 & 1 & 3 & 12 & 20 & 4 & 10 & 13 & 7 & 2 & 15 & 6 \\ L_5: & 0 & 18 & 8 & 9 & 11 & 14 & 17 & 16 & 5 & 1 & 19 & 12 & 20 & 4 & 10 & 3 & 7 & 2 & 15 & 6 & 13 \end{array}$$

2.42 Theorem [1, 5] $N(22) \geq 3$. Consider the following arrays over $\mathbb{Z}_{21}$

$$A_1 = \begin{pmatrix} 1 & 13 & 18 & 3 & 16 & 19 & - \\ 16 & 19 & 1 & 13 & 18 & 3 & 0 \\ 18 & 3 & 16 & 19 & 1 & 13 & 0 \\ 6 & 15 & 6 & 15 & 6 & 15 & 0 \\ 12 & 9 & 19 & 16 & 5 & 2 & 0 \end{pmatrix} \qquad A_2 = \begin{pmatrix} 0 & 0 \\ 7 & 14 \\ 14 & 7 \\ - & 0 \\ 0 & - \end{pmatrix}.$$

Replace each column (a,b,c,d,e) of A_1 by 3 columns (a,b,c,d,e), $(16c, 16a, 16b, 16d+7, 16e+14)$, and $(4b, 4c, 4a, 4d+14, 4e+7)$; then append the columns of A_2. This gives a (21,1;1,1;5) quasi-difference matrix.

2.43 Theorem [3, 26] $N(24) \geq 5$. Consider the matrix over $\mathbb{Z}_6 \times \mathbb{Z}_2 \times \mathbb{Z}_2$

$$\begin{pmatrix} 000 & 010 & 100 & 110 & 200 & 210 & 300 & 310 & 400 & 410 & 500 & 510 \\ 000 & 110 & 010 & 500 & 400 & 510 & 111 & 001 & 501 & 011 & 511 & 401 \\ 000 & 101 & 210 & 301 & 500 & 411 & 110 & 011 & 310 & 201 & 400 & 511 \\ 000 & 201 & 400 & 501 & 300 & 111 & 500 & 301 & 200 & 101 & 410 & 001 \\ 000 & 500 & 511 & 101 & 100 & 310 & 211 & 311 & 410 & 200 & 001 & 411 \end{pmatrix}.$$

Replace each column $(a,b,c,d,e)^T$ by $(a,b,c,d,e)^T$, and $(a+001, b+300, c+010, d+011, e+301)^T$, and add a row consisting of all entries equal to 000. The result is a difference matrix with six rows over $\mathbb{Z}_6 \times \mathbb{Z}_2 \times \mathbb{Z}_2$.

2.44 Theorem [13] $N(26) \geq 4$. Consider the matrix

$$\begin{pmatrix} - & - & - & - & - \\ 0 & 0 & 0 & 0 & 0 \\ 1 & 6 & 7 & 8 & 14 \\ 3 & 11 & 20 & 18 & 10 \\ 6 & 10 & 14 & 1 & 5 \\ 4 & 19 & 5 & 12 & 2 \end{pmatrix}.$$

Replace each column by six columns which are the six cyclic shifts of the column. The result is a quasi-difference matrix with $n = 21$, $u = 5$, $\lambda = 1$ and $\mu = 0$.

2.45 Theorem [2] $N(28) \geq 5$. Let z be a primitive element in $\mathbb{F}_4$. Consider the following array A over $\mathbb{F}_4 \times \mathbb{Z}_7$

$$\begin{pmatrix} (0,0) & (z{+}1,6) & (1,1) & (1,1) & (1,3) & (1,4) & (0,0) & (1,4) & (z,5) \\ (z,2) & (0,0) & (1,5) & (z,1) & (z,2) & (z,6) & (z{+}1,3) & (0,0) & (z,1) \\ (z,3) & (z{+}1,4) & (0,0) & (z{+}1,5) & (z{+}1,2) & (z{+}1,4) & (z{+}1,2) & (1,6) & (0,0) \\ (0,5) & (z,6) & (0,5) & (0,6) & (z,3) & (0,0) & (0,4) & (1,5) & (z{+}1,4) \\ (0,3) & (0,3) & (z{+}1,5) & (0,0) & (0,5) & (z{+}1,6) & (1,1) & (0,1) & (z,3) \\ (1,3) & (0,6) & (0,6) & (1,5) & (0,0) & (0,3) & (z{+}1,6) & (z,2) & (0,2) \end{pmatrix}$$

Replace each column $(a,b,c,d,e,f)^T$ by three columns $(a,b,c,d,e,f)^T$, $(b,c,a,f,d,e)^T$, and $(c,a,b,e,f,d)^T$. Then append a column whose entries all equal (0,0). This gives a (28,6,1) difference matrix.

2.46 Theorem [4] $N(30) \geq 4$. Consider the matrix over $\mathbb{Z}_5 \times \mathbb{Z}_5$

$$\begin{pmatrix} 00 & - & 00 & 00 & 00 & 00 & 00 \\ 00 & 00 & - & 04 & 02 & 03 & 01 \\ 00 & 31 & 30 & - & 40 & 10 & 20 \\ 00 & 30 & 02 & 12 & - & 01 & 03 \\ 00 & 33 & 12 & 42 & 20 & - & 04 \\ 00 & 42 & 24 & 03 & 23 & 32 & - \end{pmatrix}.$$

Replace each column $(a,b,c,d,e,f)^T$ by the five columns $(a+(i,i), b+(2i,i), c+(i,0), d+(4i,0), e+(3i,4i), f+(4i,4i))^T$ for $0 \leq i \leq 4$. The result is a quasi-difference matrix over $\mathbb{Z}_5 \times \mathbb{Z}_5$ with $n = 25$, $u = 5$, $\lambda = 1$, and $\mu = 1$.

2.47 Theorem [2] $N(33) \geq 5$. Consider the matrices

$$A_1 = \begin{pmatrix} 0 & 0 & 0 & 0 & 0 & 0 \\ 15 & 11 & 22 & 4 & 17 & 8 \\ 19 & 7 & 14 & 32 & 22 & 18 \\ 22 & 19 & 8 & 24 & 21 & 6 \\ 9 & 12 & 15 & 7 & 26 & 14 \\ 14 & 28 & 23 & 2 & 19 & 3 \end{pmatrix} \qquad A_2 = \begin{pmatrix} 0 & 1 & 7 \\ 0 & 4 & 28 \\ 0 & 16 & 13 \\ 0 & 31 & 19 \\ 0 & 25 & 10 \\ 0 & 22 & 0 \end{pmatrix}.$$

Replace each column $C = (a,b,c,d,e,f)^T$ of A_1 by the five columns $t^i(C)$, $0 \leq i \leq 4$, where $t(C) = (4e, 4a, 4b, 4c, 4d, 4f)^T$. Then append the columns of A_2. The result is a difference matrix over $\mathbb{Z}_{33}$.

2.48 Theorem [2] $N(34) \geq 4$. Consider the matrices

$$A_1 = \begin{pmatrix} - & 0 & 0 & 0 & 0 & 0 \\ 30 & 17 & 10 & 25 & 23 & 8 \\ 22 & 4 & 32 & 29 & 28 & 22 \\ 25 & 10 & 20 & 15 & 21 & 16 \\ 0 & 12 & 15 & 16 & 32 & 23 \\ 6 & 11 & 18 & 14 & 9 & 20 \end{pmatrix} \quad A_2 = \begin{pmatrix} 1 & 3 & 10 & 5 \\ 4 & 12 & 7 & 20 \\ 16 & 15 & 28 & 14 \\ 31 & 27 & 13 & 23 \\ 25 & 9 & 19 & 26 \\ 11 & 11 & 0 & - \end{pmatrix}.$$

Replace each column $C = (a, b, c, d, e, f)^T$ of A_1 by the five columns $t^i(C)$, $0 \leq i \leq 4$, where $t(C) = (4e, 4a, 4b, 4c, 4d, 4f)^T$. Then append the columns of A_2. The result is a (33,6;1,0;1) quasi-difference matrix.

2.49 Theorem (Wojtas, alternate construction to [33]) $N(35) \geq 5$. Append a row of all zeroes to the following array to obtain a (35,6,1) difference matrix over $\mathbb{Z}_{35}$:

$$\left(\begin{array}{ccccccccccccccccccccccccccccccccccc} 0&15&30&10&25&1&16&31&11&26&2&17&32&12&6&3&18&33&27&21&4&19&13&7&22&5&34&28&8&23&20&14&29&9&24 \\ 0&22&16&3&4&9&10&32&26&13&18&5&27&14&15&20&7&1&23&31&29&2&24&11&19&17&25&12&6&28&33&34&21&8&30 \\ 0&29&2&31&18&10&32&26&34&28&27&21&15&9&17&30&3&4&5&20&12&6&14&22&16&8&23&24&25&33&11&19&13&7&1 \\ 0&8&9&17&11&25&19&27&28&1&15&23&31&4&26&12&6&14&29&16&2&3&18&33&34&20&7&22&30&24&10&32&5&13&21 \\ 0&1&23&24&32&33&6&7&29&30&10&11&12&13&28&8&9&31&4&5&27&14&15&16&3&25&26&34&21&22&2&17&18&19&20 \end{array}\right)$$

2.50 Theorem [27] $N(36) \geq 5$. Consider the following array A over $\mathbb{Z}_2 \times \mathbb{Z}_2 \times \mathbb{Z}_3 \times \mathbb{Z}_3$. Replace each column $[(a_1, b_1, c_1, d_1), (a_2, b_2, c_2, d_2) \ldots (a_6, b_6, c_6, d_6)]^T$ of A by the three columns $[(a_1, b_1, c_1, d_1), (a_2, b_2, c_2, d_2{+}i), (a_3, b_3, c_3, d_3{+}2i)\ (a_4, b_4, c_4{+}i, d_4)(a_5, b_5, c_5{+}i, d_5 + i)(a_6, b_6, c_6 + i, d_6 + 2i)]^T$ for $i = 0, 1, 2$. This gives a $(36, 6, 1)$ difference matrix.

$$A = \begin{pmatrix} 0000 & 0000 & 0000 & 0000 & 0000 & 0000 & 0000 & 0000 & 0000 & 0000 & 0000 & 0000 \\ 0000 & 0010 & 0020 & 0100 & 0110 & 0120 & 1000 & 1010 & 1020 & 1100 & 1110 & 1120 \\ 0000 & 0020 & 0011 & 1000 & 1020 & 1010 & 1111 & 1101 & 1121 & 0111 & 0101 & 0121 \\ 0000 & 0102 & 1101 & 1100 & 1002 & 0001 & 0100 & 1011 & 0022 & 1000 & 0111 & 1122 \\ 0000 & 0110 & 1120 & 0012 & 1121 & 1011 & 0010 & 0112 & 1012 & 0122 & 1111 & 1021 \\ 0000 & 1010 & 0100 & 1002 & 1111 & 0122 & 0120 & 0001 & 1112 & 1012 & 1122 & 0020 \end{pmatrix}.$$

2.51 Theorem [4] $N(38) \geq 4$. Consider the matrix

$$\begin{pmatrix} - & 10 & 1 & 2 & 6 & 3 & 22 & 5 & 7 & 9 & 14 & 18 & 28 \\ 0 & 1 & 10 & 20 & 23 & 30 & 35 & 13 & 33 & 16 & 29 & 32 & 21 \\ 0 & 26 & 26 & 15 & 8 & 4 & 17 & 19 & 34 & 12 & 31 & 24 & 25 \\ 10 & - & 10 & 6 & 2 & 22 & 3 & 7 & 5 & 14 & 9 & 28 & 18 \\ 1 & 0 & 26 & 23 & 20 & 35 & 30 & 33 & 13 & 29 & 16 & 21 & 32 \\ 26 & 0 & 1 & 8 & 15 & 17 & 4 & 34 & 19 & 31 & 12 & 25 & 24 \end{pmatrix}.$$

Each column $(a, b, c, d, e, f)^T$ is replaced by columns $(a, b, c, d, e, f)^T$, $(b, c, a, f, d, e)^T$, and $(c, a, b, e, f, d)^T$ to obtain a quasi-difference matrix with $n = 37$, $u = 1$, $\lambda = 1$, and $\mu = 1$.

2.52 Theorem [2, 22] $N(39) \geq 4$. Consider the array

$$\begin{pmatrix} 0 & 1 & 5 & 1 & 0 & 27 & 23 & 0 & 5 & 36 & 0 & 9 & 8 \\ 19 & 11 & 35 & 7 & 9 & 0 & 37 & 24 & 0 & 35 & 18 & 0 & 25 \\ - & 7 & 17 & 11 & 31 & 25 & 0 & 17 & 16 & 0 & 23 & 12 & 0 \\ 1 & 0 & 30 & 5 & 34 & 5 & 2 & 32 & 7 & 25 & 20 & 1 & 6 \\ 11 & 19 & 20 & 35 & 14 & 10 & 35 & 23 & 34 & 11 & 4 & 26 & 7 \\ 7 & - & 26 & 17 & 17 & 22 & 32 & 1 & 9 & 10 & 11 & 28 & 30 \end{pmatrix}.$$

Replace each column $(a, b, c, d, e, f)^T$ by three columns $(a, b, c, d, e, f)^T$, $(c, a, b, e, f, d)^T$, and $(b, c, a, f, d, e)^T$. This gives a (38,6;1,1;1) quasi-difference matrix.

2.53 Theorem [3] $N(40) \geq 7$. Let ω be a primitive root in $\mathbb{F}_{2^3}$ satisfying $\omega^3 = \omega + 1$. Let

$$A = \begin{pmatrix} (0,\emptyset) & (0,\emptyset) & (0,\emptyset) & (0,\emptyset) & (0,\emptyset) & (0,\emptyset) & (0,\emptyset) & (0,\emptyset) & (0,\emptyset) & (0,\emptyset) \\ (0,\emptyset) & (1,\emptyset) & (2,2) & (3,2) & (4,2) & (2,\emptyset) & (3,\emptyset) & (4,\emptyset) & (0,2) & (1,2) \\ (0,\emptyset) & (2,5) & (4,5) & (1,2) & (3,6) & (3,4) & (0,0) & (2,1) & (4,1) & (1,6) \\ (0,\emptyset) & (3,4) & (1,4) & (4,0) & (2,5) & (3,\emptyset) & (1,0) & (4,1) & (2,2) & (0,3) \\ (0,\emptyset) & (4,6) & (3,\emptyset) & (2,3) & (1,4) & (2,1) & (1,\emptyset) & (0,4) & (4,0) & (3,2) \\ (0,\emptyset) & (1,2) & (4,6) & (4,4) & (1,0) & (0,6) & (2,3) & (3,6) & (3,5) & (2,5) \\ (1,\emptyset) & (0,3) & (1,2) & (4,5) & (4,\emptyset) & (2,3) & (0,0) & (2,2) & (3,0) & (3,\emptyset) \\ (4,\emptyset) & (1,3) & (0,0) & (1,1) & (4,0) & (3,1) & (2,5) & (0,\emptyset) & (2,1) & (3,\emptyset) \end{pmatrix},$$

$$Y = \begin{pmatrix} \emptyset & 0 & 1 & 6 & 5 & 4 & 3 & 2 \\ \emptyset & 1 & 2 & 0 & 6 & 5 & 4 & 3 \end{pmatrix},$$

where the second coordinates of entries in A, and the entries in Y are interpreted as follows. Entry i denotes ω^i, and entry $\emptyset$ denotes the zero element of $\mathbb{F}_{2^3}$. Now if $A(x,y) = (a,b)$, let $\tau_i(A(x,y)) = (a, b + Y(i,x))$ for $i = 1, 2$. The group generated by τ_1 and τ_2 has order 4. Developing columns of A under the action of this group yields 40 columns, which form a difference matrix over $\mathbb{F}_5 \times \mathbb{F}_{2^3}$.

2.54 Theorem [2] $N(42) \geq 5$. Consider the matrices

$$A_1 = \begin{pmatrix} - & - & - \\ 0 & 0 & 0 \\ 18 & 11 & 5 \\ 26 & 10 & 30 \\ 20 & 3 & 33 \\ 5 & 25 & 24 \\ 17 & 4 & 22 \end{pmatrix} \qquad A_2 = \begin{pmatrix} - \\ 0 \\ 4 \\ 23 \\ 23 \\ 4 \\ 0 \end{pmatrix}.$$

The seven cyclic shifts of the columns of $(A_1| - A_1|A_2)$ give a $(35, 7; 1, 1; 7)$ QDM.

2.55 Theorem [2] $N(44) \geq 5$. Consider the following array over $\mathbb{Z}_2 \times \mathbb{Z}_2 \times \mathbb{Z}_{11}$

$$\begin{pmatrix} 000 & 000 & 000 & 000 & 000 & 000 & 000 & 000 \\ 114 & 014 & 117 & 106 & 119 & 012 & 015 & 011 \\ 106 & 013 & 100 & 019 & 111 & 014 & 119 & 109 \\ 116 & 119 & 012 & 110 & 010 & 115 & 004 & 009 \\ 109 & 002 & 001 & 102 & 007 & 116 & 110 & 107 \\ 101 & 106 & 113 & 015 & 005 & 013 & 010 & 110 \end{pmatrix}.$$

Replace each column $C = [(x_1, y_1, z_1), (x_2, y_2, z_2), \ldots, (x_5, y_5, z_5), (x_6, y_6, z_6)]^T$ by the five columns $t^i(C)$, where

$t(C) = [(x_1, y_1, 5z_5), (x_1, y_1, 5z_1), (x_2, y_2, 5z_2), (x_3, y_3, 5z_3), (x_4, y_4, 5z_4), (x_6, y_6, 5z_6)]^T$. Append the four columns $[(x,y,z), (x,y,5z), (x,y,3z), (x,y,4z), (x,y,9z), (0,0,0)]^T$ for $(x,y,z) =$ (0,0,0), (1,0,1), (1,1,2) and (0,0,8), giving a (44,6,1) difference matrix.

2.56 Theorem [2] $N(45) \geq 6$. Over $\mathbb{Z}_5 \times \mathbb{Z}_3 \times \mathbb{Z}_3$, consider the array

$$\begin{pmatrix} 000 & 000 & 100 & 400 & 200 & 200 & 400 & 100 & 000 & 200 & 300 & 400 & 400 & 300 & 200 \\ 000 & 110 & 021 & 102 & 411 & 212 & 210 & 402 & 220 & 022 & 202 & 311 & 421 & 402 & 322 \\ 000 & 412 & 101 & 022 & 120 & 410 & 220 & 211 & 301 & 212 & 021 & 202 & 301 & 410 & 422 \\ 000 & 202 & 402 & 111 & 012 & 102 & 420 & 220 & 401 & 322 & 200 & 010 & 200 & 321 & 412 \\ 000 & 201 & 210 & 410 & 101 & 021 & 122 & 401 & 402 & 421 & 312 & 200 & 020 & 200 & 311 \\ 000 & 421 & 222 & 210 & 420 & 110 & 011 & 102 & 302 & 411 & 420 & 302 & 201 & 012 & 221 \\ 000 & 120 & 401 & 220 & 221 & 422 & 101 & 012 & 210 & 311 & 401 & 412 & 322 & 201 & 011 \end{pmatrix}$$

Replace each column $[(x_1, y_1, z_1), (x_2, y_2, z_2), \ldots (x_7, y_7, z_7)]^T$ by the three columns $[(x_1, y_1, z_1), (x_2, y_2, z_2+i), (x_3, y_3+i, z_3), (x_4, y_4+i, z_4+i), (x_5, y_5+2i, z_5+2i), (x_6, y_6+2i, z_i), (x_7, y_7, z_7+2i)]^T$. This gives a (45,7,1) difference matrix.

2.57 Theorem $N(46) \geq 4$, from the existence of a $V(4,9)$ vector (Table IV.11.55).

2.58 Theorem [3, 27] $N(48) \geq 5$. Let ω be a primitive root in $\mathbb{F}_{2^4}$ satisfying $\omega^4 = \omega + 1$. Let

$$A = \begin{pmatrix} (0,\emptyset) & (0,\emptyset) & (0,\emptyset) & (0,\emptyset) & (0,\emptyset) & (0,\emptyset) \\ (0,\emptyset) & (1,\emptyset) & (2,\emptyset) & (2,3) & (0,3) & (1,3) \\ (0,\emptyset) & (2,0) & (1,5) & (2,6) & (1,10) & (0,9) \\ (0,\emptyset) & (1,9) & (1,4) & (0,2) & (2,3) & (2,\emptyset) \\ (1,\emptyset) & (0,9) & (1,5) & (2,9) & (0,5) & (2,2) \\ (1,\emptyset) & (1,3) & (0,4) & (2,5) & (2,8) & (0,1) \end{pmatrix} \qquad Y = \begin{pmatrix} \emptyset\,0\,2\,5\,14\,6 \\ \emptyset\,1\,3\,6\;\,0\;\,7 \\ \emptyset\,2\,4\,7\;\,1\;\,8 \end{pmatrix},$$

where the second coordinates of entries in A, and the entries in Y are interpreted in $\mathbb{F}_{2^4}$ as follows. Entry i denotes ω^i, and entry $\emptyset$ denotes the zero element of $\mathbb{F}_{2^4}$. Now if $A(x,y) = (a,b)$, let $\tau_i(A(x,y)) = (a, b+Y(i,x))$ for $i = 1,2,3$. The group generated by τ_1, τ_2, and τ_3 has order 8. Developing columns of A under the action of this group yields 48 columns, which form a difference matrix over $\mathbb{F}_3 \times \mathbb{F}_{2^4}$.

2.59 Theorem $N(50) \geq 6$. This follows from the existence of a $V(6,7)$ vector (Table IV.11.55).

2.60 Theorem [2] $N(51) \geq 5$. Consider the following array with entries from $\mathbb{Z}_{51}$

$$A = \begin{pmatrix} 5 & 33 & 29 & 30 & 1 \\ 8 & 3 & 47 & 10 & 13 \\ 14 & 27 & 6 & 12 & 28 \\ 9 & 16 & 44 & 49 & 11 \\ 34 & 32 & 36 & 26 & 20 \end{pmatrix}.$$

Replace each column of $[A| - A]$ by its five cyclic shifts. Then append a row and column of zeros. The result is a (51,6,1) difference matrix.

2.61 Theorem [2] $N(52) \geq 5$. Consider arrays over $\mathbb{F}_4 \times \mathbb{Z}_{13}$

$$A_1 = \begin{pmatrix} (0,0) & (0,0) & (0,0) & (0,0) & (0,0) \\ (z^2,10) & (0,7) & (1,10) & (z,10) & (z^2,3) \\ (z,10) & (z^2,2) & (1,11) & (z,2) & (z^2,7) \\ (z,8) & (z^2,12) & (0,10) & (z^2,11) & (z^2,6) \\ (1,2) & (0,2) & (z^2,8) & (z,3) & (z,7) \\ (1,6) & (z,12) & (0,7) & (z^2,6) & (z,2) \end{pmatrix} \qquad A_2 = \begin{pmatrix} (1,1) & (z^2,11) \\ (z,3) & (1,7) \\ (z^2,9) & (z,8) \\ (1,4) & (z^2,3) \\ (z,12) & (1,9) \\ (z^2,10) & (z,1) \end{pmatrix}.$$

For a column $C = ((x_1,y_1),(x_2,y_2)\dots(x_6,y_6))^T$, let $t_1(C) = ((z\cdot x_3, 3\cdot y_3),(z\cdot x_1, 3\cdot y_1),(z\cdot x_2,3\cdot y_2),(z\cdot x_6,3\cdot y_6),(z\cdot x_4,3\cdot y_4),(z\cdot x_5,3\cdot y_5))^T$, and $t_2(C) = ((x_3,y_3),(x_1,y_1),(x_2,y_2),(x_5,y_5),(x_6,y_6),(x_4,y_4))^T$. (Note that $z^2 = z+1$.) Apply the group of order 9 generated by t_1, t_2 to the columns of A_1 to give 45 columns; then apply the group of order 3 generated by t_2 to the columns of A_2 to give another 6 columns. Finally append a column whose entries all equal (0,0). This gives a (52,6,1) difference matrix.

2.62 Theorem [22] $N(55) \geq 5$. Consider the following array over $\mathbb{Z}_{11} \times \mathbb{Z}_5$

$$\begin{pmatrix} (0,0) & (1,1) & (2,0) & (3,1) & (4,1) & (5,0) & (6,2) & (7,0) & (8,4) & (9,3) & (10,4) \\ (0,1) & (4,2) & (8,1) & (1,2) & (5,2) & (9,3) & (2,3) & (6,3) & (10,3) & (3,3) & (7,2) \\ (0,4) & (5,0) & (10,4) & (4,0) & (9,1) & (3,3) & (8,0) & (2,3) & (7,4) & (1,4) & (6,1) \\ (0,4) & (9,0) & (7,4) & (5,0) & (3,3) & (1,0) & (10,3) & (8,0) & (6,2) & (4,1) & (2,1) \\ (0,1) & (3,2) & (6,1) & (9,2) & (1,3) & (4,4) & (7,2) & (10,4) & (2,2) & (5,4) & (8,2) \\ (1,0) & (1,0) & (1,0) & (1,0) & (1,0) & (1,0) & (1,0) & (1,0) & (1,0) & (1,0) & (1,0) \end{pmatrix}.$$

Replace each column $C = ((x_1,y_1)(x_2,y_2)(x_3,y_3)(x_4,y_4)(x_5,y_5)(x_6,y_6))^T$ by $t^i(C)$ for $0 \leq i \leq 4$, where $t(C) = [(4x_5,y_5)(4x_1,y_1)(4x_2,y_2)(4x_3,y_3)(4x_4,y_4)(4x_6,y_6)]^T$. The result is a (55,6,1) difference matrix.

2.63 Theorem [1, 22] $N(56) \geq 7$. Let ω be a primitive element of $\mathbb{F}_8$ satisfying $\omega^3 = \omega + 1$. Consider the following 8×8 array over $\mathbb{F}_8 \times \mathbb{Z}_7$

$$\begin{pmatrix}
(0,0) & (\omega^0,0) & (\omega^1,0) & (\omega^2,0) & (\omega^3,0) & (\omega^4,0) & (\omega^5,0) & (\omega^6,0) \\
(0,1) & (\omega^1,6) & (\omega^2,1) & (\omega^3,1) & (\omega^4,6) & (\omega^5,1) & (\omega^6,6) & (\omega^0,6) \\
(0,4) & (\omega^2,3) & (\omega^3,4) & (\omega^4,4) & (\omega^5,3) & (\omega^6,4) & (\omega^0,3) & (\omega^1,3) \\
(0,2) & (\omega^3,5) & (\omega^4,2) & (\omega^5,2) & (\omega^6,5) & (\omega^0,2) & (\omega^1,5) & (\omega^2,5) \\
(0,2) & (\omega^4,5) & (\omega^5,2) & (\omega^6,2) & (\omega^0,5) & (\omega^1,2) & (\omega^2,5) & (\omega^3,5) \\
(0,4) & (\omega^5,3) & (\omega^6,4) & (\omega^0,4) & (\omega^1,3) & (\omega^2,4) & (\omega^3,3) & (\omega^4,3) \\
(0,1) & (\omega^6,6) & (\omega^0,1) & (\omega^1,1) & (\omega^2,6) & (\omega^3,1) & (\omega^4,6) & (\omega^5,6) \\
(1,0) & (1,0) & (1,0) & (1,0) & (1,0) & (1,0) & (1,0) & (1,0)
\end{pmatrix}.$$

Each column $C = [(x_1,y_1)(x_2,y_2)(x_3,y_3)(x_4,y_4)(x_5,y_5)(x_6,y_6)(x_7,y_7)(x_8,y_8)]^T$ is replaced by the 7 columns $t^i(C)$ $(0 \leq i \leq 6)$ where

$$t(C) = [(\omega x_7,y_7)(\omega x_1,y_1)(\omega x_2,y_2)(\omega x_3,y_3)(\omega x_4,y_4)(\omega x_5,y_5)(\omega x_6,y_6)(\omega x_8,y_8)]^T.$$

The result is a (56,8,1) difference matrix over $\mathbb{F}_8 \times \mathbb{Z}_7$.

2.64 Remark An easy way to obtain sets of MOLS of side q for q a prime power is by use of MAPLE V. Using that software, first type `readlib(MOLS):` then type `MOLS(p,e,k);` to get k MOLS of side p^e as output.

2.5 Complete Sets of Order 9

2.65 Remarks Owens and Preece [25] determined the 19 equivalence classes of complete sets of MOLS of order 9, using a definition of equivalence that permits reordering the rows and columns of all squares together, and the symbols of each square individually. There are four nonisomorphic projective planes: the desarguesian plane Φ, the translation plane Ω, the dual translation plane Ω^D, and the Hughes plane Ψ. See §VI.8.

2.66 Table The 19 inequivalent complete sets of order 9. For each, the eight orthogonal latin squares are given, and under each is provided a name for the isotopy class to which the square belongs (see [25]), and the number of intercalates in the square.

For each set, the plane from which it arises is given along with some information about the manner in which the complete set is formed. Here, ℓ_∞ is the line at infinity, t is the translation line of Ω, and T is the translation point of Ω^D. In the Hughes plane, lines are classified as real or complex [28]. Details are given in [25].

Set 1 (Plane Φ)							
123456789	123456789	123456789	123456789	123456789	123456789	123456789	123456789
231564897	456789123	645978312	897231564	312645978	564897231	789123456	978312645
312645978	789123456	897231564	645978312	231564897	978312645	456789123	564897231
456789123	312645978	978312645	564897231	789123456	645978312	231564897	897231564
564897231	645978312	231564897	312645978	978312645	789123456	897231564	456789123
645978312	978312645	456789123	789123456	897231564	231564897	564897231	312645978
789123456	231564897	564897231	978312645	456789123	897231564	312645978	645978312
897231564	564897231	789123456	456789123	645978312	312645978	978312645	231564897
978312645	897231564	312645978	231564897	564897231	456789123	645978312	789123456
a,0	a,0	a,0	a,0	a,0	a,0	a,0	a,0
Set 2 (Plane Ω, $t = \ell_\infty$)							
123456789	123456789	123456789	123456789	123456789	123456789	123456789	123456789
231564897	456789123	645978312	564897231	312645978	789123456	897231564	978312645
312645978	789123456	897231564	978312645	231564897	456789123	645978312	564897231
456789123	312645978	564897231	897231564	789123456	231564897	978312645	645978312
564897231	645978312	789123456	312645978	978312645	897231564	456789123	231564897
645978312	978312645	312645978	456789123	897231564	564897231	231564897	789123456
789123456	231564897	978312645	645978312	456789123	312645978	564897231	897231564
897231564	564897231	231564897	789123456	645978312	978312645	312645978	456789123
978312645	897231564	456789123	231564897	564897231	645978312	789123456	312645978
a,0	a,0	a,0	a,0	a,0	a,0	a,0	a,0

Set 3 (Plane Ω, $t = \ell_\infty$)							
123456789	123456789	123456789	123456789	123456789	123456789	123456789	123456789
312645978	645978312	564897231	456789123	231564897	978312645	789123456	897231564
231564897	897231564	978312645	789123456	312645978	564897231	456789123	645978312
789123456	978312645	231564897	564897231	456789123	897231564	645978312	312645978
978312645	231564897	645978312	897231564	564897231	456789123	312645978	789123456
897231564	456789123	789123456	231564897	645978312	312645978	978312645	564897231
456789123	564897231	312645978	978312645	789123456	645978312	897231564	231564897
645978312	789123456	456789123	312645978	897231564	231564897	564897231	978312645
564897231	312645978	897231564	645978312	978312645	789123456	231564897	456789123
a,0	a,0	a,0	a,0	a,0	a,0	a,0	a,0
Set 4 (Plane Ω, $t \neq \ell_\infty$)							
123456789	123456789	123456789	123456789	123456789	123456789	123456789	123456789
231564897	456789123	645978312	564897231	312645978	789123456	897231564	978312645
312645978	789123456	897231564	978312645	231564897	456789123	645978312	564897231
498732156	312867594	576394821	849521367	765189432	231975648	984613275	657248913
587293641	694538217	758162493	312974856	946817325	875341962	469725138	231689574
679328514	967214835	312589647	485763192	854971263	548692371	231847956	796135428
765189432	231975648	984613275	657248913	498732156	312867594	576394821	849521367
854971263	548692371	231847956	796135428	679328514	967214835	312589647	485763192
946817325	875341962	469725138	231689574	587293641	694538217	758162493	312974856
b_R,48	b_R,48	b_R,48	b_R,48	b_R,48	b_R,48	b_R,48	b_R,48
Set 5 (Plane Ω, $t \neq \ell_\infty$)							
123456789	123456789	123456789	123456789	123456789	123456789	123456789	123456789
231987654	789165423	897541362	978614235	312879546	456723198	645238917	564392871
312879546	456723198	645238917	564392871	231987654	789165423	897541362	978614235
456723198	231987654	978614235	645238917	789165423	312879546	564392871	897541362
564392871	897541362	456723198	231987654	978614235	645238917	789165423	312879546
645238917	564392871	231987654	789165423	897541362	978614235	312879546	456723198
789165423	312879546	564392871	897541362	456723198	231987654	978614235	645238917
897541362	978614235	312879546	456723198	645238917	564392871	231987654	789165423
978614235	645238917	789165423	312879546	564392871	897541362	456723198	231987654
b_C,48	b_C,48	b_C,48	b_C,48	b_C,48	b_C,48	b_C,48	b_C,48
Set 6 (Plane Ω, $t \neq \ell_\infty$)							
123456789	123456789	123456789	123456789	123456789	123456789	123456789	123456789
312987654	978324561	789263145	897532416	231645978	645891327	564719832	456178293
231879546	564918273	456197328	645781932	312564897	897243615	978632451	789325164
756123498	687231945	214978653	938645271	465789132	579312864	892564317	341897526
964312875	215789634	837645912	479123568	546897321	782564193	351978246	698231457
845231967	739645128	598312476	216897354	654978213	361789542	487123695	972564831
489765123	846597312	371829564	562314897	798132456	954678231	635241978	217983645
697548312	452163897	965784231	381279645	879321564	218937456	746895123	534612978
578694231	391872456	642531897	754968123	987213645	436125978	219387564	865749312
b_S,48	c,32	c,32	c,32	a,0	c,32	c,32	c,32
Set 7 (Plane Ω^D, $T \in \ell_\infty$)							
123456789	123456789	123456789	123456789	123456789	123456789	123456789	123456789
231564897	456789123	645978312	564897231	312645978	789123456	897231564	978312645
312645978	789123456	897231564	978312645	231564897	456789123	645978312	564897231
456789123	312645978	978312645	645978312	789123456	231564897	564897231	897231564
564897231	897231564	456789123	312645978	978312645	645978312	789123456	231564897
645978312	564897231	312645978	789123456	897231564	978312645	231564897	456789123
789123456	231564897	564897231	897231564	456789123	312645978	978312645	645978312
897231564	978312645	231564897	456789123	645978312	564897231	312645978	789123456
978312645	645978312	789123456	231564897	564897231	897231564	456789123	312645978
a,0	a,0	a,0	a,0	a,0	a,0	a,0	a,0
Set 8 (Plane Ω^D, $T \in \ell_\infty$)							
123456789	123456789	123456789	123456789	123456789	123456789	123456789	123456789
231564897	765189432	854971263	946817325	312645978	498732156	679328514	587293641
312645978	498732156	679328514	587293641	231564897	765189432	854971263	946817325
456789123	231975648	548692371	875341962	789123456	312867594	967214835	694538217
564897231	657248913	796135428	231689574	978312645	849521367	485763192	312974856
645978312	984613275	231847956	469725138	897231564	576394821	312589647	758162493
789123456	312867594	967214835	694538217	456789123	231975648	548692371	875341962
897231564	576394821	312589647	758162493	645978312	984613275	231847956	469725138
978312645	849521367	485763192	312974856	564897231	657248913	796135428	231689574
a,0	a,0	a,0	a,0	a,0	a,0	a,0	a,0

Set 9 (Plane Ω^D, $T \in \ell_\infty$)							
123456789	123456789	123456789	123456789	123456789	123456789	123456789	123456789
312645978	576918243	485197326	694781532	231564897	849273615	967832451	758329164
231564897	984327561	796283145	875932416	312645978	657891324	548719632	469178253
789123456	648231975	371548692	962875341	456789123	594312867	835967214	217694538
978312645	391765824	842679513	457123968	564897231	736984152	219548376	685231497
897231564	752849136	569312478	381694257	645978312	218765943	476123895	934587621
456789123	867594312	214835967	538217694	789123456	975648231	692371548	341962875
645978312	439182657	958764231	216349875	897231564	361527498	784695123	572813946
564897231	215673498	637921854	749568123	978312645	482139576	351284967	896745312
a,0	d,72	d,72	d,72	a,0	d,72	d,72	d,72
Set 10 (Plane Ω^D, $T \notin \ell_\infty$)							
123456789	123456789	123456789	123456789	123456789	123456789	123456789	123456789
231879546	756123498	845231967	964312875	312987654	489765123	697548312	578694231
312987654	489765123	697548312	578694231	231879546	756123498	845231967	964312875
489765123	231879546	578694231	845231967	756123498	312987654	964312875	697548312
578694231	697548312	756123498	231879546	964312875	845231967	489765123	312987654
697548312	964312875	231879546	489765123	845231967	578694231	312987654	756123498
756123498	312987654	964312875	697548312	489765123	231879546	578694231	845231967
845231967	578694231	312987654	756123498	697548312	964312875	231879546	489765123
964312875	845231967	489765123	312987654	578694231	697548312	756123498	231879546
b_S,48	b_S,48	b_S,48	b_S,48	b_S,48	b_S,48	b_S,48	b_S,48
Set 11 (Plane Ω^D, $T \notin \ell_\infty$)							
123456789	123456789	123456789	123456789	123456789	123456789	123456789	123456789
312879546	675948213	584697321	496785132	231564897	948213675	769132458	857321964
231987654	894321567	976213845	785132496	312645978	567894321	458769132	649578213
789165423	546297831	231549678	967813245	498732156	654378912	875921364	312684597
978614235	231865974	849371562	654928317	587293641	796182453	312547896	465739128
897541362	758139426	465782193	231674958	679328514	312965847	546893271	984217635
456723198	987612345	312864957	548397621	765189432	879531264	694278513	231945876
645238917	469783152	798125436	312569874	854971263	231647598	987314625	576892341
564392871	312574698	657938214	879241563	946817325	485729136	231685948	798163452
b_C,48	e,24	e,24	e,24	b_R,48	e,24	e,24	e,24
Set 12 (Plane Ψ, real ℓ_∞)							
123456789	123456789	123456789	123456789	123456789	123456789	123456789	123456789
231564897	456789123	564897231	645978312	312645978	789123456	978312645	897231564
312645978	789123456	978312645	897231564	231564897	456789123	564897231	645978312
456789123	312897564	897564312	564312897	789123456	231978645	645231978	978645231
564897231	645231897	312978456	789564123	978312645	897645312	231789564	456123978
645978312	978564231	456123897	312789645	897231564	564312978	789645123	231897456
789123456	231978645	645231978	978645231	456789123	312897564	897564312	564312897
897231564	564312978	789645123	231897456	645978312	978564231	456123897	312789645
978312645	897645312	231789564	456123978	564897231	645231897	312978456	789564123
a,0	f_S,36	f_S,36	f_S,36	a,0	f_S,36	f_S,36	f_S,36
Set 13 (Plane Ψ, real ℓ_∞)							
123456789	123456789	123456789	123456789	123456789	123456789	123456789	123456789
789123456	897231564	978312645	231564897	456789123	645978312	564897231	312645978
456789123	645978312	564897231	312645978	789123456	897231564	978312645	231564897
231978645	564312978	897645312	978231456	312564897	456897123	789123564	645789231
645231978	789645123	231789564	897123645	564978231	978312456	312564897	456897312
978645231	231897456	456123978	789312564	897231645	312564897	645789312	564978123
312897564	978564231	645231897	564789312	231645978	789123645	456978123	897312456
564312897	312789645	789564123	456978231	645897312	231645978	897231456	978123564
897564312	456123897	312978456	645897123	978312564	564789231	231645978	789231645
f_C,36	f_C,36	f_C,36	g_R,0	h_R,0	h_R,0	h_R,0	g_R,0
Set 14 (Plane Ψ, real ℓ_∞)							
123456789	123456789	123456789	123456789	123456789	123456789	123456789	123456789
498732156	679328514	587293641	231645978	756189423	845971362	964817235	312564897
765189432	854971263	946817325	312564897	489723156	697238514	578392641	231645978
312645978	967832451	694781532	879213645	231597864	458369127	785124396	546978213
849273615	485197326	312645978	798321564	567834291	976512843	231968457	654789132
576918243	312645978	758329164	987132456	894261537	231784695	649573812	465897321
231564897	548719632	875932416	654897321	312678945	769143258	496285173	987321564
984327561	231564897	469178253	546789132	675942318	312895476	857631924	798213645
657891324	796283145	231564897	465978213	948315672	584627931	312749568	879132456
f_R,36	f_R,36	f_R,36	g_C,0	h_C,0	h_C,0	h_C,0	g_C,0

Set 15 (Plane Ψ, real ℓ_∞)							
123456789	123456789	123456789	123456789	123456789	123456789	123456789	123456789
312564897	849327561	967283145	758932416	231645978	675891234	584719623	496178352
231645978	657918243	548197326	469781532	312564897	894372615	976823451	785239164
546897213	465789132	382514697	871325964	798132456	957648321	639271548	214963875
654978132	372561894	791632458	546897321	987213645	438129576	215384967	869745213
465789321	981243675	654978213	392164857	879321564	216537498	748695132	537812946
987321645	798132456	236845971	614578293	465789132	549213867	852967314	371694528
798213456	516894327	879321564	235649178	654978213	381765942	467132895	942587631
879132564	234675918	415769832	987213645	546897321	762984153	391548276	658321497
g_S,0	g_S,0	g_S,0	g_S,0	a,0	d,72	d,72	d,72
Set 16 (Plane Ψ, real ℓ_∞)							
123456789	123456789	123456789	123456789	123456789	123456789	123456789	123456789
489723156	314579628	678931245	962345871	756198432	241687593	597812364	835264917
756189423	271863954	864397512	538614297	489732165	317925846	945278631	692541378
312948675	796185243	987524361	845732916	231867594	469273158	654391827	578619432
975612348	639748512	452179836	381297654	567384921	894531267	218965473	746823195
648375912	567294831	215743698	479168523	894621357	936812475	782539146	351987264
231894567	485327196	546218973	697581432	312975648	758169324	879643215	964732851
867531294	948612375	391865427	754923168	675249813	582794631	436187952	219378546
594267831	852931467	739682154	216879345	948513276	675348912	361724598	487195623
h_S,0	h_S,0	d,72	d,72	h_S,0	h_S,0	d,72	d,72
Set 17 (Plane Ψ, complex ℓ_∞)							
123456789	123456789	123456789	123456789	123456789	123456789	123456789	123456789
231645978	879213564	798321456	987132645	312564897	654978321	465789213	546897132
312564897	546897213	654978132	465789321	231645978	987321645	879132564	798213456
489723156	312948675	975612348	648375912	756189423	231894567	594267831	867531294
578392641	457639128	312785964	896241573	964817235	749163852	231578496	685924317
697238514	784125936	569843271	312697458	845971362	476582193	958314627	231769845
765189432	231564897	846297513	579813264	498732156	312645978	687921345	954378621
854971263	965782341	487139625	231564897	679328514	598217436	746893152	312645978
946817325	698371452	231564897	754928136	587293641	865739214	312645978	479182563
i_R,24	i_R,24	i_R,24	i_R,24	i_R,24	i_R,24	i_R,24	i_R,24
Set 18 (Plane Ψ, complex ℓ_∞)							
123456789	123456789	123456789	123456789	123456789	123456789	123456789	123456789
231879654	765132498	946213875	854321967	312987546	489765132	578694321	697548213
312987546	489765132	578694321	697548213	231879654	765132498	946213875	854321967
465732198	572891364	854327916	319684572	798165423	637918245	281549637	946273851
546293871	318547926	692178453	765932148	987614235	854329617	439781562	271865394
654328917	946283571	317965248	482719635	879541362	291674853	765832194	538197426
798165423	637918245	281549637	946273851	465732198	572891364	854327916	319684572
879541362	291674853	765832194	538197426	654328917	946283571	317965248	482719635
987614235	854329617	439781562	271865394	546293871	318547926	692178453	765932148
i_C,24	i_C,24	i_C,24	i_C,24	i_C,24	i_C,24	i_C,24	i_C,24
Set 19 (Plane Ψ, complex ℓ_∞)							
123456789	123456789	123456789	123456789	123456789	123456789	123456789	123456789
312987546	576894312	694578123	485769231	231645897	948231675	857123964	769312458
231879654	894312567	785231496	976123845	312564978	657948231	469785312	548697123
765132498	319527846	972845361	548391627	456789132	281673954	694218573	837964215
946213875	657138294	318962547	892675413	564897321	739584162	271349658	485721936
854321967	782649135	469713852	317284596	645978213	576192348	938567421	291835674
489765132	945281673	537694218	251837964	798123456	864319527	316972845	672548391
697548213	438765921	241389675	764912358	879231564	315827496	582694137	956173842
578694321	261973458	856127934	639548172	987312645	492765813	745831296	314289567
i_S,24	j,24	j,24	j,24	k,0	j,24	j,24	j,24

2.6 MOLS of Side n $\leq$ 10,000

2.67 **Remark** Table 2.72 updates a similar table by Brouwer [10] from 1979. This table has been produced using a new program that attempts to incorporate all existing constructions for sets of MOLS. The constructions used and the program developed are described in some detail in [14]. Tables 2.68, 2.69, 2.71, and 2.73 are derived from information in Table 2.72.

n_k is the largest order for which the existence of k MOLS is unknown.

2.68 Table n_k for $k \leq 15$ and $k = 30$.

k	n_k	k	n_k	k	n_k	k	n_k	k	n_k
2	6	3	10	4	22	5	62	6	75
7	780	8	2774	9	3678	10	5804	11	7222
12	7286	13	7288	14	7874	15	8360	30	52502

i_k is the largest order for which the existence of k idempotent MOLS is unknown.

2.69 Table i_k for $k \leq 15$.

k	i_k	k	i_k	k	i_k	k	i_k	k	i_k
2	6	3	10	4	60	5	62	6	98
7	780	8	2774	9	3678	10	5804	11	7222
12	7288	13	7618	14	7874	15	10618		

2.70 Remark [1] The only orders greater than 9718 for which 15 idempotent MOLS are not known are 10606 and 10618.

o_k is the largest *odd* order for which the existence of k MOLS is unknown.

2.71 Table o_k for $k \leq 15$.

k	o_k	k	o_k	k	o_k	k	o_k	k	o_k
2	–	3	–	4	–	5	39	6	75
7	335	8	1935	9	2607	10	2607	11	3471
12	3565	13	3565	14	3603	15	3603		

2.72 Table Lower bounds on $N(n)$, the number of MOLS of side n. Italicized entries are lower bounds on $N(1^n)$ (all the MOLS are idempotent), when the best current bound matches the lower bound on $N(n)$.

	0	1	2	3	4	5	6	7	8	9	10	11	12	13	14	15	16	17	18	19
0	∞	∞	1	2	3	4	*1*	6	7	8	*2*	10	5	12	*3*	4	15	16	*3*	18
20	4	5	*3*	22	5	24	4	26	5	28	4	30	31	5	*4*	5	5	36	*4*	*4*
40	7	40	5	42	5	6	*4*	46	5	48	6	5	5	52	*4*	5	7	7	*5*	58
60	4	60	*4*	*6*	63	7	*5*	66	*5*	6	6	70	7	72	*5*	*5*	6	6	*6*	78
80	9	80	8	82	6	*6*	*6*	*6*	7	88	*6*	*7*	*6*	*6*	*6*	*6*	7	96	6	8
100	*8*	100	*6*	102	7	*7*	*6*	106	*6*	108	*6*	*6*	13	112	*6*	*7*	*6*	8	*6*	*6*
120	7	120	*6*	*6*	*6*	124	*6*	126	127	*7*	*6*	130	*6*	*7*	*6*	*7*	*7*	136	*6*	138
140	*6*	*7*	*6*	10	*10*	*7*	*6*	*7*	*6*	148	*6*	150	*7*	8	*8*	*7*	*6*	156	7	*6*
160	9	*7*	*6*	162	*6*	*7*	*6*	166	*7*	168	*6*	8	*6*	172	*6*	*6*	14	*9*	*6*	178
180	*6*	180	*6*	*6*	*7*	*9*	*6*	10	*6*	8	*6*	190	*7*	192	*6*	*7*	*6*	196	*6*	198
200	*7*	*8*	*6*	*7*	*6*	8	*6*	8	14	*11*	*10*	210	*6*	*7*	*6*	*7*	*7*	*8*	*6*	*10*
220	*6*	12	*6*	222	13	*8*	*6*	226	*6*	228	*6*	*7*	*7*	232	*6*	*7*	*6*	*7*	*6*	238
240	*7*	240	*6*	242	*6*	*7*	*6*	12	*7*	*7*	*6*	250	*6*	12	*9*	*7*	255	256	*6*	*12*
260	*6*	8	*8*	262	*7*	*8*	*6*	*10*	*6*	268	*6*	270	15	16	*6*	*13*	*10*	276	*6*	*9*
280	*7*	280	*6*	282	*6*	*12*	*6*	*7*	*15*	288	*6*	*6*	*6*	292	*6*	*6*	*7*	10	*10*	12
300	*6*	*7*	*6*	*6*	15	*15*	*6*	306	*6*	*7*	*6*	310	*7*	312	*6*	*10*	*7*	316	*6*	10
320	15	*15*	*6*	16	*6*	12	*6*	*7*	*7*	*9*	*6*	330	*6*	8	*6*	*6*	*8*	336	*6*	*7*
340	*6*	10	*10*	342	*7*	*7*	*6*	346	*6*	348	*8*	12	18	352	*6*	*9*	*6*	*9*	*6*	358
360	*8*	360	*6*	*7*	*6*	*10*	*6*	366	15	*15*	*6*	*15*	*6*	372	*6*	*15*	*7*	*13*	*6*	378
380	*6*	*12*	*6*	382	15	*15*	*6*	*15*	*6*	388	*6*	16	*7*	*8*	*6*	*7*	*6*	396	*6*	*7*
400	15	400	*7*	*15*	*11*	*7*	*6*	*15*	8	408	*6*	*13*	*8*	*12*	*10*	*9*	18	*15*	*6*	418

	0	1	2	3	4	5	6	7	8	9	10	11	12	13	14	15	16	17	18	19
420	6	420	6	15	7	16	6	7	6	10	6	430	15	432	6	15	6	18	6	438
440	7	15	6	442	6	13	6	11	15	448	6	15	6	7	6	15	7	456	6	16
460	6	460	6	462	15	15	6	466	6	8	6	15	7	15	10	18	6	15	6	478
480	15	15	6	15	6	7	6	486	7	15	6	490	6	16	6	7	15	15	6	498
500	6	12	9	502	7	15	6	15	6	508	6	15	511	18	7	15	6	12	6	15
520	8	520	10	522	12	15	6	16	15	528	6	15	6	12	6	15	8	15	10	15
540	12	540	6	15	18	7	6	546	6	8	6	18	7	7	6	7	6	556	6	12
560	15	7	6	562	6	7	6	7	7	568	6	570	6	7	15	22	8	576	6	7
580	6	7	6	10	7	8	7	586	6	18	17	7	15	592	6	15	7	7	8	598
600	14	600	12	15	7	15	16	606	18	15	6	15	7	612	6	15	7	616	6	618
620	7	22	6	15	15	624	6	7	7	16	6	630	7	7	6	7	6	12	6	8
640	9	640	6	642	6	7	6	646	7	10	7	7	7	652	6	7	15	15	7	658
660	6	660	7	15	7	15	7	22	7	15	7	15	15	672	7	24	7	676	6	15
680	7	15	7	682	7	15	7	15	15	15	6	690	7	15	7	15	7	16	7	15
700	7	700	6	18	15	15	7	15	6	708	7	15	7	22	21	15	7	15	7	718
720	15	9	8	12	10	24	12	726	7	728	16	16	18	732	6	7	22	10	8	738
740	7	7	7	742	7	15	7	8	7	10	6	750	15	15	7	15	7	756	7	15
760	7	760	6	15	6	15	7	15	15	768	7	15	7	772	6	24	23	15	7	18
780	6	18	7	26	15	15	10	786	12	15	7	15	20	15	18	15	7	796	22	16
800	24	15	7	15	7	15	7	15	7	808	7	810	8	15	7	15	15	18	7	8
820	7	820	7	822	7	15	7	826	7	828	8	15	12	16	7	8	7	26	25	838
840	7	840	7	20	7	10	8	16	15	15	12	22	7	852	16	15	22	856	7	858
860	22	15	24	862	26	15	7	15	7	15	9	15	7	15	7	15	7	876	8	15
880	15	880	8	882	8	15	7	886	7	15	8	15	10	18	7	15	13	15	8	28
900	27	16	8	8	7	22	7	906	8	18	10	910	15	14	7	15	16	10	18	918
920	7	8	22	12	24	24	26	8	28	928	7	18	7	7	7	14	7	936	7	15
940	7	940	7	22	15	15	7	946	7	12	7	15	7	952	7	15	7	15	8	15
960	15	960	29	15	8	15	7	966	7	15	8	970	10	18	12	15	15	976	16	18
980	18	15	7	982	27	15	24	15	26	22	28	990	31	31	7	15	7	996	25	26
1000	7	15	21	16	19	15	7	18	15	1008	13	18	7	1012	9	22	7	28	7	1018
1020	7	1020	7	30	1023	24	7	15	9	15	9	1030	7	1032	7	15	8	16	9	1038
1040	15	15	8	15	8	15	8	15	8	1048	8	1050	8	15	8	15	15	16	8	8
1060	8	1060	8	1062	8	15	8	15	10	1068	7	15	15	28	7	24	7	15	8	15
1080	12	22	7	15	7	15	7	1086	16	15	8	1090	7	1092	7	15	7	1096	8	15
1100	8	15	7	1102	15	15	7	26	7	1108	7	18	7	15	7	15	7	1116	7	15
1120	16	18	7	1122	7	15	7	22	7	1128	7	15	7	15	10	9	15	15	7	16
1140	7	7	7	15	7	15	7	30	30	15	7	1150	15	1152	7	15	8	26	10	24
1160	12	26	7	1162	16	18	18	15	15	15	22	1170	24	15	26	24	28	15	30	30
1180	8	1180	7	15	31	15	7	1186	7	28	7	15	7	1192	7	15	7	15	7	15
1200	15	1200	7	15	8	15	7	16	7	15	7	15	7	1212	7	15	18	1216	7	22
1220	7	15	8	1222	7	24	7	15	7	1228	7	1230	15	9	8	15	7	1236	7	15
1240	7	16	8	10	8	7	8	28	8	1248	8	8	7	7	7	8	8	8	7	1258
1260	7	12	23	7	15	15	9	15	9	26	9	30	30	23	8	15	9	1276	9	1278
1280	15	30	10	1282	12	15	9	24	16	1288	18	1290	8	18	22	15	24	1296	26	15
1300	28	1300	30	1302	8	15	8	1306	30	15	8	15	31	15	8	15	8	15	8	1318
1320	8	1320	8	26	8	24	7	1326	15	15	8	1330	7	30	30	15	8	15	9	30
1340	9	15	8	30	15	30	12	15	9	26	16	24	18	15	9	20	22	22	24	15
1360	26	1360	28	28	30	30	9	1366	28	1368	30	15	9	1372	30	15	31	16	8	15
1380	8	1380	8	15	8	15	8	18	8	15	8	15	15	15	8	15	8	10	9	1398
1400	10	15	8	22	8	8	8	15	10	1408	8	16	7	9	9	22	9	12	7	7

	0	1	2	3	4	5	6	7	8	9	10	11	12	13	14	15	16	17	18	19
1420	9	28	7 1422		15	24	9 1426		9 1428		7	26	7 1432		9	15	7	15	7 1438	
1440	15	15	7	15	9	15	9 1446		7	15	7 1450		7 1452		9	15	15	30	30 1458	
1460	7	15	7	30	7	15	8	30	10	30	12 1470		22	30	16	28	18	15	7	24
1480	22 1480		24 1482		26	18	28 1486		30 1488		13	15	7 1492		30	15	8	15	30 1498	
1500	30	18	9	15	31	15	9	15	9	14	9 1510		9	24	9	9	9	36	9	30
1520	30	9	9 1522		9	30	9	9	9	30	10 1530		12	9	9	30	16	30	18	18
1540	8	26	22 1542		24	8	26	20	28 1548		30	30	15 1552		8	15	30	8	8 1558	
1560	30	15	30	15	8	15	30 1566		31	15	8 1570		8	15	9	15	7	18	8 1578	
1580	7	15	8 1582		15	24	8	8	8	15	8	36	7	26	8	15	8 1596		8	15
1600	24 1600		8	15	8	15	8 1606		8 1608		8	15	8 1612		7	15	15	15	8 1618	
1620	7 1620		7	15	7	15	7 1626		7	15	7	15	24	22	8	15	8 1636		7	15
1640	7	15	7	30	30	15	7	26	15	30	7	15	11	30	10	30	12 1656		7	30
1660	16	30	18 1662		15	30	22 1666		24 1668		26	24	28	22	30	30	19	15	7	22
1680	30 1680		9	15	30	15	30	15	9	15	30	18	30 1692		9	15	31 1696		9 1698	
1700	9	15	8	15	8	15	8	15	8 1708		21	28	15	15	8	15	10	16	7	15
1720	7 1720		9 1722		9	15	7	15	26	21	8	15	8 1732		7	15	7	15	7	36
1740	9 1740		7	15	15	15	8 1746		8	15	8	16	9 1752		9	15	9	15	7 1758	
1760	26	15	7	40	9	15	7	15	8	28	7	27	7	15	7	24	15 1776		9	15
1780	9	15	7 1782		7	15	7 1786		7 1788		8	15	15	15	9	15	8	15	8	15
1800	7 1800		7	15	9	15	7	30	15	26	8 1810		8	36	7	15	9	22	9	16
1820	9	15	9 1822		26	24	9	15	9	30	30 1830		9	15	9	30	9	15	9	30
1840	15	30	12	18	9	30	16 1846		18 1848		9	30	22	16	24	15	28	30	28	28
1860	30 1860		25	22	8	22	30 1866		8	18	30 1870		30 1872		8	15	30 1876		30 1878	
1880	8	15	30	8	8	8	8	15	31 1888		8	30	30	15	8	15	8	30	8	15
1900	8 1900		10	30	15	15	8 1906		16	30	18	15	8 1912		22	15	24	26	26	30
1920	28	30	30	30	27	9	7	40	30	9	7 1930		30 1932		30	7	15	15	30	15
1940	30	10	8	28	30	15	7	15	7 1948		30 1950		31	15	7	15	7	18	7	15
1960	8	36	7	15	7	15	7	15	15	15	7	26	8 1972		7	24	9	15	9 1978	
1980	9	15	9	15	30	30	9 1986		9	15	30	15	10 1992		30	15	30 1996		9 1998	
2000	30	16	30 2002		9	9	30	22	9	40	9 2010		30	28	30	30	31 2016		8	15
2020	27	42	8	15	23	30	21 2026		8 2028		8	30	15	30	13	15	11	30	8 2038	
2040	8	15	8	30	9	30	8	22	2047	15	8	15	8 2052		8	15	8	16	10	28
2060	8	15	9 2062		15	15	8	15	7 2068		8	18	7	15	9	24	8	30	30	15
2080	30 2080		8 2082		8	15	8 2086		10 2088		12	15	7	30	16	30	18	15	8 2098	
2100	22	36	24	15	26	30	28	42	30	30	30 2110		15 2112		30	15	9	28	30	24
2120	30	15	10	15	30	18	30	16	15 2128		30 2130		8	26	9	15	30 2136		30	15
2140	9 2140		9 2142		31	15	9	18	9	15	9	15	9 2152		10	15	12	15	9	16
2160	15 2160		9	15	9	14	9	15	9	15	10	14	12	40	9	15	16	15	9 2178	
2180	8	15	9	36	9	15	9 2186		9	15	9	23	15	15	8	15	9 2196		8	15
2200	9	30	30 2202		8	15	9 2206		15 2208		8	30	10 2212		12	15	8	30	16	30
2220	18 2220		8	30	22	24	24	16	26	30	28	30	30	30	30	15	10 2236		30 2238	
2240	16	30	30 2242		30	15	7	15	30	22	30 2250		8	18	30	15	15	36	8	15
2260	30	15	30	30	30	30	9 2266		30 2268		8	15	31 2272		10	30	12	15	8	42
2280	16 2280		18	15	8	30	22 2286		24	15	26	30	28 2292		30	30	30 2296		9	30
2300	30	15	10	46	30	30	30	15	9 2308		30 2310		30	22	9	20	30	15	9	15
2320	15	15	30	22	30	15	9	30	28	16	30	15	9 2332		30	15	31	15	9 2338	
2340	8 2340		8	10	8	15	8 2346		8	28	8 2350		15	12	8	15	9 2356		9	10
2360	8	15	8	16	8	9	8	10	36	22	10 2370		8	10	8	18	26 2376		8	10
2380	8 2380		8 2382		15	15	8	15	8 2388		8	15	8 2392		8	42	10	15	13 2398	
2400	15 2400		8	26	8	15	7	28	7	15	7 2410		8	18	17	15	15 2416		7	40

	0	1	2	3	4	5	6	7	8	9	10	11	12	13	14	15	16	17	18	19
2420	8	15	8_{2422}		14	24	12	15	7	15	16	15	18	15	8	15	9_{2436}		9	15
2440	9_{2440}		10	15	10	15	10_{2446}		15	30	30	15	9	27	9	30	9	15	9_{2458}	
2460	10	30	12	15	15	30	16_{2466}		18	15	9	30	22_{2472}		24	15	26_{2476}		28	36
2480	30	30	30	15	9	30	30	15	9	30	30	46	30	15	9	15	30	30	30	28
2500	8	40	30_{2502}		8	15	9	22	30	15	30	30	30	30	9	15	30	30	8	15
2520	30_{2520}		30	30	12	24	9	30	31	30	18_{2530}		9	30	22	15	24	42	26_{2538}	
2540	28	30	30_{2542}		30	15	9	30	30_{2548}		9_{2550}		30	30	30	15	9_{2556}		30	30
2560	30	30	9	28	30	15	10	16	9	23	30	15	30	30	30	30	15	15	30_{2578}	
2580	9	28	30	30	30	30	12	12	9	30	30_{2590}		31_{2592}		8	30	22	48	24	22
2600	26	30	28	30	30	30	30	8	15_{2608}		30	15	9	30	30	30	30_{2616}		8	26
2620	30_{2620}		30	42	40	30	30	36	8	15	8	24	30_{2632}		30	15	8	30	8	16
2640	30	18	8	15	30	15	30_{2646}		8	15	8	15	30	15	30	15	31_{2656}		10_{2658}	
2660	8	15	9_{2662}		9	15	9	15	7	16	9_{2670}		15	15	8	24	8_{2676}		8	15
2680	9	15	8_{2682}		9	15	8_{2686}		15_{2688}		8	15	10_{2692}		8	15	8	15	9_{2698}	
2700	9	36	8	15	15	15	10_{2706}		8	15	10_{2710}		9_{2712}		8	15	10	15	10_{2718}	
2720	31	15	9	15	9	24	10	26	10_{2728}		10_{2730}		9	15	10	15	15	15	8	15
2740	8_{2740}		8	15	10	15	8	40	9_{2748}		8	15	42_{2752}		9	15	8	15	9	30
2760	30	15	8	15	9	30	7_{2766}		15	30	10	30	12	46	7	30	16_{2776}		18	15
2780	9	30	22	22	24	15	26	30	28_{2788}		30_{2790}		30	15	9	30	30_{2796}		9	30
2800	30_{2800}		30_{2802}		9	15	30	30	30_{2808}		8	30	30	28	9	15	15	30	30_{2818}	
2820	30	15	8	30	9	24	30	15	8	18	30	18	30_{2832}		9	15	9_{2836}		30	23
2840	30	15	30_{2842}		8	15	8	15	31	15	10_{2850}		8	15	9	15	8_{2856}		8	15
2860	8_{2860}		29	13	29	15	9	46	29	18	29	15	8	16	29	22	8	15	8_{2878}	
2880	29	42	29	15	9	29	9_{2886}		29	26	8	48	29	15	29	15	15_{2896}		9	15
2900	29	15	30_{2902}		8	15	8	15	8_{2908}		10	40	31	15	9	15	8_{2916}		9	15
2920	8	22	21	36	9	18	9_{2926}		15	28	9	15	10	15	12	15	9	15	16_{2938}	
2940	18	16	9	26	22	15	9	15	9	15	9	15	9_{2952}		9	15	9_{2956}		9	15
2960	15	15	9_{2962}		9	15	9	15	9_{2968}		9_{2970}		9	15	10	15	15	15	9	15
2980	9	15	10	18	9	15	9	28	9	48	8	15	15	40	9	15	9	36	9_{2998}	
3000	9_{3000}		9	15	10	15	9	30	46	15	9_{3010}		9	30	8	15	10	30	10_{3018}	
3020	12	15	9_{3022}		16	30	18	15	9	30	22	15	24	15	26	30	28_{3036}		30	30
3040	31_{3040}		9	30	30	15	9	30	30_{3048}		30	26	9	42	30	30	30	30	9	30
3060	30_{3060}		9	15	9	30	30_{3066}		30	15	9	36	15	30	30	15	8	26	30_{3078}	
3080	30	15	9_{3082}		9	18	30	15	30_{3088}		30	15	9	15	10	15	30	18	9	15
3100	9	15	9	28	31	15	9	25	9_{3108}		8	15	9	15	8	15	8	15	9_{3118}	
3120	15_{3120}		8	15	9_{3124}		8	52	8	15	9	30	30	15	8	15	48_{3136}		9	42
3140	8	30	10	30	12	15	9	30	16	46	18	22	15	30	22	15	24	15	26	30
3160	28	30	30_{3162}		30	15	8_{3166}		30_{3168}		9	30	30	30	30	24	23	15	30	30
3180	30_{3180}		10	30	30	15	10_{3186}		12	30	30_{3190}		30	30	30	30	8	30	30	30
3200	24	30	30_{3202}		30	30	12	24	9_{3208}		30	30	30	18	30	30	22_{3216}		24	15
3220	30_{3220}		28	30	30	30	30	15	30_{3228}		30	15	31	52	30	30	30	15	9	40
3240	30	30	30	30	9	30	30	16	15	15	10_{3250}		30_{3252}		30	15	9_{3256}		9_{3258}	
3260	30	15	10	30	30	30	30	26	10	26	9_{3270}		30	15	30	24	30	28	10	15
3280	15	16	30	48	9	15	10	18	28	15	30	15	9	36	30	15	31	15	9_{3298}	
3300	10_{3300}		10	15	10	15	10_{3306}		10	15	9	15	15_{3312}		9	15	9	30	30_{3318}	
3320	9	40	9_{3322}		9	23	9	30	14_{3328}		12_{3330}		10	30	16	30	18	46	9	30
3340	22	15	24_{3342}		26	30	28_{3346}		30	30	30	15	10	30	30	13	10	30	30_{3358}	
3360	30_{3360}		10	15	30	30	30	30	10	30	30_{3370}		10_{3372}		10	30	30	15	30	30
3380	30	30	9	30	30	30	9	30	30_{3388}		30_{3390}		52	30	10	30	30	42	30	24
3400	30	39	22	40	24	23	30_{3406}		28	30	30	30	30_{3412}		30	30	30	15	30	30

	0	1	2	3	4	5	6	7	8	9	10	11	12	13	14	15	16	17	18	19
3420	30	30	30	15	31	24	30	30	30	30	25	46	30_{3432}		9	15	10	30	30	18
3440	30	15	12	30	10	30	30	15	22_{3448}		30	30	30	13	24	30	26_{3456}		30	15
3460	30_{3460}		30_{3462}		9	15	9_{3466}		30_{3468}		9	10	15	22	10	24	30	18	9	48
3480	30_{3480}		30	42	10	15	30	39	31	15	9_{3490}		9	11	10	13	10	15	9_{3498}	
3500	9	15	10	30	30	11	10	15	9	30	10_{3510}		9	30	10	30	12_{3516}		9	30
3520	26	30	18	15	11	30	22_{3526}		24_{3528}		26	30	28_{3532}		30	30	30	26	10_{3538}	
3540	30_{3540}		11	30	30	30	30_{3546}		11	15	30	52	30	30	10	30	30_{3556}		10_{3558}	
3560	11	30	30	11	30	11	11	30	15	42	30_{3570}		11	30	30	30	30	48	10	30
3580	10_{3580}		30_{3582}		30	30	30	16	13	36	10	15	30_{3592}		11	15	11	18	10	58
3600	30	15	10	13	30	15	30_{3606}		9	15	30	22	30_{3612}		9	15	31_{3616}		9	15
3620	10	15	9_{3622}		10	28	11	15	9	25	11_{3630}		15	15	10	15	10_{3636}		9	15
3640	9	15	8_{3642}		11	15	9	15	26	40	10	15	9	15	10	15	9	25	9_{3658}	
3660	11	15	11	15	15	15	10	18	9	15	9_{3670}		10_{3672}		10	15	9_{3676}		8	21
3680	15	15	10	28	27	15	10	24	9	15	10_{3690}		10	18	10	16	15_{3696}		10	26
3700	16_{3700}		18	15	10	15	22	15	24_{3708}		26	15	28	46	10	15	10	15	9_{3718}	
3720	10_{3720}		10	15	11	24	9_{3726}		15	15	9	15	10_{3732}		10	15	9	36	9_{3738}	
3740	9	15	10	18	15	39	10	15	9	22	10	30	30	26	9	15	10	30	10	15
3760	15_{3760}		10	52	12	15	10_{3766}		16_{3768}		18	15	9	30	22	24	58	15	26_{3778}	
3780	28	30	30	30	30	15	10	30	30	15	10	30	30_{3792}		30	15	10_{3796}		30	30
3800	30	30	10_{3802}		30	15	9	46	31	30	30	36	30	15	9	30	9	30	30	15
3820	10_{3820}		30_{3822}		30	15	10	42	10	30	30	15	30_{3832}		30	15	10	15	10	15
3840	30	30	10	15	10	26	10_{3846}		30	15	9_{3850}		30_{3852}		30	15	15	15	30	16
3860	30	15	10_{3862}		30	15	9	15	9	52	9	48	31	15	10	30	30_{3876}		10	15
3880	10_{3880}		9	15	10	30	10	30	15_{3888}		10	30	16	30	18	15	9	30	22	15
3900	24	46	26	30	60	30	30_{3906}		30	15	9_{3910}		30	15	11	30	30_{3916}		30_{3918}	
3920	15	15	30_{3922}		30	30	9	30	30_{3928}		9_{3930}		9	30	30	15	30	30	30	30
3940	9	30	30_{3942}		9	30	30_{3946}		30	30	12	30	15	58	30	30	30	30	30	36
3960	26	40	24	15	30	30	28_{3966}		30	48	30	28	30	36	30	24	30	40	30	30
3980	30	18	30	16	30	30	30	30	30_{3988}		30	22	10	24	10	30	30	28	30	30
4000	31_{4000}		10_{4002}		30	30	10_{4006}		30	30	30	30	9_{4012}		9	30	30	30	30_{4018}	
4020	30_{4020}		9	26	9	24	30_{4026}		9	30	9	30	30	36	30	30	9	26	30	30
4040	30	30	10	20	30	18	30	30	15_{4048}		30_{4050}		18	15	9	15	28_{4056}		30	15
4060	26	30	30	16	31	30	10	48	27	30	9	30	10_{4072}		21	30	19	30	16_{4078}	
4080	18	30	11	30	22	15	24	60	26	30	28_{4090}		30_{4092}		30	30_{4095}		30	30_{4098}	
4100	9	30	30	30	30	15	9	15	30	30	30_{4110}		15	30	30	15	10	22	9	30
4120	30	15	30	15	11	30	9_{4126}		30_{4128}		10	30	30_{4132}		30	15	11	30	10_{4138}	
4140	30	40	30	30	30	15	10	15	10	15	30	30	9_{4152}		11	30	9_{4156}		30_{4158}	
4160	15	30	30	22	30	15	29	24	30	22	30	42	29	15	30	24	29_{4176}		10	15
4180	10	36	30	46	29	15	30	52	30	58	29	15	31	15	29	15	29	15	10	15
4200	10_{4200}		29	15	29	15	29	15	15	15	10_{4210}		29	15	9	15	9_{4216}		10_{4218}	
4220	29	15	9	40	29	24	29	15	10_{4228}		29_{4230}		29	15	10	15	29	18	13	26
4240	15_{4240}		10_{4242}		30	15	9	30	30	15	12	15	9_{4252}		13	15	31	30	10_{4258}	
4260	12_{4260}		10	30	16	30	18	42	9	30	22_{4270}		24_{4272}		26	30	28	30	30	30
4280	30	15	10_{4282}		30	15	9	30	63_{4288}		30	15	10	52	30	30	30_{4296}		9	30
4300	30	15	10	15	15	30	30	58	30	30	30	30	9	30	30	30	9	30	30	30
4320	30	30	12	30	10	30	30_{4326}		30	30	30	60	22	15	24	15	30_{4336}		28_{4338}	
4340	30	30	30	42	30	30	30	30	30_{4348}		30	30	30	30	30	28	30_{4356}		30	30
4360	30	48	30_{4362}		13	18	13	30	30	16	30	15	30_{4372}		30	30	30	15	30	30
4380	30	30	30	15	31	30	12	40	30	15	30_{4390}		30	22	13	15	13_{4396}		30	52
4400	15	26	12	30	13	30	30	15	13_{4408}		30	15	30	15	10	30	30	30	30	15

	0	1	2	3	4	5	6	7	8	9	10	11	12	13	14	15	16	17	18	19
4420	12_{4420}		30_{4422}		13	15	13	20	12	42	30	15	15	15	30	15	30	30	12	22
4440	30_{4440}		11	15	13	15	30_{4446}		31	15	10_{4450}		12	60	13	15	12_{4456}		13	15
4460	13	15	13_{4462}		15	15	11	15	10	40	12	16	13	15	13	24	13	36	13	15
4480	16_{4480}		10_{4482}		10	15	12	15	13_{4488}		13	15	10_{4492}		13	15	15	15	10	25
4500	10	15	10	15	10	15	12_{4506}		13	26	13	15	15_{4512}		10	15	10_{4516}		10_{4518}	
4520	10	15	12_{4522}		13	24	10	15	15	15	11	22	10	15	10	15	10	15	10	15
4540	12	18	13	15	63	15	10_{4546}		10_{4548}		10	15	10	28	10	15	12	15	12	46
4560	15_{4560}		11	26	11	15	10_{4566}		10	15	12	15	12	16	13	15	31	22	11	18
4580	11	15	10_{4582}		10	15	10	15	12	15	13_{4590}		15	15	10	15	10_{4596}		12	15
4600	10	42	10_{4602}		12	15	10	16	15	15	13	15	13	15	13	15	9	18	13	30
4620	30_{4620}		12	15	15	36	15	15	13	30	15	30	15	40	15	30	16_{4636}		18_{4638}	
4640	31	30	22_{4642}		24	15	26	30	28_{4648}		30_{4650}		30	15	15	30	30_{4656}		12	30
4660	30	58	30_{4662}		14	15	30	30	30	30	14	30	63_{4672}		12	15	14	30	30_{4678}	
4680	30	30	30	30	14	30	30	42	15	30	30_{4690}		30	30	14	30	15	30	30	36
4700	35	30	30_{4702}		22	15	24	15	30	30	28	30	30	30	30	30	30	52	30	30
4720	30_{4720}		30_{4722}		30	30	30	30	30_{4728}		30	31	30_{4732}		30	15	36	30	12	30
4740	30	15	30	15	30	30	30	46	30	15	30_{4750}		30	48	30	15	30	66	30_{4758}	
4760	30	15	30	30	30	15	13	15	31	18	30	30	11	15	10	30	10	39	30	58
4780	15	30	30_{4782}		30	15	12_{4786}		30_{4788}		30	15	10_{4792}		30	15	14	15	15_{4798}	
4800	15_{4800}		30	15	10	15	30	24	30	30	10	16	30_{4812}		14	15	15_{4816}		30	60
4820	30	15	10	22	28	24	12	15	12	15	30_{4830}		31	26	12	15	11	23	10	15
4840	10	46	11	28	12	15	15	36	15	15	11	15	10	22	10	15	11	15	10	42
4860	11_{4860}		14	15	18	17	12	30	30	15	10_{4870}		11	30	11	15	12_{4876}		11	30
4880	15	16	10	30	16	30	18	26	10_{4888}		22	66	24	15	26	30	28	58	30	30
4900	30	28	12_{4902}		30	15	10	30	30_{4908}		30	15	15_{4912}		30	30	30	30	10_{4918}	
4920	30	15	11	15	10	30	30	15	30	15	12_{4930}		11_{4932}		30	15	11_{4936}		30	30
4940	30	60	10_{4942}		15	30	30	15	30	48	30_{4950}		10	15	10	15	30_{4956}		10	15
4960	15	40	11	30	30	15	12_{4966}		30_{4968}		30	15	10_{4972}		30	30	30	15	12	15
4980	30	16	11	15	11	30	10_{4986}		30	15	10	15	30_{4992}		30	30	11	18	30_{4998}	
5000	11	15	10_{5002}		30	18	30	15	15_{5008}		10_{5010}		11	15	10	15	30	28	30	15
5020	10_{5020}		10_{5022}		31	15	10	15	10	46	10	15	9	15	10	15	10	15	10_{5038}	
5040	15_{5040}		10	15	11	15	10	48	9	15	10_{5050}		10	30	30	15	63	15	10_{5058}	
5060	11	15	10	60	10	30	12	15	10	36	16	30	18	15	10	30	22_{5076}		24	15
5080	26_{5080}		28	30	30	30	30_{5086}		31	30	30	15	12	30	30	30	30	15	10_{5098}	
5100	30_{5100}		30	30	15	30	30_{5106}		11	15	10	30	30_{5112}		30	15	10	30	10_{5118}	
5120	30	39	10	46	30	40	30	15	10	30	13	30	30	15	30	30	30	15	10	15
5140	11	52	30	36	11	15	13_{5146}		13	45	30	15	31_{5152}		30	15	30	26	10	30
5160	30	30	30	15	12	15	30_{5166}		15	15	15_{5170}		15	30	30	15	15	30	30_{5178}	
5180	30	30	15	70	63	30	15	30	15_{5188}		30	28	30	30	16	30	18_{5196}		15	30
5200	22	15	30	42	30	30	28	40	39_{5208}		30	36	15	30	30	15	31	30	30	30
5220	30	22	15	15	30	30	30_{5226}		15	30	30_{5230}		24_{5232}		26	30	30_{5236}		30	31
5240	30	30	13	48	36	30	15	30	40	30	30	58	15	30	15	30	30	30	30	30
5260	30_{5260}		22	18	24	15	30	30	28	30	30	30	30_{5272}		30	30	30	30	30_{5278}	
5280	31_{5280}		30	30	30	30	30	30	30	30	30	30	30	66	15	30	15_{5296}		30	16
5300	30	15	30_{5302}		30	30	30	15	30_{5308}		30	46	63	30	30	30	30	30	30	26
5320	30	30	30_{5322}		15	18	30	16	30_{5328}		12	16	10_{5332}		30	30	30	15	15	30
5340	30	48	30	15	31	30	30_{5346}		30	15	10_{5350}		30	52	11	15	15	30	15	30
5360	30	15	12	30	30	30	30	30	10	30	30	40	15	30	15	42	30	30	30	30
5380	16_{5380}		18	15	14	30	22_{5386}		30	18	30	30	28_{5392}		30	30	30	15	12_{5398}	
5400	30	15	30	30	30	30	30_{5406}		31	15	30	30	30_{5412}		9	30	30_{5416}		12_{5418}	

	0	1	2	3	4	5	6	7	8	9	10	11	12	13	14	15	16	17	18	19
5420	11	30	30	15	30	15	13	66	12	60	30_{5430}		11	30	30	30	30_{5436}		15	30
5440	40_{5440}		30_{5442}		30	30	30	15	11_{5448}		11	15	30	30	15	15	15	30	10	52
5460	30	42	41	30	30	38	30	15	11	30	30_{5470}		30	30	12	16	30_{5476}		16_{5478}	
5480	18	30	10_{5482}		30	18	24	15	30	30	30	30	30	31	30	30	38	22	36	30
5500	30_{5500}		40_{5502}		42	30	10_{5506}		14	15	12	24	30	36	30	15	10	15	10_{5518}	
5520	30_{5520}		10	15	10	15	30_{5526}		10	15	30_{5530}		30	15	10	15	31	48	10	28
5540	10	15	10	25	10	15	10	15	12	30	30	15	15	15	11	30	10_{5556}		10	30
5560	10	66	12_{5562}		9	30	16	30	18_{5568}		10	30	22_{5572}		24	24	26	30	28	30
5580	30_{5580}		30	15	15	30	30	36	12	30	30_{5590}		30	15	13	15	30	30	30	30
5600	15	30	30	15	11	15	10	30	30	70	30	30	30	30	10	30	30	40	12	30
5620	30	30	30_{5622}		12	30	11	30	30	30	30	30	30	42	22	15	24	15	30_{5638}	
5640	28_{5640}		30	30	30	30	30_{5646}		30	30	30_{5650}		30_{5652}		30	30	30_{5656}		30_{5658}	
5660	30	30	30	30	30	15	10	30	10_{5668}		30	52	30	15	30	30	30	30	30	15
5680	30	30	30_{5682}		30	30	30	46	30_{5688}		30	30	30_{5692}		30	15	63	30	30	40
5700	30_{5700}		10	15	11	30	30	30	30	18	11_{5710}		30	28	30	15	30_{5716}		30	30
5720	30	15	30	58	30	24	10	15	31	30	10	30	30	15	11	15	30_{5736}		30	30
5740	11_{5740}		30_{5742}		15	15	10	30	30_{5748}		30	70	11	30	11	15	14	15	11	30
5760	30	30	30	15	11	15	11	72	30	15	11	28	11	22	30	15	15	52	30_{5778}	
5780	30	15	11_{5782}		30	15	10	15	10	15	30_{5790}		31	15	10	15	11	15	10	15
5800	11_{5800}		11	15	9	15	11_{5806}		15	39	10	15	11_{5812}		12	15	10	15	10	15
5820	10_{5820}		13	15	63	24	10_{5826}		11	15	10	16	10	18	10	15	10	15	12_{5838}	
5840	15	15	10_{5842}		10	15	10	15	11_{5848}		10_{5850}		10	15	11	15	15_{5856}		10	15
5860	10_{5860}		10	27	11	15	10_{5866}		10_{5868}		10	15	15	15	13	46	11	15	10_{5878}	
5880	10_{5880}		10	15	10	15	12	15	22	21	13	42	13	70	11	15	13_{5896}		13	16
5900	13	15	13_{5902}		15	16	13	18	13	18	13	22	13	72	13	17	13	60	13	15
5920	15	30	30_{5922}		13	15	13_{5926}		13	48	13	30	13	30	13	15	15	30	16_{5938}	
5940	18	15	10	30	22	16	24	18	26	30	28	30	30_{5952}		30	15	12	30	30	58
5960	13	30	30	66	30	15	13	15	30	46	45	30	11	42	30	24	13	42	11	36
5980	30_{5980}		30	30	30	30	16_{5986}		30	52	13	30	30	30	30	30	26	30	28	30
6000	30	31	30	30	30	30	36_{6006}		24	15	40_{6010}		42	30	30	30	46	30	30	30
6020	30	30	30	30	30	30	30	30	30_{6028}		30	45	30	30	30	30	30_{6036}		13	30
6040	13	30	30_{6042}		30	15	30_{6046}		30	30	30	15	30_{6052}		30	30	30	30	30	72
6060	30	30	30	30	30	30	30_{6066}		12	30	30	30	30_{6072}		12	24	12	58	30_{6078}	
6080	46	30	10	30	30	15	30	24	30_{6088}		30_{6090}		30	15	30	15	30	15	12	15
6100	30_{6100}		30	30	30	15	12	30	30	40	30	30	31_{6112}		30	30	11	30	10	30
6120	30_{6120}		30	30	16	48	18	15	15	45	22_{6130}		30_{6132}		30	30	28	30	30	30
6140	30	15	12_{6142}		30	30	30	30	30	30	30_{6150}		30	15	30	30	30	46	12	30
6160	30	60	30_{6162}		30	30	30	15	30	30	30	30	11_{6172}		30	30	31	45	30	36
6180	30	30	12	30	11	30	30	30	30	30	30	40	22	15	24	15	30_{6196}		28_{6198}	
6200	30	30	30_{6202}		30	30	30	30	63	30	30_{6210}		30	30	30	30	30_{6216}		30	30
6220	30_{6220}		30	48	47	30	11	44	30_{6228}		30	15	30	38	30	36	30	15	30	30
6240	30_{6240}		30	30	30	30	30_{6246}		30	30	30	30	30	36	30	31	30_{6256}		30	30
6260	36	15	15_{6262}		44	30	42	30	15_{6268}		46_{6270}		48	30	30	30	30_{6276}		30	16
6280	30	15	30	60	15	15	30_{6286}		30	30	30	26	15	15	30	30	30	30	30_{6298}	
6300	30_{6300}		15	15	31	30	30	30	30	15	15_{6310}		15	58	15	15	15_{6316}		30	70
6320	30	15	15_{6322}		15	39	30	16	15_{6328}		15	30	30	15	15	30	30_{6336}		30	15
6340	15	16	30_{6342}		15	15	15	15	30	18	30	15	15_{6352}		15	18	15	15	15_{6358}	
6360	30_{6360}		30	15	15	15	30_{6366}		31	15	15	22	15_{6372}		15	15	15	15	15_{6378}	
6380	15	15	15	15	15	15	15	15	15_{6388}		15	70	15	15	14	15	15_{6396}		15	78
6400	24	36	15	18	11	15	15	42	15	48	14	15	15	52	15	15	15	16	15	48

	0	1	2	3	4	5	6	7	8	9	10	11	12	13	14	15	16	17	18	19
6420	15_{6420}		15	22	15	24	15_{6426}		15	15	15	58	15	15	13	15	15	40	10	46
6440	15	15	15	16	15	15	15	15	15_{6448}		15_{6450}		11	26	13	16	15	15	15	15
6460	15	15	15	22	63	15	15	28	15_{6468}		12	15	15_{6472}		15	15	15	15	15	15
6480	15_{6480}		15	15	13	15	13	15	15	15	15_{6490}		15	42	15	15	31	72	13	66
6500	15	15	12	15	11	15	15	26	15	22	15	16	15	15	13	15	15	18	12	16
6520	11_{6520}		15	15	15	15	15	60	24_{6528}		13	15	15	46	11	17	13	15	15	15
6540	15	30	30	15	15	15	15_{6546}		11	15	13_{6550}		11_{6552}		15	15	15	78	16	30
6560	31_{6560}		15_{6562}		22	15	24	16	26_{6568}		28_{6570}		30	30	30	24	15_{6576}		30	15
6580	11_{6580}		30	30	30	15	11	15	30	30	30	30	63	30	30	15	11	15	11_{6598}	
6600	30	15	30	15	11	30	10_{6606}		30	15	10	30	30	30	30	16	11	30	11_{6618}	
6620	30	15	30	36	30	52	11	15	11	15	30	30	11	15	11	30	11_{6636}		30	15
6640	15	30	30	15	30	15	11	30	30	60	30	15	11_{6652}		30	15	15	15	10_{6658}	
6660	11_{6660}		30	15	11	15	30	58	30	30	12	15	30_{6672}		11	15	11	30	30_{6678}	
6680	30	15	12	40	12	15	12	15	18_{6688}		30_{6690}		30	15	12	15	12	36	30	15
6700	13_{6700}		12_{6702}		30	16	10	30	30_{6708}		30	15	11	48	30	17	11	15	12_{6718}	
6720	30	30	30	80	11	24	12	23	11	15	11	52	51_{6732}		30	48	15_{6736}		30	22
6740	30	42	11	40	12	15	11	15	16	16	18	42	31	30	22	28	24	28	26	15
6760	28_{6760}		30_{6762}		11	15	10	66	36	17	11	15	40	15	42	24	15	26	46_{6778}	
6780	48_{6780}		15	15	52	15	15	17	15	15	13_{6790}		15_{6792}		15	15	15	15	15	15
6800	15	15	15_{6802}		15	17	15	15	15	23	15	48	15	15	15	15	15	16	15	15
6820	15	18	15_{6822}		15	24	15_{6826}		15_{6828}		15	15	15_{6832}		15	15	15	15	15	15
6840	15_{6840}		15	15	15	15	15	40	63	21	15	15	15	15	15	15	15_{6856}		15_{6858}	
6860	15	15	15_{6862}		15	15	15	15	13_{6868}		15_{6870}		15	16	15	15	15	15	15	15
6880	31	15	15_{6882}		15	15	15	70	15_{6888}		15	15	15	60	15	15	15	15	15_{6898}	
6900	15	66	15	15	15	15	15_{6906}		15	15	15_{6910}		26	30	30	15	15_{6916}		15	30
6920	15	16	15	30	15	30	15	15	15	40	16	30	18	17	15	30	22	24	24	26
6940	26	30	28	52	31	30	30_{6946}		15_{6948}		30	15	15	30	30	30	30	15	15_{6958}	
6960	30_{6960}		30	30	15	30	30_{6966}		15	16	15_{6970}		30	18	30	15	63_{6976}		15	30
6980	30	15	15_{6982}		30	30	30	15	15	30	15_{6990}		30	15	30	30	30_{6996}		15	15
7000	15_{7000}		30	46	15	15	15	30	15	42	30	15	15_{7012}		30	15	30	15	15_{7018}	
7020	30	30	30	15	15	24	30_{7026}		15	15	15	78	15	30	30	15	15	30	30_{7038}	
7040	30	30	15_{7042}		30	30	15	30	15	52	30	30	30	30	16	30	18_{7056}		15	30
7060	22	30	30	30	30	30	28	36	30_{7068}		30	15	16	30	30	30	30	30	30_{7078}	
7080	30	72	30	15	30	30	30	30	15	30	30	15	30	40	30	30	30	46	30	39
7100	30	30	15_{7102}		30	30	30	30	30_{7108}		30	30	30	30	15	30	30	30	30	30
7120	30_{7120}		22	16	30	15	30_{7126}		28_{7128}		30	30	30	30	30	30	31	30	30	58
7140	30	36	30	30	30	30	30	30	30	30	30_{7150}		30	22	13	30	15	30	30_{7158}	
7160	30	16	30	30	30	30	30	15	30	66	30	70	30	30	30	30	30_{7176}		30	30
7180	30	42	30	15	15	30	30_{7186}		30	30	13	15	11_{7192}		30	30	30	30	15	30
7200	30	18	30	30	30	30	30_{7206}		30	80	30_{7210}		30_{7212}		15	15	30	30	30_{7218}	
7220	30	15	10	30	30	30	30	30	30_{7228}		30	30	63	30	30	30	30_{7236}		30	30
7240	16	30	18_{7242}		12	30	22_{7246}		30	30	30	30	28_{7252}		30	30	30	15	13	30
7260	30	52	30	30	31	30	30	42	30	15	30	30	30	30	13	30	30	18	30	30
7280	30	30	30_{7282}		30	15	11	30	12	36	30	30	30	30	30	30	30_{7296}		30	30
7300	15	48	30	66	30	30	30_{7306}		15_{7308}		30	16	30	70	13	15	13	30	15	30
7320	30_{7320}		30	30	30	24	30	16	31	30	30_{7330}		30_{7332}		15	15	30	15	15	40
7340	15	30	15	30	30	15	15	15	30_{7348}		30_{7350}		15	15	30	30	15	15	15	30
7360	30	30	30	36	15	30	15	52	15_{7368}		15	30	30	72	30	58	15	15	15	46
7380	30	60	15	30	15	30	30	82	15	30	30	30	30_{7392}		15	15	30	16	15	48
7400	24	15	30	30	30	15	15	15	15	30	30_{7410}		15	15	30	30	30_{7416}		15	30

	0	1	2	3	4	5	6	7	8	9	10	11	12	13	14	15	16	17	18	19
7420	30	40	30	15	28	30	16	30	18	16	15	30	22 $_{7432}$		30	20	26	30	28	42
7440	30	30	30	18	15	30	30	22	15	30	30 $_{7450}$		30	28	15	15	31 $_{7456}$		30 $_{7458}$	
7460	15	30	30	16	15	15	15	30	30	15	30	30	30	30	15	30	30 $_{7476}$		15	30
7480	30 $_{7480}$		30	30	15	30	15 $_{7486}$		30 $_{7488}$		30	30	30	58	57	15	24	54	30 $_{7498}$	
7500	28	30	30	48	30	46	30 $_{7506}$		30	30	30	30	30	30	30	36	30 $_{7516}$		30	72
7520	30	30	30 $_{7522}$		30	31	15	30	15 $_{7528}$		36	16	30	30	40	30	42 $_{7536}$		30	30
7540	46 $_{7540}$		48	30	30	30	52 $_{7546}$		30 $_{7548}$		30	30	58	30	30	30	22	30	30 $_{7558}$	
7560	30 $_{7560}$		28	30	30	30	30	30	30	30	30	66	30 $_{7572}$		30	30	30 $_{7576}$		30	30
7580	30	30	30 $_{7582}$		30	30	30	26	30 $_{7588}$		30 $_{7590}$		30	15	30	16	30	70	30	30
7600	30	30	30 $_{7602}$		30	30	30 $_{7606}$		30	30	30	30	30	30	30	39	58	30	13	30
7620	30 $_{7620}$		30	30	15	60	15	30	30	30	30	30	30	30	30	15	30	30	30 $_{7638}$	
7640	30	30	30 $_{7642}$		30	15	30	15	31 $_{7648}$		30	30	30	30	30	15	15	15	30	30
7660	30	46	30	78	30	30	15	30	30 $_{7668}$		30	30	30 $_{7672}$		13	30	15	15	15	30
7680	15 $_{7680}$		30	30	30	15	15 $_{7686}$		15	15	30 $_{7690}$		15	48	30	30	30	42	30 $_{7698}$	
7700	30	30	30 $_{7702}$		30	15	30	15	15	15	30	17	31	30	30	15	15 $_{7716}$		15	15
7720	13	15	13 $_{7722}$		30	30	30 $_{7726}$		15	58	30	30	30	15	15	30	15	15	13	70
7740	13 $_{7740}$		15	30	63	30	15	60	59	15	15	56	15 $_{7752}$		13	15	30 $_{7756}$		17 $_{7758}$	
7760	30	15	30	16	16	17	30	15	28	38	22	36	30	15	30	24	31	30	30	31
7780	13	30	30	42	36	15	13	30	40 $_{7788}$		42	30	15 $_{7792}$		46	15	48	30	16	30
7800	56	28	15	30	22	15	58	36	60	30	28	72	30	30	30	15	15 $_{7816}$		30	16
7820	15	30	30 $_{7822}$		30	24	13	15	30 $_{7828}$		30	40	15	30	30	16	15	16	13	30
7840	30 $_{7840}$		30	15	13	30	15	30	30	46	15	30	30 $_{7852}$		30	15	15	80	13	30
7860	30	23	30	30	30	15	15 $_{7866}$		15	15	30	30	15 $_{7872}$		13	30	15 $_{7876}$		30 $_{7878}$	
7880	15	30	30 $_{7882}$		30	15	15	30	30	30	30	15	15	15	30	15	15	52	15	30
7900	15 $_{7900}$		30	15	18	15	30 $_{7906}$		30	30	15	26	30	40	15	16	15	30	30 $_{7918}$	
7920	30 $_{7920}$		15	30	15	24	15 $_{7926}$		15	30	30	30	30 $_{7932}$		15	15	30 $_{7936}$		30	16
7940	15	30	30	46	30	15	30	30	30 $_{7948}$		30 $_{7950}$		30	16	30	18	15	72	30	22
7960	30	30	30 $_{7962}$		30	28	30	30	30	30	15	15	30	30	30	15	30	30	30	78
7980	30	22	15	30	30	30	30	48	30	30	22	60	30 $_{7992}$		30	30	28	30	30	30
8000	63	30	30	52	30	15	30	30	30 $_{8008}$		30 $_{8010}$		30	18	30	30	30 $_{8016}$		30	30
8020	30	15	30	70	30	30	30	22	30	15	30	30	31	30	30	15	30	30	30 $_{8038}$	
8040	30	15	30	30	30	30	30	15	30	30	30	82	15 $_{8052}$		30	15	30	30	15 $_{8058}$	
8060	15	30	30	30	30	15	15	30	30 $_{8068}$		30	15	30	30	30	30	30	40	30	15
8080	30 $_{8080}$		15	58	30	30	30 $_{8086}$		30 $_{8088}$		15	15	30 $_{8092}$		30	30	31	18	30	30
8100	15 $_{8100}$		30	30	30	30	30	66	15	30	15 $_{8110}$		15	25	15	30	30 $_{8116}$		30	22
8120	15	15	15 $_{8122}$		30	15	15	30	63	62	30	46	59	30	30	30	30	78	53	15
8140	51	17	15	16	30	27	30 $_{8146}$		30	28	41	22	39	30	30	26	30	28	33	40
8160	31 $_{8160}$		29	30	30	36	30 $_{8166}$		23	40	21 $_{8170}$		18	15	17	46	22	48	30 $_{8178}$	
8180	26	80	28	48	30	30	30	58	26	60	30 $_{8190}$	$_{8191}$		30	30	30	30	15	15	24
8200	30	58	30	30	15	30	30	28	30 $_{8208}$		30	30	30	42	30	22	15	30	30 $_{8218}$	
8220	30 $_{8220}$		15	30	31	30	30	18	15	30	15 $_{8230}$		30 $_{8232}$		30	30	30 $_{8236}$		15	57
8240	15	15	30 $_{8242}$		15	15	15	30	15	72	30	36	15	30	30	15	62	22	15	30
8260	30	30	30 $_{8262}$		15	15	30	18	15 $_{8268}$		15	30	15 $_{8272}$		30	24	15	15	30	30
8280	30	48	14	15	30	30	15 $_{8286}$		15	30	30 $_{8290}$		30 $_{8292}$		15	30	15 $_{8296}$		15	42
8300	15	30	30	30	30	16	15	15	15	15	30 $_{8310}$		15	30	15	30	30 $_{8316}$		15	30
8320	30	52	30	15	15	17	30	25	15 $_{8328}$		15	15	30	30	30	15	15	15	14	30
8340	30	18	15	80	30	30	30	16	15	30	30	30	30 $_{8352}$		14	30	16	60	18	15
8360	14	57	22 $_{8362}$		30	30	26	30	28 $_{8368}$		30	30	30	30	15	66	30 $_{8376}$		30	30
8380	30	30	30	82	63	30	30 $_{8386}$		30 $_{8388}$		15	30	30	22	30	16	30	30	30	36
8400	30	30	30	30	30	30	30	30	15	30	30	30	30	46	15	30	31	30	30 $_{8418}$	

	0	1	2	3	4	5	6	7	8	9	10	11	12	13	14	15	16	17	18	19
8420	30	30	30	8422	22	24	24	15	30	8428	28	8430	30	30	30	30	30	30	30	30
8440	30	30	30	8442	30	30	30	8446	30	30	30	30	30	78	30	15	15	30	15	30
8460	30	8460	30	30	30	30	30	8466	30	16	30	42	30	36	30	30	30	48	30	60
8480	30	30	30	30	30	15	22	30	30	30	30	30	28	15	29	30	30	30	30	30
8500	29	8500	30	15	30	30	30	46	30	66	65	16	30	8512	30	15	29	15	30	56
8520	30	8520	30	15	29	15	30	8526	30	30	30	44	30	42	29	30	30	8536	30	8538
8540	30	31	29	8542	29	30	36	30	15	82	40	30	42	15	15	20	46	42	48	30
8560	29	30	52	8562	30	30	30	30	58	30	60	30	30	8572	30	24	66	30	30	28
8580	30	8580	30	15	29	22	15	30	30	18	30	70	30	30	30	30	30	8596	30	8598
8600	30	15	30	30	16	30	18	15	31	8608	22	78	30	30	29	30	28	30	30	30
8620	30	36	15	8622	30	16	30	8626	30	8628	30	15	29	88	30	30	30	30	15	52
8640	30	8640	30	16	30	30	30	8646	30	15	15	40	30	30	30	16	29	30	30	30
8660	30	15	29	8662	30	30	30	80	30	8668	30	30	31	15	15	24	30	8676	15	15
8680	15	8680	15	30	30	15	15	30	30	8688	30	15	15	8692	30	30	30	15	15	8698
8700	30	18	15	16	16	30	15	8706	30	15	15	30	30	8712	30	30	15	30	30	8718
8720	15	30	15	30	30	30	30	30	16	30	18	8730	15	30	22	30	30	8736	30	30
8740	28	8740	30	30	30	15	15	8746	30	30	30	30	30	8752	30	30	30	15	30	30
8760	30	8760	15	30	30	15	30	30	63	30	30	48	30	30	30	30	15	66	30	8778
8780	30	30	30	8782	30	30	30	30	15	30	30	58	30	30	30	30	22	30	30	30
8800	31	30	28	8802	30	30	30	8806	30	30	30	30	30	30	30	30	30	30	30	8818
8820	30	8820	30	30	30	30	30	30	30	80	30	8830	30	72	30	15	30	8836	30	8838
8840	30	15	30	36	30	30	30	30	30	8848	30	52	30	30	30	30	30	16	30	30
8860	30	8860	30	8862	31	15	15	8866	30	48	30	30	15	30	30	70	30	30	30	30
8880	30	82	30	15	30	28	30	8886	15	15	30	30	30	8892	30	16	63	15	30	30
8900	30	30	30	30	30	30	15	30	30	58	30	30	30	30	15	30	15	36	15	30
8920	15	30	30	8922	30	15	15	78	15	8928	30	30	15	8932	30	30	30	30	30	30
8940	30	8940	30	30	30	40	30	22	15	30	30	8950	30	30	30	15	15	52	15	30
8960	30	30	30	8962	30	30	30	30	30	8968	30	8970	30	18	30	30	16	46	18	15
8980	30	30	30	30	30	30	26	30	30	88	30	36	31	30	15	30	30	15	30	8998
9000	30	9000	30	15	15	30	30	9006	30	30	15	9010	30	9012	30	16	30	70	69	30
9020	30	66	15	30	30	30	30	60	15	9028	30	30	30	15	16	30	30	30	30	48
9040	30	9040	30	9042	30	15	28	82	30	9048	30	15	15	34	36	30	31	30	40	9058
9060	42	16	30	24	46	30	48	9066	30	18	52	46	30	42	15	15	58	30	60	30
9080	30	63	15	30	66	30	30	30	70	60	30	9090	15	30	15	30	30	30	30	30
9100	16	30	18	9102	15	30	22	30	30	9108	30	30	28	30	30	30	30	15	15	30
9120	30	30	30	30	30	72	30	9126	30	16	30	30	30	9132	15	30	30	9136	30	30
9140	30	30	30	40	30	15	15	30	15	30	30	9150	63	80	30	30	30	9156	30	30
9160	15	9160	30	16	30	30	30	88	15	52	30	30	30	9172	15	24	15	30	15	66
9180	30	9180	30	30	30	30	30	9186	30	30	30	30	30	28	15	15	30	30	30	9198
9200	30	30	30	9202	30	30	15	15	30	9208	30	60	15	15	30	30	30	30	15	30
9220	30	9220	30	22	15	30	15	9226	30	18	30	30	30	30	30	15	30	15	15	9238
9240	30	9240	30	30	15	30	30	16	31	30	30	30	30	18	15	15	30	9256	15	46
9260	15	26	30	58	30	16	15	15	15	15	15	72	71	15	30	68	30	9276	15	30
9280	63	9280	30	9282	15	30	15	36	16	15	18	30	15	9292	30	48	24	15	26	16
9300	28	70	30	31	15	17	30	40	36	30	30	9310	40	66	42	30	30	26	46	9318
9320	48	30	30	9322	52	24	30	15	15	30	58	17	60	30	30	63	68	9336	66	30
9340	30	9340	70	9342	72	30	30	15	15	9348	15	40	30	46	30	15	16	15	15	48
9360	30	15	15	15	30	15	30	16	15	26	15	9370	30	15	15	15	31	9376	15	82
9380	15	15	15	17	15	15	15	15	15	40	15	9390	15	15	15	15	15	9396	15	15
9400	15	15	15	9402	15	15	15	22	15	9408	15	15	15	9412	15	15	15	15	15	9418

	0	1	2	3	4	5	6	7	8	9	10	11	12	13	14	15	16	17	18	19
9420	15_{9420}		15	26	15	15	15	15	15	15	15_{9430}		15_{9432}		15	15	15_{9436}		15_{9438}	
9440	15	15	15	15	15	15	15	15	15	15	15	15	15	29	15	16	15	48	15	15
9460	15_{9460}		15_{9462}		15	15	15_{9466}		15	16	15	15	36_{9472}		15	24	15	18	15_{9478}	
9480	15	18	15	15	15	23	15	52	15	16	15_{9490}		15	15	15	22	15_{9496}		15	26
9500	15	28	15	30	31	18	15	15	15	36	15_{9510}		15	17	15	15	15	30	30	15
9520	15_{9520}		15	88	15	15	15	30	15	30	15	26	15_{9532}		16	30	63	15	15_{9538}	
9540	22	15	24	15	26	30	28_{9546}		30	30	30_{9550}		15	40	30	15	15	30	30	78
9560	30	15	15	72	30	30	30	30	31	30	30	17	15	15	15	30	30	60	30	15
9580	15	30	15	30	30	15	15_{9586}		30	42	30	15	15	52	15	30	30	15	30	30
9600	30_{9600}		15	15	15	15	30	30	15	15	15	30	15_{9612}		30	15	15	58	30_{9618}	
9620	30	15	15_{9622}		30	30	30	15	15_{9628}		30_{9630}		31	15	15	30	15	30	30	15
9640	15	30	30_{9642}		30	30	15	30	30_{9648}		15	30	15	48	30	30	30	30	16	30
9660	18_{9660}		15	30	63	30	30	30	30	30	28	30	30	30	30	15	15_{9676}		30_{9678}	
9680	30	30	30	30	30	30	30	15	30_{9688}		30	30	15	30	30	17	30_{9696}		30	30
9700	30	88	30	30	30	30	15	30	30	30	30	30	30	30	30	30	30	30	15_{9718}	
9720	30_{9720}		30	30	30	30	22	70	30	30	30	36	28_{9732}		30	30	30	30	30_{9738}	
9740	30	30	30_{9742}		30	30	30	30	30_{9748}		30	48	30	30	30	30	30	30	30	30
9760	30	42	30	30	30	16	30_{9766}		30_{9768}		30	16	30	30	30	30	30	30	30	30
9780	30_{9780}		30	30	30	30	30_{9786}		30	30	30_{9790}		30	30	30	16	15	96	30	40
9800	30	80	15_{9802}		30	15	30	30	30	30	30_{9810}		30	15	30	15	30_{9816}		15	15
9820	30	30	30	30	31	16	15	30	30_{9828}		30	30	30_{9832}		30	30	15	30	30_{9838}	
9840	30	30	30	30	16	30	18	42	15	30	22_{9850}		30	58	30	30	28_{9856}		30_{9858}	
9860	30	30	15	30	30	30	30	30	30	70	30_{9870}		30	30	30	78	30	30	17	30
9880	30	40	30_{9882}		30	30	30_{9886}		30	15	15	30	30	30	30	30	30	30	30	30
9900	30_{9900}		30	30	30	30	30_{9906}		30	30	30	30	30	30	30	30	30	46	30	15
9920	16	30	30_{9922}		30	24	30	30	30_{9928}		30_{9930}		30	30	30	30	30	39	15	15
9940	30_{9940}		30	60	30	30	30	30	30_{9948}		17	15	31	36	30	30	15	16	30	46
9960	30	30	15	40	30	30	30_{9966}		15	30	15	58	30_{9972}		30	30	30	30	30	30
9980	30	15	15	66	30	30	30	30	15	30	30	96	30	30	30	30	30	18	15	15

2.73 Table Unknown sets of k MOLS, $k \leq 15$. Under the entry k is given the sides for which k MOLS are unknown, but $k-1$ MOLS are known.

k	Missing Orders
2	2 6
3	3 10
4	4 14 18 22
5	5 15 20 26 30 34 38 39 46 54 60 62
6	12 21 24 28 33 35 36 42 44 48 51 52 55 58 66 68 74 75
7	7 45 50 63 69 70 76 77 78 84 85 86 87 90 92 93 94 95 98 102 106 108 110 111 114 116 118 119 122 123 124 126 130 132 134 138 140 142 146 148 150 156 159 162 164 166 170 172 174 175 178 180 182 183 186 188 190 194 196 198 202 204 206 212 214 218 220 222 226 228 230 234 236 238 242 244 246 250 252 258 260 266 268 270 274 278 282 284 286 290 291 292 294 295 300 302 303 306 308 310 314 318 322 324 326 330 332 334 335 338 340 346 348 354 356 358 362 364 366 370 372 374 378 380 382 386 388 390 394 396 398 406 410 418 420 422 426 428 430 434 436 438 442 444 446 450 452 454 458 460 462 466 468 470

k	Missing Orders													
	476	478	482	484	486	490	492	494	498	500	506	508	510	516
	518	526	530	532	534	542	546	548	550	554	556	558	562	564
	566	570	572	578	580	582	588	594	610	614	618	622	626	630
	634	636	638	642	644	646	654	660	678	690	702	708	734	750
	762	764	774	780										
8	8	40	56	57	65	72	88	91	96	104	105	115	120	129
	133	135	136	141	145	147	152	155	158	161	165	168	184	192
	195	200	203	213	215	216	231	232	235	237	240	245	248	249
	255	264	280	287	296	301	309	312	316	327	328	339	344	345
	363	376	392	395	399	402	405	424	427	440	453	456	472	485
	488	495	504	514	545	552	553	555	561	565	567	568	573	579
	581	584	586	591	596	597	604	612	616	620	627	628	632	633
	635	645	648	650	651	652	655	658	662	664	666	668	670	674
	676	680	682	684	686	692	694	696	698	700	706	710	712	716
	718	728	735	740	741	742	744	746	748	754	756	758	760	766
	770	772	778	782	790	796	802	804	806	808	810	814	818	820
	822	824	826	828	834	836	840	842	844	852	858	866	868	872
	874	876	886	888	894	904	906	914	920	930	932	933	934	936
	938	940	942	946	948	950	952	954	956	966	968	982	994	996
	1000	1006	1012	1016	1018	1020	1022	1026	1032	1034	1070	1074	1076	1082
	1084	1086	1092	1094	1096	1102	1106	1108	1110	1112	1114	1116	1118	1122
	1124	1126	1128	1130	1132	1138	1140	1141	1142	1144	1146	1150	1154	1162
	1182	1186	1188	1190	1192	1194	1196	1198	1202	1206	1208	1210	1212	1214
	1218	1220	1224	1226	1228	1230	1236	1238	1240	1245	1252	1253	1254	1258
	1260	1263	1326	1332	1412	1418	1419	1422	1430	1432	1436	1438	1442	1448
	1450	1452	1460	1462	1464	1478	1492	1576	1580	1592	1614	1620	1622	1624
	1626	1628	1630	1638	1640	1642	1646	1650	1658	1678	1718	1720	1726	1734
	1736	1738	1742	1758	1762	1766	1770	1772	1774	1782	1784	1786	1788	1800
	1802	1806	1814	1926	1930	1935	1946	1948	1954	1956	1958	1962	1964	1966
	1970	1974	2068	2072	2092	2246	2406	2408	2410	2418	2428	2668	2766	2774
9	9	82	99	100	117	153	154	171	189	201	205	207	217	225
	261	262	265	333	336	350	360	393	408	412	469	520	536	549
	576	585	598	639	722	738	747	812	819	830	835	846	878	882
	884	890	898	902	903	908	921	927	958	964	970	1036	1042	1044
	1046	1048	1050	1052	1054	1058	1059	1060	1062	1064	1066	1078	1090	1098
	1100	1156	1180	1204	1222	1234	1242	1244	1246	1248	1250	1251	1255	1256
	1257	1274	1292	1304	1306	1310	1314	1316	1318	1320	1322	1324	1330	1336
	1342	1378	1380	1382	1384	1386	1388	1390	1394	1396	1402	1404	1405	1406
	1410	1466	1496	1540	1545	1554	1557	1558	1564	1570	1572	1578	1582	1586
	1587	1588	1590	1594	1596	1598	1602	1604	1606	1608	1610	1612	1618	1634
	1636	1702	1704	1706	1708	1714	1730	1732	1746	1748	1750	1768	1790	1796
	1798	1810	1812	1864	1868	1874	1880	1883	1884	1885	1886	1890	1894	1896
	1898	1900	1906	1912	1942	1960	1972	2018	2022	2028	2030	2038	2040	2042
	2046	2050	2052	2054	2056	2060	2066	2070	2076	2082	2084	2086	2098	2132
	2180	2194	2198	2204	2210	2216	2222	2252	2258	2270	2278	2284	2340	2342
	2344	2346	2348	2350	2354	2360	2362	2364	2366	2372	2374	2378	2380	2382
	2386	2388	2390	2392	2394	2402	2404	2412	2420	2422	2434	2500	2504	2518
	2594	2607	2618	2628	2630	2636	2638	2642	2648	2650	2660	2674	2676	2678
	2682	2686	2690	2694	2696	2702	2708	2714	2738	2740	2742	2746	2750	2756
	2762	2810	2822	2828	2844	2846	2852	2856	2858	2860	2872	2876	2878	2890

k	Missing Orders
	2904 2906 2908 2916 2920 2990 3014 3076 3110 3114 3116 3122 3126 3128
	3134 3140 3166 3196 3642 3678
10	80 160 177 185 254 279 329 355 357 415 502 640 721 870
	1014 1028 1030 1038 1135 1233 1266 1268 1270 1276 1278 1286 1338 1340
	1348 1354 1366 1372 1398 1413 1414 1416 1420 1426 1428 1434 1444 1446
	1454 1502 1506 1508 1510 1512 1514 1515 1516 1518 1521 1522 1524 1526
	1527 1528 1533 1534 1574 1682 1688 1694 1698 1700 1722 1724 1740 1752
	1754 1756 1764 1778 1780 1794 1804 1816 1818 1820 1822 1826 1828 1832
	1834 1836 1838 1844 1850 1925 1929 1976 1978 1980 1982 1986 1988 1998
	2004 2005 2008 2010 2044 2062 2074 2116 2134 2140 2142 2146 2148 2150
	2152 2158 2162 2164 2166 2168 2174 2178 2182 2184 2186 2188 2190 2196
	2200 2206 2266 2298 2308 2314 2318 2326 2332 2338 2356 2358 2365 2436
	2438 2440 2452 2454 2456 2458 2470 2484 2488 2494 2506 2514 2526 2532
	2546 2550 2556 2562 2568 2580 2588 2612 2662 2664 2666 2670 2680 2684
	2698 2700 2712 2722 2724 2732 2748 2754 2758 2764 2780 2794 2798 2804
	2814 2824 2834 2836 2854 2866 2884 2886 2898 2914 2918 2924 2926 2930
	2936 2942 2946 2948 2950 2952 2954 2956 2958 2962 2964 2966 2968 2970
	2972 2978 2980 2984 2986 2988 2994 2996 2998 3000 3002 3006 3010 3012
	3022 3028 3042 3046 3052 3058 3062 3064 3070 3082 3084 3092 3098 3100
	3102 3106 3108 3112 3118 3124 3130 3138 3146 3170 3208 3238 3244 3256
	3258 3270 3284 3292 3298 3310 3314 3316 3320 3322 3324 3326 3338 3382
	3386 3434 3464 3466 3470 3478 3490 3492 3498 3500 3508 3512 3518 3608
	3614 3618 3622 3628 3638 3640 3646 3652 3656 3658 3668 3670 3676 3688
	3718 3726 3730 3736 3738 3740 3748 3754 3772 3806 3814 3816 3850 3866
	3868 3870 3882 3896 3910 3926 3930 3932 3940 3944 4012 4014 4022 4024
	4028 4030 4036 4054 4070 4100 4106 4118 4126 4152 4156 4214 4216 4222
	4246 4252 4268 4286 4298 4312 4316 4616 5032 5048 5414 5564 5804
11	11 143 144 187 210 219 267 276 297 298 315 319 341 342
	365 414 429 474 522 538 583 649 724 737 749 786 845 892
	910 917 972 1068 1134 1158 1243 1282 1397 1400 1408 1468 1530 1654
	1716 1902 1941 1992 2058 2088 2122 2154 2170 2212 2236 2274 2302 2343
	2359 2367 2370 2373 2379 2396 2442 2444 2446 2460 2566 2658 2692 2706
	2710 2716 2718 2726 2728 2730 2734 2744 2770 2850 2910 2932 2974 2982
	3004 3016 3018 3094 3142 3182 3186 3250 3262 3268 3278 3286 3300 3302
	3304 3306 3308 3332 3352 3356 3362 3368 3372 3374 3394 3436 3444 3471
	3474 3484 3494 3496 3502 3506 3510 3514 3538 3554 3558 3578 3580 3590
	3598 3602 3620 3624 3634 3636 3650 3654 3666 3672 3674 3682 3686 3690
	3692 3694 3698 3704 3714 3716 3720 3722 3732 3734 3742 3746 3750 3756
	3758 3762 3766 3786 3790 3796 3802 3820 3826 3828 3836 3838 3842 3844
	3846 3862 3874 3878 3880 3884 3886 3890 3992 3994 4002 4006 4042 4066
	4072 4116 4130 4138 4146 4148 4178 4180 4198 4200 4210 4218 4228 4234
	4242 4258 4262 4282 4292 4302 4324 4414 4450 4468 4482 4484 4492 4498
	4500 4502 4504 4514 4516 4518 4520 4526 4532 4534 4536 4538 4546 4548
	4550 4552 4554 4566 4568 4582 4584 4586 4594 4596 4600 4602 4606 4774
	4776 4792 4804 4810 4822 4838 4840 4852 4854 4858 4870 4882 4888 4906
	4918 4924 4942 4952 4954 4958 4972 4986 4990 5002 5010 5014 5020 5022
	5026 5028 5030 5034 5036 5038 5042 5046 5050 5052 5058 5062 5064 5068
	5074 5098 5110 5116 5118 5122 5128 5138 5158 5332 5350 5368 5458 5482
	5506 5516 5518 5522 5524 5528 5534 5538 5540 5542 5544 5546 5556 5558
	5560 5570 5606 5614 5666 5668 5702 5726 5730 5746 5786 5788 5794 5798

k	Missing Orders
	5810 5816 5818 5820 5826 5830 5832 5834 5836 5842 5844 5846 5850 5852 5858 5860 5862 5866 5868 5870 5878 5880 5882 5884 5942 6082 6118 6438 6606 6610 6658 6706 6766 7222
12	209 404 447 1652 2036 3493 3505 3524 3542 3548 3560 3563 3565 3566 3572 3594 3596 3626 3630 3644 3660 3662 3724 3914 4082 4124 4136 4154 4442 4466 4530 4562 4564 4578 4580 4772 4836 4842 4850 4856 4860 4872 4874 4878 4922 4932 4936 4962 4982 4984 4996 5000 5012 5044 5060 5108 5140 5144 5354 5420 5432 5448 5450 5468 5554 5604 5626 5704 5710 5734 5740 5752 5754 5758 5764 5766 5770 5772 5782 5796 5800 5802 5806 5812 5828 5848 5854 5864 5876 5894 5972 5978 6116 6172 6184 6226 6404 6452 6504 6520 6534 6548 6552 6580 6586 6596 6598 6604 6616 6618 6626 6628 6632 6634 6636 6646 6652 6660 6664 6674 6676 6712 6716 6724 6728 6730 6742 6746 6764 6770 7192 7286
13	13 221 247 253 259 285 299 325 351 381 413 501 517 524 533 540 559 602 637 723 726 788 832 850 923 949 974 1080 1160 1261 1284 1346 1417 1470 1532 1656 1842 2090 2156 2172 2214 2276 2353 2426 2462 2524 2586 2587 2772 2934 3020 3144 3188 3206 3330 3442 3516 3764 3950 4250 4260 4322 4386 4402 4420 4428 4438 4452 4456 4470 4486 4506 4522 4540 4556 4558 4570 4572 4588 4598 4604 4622 4658 4674 4738 4786 4826 4828 4834 4844 4866 4876 4902 4930 4966 4978 5066 5092 5164 5330 5362 5398 5418 5428 5474 5510 5548 5562 5588 5618 5624 5814 5838 5886 5956 6068 6074 6076 6098 6106 6142 6158 6182 6470 6502 6518 6670 6682 6684 6686 6694 6696 6702 6718 6726 6744 7244 7288
14	112 224 275 377 411 445 896 1010 1490 2034 2398 2863 3355 3453 3495 3588 3603 4238 4254 4364 4366 4394 4396 4404 4408 4424 4426 4444 4454 4458 4460 4462 4472 4474 4476 4478 4488 4490 4494 4508 4510 4524 4542 4574 4590 4610 4612 4614 4618 4628 4766 5130 5146 5148 5242 5426 5594 5822 5874 5890 5892 5896 5898 5900 5902 5906 5908 5910 5912 5914 5916 5918 5924 5926 5928 5930 5932 5934 5960 5966 5976 5990 6038 6040 6434 6454 6484 6486 6498 6514 6530 6536 6550 6700 6790 6868 7154 7190 7258 7274 7314 7316 7618 7674 7720 7722 7738 7740 7754 7780 7786 7826 7838 7844 7858 7874
15	176 208 600 913 935 1509 2165 2171 2424 3328 4664 4670 4676 4684 4694 4796 4814 4862 5384 5508 5756 6394 6410 8282 8338 8354 8360

2.7 Related Problems

Two latin squares on the same set of symbols are *r-orthogonal* if, when superimposed, there are exactly r distinct ordered pairs.

2.74 Theorem [15] For n a positive integer, two r-orthogonal latin squares of side n exist if and only if $r \in \{n, n^2\}$ or $n+2 \le r \le n^2-2$, except when

1. $n = 2$ and $r = 4$;
2. $n = 3$ and $r \in \{5, 6, 7\}$;
3. $n = 4$ and $r \in \{7, 10, 11, 13, 14\}$;
4. $n = 5$ and $r \in \{8, 9, 20, 22, 23\}$;
5. $n = 6$ and $r = 36$;

and possibly when $r = n^2 - 3$ and

$$n \in \{6, 7, 8, 10, 11, 13, 14, 16, 17, 18, 19, 20, 22, 23, 25, 26\}.$$

An *orthogonal latin square graph* of *order n* (OLSG(n)) is a graph with vertex set V and edge set E, where V corresponds to a set of latin squares of side n, and an edge connects two vertices exactly when the corresponding latin squares are orthogonal. A graph is *n-realizable* when it isomorphic to an OLSG(n).

2.75 Theorem (See [16]) Let G be a graph. There exists an integer n_G depending only upon G, such that G is n-realizable for all $n > n_G$.

2.76 Theorem (See [16]) The 6-cycle is n-realizable by the six conjugates of a single latin square of side n for all $n \geq 43$. The complete graph on 6 vertices is n-realizable by the six conjugates of a single latin square of side n for all $n \geq 2207$.

2.77 Theorem [11] Two orthogonal diagonal latin squares of side s exist if and only if $s \neq 2$, 3, or 6. (See §II.1.7.)

2.78 Theorem [17] Three orthogonal diagonal latin squares of side s exist except when $s \in \{2, 3, 4, 5, 6\}$ and possibly when $s \in \{10, 14, 15, 18, 21, 22, 26, 30, 33, 34, 46\}$.

Suppose L and M are idempotent, symmetric latin squares of the same order. Then L and M are *orthogonal-symmetric* latin squares if, for any two elements x and y, there exists at most one ordered pair (i, j) with $i < j$ such that $L(i, j) = x$ and $M(i, j) = y$.

2.79 Remark Note that orthogonal-symmetric latin squares are not orthogonal latin squares, but they are as "orthogonal" as possible, given that they are symmetric.

2.80 Remark The existence of a pair of orthogonal-symmetric latin squares of order n is equivalent to the existence of a Room square of side n (§IV.39) and to the existence of two pairwise orthogonal 1-factorizations of K_{n+1} (§VI.4).

A *magic square* is an $n \times n$ array of integers with the property that the sum (*magic sum*) of the numbers in each row, each column and the main and back diagonal is the same. The square is *nth order* if the integers are the consecutive integers from 1 to n^2.

2.81 Remark The magic sum of an nth order magic square is $\frac{1}{2}n(n^2 + 1)$.

2.82 Example A 4th order magic square and a 6th order magic square.

16	3	2	13
5	10	11	8
9	6	7	12
4	15	14	1

35	1	6	26	19	24
3	32	7	21	23	25
31	9	2	22	27	20
8	28	33	17	10	15
30	5	34	12	14	16
4	36	29	13	18	11

2.83 Construction If there exists a pair of orthogonal latin squares of order n, then there exists an nth order magic square. Let $L = (l_{ij})$ and $R = (r_{ij})$ be a pair of orthogonal latin squares of order n with L on the symbols $\{0, 1 \ldots, n-1\}$ and R on the symbols $\{1, 2, \ldots, n\}$. Construct a new square $M = (m_{ij})$ where $m_{ij} = n \times l_{ij} + r_{ij}$. It is easy to see that the properties of orthogonal latin squares ensure that M is an nth order magic square.

2.84 Remark Magic squares have long been of interest to recreational mathematicians. For a nice discussion of magic squares including construction methods and generalizations to other "magic" problems such as magic domino squares, magic hexagons and magic cubes, see Chapter 7 of [7].

2.8 See Also

§I.1, 6	BIBDs, resolvable and near-resolvable designs.
§IV.11	Difference matrices, quasi-difference matrices and $V(m,t)$ vectors are used to construct MOLS.
§IV.40	Self-orthogonal latin squares (SOLS).
§IV.27	Maximal sets of MOLS.
[14]	A lengthy survey of constructions of MOLS.
[16]	Section 15 gives more information on orthogonal latin square graphs.
[21]	Describes the use of MOLS in compiler testing.
[24]	Proposes existence of projective planes and MOLS as a successor to the Fermat problem.

References

[1] R.J.R. Abel, On the Existence of Balanced Incomplete Block Designs and Transversal Designs, Ph.D. thesis, University of New South Wales, Australia, 1995.

[2] R.J.R. Abel, previously unpublished.

[3] R.J.R. Abel and Y.W. Cheng, Some new MOLS of order $2^n p$ for p a prime power, *Australas. J. Combin.* 10 (1994), 175-186.

[4] R.J.R. Abel and D.T. Todorov, Four MOLS of order 20, 30, 38 and 44, *J. Combin. Theory A* 64 (1993), 144-148.

[5] R.J.R. Abel, H. Zhang and X. Zhang, Three mutually orthogonal idempotent latin squares of orders 22 and 26, *J. Stat. Plann. Infer.*, to appear.

[6] I. Anderson, *Combinatorial Designs: Construction Methods*, Ellis Horwood, Chichester, England, 1990.

[7] W.W. Rouse Ball and H.S.M. Coxeter, *Mathematical Recreations and Essays* (twelfth edition), University of Toronto Press, Toronto, 1974.

[8] T. Beth, Eine Bemerkung zur Abschätzung der Anzahl orthogonaler lateinischer Quadrate mittels siebverfahren, *Abh. Math. Sem. Hamburg* 53 (1983), 284-288.

[9] T. Beth, D. Jungnickel and H. Lenz, *Design Theory*, Cambridge University Press, London, 1986.

[10] A.E. Brouwer, The number of mutually orthogonal latin squares — a table up to order 10000, ZW 123/79 Report, Mathematisch Centrum, Amsterdam, 1979.

[11] J.W. Brown, F. Cherry, L. Most, M. Most, E.T. Parker and W.D. Wallis, The spectrum of orthogonal diagonal Latin squares, in: *Graphs, Matrices and Designs* (R. Rees, ed.), Marcel Dekker, New York, 1993, 43-49.

[12] S. Chowla, P. Erdös and E.G. Straus, On the maximal number of pairwise orthogonal latin squares of a given order, *Canad. J. Math.* 13 (1960), 204-208.

[13] C.J. Colbourn, Four MOLS of order 26, *J. Combin. Math. Combin. Comput.* 17 (1995), 147-148.

[14] C.J. Colbourn and J.H. Dinitz, Making the MOLS table, in: *Computational and Constructive Design Theory*, to appear.

[15] C.J. Colbourn and L. Zhu, The spectrum of r-orthogonal latin squares, in: *Combinatorics Advances*, Kluwer Academic Press, New York, 1995, 49-75.

[16] J.H. Dinitz and D.R. Stinson, Room squares and related designs, in: *Contemporary Design Theory: A Collection of Surveys* (J.H. Dinitz, D.R. Stinson; eds.) John Wiley & Sons, New York, 1992, 137-204.

[17] B. Du, Some constructions of pairwise orthogonal diagonal latin squares, *J. Combin. Math. Combin. Comput.* 9 (1991), 97-106.

[18] D.M. Johnson, A.L. Dulmage and N.S. Mendelsohn, Orthomorphisms of groups of orthogonal latin squares, *Canad. J. Math.* 13 (1961), 356-372.
[19] H.F. MacNeish, Euler squares, *Ann. Math. (NY)* 23 (1922), 221-227.
[20] K. Metsch, Improvement of Bruck's completion theorem, *Des. Codes Crypt.* 1 (1991) 99-116.
[21] R. Mandl, Orthogonal latin squares: An application of experiment design to compiler testing, *Comm. ACM* 28 (1985), 1054-1058.
[22] W.H. Mills, Some mutually orthogonal latin squares, *Congr. Numer.* 19 (1977), 473-487.
[23] A.V. Nazarok, Five pairwise orthogonal latin squares of order 21, *Issled. oper. i ASU*, 1991, 54-56.
[24] G.L. Mullen, A candidate for the "next Fermat problem", *Math. Intelligencer* 17 (1995), 18-22.
[25] P.J. Owens and D.A. Preece, Complete sets of pairwise orthogonal latin squares of order 9, *J. Combin. Math. Combin. Comput.* 18 (1995), 83-96.
[26] C.E. Roberts Jr., Sets of mutually orthogonal latin squares with 'like subsquares', *J. Combin. Theory A* 61 (1992) 50-63.
[27] C.E. Roberts Jr., Sets of mutually orthogonal latin squares with 'like subsquares' II, preprint.
[28] T.G. Room and P.B. Kirkpatrick, *Miniquaternion Geometry*, Cambridge University Press, London, 1971.
[29] P.J. Schellenberg, G.H.J. van Rees and S.A. Vanstone, Four pairwise orthogonal latin squares of order 15, *Ars Combin.* 6 (1978), 141-150.
[30] D.T. Todorov, Three mutually orthogonal latin squares of order 14, *Ars Combin.* 20 (1985), 45-48.
[31] D.T. Todorov, Four mutually orthogonal latin squares of order 20, *Ars Combin.* 27 (1989), 63-65.
[32] S.P. Wang, On Self-Orthogonal Latin Squares and Partial Transversals of Latin Squares, Ph.D. thesis, Ohio State University, 1978.
[33] M. Wojtas, Five mutually orthogonal latin squares of order 35, *J. Combin. Des.*, to appear.

3 Incomplete MOLS

R. Julian R. Abel
Charles J. Colbourn
Jeffrey H. Dinitz

3.1 One Hole

Two incomplete latin squares (ILS$(n; b_1, b_2, \ldots, b_s)$) are *orthogonal* if upon superimposition all ordered pairs in $(B \times B) \setminus \cup_{i=1}^{k}(B_i \times B_i)$ result. Two such squares are denoted IMOLS$(n; b_1, b_2, \ldots, b_s)$ (or IPOLS$(n; b_1, b_2, \ldots, b_s)$. Similarly, r-IMOLS$(n; b_1, b_2, \ldots, b_s)$ denotes a set of r ILS$(n; b_1, b_2, \ldots, b_s)$ which are pairwise orthogonal.

$N(n; b_1, b_2, \ldots, b_s)$ denotes the maximum r for which there exist a set of r-IMOLS$(n; b_1, b_2, \ldots, b_s)$.

3.1 Theorem If there exists a set of r orthogonal ILS$(n; k)$, then $n \geq (r+1)k$.

3.2 Theorem (See [7]) For $k \geq 1$, $N(n; k) \geq 2$ if and only if $n \geq 3k$, and $(n, k) \neq (6, 1)$.

3.3 Remark There are several interesting small examples of MOLS with one hole which satisfy the bound in Theorem 3.1 with equality. Example 3.4 is due to Euler [6], Example 3.5 is due to Stinson [9], and Example 3.6 is due to Brouwer [3].

3.4 Example $N(6; 2) = 2$. (The missing subsquares are on the symbols 5 and 6.)

$$\begin{array}{|cccccc|}\hline 5&6&3&4&1&2\\ 2&1&6&5&3&4\\ 6&5&1&2&4&3\\ 4&3&5&6&2&1\\ 1&4&2&3&&\\ 3&2&4&1&&\\ \hline\end{array} \qquad \begin{array}{|cccccc|}\hline 1&2&5&6&3&4\\ 6&5&1&2&4&3\\ 4&3&6&5&1&2\\ 5&6&4&3&2&1\\ 2&4&3&1&&\\ 3&1&2&4&&\\ \hline\end{array}$$

3.5 Example $N(8; 2) = 3$. Consider the following matrix

$$\begin{pmatrix} \infty_0 & \infty_0 & \infty_1 & \infty_1 \\ (0,1) & (2,0) & (1,0) & (2,1) \\ (2,0) & (2,1) & (2,1) & (0,0) \\ (2,0) & (1,1) & (2,0) & (2,1) \\ (1,0) & (1,1) & (0,0) & (0,1) \end{pmatrix}.$$

Replace each column by the five (columnwise) cyclic shifts of the column, and develop each of the resulting columns modulo $(3, -)$. The result is an incomplete orthogonal array with a hole on $\{\infty_0, \infty_1\}$.

3.6 Example $N(10, 2) = 4$. Consider the matrix on the right. Each column $(a, b, c, d, e, f)^T$ is replaced by $(a, b, c, d, e, f)^T$, $(c, a, b, f, d, e)^T$, and $(b, c, a, e, f, d)^T$. Develop each column modulo (2,4) to obtain an incomplete orthogonal array with a hole on $\{\infty_0, \infty_1\}$.

$$\begin{pmatrix} (0,0) & (0,0) & (0,0) & (0,0) \\ (0,2) & (0,1) & (0,0) & (0,3) \\ (1,0) & \infty_0 & (1,1) & \infty_1 \\ (1,1) & (0,2) & (0,0) & (1,3) \\ (1,2) & (0,2) & (0,3) & (1,1) \\ \infty_0 & (1,1) & \infty_1 & (0,3) \end{pmatrix}.$$

3.2 Equivalent Objects

An *incomplete transversal design* of *order* or *groupsize* n, *blocksize* k, *index* λ, and *holesizes* $b_1, \ldots, b_s$, denoted ITD$_\lambda(k, n; b_1, \ldots, b_s)$, is a quadruple $(V, \mathcal{G}, \mathcal{H}, \mathcal{B})$, where

1. V is a set of kn elements;
2. $\mathcal{G}$ is a partition of V into k classes (*groups*), each of size n;
3. $\mathcal{H}$ is a set of disjoint subsets $H_1, \ldots, H_s$ of V, with the property that, for each $1 \leq i \leq s$ and each $G \in \mathcal{G}$, $|G \cap H_i| = b_i$;
4. $\mathcal{B}$ is a collection of k-subsets of V (*blocks*);
5. every unordered pair of elements from V is
   - contained in a hole, and contained in no blocks; or
   - contained in a group, and contained in no blocks; or
   - contained in neither a hole nor a group, and contained in λ blocks.

Again, when $\lambda = 1$, it can be omitted from the notation. An ITD$(k, n; b_1, \ldots, b_s)$ is also denoted as a TD$(k, n) - \sum_{i=1}^{s}$ TD(k, b_i).

3.7 Theorem r-IMOLS$(n; b_1, \ldots, b_s)$ is equivalent to

1. an incomplete transversal design ITD$(r+2, n; b_1, \ldots, b_s)$;
2. an incomplete orthogonal array IOA$(r+2, n; b_1, \ldots, b_s)$.

3.8 Remark r-IMOLS$(n; 1)$ and r-IMOLS$(n; 0)$ are equivalent to r MOLS of side n.

3.3 Existence

3.9 Remark Table 3.10 of lower bounds for $N(n; k)$ was produced using the same program [4] as Table 2.72, the table of lower bounds for $N(n)$.

3.10 Table Lower bounds for $N(n; k)$ for $1 \le n \le 1000$ and $0 \le k \le 50$. (* denotes $n-1$ when $n > 100$.)

$h \rightarrow$	0									1										2										3										4										5
$n \downarrow$	1	2	3	4	5	6	7	8	9	0	1	2	3	4	5	6	7	8	9	0	1	2	3	4	5	6	7	8	9	0	1	2	3	4	5	6	7	8	9	0	1	2	3	4	5	6	7	8	9	0
7	6	2																																																
8	7	3																																																
9	8	3	2																																														
10	2	4	2																																														
11	10	3	2																																														
12	5	3	3	2																																													
13	12	3	3	2																																													
14	3	3	3	2																																													
15	4	4	4	2	2																																											
16	15	3	4	3	2																																											
17	16	3	3	3	2																																											
18	3	3	3	3	2	2																																										
19	18	4	3	3	2	2																																										
20	4	3	3	3	3	2																																										
21	5	3	3	4	3	2	2																																									
22	3	3	5	3	3	2	2																																									
23	22	3	3	4	3	2	2																																									
24	5	3	3	3	3	3	2	2																																								
25	24	3	4	3	4	3	2	2																																								
26	4	3	3	4	4	3	2	2																																								
27	26	3	3	4	4	3	2	2	2																																							
28	5	4	4	3	3	3	3	2	2																																							
29	28	3	3	5	3	3	3	2	2																																							
30	4	3	3	4	4	3	3	2	2	2																																						
31	30	3	3	3	3	4	3	2	2	2																																						
32	31	3	3	3	3	3	3	3	2	2																																						
33	5	4	3	4	3	3	3	3	2	2	2																																					
34	4	3	4	3	3	3	3	3	2	2	2																																					
35	5	4	3	4	4	3	4	3	2	2	2																																					
36	5	3	4	3	6	3	4	3	3	2	2	2																																				
37	36	3	3	4	3	5	3	3	3	2	2	2																																				
38	4	3	3	3	4	3	4	3	3	2	2	2																																				
39	4	3	3	5	3	4	3	3	3	2	2	2	2																																			
40	7	4	4	3	4	3	4	4	3	3	2	2	2																																			
41	40	3	3	7	3	3	3	4	3	3	2	2	2																																			
42	5	3	4	4	3	3	5	3	3	3	2	2	2	2																																		

$h \rightarrow$ $n \downarrow$	0 1 2 3 4 5 6 7 8 9	1 0 1 2 3 4 5 6 7 8 9	2 0 1 2 3 4 5 6 7 8 9	3 0 1 2 3 4 5 6 7 8 9	4 0 1 2 3 4 5 6 7 8 9	5 0
43	$_{42}$ 5 3 3 4 6 3 3 3	3 2 2 2 2				
44	5 3 3 3 4 3 4 3 3	3 3 2 2 2				
45	6 3 4 5 4 4 4 3 4	3 3 2 2 2 2				
46	4 3 4 3 3 3 3 4 4	3 3 2 2 2 2				
47	$_{46}$ 3 3 4 3 5 3 4 3	3 3 2 2 2 2				
48	5 3 3 3 3 4 3 4 4	3 3 3 2 2 2 2				
49	$_{48}$ 3 3 3 3 7 6 5 4	3 3 3 2 2 2 2				
50	6 4 3 3 3 3 6 5 3	4 3 3 2 2 2 2				
51	5 5 3 3 3 3 4 4 4	4 3 3 2 2 2 2 2				
52	5 4 5 3 3 2 3 3 4	3 3 3 3 2 2 2 2				
53	$_{52}$ 4 4 5 3 3 3 3 3	3 3 3 3 2 2 2 2				
54	4 3 3 4 5 3 4 4 3	3 3 3 3 2 2 2 2 2				
55	5 4 3 3 4 5 4 4 4	3 4 3 3 2 2 2 2 2				
56	7 4 4 3 3 4 6 6 3	3 3 3 3 3 2 2 2 2				
57	7 4 4 4 3 3 6 6 5	3 3 3 3 3 2 2 2 2 2				
58	5 6 4 4 4 3 5 5 5	4 3 3 3 3 2 2 2 2 2				
59	$_{58}$ 5 6 4 4 4 4 4 4	4 4 3 3 3 2 2 2 2 2				
60	4 4 5 6 4 4 4 4 3	3 3 4 3 3 3 2 2 2 2	2			
61	$_{60}$ 4 4 5 6 4 4 4 4	4 3 3 3 3 3 2 2 2 2	2			
62	4 4 3 4 5 6 4 4 4	3 3 3 3 3 3 2 2 2 2	2			
63	6 4 5 3 4 5 6 6 6	4 3 3 3 3 3 2 2 2 2	2 2			
64	$_{63}$ 5 5 5 4 4 6 7 6	4 3 3 3 3 3 3 2 2 2	2 2			
65	7 6 5 5 5 4 6 7 6	4 3 3 4 3 3 3 2 2 2	2 2			
66	5 6 6 5 5 5 5 5 5	5 3 3 4 3 3 3 2 2 2	2 2 2			
67	$_{66}$ 5 6 6 5 5 5 5 5	5 4 3 3 3 3 3 2 2 2	2 2 2			
68	5 5 5 6 6 5 5 5 5	4 4 4 3 3 3 3 3 2 2	2 2 2			
69	6 4 5 5 6 6 6 6 6	4 3 4 4 3 3 3 3 2 2	2 2 2 2			
70	6 4 3 5 5 6 6 6 6	4 3 3 3 4 3 3 3 2 2	2 2 2 2			
71	$_{70}$ 5 5 4 5 5 6 6 6	6 4 3 3 4 3 3 3 2 2	2 2 2 2			
72	7 6 5 5 4 5 6 7 7	5 3 4 3 3 3 3 3 3 2	2 2 2 2 2			
73	$_{72}$ 6 6 5 5 5 6 7 7	5 4 5 4 3 3 3 3 3 2	2 2 2 2 2			
74	5 7 6 6 5 5 5 5 5	6 4 3 3 3 3 3 3 3 2	2 2 2 2 2			
75	5 4 7 6 6 5 5 5 5	4 5 3 3 3 4 3 3 3 2	2 2 2 2 2 2			
76	6 4 4 7 6 6 5 5 5	4 3 4 3 3 4 3 3 3 3	2 2 2 2 2 2			
77	6 4 3 3 7 6 6 6 6	4 6 4 4 3 3 3 3 3 3	2 2 2 2 2 2			
78	6 3 3 3 3 7 6 6 6	3 6 5 3 4 3 3 3 3 3	2 2 2 2 2 2 2			
79	$_{78}$ 6 4 3 3 3 7 6 6	4 6 4 4 3 4 3 3 3 3	2 2 2 2 2 2 2			
80	9 6 6 3 4 3 6 7 7	6 6 3 3 3 4 4 3 3 3	3 2 2 2 2 2 2			
81	$_{80}$ 6 6 6 3 3 6 7 8	7 6 3 3 3 3 3 3 3 3	3 2 2 2 2 2 2 2			
82	8 6 6 6 6 3 5 5 8	5 6 5 3 3 3 3 3 3 3	3 2 2 2 2 2 2 2			
83	$_{82}$ 5 6 6 6 6 5 5 5	5 5 4 4 3 3 3 3 3 3	3 2 2 2 2 2 2 2			
84	6 5 5 6 6 6 6 6 5	5 6 5 3 3 3 3 3 3 3	3 3 2 2 2 2 2 2 2			
85	6 5 5 5 6 6 6 6 6	5 6 5 5 5 3 3 4 3 3	3 3 2 2 2 2 2 2 2			
86	6 6 5 5 5 6 6 6 6	6 6 5 3 4 3 4 3 3 3	3 3 2 2 2 2 2 2 2			
87	6 6 6 6 6 6 6 6 6	6 6 4 4 3 4 3 3 3 3	3 3 2 2 2 2 2 2 2 2			
88	7 6 6 6 6 6 6 7 6	6 7 5 4 3 4 4 3 3 3	3 3 3 2 2 2 2 2 2 2			
89	$_{88}$ 6 6 6 6 6 6 7 7	6 7 5 4 3 4 3 4 3 3	3 3 3 2 2 2 2 2 2 2			
90	6 7 5 6 6 6 6 6 6	7 6 5 5 3 3 3 3 4 3	3 3 3 2 2 2 2 2 2 2	2		
91	7 5 7 5 6 6 6 7 7	6 7 5 6 4 4 3 3 4 3	3 3 3 2 2 2 2 2 2 2	2		
92	6 5 5 7 5 6 6 6 6	6 6 6 6 4 4 4 3 3 3	3 3 3 3 2 2 2 2 2 2	2		
93	6 6 5 5 7 6 6 6 6	6 6 5 6 3 4 3 3 3 3	3 3 3 3 2 2 2 2 2 2	2 2		

$h \to$	0	1	2	3	4	5
$n \downarrow$	1 2 3 4 5 6 7 8 9	0 1 2 3 4 5 6 7 8 9	0 1 2 3 4 5 6 7 8 9	0 1 2 3 4 5 6 7 8 9	0 1 2 3 4 5 6 7 8 9	0
94	6 6 6 5 5 7 6 6 6	6 6 5 6 4 4 4 3 3 3	3 3 3 3 2 2 2 2 2 2	2 2		
95	6 6 6 6 5 5 7 6 6	6 6 5 6 4 4 3 4 3 4	3 3 3 3 2 2 2 2 2 2	2 2		
96	7 6 6 6 6 5 6 7 7	6 7 6 6 3 4 4 3 4 3	3 3 3 3 3 2 2 2 2 2	2 2 2		
97	96 6 6 6 6 6 6 7 7	6 7 6 6 4 4 5 4 3 4	3 3 3 3 3 2 2 2 2 2	2 2 2		
98	6 6 6 6 6 6 6 6 5	7 5 5 6 5 4 4 3 4 3	3 3 3 3 3 2 2 2 2 2	2 2 2		
99	8 5 6 6 6 6 6 6 8	4 8 5 6 4 4 3 3 3 4	3 3 3 3 3 2 2 2 2 2	2 2 2 2		
100	8 6 5 6 6 6 6 6 6	6 8 6 6 3 4 4 3 3 3	4 3 3 3 3 3 2 2 2 2	2 2 2 2		
101	* 6 6 5 6 6 6 6 6	9 6 5 6 4 4 4 4 3 3	3 3 3 3 3 3 2 2 2 2	2 2 2 2		
102	6 6 6 6 5 6 6 6 6	6 6 6 6 5 3 3 3 4 3	3 3 3 3 3 3 2 2 2 2	2 2 2 2 2		
103	* 6 7 6 6 6 6 6 6	6 6 6 6 4 4 3 3 4 4	3 3 3 3 3 3 2 2 2 2	2 2 2 2 2		
104	7 6 6 6 6 6 6 7 6	6 6 6 7 5 4 4 3 3 3	3 3 3 3 3 3 3 2 2 2	2 2 2 2 2		
105	7 6 6 6 6 6 6 7 7	6 6 6 7 6 4 3 4 3 3	4 4 3 3 3 3 3 2 2 2	2 2 2 2 2 2		
106	6 7 5 6 6 6 6 6 6	7 6 6 6 5 6 4 3 4 3	3 4 3 3 3 3 3 2 2 2	2 2 2 2 2 2		
107	* 5 7 5 6 6 6 7 7	6 7 6 7 4 4 4 4 4 4	3 3 3 3 3 3 3 2 2 2	2 2 2 2 2 2		
108	6 5 5 7 5 6 6 6 6	6 6 7 6 5 4 3 4 3 3	4 3 3 3 3 3 3 3 2 2	2 2 2 2 2 2 2		
109	* 5 5 5 7 6 6 7 7	6 6 6 7 5 4 4 3 4 4	3 4 3 3 3 3 3 3 2 2	2 2 2 2 2 2 2		
110	6 5 5 5 5 7 6 6 6	6 6 6 6 6 4 4 3 3 4	3 4 4 3 3 3 3 3 2 2	2 2 2 2 2 2 2		
111	6 4 5 5 5 6 7 6 6	5 6 6 6 4 5 4 4 3 3	4 4 4 3 3 3 3 3 2 2	2 2 2 2 2 2 2 2		
112	13 4 4 5 5 5 6 7 7	6 5 6 7 3 4 6 3 3 3	3 4 3 3 3 3 3 3 3 2	2 2 2 2 2 2 2 2		
113	* 4 4 4 5 5 6 7 7	5 6 5 7 4 4 6 5 3 3	3 3 4 3 3 3 3 3 3 2	2 2 2 2 2 2 2 2		
114	6 6 4 4 4 5 6 6 6	7 5 6 5 4 4 6 3 4 3	3 3 3 3 3 3 3 3 3 2	2 2 2 2 2 2 2 2 2		
115	7 6 6 4 4 4 6 7 7	6 7 5 7 4 4 6 4 4 4	4 3 3 4 3 3 3 3 3 2	2 2 2 2 2 2 2 2 2		
116	6 6 6 6 4 4 6 6 6	6 6 7 5 4 4 6 3 3 3	4 3 3 3 3 3 3 3 3 3	2 2 2 2 2 2 2 2 2		
117	8 6 6 6 6 4 6 6 8	6 6 6 8 4 4 6 4 3 4	3 4 3 3 3 3 3 3 3 3	2 2 2 2 2 2 2 2 2 2		
118	6 6 6 6 6 6 6 6 6	6 6 6 6 4 4 6 3 3 3	3 4 4 3 3 3 3 3 3 3	2 2 2 2 2 2 2 2 2 2		
119	6 6 6 6 6 6 6 6 6	6 6 6 6 6 4 6 6 4 4	3 3 3 4 3 3 3 3 3 3	2 2 2 2 2 2 2 2 2 2		
120	7 6 6 6 6 6 6 7 6	6 6 6 6 6 6 6 6 3 3	4 3 4 3 4 3 3 3 3 3	3 2 2 2 2 2 2 2 2 2	2	
121	* 6 6 6 6 6 6 6 6	6 10 9 6 6 6 6 6 4 4	5 4 3 3 4 3 3 3 3 3	3 2 2 2 2 2 2 2 2 2	2	
122	6 6 6 6 6 6 6 6 6	6 6 5 5 5 5 6 6 4 4	3 4 4 3 3 3 3 3 3 3	3 2 2 2 2 2 2 2 2 2	2	
123	6 6 6 6 6 6 6 6 6	6 6 6 6 5 4 6 6 4 4	3 4 3 3 3 3 3 3 3 3	3 2 2 2 2 2 2 2 2 2	2 2	
124	6 6 6 6 6 6 6 6 6	6 6 6 6 6 5 6 6 3 3	4 4 4 3 3 3 3 3 3 3	3 3 2 2 2 2 2 2 2 2	2 2	
125	* 6 6 6 7 6 6 6 6	6 6 6 6 6 6 6 6 3 3	4 4 4 4 4 4 3 3 3 3	3 3 2 2 2 2 2 2 2 2	2 2	
126	6 6 6 6 6 6 6 6 6	6 6 6 6 6 6 6 6 4 3	3 3 4 3 3 4 3 3 3 3	3 3 2 2 2 2 2 2 2 2	2 2 2	
127	* 6 6 6 6 6 6 6 6	6 6 6 6 6 6 6 6 6 4	3 3 3 4 4 4 3 3 3 3	3 3 2 2 2 2 2 2 2 2	2 2 2	
128	* 6 6 6 6 6 6 7 6	6 6 6 6 6 6 7 6 3 3	4 3 3 3 4 3 3 3 3 3	3 3 3 2 2 2 2 2 2 2	2 2 2	
129	7 6 6 6 6 6 6 7 7	6 6 6 6 6 6 7 6 4 4	3 4 3 3 3 4 3 3 3 3	3 3 3 2 2 2 2 2 2 2	2 2 2 2	
130	6 7 6 6 6 6 6 6 6	7 6 6 6 6 6 6 6 5 4	4 4 4 3 3 4 4 3 3 3	3 3 3 2 2 2 2 2 2 2	2 2 2 2	
131	* 6 7 6 6 6 6 7 7	6 7 6 6 6 6 7 6 4 4	3 3 3 4 3 3 3 3 3 3	3 3 3 2 2 2 2 2 2 2	2 2 2 2	
132	6 6 6 7 6 6 6 6 6	6 6 7 6 6 6 6 6 4 3	3 4 4 3 4 3 3 3 3 3	3 3 3 3 2 2 2 2 2 2	2 2 2 2 2	
133	7 6 6 6 7 6 6 7 7	6 6 6 7 6 6 7 6 4 6	3 3 4 3 4 4 3 3 3 3	3 3 3 3 2 2 2 2 2 2	2 2 2 2 2	
134	6 6 6 6 6 7 6 6 6	6 6 6 6 7 6 6 6 4 6	4 4 3 3 3 3 3 3 3 3	3 3 3 3 2 2 2 2 2 2	2 2 2 2 2	
135	7 6 6 6 6 6 7 7 7	6 7 6 6 6 7 6 6 4 6	4 4 3 3 4 4 3 4 3 3	3 3 3 3 2 2 2 2 2 2	2 2 2 2 2 2	
136	7 6 6 6 6 6 6 7 7	6 6 6 6 6 6 7 7 5 6	3 4 5 3 4 4 4 4 3 3	3 3 3 3 3 2 2 2 2 2	2 2 2 2 2 2	
137	* 6 6 6 6 6 6 7 7	6 6 6 6 6 6 7 7 5 6	4 4 3 4 4 4 3 3 3 3	3 3 3 3 3 2 2 2 2 2	2 2 2 2 2 2	
138	6 7 6 6 6 6 6 6 6	7 6 6 6 6 6 6 6 5 6	4 4 4 3 4 3 3 3 3 3	3 3 3 3 3 2 2 2 2 2	2 2 2 2 2 2 2	
139	* 6 7 6 6 6 6 7 7	6 7 6 6 6 6 7 7 5 6	4 4 4 3 4 4 3 3 3 3	3 3 3 3 3 2 2 2 2 2	2 2 2 2 2 2 2	
140	6 6 6 7 6 6 6 6 6	6 6 7 6 6 6 6 6 5 6	4 4 4 4 4 3 4 4 4 3	3 3 3 3 3 3 2 2 2 2	2 2 2 2 2 2 2	
141	7 6 6 6 7 6 6 7 7	6 6 6 7 9 6 7 7 5 6	4 4 4 3 4 3 3 4 4 3	3 3 3 3 3 3 2 2 2 2	2 2 2 2 2 2 2 2	
142	6 6 6 6 6 7 6 6 6	6 6 6 6 7 6 6 6 5 6	4 4 4 4 3 4 3 4 3 3	3 3 3 3 3 3 2 2 2 2	2 2 2 2 2 2 2 2	
143	10 6 6 6 6 6 7 6 6	6 10 6 10 6 7 6 6 5 6	4 4 3 4 4 3 4 4 3 3	3 3 3 3 3 3 2 2 2 2	2 2 2 2 2 2 2 2	
144	10 6 6 6 6 6 6 7 8	6 6 6 10 6 6 8 7 6 6	4 4 4 3 4 4 3 4 3 3	3 3 3 3 3 3 3 2 2 2	2 2 2 2 2 2 2 2 2	

$h \rightarrow$ / $n \downarrow$	0: 1 2 3 4 5 6 7 8 9	1: 0 1 2 3 4 5 6 7 8 9	2: 0 1 2 3 4 5 6 7 8 9	3: 0 1 2 3 4 5 6 7 8 9	4: 0 1 2 3 4 5 6 7 8 9	5: 0
145	7 6 6 6 6 6 6 7 7	6 6 11 6 6 6 7 7 7 6	4 4 4 3 5 4 4 3 3 4	3 3 3 3 3 3 3 2 2 2	2 2 2 2 2 2 2 2 2	
146	6 7 6 6 6 6 6 6 6	7 6 6 6 6 6 6 6 6 6	5 4 4 3 3 4 4 4 3 3	3 3 3 3 3 3 3 2 2 2	2 2 2 2 2 2 2 2 2	
147	7 6 9 6 6 6 6 7 7	6 7 6 6 6 6 6 7 4 6	4 5 3 3 3 4 3 4 4 3	3 3 3 3 3 3 3 2 2 2	2 2 2 2 2 2 2 2 2 2	
148	6 6 6 7 6 6 6 6 6	6 6 7 6 6 6 6 6 4 6	4 6 4 3 3 4 3 4 4 4	3 3 3 3 3 3 3 3 2 2	2 2 2 2 2 2 2 2 2 2	
149	* 6 6 6 7 6 6 7 7	6 6 6 7 6 6 6 7 4 6	4 4 4 3 4 4 3 3 3 4	3 3 3 3 3 3 3 3 2 2	2 2 2 2 2 2 2 2 2 2	
150	6 6 6 6 6 7 6 6 6	6 6 6 6 7 6 6 6 6 6	4 4 4 3 3 4 4 3 3 3	4 3 3 3 3 3 3 3 2 2	2 2 2 2 2 2 2 2 2 2	2
151	* 6 6 6 6 6 7 6 6	6 6 6 6 6 7 6 6 6 6	4 4 3 4 4 5 3 3 3 3	3 3 3 3 3 3 3 3 2 2	2 2 2 2 2 2 2 2 2 2	2
152	7 6 6 6 6 6 6 7 7	6 7 6 6 6 6 7 7 6 7	5 4 4 3 4 4 4 4 4 3	3 3 3 3 3 3 3 3 3 2	2 2 2 2 2 2 2 2 2 2	2
153	8 6 6 6 6 6 6 7 8	6 7 6 6 6 6 6 8 6 7	5 4 3 4 4 4 3 3 3 3	3 3 3 3 3 3 3 3 3 2	2 2 2 2 2 2 2 2 2 2	2
154	8 7 5 6 6 6 6 6 6	7 6 6 6 6 6 6 8 6 6	5 5 4 4 4 4 4 4 4 3	3 3 3 3 3 3 3 3 3 2	2 2 2 2 2 2 2 2 2 2	2
155	7 5 7 5 6 6 6 7 7	6 7 6 6 6 6 6 6 6 7	4 4 5 4 3 4 3 4 4 4	4 4 3 3 3 3 3 3 3 2	2 2 2 2 2 2 2 2 2 2	2
156	6 5 5 7 5 6 6 6 6	6 6 7 6 6 6 6 6 6 6	5 4 4 4 4 4 4 4 4 3	3 4 3 3 3 3 3 3 3 3	2 2 2 2 2 2 2 2 2 2	2
157	* 5 5 5 7 6 6 7 7	6 6 6 7 6 6 6 6 6 7	5 4 4 3 4 4 5 4 3 4	3 4 3 3 3 3 3 3 3 3	2 2 2 2 2 2 2 2 2 2	2
158	7 5 5 5 5 7 6 7 7	6 7 6 6 7 6 7 6 6 6	5 4 4 3 3 5 4 3 4 3	4 3 3 3 3 3 3 3 3 3	2 2 2 2 2 2 2 2 2 2	2
159	6 4 5 5 5 6 7 6 6	6 6 6 6 6 7 6 6 6 6	4 4 4 4 4 4 4 4 3 4	3 4 3 3 3 3 3 3 3 3	2 2 2 2 2 2 2 2 2 2	2
160	9 4 4 5 5 6 6 7 7	6 6 6 6 6 6 7 6 6 7	4 4 4 4 4 4 4 4 4 3	4 3 4 3 3 3 3 3 3 3	3 2 2 2 2 2 2 2 2 2	2
161	7 4 5 5 7 6 6 7 7	5 6 6 6 6 6 6 7 6 7	5 4 5 6 4 4 4 3 4 4	3 4 3 3 3 3 3 3 3 3	3 2 2 2 2 2 2 2 2 2	2
162	6 4 4 4 5 6 6 6 6	7 5 6 6 6 6 6 6 7 6	5 4 4 6 5 3 4 4 4 4	4 3 3 3 3 3 3 3 3 3	3 2 2 2 2 2 2 2 2 2	2
163	* 6 4 4 5 6 6 7 7	5 7 5 6 6 6 6 6 6 7	4 4 4 6 4 4 3 4 4 3	4 4 3 3 3 3 3 3 3 3	3 2 2 2 2 2 2 2 2 2	2
164	6 6 6 4 4 6 6 6 6	6 5 7 5 6 6 6 6 6 6	6 4 5 6 4 3 3 4 4 3	3 4 3 3 3 3 3 3 3 3	3 3 2 2 2 2 2 2 2 2	2
165	7 6 6 6 4 6 7 7 7	6 6 5 7 4 6 6 6 6 7	4 5 4 6 4 4 4 4 3 4	4 4 3 4 3 3 3 3 3 3	3 3 2 2 2 2 2 2 2 2	2
166	6 6 6 6 6 4 6 6 6	6 6 6 5 7 4 6 6 6 6	4 4 4 6 4 4 4 4 3 3	4 4 3 4 3 3 3 3 3 3	3 3 2 2 2 2 2 2 2 2	2
167	* 6 6 6 6 6 6 6 6	6 6 6 6 4 7 4 6 6 6	4 4 4 6 4 4 3 4 3 3	4 4 3 3 3 3 3 3 3 3	3 3 2 2 2 2 2 2 2 2	2
168	7 6 6 6 6 6 6 7 7	6 7 6 6 6 4 7 4 6 7	4 5 4 6 5 4 4 4 4 3	3 3 4 3 3 3 3 3 3 3	3 3 3 2 2 2 2 2 2 2	2
169	* 6 6 6 6 6 6 7 7	6 7 6 12 6 6 4 7 4 7	4 4 4 6 5 5 4 3 3 4	4 3 3 4 3 3 3 3 3 3	3 3 3 2 2 2 2 2 2 2	2
170	6 6 6 6 6 6 6 6 6	6 6 6 6 6 6 6 4 7 4	4 4 4 6 5 4 4 4 3 3	4 3 3 4 4 3 3 3 3 3	3 3 3 2 2 2 2 2 2 2	2
171	8 6 6 6 6 6 6 6 8	6 7 6 6 6 6 6 7 4 8	4 4 4 6 4 4 4 4 4 3	4 4 3 3 4 3 3 3 3 3	3 3 3 2 2 2 2 2 2 2	2
172	6 6 6 6 6 6 6 6 6	6 6 6 6 6 6 6 6 6 4	5 4 4 6 5 4 4 4 4 4	3 3 3 4 3 3 3 3 3 3	3 3 3 3 2 2 2 2 2 2	2
173	* 6 6 6 6 6 6 7 6	6 6 6 7 6 6 6 6 6 6	4 5 5 6 5 4 4 3 4 4	4 3 3 4 3 3 3 3 3 3	3 3 3 3 2 2 2 2 2 2	2
174	6 6 6 6 6 6 6 6 6	6 6 6 6 6 6 6 6 6 6	6 4 5 6 5 4 4 4 4 3	4 4 3 4 3 3 3 3 3 3	3 3 3 3 2 2 2 2 2 2	2
175	6 6 6 6 6 6 6 6 6	6 6 6 6 6 6 6 6 6 6	6 6 4 6 5 6 4 4 3 3	4 4 4 3 3 4 3 3 3 3	3 3 3 3 2 2 2 2 2 2	2
176	14 6 6 6 6 6 6 6 6	6 10 6 6 6 6 10 6 6 6	6 6 6 6 5 6 4 4 4 3	3 3 4 4 3 4 3 3 3 3	3 3 3 3 3 2 2 2 2 2	2
177	9 6 6 6 6 6 6 6 6	6 9 6 6 6 6 9 9 6 6	6 6 6 6 5 6 4 4 3 3	4 4 3 4 4 3 3 3 3 3	3 3 3 3 3 2 2 2 2 2	2
178	6 9 6 6 6 6 6 6 6	6 6 6 6 6 6 6 6 8 5	5 5 6 6 5 6 4 4 4 3	3 4 3 4 4 3 3 3 3 3	3 3 3 3 3 2 2 2 2 2	2
179	* 6 9 6 6 6 6 6 6	6 7 6 6 6 6 7 7 6 7	4 4 4 6 5 6 4 4 4 5	4 4 3 3 3 4 3 3 3 3	3 3 3 3 3 2 2 2 2 2	2
180	6 6 6 9 6 6 6 6 6	6 6 6 6 6 6 6 6 6 6	6 6 5 6 5 6 4 4 5 4	4 3 3 3 3 4 4 3 3 3	3 3 3 3 3 3 2 2 2 2	2
181	* 6 6 6 9 6 6 6 6	6 6 6 6 6 6 6 6 9 6	6 6 6 6 5 6 4 4 4 4	5 4 3 3 3 3 3 3 3 3	3 3 3 3 3 3 2 2 2 2	2
182	6 6 6 6 6 9 6 6 6	6 6 6 6 6 6 6 6 6 6	6 6 6 6 5 6 4 4 4 4	4 4 4 4 4 4 3 3 3 3	3 3 3 3 3 3 2 2 2 2	2
183	6 6 6 6 6 6 9 6 6	6 6 6 6 6 6 6 6 6 6	6 6 6 6 5 6 6 4 4 4	4 4 3 4 3 4 4 3 3 3	3 3 3 3 3 3 2 2 2 2	2
184	7 6 6 6 6 6 6 9 6	6 7 6 7 6 6 7 7 6 6	6 6 6 7 5 6 4 4 4 4	4 3 4 4 4 3 3 3 3 3	3 3 3 3 3 3 3 2 2 2	2
185	9 6 6 6 6 6 6 7 9	6 9 6 9 6 6 8 9 6 6	6 6 6 7 5 6 4 4 3 4	4 4 4 4 4 4 4 4 3 3	3 3 3 3 3 3 3 2 2 2	2
186	6 7 6 6 6 9 6 6 6	9 6 6 6 6 6 6 6 6 6	6 6 6 6 5 6 4 4 4 4	4 4 4 4 4 3 4 4 3 3	3 3 3 3 3 3 3 2 2 2	2
187	10 6 7 6 6 6 6 7 7	6 10 6 7 6 6 6 10 6 6	6 6 6 7 5 6 4 4 4 4	4 5 3 4 4 4 4 4 3 3	3 3 3 3 3 3 3 2 2 2	2
188	6 6 6 7 6 6 6 6 6	6 6 7 6 6 6 6 6 6 6	6 6 6 6 5 6 4 4 4 4	4 3 4 4 4 3 4 3 3 3	3 3 3 3 3 3 3 3 2 2	2
189	8 6 6 6 7 6 6 7 7	6 6 6 7 6 6 6 6 6 6	6 6 6 7 5 6 4 6 4 4	4 4 3 4 4 4 4 4 3 3	3 3 3 3 3 3 3 3 2 2	2
190	6 6 6 6 6 7 6 6 6	6 6 6 6 7 6 6 6 6 6	6 6 6 6 5 6 4 6 5 4	4 4 3 4 4 4 4 3 4 3	3 3 3 3 3 3 3 3 2 2	2
191	* 6 6 6 6 6 7 6 7	6 7 6 6 6 7 6 6 6 7	6 6 6 6 5 6 4 6 4 4	4 4 3 3 3 4 4 3 3 3	3 3 3 3 3 3 3 3 2 2	2
192	7 6 6 6 6 6 6 7 7	6 6 6 6 6 6 7 6 6 6	6 6 6 7 6 6 4 6 4 4	4 3 4 4 4 3 4 3 3 3	3 3 3 3 3 3 3 3 3 2	2
193	* 6 6 6 6 6 6 7 7	6 6 6 6 6 6 6 7 6 6	6 6 6 7 6 6 4 6 4 4	4 4 3 4 4 3 4 4 3 3	3 3 3 3 3 3 3 3 3 2	2
194	6 6 6 6 6 6 6 6 6	7 6 6 6 6 6 6 6 7 6	6 6 6 6 6 6 5 6 4 4	4 4 3 4 4 3 3 3 4 3	3 3 3 3 3 3 3 3 3 2	2
195	7 6 6 6 6 6 6 7 7	6 7 6 6 6 6 6 6 6 7	6 6 6 7 6 6 4 6 4 4	4 4 3 3 3 4 4 4 3 4	3 3 3 3 3 3 3 3 3 2	2

$h \rightarrow$	0	1	2	3	4	5
$n \downarrow$	1 2 3 4 5 6 7 8 9	0 1 2 3 4 5 6 7 8 9	0 1 2 3 4 5 6 7 8 9	0 1 2 3 4 5 6 7 8 9	0 1 2 3 4 5 6 7 8 9	0
196	6 6 6 6 6 6 6 6 6	6 6 7 6 6 6 6 6 6 6	7 6 6 6 6 6 4 6 5 4	4 4 3 4 4 4 4 4 3 4	3 3 3 3 3 3 3 3 3 3	2
197	* 6 6 6 6 6 6 7 7	6 6 6 7 6 6 6 6 6 6	6 7 6 7 5 6 4 6 6 5	4 4 3 4 4 4 4 4 3 3	3 3 3 3 3 3 3 3 3 3	2
198	6 6 6 6 6 9 6 6 6	6 6 6 6 7 6 6 6 6 6	6 6 7 6 6 6 4 6 5 4	4 4 3 4 3 3 4 3 4 3	3 3 3 3 3 3 3 3 3 3	2
199	* 6 6 6 6 6 9 7 7	6 6 6 6 6 7 6 6 6 6	6 6 6 7 6 6 4 6 4 4	4 4 3 3 3 3 4 4 3 4	3 3 3 3 3 3 3 3 3 3	2
200	7 6 6 6 6 6 6 7 7	6 7 6 7 6 6 7 6 6 6	6 6 6 7 6 7 5 6 5 4	4 4 4 4 3 4 4 4 3 4	4 3 3 3 3 3 3 3 3 3	3
201	8 6 6 6 6 6 6 7 8	6 7 6 7 6 6 6 8 6 6	9 6 6 7 6 7 5 6 5 4	4 4 4 4 3 3 4 3 3 3	4 3 3 3 3 3 3 3 3 3	3
202	6 7 6 6 6 6 6 6 6	7 6 6 6 6 6 6 6 7 6	6 6 6 6 6 6 5 6 5 4	4 4 3 4 4 4 4 3 3 4	4 3 3 3 3 3 3 3 3 3	3
203	7 6 7 6 6 6 6 7 7	6 7 6 6 6 6 6 6 6 7	6 6 6 7 6 7 4 6 5 6	4 3 4 4 4 4 4 3 3 4	3 3 3 3 3 3 3 3 3 3	3
204	6 6 6 7 6 6 6 6 6	6 6 7 6 6 6 6 6 6 6	7 6 6 6 6 6 5 6 5 6	4 3 3 4 3 4 4 3 3 4	3 3 3 3 3 3 3 3 3 3	3
205	8 6 6 6 7 6 6 7 8	6 6 6 8 6 6 6 8 6 6	6 7 6 6 6 7 5 6 5 6	4 4 3 3 4 4 4 4 3 3	4 4 3 3 3 3 3 3 3 3	3
206	6 6 6 6 6 7 6 6 6	6 6 6 6 7 6 6 6 6 6	6 6 7 6 6 6 5 6 5 6	4 4 3 3 3 4 4 4 4 4	3 3 3 3 3 3 3 3 3 3	3
207	8 6 6 6 6 6 7 6 8	6 6 6 6 6 8 6 6 6 6	6 6 6 8 6 6 4 6 5 6	4 4 4 4 3 4 4 4 4 4	4 3 3 3 3 3 3 3 3 3	3
208	14 6 6 6 6 6 6 7 7	6 6 6 12 6 6 12 6 6 6	6 6 6 6 6 7 6 6 5 6	4 3 4 5 3 4 4 3 3 4	4 3 3 3 3 3 3 3 3 3	3
209	11 6 6 6 6 6 6 7 7	6 10 6 11 6 6 11 11 6 10	6 6 6 6 6 7 6 6 5 6	4 4 4 4 4 4 4 3 3 4	4 4 3 3 3 3 3 3 3 3	3
210	10 11 6 6 6 6 6 6 6	7 6 6 6 6 6 6 6 10 10	6 6 6 6 6 6 5 6 5 6	4 4 3 4 4 4 4 3 3 3	3 4 4 3 3 3 3 3 3 3	3
211	* 6 11 6 6 6 6 7 7	6 7 6 9 6 6 9 9 6 9	6 6 6 6 6 7 4 6 5 6	6 4 3 4 3 4 4 4 3 3	3 4 4 3 3 3 3 3 3 3	3
212	6 6 6 11 6 6 6 6 6	6 6 7 6 6 6 6 6 6 6	8 6 6 6 6 6 6 6 5 6	4 4 4 4 3 4 4 4 3 3	3 3 3 3 3 3 3 3 3 3	3
213	7 6 6 6 11 6 6 7 7	6 6 6 7 6 6 6 6 6 6	6 7 6 6 6 7 6 6 5 6	4 4 4 4 4 4 4 4 4 3	4 4 3 3 3 3 3 3 3 3	3
214	6 6 6 6 6 11 6 6 6	6 6 6 6 7 6 6 6 6 6	6 6 7 6 6 6 6 6 5 6	4 4 4 4 4 4 4 3 4 4	3 3 3 3 3 3 3 3 3 3	3
215	7 6 6 6 6 6 11 7 7	6 6 6 6 6 7 6 6 6 6	6 6 6 7 6 7 6 6 5 6	4 4 4 4 4 4 3 3 3 4	4 4 3 4 3 3 3 3 3 3	3
216	7 6 6 6 6 6 6 11 7	6 6 6 7 6 6 7 7 6 6	6 6 6 6 7 7 6 7 5 6	4 4 4 4 4 4 4 3 3 3	4 4 3 4 3 3 3 3 3 3	3
217	8 6 6 6 6 6 6 7 11	6 6 6 8 6 6 8 8 9 6	6 6 6 6 6 7 6 7 5 6	4 6 4 4 4 4 5 4 3 3	3 4 4 4 3 3 3 3 3 3	3
218	6 7 6 6 6 6 6 6 6	11 6 6 6 6 6 6 6 7 6	6 6 6 6 6 6 6 6 5 6	4 6 5 4 4 4 4 3 4 3	3 3 4 4 3 3 3 3 3 3	3
219	10 6 7 6 6 6 6 7 7	6 11 6 10 6 6 10 10 6 7	6 6 6 6 6 7 6 7 5 6	4 6 4 4 3 4 4 4 3 4	3 3 3 4 3 3 3 3 3 3	3
220	6 6 6 7 6 6 6 6 6	6 6 11 6 6 6 6 6 6 6	7 6 6 6 6 6 6 6 5 6	4 6 4 4 4 4 4 3 4 4	4 3 3 3 4 3 3 3 3 3	3
221	12 6 6 6 7 6 6 7 7	6 6 6 12 6 6 6 12 6 6	6 7 9 6 6 6 6 7 5 6	4 6 4 4 4 4 4 4 3 4	4 4 3 3 3 3 3 3 3 3	3
222	6 6 6 6 6 7 6 6 6	6 6 6 6 7 6 6 6 6 6	6 6 7 6 6 6 6 6 5 6	4 6 4 4 4 4 5 3 4 3	4 3 4 3 3 3 3 3 3 3	3
223	* 6 6 6 6 6 7 7 7	6 6 6 6 6 7 6 6 6 6	6 6 6 7 6 7 6 6 5 6	4 6 4 4 4 4 4 5 3 4	4 4 4 4 3 3 3 3 3 3	3
224	13 6 6 6 6 6 6 7 7	6 6 6 6 6 6 7 6 6 6	6 6 6 6 7 6 6 7 6 6	4 6 6 4 5 4 4 4 4 4	4 4 3 3 3 3 3 3 3 3	3
225	8 6 6 6 6 6 6 7 9	6 8 6 8 6 6 6 7 6 8	6 6 6 6 6 8 6 7 7 6	4 6 6 5 4 4 4 4 4 4	4 4 3 3 3 4 3 3 3 3	3
226	6 6 6 6 6 6 6 6 6	7 6 6 6 6 6 6 6 7 6	6 6 6 6 6 6 6 6 6 6	5 6 6 4 4 4 4 4 3 4	4 3 4 3 3 4 3 3 3 3	3
227	* 6 6 6 6 6 6 7 7	6 7 6 6 6 6 6 6 6 7	6 6 6 6 6 6 6 7 5 6	4 6 6 4 4 4 4 4 3 4	4 4 4 4 3 4 3 3 3 3	3
228	6 6 6 6 6 6 6 6 6	6 6 7 6 6 6 6 6 6 6	7 6 6 6 6 6 6 6 6 6	4 6 6 4 4 4 4 4 4 4	4 4 4 3 4 3 3 3 3 3	3
229	* 6 6 6 6 6 6 7 7	6 6 6 7 6 6 6 6 6 6	6 7 6 6 6 6 6 7 6 6	4 6 6 4 4 4 4 4 5 4	4 3 4 4 3 4 3 3 3 3	3
230	6 6 6 6 6 6 6 6 6	6 6 6 6 7 6 6 6 6 6	6 6 7 6 6 6 6 6 6 6	4 6 6 4 4 4 4 4 4 4	4 3 4 4 4 4 4 3 3 3	3
231	7 6 6 6 6 6 6 7 7	6 6 6 6 6 7 6 6 6 6	6 6 6 7 6 6 6 7 6 6	4 6 6 5 4 4 4 4 3 4	4 4 4 4 4 4 3 3 3 3	3
232	7 6 6 6 6 6 6 7 7	6 6 6 6 6 6 7 6 6 6	6 10 6 6 7 6 6 7 6 7	5 6 6 6 4 4 4 4 3 4	4 3 4 3 3 4 3 3 3 3	3
233	* 6 6 6 6 6 6 7 7	6 6 6 6 6 6 6 7 6 6	6 6 6 6 6 7 6 7 6 7	5 6 6 5 4 4 4 4 4 3	4 4 4 4 3 4 3 3 3 3	3
234	6 7 6 6 6 6 6 6 6	7 6 6 6 6 6 6 6 7 6	6 6 6 6 6 6 7 6 6 6	5 6 6 4 4 4 4 4 3 4	4 4 4 4 3 4 3 3 3 3	3
235	7 6 7 6 6 6 6 7 7	6 7 6 6 6 7 6 6 6 7	6 6 6 6 6 6 6 7 6 7	4 6 6 5 4 4 4 4 3 4	4 3 4 4 3 4 3 4 3 3	3
236	6 6 6 7 6 6 6 6 6	6 6 7 6 6 6 6 6 6 6	7 6 6 6 6 6 6 6 6 6	5 6 6 5 4 4 4 4 4 4	4 4 3 3 4 3 3 3 3 3	3
237	7 6 6 6 7 6 6 7 7	6 6 6 7 6 6 6 6 6 6	6 7 6 6 6 6 6 6 6 7	5 6 6 5 4 4 4 5 3 4	4 4 4 4 3 4 3 3 3 3	3
238	6 6 6 6 6 7 6 6 6	6 6 6 6 7 6 6 6 6 6	6 6 7 6 6 6 6 6 6 6	5 6 6 5 4 4 4 4 4 4	4 4 4 4 3 4 4 3 3 3	3
239	* 6 6 6 6 6 7 7 7	6 6 6 6 6 7 6 6 6 6	6 6 6 7 6 6 6 7 6 6	5 6 6 5 4 4 4 4 3 4	4 4 4 4 4 4 3 4 3 3	3
240	7 6 6 6 6 6 6 7 7	6 6 6 6 6 6 7 6 6 6	6 6 6 6 7 6 6 6 6 7	6 6 6 5 4 4 4 4 4 4	4 4 4 4 3 4 4 3 4 3	3
241	* 6 6 6 6 6 6 7 7	6 6 6 6 6 6 6 7 6 6	6 6 6 6 6 7 6 7 6 7	7 6 6 5 4 4 4 4 4 4	4 4 4 4 3 4 4 3 4 3	3
242	6 6 6 6 6 6 6 6 6	7 6 6 6 6 6 6 6 7 6	6 6 6 6 6 6 7 6 6 6	6 6 6 5 4 4 4 4 4 4	4 4 4 4 4 4 4 3 3 3	3
243	* 6 6 6 6 6 6 7 8	6 7 6 6 6 6 6 6 6 7	6 6 6 6 6 6 6 8 6 7	4 6 6 5 5 4 4 4 4 4	4 4 4 4 3 4 3 3 4 3	3
244	6 6 6 6 6 6 6 6 6	6 6 7 6 6 6 6 6 6 6	7 6 6 6 6 6 6 6 6 6	6 6 6 5 4 4 4 4 4 4	4 4 4 4 4 4 4 3 4 3	3
245	7 6 6 6 6 6 6 7 7	6 6 6 7 6 6 6 6 6 6	6 7 6 6 6 6 6 6 6 7	5 6 6 5 4 5 4 4 4 4	4 4 4 4 3 4 4 4 4 4	3
246	6 6 6 6 6 6 6 6 6	6 6 6 6 7 9 6 6 6 6	6 6 7 6 6 6 6 6 6 6	6 6 6 5 4 6 5 4 4 4	4 4 4 4 3 4 4 3 4 4	3

$h \rightarrow$	0	1	2	3	4	5
$n \downarrow$	1 2 3 4 5 6 7 8 9	0 1 2 3 4 5 6 7 8 9	0 1 2 3 4 5 6 7 8 9	0 1 2 3 4 5 6 7 8 9	0 1 2 3 4 5 6 7 8 9	0
247	12 6 6 6 6 6 6 7 7	6 6 6 12 6 7 6 6 6 12	6 6 6 7 6 6 6 6 6 7	6 6 6 5 4 4 4 4 4 4	4 4 4 4 4 4 3 4 4 4	3
248	7 6 6 6 6 6 6 7 7	6 6 6 6 6 6 7 6 6 6	6 6 6 6 7 6 6 6 6 7	6 7 6 5 4 4 4 4 3 4	4 4 3 3 4 4 4 3 4 3	3
249	7 6 6 6 6 6 6 7 7	6 6 6 6 6 6 6 7 6 6	6 6 6 6 6 7 6 6 6 7	6 7 6 5 4 4 4 4 3 5	4 4 4 4 3 4 4 3 4 4	3
250	6 7 6 6 6 6 6 6 6	7 6 6 6 6 6 6 6 7 6	6 6 6 6 6 6 7 6 6 6	6 6 6 5 4 4 4 4 4 4	4 4 4 4 3 4 4 4 4 4	4
251	* 6 7 6 6 6 6 7 7	6 7 6 6 6 6 6 6 6 7	6 6 6 6 6 6 6 7 6 7	6 7 6 5 5 4 4 4 3 4	4 4 4 4 3 3 3 4 4 3	3
252	6 6 6 7 6 6 6 6 6	6 6 7 6 6 6 6 6 6 6	7 6 6 6 6 6 6 6 7 6	6 6 6 5 4 6 5 4 4 4	4 4 5 4 4 4 4 3 4 4	3
253	12 6 6 6 7 6 6 7 7	6 10 6 12 6 6 11 6 6 6	6 7 6 10 6 6 6 6 6 7	6 7 6 5 4 5 6 5 4 4	4 4 5 4 4 4 4 3 4 4	4
254	9 6 6 6 6 7 6 6 6	6 9 6 9 7 6 9 6 6 6	6 6 7 6 6 6 6 6 6 6	6 6 6 5 4 5 5 5 4 4	4 4 4 4 4 4 4 3 3 3	3
255	7 6 6 6 6 6 7 7 7	6 6 6 6 6 7 6 6 6 6	6 6 6 7 6 6 6 6 6 7	6 6 6 5 4 4 5 4 5 4	4 4 4 4 3 4 3 4 4 4	4
256	* 6 6 6 6 6 6 7 7	6 6 6 6 6 6 15 6 6 6	6 6 6 6 7 6 6 6 6 6	6 7 7 5 4 5 5 4 4 5	4 4 4 3 3 4 4 3 4 4	3
257	* 6 6 6 6 6 6 7 7	6 7 6 11 6 6 14 14 6 6	6 6 6 6 6 7 6 6 6 7	6 7 7 5 4 5 5 4 4 4	5 4 4 4 3 4 4 4 4 4	3
258	6 14 6 6 6 6 6 6 6	7 6 6 6 6 6 6 6 13 6	6 6 6 6 6 6 7 6 6 6	6 6 6 5 5 5 5 4 4 4	4 5 4 4 3 4 4 3 4 3	4
259	12 5 14 6 6 6 6 7 7	6 7 6 6 6 6 12 12 6 12	6 6 6 6 6 6 6 7 6 7	6 7 7 5 4 5 5 6 4 4	4 4 5 4 3 4 3 3 4 4	3
260	6 5 5 14 6 6 6 6 6	6 6 7 6 6 6 6 6 6 6	11 6 6 6 6 6 6 6 7 6	6 6 6 5 4 5 5 6 4 4	4 4 4 5 4 4 4 3 4 4	4
261	8 6 5 5 14 6 6 7 8	6 6 6 7 6 6 6 6 6 6	6 10 6 6 6 6 6 6 6 8	6 7 7 5 4 5 5 6 4 4	4 4 4 4 4 4 4 3 4 4	3
262	8 6 6 5 5 14 6 6 6	6 6 6 6 7 6 6 6 6 6	6 6 9 6 6 6 6 6 6 8	6 6 6 5 5 5 5 6 4 4	4 4 4 4 3 4 4 3 4 3	3
263	* 6 6 6 5 6 14 7 7	6 7 6 6 6 7 8 8 6 8	6 6 6 8 6 6 6 6 6 6	6 7 6 5 4 5 5 6 4 4	4 5 4 4 3 4 4 4 4 4	3
264	7 6 6 6 6 6 6 14 7	6 6 6 6 6 6 7 7 6 6	6 6 6 6 7 6 6 6 6 6	6 7 7 5 4 5 5 6 4 4	4 4 4 4 4 4 4 4 4 4	3
265	8 6 6 6 6 6 6 7 14	6 6 6 6 6 6 8 8 6 6	6 6 6 6 6 7 6 6 6 6	6 7 7 7 4 5 5 6 4 4	4 4 4 5 4 4 4 3 4 4	4
266	6 6 6 6 6 6 6 6 6	14 6 6 6 6 6 6 6 7 6	6 6 6 6 6 6 7 6 6 6	6 6 6 5 4 5 5 6 4 4	4 4 4 4 4 4 4 3 4 4	4
267	10 6 6 6 6 6 6 7 7	6 14 6 6 6 6 10 10 6 7	6 6 6 6 6 6 6 7 6 6	6 7 7 5 4 4 5 6 6 4	4 4 4 4 4 4 4 4 4 3	4
268	6 6 6 6 6 6 6 6 6	6 6 14 6 6 6 6 6 6 6	7 6 6 6 6 6 6 6 7 6	6 6 6 5 4 5 5 6 4 4	5 4 4 4 4 4 4 4 4 3	3
269	* 6 6 6 6 6 6 7 7	6 6 6 14 6 6 12 12 6 6	6 7 6 6 6 6 6 6 6 7	6 7 7 5 4 5 5 6 4 4	4 4 4 4 4 4 4 4 4 4	3
270	6 6 6 6 6 6 6 6 6	6 6 6 6 14 6 6 6 6 6	6 6 7 6 6 6 6 6 6 6	7 6 6 5 4 5 5 6 4 4	4 4 4 4 4 4 4 3 4 4	4
271	* 6 6 6 6 6 6 7 7	6 6 6 6 6 14 6 6 6 6	6 6 6 7 6 6 6 6 6 6	6 7 7 5 4 5 5 6 4 4	4 4 4 4 4 4 4 4 4 4	4
272	15 6 6 6 6 6 6 7 7	6 6 6 6 6 6 15 15 6 6	6 6 6 6 7 6 6 6 6 6	6 6 7 5 4 5 5 6 4 4	4 4 4 4 4 4 4 4 4 4	4
273	16 6 6 6 6 6 6 7 7	6 6 6 7 6 6 15 15 6 6	6 6 6 6 6 7 6 6 6 6	6 7 7 5 7 5 5 6 4 4	4 4 5 4 4 4 4 3 4 4	4
274	6 15 6 6 6 6 6 6 6	6 6 6 6 6 6 6 6 14 6	6 6 6 6 6 6 7 6 6 6	6 6 6 5 4 5 5 6 5 5	4 4 4 4 4 4 4 3 4 4	4
275	13 6 15 6 6 6 6 7 7	6 10 6 6 6 6 13 13 6 13	6 6 6 6 6 10 6 7 6 6	6 7 7 5 4 5 5 6 4 4	4 4 4 4 4 4 4 4 4 4	4
276	10 6 6 15 6 6 6 6 6	6 6 6 6 6 6 6 6 6 6	12 6 6 6 6 10 6 6 7 6	6 6 6 5 4 5 5 6 4 4	4 4 4 4 4 4 4 4 4 4	4
277	* 6 6 6 15 6 6 7 7	6 9 6 9 6 6 6 6 6 6	6 11 6 9 6 6 6 6 6 7	6 7 7 5 4 5 5 6 4 4	4 4 4 4 4 4 4 3 4 4	4
278	6 6 6 6 6 15 6 6 6	6 6 6 6 6 6 6 6 6 6	6 6 10 6 6 6 6 6 6 6	7 6 6 5 4 5 5 6 4 4	4 4 4 4 4 4 4 4 3 4	4
279	9 6 6 6 6 6 15 7 8	6 7 6 6 6 6 9 9 6 7	6 6 6 9 6 6 6 7 6 6	6 8 7 5 5 5 5 6 4 4	4 4 4 4 4 4 4 4 4 4	4
280	7 6 6 6 6 6 6 15 7	6 7 6 6 6 6 7 7 6 7	6 6 6 6 8 6 6 6 6 6	6 7 7 5 4 5 5 6 4 4	6 4 4 4 4 4 4 4 4 4	4
281	* 7 6 6 6 6 6 7 15	6 6 6 6 6 6 8 8 6 7	6 6 6 6 6 7 6 6 6 6	6 6 7 6 4 5 5 6 4 5	6 5 4 4 4 4 4 4 4 4	4
282	6 6 7 6 6 6 6 6 6	15 6 6 6 6 6 6 6 6 6	6 6 6 6 6 6 7 6 6 6	6 6 6 5 6 5 5 6 4 4	5 4 4 4 4 4 4 4 4 3	4
283	* 6 6 7 6 6 6 7 7	6 15 6 6 6 6 10 10 6 6	6 6 6 6 6 6 6 7 6 6	6 6 7 5 4 6 5 6 5 4	4 4 4 4 4 4 4 4 4 4	3
284	6 6 6 6 7 6 6 6 6	6 6 15 6 6 6 6 6 6 6	6 6 6 6 6 6 6 6 7 6	6 6 6 5 4 4 6 6 4 5	4 4 4 4 4 4 4 4 4 4	4
285	12 6 6 6 6 7 6 7 7	6 6 6 15 6 6 12 12 6 6	6 6 6 6 6 6 6 6 6 7	6 6 7 6 4 5 5 6 4 4	5 4 4 4 4 4 4 4 4 4	3
286	6 6 6 6 6 6 7 6 6	6 6 6 6 15 6 6 6 6 6	6 6 6 6 6 6 6 6 6 6	7 6 6 6 6 4 5 6 5 5	4 5 4 4 4 4 4 4 4 4	3
287	7 6 6 6 6 6 6 7 7	6 6 6 6 6 15 6 6 6 6	6 6 6 6 6 6 6 6 6 6	6 7 7 6 6 6 5 6 4 5	6 6 5 4 4 4 4 4 4 4	3
288	15 6 6 6 6 6 6 7 8	6 6 6 6 6 6 15 15 6 6	6 6 6 6 6 6 6 6 6 6	6 6 8 6 6 6 6 6 4 4	6 6 5 5 4 4 4 4 4 4	3
289	* 6 6 6 6 6 6 6 6	6 6 6 6 6 6 6 16 6 6	6 6 6 6 6 6 6 6 6 6	6 6 6 6 6 6 6 6 4 4	6 6 5 5 4 4 4 4 5 4	3
290	6 6 6 6 6 6 6 6 6	6 6 6 6 6 6 6 6 6 6	6 6 6 6 6 6 6 6 6 6	6 6 6 5 6 5 6 6 5 4	5 6 5 4 4 4 4 4 4 4	4
291	6 6 6 6 6 6 6 6 6	6 6 6 6 6 6 6 6 6 6	6 6 6 6 6 6 6 6 6 6	6 6 6 6 4 5 6 6 4 4	6 6 5 5 4 5 5 4 4 4	4
292	6 6 6 6 6 6 6 6 6	6 6 6 6 6 6 6 6 6 6	6 6 6 6 6 6 6 6 6 6	6 6 6 6 6 6 6 6 4 4	5 6 5 4 5 4 4 4 4 4	4
293	* 6 6 6 6 6 6 6 6	6 6 6 6 6 6 6 6 6 6	6 6 6 6 6 6 6 6 6 6	6 6 6 6 6 6 6 6 4 4	6 6 5 4 4 5 4 5 4 4	4
294	6 6 6 6 6 6 6 6 6	6 6 6 6 6 6 6 6 6 6	6 6 6 6 6 6 6 6 6 6	6 6 6 6 6 6 6 6 4 4	5 6 5 4 4 4 5 4 4 5	4
295	6 6 6 6 6 6 6 6 6	6 6 6 6 6 6 6 6 6 6	6 6 6 6 6 6 6 6 6 6	6 6 6 6 6 6 6 6 4 4	5 6 5 5 4 5 4 5 4 5	4
296	7 6 6 6 6 6 6 7 6	6 6 6 6 6 6 7 7 6 7	6 6 6 6 6 6 6 6 6 6	6 6 6 6 6 6 6 7 5 4	6 6 5 5 4 4 4 4 5 4	4
297	10 6 6 6 6 6 6 7 8	6 10 6 6 6 6 6 8 6 6	6 6 6 6 6 6 6 10 6 6	6 6 6 6 6 6 6 7 5 4	6 6 5 4 4 4 4 4 4 5	4

h →	0									1										2										3										4										5
n ↓	1	2	3	4	5	6	7	8	9	0	1	2	3	4	5	6	7	8	9	0	1	2	3	4	5	6	7	8	9	0	1	2	3	4	5	6	7	8	9	0	1	2	3	4	5	6	7	8	9	0
298	10	7	6	6	6	6	6	6	6	7	6	6	6	6	6	6	6	6	6	6	6	6	6	6	6	6	10	6	6	6	6	6	6	6	6	6	6	5	5	5	6	5	4	4	4	4	4	4	4	4
299	12	6	7	6	6	6	6	7	7	6	7	6	12	6	6	6	6	6	6	6	6	6	12	6	6	6	6	6	6	6	6	6	6	6	6	6	7	4	4	6	6	5	4	4	4	4	4	4	4	4
300	6	6	6	7	6	6	6	6	6	6	6	7	6	6	6	6	6	6	6	6	6	6	6	6	6	6	6	6	6	6	6	6	6	6	6	6	6	5	4	5	6	5	4	4	4	4	4	4	4	4
301	7	6	6	6	7	6	6	7	7	6	6	6	7	6	6	6	6	6	6	6	7	6	6	6	6	6	6	6	6	9	6	6	6	6	6	6	7	5	4	5	6	6	6	4	4	4	4	4	5	5
302	6	6	6	6	6	7	6	6	6	6	6	6	6	7	6	6	6	6	6	6	6	6	6	6	6	6	6	6	6	6	6	6	6	6	6	6	6	5	4	5	6	5	6	5	4	4	4	4	5	5
303	6	6	6	6	6	6	7	6	6	6	6	6	6	6	9	6	6	6	6	6	6	6	6	6	6	6	6	6	6	6	6	6	6	6	6	6	6	4	4	6	6	5	6	5	4	4	4	4	4	4
304	15	6	6	6	6	6	6	7	8	6	8	6	6	6	6	15	6	6	15	6	6	6	6	6	6	6	6	6	6	6	6	7	6	6	6	6	7	4	4	5	6	5	6	4	5	4	4	4	4	4
305	15	6	6	6	6	6	6	7	7	6	6	6	6	6	6	15	15	6	15	6	6	6	6	6	6	6	6	6	6	6	6	6	6	6	6	6	7	5	4	6	6	5	6	4	4	5	4	4	4	4
306	6	15	6	6	6	6	6	6	6	7	6	6	6	6	6	6	6	15	6	6	6	6	6	6	6	6	6	6	6	6	6	6	6	6	6	6	6	5	4	5	6	5	6	4	4	4	5	4	4	4
307	*	6	15	6	6	6	6	7	7	6	7	6	6	6	6	15	15	6	15	6	6	6	6	6	6	6	6	6	6	6	6	6	6	6	6	6	7	4	4	6	6	5	6	4	4	4	4	5	5	4
308	6	6	6	15	6	6	6	6	6	6	6	7	6	6	6	6	6	6	6	14	6	6	6	6	6	6	6	6	6	6	6	6	6	6	6	6	6	5	4	5	6	5	6	5	4	4	4	4	5	5
309	7	6	6	6	15	6	6	7	7	6	6	6	7	6	6	6	6	6	6	6	13	6	6	6	6	6	6	6	6	6	6	6	6	6	6	6	7	4	5	6	6	5	6	5	5	4	4	4	4	4
310	6	6	6	6	6	15	6	6	6	6	6	6	6	7	6	6	6	6	6	6	6	12	6	6	6	6	6	6	6	6	6	6	6	6	6	6	6	5	4	5	6	5	6	5	4	4	4	4	4	4
311	*	6	6	6	6	6	15	7	7	6	6	6	6	6	7	11	11	6	11	6	6	6	11	6	6	6	6	6	6	6	6	6	6	6	6	6	7	4	5	6	6	5	6	4	4	4	4	4	4	4
312	7	6	6	6	6	6	6	15	7	6	6	6	6	6	6	7	7	6	7	6	6	6	6	10	6	6	6	6	6	6	6	6	6	6	6	6	7	5	4	6	6	5	6	5	4	4	4	5	4	4
313	*	6	6	6	6	6	6	7	15	6	6	6	6	6	6	9	9	6	9	6	6	6	6	6	9	6	6	6	6	6	6	6	6	6	6	6	7	5	4	5	6	5	6	5	4	4	4	4	5	4
314	6	6	6	6	6	6	6	6	6	15	6	6	6	6	6	6	6	7	6	6	6	6	6	6	6	8	6	6	6	6	6	6	6	6	6	6	6	5	4	5	6	5	6	5	4	4	4	4	4	4
315	10	6	6	6	6	6	6	7	7	6	15	6	6	6	6	10	10	6	10	6	6	6	6	6	6	6	7	6	6	6	6	6	6	6	6	6	7	4	4	5	6	5	6	5	6	4	4	4	5	4
316	7	6	6	6	6	6	6	6	7	6	6	15	7	6	6	7	6	6	6	7	6	6	6	6	6	6	6	7	6	6	6	6	6	6	8	6	6	5	4	5	6	5	6	5	6	4	4	4	4	4
317	*	6	6	6	6	6	6	7	7	6	6	6	15	6	6	12	12	6	12	6	7	6	6	6	6	6	6	6	7	6	6	6	6	6	6	6	7	5	4	6	6	5	6	5	4	4	4	4	4	4
318	6	6	6	6	6	6	6	6	6	6	6	6	6	15	6	6	6	6	6	6	6	7	6	6	6	6	6	6	6	7	6	6	6	6	6	6	6	6	5	5	6	5	6	5	4	4	4	4	4	4
319	10	6	6	6	6	6	6	7	7	6	10	6	6	6	15	6	6	6	6	6	6	6	7	6	6	6	6	6	10	6	7	6	6	6	6	6	7	6	6	6	6	5	6	5	4	4	4	4	4	4
320	15	6	6	6	6	6	6	7	7	6	6	6	6	6	6	15	15	6	15	6	6	6	6	7	6	6	6	6	6	6	6	7	6	6	6	6	7	6	6	7	6	5	6	5	5	4	4	4	4	4
321	15	6	6	6	6	6	6	7	7	6	6	6	6	6	6	15	15	6	15	6	6	6	6	6	7	6	6	6	6	6	6	6	7	6	6	6	7	6	6	7	6	5	6	5	4	4	4	4	4	4
322	6	6	6	6	6	6	6	6	6	6	6	6	6	6	6	6	6	15	6	6	6	6	6	6	6	7	6	6	6	6	6	6	6	7	6	6	6	6	5	6	6	5	6	5	5	4	4	4	5	4
323	16	6	6	6	6	6	6	7	7	6	7	6	6	6	6	6	16	6	16	6	6	6	7	6	6	6	7	6	6	6	7	6	6	6	7	6	7	6	6	6	6	5	6	5	5	6	4	4	4	4
324	6	6	6	6	6	6	6	6	6	6	6	6	6	6	6	6	6	6	6	6	6	6	6	6	6	6	6	7	6	6	6	6	6	6	6	7	6	6	6	6	6	5	6	5	6	4	5	4	4	4
325	12	6	6	6	6	6	6	7	7	6	6	6	12	6	6	6	6	17	6	6	6	6	6	6	12	6	6	6	7	6	6	6	6	6	6	6	7	6	6	6	6	5	6	5	5	5	4	4	4	4
326	6	6	6	6	6	6	6	6	6	6	6	6	6	6	6	6	6	6	6	6	6	6	6	6	6	6	6	6	6	7	6	6	6	6	6	6	6	6	6	6	6	5	6	5	5	5	5	4	4	4
327	7	6	15	6	6	6	6	7	7	6	6	6	6	6	6	6	6	6	6	6	6	6	6	6	6	6	6	6	6	6	7	6	6	6	6	6	7	6	6	6	6	5	6	5	5	4	5	5	4	4
328	7	6	6	7	6	6	6	7	7	6	6	6	6	6	6	6	6	6	6	6	6	6	6	6	6	6	6	6	6	6	6	7	6	6	6	6	7	6	6	7	7	5	6	5	5	4	4	4	5	4
329	9	6	6	7	6	6	6	7	7	6	9	6	9	6	6	6	9	6	6	6	6	6	6	6	6	6	9	6	6	6	6	6	7	6	6	6	6	6	6	6	7	5	6	5	4	4	6	4	5	5
330	6	7	6	6	6	6	6	6	6	7	6	6	6	6	6	6	6	6	6	6	6	6	6	6	6	6	6	6	6	6	6	6	6	7	6	6	6	6	6	6	6	5	6	5	4	5	6	5	4	5
331	*	6	7	7	6	6	6	7	7	6	7	6	6	6	6	6	6	6	6	6	6	6	6	6	6	6	6	6	6	6	6	6	6	6	7	6	6	6	6	6	7	5	6	5	5	5	6	4	4	4
332	6	6	6	7	6	6	6	6	6	6	6	7	6	6	6	6	6	6	6	6	6	6	6	6	6	6	6	6	6	6	6	6	6	6	6	7	6	6	6	6	6	5	6	5	5	4	6	4	4	4
333	8	6	6	7	7	6	6	7	8	6	7	6	7	6	6	6	6	6	6	6	6	6	7	6	6	6	6	6	6	6	7	6	6	6	6	6	8	6	6	6	7	5	6	5	4	5	6	4	4	4
334	6	6	6	6	6	7	6	6	6	6	6	6	6	7	6	6	6	6	6	6	6	6	6	7	6	6	6	6	6	6	6	6	6	6	6	6	6	6	6	6	6	5	6	5	5	4	6	4	5	4
335	6	6	6	6	6	6	7	6	6	6	6	6	6	6	7	6	6	6	6	6	6	6	6	6	6	6	6	6	6	6	6	6	6	6	6	6	6	6	6	6	6	5	6	5	5	4	6	5	4	4
336	8	6	6	7	6	6	6	7	8	6	8	6	6	6	6	8	6	6	6	6	6	6	6	6	6	9	6	6	6	6	6	7	6	6	6	6	6	6	6	6	7	6	6	5	5	4	6	5	5	4
337	*	6	6	7	6	6	6	7	7	6	6	6	6	6	6	6	7	6	6	6	6	6	6	6	6	6	6	6	6	6	6	6	6	6	6	6	7	6	6	6	7	6	6	5	5	4	6	5	5	5
338	6	6	6	6	6	6	6	6	6	7	6	6	6	6	6	6	6	7	6	6	6	6	6	6	6	6	6	6	6	6	6	6	6	6	6	6	6	6	6	6	6	6	6	5	5	4	6	5	5	5
339	7	6	6	7	6	6	6	7	7	6	7	6	6	6	6	6	6	6	7	6	6	6	6	6	6	6	6	6	6	6	6	6	6	6	6	6	6	6	6	6	7	6	6	5	5	4	6	4	4	4
340	6	6	6	6	6	6	6	6	6	6	6	7	6	6	6	6	6	6	6	7	6	6	6	6	6	6	6	6	6	6	6	6	6	6	6	6	6	6	6	6	6	6	6	6	5	5	6	5	5	5
341	10	6	6	7	6	6	6	7	7	6	10	6	7	6	6	9	6	6	6	6	7	6	6	6	6	6	6	6	6	6	10	6	6	9	6	6	7	6	6	6	7	6	6	5	6	4	6	5	5	5
342	10	6	6	6	6	6	6	6	6	6	9	6	6	7	6	6	6	6	6	6	6	7	6	6	6	6	6	6	6	6	10	9	6	6	6	6	6	6	6	6	6	6	6	5	5	6	6	5	5	5
343	*	9	6	7	6	6	6	7	7	6	6	6	6	6	7	6	6	6	6	6	6	6	7	6	6	6	6	6	6	6	6	6	8	6	6	6	6	6	6	6	7	6	6	5	5	4	6	5	6	5
344	7	6	9	7	6	6	6	7	7	6	6	6	6	6	6	7	6	6	6	6	6	6	6	7	6	6	6	6	6	6	6	6	6	7	6	6	6	6	6	6	7	6	7	5	5	4	6	6	6	5
345	7	6	6	9	6	6	6	7	7	6	6	7	6	6	6	6	7	6	6	6	6	6	6	6	7	6	6	6	6	6	6	6	6	6	6	6	6	6	6	6	7	6	7	6	5	4	6	5	6	5
346	6	7	6	6	9	6	6	6	6	7	6	6	6	6	6	6	6	7	6	6	6	6	6	6	6	7	6	6	6	6	6	6	6	6	6	6	6	6	6	6	6	6	6	5	6	4	6	5	6	5
347	*	6	7	7	6	9	6	7	7	6	7	6	6	6	6	6	6	6	7	6	6	6	6	6	6	6	7	6	6	6	6	6	6	6	6	6	6	6	6	6	7	6	7	5	5	6	6	5	6	5
348	6	6	6	7	6	6	9	6	6	6	6	7	6	6	6	6	6	6	6	7	6	6	6	6	6	6	6	7	6	6	6	6	6	6	6	6	6	6	6	6	6	6	6	5	5	4	6	5	6	5

h →	0	1	2	3	4	5
n ↓	1 2 3 4 5 6 7 8 9	0 1 2 3 4 5 6 7 8 9	0 1 2 3 4 5 6 7 8 9	0 1 2 3 4 5 6 7 8 9	0 1 2 3 4 5 6 7 8 9	0
349	* 6 6 7 7 6 6 9 7	6 7 6 7 6 6 7 6 6 6	6 7 6 6 6 6 6 6 6 7	6 7 7 6 6 6 6 7 6 6	6 7 6 7 5 5 4 6 6 6	5
350	8 6 6 6 6 7 6 6 9	6 8 6 6 7 6 6 6 6 6	6 6 7 6 6 6 6 6 6 6	7 8 8 6 6 6 6 6 6 6	6 6 6 6 5 5 5 6 5 6	6
351	12 6 6 7 6 6 7 7 7	9 6 6 12 6 7 6 6 6 6	6 6 6 7 6 6 6 12 6 6	6 7 6 6 6 6 6 6 6 6	6 7 6 6 5 5 4 6 5 6	6
352	18 6 6 7 6 6 6 7 7	6 10 6 6 6 6 7 6 6 6	6 6 6 6 7 6 6 6 6 6	6 6 10 6 6 6 6 6 6 8	6 7 6 7 5 5 4 6 5 6	5
353	* 6 6 7 6 6 6 7 7	6 6 6 6 6 6 6 7 6 6	7 6 6 6 6 7 6 6 6 6	6 6 6 7 6 6 6 6 6 6	6 7 6 7 5 5 4 6 5 6	5
354	6 6 6 6 6 6 6 6 6	7 6 6 6 6 6 6 6 7 6	6 6 6 6 6 6 7 6 6 6	6 6 6 6 7 6 6 6 6 6	6 6 6 6 5 4 5 6 5 6	5
355	9 6 6 7 6 6 6 7 7	6 9 6 9 6 6 6 6 6 9	6 6 6 6 6 6 6 7 6 9	6 6 6 6 6 7 6 6 6 6	6 7 6 7 5 5 4 6 5 6	5
356	6 6 6 6 6 6 6 6 6	6 6 7 6 6 6 6 6 6 6	7 6 6 6 6 6 6 6 7 6	6 6 6 6 6 6 7 6 6 6	6 6 6 6 5 5 6 6 5 6	5
357	9 6 6 7 6 6 6 7 7	6 9 6 7 6 6 9 6 6 6	6 7 6 6 7 6 6 6 6 7	6 6 6 6 6 6 6 7 6 6	6 7 6 7 5 5 4 6 5 6	5
358	6 6 6 6 6 6 6 6 6	6 6 6 6 7 6 6 6 6 6	6 6 7 6 6 6 6 6 6 6	7 6 6 6 6 6 6 6 7 6	6 6 6 6 5 5 5 6 5 6	5
359	* 6 6 7 6 6 6 7 7	6 6 6 6 6 7 6 6 6 6	6 6 6 7 6 6 6 6 6 6	6 7 6 6 6 6 6 6 6 7	6 7 6 7 5 5 5 6 5 6	5
360	8 6 6 7 6 6 6 7 8	6 6 6 8 6 6 8 6 6 6	6 6 6 6 7 6 6 8 6 6	6 6 7 6 6 6 6 6 6 6	7 7 6 7 5 6 5 6 5 6	5
361	* 6 6 7 6 6 6 7 7	6 6 6 6 6 6 6 7 6 18	6 6 6 6 6 7 6 6 7 6	9 6 6 7 6 6 6 6 6 6	7 7 6 7 5 5 5 6 5 6	5
362	6 7 6 6 6 6 6 6 6	6 6 6 6 6 6 6 6 7 6	6 6 6 6 6 6 7 6 6 6	6 6 6 6 7 6 6 6 6 6	6 6 6 6 5 4 4 6 5 6	5
363	7 6 7 6 6 6 6 7 7	6 6 6 6 6 6 6 6 6 7	6 6 6 6 6 6 6 7 6 6	6 6 6 6 6 7 6 6 6 6	6 6 6 7 5 5 4 6 5 6	5
364	6 6 6 7 6 6 6 6 6	6 6 6 6 6 6 6 6 6 6	7 6 6 6 6 6 6 6 7 6	6 6 6 10 6 6 7 6 6 6	6 6 6 6 5 4 5 6 5 6	5
365	10 6 6 7 7 6 6 7 7	6 6 6 10 6 6 6 10 6 6	6 7 6 6 6 6 6 6 6 7	6 6 7 6 6 6 6 7 6 6	6 7 6 7 5 5 5 6 5 6	5
366	6 6 6 6 6 7 6 6 6	6 6 6 6 6 6 6 6 6 6	6 6 7 6 6 6 6 6 6 6	7 6 6 6 6 6 6 6 7 6	6 6 6 6 6 4 5 6 5 6	5
367	* 6 6 6 6 6 7 7 7	6 6 6 6 6 6 6 6 6 6	6 6 6 7 6 6 6 6 6 6	6 7 6 6 6 6 6 6 6 7	6 6 6 7 6 6 5 6 5 6	5
368	15 6 6 7 6 6 6 7 7	6 10 6 10 6 6 15 6 6 6	6 6 6 15 7 6 6 6 6 6	6 6 9 6 6 6 6 6 6 6	7 7 6 7 6 6 6 6 5 6	5
369	15 6 6 7 6 6 6 7 8	6 6 6 6 6 6 15 15 6 6	6 6 6 15 6 7 6 6 6 6	6 6 6 7 6 6 7 6 6 6	6 8 6 7 6 6 6 6 5 6	5
370	6 15 6 6 6 6 6 6 6	6 6 6 6 6 6 6 6 15 6	6 6 6 6 6 6 7 6 6 6	6 6 6 6 7 6 6 6 6 6	6 6 6 6 5 5 5 6 5 6	5
371	15 6 15 6 6 6 6 7 7	6 6 6 6 6 6 15 15 6 15	6 6 6 15 6 6 6 7 6 6	6 6 6 6 6 7 6 6 6 6	6 6 6 7 5 6 5 6 5 6	5
372	6 6 6 15 6 6 6 6 6	6 6 6 6 6 6 6 6 6 6	15 6 6 6 6 6 6 6 7 6	6 6 6 6 6 6 7 6 6 6	6 6 6 6 6 5 6 6 5 6	5
373	* 6 6 6 15 6 6 7 7	6 6 6 6 6 6 6 6 6 6	6 15 6 6 6 6 6 6 6 7	6 6 6 6 6 6 6 7 6 6	6 6 6 7 6 6 5 6 5 6	5
374	6 6 6 6 6 15 6 6 6	6 6 6 6 6 6 6 6 6 6	6 6 15 6 6 6 6 6 6 6	7 6 6 6 6 6 6 6 7 6	6 6 6 6 6 6 6 6 5 6	5
375	15 6 6 6 6 6 15 7 7	6 6 6 6 6 6 15 15 6 6	6 6 6 15 6 6 6 6 6 6	6 7 6 6 6 6 6 6 6 7	6 6 6 7 6 6 6 6 5 6	5
376	7 6 6 6 6 6 6 15 7	6 6 6 6 6 6 7 7 6 6	6 6 6 7 14 7 6 6 6 6	6 6 7 6 6 6 6 6 6 6	7 6 6 7 6 6 6 7 5 6	5
377	13 6 6 7 6 6 6 7 15	6 6 6 12 6 6 13 13 6 13	6 6 6 13 6 13 6 6 6 12	6 6 6 7 6 6 6 6 6 6	6 7 6 7 6 6 6 7 5 6	5
378	6 7 6 6 6 6 6 6 6	15 6 6 6 6 6 6 6 6 6	6 6 6 6 6 6 12 6 6 6	6 6 6 6 7 6 6 6 6 6	6 6 7 6 6 6 6 6 5 6	5
379	* 6 7 6 6 6 6 7 7	6 15 6 6 6 6 11 11 6 6	6 6 6 11 6 6 6 11 6 6	6 6 6 6 6 7 6 6 6 6	6 6 6 7 6 6 6 7 5 6	5
380	6 6 6 7 6 6 6 6 6	6 6 15 6 6 6 6 6 6 6	6 6 6 6 6 6 6 6 10 6	6 6 6 6 6 6 7 6 6 6	6 6 6 6 6 6 6 6 5 6	5
381	12 6 6 6 7 6 6 7 11	6 6 6 15 6 6 12 12 6 6	6 9 6 12 6 6 6 6 6 9	6 6 6 6 6 6 6 7 6 6	6 6 6 7 6 6 6 7 6 6	5
382	6 6 6 6 6 7 6 6 6	6 6 6 6 15 6 6 6 6 6	6 6 6 6 6 6 6 6 6 6	8 6 6 6 6 6 6 6 7 6	6 6 6 6 6 6 6 6 5 6	5
383	* 6 6 6 6 6 7 6 6	6 6 6 6 6 15 7 7 6 6	6 6 6 7 6 6 6 6 6 6	6 7 6 6 6 6 6 6 6 7	6 6 6 6 6 6 6 6 5 6	6
384	15 6 6 6 6 6 6 7 7	6 6 6 6 6 6 15 15 6 6	6 6 6 15 6 6 6 6 6 6	6 6 6 6 6 6 6 6 6 6	7 6 6 7 6 6 6 7 6 6	5
385	15 6 6 7 6 6 6 7 7	6 6 6 6 6 6 15 15 6 6	6 6 6 15 6 6 6 6 6 6	6 6 9 6 6 6 6 6 6 6	6 7 6 7 6 6 6 7 7 6	5
386	6 6 6 6 6 6 6 6 6	7 6 6 6 6 6 6 6 15 6	6 6 6 6 6 6 6 6 6 6	6 6 6 6 6 6 6 6 6 6	6 6 7 6 6 6 6 6 6 6	6
387	15 6 6 6 6 6 6 7 8	6 7 6 6 6 6 15 15 6 15	6 6 6 15 6 6 6 6 6 6	6 6 6 6 6 6 6 6 6 6	6 6 6 8 6 6 6 7 6 6	5
388	6 6 6 6 6 6 6 6 6	6 6 7 6 6 6 6 6 6 6	15 6 6 6 6 6 6 6 6 6	6 6 6 6 6 6 6 6 6 6	6 6 6 6 6 6 6 6 6 6	5
389	* 6 6 6 6 6 6 7 7	6 7 6 7 6 6 6 6 6 6	6 15 6 6 6 6 6 6 6 7	6 6 6 6 6 6 6 7 6 6	6 6 6 6 6 6 6 7 5 6	5
390	6 6 6 6 6 6 6 6 6	6 6 6 6 7 6 6 6 6 6	6 6 15 6 6 6 6 6 6 6	6 6 6 6 6 6 6 6 6 6	6 6 6 6 6 6 6 6 6 6	5
391	16 6 6 6 6 6 6 7 7	6 6 6 6 6 7 6 16 6 6	6 6 6 16 6 6 6 6 6 6	6 6 6 6 6 6 6 6 6 6	6 6 6 6 6 6 6 7 6 6	6
392	7 6 6 6 6 7 6 7 7	6 6 6 6 6 6 7 6 6 6	6 6 6 7 7 7 6 6 6 6	6 6 6 6 6 6 6 6 6 6	6 6 6 6 6 6 6 7 7 7	6
393	8 6 6 6 6 7 6 7 8	6 6 6 6 6 6 6 8 6 6	6 9 6 6 6 7 6 6 6 6	6 6 6 6 6 6 6 6 6 6	6 6 6 7 6 6 6 7 6 7	6
394	6 7 6 6 6 6 6 6 6	7 6 6 6 6 6 6 6 7 6	6 6 9 6 6 6 7 6 6 6	6 6 6 6 6 6 6 6 6 6	6 6 6 6 6 6 6 6 6 6	6
395	7 6 7 6 6 7 6 7 7	6 7 6 6 6 6 6 6 6 7	6 6 6 6 6 6 6 7 6 6	6 6 6 6 6 6 6 6 6 6	6 6 6 6 6 6 6 7 6 7	5
396	6 6 6 7 6 6 6 6 6	6 6 7 6 6 6 6 6 6 6	7 6 6 6 6 6 6 6 7 6	6 6 6 6 6 6 6 6 6 6	6 6 6 6 6 6 6 6 6 6	5
397	* 6 6 6 7 7 6 7 7	6 6 6 7 6 6 6 6 6 6	6 7 6 6 6 6 6 6 6 7	6 6 6 6 6 6 6 6 6 6	6 6 6 6 6 6 6 7 6 7	5
398	6 6 6 6 6 7 6 6 6	6 6 6 6 7 6 6 6 6 6	6 6 7 6 6 6 6 6 6 6	7 6 6 6 6 6 6 6 6 6	6 6 6 6 6 6 6 6 6 6	5
399	7 6 6 6 6 6 7 7 7	6 6 7 6 6 7 6 6 6 6	6 6 6 7 6 6 6 6 6 6	6 7 6 6 6 6 6 6 6 6	6 6 6 6 6 6 6 7 6 6	6

h →	0	1	2	3	4	5
n ↓	1 2 3 4 5 6 7 8 9	0 1 2 3 4 5 6 7 8 9	0 1 2 3 4 5 6 7 8 9	0 1 2 3 4 5 6 7 8 9	0 1 2 3 4 5 6 7 8 9	0
400	15 6 6 6 6 7 6 7 7	6 10 6 10 6 6 15 6 6 6	6 6 6 6 7 15 6 6 6 6	6 6 9 6 6 6 6 6 6 6	6 6 6 6 6 6 6 7 6 7	6
401	* 6 6 6 6 7 6 7 7	6 6 6 6 6 6 15 15 6 6	6 6 6 6 6 15 6 6 6 6	6 6 6 7 6 6 6 6 6 6	6 6 6 6 6 6 6 7 6 7	6
402	7 15 6 6 6 6 6 6 7	7 7 6 7 6 6 7 7 15 6	6 6 6 6 6 6 7 6 6 6	6 6 6 6 7 6 6 6 6 6	6 6 6 6 6 6 6 6 6 6	5
403	15 6 15 6 6 7 15 7 7	6 7 6 12 6 6 15 15 6 15	6 6 6 6 6 15 6 7 6 6	6 12 6 6 6 7 6 6 6 6	6 6 6 6 6 6 6 7 6 7	6
404	11 6 6 15 6 6 6 6 6	6 6 7 11 6 6 6 6 6 6	15 6 6 6 6 6 6 6 7 6	6 11 11 6 6 6 7 6 6 6	6 6 6 6 6 6 6 6 6 6	5
405	7 11 6 6 15 7 6 7 7	6 6 6 7 6 6 6 6 7 6	6 15 6 6 6 6 6 6 6 7	6 6 6 10 6 6 6 7 6 6	6 6 6 6 6 6 6 7 6 7	6
406	6 6 11 6 6 15 6 6 6	6 6 6 6 7 6 6 6 6 6	6 6 15 6 6 6 6 6 6 6	7 6 6 6 9 6 6 6 7 6	6 6 6 6 6 6 6 6 6 6	5
407	15 6 6 11 6 7 15 7 7	6 10 6 6 6 7 15 15 6 6	6 6 6 15 6 15 6 6 6 6	6 7 6 6 6 8 6 10 6 7	6 6 6 6 6 6 6 7 6 7	6
408	7 6 6 6 11 7 6 15 7	6 6 6 6 6 6 7 7 6 6	6 6 6 6 15 7 6 7 6 6	6 6 7 6 6 6 7 6 6 6	7 6 6 6 6 6 6 7 6 7	6
409	* 6 6 7 6 11 6 7 15	6 6 6 6 6 6 15 15 6 15	6 6 6 6 6 15 6 6 6 6	6 6 6 7 6 6 6 6 6 6	6 7 6 6 6 6 6 7 6 7	6
410	6 6 6 6 6 6 11 6 6	15 6 6 6 6 6 6 6 7 6	6 6 6 6 6 6 14 6 6 6	6 6 6 6 7 6 6 6 6 6	6 6 7 6 6 6 6 6 6 6	5
411	13 6 6 7 6 7 6 11 7	6 15 6 7 6 6 13 13 6 7	6 6 6 6 7 13 6 13 6 6	6 7 7 6 6 7 6 7 6 6	6 7 6 7 6 6 6 7 6 7	6
412	8 6 6 6 6 6 6 6 11	6 6 15 8 6 6 6 6 6 6	7 6 6 6 6 6 6 6 12 6	6 8 8 6 6 6 7 6 6 6	6 6 6 6 7 6 6 6 6 6	6
413	12 6 6 6 6 7 6 7 7	11 6 6 15 6 6 12 12 6 6	6 7 6 6 6 12 6 6 6 11	6 6 6 6 6 6 6 7 6 6	6 6 6 6 6 7 6 7 6 7	5
414	10 6 6 6 6 6 6 6 6	6 11 6 10 15 6 6 6 6 6	6 6 7 6 6 6 6 6 6 6	10 10 10 6 6 6 6 6 7 6	6 6 6 6 6 6 7 6 6 6	6
415	9 6 6 7 6 7 6 7 7	6 7 11 6 6 15 9 9 6 6	6 6 6 7 6 9 6 6 6 6	6 9 6 6 6 6 6 7 6 7	6 7 6 6 6 7 6 7 6 7	6
416	18 6 6 6 6 7 6 7 7	6 6 6 12 6 6 15 15 6 6	6 6 6 6 7 15 6 6 6 6	6 6 12 6 6 6 6 6 6 6	7 6 6 6 6 6 6 7 6 7	6
417	15 6 6 7 6 7 6 7 7	6 6 6 6 6 6 15 15 6 6	6 6 6 6 6 15 6 6 6 6	7 6 6 7 6 6 6 6 6 6	6 7 6 6 6 6 6 7 6 7	6
418	6 6 6 6 6 6 6 6 6	6 6 6 6 6 6 6 6 15 6	6 6 6 6 6 6 7 6 6 6	6 6 6 6 7 6 6 6 6 6	6 6 7 6 6 6 6 6 6 6	6
419	* 6 6 6 6 7 6 7 7	6 7 7 6 6 6 15 15 6 15	6 6 6 6 6 15 6 7 6 6	6 6 6 6 6 7 6 7 6 6	6 6 6 7 6 6 6 7 6 7	6
420	6 6 6 6 6 6 6 6 6	6 6 6 6 6 6 6 6 6 6	15 6 6 6 6 6 6 6 7 6	6 6 6 6 6 6 7 6 6 6	6 6 6 6 7 6 6 6 6 6	6
421	* 6 6 6 6 7 6 7 7	6 6 6 6 6 6 6 6 11 6	6 15 6 6 6 6 6 6 6 7	6 6 6 6 6 6 6 7 6 6	6 6 9 6 6 7 6 6 6 7	6
422	6 6 6 6 6 6 6 6 6	6 6 6 6 6 6 6 6 6 6	6 6 15 6 6 6 6 6 6 6	7 6 6 6 6 6 6 6 7 6	6 6 6 6 6 6 7 6 6 6	6
423	15 6 6 7 6 7 6 7 8	6 7 6 6 6 6 15 15 6 6	6 6 6 15 6 15 6 6 6 6	6 7 6 6 6 6 7 7 6 7	6 7 6 6 6 6 6 8 6 7	6
424	7 6 6 6 6 7 6 7 7	6 6 6 6 6 6 7 6 6 6	6 6 6 6 15 7 6 7 6 6	6 6 7 6 6 6 6 6 6 6	7 6 6 6 6 6 6 6 6 7	6
425	16 6 6 7 6 7 6 7 8	6 6 6 6 6 6 6 16 6 6	6 6 6 6 6 16 6 6 6 6	6 6 6 7 6 6 6 6 6 6	6 7 6 6 6 6 6 6 6 7	6
426	6 7 6 6 6 6 6 6 6	7 6 6 6 6 6 6 6 6 6	6 6 6 6 6 6 6 6 6 6	6 6 6 6 7 6 6 6 6 6	6 6 7 6 6 6 6 6 6 6	6
427	7 6 7 6 6 7 6 7 7	6 7 6 6 6 6 6 6 6 6	7 6 6 6 6 6 6 6 6 6	6 6 6 6 6 7 6 6 6 6	6 6 6 7 6 6 6 6 6 7	6
428	6 6 6 7 6 6 6 6 6	6 6 7 6 6 6 6 6 6 6	6 6 6 6 6 6 6 6 6 6	6 6 6 6 6 6 7 6 6 6	6 6 6 6 7 6 6 6 6 6	6
429	10 6 6 6 7 7 6 7 7	6 6 6 10 6 6 6 10 6 6	6 6 6 6 6 6 6 6 6 6	6 6 6 6 6 6 6 7 6 6	6 6 7 6 6 7 6 6 6 7	6
430	6 6 6 6 6 7 6 6 6	6 6 6 6 7 6 6 6 6 6	6 6 6 6 6 6 6 6 6 6	6 6 6 6 6 6 6 6 7 6	6 6 6 6 6 6 7 6 6 6	6
431	* 6 6 6 6 7 7 7 7	6 6 6 6 6 7 6 6 6 6	6 6 6 6 7 6 6 6 6 6	6 6 6 6 6 6 6 6 6 7	6 6 6 6 6 6 6 7 6 7	6
432	15 6 6 6 6 7 6 7 7	6 7 6 6 6 6 15 6 6 6	6 6 6 6 6 6 6 15 6 6	6 6 7 6 6 6 6 6 6 6	7 6 6 6 6 6 6 6 7 7	6
433	* 6 6 7 6 7 6 7 7	6 6 6 6 6 6 15 15 6 6	6 6 6 6 6 6 6 15 6 6	6 6 6 7 6 6 9 6 6 6	6 7 6 6 6 6 6 6 6 7	6
434	6 15 6 6 6 6 6 6 6	7 6 6 6 6 6 6 6 15 6	6 6 6 6 6 6 6 6 6 6	6 6 6 6 6 6 6 6 6 6	6 6 7 6 6 6 6 6 6 6	6
435	15 6 15 6 6 7 6 7 7	6 7 6 6 6 6 15 15 6 15	6 6 6 6 6 6 6 15 7 6	6 6 6 6 6 6 6 6 6 6	6 6 6 7 6 6 6 6 7 7	6
436	6 6 6 15 6 6 6 6 6	6 6 7 6 6 6 6 6 6 6	15 6 6 6 6 6 6 6 6 6	6 6 6 6 6 6 6 6 6 6	6 6 6 6 7 6 6 6 6 6	6
437	18 6 6 6 15 6 6 7 7	6 6 6 7 6 6 6 6 6 18	6 15 6 18 6 6 6 6 6 6	6 6 6 6 6 6 6 6 6 6	6 6 6 6 6 7 6 6 6 6	6
438	6 6 6 6 6 15 6 6 6	6 6 6 6 7 6 6 6 6 6	6 6 15 6 6 6 6 6 6 6	6 6 6 6 6 6 6 6 6 6	6 6 6 6 6 6 7 6 6 6	6
439	* 6 6 7 6 7 15 7 7	6 7 6 6 6 7 15 15 6 6	6 6 6 15 6 6 6 15 6 6	6 6 7 6 6 6 6 7 6 6	6 7 6 6 6 6 6 7 6 7	6
440	7 6 6 6 6 6 6 15 7	6 7 6 6 6 6 7 7 6 6	6 6 6 6 15 6 6 7 6 7	6 6 6 6 6 6 6 6 6 6	7 6 6 6 6 6 6 6 7 6	6
441	15 6 6 7 6 7 6 7 15	6 7 6 6 6 6 15 15 6 15	6 6 6 6 6 15 6 15 6 6	6 6 6 6 6 6 6 6 6 6	7 7 6 6 9 6 6 6 6 8	6
442	6 7 6 6 6 6 6 6 6	15 6 6 6 6 6 6 6 7 6	6 6 6 6 6 6 15 6 6 6	6 6 6 6 6 6 6 6 6 6	6 6 7 6 6 6 6 6 6 6	6
443	* 6 7 7 6 6 6 7 7	6 15 6 6 6 6 15 15 6 7	6 6 6 6 6 6 6 15 6 6	6 6 6 6 6 6 7 6 6 6	7 7 6 7 6 6 6 6 6 6	6
444	6 6 6 7 6 6 6 6 6	6 6 15 6 6 6 6 6 6 6	7 6 6 6 6 6 6 6 14 6	6 6 6 6 6 6 6 6 6 6	6 6 6 6 6 6 6 6 6 6	6
445	13 6 6 6 7 6 6 7 7	6 6 6 15 6 6 13 13 6 6	6 7 6 6 6 6 6 13 6 13	6 6 6 6 6 6 6 6 6 6	6 6 6 6 6 6 6 6 6 6	6
446	6 6 6 6 6 7 6 6 6	6 6 6 6 15 6 6 6 6 6	6 6 7 6 6 6 6 6 6 6	12 6 6 6 6 6 6 6 6 6	6 6 6 6 6 6 6 6 6 6	6
447	11 6 6 7 6 6 7 7 7	6 7 6 6 6 15 11 11 6 6	6 6 6 7 6 6 6 11 6 6	6 11 6 6 6 6 6 7 6 6	7 7 6 6 6 6 6 6 6 6	6
448	15 6 6 7 6 6 6 7 7	6 7 6 6 6 6 15 15 6 6	6 6 6 6 7 6 6 15 6 6	6 6 10 6 6 6 6 6 6 6	7 7 6 6 6 6 6 6 6 6	6
449	* 6 6 7 6 6 6 7 7	6 7 6 6 6 6 15 15 6 6	6 6 6 6 6 7 6 15 6 6	6 6 6 9 6 6 6 6 6 6	7 7 6 6 6 6 6 6 6 6	6
450	6 6 6 6 6 6 6 6 6	7 6 6 6 6 6 6 6 15 6	6 6 6 6 6 6 7 6 6 6	6 6 6 6 8 6 6 6 6 6	6 6 6 6 6 6 6 6 6 6	6

n ↓ \ h →	1	2	3	4	5	6	7	8	9	10	11	12	13	14	15	16	17	18	19
451	15	6	6	7	6	6	6	7	7	6	10	6	6	6	6	15	15	6	15
452	6	6	6	6	6	6	6	6	6	6	6	6	6	6	6	6	6	6	6
453	7	6	6	6	6	6	6	7	7	6	6	6	6	6	6	6	6	6	6
454	6	6	6	6	6	6	6	6	6	6	6	6	6	6	6	6	6	6	6
455	15	6	6	6	6	6	6	7	7	6	6	6	6	6	6	15	15	6	6
456	7	6	6	6	6	6	6	7	7	6	6	6	6	6	6	7	6	6	6
457	*	6	6	7	6	6	6	7	8	6	6	6	6	6	6	15	15	6	6
458	6	6	6	6	6	6	6	6	6	6	6	6	6	6	6	6	6	6	6
459	16	6	6	6	6	6	6	7	8	6	9	6	9	6	6	8	16	6	8
460	6	6	6	6	6	6	6	6	6	6	6	6	6	6	6	6	6	6	6
461	*	6	6	6	6	6	6	7	7	6	6	6	10	6	6	6	10	6	6
462	6	6	6	6	6	6	6	6	6	6	6	6	6	6	6	6	6	6	6
463	*	6	6	6	6	6	6	7	7	6	6	6	6	6	6	6	6	6	6
464	15	6	6	6	6	6	6	7	7	6	6	6	6	6	6	15	6	6	6
465	15	6	6	7	6	7	6	7	7	6	6	6	6	6	6	15	15	6	6
466	6	15	6	6	6	6	6	6	6	6	6	6	6	6	6	6	6	15	6
467	*	6	15	6	6	6	6	7	7	6	6	6	6	6	6	15	15	6	15
468	6	6	6	15	6	6	6	6	6	6	6	6	6	6	6	6	6	6	6
469	8	6	6	6	15	6	6	7	7	6	6	6	6	6	6	6	6	6	6
470	6	6	6	6	6	15	6	6	6	6	6	6	6	6	6	6	6	6	6
471	15	6	6	6	6	6	15	7	7	6	6	6	6	6	6	15	15	6	6
472	7	6	6	6	6	6	6	15	6	6	6	6	6	6	6	7	7	6	6
473	15	6	6	6	6	7	6	7	15	6	10	6	6	6	6	15	15	6	15
474	10	7	6	6	6	6	6	6	6	15	6	6	6	6	6	6	6	6	6
475	18	6	7	6	6	6	6	7	7	6	15	6	6	6	6	15	15	6	18
476	6	6	6	7	6	6	6	6	6	6	6	15	6	6	6	6	6	6	6
477	15	6	6	6	7	6	6	7	8	6	6	6	15	6	6	15	15	6	6
478	6	6	6	6	6	7	6	6	6	6	6	6	6	15	6	6	6	6	6
479	*	6	6	6	6	7	7	6	6	6	15	6	6	6	15	13	13	6	10
480	15	6	6	6	6	6	6	7	7	6	6	6	6	6	6	15	15	6	6
481	15	6	6	6	6	6	6	7	7	6	6	6	12	6	6	15	15	6	6
482	6	6	6	6	6	6	6	6	6	7	6	6	6	6	6	6	6	15	6
483	15	6	6	6	6	6	6	7	7	6	7	6	6	6	6	15	15	6	15
484	6	6	6	6	6	6	6	6	6	6	6	7	6	6	6	6	6	6	6
485	7	6	6	7	6	6	6	7	7	6	6	7	7	6	6	7	7	6	6
486	6	6	6	6	6	6	6	6	6	6	6	6	6	7	6	6	6	6	6
487	*	6	6	6	6	6	6	7	7	6	6	6	6	6	7	15	15	6	6
488	7	6	6	6	6	6	6	7	7	6	6	6	6	6	6	7	6	6	6
489	15	6	6	7	6	6	6	7	8	6	6	6	7	6	6	15	15	6	6
490	6	7	6	6	6	6	6	6	6	7	6	6	6	6	6	6	6	7	6
491	*	6	7	6	6	6	6	7	7	6	7	6	6	6	6	15	15	7	7
492	6	6	6	7	6	6	6	6	6	6	6	7	6	6	6	6	6	6	6
493	16	6	6	6	7	7	6	7	7	6	6	7	10	6	6	6	16	6	6
494	6	6	6	6	6	7	6	6	6	6	6	6	6	7	6	6	6	6	6
495	7	6	6	6	6	6	7	7	7	6	6	6	6	6	7	6	6	6	6
496	15	6	6	6	6	6	6	7	7	6	6	6	6	6	6	15	14	6	6
497	15	6	6	7	6	6	6	7	7	6	6	6	7	6	6	15	15	6	6
498	6	15	6	6	6	6	6	6	6	7	6	6	6	6	6	6	6	15	6
499	*	6	15	6	6	6	6	7	7	6	7	6	6	6	6	15	15	6	15
500	6	6	6	15	6	6	6	6	6	6	6	7	6	6	6	6	6	6	6
501	12	6	6	6	15	6	6	7	7	6	6	6	12	6	6	6	12	6	6

n ↓ \ h →	20	21	22	23	24	25	26	27	28	29	30	31	32	33	34	35	36	37	38	39
451	6	6	6	6	6	6	6	15	6	6	6	6	6	6	6	7	6	6	6	6
452	15	6	6	6	6	6	6	6	7	6	6	6	6	6	6	6	7	6	6	6
453	6	15	6	6	6	6	6	6	6	7	6	6	6	6	6	6	6	7	6	6
454	6	6	15	6	6	6	6	6	6	6	7	6	6	6	6	6	6	6	7	6
455	6	6	6	15	6	6	6	15	6	6	6	7	6	6	6	7	6	6	6	7
456	6	6	6	6	15	6	6	7	6	7	6	6	7	6	6	6	6	6	6	6
457	6	6	6	6	6	15	6	15	6	6	6	6	6	7	6	6	6	6	9	6
458	6	6	6	6	6	6	15	6	6	6	6	6	6	6	7	6	6	6	6	6
459	6	6	6	6	6	6	6	16	6	6	6	6	6	6	6	7	6	9	6	6
460	6	6	6	6	6	6	6	6	14	6	6	6	6	6	6	6	7	6	6	6
461	6	6	6	6	6	6	6	6	6	6	6	6	6	6	6	6	6	7	6	6
462	6	6	6	6	6	6	6	6	6	6	6	6	6	6	6	6	6	6	7	6
463	6	6	6	6	6	6	6	6	6	6	6	6	6	6	6	6	6	6	6	7
464	6	6	6	6	6	6	6	6	6	15	6	6	6	6	6	6	6	6	6	6
465	6	6	6	6	6	6	6	6	6	15	6	6	6	6	6	6	6	6	6	6
466	6	6	6	6	6	6	6	6	6	6	6	6	6	6	6	6	6	6	6	6
467	6	6	6	6	6	6	6	6	6	15	6	6	6	6	6	6	6	6	6	6
468	15	6	6	6	6	6	6	6	6	6	6	6	6	6	6	6	6	6	6	6
469	6	15	6	6	6	6	6	6	6	6	6	6	6	6	6	6	6	6	6	6
470	6	6	15	6	6	6	6	6	6	6	6	6	6	6	6	6	6	6	6	6
471	6	6	6	15	6	6	6	6	6	15	6	6	6	6	6	6	6	6	6	6
472	6	6	6	6	15	6	6	6	6	7	6	7	6	6	6	6	6	6	6	6
473	6	6	6	6	6	15	6	6	6	15	6	6	6	6	6	6	6	6	6	6
474	6	6	6	6	6	6	15	6	6	6	6	6	6	6	6	6	6	6	6	6
475	6	6	6	6	6	18	6	15	6	15	6	6	6	6	6	6	6	6	6	6
476	6	6	6	6	6	6	6	6	15	6	6	6	6	6	6	6	6	6	6	6
477	6	6	6	6	6	6	6	6	6	15	6	6	6	6	6	6	6	6	6	6
478	6	6	6	6	6	6	6	6	6	6	14	6	6	6	6	6	6	6	6	6
479	6	6	6	6	6	6	6	10	6	13	6	13	6	6	6	6	6	6	6	6
480	6	6	6	6	6	6	6	6	6	15	6	6	12	6	6	6	6	6	6	6
481	6	6	6	6	6	6	6	6	6	15	6	6	6	11	6	6	6	12	6	6
482	6	6	6	6	6	6	6	6	6	6	6	6	6	6	10	6	6	6	6	6
483	6	6	6	6	6	6	6	6	6	15	6	6	6	6	6	9	6	6	6	6
484	15	6	6	6	6	6	6	6	6	6	6	6	6	6	6	6	8	6	6	6
485	6	15	6	6	6	6	6	6	6	7	6	6	6	6	6	6	6	7	6	6
486	6	6	15	6	6	6	6	6	6	6	6	6	6	6	6	6	6	6	6	6
487	6	6	6	15	6	6	6	6	6	15	6	6	6	6	6	6	6	6	6	6
488	6	6	6	6	15	6	6	6	6	7	6	7	6	6	6	6	6	6	6	6
489	6	6	6	6	6	15	6	6	6	15	6	6	6	6	6	6	6	7	6	6
490	6	6	6	6	6	6	15	6	6	6	6	6	6	6	6	6	6	6	6	6
491	6	6	6	6	6	6	6	15	6	15	6	6	6	6	6	6	6	6	6	6
492	7	6	6	6	6	6	6	6	15	6	6	6	6	6	6	6	6	6	6	6
493	6	7	6	6	6	6	6	6	6	16	6	6	6	6	6	6	6	7	6	6
494	6	6	7	6	6	6	6	6	6	6	7	6	6	6	6	6	6	6	6	6
495	6	6	6	7	6	6	6	6	6	6	6	7	6	6	6	6	6	6	6	6
496	6	6	6	6	7	6	6	6	6	6	6	15	7	6	6	6	6	6	6	6
497	6	6	6	6	7	7	6	6	6	6	6	15	14	7	6	6	6	7	6	6
498	6	6	6	6	6	6	7	6	6	6	6	6	6	13	7	6	6	6	6	6
499	6	6	6	6	6	6	6	7	6	6	6	15	6	6	12	7	6	6	6	7
500	15	6	6	6	6	6	6	6	7	6	6	6	6	6	6	11	7	6	6	6
501	7	15	6	6	6	6	6	6	6	7	6	6	6	6	6	6	10	7	6	6

n ↓ \ h →	40	41	42	43	44	45	46	47	48	49	50
451	6	10	6	6	7	6	6	6	6	6	6
452	6	6	6	6	6	6	6	6	6	6	6
453	6	6	6	6	6	6	6	6	6	6	6
454	6	6	6	6	6	6	6	6	6	6	6
455	6	6	6	6	6	6	6	6	6	6	6
456	7	6	6	6	6	6	6	6	6	6	6
457	6	7	6	6	6	6	6	6	6	6	6
458	6	6	7	6	6	6	6	6	6	6	6
459	6	6	6	7	6	6	6	6	6	6	6
460	6	6	6	6	7	6	6	6	6	6	6
461	6	6	6	6	6	7	6	6	6	6	6
462	6	6	6	6	6	6	7	6	6	6	6
463	6	6	6	6	6	6	6	7	6	6	6
464	7	6	6	6	6	6	6	6	7	6	6
465	6	7	6	6	6	6	6	6	6	7	6
466	6	6	7	6	6	6	6	6	6	6	7
467	6	6	6	7	6	6	6	6	6	6	6
468	6	6	6	6	7	6	6	6	6	6	6
469	6	6	6	6	6	7	6	6	6	6	6
470	6	6	6	6	6	6	7	6	6	6	6
471	6	6	6	6	6	6	6	7	6	6	6
472	6	6	6	6	6	6	6	6	7	6	6
473	6	6	6	10	6	6	6	6	6	7	6
474	6	6	6	10	6	6	6	6	6	6	7
475	6	6	6	6	6	6	6	6	6	6	6
476	6	6	6	6	6	6	6	6	6	6	6
477	6	6	6	6	6	6	6	6	6	6	6
478	6	6	6	6	6	6	6	6	6	6	6
479	6	6	6	7	6	6	6	6	6	7	6
480	6	6	6	6	6	6	6	6	6	6	6
481	6	6	6	6	6	6	6	6	9	6	6
482	6	6	6	6	6	6	6	6	6	6	6
483	6	6	6	6	6	6	6	6	6	6	6
484	6	6	6	6	6	6	6	6	6	6	6
485	6	7	6	6	6	6	6	6	6	6	6
486	6	6	6	6	6	6	6	6	6	6	6
487	6	6	6	6	6	6	6	6	6	6	6
488	6	6	6	6	6	6	6	6	6	6	6
489	6	7	6	6	6	7	6	6	6	6	6
490	6	6	6	6	6	6	6	6	6	6	6
491	6	6	6	7	6	6	6	6	6	6	6
492	6	6	6	6	6	6	6	6	6	6	6
493	6	6	6	6	6	6	6	6	6	7	6
494	6	6	6	6	6	6	6	6	6	6	6
495	6	6	6	6	6	6	6	6	6	6	6
496	6	6	6	6	6	6	6	6	6	6	6
497	6	7	6	6	6	6	6	6	6	6	6
498	6	6	6	6	6	6	6	6	6	6	6
499	6	6	6	6	6	6	6	6	6	6	6
500	6	6	6	6	6	6	6	6	6	6	6
501	6	6	6	6	6	6	6	6	6	6	6

$h \rightarrow$	0	1	2	3	4	5
$n \downarrow$	1 2 3 4 5 6 7 8 9	0 1 2 3 4 5 6 7 8 9	0 1 2 3 4 5 6 7 8 9	0 1 2 3 4 5 6 7 8 9	0 1 2 3 4 5 6 7 8 9	0
502	9 6 6 6 6 15 6 6 6	6 6 6 6 7 6 9 6 6 6	6 6 15 6 6 6 6 6 6 6	7 9 9 6 6 6 6 9 7 6	6 6 6 6 6 6 6 6 6 6	6
503	* 6 6 6 6 6 15 7 7	6 6 6 6 6 7 15 15 6 6	6 6 6 15 6 6 6 6 6 6	7 15 6 6 6 6 6 6 8 7	6 6 6 6 6 6 6 6 6 6	6
504	7 6 6 6 6 6 6 15 7	6 6 6 6 6 6 7 7 6 6	6 6 6 6 15 6 6 6 6 6	6 7 7 6 6 6 6 6 6 7	7 6 6 6 6 6 6 6 6 6	6
505	15 6 6 7 6 7 6 7 15	6 6 6 6 6 6 15 15 6 15	6 6 6 6 7 15 6 6 6 6	6 15 8 7 6 6 6 6 6 6	6 7 9 6 6 6 6 6 6 7	6
506	6 7 6 6 6 6 6 6 6	15 6 6 6 6 6 6 6 7 6	6 6 6 6 6 6 15 6 6 6	6 6 6 6 7 6 6 6 6 6	6 6 7 6 6 6 6 6 6 6	7
507	15 6 7 6 6 6 6 7 7	6 15 6 6 6 6 15 15 6 7	6 6 6 6 6 6 6 15 6 6	6 15 10 6 6 7 6 6 6 6	6 6 6 7 6 6 6 6 6 6	6
508	6 6 6 7 6 6 6 6 6	6 6 15 6 6 6 6 6 6 6	7 6 6 6 6 6 6 6 15 6	6 6 6 6 6 6 7 6 6 6	6 6 6 6 7 6 6 6 6 6	6
509	* 6 6 6 7 6 6 7 7	6 6 6 15 6 6 15 15 6 6	6 7 6 6 6 6 6 6 7 15	6 15 12 6 6 6 7 7 6 6	6 6 6 6 6 7 6 6 6 6	6
510	6 6 6 6 6 7 6 6 6	6 6 6 6 15 6 6 6 6 6	6 6 7 6 6 6 6 6 6 6	15 6 6 6 6 6 6 6 7 6	6 6 6 6 6 6 7 6 6 6	6
511	15 6 6 7 6 6 7 7 7	6 9 6 9 6 15 15 15 6 6	6 6 6 7 6 6 6 6 6 6	6 15 6 6 6 6 6 6 6 7	6 9 6 6 6 6 6 7 6 6	6
512	* 6 6 6 6 6 6 7 7	6 6 6 6 6 6 15 15 6 6	6 6 6 6 7 6 6 6 6 6	6 15 15 6 6 6 6 6 6 6	7 6 6 6 6 6 6 6 7 6	6
513	18 14 6 7 6 7 6 7 8	6 6 6 7 6 6 15 15 6 18	6 6 6 6 6 7 6 18 6 6	6 15 15 14 6 6 6 7 6 6	6 7 6 6 6 6 6 6 6 7	6
514	7 15 14 6 6 6 6 6 6	7 6 6 6 6 6 6 6 15 6	6 6 6 6 6 6 7 6 6 6	6 6 6 6 12 6 6 6 6 6	6 6 7 6 6 6 6 6 6 6	7
515	15 6 15 14 6 6 6 7 7	6 7 6 6 6 6 15 15 6 15	6 6 6 6 6 6 6 7 6 6	6 15 15 6 6 11 6 6 6 6	6 6 7 7 6 6 6 6 6 6	6
516	6 6 6 15 14 6 6 6 6	6 6 7 6 6 6 6 6 6 6	15 6 6 6 6 6 6 6 7 6	6 6 6 6 6 6 10 6 6 6	6 6 6 6 7 6 6 6 6 6	6
517	12 6 6 6 15 14 6 7 7	6 10 6 15 6 6 9 12 6 12	6 15 6 6 6 6 6 6 6 12	6 9 6 6 6 6 7 9 6 6	6 6 6 6 6 7 6 10 6 6	6
518	6 6 6 6 6 15 14 6 6	6 6 6 6 7 6 6 6 6 6	6 6 15 6 6 6 6 6 6 6	7 6 6 6 6 6 6 6 8 6	6 6 6 6 6 6 7 6 6 6	6
519	15 6 6 6 6 6 15 14 7	6 7 6 6 6 7 15 15 6 15	6 6 6 15 6 6 6 6 6 6	6 15 15 6 6 6 6 6 6 7	6 6 6 6 6 6 6 7 6 6	6
520	8 6 6 6 6 6 6 15 14	6 6 6 7 6 6 7 7 6 6	6 6 6 6 15 6 6 6 6 6	6 8 8 6 6 6 6 6 6 6	7 6 6 6 6 6 6 6 7 6	6
521	* 6 6 7 6 7 6 7 15	14 7 6 7 6 6 15 15 6 15	6 6 6 6 6 15 6 6 6 6	6 15 15 7 6 6 6 7 6 6	7 7 6 6 6 6 6 6 7 7	6
522	10 7 6 6 6 6 6 6 6	15 14 6 6 6 6 9 9 7 6	6 6 6 6 6 6 15 6 6 6	6 10 10 6 7 6 6 6 6 6	6 6 7 6 6 6 6 6 6 6	7
523	* 6 7 7 6 6 6 7 7	6 15 14 7 6 6 15 15 6 7	6 6 6 6 6 6 6 15 6 6	6 15 15 6 6 7 6 6 6 6	7 7 6 7 6 6 6 6 6 6	6
524	12 6 6 7 6 6 6 6 6	6 6 15 14 6 6 11 11 6 6	7 6 6 6 6 6 6 6 15 6	6 12 12 6 6 6 7 6 6 6	6 6 6 6 7 6 6 6 6 6	6
525	15 6 6 6 7 6 6 7 7	6 6 6 15 14 6 15 15 6 6	6 7 6 6 6 6 6 6 6 15	6 15 15 6 6 6 6 7 6 6	6 6 6 6 7 7 6 6 6 6	6
526	6 6 6 6 6 7 6 6 6	6 6 6 6 15 14 6 6 6 6	6 6 7 6 6 6 6 6 6 6	15 6 6 6 6 6 6 6 7 6	6 6 6 6 6 6 7 6 6 6	6
527	16 6 6 6 6 6 7 7 7	6 6 6 6 6 15 15 16 6 6	6 6 6 7 6 6 6 6 6 6	6 16 15 6 6 6 6 6 6 7	6 6 6 6 6 6 6 7 6 6	6
528	15 6 6 7 6 6 6 7 7	6 6 6 7 6 6 15 15 6 6	6 6 6 6 7 6 6 6 6 6	6 15 15 6 6 6 6 6 6 6	7 7 6 6 6 6 6 6 7 6	6
529	* 15 6 7 6 7 6 7 7	6 6 6 7 6 6 15 15 6 6	6 6 6 22 6 7 6 6 6 6	6 6 15 15 6 6 6 6 6 6	7 7 6 6 6 6 6 6 7 7	6
530	6 6 15 6 6 6 6 6 6	7 6 6 6 6 6 6 6 15 6	6 6 6 6 6 6 7 6 6 6	6 6 6 6 13 6 6 6 6 6	6 6 7 6 6 6 6 6 6 6	7
531	15 6 6 15 6 6 6 7 8	6 7 6 7 6 6 15 15 6 15	6 6 6 6 6 6 6 7 6 6	6 6 15 6 6 12 6 6 6 6	7 7 6 7 6 6 6 6 6 6	6
532	6 6 6 6 15 6 6 6 6	6 6 7 6 6 6 6 6 6 6	15 6 6 6 6 6 6 6 7 6	6 6 6 6 6 6 11 6 6 6	6 6 6 6 7 6 6 6 6 6	6
533	12 6 6 7 6 15 6 7 7	6 6 6 12 6 6 10 10 6 6	6 15 6 6 6 6 6 6 6 7	6 10 10 6 6 6 6 10 6 6	6 12 6 6 6 7 6 6 6 6	6
534	6 6 6 6 6 6 15 6 6	6 6 6 6 6 6 6 6 6 6	6 6 15 6 6 6 6 6 6 6	7 6 6 6 6 6 6 6 9 6	6 6 6 6 6 6 7 6 6 6	6
535	15 6 6 6 6 6 6 15 7	6 6 6 6 6 6 15 15 6 6	6 6 6 15 6 6 6 6 6 6	6 7 15 6 6 6 6 6 6 8	6 6 6 6 6 6 6 7 6 6	6
536	8 6 6 6 6 6 6 7 15	6 7 6 6 6 6 8 8 6 7	6 6 6 6 15 6 6 6 6 6	6 8 8 6 6 6 6 6 6 6	7 6 6 6 6 6 6 6 7 6	6
537	15 6 6 6 6 7 6 7 7	15 6 6 6 6 6 15 15 6 6	6 6 6 6 7 15 6 6 6 6	6 6 15 9 6 6 6 6 6 6	6 6 6 6 6 6 6 6 6 7	6
538	10 7 6 6 6 6 6 6 6	7 15 6 6 6 6 10 10 6 6	6 6 6 6 6 6 15 6 6 6	6 10 10 6 7 6 6 6 6 6	6 6 6 6 6 6 6 6 6 6	7
539	15 6 7 6 6 7 6 7 7	6 10 15 6 6 6 15 15 6 6	6 6 6 6 6 6 6 15 6 6	6 6 15 6 6 7 6 6 6 6	6 6 6 6 6 6 6 6 6 10	6
540	12 6 6 7 6 7 6 6 6	6 6 7 15 6 6 12 12 6 6	6 6 6 6 6 6 6 6 15 6	6 12 12 6 6 6 7 6 6 6	6 6 6 6 6 6 6 6 6 10	6
541	* 6 6 6 7 6 6 7 7	6 6 6 7 15 6 15 15 6 6	6 6 6 6 6 6 6 6 6 15	6 6 15 6 6 6 6 7 6 6	6 6 6 6 6 6 6 6 6 6	6
542	6 6 6 6 6 7 6 6 6	6 6 6 6 7 15 6 6 6 6	6 6 6 6 6 6 6 6 6 6	15 6 6 6 6 6 6 6 7 6	6 6 6 6 6 6 6 6 6 6	6
543	15 6 6 6 6 6 7 7 7	6 6 6 6 6 7 15 15 6 6	6 6 6 6 6 6 6 6 6 6	6 15 15 6 6 6 6 6 6 7	6 6 6 6 6 6 6 6 6 6	6
544	18 6 6 6 6 6 6 7 7	6 6 6 6 6 6 14 16 6 6	6 6 6 6 6 6 6 6 6 6	6 6 16 6 6 6 6 6 6 6	7 6 6 6 6 6 6 6 6 6	6
545	7 6 6 7 6 6 6 7 7	6 6 6 6 6 6 6 7 6 6	6 6 6 6 6 6 6 6 6 6	6 6 7 7 6 6 6 6 6 6	6 7 6 6 6 6 6 6 6 6	6
546	6 6 6 6 6 6 6 6 6	7 6 6 6 6 6 6 6 7 6	6 6 6 6 6 6 6 6 6 6	6 6 6 6 7 6 6 6 6 6	6 6 7 6 6 6 6 6 6 6	6
547	* 6 6 6 6 6 6 7 7	6 7 6 6 6 6 6 6 6 7	6 6 6 6 6 6 6 6 6 6	6 6 6 6 6 7 6 6 6 6	6 6 6 7 6 6 6 6 6 6	6
548	6 6 6 6 6 6 6 6 6	6 6 7 6 6 6 6 6 6 6	7 6 6 6 6 6 6 6 6 6	6 6 6 6 6 6 7 6 6 6	6 6 6 6 7 6 6 6 6 6	6
549	8 6 6 6 6 6 6 7 8	6 6 6 7 6 6 6 6 6 6	6 7 15 6 6 6 6 6 6 6	6 6 6 6 6 6 6 7 6 6	6 6 6 6 6 7 6 6 6 6	6
550	6 6 6 6 6 6 6 6 6	6 6 6 6 7 6 6 6 6 6	6 6 7 6 6 6 6 6 6 6	6 6 6 6 6 6 6 6 7 6	6 6 6 6 6 6 7 6 6 6	6
551	18 6 6 6 6 6 6 7 7	6 6 6 6 6 7 6 6 6 18	6 6 6 7 6 6 6 6 6 18	6 6 6 6 6 6 6 6 6 7	6 6 6 6 6 6 6 7 6 6	6
552	7 6 6 6 6 6 6 7 7	6 6 6 7 6 6 7 6 6 6	6 6 6 6 7 6 6 6 6 6	6 6 6 6 6 6 6 6 6 6	7 6 6 6 6 6 6 6 7 6	6

h →	0									1										2										3										4										5
n ↓	1	2	3	4	5	6	7	8	9	0	1	2	3	4	5	6	7	8	9	0	1	2	3	4	5	6	7	8	9	0	1	2	3	4	5	6	7	8	9	0	1	2	3	4	5	6	7	8	9	0
553	7	6	6	7	6	7	6	7	7	6	6	6	6	6	6	6	7	6	6	6	6	6	6	6	7	6	6	6	6	6	6	6	6	6	6	6	6	6	6	7	7	6	6	6	6	6	6	6	7	6
554	6	6	6	6	6	6	6	6	6	6	6	6	6	6	6	6	6	7	6	6	6	6	6	6	6	7	6	6	6	6	6	6	6	6	6	6	6	6	6	6	6	7	6	6	6	6	6	6	6	7
555	7	6	6	6	6	6	6	7	7	6	6	6	6	6	15	6	6	6	7	6	6	6	6	6	6	6	7	6	6	6	6	6	6	6	6	6	6	6	6	6	6	6	7	6	6	6	6	6	6	6
556	6	6	6	6	6	6	6	6	6	6	6	6	6	6	6	6	6	6	6	7	6	6	6	6	6	6	6	7	6	6	6	6	6	6	6	6	6	6	6	6	6	6	6	7	6	6	6	6	6	6
557	*	6	6	6	6	6	6	7	7	6	6	6	6	6	6	6	6	6	6	6	7	6	6	6	6	6	6	6	7	6	6	6	6	6	6	6	6	6	6	6	6	6	6	6	7	6	6	6	6	6
558	6	6	6	6	6	6	6	6	6	6	6	6	6	6	6	6	6	6	6	6	6	7	6	6	6	6	6	6	6	7	6	6	6	6	6	6	6	6	6	6	6	6	6	6	6	7	6	6	6	6
559	12	6	6	6	6	6	6	7	7	6	6	6	12	6	6	6	6	6	6	6	6	6	7	6	6	6	6	6	6	6	7	6	6	6	6	6	6	6	6	6	6	6	12	6	6	6	7	6	6	6
560	15	6	6	6	6	6	6	7	7	6	7	6	6	6	6	15	15	6	15	6	6	6	6	7	6	6	6	6	6	6	6	15	6	6	6	6	6	6	6	6	6	6	6	6	6	6	6	7	6	6
561	7	6	17	6	6	7	6	7	7	6	6	6	6	6	6	6	6	6	6	6	6	6	6	6	7	6	6	6	6	6	6	6	7	6	6	6	6	6	6	6	6	6	6	6	6	6	6	7	7	6
562	6	6	6	6	6	6	6	6	6	6	6	6	6	6	6	6	6	6	6	6	6	6	6	6	6	7	6	6	6	6	6	6	6	7	6	6	6	6	6	6	6	6	6	6	6	6	6	6	6	7
563	*	6	6	6	6	6	6	7	7	6	6	6	6	6	6	6	6	6	6	6	6	6	6	6	6	6	7	6	6	6	6	6	6	6	7	6	6	6	6	6	6	6	6	6	6	6	6	6	6	6
564	6	6	6	6	6	6	6	6	6	6	6	6	6	6	6	6	6	6	6	6	6	6	6	6	6	6	6	7	6	6	6	6	6	6	6	7	6	6	6	6	6	6	6	6	6	6	6	6	6	6
565	7	6	6	6	6	7	6	7	7	6	6	6	7	6	6	6	6	6	6	6	6	6	6	6	6	6	6	6	7	6	6	6	6	6	6	6	7	6	6	6	6	6	7	6	7	6	6	6	7	6
566	6	6	6	6	6	6	6	6	6	6	6	6	6	6	6	6	6	6	6	6	6	6	6	6	6	6	6	6	6	7	6	6	6	6	6	6	6	7	6	6	6	6	6	6	6	6	6	6	6	6
567	7	6	6	6	6	6	6	7	7	6	6	6	6	6	6	6	6	6	6	6	6	6	6	6	6	6	6	6	6	6	7	6	6	6	6	6	6	6	7	6	6	6	6	6	6	6	6	6	6	6
568	7	6	6	6	6	6	6	7	7	6	6	6	6	6	6	6	6	6	6	6	6	6	6	6	6	6	6	6	6	6	6	7	6	6	6	6	6	6	6	7	6	6	6	6	6	6	6	6	6	6
569	*	6	6	7	6	6	6	7	7	6	6	6	6	6	6	6	6	6	6	6	6	6	6	6	6	6	6	6	6	6	6	6	7	6	6	6	6	6	6	6	7	6	6	6	6	6	6	6	6	6
570	6	7	6	6	6	6	6	6	6	7	6	6	6	6	6	6	6	6	6	6	6	6	6	6	6	6	6	6	6	6	6	6	6	7	6	6	6	6	6	6	6	7	6	6	6	6	6	6	6	6
571	*	6	7	6	6	6	6	7	7	6	7	7	6	6	6	6	6	6	6	6	6	6	6	6	6	6	6	6	6	6	6	6	6	6	7	6	6	6	6	6	6	6	7	6	6	6	6	6	6	6
572	6	6	6	7	6	6	6	6	6	6	6	7	6	6	6	6	6	6	6	6	6	6	6	6	6	6	6	6	6	6	6	6	6	6	6	7	6	6	6	6	6	6	6	7	6	6	6	6	6	6
573	7	6	6	6	7	6	6	7	7	6	6	6	7	6	15	6	6	6	6	6	6	6	6	6	6	6	6	6	6	6	6	6	6	6	6	6	7	6	6	6	6	6	6	6	7	6	6	6	6	6
574	15	6	6	6	6	7	6	6	6	6	6	6	6	7	6	15	15	6	15	6	6	6	6	6	6	6	6	6	6	6	6	15	6	6	6	6	6	7	6	6	6	6	6	6	6	7	6	6	6	6
575	22	6	6	6	6	6	7	7	7	6	6	6	6	6	7	6	6	6	6	6	6	6	22	6	22	6	6	6	6	6	6	6	6	6	6	6	6	6	7	6	6	6	6	6	6	6	7	6	6	6
576	8	6	6	6	6	6	6	7	8	6	6	6	6	6	6	7	6	6	6	6	6	6	6	6	6	6	6	6	6	6	6	6	6	6	6	6	6	6	6	7	6	6	6	6	6	6	6	7	6	6
577	*	6	6	7	6	7	6	7	7	6	6	6	7	6	6	6	7	7	6	6	6	6	6	23	6	6	6	6	6	6	6	6	6	6	6	6	6	6	6	6	7	6	7	6	6	6	6	6	7	6
578	6	7	6	6	6	6	6	6	6	7	6	6	6	6	6	6	6	7	6	6	6	6	6	6	6	6	6	6	6	6	6	6	6	6	6	6	6	6	6	6	6	7	6	6	6	6	6	6	6	7
579	7	6	21	6	6	6	6	7	7	6	7	6	6	6	6	6	6	6	7	6	6	6	6	6	6	6	6	6	6	6	6	6	6	6	6	6	6	6	6	6	6	6	7	6	6	6	6	6	6	6
580	6	6	6	7	6	6	6	6	6	6	6	7	6	6	6	6	6	6	6	7	6	6	6	6	6	6	6	6	6	6	6	6	6	6	6	6	6	6	6	6	6	6	6	7	6	6	6	6	6	6
581	7	6	6	6	7	6	6	7	7	6	6	6	7	6	6	6	6	6	6	6	7	6	6	6	6	6	6	6	6	6	6	6	6	6	6	6	6	6	6	6	6	6	6	6	7	6	6	6	6	6
582	6	6	6	6	6	7	6	6	6	6	6	6	6	7	6	6	6	6	6	6	6	7	6	6	6	6	6	6	6	6	6	6	6	6	6	6	6	6	6	6	6	6	6	6	6	7	6	6	6	6
583	10	6	6	6	6	6	7	7	7	6	10	6	7	6	7	6	6	6	6	6	6	6	7	7	6	6	6	6	6	6	6	6	6	6	6	6	6	6	6	6	6	6	7	6	6	6	7	6	6	6
584	7	6	6	6	6	6	6	7	7	6	6	6	6	6	6	7	7	6	6	6	6	6	6	7	6	6	6	6	6	6	6	6	6	6	6	6	6	6	6	6	6	6	6	6	6	6	6	7	6	6
585	8	6	6	6	6	7	6	7	8	6	6	6	6	6	6	6	7	6	6	6	6	6	6	6	7	6	6	6	6	6	6	6	6	6	6	6	6	6	6	6	6	6	6	6	6	6	6	6	7	6
586	7	7	6	6	6	6	6	7	6	7	6	6	6	6	6	6	6	7	6	6	6	6	6	6	6	7	6	6	6	6	6	6	6	6	6	6	6	6	6	6	6	6	6	6	6	6	6	6	6	7
587	*	6	7	6	6	6	6	7	7	6	7	6	6	6	6	6	6	6	7	6	6	6	6	6	6	6	7	6	6	6	6	6	6	6	6	6	6	6	6	6	6	6	6	6	6	6	6	6	6	6
588	6	6	6	7	6	6	6	6	6	6	6	7	6	6	6	6	6	6	6	7	6	6	6	6	6	6	6	7	6	6	6	6	6	6	6	6	6	6	6	6	6	6	6	6	6	6	6	6	6	6
589	18	6	6	6	7	6	6	7	7	6	9	6	9	6	6	6	6	6	18	6	7	6	6	6	6	6	6	6	7	7	18	6	6	6	6	6	9	6	6	6	6	6	7	6	6	6	9	6	6	6
590	17	6	6	6	6	7	6	6	6	6	6	6	6	7	6	6	6	6	17	6	6	7	6	6	6	6	6	6	6	7	17	17	6	6	6	6	6	6	6	6	6	6	6	6	6	6	6	6	6	6
591	7	17	6	6	6	6	7	7	7	6	6	6	6	6	7	6	6	6	6	6	6	6	7	6	6	6	6	6	6	6	7	6	16	6	6	6	6	6	6	6	6	6	6	6	6	6	6	6	6	6
592	15	6	17	6	6	6	6	7	8	6	6	6	8	6	6	15	15	6	15	6	6	6	6	7	7	6	6	6	6	6	6	15	6	15	6	6	15	6	6	6	6	6	6	6	6	6	6	6	6	6
593	*	6	6	17	6	6	6	7	7	6	6	6	6	6	6	15	15	6	6	6	6	6	6	6	7	6	6	6	6	6	6	6	7	6	14	6	15	6	6	6	6	6	6	6	6	6	6	6	6	6
594	6	15	6	6	17	6	6	6	6	7	6	6	6	6	6	6	6	15	6	6	6	6	6	6	6	7	6	6	6	6	6	6	6	7	6	13	6	6	6	6	6	6	6	6	6	6	6	6	6	6
595	15	6	15	6	6	17	6	7	7	6	7	6	6	6	6	15	15	6	15	6	6	6	6	6	6	6	7	6	6	6	12	12	6	6	7	7	15	6	6	6	6	6	6	6	6	6	6	6	6	6
596	7	6	6	15	6	6	17	6	6	6	6	7	6	6	6	7	6	6	6	15	6	6	6	6	6	6	6	7	6	6	6	6	6	6	6	7	7	11	6	6	7	6	6	6	6	6	6	6	6	6
597	7	6	6	6	15	6	6	17	7	6	6	6	7	6	6	6	6	6	7	6	15	6	6	6	6	6	6	6	7	6	7	7	6	6	6	6	7	6	10	6	6	6	6	6	6	6	6	6	6	6
598	8	6	6	6	6	15	6	6	17	6	6	6	6	7	6	6	6	6	8	6	6	15	6	6	6	6	6	6	6	7	8	8	6	6	6	6	6	7	6	9	6	6	6	6	6	6	6	6	6	6
599	*	6	6	7	6	6	15	7	7	17	6	6	6	6	7	15	15	6	8	6	6	6	15	6	6	6	6	6	6	6	8	8	6	6	6	6	15	6	7	6	8	6	6	6	6	6	6	6	6	6
600	10	6	6	7	6	6	6	15	7	6	17	6	6	6	6	7	7	6	10	6	6	6	6	15	6	6	6	6	6	6	10	10	7	6	6	6	7	6	6	7	7	7	6	6	7	6	6	6	6	6
601	*	6	6	7	6	6	6	7	15	6	6	17	6	6	6	15	15	6	15	6	6	6	6	6	15	6	6	6	6	6	6	6	7	6	6	6	15	6	6	6	7	7	6	6	6	6	6	6	6	6
602	12	6	6	6	6	6	6	6	6	15	6	6	17	6	6	6	6	7	12	6	6	6	6	6	6	15	6	6	6	6	12	12	6	7	6	6	6	6	6	6	6	7	6	6	6	6	6	6	6	6
603	15	6	6	6	6	6	6	7	8	6	15	6	6	17	6	15	15	6	7	6	6	6	6	6	6	6	15	6	6	6	6	6	9	6	7	6	15	6	6	6	6	6	7	6	6	6	6	6	6	6

$h \to$ / $n \downarrow$	0: 1 2 3 4 5 6 7 8 9	1: 0 1 2 3 4 5 6 7 8 9	2: 0 1 2 3 4 5 6 7 8 9	3: 0 1 2 3 4 5 6 7 8 9	4: 0 1 2 3 4 5 6 7 8 9	5: 0
604	7 6 6 6 6 7 6 6 6	6 6 15 6 6 17 7 6 6 6	7 6 6 6 6 6 6 6 15 6	6 6 6 6 6 6 7 7 6 6	6 6 6 6 7 6 6 6 6 7	6
605	15 6 6 6 6 6 6 7 7	6 6 6 15 6 6 17 15 6 15	6 7 6 6 6 6 6 6 6 15	6 15 15 6 6 6 6 15 6 6	6 6 6 6 6 7 6 6 6 6	6
606	16 6 6 6 6 6 6 6 6	6 6 6 6 15 6 6 17 6 16	6 6 7 6 6 6 6 6 6 6	15 16 16 6 6 6 6 6 7 6	6 6 6 6 6 6 7 6 6 6	6
607	* 6 6 6 6 6 6 7 7	6 6 6 6 6 15 15 15 17 6	6 6 6 7 6 6 6 6 6 6	6 15 6 6 6 6 6 15 6 7	6 6 6 6 6 6 6 7 7 6	6
608	18 6 6 7 6 6 6 7 8	6 8 6 6 6 6 15 15 6 18	6 6 6 6 7 6 6 6 6 6	6 6 18 6 6 6 6 15 6 6	7 7 6 6 6 6 6 6 7 6	6
609	15 6 6 7 6 7 6 7 7	6 6 6 6 6 6 15 15 6 6	6 6 6 6 7 7 6 6 6 6	6 6 6 15 6 6 6 15 6 6	6 7 6 6 6 6 6 6 6 7	6
610	6 6 6 6 6 6 6 6 6	6 6 6 6 6 6 6 6 15 6	6 6 6 6 6 6 7 6 6 6	6 6 6 6 15 6 6 6 6 6	6 6 7 6 6 6 6 6 6 6	7
611	15 6 6 6 6 6 6 7 7	6 6 6 12 6 6 15 15 6 15	6 6 6 6 6 6 6 7 6 6	6 6 6 6 6 15 6 15 6 6	6 6 6 7 6 6 6 12 6 6	6
612	7 6 6 6 6 6 6 6 6	6 6 6 6 6 6 7 6 6 6	15 6 6 6 6 6 6 6 7 6	6 6 6 6 6 6 15 7 6 6	6 6 6 6 7 6 6 6 6 6	6
613	* 6 6 6 6 6 6 7 7	6 6 6 6 6 6 15 15 6 6	6 15 6 6 17 6 6 6 6 7	6 6 6 6 6 6 6 15 6 6	6 6 6 6 6 7 6 6 6 6	6
614	6 6 6 6 6 6 6 6 6	6 6 6 6 6 6 6 6 6 6	6 6 15 6 6 6 6 6 6 6	7 6 6 6 6 6 6 6 14 6	6 6 6 6 6 6 7 6 6 6	6
615	15 6 6 6 6 6 6 7 7	6 6 6 6 6 6 15 15 6 6	6 6 6 15 6 6 6 6 6 6	6 7 6 6 6 6 6 15 6 13	6 6 6 6 6 6 6 7 6 6	6
616	7 6 6 6 6 7 6 7 7	6 7 6 6 6 6 7 7 6 6	6 6 6 6 15 6 6 6 6 6	6 6 7 6 6 6 6 7 6 6	12 6 6 6 6 6 6 6 7 7	6
617	* 6 6 7 6 7 6 7 7	6 7 6 6 6 6 15 15 6 6	6 6 6 6 6 15 6 6 6 6	6 6 7 7 6 6 6 15 6 6	6 11 6 6 6 6 6 6 6 7	6
618	6 7 6 6 6 6 6 6 6	6 6 6 6 6 6 6 6 6 6	6 6 6 6 6 6 15 6 6 6	6 6 6 6 7 6 6 6 6 6	6 6 10 6 6 6 6 6 6 6	7
619	* 6 7 6 6 6 6 7 7	6 7 6 6 6 6 15 15 6 6	6 6 6 6 6 6 6 15 6 6	6 6 6 6 6 7 6 15 6 6	6 6 6 9 6 6 6 6 6 6	6
620	7 6 6 7 6 6 6 7 6	6 6 6 6 6 6 7 6 6 6	6 6 6 6 6 6 6 6 15 6	6 6 6 6 6 6 7 7 6 6	6 6 6 6 8 6 6 6 6 6	6
621	22 6 6 6 7 6 6 7 7	6 6 6 6 6 6 15 15 6 6	6 6 6 22 6 6 6 22 6 15	6 6 6 6 6 6 6 15 6 6	6 6 6 6 6 7 6 6 6 6	6
622	6 6 6 6 6 7 6 6 6	6 6 6 6 6 6 6 6 6 6	6 6 6 6 6 6 6 6 6 6	15 6 6 6 6 6 6 6 7 6	6 6 6 6 6 6 7 6 6 6	6
623	15 6 6 6 6 6 7 7 7	6 6 6 6 6 6 15 15 6 6	6 6 6 6 6 6 6 6 6 6	6 15 6 6 6 6 6 15 6 7	6 6 6 6 6 6 6 7 6 6	6
624	15 6 6 7 6 6 6 7 7	6 7 6 6 6 6 15 15 6 6	6 6 6 6 6 6 6 6 6 6	6 6 15 6 6 6 6 15 6 6	7 7 6 6 6 6 6 6 7 6	6
625	* 6 6 7 6 7 6 7 7	6 7 6 6 6 6 6 6 6 6	6 6 6 6 6 24 6 6 6 6	6 6 6 15 6 6 6 6 6 6	7 7 6 6 6 6 6 6 6 7	6
626	6 7 6 6 6 6 6 6 6	7 6 6 6 6 6 6 6 6 6	6 6 6 6 6 6 6 6 6 6	6 6 6 6 15 6 6 6 6 6	6 6 7 6 6 6 6 6 6 6	7
627	7 6 7 6 6 6 6 7 7	6 7 6 6 6 6 6 6 6 6	6 6 6 6 6 6 6 6 6 6	6 6 6 6 6 15 6 6 6 6	6 6 6 7 6 6 6 6 6 6	6
628	7 6 6 7 6 6 6 6 6	6 6 7 6 6 6 6 6 6 6	6 6 6 6 6 6 6 6 6 6	6 6 6 6 6 6 15 6 6 6	6 6 6 6 7 6 6 6 6 6	6
629	16 6 6 6 21 6 6 7 7	6 6 6 7 6 6 6 16 6 6	6 6 6 6 6 6 6 6 6 6	6 6 6 6 6 6 6 16 6 6	6 6 6 6 6 7 6 6 6 6	6
630	6 6 6 6 6 7 6 6 6	6 6 6 6 7 6 6 6 6 6	6 6 6 6 6 6 6 6 6 6	6 6 6 6 6 6 6 6 7 6	6 6 6 6 6 6 7 6 6 6	6
631	* 6 6 6 6 6 7 7 7	6 6 6 7 6 7 6 6 6 6	6 6 6 6 6 6 6 6 6 6	6 6 6 6 6 6 6 6 6 7	6 6 6 7 6 6 6 7 6 6	6
632	7 6 6 7 6 6 7 7 7	6 7 6 6 6 6 7 6 6 6	6 6 6 6 6 6 6 6 6 6	6 6 6 6 6 6 6 7 6 6	7 7 6 6 6 6 6 6 7 6	6
633	7 6 6 7 6 7 7 7 7	6 7 6 6 6 6 7 7 6 6	6 6 6 6 6 6 6 6 6 6	6 6 6 6 6 6 6 7 6 6	6 7 6 6 6 6 6 6 7 7	6
634	6 7 6 6 6 6 6 6 6	7 6 6 6 6 6 6 6 7 6	6 6 6 6 6 6 6 6 6 6	6 6 6 6 6 6 6 6 6 6	6 6 6 6 6 6 6 6 6 6	7
635	7 6 7 6 6 6 7 7 7	6 7 6 6 6 6 6 7 6 7	6 6 6 6 6 6 6 6 6 6	6 6 6 6 6 6 6 6 6 6	6 6 6 6 6 6 6 6 6 6	6
636	6 6 6 7 6 6 6 6 6	6 6 7 6 6 6 6 6 6 6	7 6 6 6 6 6 6 6 6 6	6 6 6 6 6 6 6 6 6 6	6 6 6 6 7 6 6 6 6 6	6
637	12 6 6 7 7 7 7 7 7	6 6 6 12 6 6 6 7 6 6	6 7 6 6 6 6 6 6 6 6	6 6 6 6 6 6 6 7 6 6	6 7 6 6 6 7 6 6 6 12	6
638	6 6 6 6 6 7 6 6 6	6 6 6 6 7 6 6 6 6 6	6 6 7 6 6 6 6 6 6 6	6 6 6 6 6 6 6 6 6 6	6 6 6 6 6 6 6 6 6 6	6
639	8 6 6 6 6 6 7 7 8	6 6 6 6 6 7 6 6 6 6	6 6 6 7 6 6 6 6 6 6	6 6 6 6 6 6 6 6 6 6	6 6 6 6 6 6 6 6 6 6	6
640	9 6 6 6 6 6 7 7 7	6 7 6 6 6 6 7 7 6 6	6 6 6 6 7 6 6 6 6 6	6 6 6 6 6 6 6 6 6 6	7 6 6 6 6 6 6 6 7 6	6
641	* 7 6 7 6 7 7 7 7	6 6 7 6 6 6 7 7 6 6	6 6 6 6 7 7 6 6 6 6	6 6 6 6 6 6 6 7 6 6	7 7 6 6 6 6 6 6 6 7	6
642	6 7 7 6 6 6 6 6 6	7 6 6 6 6 6 6 6 7 6	6 6 6 6 6 6 7 6 6 6	6 6 6 6 6 6 6 6 6 6	6 6 7 6 6 6 6 6 6 6	6
643	* 6 7 7 6 6 7 7 7	6 7 6 6 6 6 7 6 6 7	6 6 6 6 6 7 6 7 6 6	6 6 6 6 6 6 6 6 6 6	7 7 6 7 6 6 6 6 6 6	6
644	6 6 6 7 7 6 6 6 6	6 6 7 6 6 6 6 6 6 6	7 6 6 6 6 6 6 6 7 6	6 6 6 6 6 6 6 6 6 6	6 6 6 6 7 6 6 6 6 6	6
645	7 6 6 7 7 7 7 7 7	6 6 6 7 6 6 7 7 9 6	6 7 6 6 6 6 6 6 6 7	6 6 6 6 6 6 6 7 6 6	6 7 6 6 6 7 6 6 6 6	6
646	6 6 6 6 6 7 7 6 6	6 6 6 6 7 6 6 6 6 6	6 6 7 6 6 6 6 6 6 6	7 6 6 6 6 6 6 6 6 6	6 6 6 6 6 6 7 6 6 6	6
647	* 6 6 7 6 6 7 7 7	6 6 6 6 6 7 7 6 6 6	6 6 6 7 6 6 6 6 6 6	6 7 6 6 6 6 6 6 6 6	7 7 6 7 6 6 6 7 6 6	6
648	7 6 6 7 6 6 7 7 7	7 6 6 6 6 6 7 6 6 6	6 6 6 6 7 6 6 6 6 6	6 6 7 6 6 6 6 7 6 6	7 7 6 6 6 6 6 6 7 6	6
649	10 6 6 7 6 6 7 7 7	7 10 6 6 6 6 7 7 6 6	7 6 6 6 7 7 6 6 6 6	6 6 7 7 6 6 6 7 6 6	7 7 6 6 6 6 6 6 6 6	6
650	7 7 6 6 6 6 6 7 7	7 7 6 6 6 6 6 6 7 6	6 6 6 6 6 6 7 6 6 6	6 6 6 6 7 6 6 6 6 6	6 6 6 6 6 6 6 6 6 6	6
651	7 6 7 7 6 6 7 7 7	7 7 7 6 6 6 7 6 6 7	6 6 6 6 7 6 6 7 6 6	6 6 6 7 6 7 6 6 6 6	7 7 6 6 6 6 6 6 6 6	6
652	7 6 6 7 6 6 6 7 7	6 6 7 7 6 6 6 6 6 6	7 6 6 6 6 6 6 6 7 6	6 6 6 6 6 6 7 6 6 6	6 6 6 6 6 6 6 6 6 6	6
653	* 6 6 7 7 6 7 7 7	7 7 6 10 7 6 7 10 6 6	6 7 6 6 7 6 6 6 6 7	6 6 6 6 6 6 6 7 6 6	7 7 6 6 6 6 6 6 6 6	6
654	6 6 6 6 6 7 6 6 6	6 6 6 6 7 7 6 6 6 6	6 6 7 6 6 6 6 6 6 6	7 6 6 6 6 6 6 6 7 6	6 6 6 6 6 6 6 6 6 6	6

h → n ↓	0 1 2 3 4 5 6 7 8 9	1 0 1 2 3 4 5 6 7 8 9	2 0 1 2 3 4 5 6 7 8 9	3 0 1 2 3 4 5 6 7 8 9	4 0 1 2 3 4 5 6 7 8 9	5 0
655	7 6 6 6 6 6 7 7 7	6 6 6 6 6 7 7 6 6 6	6 6 6 7 6 6 6 6 6 6	6 7 6 6 6 6 6 6 6 7	6 6 6 6 6 6 6 6 6 6	6
656	15 6 6 7 6 6 7 7 7	7 6 6 6 6 15 7 6 6	6 6 6 6 7 6 6 6 6 6	6 6 7 6 6 6 6 6 6 6	7 15 6 6 6 6 6 6 6 6	6
657	15 6 6 7 6 6 7 7 8	7 6 6 6 6 15 15 7 6	6 6 6 6 7 7 6 6 7 6	6 6 7 7 6 6 6 7 6 6	7 15 6 6 6 6 6 6 6 6	6
658	7 15 6 6 6 6 6 7 7	7 6 6 6 6 6 6 6 15 7	6 6 6 6 6 6 7 6 6 6	6 6 6 6 7 6 6 6 6 6	6 6 7 6 6 6 6 6 6 6	6
659	* 6 15 7 6 6 7 7 7	7 7 6 6 6 6 15 15 6 15	7 6 6 6 7 6 6 7 6 6	6 6 7 6 6 7 6 6 6 6	6 15 6 7 6 6 6 6 6 6	6
660	6 6 6 15 6 6 6 6 6	6 6 7 6 6 6 6 6 6 6	15 7 6 6 6 6 6 6 7 6	6 6 6 6 6 6 7 6 6 6	6 6 6 6 7 6 6 6 6 6	6
661	* 6 6 7 15 6 7 7 7	7 6 6 7 6 6 7 7 6 6	6 15 7 6 6 6 6 6 6 7	6 6 7 6 6 6 6 7 6 6	6 7 6 6 6 7 6 6 6 6	6
662	7 6 6 6 6 15 6 7 7	6 6 6 6 7 6 6 6 6 6	6 6 15 7 6 6 6 6 6 6	7 6 6 6 6 6 6 6 7 6	6 6 6 6 6 6 7 6 6 6	6
663	15 6 6 7 6 6 15 7 7	7 6 6 6 6 7 15 15 6 6	6 6 6 15 7 6 6 6 6 6	6 7 6 6 6 6 6 6 6 7	6 15 6 6 6 6 6 6 6 6	6
664	7 6 6 7 6 6 7 15 7	7 6 6 6 6 6 7 7 6 6	6 6 6 6 15 7 6 6 6 6	6 6 7 6 6 6 6 6 6 6	7 7 6 7 6 6 6 6 7 6	6
665	15 6 6 7 6 7 7 7 15	7 6 6 6 6 6 15 15 6 15	6 6 6 6 7 15 7 6 6 6	6 6 7 7 6 6 7 7 6 6	7 15 6 6 6 6 6 6 7 7	6
666	7 7 6 6 6 6 6 7 7	15 6 6 6 6 6 6 6 7 6	6 6 6 6 6 6 15 7 6 6	6 6 6 6 7 6 6 6 6 6	6 6 7 6 6 6 6 6 6 6	7
667	22 6 7 7 6 7 7 7 7	7 15 6 9 6 6 15 15 6 7	6 6 6 22 7 6 6 15 7 22	6 6 7 6 6 7 6 6 6 6	7 15 6 9 6 6 6 6 6 7	6
668	7 6 6 7 6 6 6 7 7	6 6 15 6 6 6 6 6 6 6	7 6 6 6 6 6 6 6 15 7	6 6 6 6 6 6 7 6 6 6	6 6 6 6 7 6 6 6 6 6	6
669	15 6 6 7 7 6 7 7 7	7 6 6 15 6 6 15 15 6 6	6 15 6 6 7 6 6 6 6 15	7 6 6 6 6 6 6 7 6 6	7 15 6 6 6 7 6 6 6 6	6
670	7 6 6 6 6 7 6 7 7	6 6 6 6 15 6 6 6 6 6	6 6 7 6 6 6 6 6 6 6	15 7 6 6 6 6 6 6 7 6	6 6 6 6 6 6 7 6 6 6	6
671	15 6 6 7 6 6 7 7 7	7 10 6 6 6 15 15 15 6 6	6 6 6 7 7 6 6 6 6 6	6 15 7 6 6 6 6 6 6 7	6 15 6 6 6 6 6 7 6 6	6
672	15 6 6 7 6 6 7 7 8	7 8 6 6 6 6 15 15 6 6	6 6 6 6 7 6 6 6 6 6	6 6 15 7 6 6 6 6 6 6	7 15 6 6 6 6 6 6 7 6	6
673	* 6 6 7 6 7 7 7 7	7 6 6 6 6 6 15 15 6 18	6 6 6 6 6 18 6 17 6 6	6 6 7 15 7 6 6 6 6 6	7 15 6 6 7 6 6 6 7 7	6
674	7 7 6 6 6 6 6 7 7	7 6 6 6 6 6 6 7 15 6	6 6 6 6 6 6 7 6 6 6	6 6 6 6 15 7 6 6 6 6	6 6 7 6 6 6 6 6 6 6	7
675	24 6 7 7 6 6 7 7 7	7 7 6 6 6 6 15 15 7 15	6 6 6 6 6 24 6 24 6 6	6 6 7 6 6 15 7 6 6 6	7 15 6 7 6 6 6 6 7 6	6
676	7 6 7 7 6 6 6 7 7	6 6 7 6 6 6 6 6 6 6	15 6 6 6 6 6 6 6 7 6	6 6 6 6 6 6 15 7 6 6	6 6 6 6 7 6 6 6 6 6	6
677	* 6 6 7 7 6 7 7 7	7 6 6 7 6 6 15 15 6 6	6 15 6 6 7 6 25 6 6 7	6 6 7 6 6 6 6 15 7 6	6 15 6 6 6 7 6 6 7 6	6
678	6 6 6 6 6 7 6 6 6	6 6 6 6 7 6 6 6 6 6	6 6 15 6 6 6 6 6 6 6	7 6 6 6 6 6 6 6 15 7	6 6 6 6 6 6 7 6 6 6	6
679	15 6 23 7 6 6 21 7 7	7 6 6 6 6 7 15 15 6 6	6 6 6 15 6 6 6 6 6 6	6 7 7 6 6 6 6 6 6 15	7 15 6 6 6 6 6 7 6 6	6
680	7 6 7 7 6 6 7 7 7	7 6 6 6 6 6 7 7 6 6	6 6 6 6 15 6 6 6 6 6	6 6 7 6 6 6 6 6 6 6	15 7 6 7 6 6 6 6 7 6	6
681	15 6 6 7 6 7 7 7 8	7 6 6 6 6 6 15 15 6 6	6 6 6 6 7 15 6 6 6 6	6 6 6 7 6 6 6 6 6 6	7 15 7 6 6 6 6 6 7 7	6
682	7 7 6 6 6 6 6 7 7	7 6 6 6 6 6 6 6 7 6	6 6 6 6 6 7 15 6 6 6	6 6 6 6 7 6 6 6 6 6	6 6 14 7 6 6 6 6 6 6	7
683	* 6 7 7 6 6 7 7 7	7 7 6 6 6 6 15 15 6 7	6 6 6 6 7 6 7 15 6 6	6 6 6 6 6 7 6 6 6 6	7 15 6 13 7 6 6 6 7 6	6
684	7 6 6 7 6 6 6 7 7	6 6 7 6 6 6 6 6 6 6	7 6 6 6 6 6 6 7 15 6	6 6 6 6 6 6 7 6 6 6	6 6 6 6 12 7 6 6 6 6	6
685	15 6 6 7 7 6 7 7 7	7 6 6 7 6 6 15 15 6 6	6 7 6 6 6 6 6 6 6 15	6 6 7 6 6 6 6 7 6 6	7 15 6 6 6 11 7 6 6 6	6
686	7 6 6 6 6 7 6 7 7	6 7 6 6 7 6 6 6 6 6	6 6 7 6 6 6 6 6 6 6	15 6 6 6 6 6 6 6 7 6	6 6 6 6 6 6 10 7 6 6	6
687	15 6 6 7 6 6 7 7 7	7 6 6 6 6 7 15 15 6 6	6 6 6 7 6 6 6 6 6 6	6 15 6 6 6 6 6 6 6 7	7 15 6 6 6 6 6 9 7 6	6
688	15 6 6 7 6 7 7 7 7	7 6 6 6 6 6 15 15 6 6	6 6 6 6 7 6 6 6 6 6	6 6 15 6 6 6 6 6 6 6	7 15 6 15 6 6 6 6 8 7	6
689	15 6 6 7 6 7 7 7 7	7 6 6 12 6 6 15 15 6 6	6 6 6 6 7 7 6 6 6 6	6 6 7 15 6 6 6 6 6 6	7 7 6 15 6 6 6 6 7 7	7
690	6 15 6 6 6 6 6 6 6	7 6 6 6 6 6 6 6 15 6	6 6 6 6 6 6 7 6 6 6	6 6 6 7 15 6 6 6 6 6	6 6 7 6 6 6 6 6 6 6	7
691	* 6 15 7 6 6 7 7 7	7 7 6 6 6 6 15 15 6 15	6 6 6 6 6 6 6 7 6 6	6 6 7 6 7 15 6 6 6 6	7 7 6 15 6 6 6 6 7 6	6
692	7 6 6 15 6 6 6 7 7	6 6 7 6 6 6 6 6 6 6	15 6 6 6 6 6 6 6 7 6	6 6 6 6 6 7 15 6 6 6	6 6 6 6 7 6 6 6 6 6	6
693	15 6 6 7 15 6 7 7 7	7 6 6 7 6 6 15 15 6 6	6 15 6 6 7 6 6 6 6 7	6 6 6 6 6 6 7 15 6 6	7 15 6 6 6 7 6 6 7 6	6
694	7 6 6 6 7 15 6 7 6	6 6 6 6 7 6 7 6 6 6	6 6 15 6 6 6 6 6 6 6	7 6 6 6 6 6 6 6 15 6	6 6 6 7 6 6 7 6 6 7	6
695	15 6 6 6 6 6 15 7 7	7 6 6 6 6 7 15 15 6 6	6 6 6 15 6 6 6 6 6 6	6 7 6 6 6 6 6 6 6 15	6 6 6 15 6 6 6 7 7 6	6
696	7 6 6 7 6 6 7 15 7	7 6 6 6 6 6 7 7 6 6	6 6 6 6 15 6 6 6 6 6	6 6 7 6 6 6 6 6 6 6	15 7 6 7 6 6 6 6 7 6	6
697	16 6 6 7 6 7 7 7 15	7 6 6 6 6 6 15 16 6 15	6 6 6 6 7 15 6 6 6 6	6 6 7 7 6 6 6 6 6 6	7 16 6 15 6 6 6 6 6 7	6
698	7 7 6 7 6 6 6 7 7	15 6 6 6 6 6 6 6 7 6	6 6 6 6 6 6 15 6 6 6	6 6 6 6 7 6 6 6 6 6	6 7 7 6 6 6 6 6 6 6	7
699	15 6 7 6 6 6 7 7 7	7 15 6 6 6 6 15 15 6 7	6 6 6 6 7 6 6 15 6 6	6 6 6 6 6 7 6 6 6 6	7 6 7 15 6 6 6 6 6 6	6
700	7 6 6 7 6 6 6 7 7	6 6 15 6 6 6 6 6 6 6	7 6 6 6 6 6 6 6 15 6	6 6 6 6 6 6 7 6 6 6	6 6 6 7 7 6 6 6 6 6	6
701	* 6 6 6 7 6 6 7 7	7 6 6 15 6 6 15 15 6 6	6 7 6 6 6 6 6 6 6 15	6 6 7 6 6 6 6 7 6 6	6 6 6 15 7 7 6 6 7 6	6
702	6 6 6 6 6 7 6 6 6	6 6 6 6 15 6 6 6 6 6	6 6 7 6 6 6 6 6 6 6	15 6 6 6 6 6 6 6 7 6	6 6 6 6 6 7 7 6 6 6	6
703	18 6 6 6 6 6 7 7 7	6 6 6 6 6 15 15 15 6 18	6 6 6 7 6 6 6 6 6 6	6 15 6 6 6 6 6 18 6 7	6 6 6 15 6 6 6 7 6 6	6
704	15 6 6 6 6 6 7 7 7	7 10 6 6 6 6 15 15 6 6	6 6 6 6 7 6 6 6 6 6	6 6 15 6 6 6 6 6 6 6	7 6 6 15 6 6 6 6 7 6	6
705	15 6 6 7 6 7 7 7 7	7 7 6 6 6 6 15 15 6 6	6 6 6 6 7 7 6 6 6 6	6 6 7 15 6 6 6 6 6 6	7 7 6 15 6 6 6 6 7 7	6

$n \downarrow$ \ $h \rightarrow$	1	2	3	4	5	6	7	8	9	10	11	12	13	14	15	16	17	18	19	20	21	22	23	24	25	26	27	28	29	30	31	32	33	34	35	36	37	38	39	40	41	42	43	44	45	46	47	48	49	50
706	7	7	6	6	6	7	6	7	7	7	6	6	6	6	6	7	6	$_{15}$	6	6	6	6	6	6	6	7	6	6	6	6	6	6	6	$_{15}$	6	6	6	6	6	6	6	7	7	6	6	6	6	6	7	7
707	$_{15}$	6	7	7	6	6	6	7	7	7	7	6	6	6	6	$_{15}$	$_{15}$	6	$_{15}$	6	6	6	6	7	6	6	7	6	6	6	6	7	6	6	$_{15}$	6	7	6	6	6	7	6	$_{15}$	6	6	6	6	7	6	7
708	6	6	6	7	6	6	6	6	6	6	6	7	6	6	6	6	6	6	6	$_{15}$	6	6	6	6	6	6	6	7	6	6	6	6	6	6	6	$_{15}$	6	6	6	6	6	6	6	7	6	6	6	6	6	6
709	*	6	6	6	7	6	6	7	7	7	6	6	7	6	6	$_{15}$	$_{15}$	6	6	6	$_{15}$	6	6	6	6	6	6	6	7	6	6	6	6	6	6	6	$_{15}$	6	6	7	6	6	$_{15}$	6	7	6	6	6	6	6
710	7	6	6	6	6	7	6	7	7	6	6	6	6	7	6	6	6	6	6	6	6	$_{15}$	6	6	6	6	6	6	6	7	6	6	6	6	6	6	6	$_{15}$	6	6	6	6	6	6	6	7	6	6	6	6
711	$_{15}$	6	6	7	6	6	7	7	8	7	6	6	6	6	7	$_{15}$	$_{15}$	6	7	6	6	6	$_{15}$	6	6	6	6	6	6	6	7	6	6	6	6	6	7	6	$_{15}$	6	7	6	$_{15}$	6	7	6	7	6	6	6
712	7	6	6	6	6	6	7	7	7	7	7	6	6	6	6	7	7	6	7	6	6	6	6	$_{15}$	6	6	6	6	6	6	6	7	6	6	6	6	6	6	6	$_{15}$	6	6	7	6	6	6	6	7	6	6
713	$_{22}$	7	6	7	6	7	6	7	7	7	7	6	6	6	6	$_{15}$	$_{15}$	6	7	6	6	6	$_{22}$	7	$_{15}$	6	6	6	6	6	$_{22}$	7	7	6	6	6	6	6	6	7	$_{15}$	6	$_{15}$	6	6	6	6	7	7	6
714	$_{21}$	7	7	6	6	6	6	7	7	7	6	6	6	6	6	6	6	7	6	6	6	6	$_{21}$	6	6	$_{15}$	6	6	6	6	$_{21}$	$_{21}$	6	7	6	6	6	6	6	6	6	$_{15}$	6	6	6	6	6	6	6	7
715	$_{15}$	$_{21}$	7	7	6	7	6	7	7	7	7	7	$_{11}$	6	6	$_{15}$	$_{15}$	6	7	6	6	6	6	6	6	6	$_{15}$	6	6	6	$_{11}$	7	$_{20}$	6	7	6	7	6	6	7	6	6	$_{15}$	6	6	6	6	6	7	6
716	7	6	$_{21}$	7	7	6	6	7	7	6	6	7	6	6	6	6	6	6	6	7	6	6	6	6	6	6	6	$_{15}$	6	6	6	6	6	$_{19}$	6	7	6	6	6	6	6	6	6	$_{14}$	6	6	6	6	6	6
717	$_{15}$	6	6	$_{21}$	7	7	6	7	7	6	7	6	7	6	6	$_{15}$	$_{15}$	6	6	6	7	6	6	6	6	6	6	6	$_{15}$	6	6	6	6	6	$_{18}$	6	7	6	6	6	6	6	$_{15}$	6	$_{13}$	6	6	7	6	6
718	7	6	6	6	$_{21}$	7	7	7	7	6	6	6	6	7	6	7	6	6	6	6	6	7	6	6	6	6	6	6	6	$_{15}$	6	6	6	6	6	$_{17}$	6	7	6	6	6	6	7	6	6	$_{12}$	6	6	6	6
719	*	6	6	7	6	$_{21}$	7	7	7	7	7	6	6	6	7	$_{15}$	$_{15}$	6	7	6	6	6	$_{16}$	6	6	6	6	6	6	6	$_{16}$	$_{16}$	6	6	6	6	$_{16}$	6	7	6	7	6	$_{15}$	6	6	6	$_{11}$	6	6	6
720	$_{15}$	6	6	6	6	6	$_{21}$	7	8	7	7	6	6	6	6	$_{15}$	$_{15}$	6	7	6	6	6	6	7	6	6	7	6	6	6	6	$_{15}$	6	6	6	6	6	$_{15}$	6	7	6	6	$_{15}$	6	6	6	6	$_{10}$	6	6
721	9	7	6	7	6	7	7	$_{21}$	7	7	7	6	6	6	6	9	9	6	7	6	6	6	7	7	7	6	7	6	6	6	7	7	$_{15}$	6	6	6	6	6	$_{14}$	7	7	6	9	6	6	6	6	7	9	6
722	8	7	6	6	6	6	6	7	$_{21}$	7	7	6	6	6	6	6	6	7	6	6	6	6	8	6	6	7	6	6	6	6	8	8	6	$_{15}$	6	6	6	6	6	$_{13}$	6	7	6	6	6	6	6	6	6	8
723	$_{12}$	6	7	7	6	6	6	7	7	$_{21}$	7	6	6	6	6	6	7	6	7	7	6	6	$_{12}$	6	6	6	7	6	6	6	$_{12}$	$_{12}$	6	6	$_{15}$	6	7	6	6	7	$_{12}$	6	7	6	6	6	6	7	6	6
724	$_{10}$	6	6	7	6	6	7	7	7	6	$_{21}$	7	6	6	6	7	6	6	6	7	6	6	$_{10}$	6	6	6	6	7	6	6	$_{10}$	$_{10}$	6	6	6	$_{15}$	6	6	6	6	6	$_{11}$	7	7	6	6	6	6	6	6
725	$_{24}$	6	6	6	7	6	6	7	7	6	7	$_{21}$	7	6	6	$_{15}$	$_{15}$	6	6	6	7	6	$_{10}$	6	$_{24}$	6	6	6	$_{24}$	6	$_{10}$	$_{10}$	6	6	6	6	$_{15}$	6	6	6	6	6	$_{15}$	6	7	6	6	6	6	6
726	$_{12}$	6	6	6	6	7	6	6	6	6	6	6	$_{21}$	7	6	6	6	6	6	6	6	7	$_{12}$	6	6	6	6	6	6	7	$_{12}$	$_{12}$	6	6	6	6	6	$_{15}$	6	6	6	6	6	9	6	7	6	6	6	6
727	*	6	6	6	6	6	7	7	7	7	7	6	6	$_{21}$	7	6	6	6	7	6	6	6	7	7	6	6	6	6	6	6	7	6	6	6	6	6	7	6	$_{15}$	6	6	6	6	6	8	6	6	6	6	6
728	7	6	7	6	6	6	6	7	7	7	7	6	7	6	$_{21}$	7	7	6	7	6	6	6	6	7	6	6	7	6	6	6	6	7	6	6	7	6	6	6	6	$_{15}$	6	6	7	6	6	7	6	7	6	6
729	*	7	6	7	6	7	6	7	$_{21}$	7	7	6	7	6	6	$_{21}$	$_{15}$	6	7	6	6	6	$_{15}$	7	8	6	$_{26}$	6	6	6	$_{15}$	$_{15}$	7	6	7	6	6	6	6	7	$_{15}$	6	$_{15}$	6	6	6	6	7	7	6
730	$_{16}$	7	6	6	6	6	6	7	7	7	6	6	6	6	6	6	$_{21}$	7	7	6	6	6	$_{16}$	6	6	7	6	6	6	6	$_{16}$	$_{16}$	6	7	6	6	6	6	6	6	6	$_{15}$	6	6	6	6	6	6	6	7
731	$_{16}$	7	7	6	6	6	7	7	7	7	7	6	7	6	6	7	$_{16}$	$_{21}$	7	6	6	6	6	6	7	6	7	7	7	6	6	6	6	6	7	6	7	6	6	6	6	6	$_{16}$	6	6	6	6	7	6	6
732	$_{18}$	6	7	7	6	6	6	6	6	6	6	7	6	6	6	6	6	6	$_{21}$	7	6	6	$_{18}$	6	6	6	6	7	6	6	$_{18}$	$_{18}$	6	6	6	7	6	6	6	6	6	6	6	7	6	6	6	6	6	6
733	*	6	6	7	$_{23}$	6	6	7	7	7	7	6	7	6	6	7	6	6	6	$_{21}$	7	6	6	6	6	6	6	6	7	6	6	6	6	6	6	6	7	6	6	6	6	6	6	6	7	6	6	6	6	6
734	6	6	6	6	7	7	6	6	6	6	6	6	6	7	6	6	6	6	6	6	$_{21}$	7	6	6	6	6	6	6	6	7	6	6	6	6	6	6	6	7	6	6	6	6	6	6	6	7	6	6	6	6
735	7	6	6	7	6	7	7	7	7	6	7	6	6	6	7	6	6	6	7	6	6	$_{21}$	7	6	6	6	6	6	6	6	7	7	6	6	6	6	7	6	7	6	7	6	6	6	6	6	7	6	6	6
736	$_{22}$	6	7	6	6	6	7	7	7	7	$_{10}$	6	$_{10}$	6	6	7	6	6	6	6	6	6	$_{22}$	7	6	6	7	6	6	6	6	$_{22}$	6	6	7	6	6	6	6	7	6	6	7	6	6	6	6	7	6	6
737	$_{10}$	6	6	7	6	7	6	7	7	7	$_{10}$	6	7	6	6	7	7	6	7	6	6	6	6	7	7	6	7	6	6	6	6	7	7	6	7	6	6	6	6	7	7	6	7	6	6	6	6	7	7	6
738	8	7	6	6	6	6	6	7	8	7	6	6	6	6	6	6	6	7	6	6	6	6	6	6	6	7	7	6	6	6	6	6	6	7	6	6	6	6	6	6	6	7	6	6	6	6	6	6	6	7
739	*	6	7	6	6	6	6	7	7	7	7	6	7	6	6	7	6	6	7	6	6	6	6	7	6	6	7	6	6	6	6	6	7	6	7	7	7	6	6	6	6	6	7	6	6	6	6	6	6	6
740	7	7	6	7	6	6	6	7	7	7	7	7	6	6	6	6	6	6	6	7	6	6	6	6	6	6	6	7	6	6	6	6	6	6	6	7	6	6	6	6	6	6	7	6	6	6	6	6	6	6
741	7	6	7	6	7	6	6	7	7	6	7	7	7	6	6	6	6	6	6	6	7	6	6	7	6	6	6	6	7	6	6	6	6	6	6	6	7	6	7	6	6	6	6	6	7	6	6	6	6	6
742	7	6	6	7	6	7	6	7	7	7	6	6	7	7	6	6	6	6	6	6	6	7	6	6	6	6	6	6	6	7	6	6	6	6	6	6	6	7	6	6	6	6	6	6	6	7	6	6	6	6
743	*	6	6	7	7	6	7	7	7	7	7	7	6	7	7	7	6	6	7	6	6	6	7	6	6	6	6	6	6	6	7	6	6	6	6	6	7	6	7	7	7	6	6	6	6	6	6	6	6	6
744	7	6	6	6	6	7	6	7	7	6	7	6	7	6	7	7	6	6	7	6	6	6	6	7	6	6	6	6	6	6	6	7	6	6	7	6	6	6	6	7	6	6	7	6	6	6	6	7	6	6
745	$_{15}$	6	6	7	6	7	7	7	7	7	9	6	9	6	6	7	$_{15}$	6	$_{15}$	6	6	6	6	7	$_{15}$	6	7	6	6	6	6	7	7	6	7	6	6	6	6	7	$_{15}$	6	7	6	6	6	6	7	9	6
746	7	7	6	6	6	6	6	7	7	7	6	6	6	6	6	6	7	7	6	6	6	6	6	6	6	7	6	6	6	6	6	6	6	7	7	6	6	6	6	6	6	7	6	6	6	6	6	6	6	7
747	8	6	7	7	6	6	6	7	8	7	7	6	6	6	6	7	7	7	7	6	6	6	6	7	6	6	7	6	6	6	6	7	6	6	7	6	7	6	6	6	7	6	7	7	6	6	6	6	6	6
748	7	6	6	7	6	6	6	7	7	7	6	7	6	6	6	6	6	6	7	7	6	6	6	6	6	6	6	7	6	6	6	6	6	6	6	7	6	6	6	6	6	6	6	7	6	6	6	6	6	6
749	$_{10}$	6	6	6	7	6	6	7	7	7	7	6	$_{10}$	6	6	7	$_{10}$	7	6	7	7	6	6	6	6	6	6	6	7	6	6	7	6	6	6	6	7	6	6	6	6	6	7	6	7	6	6	6	6	6
750	6	6	6	6	6	7	6	6	6	6	6	7	6	7	6	6	6	6	6	6	7	7	6	6	6	6	6	6	6	7	6	6	6	6	6	6	6	7	6	6	6	6	6	6	6	7	6	6	6	6
751	*	6	6	6	6	7	7	7	7	7	7	6	7	6	7	7	6	6	6	6	6	7	7	7	6	6	6	6	6	6	7	6	6	6	6	6	6	6	7	6	6	6	6	6	6	6	7	7	7	6
752	$_{15}$	6	6	6	6	6	6	7	7	7	7	6	7	7	6	$_{15}$	7	6	7	6	6	6	$_{15}$	7	$_{15}$	6	7	6	6	6	6	$_{15}$	6	6	6	6	6	6	6	7	6	6	7	6	6	6	$_{15}$	7	6	6
753	$_{15}$	6	6	7	6	7	6	7	$_{21}$	7	7	6	7	6	7	$_{15}$	$_{15}$	6	7	6	6	6	8	7	8	6	6	6	6	6	8	7	7	6	7	6	6	6	6	7	7	6	7	6	6	6	$_{15}$	7	7	6
754	7	$_{15}$	6	6	6	6	6	7	7	$_{21}$	6	6	7	6	6	7	6	$_{15}$	6	6	6	6	6	6	7	7	6	6	6	6	6	6	6	7	6	6	6	6	6	6	6	7	7	6	6	6	6	6	6	7
755	$_{15}$	6	$_{15}$	6	6	7	6	7	7	6	7	6	6	6	6	$_{15}$	$_{15}$	6	$_{15}$	6	6	6	6	7	7	7	7	6	6	6	6	7	6	6	7	6	6	6	6	7	6	6	7	6	6	6	$_{15}$	6	7	6
756	7	6	6	$_{15}$	6	6	6	7	7	7	6	7	6	6	6	6	6	7	6	$_{15}$	6	6	6	6	6	6	7	7	6	6	6	6	6	6	6	7	6	6	6	6	6	6	6	7	6	6	6	6	6	6

$h \rightarrow$	0	1	2	3	4	5
$n \downarrow$	1 2 3 4 5 6 7 8 9	0 1 2 3 4 5 6 7 8 9	0 1 2 3 4 5 6 7 8 9	0 1 2 3 4 5 6 7 8 9	0 1 2 3 4 5 6 7 8 9	0
757	* 6 6 6 $_{15}$ 6 6 7 7	7 7 6 7 6 6 7 6 6 7	6 $_{15}$ 6 6 7 7 6 6 7 7	6 6 6 6 6 6 6 7 6 6	7 6 6 6 6 7 6 6 6 6	6
758	7 6 6 6 6 $_{15}$ 6 7 7	7 6 6 6 7 6 6 6 6 6	7 6 $_{15}$ 6 6 6 6 6 6 7	7 6 6 6 6 6 6 6 7 6	6 6 6 6 6 6 7 6 6 6	6
759	$_{15}$ 6 6 7 6 6 $_{15}$ 7 7	6 7 6 6 6 7 $_{15}$ $_{15}$ 6 7	6 7 6 $_{15}$ 7 6 6 6 6 6	7 7 7 6 6 6 6 7 6 7	6 7 6 6 6 6 6 $_{15}$ 6 6	6
760	7 6 6 6 6 7 6 $_{15}$ 7	7 7 6 7 6 6 7 7 6 7	6 6 7 6 $_{15}$ 6 6 7 6 6	6 6 7 6 6 7 6 6 6 6	7 6 6 7 6 6 6 7 7 7	6
761	* 6 6 7 6 7 7 7 $_{15}$	7 7 6 7 6 6 $_{15}$ $_{15}$ 6 $_{15}$	6 6 6 7 7 $_{15}$ 6 7 6 6	7 6 7 7 6 6 6 6 6 6	7 7 6 7 6 6 6 $_{15}$ 7 7	6
762	6 7 6 6 6 6 6 6 6	$_{15}$ 6 6 6 6 6 6 6 7 6	6 6 6 6 7 6 $_{15}$ 6 6 6	6 6 6 7 7 6 6 6 6 6	6 6 7 6 6 6 6 6 6 6	7
763	$_{15}$ 6 7 7 6 6 6 7 7	7 $_{15}$ 6 6 6 6 $_{15}$ $_{15}$ 6 7	6 6 6 6 7 7 6 $_{15}$ 6 6	6 6 7 7 7 7 6 6 6 6	7 7 6 7 6 6 6 $_{15}$ 7 6	6
764	6 6 6 7 6 6 6 6 6	6 6 $_{15}$ 6 6 6 6 6 6 6	7 6 6 6 6 6 6 6 $_{15}$ 6	6 6 6 6 6 7 7 6 6 6	6 6 6 6 7 6 6 6 6 6	6
765	$_{15}$ 6 6 6 7 6 6 7 7	7 7 6 $_{15}$ 6 6 $_{15}$ $_{15}$ 6 7	6 7 6 6 7 6 6 7 6 $_{15}$	6 6 7 7 6 6 7 7 6 6	6 6 6 6 6 7 6 $_{15}$ 7 6	6
766	7 6 6 6 6 7 6 6 7	6 6 6 6 $_{15}$ 6 6 6 6 6	6 6 7 6 6 6 6 6 7 6	$_{15}$ 6 6 6 6 6 6 7 7 6	6 6 6 6 6 6 7 6 6 7	6
767	$_{15}$ 6 6 7 6 6 7 7 7	7 6 6 $_{12}$ 6 $_{15}$ $_{15}$ $_{15}$ 6 7	6 6 6 7 6 6 6 6 6 7	6 $_{15}$ 7 6 6 6 7 7 6 7	7 7 6 7 6 6 6 $_{15}$ 6 6	6
768	$_{15}$ 6 7 7 6 6 6 7 7	6 7 6 7 6 6 $_{15}$ $_{15}$ 6 7	6 6 6 6 7 6 6 6 6 6	7 6 $_{15}$ 6 6 7 6 6 6 6	7 7 6 7 6 6 6 $_{15}$ 7 6	6
769	* 6 7 7 6 7 6 7 7	7 7 6 6 6 6 $_{15}$ $_{15}$ 6 7	6 6 6 6 7 7 6 7 6 6	6 7 7 $_{15}$ 6 7 6 6 6 6	7 7 6 7 6 6 6 $_{15}$ 7 7	6
770	7 7 6 7 6 6 6 7 7	7 6 6 6 6 6 6 6 $_{15}$ 6	6 6 6 6 6 6 7 6 6 6	6 6 7 6 $_{15}$ 6 6 6 6 6	6 7 7 6 6 6 6 6 6 7	7
771	$_{15}$ 6 7 7 6 6 6 7 7	6 7 6 6 6 6 $_{15}$ $_{15}$ 6 $_{15}$	6 6 6 6 7 6 6 7 6 6	6 6 7 7 6 $_{15}$ 6 6 6 6	7 7 7 7 6 6 6 $_{15}$ 7 6	6
772	7 6 6 7 6 6 6 7 7	7 6 7 6 6 6 6 6 6 6	$_{15}$ 6 6 6 6 6 6 6 7 6	6 6 6 6 6 6 $_{15}$ 6 6 6	6 6 6 7 7 6 6 6 6 6	6
773	* 6 6 7 7 6 6 7 7	7 7 6 7 6 6 $_{15}$ $_{15}$ 6 7	6 $_{15}$ 6 6 6 6 6 7 6 7	6 6 7 6 6 6 6 $_{15}$ 6 6	7 7 7 6 7 7 6 $_{15}$ 6 6	6
774	6 6 6 6 6 7 6 6 6	6 6 6 6 7 6 6 6 6 6	6 6 $_{15}$ 6 6 6 6 6 6 6	7 6 6 6 6 6 7 6 $_{15}$ 6	6 6 6 6 6 6 7 6 6 6	6
775	$_{24}$ 6 6 7 6 6 7 7 7	6 7 6 6 6 7 $_{15}$ $_{15}$ 6 7	6 6 6 $_{15}$ 7 $_{24}$ 6 6 6 6	6 $_{24}$ 6 6 6 6 6 7 6 $_{15}$	7 7 6 6 6 6 6 $_{15}$ 7 6	6
776	$_{23}$ 6 7 7 6 7 6 7 7	7 7 6 6 6 6 7 7 6 7	6 6 6 6 $_{15}$ $_{23}$ 6 7 6 6	6 $_{23}$ $_{23}$ 6 6 6 6 6 7 6	$_{15}$ 7 6 7 6 6 6 7 7 7	6
777	$_{15}$ $_{23}$ 7 7 6 7 6 7 8	7 7 6 6 6 6 $_{15}$ $_{15}$ 6 7	6 6 6 6 7 $_{15}$ 6 7 6 6	6 6 7 $_{22}$ 6 7 6 6 6 7	7 $_{15}$ 6 7 6 6 6 $_{15}$ 7 7	6
778	7 7 $_{23}$ 6 6 7 6 7 7	7 6 6 6 6 6 6 6 7 6	6 6 6 6 6 6 $_{15}$ 6 6 6	6 6 6 6 $_{21}$ 6 6 6 6 6	7 6 $_{15}$ 6 6 6 6 6 6 7	7
779	$_{18}$ 6 7 $_{23}$ 6 7 6 7 7	7 7 6 6 6 6 $_{15}$ $_{15}$ 6 $_{18}$	6 6 6 6 7 6 6 $_{15}$ 6 6	6 6 7 6 6 $_{20}$ 6 6 6 6	7 $_{18}$ 6 $_{15}$ 6 6 6 $_{15}$ 7 7	7
780	6 6 6 7 $_{23}$ 6 6 6 6	6 6 7 6 6 6 6 6 6 6	7 6 6 6 6 6 6 6 $_{15}$ 6	6 6 6 6 6 6 $_{19}$ 6 6 6	6 6 6 6 $_{15}$ 6 6 6 6 6	6
781	$_{18}$ 6 6 6 7 $_{23}$ 7 7 7	7 $_{10}$ 6 7 6 6 $_{15}$ $_{15}$ 6 6	6 7 6 6 7 $_{18}$ 6 7 6 $_{15}$	6 $_{18}$ $_{18}$ 6 6 7 6 $_{18}$ 6 6	7 6 6 6 6 $_{15}$ 6 $_{15}$ 7 7	6
782	7 6 6 6 6 7 $_{23}$ 7 7	6 7 6 6 7 6 6 6 6 6	6 6 7 6 6 6 6 6 6 6	$_{15}$ 6 6 6 6 6 6 6 $_{17}$ 6	6 6 6 6 6 6 $_{15}$ 6 6 6	6
783	$_{26}$ 7 6 6 6 6 7 $_{23}$ 7	6 7 6 6 6 7 $_{15}$ $_{15}$ 6 $_{15}$	6 6 6 7 7 7 6 $_{26}$ 6 $_{26}$	6 $_{15}$ 7 6 6 6 6 6 6 $_{16}$	6 6 6 $_{15}$ 6 7 6 $_{15}$ 7 6	6
784	$_{15}$ 7 7 6 6 7 6 7 $_{23}$	7 7 6 6 6 6 $_{15}$ $_{15}$ 6 7	6 6 6 $_{15}$ 7 $_{15}$ 6 7 6 6	6 8 $_{15}$ 6 6 7 6 6 6 6	$_{15}$ 6 6 6 6 6 7 $_{15}$ $_{14}$ $_{15}$	6
785	$_{15}$ 7 7 7 6 7 6 7 7	$_{23}$ 7 6 6 6 6 $_{15}$ $_{15}$ 6 7	6 6 6 6 7 $_{14}$ 6 7 $_{27}$ 6	6 $_{14}$ $_{14}$ $_{15}$ 6 7 6 6 6 6	7 $_{14}$ 6 7 6 6 6 $_{13}$ 7 $_{15}$	6
786	$_{10}$ $_{15}$ 7 7 7 6 6 7 7	7 $_{23}$ 6 6 6 6 6 6 $_{15}$ 6	6 6 6 6 6 $_{10}$ 7 6 6 6	6 $_{10}$ $_{10}$ 6 $_{15}$ 6 6 6 6 6	6 6 $_{13}$ 6 6 6 6 6 7 6	$_{12}$
787	* 6 $_{25}$ 7 7 7 $_{23}$ 7 7	7 7 $_{23}$ 6 6 6 $_{15}$ $_{15}$ 6 $_{15}$	6 6 6 6 7 $_{12}$ 6 7 6 6	6 $_{12}$ $_{12}$ 6 6 $_{15}$ 6 6 6 6	7 6 6 $_{12}$ 6 6 6 6 7 $_{15}$	6
788	$_{12}$ 6 6 $_{15}$ 7 7 7 6 6	6 6 7 $_{23}$ 6 6 6 6 6 6	$_{15}$ 6 6 6 6 $_{12}$ 6 6 7 6	6 $_{12}$ $_{12}$ 6 6 6 $_{15}$ 6 6 6	6 6 6 6 $_{11}$ 6 6 6 6 6	6
789	$_{15}$ 6 6 6 $_{15}$ 7 7 7 7	6 7 6 7 $_{23}$ 6 $_{15}$ $_{15}$ 6 6	6 $_{15}$ 6 6 7 6 6 6 6 7	6 6 7 6 6 7 6 $_{15}$ 6 6	6 6 6 7 6 $_{10}$ 6 $_{15}$ 7 6	6
790	7 6 6 6 6 $_{15}$ 7 7 7	6 7 6 6 7 $_{23}$ 6 6 6 6	6 6 $_{15}$ 6 6 6 6 6 6 6	7 6 6 6 6 6 6 6 $_{15}$ 6	6 6 6 6 6 6 9 6 6 6	6
791	$_{15}$ 7 7 7 7 7 $_{15}$ 7 7	7 7 6 6 6 7 $_{23}$ $_{15}$ 6 7	6 6 6 $_{15}$ 6 $_{15}$ 6 7 6 6	6 $_{15}$ $_{15}$ 6 6 6 6 6 6 $_{15}$	7 6 6 6 6 6 6 8 6 $_{15}$	6
792	$_{16}$ 7 7 7 7 7 6 $_{15}$ 8	7 7 6 6 6 6 7 $_{23}$ 6 7	6 6 6 6 $_{15}$ $_{16}$ 6 8 6 8	6 $_{16}$ $_{16}$ 6 6 7 6 6 6 6	$_{15}$ 6 6 7 6 6 6 7 7 7	6
793	$_{15}$ 7 7 7 7 7 7 7 $_{15}$	7 7 7 $_{12}$ 6 6 $_{15}$ $_{15}$ $_{23}$ $_{15}$	6 6 6 6 7 $_{15}$ 6 7 6 6	6 6 7 7 6 7 6 6 6 6	7 $_{15}$ 6 7 6 6 6 $_{15}$ 7 $_{15}$	6
794	$_{18}$ 7 6 7 7 7 7 7 7	$_{15}$ 7 7 7 6 6 6 6 7 $_{23}$	6 6 6 6 6 $_{18}$ $_{15}$ 6 6 6	6 $_{18}$ $_{18}$ 6 7 6 6 6 6 6	6 6 $_{15}$ 6 6 6 6 6 6 6	7
795	$_{15}$ 6 7 6 7 7 7 7 7	7 $_{15}$ 7 7 7 6 $_{15}$ $_{15}$ 6 7	$_{23}$ 6 6 6 7 6 6 $_{15}$ 6 6	6 6 7 6 6 7 6 6 6 6	7 6 6 $_{15}$ 6 6 6 $_{15}$ 7 $_{15}$	6
796	7 7 6 7 6 7 7 7 7	7 6 $_{15}$ 7 7 7 6 6 6 6	7 $_{23}$ 6 6 6 6 6 6 $_{15}$ 6	6 6 6 6 6 6 7 6 6 6	6 6 6 6 $_{15}$ 6 6 6 6 6	6
797	* 6 7 6 7 7 7 7 7	7 7 6 $_{15}$ 7 7 $_{15}$ $_{15}$ 6 6	6 7 $_{23}$ 6 6 6 6 6 6 $_{15}$	6 6 7 6 6 6 6 7 6 6	7 6 6 7 6 $_{15}$ 6 6 6 $_{15}$	6
798	$_{22}$ 6 6 7 6 7 6 7 7	7 7 6 6 $_{15}$ 7 7 7 6 6	6 6 7 $_{23}$ 6 $_{22}$ 6 6 6 6	$_{15}$ $_{22}$ $_{22}$ 6 6 6 6 6 7 6	6 6 6 6 6 6 $_{15}$ 6 6 6	6
799	$_{16}$ 7 6 6 7 7 7 7 7	7 7 6 6 6 $_{15}$ $_{15}$ $_{16}$ 7 6	6 6 6 7 $_{23}$ 6 6 7 6 6	6 $_{15}$ 6 6 6 7 6 6 6 7	7 6 6 6 6 6 6 $_{16}$ 7 $_{15}$	6
800	$_{24}$ 7 7 7 6 7 6 7 7	7 $_{10}$ 6 $_{10}$ 6 6 $_{15}$ $_{15}$ 7 7	6 6 6 6 7 $_{24}$ 6 7 6 6	6 6 $_{24}$ 6 6 7 6 6 6 6	7 6 6 7 6 6 6 6 7 $_{15}$	6
801	$_{15}$ 7 7 7 6 7 7 7 8	7 7 7 6 6 6 $_{15}$ $_{15}$ 7 7	7 6 6 6 7 7 6 7 6 6	6 6 7 $_{15}$ 6 7 6 6 6 6	7 7 6 7 6 6 6 6 7 $_{15}$	6
802	7 7 7 7 7 6 6 7 7	7 7 7 7 6 6 6 6 $_{15}$ 7	7 7 6 6 6 6 7 7 6 6	6 6 6 6 $_{15}$ 6 6 6 6 6	6 6 7 6 6 6 6 6 6 7	7
803	$_{15}$ 7 7 7 7 7 6 7 7	7 $_{10}$ 6 7 7 6 $_{15}$ $_{15}$ 6 $_{15}$	7 7 7 6 6 6 6 7 6 6	6 6 7 6 6 $_{15}$ 6 6 6 6	7 6 6 7 6 6 6 6 7 $_{15}$	6
804	7 7 7 7 6 7 7 7 7	7 6 7 6 7 7 6 6 6 6	$_{15}$ 7 7 7 6 6 6 6 7 6	6 6 6 6 6 6 $_{15}$ 6 6 6	6 6 6 6 7 6 6 6 6 6	6
805	$_{15}$ 7 7 7 7 7 7 7 7	7 7 6 7 6 7 $_{15}$ $_{15}$ 6 6	6 $_{15}$ 7 7 7 6 6 6 6 7	6 6 6 6 6 6 6 $_{15}$ 6 6	7 6 6 6 6 7 6 6 7 $_{15}$	6
806	7 7 6 6 7 7 6 7 7	7 6 6 6 7 6 7 7 6 6	6 6 $_{15}$ 7 7 7 6 6 6 6	7 6 6 6 6 6 6 6 $_{15}$ 6	6 6 6 6 6 6 7 6 6 6	6
807	$_{15}$ 7 7 6 6 7 7 7 7	7 7 6 6 6 7 $_{15}$ $_{15}$ 7 6	6 6 6 $_{15}$ 7 7 7 6 6 6	6 7 7 6 6 7 6 6 6 $_{15}$	6 6 6 7 6 6 6 7 7 $_{15}$	6

n ↓ \ h →	1	2	3	4	5	6	7	8	9	10	11	12	13	14	15	16	17	18	19	20	21	22	23	24	25	26	27	28	29	30	31	32	33	34	35	36	37	38	39	40	41	42	43	44	45	46	47	48	49	50
808	7	7	7	7	6	7	7	7	7	7	7	6	6	6	6	7	7	7	7	6	6	6	6	15	7	7	7	6	6	6	6	7	6	6	7	6	6	6	6	15	6	6	7	6	6	6	6	7	7	6
809	*	7	7	7	7	7	6	7	7	7	7	6	6	6	6	15	15	6	7	7	6	6	6	7	15	7	7	7	6	6	6	7	7	6	7	6	6	6	6	7	15	6	7	6	6	6	6	7	15	6
810	7	7	7	7	6	7	6	7	7	7	7	6	6	6	6	6	7	7	6	7	7	6	6	6	6	15	7	7	7	6	6	6	6	7	7	6	6	6	6	6	6	15	6	6	6	6	6	6	6	7
811	*	7	7	7	7	7	7	7	8	7	9	6	6	6	6	15	15	7	8	6	7	7	6	7	6	6	15	7	7	7	6	6	6	6	7	6	6	6	6	7	6	6	15	6	6	6	6	7	15	6
812	8	7	7	7	6	7	6	7	8	7	8	7	6	6	6	6	6	6	6	7	6	7	7	6	6	6	6	15	7	7	7	6	6	6	6	7	6	6	6	6	6	6	6	15	6	6	6	6	6	6
813	15	6	7	7	7	7	23	7	7	7	7	7	7	6	6	15	15	6	6	6	7	6	7	7	6	6	6	6	15	7	7	7	6	6	6	6	7	6	6	6	6	6	6	6	15	6	6	7	15	6
814	7	7	6	7	7	7	6	23	7	7	7	6	7	7	6	6	6	6	6	6	6	7	6	7	7	6	7	6	6	15	7	7	7	6	6	6	6	7	6	6	6	6	6	6	6	15	6	6	6	6
815	15	7	7	7	7	7	7	7	7	7	7	6	6	7	7	15	15	6	6	6	6	6	7	6	7	7	6	6	6	6	15	7	7	7	6	6	6	6	7	7	6	6	7	6	6	6	15	6	15	6
816	15	6	7	7	6	7	7	7	7	7	7	6	6	6	7	15	15	6	6	6	6	6	6	7	15	7	15	6	6	6	6	15	7	7	7	6	6	6	6	7	6	6	7	6	6	6	6	15	15	6
817	18	7	7	7	23	7	7	7	7	7	7	6	6	6	6	15	15	6	18	6	6	6	6	7	7	6	7	7	6	6	6	7	15	7	7	7	6	6	6	7	7	6	18	6	6	6	6	7	15	6
818	7	7	7	6	7	7	6	7	7	7	7	6	6	6	6	7	7	7	6	6	6	6	6	6	7	7	6	7	7	6	6	6	6	15	7	7	7	6	6	6	6	7	7	6	6	6	6	6	6	14
819	8	7	7	7	6	7	7	7	8	7	7	6	6	6	6	7	6	7	7	6	6	6	6	7	6	7	8	6	7	7	6	7	6	6	15	7	7	7	6	6	6	6	7	6	6	6	6	7	6	6
820	7	7	7	7	6	6	7	7	7	7	7	7	6	6	6	6	7	6	7	7	6	6	6	6	6	6	7	7	6	7	7	6	6	6	6	15	7	7	7	6	6	6	6	7	6	6	6	6	6	6
821	*	7	7	7	7	7	23	7	8	7	7	7	7	6	6	15	15	8	6	7	7	6	6	6	6	6	6	6	7	6	7	7	6	6	6	6	15	7	7	7	6	6	6	6	7	6	6	6	15	6
822	7	7	7	7	7	7	6	7	7	7	7	7	7	7	6	6	7	6	6	6	7	7	6	6	6	6	6	6	6	7	6	7	7	6	6	6	6	15	7	7	7	6	6	6	6	7	6	6	6	6
823	*	7	7	7	7	7	7	7	7	7	7	6	7	7	7	6	6	7	7	6	6	7	7	6	6	6	6	6	6	6	7	6	7	7	6	6	6	6	15	7	7	7	7	6	6	6	7	7	7	6
824	7	7	7	7	7	7	7	7	7	7	7	6	6	7	7	7	7	6	6	6	6	6	7	7	6	6	6	6	6	6	6	7	6	7	7	6	6	6	6	15	7	7	7	6	6	6	6	7	7	6
825	15	7	7	7	6	7	7	7	7	7	7	6	6	6	7	15	15	6	6	6	6	6	6	7	7	6	6	6	6	6	6	7	7	6	7	7	6	6	6	7	15	7	7	7	6	6	6	7	15	6
826	7	7	7	7	7	6	7	7	7	7	7	6	6	6	6	7	7	7	6	6	6	6	6	7	7	7	7	6	6	6	6	6	7	7	6	7	7	6	6	6	6	15	7	6	7	6	6	6	6	7
827	*	7	7	6	7	7	6	7	7	7	10	6	6	6	6	15	15	7	7	6	6	6	6	7	6	7	7	6	6	6	6	7	6	7	8	6	7	7	6	7	6	6	15	6	7	7	6	6	15	6
828	7	7	7	7	6	7	7	7	7	7	7	7	6	6	6	7	7	7	7	7	6	6	6	6	7	6	7	7	6	6	6	6	6	6	7	7	6	7	7	6	6	6	6	15	7	7	7	6	6	6
829	*	7	7	7	7	7	7	7	7	7	7	7	7	6	6	7	6	7	7	7	7	6	6	7	6	7	6	7	7	6	6	6	6	6	6	7	7	6	7	7	6	6	6	6	15	7	7	7	6	6
830	8	7	7	7	7	7	6	7	7	7	8	6	7	7	6	7	7	6	6	7	7	7	6	6	7	6	8	6	7	7	6	6	6	6	6	6	6	7	6	7	7	6	6	6	6	15	7	7	7	6
831	15	6	7	7	6	7	7	7	7	7	7	7	6	7	7	15	15	7	6	6	7	7	7	6	6	7	6	6	6	7	7	6	6	6	6	6	6	6	7	6	7	7	6	6	6	6	15	7	15	7
832	12	7	7	7	7	7	7	7	7	7	7	6	12	6	7	7	7	6	6	6	6	7	7	7	6	6	7	6	6	6	7	7	6	6	6	6	6	6	6	7	6	7	7	6	6	6	6	15	7	7
833	16	7	7	7	7	7	6	7	7	7	7	6	7	7	6	7	16	7	6	6	6	6	7	7	7	6	6	6	6	6	6	7	7	6	6	6	6	6	6	7	7	6	7	6	6	6	6	7	16	7
834	7	7	6	7	6	7	7	7	7	7	7	6	6	6	7	7	7	7	6	6	6	6	6	7	7	7	6	6	6	6	6	7	7	7	6	7	6	6	6	6	7	7	6	6	7	6	6	6	6	7
835	8	7	7	6	7	7	7	7	8	7	7	6	7	6	6	7	6	7	7	6	6	6	6	7	7	7	7	6	6	6	6	7	6	7	7	7	6	6	6	7	6	7	8	6	7	7	6	7	6	6
836	7	7	7	7	6	7	6	7	7	7	7	7	6	6	6	7	7	6	7	7	6	6	6	7	7	7	7	7	6	6	6	6	7	6	7	7	6	6	6	6	6	6	7	7	6	7	7	6	6	6
837	26	7	7	7	7	6	7	7	7	7	7	7	7	6	6	7	6	7	6	7	7	6	6	7	6	7	26	7	7	6	26	7	6	7	6	7	7	6	6	6	6	6	6	7	7	6	7	7	6	6
838	25	7	7	6	7	7	6	7	7	7	7	6	7	7	6	7	7	6	7	6	7	7	6	7	7	6	25	7	7	7	25	25	7	6	7	6	7	7	6	6	6	6	6	6	7	7	6	7	7	6
839	*	25	7	7	6	7	7	7	7	7	7	7	6	7	7	7	6	7	6	7	6	7	7	7	6	7	6	6	7	7	7	6	24	7	6	8	6	7	7	6	6	6	6	6	6	6	7	6	7	7
840	7	7	25	6	7	6	7	7	7	7	7	6	7	6	7	7	7	6	6	6	7	6	7	7	7	6	7	6	6	7	7	7	6	23	7	6	6	6	7	7	6	6	6	6	6	6	6	7	6	7
841	*	7	7	25	6	7	6	7	7	7	7	6	7	7	6	7	7	7	7	6	7	7	6	7	7	7	6	6	28	6	7	7	7	6	22	7	6	6	6	7	7	6	7	6	6	6	6	7	7	6
842	7	7	7	7	25	7	7	7	7	7	7	6	6	6	7	7	7	7	6	6	6	6	7	7	7	7	7	6	6	6	6	7	7	7	6	21	6	6	6	7	7	7	6	6	7	6	6	6	7	7
843	20	7	7	6	7	25	6	7	7	7	7	6	7	6	6	7	6	7	7	6	6	6	6	7	6	7	20	6	6	6	20	20	7	7	7	6	20	6	6	7	6	7	7	6	7	6	6	7	6	7
844	7	6	7	7	6	7	25	7	7	7	7	7	6	6	6	7	7	6	7	7	6	6	6	7	7	6	7	7	6	6	6	7	7	7	7	7	6	19	6	6	7	6	7	7	7	6	6	6	6	6
845	10	7	7	7	25	6	7	25	7	7	7	7	10	6	6	7	10	7	6	7	7	6	6	7	6	7	7	7	7	6	7	7	6	7	7	7	7	6	18	7	6	7	6	7	7	6	6	6	6	6
846	8	7	7	7	7	7	7	7	25	7	7	6	7	7	6	7	7	6	7	6	7	7	6	7	7	6	8	6	7	7	8	8	7	6	7	7	7	7	6	17	7	6	7	6	7	7	6	6	6	6
847	16	7	7	7	6	7	7	7	7	25	7	7	6	7	7	7	6	7	7	7	6	7	7	7	6	7	16	7	6	7	16	16	6	7	6	7	7	7	7	6	16	7	7	7	6	7	7	6	6	6
848	15	7	7	7	7	7	7	7	7	7	25	6	12	6	7	15	12	6	6	6	7	6	7	7	15	6	15	6	7	6	10	15	7	6	7	6	6	7	7	7	7	15	7	6	8	6	6	7	6	6
849	15	7	7	7	6	7	6	7	7	7	7	25	7	7	6	15	15	7	6	6	7	7	6	7	7	7	14	6	7	7	14	14	7	7	6	7	6	6	7	7	7	6	14	7	6	6	6	7	7	6
850	12	15	7	7	7	7	7	7	7	7	7	6	25	6	7	7	7	15	6	6	6	6	7	7	7	7	12	6	6	6	12	12	7	7	7	6	6	6	6	7	7	7	6	13	7	6	6	7	7	7
851	22	7	15	7	7	7	7	7	8	7	7	6	7	25	6	15	15	7	15	6	7	6	22	7	6	7	7	6	6	6	6	7	6	7	7	7	22	6	6	7	7	7	7	6	12	6	6	7	6	7
852	7	7	7	15	6	7	7	7	7	7	7	7	6	6	25	7	7	6	7	15	6	6	6	7	7	6	7	7	6	6	6	7	7	6	7	7	6	6	6	7	7	7	7	7	6	11	6	6	7	6
853	*	7	7	7	15	7	7	7	7	7	7	7	7	6	6	25	6	7	7	7	15	6	6	7	6	7	15	7	7	6	15	15	6	7	6	7	7	6	6	7	6	7	7	7	7	6	10	7	6	7
854	16	7	7	7	7	15	7	7	7	7	7	6	7	7	6	7	25	6	7	6	7	15	6	7	7	6	16	6	7	7	16	16	7	6	7	6	7	7	6	7	7	6	7	7	7	7	6	9	7	6
855	15	7	7	7	6	7	15	7	7	7	7	7	7	7	7	15	15	25	6	7	6	7	15	7	6	7	8	7	6	7	8	8	6	7	6	7	7	7	7	7	7	7	6	7	7	7	7	6	8	7
856	18	7	7	7	7	6	7	15	7	7	7	6	7	6	7	7	7	6	25	6	7	6	7	15	7	6	18	6	7	6	18	18	7	6	7	6	7	6	7	7	7	6	7	6	7	7	6	7	6	7
857	*	7	7	7	6	7	6	7	15	7	7	7	7	7	6	15	15	7	15	25	7	7	6	7	15	7	6	6	7	7	6	7	7	7	6	7	7	7	6	7	7	7	6	7	6	6	7	7	7	6
858	7	7	7	7	7	7	7	7	7	15	7	6	7	6	7	7	7	7	6	6	25	6	7	7	7	15	7	6	6	6	7	7	7	7	7	6	6	6	7	7	7	7	7	6	7	6	6	7	7	7

h →	0									1										2										3										4										5
n ↓	1	2	3	4	5	6	7	8	9	0	1	2	3	4	5	6	7	8	9	0	1	2	3	4	5	6	7	8	9	0	1	2	3	4	5	6	7	8	9	0	1	2	3	4	5	6	7	8	9	0
859	*	7	7	7	7	7	6	7	8	7	15	6	7	7	6	15	15	7	15	6	6	25	7	7	7	7	15	6	7	6	15	7	6	7	7	7	7	6	6	7	7	7	7	7	7	6	15	7	7	7
860	22	7	7	7	6	7	7	7	7	7	7	15	6	6	7	7	7	6	7	7	6	6	25	7	7	6	22	15	6	6	22	22	7	6	7	7	6	6	6	7	7	6	7	7	7	6	6	7	7	7
861	15	7	7	7	7	6	7	7	7	7	7	7	15	6	6	15	15	7	6	7	7	6	6	25	6	7	6	7	15	6	6	7	6	7	6	7	7	6	6	7	6	7	6	7	7	6	6	7	6	7
862	24	7	7	7	7	7	7	7	7	7	7	6	7	15	6	7	7	6	7	6	7	7	6	7	25	6	24	6	7	15	24	24	7	6	7	6	7	7	6	7	7	6	7	6	7	7	6	7	7	6
863	*	7	7	7	6	7	7	7	7	7	7	7	7	7	15	15	15	7	6	7	6	7	7	7	6	25	6	7	6	7	15	7	6	7	6	7	7	7	7	7	6	7	6	7	6	7	7	7	7	7
864	26	7	7	7	7	7	7	7	7	7	7	6	7	6	7	15	15	6	7	6	7	6	7	7	7	6	26	6	7	6	7	26	7	6	7	6	7	6	7	7	7	6	7	6	7	6	6	7	7	6
865	15	7	7	7	6	7	6	7	7	7	7	7	7	7	6	15	15	6	7	7	7	7	6	7	7	7	6	6	7	7	6	7	15	7	6	7	7	7	6	7	7	7	7	7	7	7	6	7	7	7
866	7	7	7	7	7	7	7	7	7	7	7	6	7	6	7	7	7	15	6	6	7	6	7	7	7	7	7	6	7	6	7	7	7	15	7	6	6	6	7	7	7	7	7	6	7	6	6	7	7	7
867	15	7	7	7	7	7	6	7	7	7	7	6	7	7	6	15	15	7	15	6	7	7	7	7	6	7	7	6	6	6	6	7	7	7	15	7	7	6	6	7	7	7	7	7	6	6	6	7	6	7
868	7	7	7	7	7	7	7	7	7	7	7	7	7	6	7	7	7	6	7	15	6	6	7	7	7	6	7	7	6	6	6	7	7	6	7	15	6	6	6	7	7	6	7	7	7	6	6	7	7	6
869	15	7	7	7	7	7	7	7	7	7	10	7	7	6	6	15	15	7	6	7	15	6	6	7	6	7	6	7	7	6	6	7	6	7	6	7	15	6	6	7	6	7	6	7	7	6	6	7	6	7
870	9	6	7	7	7	7	7	7	7	7	9	6	7	7	6	7	7	6	7	6	7	15	6	7	7	6	7	6	7	7	6	7	7	6	7	6	7	15	6	7	7	6	7	6	7	7	6	7	7	6
871	15	9	7	7	6	7	7	7	8	7	8	7	12	7	7	15	15	7	7	7	6	7	15	7	6	7	6	7	6	7	7	7	6	7	6	7	7	7	15	7	6	7	7	7	6	7	7	7	6	7
872	7	8	9	7	7	7	7	7	7	7	7	6	7	6	7	7	7	6	7	6	7	6	7	15	7	6	7	6	7	6	7	7	7	6	7	6	7	6	7	15	7	6	7	6	7	6	6	7	7	6
873	15	7	8	9	25	7	6	7	8	7	7	7	7	7	6	15	15	6	6	7	7	7	6	7	15	6	6	7	7	7	6	7	25	7	6	7	7	7	6	7	15	7	6	7	7	7	6	7	7	7
874	7	7	7	8	9	25	7	7	7	7	7	6	7	6	7	7	7	7	6	6	7	6	7	7	7	15	6	6	7	6	7	7	7	7	7	6	7	6	7	7	7	15	7	6	7	6	6	7	7	7
875	15	7	7	7	8	9	6	7	8	7	7	7	7	7	6	15	15	7	7	6	7	7	7	7	6	7	15	6	7	7	6	7	6	7	7	6	7	6	6	7	7	7	15	7	7	6	6	7	6	7
876	7	7	7	7	7	8	9	7	7	7	7	7	7	6	6	7	7	6	7	7	7	6	7	7	7	6	7	15	6	6	7	7	7	6	7	7	6	6	6	7	7	6	7	15	6	6	6	7	7	6
877	*	7	7	7	7	7	8	9	7	7	7	7	7	7	6	15	15	7	7	7	7	6	6	7	6	7	6	7	15	6	6	7	6	7	6	7	7	6	6	7	6	7	7	7	15	6	6	7	6	7
878	8	6	7	7	7	7	7	8	9	7	8	6	7	7	7	7	7	6	7	6	7	7	6	7	7	6	7	6	7	15	6	7	7	6	7	6	7	7	6	7	7	6	7	6	7	15	6	7	7	6
879	15	7	7	7	7	7	7	7	8	9	8	7	7	7	21	15	15	7	6	7	7	7	7	7	6	7	6	7	6	7	15	7	6	7	6	7	7	7	7	7	6	7	6	7	6	7	15	7	6	7
880	15	8	7	7	7	7	7	7	7	8	9	6	7	6	7	15	15	6	7	6	7	6	7	7	7	6	15	6	15	6	7	15	7	6	7	6	7	6	7	7	7	6	7	6	7	6	6	15	7	6
881	*	7	8	7	7	7	7	7	8	7	9	7	7	7	6	15	15	7	6	7	7	7	6	7	7	6	6	7	7	7	6	7	15	6	6	7	7	7	6	7	7	7	6	7	7	7	6	7	15	7
882	8	9	7	8	7	7	7	7	8	7	8	6	7	6	7	7	7	7	7	6	7	6	7	7	7	7	6	6	7	6	7	7	7	15	6	6	7	6	7	7	7	7	7	6	7	6	6	7	7	15
883	*	8	9	7	8	7	7	7	7	7	7	7	7	6	6	7	6	7	7	7	7	7	7	7	6	7	7	6	7	7	6	7	6	7	15	6	7	7	6	7	7	7	7	6	6	6	6	7	7	7
884	8	7	8	9	7	8	7	7	8	7	8	7	7	6	6	7	7	6	7	7	7	6	6	7	7	6	7	7	7	6	7	7	7	6	7	15	6	6	7	7	7	6	7	7	6	6	6	7	7	6
885	15	7	7	8	9	7	8	7	7	7	7	7	7	7	6	15	15	7	6	7	7	7	6	7	6	7	6	7	7	6	6	7	6	7	6	7	15	6	6	7	6	7	6	7	7	6	6	7	6	7
886	7	6	7	7	8	9	7	8	7	7	7	7	7	7	7	7	7	6	7	6	7	7	7	7	7	6	7	6	7	7	6	7	7	6	7	6	7	15	6	7	7	6	7	6	7	7	6	7	7	6
887	*	7	7	7	7	8	9	7	8	7	8	7	7	7	7	7	6	7	6	7	6	7	7	7	6	7	6	7	6	7	7	7	6	7	6	7	7	7	15	7	6	7	6	7	6	7	7	7	6	7
888	7	7	7	7	7	7	8	9	7	8	7	6	7	7	7	7	7	6	7	6	7	6	7	7	7	6	7	6	7	6	7	7	7	6	7	6	7	6	7	15	7	6	7	6	7	6	6	7	7	6
889	15	7	7	7	7	7	7	8	9	7	8	7	7	7	7	15	15	7	7	7	7	7	6	7	7	7	6	7	7	7	7	7	7	6	6	7	7	7	6	7	15	6	7	7	7	7	6	7	7	7
890	8	7	7	7	7	7	7	7	8	9	8	6	7	6	7	7	7	7	7	6	7	6	7	7	7	7	7	6	7	6	7	7	7	7	6	6	7	6	7	7	7	15	6	6	7	6	6	7	7	7
891	15	8	7	7	7	7	6	7	8	8	10	7	7	6	6	15	15	7	7	7	6	6	7	7	6	7	7	7	7	7	6	7	6	7	7	6	7	7	6	7	7	7	15	6	7	7	6	7	6	7
892	10	7	8	7	7	7	7	7	8	7	8	7	6	6	6	7	7	7	7	7	7	6	6	7	7	6	7	7	7	6	6	7	7	6	7	7	7	6	7	7	7	6	7	15	6	6	6	7	7	6
893	18	8	7	8	7	7	7	7	8	7	8	7	7	6	6	7	7	7	18	7	7	7	6	7	6	7	6	7	7	7	6	7	6	7	6	7	7	6	6	7	6	7	6	7	15	6	18	7	6	7
894	7	8	8	7	8	7	7	7	7	7	7	7	7	7	6	7	7	6	7	7	7	7	7	7	7	6	7	6	7	7	7	7	7	6	7	6	7	7	6	7	7	6	7	6	7	15	6	7	7	6
895	15	7	8	8	7	8	7	7	7	7	7	7	7	7	7	15	15	7	6	7	7	7	7	7	6	7	6	7	6	7	7	7	6	7	6	7	7	7	7	7	6	7	6	7	6	7	15	7	7	7
896	13	7	7	8	8	7	8	7	7	7	7	6	7	7	7	7	7	6	7	6	7	7	7	7	7	6	7	6	7	6	7	7	7	6	7	6	7	6	7	7	7	6	7	6	7	6	6	15	7	6
897	15	7	7	7	8	8	7	8	7	7	7	6	7	7	7	15	15	7	6	6	7	7	7	7	7	7	6	7	7	7	6	7	15	7	6	7	7	7	6	7	7	6	6	7	7	7	6	7	15	6
898	8	7	7	7	7	8	8	7	8	7	8	6	6	6	7	7	7	7	7	6	6	6	7	7	7	7	7	6	7	6	7	7	7	7	7	6	7	6	7	7	7	7	6	6	7	6	6	7	7	15
899	28	7	7	7	7	7	8	8	7	8	7	6	7	6	6	7	7	7	7	7	7	6	6	7	7	7	7	7	28	6	28	7	6	7	7	7	7	7	6	7	7	7	7	6	7	7	6	7	6	7
900	27	7	7	7	7	7	7	8	8	7	8	7	6	6	6	7	7	7	7	7	6	6	6	7	7	7	7	7	27	6	27	27	7	6	7	7	7	6	6	7	7	6	7	7	7	6	6	7	7	6
901	16	27	7	7	7	7	6	7	8	8	9	7	9	6	6	7	16	7	7	7	7	6	6	7	6	7	7	7	7	29	6	7	26	7	6	7	7	7	6	7	6	7	6	7	7	6	6	7	6	7
902	8	8	27	7	7	7	7	7	8	8	8	7	7	7	6	7	7	6	7	7	7	7	6	7	7	6	7	7	7	7	7	7	7	25	7	6	7	7	7	7	7	6	7	6	7	7	6	7	7	6
903	8	7	27	27	7	7	7	7	8	7	8	7	7	7	7	7	6	7	6	7	7	7	7	7	6	7	6	7	7	7	7	7	6	7	24	7	7	7	7	7	6	7	6	7	6	7	7	7	6	7
904	7	8	7	8	27	7	7	7	7	7	7	6	7	7	7	7	7	6	7	6	7	7	7	7	7	6	7	6	7	7	7	7	7	6	7	23	7	6	7	7	7	6	7	6	7	6	6	7	7	6
905	22	7	8	7	8	27	7	7	7	7	7	6	7	7	7	7	7	7	6	6	7	7	7	7	7	7	6	6	22	7	22	22	7	7	6	7	22	7	6	7	7	7	6	7	7	7	6	7	7	14
906	7	7	7	8	7	8	27	7	7	7	7	6	7	6	7	7	7	7	7	6	6	6	7	7	7	7	7	6	6	6	7	7	7	7	7	6	7	21	7	7	7	7	7	6	7	6	6	7	7	7
907	*	7	7	7	8	7	8	27	7	7	7	6	7	7	6	7	7	7	7	6	7	6	7	7	7	7	7	7	7	6	7	7	7	7	7	7	7	6	20	7	7	7	7	7	7	7	6	7	6	7
908	8	7	7	7	7	8	7	8	27	7	7	7	6	6	7	7	7	7	7	7	6	6	6	7	7	7	7	7	8	6	8	8	7	7	7	7	7	6	6	19	7	6	7	7	7	6	6	7	7	6
909	18	7	7	7	7	7	8	7	8	27	8	7	7	6	6	7	6	7	7	7	7	6	6	7	6	7	7	7	18	6	18	18	6	7	7	7	7	7	6	7	18	7	6	7	7	7	6	7	6	7

$h \rightarrow$	0	1	2	3	4	5
$n \downarrow$	1 2 3 4 5 6 7 8 9	0 1 2 3 4 5 6 7 8 9	0 1 2 3 4 5 6 7 8 9	0 1 2 3 4 5 6 7 8 9	0 1 2 3 4 5 6 7 8 9	0
910	10 6 7 7 7 7 7 8 7	8 27 6 7 7 6 7 7 6 7	7 7 7 6 7 7 6 7 7 10	7 10 10 7 6 7 7 7 7 7	7 7 17 7 6 7 7 6 7 7	6
911	* 7 7 7 7 7 7 7 8	7 8 27 7 7 7 7 6 7 6	7 7 7 7 7 6 7 6 7 16	7 16 16 6 7 6 7 7 7 7	7 6 7 16 7 6 7 7 7 6	7
912	15 7 7 7 7 7 7 7 7	8 7 7 27 6 7 15 7 6 6	6 7 7 14 7 7 6 15 6 15	7 12 15 7 6 7 6 7 7 7	7 7 6 7 15 7 6 6 7 7	6
913	14 7 7 7 7 7 6 7 7	7 10 6 7 27 6 14 14 6 7	6 7 7 14 7 7 7 6 6 7	7 7 7 7 7 6 6 7 7 7	7 7 7 7 7 14 7 6 7 7	7
914	7 14 7 7 7 7 7 7 7	7 7 7 7 7 27 7 7 7 6	6 6 6 7 7 7 7 7 6 6	6 7 7 7 7 7 6 6 6 7	7 7 7 7 6 7 13 6 7 7	7
915	15 7 14 7 7 7 6 7 7	7 7 6 7 6 7 27 7 7 7	6 6 6 7 7 7 7 7 6 15	6 15 15 7 7 7 7 7 6 6	7 7 7 7 7 6 6 12 7 6	7
916	16 7 7 14 7 7 7 7 7	7 7 7 6 7 6 7 27 6 7	7 6 6 6 7 7 7 7 7 16	6 16 16 7 7 7 7 6 6 6	7 7 7 7 7 7 6 6 11 7	6
917	10 7 7 7 14 7 6 7 7	7 7 7 7 6 7 7 7 27 6	7 7 6 6 7 6 7 7 7 10	6 10 10 6 7 7 7 7 6 6	7 6 7 7 7 7 7 6 7 10	7
918	18 6 7 7 7 14 7 7 7	7 7 7 7 7 6 7 6 7 27	6 7 7 6 7 7 6 7 7 18	7 18 18 7 6 7 7 7 7 6	7 7 6 7 7 7 7 6 7 7	9
919	* 7 7 7 7 7 14 7 7	7 7 7 7 7 7 7 7 6 7	27 7 7 7 7 6 7 6 7 7	7 7 7 6 7 6 7 7 7 7	7 6 7 6 7 7 7 7 7 6	7
920	7 7 7 7 7 7 7 14 7	7 7 6 7 7 7 7 7 7 6	7 27 6 7 7 7 6 6 6 7	7 7 7 7 6 7 6 7 7 7	7 7 6 7 6 7 7 6 7 7	6
921	8 7 7 7 7 7 6 7 14	7 7 6 7 6 6 8 8 6 7	6 7 27 7 7 7 6 6 6 7	7 7 7 7 7 6 6 7 7 7	7 7 7 6 6 7 7 6 7 7	7
922	22 7 7 7 7 7 7 7 7	14 7 6 6 6 6 7 7 7 6	7 7 7 27 7 7 7 6 6 22	6 22 22 7 7 7 6 6 6 7	7 7 7 7 6 6 6 6 7 7	7
923	12 7 7 7 7 7 6 7 7	7 14 6 12 6 6 10 10 7 7	6 7 6 7 27 7 7 7 6 6	6 6 7 7 7 7 6 7 6 6	7 7 7 7 7 7 6 6 7 7	7
924	24 6 7 7 6 7 7 7 7	7 7 14 7 6 6 7 7 7 7	7 6 7 6 7 27 6 7 7 24	6 24 24 7 7 7 7 6 6 6	7 7 7 7 7 6 6 6 7 7	7
925	24 7 7 7 7 7 6 7 7	7 7 7 14 6 6 12 12 7 11	7 7 6 7 7 24 27 6 7 7	6 6 7 6 7 7 7 24 6 6	7 6 7 12 7 7 6 6 7 11	7
926	26 7 7 7 7 7 6 7 7	7 7 7 7 14 6 7 6 6 7	7 7 7 6 7 6 7 27 6 26	7 26 26 7 6 7 7 7 7 6	7 7 6 7 7 7 7 6 7 7	6
927	8 7 7 7 6 7 7 7 8	7 7 7 7 7 14 7 6 6 6	7 7 7 7 7 7 6 7 27 7	7 7 7 6 7 6 7 7 7 7	7 7 7 6 7 7 7 7 7 6	7
928	28 7 7 7 7 7 7 7 7	7 7 6 7 6 7 14 14 6 6	6 7 7 7 7 7 7 6 7 28	6 7 28 7 6 6 6 7 7 7	7 7 6 7 6 7 7 6 7 7	6
929	* 7 7 7 7 7 7 7 7	7 7 6 7 6 6 14 14 6 7	6 7 6 7 7 7 6 7 6 7	7 6 7 7 6 6 6 7 7 7	7 7 7 7 6 7 7 7 7 7	7
930	7 7 7 7 7 7 7 7 7	7 7 6 7 6 6 7 7 14 6	6 6 6 6 7 7 7 6 7 7	7 7 7 7 7 6 6 6 6 7	7 7 7 7 6 6 6 6 7 7	7
931	18 7 7 7 7 7 7 7 7	7 7 6 7 6 6 14 14 7 18	6 7 6 7 7 7 7 7 6 7	6 7 7 7 7 7 6 7 6 6	7 7 7 7 6 7 6 6 7 18	7
932	7 7 7 7 7 7 7 7 7	7 7 7 7 6 6 7 7 7 7	14 6 6 6 7 7 7 7 7 6	7 6 7 7 6 7 7 6 6 6	7 7 7 7 7 6 6 6 7 7	7
933	7 27 7 7 7 7 7 7 7	7 7 7 7 6 6 7 6 7 7	7 14 6 7 7 7 7 7 7 7	6 7 7 7 7 6 7 7 6 6	7 7 7 7 7 7 6 6 7 6	7
934	7 7 27 7 7 7 7 7 7	7 7 6 7 7 6 7 6 6 7	7 7 14 6 7 6 6 7 7 7	7 6 7 6 7 7 6 7 7 6	7 7 6 7 7 7 7 6 7 7	6
935	14 7 7 7 7 7 7 7 7	7 7 7 7 7 7 14 14 6 6	7 7 7 14 7 6 6 6 7 7	7 7 7 7 6 7 7 7 7 7	7 6 7 6 7 7 7 7 7 6	7
936	7 7 7 7 7 7 7 7 7	7 7 7 7 6 6 7 6 6 6	6 7 6 7 14 6 6 6 6 7	7 7 7 7 7 6 7 7 6 7	7 7 6 6 6 7 7 6 7 7	6
937	* 7 7 7 7 7 7 7 7	7 7 7 7 7 6 14 14 6 7	6 7 6 7 7 14 6 7 6 7	6 6 7 7 6 7 6 7 7 6	7 7 6 6 6 7 7 6 7 7	7
938	7 7 7 7 7 7 7 7 7	7 7 7 7 7 7 7 7 7 6	6 6 6 6 7 7 15 6 6 6	6 6 7 7 7 6 7 7 7 7	7 7 7 6 6 6 6 7 7 7	7
939	15 7 7 7 7 7 7 7 7	7 7 7 7 7 7 7 15 7 7	6 7 6 7 7 7 7 15 6 7	6 6 7 7 7 7 6 7 6 7	7 7 7 7 6 6 6 6 7 7	7
940	7 7 7 7 6 7 7 7 7	7 7 7 7 7 7 7 7 6 7	7 7 6 6 7 7 7 7 7 6	6 6 7 7 7 7 7 6 7 6	7 7 6 7 7 6 6 6 7 7	7
941	* 7 6 7 7 6 7 7 7	7 7 7 7 7 7 7 7 7 6	7 7 6 6 7 7 7 7 7 7	6 6 7 6 7 7 7 7 6 7	7 7 7 6 7 7 6 6 7 6	7
942	7 7 7 6 7 7 6 7 7	7 7 7 7 7 7 7 7 7 7	6 7 7 6 7 6 6 7 7 7	7 6 7 6 6 7 7 7 7 6	7 6 7 7 6 7 7 6 7 7	6
943	22 7 7 7 7 7 7 7 7	7 7 7 7 6 7 7 7 7 7	7 7 7 22 7 6 6 6 7 7	7 7 7 6 6 6 7 7 7 7	7 22 6 7 7 7 7 7 7 6	7
944	15 7 7 7 7 7 7 7 7	7 7 7 7 6 6 15 7 7 7	7 7 6 6 7 6 6 6 6 15	6 15 15 6 6 6 6 7 7 7	7 7 7 6 7 7 6 6 7 7	6
945	15 7 7 7 7 7 7 7 7	7 7 7 7 7 6 15 15 7 7	7 7 7 6 7 7 6 7 6 7	6 6 7 7 6 7 6 7 6 6	7 7 6 7 6 7 7 6 7 7	6
946	7 15 7 7 7 7 7 7 7	7 7 7 7 7 7 7 7 15 7	7 7 7 7 7 7 7 6 6 6	6 6 7 7 7 6 6 6 6 6	7 7 7 6 6 7 7 6 7 7	7
947	* 7 15 7 7 7 7 7 7	7 7 7 7 6 7 15 15 7 15	7 7 7 7 7 6 7 7 6 7	6 6 7 7 7 7 6 7 6 6	7 7 7 7 6 7 6 7 7 7	7
948	7 7 7 15 7 7 7 7 7	7 7 7 7 7 6 7 7 7 7	15 7 7 7 7 7 6 7 7 7	6 6 7 7 7 7 7 6 6 6	7 7 7 7 7 6 7 6 7 7	6
949	12 7 7 7 15 7 7 7 7	7 7 7 23 6 7 7 7 7 7	7 15 7 7 7 12 7 12 7 7	6 6 7 6 7 7 7 12 6 6	7 6 7 7 7 7 6 7 7 7	7
950	7 7 7 7 7 15 7 7 7	7 7 7 7 7 6 7 6 7 7	6 6 15 7 7 7 7 7 6 7	7 6 7 6 6 7 7 7 7 6	7 6 6 7 7 7 7 6 7 6	7
951	15 7 7 7 7 7 15 7 7	7 7 7 7 6 7 15 15 7 7	7 6 6 15 7 7 7 7 7 7	7 7 7 7 6 6 7 7 7 7	7 6 6 6 7 7 7 7 7 7	6
952	7 7 7 7 7 7 7 15 7	7 7 7 7 6 6 7 7 7 6	7 7 6 6 15 7 7 7 7 7	6 6 7 6 6 6 6 7 6 7	7 6 6 6 6 7 7 6 7 7	7
953	* 7 7 7 7 7 7 7 15	7 7 7 7 6 6 15 15 7 15	6 7 7 6 7 15 6 7 7 7	7 6 7 7 6 7 6 7 6 6	7 7 6 7 6 7 6 6 7 7	6
954	7 7 7 7 7 7 7 7 7	15 7 7 7 6 6 7 7 7 6	7 6 7 7 7 7 15 7 7 7	7 7 7 7 7 6 6 6 6 6	7 7 7 6 6 6 6 6 7 7	7
955	15 7 7 7 7 7 7 7 7	7 15 7 7 6 6 15 15 7 7	6 7 6 7 7 7 7 15 7 7	7 7 7 6 7 7 6 7 6 6	7 7 7 7 6 7 6 6 7 7	7
956	7 7 7 7 7 7 6 7 7	7 7 15 7 6 6 7 7 7 7	7 6 7 6 7 7 7 7 15 7	7 7 7 7 6 7 7 7 6 6	7 7 7 7 7 6 6 6 7 7	7
957	15 7 7 7 7 7 7 7 7	7 7 7 15 6 6 15 15 7 6	7 8 6 7 7 7 7 7 7 15	7 7 7 7 7 6 7 7 6 6	7 7 7 7 7 7 6 6 7 6	7
958	8 7 7 7 7 7 7 7 8	7 7 7 8 15 6 7 6 7 6	6 6 7 6 7 6 7 7 6 6	15 7 7 7 7 7 6 7 7 6	7 6 6 7 7 7 7 6 7 6	6
959	15 7 7 7 7 7 7 7 7	7 7 6 7 7 15 15 15 7 6	6 6 6 7 7 7 6 7 7 6	6 15 7 7 7 7 7 7 7 7	7 6 6 6 7 7 7 7 7 6	6
960	15 7 7 7 7 7 7 7 7	7 7 7 7 6 7 15 15 7 6	6 6 6 6 7 6 7 7 7 7	6 6 15 7 7 7 7 7 6 6	7 6 6 6 6 7 6 6 7 6	6

h →	0									1										2										3										4										5
n ↓	1	2	3	4	5	6	7	8	9	0	1	2	3	4	5	6	7	8	9	0	1	2	3	4	5	6	7	8	9	0	1	2	3	4	5	6	7	8	9	0	1	2	3	4	5	6	7	8	9	0
961	*	7	7	7	7	7	7	7	7	7	7	6	7	6	6	15	15	7	6	6	6	6	6	7	7	6	7	6	7	7	30	7	15	6	6	7	7	7	6	7	7	6	7	6	7	6	6	7	7	6
962	29	7	7	7	7	7	7	7	7	7	7	7	7	6	6	7	7	15	6	6	6	6	6	7	7	7	7	7	6	7	29	29	7	15	6	7	7	7	7	7	7	7	6	6	6	6	6	7	7	7
963	15	29	7	7	7	7	7	7	8	7	7	7	7	6	6	15	15	7	15	6	6	6	6	7	7	7	7	6	7	6	7	7	28	7	15	7	7	7	7	7	6	7	7	6	7	6	6	7	7	7
964	8	7	29	7	7	7	7	7	7	7	7	7	7	6	6	7	7	7	7	15	6	6	6	7	7	7	7	7	6	7	6	7	7	27	7	15	7	7	7	7	7	6	7	7	7	6	6	7	7	7
965	15	7	7	29	7	7	7	7	8	7	7	7	7	6	6	15	15	7	7	7	15	6	6	7	7	7	7	7	8	6	7	7	7	7	26	7	15	7	7	7	7	7	6	7	7	6	6	7	6	7
966	7	7	7	7	29	7	7	7	7	7	7	7	7	7	6	7	7	6	6	6	7	15	6	7	6	6	7	6	6	7	6	7	6	7	7	25	6	15	7	7	7	7	7	6	7	7	6	7	6	6
967	*	7	7	7	7	29	7	7	7	7	7	7	7	7	7	15	15	8	6	6	6	7	15	7	6	6	7	6	6	6	24	24	7	6	7	7	24	6	15	7	7	7	7	7	7	7	7	7	6	6
968	7	7	7	7	7	7	29	7	7	7	7	7	7	7	7	7	7	6	6	6	6	6	7	15	6	6	7	6	6	6	6	7	6	7	6	7	7	23	6	15	7	7	7	7	7	6	6	7	6	6
969	15	7	7	7	7	7	7	29	8	7	7	7	7	6	7	15	15	6	7	6	6	6	6	7	15	6	7	6	6	6	7	7	7	6	7	7	7	7	22	7	15	6	6	7	7	7	6	7	7	6
970	8	7	7	7	6	7	7	7	29	7	7	7	7	6	6	7	7	7	6	6	6	6	6	7	7	15	7	6	6	6	8	8	7	7	6	7	6	7	7	21	7	15	6	6	7	7	7	7	7	7
971	*	7	7	7	7	6	7	7	7	29	7	7	7	6	6	15	15	7	7	6	6	6	6	7	7	7	15	6	6	6	20	20	7	7	7	7	7	6	7	7	20	7	15	6	7	7	7	7	6	7
972	10	7	7	7	7	7	6	7	7	7	29	7	7	6	6	7	7	7	7	7	6	6	6	7	7	7	7	15	6	6	10	10	7	7	7	7	6	7	6	7	7	19	7	15	7	7	7	7	7	6
973	18	7	7	7	7	7	7	7	8	7	7	29	7	6	6	15	15	7	15	7	7	6	6	7	6	7	7	7	15	6	18	18	6	7	6	7	15	6	7	7	7	7	18	7	15	7	7	7	7	7
974	12	7	7	7	7	7	7	7	7	7	7	7	29	7	6	7	7	6	6	7	7	7	6	7	7	6	7	7	7	15	12	12	6	6	6	7	6	7	6	7	6	7	7	17	6	15	7	7	7	7
975	15	7	7	7	7	7	7	7	7	7	7	7	7	29	7	15	15	7	6	6	7	7	7	7	6	7	6	6	6	7	15	7	6	6	6	7	6	6	7	7	7	6	7	7	16	6	15	7	7	7
976	15	7	7	7	7	7	7	7	7	7	8	7	8	7	29	15	15	6	7	6	6	7	7	7	6	6	8	6	15	6	15	15	6	6	6	7	6	6	6	7	6	7	6	7	7	15	6	15	7	7
977	*	7	7	7	7	7	7	7	7	7	7	7	7	7	7	29	15	7	7	6	6	6	7	7	7	6	7	6	6	6	15	15	15	6	6	7	7	6	6	7	7	6	7	6	7	7	14	7	15	6
978	16	15	7	7	7	7	7	7	7	7	7	7	7	6	7	7	29	15	6	6	6	6	6	7	7	7	6	6	6	6	16	16	7	15	6	7	6	6	6	7	7	7	6	7	7	7	7	13	7	15
979	18	7	15	7	7	7	7	7	7	7	10	7	7	6	6	15	15	29	21	6	6	6	18	7	18	7	7	6	6	6	12	12	7	7	15	7	6	6	6	7	18	7	7	6	7	6	7	7	12	7
980	18	7	7	15	7	7	6	7	7	7	7	7	7	6	6	7	7	7	29	15	6	6	6	7	7	7	7	7	6	6	18	18	7	7	7	15	6	6	6	7	7	7	7	7	7	7	6	7	7	11
981	15	7	7	7	15	7	6	7	8	7	7	7	7	6	6	15	15	7	7	29	15	6	6	7	7	7	7	7	7	6	6	7	6	7	7	7	15	6	6	7	7	7	6	7	8	6	7	7	7	7
982	7	7	7	7	7	15	6	7	7	7	7	7	7	7	6	7	7	6	7	7	29	15	6	7	7	6	7	7	7	7	6	7	7	6	7	7	7	15	6	7	6	6	6	6	7	7	6	7	6	7
983	*	7	7	7	7	7	15	7	7	7	7	7	7	7	7	15	15	7	6	7	7	29	15	7	6	7	6	6	7	7	8	8	6	7	6	7	7	7	15	7	6	6	6	6	7	6	7	7	7	6
984	27	7	7	7	7	6	7	15	7	7	7	7	7	7	7	7	7	6	7	6	7	7	29	15	7	6	7	6	27	7	27	27	6	6	7	6	6	6	7	15	6	6	6	6	7	6	6	7	6	7
985	15	7	7	7	7	7	6	7	15	7	7	7	8	7	7	15	15	7	15	6	6	7	7	29	15	7	7	6	6	6	7	7	7	6	7	8	7	6	6	7	15	6	6	6	7	6	6	7	7	6
986	24	7	7	7	6	7	7	7	7	15	7	6	7	7	7	7	7	7	6	6	6	6	7	7	29	15	7	6	6	6	24	24	7	7	6	6	6	6	6	7	7	15	6	6	7	6	6	7	7	7
987	15	7	7	7	7	7	7	7	7	7	15	7	7	6	7	15	15	7	7	6	6	6	6	7	7	29	15	6	6	6	6	7	7	7	7	6	6	6	6	7	7	7	15	6	7	6	6	7	7	7
988	26	7	7	7	7	7	7	7	7	7	7	15	7	6	6	7	7	7	7	7	6	6	6	7	7	7	29	15	6	6	26	26	7	7	7	7	6	6	6	7	7	7	7	15	7	6	6	7	7	7
989	22	7	7	7	7	7	7	7	8	7	7	7	15	6	6	15	15	7	7	7	8	6	22	7	7	7	7	29	15	6	6	7	7	7	7	7	7	6	6	7	7	7	22	7	15	6	6	7	6	7
990	28	7	7	7	7	7	6	7	7	7	8	7	7	15	6	7	7	7	7	7	7	7	6	7	7	6	7	7	29	15	28	28	7	6	7	7	7	7	6	7	7	6	7	7	7	15	6	7	6	6
991	*	7	7	7	7	7	7	7	7	7	7	7	7	7	15	15	15	7	7	7	7	7	7	7	6	7	6	7	7	29	15	7	6	7	6	7	7	7	7	7	6	7	6	7	7	7	15	7	6	6
992	31	7	7	7	7	7	7	7	7	7	7	7	7	6	7	15	15	6	6	7	7	7	7	7	7	6	7	6	7	7	30	30	7	6	7	6	6	7	7	7	6	6	7	6	7	7	7	15	6	6
993	31	7	7	7	7	7	6	7	7	7	7	7	7	7	6	15	15	7	7	6	7	7	7	15	7	7	7	6	6	7	30	30	15	7	7	7	7	6	7	7	7	6	7	7	6	6	6	7	15	6
994	7	30	7	7	7	7	7	7	7	7	7	7	7	6	7	7	7	15	6	6	6	7	7	7	7	7	7	6	6	6	7	7	29	15	7	6	6	6	6	7	7	7	6	6	8	6	6	7	7	15
995	15	7	30	7	7	7	7	7	7	7	7	7	7	6	6	15	15	7	15	6	6	6	7	7	7	7	7	6	6	6	6	7	7	28	15	7	6	6	6	7	7	7	7	6	6	6	6	7	7	7
996	7	7	7	30	7	7	7	7	7	7	7	7	7	6	6	7	7	6	7	15	6	6	6	7	7	7	7	7	6	6	6	7	7	7	27	15	6	6	6	7	7	7	7	7	6	6	6	7	7	7
997	*	7	7	7	30	7	7	7	8	7	7	7	7	6	6	15	15	7	6	7	15	6	6	7	7	7	7	7	7	6	6	7	6	7	7	26	15	6	6	7	7	7	7	7	7	6	6	7	6	7
998	25	7	7	7	7	30	7	7	7	7	7	7	7	7	6	7	7	6	7	6	7	15	6	7	7	7	7	7	7	7	25	25	7	6	7	7	25	15	6	7	7	6	7	7	7	7	6	7	7	6
999	26	7	7	7	7	7	30	7	8	7	8	7	7	7	7	15	15	7	6	7	6	7	15	7	6	7	26	7	7	7	8	7	6	7	6	7	26	24	15	7	6	7	6	7	7	7	7	7	6	7
1000	7	7	7	7	7	6	7	30	7	7	7	7	7	6	7	7	7	6	7	6	7	6	7	15	7	6	7	7	7	7	7	7	7	6	7	6	7	7	23	15	7	6	7	6	7	7	7	7	6	6

3.11 Remark Table 3.12 is derived from Table 3.10. For each s, $1 \leq s \leq 50$, it gives all orders n for which it is not known that there exists r-IMOLS$(n; s)$ for $r = 2, 3, \ldots, 6$.

3.12 Table r-IMOLS$(n; s)$ not known for $r = 2, 3, \ldots, 6$, $1 \leq s \leq 50$.

	Number of Incomplete Squares Not Known				
Hole	2	3	4	5	6
1	6	10	14 18 22	15 20 26 30 34 38 39 46 54 60 62	12 21 24 28 33 35 36 42 44 48 51 52 55 58 66 68 74 75

	Number of Incomplete Squares Not Known				
Hole	2	3	4	5	6
2			11 12 13 14 16 17 18 20 21 22 23 24 25 26 27 29 30 31 32 34 36 37 38 39 41 42 44 45 46 47 48 49 54 78	15 19 28 33 35 40 50 52 53 55 56 57 60 61 62 63 69 70 75 76 77 111 112 113 159 160 161 162	43 51 59 64 67 68 71 83 84 85 91 92 99 107 108 109 110 155 156 157 158 259 260
3			17 18 19 20 21 23 24 26 27 29 30 31 32 33 35 37 38 39 41 43 44 47 48 49 50 51 54 55 62 70 77 78	25 28 34 36 40 42 45 46 53 56 57 58 61 76 79 112 113 114 160 162 163	22 52 60 63 64 65 68 69 71 72 84 85 86 90 92 93 100 106 108 109 110 111 154 156 157 158 159 161 260 261
4			20 22 24 25 28 31 32 34 36 38 40 43 44 46 48 49 50 51 52 55 56 63 77 78 79 80	26 27 30 33 35 37 42 47 54 57 58 59 62 71 113 114 115 162 163 164	29 39 45 53 61 64 65 66 69 70 72 73 85 86 91 93 94 101 107 109 110 111 112 155 157 158 159 160 161 261 262
5			28 29 31 32 33 34 37 39 41 42 46 47 48 49 50 51 52 53 56 57 78 79 81	30 35 38 40 43 44 45 55 58 59 60 63 64 72 80 114 115 116 164 165	54 62 65 66 67 70 71 73 74 86 92 94 95 102 108 110 111 112 113 156 158 159 160 162 163 262 263
6		52	30 32 33 34 35 36 38 40 41 42 44 46 50 51 53 54 57 58 79 80 81 82	39 45 48 56 59 60 61 64 65 115 116 117 166	47 55 63 66 67 68 71 72 73 74 75 95 96 112 113 114
7			37 39 41 43 46 47 48 52 53	44 45 51 54 55 59 60 61 62	58 66 67 68 74 75 76 82 83
8			42 43 44 45 52 53	48 51 54 55 59 60 61 62	58 66 67 68 74 75 76 82 83
9			47 50 53 54 56 60	55 59 61 62	66 67 68 74 75 76 83 84 98
10			52 53 54 55 56 57 60 62 78	61 63 64 65 68 69 70 75 76 77 79 99	72 73 82 83 84 85 111 113 161 163
11			56 57 58 60 61 62 63 64 65 66 69 70 72 76	67 68 71 73 74	83 98 112 114 162 164
12			61 62 63 64 65 66 67 70 71 74 75 80 81	72 76 77 79 83 87	84 85 86 88 89 90 91 93 94 95 98 99 101 113 115 122 163 165
13			67 68 70 71 72 74 75 76 78 80 81 82 84 86	79 83 87 88 89	114 116 122 164 166
14			72 73 74 75 76 77 79 80 81 82 83 84 87 88 89 90 93 96 100 112	86 91 92 94 95 97 99 101 103 107 111 113 114 115 116 117 118 165 167	98 102 104 106 108 109 122 123
15			77 78 81 82 83 84 85 86 90 102	91 92 93 94 95 96 97 98 99 100 101 103 104 105 107 108 109 110 112 113 114 115 116 117 118 119 123 166 168	111 122 124
16			81 82 83 84 85 87 89 90 91 93 95 99 102 103 105 108	96 98 100 101 104 106 107 109 110 111 167 169	

	Number of Incomplete Squares Not Known				
Hole	2	3	4	5	6
17			86 87 88 90 91 92 93 94 96 98 99 100 102 103 104 106 109 110 112 114 116 118	105 107 108 111 115 117 168 170	
18			92 93 94 95 97 99 100 101 104 105 108 110 111 112 113 116 117 118 120 124 125 128	109 114 115 119 121 122 123 126 129 131 132 133 134 135 147 148 149 169 171	130 136 137 138 139 140 141 142 143
19			96 98 100 101 102 104 105 106 108 111 112 113 114 116 118 120 124 125 126 128 132	115 117 119 121 122 123 127 129 130 131 170 172	178
20			101 102 103 104 106 107 109 110 112 113 114 117 118 119 122 123 126 127 129 131 132 133 136	120 124 125 128 130 134 135 137 138 139 140 141 142 143 144 145 147 148 149 150 151 155 159 160 163 165 166 167 168 169 170 171 173 179	146 152 153 154 156 157 158 161 162 172 178
21			107 108 113 114 115 116 119 120 126 127 128 131 133	129 130 132 134 135 136 137 138 139 140 141 142 143 144 145 146 149 150 151 152 153 155 156 157 158 159 160 161 162 163 164 166 167 169 170 171 172 174 179	147 154 165 168 173 178
22			112 114 115 116 117 119 121 123 127 128 129 131 134 135 137 143 147 151 153	132 133 138 139 140 141 142 144 145 146 148 149 150 152 154 156 157 158 159 160 162 163 165 166 167 168 169 170 171 172 175 179	155 161 164 173 174 180
23			116 117 118 120 121 122 123 124 126 128 129 130 132 133 134 135 136 138 139 141 144 145 146 147 148 149 150 152 157 158	140 142 143 151 153 154 155 156 159 160	
24			122 123 124 126 129 130 131 134 142 146 147 148 150 155 158	144 149 151 152 153 154 156 157 159 160 161 163 164 165 166 167 171	168 169 170 172 173 174 175 176 177 178 179 180 181 182 183 184 185 186 187 188 189 190 191 197
25			128 131 132 134 138 140 141 143 162 164	150 152 153 154 155 156 157 159 160 161 163 165 166 167 168 170 171 172 173 174	
26			131 132 133 134 135 137 138 139 141 142 144 147 148 149 151 153 155 163 164 167	156 158 159 160 161 162 165 166 168 169 170 171 172 173 174 175 176 177 178 179 180 181 182 184 185 186 187 188 189 190 191 192 193 195 196 197 198 199 203 207 211	194 200 201 202 204 205 206 210

	Number of Incomplete Squares Not Known				
Hole	2	3	4	5	6
27			137 138 139 145 149 150 151 153 158 161 169 173	162 163 164 165 166 167 168 170 171 172 174 175 176 177 178 179 180 181 182 183 184 185 186 187 188	
28			142 143 144 145 146 149 150 151 153 157 159 165 166 167 169 170 175 177 185	168 171 172 173 174 176 178 179 181 182 183 184 186 187 188 189 191 192 193 194 195 199	196 198 200 201 202 203 204 205 206 207 208 209 210 211 212 213 214 215 216 217 218 219 220 221 222 223 227
29			146 147 150 151 152 153 154 156 158 160 163 164 166 167 168 170 171 174 175 176 177 178	180 181 182 183 184 185 186 187 188 189 190 191 192 193 194 195 196 198 199 200 201 202	
30			151 152 153 154 156 157 159 161 164 168 172 176 178	180 182 183 184 185 186 187 188 189 190 191 192 193 194 195 196 197 198 199 200 201 202 203 204 205 206 207 208 209 210 212 213 214 215 216 217 218 219 220 221 222 223 224 225 227 228 229 230 231 235 243	226 232 233 234 236 237 238 239 245
31			158 160 162 168 169 170 172 173 176 180 184 188 192 203 204 208	186 189 190 191 193 194 195 196 197 198 199 200 201 202 205 206 207 209 210 211 212 213 214 215 216	
32			161 162 163 164 165 166 167 169 170 171 172 173 174 177 178 179 180 181 183 187 189 190 191 193 194 195 196 197 198 199 202 204 205 206 210 211	192 200 201 203 207 208 209 212 213 214 215 216 217 219 220 221 222 223	
33			167 168 171 175 179 180 181 191 195 199 205 206	198 200 201 202 203 204 207 209 210 211 212 213 214 215 216 217 218 219 220 221 222 223 224 226 227 228 229 230 234	231 233 235 236 237 238 239 240 241 242 243 244 245 246 247 248 249 250 251 252 253 254 255 256 257 258 259 260 261 262 263 264 266 267 268 269 270 271 272 273 274 275 276 277 278 279 280 282 283 284 290
34			172 173 174 175 176 179 180 181 183 191 195 198 199 200 201 204 206 207 208 211 212 219	205 209 210 213 214 215 216 217 218 220 221 222 223 225 226 227 228 229 230 231 232 233 234 235 236 237 238 239 240 241	243 251 258 262 279

	Number of Incomplete Squares Not Known				
Hole	2	3	4	5	6
				242 244 245 246 247 248 249 250 252 253 254 255 256 257 259 260 261 263 264 265 266 267 268 269 270 271 272 274 275 276 277 278 280 281 283 284 285 291	
35			177 178 181 184 186 188 192 193 194 198 199 201	210 211 212 213 214 215 216 217 218 219 220 221 222 223 224 225 226 227 228 229 230 231 232 233 234 235 236 237 238 239 240 241 242 243 244 247 248 249 250 251 255 267 284 286	245 253 254 256 257 258 259 260 261 262 263 264 265 266 268 269 270 271 272 273 274 275 276 277 278 279 280 281 282 285 290 291
36			181 182 184 194 215	216 218 219 220 221 223 224 225 226 227 228 229 230 231 232 233 234 235 236 237 238 239 240 241 242 243 244 245 247 248 249 250 251	252 254 255 256 257 258 259 260 261 262 263 264 265 266 267 268 269 270 271 272 273 274 275 276 277 278 279 280 281 282 283 285 286 287
37			188 190 191 192 194 198 201 202 203 204 208 209 210 214 215 216 218 220 222	224 225 226 227 228 229 230 231 232 233 234 235 236 238 239 240 241 242 243 244 245 246 247 248 249 250 251 252 255 256 257 258	
38			191 192 193 195 196 197 199 200 201 202 203 204 205 208 209 210 211 212 215 216 217 219 221 223 226 227 231 232 234 235 237 239 248 249 251	228 230 233 236 238 240 241 242 243 244 245 246 247 250 252 253 254 256 257 258 259 260 261 262 263 264 265 266 268 269 270 271 272 273 275 276 277 278 279 280 281 282 284 285 287 288 289 291 292 293 294 295 299 303 304 307 309 311 315	274 283 286 290 296 297 298 300 301 302 305 306 308 310 312 313 314 316 317
39			197 198 201 205 210 211 212 213 216 217 218 222 233	234 235 236 237 238 239 240 241 242 243 244 245 246 247 248 250 251 252 253 254 255 257 258 259 260 261 262 263 264 265 266 267 268 269 270 271 272 273 275 276 277 278 279 280 282 283 285 288 289 290 291 292 293 294 295 296 297 299 300 301 302 303 304 305 306 307 308 310 312 313 314 315 316 317	274 281 284 286 287 298 309 311 318 322

	Number of Incomplete Squares Not Known				
Hole	2	3	4	5	6
40			203 204 206 210 211 212 214 217 218 219	240 241 242 243 244 245 246 247 248 249 250 251 252 253 254 255 256 258 259 260 261 262 263 264 265 266 267 269 270 271 272 273 274 275 276 277 278 279 283 284 286	282 285 290 292 294 295 298 300 301 302 304 306 308 310 313 314 315 316 318
41			206 207 208 212 214 218 219 220 222 226 229 230 232 235	246 247 248 249 250 251 252 253 254 255 256 257 259 260 261 262 264 265 266 267 268 269 270 271 272 273 274 275 276 277 278 279 280 282 283 284 285	
42			212 213 214 215 216 219 220 221 224 225 236 248	254 255 256 257 258 260 261 262 263 264 265 266 267 268 269 270 271 272 274 275 276 277 278 279 280 281 282 283 284 285 286	294 295 296 297 298 299 300 302 303 304 305 306 307 308 309 310 311 312 313 314 315 316 317 318 319 320 321 322 323 324 325 326 327 328 329 330 331 332 333 334 335
43			220 221 222 224 225 226 228 232 236 248 256	258 259 261 262 263 264 266 267 268 269 270 271 272 273 274 275 276 277 278 279 280 281 282 283 284 285 286 287 290 292 293 294 297 298 299 300	
44			221 222 223 224 225 226 227 229 232 233 234 235 237 238 240 241 243 245 246 249 250 251 255 256 257 258 259 262 263	264 265 266 267 268 269 270 271 272 273 274 275 276 277 278 279 280 281 282 283 284 285 286 287 288 289 290 291 293 294 295 296 297 298 299 300 301 304 305 306 307 311	308 309 310 312 313 314 315 316 317 318 319 320 321 322 323 324 325 326 327 328 329 330 331 332 333 334 335 336 337 338 339 341 342 343 344 346 347 348 349 350 351 352 353 354 355 356 357 358 359 360 361 362 363 364 365 370 371
45			228 236 251	270 271 272 273 274 275 276 277 278 279 280 281 282 283 284 285 286 287 288 289 290 292 294 296 297 298 299 300 301 302 303 305 306 307 308 310 311 312 313 314 317 318 319 321 329 330 333 354 362 364 366	320 322 323 325 326 327 328 331 332 334 335 336 337 338 339 340 342 343 344 345 347 348 349 350 351 352 353 355 356 357 358 359 361 363 365 370 372

	Number of Incomplete Squares Not Known				
Hole	2	3	4	5	6
46			231 232 233 234 235 236 237 239 243 247 251 255 259	276 277 278 279 280 281 282 283 284 285 286 287 288 289 290 292 293 295 296 297 298 299 300 301 302 303 304 306 307 308 309 310 311 312 313 314 315 316 317 318 319 320 321 322 324 327 328 329 332 334 335 336 337 338 339 341 343 344 345 346 348 349 351 352 353 355 357 362 363	325 326 330 331 333 340 350 354 358 359 360 361 364 365 366 367 370 371 373
47			236 237 238 240 241 242 243 244 246 248 249 252 253 254 256 258 259 260 261 262 265 266 270 273 274 277	282 283 284 285 286 287 288 289 290 291 292 294 296 297 298 299 300 301 302 303 304 305 307 308 309 310 311 312 313 314 315 316 317 318 319 320 321 322 323 325 328	
48			242 254 278	288 290 291 292 293 294 295 297 298 299 300 301 302 303 304 305 306 308 309 310 311 313 314 315 316 317 318 319 320 321 322 323 324 325 326 328 329 331 332 333 334 339	336 337 338 340 341 342 343 345 346 347 348 350 351 352 353 354 355 356 357 358 359 360 361 362 363 364 365 366 367 368 369 370 371 372 373 374 375 376 377 378 379 380 382 383 389
49			248 251 254 258 262 267 268 282	296 298 299 300 302 303 304 305 306 309 310 311 312 314 316 317 318 319 320 321 323 324 325 326 327 330 331 332 333 335 339	
50			251 252 254 256 257 259 261 262 263 264 268 269 283 285 286 287 288 289	300 303 304 305 306 307 309 310 311 312 313 314 315 316 317 318 319 320 321 322 323 324 325 326 327 328 331 332 333 334 335 336 339	352 353 354 355 356 357 358 359 360 361 362 363 364 365 366 367 368 369 370 371 372 373 374 375 376 377 378 379 380 381 382 384 385 387 388 389 390 395 396 397 398 402 404 406 410 413

3.13 Theorem [8] For $n > h > 0$, an ITD$(4, n; h)$ exists if and only if $n \geq 3h$, $(n, h) \neq (6, 1)$.

3.14 Theorem [1] For $n > h > 0$, an ITD$(5, n; h)$ exists if and only if $n \geq 4h$ except when $(n, h) = (6, 1)$, and possibly when $h = 1$ and $n = 10$; or $h = 6$ and $n = 52$.

3.4 MOLS with Holes

If $b_1 = b_2 = \cdots = b_s = b$ and if $b_1 + b_2 + \cdots + b_s = n$ then an ILS$(n; b_1, b_2, \ldots, b_s)$ (or a set of IMOLS$(n; b_1, b_2, \ldots, b_s)$) is *uniform* of *type* b^s. In this case $N(b^s)$ denotes $N(n; b_1, b_2, \ldots, b_s)$.

3.15 Remark Incomplete latin squares of type 1^n are equivalent to idempotent latin squares of side n. So the best lower bounds for $N(1^n)$ are found in Table 2.72 by using Theorem 3.16, when the entry in that table is not italicized.

3.16 Theorem $N(n) - 1 \le N(1^n) \le N(n)$.

3.17 Theorem [10] $N(h^n) \ge 2$ if and only if $h \ge 1$, $n \ge 4$, and $(h, n) \ne (1, 6)$.

3.18 Theorem [2] $N(h^n) \ge 3$ if and only if $h \ge 1$, $n \ge 5$, and $(h, n) \ne (1, 6)$, except possibly for $(h, n) = (1, 10), (3, 6), (3, 18), (3, 28), (3, 34), (6, 18), (6, 19)$, and $(6, 23)$.

3.19 Remark Theorem 3.20 is an analog to MacNeish's Theorem 2.18 for the case of uniform IMOLS.

3.20 Theorem For any divisor a of h, $N(h^n) \ge \min\{N(a^n), N(\frac{h}{a})\}$.

3.21 Table Lower bounds for $N(h^n)$ for $1 \le h \le 20$ and $4 \le n \le 50$.

	Size h of Groups																			
n	1	2	3	4	5	6	7	8	9	10	11	12	13	14	15	16	17	18	19	20
4	2	2	2	2	2	2	2	2	2	2	2	2	2	2	2	2	2	2	2	2
5	3	3	3	3	3	3	3	3	3	3	3	3	3	3	3	3	3	3	3	3
6	1	3	2	4	3	3	3	4	3	3	3	4	3	3	3	3	3	3	3	4
7	5	3	4	3	4	3	5	5	5	3	5	5	5	3	4	5	5	3	5	4
8	6	4	4	3	4	6	6	6	6	5	6	5	6	4	5	6	6	5	6	4
9	7	4	3	3	4	3	7	7	7	6	7	6	7	4	6	7	7	6	7	4
10	2	3	3	3	3	3	6	8	3	3	3	4	4	3	3	4	3	3	4	4
11	9	3	4	3	4	5	6	7	8	3	9	5	9	3	4	9	9	3	9	4
12	4	3	3	3	4	6	6	7	4	10	4	4	4	3	4	4	4	4	4	4
13	11	3	4	3	4	7	6	7	8	3	10	5	11	3	4	11	11	4	11	4
14	3	3	4	3	3	3	6	7	4	3	5	12	4	3	4	4	4	3	4	4
15	3	3	5	3	3	5	3	6	3	3	3	3	3	3	4	3	3	3	3	3
16	14	3	3	3	4	3	6	7	8	3	10	5	12	3	4	14	14	3	14	4
17	15	4	4	3	4	3	6	7	8	4	10	5	12	4	15	15	15	14	15	4
18	3	3	2	3	3	2	6	7	3	3	5	5	4	3	4	16	3	3	4	4
19	17	3	4	3	4	2	6	7	8	7	10	5	12	3	4	15	16	3	17	4
20	4	3	4	3	4	3	6	7	5	3	5	5	4	3	4	15	4	18	4	4
21	4	3	4	3	4	3	5	4	7	3	4	14	4	3	4	4	4	3	4	4
22	3	3	4	3	4	3	4	4	3	3	4	4	4	3	4	4	3	3	4	4
23	21	3	3	3	4	2	6	7	8	3	10	9	12	3	4	15	16	3	18	4
24	4	3	3	3	4	3	6	7	4	3	5	5	4	3	4	15	4	3	4	4
25	23	6	3	3	4	3	6	7	8	4	10	5	12	6	5	15	16	6	18	4
26	3	3	3	3	3	3	6	7	4	3	5	5	4	3	4	15	3	3	4	4
27	25	3	3	3	4	3	6	7	8	3	10	5	12	3	4	15	16	3	18	4
28	4	3	2	3	4	5	6	7	4	3	5	5	4	3	4	15	4	3	4	4
29	27	6	3	3	4	3	6	7	8	4	10	5	12	6	4	15	16	6	18	4
30	3	3	3	3	3	3	6	7	3	3	5	5	4	3	4	15	3	3	4	4

	Size h of Groups																			
n	1	2	3	4	5	6	7	8	9	10	11	12	13	14	15	16	17	18	19	20
31	29	3	7	4	4	3	6	7	8	3	10	5	12	3	4	15	16	3	18	23
32	30	3	6	3	4	3	6	7	8	3	10	5	12	3	4	15	16	3	18	4
33	4	3	7	3	4	3	6	7	4	3	5	5	4	3	14	15	4	3	4	4
34	4	3	2	3	4	3	4	4	4	3	4	4	4	3	4	4	4	3	4	4
35	4	3	3	3	4	3	4	4	4	3	4	4	4	3	4	4	4	3	4	4
36	4	3	3	4	4	5	4	6	4	3	4	4	4	3	4	4	4	3	4	4
37	35	6	7	3	4	3	6	7	8	4	10	5	12	6	4	15	16	6	18	4
38	4	3	3	3	4	3	6	7	4	3	5	5	4	3	4	15	4	3	4	4
39	4	3	3	3	4	3	4	4	4	3	4	4	4	3	4	4	4	3	4	4
40	6	3	3	3	4	3	6	6	6	3	6	5	6	3	4	6	6	3	6	4
41	39	6	3	4	4	3	6	7	8	4	10	5	12	6	4	15	16	6	18	4
42	4	3	3	4	4	3	6	7	4	3	5	5	4	3	4	15	3	3	4	4
43	41	3	7	4	4	3	6	7	8	3	10	5	12	3	4	15	16	3	18	4
44	4	3	3	3	4	3	6	7	4	3	5	4	6	3	4	15	4	3	6	4
45	5	3	3	3	4	3	5	5	5	3	5	5	5	3	4	5	5	3	5	4
46	4	3	3	4	4	3	5	5	4	3	4	4	4	3	4	4	4	3	4	4
47	45	3	3	3	4	3	6	7	8	3	10	5	12	3	4	15	16	3	18	4
48	4	3	3	4	4	4	6	7	4	3	4	4	4	3	4	15	4	4	4	4
49	47	6	4	4	4	5	6	7	8	4	10	5	12	6	4	15	16	6	18	4
50	5	3	4	3	4	5	6	7	5	5	5	5	5	3	5	15	5	5	5	4

3.5 See Also

§II.2	Incomplete MOLS are used to construct MOLS.
§II.4	Orthogonal arrays of higher index are analogues of transversal designs and OAs of index one.
§III.1, III.2	TDs and ITDs provide numerous ingredients for the construction of PBDs and GDDs.
§IV.11	Difference matrices, quasi-difference matrices, and $V(m,t)$ vectors are used to construct MOLS and IMOLS.
[7]	Contains a nice section on MOLS with holes.
[5]	The first paper on MOLS with holes.
[4]	Contains a discussion of all known construction methods for IMOLS.

References

[1] R.J.R. Abel, C.J. Colbourn, J.X. Yin and H. Zhang, Existence of transversal designs with block size five and any index λ, preprint.

[2] F.E. Bennett, C.J. Colbourn and L. Zhu, Existence of certain types of three HMOLS, *Discrete Math.*, to appear.

[3] A.E. Brouwer, Four MOLS of order 10 with a hole of order 2, *J. Stat. Plann. Infer.* 10 (1984), 203-205.

[4] C.J. Colbourn and J.H. Dinitz, Making the MOLS table, in: *Computational and Constructive Design Theory*, to appear.

[5] J.H. Dinitz and D.R. Stinson, MOLS with holes, *Discrete Math.* 44 (1983), 145-154.

[6] L. Euler, Recherches sur une nouvelle espèce de quarrés magiques, *Verh. Zeeuw. Gen. Weten. Vlissengen* 9 (1782), 85-239.

[7] K.E. Heinrich, Latin squares with and without subsquares, in: *Latin Squares: New Developments in the Theory and Applications* (J. Dénes, A.D. Keedwell; eds.) North-Holland, Amsterdam, 1991.
[8] K.E. Heinrich and L. Zhu, Existence of orthogonal latin squares with aligned subsquares, *Discrete Math.* 59 (1984), 241-248.
[9] D.R. Stinson, The equivalence of certain incomplete transversal designs and frames, *Ars Combin.* 22 (1986), 81-87.
[10] D.R. Stinson and L. Zhu, On the existence of MOLS with equal-sized holes, *Aequat. Math.* 33 (1987), 96-105.

4 Orthogonal Arrays of Index More Than One

CHARLES J. COLBOURN

4.1 Definition and Equivalent Objects

An *orthogonal array* of *size* N, with k *constraints* (or of *degree* k), s *levels* (or of *order* s), and *strength* t, denoted $\mathrm{OA}(N,k,s,t)$, is a $k \times N$ array with entries from a set of $s \geq 2$ symbols, having the property that in every $t \times N$ submatrix, every $t \times 1$ column vector appears the same number $\lambda = \frac{N}{s^t}$ times. The parameter λ is the *index* of the orthogonal array. An $\mathrm{OA}(N,k,s,t)$ is also denoted by $\mathrm{OA}_\lambda(t,k,s)$; in this notation, if t is omitted it is understood to be 2, and if λ is omitted it is understood to be 1.

4.1 **Remark** In this section, only OAs of strength two and arbitrary index are treated.

An $(n,k;\lambda)$-*net* is a set X of λn^2 points together with a set $\mathcal{D}$ of kn subsets of X (*blocks*) each of size λn. The set of all blocks is partitioned into k parallel classes, each containing n disjoint blocks. Every two nonparallel blocks intersect in λ points.

4.2 **Remark** Alternative names for a $(n,k;\lambda)$-net are *affine resolvable 1-design* or *affine 1-design*. Here the adjective 'affine' is used to record that every two nonparallel blocks intersect in the same number of points.

4.3 **Theorem** An $\mathrm{OA}_\lambda(k,n)$ is equivalent to

1. a $\mathrm{TD}_\lambda(k,n)$;
2. an $(n,k;\lambda)$-net (the dual structure).

An $\mathrm{OA}_\lambda(k,n)$ is *class-regular* or *regular* if some group Γ of order n acts regularly on the symbols of the array.

4.4 **Theorem** A class regular $\mathrm{OA}_\lambda(k,n)$ is resolvable and hence gives an $\mathrm{OA}_\lambda(k+1,n)$.

4.2 Existence

4.5 **Theorem** (*Bose-Bush bound*) An $\mathrm{OA}_\lambda(k,v)$ exists only if $k \leq \left\lfloor \frac{\lambda v^2 - 1}{v-1} \right\rfloor$. Moreover, if $\lambda - 1 \equiv b \pmod{v-1}$ and $1 < b < v-1$, then

$$k \leq \left\lfloor \frac{\lambda v^2 - 1}{v-1} \right\rfloor - \left\lfloor \frac{\sqrt{1+4v(v-1-b)} - (2v-2b-1)}{2} \right\rfloor - 1.$$

4.6 Theorem A Hadamard matrix $H(4n)$ is equivalent to a $\mathrm{TD}_n(4n-1,2)$.

4.7 Theorem [11] An affine resolvable 2-design with parameters 2-(v,k,λ) exists only if there are integers $n \geq 2$ and $\mu \geq 0$ for which $v = \mu n^2$, $k = \mu n$ and $\lambda = \frac{\mu n - 1}{n-1}$. Such an *affine design* is denoted $\mathrm{AD}(n,\mu)$; every two nonparallel blocks intersect in μ points. An $\mathrm{AD}(n, \frac{\sigma(n-1)+1}{n})$ is equivalent to an $\mathrm{OA}_\sigma(\frac{\sigma n^2 - 1}{n-1}, n)$.

4.8 Remarks Theorem 4.7 characterizes, in one sense, the upper bound in Theorem 4.5. Only two families of $\mathrm{AD}(n,\mu)$ are known:

- $\mathrm{AD}(q, q^n)$ for q a prime power, $n \geq 0$, are the affine geometries (see §VI.7.6).
- If a Hadamard matrix of order $4n$ exists, its Hadamard 3-design is an $\mathrm{AD}(2,n)$.

Shrikhande [11] conjectures that no others exist.

An *α-parallel class* in an $\mathrm{OA}_\lambda(k,n)$ is a set of αn columns, so that each of the n symbols occurs in each row α times within these columns. An $\mathrm{OA}_\lambda(k,n)$ is *α-resolvable* if its columns can be partitioned into α-parallel classes.

4.9 Theorem If an α-resolvable $\mathrm{OA}_\lambda(k,n)$ exists and a β-resolvable $\mathrm{OA}_\mu(k',n)$ exists for which $\mu = \frac{\lambda}{\alpha n}$, then an $\alpha\beta n$-resolvable $\mathrm{OA}_\lambda(k+k',n)$ exists.

4.10 Theorem Let p be a prime, and $u \geq 0$, $v \geq 1$ be integers. Let $s = p^v$ and $d = \lfloor \frac{u}{v} \rfloor$. Then an $\mathrm{OA}_{p^u}(p^u \frac{s^{d+1}-1}{s^d - s^{d-1}} + 1, s)$ exists.

4.11 Theorem Let q be a prime power and $n \geq 0$ be an integer. Then there exists an $\mathrm{OA}_{2q^n}(2\frac{q^{n+1}-1}{q-1} - 1, q)$.

4.12 Theorem [8] If there exists an $\mathrm{OA}_\lambda(k_1,s)$ and a $(s,k_2;\mu)$-difference matrix, then there exists an $\mathrm{OA}_{\lambda\mu s}(k_1k_2+1,s)$. Moreover, if the $\mathrm{OA}_\lambda(k_1,s)$ is completely resolvable, and an $\mathrm{OA}_{\lambda\mu}(k_3,s)$ exists, then there exists an $\mathrm{OA}_{\lambda\mu s}(k_1k_2+k_3,s)$.

4.13 Theorem [8] If an $\mathrm{OA}_\lambda(k_1,s)$ and an $\mathrm{OA}_\mu(k_2,s)$ both exist, then an $\mathrm{OA}_{\lambda\mu s^2}((k_1+1)(k_2+1)-1,s)$ exists.

4.14 Theorem [5] A $\mathrm{TD}_\lambda(7,n)$ exists for all $\lambda \geq 2$ and $n \geq 2$.

4.15 Theorem [4] For $\lambda \geq 2$, a $\mathrm{TD}_\lambda(8,n)$ exists except when $\lambda = 2$ and $n \in \{2,3\}$, and possibly when $\lambda = 2$ and $n \in \{6,14,21,22,34\}$.

4.16 Theorem [4] For $\lambda \geq 2$, a $\mathrm{TD}_\lambda(9,n)$ exists except when $\lambda = 2$ and $n \in \{2,3\}$, and possibly when

1. $\lambda = 2$ and $n \in \{6,14,15,21,22,33,34,35,38,39,50,51,54,62,74,75\}$;
2. $\lambda = 3$ and $n \in \{5,6,28,45,60\}$;
3. $\lambda = 4$ and $n \in \{6,15,21\}$;
4. $\lambda = 5$ and $n \in \{6,14,21,28,35\}$;
5. $\lambda = 7$ and $n \in \{15,21\}$;
6. $\lambda = 11$ and $n = 6$.

4.17 Remark Abel [1] found $\mathrm{TD}_3(9,n)$ for $n = 14$ and 20, which eliminates a number of the exceptions in Theorems 4.15 and 4.16 from [4].

4.18 Theorem [10] For fixed positive integers k and n, there exists an integer λ_0 such that for all $\lambda \geq \lambda_0$, a $\mathrm{TD}_\lambda(k,n)$ exists.

4.3 Tables for Existence

4.19 Table Upper and lower bounds on k in $\mathrm{OA}_\lambda(k,n)$ for $1 \le \lambda \le 18$ and $2 \le n \le 50$. Entries for which the upper and lower bounds match are shown in bold face.

$n \backslash \lambda$	1	2	3	4	5	6	7	8	9	10	11	12	13	14	15	16	17	18
2	**3**	**7**	**11**	**15**	**19**	**23**	**27**	**31**	**35**	**39**	**43**	**47**	**51**	**55**	**59**	**63**	**67**	**71**
3	**4**	**7**	**13**	16	22	**25**	31	34	**40**	43	49	52	58	61	67	70	76	**79**
				13	10		13	25		31	13	49	13	25	37	49	25	
4	**5**	**9**	14	**21**	25	30	37	**41**	46	53	57	62	69	73	78	**85**	89	94
			13		10	13	13		37	21	13	61	21	57	21		37	37
5	**6**	**11**	17	23	**31**	36	42	48	56	**61**	67	73	81	86	92	98	106	111
			8	21		16	18	21	21		18	26	21	21	43	81	36	91
6	4	13	20	27	34	43	49	56	63	70	79	85	92	99	106	115	121	128
	3	7	8	8	8	12	9	12	11	12	8	19	9	17	13	23	11	18
7	**8**	**15**	23	30	38	46	**57**	64	72	79	87	95	106	**113**	121	128	136	144
			9	29	12	16		29	29	16	29	37	29		36	38	29	64
8	**9**	**17**	26	34	43	52	61	**73**	81	90	98	107	116	125	137	**145**	154	162
			9	33	9	17	57		22	41	33	33	22	57	57		41	73
9	**10**	**19**	29	38	48	58	68	78	**91**	100	110	119	129	139	149	159	172	**181**
			28	37	19	55	28	73		37	28	109	37	55	55	73	73	
10	11	21	32	42	53	64	75	86	97	111	121	132	142	153	164	175	186	197
	4	10	12	10	10	20	12	11	12	12	11	28	12	12	12	19	12	30
11	**12**	**23**	35	46	58	70	81	93	105	118	**133**	144	156	167	179	191	202	214
			12	45	12	23	12	45	20	23		45	23	23	45	45	38	23
12	13	25	38	50	63	76	88	101	114	127	141	157	169	182	194	207	220	232
	7	12	14	12	12	24	12	24	14	12	14	43	12	24	14	47	14	24
13	**14**	**27**	41	54	68	81	95	109	123	137	151	165	**183**	196	210	223	237	250
			14	53	14	27	14	53	20	27	24	53		28	66	53	53	40
14	12	29	44	58	73	87	102	117	132	147	162	177	192	211	225	240	254	269
	5	7	9	14	8	27	8	28	9	15	8	27	9	27	11	55	9	42
15	16	31	47	62	78	93	109	125	141	156	172	188	205	221	241	256	272	287
	6	8	17	8	17	30	8	30	17	17	17	32	17	30	31	59	17	30
16	**17**	**33**	50	66	83	99	116	133	149	166	183	200	217	234	252	**273**	289	306
			18	65	18	33	18	129	18	33	18	65	18	33	66		34	145
17	**18**	**35**	53	70	88	105	123	141	158	176	194	212	230	248	266	285	**307**	324
			18	69	18	35	18	69	20	35	24	69	28	35	32	273		40
18	19	37	56	74	93	111	130	149	167	186	205	224	243	262	281	300	320	343
	5	18	10	18	10	35	10	18	11	19	11	38	11	19	19	19	18	35
19	**20**	**39**	59	78	98	117	137	157	176	196	216	235	255	275	295	315	336	357
			20	77	20	39	20	77	20	39	24	77	28	39	32	77	32	40
20	21	41	62	82	103	123	144	165	185	206	227	247	268	289	310	331	352	374
	6	9	9	11	9	39	9	11	11	11	9	42	11	11	13	21	11	58
21	19	43	65	86	108	129	151	172	194	216	238	259	281	303	325	347	369	391
	7	7	11	8	8	13	8	42	11	10	11	44	10	13	12	83	11	15
22	20	45	68	90	113	135	158	180	203	226	248	271	294	317	340	363	386	409
	5	7	24	12	24	44	12	24	24	24	24	44	24	24	24	24	24	44
23	**24**	**47**	71	94	118	141	165	188	212	236	259	283	307	331	354	378	402	426
			24	93	24	47	24	185	24	47	24	93	28	47	32	185	32	47
24	25	49	74	98	123	147	172	196	221	246	270	295	320	345	369	394	419	444
	7	24	26	24	24	48	24	48	26	24	26	48	24	48	26	95	26	70
25	**26**	**51**	77	102	128	153	179	204	230	256	281	307	333	358	384	410	436	462
			26	101	126	51	51	201	101	126	51	101	126	101	126	201	101	126

$n\backslash\lambda$	1	2	3	4	5	6	7	8	9	10	11	12	13	14	15	16	17	18
26	27	53	80	106	133	159	186	212	239	266	292	319	346	372	399	426	453	480
	6	26	11	26	14	51	28	26	26	26	26	51	28	28	26	27	26	51
27	**28**	**55**	83	110	138	165	193	220	248	276	303	331	359	386	414	442	470	497
			82	109	55	82	82	109	244	82	82	109	109	82	82	109	109	244
28	29	57	86	114	143	171	200	228	257	286	314	343	372	400	429	458	486	515
	6	28	8	28	8	28	8	28	9	28	9	30	9	28	12	28	12	28
29	**30**	**59**	89	118	148	177	207	236	266	296	325	355	384	414	444	474	503	533
			30	117	30	59	30	233	30	59	30	117	30	59	32	233	32	59
30	27	61	92	122	153	183	214	244	275	306	336	367	397	428	459	490	520	551
	6	30	32	30	30	59	30	58	32	30	32	59	30	58	32	58	32	59
31	**32**	**63**	95	126	158	189	221	252	284	315	347	379	410	442	474	505	537	569
			33	125	33	63	33	249	33	63	33	125	33	63	33	249	33	63
32	**33**	**65**	98	130	163	195	228	260	293	325	358	391	423	456	489	521	554	587
			33	129	33	65	33	257	33	65	33	129	33	65	33	513	33	65
33	30	67	101	134	168	201	235	268	302	335	369	403	436	470	504	537	571	605
	7	8	12	14	17	65	17	17	17	17	17	65	17	17	17	23	17	65
34	35	69	104	138	173	207	242	276	311	345	380	415	449	484	519	553	588	623
	6	7	11	15	18	18	18	18	18	19	18	23	18	27	18	31	18	100
35	36	71	107	142	178	213	249	284	320	355	391	427	462	498	534	569	605	641
	6	8	9	11	9	12	9	15	9	18	9	18	9	18	9	21	9	18
36	37	73	110	146	183	219	256	292	329	365	402	439	475	512	549	585	622	659
	7	9	38	14	10	70	14	70	38	18	38	70	18	70	38	70	38	106
37	**38**	**75**	113	150	188	225	263	300	338	375	413	451	488	526	563	601	639	676
			38	149	38	75	38	297	38	75	38	149	38	75	38	297	38	75
38	35	77	116	154	193	231	270	308	347	385	424	463	501	540	578	617	656	694
	6	8	11	15	19	20	20	20	20	20	20	23	20	27	20	31	20	35
39	40	79	119	158	198	237	277	316	356	395	435	475	514	554	593	633	673	712
	6	8	13	13	10	14	13	14	14	14	13	25	13	14	14	25	14	27
40	41	81	122	162	203	243	284	324	365	405	446	487	527	568	608	649	690	730
	9	11	42	13	11	78	13	78	42	17	42	78	17	78	42	78	42	78
41	**42**	**83**	125	166	208	249	291	332	374	415	457	499	540	582	623	665	707	748
			42	165	42	83	42	329	42	83	42	165	42	83	42	329	42	83
42	39	85	128	170	213	255	298	340	383	425	468	511	553	596	638	681	724	766
	6	42	44	42	42	82	42	42	44	42	42	82	42	42	44	42	42	124
43	**44**	**87**	131	174	218	261	305	348	392	435	479	522	566	610	653	697	741	784
			44	173	44	87	44	345	44	87	44	173	44	87	44	345	44	87
44	45	89	134	178	223	267	312	356	401	445	490	534	579	624	668	713	758	802
	7	9	12	12	10	87	12	21	12	12	12	87	12	21	19	23	12	87
45	46	91	137	182	228	273	319	364	410	455	501	546	592	638	683	729	775	820
	8	10	8	11	10	12	14	19	13	19	11	21	12	18	28	21	14	19
46	43	93	140	186	233	279	326	372	419	465	512	558	605	652	698	745	791	838
	6	10	11	15	19	23	24	24	24	24	24	48	24	27	24	31	24	136
47	**48**	**95**	143	190	238	285	333	380	428	475	523	570	618	666	713	761	808	856
			48	189	48	95	48	377	48	95	48	189	48	95	48	377	48	95
48	49	97	146	194	243	291	340	388	437	485	534	582	631	680	728	777	825	874
	7	48	13	48	13	95	13	48	17	48	17	95	17	48	19	48	19	95
49	**50**	**99**	149	198	248	297	347	396	446	495	545	594	644	694	743	793	842	892
			50	197	50	99	344	393	99	99	197	197	99	344	344	393	99	197
50	51	101	152	202	253	303	354	404	455	505	556	606	657	708	758	809	859	910
	8	8	11	15	19	99	26	26	26	26	26	99	26	27	26	31	26	148

4.20 Table Lower bounds on k in $OA_\lambda(k, n)$ for $19 \le \lambda \le 36$ and $2 \le n \le 50$. Upper and lower entries for which the upper and lower bounds match are shown in bold face.

$n\backslash\lambda$	19	20	21	22	23	24	25	26	27	28	29	30	31	32	33	34	35	36
2	**75**	**79**	**83**	**87**	**91**	**95**	**99**	**103**	**107**	**111**	**115**	**119**	**123**	**127**	**131**	**135**	**139**	**143**
3	31	31	46	31	25	97	40	31	**121**	49	31	121	37	97	46	52	31	157
4	21	46	37	41	37	109	37	57	109	61	37	61	37	**169**	133	46	41	181
5	21	121	31	31	31	61	**156**	61	36	61	31	91	43	81	43	86	103	91
6	11	26	13	12	12	44	12	32	13	19	13	31	13	24	13	18	13	49
7	37	29	64	36	38	37	57	183	64	225	36	92	37	64	57	57	89	127
8	33	81	57	57	57	73	57	105	57	73	57	100	57	289	57	73	57	145
9	37	82	91	55	55	109	73	73	271	73	82	109	55	82	91	307	73	361
10	12	37	12	19	12	39	19	20	33	21	12	28	19	37	20	38	27	59
11	45	45	38	**265**	45	122	56	45	45	62	45	45	45	45	133	45	122	67
12	14	48	14	24	14	73	14	24	38	43	14	24	14	48	24	24	24	85
13	53	53	53	40	53	67	56	**365**	53	170	53	66	68	53	53	74	53	67
14	11	27	11	27	11	28	12	28	44	27	12	82	14	55	27	42	28	83
15	17	58	30	30	30	59	17	30	31	32	30	43	31	59	30	30	31	58
16	65	65	46	65	66	129	46	129	65	65	46	66	66	**545**	66	145	65	289
17	35	86	69	40	35	69	69	40	35	69	69	40	35	273	273	**613**	40	97
18	18	70	27	19	18	70	19	35	28	27	19	35	19	70	28	27	19	38
19	**381**	86	46	40	77	77	46	40	77	77	62	40	77	77	62	40	77	97
20	11	77	13	21	11	78	11	39	13	21	11	42	13	78	13	21	13	58
21	11	42	43	13	12	82	11	15	13	44	42	15	12	83	43	15	13	44
22	24	24	24	24	24	87	24	24	24	27	24	44	24	31	24	27	24	44
23	32	93	44	47	**553**	185	47	47	93	93	47	64	185	185	70	64	93	93
24	26	93	26	48	26	95	26	48	26	48	26	48	26	95	26	70	26	93
25	101	126	126	101	126	201	**651**	126	101	126	126	126	126	201	201	126	126	126
26	28	102	28	27	27	51	28	51	28	28	27	126	28	51	28	28	28	51
27	82	109	109	109	82	109	109	109	**757**	109	109	157	109	109	109	109	109	244
28	12	109	13	28	13	30	13	28	13	31	13	30	13	30	13	28	13	30
29	32	117	44	59	48	233	48	59	48	117	**871**	157	65	233	117	117	65	117
30	32	117	32	58	32	59	32	59	32	58	32	59	32	59	32	58	32	59
31	33	125	44	63	48	249	48	63	48	125	60	157	**993**	497	497	125	125	187
32	33	129	33	65	33	257	33	65	33	129	33	65	993	**1057**	70	65	129	187
33	17	23	17	23	17	65	17	23	17	23	17	65	33	25	43	23	17	65
34	18	35	18	35	18	35	18	35	18	35	18	35	18	35	18	35	18	100
35	9	29	9	18	9	18	12	18	9	21	11	18	12	21	12	18	31	21
36	38	70	38	70	38	70	38	70	38	70	38	70	38	70	38	70	38	106
37	38	149	44	75	48	297	48	75	48	149	60	75	60	297	68	75	72	187
38	20	39	20	39	20	39	20	39	20	39	20	39	20	39	20	39	20	39
39	14	27	14	14	14	27	14	14	14	25	14	27	14	27	14	25	14	40
40	42	78	42	78	42	78	42	78	42	78	42	78	42	78	42	78	42	78
41	42	165	44	83	48	329	48	83	48	165	60	83	60	329	68	83	72	165
42	42	166	44	42	44	82	42	82	44	42	44	82	42	166	44	42	44	124
43	44	173	44	87	48	345	48	87	48	173	60	87	60	345	68	87	72	173
44	12	21	24	23	19	87	19	21	24	23	21	87	19	41	24	23	24	87
45	14	31	18	19	19	21	19	19	21	21	18	31	21	178	19	19	28	26
46	24	39	24	43	24	48	24	47	24	47	24	48	24	47	24	47	24	136
47	48	189	48	95	48	377	48	95	48	189	60	95	60	377	68	95	72	189
48	19	190	26	48	26	95	26	95	26	48	26	95	26	190	26	48	26	95
49	197	197	344	344	344	393	197	197	197	344	344	344	344	393	197	197	344	344
50	26	39	26	43	26	99	26	51	26	51	26	99	26	51	26	51	27	148

4.4 Hypercubes

A *hypercube of dimension* d, *size* n, and *order* m, denoted $\mathrm{HC}(m; n, d)$, is an $n \times \ldots \times n$ array of dimension d, in which each cell contains a single element from an m-set of symbols. When $m = n$, the simpler notation $\mathrm{HC}(n, d)$ is used.

An $\mathrm{HC}(m; n, d)$ has *type* j and is denoted by $\mathrm{HC}^j(m; n, d)$, if, whenever j of the coordinates are fixed, the corresponding subcube of dimension $d - j$ contains each symbol in exactly n^{d-j}/m cells.

A hypercube $\mathrm{HC}(m; n, d)$ of type 1 is a *Youden frequency hypercube*, $\mathrm{YFHC}(m; n, d)$. Such a hypercube is *latin* if $m = n$.

Two $\mathrm{HC}(m; n, d)$'s are *orthogonal* if, when superimposed, each of the m^2 ordered pairs of elements occurs precisely $\frac{n^d}{m^2}$ times. A set of mutually orthogonal $\mathrm{HC}^j(m; n, d)$'s is denoted by $\mathrm{MOHC}^j(m; n, d)$.

4.21 Theorem An $\mathrm{OA}_{n^d/m^2}(k, m)$ is equivalent to k $\mathrm{MOHC}^0(m; n, d)$. In particular, when $n = m$, an $\mathrm{OA}_{n^{d-2}}(k, n)$ is equivalent to k $\mathrm{MOHC}^0(n, d)$.

4.22 Theorem [3] There are at most $\frac{n^d - dn + d - 1}{m - 1}$ mutually orthogonal $\mathrm{YFHC}(m; n, d)$.

4.23 Theorem [7] For $0 \le j \le d - 1$ and $n \ge 2$, the number of $\mathrm{MOHC}^j(n, d)$ is at most $(n-1)^{d-1} + \sum_{i=j}^{d-2} \binom{d}{i+1}(n-1)^i$.

A set of $\mathrm{MOHC}^j(n, d)$ is *complete* if the number of hypercubes in the set meets the upper bound of Theorem 4.23.

4.24 Theorem [7] If h $\mathrm{MOHC}^1(n, d)$ exist and ℓ $MOLS(n)$ exist, then $h(\ell + 1) + \ell d$ $\mathrm{MOHC}^1(n, d + 1)$ exist. In particular, if both sets are complete, the resulting set is also complete.

4.25 Corollary There exists a complete set of $\mathrm{MOHC}^1(n, d)$ whenever n is a prime power and $d \ge 2$.

4.26 Example [2] Three $\mathrm{MOHC}^2(6, 3)$.

012345	201534	120453	534012	453201	345120
130452	013245	301524	245130	524013	452301
341520	134052	413205	052341	205134	520413
453201	345120	534012	120453	012345	201534
524013	452301	245130	301524	130452	013245
205134	520413	052341	413205	341520	134052

012345	450123	531204	345012	123450	204531
425013	134250	302541	013425	250134	541302
153420	201534	045312	420153	534201	312045
230154	542301	413025	154230	301542	025413
504231	315042	120453	231504	042315	453120
341502	023415	254130	502341	415023	130254

012345	201534	120453	453120	345012	534201
324501	432150	243015	015243	501324	150432
540132	054213	405321	321405	132540	213054
103254	310425	031542	542031	254103	425310
235410	523041	352104	104352	410235	041523
451023	145302	514230	230514	023451	302145

4.27 Theorem For $d > j > 0$, a set of k MOHC$^j(n,d)$ gives $k + \binom{d}{j}$ MOHC$^{j-1}(n,d)$.

4.28 Remark When $d = 2$, the converse to Theorem 4.27 holds.

4.29 Example Adding the three squares defined by $L_4(i,j,k) = i+j \pmod 6$, $L_5(i,j,k) = i+k \pmod 6$, and $L_6(i,j,k) = j+k \pmod 6$ to the set of three MOHC$^2(6,3)$ in Example 4.26 gives a set of six MOHC$^1(6,3)$. Further adding the three squares defined by $L_7(i,j,k) = i$, $L_8(i,j,k) = j$, and $L_9(i,j,k) = k$ gives a set of nine MOHC$^0(6,3)$. From Table 4.19, 12 MOHC$^0(6,3)$ can be constructed; however, they are not obtained by using Theorem 4.27.

4.5 See Also

§I.6	More on affine resolvable designs.
§II.2	Orthogonal arrays of index one and strength two.
§II.5	Orthogonal arrays of strength more than two.
§III.5	Transversal designs of higher index are used to construct PBDs of index more than one.
§IV.11	Difference matrices are extensively used in the direct construction of orthogonal arrays.
§IV.19	Orthogonal arrays are used in the construction of factorial designs.
§IV.20	Mutually orthogonal frequency squares are constructed using OAs of higher index.
§IV.32	Orthogonal arrays are types of orthogonal main effect plans.
[6]	A new textbook on orthogonal arrays.
[9]	An old textbook, valuable for the early results.

References

[1] R.J.R. Abel, private communication.

[2] J. Arkin, P. Smith and E.G. Straus, Euler's 36 officer problem in three dimensions — solved, *J. Recreational Math.* 15 (1983), 81-84.

[3] C.-S. Cheng, Orthogonal arrays with a variable number of symbols, *Ann. Statist.* 8 (1980), 447-453.

[4] C.J. Colbourn, Transversal designs with block size eight and nine, *Europ. J. Combin.*, to appear.

[5] H. Hanani, On transversal designs, in: *Combinatorics*, Mathematisch Centrum, Amsterdam, 1974, 42-52.

[6] A.S. Hedayat, N.J.A. Sloane and J. Stufken, *Orthogonal Arrays*, to appear.

[7] C.F. Laywine, G.L. Mullen and G. Whittle, D-dimensional hypercubes and the Euler and MacNeish conjectures, *Monatsh. Math.*, to appear.

[8] A.C. Mukhopadhyay, Construction of some series of orthogonal arrays, *Sankhya* B43 (1981), 81-92.

[9] D. Raghavarao, *Constructions and Combinatorial Problems in Design of Experiments*, Dover, New York, 1988.

[10] D.K. Ray-Chaudhuri and N.M. Singhi, On the existence and number of orthogonal arrays, *J. Combin. Theory A* 47 (1988), 28-36.

[11] S.S. Shrikhande, Affine resolvable balanced incomplete block designs: A survey, *Aequat. Math.* 14 (1976), 251-269.

5 Orthogonal Arrays of Strength More Than Two

Jürgen Bierbrauer
Charles J. Colbourn

5.1 Examples

5.1 Example An $\mathrm{OA}_4(3, 16, 2)$.

$$\begin{pmatrix}
0&0&0&0&0&0&0&0&0&0&0&0&0&0&0&0&1&1&1&1&1&1&1&1&1&1&1&1&1&1&1&1\\
0&1&0&1&0&1&0&1&0&1&0&1&0&1&0&1&1&0&1&0&1&0&1&0&1&0&1&0&1&0&1&0\\
0&0&1&1&0&0&1&1&0&0&1&1&0&0&1&1&1&1&0&0&1&1&0&0&1&1&0&0&1&1&0&0\\
0&1&1&0&0&1&1&0&0&1&1&0&0&1&1&0&1&0&0&1&1&0&0&1&1&0&0&1&1&0&0&1\\
0&0&0&0&1&1&1&1&0&0&0&0&1&1&1&1&1&1&1&1&0&0&0&0&1&1&1&1&0&0&0&0\\
0&1&0&1&1&0&1&0&0&1&0&1&1&0&1&0&1&0&1&0&0&1&0&1&1&0&1&0&0&1&0&1\\
0&0&1&1&1&1&0&0&0&0&1&1&1&1&0&0&1&1&0&0&0&0&1&1&1&1&0&0&0&0&1&1\\
0&1&1&0&1&0&0&1&0&1&1&0&1&0&0&1&1&0&0&1&0&1&1&0&1&0&0&1&0&1&1&0\\
0&0&0&0&0&0&0&0&1&1&1&1&1&1&1&1&1&1&1&1&1&1&1&1&0&0&0&0&0&0&0&0\\
0&1&0&1&0&1&0&1&1&0&1&0&1&0&1&0&1&0&1&0&1&0&1&0&0&1&0&1&0&1&0&1\\
0&0&1&1&0&0&1&1&1&1&0&0&1&1&0&0&1&1&0&0&1&1&0&0&0&0&1&1&0&0&1&1\\
0&1&1&0&0&1&1&0&1&0&0&1&1&0&0&1&1&0&0&1&1&0&0&1&0&1&1&0&0&1&1&0\\
0&0&0&0&1&1&1&1&1&1&1&1&0&0&0&0&1&1&1&1&0&0&0&0&0&0&0&0&1&1&1&1\\
0&1&0&1&1&0&1&0&1&0&1&0&0&1&0&1&1&0&1&0&0&1&0&1&0&1&0&1&1&0&1&0\\
0&0&1&1&1&1&0&0&1&1&0&0&0&0&1&1&1&1&0&0&0&0&1&1&0&0&1&1&1&1&0&0\\
0&1&1&0&1&0&0&1&1&0&0&1&0&1&1&0&1&0&0&1&0&1&1&0&0&1&1&0&1&0&0&1
\end{pmatrix}$$

5.2 Example An $\mathrm{OA}_1(6, 7, 2)$.

$$\begin{pmatrix}
0100000111110000000000011111111110000000000011111111110000011111 01\\
0010000100001111000000011110000000111111000011111100001111011110 11\\
0001000010001000111000100011100011100011101110001110111011110111\\
0000100001000100100110010010011010011011011001101101110111101111\\
0000010000100010010101001001010101011011010101101110111110111111\\
0000001000010001001011000100101100101101110010110111011110111111\\
0111111000000000000000011111111111111111111000000000000000111111 0
\end{pmatrix}$$

5.3 Example An $\mathrm{OA}_1(3, 6, 4)$.

$$\begin{pmatrix}
0000000000000000111111111111111122222222222222223333333333333333\\
0000111122223333000011112222333300001111222233330000111122223333\\
0123012301230123012301230123012301230123012301230123012301230123\\
0123103223013210103201233210230123013210012310323210230110320123\\
0123230132101032321010320123230110323210230101232301012310323210\\
0123321010322301230110323210012332100123230110321032230101233210
\end{pmatrix}$$

5.4 Example An $\mathrm{OA}_1(3, 4, 3)$.

$$\begin{pmatrix}
0&0&0&0&0&0&0&0&0&1&1&1&1&1&1&1&1&1&2&2&2&2&2&2&2&2&2\\
0&1&2&0&1&2&0&1&2&0&1&2&0&1&2&0&1&2&0&1&2&0&1&2&0&1&2\\
0&1&2&1&2&0&2&0&1&1&2&0&2&0&1&0&1&2&2&0&1&0&1&2&1&2&0\\
0&1&2&2&0&1&1&2&0&1&2&0&0&1&2&2&0&1&2&0&1&1&2&0&0&1&2
\end{pmatrix}$$

5.5 **Remark** See §II.4.1 for required definitions. Only the cases when $t \geq 3$ are considered in this chapter.

5.2 Equivalent and Related Objects

5.6 **Remark** (The link with codes) Let $\mathcal{C}$ be a v-ary code of length n, equivalently a set of n-tuples with entries in a fixed v-set, the *alphabet*. The *distance distribution* of $\mathcal{C}$ is the $(n+1)$-tuple of nonnegative rational numbers $A_i, i = 0, 1, \ldots, n$, where

$$A_i = (\#(x, y) \mid x, y \in \mathcal{C}, d(x, y) = i) / \mid \mathcal{C} \mid .$$

Here the *Hamming distance* $d(x, y)$ is the number of coordinates where x and y differ. The *minimum distance* d of $\mathcal{C}$ is the smallest positive integer such that $A_d > 0$. The *dual distance distribution* of $\mathcal{C}$ is the $(n+1)$-tuple of rational numbers $A'_k, k = 0, 1, \ldots, n$, where A'_k is defined via the *Kravchouk transform*

$$A'_k = \frac{1}{|\mathcal{C}|} \sum_{i=0}^{n} A_i K_k(i).$$

The numbers $K_k(i)$ are evaluations of the *Kravchouk polynomials*

$$K_k(X) = \sum_{j=0}^{k} (-1)^j \binom{X}{j} \binom{n-X}{k-j} (v-1)^{k-j}.$$

The A'_k are nonnegative and $A'_0 = 1$. The *dual distance* d' of $\mathcal{C}$ is the smallest integer greater than 0 such that $A'_{d'} > 0$.

It is a theorem of Delsarte's that $\mathcal{C}$ is an orthogonal array of strength $d' - 1$. If $v = q$ is a prime power and $\mathcal{C}$ is a linear code, then d' is the minimum distance of the dual code $\mathcal{C}^{\perp}$. This theory was first developed in various papers of Delsarte; see [6].

5.7 **Remark** $\mathrm{OA}_1(t, k, s)$ are known as *MDS codes* in coding theory. The letters stand for *maximum distance separable*.

5.8 **Theorem** An $\mathrm{OA}_\lambda(t, k, s)$ is an $\mathrm{OA}_{s\lambda}(t-1, k, s)$.

5.9 **Remark** Suppose there is an $\mathrm{OA}_\lambda(t, k, s)$. Fixing one row i and one symbol σ, then selecting all columns that contain σ in row i, and finally deleting row i, one obtains an $\mathrm{OA}_\lambda(t-1, k-1, s)$, the *derived OA*.

5.3 Existence

5.10 **Theorem** (*Rao bound*) An $\mathrm{OA}_\lambda(t, k, s)$ exists only if

$$\lambda s^t \geq 1 + \begin{cases} \sum_{i=1}^{t/2} \binom{k}{i} (s-1)^i & \text{if } t \text{ even}, \\ \sum_{i=1}^{(t-1)/2} \binom{k}{i} (s-1)^i + \binom{k-1}{(t-1)/2} (s-1)^{(t+1)/2} & \text{if } t \text{ odd}. \end{cases}$$

5.11 **Theorem** [9] The Rao bound is met with equality

1. when $t = 3$, only for $\mathrm{OA}_n(3, 4n, 2)$ with n integral, and for $\mathrm{OA}_q(3, q+2, q)$ with q even;
2. when $t = 4$, only for $\mathrm{OA}_1(4, 5, 2)$, $\mathrm{OA}_3(4, 11, 3)$, and $\mathrm{OA}_\lambda(4, k, 6)$ with $\lambda = \frac{a^2(9a^2-1)}{288}$, $k = \frac{9a^2+1}{5}$, and $a \equiv 21, 69 \pmod{240}$.
3. when $t = 5$, only for $\mathrm{OA}_1(5, 6, 2)$ and $\mathrm{OA}_3(5, 12, 3)$.

5.12 **Theorem** (*Bush bound*) An $\mathrm{OA}_1(t, k, s)$ with $t > 1$ exists only if

$$k \leq \begin{cases} s + t - 1 & \text{if } s \text{ even and } t \leq s, \\ s + t - 2 & \text{if } s \text{ odd and } 3 \leq t \leq s, \\ t + 1 & \text{if } t \geq s. \end{cases}$$

5.13 Theorem (*Bose-Bush bound for* $t = 3$) An $OA_\lambda(3, k, s)$ exists only if $k \le \left\lfloor \frac{\lambda s^2 - 1}{s-1} \right\rfloor + 1$. Moreover, if $\lambda - 1 \equiv b \pmod{s-1}$ and $1 \le b < s - 1$, then

$$k \le \left\lfloor \frac{\lambda s^2 - 1}{s - 1} \right\rfloor - \left\lfloor \frac{\sqrt{1 + 4s(s-1-b)} - (2s - 2b - 1)}{2} \right\rfloor.$$

5.14 Theorem [5] If s is a prime power and $t < s$, then an $OA_1(t, s+1, s)$ exists. Moreover, if $s \ge 4$ is a power of 2, an $OA_1(3, s+2, s)$ exists.

5.15 Remark Theorem 5.16, due to Bush, is a generalization of Theorem 2.18, due to MacNeish.

5.16 Theorem [4] If $OA_{\lambda_i}(t, k, s_i)$ exist for $1 \le i \le m$, then an $OA_{\prod_{i=1}^m \lambda_i}(t, k, \prod_{i=1}^m s_i)$ exists.

5.17 Theorem [1] If t is even, any $OA_\lambda(t, k, 2)$ yields an $OA_\lambda(t + 1, k + 1, 2)$. Hence, whenever a Hadamard matrix of order $4n$ exists, so also does a $OA_n(3, 4n, 2)$.

5.18 Theorem [8, 2] If s and σ are powers of the same prime, then an $OA_{\sigma^{t-1}}(t, \sigma s + 1, s)$ exists for all $t \ge 3$ and an $OA_{\sigma^{t-2}/s}(t, \sigma s, s)$ exists for all $t > s$.

5.19 Theorem [8] If s and σ are powers of the same prime, then if an $OA_\lambda(3, r, s)$ exists, so also does an $OA_{\lambda s^2 \sigma^2}(3, r\sigma s, s)$.

5.20 Theorem [8] If an $OA_\lambda(3, r, s)$ exists, so does an $OA_{s\lambda}(3, 2r, s)$.

5.21 Theorem [3] For any t, k with $2 \le t \le k$, there exists a number $e_0 = e_0(t, k)$ such that for any positive v and any prime power q there is an orthogonal array $OA(t, k, v \cdot q^e)$ for all $e \ge e_0$.

5.22 Theorem [11] For $j \ge 3$ an odd integer, there exists an $OA_{2^{2j-3}}(5, 2^{j+1}, 2)$.

5.23 Remark The OAs from Theorem 5.22 have the largest number of rows permitted by the Rao bound. The Rao bound is not met "with equality", however; compare Theorem 5.11.

5.4 See Also

§II.1-4	MOLS and transversal designs arise in the case $t = 2$.
§IV.19	Orthogonal arrays are used to construct factorial designs.
§IV.24	Hadamard matrices correspond to certain OAs.
§IV.47	(t,m,s)-nets are constructed using orthogonal arrays.
§V.4	OAs are useful in derandomization of randomized algorithms.
§VI.1	Existence results on codes.
[7]	A new textbook on orthogonal arrays.
[10]	An old textbook, valuable for the early results.

References

[1] T. Atsumi, A study of orthogonal arrays from the point of view of design theory, *J. Combin. Theory A* 35 (1983), 241-251.

[2] J. Bierbrauer, Construction of orthogonal arrays, *J. Stat. Plann. Infer.*, to appear.

[3] J.L. Blanchard, A construction for orthogonal arrays with strength $t \ge 3$, *Discrete Math.* 137 (1995), 35-44.

[4] K.A. Bush, A generalization of the theorem due to MacNeish, *Ann. Math. Stat.* 23 (1952), 293-295.
[5] K.A. Bush, Orthogonal arrays of index unity, *Ann. Math. Stat.* 23 (1952), 426-434.
[6] P. Delsarte, Four fundamental parameters of a code and their combinatorial significance, *Inf. Control* 23 (1973), 407-438.
[7] A.S. Hedayat, N.J.A. Sloane and J. Stufken, *Orthogonal Arrays*, to appear.
[8] A.C. Mukhopadhyay, Construction of some series of orthogonal arrays, *Sankhya* B43 (1981), 81-92.
[9] R. Noda, On orthogonal arrays of strength 3 and 5 achieving Rao's bound, *Graphs Combin.* 2 (1986), 277-282.
[10] D. Raghavarao, *Constructions and Combinatorial Problems in Design of Experiments*, Dover, New York, 1988.
[11] D.R. Stinson and J.L. Massey, An infinite class of counterexamples to a conjecture concerning nonlinear resilient functions, *J. Cryptology*, to appear.

PART III

PAIRWISE BALANCED DESIGNS

1 PBDs and GDDs: The Basics

Ronald C. Mullin
Hans-Dietrich O.F. Gronau

1.1 PBDs and GDDs

Let K be a subset of positive integers and let λ be a positive integer. A *pairwise balanced design* (PBD$(v, K; \lambda)$ or (K, λ)-PBD) of order v with block sizes from K is a pair $(\mathcal{V}, \mathcal{B})$, where $\mathcal{V}$ is a finite set (the *point set*) of cardinality v and $\mathcal{B}$ is a family of subsets (*blocks*) of $\mathcal{V}$ which satisfy the properties:

1. If $B \in \mathcal{B}$, then $|B| \in K$.
2. Every pair of distinct elements of $\mathcal{V}$ occurs in exactly λ blocks of $\mathcal{B}$.

The integer λ is the *index* of the PBD. PBD(v, K) is often used when $\lambda = 1$.

1.1 **Example** A PBD$(10, \{3, 4\})$. The blocks are listed columnwise in this entire part.

1	1	1	2	2	2	3	3	3	4	4	4
2	5	8	5	6	7	5	6	7	5	6	7
3	6	9	8	9	10	10	8	9	9	10	8
4	7	10									

Let K and G be sets of positive integers and let λ be a positive integer. *A group divisible design of index λ and order v* ((K, λ)-GDD) is a triple $(\mathcal{V}, \mathcal{G}, \mathcal{B})$, where $\mathcal{V}$ is a finite set of cardinality v, $\mathcal{G}$ is a partition of $\mathcal{V}$ into parts (*groups*) whose sizes lie in G, and $\mathcal{B}$ is a family of subsets (*blocks*) of $\mathcal{V}$ which satisfy the properties:

1. If $B \in \mathcal{B}$ then $|B| \in K$.
2. Every pair of distinct elements of $\mathcal{V}$ occurs in exactly λ blocks or one group, but not both.
3. $|\mathcal{G}| > 1$.

If $v = a_1 g_1 + a_2 g_2 + \ldots + a_s g_s$, and if there are a_i groups of size g_i, $i = 1, 2, \ldots, s$, then the (K, λ)-GDD is of type $g_1^{a_1} g_2^{a_2} \ldots g_s^{a_s}$. This is *exponential notation* for the group type. Alternatively, if the GDD has groups $G_1, G_2, \ldots, G_t$, then the list $T = [|G_i| : i = 1, 2, \ldots, t]$ is the *type* of the GDD when more convenient.

If $K = \{k\}$, then the (K, λ)-GDD is a (k, λ)-GDD. If $\lambda = 1$, the GDD is a K-GDD. Furthermore, a $(\{k\}, 1)$-GDD is a k-GDD.

1.2 **Remark** By convention, factors of type 0^a can be included in the exponential form of the type to accommodate null groups when needed.

1.3 **Example** A {3,4}-GDD of type $1^1 3^3$ or type [3,3,3,1] (the groups are in boldface).

1	**4**	**7**	**10**	1	2	3	1	2	3	1	2	3
2	**5**	**8**		4	5	6	5	6	4	6	4	5
3	**6**	**9**		7	8	9	9	7	8	8	9	7
				10	10	10						

1.4 **Example** A 4-GDD of type 3^9. Let $V = \mathbb{Z}_3^3$. The orbits of {000,020,111,211} and {000,102,012,110} (writing $(x, y, z) \in \mathbb{Z}_3^3$ as xyz) form the blocks. The groups are the orbit of {000,001,002}. Adding a new symbol ∞ to each group gives a (28,4,1) design.

A GDD D_1 is *derivable* if it can be obtained from another GDD D_2 by replacing a group G of D_2 by the groups and blocks of a third GDD D_3 defined on the point-set of G. A GDD is *primitive* if it is not derivable.

A GDD is *uniform* if all groups have the same size, that is, if it is of type m^u for some positive integer u.

1.5 **Theorem** [21] The necessary conditions for the existence of a uniform (k, λ)-GDD of type m^u are

1. $u \geq k$,
2. $\lambda(u-1)m \equiv 0 \pmod{k-1}$,
3. $\lambda u(u-1)m^2 \equiv 0 \pmod{k(k-1)}$.

1.6 **Remarks**

1. A K-GDD of order v with group sizes in a set G can be viewed as a PBD$(v, K \cup G)$ by considering the groups of the GDD to be blocks of the PBD also.
2. A K-GDD of order v with group sizes in a set G can be used to create a PBD$(v+1, K \cup (G+1))$ (where $G+1 = \{g+1 : g \in G\}$), by adjoining a new point (say ∞) to each group and considering the resulting subsets as blocks. Conversely, a GDD can be obtained from a PBD by deleting a point.
3. (See §II.2.) A *transversal design* TD(k, n) is a k-GDD of type n^k. A GDD is a transversal design if and only if each block meets every group in exactly one point.
4. (See §I.1.) A BIBD$(v, k, 1)$ is equivalent to a PBD$(v, \{k\})$ or a k-GDD of type 1^v.

Let $\mathcal{B}$ be a set of blocks in a PBD, GDD, or BIBD. A *parallel class* or *resolution class* is a collection of blocks which partition the point-set of the design.

A PBD, GDD, or BIBD is *resolvable* if the blocks of the design can be partitioned into parallel classes. In the case of a resolvable BIBD(v, k, λ), RBIBD(v, k, λ) or RB(v, k, λ) is used. A resolvable GDD is denoted by RGDD.

1.7 **Theorem** [21] The necessary conditions for the existence of a uniform (k, λ)-RGDD of type m^u are

1. $u \geq k$,
2. $um \equiv 0 \pmod{k}$,
3. $\lambda(u-1)m \equiv 0 \pmod{k-1}$.

Let K be a set of positive integers, and let k be a positive integer. PBD$(v, K \cup k^\star)$ denotes a PBD containing a block of size k. If $k \notin K$, this indicates that there is only one block of size k in the PBD. On the other hand, if $k \in K$, then there is at least one block of size k in the PBD.

1.8 **Theorem** Let D be a BIBD$(v, k, 1)$. Let $r = (v-1)/(k-1)$ and suppose $n = v/k$ is an integer. Suppose that D contains $s \geq 1$ mutually block disjoint parallel classes, and let t be an integer. By adjoining new points, one obtains:

1. For $0 \leq t \leq s-1$, there exists a $\{k, k+1\}$-GDD of type $k^n t^1$.
2. For $0 \leq t \leq s$, there exists a PBD$(nk+t, \{k, k+1, t^\star\})$.
3. If $t = s = r$, then there exists a PBD$(nk+r, \{k+1, r^\star\})$. In this case, the operation of creating the design is the *completion* of the (resolvable) BIBD.

1.9 **Example** D is a resolvable PBD(9,{3}), a 3-GDD of type 3^3, and a BIBD(9,3,1).

$$D = \begin{array}{cccc} 147 & 123 & 123 & 123 \\ 258 & 456 & 564 & 645 \\ 369 & 789 & 978 & 897 \end{array}$$

The blocks are written as columns with space between parallel classes of blocks. If {1,2,3}, {4,5,6}, {7,8,9} are taken as groups of a GDD, there are three parallel classes of the remaining blocks.

Let K be a set of positive integers, and suppose that $k \in K$. A pairwise balanced design of order v and with block sizes from K has a *mandatory block* of size k if it contains a block of size k.

1.10 **Example** A PBD$(11, \{3, 5^\star\})$. (The distinguished block is in boldface.)

$$\begin{array}{cccccccccccccccc} \mathbf{1} & 1 & 1 & 1 & 2 & 2 & 2 & 3 & 3 & 3 & 4 & 4 & 4 & 5 & 5 & 5 \\ \mathbf{2} & 6 & 8 & 9 & 6 & 7 & 10 & 6 & 7 & 8 & 6 & 7 & 9 & 6 & 7 & 8 \\ \mathbf{3} & 7 & 11 & 10 & 8 & 9 & 11 & 9 & 11 & 10 & 10 & 8 & 11 & 11 & 10 & 9 \\ \mathbf{4} \\ \mathbf{5} \end{array}$$

1.11 **Theorem** [15, 6] Let $N_{\geq 3} = \{n \in \mathbb{Z} : n \geq 3\}$. Then a PBD$(v, N_{\geq 3} \cup \{k^\star\})$ exists if and only if $v \geq 2k + 1$ except when

1. $v = 2k + 1$ and $k \equiv 0 \pmod 2$;
2. $v = 2k + 2$ and $k \not\equiv 4 \pmod 6$, $k > 1$;
3. $v = 2k + 3$ and $k \equiv 0 \pmod 2$, $k > 6$;
4. $(v, k) \in \{(7, 2), (8, 2), (9, 2), (10, 2), (11, 4), (12, 2), (13, 2), (17, 6)\}$.

1.12 **Theorem** (Colbourn, Haddad, Linek [9]) Let N_{odd} denote the set of all odd positive integers. Let v and k be odd positive integers which satisfy $v \geq 2k + 1$. Then there exists a PBD$(v, N_{odd} \cup \{k^\star\})$.

1.13 **Theorem** [18] There exists a PBD$(v, \{4, w^\star\})$ if and only if $v \geq 3w + 1$, and

1. $v \equiv 1$ or $4 \pmod{12}$ and $w \equiv 1$ or $4 \pmod{12}$, or
2. $v \equiv 7$ or $10 \pmod{12}$ and $w \equiv 7$ or $10 \pmod{12}$.

1.2 Flats and Holes

Let $D = (\mathcal{V}, \mathcal{B})$ be a PBD(v, K). A *flat* F in D is a pair $(\mathcal{F}, \mathcal{C})$ where $\mathcal{F} \subset \mathcal{V}$ and $\mathcal{C} \subset \mathcal{B}$ which has the property that $(\mathcal{F}, \mathcal{C})$ is also a PBD. The order of the flat is $|\mathcal{F}|$. In general, the degenerate PBDs of orders 0 and 1 are considered to be flats of all PBDs of order $v > 1$. Such flats are *trivial.*

1.14 **Example** A PBD$(7, \{2, 3\})$ which contains a flat of order 3. (The point-set of the flat is {1,2,3}, the blocks of the flat are in boldface.)

$$\begin{array}{ccccccccc} \mathbf{1} & \mathbf{1} & \mathbf{2} & 1 & 1 & 2 & 2 & 3 & 3 \\ \mathbf{2} & \mathbf{3} & \mathbf{3} & 4 & 6 & 4 & 5 & 4 & 5 \\ & & & 5 & 7 & 6 & 7 & 7 & 6 \end{array}$$

1.15 **Remark** In any PBD(v, K), any block can be considered to be a flat. Conversely the blocks of any flat of order w can be replaced by a block of size w containing these points to obtain a PBD$(v, K \cup \{w\})$.

Let K be a set of positive integers and let λ be a positive integer. An *incomplete* PBD of order v and index λ with a hole of order h ($(v, h; K, \lambda)$-IPBD) is a triple $(\mathcal{V}, \mathcal{H}, \mathcal{B})$ where $\mathcal{V}$ is a set of cardinality v, $\mathcal{H}$ is a subset of $\mathcal{V}$ which contains h elements, and $\mathcal{B}$ is a family of subsets (*blocks*) of $\mathcal{V}$ which satisfy the properties:

1. If $B \in \mathcal{B}$, then $|B| \in K$;
2. Every pair of distinct elements $\{x, y\}$ from $\mathcal{V}$ with the exception of those in which both x and y lie in $\mathcal{H}$ occur in precisely λ blocks;
3. No pair of distinct elements of $\mathcal{H}$ occur in any block.

If $\lambda = 1$, the design is a $(v, h; K)$-IPBD, or simply a K-IPBD of order v with a hole of size h.

1.16 Remark An incomplete PBD of order v and index 1 with a hole of size h is equivalent to a PBD(v, K) from which a block of size h has been removed.

Let K be a set of positive integers, and let λ be a positive integer. An *incomplete group divisible design* with block-sizes from K and index λ ((K, λ)-IGDD) is a quadruple $(\mathcal{V}, \mathcal{G}, \mathcal{H}, \mathcal{B})$ where $\mathcal{V}$ is a finite set of cardinality v, $\mathcal{G} = (G_1, G_2, \dots, G_s)$ is a partition of $\mathcal{V}$, $\mathcal{H} = \{H_1, H_2, \dots, H_s\}$ is a collection of subsets of V with the property $H_i \subseteq G_i$, $i = 1, 2, \dots, s$ (the G_i are *groups* and the H_i are *holes*), and $\mathcal{B}$ is a family of subsets of V (*blocks*) which satisfy the properties:

1. Any pair of distinct elements of $\mathcal{V}$ which occurs in a group does not occur in any block.
2. If a pair of distinct elements from V comes from distinct groups and each element occurs in the hole of its respective group, then that pair occurs in no block of $\mathcal{B}$. Otherwise; it occurs in exactly λ blocks of $\mathcal{B}$.

An IGDD is of *type* $(g_1, h_1)^{a_1} (g_2, h_2)^{a_2} \dots (g_t, h_t)^{a_t}$ if there are a_i groups of size g_i which contains a hole of size h_i, $i = 1, 2, \dots, t$. An IGDD is a K-IGDD if $\lambda = 1$. Further, it is a k-IGDD if $K = \{k\}$.

1.17 Example A {3,4}-IGDD of type $(3, 1)^3 (1, 1)^1$. (The groups are printed in boldface; the holes are overscored.)

$$\begin{array}{cccc|cccccccc}
\mathbf{1} & \mathbf{4} & \mathbf{7} & \overline{\mathbf{10}} & 1 & 2 & 1 & 2 & 3 & 1 & 2 & 3 \\
\mathbf{2} & \mathbf{5} & \mathbf{8} & & 4 & 5 & 5 & 6 & 4 & 6 & 4 & 5 \\
\overline{\mathbf{3}} & \overline{\mathbf{6}} & \overline{\mathbf{9}} & & 7 & 8 & 9 & 7 & 8 & 8 & 9 & 7 \\
 & & & & 10 & 10 & & & & & &
\end{array}$$

1.18 Theorem [21] Necessary conditions for the existence of a (k, λ)-IGDD of type $(m, h)^u$ are

1. $m \geq (k - 1)h$,
2. $\lambda m(u - 1) \equiv 0 \pmod{k - 1}$,
3. $\lambda(m - h)(u - 1) \equiv 0 \pmod{k - 1}$,
4. $\lambda u(u - 1)(m^2 - h^2) \equiv 0 \pmod{k(k - 1)}$.

1.19 Example Incomplete MOLS with one hole (§II.3) provide examples of IGDD. In particular, an ITD$(k, m; h)$ (other notation: a TD(k, m) − TD(k, h)) is a k-IGDD of type $(m, h)^k$.

1.20 Theorem [21] There exists a 3-IGDD of type $(m, h)^u$ if and only if the following conditions hold:

1. $m \geq 2h$;
2. $m(u - 1) \equiv 0 \pmod 2$;
3. $(m - h)(u - 1) \equiv 0 \pmod 2$;
4. $u(u - 1)(m^2 - h^2) \equiv 0 \pmod 6$.

1.21 Theorem [20] There exists a 4-IGDD of type $(m,h)^u$ where $u \equiv 1 \pmod 3$, $u \notin \{10, 19\}$, and $(m,h) \neq (6,1)$ if and only if

1. $m \geq 3h$ for $u \equiv 1$ or $4 \pmod{12}$, or
2. $m \geq 3h$ and $m-h$ is even for $u \equiv 7$ or $10 \pmod{12}$.

Let K be a set of positive integers, let v, w_1, w_2, and λ be positive integers, and let w_3 be a nonnegative integer. Then a $(v; w_1, w_2; w_3; K, \lambda)$-$\diamondsuit$-IPBD *(incomplete-$\diamondsuit$-PBD)* is a quadruple $(\mathcal{V}, \mathcal{Y}_1, \mathcal{Y}_2, \mathcal{B})$ where $\mathcal{V}$ is a set of cardinality v, $\mathcal{Y}_1$ and $\mathcal{Y}_2$ are subsets of $\mathcal{V}$ of cardinalities w_1 and w_2 respectively, $|\mathcal{Y}_1 \cap \mathcal{Y}_2| = w_3$, and $\mathcal{B}$ is a collection of subsets (*blocks*) of $\mathcal{V}$ which satisfy the properties:

1. If $B \in \mathcal{B}$, then $|B| \in K$;
2. If $\{x, y\}$ is a pair of distinct elements of $\mathcal{V}$, then $\{x, y\}$ occurs in precisely λ blocks of $\mathcal{B}$ unless $\{x, y\} \subset \mathcal{Y}_1$ or $\{x, y\} \subset \mathcal{Y}_2$, in which case the pair occurs in no block.

If $\lambda = 1$, then the design is a $(v; w_1, w_2; w_3; K)$-$\diamondsuit$-IPBD.

1.3 Some Families of Group Divisible Designs

1.22 Theorem [21] Let u and t be positive integers. There exists a 3-GDD of type t^u if and only if the conditions in the following table are satisfied.

t	u	Necessary and Sufficient Conditions
$1, 5 \pmod 6$	$1, 3 \pmod 6$	$u \geq 3$
$2, 4 \pmod 6$	$0, 1 \pmod 3$	$u \geq 3$
$3 \pmod 6$	$1 \pmod 2$	$u \geq 3$
$0 \pmod 6$	no constraint	$u \geq 3$

1.23 Table (Colbourn, [7]) Every primitive 3-GDD on at most 30 points has one of the following types. (The number of points is given in boldface.)

3: 1^3 **6:** 2^3 **7:** $3^1 1^4$ **8:** 2^4 **9:** 3^3 **10:** $4^1 2^3$
11: $5^1 1^6$ **12:** 2^6, 4^3 **13:** $3^4 1^1$ **14:** $4^3 2^1$, $6^1 2^4$ **15:** 3^5, 5^3, $7^1 1^8$
16: 4^4, $6^1 4^1 2^3$ **17:** $5^1 3^4$ **18:** $6^1 4^3$, 6^3, $8^1 2^5$ **19:** $3^6 1^1$, $5^3 3^1 1^1$, $7^1 3^4$, $9^1 1^{10}$
20: $6^1 4^3 2^1$, $6^3 2^1$, $8^1 2^6$, $8^1 4^3$ **21:** $5^3 3^2$, $7^1 3^4 1^2$, 7^3, $9^1 3^4$
22: $6^1 4^4$, $6^3 4^1$, $8^1 6^1 4^1 2^2$, $10^1 2^6$
23: $5^4 3^1$, $7^1 5^1 3^3 1^2$, $7^2 5^1 3^1 1^1$, $9^1 5^1 3^3$, $11^1 1^{12}$
24: 4^6, $6^2 4^3$, 6^4, $8^1 4^3 2^2$, $8^1 6^2 2^2$, 8^3, $10^1 4^2 2^3$
25: $7^1 5^3 3^1$, $7^2 3^3 1^2$, $7^3 3^1 1^1$, $9^1 3^5 1^1$, $9^1 5^3 1^1$, $9^1 7^1 3^3$
26: $4^6 2^1$, $6^2 4^3 2^1$, $6^4 2^1$, $8^1 6^1 4^3$, $8^1 6^3$, $8^2 6^1 2^2$, $8^3 2^1$, $10^1 6^1 4^2 2^1$, $12^1 2^7$
27: $7^1 5^3 3^1 1^2$, $7^2 3^4 1^1$, $7^3 3^2$, $9^1 5^3 3^1$, $9^1 7^1 3^3 1^2$, $9^1 7^2 1^4$, 9^3, $11^1 5^2 3^1 1^3$, $13^1 1^{14}$
28: $6^2 4^4$, $6^4 4^1$, $8^1 4^4 2^2$, $8^1 6^2 4^1 2^2$, $8^2 6^1 4^1 2^1$, $8^3 4^1$, $10^1 6^1 4^3$, $10^1 6^3$, $10^1 8^1 2^5$, $12^1 4^1 2^6$, $12^1 4^4$
29: $7^1 5^4 1^2$, $7^3 5^1 3^1$, $9^1 5^4$, $9^1 7^1 5^1 3^2 1^2$, $9^1 7^2 5^1 1^1$, $9^2 5^1 3^2$, $11^1 3^6$, $11^1 5^3 3^1$, $11^1 7^1 3^3 1^2$
30: 6^5, $8^1 6^1 4^3 2^2$, $8^1 6^3 2^2$, $8^2 4^3 2^1$, $8^2 6^2 2^1$, $8^3 6^1$, $10^1 4^5$, $10^1 6^2 4^2$, $10^1 8^1 4^2 2^2$, 10^3, $12^1 6^1 4^3$, $12^1 6^3$, $14^1 2^8$

1.24 Theorem [10] Let g, t, and u be nonnegative integers. There exists a 3-GDD of the type $g^t u^1$ if and only if the following conditions are all satisfied:

1. if $g > 0$, then $t \geq 3$, or $t = 2$ and $u = g$, or $t = 1$ and $u = 0$, or $t = 0$;

2. $u \le g(t-1)$ or $gt = 0$;
3. $g(t-1) + u \equiv 0 \pmod 2$ or $gt = 0$;
4. $gt \equiv 0 \pmod 2$ or $u = 0$;
5. $\frac{1}{2}g^2t(t-1) + gtu \equiv 0 \pmod 3$.

1.25 Theorem [8] Let u, r, and t be positive integers. Then there exists a 3-GDD of type $u^r 1^t$ if and only if the following conditions are all satisfied:

1. $u \equiv 1 \pmod 2$;
2. $r + t \equiv 1 \pmod 2$;
3. if $r = 1, t \ge u + 1$;
4. if $r = 2, t \ge u$;
5. $\binom{t}{2} + rut + \binom{r}{2}u^2 \equiv 0 \pmod 3$.

1.26 Theorem [8] Let u and v be positive integers with $v \le u$. Then a 3-GDD of type $u^1v^11^u$ exists if and only if $(u, v) \equiv (1,1), (3,1), (3,3), (3,5), (5,1)$ modulo (6,6).

1.27 Theorem [16] Let u and t be positive integers. Then there exists a 4-GDD of type t^u if and only if the conditions in the following table are satisfied.

Existence of 4-GDDs of Type t^u		
t	u	Necessary and Sufficient Conditions
$1, 5 \pmod 6$	$1, 4 \pmod{12}$	$u \ge 4$
$2, 4 \pmod 6$	$1 \pmod 3$	$u \ge 4,\ (t, u) \ne (2, 4)$
$3 \pmod 6$	$0, 1 \pmod 4$	$u \ge 4$
$0 \pmod 6$	no constraint	$u \ge 4, (t, u) \ne (6, 4)$

1.28 Theorem [16] Let t and u be positive integers. Then there exists a 4-GDD of type t^11^u or t^13^u if and only if the conditions in the following table are satisfied.

Existence of 4-GDDs of Types t^11^u and t^13^u			
t	u	type	Necessary and Sufficient Conditions
$1, 7 \pmod{12}$	$0, 3 \pmod{12}$	t^11^u	$u \ge 2t + 1$
$4, 10 \pmod{12}$	$0, 9 \pmod{12}$	t^11^u	$u \ge 2t + 1$
$0, 6 \pmod{12}$	$0, 1 \pmod 4$	t^13^u	$u \ge (2t + 3)/3$
$3, 9 \pmod{12}$	$0, 3 \pmod 4$	t^13^u	$u \ge (2t + 3)/3$

1.29 Theorem [16] Let t and a be positive integers. Then there exists a 4-GDD of type $t^u(t(u-1)/2)^1$ if and only if the conditions in the following table are satisfied.

Existence of 4-GDDs of Type $t^u(t(u-1)/2)^1$		
t	u	Necessary and Sufficient Conditions
$1, 5 \pmod 6$	$3 \pmod 6$	$u \ge 3$
$2, 4 \pmod 6$	$0 \pmod 3$	$u \ge 3,\ (t, u) \ne (2, 3), (2, 6)$
$3 \pmod 6$	$1 \pmod 2$	$u \ge 3$
$0 \pmod 6$	no constraint	$u \ge 3, (t, u) \ne (6, 3)$

1.30 Remark The latter class of 4-GDDs in Theorem 1.29 is intimately related to the class of resolvable 3-GDDs of type t^u. Such a resolvable 3-GDD exists if and only if $t(u-1)$ is even, $tu \equiv 0 \pmod 3$, $u \ge 3$, $(t, u) \ne (2, 3), (6, 3)$ or (2,6). The 4-GDDs are obtained by adjoining a new point to each resolution class, then adjoining the set of new points as a new group.

1.31 Theorem (Shen; Rees-Stinson [18]) A resolvable 4-GDD of type 3^u exists if and only if $u \geq 8$ and $u \equiv 0 \pmod 4$, with the possible exception of

$$u \in \{28, 44, 88, 152, 184, 220, 284, 288\}.$$

1.32 Table Existence of 4-GDDs of small order [16].

v	Group Type	Exist
4	1^4	YES
8	2^4	NO
12	3^4	YES
13	1^{13}	YES
	$1^9 4^1$	YES
14	2^7	YES
15	3^5	YES
16	1^{16}	YES
	$1^{12} 4^1$	YES
	$1^8 4^2$	YES
	$1^4 4^3$	YES
	4^4	YES
17	$2^6 5^1$	NO
20	2^{10}	YES
	5^4	YES
21	$3^5 6^1$	YES
22	$1^{15} 7^1$	YES
23	$2^9 5^1$	YES
24	3^8	YES
	$3^4 6^2$	YES
	6^4	NO
25	1^{25}	YES
	$1^{21} 4^1$	YES
	$1^{17} 4^2$	YES
	$1^{13} 4^3$	YES
	$1^9 4^4$	YES
	$1^5 4^5$	YES
	$1^1 4^6$	YES
26	2^{13}	YES
	$2^3 5^4$	?
	$2^9 8^1$	YES
27	3^9	YES
	$3^5 6^2$	?
	$3^1 6^4$	YES
28	1^{28}	YES
	$1^{24} 4^1$	YES
28	$1^{20} 4^2$	YES
	$1^{16} 4^3$	YES
	$1^{12} 4^4$	YES
	$1^8 4^5$	YES
	$1^4 4^6$	YES
	4^7	YES
	$1^{14} 7^2$	YES
	$1^{10} 4^1 7^2$	YES
	$1^6 4^2 7^2$	YES
	$1^2 4^3 7^2$	YES
	7^4	YES
29	$2^{12} 5^1$	YES
	$2^2 5^5$	?
	$2^8 5^1 8^1$	YES
30	$3^8 6^1$	YES
	$3^4 6^3$	YES
	6^5	YES
	$3^7 9^1$	YES

1.33 Theorem [5] A 4-GDD of type $2^u 5^1$ exists if and only if $u \equiv 0 \pmod 3$, $u \geq 9$.

1.34 Theorem

1. [19] There exists a 4-GDD of type $6^5 9^1$.
2. There exists a 5-GDD of type 5^9 (a BIBD (45,5,1) with a parallel class) (see Example IV.10.57).
3. [14] There exists a 7-GDD of type 7^{13} (a BIBD (91,7,1) with a parallel class).

1.35 Theorem (*elliptic semiplanes*) There exists a 7-GDD of type 3^{15} [1] and a 9-GDD of type 3^{33} [17].

1.36 Theorem If there exists a $\mathrm{TD}(k,n) - \mathrm{TD}(k,h)$ with $(k-1)h = n$, then there exists a $(k-1)$-GDD of type $(n-h)^k$.

1.37 Example By Theorem II.3.6, there is a $\mathrm{TD}(6,10) - \mathrm{TD}(6,2)$. So there exists a 5-GDD of type 8^6. Another construction follows: The point set is $(I_2 \times \mathbb{Z}_3) \times (\mathbb{Z}_2 \times \mathbb{Z}_4)$ where $I_2 = \{0,1\}$. Let the groups be $g \times (\mathbb{Z}_2 \times \mathbb{Z}_4)$ for $g \in \mathbf{I}_2 \times \mathbb{Z}_3$. For blocks take

$$\begin{array}{l} \{(0,0,0,0), (0,1,0,2), (0,2,1,0), (1,0,1,1), (1,1,1,2)\}, \\ \{(0,0,0,0), (0,1,0,0), (0,2,1,1), (1,0,0,0), (1,1,0,3)\}, \\ \{(0,0,0,0), (0,1,0,1), (1,0,0,2), (1,1,0,2), (1,2,1,1)\}, \\ \{(0,0,0,0), (0,1,0,3), (1,0,1,3), (1,1,1,1), (1,2,0,3)\}, \end{array} \quad \text{all mod } (\text{-},3,2,4).$$

1.4 Some Families of PBDs with Holes

1.38 Table Some results on specific families of PBDs with a specified hole size. See also Theorem 1.13.

Some Families of PBDs with a Given Hole Size					
$K \cup \{k^\star\}$	Congruence Classes	Range For v	Known Exceptions	Possible Exceptions	Ref.
$3, 4, 6^\star$	0,1 (mod 3)	≥ 6	7,9,10,12,13		[12]
$3, 5^\star$	5 (mod 6)	≥ 5	-	-	[11]
$4, 5, 11^\star$	2,3 (mod 4)	≥ 11	14, 15, 18, 19, 22, 23, 26, 27, 30, 31, 34	38, 42, 43, 46, 50, 54, 55, 58, 62, 66, 67, 70, 74, 78, 82, 90, 94, 98, 102, 106, 114, 115, 118, 126, 139, 142, 154, 166, 178, 190, 202, 214, 226	[2]
$4, 7^\star$	7,10 (mod 12)	≥ 7	10,19	-	[4]
$5, 9^\star$	9,17 (mod 20)	≥ 9	17,29	49	[13]
$5, 13^\star$	13 (mod 20)	≥ 13	33	-	[13]
$5, 17^\star$	9, 17 (mod 20)	≥ 17	29, 37, 49, 57	77, 89, 109, 129, 137, 149, 169, 189, 209, 229, 249, 269, 289	[3]
$5, 21^\star$	1,5 (mod 20)	≥ 21	25,41,45,61,65,81	125	[3]
$5, 25^\star$	1,5 (mod 20)	≥ 25	41,45,61,65,81,85	141	[3]
$5, 29^\star$	9, 17 (mod 20)	≥ 29	37, 49, 57, 69, 77, 89, 97, 109	137, 157, 177, 217, 237, 277, 337, 397, 417	[3]
$5, 33^\star$	13 (mod 20)	≥ 33	53,73,93,113	153	[3]

(Breaking up blocks). Let D be a K-PBD of order v which contains a flat F of order f. Suppose that there exists a PBD(F, K'), say D'. Then by replacing the blocks of F by the blocks of D' defined on the point-set of F, a $K \cup K'$-PBD of order v is obtained. In particular, if the flat F is a block B of D, then the design D' is *obtained by breaking up the block B*.

1.39 Remark This operation is useful for reducing the set of block sizes of a PBD. For example, let $K' = K \setminus \{k\}$, where $k \in K$. If there exists a K'-PBD of order k, and a K-PBD of order v, say D, then a K'-PBD of order v can be obtained by breaking up the blocks of size k in D.

1.5 See Also

§I.1	BIBDs are PBDs with one block size.
§II.2	TDs are special classes of GDDs.
§II.3	ITDs are special classes of IGDDs.
[14]	A large collection of PBDs and GDDs.

References

[1] R.D. Baker, An elliptic semiplane, *J. Combin. Theory A* 25 (1978), 193-195.

[2] F.E. Bennett, Concerning pairwise balanced designs with prime power block sizes, *Contemp. Math.* 3 (1990), 23-37.

[3] F.E. Bennett and J. Yin, Some results on $(v, \{5, w^\star\})$-PBDs, *J. Combin. Des.* 3 (1995), 455-468.

[4] A.E. Brouwer, Optimal packings of K_4's into K_n, *J. Combin. Theory A* 26 (1979), 278-297.

[5] A.E. Brouwer, A. Schrijver and H. Hanani, Group divisible designs with block size 4, *Discrete Math.* 20 (1977), 1-10.

[6] Y.M. Chee, C.J. Colbourn, R.P. Gallant and A.C.H. Ling, On a problem of Hartman and Heinrich concerning pairwise balanced designs with holes, *J. Combin. Math. Combin. Comput.*, to appear.

[7] C.J. Colbourn, Small group divisible designs with block size 3, *J. Combin. Math. Combin. Comput.* 14 (1993), 151-171.

[8] C.J. Colbourn, C.A. Cusack and D.L. Kreher, Partial Steiner triple systems with equal-sized holes, *J. Combin. Theory A* 70 (1995), 56-65.

[9] C.J. Colbourn, L. Haddad and V. Linek, Equitable embeddings of Steiner triple systems, *J. Combin. Theory A*, to appear.

[10] C.J. Colbourn, D.G. Hoffman, R. Rees, A new class of group divisible designs with block size three, *J. Combin. Theory A* 59 (1992), 73-89.

[11] M.K. Fort Jr. and G.A. Hedlund, Minimal coverings of pairs by triples, *Pacific J. Math.* 8 (1958), 709-719.

[12] R. Gallant and C. Pietsch, private communication.

[13] A.M. Hamel, W.H. Mills, R.C. Mullin, R. Rees, D.R. Stinson and J. Yin, The spectrum of PBD($\{5, k^\star\}, v$) for $k = 9$, 13, *Ars Combin.* 36 (1993), 7-26.

[14] H. Hanani, Balanced incomplete block designs and related designs, *Discrete Math.* 11 (1975), 255-369.

[15] A. Hartman and K. Heinrich, Pairwise balanced designs with holes, in: *Graphs, Matrices and Designs*, (R.S. Rees, ed.), Marcel Dekker, New York, 1993, 171-204.

[16] D.L. Kreher and D.R. Stinson, Small group divisible designs with block size 4, *J. Stat. Plann. Infer.*, to appear.

[17] R. Mathon, private communication.

[18] R. Rees and D.R. Stinson, On the existence of incomplete designs of block size four having one hole, *Utilitas Math.* 35 (1989), 119-152.

[19] R. Rees and D.R. Stinson, On resolvable group divisible designs with block size 3, *Ars Combin.* 23 (1987), 107-120.

[20] Wu Dianhu and Wei Ruizhong, A note on IGDDs, *J. Suzhou Univ.* 8 (1992), 142-145.

[21] L. Zhu, Some recent developments on BIBDs and related designs, *Discrete Math.* 123 (1993), 189-214.

2 PBDs: Recursive Constructions

RONALD C. MULLIN
HANS-DIETRICH O.F. GRONAU

2.1 Some Constructions

2.1 Theorem (*singular direct product* [11]) If there exists a PBD(v, K) which contains a flat of order f, and if there exists a K-GDD of type $(v-f)^m$, then there exists a PBD($m(v-f)+f, K$) which contains flats of orders v and f. (If $f = 0$, this is a product construction.)

2.2 Examples The construction of BIBDs.

1. A TD$(5,21)$ is a 5-GDD of type 21^5, and a PBD$(21,\{5\})$ exists, so there exists a PBD(105,{5}). This is a (105,5,1) design.
2. A TD$(4,21)$ is a 4-GDD of type 21^4, and a PBD$(25,\{4\})$ having a flat of size four exists, so a PBD(88,{4}) exists (an (88,4,1) design).
3. A TD$(4,25)$ and a (25,4,1) design give a (100,4,1) design.
4. A TD$(5,25)$ and a (25,5,1) design give a (125,5,1) design.
5. A TD$(4,28)$ and a (28,4,1) design give a (112,4,1) design.
6. A TD$(6,25)$ and a (31,6,1) design with flats of size 0 and 6 exist. So there are (156,6,1) and (186,6,1) designs.

2.3 Theorem (*singular direct product*, generalized from [11]) Suppose that there exists a K-GDD of type $g_1^{a_1}g_2^{a_2}\ldots g_s^{a_s}$.

1. If for each $g_i, i=1,2,\ldots s$, there exists a PBD(g_i+f,K) which contains a flat of order f, then there exists a PBD$((\sum_{i=1}^{s} a_i g_i)+f,K)$ which contains flats of order $g_i+f, i=1,2,\ldots,s$ and a flat of order f.
2. If for some j there exists exactly one group of size g_j in the GDD (that is, $a_j=1$), and if for all $i=1,2,\ldots,s$, $i\neq j$, there exists a PBD(g_i+f,K) which contains a flat of order f, and if there exists PBD(g_j+f,K), then there exists a PBD$((\sum_{i=1}^{s} a_i g_i)+f,K\cup\{g_j+f\})$.

2.4 Remark Theorems 2.1 and 2.3 are examples of constructions known as "filling in holes". The groups are considered as "holes" to be replaced by other objects. They require group divisible designs. The most important recursive construction for group divisible designs is Wilson's Fundamental Construction.

2.5 Theorem (*Wilson's Fundamental Construction* [11]) Let $(V,\mathcal{G},\mathcal{B})$ be a GDD (the *master* GDD) with groups $G_1,G_2\ldots G_t$. Suppose there exists a function $w: V\to\mathbb{Z}^+\cup\{0\}$ (*a weight function*) which has the property that for each block $B=\{x_1,x_2,\ldots,x_k\}\in\mathcal{B}$ there exists a K-GDD of type $[w(x_1),w(x_2)\ldots,w(x_k)]$ (such a GDD is an "ingredient" GDD). Then there exists a K-GDD of type

$$\left[\sum_{x\in G_1} w(x),\ \sum_{x\in G_2} w(x),\ldots,\ \sum_{x\in G_t} w(x)\right].$$

2.6 Remarks The block sizes of the master GDD do not necessarily occur among the block sizes of the constructed GDD. Most recursive constructions for pairwise balanced designs employ some application or variant of Theorem 2.5, often in imaginative ways.

2.7 Examples Wilson's Fundamental Construction for BIBDs.

1. Use a (41,5,1) design as the master design, and a {5}-GDD of type 4^5 as the ingredient GDD to produce a {5}-GDD of type 4^{41}. Fill the holes using one additional point, and a (5,5,1) design to get a (165,5,1) design.
2. Start with TD$(5,9)$ and delete four points from a group to get a $\{4,5\}$-GDD of type $9^4 5^1$. Use weight three on each point, and ingredient {4}-GDDs of types 4^4 and 4^5 to get a {4}-GDD of type $27^4 15^1$. Fill the holes using (28,4,1) and (16,4,1) designs to get a (124,4,1) design.

2.8 Remark In order to apply Wilson's Fundamental Construction, group divisible designs are required both as master designs and ingredients. Theorems 2.9, 2.10, 2.2, and 2.11 are used to construct group divisible designs, which are special types of pairwise balanced designs. Often the results are viewed directly as pairwise balanced designs.

2.9 Theorem If there exists a BIBD$(v, k, 1)$, then there exists a k-GDD of type $(k-1)^r$, where $r = (v-1)/(k-1)$.

2.10 Theorem Suppose that there exists a BIBD$(v, k, 1)$ which contains u mutually disjoint parallel classes. Then there exists a $\{k, k+1\}$-GDD of type $k^n w^1$ for $0 \le w \le u-1$, where $n = v/k$. Further if $u = (v-1)/(k-1)$ (that is, if the BIBD$(v, k, 1)$ is resolvable), then there exists a $(k+1)$-GDD of type $k^n (u-1)^1$.

2.2 PBDs Obtained by Truncation of TDs

2.11 Theorem (*truncation of groups of a transversal design* [8]) Let k be an integer, $k \ge 2$. Let $K = \{k, k+1, \ldots, k+s\}$. Suppose that there exists a TD$(k+s, m)$. Let $g_1, g_2, \ldots, g_s$ be integers satisfying $0 \le g_i \le m$, $i = 1, 2, \ldots, s$. Then there exists a K-GDD of type $m^k g_1 g_2 \cdots g_s$.

2.12 Remark The GDD in Theorem 2.11 is obtained by deleting points from a transversal design. Because of this, it is possible to see how the blocks of the GDD relate to its groups, a necessary requirement for the application of Wilson's Fundamental Construction. An example of a typical application is given in Example 2.13.

2.13 Example An application of Wilson's Fundamental Construction. There are $\{3,4\}$-GDDs of type 3^5 and $3^5 5^1$ respectively. (These are to be used as ingredient GDDs). Suppose that there exists a TD$(6, m)$. Let its groups be $G_1, G_2, \ldots, G_6$. Let t be an integer satisfying $0 \le t \le m$, and delete all but t points from group G_6 to obtain a master $\{5,6\}$-GDD of type $5^m t^1$. Now assign a weight of 5 to each point of the group of size t obtained from G_6, and assign a weight of 3 to all other points of the master GDD. Let $B = \{x_1, x_2, \ldots, x_5\}$ be a block of size 5 of the master GDD. It meets only the groups $G_1, G_2, \ldots G_5$. So the corresponding list of weights on the points of B is $[3,3,3,3,3]$, and there is an ingredient $\{3,4\}$-GDD of this type. Similarly, a block of size 6 of the master GDD meets all groups and therefore has a list of weights $[3,3,3,3,3,5]$, and again there is an ingredient $\{3,4\}$-GDD of this type. Thus the conditions required to apply Wilson's Fundamental Construction are satisfied. It yields a $\{3,4\}$-GDD of type $(3m)^5 (5t)^1$ in this case.

2.14 Theorem (*spike construction*) Let k be an integer, $k \ge 2$. Let $K = \{k, k+1, \ldots, k+s+1\}$. Suppose there exists a TD$(k+s+n, m)$ where n is a nonnegative integer. Let $g_1, g_2, \ldots, g_s$ be integers satisfying $1 \le g_i \le m$, $i = 1, 2, \ldots, s$. Then there exists a $K \cup \{k+s+n\}$-GDD of type $m^k g_1 g_2 \cdots g_s 1^n$.

2.15 Remarks Theorem 2.14 is similar to Theorem 2.11, except that certain points on a particular block are not deleted, hence the name "spike". Some of these points remain as groups of size one.

Theorem 2.14 is usually applied with $s = 0$ or 1. For $s = 1$, this yields the statement: If there exists a TD$(k+1+n, m)$ where n is a nonnegative integer, and if g satisfies $1 \le g \le m$, then there exists a $\{k, k+1, k+2, k+n+1\}$-GDD of type $m^k g^1 1^n$. For $s = 0$, if there exists a TD$(k+n, m)$, then there exists a $\{k, k+1, k+n\}$-GDD of type $m^k 1^n$.

2.16 Example The spike construction for $s = 1$. There exists a TD$(8, 9)$. Take $k = 6$. Then there exist $\{6,7,8\}$-GDD of type $9^6 g^1 1^1$, $1 \le g \le 9$.

2.17 Theorem [6] Let k be an integer, $k \ge 3$, and let $K = \{k-1, k, k+1, \ldots, k+s+1\}$. Suppose that there exists a TD$(k+s+n, m)$, where n is a positive integer. Let

$g_1, g_2, \ldots, g_s$ be integers satisfying $0 \le g_i \le m-1$, $i = 1, 2, \ldots, s$. Then there exists a K-GDD of type $(m-1)^k g_1^1 g_2^1 \cdots g_s^1 n^1$.

2.18 Remark Theorem 2.17 is usually applied in the case $s = 0$. In this case the result states that there exists a $\{k-1, k, k+1\}$-GDD of type $(m-1)^k n^1$. Further if $k+n = m+1$, then there exists a $\{k-1, k+1\}$-GDD of type $(m-1)^k n^1$.

2.19 Example Application of Theorem 2.17 with $n + k = m + 1$, and $s = 0$. Type is the group type $(m-1)^k (m-k+1)^1$.

k	m	$\{k-1, k+1\}$	Type
5	5	$\{4, 6\}$	$4^5 1^1$
	7		$6^5 3^1$
	8		$7^5 4^1$
	9		$8^5 5^1$
	11		$10^5 7^1$
6	7	$\{5, 7\}$	$6^6 2^1$
	8		$7^6 3^1$
	9		$8^6 4^1$
	11		$10^6 6^1$

k	m	$\{k-1, k+1\}$	Type
7	7	$\{6, 8\}$	$6^7 1^1$
	8		$7^7 2^1$
	9		$8^7 3^1$
	11		$10^7 5^1$
8	8	$\{7, 9\}$	$7^8 1^1$
	9		$8^8 2^1$
	11		$10^8 4^1$
9	9	$\{8, 10\}$	$8^9 1^1$
	11		$10^9 3^1$

2.20 Theorem [5] Suppose that there exists a TD(m, m). Let k be an integer satisfying $1 \le k \le m-1$. Then there exists an $\{m-1, k-1, k+1\}$-GDD of type $k^{m-1}(m-k)^1$.

2.21 Example Application of Theorem 2.20.

k	m	$\{m-1, k-1, k+1\}$	Group Types $k^{m-1}(m-k)^1$
5	7	$\{6, 4, 6\}$	$5^6 2^1$
	8	$\{7, 4, 6\}$	$5^7 3^1$
	9	$\{8, 4, 6\}$	$5^8 4^1$
	11	$\{10, 4, 6\}$	$5^{10} 6^1$
6	7	$\{6, 5, 7\}$	$6^6 1^1$
	8	$\{7, 5, 7\}$	$6^7 2^1$
	9	$\{8, 5, 7\}$	$6^8 3^1$
	11	$\{10, 5, 7\}$	$6^{10} 5^1$
7	8	$\{7, 6, 8\}$	$7^7 1^1$
	9	$\{8, 6, 8\}$	$7^8 2^1$
	11	$\{10, 6, 8\}$	$7^{10} 4^1$
8	9	$\{8, 7, 9\}$	$8^8 1^1$
	11	$\{10, 7, 9\}$	$8^{10} 3^1$

2.22 Theorem [5] Let k be a positive integer and let s be an integer satisfying $0 \le s \le k-1$. Suppose that there exists a TD$(k+n, m)$. Then there exists a $\{k-1, k, k+1, m-1, m\}$-GDD of type $k^{m-1}(n+s)^1$.

2.3 Constructions from Projective Planes and Geometries

Let A be a set of nonnegative integers, and let D be a PBD. Then a $\{w; A\}$ *arc*, (or A-arc of order w) in a PBD(v, K), say D, is a set of w points S of D such that if B is a block, then $|B \cap S| \in A$.

2.23 Remarks Many PBDs and GDDs can be obtained by deleting judiciously chosen sets of points from other PBDs. Most frequently these are sets of points in projective or affine spaces (See §VI.7). Where the results are obtained from deletion of points from desarguesian geometries, the results are stated in terms of q for q a prime or prime power.

Suppose that a projective plane of order n contains a $\{w; A\}$ arc. Then it also contains a complementary $\{n^2+n+1-w; n+1-A\}$ arc, where $n+1-A = \{n+1-a : a \in A\}$.

2.24 Example Let π be PG$(2,n)$ where $n = 2^k$. Then π contains a set $\mathcal{A}$ of $n+2$ points, no two collinear (an oval or hyperoval). $\mathcal{A}$ is a $\{0,2\}$-arc of order $n+2$ in π.

2.25 Example Let π be PG$(2,q)$ where q is odd. Then π contains a set $\mathcal{A}$ on $n+1$ points, no two collinear (an oval or conic). $\mathcal{A}$ is a $\{0,1,2\}$-arc of order $n+1$ in π.

2.26 Remark Let A be a set of positive integers, and let k be a positive integer. As in the case of PBDs, $\{w, A \cup k^\star\}$ arc is used to designate the case of an arc which has the property that at least one block of the PBD meets the arc in exactly k points. If $k \notin A$, then there is exactly one block with this property. In the case in which the set is displayed explicitly, if the displayed set contains entries k and $k^\star$, this indicates that at least one block meets the arc in exactly k points. If the symbol $k^\star$ appears without k, there is exactly one such block.

2.27 Theorem [6] Let q be an odd prime power. Then
1. there exists a $\{(q+1)/2, (q-1)/2\}$-GDD of type $((q-1)/2)^q$,
2. there exists a $\{(q+1)/2, (q+3)/2\}$-GDD of type $((q+1)/2)^q$.

2.28 Theorem (*subfield subplanes*) Let q be a prime power. Let a and b be positive integers such that b divides a. Then PG$(2,q^a)$ contains a subplane PG$(2,q^b)$. (If $a = 2b$, then the subplane is a *Baer subplane*; see §VI.7.7).

2.29 Remarks Subfield subplanes give rise to $\{q^{2b}+q^b+1; \{0,1,q^b+1\}\}$ arcs in PG$(2,q^a)$ when $b \mid a$. When $a = 2b$, the arc has no lines of size zero. These Baer subplanes are the source of several constructions, including Theorems 2.30 and 2.31.

2.30 Theorem (Wallis, see [10]) If q is a prime power, then there exists a $(q+1)$-GDD of type $(q^2-q)^{q+1}(q^2)^1$.

2.31 Theorem (*Brouwer construction* [2]) Let q be a prime power, and let t be an integer satisfying $0 < t < q^2-q+1$. Then there exists a $\{t, q+t\}$-GDD of type t^{q^2+q+1}; that is, there exists a PBD$(t(q^2+q+1), \{t, t+q\})$.

2.32 Remark In the PBD constructed in Theorem 2.31, the blocks of size t can be partitioned into $q^2-q+1-t$ parallel classes. By adjoining new points to these classes, Theorem 2.31 extends to Theorem 2.33.

2.33 Theorem (*extended Brouwer construction* [2]) Let q be a prime power, and t be an integer satisfying $0 < t < q^2-q+1$. If u is an integer satisfying $0 \le u \le q^2-q-t$, then there exists a $\{t, t+1, q+t\}$-GDD of type $t^{q^2+q+1}\, u^1$. Therefore, for $0 \le u \le q^2-q+1-t$, there exists an incomplete PBD of order $t(q^2+q+1)+u$ and blocks from the set $\{t, t+1, t+q\}$ having a hole of size u. If $u = q^2-q+1-t$, then no block of size t occurs in the incomplete PBD. If $u = q^2-q-t$, then no block of size t occurs in the GDD.

2.34 Corollary If q is a prime power, then there exists a $\{q^2, q^2-q\}$-GDD of type $(q^2-q)^{q^2+q+1}$. This corresponds to the case of $t = q^2-q$ (and $u = 0$) in Theorem 2.33.

2.35 Example Applications of the extended Brouwer construction of a $\{t, t+1, q+t\}$-GDD of type $t^{q^2+q+1} u^1$ with $u = q^2 - q - t$.

q	q^2-q	t	q^2-q-t	$\{t+1, q+t\}$	GDD Type
3	6	6	0	$\{7,9\}$	6^{13}
3	6	5	1	$\{6,8\}$	$5^{13}1^1$
3	6	4	2	$\{5,7\}$	$4^{13}2^1$
3	6	3	3	$\{4,6\}$	$3^{13}3^1$
3	6	2	4	$\{3,5\}$	$2^{13}4^1$
3	6	1	5	$\{2,4\}$	$1^{13}5^1$
4	12	12	0	$\{13,16\}$	12^{21}
4	12	11	1	$\{12,15\}$	$11^{21}1^1$
4	12	10	2	$\{11,14\}$	$10^{21}2^1$
4	12	9	3	$\{10,13\}$	$9^{21}3^1$
4	12	8	4	$\{9,12\}$	$8^{21}4^1$
4	12	7	5	$\{8,11\}$	$7^{21}5^1$
4	12	6	6	$\{7,10\}$	$6^{21}6^1$
4	12	5	7	$\{6,9\}$	$5^{21}7^1$
4	12	4	8	$\{5,8\}$	$4^{21}8^1$
4	12	3	9	$\{4,7\}$	$3^{21}9^1$
4	12	2	10	$\{3,6\}$	$2^{21}10^1$
4	12	1	11	$\{2,5\}$	$1^{21}11^1$
5	20	20	0	$\{21,25\}$	20^{31}
5	20	19	1	$\{20,24\}$	$19^{31}1^1$
5	20	18	2	$\{19,23\}$	$18^{31}2^1$
5	20	17	3	$\{18,22\}$	$17^{31}3^1$
5	20	16	4	$\{17,21\}$	$16^{31}4^1$
5	20	15	5	$\{16,20\}$	$15^{31}5^1$
5	20	14	6	$\{15,19\}$	$14^{31}6^1$
5	20	13	7	$\{14,18\}$	$13^{31}7^1$
5	20	12	8	$\{13,17\}$	$12^{31}8^1$
5	20	11	9	$\{12,16\}$	$11^{31}9^1$
5	20	10	10	$\{11,15\}$	$10^{31}10^1$
5	20	9	11	$\{10,14\}$	$9^{31}11^1$
5	20	8	12	$\{9,13\}$	$8^{31}12^1$
5	20	7	13	$\{8,12\}$	$7^{31}13^1$
5	20	6	14	$\{7,11\}$	$6^{31}14^1$
5	20	5	15	$\{6,10\}$	$5^{31}15^1$
5	20	4	16	$\{5,9\}$	$4^{31}16^1$
5	20	3	17	$\{4,8\}$	$3^{31}17^1$
5	20	2	18	$\{3,7\}$	$2^{31}18^1$
5	20	1	19	$\{2,6\}$	$1^{31}19^1$

2.36 Theorem (*Brouwer's Singer set construction*) [2] Let D be a Singer difference set for a projective plane π of order q (see §IV.12.2). Let $v = q^2+q+1$, and let $S = \{0, 1, \cdots, v-1\}$ be the representatives of $\mathbb{Z}_v$ used to represent and develop the difference set. Let u be a proper divisor of v, say $v = um$. Let $X = \{0, m, 2m, \cdots, v-m\}$; so that $|X| = u$. Then X is a $\{m; A\}$ arc in π for some A, where clearly A contains no more than m distinct integers.

2.37 Remarks Theorem 2.36 has been extended by Greig [6] to multiple divisors of v. Theorem 2.38 is a specific application of Theorem 2.36.

2.38 Theorem (*Brouwer's 1 mod 3 construction*) [2] Let $q \equiv 1 \pmod 3$ be a prime power. Let $u = (q^2 + q + 1)/3$. Then there exists a PBD$(u, \{k_0, k_1, k_2\})$ where k_0, k_1, and k_2 are a solution of $k_0 + k_1 + k_2 = q + 1$ and $k_0^2 + k_1^2 + k_2^2 = q + u$.

2.39 Example PBD$(u, \{k_0, k_1, k_2\})$ via applications of Brouwer's 1 mod 3 construction.

q	7	13	16	19	25	31	37
u	19	61	91	127	217	331	469
$\{k_0, k_1, k_2\}$	$\{1,3,4\}$	$\{3,4,7\}$	$\{3,7,7\}$	$\{4,7,9\}$	$\{7,7,12\}$	$\{7,12,13\}$	$\{9,13,16\}$

2.40 Remark The line flip method, in which the points of a block transverse to an arc in a projective plane or other design are replaced by the complementary set, or vice versa, is a powerful technique for creating new PBDs.

2.41 Theorem [6] If a projective plane of order q contains a $(v; K \cup k^\star)$ arc, and if t is an integer satisfying $0 \le t \le q+1-k$, then the plane contains a $\{v+t; K \cup (K+1) \cup (k+t)^\star\}$ arc, where $K + 1 = \{k + 1 : k \in K\}$.

2.42 Theorem [6] If a projective plane of order q contains a $\{v; K \cup k^\star\}$ arc, then the plane contains a $\{v + q + 1 - 2k;\ (K - 1) \cup (K + 1) \cup (q + 1 - k)^\star\}$ arc, where $K + 1 = \{k + 1 : k \in K\}$, and $K - 1 = \{k - 1 : k \in K, k > 0\}$.

2.43 Theorem [6] Suppose that a projective plane of order q contains a $\{0, a\}$ arc of order w. Let $s = w/a$. Then

1. (secant line flip) there exists a $\{a - 1, a + 1\}$-GDD of type $a^{s-1}(q - a)^1$,
2. (external line flip) there exists an $(a + 1)$-GDD of type $a^s q^1$.

2.44 Theorem (Denniston) [6] There exists a $\{0, 2^n\}$ arc of order $w = 2^{2n+m} - 2^{n+m} + 2^n$ in PG$(2, 2^{n+m})$. The $\{0, 2^n\}$ arc contains a $\{0, 2^\ell\}$ Denniston arc for all $\ell \le n$.

2.45 Corollary [6] Let m and n be positive integers.

1. There exists a $(2^n - 1, 2^n + 1)$-GDD of type $(2^n)^{2^m(2^n-1)}(2^{m+n} - 2^n)^1$.
2. There exists a $(2^n + 1)$-GDD of type $(2^n)^{2^{m+n}-2^m+1}(2^{m+n})^1$.

2.46 Theorem (*the $q - x$ construction*) [6] If there is a projective plane of order q, and if x is an integer satisfying $0 \le x \le q - 1$, then there exists a $\{q - x - 1, q - x + 1\}$-GDD of type $(q - 1)^{q-x}(x + 1)^1$.

2.47 Example Applications of the $q - x$ construction. $K = \{q - x - 1, q - x + 1\}$.

q	x	K	Type	q	x	K	Type	q	x	K	Type
7	1	$\{5,7\}$	$6^6 2^1$	7	2	$\{4,6\}$	$6^5 3^1$				
8	1	$\{6,8\}$	$7^7 2^1$	8	2	$\{5,7\}$	$7^6 3^1$	8	3	$\{4,6\}$	$7^5 4^1$
9	1	$\{7,9\}$	$8^8 2^1$	9	2	$\{6,8\}$	$8^7 3^1$	9	3	$\{5,7\}$	$8^6 4^1$
9	4	$\{4,6\}$	$8^5 5^1$								
11	1	$\{9,11\}$	$10^{10} 2^1$	11	2	$\{8,10\}$	$10^9 3^1$	11	3	$\{7,9\}$	$10^8 4^1$
11	4	$\{6,8\}$	$10^7 5^1$	11	5	$\{5,7\}$	$10^6 6^1$	11	6	$\{4,6\}$	$10^5 7^1$
13	1	$\{11,13\}$	$12^{12} 2^1$	13	2	$\{10,12\}$	$12^{11} 3^1$	13	3	$\{9,11\}$	$12^{10} 4^1$
13	4	$\{8,10\}$	$12^9 5^1$	13	5	$\{7,9\}$	$12^8 6^1$	13	6	$\{6,8\}$	$12^7 7^1$
13	7	$\{5,7\}$	$12^6 8^1$	13	8	$\{4,6\}$	$12^5 9^1$				
16	1	$\{14,16\}$	$15^{15} 2^1$	16	2	$\{13,15\}$	$15^{14} 3^1$	16	3	$\{12,14\}$	$15^{13} 4^1$
16	4	$\{11,13\}$	$15^{12} 5^1$	16	5	$\{10,12\}$	$15^{11} 6^1$	16	6	$\{9,11\}$	$15^{10} 7^1$
16	7	$\{8,10\}$	$15^9 8^1$	16	8	$\{7,9\}$	$15^8 9^1$	16	9	$\{6,8\}$	$15^7 10^1$
16	10	$\{5,7\}$	$15^6 11^1$	16	11	$\{4,6\}$	$15^5 12^1$				

Let p be a prime. Then a *(p,a,b)-Baer configuration* in $\pi = PG(2,p^a)$ is a set X of $p^a + p^b + 1$ points which has the properties that, restricting the lines of π to the points of X,

1. one point of X, the *focus*, lies on $p^{a-b}+1$ lines each of size p^b+1, and
2. the remaining points each lie on one line of size p^b+1 and p^b lines each of size $p^{a-b}+1$.

2.48 Remark By removing the focus, a $\mathrm{TD}(p^{a-b}+1, p^b)$ is obtained. This is of itself not remarkable, since p is prime. However the embedding of the configuration into $\mathrm{PG}(2,p^a)$ (and in particular, an associated transversal design) is remarkable. This is partly due to the fact that every line of $\mathrm{PG}(2,p^a)$ meets the Baer configuration in at least one point, so the configuration is a $\{1, p^b+1, p^{a-b}+1\}$ arc in the plane.

2.49 Theorem (Ostrom) [3] The plane $\mathrm{PG}(2,p^a)$ contains a (p,a,b)-Baer configuration if p is prime and $a-b$ is a divisor of a.

2.50 Remark To discuss the embedding of the configuration in the TD (in Theorem 2.49), the notion of a thwart is required.

Let k and x be nonnegative integers, and let $\mathcal{I} = \{i_1, i_2, \ldots, i_s\}$ be a set of integers satisfying $0 \le i_1 < i_2 < \cdots < i_s \le x$. Further suppose that there are integers s_i satisfying $0 \le s_1 \le s_2 \le \cdots \le s_x \le n$. Let $(\mathcal{X}, \mathcal{G}, B)$ be a $\mathrm{TD}(k+x, n)$ with $\mathcal{G} = \{G_1, G_2, \cdots, G_k, H_1, H_2, \cdots, H_x\}$. Then an $(x, \mathcal{I}, s_1, s_2, \ldots, s_x)$-*thwart* is a set $S = \cup_{j=1}^{x} S_j$, where $S_j \subseteq H_j$ with $|S_j| = s_j$ for $1 \le j \le x$, such that for every $B \in \mathcal{B}$, $|B \cap S| \in \mathcal{I}$.

2.51 Theorem [3] Let p be a prime, and let a and b be integers satisfying $a \ge b \ge 1$ and $(a-b)|a$. Then there is a $\mathrm{TD}(p^a+1, p^a)$

1. containing a $(p^{a-b}+1, \{1, p^{a-b}+1\}, p^b, \ldots p^b)$-thwart and
2. a $(p^{a-b}+1, \{0, p^{a-b}\}, p^a - p^b, \ldots, p^a - p^b)$-thwart.

(These arise by deleting the focus of the Baer configuration from the plane.)

2.52 Example A PBD(53,{9,8,7,5}) can be obtained from part 1 of Theorem 2.51 as follows. Take $a = 3$, $b = 2$, and $p = 2$ to obtain a $\mathrm{TD}(9,8)$ containing (3, {1,3}, 4,4,4)-thwart. Delete a group which does not meet the thwart to obtain a $\mathrm{TD}(8,8)$ containing a $(3, \{1,3\}, 4, 4, 4)$-thwart, then delete the points of the thwart to obtain a {8,7,5}-GDD of type $8^5 4^3$. Now adjoin a new point, say ∞, to the groups of this GDD to obtain the PBD(53,{9,8,7,5}).

2.53 Example An application of thwarts employing Wilson's Fundamental Construction. Let D be the $\mathrm{TD}(9,8)$ containing a (3,{1,3},4,4,4)-thwart, and let S denote the points of the thwart. Let the groups of D be G_i, $i = 1, 2, \ldots, 9$, and let $G_1, G_2, \ldots, G_6$ be the groups which do not meet S. Let $G_i = N_i \cup T_i$, $i = 7, 8, 9$ where $T_i = G_i \cap S$, so $|T_i| = 4, i = 7, 8, 9$. For ingredient GDDs, by the table of 4-GDDs there exist 4-GDDs of type $3^8 0^1$ and 3^9. There exists a resolvable BIBD (15,3,1) and BIBD (21,3,1), respectively. By completing the former by adjoining 7 new points, then deleting one of the old points, a $\{4,7\}$-GDD of type $3^7 0^2$ is obtained, whereas by adjoining 9 new points to the latter, a 4-GDD of type $3^7 9^1 0^1$ is obtained. Further by deleting a point from a $\mathrm{TD}(4,7)$, a $\{4,7\}$-GDD of type $3^7 6^1 0^1$ is obtained, and by adjoining a new point to the groups of a $\mathrm{TD}(4,7)$ then deleting another point, a $\{4,8\}$-GDD of type $3^7 7^1 0^1$ is obtained.

Now assign weights to the points of D as follows. To each point in G_i, $i = 1, 2, \ldots, 6$, assign a weight of 3. To each point of $N_i, i = 7, 8$, also assign a weight of 3 and assign a weight of 0 to each point $T_i, i = 7, 8$. To each point of T_9 also assign a weight of 3. Let x_1, x_2, x_3, and x_4 be nonnegative integers. To x_1, x_2, x_3, and x_4 points of N_9, assign weights of 0, 3, 7, and 9 respectively.

A block of D is a *tangent* if it meets the thwart in one point, and a *transversal* if it meets the thwart in three points. Each transversal has 2 points of weight 0 and 7 points of weight 3, (that is, it has type $3^7 0^2$). If the block is a tangent meeting T_9 it is of type 3^9, otherwise it is of type $3^7 w^1 0^1$, where w is the weight of the point in which it meets N_9, that is, $w \in \{0, 6, 7, 9\}$. Since there is a {4,7,8}-GDD of each of these types, then one can apply Wilson's Fundamental Construction to obtain a {4,7,8}-GDD of type $24^6 12^2 (12 + 6x_2 + 7x_3 + 9x_4)$ where $0 \leq x_2 + x_3 + x_4 \leq 4$. Since there exists a BIBD$(v, 4, 1)$ for $v \equiv 1$ or $4 \pmod{12}$, then there exists a PBD$(v + f, \{4\})$ with a flat of order f, for $v \in \{12, 24\}$ and $f \in \{1, 4\}$. Thus by applying the generalized singular product, one obtains PBD$(180{+}6x_2{+}7x_3{+}9x_4{+}f, \{4, 7, 8, (12{+}6x_2{+}7x_3{+}9x_4{+}f)^\star\})$ for any collection of x_2, x_3, x_4 satisfying $0 \leq x_2 + x_3 + x_4 \leq 4$ and $f \in \{1, 4\}$. If in addition $w = 12 + 6_2 + 7x_3 + 9x_4 + f$, and if there exists a PBD$(w, \{4, 7, 8\})$, then there exists a PBD$(168 + w, \{4, 7, 8\})$.

2.54 Remark A (p, a, b)-Baer configuration induces an arc in the dual plane. The corresponding PBD is noted in Theorem 2.55.

2.55 Theorem [3] Let p be a prime, and let a and b be integers satisfying $a \geq b \geq 1$ and $(a - b)|a$. Then PG$(2, p^a)$ contains a $\{p^a, p^a - p^{2b-a} + 1\}$-GDD of type $(p^a - p^b)^{p^a+p^b}$ $(p^a - p^{a-b})^1$.

2.56 Remark Interesting thwarts also arise from other structures in planes such as ovals, hyperovals, and Denniston arcs. See [3] and [4]. The idea of a thwart extends naturally to configurations of GDDs.

2.4 Constructions Involving Incomplete Objects

2.57 Theorem (*Wilson-type theorem for incomplete GDDs*) [7] Let K and L be sets of positive integers and let there exist an L-IGDD of type $(g_1, h_1)^{a_1} (g_2, h_2)^{a_2} \ldots (g_s, h_s)^{a_s}$. For some positive integer n and for each $\ell \in L$, suppose there exists a K-GDD of type n^ℓ. Then there exists a K-IGDD of type $(ng_1, nh_1)^{a_1} (ng_2, nh_2)^{a_2} \ldots (ng_s, nh_s)^{a_s}$.

2.58 Remarks Theorem 2.57 can be generalized to use arbitrary weights on the master GDD. However the form given appears to be the most useful. Theorems 2.59 and 2.60 are further examples of "filling in holes". By making various changes, a host of variations is possible.

2.59 Theorem [9] Let K be a set of positive integers and suppose that there exists a K-IGDD of type $(g_1, h_1)^{a_1} (g_2, h_2)^{a_2} \ldots (g_s, h_s)^{a_s}$. Let q be a positive integer. Suppose that for $i = 1, 2, \ldots, s$, there exists a K-IPBD of order $g_i + q$ with a hole of $h_i + q$. Then there exists a K-IPBD of order $v = (\sum_{i=1}^{s} a_i g_i) + q$ with a hole of order $h = (\sum_{i=1}^{s} a_i h_i) + q$, that is, a PBD$(v, K \cup \{h^\star\})$.

2.60 Theorem (*Stinson's diamond construction*) [12] Let K be a set of positive integers. Let a and b be integers satisfying $b \geq a \geq 0$. Suppose that there exist

1. a K-IGDD of type $(t_1, u_1)(t_2, u_2), \ldots, (t_n, u_n)$,
2. a $(t_i + b; u_i + a, b; a; K) - \diamondsuit$-IPBD for $1 \leq i \leq n - 1$; (§III.1.2), and
3. a $(t_n + b, u_n + a; K)$-IPBD.

Then there exists a $(t + b, u + a; K)$-IPBD, where $t = \sum_{i=1}^{n} t_i$ and $u = \sum_{i=1}^{n} u_i$.

2.5 Other Constructions

Let m, k, λ, and u be positive integers. A *modified group divisible design* is a quadruple $(V, \mathcal{G}, \mathcal{R}, \mathcal{B})$ where V is a set of cardinality μ, $\mathcal{G}$ is a partition of V into u parts (*groups*) of size m, $\mathcal{R}$ is a partition of V into m parts (*rows*) of size u, and $\mathcal{B}$ is a family of subsets of V which satisfy the properties:

1. If $B \in \mathcal{B}$, then $|B| = k$;
2. Each row meets every group in precisely one element of V;
3. Let $\{x, y\}$ be a pair of distinct elements of V. If x and y occur in the same group or same row, then $\{x, y\}$ occurs in no block of $\mathcal{B}$; otherwise, $\{x, y\}$ occurs in λ blocks of $\mathcal{B}$.

A modified group divisible design D is *resolvable* if the blocks of D can be partitioned into parallel classes containing each point of V exactly once.

2.61 Remarks Modified group divisible designs and resolvable group divisible designs have been used to create group divisible designs, pairwise balanced designs, and other designs. See [1].

A very general construction for GDDs and IGDDs of general index λ was originated by Stinson [9] and used in the form of double group divisible designs (DGDD). A DGDD can be thought of as a GDD with missing sub-GDDs, and is therefore an extension of an IGDD. (An example is the configuration used in Stinson's diamond construction). See Zhu [12].

2.6 See Also

§I.1	BIBDs are PBDs with one block size.
§II.2	TDs are special classes of GDDs.
§II.3	ITDs are special classes of IGDDs.

References

[1] A.M. Assaf, An application of modified group divisible designs, *J. Combin. Theory A* 68 (1994), 152-168.

[2] A.E. Brouwer, A series of separable designs with application to pairwise orthogonal latin squares, *Europ. J. Combin.* 1 (1990), 39-41.

[3] C.J. Colbourn, J.H. Dinitz and D.R. Stinson, More thwarts in transversal designs, *Finite Fields Appl.*, to appear.

[4] C.J. Colbourn, J.H. Dinitz and M. Wojtas, Thwarts in transversal designs, *Des. Codes Crypt.* 5 (1995), 189-197.

[5] M. Greig, Some balanced incomplete block design constructions, *Congr. Numer.* 77 (1990), 121-134.

[6] M. Greig, Designs from projective planes and PBD bases, unpublished.

[7] A.M. Hamel, W.H. Mills, R.C. Mullin, R. Rees, D.R. Stinson and J. Yin, The spectrum of PBD($\{5, k^\star\}, v$) for $k = 9$, 13, *Ars Combin.* 36 (1993), 7-26.

[8] H. Hanani, Balanced incomplete block designs and related designs, *Discrete Math.* 11 (1975), 255-369.

[9] D.R. Stinson, A general construction for group divisible designs, *Discrete Math.* 33 (1981), 89-94.

[10] W.D. Wallis, Configurations arising from maximal arcs, *J. Combin. Theory A* 15 (1973), 115-119.

[11] R.M. Wilson, An existence theory for pairwise balanced designs I: Composition theorems and morphisms, *J. Combin. Theory A* 13 (1971), 220-245.
[12] L. Zhu, Some recent developments on BIBDs and related designs, *Discrete Math.* 123 (1993), 189-214.

3 PBD-Closure

Frank E. Bennett
Hans-Dietrich O.F. Gronau
Alan C.H. Ling
Ronald C. Mullin

3.1 Necessary Conditions

3.1 **Theorem** (Drake and Larson [10]) Let K be a set of positive integers and let m denote the smallest integer in K. Suppose that there exists a PBD(v, K) which contains blocks B_h and B_k of sizes h and k, respectively. Then

1. $v \geq (m-1)k + h - m + 1$; hence
2. $v \geq (m-1)k+1$, with equality if and only if there exists a resolvable BIBD$(k(m-2)+1, m-1, 1)$;
3. if B_h and B_k do not intersect, then $v \geq (m-1)k + h$.

A set S of positive integers is *PBD-closed* if the existence of a PBD(v, S) implies that v belongs to S.

Let K be a set of positive integers and let $B(K) = \{v : \exists \text{PBD}(v, K)\}$. Then $B(K)$ is the PBD-*closure* of K. If $K = \{k\}$, the notation $B(k)$ is used.

Let K be a set of positive integers. Define $\alpha(K) = gcd\{k-1 : k \in K\}$ and $\beta(K) = gcd\{k(k-1) : k \in K\}$.

3.2 **Lemma** [26] Necessary conditions for the existence of a PBD(v, K) are

$$v - 1 \equiv 0 \pmod{\alpha(K)}, \qquad \text{and} \qquad v(v-1) \equiv 0 \pmod{\beta(K)}.$$

3.3 **Theorem** [26] The necessary conditions for the existence of a PBD(v, K) are asymptotically sufficient. More precisely, let K be a PBD-closed set. Then there exists a constant $v_0 = v_0(K)$ such that if $v > v_0$ and v satisfies the conditions

$$v - 1 \equiv 0 \pmod{\alpha(K)}, \qquad \text{and} \qquad v(v-1) \equiv 0 \pmod{\beta(K)},$$

then there exists a PBD(v, K).

3.2 Closures

3.4 **Theorem** [28] The asymptotic sufficiency of the necessary conditions for the existence of PBD(v, K) can be restated in terms of PBD-closed sets as follows. Let K be a PBD-closed set. Then K is *eventually periodic* with period $\beta(K)$; that is, there exists a constant $C(K)$ such that for every $k \in K$, $\{v : v \geq C(K), v \equiv k \pmod{\beta(K)}\} \subseteq K$.

3.5 Remark Theorem 3.4 makes the theory of PBD-closure useful. A few examples of a certain set of objects can prove sufficient to establish the existence of an entire set of objects, as is illustrated in Example 3.6.

3.6 Example By breaking up blocks, the set $S = \{v : \exists \text{ a BIBD}(v, 4, 2)\}$ is PBD-closed. A necessary condition for the existence of a BIBD$(v, 4, 2)$ is $v \equiv 1 \pmod 3$. A BIBD(4,4,2) can be obtained by taking the block {1,2,3,4} twice. (It is degenerate since $v = k$, but it is sufficient for our purposes here.) A BIBD(7,4,2) is obtained by developing the set $\{0, 1, 2, 4\}$ modulo 7. Now $\gcd(7 \cdot 6, 4 \cdot 3) = 6$, so $\beta(S)|6$. Also $\{4, 7\} \subset S$. Therefore S is eventually periodic, with period 3, so for sufficiently large $v \equiv 1 \pmod 3$, there exists a BIBD$(v, 4, 2)$, simply by the closure theorem. However, it is known that $B(\{4, 7\}) = \{v : v \equiv 1 \pmod 3), v \neq 10, 19\}$, so that there exists a BIBD$(v, 4, 2)$ for all positive $v \equiv 1 \pmod 3$ with the possible exception of $v = 10$ or $v = 19$. (It is for this reason that the study of closures is important). Since it is known that there exists a BIBD$(v, 4, 2)$ for $v = 10$ and $v = 19$, then $S = \{v : v \equiv 1 \pmod 3)\}$.

3.7 Theorem [26] The primitive (or minimal) eventual period of a PBD-closed set is either $\beta(K)$ or $\frac{1}{2}\beta(K)$, and the latter is possible only if $\beta(K) \equiv 2 \pmod 4$.

3.8 Remark Every PBD-closed set $K \neq \{1\}$ is infinite.

3.9 Examples Examples of PBD-closed sets.

- $B(k, \lambda) = \{v : \exists \text{ a BIBD}(v, k, \lambda)\}$.
- $R(k) = \{r : \exists \text{ a BIBD}(v, k, 1) \text{ with replication number } r\}$.
- $R^*(k) = \{r : \exists \text{ a resolvable BIBD}(v, k, 1) \text{ with replication number } r\}$.
- $U(K, u) = \{t : \exists \text{ a } K\text{-GDD of type } u^t\}$, where K is a set of positive integers, and u is a fixed integer such that there exists a K-GDD of type u^t.
- $S = \{s : \exists \text{ a Room square of side } s\}$. (See §IV.39).
- $S = \{s : \exists \text{ a self-orthogonal latin square of order } s\}$. (See §IV.40.)

3.10 Example A set which is not PBD-closed: $RB(k) = \{v : \exists \text{ a resolvable BIBD}(v, k, 1)\}$.

3.11 Theorem [11] Let n be an integer, $n \geq 3$. Let $P_{1,n} = \{s : s \equiv 1 \pmod n), s = p^t$ for some prime p and positive integer $t\}$. Then $B(P_{1,n})$ is eventually periodic with minimum period n; that is, there exists a constant C_n such that if $v > C_n$ and $v \equiv 1 \pmod n$, then $v \in B(P_{1,n})$.

3.12 Example [11] The use of Theorem 3.11. A *Hering configuration* of type n and order $v = sn + 1$ is a decomposition of the complete directed graph DK_{sn+1} into a family $H = \{H_1, H_2, \ldots, H_{sn+1}\}$ of spanning directed subgraphs, each consisting of s directed n-cycles and an isolated vertex, which has the property that any pair H_i, $H_j, i \neq j$ share precisely one edge in common when viewed as undirected graphs, but that the edge is oriented in opposite directions in H_i and H_j. Let $S_n = \{v : \exists$ a Hering configuration of type n and order $v\}$. A necessary condition for $v \in S_n$ is that $v \equiv 1 \pmod n$. It is known that S_n is PBD-closed. It is also known that if $q = sn+1$ is a prime power, then there exists a Hering configuration of type n and of order q. In view of Dirichlet's theorem, which guarantees the existence of a prime $p \equiv 1 \pmod n$, and Theorem 3.11, it follows that for all fixed n and sufficiently large $v \equiv 1 \pmod n$, there exists a Hering configuration of type n and order v. In fact, as shown in Table 3.17, $B(\{4, 7\}) = \{v : v \equiv 1 \pmod 3), v \neq 10, 19\}$. Since 19 is a prime, there is a Hering configuration of type 3 and order 19. Further, it can be shown that there is no Hering configuration of type 3 and order 10. Hence, there is a Hering configuration of type 3 and order $v \equiv 1 \pmod 3$ if and only if $v \geq 4$ and $v \neq 10$. Similarly it is shown in Table 3.17 that $B(\{5,9,13,17,29,33\}) = \{n : n \geq 5,\ n \equiv 1 \pmod 4\}$. Since $\{5,9,13,17,29\}$ is a set of prime powers and since a Hering configuration of type 4 and order 33 can be constructed, there is a Hering configuration for all $v \equiv 1 \pmod 4$, $v \geq 5$.

Let K be a PBD-closed set. Let G be a set of positive integers such that $K = B(G)$. Then G is a *generating set* for K.

Let K be a PBD-closed set. An element $x \in K$ is *essential* in K if and only if $x \notin B(K \backslash \{x\})$, that is; if and only if there does not exist a PBD$(x, K\backslash\{x\})$.

3.13 Theorem [26] Let E_K denote the set of all essential elements in a PBD-closed set K. Then E_K is the unique minimal generating set for K. Moreover, E_K is finite.

3.14 Theorem [26] Let K be a PBD-closed set. Suppose that $G \subseteq K$. Then G is a generating set for K if and only if $E_K \subseteq G$.

3.15 Remark There is some confusion in the use of language of bases and generating sets. Some authors refer to any generating set for a set K as a basis, and speak of E_K as the minimal basis. Others use the term "generating set" as defined and speak of E_K as being the minimal generating set. Still others use the term "generating set" as defined, but refer to the minimal set E_K as a basis. Here, basis refers to the set E_K.

3.16 Remark There are two approaches to PBD-closed sets. One approach is to be given a set of positive integers K and to determine $B(K)$ as accurately as possible. Alternatively one can select a PBD-closed set, and determine as small a generating set as possible. Tables 3.17, 3.18, and 3.19 reflect both approaches.

3.17 Table Closures of subsets. Let K be the set whose closure is being tabulated. Let $m = \min(K)$. In the exceptions, nonnegative integers less than m are omitted, since 0 and 1 are always present and the remaining integers are always absent. Braces and commas are omitted from sets. $\mathbb{N}$ is used when the closure contains all sufficiently large integers.

The precise closure of all subsets of $\{3,4,\ldots,10\}$ which contain the element 3 are exhibited first [14]. Since $\{7,9\} \subset B(3)$, it is not necessary to include 7 or 9 in the list of sets whose closures are given. Similarly since $10 \in B(\{3,4\})$, it is not necessary to include 10 when 3 and 4 are both present.

Subset	Closure	**Genuine Exceptions**
3	1 3 mod 6	–
3 4	0 1 mod 3	**6**
3 5	1 mod 2	–
3 6	0 1 mod 3	**4 10 12 22**
3 8	$\mathbb{N}$	**4 5 6 10 11 12 14 16 17 18 20 23 26 28 29 30 34 35 36 38**
3 10	0 1 mod 3	**4 6 12 16 18 22 24 34 36 42**
3 4 5	$\mathbb{N}$	**6 8**
3 4 6	0 1 mod 3	–
3 4 8	$\mathbb{N}$	**5 6 11 14 17**
3 5 6	$\mathbb{N}$	**4 8 10 12 14 20 22**

3 5 8	$\mathbb{N}$	**4 6 10 12 14 16 18 20 26 28 30 34**
3 5 10	$\mathbb{N}$	**4 6 8 12 14 16 18 20 22 24 26 32 34 36 38 42 44**
3 6 8	$\mathbb{N}$	**4 5 10 11 12 14 17 20 23**
3 6 10	0 1 mod 3	**4 12 22**
3 8 10	$\mathbb{N}$	**4 5 6 11 12 14 16 17 18 20 23 26 29 35 36**
3 4 5 6	$\mathbb{N}$	**8**
3 4 5 8	$\mathbb{N}$	**6**
3 4 6 8	$\mathbb{N}$	**5 11 14 17**
3 5 6 8	$\mathbb{N}$	**4 10 12 14 20**
3 5 6 10	$\mathbb{N}$	**4 8 12 14 20 22**
3 5 8 10	$\mathbb{N}$	**4 6 12 14 16 18 20 26**
3 6 8 10	$\mathbb{N}$	**4 5 11 12 14 17 20 23**
3 4 5 6 8	$\mathbb{N}$	–
3 5 6 8 10	$\mathbb{N}$	**4 12 14 20**

Subset	Closure	Ref(s)	**Genuine** and Possible Exceptions
4	1 4 mod 12	[17]	–
4 5	0 1 mod 4	[8]	**8 9 12**
4 6	0 1 mod 3	[25]	**7 9 10 12 15 18 19 22 24 27** 33 34 39 45 46 51 55 75 87
4 7	1 mod 3	[9]	**10 19**
4 8	0 1 mod 4	[2, 25]	**5 9 12 17 20 21 24 33 41 44 45** 48 53 60 65 69 77 89 101 161 164 173
4 9	0 1 4 9 mod 12	[25]	**12 21 24 48** 60 69 84 93 96 192
4 5 6	$\mathbb{N}$	[18]	**7 8 9 10 11 12 14 15 18 19 23** 47
4 5 7	$\mathbb{N}$	[13, 18]	**6 8 9 10 11 12 14 15 18 19 23 26 27** 30 39 42 50 51 54 62 63 66 74 78
4 5 8	0 1 mod 4	[8]	**9 12**
4 5 9	0 1 mod 4	[8]	**8 12**
4 5 11	$\mathbb{N}$	[3]	**6 7 8 9 10 12 14 15 18 19 22 23 26 27 30 31 34** 38 42 43 46 50 54 58 62 66 67 70 74 78 82 90 94 98 102 106 114 115 118 126 130 142 154
4 6 7	0 1 mod 3	[25]	**9 10 12 15 18 19 24 27** 33 45 75 87
4 6 8	$\mathbb{N}$	[25]	**5 7 9 10 11 12 14 15 17 18 19 20 22 23 24 26 27** 33 34 35 39 41 47 50 51 53 55 59 62 65 71 74 75 77 87 89 95 98 101 110 122 131 161 170 182 194 242 266 290
4 6 9	0 1 mod 3	[25]	**7 10 12 15 18 19 22 24 27** 34 55 75 87
4 7 8	$\mathbb{N}$	[25]	**5 6 9 10 11 12 14 15 17 18 19 20 21 23 24 26 27 30 33 35 38 39 41 42** 44 45 47 48 51 54 59 62 65 66 69 74 75 77 78 83 86 87 89 90 93 101 102 107 110 111 114 122 123 126 129 131 135 138 143 146 150 158 159 161 162 164 165 167 170 171 173 174 186 194 195 198

4 7 9	0 1 mod 3	[25]	**6 10 12 15 18 19 21 24 27 30 39 42** 48 51 54 60 66 69 75 78 84 87 90 93 96 99 102 111 114 126 138 147 150 159 162 171 174 183 186 195 210 219 234
4 8 9	0 1 mod 4	[25]	**5 12 17 20 21 24 41 44** 48 53 60 69 77 101 161 164 173
4 5 6 7	$\mathbb{N}$	[18]	**8 9 10 11 12 14 15 18 19 23**
4 5 6 8	$\mathbb{N}$	[18]	**7 9 10 11 12 14 15 18 19 23**
4 5 6 9	$\mathbb{N}$	[18]	**7 8 10 11 12 14 15 18 19 23**
4 5 7 8	$\mathbb{N}$	[13, 18]	**6 9 10 11 12 14 15 18 19 23 26 27** 30 38 42 51 62 66 74 78
4 5 7 9	$\mathbb{N}$	[13, 18]	**6 8 10 11 12 14 15 18 19 23 26 27** 30 50 51 54 62 66 74 78
4 5 7 11	$\mathbb{N}$	[3]	**6 8 9 10 12 14 15 18 19 23 26 27** 30 42 50 54 62 66 74 78
4 5 8 9	0 1 mod 4	[8]	**12**
4 6 7 8	$\mathbb{N}$	[25]	**5 9 10 11 12 14 15 17 18 19 20 23 24 26 27** 33 35 41 65 74 75 77 123 131 143
4 6 7 9	0 1 mod 3	[25]	**10 12 15 18 19 24 27** 75
4 6 8 9	$\mathbb{N}$	[25]	**5 7 10 11 12 14 15 17 18 19 20 22 23 24 26 27** 34 35 41 47 50 53 55 59 62 71 74 75 77 87 95 98 101 122 131 161 170 182 194 242 266
4 7 8 9	$\mathbb{N}$	[25]	**5 6 10 11 12 14 15 17 18 19 20 21 23 24 26 27 30** 35 38 39 41 42 44 47 48 51 54 59 62 110 143 146 147 150 158 159 161 162 164 167 170 171 173 174
4 5 6 7 8	$\mathbb{N}$	[13]	**9 10 11 12 14 15 18 19 23**
4 5 6 7 9	$\mathbb{N}$	[13]	**8 10 11 12 14 15 18 19 23**
4 5 6 8 9	$\mathbb{N}$	[13]	**7 10 11 12 14 15 18 19 23**
4 5 7 8 9	$\mathbb{N}$	[13]	**6 10 11 12 14 15 18 19 23 26 27** 30 51
4 6 7 8 9	$\mathbb{N}$	[25]	**5 10 11 12 14 15 17 18 19 20 23 24 26 27** 35 41
4 5 6 7 8 9	$\mathbb{N}$	[13]	**10 11 12 14 15 18 19 23**
4 5 7 9 10 11	$\mathbb{N}$	[15]	**6 8 12 14 15 18 19 23 26 27** 30
4 5 9 11 19 31	$\mathbb{N}$	[3]	**6 7 8 10 12 14 15 18 22 23 26 27 30 34** 38 42 43 46 50 54 62 66 74 78 90 98 102 114 126

5	1 5 mod 20	[17]	–
5 6	0 1 mod 5	[4]	**10 11 15 16 20 35** 40 50 51 80
5 7	1 mod 2	[4]	**9 11 13 15 17 19 23 27 29 31 33 37 39** 43 47 51 53 55 57 59 63 69 71 73 75 77 79 83 87 89 93 95 97 99 107 109 111 113 115 119 129 131 133 135 137 139 143 149 153 157 159 167 173 177 179 183 189 191 193 195 199 207 209 211 219 227 229 231 233 235 237 239 243 253 255 259 263 269 271 275 277 279 283 291 299 303 311 323 327 329 347 351 353 355 359 363 373 383 395 399 403 407 411 413 415 419 423 431 435 437 439 443 447 453 457 459 471 473 475 477 479 483 491 499 503 507 509 519 523 531 533 535 539 543 559 563 619 639 699

709 719 743 779 839 899 1059 1319

5 8 | 0 1 mod 4 | [4] | **9 12 13 16 17 20 24 28 29 32 33 37 44 49 52** 53 60 68 69 72 73 76 77 84 89 92 93 96 97 100 104 108 109 112 113 116 124 129 132 136 140 149 152 153 156 157 164 169 172 173 177 184 189 192 193 204 209 212 213 217 224 228 233 237 244 252 256 268 269 272 273 284 292 300 308 309 312 313 324 329 332 333 348 352 353 364 369 373 404 409 412 429 444 452 468 492 508 524 532 544 548 552 564 572 580 588 628 644 709 724 732 772 812 852 909 912 1012 1052 1292

5 9 | 1 mod 4 | [4] | **13 17 29 33** 93 113 173

5 6 7 | $\mathbb{N}$ | [21] | **8 9 10 11 12 13 14 15 16 17 18 19 20 22 23 24 27 28 29 32 33 34** 68 69 93 94 98 99 104 108 109 114 124

5 6 8 | $\mathbb{N}$ | [4] | **7 9 10 11 12 13 14 15 16 17 18 19 20 22 23 24 27 28 29 32 33 34 35 37 38 39** 42 44 47 49 50 51 52 53 58 59 62 63 67 68 69 72 74 77 78 79 82 83 84 87 89 92 94 97 98 99 103 104 107 109 112 114 118 119 122 123 124 127 129 132 139 142 147 149 152 154 157 159 162 167 169 172 174 177 179 182 184 187 189 192 194 199 202 204 207 212 214 217 219 222 234 237 239 242 244 247 252 259 262 267 269 272 274 307 312

5 6 9 | $\mathbb{N}$ | [4] | **7 8 10 11 12 13 14 15 16 17 18 19 20 22 23 24 27 28 29 32 33 34 35 38 39** 40 42 43 44 47 48 52 58 59 62 64 67 68 72 79 80 82 83 84 87 88 92 93 98 99 102 103 104 107 108 112 113 118 119 122 123 124 127 128 132 138 142 143 147 148 152 158 162 163 167 168 172 173 178 179 182 183 188 198 199 203 207 208 218 219 223 227 228 238 243 248 258 283 288

5 7 8 | $\mathbb{N}$ | [4] | **6 9 10 11 12 13 14 15 16 17 18 19 20 22 23 24 26 27 28 29 30 31 32 33 34 37 38 39** 42 43 44 46 47 51 52 53 58 59 60 62 66 68 69 70 71 72 73 74 75 76 77 78 79 82 83 84 86 87 89 90 93 94 95 96 97 98 99 100 102 104 106 107 108 109 110 111 114 115 116 118 122 124 126 129 130 131 132 134 135 136 138 139 140 142 143 146 149 150 152 153 154 156 157 158 159 162 164 166 167 170 172 173 174 177 178 179 184 186 190 191 192 194 195 198 199 202 206 207 209 210 211 214 219 226 227 228 230 233 234 237 239 244 256 258 262 268 269 272 274 278 279 282 284 298 300 303 324 338 359 422 443 471 478 492 499 562 619

5 7 9 | 1 mod 2 | [4] | **11 13 15 17 19 23 27 29 31 33 39** 43 51 59 71 75 83 87 93 95 99 107 111 113 115 119 131 135 139 143 167 173 179 183 191 195 243 283 411 563

5 8 9 | 0 1 mod 4 | [4] | **12 13 16 17 20 24 28 29 32 33 44** 52 60 68 84 92 96 100 104 108 112 113 116 124 132 140 156 172 173 192 204 212 228 244 252 268 272 300 308 312

5 9 13 | 1 mod 4 | [16] | **17 29 33**

5 6 7 8 | $\mathbb{N}$ | [21] | **9 10 11 12 13 14 15 16 17 18 19 20 22 23 24 27 28 29 32 33 34** 94 124

5 6 7 9 | $\mathbb{N}$ | [21] | **8 10 11 12 13 14 15 16 17 18 19 20 22 23 24 27 28 29 32 33 34** 68 93 98 99 104 108 124

5 6 8 9 | $\mathbb{N}$ | [4] | **7 10 11 12 13 14 15 16 17 18 19 20 22 23 24 27 28 29 32 33 34 35 38 39** 42 47 52 67 79 82 83 84 87 92 98 99 103 107 118 119 122 123 124 127 132 142 147 162 167 172 174 182

5 7 8 9 | $\mathbb{N}$ | [21] | **6 10 11 12 13 14 15 16 17 18 19 20 22 23 24 26 27 28 29 30 31 32 33 34 38 39** 42 43 44 46 51 52 60 94 95 96 98 99 100 102 104 106 107 108 110 111 116 138 139 140 142 143 146 150 154 156 158 162 163

166 167 170 172 173 174 206 228 243

5 6 7 8 9 | $\mathbb{N}$ | [21] **10 11 12 13 14 15 16 17 18 19 20 22 23 24 27 28 29 32 33 34**

---

6 | 1 6 mod 15 | Table I.2.5

6 7 8 | $\mathbb{N}$ | [21] **9 10 11 12 13 14 15 16 17 18 19 20 21 22 23 24 25 26 27 28 29 30 32 33 34 35 36 37 38 39 40 41 45** 46 **47** 65 68 69 70 71 73 74 75 77 93 94 95 98 99 100 101 122 123 124 125 128 129 130 131 135 137 138 139 140 141 142 143 144 145 146 147 148 149 150 152 153 154 155 159 160 161 165 166 167 170 172 173 174 175 176 177 178 179 180 184 185 233 235 236 237 238 239 240 242 243 244 245

6 7 8 9 | $\mathbb{N}$ | [21] **10 11 12 13 14 15 16 17 18 19 20 21 22 23 24 25 26 27 28 29 30 32 33 34 35 36 37 38 39 40 41 45** 46 **47** 93 94 95 98 99 100 101 137 138 139 140 141 142 143 144 145 146 147 148 149 150 152 153 154 155 160 161 166 167 170 172 173 174 175 176 177 178 179 180 185

---

7 | 1 7 mod 42 | Tables I.2.6, I.2.7

7 8 9 | $\mathbb{N}$ | [21] **10 11 12 13 14 15 16 17 18 19 20 21 22 23 24 25 26 27 28 29 30 31 32 33 34 35 36 37 38 39 40 41 42 43 44 45 46 47 48 51 52 53 54 55 59 60 61 62** 93 94 95 96 97 98 99 100 101 102 103 104 105 106 107 108 109 110 111 116 117 118 132 138 139 140 141 142 143 144 145 146 147 148 149 150 151 152 153 154 155 156 157 158 159 160 161 162 163 164 165 166 167 168 170 171 172 173 174 175 176 177 178 179 180 181 182 183 184 185 186 187 188 189 190 191 192 193 194 195 196 197 198 199 200 201 202 203 204 205 206 207 208 209 210 211 212 213 214 215 216 219 220 221 222 223 228 229 230 242 243 244 245 246 247 248 249 250 251 252 253 254 255 256 257 258 259 260 261 262 263 264 265 266 267 268 269 270 271 272 273 274 275 276 277 278 279 283 284 285 286 298 299 300 303 304 305 306 307 311 312 313 314 315 316 317 318 319 320 321 322 323 324 325 326 327 328 329 330 331 332 333 334 335 339 340 341 342

---

8 | 1 8 mod 56 | Tables I.2.9, I.2.10

8 9 10 | $\mathbb{N}$ | [19] **11 12 13 14 15 16 17 18 19 20 21 22 23 24 25 26 27 28 29 30 31 32 33 34 35 36 37 38 39 40 41 42 43 44 45 46 47 48 49 50 51 52 53 54 55 56 58 59 60 61 62 63 66 67 68 69 70 71 75 76 77 78 79** 101 102 103 104 105 106 107 108 109 111 112 113 115 116 117 118 119 126 127 133 134 135 155 156 157 158 159 160 166 167 173 174 175 176 177 178 179 180 181 182 183 184 185 186 187 188 189 190 191 192 193 194 195 196 197 198 199 200 201 202 203 204 205 206 207 208 209 210 211 212 213 214 215 216 217 218 219 220 221 222 223 224 225 226 227 228 229 230 231 239 247 248 249 250 251 252 253 254 255 256 257 258 259 260 261 262 263 264 265 266 267 268 269 270 271 272 273 274 275 276 277 278 279 280 281 282 283 284 285 286 287 290 291 292 293 294 295 299 300 301 302 303 304 305 306 307 308 309 310 311 312 313 314 315 316 317 318 319 320 321 322 323 324 325 326 327 328 329 330 331 332 333 334 335 336 337 338 339 340 341 342 343 346 347 348 349 350 351 355 356 357 358 359 360 361 362 363 364 365 366 367 368 369 370 371 372 373 374 375 376 377 378 379 380 381 382 383 384 385 386 387 388 389 390 391 392 393 394 395 396 397 398 399 403 404 405 406 407 411 412 413 414 415 416 417 418 419 420 421 422 423 426 427 428 429 430 431 435 436 437 438 439 443 444 445 446 447 448 452 453 454 455 472 473 474 475 476 477 478 479 480 481 482 483 484 485 486 487 488 489 490 491 492 493 494 495 496 497 499 500 501 502 503 507 508 509 510 511 580 581 582

3.18 Table More Closures. Let n be an integer ≥ 3.

$$Q_{\geq n} = \{v : v \text{ is a prime power; } v \geq n\}$$
$$OQ_{\geq n} = \{v : v \text{ is a prime power; } v \geq n,\ v \text{ is odd}\}$$
$$Q_{1 \bmod n} = \{v : v \text{ is a prime power, } v \equiv 1 \bmod n\}$$
$$Q_{0,1 \bmod n} = \{v : v \text{ is a prime power, } v \equiv 0, 1 \bmod n\}$$

(Unless n is prime power, $Q_{0,1 \bmod n}$ is identical to $Q_{1 \bmod n}$.)

Subset	Closure	Ref(s)	**Genuine** and Possible Exceptions
$OQ_{\geq 3}$	1 mod 2		–
$OQ_{\geq 5}$	1 mod 2	[3]	**15 33 39** 51 75 87 183 195
$OQ_{\geq 7}$	1 mod 2	[13]	**15 21 33 35 39 45 51 55 65** 69 75 87 93 95 105 111 115 123 129 135 141 155 159 183 195 213 215 219 231 235 237 245 255 267 285 291 303 305 309 315 321 327 335 339 345 363 365 375 395 415 445 447 453 455 465 471 483 485 501 507 519 525 543 573 597 605 615 651 699 717 735 843 861 1047
$Q_{\geq 3}$	$\mathbb{N}$		**6**
$Q_{\geq 4}$	$\mathbb{N}$	[13]	**6 10 11 12 14 15 18 19 23 26 27** 30 51
$Q_{\geq 5}$	$\mathbb{N}$	[7]	**6 10 12 14 15 18 20 22 24 26 28 30 33 34 38 39** 42 44 46 51 52 60
$Q_{\geq 7}$	$\mathbb{N}$	[1]	**10 12 14 15 18 20 21 22 24 26 28 30 33 34 35 36 38 39 40 42 44 45 46 48 51 52 54 55 60 62**
$Q_{\geq 8}$	$\mathbb{N}$	[7, 12]	**10 12 14 15 18 20 21 22 24 26 28 30 33 34 35 36 38 39 40 42 44 45 46 48 50 51 52 54 55 56 58 60 62 63 66 68 69 70 74 75 76 77 78 82 84 85 86 87** 90 92 93 94 95 98 100 102 106 108 110 111 114 116 118 119 122 124 126 130 132 134 138 140 142 146 148 150 154 159 162 164 166 170 172 174 175 178 180 182 183 186 188 190 194 196 198 202 204 206 210 212 214 218 220 222 226 228 230 234 236 238 242 244 246 250 252 258 260 262 266 268 270 274 276 278 282 284 286 290 292 294 298 300 302 303 306 308 310 314 316 318 322 324 326 330 332 334 335 338 340 346 348 350 354 356 358 362 364 366 370 372 374 378 380 382 386 388 390 394 396 398 402 406 410 412 414 418 420 422 426 428 430 434 436 438 442 444 446 450 452 454 458 460 462 466 468 470 474 476 478 482 484 486 490 492 494 498 500 502 506 508 510 514 516 518 526 530 532 534 542 546 548 550 554 556 558 562 564 566 570 572 574 578 580 582 586 588 594 596 598 602 604 606 610 612 614 618 620 622 626 628 630 634 636 638 642 644 646 650 652 654 658 660 662 666 668 670 674 676 678 682 684 686 690 692 694 698 700 702 706 708 710 716 718 722 724 726 730 732 734 738 740 742 746 748 750 754 756 758 762 764 766 770 772 774 778 780 786 788 794 796 798 802 804 806 810 812 814 818 820 822 826 828 830 834 836 842 844 846 850 852 854 858 860 862 866 868 874 876 878 882 884 886 890 892 894 898 902 906 908 910 914 916 918 922 926 930 934 938 940 942 946 948 950 954 956 958 964 966 982 994 996 998 1002 1004 1006 1010 1012 1014 1018 1020 1022 1026 1030 1034 1036 1038 1042 1044 1046 1050 1052 1054 1058 1060 1062 1066 1068 1070 1074 1076 1078 1082 1084 1086 1090 1092 1094 1098 1100 1106 1108 1110 1114 1116 1118 1122 1124 1126 1130 1132 1138 1140 1146 1150 1154 1162 1180 1186 1188 1190 1194 1196 1198 1206 1210 1212 1214 1218 1226 1228 1230 1234 1236 1238 1242 1244 1246 1250 1252 1254 1258 1260 1262 1266 1270 1274 1306 1314 1318 1322 1330 1354 1378 1386 1394 1402 1410 1418 1426 1434 1442 1450 1490 1506 1514 1522 1554 1570 1586 1594 1602 1610 1618 1650 1658 1682 1698 1706 1714 1722 1730

1738 1754 1762 1770 1778 1786 1794 1802 1810 1826 1834 1850 1874 1890 1898 1906 1930 1954 1962 1970 1974 1978 1986 2058 2066 2074 2082 2098 2122 2146 2186

$Q_{1 \bmod 3}$	1 mod 3	[9]	**10**
$Q_{1 \bmod 4}$	1 mod 4	[16]	**33**
$Q_{1 \bmod 5} \cup \{6\}$	1 mod 5	[22]	**21 26 36** 46 51 56 86 116 146 161 166 196 221 226 231 236 261 266 276 286 291 296 316 321 326 351 376 386 411 416 441 446 471 476 501 536 561 566 596 626 651 686 716 741 746 771 776 801 806 866 896 926 946 986 1046
$Q_{1 \bmod 6}$	1 mod 6	[13]	**55** 115 145 205 235 265 319 355 391 415 445 451 493 649 667 685 697 745 781 799 805 1315
$Q_{1 \bmod 8}$	1 mod 8	[12]	**33 57 65 105 129** 161 177 185 201 209 217 249 265 297 305 321 345 377 385 417 465 473 481 489 497 505 537 545 553 561 609 633 681 705 713 737 745 753 785 825 849 897 913 921 945 985 993 1001 1041 1057 1065 1073 1113 1121 1137 1145 1185 1209 1257 1281 1329 1401 1425 1473 1497 1529 1537 1561 1569 1577 1593 1617 1625 1633 1641 1665 1673 1689 1705 1713 1745 1761 1769 1785 1793 1809 1817 1833 1857 1905 1977 2001 2049 2073 2121 2145 2913 2937 2945 2985 3009 3057 3081 3129 3201 3393 3417 3505 3705 4857 4985 5025 5097 5145 5169 5361 6585 6657 8193 8313 8385 8673
$Q_{0,1 \bmod 7}$	0 1 mod 7	[5, 6]	**14 15 21 22 28 35 36 42** 70 77 78 84 85 98 99 105 106 126 133 134 140 141 147 148 154 155 161 162 168 175 176 182 183 189 190 196 238 245 246 252 253 259 260 266 267 273 274 294 315 316 322 329 364 371 378 420 574 581 588 623 630 1078 1085 1092 1106 1107 1113 1114 1120 1127 1134 1162 1169 1170 1218 1260 1372 1407

3.19 Table Generating sets. Here the opposite view is taken: A PBD-closed set is specified, and a generating set is presented for it. The notation used is

$$\begin{aligned} K_n &= \{v : v > n\}, \\ H_{1(a)} &= \{v : v > a,\ v \equiv 1 \pmod{a}\}, \\ H_{0,1(a)} &= \{v : v > a,\ v \equiv 0, 1 \pmod{a}\}. \end{aligned}$$

Essential elements are shown in bold, and elements not known to be essential are shown in normal typeface.

Closed Set	Generating Set	Ref.
K_3	**3 4 5 6 8**	[17]
K_4	**4 5 6 7 8 9 10 11 12 14 15 18 19 23**	[17, 20]
K_5	**5 6 ... 20 22 23 24 27 28 29 32 33 34**	[17, 20]
K_6	**6 7 ... 41 45** 46 **47**	[17, 20]
K_7	**7 8 ... 48 51 52 53 54 55 59 60 61 62**	[20]
K_8	**8 9 ... 56 58 59 60 61 62 63 66 67 68 69 70 71 75 76 77 78 79**	[20]
$H_{1(2)}$	**3 5**	[27]
$H_{1(3)}$	**4 7 10 19**	[27]
$H_{0,1(3)}$	**3 4 6**	[27]
$H_{0,1(3)} \setminus \{3\}$	**4 6 7 9 10 12 15 18 19 24 27**	[23]
$H_{0,1(3)} \setminus \{3, 4\}$	**6 7 9 10 12 13 15 16 18 19 21 22 24 25 27 28 30 33 34 36 37 39 40 45** 46 51 **52** 69 93 94	[24]
$H_{1(4)}$	**5 9 13 17 29 33**	[16]
$H_{0,1(4)}$	**4 5 8 9 12**	[8]

Closed Set	Generating Set	Ref.
$H_{1(5)}$	**6 11 16 21 26 36 41** 46 51 56 61 71 86 101 116 131 146 166 191 196 221 226 231 236 251 261 266 286 291 296 311 316 321 326 351 356	[13]
$H_{0,1(5)}$	**5 6 10 11 16 20 35** 40	[27]
$H_{1(6)}$	**7 13 19 25 31 37 43 55 61 67 73** 79 97 103 109 115 121 127 139 145 157 163 181 193 199 205 211 229 235 241 265 271 277 283 289 313 319 331 349 355 373 391 397 409 415 445 451 457 487 493 499 643 649 661 667 685 691 697 709 733 739 745 751 781 787 811 1063 1069 1315 1321 1327 1567 1579	[13]
$H_{1(7)}$	**8 15 22 29 36 43 50 71 78 85 92 99** 106 113 127 134 141 148 155 162 169 176 183 190 197 204 211 218 225 239 246 253 260 267 274 281 295 302 309 316 323 330 337 351 358 365 372 379 386 414 421 428 442 575 582 589 596 603 610 701 708 715 722 827 834 1205 1212 1219 1226 1261 1268 1275 1282	[13]
$H_{1(8)}$	**9 17 25 33 41 49 57 65 89 97 105 113 121 129** 161 169 177 185 193 201 209 217 233 241 249 257 281 305 313 321 337 345 353 377 385 401 409 417 449 465 473 481 489 497 537 553 569 601 609 617 633 641 737 745 753 761 769 785 897 913 921 929 1049 1065 1073 1113 1121 1129 1145 1529 1537 1553 1561 1569 1577 1625 1633 1641	[12]

3.3 See Also

§I.6	Replication numbers of resolvable BIBDs with index one form PBD-closed sets.
§I.1	Orders of BIBD(v, k, λ) form PBD-closed sets.
§IV.39	The set of sides of Room squares is PBD-closed.

References

[1] F.E. Bennett, On r-fold perfect Mendelsohn designs, *Ars Combin.* 23 (1987), 57-68.

[2] F.E. Bennett, PBDs with block sizes 4 and 8, *Australas. J. Combin.* 2 (1990), 77-100.

[3] F.E. Bennett, Concerning pairwise balanced designs with prime power block sizes, *Contemp. Math.* 111 (1990), 23-37.

[4] F.E. Bennett, C.J. Colbourn and R.C. Mullin, Quintessential pairwise balanced designs, preprint.

[5] F.E. Bennett, B. Du and L. Zhu, On the existence of $(v, 7, 1)$-perfect Mendelsohn designs, *Discrete Math.* 84 (1990), 221-239.

[6] F.E. Bennett, J. Yin and L. Zhu, On the existence of perfect Mendelsohn designs with k = 7 and lambda even, *Discrete Math.* 113 (1993), 7-25.

[7] F.E. Bennett and L. Zhu, Conjugate-orthogonal latin squares and related structures, in: *Contemporary Design Theory: A Collection of Surveys* (J.H. Dinitz, D.R. Stinson; eds.) John Wiley & Sons, New York, 1992, 41-96.

[8] T. Beth, D. Jungnickel and H. Lenz, *Design Theory,* Cambridge University Press, London, 1986.

[9] A.E. Brouwer, Optimal packings of K_4's into K_n, *J. Combin. Theory A* 26 (1979), 278-297.

[10] D.A. Drake and J.A. Larson, Pairwise balanced designs whose line sizes do not divide 6, *J. Combin. Theory A* 34 (1983), 266-300.
[11] A. Granville, H.-D. O.F. Gronau and R.C. Mullin, On a problem of Hering concerning orthogonal covers of $\vec{K}_n$, *J. Combin. Theory A* 72 (1995), 345-350.
[12] M. Greig, private communication.
[13] M. Greig, Designs from projective planes and PBD bases, unpublished.
[14] H.-D. Gronau, R.C. Mullin and C. Pietsch, The closure of all subsets of $\{3, 4, \ldots, 10\}$ which include 3, *Ars Combin.*, to appear.
[15] H.-D.O.F. Gronau, R.C. Mullin and P.J. Schellenberg, On orthogonal double coverings of $K_n^{\rightarrow}$, preprint.
[16] A.M. Hamel, W.H. Mills, R.C. Mullin, R. Rees, D.R. Stinson and J. Yin, The spectrum of PBD($\{5, k^\star\}, v$) for $k = 9,\ 13$, *Ars Combin.* 36 (1993), 7-26.
[17] H. Hanani, Balanced incomplete block designs and related designs, *Discrete Math.* 11 (1975), 255-369.
[18] H. Lenz, Some remarks on pairwise balanced designs, *Mitt. Math. Sem. Giessen* 165 (1984), 49-62.
[19] A.C.H. Ling and C.J. Colbourn, Pairwise balanced designs with block sizes 8,9 and 10, preprint.
[20] A.C.H. Ling and C.J. Colbourn, Concerning the generating set of $\mathbb{Z}_{\geq n}$, preprint.
[21] A.C.H. Ling, X.J. Zhu, C.J. Colbourn and R.C. Mullin, Pairwise balanced designs with consecutive block sizes, preprint.
[22] Y. Miao and L. Zhu, Pairwise balanced designs with block sizes 5t+1, *Ars Combin.* 32 (1991), 239-251.
[23] R.C. Mullin, On a PBD generating set problem of Rosa, *Ars Combin.* 27 (1989), 75-84.
[24] R.C. Mullin, B. Gardner, K. Metsch and G.H.J. van Rees, Some properties of finite bases for the Rosa set, *Utilitas Math.* 38 (1990), 199-215.
[25] R.C. Mullin, A.C.H. Ling, R.J.R. Abel and F.E. Bennett, On the closure of subsets of {4,5,. . .,9} which contain 4, preprint.
[26] R.M. Wilson, An existence theory for pairwise balanced designs II. The structure of PBD-closed sets and the existence conjectures, *J. Combin. Theory A* 13 (1972), 246-273.
[27] R.M. Wilson, Constructions and uses of pairwise balanced designs, *Mathematisch Centre Tracts* 55, Amsterdam (1974), 18-41.
[28] R.M. Wilson, An existence theory for pairwise balanced designs III: A proof of the existence conjectures, *J. Combin. Theory A* 18 (1975), 71-79.

4 Pairwise Balanced Designs as Linear Spaces

Hans-Dietrich O.F. Gronau
Ronald C. Mullin
Christian Pietsch

4.1 Linear Spaces

A PBD on v points is also a *linear space* of order v if all blocks have size at least 2. If all blocks have size at least 3, then such a linear space is *proper*. It is *nontrivial* if it contains more than one block.

4.1 Remarks It is customary to refer to elements as *points* and to blocks as *lines* when dealing with linear spaces, and theorems are often stated in this geometric language. In geometric terms, a linear space is a pair $(\mathcal{P}, \mathcal{B})$ where $\mathcal{P}$ is a set of *points* and $\mathcal{B}$ is a collection of subsets of $\mathcal{P}$ (the *lines*) satisfying:

1. every pair of points is contained on exactly one line, and
2. every line contains at least two points.

Let L be a finite linear space. The *degree of a line* is the number of points on it, and the *degree of a point* is the number of lines that pass through it.

A finite linear space with $n \geq 3$ points is a *near pencil* if it contains one line consisting of $n-1$ points, and the remaining lines contain two points.

4.2 Example The lines of a near pencil on the 5 points $\{1,2,3,4,5\}$ are

$$\{1,2,3,4\}, \{1,5\}, \{2,5\}, \{3,5\}, \{4,5\}.$$

4.3 Theorem (de Bruijn and Erdös) [3] Let L be a nontrivial finite linear space. Let v denote the number of points in L, and b denote the number of lines. Then $b \geq v$, with equality if and only if L is a finite projective plane, or L is a near-pencil.

4.4 Remark The inequality of Theorem 4.3 holds for all PBDs with more than one block, since adjoining additional blocks of sizes 0 or 1 only strengthens the inequality.

A finite linear space is *nondegenerate* if it is nontrivial and is not a near-pencil.

4.5 Theorem (See [3]) Let L be a nondegenerate linear space having v points and b lines. If n is the unique integer satisfying $n^2-n+1 = (n-1)^2+(n-1)+1 < v \leq n^2+n+1$, then $b \geq B(v)$ where

$$B(v) = \begin{cases} n^2+n+1 & \text{if } \; n^2+2 \leq v \leq n^2+n+1 \\ n^2+n & \text{if } \; n^2-n+3 \leq v \leq n^2+1 \quad \text{or} \quad v=4 \\ n^2+n-1 & \text{if } \; v = n^2-n+2 \neq 4 \end{cases}$$

This is best possible inasmuch as the bounds are attained infinitely often.

Let L be a finite linear space. Then L is *embeddable* in a projective plane of order n if there exists a projective plane π of order n with the property that L can be obtained by deleting points (and any resulting lines of length 0 and 1) from π.

4.6 Remark If a linear space is embeddable in a projective plane, this yields a method of generating the linear space when n is a prime power and may be an indicator of the difficulty of constructing the linear space otherwise.

To discuss the embeddability of finite linear spaces whose numbers of points and lines meet the bound in Theorem 4.3, Example 4.7 is required.

4.7 Example A special finite linear space $E_1 : v = 8, b = 11$. The lines are listed as columns.

1	1	2	3	1	2	3	4	5	6	7
2	4	4	4	6	5	5	8	8	8	8
3	5	6	7	7	7	6				
8										

4.8 Remark The configuration of Example 4.7 is not embeddable in a projective plane of order 3.

4.9 Theorem [3] Let L be a nondegenerate finite linear space with v points and b lines. Suppose that n is the unique integer such that $n^2-n+2 \leq v \leq n^2+n+1$. If $b = B(v)$, then either L can be embedded in a projective plane of order n, or $v = 8$, $b = 11$, and L is the special linear space E_1 given in Example 4.7.

4.10 Remark The case of finite linear spaces L with v points and b lines where $n^2 - n + 2 \le v \le b \le B(v) + 1$ is also of interest. Such spaces have been classified [3], but infinitely many cannot be embedded in a finite projective plane of order n. However if such a linear space contains a point of degree at most n, then, apart from a finite number of exceptions, it can be embedded in a projective plane of order n. For this and other results an expanded set of special linear spaces is required.

4.11 Example More special finite linear spaces.

$E_2 :\ v = 8,\ b = 12.$
1 3 1 1 2 2 4 4 5 5 6 6
2 7 7 8 7 8 7 8 7 8 7 8
3 8
4
5
6

$E_3 :\ v = 8,\ b = 13.$
1 1 3 5 1 2 2 2 3 4 4 5 4
2 6 7 6 8 6 7 8 6 6 7 7 8
3 7 8 8
4
5

$E_4 :\ v = 8,\ b = 13.$
1 1 1 2 3 4 2 2 3 3 4 4 6
2 5 7 5 5 6 6 7 6 8 5 7 7
3 6 8 8 7 8
4

$E_5 :\ v = 6,\ b = 8.$
1 2 1 1 3 3 4 4
2 5 5 6 5 6 5 6
3 6
4

$E_6 :\ v = 7,\ b = 10.$
1 3 1 1 2 2 4 4 5 5
2 6 6 7 6 7 6 7 6 7
3 7
4
5

$E_7 :\ v = 7,\ b = 10.$
1 1 4 3 1 2 2 2 3 4
2 5 5 6 6 5 6 7 5 7
3 7 6 7
4

4.12 Theorem Let L be a nondegenerate linear space with at least $n^2 - n + 2$ points and at most $n^2 + n + 1$ lines for some integer $n \ge 2$. If L has a point of degree at most n, then one of the following cases occurs.

1. L can be embedded into a projective plane of order n.
2. L is one of the linear spaces E_1, E_2, E_3, or E_4; in particular, $n = 3$ and $v = n^2 - n + 2 = 8$.
3. $v = n^2 - n + 2$ and $b = n^2 + n + 1$, L has a point Q of degree n which lies on one line l of degree $n + 1$ and on $n - 1$ lines of degree n. The n points different from Q on l do not all have degree $n + 1$.

4.13 Corollary In a nondegenerate linear space with $n^2 + n + 1$ lines and at least $n^2 - n + 3$ points, every point has degree at least $n + 1$.

4.14 Theorem [3] If L is a nondegenerate linear space with $v = n^2 - n + 2$ points and $b = B(v) + 1 = n^2 + n$ lines for some integer $n \ge 2$, then either L can be embedded in a projective plane of order n, or $v = 8, b = 12$, and L is the linear space E_2.

4.15 Remark Totten studied *restricted* linear spaces, those that satisfy $(b - v)^2 \le v$, where v and b are the number of points and lines of the space respectively. Theorem 4.16 yields useful information about restricted linear spaces.

4.16 Theorem [3] Let L be a nondegenerate linear space with v points and b lines. Let n be the unique integer with $n^2 - n + 2 \le v \le n^2 + n + 1$.

1. If $(b - v)^2 \le v$, then $b \le v + n$.
2. If $v < (b - v)^2 \le b$, then $v = 8$ and $b = 11$, or $v \in \{n^2 + n, n^2 + n + 1\}$ and $b = v + n + 1$.
3. If $(b - v)^2 \le b$ and if L has maximal line degree $k \ge n + 2$, then L is one of the linear spaces E_5, E_6 or E_7.

A *punctured* affine plane is an affine plane with one of its points deleted.

4.17 Theorem (Totten [3]) Every nondegenerate finite linear space with v points and b lines, where $(b-v)^2 \leq v$, is one of the following structures.

1. A projective plane with at most n points deleted but not more than $n-1$ being deleted from the same line.
2. An affine plane π of order n, or an affine plane π of order n with a new point, say ∞, which has been adjoined to all lines of one of the parallel classes of π.
3. An affine plane π of order n from which a point has been deleted, and to which a new point, say ∞, has been adjoined to all lines which are the remnants of one of the original parallel classes of π.
4. The configuration obtained from a projective plane π of order n by deleting s points from a line ℓ, and replacing the remaining $n-s+1$ points of ℓ by either a projective plane of order $n-s$ or a near pencil on $n-s+1$ points.
5. The linear space E_5 $(v=6, b=8)$.

4.18 Remarks

1. Theorems 4.12, 4.14, 4.16, and 4.17 are useful in ruling out the existence of certain linear spaces as well as yielding information about the types of linear spaces obtained by deleting relatively few points from a projective plane.
2. Good general references for properties of finite linear spaces are [3] and [1].
3. A complete catalogue of finite linear spaces has been generated in those cases in which the total number of points is relatively small.

4.2 Enumeration

LIN(v) is the number of nonisomorphic linear spaces of order v.

PLIN(v) is the number of proper nontrivial nonisomorphic linear spaces of order v.

4.19 Table The number of nonisomorphic linear and proper linear spaces of order $v \leq 16$. [2, 4]

v	3	4	5	6	7	8	9	10	11	12	13	14	15	16
LIN(v)	2	3	5	10	24	69	384	5250	232929	?	?	?	?	?
PLIN(v)	0	0	0	0	1	0	1	1	1	3	7	1	120	398

Let $K = \{k_1, k_2, \ldots, k_n\}$ be a set of positive integers and $P = (V, \mathcal{B})$ be a PBD(v, K). Let β_i denote the number of blocks in P with size $k_i, i = 1, \ldots, n$. Then the vector $\vec{\beta} = (\beta_1, \beta_2, \ldots, \beta_n)$ is the *block type* of P. For a fixed point $x \in V$ let γ_i denote the number of blocks of size k_i through x, $i = 1, \ldots, n$. Then $\vec{\gamma} = (\gamma_1, \gamma_2, \ldots, \gamma_n)$ is the *point type* of x in P.

4.20 Table List of proper nontrivial linear spaces of order $v \leq 15$. Steiner triple systems of orders 7, 9, 13 and 15 are not listed here; see §I.1.

Blocks are shown vertically, using letters as points. Each design is labeled by $a.b$, where a is the number of points and b is a sequence number; the order of the automorphism group is given under the label.

10.1 (108)
```
aaabbbcccddd
behefgefgefg
cfihijjhiijh
dgj
```

11.1 (120)
```
aaaabbbcccdddeee
bfhjfgifgifghfgh
cgikhjkkhjikjjik
d
e
```

12.1 (432)
```
aaabbccddabcd
behefegfgkgfe
cfihijijhljhi
dgjkllkkl
```

12.2 (576)
```
aeiaaaabbbbccccdddd
bfjefghefghefghefgh
cgkijkljilkklijlkji
dhl
```

12.3 (192)
```
aeiaaaabbbbccccdddd
bfjefghefghefghefgh
cgkijkljilklkijklji
dhl
```

13.1 (576)
```
aaabbbbccccddddeeee
bfjfghifghifghifghi
cgkjklmkjmllmjkmlkj
dhl
eim
```

13.2 (192)
```
aaabbbbccccddddeeee
bfjfghifghifghifghi
cgkjklmkjmlmljklmkj
dhl
eim
```

13.3 (5616)
```
aaaabbbcccddd
behkefgefgefg
cfilhijjhiijh
dgjmklmlmkmkl
```

13.4 (72)
```
aaaabcdbbbcccdddefg
behkefgfgiegjefhjhi
cfilhijjlmmhkikllmk
dgjmklm
```

13.5 (48)
```
aabcdgaaabbbcccdddef
beefhihilfgkegiefgjh
cfhjjkkjmljmlhmkimmm
dgikll
```

14.1
```
aaaabbbcdecccdddeee
bfilfghfghghjfhkfgi
cgjmijkkijlinjmlnkm
dhknlmnmnl
e
```

15.1 (1344)
```
aaaaabbbbccccddddeeeeffffgggg
bhjlnhilmhilmhijkhijkhijkhijk
cikmojknokjonlmnomlonnolmonml
d
e
f
g
```

15.2 (64)
```
aaaaabbbbccccddddeeeeffffgggg
bhjlnhilmhilmhijkhijkhijkhijk
cikmojknokjonlmnomlononlmnoml
d
e
f
g
```

15.3 (16)
```
aaaaabbbbccccddddeeeeffffgggg
bhjlnhilmhilmhijkhijkhijkhijk
cikmojknokjonlmonnlmoonlmmonl
d
e
f
g
```

15.4 (96)
```
aaaaabbbbccccddddeeeeffffgggg
bhjlnhilmhilmhijkhijkhijkhijk
cikmojknokjonlomnolnmmnlonmol
d
e
f
g
```

15.5 (24)
```
aaaaabbbbccccddddeeeeffffgggg
bhjlnhilmhikmhijlhijkhijkhijk
cikmojknoljonknmoolnmmolnnmol
d
e
f
g
```

15.6 (42)
```
aaaaabbbbccccddddeeeeffffgggg
bhjlnhikmhikmhijlhijlhijkhijk
cikmojlnoojlnkmonnkmolonmmnlo
d
e
f
g
```

15.7 (36)
```
aabcdefaaabbbcccdddeeefff
bgghijkjkmhilgikghjghighi
chjlnlmolnnmoojnlkmmoknjl
dikmono
e
f
```

15.8 (720)
```
aabcdeaaabbbcccdddeee
bffghijklghifhifgifgh
cgjjklonmonmolknljmkj
dhkmmn
eilnoo
```

15.9 (48)
```
aaabbccddeeabbccddee
bfjfgfghihinhihifgfg
cgkjklmjklmomlkjmlkj
dhlnoononno
eim
```

Design	Blocks	Design	Blocks
15.10 (6)	aaaabbbbbcccccddddde eeeefghi bfjnfghimfghilfghikfghijmlkj cgkojklnokjnmoomjlnlomknnnoo dhl eim	15.11 (12)	aaaabbbbbcccccdddddeeeeefghi bfjnfghimfghilfghikfghijklmj cgkojklnonjkmomojlnlmoknonno dhl eim
15.12 (6)	aabaaabbbcccccdddddeeeeeghij bffjklghifghilfghimfghikkmln cgjmnomonmjkonnljkoonljmonmo dhk eil	15.13 (18)	aabaaabbbcccccdddddeeeeeghim bffjklghifghilfghikfghijkljn cgjmnomnomjoknonjlmnlkmoomno dhk eil
15.14 (72)	afkaaaaabbbbbcccccdddddeeeee bglfghijfghijfghijfghijfghij chmklmnolkomnonklmmonklnmlok din ejo	15.15 (600)	afkaaaaabbbbbcccccdddddeeeee bglfghijfghijfghijfghijfghij chmklmnooklmnnoklmmnokllmnok din ejo
15.16 (360)	aaaabbbcccdddfgabcde behkefgefhegijhnjgfi cfilhlijilmjkkmomkhl dgjmknonmoolnon	15.17 (336)	aaaabbbcccdddgabcdefh behkefgefgefjinmigkjl cfilhijjlhihkmoooooo dgjmnklmnklmnn
15.18 (4)	aaaabbbcdgabccccddddeffgh behkefjeejngfgilfghiihmim cfilhilkjkomjhmnolknllnoo dgjmnkoomn	15.19 (20)	aaaabbbcdgabccccddddeffgi behkefjeejngfghlfghiihmhm cfilhilkjkomjimnolknllnoo dgjmnkoomn
15.20 (4)	aaaabcdabbbbccccddddeeffgghh behkefgnfgijegjmefilijjkiklm cfilhijomlnonhkoohknmlnoonon dgjmklm	15.21 (6)	aaaabcdabbbbccccddddeeffgghh behkefgnfgijegjmefikijjkillm cfilhijomnolohknlhnomnonkono dgjmklm
15.22 (6)	aabcdgaaaabbbbccccddddeeffghi beefhihiklfgjkegilefgmjlhmjkn cfhjjkmjnolnomohmnkionnmnomoo dgikll	15.23 (24)	aabcdgaaaabbbbccccddddeeffghm beefhihiklfgjkegilefgijlhijkn cfhjjkmjnolnomohmnkmonnmnomoo dgikll
15.24 (120)	aabcdaaaabbbbccccddddeeffgghij beefghijnfgjmegilefhkjmilhklkk cfhhiklmoklnokmnolmnoonononmml dgijj	15.25 (4)	aabcdaaaabbbbccccddddeeffgghij beefghijnfgjmegilefhkjmilhklkk cfhhiklmoklnokmnolmononononnomml dgijj
15.26 (1)	aabcdaaaabbbbccccddddeeffgghij beefghijnfgjmegilefhkjmilhklkk cfhhiklmoklnokmonlomnonnmnooml dgijj	15.27 (6)	aabcdaaaabbbbccccddddeeffgghij beefghijnfgjmegilefhkjmilhklkk cfhhiklmoklnokomnlomnonnmnmool dgijj
15.28 (2)	aabcdaaaabbbbccccddddeeffgghij beefghijnfgjmegilefhkjlimhklkk cfhhiklmoklnokmnomlonononnomml dgijj	15.29 (3)	aabcdaaaabbbbccccddddeeffgghij beefghijnfgjmegilefhkjlimhklkk cfhhiklmoklnokomnnlmoomonnmonl dgijj
15.30 (1)	aabcdaaaabbbbccccddddeeffgghij beefghijnfgjmegilefhkjlimhklkk cfhhiklmoklnoknmonlomomonmonnl dgijj	15.31 (8)	aabcdaaaabbbbccccddddeeffgghij beefghijnfgjmegilefhkjmilhklkk cfhhiklmoklnoknmoomnllnonommno dgijj
15.32 (12)	aabcdaaaabbbbccccddddeeffgghij beefghijnfgjmegimefhkjlilhklkk cfhhiklmoknloklonmonlnomnommno dgijj	15.33 (3)	aabcdaaaabbbbccccddddeeffgghij beefghijnfgjmegimefhkjlilhklkk cfhhiklmoknloklonmonlonnmmoomn dgijj

15.34 (3)
```
aabcdaaaabbbbccccddddeeffgghij
beefghijnfgjmegimefhkjlilhklkk
cfhhiklmoknloklononmlnmmoomnno
dgijj
```

15.35 (1)
```
aabcdaaaabbbbccccddddeeffgghij
beefghijnfgjmegilefhkjkilhmlkk
cfhhiklmoklnomkonomnllnnoonmmo
dgijj
```

15.36 (2)
```
aabcdaaaabbbbccccddddeeffgghij
beefghijnfgjmegilefhkjkilhmlkk
cfhhiklmoklnonkmoomnllmononmno
dgijj
```

15.37 (1)
```
aabcdaaaabbbbccccddddeeffgghij
beefghijnfgjmegilefhkjkilhmlkk
cfhhiklmoklnonkomonmllmmoonnno
dgijj
```

15.38 (1)
```
aabcdaaaabbbbccccddddeeffgghij
beefghijnfgjmegimefhkjkilhllkk
cfhhiklmoknlolkononmlnmmoomnno
dgijj
```

15.39 (5)
```
aabcdaaaabbbbccccddddeeffgghij
beefghijnfgjmegilefhkjkikhllmk
cfhhiklmolknomoknonmllnomnmono
dgijj
```

4.21 Table [2] Proper linear spaces on 16 points having a block of size at least five.

16.1 (72)
```
aaabbbbbcccccdddddeeffgghhii
belefghiefghiefghijkjkjkjkjk
cfmpolmnmlpnolmnoponnpmopllm
dgn
jho
kip
```

16.2 (600)
```
aaabbbbbcccccdddddeeffgghhii
belefghiefghiefghijkjkjkjkjk
cfmoplmnplmnolmnopmnnoopplllm
dgn
jho
kip
```

16.3 (20)
```
aaabbccddeeaabbccddeefffff
bghgjijgjhiikhighmnglghikl
clmkmlohknkjnlpmkpojonjopm
dpoonnpilpm
e
f
```

16.4 (1)
```
aaabbccaabbccddddddeeeeefffffkm
bghghgikolnklghijlghijkghijmln
cjiijhjmpopnmmpoknpnmlolknoppo
dnlkmop
e
f
```

16.5 (72)
```
aaaabbbcccdddefhbcdg
beinegkefgfglhhjfieh
cfjoijmkjlkimlimlmjk
dgkppnonopponomp
hml
```

16.6 (6)
```
aaaabcdfbbbbcccdddeeeffgghhi
bcjnehigcfijdgmefkghjhkimill
dekokjmllmpopkoljoompopjpkpo
gflpnnnn
him
```

16.7 (4)
```
aaaabcdfbbbbcccdddeeeffgghhi
bcjnehigcfijdgmefkghkhjimill
dekomkjllkpopjolmoojpopkpmpo
gflpnnnn
him
```

16.8 (120)
```
aabcdkabcdhaabbccefgi
bdefelefgijfidhdehgjj
cgghfmlnkmooklkonmmln
jhiijnppppp
mnolko
```

16.9 (3)
```
aababcddefaabbccdefgghhijk
bccdefekghijhkglgjijmillon
fednmmflijpmponphlkkpmonpp
ghionopmon
lkj
```

16.10 (2)
```
aababcdfaabbccdddeeeffggghij
bccjhgekikeldfhjlfgkhnhikonl
deimmmilpnopnokpojnpipjlopon
fhjonpmm
glk
```

16.11 (6)
```
aababcdhaabbccdddeeefffgghjn
bccjfgekikeldhfjlghkghiikilo
deimmmilpnopnokponjpjpolonnp
gfjonpmm
hlk
```

16.12 (6)
```
aababcdfiaabbccdddeeeefffgh
bccjgehklikhldfgijghjkghljk
dgionmmmnmnomonkploinlijpmp
ehjpppnoo
flk
```

16.13 (40)
```
aekabcdiaaaabbbbccccddddefgh
bflhgfejefgiefhieghifghijjjj
cgmmonklnolkmklnlkomlmnoomnk
dhnppppp
jio
```

16.14 (24)
```
aekabcdiaaaabbbbccccddddefgh
bflhgfejefgiefhieghifghijjjj
cgmmnoklonlklkommklnlmnonmok
dhnppppp
jio
```

4.3 See Also

§I.1	BIBDs of index one give linear spaces.
§VI.7	Linear spaces are simple finite geometries.
[1]	A text dedicated to finite linear spaces.

References

[1] L.M. Batten and A. Beutelspacher, *The Theory of Finite Linear Spaces*, Cambridge University Press, London, 1994.

[2] G. Heathcote, Linear spaces on 16 points, *J. Combin. Des.* 1 (1993), 359-378.

[3] K. Metsch, *Linear Spaces with Few Lines*, Springer-Verlag, Berlin, 1991.

[4] C. Pietsch, On the classification of linear spaces of order 11, *J. Combin. Des.* 3 (1995), 185-193.

5 PBDs and GDDs of Higher Index

Ronald C. Mullin
Hans-Dietrich O.F. Gronau

5.1 PBDs of Higher Index

5.1 **Remark** The construction of PBDs (and GDDs) of index $\lambda > 1$ is considered. The theory of PBD-closure provides a powerful tool to use in connection with such objects. For the most part, theorems for index 1 have natural extensions to cases of higher λ. The main exception is that the set of replication numbers of resolvable BIBDs of index $\lambda > 1$ is not PBD-closed. Resolvability is treated in §III.6.

5.2 **Theorem** Suppose that K is a set of positive integers and let λ be a positive integer. Let $B(K;\lambda) = \{v : \exists \text{ a PBD}(v,K;\lambda)\}$. Then $B(K;\lambda)$ is a PBD-closed set.

5.3 **Remark** The theory of PBD-closed sets was developed by Wilson partly to settle the asymptotic existence for BIBDs. It is trivial to show that necessary conditions for the existence of a BIBD(v,k,λ) are $\lambda(v-1) \equiv 0 \pmod{k-1}$ and $\lambda v(v-1) \equiv \pmod{k(k-1)}$. These are asymptotically sufficient, as is indicated in Theorem 5.4.

5.4 **Theorem** [3, 4] Let positive integers k and λ be given. Then there exists a constant $C(k,\lambda)$ such that if $v > C(k,\lambda)$ and if v satisfies $\lambda(v-1) \equiv 0 \pmod{k-1}$ and $\lambda v(v-1) \equiv 0 \pmod{k(k-1)}$, then there exists a BIBD(v,k,λ).

5.5 **Theorem** (*Breaking up blocks*) Let K and L be sets of positive integers, and let λ_1, λ_2 be positive integers. Suppose that there exists D, a PBD$(v,L;\lambda_1)$, and that for each ℓ which occurs as a block-size of D there exists a PBD$(\ell,K;\lambda_2)$. Then there exists a PBD$(v,K;\lambda_1\lambda_2)$.

5.6 **Remark** A (K,λ)-GDD of order v can be viewed as a PBD$(v,K\cup G;\lambda)$ by replicating each group G λ times and adjoining the resulting sets as blocks. A (K,λ)-GDD of order v with block sizes in a set G can be used to create a PBD$(v+1,K\cup(G+1);\lambda)$

(where $G+1 = \{g+1 : g \in G\}$) by adjoining a new point (say ∞) to each group, then replicating the resulting set of blocks λ times and adjoining the resulting sets as blocks.

5.7 Theorem (*Wilson's Fundamental Construction for group divisible designs of higher index*) Let $D = (\mathcal{V}, \mathcal{G}, \mathcal{B})$ be a GDD of index λ_1 (the *master* GDD). Suppose that D has groups $G_1, G_2, \ldots, G_s$. Let $w : \mathcal{V} \to \mathbb{Z}^+ \cup \{0\}$ be a mapping (weight function) from the point set of D to the nonnegative integers which has the property that for each $B = \{x_1, x_2, \ldots, x_k\}$ in $\mathcal{B}$ there exists (K, λ_2)-GDD (an *ingredient* GDD) of type $[w(x_1), w(x_2), \ldots, w(x_k)]$. Then there exists a $(K, \lambda_1\lambda_2)$-GDD of type

$$T = \left[\sum_{x \in G_1} w(x), \sum_{x \in G_2} w(x), \ldots, \sum_{x \in G_s} w(x)\right].$$

5.8 Remark For indices $\lambda > 1$, there is a greater choice of ingredients for Wilson's fundamental construction. For example if λ is a prime p, then either the master GDD can be of index p and the ingredients of index one, or vice versa. For composite λ, the number of choices is even greater.

5.9 Examples Wilson's Fundamental Construction for BIBDs.

1. As master GDD, start with a TD$(5,4)$, and remove 1 or 2 points from a group to obtain a $\{4,5\}$-GDD of type 4^4x^1 for $x \in \{2,3\}$. Give every point weight 3, using $\{4\}$-GDDs of types 3^4 and 3^5 as ingredient GDDs. The result is a $\{4\}$-GDD of type $12^4(3x)^1$. Now duplicate each block and fill the holes with (13,4,2) and $(3x+1$,4,2) designs to produce both a (55,4,2) and a (58,4,2) design.
2. The master GDD is a (16,6,2) design, which is a $(\{6\},2)$-GDD of type 1^{16}. Give weight 5 to each point, using a $\{6\}$-GDD of type 5^6 (a TD$(6,5)$ as the ingredient, to form a $(\{6\},2)$-GDD of type 5^{16}. Fill the holes using a (6,6,2) design to produce an (81,6,2) design.
3. The master GDD is an (11,6,3) design. Proceed as above, filling with (6,6,3) designs, to produce a (56,6,3) design.

5.10 Theorem [5] Necessary and sufficient conditions for the existence of a $(3,\lambda)$-IGDD of type $(m,h)^u$ are

1. $m \geq 2h$,
2. $\lambda m(u-1) \equiv 0 \pmod 2$,
3. $\lambda(m-h)(u-1) \equiv 0 \pmod 2$, and
4. $\lambda u(u-1)(m^2-h^2) \equiv 0 \pmod 3$.

5.11 Theorem [5] There exists a $(4,\lambda)$-IGDD of type $(m,h)^u$ where $u \equiv 1 \pmod 3$, $u \notin \{10, 19\}$, and $(m,h) \neq (6,1)$ if and only if

1. $m \geq 3h$ for $u \equiv 1$ or $4 \pmod{12}$ or $\lambda \equiv 0 \pmod 2$,
2. $m \geq 3h$ and $m-h \equiv 0 \pmod 2$ for $u \equiv 7$ or $10 \pmod{12}$ and $\lambda \equiv 1 \pmod 2$.

5.12 Remarks

1. A PBD$(v, K; \lambda_1\lambda_2)$ ($(v,h;K,\lambda_1\lambda_2)$-IPBD) is obtained from a PBD$(v,K;\lambda_1)$ ($(v,h;K,\lambda_1)$-IPBD, respectively) by replicating the blocks of the design λ_2 times.
2. A $(K,\lambda_1\lambda_2)$-GDD (or $(K,\lambda_1\lambda_2)$-IGDD) of type T is obtained from a (K,λ_1)-GDD (or (K,λ_1)-IGDD) by replicating its blocks λ_2 times.
3. In view of this, a rich source of PBDs, IPBDs, GDDs, and IGDDs of index $\lambda > 1$ is the class of equivalent objects of index 1.

4. GDDs and IGDDs of higher index are frequently obtained from transversal designs and incomplete transversal designs of general index. See §II.4.
5. The notion of filling in holes applies to GDDs and IGDDs of higher index. For example, the following BIBDs can be obtained:

GDD:	Fill the Hole with:	Resulting BIBD:
$TD_2(5, 11)$	(11,5,2) design	(55,5,2) design
$TD_3(4, 9)$	(9,4,3) design	(36,4,3) design
$TD_2(10, 10)$	(10,10,2) design	(100,10,2) design
$TD_2(12, 12)$	(12,12,2) design	(144,12,2) design
$TD_3(10, 10)$	(10,10,3) design	(100,10,3) design
$TD_3(11, 10)$	(11,11,3) design	(111,11,3) design
$TD_2(18, 18)$	(18,18,2) design	(324,18,2) design
$TD_3(12, 12)$	(12,12,3) design	(144,12,3) design
$TD_3(13, 12)$	(13,13,3) design	(157,13,3) design

5.13 Theorem Suppose that there exists a (K, λ)-GDD of type $g_1^{a_1} g_2^{a_2} \cdots g_s^{a_s}$.

1. If for each $g_i, i = 1, 2, \cdots, s$, there exists a $(g_i + h, h; K, \lambda)$-IPBD, then there exists a $(w + h, h; K, \lambda)$-IPBD where $w = \sum_{i=1}^{s} a_i g_i$.
2. If for some j there exists exactly one group of size g_j (that is, $a_j = 1$) and if for all $i = 1, 2, \cdots, s, i \neq j$ there exists a $(g_i + h, h; K, \lambda)$-IPBD, then there exists a PBD$(w + h, K; \lambda)$, where $w = \sum_{i=1}^{s} a_i g_i$.

5.14 Theorem (Stinson) [8] Let K be a set of positive integers and let λ be a positive integer. Further, let a and b be integers satisfying $b \geq a \geq 0$. Suppose that there exist

1. a (K, λ)-IGDD of type $(t_1, u_1)(t_2, u_2), \ldots, (t_n, u_n)$;
2. a $(t_i + b; u_i + a, b; a, K; \lambda) - \diamondsuit$-IPBD for $1 \leq i \leq n - 1$ (§III.1.2); and
3. a $(t_n + b, u_n + a; K, \lambda)$-IPBD.

Then there exists a $(t + b, u + a; K, \lambda)$-IPBD, where $t = \sum_{i=1}^{n} t_i$ and $u = \sum_{i=1}^{n} u_i$.

5.15 Theorem Suppose a (K, λ)-IGDD of type $(g_1, h_1)^{a_1} (g_2, h_2)^{a_2} \cdots (g_s, h_s)^{a_s}$ exists. Suppose that for $i = 1, 2, \ldots, s$ there exists a $(g_i + q, h_i + q; K, \lambda)$-IPBD. Then there exists a $(v, w; K, \lambda)$-IPBD where $v = (\sum_{i=1}^{s} a_i g_i) + q$ and $w = (\sum_{i=1}^{s} a_i h_i) + q$.

If in addition there exists a PBD$(w, K; \lambda)$, then there exists a PBD$(v, K; \lambda)$.

5.16 Theorems

1. [6] If $v \equiv 2, 4 \pmod 5$ and $v \geq 9$, then there exists a $(v, 2; \{5\}; 4)$-IPBD.
2. [6] If $v \equiv 3 \pmod 5$ and $v \geq 13$, then there exists a $(v, 2; \{5\}; 4)$-IPBD.
3. [7] If $v \equiv 2 \pmod 3$ and $v \geq 11$, then there exists a $(v, 2, \{6\}; 5)$-IPBD.

5.17 Remark Another method of creating IPBDs of greater index is by identifying certain points in PBDs of smaller index. The IPBDs in Theorem 5.18 are produced by this method.

5.18 Theorems [2]

1. If $v \equiv 3 \pmod{20}$ and $v \geq 23$, then there exists a $(v, 3; \{5\}, 2)$-IPBD.
2. If $v \equiv 7 \pmod{20}$ and $v \geq 47$, then there exists a $(v, 7; \{5\}, 2)$-IPBD.
3. If $v \equiv 19 \pmod{20}$ and $v \geq 39$, then there exists a $(v, 7; \{5\}, 2)$-IPBD with the possible exception of $v \in \{99, 199\}$. Further there exists a $(v, 19; \{5\}, 2)$-IPBD for $v \in \{99, 199\}$.

5.19 Remark Another powerful method of creating designs of higher index is the method of differences. See §IV.10.

5.2 GDDs of Higher Index

5.20 Theorem [8] Necessary and sufficient conditions for the existence of a $(3,\lambda)$-GDD of type m^u are

1. $u \geq 3$,
2. $\lambda(u-1)m \equiv 0 \pmod 2$,
3. $\lambda u(u-1)m^2 \equiv 0 \pmod 6$.

5.21 Theorem [8] Necessary and sufficient conditions for the existence of a $(4,\lambda)$-GDD of type m^u are

1. $u \geq 3$,
2. $\lambda(u-1)m \equiv 0 \pmod 4$,
3. $\lambda u(u-1)m^2 \equiv 0 \pmod{12}$.

with the exception of $(m,u,\lambda) \in \{(2,4,1),(6,4,1)\}$, in which case no such GDD exists.

5.3 See Also

§I.1	BIBDs of higher index are restricted classes of PBDs and are often constructed using PBDs of higher index.
§III.1	PBDs of higher index generalize PBDs of index one.
§III.2	Constructions of PBDs of index one often generalize to larger indices.
§IV.8	Coverings of pairs are often constructed using PBDs.
§IV.33	Packings for pairs are often constructed using PBDs.
[1, 8]	More on designs with higher index.

References

[1] H. Hanani, Balanced incomplete block designs and related designs, *Discrete Math.* 11 (1975), 255-369.

[2] W.H. Mills and R.C. Mullin, On λ-covers of pairs by quintuples: v odd, *J. Combin. Theory A* 67 (1994), 245-272.

[3] R.M. Wilson, An existence theory for pairwise balanced designs II. The structure of PBD-closed sets and the existence conjectures, *J. Combin. Theory A* 13 (1972), 246-273.

[4] R.M. Wilson, An existence theory for pairwise balanced designs III: A proof of the existence conjectures, *J. Combin. Theory A* 18 (1975), 71-79.

[5] Wu Dianhu and Wei Ruizhong, A note on IGDDs, *J. Suzhou Univ.* 8 (1992), 142-145.

[6] J. Yin, On λ-packings of pairs by quintuples: $\lambda \equiv 0 \pmod 4$, *Ars Combin.* 37 (1994), 288-300.

[7] J. Yin and K. Chen, A complete solution to the packing problem with block size six and index five, *Ars Combin.* 37 (1994) 257-261.

[8] L. Zhu, Some recent developments on BIBDs and related designs, *Discrete Math.* 123 (1993), 189-214.

6 PBDs, Frames, and Resolvability

Ronald C. Mullin
Hans-Dietrich O.F. Gronau

6.1 Frames and Resolvability

6.1 **Remark** While the set $R(k) = \{r : \exists$ a resolvable BIBD$(v,k,1)$ with replication number $r\}$ is PBD-closed for fixed k, the set of replication numbers of a resolvable BIBD(v,k,λ) is not in general PBD-closed for $\lambda \geq 2$. To overcome this difficulty, Hanani [3] introduced resolvable frames.

Let $\mathcal{B}$ be the block set of a PBD, GDD, or BIBD of index λ. Then an *α-resolution class* is a collection of blocks, which together contain every element of the design exactly α times.

A PBD, GDD, or BIBD of index λ is *α-resolvable* if the blocks of the design can be partitioned into α-resolution classes.

6.2 **Theorem** Let D be a PBD$(v,K;\lambda)$ that is λ-resolvable. Let n denote the number of λ-resolution classes of D. Then there exists a PBD$(v+n, K \cup (K+1) \cup \{n\}; \lambda)$, where $K+1 = \{k+1 : k \in K\}$. In particular, if there exists a λ-resolvable BIBD(v,k,λ), with $v \neq k^2$ then there exists a PBD$(v+n, \{k+1, n^\dagger\}; \lambda)$, where the symbol $n^\dagger$ indicates that there are λ identical blocks of size n in the design, and no other blocks of size n. If $v = k^2$, then one obtains a PBD$(k^2+k+1, \{k+1\}; \lambda)$, that is, a BIBD$(k^2+k+1, k+1, \lambda)$, a design which has the parameters of a λ-fold copy of a (perhaps nonexistent) projective plane of order n.

6.3 **Remark** Theorem 6.2 demonstrates that λ-resolvable PBDs of index λ are useful in creating λ-resolvable BIBDs, as well as other objects. The purpose of using α-resolvable designs of index λ is that it is possible to combine such designs with certain values for α, with those with certain combinations of λ to produce a λ'-resolvable PBD.

6.4 **Remark** An even more powerful method for creating resolvable objects is the use of frames.

Let k be a positive integer and let v and α be positive integers. Let $\mathcal{V}$ be a set of cardinality v. Let any subset of size k of $\mathcal{V}$ be a *block*. Then a *(k,α)-partial* resolution class is a collection $\mathcal{C}$ of blocks such that every element of V occurs in either exactly α or exactly zero blocks of $\mathcal{C}$. The set of elements of V not occurring in the partial resolution class is the *complement* of the class.

Let k, α, and λ be positive integers. A *$(\lambda,\alpha;k)$-frame* is a triple $(\mathcal{V},\mathcal{G},\mathcal{P})$ where $\mathcal{V}$ is a set of cardinality v, $\mathcal{G}$ is a partition of $\mathcal{V}$ into parts (*groups*), and $\mathcal{P}$ is a collection of (k,α)-partial resolution classes of $\mathcal{V}$ which satisfies the conditions:

1. The complement of each (k,α)-partial resolution class P of $\mathcal{P}$ is a group $G \in \mathcal{G}$;
2. Each unordered pair $\{x,y\}$ of $\mathcal{V}$ which does not lie in some group G of $\mathcal{G}$ lies in precisely λ blocks of $\mathcal{P}$;
3. No unordered pair $\{x,y\}$ of elements of $\mathcal{V}$ which lies in some group G of $\mathcal{G}$ also lies in a block of $\mathcal{P}$.

The *type* of the $(\lambda, \alpha; k)$-frame is the multiset $T = [|G| : G \in \mathcal{G}]$. If $\mathcal{G}$ contains a_1 groups of size g_1, a_2 groups of size g_2, etc. where $v = a_1g_1 + a_2g_2 + \ldots + a_sg_s$, then the exponential notation $g_1^{a_1}g_2^{a_2}\ldots g_s^{a_s}$ is also used. By convention, factors of the type 0^a can be included in the exponential form of the type to accommodate null groups when necessary.

6.5 **Remark** A $(\lambda, \alpha; k)$-frame is simply a (k, λ)-GDD with an additional property on the blocks of the design.

6.6 **Theorem** [2] Let k, α, and λ be positive integers. Suppose that there exists a $(\lambda, \alpha; k)$-frame of type $g_1^{a_1}g_2^{a_2}\ldots g_s^{a_s}$, and that for $i = 1, 2, \ldots, s$, there exists an α-resolvable PBD$(g_i + 1, k; \lambda)$. Then there exists an α-resolvable PBD$((\sum_{i=1}^{s} a_ig_i) + 1, k, \lambda)$.

6.7 **Theorem** [2] Let $(\mathcal{V}, \mathcal{G}, \mathcal{B})$ be a GDD of index λ_1 (the master GDD) and let w be a mapping $\mathcal{V} \to \mathbb{Z}^+ \cup \{0\}$. Suppose that for each $B = \{x_1, x_2, \ldots, x_k\}$ in $\mathcal{B}$, there exists a $(\lambda_2, \alpha; k)$-frame of type $[w(x_1), w(x_2), \ldots, w(x_k)]$. Then there exists a $(\lambda_1\lambda_2, \lambda_1\alpha; k)$-frame of type $T = \left[\sum_{x \in G} w(x) : G \in \mathcal{G}\right]$.

6.8 **Corollary** Let $F(\lambda, \alpha; k, g) = \{u : \exists \text{ a } (\lambda, \alpha; k)\text{-frame of type } g^u\}$. Then $F(\lambda, \alpha; k, g)$ is PBD-closed. (Apply Theorem 6.7 with $\lambda_1 = 1$).

6.9 **Remark** Theorem 6.7 has been used to construct classes of α-resolvable GDD of index λ and α-resolvable BIBD(v, k, λ). However, the main application lies in the area of 1-resolvability, that is, the case of resolvable designs. Within that area, the main emphasis has been on $(1, 1; k)$-frames, both because of the importance of the construction of RBIBD$(v, k, 1)$, and due to the fact that a $(\lambda, 1; k)$-frame can be obtained from a $(1, 1; k)$-frame by replicating the block set λ times, maintaining the frame structure.

A *k-frame* of type T is a $(1, 1; k)$-frame of type T. (This simplified notation is frequently used for this special case).

6.10 **Theorem** (Stinson) [6] If there exists a TD$(k+1, kw) -$ TD$(k+1, w)$, then there exists a k-frame of type $(w(k-1))^{k+1}$. Conversely if there exists a k-frame of type t^{k+1}, then $(k-1)$ divides t, say $t = w(k-1)$, and there exists a TD$(k+1, kw) -$ TD$(k+1, w)$.

6.11 **Theorem** [8] Necessary conditions for the existence of a (k, λ)-frame of type h^u are

1. $u \geq k + 1$,
2. $h(u-1) \equiv 0 \pmod{k}$,
3. $\lambda h \equiv 0 \pmod{k-1}$.

6.12 **Theorem** [5] There exists a $(2, \lambda)$-frame of type h^u if and only if $u \geq 3$ and $h(u-1) \equiv 0 \pmod{2}$.

6.13 **Theorem** [7] There exists a $(3, \lambda)$-frame of type h^u if and only if $u \geq 4$, $\lambda h \equiv 0 \pmod{2}$ and $h(u-1) \equiv 0 \pmod{3}$.

6.14 **Theorem** [1] There exists a 4-frame of type h^u if and only if $u \geq 5$, $h \equiv 0 \pmod{3}$ and $h(u-1) \equiv 0 \pmod{4}$, except possibly where

1. $h = 9$ and $u \in \{13, 17, 29, 93, 113, 153, 173, 193\}$,
2. $h \equiv 0 \pmod{12}$ and $u \in \{8, 12\}$,
   $h = 36$ and $u \in \{18, 23, 28, 38, 43, 48\}$,
3. $h \equiv 6 \pmod{12}$ and $u \in \{17, 29, 33, 563\} \cup \{n : n \equiv 3 \text{ or } 11 \pmod{12} \text{ and } n \leq 527\} \cup \{n : n \equiv 7 \pmod{12} \text{ and } n \leq 259\}$,
   $h = 18$, $u = 17$ or $u \in \{n : n \equiv 3 \pmod{4} \text{ and } n \leq 267\}$.

Further there exists a 4-frame of type h^u if

1. $h = 24$ and $u = 8$,
2. $h = 72$ and $u \geq 5$,
3. $h = 6$ and $u \in \{17, 29, 33, 563\} \cup \{n : n \equiv 3 \text{ or } 11 \pmod{12} \text{ and } n \geq 275\}$.

6.2 See Also

§I.6	More on frames and resolvable designs.
[4, 8]	Surveys of frames and resolvable structures.

References

[1] C.J. Colbourn, D.R. Stinson and L. Zhu, More frames with block size 4, *J. Combin. Math. Combin. Comput.*, to appear.

[2] S.C. Furino and R.C. Mullin, Block designs with large holes and α resolvable BIBDs, *J. Combin. Des.* 1 (1993), 101-112.

[3] H. Hanani, Balanced incomplete block designs and related designs, *Discrete Math.* 11 (1975), 255-369.

[4] Y. Miao, J. Yin and S. Furino, *Frames and Resolvable Designs*, CRC Press, Boca Raton, to appear.

[5] R. Rees and D.R. Stinson, Frames with block size four, *Canad. J. Math.* 44 (1992), 1030-1049.

[6] D.R. Stinson, The equivalence of certain incomplete transversal designs and frames, *Ars Combin.* 22 (1986), 81-87.

[7] D.R. Stinson, Frames for Kirkman triple systems, *Discrete Math.* 65 (1987), 289-300.

[8] L. Zhu, Some recent developments on BIBDs and related designs, *Discrete Math.* 123 (1993), 189-214.

PART IV

OTHER COMBINATORIAL DESIGNS

1 Association Schemes

Christopher D. Godsil

1.1 Preliminaries

An *association scheme* $\mathcal{A}$ with vertex set X and d classes is a set of symmetric $\{0,1\}$-matrices $A_0, \dots, A_d$ with rows and columns indexed by X, such that

1. $A_0 = I$,
2. $\sum_{i=0}^{d} A_i = J$,
3. A_iA_j lies in the real span of $A_0, \dots, A_d$: $A_iA_j = \sum_k p_{ij}^k A_k$.

Here J denotes the matrix with each entry equal to one (and order determined by context).

1.1 **Remarks** The third axiom implies that the product A_iA_j is symmetric, from which it follows that $A_iA_j = A_jA_i$. It also implies that the vector space formed by the real span of $A_0, \dots, A_d$ is a matrix algebra. This is the *Bose-Mesner algebra* of the association scheme. Association schemes as just defined are *symmetric.*

A *nonsymmetric* association scheme $\mathcal{A}$ with vertex set X and d classes is a set of $\{0,1\}$-matrices $A_0, \dots, A_d$ with rows and columns indexed by X such that axioms (1), (2) and (3) hold, along with:

4. if $A_i \in \mathcal{A}$, then $A_i^T \in \mathcal{A}$,
5. $A_iA_j = A_jA_i$ for all i and j.

1.2 **Remark** In the symmetric case, axiom (3) implies axiom (5).

1.3 **Example** The center of the complex group algebra of a finite group G, viewed as an algebra of $|G| \times |G|$ matrices, is a nonsymmetric scheme (that is symmetric if and only if each element of G is conjugate to its inverse).

1.4 **Remark** The matrices $A_1, \dots, A_d$ can be viewed as adjacency matrices of graphs $G_1, \dots, G_d$; thus an association scheme may be viewed as a partition of the edge set of a complete graph into d spanning subgraphs.

The *Hamming scheme* $H(n,q)$ has the set of all words of length n over an alphabet of q symbols as its vertex set. The $\alpha\beta$-entry of A_i is one if and only if the Hamming distance between α and β is i. The *Johnson scheme* $J(v,k)$ has the set of all k-subsets of a set of size v as its vertices. The $\alpha\beta$-entry of A_i is one if and only if $|\alpha \cap \beta| = k - i$.

1.5 **Lemma** Let $\mathcal{A}$ be an association scheme with d classes. Then each graph in $\mathcal{A}$ is regular, with diameter at most d. If $d = 2$ then both graphs in $\mathcal{A}$ are strongly regular (see §VI.5).

An association scheme is *metric* relative to G_i if the diameter of G_i is equal to d. An association scheme with d classes is *primitive* if each of the graphs $G_1, \dots, G_d$ is connected; otherwise, it is *imprimitive.*

1.6 **Example** The Hamming and Johnson schemes are metric, relative to G_1 in each case.

If a scheme is metric relative to G_1 and $S \subseteq X$, then the *minimum distance* of S is the minimum distance in G_1 between two distinct vertices of S. If S has minimum distance δ, then $\lfloor \frac{\delta-1}{2} \rfloor$ is the *packing radius* of S; the *covering radius* of S is the maximum distance of any vertex in X from S. If the packing radius of S is equal to its covering radius, it is a *perfect code.*

If S is a subset of the vertices of an association scheme with size b and characteristic vector x, then the vector a in $\mathbb{R}^{d+1}$ defined by $a_i = b^{-1}x^T A_i x$ is the *inner distribution* of S. The number of positive indices i such that $a_i \neq 0$ is the *degree* of S; it represents the number of distinct nonzero distances which occur among the members of S. The entries of the inner distribution of a single vertex are the *valencies* of the scheme, denoted by v_i (v_i is the row sum of A_i, a constant).

1.7 Remarks As the matrices in the Bose-Mesner algebra are normal and commute, $\mathbb{R}^X$ can be expressed as the direct sum of its eigenspaces. There are exactly $d+1$ eigenspaces and the orthogonal projections onto them are denoted by $E_0, \ldots, E_d$, where $E_0 = |X|^{-1}J$. As E_i is idempotent, its trace and its rank are equal, and both are equal to the dimension of the corresponding eigenspace.

The dimension of the i-th eigenspace is the i-th *multiplicity* of the scheme, and is denoted by f_i.

The *dual degree* of a subset S of X with characteristic vector x is one less than the size of the set $\{i : E_i x \neq 0\}$. It is denoted by s^*.

1.8 Remark Covering radius and dual degree are interesting and useful concepts in coding theory, but have not received much attention in design theory yet.

1.9 Lemma Let $\mathcal{A}$ be a metric association scheme with vertex set X, and let S be a subset of X with covering radius r and dual degree s^*. Then $r \leq s^*$.

The *outer distribution matrix* of S is the matrix with columns $A_0x, A_1x, \ldots, A_dx$. The dual degree of S is also one less than the rank of the outer distribution matrix of S. A subset S of the vertices of a metric association scheme is *completely regular* if its covering radius is one less than the number of distinct rows in its outer distribution matrix.

The *dual inner distribution* of S is the vector b in $\mathbb{R}^{d+1}$ with $b_i := \frac{|X|}{|S|}x^T E_i x$. It is also known as the *MacWilliams transform* of the inner distribution.

1.10 Example If $H(n,q)$ is the Hamming scheme with a finite field of order q as alphabet and C is a subspace of $H(n,q)$, then the dual inner distribution of C is the inner distribution of the dual code to C, scaled by $|C|$.

1.11 Theorem Let $A_0, \ldots, A_d$ form an association scheme and let $E_0, \ldots, E_d$ be its principal idempotents. Then both the matrices A_i and the matrices E_j form a basis for the Bose-Mesner algebra of the scheme. Hence there are numbers $p_i(j)$ such that $A_i = \sum_j p_i(j)E_j$ and numbers $q_j(i)$ such that $E_j = |X|^{-1}\sum_i q_j(i)A_i$.

The $p_i(j)$ are the *eigenvalues* of the scheme, while the $q_i(j)$ are the *dual eigenvalues.* Define the $(d+1)\times(d+1)$ matrix P by $(P)_{i,j} := p_j(i)$ and define the $(d+1)\times(d+1)$ matrix Q by $(Q)_{i,j} := q_j(i)$. The entries of the i-th column of P are the eigenvalues of A_i. One has $PQ = |X|I$ and also $f_j p_i(j) = v_i q_j(i)$.

1.12 Theorem In any association scheme $\sum_r v_r q_i(r) q_j(r) = \delta_{ij}|X|f_r$.

1.13 Theorem If a is the inner distribution of a subset of X, one has *Delsarte's inequalities*, also known as the *linear programming bound*: $Q^T a \geq 0$. Here $Q^T a = b$, the dual inner distribution vector.

1.14 Remark Delsarte's inequalities are very important, and provide the main reason why the entries of Q are given for $J(v,k)$ and $H(n,q)$ in §1.2. For applications in design theory and coding theory, see [6, 11, 16].

1.2 The Hamming and Johnson Schemes

1.15 Remark The vertices of the Hamming scheme can be regarded as words over an alphabet Q with $|Q| = q$. If α is a vertex of $H(n,q)$ and $I \subseteq \{1,\ldots,n\}$, let $\alpha(I)$ be the "subword" $(\alpha_i)_{i\in I}$. A subset S of the vertices of $H(n,q)$ has *strength at least* t if, for each t-subset T of $\{1,\ldots,n\}$, every sequence in Q^t occurs exactly λ_t times as a subword $\alpha(T)$ when α ranges over the elements of S.

1.16 Theorem For the Hamming scheme $H(n,q)$, $v_i = f_i = (q-1)^i\binom{n}{i}$ and

$$q_i(j) = p_i(j) = \sum_{r=0}^{i}(-1)^r(q-1)^{i-r}\binom{n-j}{i-r}\binom{j}{r}.$$

1.17 Theorem A subset of $H(n,q)$ with strength at least t is a orthogonal array of strength at least t.

1.18 Example If S is a linear code then its strength as an orthogonal array is one less than the minimum distance of its dual code.

The *strength* of a subset S of $J(v,k)$ is the largest integer t such that S is a t-design.

1.19 Theorem For the Johnson scheme, with the understanding that $\binom{v}{-1} = 0$, one has

$$v_i = \binom{k}{i}\binom{v-k}{i}, \qquad f_i = \binom{v}{i} - \binom{v}{i-1},$$

$$p_i(j) = \sum_{r=0}^{i}(-1)^{i-r}\binom{k-r}{i-r}\binom{v-k+r-j}{r}\binom{k-j}{r}, \text{ and}$$

$$q_i(j) = \binom{v-2i}{k-i}^{-1}\binom{v}{k}\sum_{r=0}^{i}(-1)^{i-r}\binom{v+1-i-r}{i-r}^{-1}\binom{k-r}{i-r}^2\binom{k-j}{r}.$$

1.20 Lemma (Delsarte) Let S be a subset of the vertices of $J(v,k)$. Then S is a t-design if and only if the entries $b_1,\ldots,b_t$ of the dual inner distribution of S are all zero.

1.21 Lemma (Delsarte) Let S be a subset of the vertices of $H(n,q)$. Then S is an orthogonal array of strength t if and only if the entries $b_1,\ldots,b_t$ of the dual inner distribution of S are all zero.

1.3 Parameters of Designs

1.22 Lemma If $\mathcal{D}$ is a t-design with degree s, then $t \leq 2s$.

1.23 Theorem (Ray-Chaudhuri and Wilson [14]) If $\mathcal{D}$ has v points and b blocks, then $\binom{v}{t/2} \leq b \leq \binom{v}{s}$. If either one of these inequalities is tight, so is the other.

Designs with $t = 2s$ are *tight designs*.

1.24 Theorem

1. [3, 7, 9, 10, 15] The only tight 4-designs are the Witt design on 23 points and its complement (see §I.4.7).
2. [13] There are no tight 6-designs.

3. [1] If $t = 2s \geq 10$, then there are at most finitely many tight t-designs.

1.25 Remark When $t = 2$, tight designs are equivalent to symmetric designs.

Let $\mathcal{D}$ be a design with degree set $i_1, \ldots, i_s$. Let A_r be the 01-matrix with rows and columns indexed by the blocks of $\mathcal{D}$ and with $\alpha\beta$-entry equal to one if and only if $|\alpha \cap \beta| = i_r$. Let A_0 be the identity matrix. $\mathcal{D}$ is *schematic* if $A_0, \ldots, A_s$ forms an association scheme, the scheme *induced* by S. A 1-design is a *partially balanced incomplete block design* (PBIBD) if and only if its dual design is schematic.

1.26 Theorem (Cameron [5]) Let $\mathcal{D}$ be a t-design with degree s. If $t \geq 2s - 2$, then $\mathcal{D}$ is schematic.

A 2-design with degree two is a *quasi-symmetric design.*

1.27 Lemma The degree set of a design with $t \geq 2s - 1$ is determined by v, k, and the number of blocks.

1.28 Remark Schematic subsets in the Hamming scheme may be defined in the same way; the condition $t \geq 2s - 2$ is again sufficient for a subset to be schematic and one always has $t \leq 2s$. (These claims, and close analogs of the Ray-Chaudhuri and Wilson bounds, hold in any Q-polynomial association scheme. See Delsarte [6]: §5.3.)

1.4 See Also

§I.5	Symmetric designs are equivalent to tight designs with $t = 2$.
§IV.35	Association schemes underlie PBIBDs.
§IV.37	Quasi-symmetric designs are 2-designs of degree 2.
[2, 4]	Standard texts, concentrating on distance-regular graphs.
[12]	Contains a brief introduction to association schemes.
[8]	A more detailed introduction to association schemes.
[6]	The most influential work, especially Delsarte's inequalities.
[11, 16, 17]	Contains applications of Delsarte's inequalities.

References

[1] E. Bannai, On tight designs, *Quart. J. Math. Oxford* 28 (1977), 433-448.
[2] E. Bannai and T. Ito, *Algebraic Combinatorics I: Association Schemes*, Benjamin/ Cummings, London, 1984.
[3] A. Bremner, A diophantine equation arising from tight 4-designs, *Osaka J. Math.* 16 (1979), 352-356.
[4] A.E. Brouwer, A.M. Cohen and A. Neumaier, *Distance-Regular Graphs*, Springer-Verlag, Berlin, 1989.
[5] P.J. Cameron, Near regularity conditions for designs, *Geom. Ded.* 2 (1973), 213-223.
[6] P. Delsarte, An algebraic approach to the association schemes of coding theory, *Philips Research Reports Supplements* 1973, No. 10.
[7] H. Enomoto, N. Ito and R. Noda, Tight 4-designs, *Osaka J. Math.* 16 (1979), 39-43.
[8] C.D. Godsil, *Algebraic Combinatorics*, Chapman and Hall, New York, 1993.
[9] N. Ito, On tight 4-designs, *Osaka J. Math.* 12 (1975), 493-522.
[10] N. Ito, Corrections and supplements to 'On tight 4-designs', *Osaka J. Math.* 15 (1978), 693-697.
[11] F.J. MacWilliams and N.J.A. Sloane, *The Theory of Error-Correcting Codes*, North-Holland, Amsterdam, 1977.

[12] J.H. van Lint and R.M. Wilson, *A Course in Combinatorics*, Cambridge University Press, London, 1992.
[13] C. Peterson, Tight 6-designs, *Osaka J. Math.* 14 (1977), 417-435.
[14] D.K. Ray-Chaudhuri and R.M. Wilson, On t-designs, *Osaka J. Math.* 12 (1975), 737-744.
[15] R.J. Stroeker, On the diophantine equation $(2y^2-3)^2 = x^2(3x^2-2)$ in connection with the existence of nontrivial tight 4-designs, *Indag. Math.* 43 (1981), 353-358.
[16] R.M. Wilson, On the theory of t-designs, in: *Enumeration and Designs* (D.M. Jackson and S.A. Vanstone; eds.), Academic Press, Toronto, 1984, 19-49.
[17] R.M. Wilson, The exact bound in the Erdős-Ko-Rado theorem, *Combinatorica* 4 (1984), 247-257.

2 Balanced (Part) Ternary Designs

Thomas Kunkle
Dinesh G. Sarvate

2.1 Definitions and Example

A *balanced ternary design* BTD$(V, B; \rho_1, \rho_2, R; K, \Lambda)$, is an arrangement of V elements into B multisets, or *blocks*, each of cardinality K $(K \leq V)$, satisfying:

1. Each element appears $R = \rho_1 + 2\rho_2$ times altogether, with multiplicity one in exactly ρ_1 blocks, with multiplicity two in exactly ρ_2 blocks.
2. Every pair of distinct elements appears Λ times; i.e., if m_{vb} is the multiplicity of the vth element in the bth block, then for every for every pair of distinct elements v and w, $\sum_{b=1}^{B} m_{vb} m_{wb} = \Lambda$.

A *balanced part ternary design* BPTD$(V; b_1, b_2, B; \rho_1, \rho_2, R; K, \Lambda)$ is a BTD that also satisfies:

3. There exist exactly $b_2 = B - b_1$ blocks each containing at least one element of multiplicity two.

Hence, every BTD is a BPTD for some choice of b_1 and b_2.

2.1 **Example** A BPTD(4; 1,8,9; 3,3,9; 4,7).

1	1	1	2	2	3	3	3	1
1	1	1	2	2	3	3	3	2
2	3	4	3	4	1	1	4	3
2	4	4	4	4	2	2	4	4

2.2 A BPTD Parameter List

2.2 **Table** The known and possible values of b_2 for each $V; B; \rho_1, \rho_2, R; K, \Lambda$, based on the list in [3], Only parameters for which $\min(K, \Lambda) > 3$ are given, since otherwise $b_2 = V\rho_2$. Column 1 shows, in parentheses, the row number of these parameters in [3]. Column 9 contains the values of b_2 for which a BPTD is known to exist. Column 10 shows the values of b_2 for which existence is in question. See also [3] and [6] throughout.

# ([3])	V	B	ρ_1	ρ_2	R	K	Λ	Known b_2's	Open b_2's	Ref.
1 (7)	5	5	1	2	5	5	4	5	none	[6]
2 (11)	8	8	4	1	6	6	4	8	none	[7]
3 (14)	5	6	2	2	6	5	5	5	none	[6]
4 (18)	11	11	5	1	7	7	4	11	none	[7]
5 (22)	5	7	3	2	7	5	6	5	none	[6]
6 (23)	4	7	1	3	7	4	5	6	none	[6]
7 (24)	10	10	1	3	7	7	4	10	none	[7]
8 (25)	7	7	1	3	7	7	6	7	none	[7]
9 (30)	10	10	6	1	8	8	6	10	none	[7]
10 (33)	6	12	4	2	8	4	4	8, 10...12	none	[7]
11 (34)	5	10	4	2	8	4	5	9, 10	none	[7]
12 (35)	5	8	4	2	8	5	7	5	none	[6]
13 (37)	4	8	2	3	8	4	6	6, 8	none	[6]
14 (38)	7	8	2	3	8	7	7	7	none	[7]
15 (43)	14	18	7	1	9	7	4	none	14	[7, 4]
16 (45)	15	15	7	1	9	9	5	15	none	[7]
17 (46)	11	11	7	1	9	9	7	11	none	[7]
18 (49)	5	9	5	2	9	5	8	5, 9	none	[6]
19 (50)	18	18	5	2	9	9	4	18	none	[7]
20 (55)	4	9	3	3	9	4	7	6, 8, 9	none	[6]
21 (56)	7	9	3	3	9	7	8	7	9	[6, 7]
22 (57)	12	12	3	3	9	9	6	12	none	[7, 1]
23 (59)	9	9	1	4	9	9	8	9	none	[7]
24 (61)	8	20	8	1	10	4	4	4	5...8	
25 (63)	9	15	8	1	10	6	6	9	none	[4]
26 (66)	23	23	8	1	10	10	4	23	none	[7, 1]
27 (67)	12	12	8	1	10	10	8	12	none	[7]
28 (73)	10	20	6	2	10	5	4	10	11...20	
29 (74)	7	14	6	2	10	5	6	14	7...13	
30 (75)	5	10	6	2	10	5	9	5, 9, 10	none	[6]
31 (76)	12	15	6	2	10	8	6	none	6...15	
32 (78)	4	10	4	3	10	4	8	6, 8...10	none	[6]
33 (79)	12	20	4	3	10	6	4	none	12...20	
34 (80)	7	10	4	3	10	7	9	7	none	[6, 4]
35 (83)	13	13	4	3	10	10	7	13	none	[1]
36 (84)	9	18	2	4	10	5	4	18	none	
37 (85)	5	10	2	4	10	5	8	10	none	
38 (87)	9	10	2	4	10	9	9	9	none	[7]
39 (91)	16	22	9	1	11	8	5	none	4...16	
40 (94)	28	28	9	1	11	11	4	none	28	[7, 1]
41 (95)	19	19	9	1	11	11	6	19	none	[7]
42 (96)	13	13	9	1	11	11	9	13	none	[7]
43 (98)	5	11	7	2	11	5	10	5, 9, 10	none	[6]
44 (99)	20	22	7	2	11	10	5	none	8...22	
45 (102)	4	11	5	3	11	4	9	6, 8...11	none	[6]
46 (103)	7	11	5	3	11	7	10	7	11	[6, 4]
47 (104)	27	27	5	3	11	11	4	none	17...27	
48 (105)	14	14	5	3	11	11	8	14	none	[1]
49 (106)	10	22	3	4	11	5	4	none	20...22	
50 (107)	5	11	3	4	11	5	9	10	none	[7]
51 (108)	9	11	3	4	11	9	10	9	10, 11	

# ([3])	V	B	ρ_1	ρ_2	R	K	Λ	Known b_2's	Open b_2's	Ref.
52 (110)	6	11	1	5	11	6	9	10	none	[7]
53 (113)	11	11	1	5	11	11	10	11	none	[7]
54 (121)	27	27	10	1	12	12	5	none	27	[1]
55 (122)	14	14	10	1	12	12	10	14	none	[7]
56 (128)	9	27	8	2	12	4	4	18	9...17	
57 (129)	5	15	8	2	12	4	8	7...10	none	[7]
58 (130)	5	12	8	2	12	5	11	5, 9, 10	none	[6]
59 (132)	15	30	8	2	12	6	4	30	10...29	
60 (133)	9	18	8	2	12	6	7	18	6...17	
61 (134)	8	16	8	2	12	6	8	16	8...15	[4]
62 (135)	24	32	8	2	12	9	4	none	12...32	
63 (136)	33	33	8	2	12	12	4	none	33	[7, 1]
64 (137)	17	17	8	2	12	12	8	none	17	[1]
65 (144)	7	21	6	3	12	4	5	21	11...20	[7]
66 (145)	6	18	6	3	12	4	6	18	10...17	[7]
67 (146)	4	12	6	3	12	4	10	6, 8...12	none	[6]
68 (147)	10	20	6	3	12	6	6	20	10...19	
69 (148)	7	14	6	3	12	6	9	7	8...14	[6]
70 (149)	7	12	6	3	12	7	11	7	11, 12	[6, 4]
71 (150)	14	21	6	3	12	8	6	none	11...21	
72 (152)	19	19	6	3	12	12	7	19	none	[1]
73 (153)	15	15	6	3	12	12	9	15	none	[5, 1]
74 (157)	8	24	4	4	12	4	4	24	16...23	
75 (158)	5	15	4	4	12	4	7	10, 12...15	none	[6, 4]
76 (159)	5	12	4	4	12	5	10	10	11...12	
77 (160)	14	28	4	4	12	6	4	28	19...27	
78 (161)	20	30	4	4	12	8	4	none	20...30	
79 (162)	12	16	4	4	12	9	8	none	12...16	
80 (163)	9	12	4	4	12	9	11	9, 12	10, 11	
81 (164)	32	32	4	4	12	12	4	none	22...32	
82 (166)	6	12	2	5	12	6	10	10, 12	none	[7]
83 (168)	11	12	2	5	12	11	11	11	none	[7]
84 (171)	18	26	11	1	13	9	6	none	5...18	
85 (172)	33	39	11	1	13	11	4	none	7...33	
86 (175)	23	23	11	1	13	13	7	23	none	[7]
87 (176)	15	15	11	1	13	13	11	15	none	[7]
88 (179)	8	26	9	2	13	4	5	13	8...12, 14...16	
89 (181)	5	13	9	2	13	5	12	5, 9, 10	none	[6]
90 (182)	22	26	9	2	13	11	6	none	9...26	
91 (183)	39	39	9	2	13	13	4	none	39	[1]
92 (184)	20	20	9	2	13	13	8	20	none	[1]
93 (188)	4	13	7	3	13	4	11	6, 8...12	none	[6]
94 (189)	7	13	7	3	13	7	12	7	11...13	[6, 4]
95 (190)	31	31	7	3	13	13	5	none	31	[1]
96 (192)	16	16	7	3	13	13	10	none	16	[1]
97 (194)	5	13	5	4	13	5	11	10	11...13	
98 (195)	9	13	5	4	13	9	12	9	10...13	
99 (197)	12	26	3	5	13	6	5	none	20...26	
100 (198)	6	13	3	5	13	6	11	10, 12	none	[7, 4]
101 (199)	11	13	3	5	13	11	12	11	12, 13	
102 (200)	4	13	1	6	13	4	9	12	none	[7, 6]

# ([3])	V	B	ρ_1	ρ_2	R	K	Λ	Known b_2's	Open b_2's	Ref.
103 (203)	19	19	1	6	13	13	8	19	none	[7]
104 (205)	13	13	1	6	13	13	12	13	none	[7]
105 (207)	6	21	12	1	14	4	8	6	5	[7]
106 (208)	10	28	12	1	14	5	6	9	5...8, 10	
107 (209)	18	42	12	1	14	6	4	6	7...18	[8]
108 (214)	37	37	12	1	14	14	5	none	37	[1]
109 (215)	31	31	12	1	14	14	6	31	none	[1]
110 (216)	21	21	12	1	14	14	9	none	21	[1]
111 (218)	16	16	12	1	14	14	12	16	none	[7]
112 (222)	5	14	10	2	14	5	13	5, 9, 10	none	[6]
113 (223)	12	28	10	2	14	6	6	none	8...24	
114 (225)	21	42	10	2	14	7	4	42	14...41	
115 (226)	17	34	10	2	14	7	5	34	12...33	
116 (227)	11	22	10	2	14	7	8	22	8...21	
117 (228)	9	18	10	2	14	7	10	18	6...17	
118 (231)	10	35	8	3	14	4	4	25	15...24, 26...30	
119 (232)	4	14	8	3	14	4	12	6, 8...12	none	[6]
120 (233)	9	21	8	3	14	6	8	16	9...15, 17...21	
121 (234)	14	28	8	3	14	7	6	none	14...28	
122 (235)	7	14	8	3	14	7	13	7	11...14	[6, 4]
123 (236)	24	42	8	3	14	8	4	none	18...42	
124 (237)	25	35	8	3	14	10	5	none	15...35	
125 (238)	45	45	8	3	14	14	4	none	45	[1]
126 (239)	23	23	8	3	14	14	8	none	23	[1]
127 (240)	17	17	8	3	14	14	11	none	17	[1]
128 (243)	5	14	6	4	14	5	12	10	11...14	
129 (244)	20	40	6	4	14	7	4	40	27...39	
130 (245)	16	28	6	4	14	8	6	16	17...28	[8]
131 (246)	27	42	6	4	14	9	4	none	27...42	
132 (247)	9	14	6	4	14	9	13	9	10...14	
133 (248)	30	30	6	4	14	14	6	none	30	[1]
134 (249)	6	14	4	5	14	6	12	10, 12, 14	none	[7, 4]
135 (250)	12	21	4	5	14	8	8	none	15...21	
136 (251)	18	28	4	5	14	9	6	none	23...28	
137 (252)	30	42	4	5	14	10	4	none	30...42	
138 (253)	11	14	4	5	14	11	13	11	12...14	
139 (255)	6	21	2	6	14	4	6	21	18...20	[3, 6]
140 (256)	4	14	2	6	14	4	10	12	13...14	
141 (257)	19	38	2	6	14	7	4	none	none	[4]
142 (258)	13	26	2	6	14	7	6	26	none	
143 (259)	10	20	2	6	14	7	8	20	none	
144 (261)	7	14	2	6	14	7	12	14	none	
145 (262)	20	28	2	6	14	10	6	none	24...28	
146 (264)	13	14	2	6	14	13	13	13	none	[7]
147 (266)	18	18	2	6	14	14	10	none	18	[1]
148 (273)	20	30	13	1	15	10	7	none	4...20	
149 (275)	53	53	13	1	15	15	4	53	none	[1]
150 (276)	27	27	13	1	15	15	8	27	none	[7]
151 (277)	17	17	13	1	15	15	13	17	none	[7]
152 (281)	15	45	11	2	15	5	4	30	15...29	
153 (282)	9	27	11	2	15	5	7	9	10...18	

# ([3])	V	B	ρ_1	ρ_2	R	K	Λ	Known b_2's	Open b_2's	Ref.
154 (283)	8	24	11	2	15	5	8	16	8...15	
155 (284)	5	15	11	2	15	5	14	5, 9, 10	none	[6]
156 (285)	30	50	11	2	15	9	4	none	15...50	
157 (286)	24	30	11	2	15	12	7	none	8...30	
158 (293)	4	15	9	3	15	4	13	6, 8...12	none	[6]
159 (296)	10	30	9	3	15	5	6	none	15...30	
160 (297)	7	21	9	3	15	5	9	21	11...20	
161 (298)	7	15	9	3	15	7	14	7	11...15	[6, 4]
162 (299)	28	30	9	3	15	14	7	none	12...30	
163 (301)	35	35	9	3	15	15	6	none	35	[1]
164 (302)	18	18	9	3	15	15	12	18	none	[5, 1]
165 (305)	14	42	7	4	15	5	4	42	28...41	
166 (306)	5	15	7	4	15	5	13	10	11...15	
167 (307)	15	25	7	4	15	9	8	none	15...25	
168 (308)	9	15	7	4	15	9	14	9	10...15	
169 (313)	8	30	5	5	15	4	5	27	20...26, 28	[7]
170 (315)	6	18	5	5	15	5	10	18	15...17	
171 (316)	14	35	5	5	15	6	5	none	24...35	
172 (317)	6	15	5	5	15	6	13	10, 12, 14, 15	none	[7, 4]
173 (318)	21	45	5	5	15	7	4	none	35...45	
174 (319)	12	20	5	5	15	9	10	none	15...20	
175 (320)	26	39	5	5	15	10	5	none	26...39	
176 (321)	11	15	5	5	15	11	14	11	12...15	
177 (322)	32	40	5	5	15	12	5	none	27...40	
178 (326)	21	21	5	5	15	15	10	none	21	[1]
179 (327)	4	15	3	6	15	4	11	12	13...15	
180 (328)	13	39	3	6	15	5	4	39	none	
181 (329)	9	27	3	6	15	5	6	27	none	
182 (330)	7	21	3	6	15	5	8	21	none	
183 (331)	5	15	3	6	15	5	12	15	none	
184 (332)	10	25	3	6	15	6	7	none	none	[2]
185 (334)	14	30	3	6	15	7	6	none	28...30	
186 (335)	7	15	3	6	15	7	13	14	15	
187 (336)	13	15	3	6	15	13	14	13	14, 15	
188 (340)	8	15	1	7	15	8	13	14	15	[6]
189 (342)	50	50	1	7	15	15	4	50	none	[7]
190 (344)	15	15	1	7	15	15	14	15	none	[7]

2.3 See Also

[1, 2, 3]	For excellent surveys and the original BTD parameter list. An in-progress paper of Billington is to update that list.
[6, 7]	For some BPTD results and constructions.

References

[1] E.J. Billington, Balanced n-ary designs: A combinatorial survey and some new results, *Ars Combin.* 17A (1984), 37-72.

[2] E.J. Billington, Designs with repeated elements in blocks: A survey and some recent results, *Congr. Numer.* 68 (1989), 123-146.

[3] E.J. Billington and P.J. Robinson, A list of balanced ternary designs with $R \leq 15$, and some necessary existence conditions, *Ars Combin.* 16 (1983), 235-258.
[4] P. Chawathe and D.G. Sarvate, Improved entries in a list of BPTDs, preprint.
[5] J.F. Dillon and M.A. Wertheimer, Balanced ternary designs derived from other combinatorial designs, *Congr. Numer.* 47 (1985), 285-298.
[6] T. Kunkle and D.G. Sarvate, On ternary designs with a specified number of blocks with repeated elements, *Ars Combin.*, to appear.
[7] T. Kunkle and D.G. Sarvate, designs: Some new results, *J. Combin. Math. Combin. Comput.*, to appear.
[8] K. Sinha, *Ars Combin.* 33 (1992), 276-278.

3 Balanced Tournament Designs

Esther R. Lamken

3.1 Definitions, Examples, and Existence

A *balanced tournament design* of order n, BTD(n), defined on a $2n$-set V is an arrangement of the $\binom{2n}{2}$ distinct unordered pairs of the elements of V into an $n \times 2n-1$ array such that

1. every element of V is contained in precisely one cell of each column, and
2. every element of V is contained in at most two cells of any row.

A *factored balanced tournament design* of order n, FBTD(n), is a BTD(n) with the property that in each row there exist n cells, a *factor*, which contain all $2n$ elements of V.

3.1 **Remark** The columns of a BTD(n) provide a 1-factorization of the complete graph on $2n$ vertices, K_{2n}.

3.2 **Example** A balanced tournament design of order 3.

1 6	3 5	2 3	4 5	2 4
2 5	4 6	1 4	1 3	3 6
3 4	1 2	5 6	2 6	1 5

3.3 **Example** An FBTD(6). The factors are underlined.

$\alpha\ 8$	$\underline{4\ 6}$	$6\ 2$	$\underline{\infty\ 0}$	$0\ 3$	$\underline{\alpha\ 3}$	$\underline{9\ 1}$	$1\ 7$	$\infty\ 5$	$\underline{5\ 8}$	$\underline{2\ 7}$
$\underline{6\ 9}$	$\alpha\ 9$	$\underline{5\ 7}$	$7\ 3$	$\underline{\infty\ 1}$	$1\ 4$	$\underline{\alpha\ 4}$	$\underline{0\ 2}$	$2\ 8$	$\infty\ 6$	$\underline{3\ 8}$
$\infty\ 7$	$\underline{7\ 0}$	$\alpha\ 0$	$\underline{6\ 8}$	$8\ 4$	$\underline{\infty\ 2}$	$2\ 5$	$\underline{\alpha\ 5}$	$\underline{1\ 3}$	$3\ 9$	$\underline{4\ 9}$
$4\ 0$	$\infty\ 8$	$\underline{8\ 1}$	$\alpha\ 1$	$\underline{7\ 9}$	$9\ 5$	$\underline{\infty\ 3}$	$3\ 6$	$\underline{\alpha\ 6}$	$\underline{4\ 2}$	$\underline{5\ 0}$
$\underline{5\ 3}$	$5\ 1$	$\infty\ 9$	$\underline{9\ 2}$	$\alpha\ 2$	$\underline{8\ 0}$	$0\ 6$	$\underline{\infty\ 4}$	$4\ 7$	$\underline{\alpha\ 7}$	$\underline{6\ 1}$
$\underline{1\ 2}$	$2\ 3$	$\underline{3\ 4}$	$4\ 5$	$\underline{5\ 6}$	$6\ 7$	$\underline{7\ 8}$	$8\ 9$	$\underline{9\ 0}$	$0\ 1$	$\underline{\infty\ \alpha}$

3.4 **Remark** Balanced tournament designs can be used to schedule round robin tournaments: the rows represent the locations, the columns the rounds of play and the elements the players or the teams. The properties of the array provide a schedule of play with the following properties:

1. every team plays every other team precisely once during the tournament,
2. each team plays in each round, and
3. no team plays more than twice at any location.

See §V.7 for more on the scheduling of tournaments.

3.5 Theorem There exists a (factored) BTD(n) if and only if n is a positive integer, $n \neq 2$.

3.6 Remark Very little is known about the enumeration of balanced tournament designs. Up to isomorphism, there is precisely one BTD(3). There are 47 nonisomorphic BTD(4)s; 29 of these are FBTD(4)s. Each of the nonisomorphic 1-factorizations of K_{10} gives rise to at least one BTD(5) [1].

3.2 Partitioned Balanced Tournament Designs

A *partitioned balanced tournament design*, PBTD(n), is a BTD(n) which also satisfies the following properties:

3. each row contains a factor in the first n columns, and
4. each row contains a factor in the last n columns.

3.7 Example A partitioned balanced tournament design of order 5.

α 4	∞ 2	1 3	5 7	0 6	2 3	4 5	∞ 7	α 1
∞ 3	α 5	4 6	0 2	1 7	∞ 4	α 2	0 5	6 3
5 6	0 3	α 7	∞ 1	4 2	6 7	0 1	α 3	∞ 5
1 2	4 7	∞ 0	α 6	5 3	α 0	∞ 6	1 4	7 2
0 7	1 6	2 5	4 3	α ∞	1 5	3 7	2 6	0 4

3.8 Theorem There exists a PBTD(n) for n a positive integer, $n \geq 5$, except possibly for $n \in \{9, 11, 15\}$.[4]

3.9 Theorem (See [7]) For n a positive integer, the existence of the following designs are equivalent:

1. a PBTD(n);
2. a maximum empty subarray Room square of side $2n-1$, a MESRS($2n-1$);
3. a pair of almost disjoint $H(n, 2n)$s (Howell designs); and
4. a partitionable house of order n.

3.10 Remark In terms of tournament scheduling, a PBTD(n) represents a tournament where each team plays precisely once at each location in the first n rounds of play and each team plays precisely once at each location in the last n rounds of the tournament.

3.3 Odd and Hamiltonian Balanced Tournament Designs

An *odd balanced tournament design*, OBTD(n), defined on a $2n+1$-set V is an arrangement of the $\binom{2n+1}{2}$ distinct unordered pairs of the elements of V into an $n \times (2n+1)$ array such that

1. each column of the array contains $2n$ distinct elements of V, and
2. each element of V occurs precisely twice in each row.

If each row of the OBTD(n) corresponds to a hamiltonian cycle, then the OBTD(n) is a *Kotzig factorization* of order $2n+1$.

3.11 Example A Kotzig factorization of order 7.

1 6	2 7	3 1	4 2	5 3	6 4	7 5
2 5	3 6	4 7	5 1	6 2	7 3	1 4
3 4	4 5	5 6	6 7	7 1	1 2	2 3

3.12 Remark Starters can be used to construct OBTDs. (See §IV.45 and Remark V.7.6)

3.13 Theorem For n a positive integer, there exists an OBTD(n).

3.14 Theorem For each positive integer n, there exists a Kotzig factorization of order $2n+1$.

If each row of a BTD(n) gives a hamiltonian path in K_{2n}, the BTD(n) is a *hamiltonian balanced tournament design*, HBTD(n).

3.15 Remark An HBTD is the analogue of a Kotzig factorization.

3.16 Example A hamiltonian balanced tournament design of side 4.

1 2	1 3	5 8	3 7	4 7	2 8	4 6
3 4	5 7	1 4	2 6	2 5	3 6	1 8
5 6	2 4	6 7	4 8	3 8	1 7	3 5
7 8	6 8	2 3	1 5	1 6	4 5	2 7

3.17 Theorem The following existence results are known for HBTD(n).

1. There exist HBTD(n) for $n = 1, 4, 5$. There do not exist HBTD(n) for $n = 2, 3$.
2. There exist HBTD(n) for all positive integers n not divisible by 2, 3 or 5. [3]

3.4 Generalized Balanced Tournament Designs

A *generalized balanced tournament design*, GBTD(n, k), defined on a kn-set V, is an arrangement of the blocks of a $(kn, k, k-1)$ BIBD defined on V into an $n \times (kn-1)$ array such that

1. every element of V is contained in precisely one cell of each column, and
2. every element of V is contained in at most k cells of each row.

3.18 Example A GBTD(3, 3).

1 2 9	3 4 9	5 6 9	1 4 5	3 5 7	1 7 8	2 3 8	2 6 7
3 5 7	1 6 7	1 3 8	2 3 6	4 6 8	2 4 5	7 4 9	5 8 9
4 6 8	2 5 8	2 4 7	7 8 9	1 2 9	3 6 9	1 6 5	1 3 4

3.19 Remarks

1. A GBTD(n, 2) is a BTD(n).
2. Factored GBTDs, partitioned GBTDs, and near GBTDs are the generalizations of FBTDs, PBTDs, and OBTDs, respectively, to larger block size k [5, 6].
3. GBTDs are also equivalent to other combinatorial designs [5].

3.20 Remark Generalized balanced tournament designs can be used to construct several other types of combinatorial designs including resolvable, near resolvable, doubly resolvable, and doubly near resolvable balanced incomplete block designs [5].

3.5 See Also

§I.6	GBTDs and near GBTDs are RBIBDs and near RBIBDs.
§IV.22	Kotzig factorizations and HBTDs are graph decompositions.
§IV.26	A PBTD(n) can also be defined in terms of Howell designs.
§IV.39	A PBTD(n) is equivalent to a MESRS($2n-1$).
§IV.45	Starters can be used to construct BTDs and OBTDs.
§V.7	A section on scheduling tournaments.
[7]	A survey of balanced tournament designs.
[5, 6]	Further information and references on GBTDs.
[2]	Applications of tournament designs to scheduling softball leagues.

References

[1] J. Corriveau, Enumeration of balanced tournament designs, *Ars Combin.* 25 (1988), 93-105.

[2] N.J. Finizio, Tournament designs balanced with respect to several bias categories, *Bull. ICA* 9 (1993), 69-95.

[3] J.D. Horton, Hamilton path tournament designs, *Ars Combin.* 27 (1989), 69-74.

[4] E.R. Lamken, A few more partitioned balanced tournament designs, *Ars Combin.*, to appear.

[5] E.R. Lamken, Generalized balanced tournament designs, *Trans. Amer. Math. Soc.* 318 (1990), 473-490.

[6] E.R. Lamken, Existence results for generalized balanced tournament designs with block size 3, *Des. Codes Crypt.* 3 (1992), 33-61.

[7] E.R. Lamken and S.A. Vanstone, Balanced tournament designs and related topics, *Discrete Math.* 77 (1989), 159-176.

4 Bhaskar Rao Designs

Warwick de Launey

4.1 GBRDs: Definition and Example

A *generalized Bhaskar Rao design* GBRD($v, b, r, k, \lambda; G$) is a v by b array whose nonzero entries are drawn from the group G of finite order g so that

1. each row contains precisely r nonzero entries,
2. each column contains precisely k nonzero entries, and
3. for any pair of distinct rows $(x_1, x_2, \cdots, x_b)$ and $(y_1, y_2, \cdots, y_b)$, the list $x_1y_1^{g-1}, x_2y_2^{g-1}, \cdots, x_by_b^{g-1}$ contains each nonzero group element exactly λ/g times.

4.1 **Remarks** Replacing the nonzero entries by 1 produces a balanced block design; consequently, $r = 1 + \lambda(v-1)/k$ and $b = rv/k$ and so GBRD($v, b, r, k, \lambda; G$) is usually just written as GBRD($v, k, \lambda; G$). When $v > k$, the underlying $(0, 1)$ design is incomplete, and then $b \geq v$. The cases where $v = k$ or $b = v$ or $G = \mathbb{Z}_2$ are dealt with in §IV.11.3, §4.4 and §4.3, respectively.

4.2 **Example** A GBRD$(4, 3, 6; \mathbb{Z}_6)$.

1	1	1	1	1	1	1	1	1	0	0	0
1	ω^4	ω^2	ω^3	ω	ω^5	0	0	0	1	1	1
1	ω^2	ω^4	0	0	0	ω^3	ω	ω^5	ω^3	ω	ω^5
0	0	0	1	ω^2	ω^4	ω^3	ω^5	ω	1	ω^2	ω^4

4.2 Existence of Generalized Bhaskar Rao Designs

4.3 **Theorem** A GBRD$(v, k, \lambda; H)$ is obtained by applying the onto group homomorphism $\phi : G \to H$ entrywise to a GBRD$(v, k, \lambda; G)$.

4.4 **Remarks** Some general techniques [11, pp. 230-232] are known for obtaining families of designs with $k > 4$; however, all attempts to determine the spectrum of admissible parameters v, b, r, k, λ, G have been restricted to $k = 3, 4$. A construction for designs with large k is given in [22]. A family of GBRD$(k + 1, k, k^2 - k; G)$ appears in [14]. That paper also lists all the GBRD$(k + 1, k, lk - l; G)$ with $k = 3, 4, 5$ and l small. Theorem 4.5 is a complete asymptotic result for odd order groups and large k.

4.5 **Theorem** [8] Let G be any group of odd order g. For all $k, t > 0$, there is an integer M such that for all $v > M$ there exists a GBRD$(v, k, tg; G)$ if and only if $tg(v - 1) \equiv 0 \pmod{k - 1}$ and $tgv(v - 1) \equiv 0 \pmod{k(k - 1)}$.

Let $EA(g)$ denote the order g direct product of prime order cyclic groups.

4.6 **Theorem** [21] A GBRD$(v, 3, tg; EA(g))$ exists if and only if $tg(v - 1)$ is even and $tgv(v - 1) \equiv 0 \pmod 6$ for g odd and $tgv(v - 1) \equiv 0 \pmod{24}$ for g even.

4.7 **Remark** Necessary and sufficient conditions are also known for any nilpotent odd order group [18] and for $S_3, D_4, Q_8, \mathbb{Z}_4 \times \mathbb{Z}_2$ and D_6 [17]. The group $\mathbb{Z}_4$ is studied in [10], and partial results for some other even order cyclic groups appear in [6].

4.8 **Theorem** [11] Let $v > 50$ and $t > 1$ be integers. A GBRD$(v, 4, tg; EA(g))$ exists if and only if $tg(v - 1)$ is even and $tgv(v - 1) \equiv 0 \pmod{12}$.

4.9 **Remark** Many examples [11] are known for $k = 4$ and $v < 50$ or $t = 1$. There is no GBRD$(5, 4, 6; \mathbb{Z}_6)$ or GBRD$(7, 4, 4; \mathbb{Z}_2 \times \mathbb{Z}_2)$ [9].

4.3 Bhaskar Rao Designs

A *Bhaskar Rao design* BRD(v, k, λ) is a GBRD$(v, k, \lambda; G)$ where $v > k$ and $G = \mathbb{Z}_2$.

4.10 **Remarks** Bhaskar Rao [4, 5] called these designs *balanced orthogonal designs*. Bhaskar Rao designs are often defined to include a wider class of $(0, \pm 1)$-matrices where the underlying design is a PBIBD instead of a BIBD, and the between class inner products are fully specified but not necessarily zero. Bhaskar Rao designs have been studied at this level of generality in [23, 24].

4.11 **Theorem** [5] A BRD(v, k, λ) can exist only if the following equations have nonnegative integer solutions:

$$x_3 + 3x_5 + 6x_7 + \cdots + \tfrac{(k^2-1)}{8} x_k = b\tfrac{(k-1)}{8} \quad \text{for } k \text{ odd},$$
$$-x_0 + 3x_2 + 8x_4 + \cdots + \tfrac{(k^2-4)}{4} x_k = b\tfrac{(k-4)}{4} \quad \text{for } k \text{ even}.$$

4.12 **Theorem** [20] A BRD$(v, 3, 2t)$ exists for $v \geq 3$ whenever $tv(v - 1) \equiv 0 \pmod{12}$.

4.13 Example A BRD(7, 4, 4), with "–" representing −1.

$$\begin{array}{|cccccccccccccc|}\hline 1 & 1 & 1 & 1 & 1 & 1 & 1 & 1 & 0 & 0 & 0 & 0 & 0 & 0 \\ 1 & 1 & - & 0 & 0 & - & 0 & 0 & 0 & - & 1 & 0 & - & 1 \\ 1 & 1 & 0 & - & 0 & 0 & - & 0 & 1 & 0 & - & 1 & 0 & - \\ 1 & 1 & 0 & 0 & - & 0 & 0 & - & - & 1 & 0 & - & 1 & 0 \\ 0 & 0 & 0 & - & 1 & 0 & 1 & - & 0 & 1 & 1 & 0 & - & - \\ 0 & 0 & 1 & 0 & - & - & 0 & 1 & 1 & 0 & 1 & - & 0 & - \\ 0 & 0 & - & 1 & 0 & 1 & - & 0 & 1 & 1 & 0 & - & - & 0 \\ \hline\end{array}$$

4.14 Theorem [11, pp. 251-252] A BRD$(v, 4, 2)$ exists whenever $v \equiv 1 \pmod 6$ is a prime power. A BRD$(v, 4, 2)$ can exist only if $v \equiv 1, 4 \pmod 6$.

4.15 Remark There is no BRD(4, 4, 2)[11, Remark 4.1.4] or BRD(10, 4, 2)[9].

4.16 Theorem [11, Theorem 6.6.2] A BRD$(v, 4, 2t)$ exists for $v > 4$ and $t > 1$ if and only if $t(v-1) \equiv 0 \pmod 3$ except possibly when $t = 3, 5, 7$ and $v = 28, 34, 39$ or when $t = 3$ and $v = 39$. A BRD$(4, 4, 2t)$ exists for $t > 0$ if and only if t is even.

4.17 Theorem [8, Theorem 5.1.1(ii)] For all $k, t > 0$, there is an integer M such that there is a BRD$(v, k, 2t; \mathbb{Z}_2)$ whenever $2t(v-1)/(k-1)$ and $2tv(v-1)/k(k-1)$ are integers and $v > M$. When $k \equiv 2, 3 \pmod 4$, it is necessary that $2t(v-1)/(k-1)$ and $2tv(v-1)/k(k-1)$ be integers.

4.4 Balanced Generalized Weighing Matrices

A *balanced generalized weighing matrix* BGW$(v, k, \lambda; G)$ is a GBRD$(v, b, r, k, \lambda; G)$ where $b = v$, $r = k$ and $k < v$.

4.18 Remarks Replacing the nonzero entries by 1 yields the incidence matrix of an symmetric BIBD(v, k, λ). Any BGW with $G = \mathbb{Z}_2$ is a weighing matrix with additional structure. These balanced weighing matrices were studied in [16] and related papers. See also §IV.52.

4.19 Examples A w-cyclic BGW$(5, 4, \mathbb{Z}_3)$, a normalized BGW$(6, 5, 4; \mathbb{Z}_4)$ with circulant core, and a circulant BGW$(7, 4, 2; \mathbb{Z}_2)$.

$$\begin{array}{|ccccc|}\hline 0 & w^2 & w & w^2 & w^2 \\ e & 0 & w^2 & w & w^2 \\ e & e & 0 & w^2 & w \\ w^2 & e & e & 0 & w^2 \\ e & w^2 & e & e & 0 \\ \hline\end{array} \qquad \begin{array}{|cccccc|}\hline 0 & 1 & 1 & 1 & 1 & 1 \\ 1 & 0 & c & c^2 & c^4 & c^3 \\ 1 & c^3 & 0 & c & c^2 & c^4 \\ 1 & c^4 & c^3 & 0 & c & c^2 \\ 1 & c^2 & c^4 & c^3 & 0 & c \\ 1 & c & c^2 & c^4 & c^3 & 0 \\ \hline\end{array} \qquad \begin{array}{|ccccccc|}\hline - & 1 & 1 & 0 & 1 & 0 & 0 \\ 0 & - & 1 & 1 & 0 & 1 & 0 \\ 0 & 0 & - & 1 & 1 & 0 & 1 \\ 1 & 0 & 0 & - & 1 & 1 & 0 \\ 0 & 1 & 0 & 0 & - & 1 & 1 \\ 1 & 0 & 1 & 0 & 0 & - & 1 \\ 1 & 1 & 0 & 1 & 0 & 0 & - \\ \hline\end{array}$$

4.20 Example A circulant BGW$(31, 16, 8; \mathbb{Z}_2)$ (Eades, see [12]). Its first row is $-0000-0+0--+0++000-+-++00++0+00$.

4.21 Remark The only known families of balanced generalized weighing matrices are listed in Theorem 4.22. They are strongly related to finite geometries. Three sporadic examples are known: a circulant BGW$(13, 9, 6; S_3)$ [19], a BGW$(15, 7, 3; \mathbb{Z}_3)$ [2], and a BGW$(19, 9, 4; \mathbb{Z}_2)$ [13, 9]. The first design has first row equal to

$$0, a, e, ab, 0, e, e, ab^2, a, ab^2, 0, ab, 0$$

where $a = (1\ 5)(2\ 4)(3\ 6)$ and $b = (1\ 2\ 3)(4\ 5\ 6)$ generate the symmetric group S_3. Here $ab = (1\ 4)(2\ 6)(3\ 5)$. The other two designs are shown here. Existence has been settled for $v \leq 19$ [13]. The techniques of Gibbons and Mathon [15] give profound insight into the use of signings of designs leading systematically to biplanes and projective planes.

0	1	1	1	1	1	1	1	1	1	0	0	0	0	0	0	0	0	0
1	0	0	0	1	0	−	1	−	0	0	−	−	0	1	0	0	0	1
1	0	0	0	−	1	0	0	1	−	−	0	−	0	0	1	1	0	0
1	0	0	0	0	−	1	−	0	1	−	−	0	1	0	0	0	1	0
1	1	−	0	0	0	0	1	0	−	0	0	1	0	−	−	0	1	0
1	0	1	−	0	0	0	−	1	0	1	0	0	−	0	−	0	0	1
1	−	0	1	0	0	0	0	−	1	0	1	0	−	−	0	1	0	0
1	1	0	−	1	−	0	0	0	0	0	1	0	0	0	1	0	−	−
1	−	1	0	0	1	−	0	0	0	0	0	1	1	0	0	−	0	−
1	0	−	1	−	0	1	0	0	0	1	0	0	0	1	0	−	−	0
0	0	−	−	0	1	0	0	0	1	−	1	1	0	1	0	0	0	1
0	−	0	−	0	0	1	1	0	0	1	−	1	0	0	1	1	0	0
0	−	−	0	1	0	0	0	1	0	1	1	−	1	0	0	0	1	0
0	0	0	1	0	−	−	0	1	0	0	0	1	−	1	1	0	1	0
0	1	0	0	−	0	−	0	0	1	1	0	0	1	−	1	0	0	1
0	0	1	0	−	−	0	1	0	0	0	1	0	1	1	−	1	0	0
0	0	1	0	0	0	1	0	−	−	0	1	0	0	0	1	−	1	1
0	0	0	1	1	0	0	−	0	−	0	0	1	1	0	0	1	−	1
0	1	0	0	0	1	0	−	−	0	1	0	0	0	1	0	1	1	−

0	e	e	e	e	e	e	e	0	0	0	0	0	0	0
e	e	w	0	w^2	0	0	0	e	w^2	0	w	0	0	0
e	0	e	w	0	w^2	0	0	0	e	w^2	0	w	0	0
e	0	0	e	w	0	w^2	0	0	0	e	w^2	0	w	0
e	0	0	0	e	w	0	w^2	0	0	0	e	w^2	0	w
e	w^2	0	0	0	e	w	0	w	0	0	0	e	w^2	0
e	0	w^2	0	0	0	e	w	0	w	0	0	0	e	w^2
e	w	0	w^2	0	0	0	e	w^2	0	w	0	0	0	e
0	e	w^2	0	w	0	0	0	0	0	w^2	0	w^2	w^2	e
0	0	e	w^2	0	w	0	0	e	0	0	w^2	0	w^2	w^2
0	0	0	e	w^2	0	w	0	w^2	e	0	0	w^2	0	w^2
0	0	0	0	e	w^2	0	w	w^2	w^2	e	0	0	w^2	0
0	w	0	0	0	e	w^2	0	0	w^2	w^2	e	0	0	w^2
0	0	w	0	0	0	e	w^2	w^2	0	w^2	w^2	e	0	0
0	w^2	0	w	0	0	0	e	0	w^2	0	w^2	w^2	e	0

4.22 Theorem Let q be a prime power. A BGW(v, k, λ) exists for the following parameter sets:

1. $v = (q^{t+1} - 1)/(q - 1)$, $k = q^t$, $\lambda = q^{t-1}(q - 1)$ and $G = \mathbb{Z}_m$ where m divides $q - 1$ and $t \geq 0$;
2. $v = (q^{2(t+1)} - 1)/(q^2 - 1)$, $k = q^{2t}$, $\lambda = q^{2(t-1)}(q^2 - 1)$ and G any group of order $q + 1$;
3. $v = k + 1$, $k = g(2g - 1)$, $\lambda = k - 1$ and $G = \mathbb{Z}_g$ where $g = 2^{m-1} - 1$ and $m \geq 3$.

4.23 Theorem

1. If v is odd and g is even, then k must be a square.
2. [7] Suppose there is an onto homomorphism $\phi\colon G \to \mathbb{Z}_p$ where p is an odd prime, and suppose there exists a BGW$(v, k, \lambda; G)$ where v is odd. If $m \not\equiv 0 \pmod{p}$ is an integer dividing the squarefree part of k, then the order of m modulo p is odd.

4.24 Remark Circulant balanced generalized weighing matrices are given in [3]. Apart from the requirement that an underlying symmetric BIBD(v, k, λ) exist, two nonexistence results are given in Theorem 4.23. See also [1].

4.5 See Also

§III.1	A GBRD$(v, k, \lambda; G)$ is equivalent to a group divisible design with G acting regularly on the points and semi-regularly on the lines.
§IV.11	A GH(g, λ) over G is a GBRD$(\lambda g, \lambda g, \lambda g; G)$. A $(g, k; \lambda)$-difference matrix over G is a GBRD$(k, k, g\lambda; G)$.

References

[1] K.T. Arasu and J.R. Seberry, Circulant weighing matrices, preprint.

[2] R.D. Baker, Elliptic semi-planes I: Existence and classification, *Congr. Numer.* 19 (1977), 61-73.

[3] G. Berman, Families of generalized weighing matrices, *Canad. J. Math.* 30 (1978), 1016-1028.

[4] M. Bhaskar Rao, Group divisible family of PBIB designs, *J. Indian Stat. Assoc.* 4 (1966), 14-28.

[5] M. Bhaskar Rao, Balanced orthogonal designs and their application in the construction of some BIB and group divisible designs, *Sankhya (A)* 32 (1970), 439-448.

[6] A. Bowler, K. Quinn and J. Seberry, Generalized Bhaskar Rao designs with elements from cyclic groups of even order, *Australas. J. Combin.* 3 (1991), 5-13.

[7] W. de Launey, On the non-existence of certain generalized weighing matrices, *Ars Combin.* 17A (1984), 117-132.

[8] W. de Launey, $(0, G)$-Designs and Applications, Ph.D. thesis, University of Sydney, 1987.

[9] W. de Launey and D.G. Sarvate, Non-existence of certain GBRDs, *Ars Combin.* 18 (1983), 5-20.

[10] W. de Launey, D.G. Sarvate and J. Seberry, Generalized Bhaskar Rao designs with blocksize 3 over $\mathbb{Z}_4$, *Ars Combin.* A19 (1985), 273-285.

[11] W. de Launey and J. Seberry, On generalized Bhaskar Rao designs of block size four, *Congr. Numer.* 41 (1984), 229-294.

[12] A. Geramita and J. Seberry, *Quadratic Forms, Orthogonal Designs, and Hadamard Matrices*, Marcel Dekker, New York, 1979.

[13] P.B. Gibbons and R.A. Mathon, Group signings of symmetric balanced incomplete block designs, *Ars Combin.* A23 (1987), 123-134.

[14] P.B. Gibbons and R.A. Mathon, Construction methods for Bhaskar Rao and related designs, *J. Austral. Math. Soc. (A)* 42 (1987), 5-30.

[15] P.B.Gibbons and R. Mathon, Signings of group divisible designs and projective planes, *Australas. J. Combin.* 11 (1995), 79-104.

[16] R.C. Mullin and R.G. Stanton, Group matrices and balanced weighing designs, *Utilitas Math.* 8 (1975), 277-301.

[17] W. Palmer and J. Seberry, Bhaskar Rao designs over small groups, *Ars Combin.* A26 (1988), 125-148.

[18] W. Palmer, A composition theorem for generalized Bhaskar Rao designs, *Australas. J. Combin.* 6 (1992), 221-228.

[19] J. Seberry, Some remarks on generalized Hadamard matrices and theorems of Rajkundlia on SBIBDs, *Combin. Math. VI*, Lecture Notes in Math. 784 (1979), 154-164.

[20] J. Seberry, Regular group divisible designs and Bhaskar Rao designs with block size three, *J. Stat. Plann. Infer.* 10 (1984), 69-82.
[21] J. Seberry, Generalized Bhaskar Rao designs of block size three, *J. Stat. Plann. Infer.* 11 (1985), 373-379.
[22] J. Seberry, Vanstone's construction applied to Bhaskar Rao designs, *Congr. Numer.* 59 (1987), 265-274.
[23] D.J. Street, Bhaskar Rao designs from cyclotomy, *J. Austral. Math. Soc.* 29 (1980), 425-430.
[24] D. J. Street and C.A. Rodger, Some results on Bhaskar Rao designs, *Combin. Math. VII*, Lecture Notes in Math. 829 (1980), 238-245.

5 Complete Mappings and Sequencings of Finite Groups

DONALD KEEDWELL

5.1 Definitions, Examples, and Existence Conditions

A *complete mapping* of a group or quasigroup $(G, \odot)$ is a permutation $g \to \theta(g)$ of the elements of G such that the mapping $\phi : g \to g\theta(g)$ is again a permutation of the elements of G. The mapping $g \to \phi(g)$ is an *orthomorphism* of $(G, \odot)$.

A *near-complete* mapping of a group $(G, \odot)$ is a bijection $g \to \theta(g)$ from $G \setminus \{h\}$ to $G \setminus \{e\}$ such that the mapping $\phi : g \to g\theta(g)$ is again a one-to-one mapping from $G \setminus \{h\}$ to $G \setminus \{k\}$ say. Here $h \neq e$ and k are fixed elements of G and e is the identity element. The mapping $g \to \phi(g)$ is a *near-orthomorphism* of $(G, \odot)$.

A complete mapping of a group $(G, \odot)$ is in *canonical form* if it maps the identity element e of G to itself. A near-complete mapping of a group $(G, \odot)$ is in *canonical form* if $k = e$. In that case, h is the *exdomain element* of the mapping.

5.1 **Remark** If θ is a complete mapping of the group $(G, \odot)$ such that $\theta(e) = k$, then the mapping $\theta_k : g \to \theta(g)k^{-1}$ is a complete mapping in canonical form.

5.2 **Theorem** If $(G, \odot)$ is a group or quasigroup with a complete mapping, then its Cayley table (a latin square) has a transversal. Conversely, if L is a latin square with a transversal, then any quasigroup which has L as its Cayley table has a complete mapping.

5.3 **Example** For Klein's Viergruppe $\mathbb{Z}_2 \times \mathbb{Z}_2 = \langle a, b \colon a^2 = b^2 = e, ab = c = ba\rangle$, a complete mapping in canonical form is $\theta = \begin{pmatrix} e & a & b & c \\ e & b & c & a \end{pmatrix}$ with corresponding orthomorphism $\phi = \begin{pmatrix} e & a & b & c \\ e & c & a & b \end{pmatrix}$. The mapping corresponds to the transversal underlined in the Cayley table on the right.

$\underline{e}$	a	b	c
a	e	$\underline{c}$	b
b	c	e	$\underline{a}$
c	$\underline{b}$	a	e

5.4 **Remarks**

1. If $(G, \odot)$ is a group (not just a quasigroup), the existence of a transversal in the latin square L formed by the Cayley table of $(G, \odot)$ is sufficient (as well as necessary) for L to possess an orthogonal mate. (See also §II.1.4.)
2. A group which has complete mappings is often called *admissible*.

5.5 Theorem A group whose Sylow 2-subgroups are cyclic (and nontrivial) has no complete mappings.

5.6 Theorem A sufficient condition for a solvable group (in particular, an abelian group) to possess complete mappings is that its Sylow 2-subgroups be noncyclic.

5.7 Theorem A necessary condition for a group to possess complete mappings is that the product of all its elements in some order be equal to the identity element. For an abelian or solvable group, this condition is also sufficient.

5.8 Remarks

1. All nonsolvable groups have the properties stated in Theorems 5.6 and 5.7 but it is not known whether all such groups are admissible.
2. For a group G of odd order, the identity map $\theta(g) = g$, for all $g \in G$, is a complete mapping.
3. A cyclic group of even order has no complete mappings (Theorem 5.5).
4. A frequently-cited sufficient condition (originally given by Paige) for a non-abelian group to have a complete mapping is that it be R-sequenceable. (See §5.3.)

5.9 Theorem If $(G, \odot)$ is a group or quasigroup with a near-complete mapping, then its Cayley table possesses a set of $n-1$ cells in distinct rows and columns and containing distinct elements (an *almost-transversal*).

5.10 Conjecture Every latin square possesses an almost-transversal.

5.11 Example For the abelian group $(\mathbb{Z}_4, +)$, a near-complete mapping in canonical form is $\theta = \begin{pmatrix} 0\ 1\ 2\ 3 \\ 1\ 2\ \cdot\ 3 \end{pmatrix}$ with corresponding near-orthomorphism $\phi = \begin{pmatrix} 0\ 1\ 2\ 3 \\ 1\ 3\ \cdot\ 2 \end{pmatrix}$. The mapping corresponds to the almost-transversal underlined in the Cayley table on the right. The exdomain element is 2.

0	$\underline{1}$	2	3
1	2	$\underline{3}$	0
2	3	0	1
3	0	1	$\underline{2}$

5.12 Theorem If a group possesses a near-complete mapping, then the product of its elements in some order is equal to the exdomain element of that near-complete mapping. (In particular, an abelian group has near-complete mappings only if it has a unique element of order two.)

5.2 Enumeration of Mappings for Groups of Small Order

5.13 Table The number of complete and near-complete mappings in canonical form for groups with up to 13 elements. Here $\mathbb{Z}_{r,s,t}$ denotes the abelian group $\mathbb{Z}_r \times \mathbb{Z}_s \times \mathbb{Z}_t$; $V_4 \cong \mathbb{Z}_2 \times \mathbb{Z}_2$; $D =$ dihedral; $Q =$ dicyclic; and $A =$ alternating. [4]

Group	$\mathbb{Z}_2$	$\mathbb{Z}_3$	$\mathbb{Z}_4$	V_4	$\mathbb{Z}_5$	$\mathbb{Z}_6$	D_3	$\mathbb{Z}_7$	$\mathbb{Z}_8$	$\mathbb{Z}_{4,2}$	$\mathbb{Z}_{2,2,2}$	D_4	Q_4	$\mathbb{Z}_9$
Complete mappings	0	1	0	2	3	0	0	19	0	48	48	48	48	225
Near-complete mappings	1	0	2	0	0	8	6	0	64	0	0	16	16	0

Group	$\mathbb{Z}_{3,3}$	$\mathbb{Z}_{10}$	D_5	$\mathbb{Z}_{11}$	$\mathbb{Z}_{12}$	$\mathbb{Z}_{2,6}$	D_6	A_4	Q_6	$\mathbb{Z}_{13}$
Complete mappings	249	0	0	3441	0	16512	6336	3840	0	79259
Near-complete mappings	0	928	820	0	17536	0	10944	12864	17280	0

5.14 Remark To obtain the total number of complete mappings, multiply the number of complete mappings in canonical form by the order of the group.

5.15 Theorem For any group G the number of left neofields whose multiplicative group is G equals the total number of canonical form complete and near-complete mappings of G.

5.3 Group Sequencings: Definitions and Examples

A group $(G, \odot)$ of order n with identity element e is *R-sequenceable* if there is an ordering $a_0 = e, a_1, a_2, \ldots, a_{n-1}$ of the elements of G such that $a_0 a_1 a_2 \cdots a_{n-1} = e$ and so that all of the partial products $b_0 = a_0 = e,\ b_1 = a_0 a_1,\ b_2 = a_0 a_1 a_2, \ldots, b_{n-2} = a_0 a_1 a_2 \cdots a_{n-2}$ are different. (One element, say c, does not occur among these partial products.)

5.16 Example For the cyclic group $(\mathbb{Z}_{4m+1}, +)$ an R-sequencing is $0, 4m, 2, 4m-2, 4, 4m-4, 6, 4m-6, \ldots, 2m-2, 4m-(2m-2), 2m, 2m-1, 2m+3, 2m-3, 2m+5, \ldots, 2m-(2m-3), 2m+(2m-1), 2m-(2m-1), 2m+1$.

For $(\mathbb{Z}_{4m+3}, +)$ an R-sequencing is $0, 4m+2, 2, 4m, 4, 4m-2, 6, 4m-4, \ldots, 2m-2, 4m-(2m-4), 2m, 2m+1, 2m+3, 2m-1, 2m+5, 2m-3, \ldots, 2m-(2m-3), 2m+(2m+1), 2m-(2m-1), 2m+2$.

A group $(G, \odot)$ of order n with identity element e is *sequenceable* if there is an ordering $a_0 = e, a_1, a_2, \ldots, a_{n-1}$ of the elements of G such that all of the partial products $b_0 = a_0 = e,\ b_1 = a_0 a_1,\ b_2 = a_0 a_1 a_2, \ldots, b_{n-1} = a_0 a_1 a_2 \cdots a_{n-1}$ are different. (Every element must occur among these partial products and $a_0 a_1 a_2 \cdots a_{n-1} \neq e$.)

5.17 Example For the cyclic group $(\mathbb{Z}_{2m}, +)$ a sequencing is

$$0, 1, 2m-2, 3, 2m-4, 5, 2m-6, 7, \ldots, 6, 2m-5, 4, 2m-3, 2, 2m-1.$$

A group $(G, \odot)$ of order n with identity element e is *harmonious* if there is an ordering $a_0 = e, a_1, a_2, \ldots, a_{n-1}$ of the elements of G such that all of the products $a_0 a_1,\ a_1 a_2,\ a_2 a_3, \ldots,\ a_{n-2} a_{n-1},\ a_{n-1} a_0$ are different.

5.18 Example For the cyclic group $(\mathbb{Z}_{2m+1}, +)$ a harmonious ordering is $0, 1, 2, \ldots, 2m$.

5.19 Remark Both R-sequenceability and harmoniousness are sufficient conditions for a group to have a complete mapping. If $(G, \odot)$ has a harmonious ordering, then $\theta = (a_0 a_1 a_2 \ldots a_{n-1})$ is a complete mapping of G expressed in cycle form. Note that $\theta(a_i) = a_{i+1}$ and so $\phi(a_i) = a_i a_{i+1}$ is a permutation of G, where suffix addition is modulo n. If, on the other hand, $(G, \odot)$ has an R-sequencing, then $\phi = (c)(b_0 b_1 b_2 ... b_{n-2})$ is an orthomorphism of G expressed in cycle form. Note that $\theta(c) = e$ and that $\theta(b_i) = a_{i+1}$ for $i = 0, 1, 2, \ldots, n-2$, where suffix addition is modulo $n-1$, so θ is a permutation of G.

5.20 Remark Sequenceability is a sufficient condition for a group to have a near-complete mapping. Observe that $\theta(b_i) = a_{i+1}$ for $i = 0, 1, 2, \ldots, n-2$ and that $\phi(b_i) = b_i \theta(b_i) = b_i a_{i+1} = b_{i+1}$ for $i = 0, 1, 2, \ldots, n-2$ are both bijective mappings from $G \setminus \{b_{n-1}\}$ to $G \setminus \{e\}$ so θ is a near-complete mapping (in canonical form) and ϕ is the corresponding near-orthomorphism.

5.21 Remarks

1. For abelian groups, R-sequenceability and sequenceability are mutually exclusive properties. An abelian group is sequenceable if and only if it has a unique element z of order 2 ([6], p. 86), in which case the product of all of its elements is z and so it has no complete mapping (Theorem 5.7).

2. A nonabelian group may have both, one, or neither of the properties of sequenceability and R-sequenceability.

5.22 Theorem The Cayley table of a group can be expressed in any one of the forms of row complete, column complete, or complete latin square if and only if the group is sequenceable ([7], pp. 45-47). (See also §IV.48.)

5.23 Theorem If G is a sequenceable group with a complete mapping, then there exist two orthogonal row complete (or column complete) latin squares based on G [5].

5.4 Groups: Harmonious, R-Sequenceable, and Sequenceable

5.24 Theorem Classes of groups that are harmonious include:

1. all groups of odd order [3];
2. all finite nontrivial abelian groups which have noncyclic Sylow 2-subgroups except the elementary abelian 2-groups [3];
3. the dihedral groups D_n (of order $2n$) [3] and the dicyclic groups Q_{2n} (of order $4n$) [12] for $n \equiv 0 \pmod 4$ or $n \equiv 0 \pmod 6$ in both cases. (Also, D_n and Q_{2n} are not harmonious if $n = 2$ or n is odd [3].)

5.25 Remark The question of which classes of abelian group are R-sequenceable is only partially solved. For a list of those classes of abelian groups which are known to be R-sequenceable, see [7], p. 68. They include all cyclic groups of odd order and all abelian groups of orders prime to 6.

5.26 Theorem Classes of nonabelian groups that are R-sequenceable include:

1. the dihedral groups D_n (of order $2n$) if and only if $n \equiv 0 \pmod 2$;
2. the dicyclic groups Q_{2n} of order $4n$ ($n > 2$) when $n \equiv 0 \pmod 4$ and when $n+1$ is a prime which has -2 as a primitive root [13];
3. the nonabelian groups of order pq, where $p < q$ are odd primes such that p has 2 as a primitive root (he last condition is sufficient but may not be necessary).

5.27 Theorem An abelian group is sequenceable if and only if it has a unique element of order two.

5.28 Conjecture The nonabelian groups D_3, D_4, and Q_4 of orders 6, 8, and 8, respectively, are not sequenceable, but it is conjectured that all nonabelian groups of orders greater than 8 are sequenceable. This has been confirmed by computer search for orders up to and including 32. It has also been shown that all dihedral groups D_n with $n \geq 5$ are sequenceable. See [7], p. 54.

5.29 Theorem Classes of nonabelian groups that are sequenceable include:

1. all finite solvable groups with a unique element of order 2 except the quaternion group Q_4 [2];
2. the dihedral groups except D_3 and D_4 [10, 11];
3. the dicyclic groups with the exception of the quaternion group Q_4 (see [7], p.96);
4. the nonabelian groups of order pq, where $p < q$ are odd primes such that p has 2 as a primitive root the last condition is sufficient but may not be necessary, see [7], pp. 48, 56-57).

5.30 Remark An extremely effective hill-climbing algorithm for sequencing nonabelian groups is given in [1].

5.31 Examples

1. An R-sequencing for the dihedral group D_4 is $e, a, b, a^3, a^2, ba^2, ba^3, ba$. The partial products are $e, a, ba^3, ba^2, b, a^2, ba, e$.
2. A sequencing for the dihedral group D_5 is $e, a, a^2, b, ba, ba^4, ba^2, ba^3, a^3, a^4$. The partial products are $e, a, a^3, ba^2, a^4, b, a^2, ba, ba^4, ba^3$. (The partial products form a directed terrace of G, cf. §5.6.)

5.5 Direct Product Constructions

5.32 Remark Almost all of the known direct product theorems require strengthened versions of harmoniousness, R-sequenceability or sequenceability.

A finite group G of odd order $2m+1$ is *symmetrically harmonious* if its elements have a harmonious ordering $a_0 = e, a_1, a_2, \ldots, a_{2m}$ such that $(a_{m+i})^{-1} = a_{m-i+1}$ for $i = 1, 2, \ldots, m$.

If a group of order n has an R-sequencing $a_0 = e, a_1, a_2, \ldots, a_{n-1}$ such that $a_i = a_{i-1}a_{i+1} = a_{i+1}a_{i-1}$ for some integer i, the sequencing is an *$R^\star$-sequencing* and the group is then *$R^\star$-sequenceable.*

A sequenceable group of even order $2m$ which has a unique element z of order two is *symmetrically sequenceable* if it has a sequencing $a_0 = e, a_1, a_2, \ldots, a_{2m-1}$ such that $a_m = z$ and $(a_{m+i})^{-1} = a_{m-i}$ for $i = 1, 2, \ldots, m-1$. A sequencing with these properties is a *symmetric sequencing.*

5.33 Theorem If G and H are both harmonious groups and H has odd order, then $G \times H$ is harmonious. [3]

5.34 Theorem The direct product of two symmetrically harmonious groups of odd order is symmetrically harmonious [15].

5.35 Theorem If G is a finite $R^\star$-sequenceable group and H is a symmetrically harmonious group, then the direct product $G \times H$ is $R^\star$-sequenceable except possibly when $|G|$ is odd and $|H|$ is divisible by 3 [15].

5.36 Corollary If G is an $R^\star$-sequenceable group of even order and H is an abelian group of odd order, the direct product $G \times H$ is $R^\star$- sequenceable. (Stronger but less easily stated corollaries can be deduced from Theorem 5.35. See [15].)

5.37 Theorem If the group G has a symmetric sequencing and B is an abelian group such that the orders of G and B are relatively prime, then the direct product $G \times B$ has a symmetric sequencing. (See [7], p. 85)

5.6 Quasi-Orderings and Quasi-Complete Mappings

A list $a_1, a_2, \ldots, a_n$ of elements of a finite group $(G, \odot)$ of order n (with identity element e) is a *quasi-ordering* of G if and only if the list contains exactly one occurrence of each element z of G such that $z^2 = e$; and, for every element g of G such that $g^2 \neq e$, there are either two occurrences of g and none of g^{-1} or one occurrence of g and one of g^{-1} or two occurrences of g^{-1} and none of g.

A *quasi-complete mapping* of a group or quasigroup $(G, \odot)$ is a mapping $g \to \theta(g)$ of the elements of G onto a quasi-ordering of G such that the mapping $\phi \to g\theta(g)$ is a permutation of the elements of G. The mapping $g \to \phi(g)$ is a *quasi-orthomorphism* of $(G, \odot)$.

A *quasi-near-complete mapping* of a group $(G, \odot)$ is a mapping $g \to \theta(g)$ of the elements of $G \setminus \{h\}$ onto a quasi-ordering of $G \setminus \{e\}$ such that the mapping $\phi \to g\theta(g)$ is a one-to-one mapping from $G \setminus \{h\}$ to $G \setminus \{k\}$ say. Here h and k are fixed elements of G and e is the identity element. The mapping $g \to \phi(g)$ is a *quasi-near-orthomorphism* of $(G, \odot)$.

Two latin squares of order n, each defined on the same symbol set S, are *quasi-orthogonal* if, when they are superimposed to form an array Λ of unordered pairs, each pair of the form $\{x, x\}$ occurs in Λ exactly once and each unordered pair of distinct elements $\{x, y\}$ occurs in Λ exactly twice.

5.38 **Theorem** [5] If L is a latin square which is the Cayley table of a finite group G (i.e., L is *based on* the group G) and if G has a quasi-complete mapping, then L has a quasi-orthogonal mate M.

5.39 **Remarks**

1. A group which has no complete mappings may nonetheless have quasi-complete mappings. The cyclic groups $\mathbb{Z}_{4n}$ are examples [5].
2. If L is a latin square based on a finite group G which has a complete mapping, then L has an orthogonal mate by Theorem 5.2 and Remark 5.4(1). The theory of quasi-orthogonal latin squares parallels that of orthogonal latin squares. See [5] for details.

If $a_0 = e, a_1, a_2, \ldots, a_{n-1}$ is a quasi-ordering of the elements of a group G of order n such that the partial products $b_0 = a_0 = e$, $b_1 = a_0a_1, b_2 = a_0a_1a_2, \ldots, b_{n-1} = a_0a_1a_2\cdots a_{n-1}$ are all distinct, then $a_0, a_1, a_2, \ldots, a_{n-1}$ is a *quasi-sequencing* of G and the sequence $b_0, b_1, b_2, \ldots, b_{n-1}$ is a *terrace* of G.

5.40 **Theorem** The Cayley table of a group G can be expressed in the form of a quasi-row complete latin square L such that each unordered pair of distinct elements occurs exactly twice in the rows of L if and only if G is quasi-sequenceable. (Compare Theorem 5.22, and see also §IV.48.)

5.41 **Remarks**

1. Directed terraces give rise to row complete latin squares, while (undirected) terraces give rise to quasi-row complete latin squares.
2. A quasi-sequencing of a group G is a quasi-near-complete mapping for which the corresponding near-orthomorphism ϕ satisfies $\phi(b_i) = b_i\theta(b_i) = b_ia_{i+1} = b_{i+1}$ for $i = 0, 1, 2, \ldots, n-2$. Compare Remark 5.20.
3. Quasi-sequencings have been called *2-sequencings* by some authors.

5.42 **Example** A quasi-sequencing for the dihedral group D_4 is $e, a, ba^3, ba, b, a^2, a, ba^2$. The corresponding terrace (sequence of partial products) is $e, a, ba^2, a^3, ba, ba^3, b, a^2$.

5.7 Other Generalizations of Complete Mappings

A complete mapping of a group or quasigroup $(G, \odot)$ is *strong* if the corresponding orthomorphism $g \to \phi(g) = g\theta(g)$ is itself a complete mapping of G (that is, if the map $g \to g\phi(g)$ is another permutation of G).

5.43 **Example** The complete mapping of $\mathbb{Z}_2 \times \mathbb{Z}_2$ given in Example 5.3 is strong.

5.44 **Theorem** Let θ be a strong complete mapping of the cyclic group $(\mathbb{Z}_n, +)$. Then the square matrix whose $(i,j)th$ cell contains $i + \phi(j)$ is a Knut Vik design (§II.1.7). Conversely, any Knut Vik design $A = (a_{ij})$ on the n symbols $0, 1, \ldots, n-1$ defines a strong complete mapping $g \to \theta(g) = \phi(g) - g$ of $(\mathbb{Z}_n, +)$ such that $\phi(j) = a_{ij} - i$ [8].

5.45 **Example** The identity mapping is a strong complete mapping of $(\mathbb{Z}_7, +)$. The corresponding orthomorphism maps h to $2h$ for all $h \in \mathbb{Z}_7$. This defines the Knut Vik design on the right.

0	2	4	6	1	3	5
1	3	5	0	2	4	6
2	4	6	1	3	5	0
3	5	0	2	4	6	1
4	6	1	3	5	0	2
5	0	2	4	6	1	3
6	1	3	5	0	2	4

A (K, λ)-*orthomorphism* of a group $(G, \odot)$, where $K = \{k_1, k_2, \ldots, k_s\}$ and the k_i are integers such that $\sum k_i = \lambda(|G| - 1)$, is an arrangement of the nonidentity elements of G (each used λ times) into s cyclic sequences (of distinct elements) of lengths $k_1, k_2, \ldots, k_s$, say $\phi = (g_{11}g_{12}\cdots g_{1k_1})(g_{21}g_{22}\cdots g_{2k_2})\cdots(g_{s1}g_{s2}\cdots g_{sk_s})$, such that the elements $g_{ij}^{-1}g_{i,j+1}$ (where $i = 1, 2, \ldots, s$; and the second suffix j is added modulo k_i) comprise the nonidentity elements of G each counted λ times.

5.46 **Remark** When $\lambda = 1$, a generalized orthomorphism is an orthomorphism.

5.47 **Theorem** If $\phi = (g_{11}\cdots g_{1k_1})(g_{21}\cdots g_{2k_2})\cdots(g_{s1}\cdots g_{sk_s})$ is a (K, λ)-orthomorphism of a group $(G, \odot)$, then the blocks of the set

$$B = \cup_{g \in G}(gg_{11}\ gg_{12}\cdots gg_{1k_1}), (gg_{21}\ gg_{22}\cdots gg_{2k_2}), \ldots, (gg_{s1}\ gg_{s2}\cdots gg_{sk_s})$$

form a (v, K, λ)-Mendelsohn design which has G as a regular group of automorphisms. (See §IV.28.)

5.48 **Remark** Generalized near-orthomorphisms also exist and define Mendelsohn designs in a similar way. See [7], pp. 38-39 and 74-75.

An *anti-symmetric permutation* of a group $(G, \odot)$ is a permutation $g \mapsto \psi(g)$ of the elements of G such that $\psi(g_1) \odot g_2 = \psi(g_2) \odot g_1$ implies $g_1 = g_2$ for all g_1, g_2 in G.

5.49 **Remarks**

1. An anti-symmetric permutation of an abelian group is equivalent to an orthomorphism, since then the property is equivalent to the requirement that the mapping $\theta : g \mapsto g^{-1}\psi(g)$ be a permutation of G.
2. Anti-symmetric permutations play an important role in the construction of decimal codes with a single check digit which detect transposition as well as single digit errors [7, pp. 310-311].
3. See [9] for details of those groups known to possess anti-symmetric permutations. It is conjectured that all nonabelian groups have such mappings.

5.8 See Also

[7] Chapters 2 and 3 contain detailed accounts of the material of this section up to mid-1990 and include proofs of many of the theorems.

[6] A more basic discussion of complete mappings and sequencing, together with proofs of the earlier theorems not given in [7], can be found in Chapters 1 and 2.

References

[1] B.A. Anderson, A fast method for sequencing low order non-abelian groups, *Ann. Discrete Math.* 34 (1987), 27-42.

[2] B.A. Anderson and E.C. Ihrig, Every finite solvable group with a unique element of order two, except the quaternion group, has a symmetric sequencing, *J. Combin. Des.* 1 (1993), 3-14.

[3] R. Beals, J.A. Gallian, P. Headley and D. Jungreis, Harmonious groups, *J. Combin. Theory A* 56 (1991), 223-238.

[4] D. Bedford, Construction of orthogonal latin squares using left neofields, *Discrete Math.* 115 (1993), 17-38.

[5] D. Bedford, Quasi-orthogonal latin squares and related designs, preprint.

[6] J. Dénes and A.D. Keedwell, *Latin Squares and their Applications*, Akadémiai Kiadó, Budapest; English Universities Press, London; Academic Press, New York, 1974.

[7] J. Dénes and A.D. Keedwell, *Latin Squares: New Developments in the Theory and Applications*, North-Holland, Amsterdam, 1991.

[8] A.B. Evans, On strong complete mappings, *Congr. Numer.* 70 (1990), 241-248.

[9] J.A. Gallian and M.D. Mullin, Groups with anti-symmetric mappings, *Arch. Math.*, to appear.

[10] J. Isbell, Sequencing certain dihedral groups, *Discrete Math.* 86 (1990), 323-328.

[11] P. Li, Sequencing the dihedral groups D_{4k}, preprint.

[12] ChengDe Wang, On the harmoniousness of dicyclic groups, *Discrete Math.* 120 (1993), 221-225.

[13] ChengDe Wang, On the R-sequenceability of dicyclic groups, *Discrete Math.* 125 (1994), 393-398.

[14] ChengDe Wang and P.A. Leonard, On $R^\star$-sequenceability and symmetric harmoniousness of groups, *J. Combin. Des.* 2 (1994), 71-78.

[15] ChengDe Wang and P.A. Leonard, On $R^\star$-sequenceability and symmetric harmoniousness of groups II, *J. Combin. Des.* 3 (1995), 313-320.

6 Configurations

HARALD GROPP

6.1 Definitions and Examples

A *configuration* (v_r, b_k) is an incidence structure of v points and b lines such that

1. each line contains k points,
2. each point lies on r lines,
3. two different points are connected by at most one line.

If $v = b$ and hence $r = k$, then configuration is *symmetric* and denoted by v_k.

A configuration (v_r, b_k) can be described by three parameters: *deficiency* d, *order* m, and *index* t, where $d = v - r(k-1) - 1, m = k - 1, t = r/k$.

The *configuration graph* of a configuration (v_r, b_k) has the v points as vertices; two vertices are joined by an edge if and only if the points are not collinear in the configuration. The configuration graph is d-regular.

6.1 Example The Fano configuration 7_3 (see Example I.5.3).

$$\{1,2,4\}, \{2,3,5\}, \{3,4,6\}, \{4,5,7\}, \{1,5,6\}, \{2,6,7\}, \{1,3,7\}.$$

6.2 Example The Pappus configuration 9_3 (see §VI.8.1).

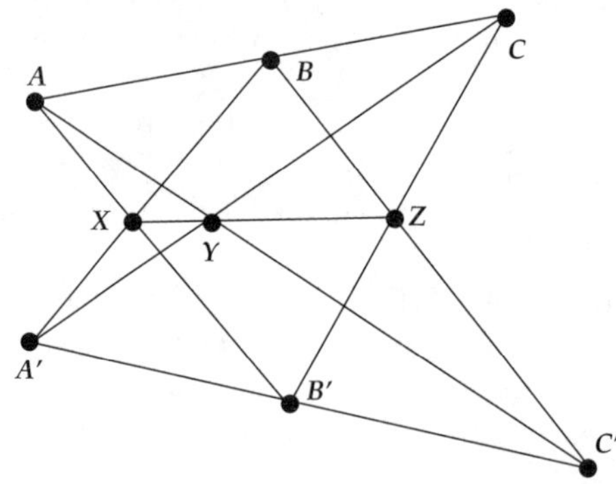

6.3 Example The Desargues configuration 10_3 is given in Example VI.8.11.

6.4 Example A configuration $(12_4, 16_3)$.

$\{1, 5,12\}$, $\{1,6,8\}$, $\{1,7,10\}$, $\{1,9,11\}$, $\{2,4,9\}$, $\{2,6,7\}$, $\{2,8,11\}$, $\{2,10,12\}$, $\{3, 4,10\}$, $\{3,5,11\}$, $\{3,7,12\}$, $\{3,8,9\}$, $\{4,5,6\}$, $\{4,11,12\}$, $\{5,7,8\}$, $\{6,9,10\}$.

6.5 Remarks

1. A configuration (v_r, b_k) is a k-uniform r-regular linear hypergraph with v vertices and b hyperedges.
2. Steiner systems are configurations with deficiency 0.
3. The point-line incidence graph of a configuration is a bipartite semiregular graph with girth at least 6.

6.2 Existence and Enumeration

6.6 Remark For convenience assume that $v \le b$. (If not, take the dual configuration by interchanging the role of points and lines).

6.7 Proposition The necessary conditions for the existence of a configuration (v_r, b_k) are that $vr = bk$ and that $v \ge r(k-1)+1$.

6.8 Theorem The necessary conditions are also sufficient for $k = 3$ [3].

6.9 Remarks

1. For $k = 4$, no nonexistence results concerning configurations (v_r, b_k) are known. All symmetric configurations $v_4, v \ge 13$ exist.
2. For $k = 5$, the necessary conditions are not sufficient. There is no configuration 22_5.

6.10 Table Existence of configurations v_k for $5 \le k \le 7, d \le 10$ [2]. (v_k means configuration exists, – means configuration does not exist, ? means existence is unknown.)

$d=$	0	1	2	3	4	5	6	7	8	9	10
$k=5$	21_5	–	23_5	24_5	25_5	26_5	27_5	28_5	29_5	30_5	31_5
6	31_6	–	?	?	35_6	36_6	37_6	38_6	39_6	40_6	41_6
7	–	–	45_7	?	?	48_7	49_7	50_7	51_7	52_7	53_7

6.11 Table The number n of nonisomorphic configurations v_3 for $v \leq 14$ [2].

v	7	8	9	10	11	12	13	14
n	1	1	3	10	31	229	2036	21399

6.12 Table The number N of nonisomorphic configurations (v_r, b_k) (with $d > 0$).

(v_r, b_k)	$(12_4, 16_3)$	$(12_5, 20_3)$	$(14_6, 28_3)$
N	574	5	787

6.3 Blocking Set Free Configurations and λ-Configurations

A subset of the point set of a configuration is a *blocking set* if it intersects all lines in at least 1 and at most $k-1$ points. A configuration is *blocking set free* if it does not contain a blocking set.

6.13 Remarks

1. There is no blocking set free configuration v_3 for $v = 8, 9, 10, 11, 12, 14$ [6].
2. There is a blocking set free configuration v_3 for $v = 7, 13, 19, 21, 22, 25$ and for $v \geq 27$. The existence is open for $v = 15, 16, 17, 18, 20, 23, 24, 26$ [6].
3. All symmetric configurations v_k with $k \geq 4$ have a blocking set [6].

A *λ-configuration* $(v_r, b_k)_\lambda$ is an incidence structure of v points and b lines such that

1. each line contains k points,
2. each point lies on r lines,
3. two different points are connected by at most λ lines.

A 2-configuration $(v_r, b_k)_2$ is also called a *spatial* configuration.

6.14 Remarks

1. All 2-configurations $(v_k)_2$, $3 \leq k \leq 6$ exist for $v \geq (k(k-1)/2) + 1$ [4].
2. $(v_7)_2$ exist for $v \geq 24$. A $(22_7)_2$ does not exist. The existence of a 2-configuration $(23_7)_2$ is still open [4].

6.4 See Also

§IV.35	Configurations are partially balanced incomplete block designs.
§IV.38	Configurations are special $(r, 1)$-designs (see [5]).
[1]	A survey on the history of configurations.

References

[1] H. Gropp, On the history of configurations, (A. Díez, J. Echeverría, and A. Ibarra, eds.), Universidad del País Vasco-Euskal Herriko Univ, Bilbao, Spain, 1990, 263-268.

[2] H. Gropp, Non-symmetric configurations with deficiencies 1 and 2, *Ann. Discrete Math.* 52 (1992), 227-239.

[3] H. Gropp, Non-symmetric configurations with natural index, *Discrete Math.* 124 (1994), 87-98.

[4] H. Gropp, On symmetric spatial configurations, *Discrete Math.* 125 (1994), 201-209.

[5] H. Gropp, Configurations and $(r, 1)$-designs, *Discrete Math.* 129 (1994), 113-137.

[6] H. Gropp, Blocking set free configurations and their relations to digraphs and hypergraphs, preprint.

7 Costas Arrays

HERBERT TAYLOR

7.1 Definitions and Examples

A *Costas array* of *order* n is an n by n array of dots and blanks which satisfies:

1. There are n dots and $n(n-1)$ blanks, with exactly one dot in each row and column.
2. All the segments between pairs of dots differ in length or in slope.

$C(n)$ denotes the number of distinct n by n Costas arrays.
$c(n)$ denotes the number of n by n Costas arrays, inequivalent under the dihedral group of rotations and reflections of the square.
$s(n)$ denotes the number of n by n Costas arrays which are symmetric across a diagonal and inequivalent under the dihedral group of rotations and reflections of the square.

7.1 Example $C(4) = 12,\quad c(4) = 2,\quad$ and $\quad s(4) = 1$.

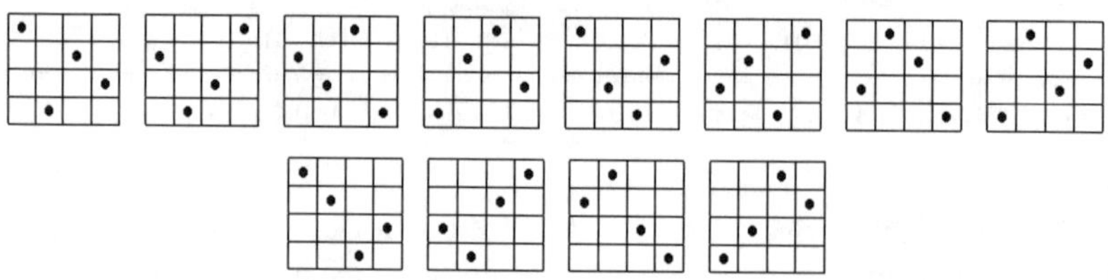

7.2 Remark For $n > 2$, if a Costas array is symmetric across a diagonal its dihedral equivalence class contains four arrays; otherwise, it contains eight arrays. Thus, for $n > 2$, $C(n) = 8 \cdot c(n) - 4 \cdot s(n)$.

Let $\pi(j) = i$ when there is a dot at (i, j). A Costas array of order n can be presented in the form:

$\pi(1)$	$\pi(2)$	$\pi(3)$	$\pi(4)$	$\cdots\cdots$	$\pi(n-1)$	$\pi(n)$

7.3 Example $c(5) = 6$. One representative of each dihedral equivalence class.

1	3	4	2	5

1	5	3	2	4

1	5	4	2	3

2	1	5	3	4

2	4	1	5	3

2	5	1	3	4

7.4 Example A Costas array of order 7 that is not obtained by any known algebraic construction.

3	6	7	1	5	2	4

7.5 Theorem [7] For $n \geq 4$, two Costas arrays of order n can not have disjoint sets of segments (with respect to length and slope) between pairs of dots.

7.2 Algebraic Constructions and Examples

7.6 Construction (*Welch construction*) Let p be prime and α be a primitive element in the field $\mathbb{F}_p$. Let $n = p-1$. A Costas array of order n is obtained by placing a dot at (i,j) if and only if $i = \alpha^j$, for $a \le j < n+a$, a a nonnegative integer, and $i = 1, \cdots, n$.

7.7 Construction Let α and β be primitive elements in the field $\mathbb{F}_q$ for q a prime power. Let $n = q-2$. Costas arrays of order n are obtained by:

1. *Lempel construction:* Put a dot at (i,j) if and only if $\alpha^i + \alpha^j = 1$, $1 \le i,j \le n$.
2. *Golomb construction:* Put a dot at (i,j) if and only if $\alpha^i + \beta^j = 1$, $1 \le i,j \le n$.

7.8 Remark $C(p-1) \ge 1$ and $C(q-2) \ge 1$ from Constructions 7.6 and 7.7, respectively. Also, if a corner dot is present in a Costas array of order n, it can be removed along with its row and column to obtain a Costas array of order $(n-1)$.

7.9 Theorem If $q > 2$ is a prime power, then there exist primitive elements α and β in $\mathbb{F}_q$ such that $\alpha + \beta = 1$.

7.10 Corollary Removing the corner dot at $(1,1)$ in the Costas array of order $(q-2)$ from Construction 7.7(2) yields $C(q-3) > 0$.

7.11 Example If there exist primitive elements α and β satisfying the conditions stated, then a Costas array of order n can be obtained by removing one or more corner dots.

Conditions	n
$\alpha^1 = 2$	$q-3$
$\alpha^1 + \beta^1 = 1$ and $\alpha^2 + \beta^2 = 1$	$q-4$
$\alpha^2 + \alpha^1 = 1$	$q-4$
$\alpha^1 + \beta^1 = 1$ and $\alpha^2 + \beta^{-1} = 1$	$q-4$
and necessarily $\alpha^{-1} + \beta^2 = 1$	$q-5$

7.12 Remark For $n \ge 30$, except for $n = 31$ and $n = 53$, the only orders for which Costas arrays are known are orders $n = p-1$ or $n = q-2$ or orders for which some algebraic condition exists which guarantees corner dots whose removal leaves a smaller Costas array.

7.13 Proposition Costas arrays of order n are known to exist for all $1 \le n \le 359$ except for $n \in \{32, 33, 43, 48, 49, 54, 63, 73, 74, 83, 84, 85, 89-93, 97, 103, 109, 113-117, 120, 121, 131-133, 139-143, 151-153, 157-159, 163, 168, 169, 173, 174, 181-186, 193, 199-207, 211-219, 223, 229, 233, 234, 242-246, 251, 257-259, 263, 271-273, 277, 283-285, 288, 289, 293-303, 313, 317-327, 331-333, 337-339, 342, 343, 349, 353, 354\}$.

7.14 Remark The only known Costas array of order 53 was found by Carbonera by adding a corner dot to one of the Costas arrays of order 52. The only known Costas array of order 31 was found in similar fashion from one of order 30.

7.3 Singly Periodic Costas Arrays

Let an n by n array be extended left and right putting a dot at $(i, j+n)$ if and only if there is a dot at (i,j). A *singly periodic Costas array* of order n is an n by n array such that every n by n window in the left right extension is a Costas array of order n.

7.15 Example A singly periodic Costas array of order 6 with $i = 3^j$ in $\mathbb{F}_7$.

$$\ldots, (1,0), (3,1), (2,2), (6,3), (4,4), (5,5), (1,6), (3,7), (2,8), \ldots.$$

7.16 Remarks The only known singly periodic Costas arrays of order n come from Construction 7.6 when $n = p-1$. A singly periodic Costas array of order n does not exist if $n > 1$ is odd, nor if $n \in \{8, 14, 20\}$.

7.17 Theorem [4] If M is a singly periodic Costas array of order n given by Construction 7.6, and $n > 4$, then the transpose of M is not a singly periodic Costas array of order n.

7.18 Remark It remains open whether for $n > 20$ a matrix M exists such that both M and its transpose are singly periodic Costas arrays of order n.

7.19 Proposition [4] A singly periodic Costas array of order n is equivalent to a vatican square of order n by the polygonal path construction. See §IV.48.1.

7.4 G-Symmetry (Glide-Reflection Symmetry)

A Costas array of order $2m$ has *G-symmetry* if for j from 1 to m there is a dot at (i,j) if and only if there is a dot at $(2m-i+1,\ m+j)$.
A Costas array of order $(2m-1)$ has *G-symmetry* if there is a dot at (m,m) and for j from 1 to $(m-1)$ there is a dot at (i,j) if and only if there is a dot at $(2m-i,\ m+j)$.

$c_G(n)$ denotes the number of Costas arrays of order n that have G-symmetry and are inequivalent under the dihedral group of rotations and reflections of the square. $c^\star_G(n) = c_G(n)$ minus the number of those obtained from Construction 7.6.

7.20 Example [4] The only Costas array of order $2m$ different from its transpose, both having G-symmetry.

1	7	6	4	8	2	3	5

1	6	7	4	8	3	2	5

7.21 Theorem [4] For $2m > 8$, if a Costas array of order $2m$ has G-symmetry, then its transpose does not have G-symmetry.

7.22 Remark All known singly periodic Costas arrays of order n have G-symmetry.

7.5 Further Examples and Enumeration

7.23 Example The only known Costas array of order 31 (all arithmetic in $\mathbb{F}_{31}$).

0	3^3	3^4	3^5		……		3^{30}	3^{31}	3^{32}

7.24 Example A Costas array of nonattacking Kings.

3	5	8	1	7	4	2	6

A *honeycomb Costas array* is a Costas array of order $(2r+1)$ whose dot positions (i,j) are such that $(i-j)$ takes each value exactly once from $-r$ to r.

7.25 Remark Honeycomb arrays are discussed in [6] and are related to the "Zero Sum Arrays" of Bennett and Potts [1].

7.26 Example A honeycomb Costas array not given by any known algebraic construction.

1	3	6	2	7	5	4

7.27 Table For each triple n, α, q below there exists a honeycomb Costas array of order n from Construction 7.7(1) with primitive α in $\mathbb{F}_q$.

n	7	9	9	15	15	21	27	45
α	$\alpha^2 = \alpha + 1$	2	7	3	5	7	3	30
q	9	11	11	17	17	23	29	47

7.28 Remark Proofs for the algebraic constructions are given in [5]. Cenkl and Rushanan report that honeycomb Costas arrays of orders 17 and 19 do not exist.

7.29 Table Enumeration of Costas arrays of order n, $1 \leq n \leq 32$.

n	$C(n)$	$s(n)$	$c(n)$	$c_G(n)$	$c^\star_G(n)$	n	$C(n)$	$s(n)$	$c(n)$	$c_G(n)$	$c^\star_G(n)$
1	1	1	1	1	0	2	2	1	1	1	0
3	4	1	1	0	0	4	12	1	2	2	0
5	40	2	6	1	1	6	116	5	17	4	1
7	200	10	30	0	0	8	444	9	60	3	3
9	760	10	100	0	0	10	2160	14	277	24	14
11	4368	18	555	0	0	12	7852	17	990	44	32
13	12828	25	1616	4	4	14	17252	23	2168	31	31
15	19612	31	2467	0	0	16	21104	20	2648	77	45
17	18276	19	2294	0	0	18	15096	10	1892	29	2
19	10240	6	1283	0	0	20	6464	4	810	3	3
21	3536	8	446	0	0	22	2052	5	259	55	0
23	872	10	114	0	0	24	≥ 1	0	≥ 1		
25	≥ 1	2	≥ 1			26	≥ 1	2	≥ 1		
27	≥ 1	7	≥ 1			28	≥ 1	0	≥ 1		
29	≥ 1	5	≥ 1			30	≥ 1	4	≥ 1		
31	≥ 1	0	≥ 1			32	?	0	?		

7.30 Remark The properties of a Costas array make it an ideal discrete waveform for Doppler sonar. Having one dot in each row and column minimizes reverberation. Distinct segments between pairs of dots give it a thumbtack ambiguity function, because, shifted left-right in time and up-down in frequency, copies of the pattern can only agree with the original in one dot, no dots, or in all n dots at once. Thus the spike of the thumbtack makes a sharp distinction between the actual shift and all the near misses. See [3, 6, 7].

7.6 See Also

§IV.48	Proposition 7.19 gives a connection between Costas arrays and vatican squares.
[7]	Chapter 8 contains a discussion of the use of Costas arrays as coded radar signals.

References

[1] B.T. Bennett and R.B. Potts, Arrays and brooks, *J. Australian Math. Soc.* 7 (1967), 23-31.

[2] C.P. Brown, M. Cenkl, R.A. Games, J.J. Rushanan, O. Moreno and Pei Pei, New enumeration results for Costas arrays, *Proceedings of the 1993 International Symposium on Information Theory*, IEEE catalog no. 92CH3230-0, p. 405.

[3] J.P. Costas, A study of a class of detection waveforms having nearly ideal range-doppler ambiguity properties, *Proc. IEEE* 72 (1984), 996-1009.
[4] T. Etzion, Combinatorial designs derived from Costas arrays, in: *Sequences* (R.M. Capocelli, ed.), Springer-Verlag, New York, 1989, 206-227.
[5] S.W. Golomb, Algebraic constructions for Costas arrays, *J. Combin. Theory A* 37 (1984), 13-21.
[6] S.W. Golomb and H. Taylor, Constructions and properties of Costas arrays, *Proc. IEEE* 72 (1984), 1143-1163.
[7] N. Levanon, *Radar Principles*, John Wiley and Sons, New York, 1988.
[8] J. Silverman, V.E. Vickers and J.M. Mooney, On the number of Costas arrays as a function of array size, *Proc. IEEE* 76 (1988), 851-853.

8 Coverings

DOUGLAS R. STINSON

8.1 Definitions and Examples

Let $v \geq m \geq k$. A t-(v,k,λ) *covering* is a pair $(X,\mathcal{B})$, where X is a v-set of elements (*points*) and $\mathcal{B}$ is a collection of k-subsets (*blocks*) of X, such that every t-subset of points occurs in at least λ blocks in $\mathcal{B}$. Repeated blocks in $\mathcal{B}$ are permitted.

The *covering number* $C_\lambda(v,k,t)$ is the minimum number of blocks in any t-(v,k,λ) covering. A t-(v,k,λ) covering $(X,\mathcal{B})$ is *optimal* if $|\mathcal{B}| = C_\lambda(v,k,t)$. If $\lambda = 1$, then write $C(v,k,t)$ for $C_1(v,k,t)$.

8.1 Examples Optimal coverings for certain parameter sets t-(v,k,λ).

t-(v,k,λ)	Covering
2-$(5,3,1)$	{1,2,3}, {1,4,5}, {2,3,4}, {2,3,5}
2-$(6,3,1)$	{0,1,3} modulo 6
2-$(8,3,1)$	{0,1,3} modulo 7, {0,1,∞}, {2,3,∞}, {4,5,∞}, {5,6,∞}
2-$(6,4,1)$	{1,2,3,4}, {1,2,5,6}, {3,4,5,6}
2-$(9,4,1)$	{1,2,3,4}, {1,2,5,6}, {1,7,8,9}, {2,4,6,8}, {2,7,8,9}, {3,5,8,9}, {3,6,7,9}, {4,5,7,9}

8.2 Remark The survey paper by Mills and Mullin [6] covers much of the material in this section. For this reason individual references are not given in general.

8.2 Equivalent Combinatorial Objects

8.3 Theorem A t-(v,k,λ) covering with $\lambda\binom{v}{t}/\binom{k}{t}$ blocks is equivalent to a t-(v,k,λ) design or an $S_\lambda(t,k,v)$ (possibly containing repeated blocks).

Let $v \geq k \geq t$. A (v,m,k) *Turán design* is a pair $(X,\mathcal{B})$, where X is a v-set of elements (*points*) and $\mathcal{B}$ is a collection of k-subsets (*blocks*) of X, such that every m-subset of points is a superset of at least one block $B \in \mathcal{B}$.

The *Turán number* $T(v,m,k)$ is the minimum number of blocks in any (v,m,k) Turán design.

8.4 Theorem $(X,\mathcal{B})$ is a (v,m,k) *Turán design* if and only if $(X,\{X\backslash B : B \in \mathcal{B}\})$ is a $(v-m)$-$(v,v-k,1)$ covering.

8.5 Corollary $T(v,m,k) = C(v,v-k,v-m)$.

8.3 Bounds

8.6 Theorem (*Schönheim bound*) $C_\lambda(v,k,t) \geq \lceil vC_\lambda(v-1,k-1,t-1)/k \rceil$. Iterating this bound yields $C_\lambda(v,k,t) \geq L_\lambda(v,k,t)$, where

$$L_\lambda(v,k,t) = \left\lceil \frac{v}{k} \left\lceil \frac{v-1}{k-1} \cdots \left\lceil \frac{\lambda(v-t+1)}{k-t+1} \right\rceil \right\rceil \right\rceil.$$

Further, if $\lambda(v-1) \equiv 0 \pmod{k-1}$ and $\lambda v(v-1) \equiv 1 \pmod{k}$, then

$$C_\lambda(v,k,2) \geq L_\lambda(v,k,2) + 1.$$

8.4 Determination of Covering Numbers

8.7 Theorem If $v \equiv 2 \pmod 3$, $\lambda \equiv 2 \pmod 3$ and $\lambda(v-1) \equiv 0 \pmod 2$, then $C_\lambda(v,3,2) = L_\lambda(v,3,2) + 1$. Otherwise, $C_\lambda(v,3,2) = L_\lambda(v,3,2)$.

8.8 Remark $C_\lambda(v,3,2) = L_\lambda(v,3,2)$ except when the final assertion of Theorem 8.6 applies.

8.9 Theorem $C(v,4,2) = L_1(v,4,2) + \epsilon$, where

$$\epsilon = \begin{cases} 1 & \text{if } v = 7, 9 \text{ or } 10 \\ 2 & \text{if } v = 19 \\ 0 & \text{otherwise.} \end{cases}$$

8.10 Theorem If $\lambda > 1$, then $C_\lambda(v,4,2) = L_\lambda(v,4,2)$.

8.11 Theorem If $v \not\equiv 7 \pmod{12}$, then $C(v,4,3) = L_1(v,4,3)$. Also, if $v \equiv 7 \pmod{12}$, $v \geq 52423$, then $C(v,4,3) = L_1(v,4,3)$.

8.12 Remark Determination of the numbers $C(v,4,3)$ for $v \equiv 7 \pmod{12}$ is made difficult by the fact that $C(7,4,3) = L_1(7,4,3) + 1 = 12$. The smallest value $v \equiv 7 \pmod{12}$ for which it is known that $C(7,4,3) = L_1(7,4,3)$ is $v = 499$.

8.13 Theorem $C(v,5,2) = L_1(v,5,2)$ in the following cases:

1. $v \equiv 2 \pmod 4$,
2. $v \equiv 3 \pmod 4, v \neq 15$,
3. $v \equiv 1, 5 \pmod{20}$,
4. $v \equiv 9 \pmod{20}, v > 3149$, and
5. $v \equiv 17 \pmod{20}, v > 757$.

8.14 Theorem $C(v,5,2) = L_1(v,5,2)+1$ when $v = 15$ and when $v \equiv 13 \pmod{20}, v > 753$. (Note that $C(v,5,2) \geq L_1(v,5,2) + 1$ for all $v \equiv 13 \pmod{20}$.)

8.15 Theorem $C_2(v,5,2) = L_2(v,5,2)$ if v is odd, $v \neq 9, 13, 15, 53, 63, 73, 83$.

8.16 Theorem If $v \equiv 0 \pmod 4$ or if $v \equiv 1, 5, 13 \pmod{20}$, then $C_3(v,5,2) = L_3(v,5,2)$; if $v \equiv 9, 17 \pmod{20}$, then $C_3(v,5,2) = L_3(v,5,2) + 1$.

8.17 Theorem (Assaf [1]) Suppose $\lambda \equiv 0 \pmod 4$ and $(v, \lambda) \neq (28, 4)$. Then $C_\lambda(v,5,2) = L_\lambda(v,5,2)+1$ if $\lambda(v-1) \equiv 0 \pmod 4$ and $\lambda v(v-1)/4 \equiv 4 \pmod 5$; and $C_\lambda(v,5,2) = L_\lambda(v,5,2)$, otherwise.

8.18 Theorem Suppose v is odd and $\lambda > 2$. Then $C_\lambda(v,5,2) = L_\lambda(v,5,2)+1$ if $\lambda(v-1) \equiv 0 \pmod 4$ and $\lambda v(v-1) \equiv 1 \pmod 5$; and $C_\lambda(v,5,2) = L_\lambda(v,5,2)$, otherwise.

8.19 Remark Various other partial results are known concerning $C_\lambda(v,5,2)$, $\lambda > 4$.

8.20 Table Known values $C(v,k,t)$ for $v \leq 20$, $2 \leq t \leq 3$ and $1 \leq k-t \leq 4$.

$t = 2$

$v\backslash k$	3	4	5	6
3	1			
4	3	1		
5	4	3	1	
6	6	3	3	1
7	7	5	3	3
8	11	6	4	3
9	12	8	5	3
10	17	9	6	4
11	19	11	7	6
12	24	12	9	6
13	26	13	10	7
14	33	18	12	7
15	35	19	13	10
16	43	20	15	10
17	46	26	16	12
18	54	27	18	12
19	57	31	19	≤ 15
20	67	35	21	16

$t = 3$

$v\backslash k$	4	5	6	7
4	1			
5	4	1		
6	6	4	1	
7	12	5	4	1
8	14	8	4	4
9	25	12	7	4
10	30	17	10	6
11	47	20	11	8
12	57	≤ 29	15	11
13	78	≤ 34	≤ 21	13
14	91	47	≤ 25	14
15	124	≤ 60	≤ 31	15
16	140	≤ 68	≤ 38	≤ 25
17	183	68	≤ 44	≤ 28
18	207	94	48	≤ 34
19	≤ 261	≤ 114	≤ 66	≤ 44
20	285	≤ 145	≤ 75	≤ 52

8.21 Theorem The values $C(v,k,2)$ are known in the following cases:

1. $C(v,k,2) = 3$ for $1 < v/k \leq 3/2$;
2. $C(v,k,2) = 4$ for $3/2 < v/k \leq 5/3$;
3. $C(v,k,2) = 5$ for $5/3 < v/k \leq 9/5$;
4. $C(v,k,2) = 6$ for $9/5 < v/k \leq 2$;
5. $C(v,k,2) = 7$ for $2 < v/k \leq 7/3$, except when $3v = 7k-1$;
6. $C(v,k,2) = 8$ for $7/3 < v/k \leq 12/5$, except when $12k - 5v = 0, 1$ and $v-k$ is odd;
7. $C(v,k,2) = 9$ for $12/5 < v/k \leq 5/2$, except when $2v = 5k$ and $v-k$ is odd;
8. $C(v,k,2) = 10$ for $5/2 < v/k \leq 8/3$, except when $8k - 3v = 0, 1$, $v-k$ is odd and $k > 2$;
9. $C(v,k,2) = 11$ for $8/3 < v/k \leq 14/5$, except when $14k - 5v = 0, 1$, $v-k$ is odd and $k > 4$;
10. $C(v,k,2) = 12$ for $14/5 < v/k \leq 3$, except when $v = 3k$, $k \not\equiv 0 \pmod 3$ and $k \not\equiv 0 \pmod 4$.

8.22 Theorem (Mills) The values $C(v,k,3)$ are known in the following cases:

1. $C(v,k,3) = 4$ for $1 < v/k \leq 4/3$;
2. $C(v,k,3) = 5$ for $4/3 < v/k \leq 7/5$;
3. $C(v,k,3) = 6$ for $7/5 < v/k \leq 3/2$, except when $2v = 3k$ and v is odd;
4. $C(v,k,3) = 7$ for $3/2 < v/k \leq 17/11$, except when $11v = 17k - 1$;
5. $C(v,k,3) = 8$ for $17/11 < v/k \leq 8/5$, except when $5v = 8k - 1$ and $k > 7$.

8.23 Theorem $C(v, v-1, t) = t+1$ for all t.

8.24 Theorem (Turán) Suppose $q = \left\lfloor \frac{v}{v-t+1} \right\rfloor$. Then $C(v, v-2, t) = qv - \binom{q+1}{2}(v-t+1)$.

8.25 Table Known values $C(v,k,t)$ for $v \le 20$, $4 \le t \le 5$ and $1 \le k-t \le 4$.

$t = 4$

$v\backslash k$	5	6	7	8
5	1			
6	5	5		
7	9	7	1	
8	20	12	5	1
9	30	20	6	5
10	≤ 51	32	10	5
11	66	≤ 41	17	9
12	113	≤ 66	≤ 24	12
13	≤ 157	≤ 87	≤ 30	≤ 19
14	≤ 235	≤ 134	≤ 44	≤ 27
15	≤ 313	≤ 178	≤ 59	≤ 30
16	≤ 437	≤ 243	≤ 90	30
17	≤ 558	≤ 258	≤ 119	≤ 55
18	≤ 732	≤ 352	≤ 157	≤ 68
19	≤ 926	≤ 456	≤ 187	≤ 98
20	≤ 1165	≤ 594	≤ 246	≤ 116

$t = 5$

$v\backslash k$	6	7	8	9
6	1			
7	6	1		
8	12	6	1	
9	30	9	6	1
10	50	20	8	6
11	≤ 100	≤ 34	16	7
12	132	≤ 59	26	12
13	245	≤ 88	≤ 43	≤ 19
14	≤ 385	≤ 154	≤ 66	≤ 36
15	≤ 620	≤ 224	≤ 108	≤ 49
16	≤ 840	≤ 358	≤ 118	≤ 79
17	≤ 1277	≤ 506	≤ 208	≤ 94
18	≤ 1791	≤ 696	≤ 296	≤ 149
19	≤ 2501	≤ 930	≤ 419	≤ 199
20	≤ 3297	≤ 1239	≤ 541	≤ 267

8.5 Structure of Optimal Coverings

Let $(X, \mathcal{B})$ be a 2-$(v,k,1)$ covering. The *excess graph* of $(X,\mathcal{B})$ is the multigraph (X, E), where each edge xy occurs with multiplicity $|\{B \in \mathcal{B} : \{x,y\} \subseteq B\}| - 1$.

8.26 Table Optimal 2-$(v,3,1)$ coverings.

$v \equiv$	$D(v,3,2)$	Excess Graph	Construction
1, 3 (mod 6)	$\frac{v^2-v}{6}$	Empty	$(v,3,1)$ BIBD
0 (mod 6)	$\frac{v^2}{6}$	$\frac{v}{2}K_2$	For $v \ge 18$, fill in each group of a $\{3\}$-GDD of type $6^{v/6}$ with an optimal covering on six points.
2, 4 (mod 6)	$\frac{v^2+2}{6}$	$K_{1,3} \cup \frac{v-4}{2}K_2$	For $v \equiv 4$ (mod 6), $v \ge 22$, fill in each group of a $\{3\}$-GDD of type $6^{(v-4)/6}4^1$ with an optimal covering on four or six points; for $v \equiv 2$ (mod 6), $v \ge 26$, fill in each group of a $\{3\}$-GDD of type $6^{(v-8)/6}8^1$ with an optimal covering on six or eight points.
5 (mod 6)	$\frac{v^2-v+4}{6}$	One edge of multiplicity 2	For $v \ge 11$, take a $(v,\{3,5^*\})$-PBD on v points and fill in the block of size 5 with an optimal covering on five points.

8.27 Table Optimal 2-$(v,4,1)$ coverings, $v \notin \{7,9,10,19\}$.

$v \equiv$	$D(v,4,2)$	Excess Graph	Construction
$1,4 \pmod{12}$	$\frac{v^2-v}{12}$	Empty	$(v,4,1)$ BIBD
$0,6 \pmod{12}$	$\frac{v^2}{12}$	$\frac{v}{2}K_2$	For $v \geq 30$, fill in each group of a $\{4\}$-GDD of type $6^{v/6}$ with an optimal covering on six points.
$3,9 \pmod{12}$	$\frac{v^2+3}{12}$	$K_{1,4} \cup \frac{v-5}{2}K_2$	For $v \geq 51$, fill in each group of a $\{4\}$-GDD of type $6^{(v-15)/6}15^1$ with an optimal covering on six or fifteen points.
$7,10 \pmod{12}$	$\frac{v^2-v+6}{12}$	One edge of multiplicity three	Take a $(v,\{4,22^*\})$-PBD and replace the block of size 22 by an optimal covering on 22 points.
$8,11 \pmod{12}$	$\frac{v^2+v}{12}$	A 2-regular multigraph on v points	Take an optimal covering on $v-1$ points in which the pair 12 occurs four times, and in which $\{1,2,3,4\}$ is a block. Then replace the block $\{1,2,3,4\}$ by the two blocks $\{1,3,4,v\}$ and $\{2,3,4,v\}$. Finally, adjoin new blocks $\{5,6,7,v\},\ldots,\{v-3,v-2,v-1,v\}$.
$2,5 \pmod{12}$	$\frac{v^2+v+6}{12}$	A multigraph on v points in which two vertices have degree 5 and the remaining $v-2$ vertices have degree 2; or one in which one vertex has degree 8 and the remaining $v-1$ vertices have degree 2	Take a $(v-1,4,1)$ BIBD and adjoin new blocks $\{1,2,3,v\}$, $\{4,5,6,v\}$, ..., $\{v-4,v-3,v-2,v\}$ and $\{v-3,v-2,v-1,v\}$.

8.6 Resolvable Coverings with $\lambda = 1$

A t-(v,k,λ) covering $(X,\mathcal{B})$ is *resolvable* if $\mathcal{B}$ can be partitioned into *parallel classes*, each of which consists of v/k disjoint blocks.

8.28 Example A resolvable 2-$(24,4,1)$ covering on $48 = L_1(24,4,2)$ blocks [4]. Let $X = \mathbb{Z}_3 \times \{0,\ldots,7\}$. Three parallel classes are formed by developing the following class modulo 3:

$$\{(0,0),(1,0),(1,1),(2,2)\}\ \{(2,0),(2,3),(0,4),(0,6)\}\ \{(0,1),(1,2),(0,3),(1,3)\}$$
$$\{(2,1),(2,2),(0,5),(0,7)\}\ \{(1,4),(1,5),(2,6),(2,7)\}\ \{(2,4),(2,5),(1,6),(1,7)\}.$$

Three more parallel classes are formed by developing the following class modulo 3:

$$\{(0,0),(2,1),(0,4),(2,4)\}\ \{(1,0),(2,3),(1,5),(2,5)\}\ \{(2,0),(1,3),(1,7),(2,7)\}$$
$$\{(0,1),(1,1),(1,6),(0,7)\}\ \{(0,2),(1,2),(1,4),(0,5)\}\ \{(2,2),(0,3),(0,6),(2,6)\}.$$

The seventh parallel class is formed by developing the following two blocks modulo 3:

$$\{(0,0),(0,1),(2,5),(2,6)\}\ \{(0,2),(0,3),(2,4),(2,7)\}.$$

Finally, the eighth parallel class is formed by developing the following two blocks modulo 3:

$$\{(0,0),(1,2),(0,6),(1,7)\}\ \{(0,1),(2,3),(2,4),(0,5)\}.$$

8.29 Theorem Suppose $v \equiv 0 \pmod{k}$. If $v-1 \equiv 0 \pmod{(k-1)}$, then a resolvable 2-$(v,k,1)$ covering with $L_1(v,k,2)$ blocks is equivalent to a resolvable $(v,k,1)$ BIBD.

8.30 Theorem There exists a resolvable 2-$(v,3,1)$ covering having $L_1(v,3,2)$ blocks for all $v \equiv 0 \pmod{6}$, $v \geq 18$.

8.31 Theorem [4] There exists a resolvable 2-$(v,4,1)$ covering having $L_1(v,4,2)$ blocks for all $v \equiv 0 \pmod{4}$, $v \neq 12$, except for possibly 10 values of v all greater than 100.

8.7 See Also

§I.1	BIBDs are coverings with void excess graph.
§II.3	Incomplete transversal designs are used extensively in various constructions for coverings.
§III	Pairwise balanced designs and group divisible designs.
§IV.49	t-wise balanced designs.
§V.8	Gives the connection between coverings, Turán designs and lottery schemes.
[6]	A general survey of coverings and packings with an extensive bibliography.
[2]	Up-to-date numerical results and tables of the best known covering numbers.
[3]	Information on resolvable coverings with $k=3$.
[8]	Computational methods for coverings.

References

[1] A.M. Assaf, Covering pairs by quintuples with index $\lambda \equiv 0 \pmod{4}$, *J. Combin. Des.* 1 (1993), 453-458.

[2] D.M. Gordon, G. Kuperberg and O. Patashnik, New constructions for covering designs, *J. Combin. Des.* 3 (1995), 269-284.

[3] E.R. Lamken and W.H. Mills, Resolvable coverings, *Congr. Numer.* 96 (1993), 21-26.

[4] E.R. Lamken, W.H. Mills and R. Rees, Resolvable minimum covers with quadruples, preprint.

[5] W.H. Mills, Covering designs I: coverings by a small number of subsets, *Ars Combin.* 8 (1979), 199-315.

[6] W.H. Mills and R.C. Mullin, Coverings and packings, in: *Contemporary Design Theory: A Collection of Surveys* (J.H. Dinitz, D.R. Stinson; eds.) John Wiley & Sons, New York, 1992, 371-399.

[7] W.H. Mills and R.C. Mullin, On λ-covers of pairs by quintuples: v odd, *J. Combin. Theory A* 67 (1994), 245-272.

[8] K.J. Nurmela and P.R. Östergård, Upper bounds for covering designs by simulated annealing, *Congr. Numer.* 96 (1993), 93-111.

9 Cycle Systems

C.A. RODGER

9.1 Definitions and Example

λK_n is the graph on n vertices in which each pair of vertices is joined by exactly λ edges. λD_n is the directed graph on n vertices in which for each pair of vertices u, v, there are λ arcs (u, v) and λ arcs (v, u).

A *(directed) m-cycle system* of a (directed) graph G is a collection C of (directed) cycles of length m whose (directed) edges partition the (directed) edges of G. A (directed) m-cycle system of *order* n and *index* λ is a (directed) m-cycle system of λK_n (λD_n).

For m-cycle systems of λK_n, n is (m, λ)*-admissible* if m divides $\lambda n(n-1)/2$, $\lambda(n-1) \equiv 0 \pmod 2$, and either $n = 1$ or $n \geq m$. For directed m-cycle systems of λD_n, n is $(m, \lambda)^*$*-admissible* if m divides $\lambda n(n-1)$ and either $n = 1$ or $n \geq m$.

9.1 **Example** A 5-cycle system of order 11 and index 1.

$$C = \{(1+i, 2+i, 5+i, 3+i, 8+i) | i \in \mathbb{Z}_{11}\}.$$

9.2 Existence and Enumeration

9.2 **Table** The existence of m-cycle systems of order n and index λ has been settled in the cases shown.

m	λ	n
≤ 50	1	all $(m, 1)$-admissible
p^e, p prime, $e \geq 1$	1	all $(m, 1)$-admissible
≤ 14, or 16	all	all (m, λ)-admissible
all	all	$\equiv 1 \pmod{2m}$
odd	all	$\equiv m \pmod{2m}$

9.3 **Theorem** [5, 10] If there exists an m-cycle system of K_n, for which $m \leq n < 3m$ if m is odd, then there exists an m-cycle system of K_v for all $v \equiv n \pmod{2m}$, $v \geq n$.

9.4 **Table** The existence of directed m-cycle systems of order n and index λ has been settled in the following cases.

m	λ	n
3	all	all $(3, \lambda)^*$-admissible except when $n = 6$ and $\lambda = 1$
4	all	all $(4, \lambda)^*$-admissible except when $n = 4$ and λ is odd
6	all	all $(6, \lambda)^*$-admissible except when $n = 6$ and $\lambda = 1$
5, 7, 8, 10, 12, 14	all	all $(m, \lambda)^*$-admissible
≥ 7	all	$\equiv 0$ or $1 \pmod m$

9.5 **Table** [4] The number N of nonisomorphic m-cycle systems of order n and index 1.

n	m	N	n	m	N	n	m	N
7	3	1	7	7	2	9	3	1
9	4	8	9	6	640	9	9	122
11	5 (cyclic)	5	11	11	3140 with a nontrivial automorphism group			

9.3 i-Perfect m-Cycle Systems

If c is a cycle, let $c(i)$ be the graph formed from c on the same vertex set by joining two vertices if and only if they are distance i apart in c. An m-cycle system C of order n and index λ is *i-perfect* if and only if $C(i) = \{c(i) | c \in C\}$ is a cycle system of order n and index λ.

9.6 **Example** The 5-cycle system in Example 9.1 is 2-perfect, since $C(2) = \{(1+i, 5+i, 8+i, 2+i, 3+i) | i \in \mathbb{Z}_{11}\}$ is also a 5-cycle system of order 11 and index 1.

9.7 **Remark** One can assume that $2 \leq i < m/2$, since i-perfect m-cycle systems are $(m-i)$-perfect.

9.8 **Table** For the given values of m, λ and i, $2 \leq i < m/2$, i-perfect m-cycle systems are known to exist for all (m, λ)-admissible integers n, except for those listed in the next to last column, and possibly except those in the last column [1, 7].

m	λ	i	Exceptions of (n, λ)	Possible Exceptions of (n, λ)
5	all	2	(15, 1)	
6	all	2	(9, 1),(6, 2(mod 4))	
7	1	all		
8	all	2	(8, 2)	(185,2)
8	1	3		
9	1	2, 4	(9, 1)	(45,1)
9	all	3		
10	1	3		
11, 12	1	all		
13	1	2,3,4,6		
15	1	2,4,5,7		(55, 1)
16	1	all		
17	1	all		
19	1	all		(57,1) for $i = 2, 9$
23	1	2		(69,1)
$m = p_1, 2m+1 = p_2^e, p_i$ prime	1	2		$3m, 4m+1, 6m+1$

9.9 **Theorem** [10] Let m be odd. There exists a 2-perfect m-cycle system of order n and index λ if and only if there exists a $\lfloor m/2 \rfloor$-perfect m-cycle system of order n and index λ.

9.10 **Remark** Cycle systems that are 2-perfect or $\lfloor m/2 \rfloor$-perfect with m odd are of particular interest since they can be used to construct quasigroups satisfying special identities. See also [6] and §IV.36.

9.4 Embedding and Incomplete m-Cycle Systems

A *partial* (directed) m-cycle system of order n and index λ is a collection C of (directed) cycles of length m such that each edge in $\lambda K_n(\lambda D_n)$ is in at most one (directed) m-cycle in C.

The partial m-cycle system C' is *embedded* in the m-cycle system C if $C' \subseteq C$.

9.11 **Table** The best known results for embedding partial cycle systems. The Best Embedding column lists an integer v such that any partial (directed) m-cycle system of

λK_n (λD_n) can be embedded in a (directed) m-cycle system of index λ and order at most v [9, 11]. For lower bounds on v, see Theorem 9.13.

Parity of m	K_n or D_n	Best Embedding	Conditions
Odd	K_n	$2n+1$	$\lambda \equiv 0 \pmod 2$ and $m=3$
·	·	$4n+1$	$\lambda \equiv 1 \pmod 2$ and $m=3$
·	·	$(2n+1)m$	$\lambda = 1$ and $m > 3$
·	·	$(4n+17)m$	$\lambda \geq 2$ and $m > 3$
Odd	D_n	$(2n+1)m$	$\lambda \geq 1$ or $m > 3$
·	·	$4n+1$	$\lambda = 1$ and $m = 3$
Even	K_n	nm	λ even
·	·	$2nm+1$	$\lambda \geq 3$ and odd, or $\lambda = 1$ and $m > 4$
·	·	$\approx 2n+\sqrt{2n}$	$\lambda = 1$ and $m = 4$
Even	D_n	nm	$\lambda \geq 1$ and $m \geq 8$
·	·	$nm+1$	$\lambda \geq 1$ and $m = 6$, or $\lambda \geq 2$ and $m = 4$
·	·	$\approx 2n+\sqrt{2n}$	$\lambda = 1$ and $m = 4$

For $n < v$, $K_v - K_n$ is the subgraph of K_v formed by removing all edges that join two vertices in some set of n vertices of K_v.

An *incomplete* m-cycle system of order v with a *hole* of size n is an m-cycle system of $K_v - K_n$.

9.12 Remark If an m-cycle system of K_n is embedded in an m-cycle system of K_v, then the embedding produces an m-cycle system of $K_v - K_n$.

9.13 Theorem If there exists an incomplete m-cycle system of order $v > n$ with a hole of size n, then (1) $v \equiv n \equiv 1 \pmod 2$; (2) $(v-n)(v+n-1) \equiv 0 \pmod{2m}$; (3) if m is odd, then $v \geq (m+1)n/(m-1)+1$ and (4) if m is even, then $(v-m)(v-1) \geq n(n-1)$.

9.14 Table There exists an incomplete m-cycle system of order v with a hole of size n for the following values of v that also satisfy conditions (1) to (4) of Theorem 9.13.

m	v
$3 \leq m \leq 7$, m odd	all
$4 \leq m \leq 14$, m even	all
$9 \leq m \leq 13$, m odd	all $(m,1)$ admissible n and v
even	all $v \equiv n \pmod{2m}$
odd	if $n \equiv 1$ or $m \pmod{2m}$, then all $v \equiv 1$ or $m \pmod{2m}$, except possibly for the smallest such v in some cases.

9.5 Resolvable m-Cycle Systems

A (directed) m-cycle system C of K_n (D_n) is *resolvable* if the (directed) cycles in C can be partitioned into 2-factors of K_n (D_n).

9.15 Remark If there exists a resolvable m-cycle of K_n, then necessarily m is odd and $n \equiv m \pmod{2m}$. In the directed case, necessarily $n \equiv 0 \pmod m$.

An *almost parallel class* of a graph G is a 2-factor of $G - v$ for some vertex v in G. A (directed) m-cycle system C of λK_n (λD_n) is *almost resolvable* if the (directed) m-cycles in C can be partitioned into almost parallel classes.

9.16 Remark If there exists an almost resolvable m-system of λK_n, then necessarily λ is even and $n \equiv 1 \pmod m$. In the directed case, then necessarily $n \equiv 1 \pmod m$.

A *nearly resolvable* m-cycle system of order n is a resolvable m-cycle system of $K_n - F$, where F is a 1-factor of K_n.

9.17 Remark If there exists a nearly resolvable m-cycle system of K_n, then necessarily $n \equiv 0 \pmod{m}$ if m is even and $n \equiv 0 \pmod{2m}$ if m is odd.

9.18 Table Existence of m-cycle systems with resolvability constraints [2, 8].

Resolvability	Graph	m	λ	$n \equiv$
resolvable	K_n	odd	1	$m \pmod{2m}$
·	D_n	3, 4	1	$0 \pmod{m}$, $(m,n) \notin \{(3,6),(4,4)\}$, and possibly $(m,n) \neq (4,12)$
almost resolvable	K_n	all	even	$1 \pmod{m}$
·	D_n	3, 5	1	$1 \pmod{m}$
·	·	even	1	$1 \pmod{m}$
·	·	$3 \pmod{6}$)	1	$1 \pmod{m}$ except possibly $n \in \{3m+1, 6m+1\}$
nearly resolvable	K_n	even	1	$0 \pmod{m}$
·	·	odd	1	$0 \pmod{2m}$, $(m,n) \notin \{(3,6),(3,12)\}$

9.6 Nestings of m-Cycle Systems

A *nesting* of an m-cycle system C of K_n with vertex set V is a function $f : C \to V$ such that $\{\{x, f(c)\} | c \in C, x \in V(c)\}$ partitions the edges of K_n (into stars).

9.19 Remark If there exists a nesting of an m-cycle system of K_n, then $n \equiv 1 \pmod{2m}$.

9.20 Table Existence of an m-cycle system of K_n with a nesting.

m	All $n = 2xm + 1$ except possibly if
2^e	$x \in \{2,3,4,7,8,12,14,18,19,23,24,33,34\}$
$\neq 2^e$	$x \in \{2,3,4,6,22,23,24,26,27,28,30,34,38\}$
$3 \leq m \leq 15$, m odd	$(m,x) \in \{(7,4),(7,6),(9,3),(11,6),(13,4),(15,3)\}$
odd	$x \in \{3,4,6\}$, providing there is an m-cycle system of K_{4m+1} with a nesting

An *x-nesting* of a directed cycle system C of D_n with vertex set V is a function $f : C \to V$ such that $\{(x_1, f(c)), (f(c), x_2) | c \in C, x_1 \in X_1, x_2 \in X_2, V(c) = X_1 \cup X_2, |X_1| = x, |X_2| = m - x\}$ partitions the directed edges of D_n (into directed stars).

9.21 Remark If there exists an x-nesting of a directed m-cycle system of D_n, then $n \equiv 1 \pmod{m}$.

9.22 Table Existence of directed m-cycle systems with an x-nesting.

m	x	n
odd	$\lfloor m/2 \rfloor$	all $n \equiv 1 \pmod{m}$, except possibly $n \in \{3m+1, 6m+1\}$
even	even	all $n \equiv 1 \pmod{m}$, except possibly $n = 2m+1$
$2 \pmod{4}$	$m/2$	all $n \equiv 1 \pmod{2m}$ except possibly $n = 4m+1$

9.7 See Also

§IV.36	2-perfect m-cycle systems give rise to quasigroups satisfying special identities.
§IV.22	Decompositions of K_n into copies of other small graphs.

[5] A general survey of cycle systems with an extensive bibliography.
[6] A survey of connections between i-perfect cycle systems and quasigroups satisfying certain identities.
[10] A survey of decompositions of K_n into copies of other graphs.
[11] A survey of embedding (partial) cycle systems.

References

[1] P. Adams and E.J. Billington, Completing some spectra for 2-perfect cycle systems, *Australas. J. Combin.* 7 (1993), 175-188.
[2] B. Alspach, P. Schellenberg, D.R. Stinson and D. Wagner, The Oberwolfach problem and factors of uniform odd length cycles, *J. Combin. Theory A* 52 (1989) 20-43.
[3] D.E. Bryant and C.A. Rodger, On the Doyen-Wilson Theory for m-cycle systems, *J. Combin. Des.* 2 (1994), 253-271.
[4] C.J. Colbourn, Hamiltonian decompositions of complete graphs, *Ars Combin.* 14 (1982), 261-269.
[5] D.G. Hoffman, C.C. Lindner and C.A. Rodger, On the construction of odd cycle systems, *J. Graph Theory* 13 (1989), 417-426.
[6] C.C. Lindner, 2-perfect m-cycle systems and quasigroup varieties: A survey, *Proc. 24th Annual Iranian Math. Conf., 1993*, to appear.
[7] C.C. Lindner and C.A. Rodger, 2-perfect m-cycle systems, *Discrete Math.* 104 (1992), 83-90.
[8] C.C. Lindner and C.A. Rodger, Decomposition into cycles II: Cycle systems, in: *Contemporary Design Theory: A Collection of Surveys* (J.H. Dinitz, D.R. Stinson; eds.) John Wiley & Sons, New York, 1992, 325-369.
[9] C.C. Lindner and C.A. Rodger, Embedding directed and undirected partial cycle systems of index $\lambda > 1$, *J. Combin. Des.* 1 (1993), 113-123.
[10] C.A. Rodger, Graph decompositions, *Le Matematiche* 45 (1990), 119-140.
[11] C.A. Rodger, Recent results on the embedding of latin squares and related structures, cycle systems and graph designs, *Le Matematiche* 47 (1992), 295-311.
[12] C.A. Rodger and D.R. Stinson, Nesting directed cycle systems of even length, *Europ. J. Combin.* 13 (1992), 213-218.

10 Difference Families

R. Julian R. Abel

10.1 Definitions and Examples

Let G be an additive abelian group of order v. Then t k-element subsets of G, $B_i = \{b_{i,1}, b_{i,2} \ldots b_{i,k}\}$ $(1 \le i \le t)$ form a (v, k, λ) *difference family* (or *difference system*) if every nonzero element of G occurs λ times among the differences $b_{i,x} - b_{i,y}$ $(i = 1, \ldots, t;\ x, y = 1, \ldots, k)$. The sets B_i are *base blocks*.

10.1 Remarks

1. If $t = 1$, then B_1 is an abelian (v, k, λ) difference set. (See §IV.12).

2. If $B_1, \ldots, B_t$ form a (v, k, λ) difference family, then the translates of the base blocks, namely $B_i + g = \{b_{i,1} + g, b_{i,2} + g \ldots b_{i,k} + g\}$ $(i = 1, \ldots, t, g \in G)$, form a (v, k, λ) BIBD. The BIBD is cyclic if $G = \mathbb{Z}_v$ and in this case the difference family is a *cyclic difference family.*

A k-subset B of an abelian group G is a *short block* if $B + g = B$ for some nonzero $g \in G$. A collection of t k-element subsets of G, B_i $(1 \leq i \leq t)$ form a (v, k, λ) *partial difference family* if the collection of distinct translates of each of the base blocks form a (v, k, λ) BIBD.

10.2 Remark Sometimes one can define multiplication as well as addition over G; including a multiplicative structure often simplifies the task of obtaining base blocks. The following definitions apply in this situation.

1. $C = \{c_1, \ldots, c_k\}$ is a *multiple* of $B = \{b_1, \ldots, b_k\}$ if, for some w, $c_i = w \cdot b_i$ for all i.
2. w is a *multiplier* (of order n) of $B = \{b_1, \ldots, b_k\}$ if
   (a) for some $g \in G$, $B = w \cdot B + g$ $(= \{w \cdot b_1 + g, w \cdot b_2 + g \ \ldots \ w \cdot b_k + g\})$
   (b) $w^n = 1$ but $w^i \neq 1$ for $0 < i < n$.
3. w is a *multiplier* of a difference family D if, for each base block B in D, $w \cdot B + g = C$ for some base block C in D and some $g \in G$.
4. If q is a prime power and D is a (q, k, λ) difference family over $\mathbb{F}_q$ in which one base block, B, has a multiplier of order k or $k-1$ and all other base blocks are multiples of B, then D is *radical.*

10.3 Examples

1. $B_1 = \{0, 1, 4\}, B_2 = \{0, 2, 8\}, B_3 = \{0, 5, 10\}$ form a partial (15,3,1) difference family over $\mathbb{Z}_{15}$ in which 2 is a multiplier of order 4 and B_3 is a short block.
2. $B_1 = \{0, 1, 3, 24\}, B_2 = \{0, 4, 9, 15\}, B_3 = \{0, 7, 17, 25\}$, form a (37,4,1) difference family over $\mathbb{Z}_{37}$ with no multiplier.
3. $B_1 = \{0, 1, 3, 24\}, B_2 = \{0, 10, 18, 30\} = 10 \cdot B_1, B_3 = \{0, 4, 26, 32\} = 26 \cdot B_1$, form a (37,4,1) difference family over $\mathbb{Z}_{37}$ in which 10 is a multiplier of order 3.

10.4 Examples $(v, 3, 1)$ cyclic difference families for $7 \leq v \leq 81$; short blocks shown *slanted.*

v	Base Blocks							
7	0 1 3							
13	0 1 4	0 2 7						
15	0 1 4	0 2 9	*0 5 10*					
19	0 1 4	0 2 9	0 5 11					
21	0 1 3	0 4 12	0 5 11	*0 7 14*				
25	0 1 3	0 4 11	0 5 13	0 6 15				
27	0 1 3	0 4 11	0 5 15	0 6 14	*0 9 18*			
31	0 1 12	0 2 24	0 3 8	0 4 17	0 6 16			
33	0 1 3	0 4 10	0 5 18	0 7 19	0 8 17	*0 11 22*		
37	0 1 3	0 4 26	0 5 14	0 6 25	0 7 17	0 8 21		
39	0 1 3	0 4 18	0 5 27	0 6 16	0 7 15	0 9 20	*0 13 26*	
43	0 1 3	0 4 9	0 6 28	0 7 23	0 8 33	0 11 30	0 12 26	
45	0 1 3	0 4 10	0 5 28	0 7 34	0 8 32	0 9 29	0 12 26	*0 15 30*
49	0 1 3	0 4 9	0 6 17	0 7 23	0 8 30	0 10 31	0 12 36	0 14 34
51	0 1 3	0 4 9	0 6 25	0 7 35	0 8 22	0 10 21	0 12 27	0 13 31
	0 17 34							
55	0 1 3	0 4 9	0 6 16	0 7 32	0 8 29	0 11 42	0 12 27	0 14 36
	0 17 37							

v	Base Blocks							
57	0 1 3	0 4 9	0 6 13	0 8 26	0 10 33	0 11 32	0 12 40	0 14 41
	0 15 35	*0 19 38*						
61	0 1 3	0 4 9	0 6 13	0 8 25	0 10 33	0 11 30	0 12 32	0 14 40
	0 15 37	0 16 34						
63	0 1 3	0 4 9	0 6 13	0 8 25	0 10 41	0 11 44	0 12 36	0 14 37
	0 15 43	0 16 34	*0 21 42*					
67	0 1 3	0 4 9	0 6 13	0 8 23	0 10 38	0 11 33	0 12 42	0 14 32
	0 16 43	0 17 36	0 20 46					
69	0 1 3	0 4 9	0 6 13	0 8 24	0 10 38	0 11 47	0 12 32	0 14 40
	0 15 50	0 17 42	0 18 39	*0 23 46*				
73	0 1 3	0 4 10	0 5 35	0 7 32	0 8 24	0 9 55	0 11 53	0 12 52
	0 13 39	0 14 29	0 17 54	0 22 45				
75	0 1 67	0 2 47	0 3 41	0 4 69	0 5 68	0 11 55	0 13 61	0 15 33
	0 16 52	0 17 43	0 19 40	0 22 51	*0 25 50*			
79	0 1 29	0 2 19	0 3 14	0 4 42	0 5 13	0 6 22	0 7 52	0 9 55
	0 10 53	0 12 59	0 15 54	0 18 48	0 21 56			
81	0 1 39	0 2 58	0 3 34	0 4 21	0 5 67	0 6 15	0 7 36	0 8 59
	0 10 63	0 11 37	0 12 61	0 13 48	0 16 40	*0 27 54*		

10.5 Examples $(v, 3, 2)$ cyclic difference families.

v	Base Blocks							
16	0 1 2	0 2 8	0 3 7	0 4 7	0 5 10			
28	0 1 12	0 2 11	0 2 12	0 3 7	0 3 13	0 4 9	0 5 13	0 6 7
	0 6 14							
40	0 1 4	0 1 16	0 2 7	0 2 9	0 3 17	0 4 17	0 5 19	0 6 16
	0 6 18	0 8 18	0 8 19	0 9 20				

10.6 Examples $(v, 4, 1)$ cyclic difference families; short blocks shown *slanted*.

v	Base Blocks						
13	0 1 3 9						
37	0 1 3 24	0 4 6 32	0 7 17 25				
40	0 1 4 13	0 2 7 24	0 6 14 25	*0 10 20 30*			
49	0 1 3 8	0 4 18 29	0 6 21 33	0 9 19 32			
52	0 1 3 7	0 5 19 35	0 8 20 31	0 9 24 34	*0 13 26 39*		
64	0 1 3 7	0 5 18 47	0 8 33 44	0 9 19 43	0 12 26 49	*0 16 32 48*	
76	0 1 7 22	0 2 11 45	0 3 59 71	0 4 32 50	0 10 37 51	0 13 36 60	*0 19 38 57*
85	0 2 41 42	0 17 32 38	0 18 27 37	0 13 29 36	0 11 31 35	0 12 26 34	0 5 30 33

10.7 Examples $(v, 4, \lambda)$ cyclic difference families; short blocks shown *slanted*.

v, λ	Base Blocks						
11,6	0 1 8 9	0 2 5 7	0 1 4 5	0 2 3 5	0 4 5 9		
15,6	0 1 2 3	0 2 4 6	0 4 8 12	0 8 1 9	3 6 9 12	0 1 5 10	0 2 5 10
19,2	0 1 3 12	0 1 5 13	0 4 6 9				
21,3	0 2 3 7	0 3 5 9	0 1 7 11	0 2 8 11	0 1 9 14		
22,2	0 4 16 17	0 12 14 21	0 14 16 19	*0 4 11 15*			
29,3	0 2 5 7	0 1 12 13	0 4 19 23	0 3 22 25	0 6 15 21	0 8 9 17	0 11 16 27
31,2	0 1 8 11	0 1 13 17	0 2 11 14	0 5 7 13	0 5 9 15		
34,2	0 1 22 24	0 1 19 25	0 2 6 29	0 4 7 20	0 5 8 20	*0 8 17 25*	
43,2	0 1 6 36	0 3 18 22	0 9 11 23	0 10 12 26	0 26 27 33	0 13 35 38	0 19 28 39
46,2	0 2 7 10	0 4 19 32	0 10 34 35	0 5 8 24	0 26 30 39	0 17 26 32	0 28 34 45
	0 2 23 25						

10.8 Examples $(v, 5, 1)$ cyclic difference families; short blocks shown *slanted*.

v	Base Blocks			
21	0 1 4 14 16			
41	0 1 4 11 29	0 2 8 17 22		
61	0 1 3 13 34	0 4 9 23 45	0 6 17 24 32	
65	0 1 3 31 45	0 4 10 19 57	0 5 16 41 48	*0 13 26 39 52*

10.9 Examples $(v, 5, \lambda)$ cyclic difference families; short blocks shown *slanted*.

v, λ	Base Blocks				
13,5	0 1 2 4 8	0 1 3 6 12	0 2 5 6 10		
15,6	0 1 2 3 6	0 2 4 7 8	0 2 4 9 10	0 3 6 10 11	*0 3 6 9 12*
17,5	0 1 4 13 16	0 3 5 12 14	0 2 8 9 15	0 6 7 10 11	
31,2	0 1 3 7 15	0 3 9 14 21	0 4 5 13 15		
33,5	0 2 3 7 25	0 3 13 14 29	0 4 5 12 13	0 2 12 16 26	0 3 12 20 31
	3 9 12 15 27	0 8 13 14 31	0 2 7 13 29		
35,2	0 2 8 12 13	0 3 18 22 27	0 17 23 32 33	*0 7 14 21 28*	*0 7 14 21 28*
51,2	0 1 14 31 35	0 1 9 23 33	0 11 16 18 42	0 7 13 36 39	0 4 10 12 15
71,2	1 5 25 54 57	3 4 15 20 29	9 12 16 45 60	27 36 38 48 64	2 10 37 43 50
	6 8 30 40 58	18 19 24 32 49			

10.10 Examples $(v, 6, \lambda)$ cyclic difference families; short blocks shown *slanted*.

v, λ	Base Blocks			
21,3	0 2 10 15 19 20	0 3 7 9 10 16		
21,5	0 2 6 12 15 16	0 3 6 7 11 19	0 7 15 16 17 18	*0 2 7 9 14 16*
22,5	0 1 2 5 10 13	0 1 5 6 8 15	0 2 3 6 16 18	*0 2 6 11 13 17*
28,5	0 4 7 8 16 21	5 7 8 9 14 20	7 12 14 16 17 25	1 4 7 13 14 24
	2 4 8 16 18 22			
33,5	0 3 12 17 18 28	0 2 3 16 28 29	0 16 20 26 28 30	0 2 3 12 16 27
	0 6 20 21 28 30	*0 4 11 15 22 26*		
39,5	0 3 4 17 19 32	0 1 5 12 30 36	0 3 8 9 25 27	0 7 10 12 17 21
	0 16 18 19 27 35	0 2 18 27 28 33	0 6 13 19 26 32	
41,3	0 1 10 16 18 37	0 6 14 17 19 26	0 2 20 32 33 36	0 11 12 28 34 38
46,2	0 1 3 11 31 35	0 1 4 10 23 29	0 2 7 15 32 41	
51,3	15 17 18 27 34 48	3 17 30 34 42 45	9 17 24 33 34 39	3 25 41 43 44 48
	3 5 25 29 43 48			
61,2	12 15 28 34 35 59	1 13 18 47 51 53	8 10 11 21 29 43	16 20 25 32 40 50
61,3	0 1 9 20 58 34	0 2 7 18 40 55	0 4 14 19 36 49	0 8 11 28 37 38
	0 13 15 16 22 56	0 26 30 32 44 51		
91,1	0 1 3 7 25 38	0 16 21 36 48 62	0 30 40 63 74 82	

10.11 Remark There are known infinite classes of difference families for $k \leq 5$ and for q a prime power. Some of these are given in the next few theorems.

10.12 Theorem There exists a $(v, 3, 1)$ difference family for every $v \equiv 1 \pmod 6$. (Construction IV.43.11 gives one explicit solution.)

10.13 Theorem [10] If $v \equiv 1, 3 \pmod{\frac{6}{\gcd(6,\lambda)}}$ and $(v, \lambda) \neq (9,1)$ or $(9,2)$, there exists a $(v, 3, \lambda)$ difference family (possibly partial) over $\mathbb{Z}_v$, except when $v \equiv 2 \pmod 4$ and $\lambda \equiv 2 \pmod 4$.

10.14 Construction When $v = 2t + 1$, $t \geq 1$, the set $\{\{0, i, 2i\} : 1 \leq i \leq t\}$ is a $(v, 3, 3)$-difference family over $\mathbb{Z}_v$.

10.15 Theorem Suppose $q \equiv 7 \pmod{12}$ is a prime power and there exists a cube root of unity ω in $\mathbb{F}_q$ such that $x = \omega - 1$ is a primitive root. Then the following base blocks form a (7q,4,1) difference family over $\mathbb{Z}_7 \times \mathbb{F}_q$:

1. $\{(0,0), (0,(x-1)x^{2t-1}), (0,\omega(x-1)x^{2t-1}), (0,\omega^2(x-1)x^{2t-1})\}$ for $1 \le t \le (q-7)/12$,
2. $\{(0,0), (1,x^{2t}), (2,\omega x^{2t}), (4,\omega^2 x^{2t})\}$ for $1 \le t \le (q-3)/2$, $x^{2t} \neq \omega$,
3. $\{(0,0), (2^t,\omega^{2t}), (2^t, x\cdot\omega^{2t}), (2^{t+2},0)\}$ for $0 \le t \le 2$.

10.16 Remark (7q,4,1) difference families are obtainable by Theorem 10.15 for q = 7, 19, 31, 43, 67, 79, 103, 127, 151, 163, 199, 211, 367, 379, 439, 463, 487, 571, but not for q = 139, 223, 271, 283, 307, 331, 523, 547.

10.17 Remark Theorem 10.18 updates some known results for $k = 4, 5$ in [16].

10.18 Theorem (12t+1,4,1) and (20t+1,5,1) difference families over $\mathbb{Z}_{12t+1}$ and $\mathbb{Z}_{20t+1}$ are known to exist for the following values of $t \le 50$:

1. (12t+1,4,1): t = 1,3-11,13-16,18-21,23,25,26,28-31,33-36,38-41,43,45,46,48,50.
2. (20t+1,5,1): t = 1-3,5-10,12-15,18,20-28,30,32-35,38-44,47-50.

10.19 Remark Buratti [7] gives constructions for $(q, 4, 1)$ difference families over $\mathbb{F}_q$ when $q = 12t + 1$ is prime and $q < 10^6$, or q is a prime power that is not a prime; and for $(q, 5, 1)$ difference families over $\mathbb{F}_q$ when $q = 20t + 1$ is a prime power and $q < 10^4$.

10.20 Theorem [18] Suppose q is a prime power, and $\lambda(q-1) \equiv 0 \pmod{k(k-1)}$. Then a (q, k, λ) difference family over $\mathbb{F}_q$ exists if one of the following holds:

1. λ is a multiple of $k/2$ or $(k-1)/2$.
2. $\lambda \ge k(k-1)$
3. $q > \binom{k}{2}^{k(k-1)}$

10.21 Examples More (v, k, λ) cyclic difference families.

v,k,λ	Base Blocks
21,10,9	{0,1,2,3,4,7,8,11,14,16}, {0,6,7,9,11,12,15,16,17,19}
21,8,14	{0,9,10,13,14,15,18,19}, {0,1,4,7,9,15,16,18}, {0,1,2,4,6,14,15,16}, {0,1,3,4,8,14,16,18}, {0,1,4,9,11,12,14,16}
22,7,4	{0,2,6,8,9,10,13}, {0,3,5,6,12,13,17}
22,8,8	{0,1,5,7,13,17,20,21}, {0,2,7,11,13,14,16,17}, {0,3,4,12,14,15,17,21}
22,8,12	{0,1,2,3,5,14,17,18}, {0,1,3,5,8,9,11,15}, {1,2,5,6,8,10,17,18}, {1,4,7,12,14,16,20,21}, {0,1,2,5,11,12,13,16}
29,7,3	{1,7,16,20,23,24,25}, {2,3,11,14,17,19,21}
29,8,4	{0,1,7,16,20,23,24,25}, {0,2,3,11,14,17,19,21}
29,8,6	{0,5,10,11,12,13,16,20}, {0,8,10,12,17,22,23,26}, {0,4,5,11,13,23,25,26}
34,12,8	{0,5,9,14,15,17,20,25,26,27,28,30}, {0,6,7,10,13,17,18,20,22,24,25,26}
37,10,5	{0,1,7,9,10,12,16,26,33,34}, {0,2,14,15,18,20,24,29,31,32}
43,7,2	{0,1,11,19,31,38,40}, {0,2,10,16,25,38,42}
43,7,3	{1,4,11,16,21,35,41}, {3,5,12,19,20,33,37}, {9,13,14,15,17,25,36}
43,8,4	{0,1,4,11,16,21,35,41}, {0,3,5,12,19,20,33,37}, {0,9,13,14,15,17,25,36}
43,15,10	{1,3,6,13,18,21,22,25,26,27,33,35,36,38,40}, {9,10,11,13,16,17,19,23,26,27,28,33,35,38,39}
46,10,8	{0,1,3,12,13,16,18,19,26,27}, {0,1,9,22,24,27,29,33,39,43}, {0,1,4,12,26,28,30,35,36,41}, {0,2,7,10,14,27,29,30,35,39}
53,13,6	{1,10,13,15,16,24,28,36,42,44,46,47,49}, {2,3,19,20,26,30,31,32,35,39,41,45,48}
53,14,7	{0,1,10,13,15,16,24,28,36,42,44,46,47,49}, {0,2,3,19,20,26,30,31,32,35,39,41,45,48}

v,k,λ	Base Blocks
55,9,4	{0,4,21,25,26,42,45,53,54}, {0,6,8,25,37,39,45,48,52}, {2,5,6,13,15,20,25,39,45}
55,10,5	{0,5,11,15,20,22,25,33,44,45}, {3,7,8,10,31,37,39,45,46,49} twice
57,7,3	{0,1,11,12,15,35,53}, {0,7,17,20,27,29,48}, {0,5,18,26,32,49,51}, {0,2,6,9,14,41,42}
61,10,6	{0,1,2,8,10,13,26,33,43,47}, {0,3,12,15,19,32,34,39,40,50}, {0,4,5,8,10,22,42,43,52,58}, {0,2,6,12,15,17,26,33,34,38}
61,15,7	{0,1,3,4,8,10,13,22,30,35,44,45,46,50,58}, {0,1,3,5,13,18,29,34,35,37,41,43,44,51,55}
67,11,5	{1,9,14,15,22,24,25,40,59,62,64}, {2,13,18,28,30,44,48,50,51,57,61}, {4,21,26,29,33,35,36,47,55,56,60}
67,12,4	{1,8,23,25,28,29,31,37,47,54,55,64}, {3,20,25,32,36,39,44,45,54,55,57,59}
67,12,6	{0,1,9,14,15,22,24,25,40,59,62,64}, {0,2,13,18,28,30,44,48,50,51,57,61}, {0,4,21,26,29,33,35,36,47,55,56,60}
71,7,3	{1,20,30,32,37,45,48}, {2,3,19,25,40,60,64}, {4,6,9,38,49,50,57}, {5,8,12,18,27,29,43}, {10,15,16,24,36,54,58}
71,8,4	{0,1,20,30,32,37,45,48}, {0,2,3,19,25,40,60,64}, {0,4,6,9,38,49,50,57}, {0,5,8,12,18,27,29,43}, {0,10,15,16,24,36,54,58}
73,10,5	{0,1,2,4,8,16,32,37,55,64}, {0,5,7,10,14,20,28,39,40,56} {0,25,27,35,49,50,54,61,67,70}, {0,11,15,21,22,30,42,44,47,60}
85,7,2	{0,1,11,20,32,35,39}, {0,2,6,16,29,50,65}, {0,3,9,27,55,72,80}, {0,5,7,30,47,48,59}
97,9,3	{1,2,25,35,46,58,61,70,90}, {3,4,8,38,43,50,69,86,87}, {6,12,16,32,53,55,57,75,82}, {9,18,24,26,31,34,37,48,64}
181,10,2	{1,7,40,42,51,59,113,125,135,151}, {19,22,31,35,36,64,74,133,154,156}, {10,15,34,47,58,65,83,87,161,164}, {12,18,32,52,77,78,142,157,165,172}
217,7,1	{0,1,37,67,88,92,149}, {0,15,18,65,78,121,137}, {0,8,53,79,85,102,107}, {0,11,86,100,120,144,190}, {0,29,64,165,198,205,207}, *{0,31,62,93,124,155,186}*

10.22 Examples Difference families over products of cyclic groups.

v,k,λ	Group	Base Blocks
25,7,7	$\mathbb{Z}_5 \times \mathbb{Z}_5$	{(0,0),(0,1),(0,4),(1,1),(1,2),(4,3),(4,4)}, {(0,0),(1,0),(1,3),(2,3),(3,2),(4,0),(4,2)}, {(0,0),(0,2),(0,3),(2,2),(2,4),(3,1),(3,3)}, {(0,0),(1,4),(2,0),(2,1),(3,0),(3,4),(4,1)}
25,8,7	$\mathbb{Z}_5 \times \mathbb{Z}_5$	{(0,1),(0,2),(0,3),(0,4),(1,2),(2,4),(3,1),(4,3)}, {(1,0),(2,0),(3,0),(4,0),(1,1),(2,2),(3,3),(4,4}, {(1,3),(2,1),(3,4),(4,2),(1,4),(2,3),(3,2),(4,1)}
45,5,1	$\mathbb{Z}_3^2 \times \mathbb{Z}_5$	{(0,1,0),(0,2,0),(1,0,2),(2,0,2),(0,0,1)}, {(2,1,0),(1,2,0),(2,2,2),(1,1,2),(0,0,1)}, *{(0,0,0),(0,0,1),(0,0,2),(0,0,3),(0,0,4)}*
49,9,6	$\mathbb{Z}_7 \times \mathbb{Z}_7$	{(0,6),(0,5),(0,3),(1,0),(1,1),(2,0),(2,2),(4,0),(4,4)}, {(0,1),(0,2),(0,4),(1,0),(1,6),(2,0),(2,5),(4,0),(4,3)}, {(1,0),(2,0),(3,4),(3,1),(4,0),(5,2),(5,4),(6,1),(6,2)}, {(1,0),(2,0),(3,3),(3,6),(4,0),(5,5),(5,3),(6,6),(6,5)}
75,5,2	$\mathbb{Z}_5 \times \mathbb{Z}_{15}$	{(0,0),(1,10),(1,2),(4,1),(4,2)}, {(0,0),(2,5),(2,10),(3,7),(3,13)}, {(0,0),(1,10),(1,8),(4,4),(4,8)}, {(0,0),(2,5),(2,10),(3,14),(3,11)}, {(0,0),(1,4),(1,5),(4,1),(4,8)}, {(0,0),(1,1),(1,5),(4,4),(4,2)}, {(0,0),(2,7),(2,13),(3,1),(3,4)}, *{(0,0),(1,0),(2,0),(3,0),(4,0)}* (twice)
81,5,1	$\mathbb{Z}_3^4$	{(0,0,0,1),(2,0,0,1),(0,0,2,1),(1,2,0,2),(0,1,1,1)}, {(0,0,1,0),(1,1,0,2),(0,2,1,0),(1,2,0,1),(1,1,1,0)}, {(2,2,1,1),(1,2,2,2),(2,0,1,2),(0,1,2,1),(1,1,0,0)}, {(0,2,0,2),(1,1,0,1),(1,2,1,2),(1,2,1,0),(0,2,1,1)}

v,k,λ	Group	Base Blocks
121,6,1	$\mathbb{Z}_{11}\times\mathbb{Z}_{11}$	{(0,0),(0,3),(0,4),(1,1),(1,7),(4,6)}, {(0,0),(0,2),(2,5),(4,7),(6,4),(8,0)}, {(0,0),(1,5),(2,0),(4,1),(6,0),(7,2)}, {(0,0),(1,0),(3,9),(4,8),(6,1),(9,5)}
121,10,3	$\mathbb{Z}_{11}\times\mathbb{Z}_{11}$	{(0,1),(0,3),(0,4),(0,5),(0,9),(1,8),(3,2),(4,10),(5,7),(9,6)}, {(1,2),(3,6),(4,8),(5,10),(9,7),(10,2),(8,6),(7,8),(6,10),(2,7)}, {(1,7),(3,10),(4,6),(5,2),(9,8),(1,4),(3,1),(4,5),(5,9),(9,3)}, {(10,10),(8,8),(7,7),(6,6),(2,2),(1,0),(3,0),(4,0),(5,0),(9,0)}
121,16,4	$\mathbb{Z}_{11}\times\mathbb{Z}_{11}$	{(0,0),(0,1),(0,3),(0,9),(0,5),(0,4),(2,1),(6,3), (7,9),(10,5),(8,4),(1,8),(3,2),(9,6),(5,7),(4,10)}, {(0,0),(8,3),(2,9),(6,5),(7,4),(10,1),(10,7),(8,10), (2,8),(6,2),(7,6),(8,1),(2,3),(6,9),(7,5),(10,4)}
153,9,2	$\mathbb{Z}_3^2\times\mathbb{Z}_{17}$	*{(0,0,0),(0,1,0),(0,2,0),(1,0,0),(1,1,0),(1,2,0),(2,0,0),(2,1,0),(2,2,0)}*, *{(0,0,0),(0,1,0),(0,2,0),(1,0,0),(1,1,0),(1,2,0),(2,0,0),(2,1,0),(2,2,0)}*, {(0,0,0),(0,1,t),(0,1,16t),(0,2,4t),(0,2,13t),(1,0,3t), (1,0,14t),(2,0,5t),(2,0,12t)} for t =1, 2, 3, and 6.
161,5,1	$\mathbb{Z}_7\times\mathbb{Z}_{23}$	{(0,0),(0,2),(0,14),(0,15),(0,18)}, {(0,3),(0,20),(1,0),(2,0),(4,0)}, {{(t,0),(2t,1),(3t,7),(4t,9),(5t,19)} : $1\le t\le 6$}

10.23 Example A (49,21,10) difference family over $\mathbb{Z}_7\times\mathbb{Z}_7$. Base blocks are the short block $\bigcup_{i=0}^{6}\{(i,1),(i,2),(i,4)\}$, and the block

$$\left(\bigcup_{i=1,2,4}\{(i,0),(i,1),(i,2),(i,4)\}\right)\cup\left(\bigcup_{i=3,5,6}\{(i,3),(i,5),(i,6)\}\right).$$

10.2 Basic Recursive Constructions

10.24 Theorem If a (v_i,k,λ_i) difference family over H_i exists for $i=1,2$, a $(v_1v_2,k,\lambda_1\lambda_2)$ difference family over $H_1\times H_2$ also exists.

10.25 Theorem Suppose that there exists a (v,k,λ_1) (possibly partial) difference family over $\mathbb{Z}_v$ having μ base blocks in the orbit of $\{0,\frac{v}{k},2\frac{v}{k},\ldots,(k-1)\frac{v}{k}\}$, and all other base blocks generating v distinct blocks under the action of $\mathbb{Z}_v$. If in addition there exist

1. a $(u,k,(\lambda_1-\mu)\lambda_2)$ (possibly partial) difference family over $\mathbb{Z}_u$,
2. a $(ku,k,\mu\lambda_2)$ partial difference family over $\mathbb{Z}_{ku}$, and
3. a (u,k,λ_2)-difference matrix over $\mathbb{Z}_u$,

then there exists a $(uv,k,\lambda_1\lambda_2)$ difference family over $\mathbb{Z}_{uv}$.

10.26 Corollary

1. Let k be a prime or prime power, and $v\equiv 1 \pmod{k(k-1)}$. If a $(v,k,1)$ difference family over $\mathbb{Z}_v$ exists and a $(u,k,1)$ (possibly partial) difference family over $\mathbb{Z}_u$ exists, then a $(uv,k,1)$ difference family over $\mathbb{Z}_{uv}$ exists.
2. Let k be prime and $v\equiv k \pmod{k(k-1)}$. If a $(v,k,1)$ partial difference family over $\mathbb{Z}_v$ exists and a $(ku,k,1)$ partial difference family over $\mathbb{Z}_{ku}$ exists, then a $(uv,k,1)$ partial difference family over $\mathbb{Z}_{uv}$ exists.

10.3 Perfect Difference Families

If $v=k(k-1)t+1$, then t blocks $B_i=\{b_{i,1},b_{i,2},\ldots,b_{i,k}\}$ form a *perfect* $(v,k,1)$ difference family over $\mathbb{Z}_v$ if the $tk(k-1)/2$ differences $b_{i,m}-b_{i,n}$ $(i=1,\ldots,t,$ $1\le m<n\le k)$ cover the set $\{1,2,\ldots,(v-1)/2\}$.

10.27 Example $\{0,1,12\}, \{0,2,8\}, \{0,5,9\}$, and $\{0,3,10\}$ form a perfect (25,3,1) difference family.

10.28 Theorem ([4], p. 318) If $v \equiv 1$ or 7 (mod 24), then a $(v,3,1)$ perfect difference family exists.

10.29 Theorem ([14, 16] and §IV.14) Perfect (12t+1,4,1) difference families exist for the following values of $t < 50$: 1,4-23,26,28,30,31,36,41.

10.30 Theorem [16] Suppose a $(12t+1,4,1)$ perfect difference family exists. Then $(q,4,1)$ perfect difference families exist for $q = 60t{+}13$, $84t{+}13$, $156t{+}13$, $228t{+}49$, $276t{+}61$, $300t{+}61$ and $300t{+}79$.

10.31 Remark [16] $(20t+1,5,1)$ perfect difference families are known for $t = 6,8,10$ but for no other small values of t.

10.32 Theorem There are no $(v,k,1)$ perfect difference families for the following values:

1. $k=3$, $v \equiv 13$ or 19 (mod 24),
2. $k=4$, $v \in \{25,37\}$,
3. $k=5$, $v \equiv 21$ (mod 40) or $v \in \{41,81\}$,
4. $k \geq 6$ [16].

10.4 Prime Power Difference Families

10.33 Remarks Here (q,k,λ) difference families are examined, for q a prime power, and all base blocks multiples of one block. For all of these, x denotes a primitive element of $\mathbb{F}_q$, B the initial base block, and n the highest order of a multiplier of B. The other base blocks are then $x^m \cdot B, x^{2m} \cdot B, \ldots, x^{(t-1)m} \cdot B$ for $t = \lambda(q-1)/(k(k-1))$ and $m = (q-1)/2t = k(k-1)/(2n\lambda)$. The initial block $B = \{b_1, b_2 \ldots b_k\}$ has the property that the $k(k-1)/2$ differences $b_i - b_j$ $(1 \leq i < j \leq k)$ are evenly spread among the multiplicative cosets H_i where $H_i = \{x^a : a \equiv i \pmod{k(k-1)/(2n\lambda)}\}$. Three main cases are considered: (1) $n = k$ or $k-1$ (radical difference families); (2) $n = 3$; and (3) $n = 1$ (no multiplier). If q is a power of a prime, x is taken to be a root of the appropriate primitive polynomial in Table 10.34.

10.34 Table Primitive polynomials for finite fields, $\mathbb{F}_q$, $q \leq 7921$.

q	Polynomial	q	Polynomial	q	Polynomial
25	x^2-x-3	49	x^2-3x-2	121	x^2-8x-3
169	x^2-4x-7	289	x^2-6x-3	361	x^2-6x-6
529	x^2-7x-4	625	$x^4+x^3+x^2+x+2$	729	x^6-2x-1
841	x^2-3x-2	961	x^2-3x-7	1369	x^2-3x-5
1681	x^2-5x-7	1849	x^2-7x-9	2197	x^3-x^2-2
2401	x^4-3x^3-x-2	2809	x^2-3x-3	5041	x^2-5x-4
5329	x^2-3x-5	6889	x^2-6x-4	7921	x^2-2x-6

▷ Radical Difference Families

10.35 Theorem Suppose q is an odd prime power. Then there exists a (q,k,λ) radical difference family if either:

1. λ is a multiple of $k/2$ and $q \equiv 1$ (mod $k-1$), or
2. λ is a multiple of $(k-1)/2$ and $q \equiv 1$ (mod k).

10.36 Remark For radical difference families with $\lambda = 1$, the multiplier must have odd order (i.e. order k if q is odd, or order $k - 1$ if q is even).

10.37 Theorem [5] Let $q = 12t + 1$ be a prime power and 2^e be the largest power of 2 dividing t. Then a $(q, 4, 1)$ radical difference family in $\mathbb{F}_q$ exists if and only if -3 is not a 2^{e+2}th power in $\mathbb{F}_q$. (This condition holds for q=13, 25, 73, 97, 109, 121, 169, 181, 193, 229, 241, 277, 289, 313, 337, 409, 421, 433, 457, 529, 541, 577, 601, 625, 673, 709, 733, 757, 769, 829, 841.)

10.38 Theorem [5] Let $q = 20t + 1$ be a prime power, and let 2^e be the largest power of 2 dividing t. Then a $(q, 5, 1)$ radical difference family in $\mathbb{F}_q$ exists if and only if $(11+5\sqrt{5})/2$ is not a 2^{e+1}th power in $\mathbb{F}_q$. (This condition holds for q=41, 61, 81, 241, 281, 401, 421, 601, 641, 661, 701, 761, 821, 881.)

10.39 Remark Buratti [6] gives necessary and sufficient conditions for a $(q, k, 1)$ radical difference family with $k \in \{6, 7\}$ to exist in $\mathbb{F}_q$; he also gives a sufficient condition for $k \geq 8$.

10.40 Theorem [6, 12, 18] $(q, k, 1)$ radical difference families exist, among others, for the following values of q and k:

1. $k = 6$: q=181, 211, 241, 631, 691, 1531, 1831, 1861, 2791, 2851, 3061.
2. $k = 7$: q=337, 421, 463, 883, 1723, 3067, 3319, 3823, 3907, 4621, 4957, 5167, 5419, 5881, 6133, 8233, 8527, 8821, 9619, 9787, 9829.
3. $k = 8$: q=449, 1009, 3137, 3697, 6217, 6329, 8233, 9869.
4. $k = 9$: q=73, 1153, 1873, 2017, 6481, 7489, 7561, 8359.

▷ Difference Families Where the Blocks Have a Multiplier of Order 3

10.41 Remark Here $q \equiv 1 \pmod 3$ and w denotes a cube root of unity in $\mathbb{F}_q$. The initial base block is a union of sets of the form $\{c_i, w \cdot c_i, w^2 \cdot c_i\}$, plus the zero element of $\mathbb{F}_q$ if $k \equiv 1 \pmod 3$. For convenience, c_1 is taken to be 1. Tables 10.42-10.45 give the other c_i's for $\lambda = 1$, $k \in \{6, 7, 9\}$ and $\lambda = 2$, $k = 9$.

10.42 Table $(q, 6, 1)$ difference families.

q	c_2	q	c_2	q	c_2	q	c_2	q	c_2
31	27	151	7	181	4	211	9	241	13
271	9	331	24	361	x^{42}	421	74	541	100
571	20	601	46	631	52	661	6	691	60
751	39	811	6	841	x^{164}	961	x^2	991	2
1021	29	1051	11	1171	112	1201	19	1231	53
1291	2	1321	11	1381	5	1471	12	1531	79
1621	8	1681	x^{112}	1741	21	1801	47	1831	63
1861	22	1951	16	2011	19	2131	84	2161	39
2221	20	2251	4	2281	37	2311	4	2341	37
2371	47	2401	x^6	2521	44	2551	5	2671	16
2731	58	2791	53	2851	111	2971	3	3001	9

10.43 Table $(q, 7, 1)$ difference families.

q	c_2	q	c_2	q	c_2	q	c_2	q	c_2
421	4	463	190	631	363	967	904	1009	115
1051	464	1093	936	1303	472	1429	1308	1723	898
1849	x^{223}	1933	1222	2017	139	2143	1063	2269	865
2311	1792	2437	929	2521	372	2647	1132	2689	156

q	c_2	q	c_2	q	c_2	q	c_2	q	c_2
2731	1504	2857	1537	3067	1052	3109	556	3319	2237
3361	2957	3529	2758	3571	3505	3613	1004	3697	370
3739	506	3823	1078	3907	1896	4159	3334	4201	3632
4243	3361	4327	145	4621	2115	4663	4658	4789	3279
4831	4251	4957	197	4999	3635	5041	x^{33}	5209	1865
5419	4060	5503	4907	5839	1501	5881	3387	5923	383
6007	2538	6133	5730	6217	4012	6301	1018	6343	5552
6427	3371	6469	3920	6637	4649	6763	391	7057	25

10.44 Table $(q, 9, 1)$ difference families.

q	c_2	c_3	q	c_2	c_3	q	c_2	c_3	q	c_2	c_3
937	82	211	1153	868	950	1297	233	810	1369	x^{30}	x^{223}
1657	11	855	1801	143	1740	1873	44	1540	2017	212	1506
2089	138	1570	2161	417	1858	2377	275	1942	2521	534	1002
2593	963	1430	2809	x^5	x^{145}	2953	572	2252	3169	989	1743
3313	464	1108	3457	322	2708	3529	564	2815	3673	735	933
3889	342	2179	4177	473	3595	4969	525	748	5041	x^2	x^{927}
5113	23	2102	5329	x^9	x^{947}	5689	21	1499	6121	39	4943
6337	10	5111	6481	49	592	6553	10	5176	6841	119	4383
7057	625	5279	7129	7	1961	7417	25	1558	7489	343	5019
7561	37	5893	7921	x^4	x^{218}	7993	51	3497	8209	143	4533
8353	25	640	8641	21	4365	8713	125	2254	8929	11	4766
9001	14	6151	9433	121	8372	9649	49	152	9721	19	4464

10.45 Table $(q, 9, 2)$ difference families.

q	c_2	c_3	q	c_2	c_3	q	c_2	c_3	q	c_2	c_3
109	6	77	181	4	143	289	x^3	x^{49}	361	x^5	x^{93}
397	5	282	541	8	235	577	5	304	613	5	246
757	6	83	829	2	653	1117	2	220	1549	4	1501
1621	4	1361	1693	5	1202	2053	5	959	2197	x^2	x^{395}
2269	2	1441	2341	7	674	2557	5	531	2917	5	1618
3061	6	2496	3637	2	2151	3709	2	577	3853	2	1639
4357	5	2297	4789	2	1562	4861	11	1636	4933	25	918
5077	8	2586	5437	5	2503	5581	19	79	5653	18	93
5869	2	4352	6229	2	4374	6301	10	4367	6373	2	1174
6661	11	600	6733	2	6649	6949	7	6541	7237	5	695
7309	19	6382	7669	19	1852	7741	7	2997	8101	29	7928
8317	15	6892	8389	36	1079	8461	6	6313	8677	5	3670
8821	10	5959	8893	20	7963	9109	10	8191	9181	2	6807
9397	15	8532	9613	5	1027	9829	11	5553	9901	2	1173

▷ Difference Families Where No Base Block has a Multiplier

10.46 Remarks These are given for $\lambda = 1$, $k \in \{4, 5, 8\}$ and $\lambda = 2$, $k = 8$. They are also given for $\lambda = 1$, $k \in \{7, 9\}$ if no difference family with a multiplier of order 3 was possible (Remarks 10.33 apply here as well). The full initial base block is given for these difference families.

10.47 Example Consider a (q, k, λ) difference family for $(q, k, \lambda) = (61, 4, 1)$. Here, $n = 1$ (no multiplier), $t = \lambda(q-1)/k(k-1) = 5$, $m = (q-1)/2t = 6$, and $x = 2$ is a primitive element of $\mathbb{F}_{61}$. Suitable base blocks are B= {0,1,5,11}, 2^6B = 3B = {0,3,15,33}, 2^{12}B = 9B = {0,9,45,38}, 2^{18}B = 27B = {0,27,13,53}, and 2^{24}B = 20B = {0,20,39,37}.

10.48 Table $(q, 4, 1)$ difference families.

q	Base Block	q	Base Block	q	Base Block	q	Base Block
25	$\{0,1,x,x^3\}$	37	{0,1,3,24}	49	$\{0,1,x,x^3\}$	61	{0,1,5,11}
73	{0,1,5,18}	97	{0,1,4,30}	109	{0,1,3,60}	121	$\{0,1,x,x^{10}\}$
157	{0,1,3,72}	181	{0,1,3,19}	229	{0,1,3,39}	241	{0,1,3,22}
277	{0,1,3,94}	289	$\{0,1,x,x^{17}\}$	337	{0,1,3,20}	349	{0,1,3,21}
361	$\{0,1,x,x^{16}\}$	373	{0,1,3,11}	397	{0,1,3,39}	409	{0,1,3,158}
421	{0,1,3,19}	457	{0,1,6,36}	529	$\{0,1,x,x^{35}\}$	541	{0,1,3,32}
613	{0,1,5,25}	661	{0,1,6,23}	709	{0,1,3,33}	733	{0,1,3,87}
757	{0,1,7,24}	829	{0,1,3,47}	841	$\{0,1,x^2,x^{58}\}$	853	{0,1,5,98}
877	{0,1,3,15}	937	{0,1,3,83}	961	$\{0,1,x,x^{52}\}$		

10.49 Table $(q, 5, 1)$ difference families.

q	Base Block	q	Base Block	q	Base Block
61	{0,1,4,11,27}	101	{0,1,4,16,41}	121	$\{0,1,x^2,x^{15},x^{118}\}$
181	{0,1,3,10,137}	241	{0,1,6,19,173}	281	{0,1,3,14,150}
361	$\{0,1,x,x^3,x^{89}\}$	401	{0,1,3,12,109}	421	{0,1,3,9,341}
461	{0,1,4,16,413}	521	{0,1,3,9,217}	541	{0,1,3,7,210}
601	{0,1,3,17,430}	641	{0,1,7,18,245}	661	{0,1,5,19,89}
701	{0,1,3,8,392}	761	{0,1,3,7,149}	821	{0,1,4,11,68}
841	$\{0,1,x,x^4,x^{25}\}$	881	{0,1,3,9,34}	941	{0,1,3,17,65}
961	$\{0,1,x,x^3,x^{414}\}$	1021	{0,1,5,21,851}	1061	{0,1,3,7,411}
1181	{0,1,3,18,483}	1201	{0,1,3,11,498}	1301	{0,1,3,13,1048}
1321	{0,1,3,14,224}	1361	{0,1,3,7,172}	1381	{0,1,3,8,372}
1481	{0,1,3,7,454}	1601	{0,1,3,7,430}	1621	{0,1,3,9,605}
1681	$\{0,1,x,x^3,x^{439}\}$	1721	{0,1,3,7,593}	1741	{0,1,3,9,57}
1801	{0,1,3,19,133}	1861	{0,1,3,7,19}	1901	{0,1,3,8,790}

10.50 Table $(q, 8, 1)$ difference families.

q	Base Block	q	Base Block
449	{0,1,3,8,61,104,332,381}	617	{0,1,3,16,133,334,371,508}
673	{0,1,3,18,218,252,462,529}	729	$\{0,1,x,x^3,x^8,x^{362},x^{543},x^{685}\}$
841	$\{0,1,x,x^3,x^{378},x^{524},x^{824},x^{834}\}$	953	{0,1,3,18,247,370,789,794}
1009	{0,1,4,14,215,272,834,959}	1289	{0,1,3,7,68,297,466,551}
1681	$\{0,1,x,x^3,x^{21},x^{198},x^{471},x^{536}\}$	1849	$\{0,1,x,x^3,x^{24},x^{386},x^{459},x^{1353}\}$
2129	{0,1,3,7,55,341,1336,1413}	2521	{0,1,3,9,75,1578,1817,2467}
2633	{0,1,3,7,48,1035,2009,2224}	2689	{0,1,3,7,32,1019,1632,2539}
2801	{0,1,3,11,54,1726,1749,1864}	2857	{0,1,6,15,85,878,1269,2329}
2969	{0,1,3,7,65,567,1379,2003}	3137	{0,1,3,7,15,269,1814,2254}
3361	{0,1,3,7,44,141,1088,2681}	3529	{0,1,4,16,69,646,1219,2053}
3697	{0,1,3,7,21,2912,3224,3557}	4201	{0,1,3,7,16,44,1547,1619}
4481	{0,1,3,7,23,890,1468,2075}	4649	{0,1,3,7,31,210,2797,3229}
4817	{0,1,3,8,19,196,2964,3847}	5041	$\{0,1,x,x^3,x^{13},x^{643},x^{2116},x^{3834}\}$
5153	{0,1,3,7,17,187,782,1140}	5209	{0,1,3,7,17,347,548,1976}
5657	{0,1,3,7,17,1354,1574,3014}	5881	{0,1,3,7,18,1710,2150,5492}
6217	{0,1,3,12,29,411,1131,4700}	6329	{0,1,3,7,19,188,1832,2927}
6553	{0,1,11,39,86,1044,1579,1756}	6833	{0,1,3,7,17,745,3503,5383}

10.51 Table $(q, 8, 2)$ difference families.

q	Base Block	q	Base Block
113	{0,1,3,9,24,50,85,97}	169	$\{0, 1, x, x^3, x^4, x^{21}, x^{81}, x^{117}\}$
281	{0,1,3,7,16,100,165,267}	337	{0,1,3,7,16,68,211,228}
421	{0,1,3,7,16,44,305,334}	701	{0,1,3,7,16,48,298,629}
757	{0,1,3,7,24,50,141,662}	1093	{0,1,3,7,16,173,495,974}
1373	{0,1,3,7,21,74,326,1310}	1429	{0,1,3,7,17,37,160,1300}
1597	{0,1,3,7,17,51,985,1356}	1709	{0,1,3,7,18,39,467,868}
1877	{0,1,3,7,17,45,848,1117}	1933	{0,1,3,7,16,36,1532,1636}
2213	{0,1,3,7,16,38,243,1853}	2269	{0,1,3,7,16,39,871,2141}
2381	{0,1,3,7,16,34,128,1040}	2437	{0,1,3,7,16,34,48,556}
2549	{0,1,3,7,18,41,929,2311}	3109	{0,1,3,7,22,52,880,1468}
3221	{0,1,3,7,16,35,258,1382}	3389	{0,1,3,7,20,46,159,961}
3557	{0,1,3,7,16,34,225,2508}	3613	{0,1,3,7,16,34,508,828}
4229	{0,1,3,7,21,100,1109,1206}	4397	{0,1,3,7,16,36,115,3723}
4621	{0,1,3,7,16,41,317,4562}	4733	{0,1,3,7,18,42,335,1141}
4789	{0,1,3,8,18,39,282,736}	4957	{0,1,3,7,16,37,3222,4913}
5237	{0,1,3,7,18,38,133,237}	5573	{0,1,3,7,16,37,72,2165}
5741	{0,1,3,7,16,34,69,4709}	6133	{0,1,3,7,16,34,142,3511}
6301	{0,1,3,7,16,45,164,4348}	6469	{0,1,3,7,16,34,70,5583}
6581	{0,1,3,7,16,39,258,3951}	6637	{0,1,3,7,16,34,106,883}

10.52 Table $(q, 7, 1)$ difference families and $(q, 9, 1)$ difference families for q=433,1009.

q	Base Block	q	Base Block
169	$\{0, 1, x, x^3, x^8, x^{51}, x^{58}\}$	337	{0,1,3,15,54,103,164}
379	{0,1,3,24,74,167,245}	547	{0,1,3,8,67,169,373}
673	{0,1,3,7,74,180,392}	757	{0,1,5,11,186,392,620}
841	$\{0, 1, x, x^{12}, x^{172}, x^{278}, x^{393}\}$	883	{0,1,3,7,242,423,505}
1471	{0,1,3,7,35,620,1126}	1597	{0,1,5,11,83,372,503}
1681	$\{0, 1, x, x^3, x^{17}, x^{1339}, x^{1441}\}$	5167	{0,1,3,7,18,2535,2976}
6091	{0,1,6,14,34,987,5380}	6553	{0,1,5,11,67,644,4144}
6679	{0,1,6,15,50,180,4391}	6889	$\{0, 1, x, x^3, x^4, x^{590}, x^{2830}\}$
433	{0,1,3,30,52,61,84,280,394}	1009	{0,1,3,38,45,409,427,516,988}

10.53 Theorem Suppose $k \leq 9$, $q < 10000$ and q is an odd prime power with $q \equiv 1 \pmod{k(k-1)/\gcd(k(k-1), \lambda)}$. Then a (q, k, λ) difference family over $\mathbb{F}_q$ exists, except possibly for the following values:

1. $k = 6$, $\lambda = 1$, q=61 (nonexistent).
2. $k = 7$, $\lambda = 1$, q=43 (nonexistent), 127 or 211.
3. $k = 8$, $\lambda = 1$, q=113, 169, 281 or 337.
4. $k = 8$, $\lambda = 2$, q=29 (nonexistent).
5. $k = 9$, $\lambda = 1$, q=289, 361 or 577.
6. $k = 9$, $\lambda = 3$, q=25 (nonexistent), or 49.

10.5 Partial Difference Families

10.54 Theorem A partial $(v, 3, 1)$ difference family over $\mathbb{Z}_v$ exists if and only if $v \neq 9$ and $v \equiv 3 \pmod 6$. (Construction IV.43.13 gives one explicit solution.)

10.55 Theorem [8] Let $p \equiv 13 \pmod{24}$ be a prime. Then for $\alpha \geq 1$, a $(4p^\alpha, 4, 1)$ partial difference family over the cyclic group exists.

10.56 Remarks For several known (v,k,λ) partial difference families, $v = kq$ where k, q are odd prime powers, $k-1 \equiv 0 \pmod{2\lambda}$, and $\lambda(q-1) \equiv 0 \pmod{k-1}$. These difference families are over $\mathbb{F}_k \times \mathbb{F}_q$. In each case there are λ identical base blocks containing the values $(i,0)$ $(i \in \mathbb{F}_k)$. There is an initial base block of the form $B = \{(0,0), (c_1, \pm z_1), (c_2, \pm z_2), \ldots (c_{(k-1)/2}, \pm z_{(k-1)/2})\}$ where the c_i's are the quadratic residues in $\mathbb{F}_k$ and the values z_i depend on q. Other base blocks are obtained by multiplying B by $(1, x^{mn})$ for $m = \frac{k-1}{2\lambda}$ and $n = 1,2,\ldots,\frac{\lambda(q-1)}{k-1} - 1$.

10.57 Examples

1. $k = 5, q = 17$: $(v,k,\lambda) = (85,5,1)$: $B=\{(0,0),(1,1),(1,16),(4,7),(4,10)\}$.
2. $k = 5, q = 29$: $(v,k,\lambda) = (145,5,1)$: $B=\{(0,0),(1,1),(1,28),(4,12),(4,17)\}$.
3. $k = 7, q = 13$: $(v,k,\lambda) = (91,7,1)$: $B=\{(0,0),(1,1),(1,12),(2,4),(2,9),(4,3),(4,10)\}$.
4. $k = 7, q = 11$: $(v,k,\lambda) = (77,7,3)$: $B=\{(0,0),(1,1),(1,10),(2,2),(2,9),(4,4),(4,7)\}$.
5. $k = 9, q = 13$: $(v,k,\lambda) = (117,9,2)$: $B=\{(0,0), (1,1), (1,12), (2,5), (2,8), (x+1,2), (x+1,11), (2x+2,3), (2x+2,10)\}$.
6. $k = 9, q = 89$: $(v,k,\lambda) = (801,9,1)$: $B=\{(0,0), (1,1), (1,88), (2,34), (2,55), (x+1,15), (x+1,74), (2x+2,24), (2x+2,65)\}$ (here x is a primitive element of $\mathbb{F}_9$ satisfying $x^2 = x + 1$).

10.58 Theorem ([4], p. 334) If v is odd and $v \not\equiv 3, 15 \pmod{18}$, then a partial $(5v, 5, 2)$ difference family exists.

10.59 Theorem [13] If k, q are odd prime powers and $q \geq k$, then a $(kq, k, (k-1)/2)$ partial difference family over $\mathbb{F}_k \times \mathbb{F}_q$ exists.

10.60 Theorem Suppose $q \equiv 1 \pmod{k-1}$ is a prime power. Then a $(kq, k, 1)$ partial difference family over $\mathbb{F}_k \times \mathbb{F}_q$ exists if one of the following holds:

1. $k \in \{3, 5\}$ (for $k = 5$, the initial block B can be taken as $\{(0,0), (1,1), (1,-1), (4,x), (4,-x)\}$ where x is any nonsquare in $\mathbb{F}_q$ such that exactly one of $x-1, x+1$ is a square).
2. $k = 7, q < 5000$, and $q \notin \{19, 25, 37\}$.
3. $k = 9, q < 5834$, and $q \notin \{17, 25, 41, 97, 113\}$.
4. $k = 11, q < 1202$, q is prime, and $q \notin [30,192]$, $[240,312]$ or $[490,492]$.

10.61 Remark The partial difference families for $k = 7, 9, 11$ in Theorem 10.60 are due to Greig, but many of them remain unpublished. Some can be found in [3], [13], or [16].

10.6 1-Rotational Designs

A (v,k,λ) design $(V, \mathcal{B})$ is *1-rotational* if $V = \{\infty\} \cup \mathbb{Z}_{v-1}$, and the mapping ϕ from i to $i+1 \pmod{v}$, fixing ∞, is an automorphism of the design. *Base blocks* for a 1-rotational design are obtained by selecting orbit representatives of the orbits of blocks under the action of ϕ.

10.62 Remark A more general definition permits the obvious extension to sets of elements $V = G \cup \{\infty\}$ for an abelian group G.

10.63 Theorem [17] A 1-rotational STS(v) exists if and only if $v \equiv 3, 9 \pmod{24}$.

10.64 Theorem [2] A 1-rotational $(v, 4, 1)$ design exists whenever $v = 1 + 3\prod_{i=1}^{k} p_i$, where $p_1, \ldots, p_k$ are primes, $p_i \equiv 1 \pmod 4$, and the largest powers of 2 dividing any two of $\{p_i - 1 : 1 \leq i \leq k\}$ are the same.

10.65 Examples 1-rotational $(v, 3, 2)$ designs. Short are blocks shown *slanted.* The designs for v = 12, 18, 24, 30, 36, and 42 are resolvable.

v	Base Blocks						
6	∞ 0 1	0 2 4					
10	∞ 0 1	0 1 4	0 2 4	*0 3 6*			
12	∞ 3 6	0 1 5	2 4 8	7 9 10			
18	∞ 13 16	0 1 5	3 4 11	8 10 14	6 12 15	2 7 9	
22	∞ 0 3	0 1 7	0 1 9	0 2 5	0 2 10	0 4 9	0 4 10
	0 7 14						
24	∞ 16 20	0 7 21	1 3 11	4 5 18	6 12 17	2 10 13	8 9 14
	15 19 22						
30	∞ 2 22	0 1 19	6 8 27	9 15 16	7 10 14	11 25 28	12 18 23
	13 21 26	4 20 24	3 5 17				
34	∞ 0 1	0 1 14	0 2 7	0 2 8	0 3 13	0 3 15	0 4 12
	0 4 14	0 5 16	0 6 15	0 7 16	*0 11 22*		
36	∞ 14 29	0 1 13	2 17 25	3 6 12	4 9 28	5 21 23	7 18 32
	8 16 22	10 19 20	11 15 33	24 26 31	27 30 34		
42	∞ 7 22	0 1 20	9 15 40	18 23 36	26 30 38	8 11 35	19 21 28
	13 16 34	6 17 33	10 32 39	5 31 37	3 14 27	2 4 12	24 25 29

10.66 Examples 1-rotational $(v, 3, 6)$ designs.

v	Base Blocks							
8	∞ 0 1	∞ 0 2	∞ 0 3	0 1 2	0 2 4	0 1 3	0 2 3	0 3 6
14	∞ 0 4	∞ 0 5	∞ 0 6	0 1 2	0 2 4	0 1 3	0 3 6	0 4 8
	0 5 10	0 6 12	0 1 4	0 3 4	0 2 7	0 5 7		

10.67 Examples Resolvable 1-rotational $(v, 4, 3)$ designs.

v	Base Blocks				
8	∞ 3 5 6	0 1 2 4			
12	∞ 3 7 9	0 1 4 10	2 5 6 8		
20	∞ 8 12 18	0 2 3 14	1 4 13 16	7 9 15 17	5 6 10 11
24	∞ 0 7 10	1 8 12 22	2 5 6 11	3 9 14 18	4 16 17 19
	13 15 20 21				
32	∞ 17 22 23	0 4 7 20	2 14 25 28	1 8 10 16	5 9 18 19
	6 11 13 21	3 12 24 30	15 26 27 29		

10.68 Example A 1-rotational (18,4,6) design.

∞ 0 1 4	∞ 0 4 5	1 4 13 16	1 4 13 16	3 5 12 14
3 5 12 14	2 8 9 15	2 8 9 15	6 7 10 11	

10.69 Examples 1-rotational $(v, 5, 4)$ designs. When $v = 20, 40$, the designs given are resolvable.

v	Base Blocks			
10	∞ 0 1 4 6	0 1 2 4 8		
15	∞ 0 1 3 8	0 1 2 5 6	0 2 4 7 10	
20	∞ 0 2 3 14	1 5 6 10 12	4 7 8 13 16	9 11 15 17 18
36	∞ 8 13 22 27	0 1 10 12 13	0 2 6 8 25	0 16 20 27 33
	0 15 22 23 26	14 15 21 25 30	12 15 25 30 33	
40	∞ 0 3 9 27	2 5 13 26 32	1 4 20 29 36	8 16 25 30 35
	10 11 12 14 22	7 15 17 31 33	6 21 28 34 38	18 19 23 24 37

10.70 Examples 1-rotational $(v, 6, \lambda)$ designs. The (24,6,5), (30,6,5), (36,6,2), (36,6,3) and (42,6,5) designs given are resolvable.

v, λ	Base Blocks			
14,15	∞ 0 1 5 7 10	∞ 0 2 3 4 8	∞ 0 6 9 11 12	1 3 4 9 10 12
	0 1 2 6 11 12	0 3 5 6 7 10	0 2 4 5 8 9	
15,5	∞ 0 3 5 7 13	1 6 9 10 11 12	1 2 4 8 9 11	
18,5	∞ 0 2 8 9 12	0 1 2 4 7 15	0 1 3 7 8 12	
24,5	∞ 0 1 7 15 20	2 3 4 5 10 14	6 11 13 17 19 22	8 9 12 16 18 21
27,5	∞ 0 7 11 13 21	0 3 6 14 17 24	0 9 16 18 20 25	0 1 2 8 22 23
	2 5 6 15 18 19			
30,5	∞ 5 16 19 21 22	0 1 6 7 11 14	2 9 13 15 17 26	3 10 12 20 23 24
	4 8 18 25 27 28			
34,5	∞ 0 3 10 12 32	0 3 16 18 24 32	0 9 15 23 28 30	0 3 4 12 31 32
	0 7 13 21 23 30	*0 4 11 15 22 26*	*0 5 11 16 22 27*	
36,2	∞ *0 7 14 21 28*	∞ *0 7 14 21 28*	4 15 16 20 24 26	2 3 12 15 18 20
36,3	2 4 8 12 17 20	1 2 3 6 12 18	3 12 14 15 25 30	∞ *0 7 14 21 28*
	∞ *0 7 14 21 28*	∞ *0 7 14 21 28*		
40,5	∞ 0 3 19 27 32	0 6 7 15 18 35	0 17 19 21 24 33	0 1 7 8 29 30
	0 5 9 19 23 25	10 12 15 16 24 27	*0 5 13 18 26 31*	*0 2 13 15 26 38*
42,5	∞ 15 22 24 27 35	0 3 7 13 29 30	8 17 18 25 34 36	9 10 14 20 23 28
	1 2 5 11 26 31	19 21 33 37 38 40	4 6 12 16 32 39	

10.71 Examples Resolvable 1-rotational (v, k, λ) designs.

v, k, λ	Base Blocks
28,7,6	{∞,0,5,14,15,24,25}, {1,10,17,20,22,23,26}, {2,6,8,9,13,19,21}, {3,4,7,11,12,16,18}
32,8,7	{∞,0,1,5,16,18,25,28}, {3,9,17,19,20,21,24,26}, {2,6,7,12,14,15,23,27}, {4,8,10,11,13,22,29,30}
40,8,7	{∞,13,17,19,23,28,31,38}, {0,1,14,16,22,26,29,35}, {3,5,8,15,24,25,36,37}, {2,6,7,9,10,11,30,33}, {4,12,18,20,21,27,32,34}
42,7,6	{∞,0,3,7,13,29,30}, {8,15,17,18,25,34,36}, {9,10,14,20,22,23,28}, {1,2,5,11,26,31,35}, {19,21,24,33,37,38,40}, {4,6,12,16,27,32,39}
63,7,3	{6,14,32,44,49,51,52}, {7,8,12,30,34,36,59}, {26,35,40,46,47,56,60}, {0,2,10,17,23,50,53}, {∞,*11,24,27,42,55,58*}
63,9,4	{4,9,20,32,36,38,39,45,56}, {24,34,37,42,44,46,57,58,61}, {12,16,18,19,28,33,41,52,60}, {∞,*0,17,22,23,31,48,53,54*}

10.72 Examples 1-rotational (v, k, λ) designs.

v, k, λ	Base Blocks
20,8,14	{∞,0,4,5,6,9,16,17}, {∞,0,1,4,6,7,9,11}, {0,1,5,6,7,9,15,17}, {0,4,5,6,7,10,11,16}, {0,1,4,9,11,13,16,17}
24,8,7	{∞,0,1,4,5,7,9,17}, {0,1,9,10,15,17,20,21}, {0,3,8,9,10,12,16,21}
30,10,9	{∞,0,1,5,7,16,19,20,23,28}, {0,2,3,6,8,10,11,18,22,23}, {0,2,3,13,14,19,22,24,26,28}
34,11,10	{∞,1,6,7,11,14,22,23,25,27,29}, {0,1,3,4,6,10,18,23,24,26,32}, {0,1,2,6,9,10,15,23,24,26,32}, {*0,3,6,9,12,15,18,21,24,27,30*}
34,12,12	{∞,0,1,3,5,9,16,17,23,25,27,28,30}, {1,3,6,9,10,11,20,22,24,27,30}, {0,2,4,7,12,13,17,23,27,28,29,30}, {∞,*0,3,6,9,12,15,18,21,24,27,30*}
36,9,8	{∞,5,15,18,20,24,30,31,32}, { 0,3,4,5,11,15,17,20,30}, {0,4,7,8,9,11,16,22,28}, {0,3,4,11,12,14,17,21,33}
36,12,11	{∞,0,1,4,6,7,11,19,26,28,29,34}, {1,3,6,7,15,17,18,20,28,29,32,34}, {1,4,5,7,9,10,11,16,20,25,28,29}
39,13,12	{∞,1,7,8,9,11,12,18,20,23,25,26,30}, {0,3,4,6,13,15,21,27,28,29,31,33,37}, {1,2,7,10,11,13,14,15,19,22,29,32,34}

v,k,λ	Base Blocks
40,10,9	{∞,0,3,7,9,13,26,27,34,37}, {0,7,12,17,23,30,34,36,37,38}, {2,3,5,9,12,13,27,30,32,36}, {2,4,5,6,10,15,18,25,26,32}
40,12,11	{∞,0,3,4,6,9,10,15,18,25,26,27}, {1,2,5,6,7,8,9,15,19,24,29,31}, {1,9,16,20,21,23,25,27,29,30,35,37}, {*0,1,3,9,13,14,16,22,26,27,29,35*}
70,7,2	{∞,*0,2,23,25,46,48*}, {0,1,8,12,14,17,41}, {0,5,14,27,49,53,59}, {0,17,20,32,50,51,58}

10.73 Example A (26,6,3) design on elements $(\mathbb{Z}_5 \times \mathbb{Z}_5) \cup \{\infty\}$. Base blocks are

$$\{(0,0),(0,x),(0,2x),(x,x),(2x,4x),(4x,0)\}$$

for $x = 1, 2$, and short blocks $\{\infty,(0,0),(1,y),(2,2y),(3,3y),(4,4y)\}$ for $y = 0,2,4$.

10.74 Example A 1-rotational (46,6,3) difference family over $(\mathbb{Z}_5 \times \mathbb{Z}_3^2) \cup \{\infty\}$. The base blocks are

{(0,0,1),(0,0,2),(1,1,1),(1,2,2),(4,1,2),(4,2,1)}, {(0,1,0),(0,2,0),(1,1,2),(1,2,1),(4,0,1),(4,0,2)}, {(0,0,1),(0,0,2),(2,1,0),(2,2,0),(3,1,1),(3,2,2)}, {(0,1,0),(0,2,0),(2,1,2),(2,2,1),(3,1,1),(3,2,2)}

and the short block (included three times) is {∞,(0,0,0),(1,0,0),(2,0,0),(3,0,0),(4,0,0)}.

10.75 Example A 1-rotational (50,8,4) difference family over $(\mathbb{Z}_7 \times \mathbb{Z}_7) \cup \{\infty\}$. The base blocks are $\{(0,0),(0,x),(0,2x),(x,0),(2x,2x),(4x,5x),(6x,6x)\}$ for $x \in \{1,2,4\}$, and short blocks $\{\infty,(0,0),(1,y),(2,2y),(3,3y),(4,4y),(5,5y),(6,6y)\}$ for $y \in \{0,2,3,6\}$.

10.76 Example A (51,6,2) design over $(\mathbb{Z}_5 \times \mathbb{Z}_{10}) \cup \{\infty\}$ which is 2-resolvable has short base blocks $\{\infty,(0,5),(1,5),(2,5),(3,5),(4,5)\}$, $\{\infty,(0,8),(1,8),(2,8),(3,8),(4,8)\}$; base blocks {(0,1),(0,3),(0,7),(1,2),(1,9),(3,4)}, {(0,2),(0,7),(2,3),(2,6),(3,0),(4,4)}, and {(0,0),(0,1),(0,9),(1,6),(2,5),(4,8)}.

10.77 Example A resolvable (126,6,1) design over $\mathbb{Z}_5^3 \cup \{\infty\}$. Element $(x,y,z) \in \mathbb{Z}_5^3$ is written as xyz. A short base block is $\{\infty, 000, 100, 200, 300, 400\}$. The other base blocks are {001,004,122,133,421,434}, {002,003,144,111,442,413}, {043,012,220,230,332,323}, {031,024,240,210,314,341}.

10.7 Enumeration

Two difference families D and D' over G are *equivalent* if there is an element $w \in G$ for which $D = w \cdot D'$.

10.78 Remark Equivalent difference families lead to isomorphic BIBD, although the converse is not true in general.

10.79 Table [9] The number of inequivalent $(v,3,\lambda)$ difference families over $\mathbb{Z}_v$, $v \le 28$.

$v\backslash\lambda$	1	2	3	4	5	6	7	8	9
5			1			1			1
6		0		2		0		3	
7	1	2	3	4	5	7	8	10	12
8						12			
9	0	0	4	3	4	24	27	30	108
10		0		21		0		360	
11			8			209			2593
12		4		180		3561		15264	
13	1	9	47	253	1224	5285	20347	70924	225900
14						0			
15	2	9	421	2981	15686	375372	?	?	?

$v\backslash\lambda$	1	2	3	4	5	6
16		71		21742		2460490
17			1550			?
18		0		320760		0
19	4	201	13316	?	?	?
21	7	175	212968	?	?	?
24		2939		?		?
25	12	19543	?	?	?	?
27	8	10841	?	?	?	?
28		609483		?		?

10.80 Table [9] The number N of inequivalent cyclic $(v, 3, 1)$ difference families, $31 \leq v \leq 57$.

v	31	33	37	39	43	45	49	51	55	57
N	80	84	820	798	9508	11616	157340	139828	3027456	2353310

10.81 Table [11] The number N of inequivalent $(v, 4, 1)$ difference families over $\mathbb{Z}_v$.

v	13	16	25	28	37	40	49	52	61	64
N	1	0	0	0	2	10	224	206	18132	12048

10.82 Table [11] The number N of inequivalent $(v, 5, 1)$ difference families over $\mathbb{Z}_v$.

v	21	25	41	45	61	65
N	1	0	1	0	10	2

10.8 See Also

§I.1	Difference families generate BIBDs.
§IV.14	Difference triangle sets are generalizations of perfect difference families.
§IV.12	Difference sets are difference families with one base block.
§IV.11	Difference matrices are used in recursive constructions of difference families.
§IV.43	Skolem sequences are used to construct difference families with block size three.
[4]	A text covering the main results of this section.
[15]	A survey of recursive constructions for difference families over the cyclic group.

References

[1] R.J.R. Abel, Some new BIBDs with $\lambda = 1$ and $6 \leq k \leq 10$, *J. Combin. Des.*, to appear.

[2] I. Anderson and N.J. Finizio, Cyclically resolvable designs and triplewhist tournaments, *J. Combin. Des.* 1 (1993), 347-358.

[3] S. Bagchi and B. Bagchi, Designs from pairs of finite fields I. A cyclic unital U(6) and other regular Steiner 2-designs, *J. Combin. Theory A* 52 (1989), 51-61.

[4] T. Beth, D. Jungnickel and H. Lenz, *Design Theory*, Cambridge University Press, London, 1986.

[5] M. Buratti, Improving two theorems of Bose on difference families, *J. Combin. Des.* 3 (1995), 15-24.

[6] M. Buratti, On simple radical difference families, *J. Combin. Des.* 3 (1995), 161-168.

[7] M. Buratti, Constructions of $(q, k, 1)$ difference families with q a prime power and $k = 4, 5$, *Discrete Math.* 138 (1995), 169-175.
[8] P.L. Check and C.J. Colbourn, A note concerning difference families with block size four, *Discrete Math.* 133 (1994), 285-289.
[9] C.J. Colbourn, private communication, 1994.
[10] C.J. Colbourn and M.J. Colbourn, Cyclic block designs with block size 3, *Europ. J. Combin.* 2 (1981), 21-26.
[11] M.J. Colbourn and R.A. Mathon, On cyclic Steiner 2-designs, *Ann. Discrete Math.* 7 (1980), 215-253.
[12] M. Greig, Some balanced incomplete block design constructions, *Congr. Numer.* 77 (1990), 121-134.
[13] H. Hanani, Balanced incomplete block designs and related designs, *Discrete Math.* 77 (1975), 255-369.
[14] J.H. Huang and S.S. Skiena, Gracefully labeling prisms, *Ars Combin.* 38 (1994), 225-242.
[15] M. Jimbo, Recursive constructions for cyclic BIB designs and their generalizations, *Discrete Math.* 116 (1993), 79-95.
[16] R. Mathon, Constructions for cyclic Steiner 2-designs, *Ann. Discrete Math.* 34 (1987), 353-362.
[17] K.T. Phelps and A. Rosa, Steiner triple systems with rotational automorphisms, *Discrete Math.* 33 (1981), 57-66.
[18] R.M. Wilson, Cyclotomy and difference families in elementary abelian groups, *J. Number Theory* 4 (1972), 17-47.

11 Difference Matrices

Charles J. Colbourn
Warwick de Launey

11.1 Definition and Examples

Let $(G, \odot)$ be a group of order g. A $(g, k; \lambda)$-*difference matrix* is a $k \times g\lambda$ matrix $D = (d_{ij})$ with entries from G, so that for each $1 \le i < j \le k$, the multiset

$$\{d_{i\ell} \odot d_{j\ell}^{-1} : 1 \le \ell \le g\lambda\}$$

contains every element of G λ times. When G is abelian, typically additive notation is used, so that differences $d_{i\ell} - d_{j\ell}$ are employed.

11.1 Examples

1. A $(3, 6; 2)$-difference matrix and a $(15, 5; 1)$-difference matrix [18]. For the latter, append a column of zeroes to $(B| - B)$ to get the required 5×15 matrix.

$$A = \begin{pmatrix} 0&0&0&0&0&0 \\ 1&2&0&2&0&1 \\ 1&0&2&2&1&0 \\ 0&2&2&0&1&1 \\ 2&2&0&1&1&0 \\ 2&0&2&1&0&1 \end{pmatrix} \qquad B = \begin{pmatrix} 0&0&0&0&0&0&0 \\ 1&2&3&4&5&6&7 \\ 2&5&7&9&12&4&1 \\ 6&3&14&10&7&13&4 \\ 10&6&1&11&2&7&12 \end{pmatrix}$$

2. Theorem II.2.35 gives (implicitly) a $(12,6;1)$-difference matrix. Numerous other difference matrices are given in §II.2.

11.2 Remark The $(15,5;1)$- and $(12,6;1)$-difference matrices of Examples 11.1 can be used to construct four MOLS(15) and five MOLS(12), respectively.

11.3 Remark By removing any row from a $(g,k;\lambda)$-difference matrix, one obtains a $(g,k-1;\lambda)$-difference matrix.

11.4 Theorems

1. [17] A $(g,k;\lambda)$-difference matrix does not exist if $k > \lambda g$.
2. [15] A $(g,3;\lambda)$-difference matrix does not exist if $g \equiv 2 \pmod 4$ and λ is odd.

11.5 Theorem [15] The multiplication table for the finite field $\mathbb{F}_q$ is a $(q,q;1)$-difference matrix over EA(q).

11.2 Related Objects

11.6 Remark A $(s,k;\lambda)$-difference matrix D over a group $(G,\odot)$ with $G=\{g_1,\ldots,g_s\}$ can be developed over G to obtain a completely resolvable $\mathrm{OA}_\lambda(k,s)$, $(g_1D\ g_2D\ \ldots\ g_sD)$.

11.7 Theorem A $(g,k;\lambda)$-difference matrix over G gives rise to a resolvable $\mathrm{OA}_\lambda(k,g)$, and hence to an $\mathrm{OA}_\lambda(k+1,g)$ and a $\mathrm{TD}_\lambda(k+1,g)$.

11.8 Theorem An $(s,3;1)$-difference matrix is equivalent to a complete mapping.

11.9 Theorem An $\mathrm{OA}_\lambda(k,g)$ can be viewed as a $(g,k;\lambda g)$-difference matrix over an arbitrary group of order g.

11.10 Remark A $(g,k;\lambda)$-difference matrix over a group G of order g is a generalized Bhaskar Rao design GBRD$(k,k,\lambda g;G)$. See §IV.4.

11.3 Generalized Hadamard Matrices

A *generalized Hadamard matrix* GH(g,λ) is a $(g,g\lambda;\lambda)$-difference matrix.

Let EA(g) denote the order g direct product of prime order cyclic groups.

11.11 Example A normalized GH(4,1) over EA(4), a circulant GH(5,1), and a GH(3,2) over $\mathbb{Z}_3$.

$$\begin{array}{|cccc|}\hline 1&1&1&1\\ 1&ab&a&b\\ 1&b&ab&a\\ 1&a&b&ab\\ \hline\end{array}\qquad
\begin{array}{|ccccc|}\hline 1&w&w^4&w^4&w\\ w&1&w&w^4&w^4\\ w^4&w&1&w&w^4\\ w^4&w^4&w&1&w\\ w&w^4&w^4&w&1\\ \hline\end{array}\qquad
\begin{array}{|cccccc|}\hline 1&1&w&w&w^2&w^2\\ 1&1&w^2&w^2&w&w\\ w&w^2&w&w^2&w&w^2\\ w&w^2&w^2&w&w^2&w\\ w^2&w&w&w^2&w^2&w\\ w^2&w&w^2&w&w&w^2\\ \hline\end{array}$$

11.12 Remark By Remark 11.10, a GH(g,λ) over the group G is a generalized Bhaskar Rao design GBRD$(g\lambda,g\lambda,g\lambda;G)$.

11.13 Remark A GH(g,λ) is a $(g,k;\lambda)$-difference matrix with $k=g\ell$ as large as possible (see Theorem 11.4).

11.14 Theorem A Hadamard matrix $H(4n)$ is a GH$(2,2n)$ over the group $(\{-1,+1\},\cdot)$.

11.15 Theorem [10] For any group G of prime power order q and any integer $t > 0$, there is a GH(q, q^{2t-1}) over G.

11.16 Theorem [3, 11] Let p be an odd prime. A circulant GH(p, p^{t-1}) exists if and only if $t = 1, 2$.

11.17 Table Direct constructions over EA(g).

g	λ	Conditions	Refs.
p^n	p^{m-n}	$m \geq n \geq 1$, p a prime	[1]
p^n	$2p^{m-n}$	$m \geq n \geq 1$, p a prime	[1]
p^n	$4p^{m-n}$	$m \geq n \geq 1$, p a prime	[7]
q	8	$19 < q < 200$, q prime power	[13]
p	8	$p > 19$ a prime	[13]
q	k	$q > ((k-2)2^{k-2})^2$, $\exists$ Hadamard matrix of order k	[14]

11.18 Theorems

1. [15] If there is a GH(g, λ_1) and a GH(g, λ_2) over the group G, then there is a GH$(g, g\lambda_1\lambda_2)$ over G.
2. [7] Let $q = g\lambda - 1$ be a prime power. If there is a GH(g, λ) over the group G, then there are GH$(g, \lambda q^t)$ over G for all $t \geq 0$.

11.19 Table The existence of GH(g, λ) matrices over EA(g) with $n = g\lambda$ for $n < 100$. The entry $3, 5^2$ at the end of the row headed by 0 means that GH$(3, 30)$ and GH$(5, 18)$ are known and that no GH$(2t, 45/t)$ can exist over EA$(2t)$; for $g = 9, 15, 45$ it is not known whether a GH$(g, 90/g)$ exists.

n	0	10	20	30	40	50	60	70	80	90
0		5^2	2, 5	3^2	2	25^2	2	2	2, 5	$3, 5^2$
1		11	3^7	31	41	3,17	61	71	81	7,13
2	2	3, 4	11^2	32	2	2, 13	31^2	2, 9	41^2	2, 23
3	3	13	23	3	43	53		73	83	31
4	4	7^2	2, 3	17^2	2, 11	27^2	64	37^2	2	47^2
5	5	3,5	25	5,7	3	5,11	5,13	5	5	
6	3^2	16	13^2	4, 9	23^2	8	2	2, 19	43^2	3, 4
7	7	17	27	37	47	19	67	11	3,29	97
8	8	9^2	2, 7	19^2	3, 4	29^2	2, 17	2	2	47^2
9	9	19	29		49	59	3	79	89	3

11.20 Theorem [6] Let G be abelian group of order g. Let p be an odd prime dividing g. A GH(g, λ) of odd order $g\lambda$ exists over G only if every positive integer $m \not\equiv 0 \pmod{p}$ which divides the squarefree part of $g\lambda$ has odd multiplicative order modulo p. In particular, for all divisors t of g, and all primes p, the Hilbert norm $(q, (-1)^{(t-1)/2}t)_p = 1$, and hence there exists a nontrivial solution for the equation $nx^2 + (-1)^{(t-1)/2}ty^2 = z^2$.

11.21 Remark Theorem 11.20 is true for nilpotent G. If $g = \prod p_i^{k_i}$ is the prime power decomposition of g, and for all pairs i and j, $p_i \not| p_j^{k_j} - 1$, then G must be nilpotent. The only odd values $g < 100$ for which this is not true are 39, 55, 63, and 75.

11.22 Theorem [2] A GH(g, λ) with $n = g\lambda$ odd exists over a group G only if a nontrivial solution in integers x, y, z to $z^2 = nx^2 + (-1)^{(t-1)/2}ty^2$ exists for every order t of a homomorphic image of G.

11.4 Shorter Difference Matrices

11.23 Example A $(15,7;2)$-difference matrix over $\mathbb{Z}_{15}$. Take $(A \mid 2A \mid -A \mid -2A)$, where A is the circulant matrix with first row $(2\ 4\ 10\ 3\ 13\ 0\ 9)$. Append two columns of zeroes.

11.24 Theorem If there is a $(g,k;\lambda_1)$-difference matrix and a $(g,k;\lambda_2)$-difference matrix over the same group G, then there is a $(g,k;\lambda_1+\lambda_2)$-difference matrix over G.

11.25 Theorem If there is a $(g,k;\lambda)$-difference matrix over G and a $(g',k;\lambda')$-difference matrix over G', then there is a $(gg',k;\lambda\lambda')$-difference matrix over $G \times G'$.

11.26 Theorem If a $(g,k;\lambda)$-difference matrix over a group G exists, and G' is a normal subgroup of G of index x, then an $(x,k;\lambda g/x)$-difference matrix over G/G' exists.

11.27 Theorem (*tensor or Kronecker product* [19]) If a $(g,k;\lambda)$-difference matrix and a $(g,k';\lambda')$-difference matrix exist over the same group G, then a $(g,kk';g\lambda\lambda')$-difference matrix exists over G.

11.28 Theorem [17] Let $s = q_1q_2\ldots q_n$ be the prime power factorization of s, and let $q = \min(q_1,q_2,\ldots,q_n)$. Then if there exists a transversal design $\mathrm{TD}(r,s)$, there exists a $(s,r(q-1)+1;s)$-difference matrix over $\mathrm{EA}(q_1)\times\ldots\times\mathrm{EA}(q_n)$.

11.29 Theorem [17] Suppose there is a (v,k,λ)-difference family over G. Then if there exists an idempotent $\mathrm{TD}(r,k)$, there exists a $(v,r;\lambda)$-difference matrix over G. Similarly, if there exists a $\mathrm{TD}_\mu(r,k)$ having a parallel class and $k(\mu-1)\le\lambda\mu$, then there exists a $(v,r;\lambda\mu)$-difference matrix over G.

A (G,s,t)-*complementary pair* is a pair of $s\times t$ $(0,G)$-matrices A and B such that $a_{ij}=0$ if and only if $b_{ij}\ne 0$ and for $1\le i<j\le t$ there is an integer λ_{ij} such that the multiset $\{a_{ik}-a_{jk}|k=1,\ldots,t\}\cup\{b_{ik}-b_{jk}|k=1,\ldots,t\}$ contains each element of G exactly λ_{ij} times.

11.30 Example A $(\mathbb{Z}_3,7,7)$-complementary pair. Let A and B be the circulant matrices with respective first rows $(w^2,w,w,0,w,0,0)$ and $(0,0,0,w^2,0,e,w)$. A (EA(12),7,28)-complementary pair is obtained by developing the rows $(a_1w^2,a_2w,a_3w,0,a_4w,0,0)$ and $(0,0,0,a_1w^2,0,a_2,a_3w)$ for $(a_1,a_2,a_3,a_4)=(e,e,e,e)$, (e,a,b,ab), (e,ab,a,b) and (e,b,ab,a).

11.31 Theorem (see [7, 9]) Let a be 0 or 1. If there are a (G,k,b)-complementary pair and a $(g,2n+a,\lambda)$-difference matrix over G, then there is a $(g,kn+a,b\lambda)$-difference matrix over G.

11.32 Theorem (see [7, 9]) Suppose g is square-free, and let a be 0 or 1. If there is a $(g,2n+a,\lambda)$-difference matrix over $\mathbb{Z}_g$, then there is a $(g,sn+a,s\lambda)$-difference matrix whenever $s=q^t(q+1), q\equiv 1 \pmod g$ is a prime power and $t=0,1$ or 2.

11.33 Remark See [7, 9] for the definition of mixed difference matrices and generalizations of Theorems 11.31 and 11.32. A (g,k,λ)-mixed difference matrix gives a resolvable $\mathrm{TD}_\lambda(k,g)$, and the generalization of Theorem 11.32 gives (g,k,λ)-mixed difference matrices for $(g,k,\lambda) =$ (4,20,40), (4,40,20), (4,52,26), (4,60,30), (9,81,20), (12,42,14), (12,78,26), (12,42,28), (16,144,18), (16,288,36), (18,20,20), (24,78,26), (28,61,30). Of these, some have many more rows than the analogous difference matrices.

Let $x\le\lambda$. A (g,v,λ,x)-*impulse matrix* over the group G is a $v\times(\lambda g-x)$ array $A=(a_{ij})$ of elements of G such that the multisets $\{a_{ik}-a_{jk}|k=1,\ldots,\lambda g-x\}$ each contain the identity element $\lambda-x$ times and each nonidentity element λ times.

11.34 Theorem Let $k > 0$ be an integer. If n and $n+k$ are prime powers, then an $(n+k, n^2, n, 1-k)$-impulse matrix over $\mathrm{EA}(n)$ exists.

11.35 Remarks

1. A (g, v, λ, x)-impulse matrix with $x = 0$ or 1 is equivalent to a (g, v, λ)-difference matrix.
2. If there is a (g, v, λ, x)-impulse matrix with $x \geq 0$, then there is a (g, v, λ)-difference matrix.
3. If there is a (g, v, λ)-difference matrix, then there is a $(g, v, \lambda x, x)$-impulse matrix.
4. If there is a (g, v, λ, t)-impulse matrix and a $(g, v, \mu, -t)$-impulse matrix, then there is a $(g, v, \lambda + \mu)$-difference matrix.

11.36 Theorem Let G be a group, and $h = g\lambda - x$. If there are a (g, v, λ, x)-impulse matrix with w zero columns over G, an (h, u, μ)-difference matrix, and a $(g, u, \mu\lambda x, y)$-impulse matrix, then there is a $(g, v(u+1), \mu\lambda h, y - \mu x^2 + w)$-impulse matrix over G.

11.37 Theorem Let q be a prime power. If there is a $\mathrm{GH}(g, (q+1)/g)$ and a $\mathrm{GH}(g, n(q+1)/g)$ over a group G, then there is a $(g, (q+1)(q+n-1), nq(q+1)/g)$-difference matrix over G.

11.38 Theorem Let $g\lambda = h + 1$.

1. If there is a $(g, v+1, \lambda)$-difference matrix and an (h, u, μ)-difference matrix, and a $(g, u+1, \mu\lambda)$-difference matrix, then there is a $(g, v(u+1), \mu\lambda h, 1-\mu)$-impulse matrix.
2. If there is a $(g, v+1, \lambda)$-difference matrix and an (h, v, μ)-difference matrix, and a $(g, u+1, \mu\lambda)$-difference matrix, then there is a $(g, u+v^2, \mu\lambda h)$-difference matrix.

11.39 Remarks See [7] for special cases of Theorems 11.36 and 11.38. The difference matrix in Theorems 11.36 and 11.38 may be replaced by the appropriate resolvable transversal design.

11.40 Theorem [8] Let H be a finite group of odd order and let $G = \mathbb{Z}_2^s \times H$, where $s \geq 0$; Let g be the order of the group G. Then there exists an integer $N(k; G)$ such that

1. if $s \neq 1$, there is a $(g, k; \lambda)$-difference matrix over G for all $\lambda > N(k; G)$;
2. if $s = 1$, there is a $(g, k; 2\lambda)$-difference matrix over G for all $\lambda > N(k; G)$.

A set of m disjoint blocks, together containing c points of a PBD on v points is a *partial parallel class* of *width* $m + v - c$. A PBD is an (ℓ, w)-*PBD* if its blocks can be partitioned into ℓ partial parallel classes of maximum width w.

11.41 Theorem [5] If an (ℓ, w)-PBD of index λ on v points exists, then for any prime power s satisfying $w \leq s \leq \lfloor \frac{\ell}{\lambda} \rfloor$, there exists a $(s, v; \ell - \lambda)$-difference matrix over $\mathrm{EA}(s)$. In fact, there is a $(s, v, \ell - \lambda, l - \lambda s)$-impulse matrix over $\mathrm{EA}(s)$.

11.42 Theorem [5] Let $v = 1 + nk$ be a prime power, with v or k even. For any odd prime power s satisfying $n + 1 \leq s \leq \lfloor \frac{v}{k-1} \rfloor$, there exists a $(s, v; \frac{v-k+1}{2})$-difference matrix.

11.43 Theorem [5] If an $\mathrm{OA}_\lambda(k, n)$ exists having λ constant columns, then, over any group G of order $n+1$, there is a $(n+1, k; \lambda(n-1))$-difference matrix.

11.44 Theorem [5] Let v be a prime power with $v = 1 + nk$ for n and k integers satisfying $n \geq k-2 \geq 0$. For any group G of order $n+1$, there is a $(n+1, v; 2+(n-1)k)$-difference matrix over G.

11.45 Table Lower bounds on k in $(s, k; \lambda)$-difference matrices with $2 \le s \le 50$ and $1 \le \lambda \le 18$.

$s \setminus \lambda$	1	2	3	4	5	6	7	8	9	10	11	12	13	14	15	16	17	18
2	2^{g}	4^{g}	2	8^{g}	2	12^{g}	2	16^{g}	2	20^{g}	2	24^{g}	2	28^{g}	2	32^{g}	2	36^{g}
3	3^{g}	6^{g}	9^{g}	12^{g}	9^{s}	18^{g}	11^{s}	24^{g}	27^{g}	30^{g}	11	36^{g}	13^{i}	21^{c}	27^{t}	48^{g}	24	54^{g}
4	4^{g}	8^{g}	12^{g}	16^{g}	9^{s}	16^{d}	12	32^{g}	36^{g}	16	12	48^{g}	16	56^{g}	21^{i}	64^{g}	32	36
5	5^{g}	10^{g}	7^{q}	20^{g}	25^{g}	15^{p}	17^{q}	25^{d}	20	50^{g}	17	25^{l}	25	20	35^{t}	80^{g}	35^{l}	90^{g}
6	2	6^{s}	2	6	2	7^{s}	2	10^{j}	2	11^{m}	2	12^{t}	2	16^{m}	2	24^{l}	2	10
7	7^{g}	14^{g}	7	28^{g}	11^{q}	18^{i}	49^{g}	28	28^{l}	18	28	49^{d}	28	98^{g}	35^{p}	37^{q}	28	63^{l}
8	8^{g}	16^{g}	8	32^{g}	8	16	56^{g}	64^{g}	21^{p}	16	32	32	25^{i}	64^{d}	56	128^{g}	25	32
9	9^{g}	18^{g}	27^{g}	36^{g}	18	54^{g}	27	72^{g}	81^{g}	36	36^{i}	108^{g}	49^{i}	54	54	81^{d}	72	162^{g}
10	2	5^{s}	2	8^{s}	2	5	2	10^{s}	2	5	2	12^{l}	2	5	2	18^{j}	2	19^{m}
11	11^{g}	22^{g}	11	44^{g}	11	22	11	44	19^{q}	30^{i}	121^{g}	44	44^{i}	30	44	44	44	30
12	6^{s}	6	6	6	6	8^{a}	6	8^{a}	9^{a}	11^{j}	30^{d}	36^{t}	6	8	9^{a}	12^{a}	8	12^{a}
13	13^{g}	26^{g}	13	52^{g}	13	26	13	52	19^{q}	26	23^{q}	84^{i}	169^{g}	27^{p}	65^{p}	52	52	39^{p}
14	2	5^{s}	2	7^{a}	2	7^{a}	2	8^{a}	2	7	2	13^{j}	2	7	2	14^{a}	2	7
15	5^{s}	7^{s}	7^{s}	7	7	9^{a}	7	10^{a}	7	9^{a}	7	10^{a}	7	12^{i}	25^{t}	12^{a}	35^{l}	10^{a}
16	16^{g}	32^{g}	16	64^{g}	16	32	16	128^{g}	16	32	16	64	16	32	90^{i}	256^{g}	33^{p}	39^{p}
17	17^{g}	34^{g}	17	68^{g}	17	34	17	68	19^{q}	34	23^{q}	68	27^{q}	34	31^{q}	272^{g}	289^{g}	39^{p}
18	2	6^{s}	2	8^{a}	2	9^{a}	2	13^{j}	2	9^{a}	2	12^{a}	2	9	2	17^{j}	2	12^{a}
19	19^{g}	38^{g}	19	76^{g}	19	38	19	76	19	38	23^{q}	76	27^{q}	38	31^{q}	76	31	54^{i}
20	4^{a}	8^{s}	5^{a}	10^{s}	5	10^{a}	5	10	7^{a}	10	7	12^{a}	7	10	12^{a}	16^{a}	8	19^{j}
21	6^{s}	6	7^{a}	7^{a}	7^{a}	9^{a}	7	12^{a}	7	9^{a}	7	14^{a}	7	11^{a}	9^{a}	14^{a}	7	14^{a}
22	2	4^{a}	2	8^{a}	2	11^{a}	2	11^{a}	2	11^{a}	2	12^{a}	2	11	2	16^{a}	2	11
23	23^{g}	46^{g}	23	92^{g}	23	46	23	184^{g}	23	46	23	92	27^{q}	46	31^{q}	184	31	46
24	6^{s}	6	8^{a}	8^{a}	8^{a}	9^{a}	8	12^{a}	8	9^{a}	8	16^{a}	8	11^{a}	8	16^{a}	8	16^{a}
25	25^{g}	50^{g}	25	100^{g}	125^{g}	50	50	200^{g}	100	125	50	100	125	100	125	200	100	125
26	2	4^{a}	2	8^{a}	2	12^{a}	2	13^{a}	2	13^{a}	2	13^{a}	2	13^{a}	2	16^{a}	2	13
27	27^{g}	54^{g}	81^{g}	108^{g}	54	81	81	216^{g}	243^{g}	81	81	108	108	81	81	216	216	243
28	5^{s}	7^{a}	7^{a}	8^{a}	7	12^{a}	7	14^{a}	7	9^{a}	7	14^{a}	7	12	11^{a}	16^{a}	7	14^{a}
29	29^{g}	58^{g}	29	116^{g}	29	58	29	232^{g}	29	58	29	116	29	58	31^{q}	232	31	58
30	2	5^{a}	2	6^{a}	2	6^{a}	2	7^{a}	2	6	2	8^{a}	2	6	2	10^{a}	2	7^{a}
31	31^{g}	62^{g}	31	124^{g}	31	62	31	248^{g}	31	62	31	124	31	62	31	248	31	62
32	32^{g}	64^{g}	32	128^{g}	32	64	32	256^{g}	32	64	32	128	32	64	32	512^{g}	32	64
33	6^{s}	6	9^{a}	11^{a}	9^{a}	11^{a}	11^{a}	12^{a}	11^{a}	11	11	18^{a}	11	11	11	22^{a}	11	22^{a}
34	2	4^{a}	2	8^{a}	2	12^{a}	2	16^{a}	2	17^{a}	2	17^{a}	2	17^{a}	2	17^{a}	2	17^{a}
35	5^{a}	7^{a}	7^{a}	10^{a}	7	7	7	14^{a}	7	14^{a}	7	14^{a}	7	14^{a}	7	20^{a}	7	14
36	6^{s}	8^{a}	9^{a}	13^{j}	9^{a}	12^{a}	9	16^{a}	12^{a}	12	9	16^{a}	12	12	12	18^{a}	12	18^{a}
37	37^{g}	74^{g}	37	148^{g}	37	74	37	296^{g}	37	74	37	148	37	74	37	296	37	74
38	2	4^{a}	2	8^{a}	2	12^{a}	2	16^{a}	2	19^{a}	2	19^{a}	2	19^{a}	2	19^{a}	2	19^{a}
39	5^{s}	6^{a}	9^{a}	12^{a}	9^{a}	13^{a}	11^{a}	13^{a}	13^{a}	13^{a}	11	18^{a}	13^{a}	13	13	24^{a}	13	26^{a}
40	8^{s}	10^{s}	8	10	8	10	8	16^{a}	8	16^{a}	8	15^{a}	8	16^{a}	8	20^{a}	8	16
41	41^{g}	82^{g}	41	164^{g}	41	82	41	328^{g}	41	82	41	164	41	82	41	328	41	82
42	2	6^{a}	2	6	2	7^{a}	2	7^{a}	2	7^{a}	2	7	2	7	2	10^{a}	2	7
43	43^{g}	86^{g}	43	172^{g}	43	86	43	344^{g}	43	86	43	172	43	86	43	344	43	86
44	4^{a}	8^{a}	11^{a}	11^{a}	9^{a}	12^{a}	11	16^{a}	11	11	11	16^{a}	11	12	11	22^{a}	11	22^{a}
45	7^{s}	9^{a}	7	10^{a}	9^{a}	10^{a}	9	18^{a}	9	18^{a}	9	20^{a}	9	17^{a}	25^{a}	20^{a}	9	18
46	2	4^{a}	2	8^{a}	2	12^{a}	2	16^{a}	2	20^{a}	2	23^{a}	2	23^{a}	2	23^{a}	2	23^{a}
47	47^{g}	94^{g}	47	188^{g}	47	94	47	376^{g}	47	94	47	188	47	94	47	376	47	94
48	6^{s}	6	9^{a}	12^{a}	9^{a}	16^{a}	11^{a}	16^{a}	16^{a}	16^{a}	11	18^{a}	13^{a}	16	16	24^{a}	16	27^{a}
49	49^{g}	98^{g}	49	196^{g}	49	98	343^{g}	392^{g}	98	98	196	196	98	343	343	392	98	196
50	2	5^{a}	2	8^{a}	2	12^{a}	2	16^{a}	2	20^{a}	2	24^{a}	2	25^{a}	2	25^{a}	2	25^{a}

11.46 Table Lower bounds on k in $(s,k;\lambda)$-difference matrices with $2 \le s \le 50$ and $19 \le \lambda \le 36$.

$s \setminus \lambda$	19	20	21	22	23	24	25	26	27	28	29	30	31	32	33	34	35	36
2	2	40[g]	2	44[g]	2	48[g]	2	52[g]	2	56[g]	2	60[g]	2	64[g]	2	68[g]	2	72[g]
3	27	36[d]	33[t]	30	24	72[g]	27	30	81[g]	42[l]	27	90[g]	30	96[g]	36[i]	51[l]	29[c]	108[g]
4	16	36[t]	36	32	36	96[g]	36	48	108[g]	64[d]	36	56	32	128[g]	132[g]	36	36	144[g]
5	20	100[g]	25	25	35[i]	60[l]	125[g]	50	35	50	25	75[t]	35	80	36[i]	85[l]	85[t]	100[d]
6	2	25[j]	2	30[d]	2	36[t]	2	31[m]	2	42[l]	2	16	2	32[l]	2	16	2	24
7	49	28	49	35	37	63[d]	49	182[g]	63[p]	196[g]	35	91[l]	49	63	49	49	77[t]	126[l]
8	32	32	56	64	56	64	32	36[i]	36[i]	80[d]	56	105[i]	56	256[g]	36	56[i]	63[i]	64
9	49	72	81	54	54	216[g]	81	72	243[g]	108[i]	72	108	54	99[d]	81	306[g]	72	324[g]
10	2	10	2	8	2	27[n]	2	10	2	12	2	12	2	36[j]	2	37[m]	2	19
11	44	81[i]	44	242[g]	44	121[l]	55[p]	44	44	61[q]	46[i]	64[i]	81	64[i]	121	55[i]	121	66[p]
12	9	22[j]	11	30	30	36	9	11	12	21[c]	12	12	22	22	30	30	30	36
13	52	52	52	39	52	121[i]	84	338[g]	65	169[l]	52	65	67[q]	52	52	73[q]	66[i]	84
14	2	14[a]	2	7	2	26[j]	2	27[m]	2	14[a]	2	30[l]	2	16[a]	2	8	2	14
15	7	12[a]	9	10	10	18[a]	10	10	10	29[m]	12	35[t]	12	65[l]	12	52[l]	12	20[a]
16	64	64	64[i]	39	90	128	64	39	64	64	128[i]	90	90	512[g]	64	64	64	96[p]
17	65[i]	85[p]	68	39	65	68	68	39	65	68	68	39	68[i]	289[d]	272	578[g]	68	96[p]
18	2	18[a]	2	9	2	18[a]	2	12	2	18[a]	2	20[a]	2	34[j]	2	12	2	24[a]
19	361[g]	85[p]	76[i]	54	76	76	76	54	76	76	76	54	76	76	76	54	76	289[i]
20	12[i]	19[r]	12[a]	10	10	16[a]	10	10	12	16[a]	10	16[a]	12	20[a]	12	16	12	38[j]
21	7	25[i]	36[t]	11	9	18[a]	9	13[a]	9	14	12	14	9	24[a]	14	14	11	27[a]
22	2	20[a]	2	11	2	24[l]	2	11	2	22[a]	2	11	2	22[a]	2	11	2	22[a]
23	31	92	43[q]	66[i]	529[g]	184	46	66	92	92	46	66	184	184	69[p]	66	92	92
24	8	16[a]	9[a]	23[j]	30[d]	36[t]	8	13[a]	9	16	9	16	12	24[a]	9	16	16	27[a]
25	100	125	125	100	125	200	625[g]	125	100	125	125	125	200[i]	200	200	125	125	125
26	2	20[a]	2	13	2	25[j]	2	13	2	28[l]	2	13	2	26[a]	2	13	2	26[a]
27	81	108	108	108	81	216	216	216	729[g]	108	108	156[p]	108	216	216	243[i]	216	243
28	8	14	12[a]	12	11	16[a]	9	27[j]	20[d]	25[t]	12	14	11	28[a]	12	14	14	28[a]
29	31	116	43[q]	58	47[q]	232	47	58	47	168[i]	841[g]	156[p]	116[i]	232	116	116	232[i]	168
30	2	10[a]	2	6	2	10[a]	2	7	2	29[j]	2	7	2	32[l]	2	7	2	10
31	31	124	43[q]	62	47[q]	248	47	62	47	124	59[q]	156[p]	961[g]	496[p]	496[l]	124	124	186[p]
32	32	128	32	64	32	256	32	64	32	128	32	64	992[g]	1024[g]	69[p]	128[i]	128	186[p]
33	11	22[a]	11	11	11	22[a]	11	13[a]	11	21[a]	11	22[a]	32[j]	30[d]	36[t]	22	11	27[a]
34	2	20[a]	2	17	2	24[a]	2	17	2	28[a]	2	17	2	32[a]	2	17	2	34[a]
35	7	25[a]	7	14	7	18[a]	11[a]	14	7	20[a]	10	18[a]	7	25[a]	11	14	25[a]	20
36	12	16	12	13	12	27[a]	12	16	27[a]	18[a]	12	18[a]	13	32[a]	12	18	30[i]	36[t]
37	37	148	43[q]	74	47[q]	296	47	74	47	148	59[q]	74	59	296	67[q]	74	71[q]	252[i]
38	2	20[a]	2	19	2	24[a]	2	19	2	28[a]	2	19	2	32[a]	2	19	2	37[j]
39	13	26[a]	13	13	13	26[a]	13	13	13	21[a]	13	26[a]	13	26[a]	13	24	13	27[a]
40	8	25[a]	8	16	8	16	8	16	8	20[a]	8	16	8	25[a]	8	16	25[a]	20
41	41	164	43[q]	82	47[q]	328	47	82	47	164	59[q]	82	59	328	67[q]	82	71[q]	164
42	2	11[a]	2	7	2	12[a]	2	7	2	14[a]	2	7	2	14[a]	2	7	2	10
43	43	172	43	86	47[q]	344	47	86	47	172	59[q]	86	59	344	67[q]	86	71[q]	172
44	11	16	11	12	11	22[a]	11	16	12[a]	22[a]	11	21[a]	11	32[a]	12	22	12	36[a]
45	10	25[a]	17[a]	18	18	25[a]	18	18	20	20	17	27[a]	20	25[a]	18	18	25	25[a]
46	2	23[a]	2	23[a]	2	24[a]	2	23	2	28[a]	2	23	2	32[a]	2	23	2	36[a]
47	47	188	47	94	47	376	47	94	47	188	59[q]	94	59	376	67[q]	94	71[q]	188
48	16	30[a]	16	16	16	32[a]	16	16	16	21[a]	16	27[a]	16	32[a]	16	24	16	32[a]
49	196	196	343	343	343	392	196	196	196	343	343	343	343	392	196	196	343	343
50	2	25[a]	2	25[a]	2	25[a]	2	25[a]	2	28[a]	2	25	2	32[a]	2	25	2	36[a]

11.47 Remark Tables 11.45 and 11.46 give lower bounds on the number k of rows in an $(s,k;\lambda)$-difference matrix for small values of s and λ. Authorities for each construction are given as superscripts on the table entries; when a superscript is omitted, the value is obtained from Theorem 11.24; when present, the superscript can be interpreted using Table 11.48. The group G carrying the difference matrix is not recorded.

11.48 Table Key for Tables 11.45 and 11.46. Sporadic examples are found in §II.2 and [1], with the exception of (3,5), (3,7), (4,5), (6,6), and (10,2) due to Jun Meng and D.L. Kreher (private communication).

a	Theorem 11.25
c	Theorem 11.31
d	Theorem 11.32
g	Generalized Hadamard matrix
h	Hadamard matrix
i	Impulse matrix
j	Theorem 11.29
l	Theorem 11.38

m	Theorem 11.44
n	Theorem 11.43
p	Theorem 11.41
q	Theorem 11.42
r	Theorem 11.28
s	Sporadic example
t	Theorem 11.27
w	Theorem 11.37

11.5 Quasi-Difference Matrices

Let G be an abelian group of order n. A $(n,k;\lambda,\mu;u)$-*quasi-difference matrix* (QDM) is a matrix $Q=(q_{ij})$ with k rows and $\lambda(n-1+2u)+\mu$ columns, with each entry either empty (usually denoted by $-$) or containing a single element of G. Each row contains exactly λu empty entries, and each column contains at most one empty entry. Furthermore, for each $1\le i<j\le k$, the multiset

$$\{q_{i\ell}-q_{j\ell}: 1\le \ell\le \lambda(n-1+2u)+\mu, \text{ with } q_{i\ell} \text{ and } q_{j\ell} \text{ not empty}\}$$

contains every nonzero element of G λ times, and contains 0 μ times.

Let q be a prime power and let $q=mt+1$ for m,t integer. Let ω be a primitive element of $\mathbb{F}_q$. A *$V(m,t)$ vector* is a vector $(a_1,\ldots,a_{m+1})$ for which, for each $1\le k<m$, the differences $\{a_{i+k}-a_i: 1\le i\le m+1, i+k\ne m+2\}$ represent the m cyclotomic classes of $\mathbb{F}_{mt+1}$ (compute subscripts modulo $m+2$). In other words, for fixed k, if $a_{i+k}-a_i=\omega^{mx+\alpha}$ and $a_{j+k}-a_j=\omega^{my+\beta}$, then $\alpha\not\equiv\beta \pmod m$.

11.49 Examples A (3,4;1,0;1)-QDM and an (11,4;1,0;3)-QDM (see [1]).

–	0	1	2
0	–	2	1
1	2	–	0
2	1	0	–

0	6	4	1	–	2	6	0	–	8	9	0	–	1	4	0
1	0	6	4	0	–	2	6	0	–	8	9	0	–	1	4
4	1	0	6	6	0	–	2	9	0	–	8	4	0	–	1
6	4	1	0	2	6	0	–	8	9	0	–	1	4	0	–

11.50 Examples $(19,7;\lambda,\lambda;1)$-QDM for $\lambda=2,3$. Let A and C be the circulant matrices with first rows (– 4 9 6 6 9 4) and (0 2 14 11 3 7 1), respectively. Then $(A\mid 2A\mid C\mid -C\mid 7C\mid -7C)$ is a $(19,7;2,2;1)$-QDM, and $(3A\mid 6A\mid 8A\mid C\mid -C\mid 3C\mid -3C\mid 4C\mid -4C)$ is a $(19,7;3,3;1)$-QDM.

11.51 Construction [20] If a $(n,k;\lambda,\mu;u)$-QDM exists and $\mu\le\lambda$, then an $\mathrm{ITD}_\lambda(k,n+u;u)$ exists. Start with a $(n,k;\lambda,\mu;u)$-QDM A over the group G. Append $\lambda-\mu$ columns of zeroes. Then select u elements $\infty_1,\ldots,\infty_u$ not in G, and replace the empty entries $(-)$, each by one of these infinite symbols, so that ∞_i appears exactly once in each row, for $1\le i\le u$. Develop the resulting matrix over the group G (leaving infinite symbols fixed), to obtain a $k\times\lambda(n^2+2nu)$ matrix T. Then T is an incomplete orthogonal array with k rows and index λ, having $n+u$ symbols and one hole of size u. The technique in Remark II.2.6 then gives an $\mathrm{ITD}_\lambda(k,n+u;u)$.

11.52 Remark $V(m,t)$ vectors are useful in the construction of quasi-difference matrices via Construction 11.53.

11.53 Construction Starting with a $V(m,t)$ vector $(a_1,\ldots,a_{m+1})$, form a single column of length $m+2$ whose first entry is empty, and whose remaining entries are $(a_1,\ldots,a_{m+1})$. Form t columns by multiplying this column by the powers of ω^m. From each of these t columns, form $m+2$ columns by taking the $m+2$ cyclic shifts of the column. The result is a $(q, m+2; 1, 0; t)$-QDM.

11.54 Theorem [4] A $V(m,t)$ vector exists if m and t are not both even,

1. whenever $m = 2$ and $mt+1$ is a prime power;
2. whenever $m = 3$ and $mt+1$ is a prime;
3. whenever $mt+1 \le 5000$, $m-1 \le t$, $m \le 10$ and $mt+1$ is prime, except when $m=9$ and $t=8$, as no $V(9,8)$ exists;
4. whenever $mt+1 \le 5000$, $m-1 \le t$, $m \le 6$ and $mt+1$ is a prime power, except when $m=3$ and $t=5$, as no $V(3,5)$ exists.

11.55 Table Some $V(m,t)$ vectors with $3 \le m \le 10$, $mt+1$ prime.

Solutions for $V(m,t)$, $m=3$, $v=mt+1$

t	v	Vector	t	v	Vector	t	v	Vector
2	7	(0 1 3 6)	4	13	(0 1 3 9)	6	19	(0 1 3 7)
10	31	(0 1 4 13)	12	37	(0 1 3 10)	14	43	(0 1 4 13)
20	61	(0 1 3 13)	22	67	(0 1 3 7)	24	73	(0 1 3 15)
26	79	(0 1 3 8)	32	97	(0 1 3 9)	34	103	(0 1 3 7)

Solutions for $V(m,t)$, $m=4$, $v=mt+1$

t	v	Vector	t	v	Vector	t	v	Vector
3	13	(0 1 3 7 2)	7	29	(0 1 3 7 19)	9	37	(0 1 3 2 8)
13	53	(0 1 3 7 19)	15	61	(0 1 3 7 5)	25	101	(0 1 3 2 31)

Solutions for $V(m,t)$, $m=5$, $v=mt+1$

t	v	Vector	t	v	Vector	t	v	Vector
6	31	(0 1 3 7 30 17)	8	41	(0 1 3 22 14 18)	12	61	(0 1 3 7 23 50)
14	71	(0 1 3 9 25 54)	20	101	(0 1 3 10 43 91)	26	131	(0 1 3 6 48 15)

Solutions for $V(m,t)$, $m=6$, $v=mt+1$

t	v	Vector	t	v	Vector
5	31	(0 1 7 30 12 21 15)	7	43	(0 1 3 16 35 26 36)
11	67	(0 1 3 14 7 24 27)	13	79	(0 1 3 7 55 47 34)
17	103	(0 1 3 2 14 99 29)	21	127	(0 1 4 13 66 93 45)

Solutions for $V(m,t)$, $m=7$, $v=mt+1$

t	v	Vector	t	v	Vector
6	43	(0 1 12 27 37 16 30 35)	10	71	(0 1 3 45 9 50 28 16)
16	113	(0 1 3 7 82 72 93 39)	18	127	(0 1 3 6 97 114 99 26)

Solutions for $V(m,t)$, $m=8$, $v=mt+1$

t	v	Vector	t	v	Vector
9	73	(0 1 20 70 23 59 3 8 19)	11	89	(0 1 6 56 22 35 47 23 60)
17	137	(0 1 3 2 133 126 47 109 74)	29	233	(0 1 4 11 94 60 85 16 198)

Solutions for $V(m,t)$, $m=9$, $v=mt+1$

t	v	Vector	t	v	Vector
8	73	none	12	109	(0 1 4 19 56 22 83 95 52 96)
14	127	(0 1 11 25 37 8 100 23 95 42)	18	163	(0 1 3 7 36 30 158 94 52 70)
20	181	(0 1 3 19 145 70 173 159 18 85)	22	199	(0 1 3 31 99 190 174 46 87 127)
30	271	(0 1 3 8 197 68 119 13 215 105)	34	307	(0 1 3 13 140 81 74 131 303 238)
42	379	(0 1 3 6 66 258 186 346 104 152)	44	397	(0 1 4 11 144 103 216 77 160 363)

Solutions for $V(m,t)$, $m=10$, $v=mt+1$

t	v	Vector	t	v	Vector
13	131	(0 1 5 10 22 6 14 9 53 129 84)	15	151	(0 1 45 146 51 97 70 137 85 133 18)
19	191	(0 1 3 96 143 156 182 142 4 189 25)	21	211	(0 1 6 188 205 39 101 113 30 32 42)
25	251	(0 1 3 85 140 178 195 22 48 179 188)	27	271	(0 1 3 82 109 241 36 112 141 263 126)
31	311	(0 1 3 57 128 247 289 239 70 271 96)	33	331	(0 1 3 67 319 44 249 146 302 282 90)
43	431	(0 1 6 29 170 207 385 290 375 32 336)	49	491	(0 1 3 8 406 72 335 197 324 383 395)

11.6 See Also

§II.2	Difference matrices generate transversal designs.
§II.3	Quasi-difference matrices generate incomplete transversal designs.
§II.4	Difference and quasi-difference matrices generate orthogonal arrays and incomplete orthogonal arrays.
§III.4	Linear spaces are used to construct difference matrices.
§IV.4	A GH(g,λ) over G is a GBRD$(\lambda g, \lambda g, \lambda g; G)$. A $(g,k;\lambda)$-difference matrix over G is a GBRD$(k,k,g\lambda;G)$.
[1]	Develops the theory of difference and quasi-difference matrices and gives numerous examples.
[16]	Studies difference matrices under the name "difference schemes" and gives connections to orthogonal arrays.

References

[1] T. Beth, D. Jungnickel and H. Lenz, *Design Theory*, Cambridge University Press, London, 1986.

[2] B.W. Brock, Hermitian congruence and the existence and completion of generalized Hadamard matrices, *J. Combin. Theory A* 49 (1988), 233-261.

[3] B.W. Brock, A new construction of circulant GH$(p^2;\mathbb{Z}_p)$s, *Discrete Math.* 112 (1993), 249-252.

[4] C.J. Colbourn, Construction techniques for mutually orthogonal latin squares, in: *Combinatorics Advances*, Kluwer Academic Press, to appear.

[5] C.J. Colbourn and D.L. Kreher, Concerning difference matrices, *Des. Codes Crypt.*, to appear.

[6] W. de Launey, On the non-existence of generalised weighing matrices, *Ars Combin.* 17A (1984), 117-132.

[7] W. de Launey, A survey of generalized Hadamard matrices and difference matrices $D(k,\lambda;G)$ with large k, *Utilitas Math.* 30 (1986), 5-29.

[8] W. de Launey, $(0,G)$-Designs and Applications, Ph.D. thesis, University of Sydney (1987).

[9] W. de Launey, On difference matrices, transversal designs, resolvable transversal designs, and large sets of mutually orthogonal F-squares, *J. Stat. Plann. Infer.* 16 (1987), 107-125.
[10] W. de Launey, Square GBRDs over non-abelian groups, *Ars Combin.* 27 (1989), 40-49.
[11] W. de Launey, Generalized Hadamard matrices which are developed modulo a group, *Discrete Math.* 104 (1992), 49-65.
[12] W. de Launey, Circulant GH($p^2; \mathbb{Z}_p$) exist for all primes p, *Graphs Combin.* 8 (1992), 317-321.
[13] W. de Launey and J.E. Dawson, A note on the construction of GH($4tq$; EA(q)) for $t = 1, 2$, *Australas. J. Combin.* 6 (1992), 177-186.
[14] W. de Launey and J.E. Dawson, An asymptotic result on the existence of generalized Hadamard matrices, *J. Combin. Theory A* 65 (1992), 158-163.
[15] D.A. Drake, Partial λ-geometries and generalized Hadamard matrices over groups, *Canad. J. Math.* 31 (1979), 617-627.
[16] A.S. Hedayat, N.J.A. Sloane and J. Stufken, *Orthogonal Arrays*, to appear.
[17] D. Jungnickel, On difference matrices, resolvable transversal designs and generalized Hadamard matrices, *Math. Zeitschr.* 167 (1979), 49-60.
[18] P.J. Schellenberg, G.H.J. van Rees and S.A. Vanstone, Four pairwise orthogonal Latin squares of order 15, *Ars Combin.* 6 (1978), 141-150.
[19] S.S. Shrikhande, Generalized Hadamard matrices and orthogonal arrays of strength 2, *Canad. J. Math.* 16 (1964), 131-141.
[20] R.M. Wilson, A few more squares, *Congr. Numer.* 10 (1974), 675-680.

12 Difference Sets: Abelian

Dieter Jungnickel
Alexander Pott

12.1 Basics

Let G be an additively written group of order v. A k-subset D of G is a $(v, k, \lambda; n)$-*difference set* of *order* $n = k - \lambda$ if every nonzero element of G has exactly λ representations as a difference $d - d'$ with elements from D. The difference set is *abelian*, *cyclic* etc. if the group G has the respective property.

12.1 Example The set $\{1, 3, 4, 5, 9\}$ is a $(11, 5, 2; 3)$-difference set in the group $\mathbb{Z}_{11}$.

12.2 Remarks

1. This definition holds for nonabelian as well as abelian groups. Only abelian groups are treated here. See §IV.13 for the nonabelian case.
2. The group G itself and $G\backslash\{g\}$ for an arbitrary $g \in G$ are $(v, v, v, 0)$- and $(v, v - 1, v - 2; 1)$-difference sets. In the following tables these *trivial* difference sets are excluded.
3. The complement of a $(v, k, \lambda; n)$-difference set is again a difference set with parameters $(v, v - k, v - 2k + \lambda; n)$. Therefore, assume $k \leq v/2$ (the case $k = v/2$ is actually impossible).

The *development* of a difference set D is the incidence structure $dev(D)$ whose points are the elements of G and whose blocks are the translates $D+g=\{d+g : d \in D\}$

12.3 Theorem [3] The existence of a $(v,k,\lambda;n)$-difference set is equivalent to the existence of a symmetric (v,k,λ)-design $\mathcal{D}$ admitting G as a point regular automorphism group; i.e., for any two points p and q, there is a unique group element g which maps p to q. The design $\mathcal{D}$ is isomorphic with dev(D). The design is cyclic when the difference set is cyclic.

12.4 Remarks

1. Many symmetric designs do not have difference set representations.
2. Since difference sets correspond to symmetric designs, the parameters v, k and λ must satisfy the trivial necessary conditions for the existence of a symmetric design ($\lambda(v-1)=k(k-1)$) as well as the Bruck-Ryser-Chowla condition.
3. The parameters of difference sets must also satisfy some further necessary conditions. See [6, 10] for a discussion of these additional conditions.

Two difference sets D_1 (in G_1) and D_2 (in G_2) are *equivalent* if there is a group isomorphism φ between G_1 and G_2 such that $D_1^{\varphi}=\{d^{\varphi} : d \in D_1\}=D_2+g$ for a suitable $g \in G$. The difference sets are *isomorphic* if the designs dev(D_1) and dev(D_2) are isomorphic.

12.5 Remarks

1. Equivalent difference sets are isomorphic; however, there are examples of isomorphic difference sets which are not equivalent.
2. In the cyclic case, in all known examples isomorphic difference sets are also equivalent.

A *multiplier* of a difference set D in G is an automorphism φ of G such that $D^{\varphi}=D+g$ for some $g \in G$. If φ is the automorphism which maps h to $t \cdot h$, then t is a *numerical* multiplier.

12.6 Theorem [3] If φ is a multiplier of the difference set D, then there is at least one translate $D+g$ of D which is fixed by φ.

12.7 Remark The content of the multiplier theorems is the assertion that certain automorphisms (integers) have to be (numerical) multipliers depending only on the parameters v, k and λ. Theorem 12.8 is the *first multiplier theorem.*

12.8 Theorem [6, 10] Let D be an abelian $(v,k,\lambda;n)$-difference set. If p is a prime which satisfies $(p,v)=1$, $p|n$ and $p>\lambda$, then p is a numerical multiplier.

12.9 Conjecture Every prime divisor p of n which is relatively prime to v is a multiplier of a $(v,k,\lambda;n)$-difference set, i.e. the condition that $p>\lambda$ in Theorem 12.8 is unnecessary.

12.10 Remarks

1. Multipliers are quite useful in constructing difference sets and in proofs of nonexistence. In particular if $\lambda=1$ it is conjectured that n must be a prime power; this holds for all values of $n \leq 2,000,000$; see [5].
2. Several attempts have been made to weaken the assumption "$p>\lambda$" in Theorem 12.8 (2nd multiplier theorem, McFarland's multiplier theorem; see [3, 6, 10]).

12.2 Difference Sets with Classical Parameters

Let α be a generator of the multiplicative group of $\mathbb{F}_{q^{d+1}}$. Then the set of integers $\{i : 0 \le i < (q^{d+1}-1)/(q-1), \mathrm{trace}_{(d+1)/1}(\alpha^i) = 0\}$ modulo $(q^{d+1}-1)/(q-1)$ form a (cyclic) difference set with parameters

$$\left(\tfrac{q^{d+1}-1}{q-1}, \tfrac{q^{d}-1}{q-1}, \tfrac{q^{d-1}-1}{q-1}; q^{d-1}\right).$$

Here the *trace* denotes the usual trace function $\mathrm{trace}_{(d+1)/1}(\beta) = \sum_{i=0}^{d} \beta^{q^i}$ from $\mathbb{F}_{q^{d+1}}$ onto $\mathbb{F}_q$. The designs corresponding to these *Singer difference sets* are the point-hyperplane designs of the projective geometry $\mathrm{PG}(d,q)$; see [6]. If s divides $d+1$ and r is relatively prime to q^s-1, then the set of integers $\{i : 0 \le i < (q^{d+1}-1)/(q-1), \mathrm{trace}_{s/1}(\mathrm{trace}_{(d+1)/s}(\alpha^i)^r) = 0\}$ is another, in general not equivalent, cyclic difference set with the same parameters as a Singer difference set. These difference sets are the *Gordon-Mills-Welch difference sets*; see [2].

12.11 Table A complete list (up to equivalence) of cyclic $(2^{d+1}-1, 2^d-1, 2^{d-1}-1)$-difference sets with $d \le 8$ (see [11]). All of these difference sets are fixed by the multiplier 2; see Theorem 12.6. The table contains only one representative from each orbit under the multiplier group generated by the multiplier 2, i.e. the first difference set is $\{1, 2\cdot 1, 2^2\cdot 1\}$. A "?" indicates that the difference set is not (yet) a member of an infinite class of difference sets. Paley and Hall difference sets are explained in Example 12.15. The difference sets in this table describe nonisomorphic designs. The cases $d \ge 9$ are still open.

d	Difference Set	Comment
2	1	Singer
3	0 1 5	Singer
4	1 3 15	Singer
	1 5 7	Paley
5	0 1 3 7 9 13 27	Singer
	0 1 3 5 9 23 27	GMW
6	1 3 7 9 15 27 29 31 47	Singer
	1 9 11 13 15 19 21 31 47	Paley
	1 3 5 7 19 23 27 47 63	Hall
	1 3 7 9 13 19 27 31 47	?
	1 3 5 9 15 19 27 29 55	?
	1 3 5 11 19 21 27 29 63	?
7	0 1 3 7 13 17 19 23 25 27 31 37 45 51 59 85 111 119 127	Singer
	0 7 11 13 17 19 23 27 39 43 47 51 53 55 63 85 95 111 119	GMW
	0 1 3 7 13 15 17 25 29 37 39 43 47 51 53 61 63 85 119	?
	0 7 9 11 15 17 19 25 29 31 37 43 51 53 61 63 85 87 119	?
8	1 7 13 17 21 23 31 35 37 39 51 53 55 59 61 75 77 79 83 85 91 95 103 109 123 183 187 219 223	Singer
	1 5 23 27 29 31 37 39 41 47 55 59 63 75 77 79 85 87 95 109 117 123 171 175 183 191 219 223 255	GMW
	1 3 5 7 13 15 19 21 23 25 27 37 39 45 55 57 59 63 77 83 91 107 117 123 183 191 219 239 255	?
	1 3 5 7 9 11 17 29 31 43 53 55 57 61 63 75 77 91 95 109 117 123 125 171 175 183 219 239 255	?
	1 3 5 7 9 11 17 29 31 43 53 55 57 61 63 75 77 91 95 109 117 123 125 171 175 183 219 239 255	?

12.12 Remark There are even more difference sets with the parameters of the Singer difference sets but which are not equivalent (isomorphic) to them; see Table 12.22. However, the only known infinite family of such difference sets are the Gordon-Mills-Welch difference sets. The other examples seem to be sporadic.

12.13 Proposition If D is a Singer difference set for $\mathrm{PG}(2,q)$, then $-D$ is an oval in the corresponding projective plane.

12.3 Other Series of Difference Sets

Let q be a prime power and d a positive integer. Let G be an abelian group of order $v = q^{d+1}(q^d + \ldots + q^2 + q + 2)$ which contains an elementary abelian subgroup E of order q^{d+1}. View E as the additive group of $\mathbb{F}_q^{d+1}$. Put $r = (q^{d+1} - 1)/(q - 1)$ and let $H_1, \ldots, H_r$ be the hyperplanes of order q^d of E. If $g_0, \ldots, g_r$ are distinct coset representatives of E in G then $D = (g_1 + H_1) \cup (g_2 + H_2) \cup \ldots \cup (g_r + H_r)$ is a *McFarland difference set* (see [6]) denoted $\mathrm{MF}(q,d)$ with parameters

$$\left(q^{d+1}(1 + \tfrac{q^{d+1}-1}{q-1}), q^d \cdot \tfrac{q^{d+1}-1}{q-1}, q^d \cdot \tfrac{q^d-1}{q-1}; q^{2d}\right)$$

Let E be the elementary abelian group of order 3^{d+1} and let G be a group of order $v = 3^{d+1}(3^{d+1}-1)/2$ containing E. Put $m = (3^{d+1}-1)/2$ and let $H_1, \ldots, H_m$ denote the subgroups of E of order 3^d. If $g_1, \ldots, g_m$ are distinct coset representatives of E in G, then $D = (g_1 + (E \backslash H_1)) \cup (g_2 + H_2) \cup (g_3 + H_3) \cup \ldots \cup (g_m + H_m)$ is a *Spence difference set* (see [6]) denoted $\mathrm{Sp}(d)$. It has parameters

$$\left(3^{d+1} \cdot \tfrac{3^{d+1}-1}{2}, 3^d \cdot \tfrac{3^{d+1}+1}{2}, 3^d \cdot \tfrac{3^d+1}{2}; 3^{2d}\right).$$

12.14 Remark There are many inequivalent McFarland and Spence difference sets.

12.15 Example The following subsets of $\mathbb{F}_q$ are difference sets in the additive subgroup of $\mathbb{F}_q$:

- $\mathbb{F}_q^{(2)} = \{x^2 : x \in \mathbb{F}_q \backslash \{0\}\}, q \equiv 3 \pmod 4$ (quadratic residues, *Paley difference sets*);
- $\mathbb{F}_q^{(4)} = \{x^4 : x \in \mathbb{F}_q \backslash \{0\}\}, q = 4t^2 + 1, t$ odd;
- $\mathbb{F}_q^{(4)} \cup \{0\}, q = 4t^2 + 9, t$ odd;
- $\mathbb{F}_q^{(8)} = \{x^8 : x \in \mathbb{F}_q \backslash \{0\}\}, q = 8t^2 + 1 = 64u^2 + 9, t, u$ odd;
- $\mathbb{F}_q^{(8)} \cup \{0\}, q = 8t^2 + 49 = 64u^2 + 441, t$ odd, u even;
- $H(q) = \{x^i : x \in \mathbb{F}_q \backslash \{0\}, i \equiv 0, 1 \text{ or } 3 \pmod 6\}$ $q = 4t^2 + 27, q \equiv 1 \pmod 6$ (*Hall difference sets*).

These are *cyclotomic difference sets.*

12.16 Remark More examples of difference sets related to cyclotomic cosets are given in [6].

Let q and $q + 2$ be prime powers. Then the set $D = \{(x, y) : x, y$ are both nonzero squares or both nonsquares or $y = 0\}$ is a *twin prime power difference set* ($\mathrm{TPP}(q)$) (see [6] Example 9.2) with parameters

$$\left(q^2 + 2q, \tfrac{q^2+2q-1}{2}, \tfrac{q^2+2q-3}{4}; \tfrac{q^2+2q+1}{4}\right)$$

in the group $(\mathbb{F}_q, +) \oplus (\mathbb{F}_{q+2}, +)$.

12.17 Remark The Paley difference sets and the twin prime power difference sets are examples of difference sets D with $v = 4n - 1$. In these cases, $\mathrm{dev}(D)$ is a Hadamard design.

A difference set with parameters $(4u^2, 2u^2 - u, u^2 - u; u^2)$ is a *Menon-Hadamard difference set.*

12.18 Theorem [4] Let $G \cong H \times EA(w^2)$ be an abelian group of order $4u^2$ with $u = 2^a 3^b w^2$ where w is the product of not necessarily distinct primes $p \equiv 3 \pmod 4$ and $EA(w^2)$ denotes the group of order w^2 which is the direct product of groups of prime order. If H is of type $(2^{a_1})(2^{a_2})\dots(2^{a_s})(3^{b_1})^2\dots(3^{b_r})^2$ with $\sum a_i = 2a+2$ $(a \geq 0,\ a_i \leq a+2)$, $\sum b_i = 2b$ $(b \geq 0)$, then G contains a Menon-Hadamard difference set of order u^2.

12.19 Theorem [4] An abelian group of order 2^{2a+2} contains a Menon-Hadamard difference set if and only if the exponent of G is at most 2^{a+2}.

12.20 Remark Many constructions of Menon-Hadamard difference sets are known. Theorem 12.21 is one such construction.

12.21 Theorem [6] Let $H = \langle a, b\colon\ a^{s+1} = b^{s+1} = 1\rangle$ be an abelian group of type $(2^{s+1})(2^{s+1})$. Let f be a mapping $\mathbb{Z}_{2^{s+1}} \to \{\pm 1\}$ satisfying $f(i + 2^s) = -f(i)$. Define a mapping $\mu : \mathbb{Z}_{2^{s+1}} \to \mathbb{Z}_{2^{s+1}}$ by $\mu(2^r i) = 2^r i^*$, where i is odd and $ii^* \equiv 1 \pmod{2^{s+1}}$. Then the set $D = \{a^i b^j : f(\mu(i)j) = -1\}$ is a Menon-Hadamard difference set with $u = 2^s$. Let $G = \langle a^2, c : c^2 = b\rangle$ be an abelian group of type $(2^s)(2^{s+2})$. If $A = D \cap \langle a^2\rangle\langle b\rangle$ and $B = a^{-1}(D \backslash A)$, then $A \cup cB$ is a Menon-Hadamard difference set with $u = 2^s$ in G. These difference sets are denoted D(s).

12.4 Tables of Abelian Difference Sets

12.22 Table Abelian difference sets of order $n \leq 30$. In this table, one difference set for each abelian group is listed. In general, there are many more examples! There are no other groups or parameters with $n \leq 30$ for which the existence of a difference set is undecided (see Table 12.26). In the column "group" the decomposition of the group is given as a product of cyclic subgroups. If the group is cyclic, the integers modulo the group order are used to describe the difference set.

n	v	k	λ	Group	Difference Set	Comment
2	7	3	1	(7)	1 2 4	PG(2, 2)
3	13	4	1	(13)	0 1 3 9	PG(2, 3)
3	11	5	2	(11)	1 3 4 5 9	Paley
4	21	5	1	(21)	3 6 7 12 14	PG(2, 4)
4	16	6	2	(8)(2)	(00) (10) (11) (20) (40) (61)	MF(2, 1)
				$(4)^2$	(00) (01) (10) (12) (20) (23)	MF(2, 1)
				$(4)(2)^2$	(000) (010) (100) (101) (200) (211)	MF(2, 1)
				$(2)^4$	(0000) (0010) (1000) (1001) (1100) (1111)	MF(2, 1)
4	15	7	3	(3)(5)	0 1 2 4 5 8 10	PG(3, 2)
5	31	6	1		1 5 11 24 25 27	PG(2, 5)
5	19	9	4		1 4 5 6 7 9 11 16 17	Paley
6	23	11	5	(23)	1 2 3 4 6 8 9 12 13 16 18	Paley
7	57	8	1	(3)(19)	1 6 7 9 19 38 42 49	PG(2, 7)
7	27	13	6	$(3)^3$	(001) (011) (021) (111) (020) (100) (112) (120) (121) (122) (201) (202) (220)	Paley
8	73	9	1	(73)	1 2 4 8 16 32 37 55 64	PG(2, 8)
8	31	15	7	(31)	1 2 3 4 6 8 12 15 16 17 23 24 27 29 30	PG(4, 2)
9	91	10	1	(7)(13)	0 1 3 9 27 49 56 61 77 81	PG(2, 9)
9	45	12	3	$(3)^2(5)$	(000) (001) (002) (003) (010) (020) (101) (112) (123) (201) (213) (222)	MF(3, 1)

n	v	k	λ	Group	Difference Set	Comment
9	40	13	4	(8)(5)	1 2 3 5 6 9 14 15 18 20 25 27 35	PG(3, 3)
9	36	15	6	$(4)(3)^2$	(010) (011) (012) (020) (021) (022) (100) (110) (120) (200) (211) (222) (300) (312) (321)	Sp(1)
				$(2)^2(3)^2$	(0010) (0011) (0012) (0020) (0021) (0022) (0100) (0110) (0120) (1000) (1011) (1022) (1100) (1112) (1121)	Sp(1)
9	35	17	8	(5)(7)	0 1 3 4 7 9 11 12 13 14 16 17 21 27 28 29 33	TPP(5)
11	133	12	1	(7)(19)	1 11 16 40 41 43 52 60 74 78 121 128	PG(2, 11)
11	43	21	10	(43)	1 4 6 9 10 11 13 14 15 16 17 21 23 24 25 31 35 36 38 40 41	Paley
12	47	23	11	(47)	1 2 3 4 6 7 8 9 12 14 16 17 18 21 24 25 27 28 32 34 36 37 42	Paley
13	183	14	1	(3)(61)	0 2 3 10 26 39 43 61 109 121 130 136 141 155	PG(2, 13)
15	59	29	14	(59)	1 3 4 5 7 9 12 15 16 17 19 20 21 22 25 26 27 28 29 35 36 41 45 46 48 49 51 53 57	Paley
16	273	17	1	(3)(7)(13)	1 2 4 8 16 32 64 91 117 128 137 182 195 205 234 239 256	PG(2, 16)
16	96	20	4	$(2)(4)^2(3)$	(0000) (0001) (0002) (0020) (0101) (0031) (0132) (0200) (0212) (0220) (0221) (0302) (1012) (1101) (1132) (1201) (1211) (1222) (1231) (1322)	[1]
				$(2)^3(4)(3)$	(00001) (00002) (00010) (00011) (00012) (00021) (00101) (00121) (01002) (01030) (01131) (01112) (10002) (10031) (10110) (10132) (11002) (11032) (11130) (11111)	MF(4, 1)
				$(2)^5(3)$	(000001) (000002) (000010) (000011) (000012) (000101) (001001) (001101) (010002) (010110) (011012) (011111) (100002) (100111) (101010) (101112) (110002) (110112) (111011) (111110)	MF(4, 1)
16	85	21	5	(5)(17)	0 1 2 4 7 8 14 16 17 23 27 28 32 34 43 46 51 54 56 64 68	PG(3, 4)
16	64	28	12	(4)(16)	(0 9) (0 11) (0 13) (0 15) (1 4) (1 5) (1 6) (1 9) (1 11) (1 12) (1 14) (1 15) (2 2) (2 3) (2 6) (2 7) (2 9) (2 10) (2 13) (2 14) (3 2) (3 3) (3 4) (3 5) (3 7) (3 9) (3 10) (3 12)	D(2)
				$(2)^2(16)$	(0 0 1) (0 0 2) (0 0 3) (0 0 4) (0 0 5) (0 0 6) (0 0 7) (0 0 10) (0 0 12) (0 0 14) (0 1 1) (0 1 4) (0 1 5) (0 1 11) (0 1 12) (0 1 15) (1 0 1) (1 0 2) (1 0 3) (1 0 10) (1 0 13) (1 0 15) (1 1 1) (1 1 6) (1 1 7) (1 1 11) (1 1 13) (1 1 14)	MF(2, 2)
				$(8)^2$	(14) (15) (16) (17) (22) (23) (26) (27) (32) (34) (35) (37) (41) (43) (45) (47) (51) (53) (54) (56) (61) (62) (65) (66) (71) (72) (73) (74)	D(2)
				$(2)^3(8)$	(0001) (0002) (0003) (0004) (0005) (0006) (0007) (0012) (0014) (0016) (0101) (0104) (0105) (0113) (0114) (0117) (1001) (1002) (1003) (1012) (1015) (1017) (1101) (1106) (1107) (1113) (1115) (1116)	MF(2, 2)

n	v	k	λ	Group	Difference Set	Comment
				$(4)^3$	(001) (010) (011) (012) (021) (033) (100) (101) (102) (110) (111) (112) (120) (121) (122) (133) (201) (210) (211) (212) (221) (233) (303) (313) (323) (330) (331) (332)	MF(2,2)
				$(2)^2(4)^2$	(0001) (0002) (0003) (0010) (0011) (0012) (0013) (0021) (0022) (0023) (0102) (0110) (0112) (0122) (0131) (0133) (1001) (1010) (1011) (1021) (1032) (1033) (1103) (1110) (1113) (1123) (1131) (1132)	MF(2,2)
				$(2)^4(4)$	(00001) (00002) (00003) (00010) (00011) (00012) (00013) (00101) (00102) (00103) (01002) (01010) (01012) (01102) (01111) (01113) (10001) (10010) (10011) (10101) (10112) (10113) (11003) (11010) (11013) (11103) (11111) (11112)	MF(2,2)
16	64	28	12	$(2)^6$	(000001) (000010) (000011) (000100) (000101) (000110) (000111) (001010) (001100) (001110) (010001) (010100) (010101) (011011) (011100) (011111) (100001) (100010) (100011) (101010) (101101) (101111) (110001) (110110) (110111) (111011) (111101) (111110)	MF(2,2)
16	63	31	15	(9)(7)	0 1 2 3 4 6 7 8 9 12 13 14 16 18 19 24 26 27 28 32 33 35 36 38 41 45 48 49 52 54 56	PG(5,2)
				$(3)^2(7)$	(000) (001) (002) (003) (004) (005) (006) (011) (012) (014) (021) (022) (024) (103) (105) (106) (113) (115) (116) (121) (122) (124) (203) (205) (206) (211) (212) (214) (223) (225) (226)	TPP(7)
17	307	18	1	(307)	0 1 3 30 37 50 55 76 98 117 129 133 157 189 199 222 293 299	PG(2,17)
17	67	33	16	(67)	1 4 6 9 10 14 15 16 17 19 21 22 23 24 25 26 29 33 35 36 37 39 40 47 49 54 55 56 59 60 62 64 65	Paley
18	71	35	17	(71)	1 2 3 4 5 6 8 9 10 12 15 16 18 19 20 24 25 27 29 30 32 36 37 38 40 43 45 48 49 50 54 57 58 60 64	Paley
19	381	20	1	(3)(127)	0 1 19 28 96 118 151 153 176 202 240 254 290 296 300 307 337 361 366 369	PG(2,19)
19	101	25	6	(101)	1 5 16 19 24 25 31 36 37 52 54 56 58 68 71 78 79 80 81 84 87 88 92 95 97	cyclotomy
20	79	39	19	(79)	1 2 4 5 8 9 10 11 13 16 18 19 20 21 22 23 25 26 31 32 36 38 40 42 44 45 46 49 50 51 52 55 62 64 65 67 72 73 76	Paley
21	109	28	7	(109)	0 1 3 5 7 9 15 16 21 22 25 26 27 35 38 45 48 49 63 66 73 75 78 80 81 89 97 105	$\mathbb{F}_{109}^{(4)}$
21	83	41	20	(83)	1 3 4 7 9 10 11 12 16 17 21 23 25 26 27 28 29 30 31 33 36 37 38 40 41 44 48 49 51 59 61 63 64 65 68 69 70 75 77 78 81	Paley

n	v	k	λ	Group	Difference Set	Comment
23	553	24	1	(7)(79)	1 23 52 90 108 120 152 163 173 178 186 223 232 272 359 407 411 431 438 512 513 515 529 548	PG(2, 23)
25	651	26	1	(3)(7)(31)	1 5 25 42 71 107 125 201 210 217 354 355 357 387 399 412 434 462 468 473 483 521 535 561 625 633	PG(2, 25)
25	175	30	5	$(5)^2(7)$	(001) (002) (003) (004) (005) (006) (011) (021) (031) (041) (106) (112) (124) (133) (145) (206) (213) (222) (235) (244) (306) (314) (325) (332) (343) (406) (415) (423) (434) (442)	MF(5, 1)
25	156	31	6	(4)(3)(13)	0 1 5 11 13 25 28 39 46 55 58 65 68 74 76 86 87 91 111 117 118 119 122 123 125 127 134 140 142 143 147	PG(3, 5)
25	133	33	8	(7)(19)	1 4 5 14 16 19 20 21 25 38 54 56 57 64 66 70 76 80 83 84 91 93 95 98 100 101 105 106 114 123 125 126 131	
25	99	49	24	$(3)^2(11)$	(0 0 0) (0 1 0) (0 1 1) (0 1 3) (0 1 4) (0 1 5) (0 1 9) (0 2 0) (0 2 1) (0 2 3) (0 2 4) (0 2 5) (0 2 9) (1 0 0) (1 0 2) (1 0 6) (1 0 7) (1 0 8) (1 0 10) (1 1 0) (1 1 2) (1 1 6) (1 1 7) (1 1 8) (1 1 10) (1 2 0) (1 2 1) (1 2 3) (1 2 4) (1 2 5) (1 2 9) (2 0 0) (2 0 2) (2 0 6) (2 0 7) (2 0 8) (2 0 10) (2 1 0) (2 1 1) (2 1 3) (2 1 4) (2 1 5) (2 1 9) (2 2 0) (2 2 2) (2 2 6) (2 2 7) (2 2 8) (2 2 10)	TPP(9)
26	103	51	25	(103)	1 2 4 7 8 9 13 14 15 16 17 18 19 23 25 26 28 29 30 32 33 34 36 38 41 46 49 50 52 55 56 58 59 60 61 63 64 66 68 72 76 79 81 82 83 91 92 93 97 98 100	Paley
27	757	28	1	(757)	0 1 3 9 27 43 81 129 173 220 243 310 387 404 409 445 455 466 470 505 519 578 608 641 653 660 673 729	PG(2, 27)
27	121	40	13	(121)	1 3 4 7 9 11 12 13 21 25 27 33 34 36 39 44 55 63 64 67 68 70 71 75 80 81 82 83 85 89 92 99 102 103 104 108 109 115 117 119	PG(4, 3)
27	107	53	26	(107)	1 3 4 9 10 11 12 13 14 16 19 23 25 27 29 30 33 34 35 36 37 39 40 41 42 44 47 48 49 52 53 56 57 61 62 64 69 75 76 79 81 83 85 86 87 89 90 92 99 100 101 102 105	Paley
29	871	30	1	(13)(67)	1 24 29 69 151 167 216 234 259 263 295 321 329 414 488 543 582 599 645 659 683 689 696 716 731 819 820 822 831 841	PG(2, 29)

12.23 Remark Table 12.22 does not contain inequivalent difference sets. In general, it is not known how many inequivalent difference sets exist. A complete list of non-cyclic difference sets with $k < 20$ is given in [8].

12.24 Table Cyclic difference sets with $\lambda = 1$ of prime power order n (i.e. cyclic $(n^2+n+1, n+1, 1)$ difference sets) for $31 \leq n \leq 97$.

n	Difference Set
31	0 1 23 31 60 66 77 84 87 195 253 257 291 331 401 416 468 473 515 590 606 618 711 713 752 761 841 867 892 912 961 980
32	1 2 4 8 16 32 55 64 110 128 139 220 256 278 299 339 349 440 453 512 529 556 598 678 698 703 755 793 880 906 925 991 1024
37	0 1 37 63 205 274 289 302 314 316 321 362 414 420 436 465 469 486 550 621 644 652 655 711 731 844 854 924 981 1098 1122 1152 1187 1230 1248 1316 1325 1369
41	1 31 41 99 152 187 232 271 313 345 361 421 508 594 613 618 638 672 679 710 761 772 773 775 879 897 988 996 1011 1017 1063 1067 1194 1207 1216 1243 1271 1542 1579 1612 1681 1707
43	0 1 5 43 147 164 215 284 308 356 375 439 457 459 537 627 642 655 715 721 744 755 777 780 807 854 981 1104 1230 1262 1338 1359 1373 1468 1507 1592 1647 1663 1673 1704 1779 1840 1849 1886
47	1 47 102 136 149 163 232 243 263 280 353 572 622 792 848 890 918 994 999 1037 1076 1112 1136 1204 1289 1324 1342 1375 1429 1481 1487 1578 1702 1710 1742 1813 1875 1876 1878 1885 1897 1901 1942 2057 2135 2150 2179 2209
49	0 1 7 49 170 223 267 343 348 382 491 508 688 750 828 886 894 977 986 989 1105 1123 1177 1190 1300 1304 1356 1561 1634 1716 1745 1747 1775 1849 1869 1892 1937 2000 2021 2101 2139 2171 2208 2269 2346 2365 2401 2411 2425 2436
53	1 21 53 72 189 225 292 345 386 417 443 451 473 561 566 575 644 904 929 953 966 999 1103 1107 1113 1161 1199 1234 1246 1347 1368 1410 1411 1413 1428 1487 1510 1700 1729 1838 1845 2060 2076 2104 2165 2233 2416 2443 2527 2639 2679 2718 2729 2809
59	1 59 63 79 102 135 176 266 348 366 373 374 376 462 595 608 629 761 820 883 938 962 1009 1120 1211 1379 1421 1530 1557 1607 1701 1713 1745 1850 1919 2187 2227 2244 2312 2342 2347 2396 2407 2471 2477 2523 2728 2747 2827 2875 2920 3198 3236 3251 3265 3302 3338 3450 3459 3481
61	0 1 61 220 239 246 268 314 400 570 572 613 641 650 720 723 756 845 950 1042 1107 1205 1216 1271 1313 1347 1404 1488 1492 1628 1681 1702 1820 1836 1871 2071 2161 2206 2289 2299 2307 2319 2366 2418 2490 2522 2724 3034 3199 3216 3230 3243 3346 3441 3490 3495 3607 3657 3663 3721 3744 3759
64	1 2 4 8 16 32 64 128 256 285 307 357 399 425 512 570 614 714 751 798 850 1024 1117 1140 1228 1387 1428 1502 1551 1596 1605 1700 1761 1847 2043 2048 2081 2223 2234 2259 2280 2293 2456 2639 2774 2856 2883 2961 3004 3102 3121 3192 3210 3227 3400 3522 3561 3641 3694 3861 3901 4011 4031 4086 4096
67	0 1 25 67 192 213 256 305 328 350 363 481 491 544 600 615 665 820 858 897 998 1150 1318 1359 1516 1519 1536 1675 1723 1760 2045 2102 2107 2153 2207 2548 2629 2658 2762 2774 2802 2857 2888 2904 2977 2984 3036 3068 3174 3285 3359 3481 3508 3542 3578 3744 3748 3750 3826 3977 3995 4021 4124 4138 4459 4470 4489 4549
71	1 71 222 260 419 423 499 510 819 829 877 897 911 1179 1361 1502 1668 1707 1732 1885 1901 1906 2003 2004 2006 2033 2143 2331 2346 2388 2435 2580 2616 2704 2803 2856 2869 2931 2950 3065 3121 3325 3369 3421 3581 3598 3635 3714 3845 3876 3989 4001 4061 4156 4162 4184 4207 4225 4233 4268 4292 4342 4375 4382 4468 4597 4719 4751 4921 4930 4976 5041
73	0 1 23 73 275 352 470 546 644 686 732 734 864 885 900 967 971 979 985 1104 1148 1228 1262 1291 1451 1470 1496 1535 1666 1679 1720 1801 1892 2037 2392 2427 2703 2728 2752 2759 2811 2820 3041 3164 3196 3266 3306 3442 3596 3606 3639 3701 3788 3866 3894 3995 4046 4084 4104 4275 4636 4653 4683 4749 4752 4809 4950 4955 5117 5172 5265 5276 5292 5329
79	0 1 14 17 43 79 255 490 578 717 784 822 993 1037 1106 1143 1182 1324 1343 1415 1479 1537 1728 1779 1803 1836 1906 1989 2375 2384 2540 2595 2712 2733 2783 2785 2881 3007 3014 3063 3246 3375 3397 3460 3594 3676 3771 4214 4229 4275 4316 4328 4709 4756 4824 4884 4916 4943 4961 5027 5047 5101 5191 5201 5226 5231 5393 5399 5427 5534 5545 5655 5802 5850 5951 5959 5982 6071 6075 6241

n	Difference Set
81	0 1 3 9 27 32 81 96 151 199 243 288 453 457 509 584 597 614 729 803 864 904 932 1003 1133 1359 1371 1493 1527 1745 1752 1791 1842 1856 2044 2187 2225 2384 2409 2419 2482 2525 2592 2712 2796 2833 2956 3009 3056 3233 3292 3399 3418 3478 3554 3611 3791 3802 4019 4077 4113 4190 4429 4479 4495 4581 4730 4763 5110 5235 5256 5373 5414 5526 5568 5588 5696 5905 5927 6132 6397 6561
83	1 83 219 242 274 275 277 293 472 591 598 691 699 823 900 1043 1056 1103 1129 1201 1544 1569 1635 1823 1906 1945 2044 2061 2072 2120 2190 2233 2300 2508 2523 2629 2638 2786 2791 2893 3037 3058 3218 3280 3400 3429 3504 3599 3677 3687 3691 3711 3741 3927 3972 4041 4077 4174 4231 4311 4496 4624 4713 4765 4792 4830 4836 4876 4917 4939 4970 5007 5183 5352 5491 5503 5552 5687 5851 5947 6140 6182 6514 6889
89	1 50 82 89 341 342 344 403 411 631 647 855 927 931 1024 1043 1084 1264 1300 1354 1382 1407 1506 1546 1572 1789 1914 2115 2232 2262 2373 2393 2598 2718 2727 2749 2833 2911 3015 3065 3148 3160 3165 3511 3546 3721 3785 3796 3823 3917 3959 3972 3982 3996 4030 4140 4185 4331 4387 4450 4535 4564 4691 4706 4828 5058 5076 5109 5646 5720 5806 5812 5858 5915 6085 6186 6316 6384 6405 6510 6583 6914 7012 7221 7298 7406 7798 7878 7921 7965
97	1 13 68 97 137 360 568 611 657 670 696 717 833 889 963 1070 1071 1073 1107 1122 1261 1378 1402 1503 1984 1989 2054 2163 2225 2301 2308 2670 2748 2793 2802 2825 2843 2896 3000 3008 3169 3186 3211 3527 3782 3929 4128 4257 4536 4594 4725 4745 4818 5209 5215 5253 5367 5371 5588 5598 5670 5790 5847 6034 6113 6124 6246 6338 6399 6426 6566 6596 6671 6687 6921 7221 7243 7561 7609 7829 7848 7862 7948 8091 8233 8296 8360 8560 8720 8817 8930 9011 9098 9126 9224 9374 9409 9440

12.25 Table Baumert's cyclic difference sets with $k \leq 100$. Baumert [2] listed all known cyclic difference sets with $k \leq 100$. The following examples are in his table but are not covered by Table 12.11 or 12.22. As in Table 12.11, only one representative of each orbit under the multiplier group is listed.

n	v	k	λ		Multipliers	Comment
11	43	21	10	1 2 3	$\langle x \to 4x \rangle$	H(43)
36	143	71	35	0 1 13	$\langle x \to 2x \rangle$	TPP(11)
37	197	49	12	1	$\langle x \to 16x \rangle$	biquadratic residues

12.26 Table Open cases with $k \leq 100$.

No.	n	v	k	λ	Group	No.	n	v	k	λ	Group
1	36	288	42	6	$(4)(8)(3)^2$	11	36	160	54	18	$(4)(8)(5)$
2					$(2)^2(8)(3)^2$	12	64	783	69	6	$(3)^3(29)$
3	36	208	46	10	$(4)^2(13)$	13	64	640	72	8	$(2)(4)^3(5)$
4					$(2)^2(4)(13)$	14					$(2)^3(4)^2(5)$
5					$(2)^4(13)$	15	64	320	88	24	$(4)^3(5)$
6	36	189	48	12	$(3)^3(7)$	16					$(2)^2(4)^2(5)$
7	36	176	50	14	$(4)^2(11)$	17					$(2)^4(4)(5)$
8					$(2)^2(4)(11)$	18					$(2)^6(5)$
9					$(2)^4(11)$	19	81	891	90	9	$(9)^2(11)$
10	36	160	54	18	$(2)(16)(5)$	20					$(3)^2(9)(11)$

12.27 Remark Table 12.26 contains the parameters of all abelian difference sets with $k \leq 100$ whose existence is not yet known or whose existence has not been ruled out in a published paper. It is an update of Lander's [10] and Kopilovich's [9] tables. The existence of some of these difference sets has been ruled out recently (numbers 3 and 7 with a computer by Arasu; numbers 1, 2, 4, 5, 8, and 9 by Iiams; numbers 10 and 11 by Jedwab, Ma and Schmidt (personal communications)).

12.28 Remark The abelian groups for which the existence of a Menon-Hadamard difference set of order u^2 with $u < 20$ is still undecided (see [4] and [11]) are $\mathbb{Z}_2^2 \times \mathbb{Z}_{16} \times \mathbb{Z}_9$, $\mathbb{Z}_4 \times \mathbb{Z}_{16} \times \mathbb{Z}_9$, $\mathbb{Z}_8^2 \times \mathbb{Z}_9$, $\mathbb{Z}_4 \times \mathbb{Z}_3^2 \times \mathbb{Z}_5^2$, and $\mathbb{Z}_2 \times \mathbb{Z}_8 \times \mathbb{Z}_3^2 \times \mathbb{Z}_9$.

12.5 See Also

§I.5	Every difference set yields a symmetric design (Theorem 12.3).
§IV.10	Difference families are a generalization of difference sets.
§IV.24	Some difference sets yield Hadamard designs; see Remark 12.17.
§VI.1	The code generated by dev(D) is an ideal in a group algebra.
§VI.2	The quadratic residues can be difference sets. Primitive polynomials have to be used to construct Singer difference sets.
§VI.7	Singer difference sets correspond to classical geometries and the cyclicity of the designs can be used to construct ovals, arcs, and the like.
[6]	This survey contains most of the material used here (non-existence and existence results).
[7]	The connection between difference sets and 2-level-autocorrelation sequences is explained in this book. This has applications in signal processing and cryptography.
[10]	This monograph develops the non-existence theory quite nicely.

References

[1] K.T. Arasu and S.K. Sehgal, Some difference sets, *J. Combin. Theory A* 69 (1995), 170-172.

[2] L.D. Baumert, *Cyclic Difference Sets*, Lect. Notes Math. 182, Springer-Verlag, Berlin, 1971.

[3] T. Beth, D. Jungnickel and H. Lenz, *Design Theory*, Cambridge University Press, London, 1986.

[4] J.A. Davis and J. Jedwab, A summary of Menon-difference sets, in: *Groups, Difference Sets and the Monster*, de Gruyter Verlag, Berlin-New York, to appear.

[5] D.M. Gordon, The prime power conjecture is true for $n \leq 2,000,000$, *Electron. J. Comb.* 1 (1994).

[6] D. Jungnickel, Difference sets, in: *Contemporary Design Theory: A Collection of Surveys* (J.H. Dinitz, D.R. Stinson; eds.) John Wiley & Sons, New York, 1992, 241-324.

[7] D. Jungnickel, *Finite Fields: Structure and Arithmetics*, BI Wissenschaftsverlag, Mannheim, 1993.

[8] R.E. Kibler, A summary of non-cyclic difference sets, $k < 20$, *J. Combin. Theory A* 25 (1978), 62-67.

[9] L.E. Kopilovich, Difference sets in non-cyclic abelian groups, *Kibernetika* 2 (1989), 20-23.

[10] E.S. Lander, *Symmetric Designs: An Algebraic Approach*, London Math. Soc. Lect Note Series 74, Cambridge University Press, London, 1983.

[11] A. Pott, *Applications of Character Theory to Problems in Finite Geometry*, Springer-Verlag, Berlin, to appear.

13 Difference Sets: Nonabelian

KEN W. SMITH

13.1 Introduction

13.1 Remark A difference set D may exist in a nonabelian group G; use the definitions of §IV.12 but use multiplicative notation for the group. Replace "$d_i - d_j$" with "$d_i d_j^{-1}$" and the blocks of $dev(D)$ are the left translates $\{gD : g \in G\}$.

13.2 Example The set $D = \{3, 6, 7, 12, 14\}$ is a difference set in $\mathbb{Z}_{21}$ (see §IV.12). Written multiplicatively, $D = x^3 + x^6 + x^7 + x^{12} + x^{14}$ in $\langle x : x^{21} = 1\rangle$. This is isomorphic to $D = a + a^2 + b + a^4 + b^2$ in $\langle a, b : a^7 = b^3 = 1\rangle$. But D, written in this form, is also a difference set in the nonabelian group $\langle a, b : a^7 = b^3 = bab^{-1}a^{-2} = 1\rangle$.

13.3 Example The set

$$D = (1 + x^{16})[1 + x^4 + x(1 + x^8) + y(1 + x^{12})] + (1 + x^8)[(x^{-2} + x^2) + (x^{13} + x^3) + y(x^{-2} + x^{18}) + y(x^{-3} + x^3)]$$

is a $(64, 28, 12)$ difference set in the nonabelian group $\langle x, y : x^{32} = y^2 = yxyx^{-17} = 1\rangle$. This group has exponent 32. An abelian group of order 64 has a difference set if and only if its exponent is 16 or smaller. This is the first example in an infinite family of nonabelian difference sets with exponent higher than allowed in abelian groups [2].

13.4 Example The set

$$\begin{aligned} D = {} & (1 + a + a^4) + (1 + a)b + (1 + a^2 + a^3 + a^4)b^2 + (1 + a + a^2 + a^3)b^3 \\ & + (1 + a^4)b^4 + [(a^2 + a^4) + a^4 b + a^3 b^2 + (1 + a^2)b^3 + (a + a^2 + a^3 + a^4)b^4]c \\ & + [a^4 + (a + a^2 + a^4)b + (a + a^4)b^2 + (1 + a^2 + a^4)b^3 + a^3 b^4]c^2 \\ & + [(a^3 + a^4) + (1 + a^4)b + a^3 b^2 + (a + a^2 + a^3 + a^4)b^3 + ab^4]c^3 \end{aligned}$$

is a $(100, 45, 20)$ difference set in the nonabelian group

$$\langle a, b, c : a^5 = b^5 = c^4 = [a, b] = cac^{-1}a^{-2} = cbc^{-1}b^{-2} = 1\rangle$$

There are no abelian difference sets with these parameters [8, 9].

13.2 Nonabelian Difference Sets from Abelian Ones

13.5 Remark A symmetric design with a large automorphism group may have both abelian and nonabelian subgroups acting regularly on the points.

13.6 Theorem (Bruck [1]) Let D be a cyclic difference set with $\lambda = 1$ and $k \equiv 2 \pmod 3$. Then there exists a nonabelian difference set with the same parameters. The associated projective planes are isomorphic.

13.7 Theorem (Dillon [3]) Let G be an abelian group with subgroup H of index 2. Let $\hat{G}$ be the dihedral extension of H; that is, $\hat{G} = \langle H, \theta\rangle$, $\theta^2 = 1$, and for all $h \in H, \theta h \theta = h^{-1}$. Then the existence of a difference set in $\hat{G}$ forces the existence of one in G.

13.8 Corollary Let G be a group of order 2^{2d+2}. If there exists a normal subgroup K of index at least 2^{d+3} such that G/K is cyclic or dihedral, then G does not have a difference set.

13.9 Remark Corollary 13.8 gives a necessary and sufficient condition for the existence of a difference set in groups of order 64. It is an open question as to whether this is true for all 2-groups.

13.10 Theorem (Dillon [3]) Let G be a group of order 2^{2d+2}. If there exists an elementary abelian subgroup of order 2^{d+1} in the center of G, then G has a difference set.

13.3 Genuinely Nonabelian Difference Sets

Of special interest are nonabelian difference sets which give a symmetric design having no abelian group acting regularly on its points. Such a difference set is *genuinely nonabelian.*

13.11 Example The set $D_1 = 1 + a + a^2 + b + c + abc$ is a difference set in the group $G_1 = \langle a, b, c : a^4 = c^2 = b^2a^2 = a^3ba^{-1}b^{-1} = [a, c] = [b, c] = 1\rangle$ of order 16. It gives a (16, 6, 2) design isomorphic to the one created by the set $D_2 = 1+a+b+ab+c+a^2c$ in the group $G_2 = \langle a, b, c : a^4 = c^4 = b^4 = a^2b^2c^2 = a^3ba^{-1}b^{-1} = [a, c] = [b, c] = 1\rangle$.

13.12 Example Let $A = (1+a+b)+(a+b+ab)ab+(1+a+a^2)c+(1+ab+a^2b^2)c^2$ and let $B = (1+b+b^2)c^3$. Then $D_9 = A+B$ and $D_{10} = A+aB$ are (36, 15, 6) difference sets in the group $G = \langle a, b, c : a^3 = b^3 = c^4 = [a, b] = [a, c] = b^2cb^{-1}c^{-1} = 1\rangle$.

13.13 Theorem (Dillon [3]) Let H_1 and H_2 be two groups with Menon-Hadamard difference sets (see §IV.12). Then any group G which can be written as the product of H_1 and H_2 also has a Menon-Hadamard difference set.

13.14 Remark The only known genuinely nonabelian difference sets are Menon-Hadamard difference sets with parameters $(4u^2, 2u^2 - u, u^2 - u)$ with u a multiple of 2, 3, or 5.

13.15 Theorem (Iiams [5]) Suppose p is a prime greater than 3 and G is a group of order $4p^2$ containing a Menon-Hadamard difference set. Then either G has an irreducible representation of degree 4 or $p \equiv 1 \pmod 4$ and G is the single group

$$G = \langle x, y, z : x^p = y^p = z^4 = [x, y] = [x, z] = zyz^{-1}y = 1\rangle.$$

13.16 Remark McFarland [8] ruled out all abelian groups of order $4p^2$.

13.4 Nonabelian Difference Sets with Small k

13.17 Remark Difference sets with $k < 20$ are listed in [6]. In addition to the examples previously cited, the following nonabelian groups have difference sets.

13.18 Examples From hyperplanes of a finite projective space.

(57, 8, 1), $G = \langle a, b : a^{19} = b^3 = a^7ba^{-1}b^{-1} = 1\rangle$

$D = (1 + a + a^3 + a^8) + (1 + a^4 + a^{13})b + a^{18}b^2$

(183, 14, 1), $G = \langle a, b : a^{61} = b^3 = a^{13}ba^{-1}b^{-1} = 1\rangle$

$D = (1+a+a^3+a^{20}+a^{26}+a^{48}+a^{57})+(1+a^8+a^{18}+a^{29})b+(1+a^{12}+a^{46})a^{32}b^2$

(273, 17, 1), $G = \langle a, b : a^{91} = b^3 = a^{16}ba^{-1}b^{-1} = 1\rangle$

$D = (1 + a^5 + a^6 + a^{20} + a^{24} + a^{47} + a^{80})a^7$

$+(1+a^{22}+a^{25}+a^{32}+a^{34}+a^{62}+a^{70})a^6b+(1+a^{26}+a^{39})a^{39}b^2$

(40, 13, 4), $G_1=\langle a,b : a^5=b^8=a^4ba^{-1}b^{-1}=1\rangle$

$D=(1+a+a^2)+(1+a^3)b+(a+a^3+a^4)b^2+ab^4+(a+a^2)b^5+ab^6+a^4b^7$

(40, 13, 4), $G_2=\langle a,b : a^8=b^{10}=a^4b^5=ab^9a^{-1}b^{-1}=1\rangle$

$D=(a+a^4)+(a+a^2)b^2+(a^3+a^6)b^4+(a+a^3+a^5+a^6+a^7)b^6+(a^2+a^3)b^8$

(40, 13, 4), $G_3=\langle a,b : a^8=b^{20}=a^{-2}b^5=ab^9a{-}1b^{-1}=1\rangle$

$D=(1+a^4)+(a+a^2)b^4+(a^3+a^6)b^8+(a+a^3+a^5+a^6+a^7)b^{12}+(a^2+a^3)b^{16}$

13.19 Examples Quadratic difference sets. The quadratic residues of a field of order $q\equiv 3 \pmod 4$ form an abelian difference set in the additive group of the field. In some cases this leads to nonabelian difference sets. The smallest example occurs in $\mathbb{F}_{3^3}$ and gives two nonabelian (27, 13, 6) difference sets. The group is $G=\langle a,b : a^9=b^3=a^4ba^{-1}b^{-1}\rangle$.

$D_1=(1+a+a^2+a^4+a^5+a^7)+(1+a+a^2+a^5)b+(a^5+a^6+a^8)b^2$.

$D_2=(1+a+a^2+a^4+a^5+a^7)+(1+a^5+a^8)b+(a^2+a^4+a^5+a^6)b^2$.

13.20 Examples (16,6,2) difference sets. There are 14 groups of order 16. The cyclic and dihedral do not have difference sets. The abelian groups have a total of eight difference sets. The set $1+a+b+c+d+abcd$ in the elementary abelian group of order 16 gives a $(16,6,2)$ design with a very large automorphism group. The nonabelian difference sets all give a design isomorphic to that one except for D_{11} and D_{14}. These give isomorphic $(16,6,2)$ designs which are not isomorphic to the other. The third $(16,6,2)$ design can be obtained from the abelian difference set D_1 (see [6]). All 27 difference sets are listed in [6]. The nonabelian ones are given and numbered as in [6]. The three designs are given in Table I.1.25.

$G_1=\langle a,b,c : a^4=c^2=b^2a^2=a^3ba^{-1}b^{-1}=[a,c]=[b,c]=1\rangle$.

$D_{11}=1+a+a^2+b+c+abc$ $\quad D_{12}=1+a+a^2+b+ac+a^2bc$

$G_2=\langle a,b,c : a^4=c^4=b^4=a^2b^2c^2=a^3ba^{-1}b^{-1}=[a,c]=[b,c]=1\rangle$.

$D_{13}=1+a+a^2+b+ac+bc$ $\quad D_{14}=1+a+b+ab+c+a^c$

$G_3=\langle a,b,c : a^4=b^2=c^2=baba=[a,c]=[b,c]=1\rangle$.

$D_9=1+a+a^2+b+ac+a^2bc$ $\quad D_{10}=1+a+b+a^2b+c+a^3c$

$G_4=\langle a,b,c : a^4=b^2=bc^2=a^3bca^{-1}c^{-1}=[a,b]=1\rangle$.

$D_{15}=1+a+a^2+ab+c+a^2bc$ $\quad D_{16}=1+a+a^2+ab+ac+a^3bc$

$D_{17}=1+a+b+a^3b+ac+a^3c$ $\quad D_{18}=1+a+a^2+a^3b+ac+abc$

$G_5=\langle a,b : a^4=b^4=a^3ba^{-1}b^{-1}=1\rangle$.

$D_{19}=1+a+a^2+b+ab^2+a^2b^3$ $\quad D_{20}=1+a+a^2+b+b^3+a^3b^2$

$D_{21}=1+a+b+a^2b+b^2+a^3b^2$

$G_6=\langle a,b : a^8=b^2=a^5ba^{-1}b^{-1}=1\rangle$.

$D_{22}=1+a+a^2+a^5+a^4b+a^2b$ $\quad D_{23}=1+a+a^3+a^4+a^3b+a^5b$

$G_7=\langle a,b : a^8=b^4=a^4b^2=a^3ba^{-1}b^{-1}=1\rangle$.

$D_{24}=1+a+a^2+a^5+b+a^2b$ $\quad D_{25}=1+a+a^3+a^4+ab+a^3b$

$G_8=\langle a,b : a^8=b^4=a^4b^2=a^7ba^{-1}b^{-1}=1\rangle$.

$D_{26}=1+a+a^2+a^5+b+a^2b$ $\quad D_{27}=1+a+a^3+a^4+b+a^2b$

13.21 Examples (36,15,6) difference sets. There are 14 groups of order 36. Five do not have difference sets. Two abelian groups have seven difference sets giving seven nonisomorphic designs. The abelian group with invariants [3, 3, 4] has four inequivalent difference sets labeled D_1, D_2, D_3, and D_4 in [6]. The abelian group with invariants [3, 3, 2, 2] has three inequivalent difference sets, labeled D_{17}, D_{18}, and D_{19}. These seven sets give seven nonisomorphic designs.

Seven nonabelian groups have 27 difference sets but 25 of these arise as automorphism groups of the seven designs constructible by abelian groups. Only two difference sets (D_9 and D_{10}) are genuinely nonabelian and are displayed in Example 13.12. All 34 difference sets are listed in [6]. The nonabelian ones are given below.

$G = \langle a,b,c : a^3 = b^3 = c^4 = [a,b] = [a,c] = b^2cb^{-1}c^{-1} = 1\rangle$.

$D_5 = (1+a+a^2)+(1+a+a^2)b+(1+ab+a^2b^2)c+(1+b+b^2)c^2+(1+a^2b+ab^2)c^3$

$D_6 = (1+a+a^2)+(1+a+a^2)b+(1+ab+a^2b^2)c+(1+b+b^2)c^2+a(1+a^2b+ab^2)c^3$

$D_7 = (1+a+b)+(ab+b^2+ab^2)+(1+ab+a^2b^2)c+(1+a+a^2)c^2+(1+a^2b+ab^2)c^3$

$D_8 = (1+a+b)+(ab+b^2+ab^2)+(1+ab+a^2b^2)c+(1+a+a^2)c^2+a(1+a^2b+ab^2)c^3$

$G = \langle a,b,c : a^3 = b^3 = c^4 = [a,b] = a^2ca^{-1}c^{-1} = b^2cb^{-1}c^{-1} = 1\rangle$.

$D_{11} = (1+a+a^2)+(1+a+a^2)b+(a^2+b+ab^2)c+(1+b+b^2)c^2+(1+a^2b+ab^2)c^3$

$G = \langle a,b,c : a^3 = b^3 = c^4 = [a,b] = bca^{-1}c^{-1} = a^2cb^{-1}c^{-1} = 1\rangle$.

$D_{12} = (1+a+a^2)+(1+a+a^2)b+(a^2+b+ab^2)c+(1+b+b^2)c^2+(1+a^2b+ab^2)c^3$

$D_{13} = (1+a+a^2)+(1+a+a^2)b+(a^2+b+ab^2)c+(1+b+b^2)c^2+a(1+a^2b+ab^2)c^3$

$D_{14} = (1+a+a^2)+(1+a+a^2)b+(a^2+b+ab^2)c+(1+b+b^2)c^2+a^2(1+a^2b+ab^2)c^3$

$D_{15} = (1+a+a^2)+(1+a+a^2)b+(a^2+b+ab^2)c+a(1+b+b^2)c^2+(1+a^2b+ab^2)c^3$

$D_{16} = (1+a+a^2)+(1+a+a^2)b+(a^2+ab+b^2)c+(1+b+b^2)c^2+(1+ab+a^2b^2)c^3$

$G = \langle a,b,c,d : a^3=b^3=c^2=d^2=[a,b]=[c,d]=[a,d]=[b,d]=1, a^2c=ca, b^2c=cb\rangle$.

$D_{20} = (1+a+a^2)+(1+a+a^2)b+(1+b+b^2)c+(1+ab+a^2b^2)d+(1+a^2b+ab^2)cd$

$G = \langle a,b,c,d : a^3=b^3=c^2=d^2=[a,b]=[c,d]=[a,d]=[b,c]=[b,d]=a^2ca^{-1}c^{-1} = 1\rangle$.

$D_{21} = (1+a+a^2)+(1+a+a^2)b+(1+ab+a^2b^2)c+(1+b+b^2)d+(1+a^2b+ab^2)cd$

$D_{22} = (1+a+a^2)+(1+a+a^2)b+(1+ab+a^2b^2)c+(1+b+b^2)d+a(1+a^2b+ab^2)cd$

$D_{23} = (1+a+b)+(ab+b^2+ab^2)+(1+ab+a^2b^2)c+(1+a+a^2)d+(1+a^2b+ab^2)cd$

$D_{24} = (1+a+b)+(ab+b^2+ab^2)+(1+ab+a^2b^2)c+(1+a+a^2)d+a(1+a^2b+ab^2)cd$

$D_{25} = (1+a+b)+a(ab+b^2+ab^2)+(1+a+a^2)c+(1+ab+a^2b^2)d+(1+b+b^2)cd$

$D_{26} = (1+a+b)+a(ab+b^2+ab^2)+(1+a+a^2)c+a^2(1+ab+a^2b^2)d+(1+b+b^2)cd$

$G = \langle a,b,c,d : a^3=b^3=c^2=d^2=[a,b]=[c,d]=[a,d]=[b,c]=1, b^2d=db, a^2c=ca\rangle$.

$D_{27} = (1+a+a^2)+(b+ab+a^2b)+(c+abc+a^2b^2c)+(d+a^2bd+ab^2d)+(cd+bcd+b^2cd)$

$D_{28} = (1+a+a^2)+(1+a+a^2)b+(1+ab+a^2b^2)c+(1+a^2b+ab^2)d+a^2(1+b+b^2)cd$

$D_{29} = (1+a+b)+(a^2b+ab^2+a^2b^2)+(c+ac+a^2c)+(d+bd+b^2d)+(cd+abcd+a^2b^2cd)$

$D_{30} = (1+a+b)+(a^2b+ab^2+a^2b^2)+(1+a+a^2)c+a^2(1+b+b^2)d+(1+ab+a^2b^2)cd$

$D_{31} = (1+a+b)+(a^2b+ab^2+a^2b^2)+(1+b+b^2)c+(1+a+a^2)d+(1+ab+a^2b^2)cd$

$D_{32} = (1+a+b)+(a^2b+ab^2+a^2b^2)+(1+b+b^2)c+(1+a+a^2)d+a(1+ab+a^2b^2)cd$

$G = \langle a,b,c,d : a^3=b^3=c^2=d^2=[a,b]=[c,d]=[b,c]=[d,b]=1, da=ac, cda=ad\rangle$.

$D_{33} = (1+c+d+cd+a+ca+da^2)+(c+ca+cda+cda^2)b+(c+ca+da+a^2)b^2$

$D_{34} = (1+c+d+cda+a+ca^2+cda^2+(a+da+ca^2+da^2)b+(a+ca+ca^2+a^2)b^2$

13.5 See Also

§I.5	Symmetric designs.
§IV.12	Abelian difference sets.
[6]	A listing of many nonabelian difference sets.

References

[1] R.H. Bruck, Difference sets in a finite group, *Trans. Amer. Math. Soc.* 78 (1955), 464-481.

[2] J.A. Davis and J. Jedwab, A survey of Menon difference sets, *Congr. Numer.* 93 (1993), 203-207.

[3] J.F. Dillon, Variations on a scheme of McFarland for noncyclic difference sets, *J. Combin. Theory A* 40 (1985), 9-21.
[4] S. Gao and W. Wei, On non-Abelian group difference sets, *Discrete Math.* 112 (1993), 93-102.
[5] J. Iiams, On difference sets in groups of order $4p^2$, *J. Combin. Theory A*, to appear.
[6] R.E. Kibler, A summary of noncyclic difference sets, $k < 20$, *J. Combin. Theory A* 25 (1978), 62-67.
[7] E.S. Lander, *Symmetric Designs: An Algebraic Approach*, Cambridge University Press, New York, 1983.
[8] R.L. McFarland, Difference sets in abelian groups of order $4p^2$, *Mitt. Math. Sem. Giessen*, 192 (1989), 1-70.
[9] K.W. Smith, Nonabelian Hadamard difference sets, *J. Combin. Theory A* 70 (1995), 144-156.

14 Difference Triangle Sets

CHARLES J. COLBOURN

14.1 Definition and Example

An (n, k)-*difference triangle set*, or (n, k)-DΔS, is a set $\mathcal{X} = \{X_1, \ldots, X_n\}$, where for $1 \le i \le n$, $X_i = \{a_{i0}, a_{i1}, \ldots, a_{ik}\}$, with a_{ij} an integer, and the differences

$$a_{i\ell} - a_{ij} : 1 \le i \le n,\ 0 \le j \ne \ell \le k$$

are all distinct and nonzero. It is *normalized* if, for every $1 \le i \le n$,

$$0 = a_{i0} < a_{i1} < \cdots < a_{ik}.$$

14.1 Example A $(3,3)$-DΔS.

$$\{\{0,3,15,19\},\{0,6,11,13\},\{0,8,17,18\}\}.$$

The "difference triangle" nomenclature arises from writing the differences obtained in triangular format; in the example here,

3		12		4		6		5		2		8		9		1
	15		16				11		7				17		10	
		19						13						18		

The *scope* of a (normalized) (n, k)-DΔS $\mathcal{X} = \{X_1, \ldots, X_n\}$ is

$$m(\mathcal{X}) = \max(\max(x : x \in X_i) : X_i \in \mathcal{X}).$$

Then m_{nk} denotes $\min(m(\mathcal{X}) : \mathcal{X}$ is a (n, k)-DΔS$)$.

14.2 Example The scope of the $(3,3)$-DΔS in Example 14.1 is 19.

14.3 Remarks Difference triangle sets have applications in the design of missile guidance codes [9] by forming sequences with low autocorrelation, in the design of radio systems to reduce intermodulation interference [1, 10], in the construction of convolutional self-orthogonal codes [13], in the spacing of radio antennae [2], in the construction of optical orthogonal codes [15], and in numerous other areas [3, 11].

In these applications, the problem of main interest is to find difference triangle sets having small scope, and hence the determination of m_{nk} is of interest.

14.2 Difference Packings

A $(v, k; n)$-*difference packing*, or n-DP(v, k), is a set $\mathcal{X} = \{X_1, \ldots, X_n\}$, where for $1 \leq i \leq n$, $X_i = \{b_{i1}, \ldots, b_{ik}\}$, and for $1 \leq i, i' \leq n$, $1 \leq j \neq \ell \leq k$ and $1 \leq j' \neq \ell' \leq k$, one has $b_{i\ell} - b_{ij} \equiv b_{i'\ell'} - b_{i'j'} \pmod{v}$ only if $i = i'$, $j = j'$ and $\ell = \ell'$.

A (v, k)-*disjoint difference set* is a 1-DP(v, k).

14.4 Theorem A $(v, k, 1)$-cyclic difference family is a $\frac{v-1}{k(k-1)}$-DP(v, k). A $(v, k, 1)$-partial cyclic difference family is a $\frac{v-k}{k(k-1)}$-DP(v, k) (See §IV.10). A difference set of order q and index one is a 1-DP$(q^2 + q + 1, q + 1)$ (See §IV.12 and §IV.13).

14.5 Construction [4] For every prime power q, there is a 1-DP$(q^2 - 1, q)$. Let μ be a primitive element of $\mathbb{F}_{q^2}$. Then, for u a fixed integer not a multiple of $q + 1$, define $d_1 = 0$ and $\mu^{d_{k+2}} = 1 + \mu^{u+k(q+1)}$ for $k = 0, \ldots, q - 2$. The set $\{d_1, \ldots, d_m\}$ is a 1-DP$(q^2 - 1, q)$. Indeed, the pairwise differences cover all elements of $\mathbb{Z}_{q^2+1}$ which are not multiples of $q + 1$, and hence $\{d_1, \ldots, d_m\}$ is a base block for a q-GDD of type $(q-1)^{q+1}$. Together with the (short) base block $\{\infty\} \cup \{a(q+1) : 0 \leq a \leq q - 2\}$, one obtains a 1-rotational presentation of the affine plane AG$(2, q)$.

14.6 Examples for Construction 14.5.

q	Block
2	1 2
3	1 6 7
4	1 3 4 12
5	1 3 16 17 20
7	1 3 10 20 31 45 46
8	1 8 13 39 41 47 60 61
9	1 12 13 46 49 54 68 75 77
11	1 3 6 44 64 70 71 79 89 93 110
13	1 9 11 12 18 33 59 72 77 97 134 146 150
16	1 2 4 8 16 27 32 54 64 99 108 128 141 177 198 216
17	1 23 24 38 101 133 140 159 161 171 214 220 244 247 255 260 264
19	1 63 116 130 154 158 193 199 215 224 226 247 297 302 305 312 331 348 349
23	1 27 36 64 152 188 194 207 273 287 305 307 354 357 364 415 446 461 469 486 490 491 502
25	1 3 57 60 67 85 94 98 115 134 166 258 266 337 352 376 381 387 477 489 522 542 564 565 611
27	1 94 103 134 137 154 192 233 245 268 275 320 342 390 391 423 427 437 452 491 496 509 515 517 652 702 718
29	1 46 47 55 74 103 171 234 236 303 323 328 335 339 345 378 409 449 462 486 500 521 544 547 562 640 692 758 807
31	1 3 23 28 144 168 218 233 246 254 345 379 388 426 467 511 518 557 573 615 625 626 629 655 674 686 692 709 779 812 888
32	1 5 32 75 81 160 169 185 286 293 303 314 316 351 354 398 406 460 465 489 546 558 716 734 766 805 841 905 968 982 983 1002
37	1 17 41 45 119 158 203 294 326 346 409 434 610 618 708 721 769 787 818 828 833 854 855 888 981 983 1052 1058 1075 1087 1174 1224 1231 1304 1313 1355 1366
41	1 22 51 55 62 65 67 145 227 279 306 325 326 428 456 519 528 541 614 648 674 710 725 749 766 784 814 837 881 1056 1064 1113 1181 1251 1309 1390 1415 1557 1588 1594 1626

q	Block
43	1 71 124 146 281 346 374 387 442 479 487 488 494 712 816 900 933 967 994 997 999 1018 1137 1162 1176 1193 1211 1253 1265 1291 1301 1348 1377 1427 1450 1537 1580 1596 1600 1658 1727 1767 1838
47	1 4 60 98 207 224 373 414 424 438 592 611 653 665 690 698 790 813 883 889 935 955 971 984 999 1039 1046 1117 1151 1160 1234 1292 1294 1361 1447 1659 1689 1690 1694 1716 1779 1797 1818 1829 1948 2001 2132
49	1 8 70 76 113 116 124 157 206 277 338 361 373 464 483 640 682 779 803 867 941 945 962 971 996 998 1018 1131 1249 1360 1419 1485 1592 1693 1694 1709 1905 1937 1965 1975 2004 2084 2089 2102 2147 2178 2230 2322 2336

14.7 Theorem An $(n, k-1)$-DΔS of scope m is an n-DP(v, k) for all $v \geq 2m+1$.

14.8 Theorem An n-DP(v, k) is an $(n, k-1)$-DΔS, whose scope is the largest element of a set of the difference packing.

14.9 Remarks In Theorem 14.8, one can often reduce the scope as follows. For the difference packing $\mathcal{X} = \{X_1, \ldots, X_n\}$, choose integers $a_1, \ldots, a_n$ and an integer m relatively prime to v. Then $\{m \cdot X_i + a_i \pmod{v} : i = 1, \ldots, n\}$ is a difference packing whose largest element may differ from that of the original. This translation can also be used in conjunction with the following generalization of Theorem 14.8.

14.10 Theorem [12] Let $\mathcal{X} = \{X_1, \ldots, X_n\}$ be an n-DP(v, k), where $X_i = \{b_{i1}, \ldots, b_{ik}\}$ and $b_{i1} < b_{i2} < \cdots < b_{ik}$. For each $1 \leq i \leq n$, let $\{B_{i1}, \ldots, B_{is_i}\}$ be a packing of pairs by $(\ell+1)$-sets on the elements $\{1, \ldots, k\}$. Let $B_{ij} = \{c_{ij0}, \ldots, c_{ij\ell}\}$ with $c_{ij0} < c_{ij1} < \cdots < c_{ij\ell}$. Then $\{\{b_{ic_{ijm}} - b_{ic_{ij0}} : 0 \leq m \leq \ell\} : 1 \leq i \leq n,\ 1 \leq j \leq s_i\}$ is a $(\sum_{i=1}^{n} s_i, \ell)$-D$\Delta$S whose scope is $\max(b_{ic_{ij\ell}} - b_{ic_{ij0}} : 1 \leq i \leq n,\ 1 \leq j \leq s_i)$.

14.11 Theorem [7] If there exists an n-DP(v, k) and a $(g, k; 1)$-difference matrix over $\mathbb{Z}_g$, then there exists a gn-DP(gv, k). If an n'-DP(g, k) also exists, then there exists a $gn + n'$-DP(gv, k).

14.3 Difference Triangle Sets: Existence

14.12 Theorem There exists an $(n, 1)$-DΔS with scope n for all $n \geq 1$. There exists an $(n, 2)$-DΔS with scope $3n$ whenever $n \equiv 0, 1 \pmod 4$, and scope $3n+1$ whenever $n \equiv 2, 3 \pmod 4$; these scopes are minimum.

14.13 Remarks $(n, 2)$-DΔS can be constructed from Skolem sequences and hooked Skolem sequences (see §IV.43). The minimum scope of (n, k)-DΔS is unknown for all $k \geq 3$, except for some values of n. Upper bounds on m_{nk} are primarily obtained by computer search, by construction of difference packings, by *shortening* (removing the largest element in each block), or by *reduction* (removing the block containing the largest element). Theorem 14.14 gives the best general lower bound on m_{nk}.

14.14 Theorem [12] Let t be an integer satisfying $1 \leq t \leq \frac{k}{2} + 1$. Then

$$m_{nk} \geq \tfrac{1}{6}((2t^2 - 2t + 3)n + (2t - 5)) + \tfrac{1}{t+1}\left(k - \tfrac{t-1}{2}\right)\left(tnk - \tbinom{t}{2}n + 1\right) + \tfrac{\alpha}{3t(t+1)},$$

$$\text{where } \alpha = \begin{cases} 0 & \text{if } t = 1 \text{ or } n = 1 \\ 6n - 6 & \text{if } t = 2 \text{ and } n > 1 \\ 18n - 24 & \text{if } t = 3 \text{ and } n > 1 \\ (9n - 12)(t - 1) & \text{if } t \geq 4 \text{ and } 2 \leq n \leq 8 \\ (10n - 20)(t - 1) & \text{if } t \geq 4 \text{ and } n \geq 9. \end{cases}$$

14.15 Table Bounds for m_{nk} [6, 7, 5, 8, 12]. Exact values are shown in **bold**; when no exact value is known, the upper bound is given above the lower bound.

$k \backslash n$	1	2	3	4	5	6	7	8	9	10	11	12	13	14	15
1	**1**	**2**	**3**	**4**	**5**	**6**	**7**	**8**	**9**	**10**	**11**	**12**	**13**	**14**	**15**
2	**3**	**7**	**10**	**12**	**15**	**19**	**22**	**24**	**27**	**31**	**34**	**36**	**39**	**43**	**46**
3	**6**	**13**	**19**	**24**	**30**	**36**	**42**	**48**	**54**	**60**	**66**	**72**	**78**	**84**	**90**
4	**11**	**22**	**32**	**41**	**51**	**60**	**71**	**80**	96 91	**100**	123 110	133 120	146 130	156 140	162 150
5	**17**	**34**	**49**	**65**	**83**	110 91	130 106	145 121	170 136	186 151	204 166	222 181	234 196	259 211	275 226
6	**25**	**51**	**77**	117 88	146 110	172 132	198 153	225 175	251 197	277 219	314 240	340 262	366 284	393 306	432 327
7	**34**	**70**	126 91	164 121	206 151	249 181	281 211	327 241	368 271	415 301	438 331	476 361	501 391	523 421	524 451
8	**44**	100 81	163 121	217 161	283 200	346 240	407 280	415 320	518 359	550 399	580 439	678 479	688 518	797 558	845 598
9	**55**	130 104	198 155	284 206	379 257	462 308	530 359	601 410	624 461	712 512	739 563	862 614	877 665	1046 716	1088 767
10	**72**	168 130	258 193	347 257	436 321	536 385	653 448	762 512	871 576	965 640	1009 703	1029 767	1195 831	1362 895	1415 958
11	**85**	195 158	313 236	416 314	567 392	676 470	782 548	948 626	1068 704	1224 782	1298 860	1300 938	1334 1016	1629 1094	1707 1172
12	**106**	250 190	370 284	520 378	664 471	809 565	935 659	1124 753	1263 846	1439 940	1550 1034	1551 1128	1721 1221	1773 1315	2234 1409
13	**127**	288 225	447 336	599 447	800 558	967 669	1138 780	1321 891	1510 1002	1665 1113	1886 1224	1906 1335	1953 1446	2400 1557	2476 1668
14	**151**	334 263	526 393	690 523	925 653	1134 783	1319 913	1544 1042	1748 1172	1946 1302	2132 1432	2373 1562	2580 1692	2874 1822	3029 1952
15	**177**	365 305	593 455	834 606	1043 757	1292 908	1517 1058	1713 1209	2005 1360	2224 1511	2461 1661	2736 1812	3031 1963	3266 2114	3395 2265
16	**199**	450 350	678 523	937 696	1210 869	1436 1042	1720 1215	1929 1388	2259 1562	2540 1735	2849 1908	3192 2081	3436 2254	3523 2428	3569 2601
17	**216**	499 398	765 595	1058 792	1306 989	1643 1186	1915 1384	2277 1581	2589 1778	2923 1975	3208 2172	3512 2369	3871 2567	4188 2764	
18	**246**	559 449	861 672	1205 895	1522 1117	1853 1340	2122 1563	2550 1786	2876 2008	3269 2231	3620 2454	3992 2677	4389 2900		
19	283 254	623 504	959 754	1288 1004	1706 1254	2075 1504	2486 1754	2810 2003	3249 2253	3656 2503	3938 2753	4301 3003	4964 3253		
20	333 283	686 562	1083 840	1467 1119	1847 1398	2275 1677	2712 1955	3140 2234	3531 2513	3996 2792	4406 3070	4983 3349			
21	356 314	768 623	1159 932	1598 1241	2081 1550	2572 1859	3015 2168	3454 2477	3914 2787	4411 3096					
22	372 346	826 687	1328 1028	1781 1369	2308 1710	2809 2051	3327 2393	3785 2734	4393 3075	4885 3416					
23	425 380	921 754	1429 1129	1976 1505	2530 1880	3073 2255	3621 2630	4154 3005	4721 3380	5303 3755					
24	480 415	991 826	1551 1237	2143 1648	2667 2059	3346 2470	3914 2881	4565 3292	5190 3703	5820 4114					
25	492 453	1076 901	1689 1350	2307 1798	2947 2247	3610 2695	4255 3143	4962 3592	5634 4040	6493 4489					

Let c and m be positive integers. A family of sets $\mathcal{B} = \{B_1, \ldots, B_m\}$ with $B_i = \{b_{i0}, \ldots, b_{iv_i}\}$, $b_{i0} = 0$, and $b_{i\ell} > b_{ik}$ when $\ell > k$, is a *perfect system of difference sets* with *threshold* c when

$$\bigcup_{i=1}^{m} \{b_{i\ell} - b_{ik} : 0 \le k < \ell \le v_i\} = \{c, c+1, \ldots, c-1+\sigma\},$$

where $\sigma = \sum_{i=1}^{m} \frac{v_i(v_i+1)}{2}$.

14.16 Remark Perfect systems of difference sets restrict to *consecutive* sets of differences, but permit *varying* block sizes. When all block sizes are equal, they are special types of difference triangle sets. There is a large literature on perfect systems of difference sets; see, for example, [14] and references therein. Perfect difference families are perfect systems of difference sets with threshold 1 and all blocks having the same size; see §IV.10.3.

A *spanning ruler* is a ruler with n marks and length L which measures every distance from 1 to L as a distance between two marks on the ruler in at most one way.
A *Golomb ruler* with n marks is a spanning ruler (with n marks) of minimum length.

14.17 Proposition A Golomb ruler with k marks is a $(1, k-1)$-DΔS of minimum scope.

14.18 Examples $(1, k)$-DΔS of minimum scope (Golomb rulers with $k+1$ marks) for $k \leq 18$.

k	Block
3	0 1 4 6
4	0 1 4 9 11
5	0 1 4 10 12 17
6	0 1 4 10 18 23 25
7	0 1 4 9 15 22 32 34
8	0 1 5 12 25 27 35 41 44
9	0 1 6 10 23 26 34 41 53 55
10	0 1 4 13 28 33 47 54 64 70 72
11	0 2 6 24 29 40 43 55 68 75 76 85
12	0 2 5 25 37 43 59 70 85 89 98 99 106
13	0 4 6 20 35 52 59 77 78 86 89 99 122 127
14	0 4 20 30 57 59 62 76 100 111 123 136 144 145 151
15	0 1 4 11 26 32 56 68 76 115 117 134 150 163 168 177
16	0 5 7 17 52 56 67 80 81 100 122 138 159 165 168 191 199
17	0 2 10 22 53 56 82 83 89 98 130 148 153 167 188 192 205 216
18	0 1 6 25 32 72 100 108 120 130 153 169 187 190 204 231 233 242 246

14.4 See Also

§IV.10	Perfect difference families are (n, k)-DΔS having scope $n\binom{k+1}{2}$, the minimum possible.
§IV.11	Difference matrices are used in the construction of difference packings.
[12]	A somewhat dated survey of difference triangle sets.
[16]	*Synch-sets* are analogues of $(1, k)$-DΔS in which each difference is permitted to occur at most some fixed number λ of times.

References

[1] W.C. Babcock, Intermodulation interference in radio systems, *Bell Syst. Tech. J.* 31 (1953), 63-73.

[2] F. Biraud, E.J. Blum and J.C. Ribes, On optimal synthetic linear arrays with applications to radioastronomy, *IEEE Trans. Antennas Propagation* AP-22 (1974), 108-109.

[3] G.S. Bloom and S.W. Golomb, Applications of numbered undirected graphs, *Proc. IEEE* 65 (1977), 562-571.

[4] R.C. Bose, An affine analogue of Singer's theorem, *J. Ind. Math. Soc.* 6 (1942), 1-15.

[5] Y.M. Chee and C.J. Colbourn, unpublished.

[6] Z. Chen, Further results on difference triangle sets, *IEEE Trans. Inform. Theory* IT-40 (1994), 1268-1270.

[7] Z. Chen, P. Fan and F. Jin, Disjoint difference sets, difference triangle sets, and related codes, *IEEE Trans. Inform. Theory* IT-38 (1992), 518-522.

[8] A. Dollas, W.T. Rankin and D. McCracken, New algorithms for Golomb ruler derivation and proof of the 19 mark ruler, preprint.

[9] A.R. Eckler, The construction of missile guidance codes resistant to random interference, *Bell Syst. Tech. J.* (1960), 973-994.

[10] R.J.A. Fang and W.A. Sandrin, Carrier frequency assignment for nonlinear repeaters, *Comsat Tech. Rev.* 7 (1977), 227-245.

[11] R. Gagliardi, J. Robbins and H. Taylor, Acquisition sequences in PPM communication, *IEEE Trans. Inform. Theory* IT-33 (1987), 738-744.

[12] T. Kløve, Bounds and construction for difference triangle sets, *IEEE Trans. Inform. Theory* IT-35 (1989), 879-886.

[13] J.P. Robinson and A.J. Bernstein, A class of binary recurrence codes with limited error propagation, *IEEE Trans. Inform. Theory* IT-13 (1967), 106-113.

[14] D.G. Rogers, On critical perfect systems of difference sets, *Discrete Math.* 135 (1994), 287-301.

[15] J.A. Salehi, F.R.K. Chung and V.K. Wei, Optical orthogonal codes: design, analysis and applications, *IEEE Trans. Inform. Theory* IT-35 (1989), 595-604.

[16] G.J. Simmons, Synch-sets: a variant of difference sets, *Congr. Numer.* 10 (1974), 625-645.

15 Directed Designs

FRANK E. BENNETT
ALIREZA MAHMOODI

15.1 Definitions and Examples

An ordered set $A = (a_1, a_2, \ldots, a_m)$ is *contained* in an ordered set $B = (b_1, b_2, \ldots, b_n)$ if $a_i = b_{\ell_i}$, $i = 1, 2, \ldots, m$ and $\ell_i < \ell_j$ for $i < j$.

A *directed t-design* of type t-(v, k, λ), denoted by t-(v, k, λ) DD, is a pair $(X, \mathcal{B})$ where X is a v-element set of points and $\mathcal{B}$ is a collection of *ordered* k-subsets of X (*blocks*) with the property that every *ordered* t-subset of X is contained in exactly λ blocks.

15.1 Example Let $X = \{0, 1, 2, 3\}$ and let $\mathcal{B}$ be the set of blocks $(0, 1, 2, 3)$, $(1, 0, 3, 2)$, $(2, 0, 3, 1)$, $(3, 0, 2, 1)$, $(2, 1, 3, 0)$, and $(3, 1, 2, 0)$. Then $(X, \mathcal{B})$ forms a 3-(4, 4, 1) DD. The block $(0, 1, 2, 3)$ contains the four ordered triples $(0, 1, 2)$, $(0, 1, 3)$, $(0, 2, 3)$, and $(1, 2, 3)$.

A t-(v, k, λ) directed design is also denoted by $DB_t(k, \lambda; v)$ or $T_\lambda(t, k, v)$ with the subscript dropped when $t = 2$ and $\lambda = 1$, respectively. A 2-(v, k, λ) DD is often called a *directed* BIBD (DBIBD for short) and is denoted by (v, k, λ) DD. A $(v, 3, \lambda)$ DD is referred to as a *directed triple system* and is denoted by DTS(v, λ).

The *converse* of a directed design $(X, \mathcal{B})$, is $(X, \mathcal{B}^{-1})$ where $\mathcal{B}^{-1} = \{(a_k, \ldots, a_2, a_1) : (a_1, a_2, \ldots, a_k) \in \mathcal{B}\}$. The design is *self-converse* if it is isomorphic to its converse.

15.2 Example The blocks $(0,1,2), (0,2,4), (0,3,1), (0,4,3)$ when developed modulo 5 form a DTS(5, 3).

A *directed group divisible design*, (k, λ)-DGDD of type $m^{v/m}$, is a $(k, 2\lambda)$-GDD of type $m^{v/m}$ whose blocks are transitively ordered so that every ordered pair not contained in the same group appear in exactly λ blocks.

15.3 Remark A 2-(v, k, λ) DD is equivalent to a decomposition of the complete symmetric directed multigraph, λK_v^*, into transitive tournaments of size k. Directed designs are sometimes called *transitive* designs.

15.2 Elementary Properties

15.4 Theorem A t-(v, k, λ) DD is also a t-$(v, k, \lambda t!)$ block design.

The corresponding t-design of a directed design is its *underlying design*.

15.5 Theorem The existence of a t-(v, k, λ) DD implies the existence of a $(t-1)$-$(v-1, k-1, t\lambda)$ DD.

15.6 Theorem Every t-(v, k, λ) DD is a $(t-1)$-(v, k, λ') DD, where $\lambda' = \frac{\lambda t(v-t+1)}{(k-t+1)}$.

15.7 Corollary Every t-(v, k, λ) DD is an s-(v, k, λ_s) DD, for $0 \le s \le t-1$ where

$$\lambda_s = \lambda(\tbinom{v-s}{t-s}t!)/(\tbinom{k-s}{t-s}s!)$$

15.8 Proposition Corollary 15.7 implies that a necessary condition for the existence of a t-(v, k, λ) DD is that λ_s must be an integer for $s = 0, 1, \ldots, t-1$.

15.3 Existence

15.9 Example Let $(X, \mathcal{B})$ be any (v, k, λ) design with some arbitrary but fixed ordering imposed on the elements of each block. Then $(X, \mathcal{B} \cup \mathcal{B}^{-1})$ is a (v, k, λ) DD.

Listed below are some small (v, k, λ) DD where a (v, k, λ) design does not exist.

(v, k, λ)	Construction
$(6,3,1)$	Develop $(0, \infty, 4)$ and $(0,1,3)$ modulo 5
$(8,3,3)$	Develop $(0,4,1)$, $(0,4,2)$, $(0,4,3)$, $(0,1,7)$, $(0,2,1)$, $(0,5,3)$, and $(0,2,5)$ modulo 8
$(7,4,1)$	Develop $(0,1,4,6)$ modulo 7
$(19,4,1)$	Develop $(0,3,12,1)$, $(13,1,0,5)$, and $(9,4,6,0)$ modulo 19
$(6,5,2)$	Blocks: $(0,1,2,3,4)$, $(0,1,2,3,5)$, $(2,1,0,4,5)$, $(3,4,5,1,0)$, $(5,4,3,2,0)$, $(5,4,3,2,1)$
$(11,5,1)$	Develop $(0,2,1,6,9)$ modulo 11
$(16,6,1)$	Develop $(0,1,x^3,x^6,x^9,x^{12})$ modulo 16, where x is a primitive root of $\mathbb{F}_{16}$, and $x^4 = x+1$

15.10 Remark Directed triple systems have been studied quite extensively and they are closely related to ordinary triple systems. See [2], where an excellent bibliography is given. Theorem 15.11 summarizes the existence results for DBIBDs (see [1] and [2]).

15.11 Theorem A 2-(v, k, λ) DD exists for $k \in \{3, 4, 5, 6\}$ if and only if $(v, k, \lambda) \neq (15, 5, 1)$ or $(21, 6, 1)$, $v(v-1)2\lambda \equiv 0 \pmod{k(k-1)}$, and $(v-1)2\lambda \equiv 0 \pmod{k-1}$.

15.12 Theorem [3] A self-converse $(v, 3, 1)$ DD exists if and only if $v \equiv 0, 1 \pmod 3$ and $v \neq 6$.

15.13 Theorem [8, 9] A (k, λ)-DGDD of type $m^{v/m}$ exists for $k = 3$ and 4 if and only if $v \geq km$, $v \equiv 0 \pmod m$, $(v-m)2\lambda \equiv 0 \pmod{k-1}$, and $v(v-m)2\lambda \equiv 0 \pmod{k(k-1)}$.

15.14 Theorem [4, 10] A 3-$(v, 4, \lambda)$ DD exists for all even v.

15.15 Theorem [4] The set of permutations of length k, S_k, can be partitioned into k systems $T(k-1, k, k)$.

15.16 Remark The only known example of a $T(t, k, k)$ when $k - t \geq 2$ is a $T(4, 6, 6)$ constructed by van Trung [11]. The existence of any other such system is in doubt.

15.17 Remark Replacing each block of an $S(k-1, k, v)$ with a $T(k-1, k, k)$ yields a $T(k-1, k, v)$, and hence $T(4, 5, v)$ and $T(5, 6, v+1)$ exist for $v = 11, 23, 47, 71, 83, 107, 131$, and 167. Also, $T(4, 6, 6)$ together with $S(4, 6, 27)$ produce a $T(4, 6, 27)$ (see §I.4).

15.4 Directability and Consequences

Since a t-(v, k, λ) DD is also a t-$(v, k, \lambda t!)$ design, it is natural to ask whether a t-(v, k, λ) DD can be constructed from a t-$(v, k, \lambda t!)$ design by ordering the elements in each block. Such a process is referred to as *directing*, and when the directing process is successful the t-design is *directable.*

15.18 Theorem (Colbourn and Harms (see [2])) Every $(v, 3, 2\lambda)$ block design is directable.

15.19 Remark Theorem 15.18 together with the existence results for $(v, 3, 2\lambda)$ designs provide an existence proof for directed triple systems. An important consequence of the directability theorem is that results on one type of design can be carried over to the other type. There are numerous results concerning triple systems which are thus applicable to directed triple systems (see [2]). Theorems 15.20 and 15.21 are applications of Theorem 15.18.

15.20 Theorem Let $v \geq 2w + 1, v, w \neq 2$. If $\lambda \equiv 0 \pmod{\gcd(v-2, 3)}$ and $\lambda \equiv 0 \pmod{\gcd(w-2, 3)}$, then a DTS$(v, \lambda)$ exists having a sub-DTS(w, λ).

15.21 Theorem For $v \geq 3$, $\lambda \equiv 0 \pmod{\gcd(v-2, 3)}$, and $\lambda \leq (v-2)/2$ there exists a DTS(v, λ) whose underlying design is simple.

15.22 Remark There are two known optimal algorithms for directing triple systems which are quite different from each other. Colbourn (see [2]) has shown a much stronger relationship between triple systems and directed triple systems. He presents an optimal algorithm for directing triple systems of index two which preserves an automorphism group of the underlying design when certain necessary conditions are met.

15.23 Remarks Computational investigations [5] have shown that none of the known biplanes (symmetric $(v, k, 2)$ designs) with $v \geq 37$ is directable, and hence the directability theorem for triple systems does not hold for BIBDs in general. The smallest known design which cannot be directed is a 2-$(28, 7, 2)$ design. Theorems 15.24 and 15.25 are consequences of the directing algorithm by Harms and Colbourn.

15.24 Theorem Every partial triple system of index 2λ underlies a partial directed triple system of index λ.

15.25 Theorem Every simple $(v, 3, 2\lambda)$ block design underlies at least four disjoint $(v, 3, \lambda)$ DD.

15.26 Conjecture (Harms and Colbourn) Every simple $(v, 3, 2\lambda)$ block design underlies six disjoint $(v, 3, \lambda)$ DD.

15.5 Automorphism, Isomorphism, and Large Sets

15.27 Theorem A DTS(v, λ) having a cyclic automorphism exists if and only if

1. $\lambda \equiv 1, 5 \pmod 6$ and $v \equiv 1, 4, 7 \pmod{12}$;
2. $\lambda \equiv 2, 4 \pmod 6$ and $v \equiv 1 \pmod 3$;
3. $\lambda \equiv 3 \pmod 6$ and $v \equiv 0, 1, 3 \pmod 4$; or
4. $\lambda \equiv 0 \pmod 6$ and $v \neq 2$.

15.28 Remark A similar theorem by Cho, Chae, and Hwang settles existence for k-rotational DTS$(v$,1)'s.

15.29 Theorem For v sufficiently large, $v \equiv 0, 1 \pmod 3$, the number of nonisomorphic DTS$(v, 1)$ is $\exp[(v^2 \ln v/3)(1 + o(1)]$.

15.30 Remark The number of nonisomorphic DTS$(v, 1)$ for $v = 3, 4, 6$, and 7 is $1, 3, 32$, and 2368 respectively.

15.31 Theorem For all $v \equiv 0, 1 \pmod 3$, there is a large set of disjoint DTS$(v, 1)$ (see [2] for references).

15.32 Remark A large set of $T(k-1, k, k)$ exists for every positive integer k (see §15.3). Also, a large set of $T(4, 6, 6)$ exists [11].

15.6 Intersection and Other Results

15.33 Theorem For $v \equiv 0, 1 \pmod 3$, there exist two DTS$(v, 1)$ on the same element set intersecting in s triples if and only if $s \in \{0, 1, \ldots, v(v-1)/3\}$, $s \neq v(v-1)/3 - 1$.

15.34 Theorem [6] For $v \equiv 1 \pmod 3$ there exist two 2-$(v, 4, 1)$ DD on the same element set intersecting in s blocks if and only if either $s \in \{0, 1, \ldots, v(v-1)/6\} \setminus \{v(v-1)/6 - 1\}$, and $v \neq 7$, or $s \in \{0, 1, 7\}$ and $v = 7$.

15.35 Theorem [7] For any even v, there exist two 3-$(v, 4, 1)$ DD on the same element set intersecting in s blocks if and only if $s \in \{0, 1, \ldots, v(v-1)(v-2)/4\} \setminus \{v(v-1)(v-2)/4 - 3, v(v-1)(v-2)/4 - 1\}$

15.7 See Also

§IV.28	Mendelsohn designs correspond to a different ordering of pairs within each block.
[2]	A survey of directed triple systems with proofs and original references.

References

[1] F. Bennett, R. Wei, J. Yin and A. Mahmoodi, Existence of DBIBDs with block size six, *Utilitas Math.* 43 (1993), 205-217.

[2] C.J. Colbourn and A. Rosa, Directed and Mendelsohn triple systems, in: *Contemporary Design Theory: A Collection of Surveys* (J.H. Dinitz, D.R. Stinson; eds.) John Wiley & Sons, New York, 1992, 97-136.

[3] Q. Kang, Y. Chang and G. Yang, The spectrum of self-converse DTS, *J. Combin. Des.* 2 (1994), 415-425.

[4] V. Levenshtein, On perfect codes in deletion and insertion metric, *Discrete Math. Appl.*, 2 (1992), 241-258.

[5] A. Mahmoodi and R. Mathon, Directability of biplanes and related results, preprint.

[6] E. Mahmoodian and N. Soltankhah, Intersections of 2-$(v, 4, 1)$ directed designs, preprint.

[7] E. Mahmoodian and N. Soltankhah, Intersections of directed quadruple systems, preprint.

[8] D.G. Sarvate, All directed GDDs with block size three, $\lambda_1 = 0$, exist, *Utilitas Math.* 26 (1984), 311-317.

[9] D.G. Sarvate, Some results on directed and cyclic designs, *Ars Combin.* 19A (1985), 311-317.

[10] N. Soltankhah, Directed quadruple systems, preprint.

[11] Tran van Trung, private communication.

16 D-Optimal Matrices

HADI KHARAGHANI

16.1 Definitions and Notation

An (α, β)-*matrix* is a square matrix with entries in the set $\{\alpha, \beta\}$. A $(1, -1)$-matrix B of order n is a *D-optimal matrix* if the determinant of B is the maximum determinant among all $(1, -1)$-matrices of order n. A D-optimal matrix is *constructible from two circulant matrices* if it can be written in the form $\begin{pmatrix} A & B \\ B^t & -A^t \end{pmatrix}$, where A and B are circulant matrices. α_n denotes the maximum determinant among all $(1, -1)$-matrices of order n. I_n denotes the identity matrix of order n and J_n is the matrix of order n with all entries 1.

16.2 Determinant Inequalities

16.1 Theorem

1. For $n \equiv 0 \pmod 4$, $\alpha_n \leq n^{n/2}$. Equality occurs if and only if there is a $(1, -1)$-matrix H of order n with $HH^t = nI_n$ (a Hadamard matrix). (See §IV.24).
2. For odd n, $\alpha_n \leq \sqrt{2n-1}\,(n-1)^{(n-1)/2}$. Equality occurs if and only if there is a $(1, -1)$-matrix A with $AA^t = A^tA = (n-1)I_n + J_n$.

3. For $n \equiv 2 \pmod 4$, $\alpha_n \le (2n-2)(n-2)^{(n-2)/2}$. Equality occurs if and only if there is a $(1,-1)$-matrix B with $BB^t = B^tB = \begin{pmatrix} M & 0 \\ 0 & M \end{pmatrix}$, where $M = (n-2)I_{n/2} + 2J_{n/2}$.

16.2 Remark Equality occurs in (2) of Theorem 16.1 only if $2n-1$ is a perfect square, hence only when $n \equiv 1 \pmod 4$. The problem of finding matrices for which the equality occurs is equivalent to finding certain symmetric balanced incomplete block designs. Equality holds in (3) of Theorem 16.1 only if $2n-2$ is a sum of two squares.

16.3 Existence Results

16.3 Theorem Let A, B be circulant $(1,-1)$-matrices of order $n/2$ with $AA^t + BB^t = (n-2)I_{n/2} + 2J_{n/2}$. Then there is a D-optimal matrix of order n constructible from two circulant matrices.

16.4 Theorem If there exists a cyclic projective plane of order q^2, then there exists two circulant $(1,-1)$-matrices A, B of order $1+q+q^2$, such that

$$AA^t + BB^t = (2q^2+2q)I_{1+q+q^2} + 2J_{1+q+q^2}.$$

16.5 Corollary There exist D-optimal matrices of order $n = 2(q^2+q+1)$ constructible from two circulant matrices, where q is a prime power.

16.6 Theorem If there is a D-optimal matrix of order $n \equiv 1 \pmod 4$, then there is a D-optimal matrix of order $2n$.

16.7 Theorem There exists a D-optimal matrix of order $n = 2q^2+2q+1$, where q is an odd prime power.

16.8 Table Known α_n not included in Theorem 16.7, for $n \equiv 1 \pmod 4$.

n	5	9	13	17	21	41
α_n	$3 \cdot 4^2$	$2^{11} \cdot 7$	$5 \cdot 12^6$	$16^7 \cdot 80$	$20^9 \cdot 116$	$9 \cdot 40^{20}$

16.9 Remark There are circulant D-optimal matrices of order 5 and 13.

16.10 Theorem There are D-optimal matrices constructible from two circulant matrices of order $n =$ 6, 10, 14, 18, 26, 30, 38, 42, 46, 50, 54, 62, 66, 74, 82, 86, 90, 98, 102, 122, 126, 186.

16.11 Theorem $\alpha_7 = 2^6 \cdot 9$, $\alpha_{11} = 5 \cdot 2^{16}$. Other $\alpha_n, n \equiv 3 \pmod 4$ are unknown.

16.12 Remark D-optimal matrices are useful in finding the weights of n objects in n weighings. A D-optimal matrix is referred to, in statistics, as an exact D-optimal first order saturated design with n observations.

16.4 See Also

§I.5	Symmetric designs.
§IV.24	Hadamard designs and matrices.
[1]	A survey of maximum determinant problems.
[4]	An extension to complex D-optimal matrices.
[3]	D-optimum weighing designs.

References

[1] J. Brenner and L. Cummings, The Hadamard maximum determinant problem, *Amer. Math. Monthly* 70 (1972), 626-630.

[2] J.H.E. Cohn, Determinants with elements ±1, II, *Bull. London. Math. Soc.* 21 (1989), 36-42.

[3] Z. Galil and J. Kiefer, D-optimum weighing designs, *Ann. Statist.* 8 1980, 1293-1306.

[4] W.H. Holzmann and H. Kharaghani, An infinite class of complex D-optimal designs, *Congr. Numer.* 99 (1994), 285-289.

[5] C. Koukouvinos, S. Kounias and J. Seberry, Supplementary difference sets and optimal designs, *Discrete Math.* 88 (1991), 49-58.

17 Embedding Partial Quasigroups

C.C. Lindner

17.1 Definitions and Examples

A (partial) *quasigroup of order n* is a pair $(Q, \circ)$, where Q is a finite set of size n and $\circ$ is a binary (partial) operation on Q such that the equations $a \circ x = b$ and $y \circ a = b$ have unique (at most one) solution for all a, $b \in Q$.

The partial quasigroup $(P, \circ_1)$ is *embedded* in the quasigroup $(Q, \circ_2)$ provided: (1) $P \subseteq Q$, and (2) $\circ_2$ agrees with $\circ_1$ on P.

17.1 Example The partial quasigroup $(P, \circ_1)$ of order 3 is embedded in the quasigroup $(Q, \circ_2)$ of order 6.

$(P, \circ_1) =$

$\circ_1$	1	2	3
1	3	2	1
2	2	1	
3	1		3

$(Q, \circ_2) =$

$\circ_2$	1	2	3	4	5	6
1	3	2	1	4	5	6
2	2	1	4	5	6	3
3	1	4	3	6	2	5
4	4	5	6	2	3	1
5	5	6	2	3	1	4
6	6	3	5	1	4	2

Let $w(x,y) = w_1(x,y)w_2(x,y)$ and $v(x,y) = v_1(x,y)v_2(x,y)$ be words in the free groupoid over the alphabet $\{x, y\}$. The partial quasigroup $(P, \circ)$ *satisfies the identity* $w(x,y) = v(x,y)$ if and only if whenever $w(a,b)$ is defined and $v_1(a,b)$ and $v_2(a,b)$ are defined, then $v_1(a,b) \circ v_2(a,b) = v(a,b)$ is defined and $v(a,b) = w(a,b)$. A similar statement holds if $v(a,b)$ is defined.

17.2 Example The identities $x^2 = x$, $xy = yx$, $x(xy) = y$, $(yx)x = y$, and $x(yx) = y$.

Identity $w(x,y) = v(xy)$	The partial quasigroup $(P, \circ)$ satisfies $w(x,y) = v(x,y)$ if and only if:
$x^2 = x$	$a \circ a$ is defined and $a \circ a = a$ for all $a \in P$.
$xy = yx$	for all a, $b \in P$, either both $a \circ b$ and $b \circ a$ are defined, and $a \circ b = b \circ a$, or neither is defined.

Identity $w(x,y) = v(xy)$	The partial quasigroup $(P, \circ)$ satisfies $w(x,y) = v(x,y)$ if and only if:
$(yx)x = y$	if $a \circ b$ is defined, then $(a \circ b) \circ b$ is defined and $(a \circ b) \circ b = a$.
$x(xy) = y$	if $a \circ b$ is defined, then $a \circ (a \circ b)$ is defined and $a \circ (a \circ b) = b$.
$x(yx) = y$	if $a \circ b$ is defined, then $b \circ (a \circ b)$ is defined and $b \circ (a \circ b) = a$.

17.3 Example In Example 17.1 the partial quasigroup $(P, \circ_1)$ as well as the quasigroup $(Q, \circ_2)$ satisfy the identity $xy = yx$. The partial $xy = yx$ quasigroup $(P, \circ_1)$ is *embedded* in the $xy = yx$ quasigroup $(Q, \circ_2)$.

17.4 Example Let $I = \{x^2 = x, xy = yx, (yx)x = y\}$. The partial I quasigroup $(P, \circ_1)$ of order 5 is embedded in the I quasigroup $(Q, \circ_2)$ of order 7.

$(P, \circ_1) =$

$\circ_1$	1	2	3	4	5
1	1	3	2	5	4
2	3	2	1		
3	2	1	3		
4	5			4	1
5	4			1	5

$(Q, \circ_2) =$

$\circ_2$	1	2	3	4	5	6	7
1	1	3	2	5	4	7	6
2	3	2	1	7	6	4	5
3	2	1	3	6	7	5	4
4	5	7	6	4	1	2	3
5	4	6	7	1	5	3	2
6	7	4	5	2	3	6	1
7	6	5	4	3	2	1	7

17.2 Best Possible Results

17.5 Table Let I be a subset of the identities $\{x^2 = x, xy = yx, (yx)x = y, x(xy) = y, x(yx) = y\}$. The best results to date (with respect to the size of the containing quasigroup) of the problem of embedding a partial I quasigroup of order n in an I quasigroup of order t:

	Partial I-Quasigroup of Order n with $I =$	Best Possible Embedding	Best Embedding to Date
1	ϕ	all $t \geq 2n$	all $t \geq 2n$ [5]
2	$xy = yx$ (commutative)	all even $t \geq 2n$	all even $t \geq 2n$ [3]
3	$x(xy) = y$		
4	$(yx)x = y$		
5	$x^2 = x, xy = yx$	all odd $t \geq 2n + 1$	all odd $t \geq 2n + 1$ [3]
6	$x^2 = x, x(xy) = y$		
7	$x^2 = x, (yx)x = y$		
8	$x^2 = x$ (idempotent)	all $t \geq 2n + 1$	all $t \geq 2n + 1$ [2]
9	$x(yx) = y$ (semi-symmetric)	all $t \geq 2n$	all $t \geq 6n$ such that $t \equiv 0$ or 3 (mod 6) [4]
10	$x(xy) = y$, $(yx)x = y$ (totally symmetric)	all $t \geq 2n$ if n is odd, and all even $t \geq 2n$ if n is even	all $t \geq 6n$ such that $t \equiv 0$ (mod 6) [4]
11	$x^2 = x$, $x(yx) = y$ (Mendelsohn quasigroup)	all $t \geq 2n + 1$ such that $t \equiv 0$ or 1 (mod 3)	all $t \geq 4n$ such that $t \equiv 0$ or 1 (mod 3) [10]
12	$x^2 = x$, $x(xy) = y$ and $(yx)x = y$ (Steiner quasigroup)	all $t \geq 2n + 1$ such that $t \equiv 1$ or 3 (mod 6)	all $t \geq 4n + 1$ such that $t \equiv 1$ or 3 (mod 6) [1]

17.6 Remark In Table 17.5 the identities (2), (3), and (4) are conjugate (see §IV.36) as well as the identities (5), (6), and (7). So the embedding in [3] for (2) takes care of (3) and (4), and the embedding in [3] for (5) takes care of (6) and (7). The column headed "Best Possible Embedding" means that with possibly one or two small examples, it is possible to construct a partial I quasigroup of order n which cannot be embedded in an I quasigroup of order less than the given bound.

17.7 Remarks There is a vast amount of literature on embedding partial quasigroups and designs. As a rule of thumb, almost any partial quasigroup or design one can think of can be finitely embedded. But in almost all of these embeddings the containing quasigroup or design is "extremely large". The partial quasigroups listed in Table 17.5 are, in the author's opinion, the most fundamental partial quasigroups for which a "small" embedding exists. Hence the restricted list. All Doyen-Wilson type embeddings (complete systems in complete systems) have been omitted as well, since they are not partial.

17.3 See Also

§I.4.5	Steiner triple systems underlie Steiner quasigroups.
§II.1	Latin squares are multiplication tables of quasigroups.
§IV.28	Mendelsohn triple systems underlie Mendelsohn quasigroups.
§IV.36	Results on various classes of quasigroups.
[6, 7, 8, 9]	An extensive introduction to the general problem of embedding (partial) quasigroups and designs into quasigroups and designs of the same type.

References

[1] L.D. Anderson, A.J.W. Hilton and E. Mendelsohn, Embedding partial Steiner triple systems, *Proc. London Math. Soc.* (3), 41 (1980), 447-576.

[2] L.D. Anderson, A.J.W. Hilton and C.A. Rodger, A solution to the problem of embedding partial idempotent latin squares, *J. London Math. Soc.* (2), 26 (1982), 21-27.

[3] A.B. Cruse, On embedding incomplete symmetric latin squares, *J. Combin. Theory A* 16 (1974), 18-27.

[4] A.B. Cruse and C.C. Lindner, Small embeddings for partial semisymmetric and totally symmetric quasigroups, *J. London Math. Soc.* (2), 12 (1976), 479-484.

[5] T. Evans, Embedding incomplete latin squares, *Amer. Math. Monthly* 67 (1960), 958-961.

[6] C.C. Lindner, A survey of embedding theorems for Steiner systems, *Ann. Discrete Math.* 7 (1980), 175-202.

[7] C.C. Lindner, Embedding theorems for partial latin squares, in: *Latin squares: New Developments in the Theory and Applications* (J. Dénes, A.D. Keedwell; eds.), North-Holland, Amsterdam, 1991, 217-265.

[8] C.C. Lindner and C.A. Rodger, Generalized embedding theorems for partial latin squares, *Bull. ICA* 5 (1992), 81-99.

[9] C.C. Lindner and C.A. Rodger, Decompositions into cycles II: Cycle Systems, in: *Contemporary Design Theory: A Collection of Surveys* (J.H. Dinitz, D.R. Stinson; eds.) John Wiley & Sons, New York, 1992, 325-369.

[10] C.A. Rodger, Embedding partial Mendelsohn triple systems, *Discrete Math.* 65 (1987), 187-196.

18 Equidistant Permutation Arrays

Alireza Mahmoodi
Paul J. Schellenberg

18.1 Definitions and Examples

Let S be a nonempty set of cardinality r. An *equidistant permutation array* (EPA) defined on the elements of S is a $v \times r$ array satisfying the following properties:

1. each row is a permutation of the symbols of S, and
2. any two rows agree in precisely λ columns (i.e. the Hamming distance between any two rows is $r - \lambda$).

Such an EPA is denoted by $A(r, \lambda; v)$. For any positive integer r and any nonnegative integer λ, define $R(r, \lambda) = \max\{v : \exists\ A(r, \lambda; v)\}$.

18.1 Example $R(7, 1) = 13$. An $A(7, 1; 13)$ and its corresponding $S(7, 1; 13)$ (see §18.2).

$A(7, 1; 13)$

1	2	3	4	5	6	7
1	3	2	5	4	7	6
1	6	7	3	2	4	5
1	7	5	2	6	3	4
2	6	5	4	3	7	1
3	2	5	1	7	4	6
3	6	1	2	4	5	7
3	7	4	5	2	6	1
4	3	5	6	2	1	7
5	2	7	6	4	3	1
7	1	5	3	4	6	2
7	3	6	2	5	4	1
7	5	1	4	2	3	6

$S(7, 1; 13)$

abcd	k	gm	f		i	ehjl
e	afj	b	dgl	chim		k
fgh	bil	a	ck	e	djm	
i		h	aem	bgjk	cfl	d
j	m	defik	bh	al	g	c
	ceg	l	ij	d	ahk	bfm
klm	dh	cj		f	be	agi

To construct this $S(7, 1; 13)$, label the rows of the EPA with symbols a,b,...,m in this order. If entry (x,j) of $A(7, 1; 13)$ is i, then place symbol x in cell (i, j) of this $S(7, 1; 13)$.

18.2 Theorems Some basic results on $R(r, \lambda)$.

1. $R(r + 1, \lambda + 1) \geq R(r, \lambda)$.
2. [1] $R(r, 0) = r$, $R(r, r - 1) = 1$, $R(r, r - 2) = 2$.
3. [1] For $r > 3$, $R(r, r - 3) = r - 1$. For $r > 9$, $R(r, r - 4) = \lfloor \frac{r}{2} \rfloor$.
4. [6] $R(r, r - 5) = \max\{10, \lfloor \frac{r-1}{2} \rfloor\}$ for $6 \leq r \leq 23$. $R(r, r - 6) = 13$ for $7 \leq r \leq 20$.

18.2 Equivalent Combinatorial Designs

Let X be a set of cardinality v. A *generalized Room square* (GRS) of *side* r and *index* λ defined on X is an $r \times r$ array F having the following properties:

1. every cell of F contains a subset (possibly empty) of X,
2. each symbol of X occurs once in each row and column of F, and
3. any two distinct symbols of X occur together in exactly λ cells of F.

Denote such a GRS by $S(r, \lambda; v)$.

18.3 Example An $S(7,1;13)$ is displayed above.

18.4 Remark Room squares are GRSs where every nonempty cell contains exactly two elements. (See §IV.39).

A *resolution* of a resolvable (r,λ)-design, $(X,\mathcal{B})$, is a partition of $\mathcal{B}$ into resolution classes $\mathcal{R} = \{\mathcal{R}_1, \mathcal{R}_2, \ldots, \mathcal{R}_r\}$. Two resolutions of $(X,\mathcal{B})$, say $\mathcal{R}$ and $\mathcal{S}$, are *orthogonal* if any resolution class of $\mathcal{R}$ intersects any resolution class of $\mathcal{S}$ in at most one block.

18.5 Remarks See §IV.38 for background on (r,λ)-designs. In the definition above, it must be understood that, even when two blocks of $\mathcal{B}$, say B_1 and B_2, are the same subset $Y \subset X$, they are distinguishable as blocks.

An (r,λ)-design is *doubly resolvable* if it has two orthogonal resolutions. A doubly resolvable (r,λ)-design of order v is denoted by $\mathrm{DR}(r,\lambda;v)$.

18.6 Theorem The existence of any one of $A(r,\lambda;v)$, $S(r,\lambda;v)$, or $\mathrm{DR}(r,\lambda;v)$ implies the existence of the others.

18.7 Remark These equivalent designs play an important role in the study of $R(r,\lambda)$.

18.3 Lower Bounds for R(r,λ)

18.8 Remark Lower bounds for $R(r,\lambda)$ are established by constructing $A(r,\lambda;v)$s.

18.9 Theorem [3] For $r = 4$ or 5, $R(r,1) = 2r-5$. For all other positive integers r, $R(r,1) \geq 2r-4$.

18.10 Theorem [4] If $p = r-3 > 6$ is a prime and $p \equiv 1$ or 3 (mod 8), then $R(r,1) \geq 2r-3$.

18.11 Theorem [4] If there are $\lambda+2$ MOLS of order $n-2$, then $R(n+\lambda,\lambda) \geq (\lambda+1)(n-2)+1$.

18.12 Corollary If n is a prime power, then $R(2n-1,n-3) \geq (n-1)^2$.

18.13 Theorem [5] If there are $\lambda+1$ MOLS of order $n-2$, then $R(n+\lambda,\lambda) \geq \lambda(n-2)+2$.

18.14 Remark The above results, together with known lower bounds for $R(n+1,1)$ and some exact values of $R(n+\lambda,\lambda)$ (see [5, 6, 7]), are sufficient to complete Table 18.15.

18.15 Table $R(n+\lambda,\lambda)$: Exact values (*in italics*) and lower bounds.

$n\backslash\lambda$	1	2	3	4	5	6	7	8	9	10	11	12	13	14
1	*1*	*1*	*1*	*1*	*1*	*1*	*1*	*1*	*1*	*1*	*1*	*1*	*1*	*1*
2	*2*	*2*	*2*	*2*	*2*	*2*	*2*	*2*	*2*	*2*	*2*	*2*	*2*	*2*
3	*3*	*4*	*5*	*6*	*7*	*8*	*9*	*10*	*11*	*12*	*13*	*14*	*15*	*16*
4	*5*	*5*	*5*	*5*	*5*	*5*	*5*	*6*	*6*	*7*	*7*	*8*	*8*	*9*
5	*10*	*10*	*10*	*10*	*10*	*10*	*10*	*10*	*10*	*10*	*10*	*10*	*10*	*10*
6	*13*	*13*	*13*	*13*	*13*	*13*	*13*	*13*	*13*	*13*	*13*	*13*	*13*	*13*
7	*16*	*16*	17	17	17	17	17	17	17	17	17	17	17	17
8	20	20	20	20	20	20	20	20	20	20	20	20	20	20
9	27	27	29	36	37	37	37	37	37	37	37	37	37	37
10	24	25	33	41	49	50	50	50	50	50	50	50	50	50
11	35	35	37	46	55	64	65	65	65	65	65	65	65	65
12	40	40	40	40	40	40	40	40	40	40	40	40	40	40
13	25	34	45	56	67	78	89	100	101	101	101	101	101	101
14	26	37	49	50	50	50	50	50	50	50	50	50	50	50

$n\backslash\lambda$	1	2	3	4	5	6	7	8	9	10	11	12	13	14
15	28	40	53	66	79	92	105	118	131	144	145	145	145	145
16	30	40	40	40	40	40	40	40	40	40	40	40	40	40
17	32	46	47	47	47	47	47	47	47	47	47	47	47	47
18	34	49	65	81	97	113	129	145	161	177	193	209	225	226
19	37	52	69	86	103	120	137	154	171	188	205	222	239	256

18.16 Remark There are, of course, many constructions for EPAs. Theorem 18.17 below gives better than linear bounds for $R(r,1)$.

18.17 Theorem [8] If q is a prime power, then $R(q^2+q+1,1) \geq q^3+q^2$.

18.4 Upper Bounds for R(r,λ) and Other Results

18.18 Theorem [9] $R(n+\lambda,\lambda) \leq \max\{R(n+1,1),\ \frac{(n+2)^2}{2},\ n^2-5n+7,\ 2+\left\lfloor \frac{\lambda}{\lceil \frac{n}{3} \rceil} \right\rfloor\}$.

18.19 Theorem [7] $R(r,1) \leq r^2-4r-\sqrt{2r}+O(1)$.

18.20 Theorem [10] $R(n+\lambda,\lambda) = 2+\left\lfloor \frac{\lambda}{\lceil \frac{n}{3} \rceil} \right\rfloor$, when $\lambda > \lceil \frac{n}{3} \rceil (n^2+n)$.

18.21 Theorem [5] If $M = \max\{\frac{1}{2}(n+2)^2, R(n+1,1), n^2-5n+7\}$, then

$$R(n+\lambda,\lambda) = 2+\left\lfloor \frac{\lambda}{\lceil \frac{n}{3} \rceil} \right\rfloor, \quad \text{when } \lambda > \lceil \tfrac{n}{3} \rceil (M-1).$$

18.22 Remark Theorems 18.20 and 18.21 indicate that for fixed n, there are only finitely many values of λ for which the exact value of $R(n+\lambda,\lambda)$ is not known.

18.23 Table [5, 6] The number of nonisomorphic $A(r,\lambda;v)$'s having $v = R(n+\lambda,\lambda)$.

$n\setminus\lambda$	1	2	3	4	5
3	2	1	1	1	1
4	1	2	2	2	2
5	1	1	1	1	
6	2	2	2		
7	10	19			

If property (2) in the definition of an EPA $A(r,\lambda;v)$ is replaced by property
2′. any two rows agree in *at most* λ columns,
the resulting array is a *permutation array* and is denoted by $A(r,\leq\lambda;v)$.

18.24 Remark The maximum integer v such that there is an $A(r,\leq\lambda;v)$ is considerably larger than $R(r,\lambda)$.

18.5 See Also

§II.2	MOLS now provide the most efficient constructions of EPAs.
§IV.38	(r,λ)-designs.
§IV.39	Room squares.
§IV.45	Many of the early constructions of EPAs employed starters.
§VI.7	Finite geometries provide the other class of efficient constructions.
[2]	A survey of permutation arrays.
[5]	This thesis contains a comprehensive survey of EPAs.

References

[1] D. W. Bolton, Problem, in: *Combinatorics*, Proc. Conf. on Combinatorial Math., (D.J.A. Welsh, D.R. Woodall; eds.), Instit. Math. Applic., Southend-on-Sea, Essex, England, 1972, 351-352.

[2] M. Deza and S.A. Vanstone, Bounds for permutation arrays, *J. Stat. Plann. Infer.* 2 (1978), 197-209.

[3] K. Heinrich and G.H.J. van Rees, Some constructions for equidistant permutation arrays of index one, *Utilitas Math.* 13 (1978), 193-200.

[4] K. Heinrich, G.H.J. van Rees and W.D. Wallis, A general construction for equidistant permutation arrays, in: *Graph Theory and Related Topics* (J.A. Bondy and U.S.R. Murty; eds.) Academic Press, New York, 1979, 247-252.

[5] A. Mahmoodi, Equidistant Permutation Arrays, M.Sc. thesis, University of Toronto, 1991.

[6] A. Mahmoodi and R. Mathon, Enumeration and analysis of small equidistant permutation arrays, *Congr. Numer.* 87, (1992), 5-25.

[7] R. Mathon, Bounds for equidistant permutation arrays of index one, in: *Combinatorial Mathematics: Proc. of the Third International Conf.* New York Academy of Sciences, New York, 1989, 303-309.

[8] R. Mathon and S. A. Vanstone, On the existence of doubly resolvable Kirkman systems and equidistant permutation arrays, *Discrete Math.* 30 (1980), 157-172.

[9] G.H.J. van Rees and S.A. Vanstone, Equidistant permutation arrays: A bound, *J. Australian Math. Soc. (A)* 33 (1982), 262-274.

[10] S.A. Vanstone, The asymptotic behaviour of equidistant permutation arrays, *Canad. J. Math.* 21 (1979), 45-48.

19 Factorial Designs

Deborah J. Street

19.1 Definitions and Examples

A *factor* is any attribute of the experimental units which may affect the response observed in the experiment. A *treatment factor* is allocated by the experimenter to the units. A *block factor* is an inherent attribute of the units. They are used to allocate the units into homogeneous sets which form the blocks. In either case, each factor may take one of several values. These values are the *levels* of the factor. If there are k treatment factors and factor i has s_i levels, $1 \le i \le k$, then it is a $s_1 \times s_2 \times \ldots \times s_k$ *factorial design*. If $s_1 = s_2$, say, then one writes ${s_1}^2 \times s_3 \times \ldots \times s_k$.

19.1 Example A typical factorial experiment. An engineer is interested in knowing which capacitors and resistors should be used in a circuit. There are three places where resistors are used and two places where capacitors are used. These are the five treatment factors. In each place there are three possible resistors or capacitors, as the case may be, so all five factors have three levels. If the experiment is conducted over seven days, then the day of manufacture is a potential source of variation but is an attribute of the manufactured unit. Hence the day of manufacture is a blocking factor with seven levels. To produce an actual layout for the experiment, one must know how many combinations can be tested each day. If all $3^5 = 243$ treatment combinations can be tested, then these are tested in a random order on each day. If 81 treatment combinations can be tested on each day, then a suitable subset to run on each day must be determined. For instance, one would not like all the observations on each day to have the same level of one of the resistors, say, since if the results on one day were better, or worse, than those on another, one would not know whether these differences represented differences between the days or differences between the levels of that resistor. Specific designs may be found in the tables in §19.8.

19.2 Remark The aim of a factorial design is to determine the effect of each of the treatment factors on the response observed and to determine whether there are any interactions between two, or more, of the treatment factors. The block factors are regarded as nuisance factors whose effects must be allowed for in the analysis but which are not of interest *per se*.

Each of the possible combinations of treatment factor levels is a *treatment combination*. These are usually represented by k-tuples, with the ith entry being from a set of s_i symbols.

19.3 Example A $2 \times 2 \times 3$ factorial design has treatment combinations

$$\{(000), (001), (002), (010), (011), (012), (100), (101), (102), (110), (111), (112)\}.$$

If all the treatment factors have the same number of levels then the factorial design is *symmetrical* (or *pure*); otherwise, the design is *asymmetrical* (or *mixed*).

19.4 Remarks

1. The factor levels can be represented by any finite set of the elements. If the number of levels is a prime or a prime power then it is usual to use the elements of a finite field of the appropriate order.
2. Symmetrical factorial designs in which the number of levels is a prime or a prime power are related to finite geometries (§VI.7) [41].
3. When the experiment is conducted a response is observed on each occasion that a treatment combination appears in the design. The responses are typically represented by y subscripted by the treatment combination and, if necessary, the appropriate blocking factor(s). The analysis of factorial designs is discussed in [35, 22, 30].

A factorial design may have each possible treatment combination appearing an equal number of times, in which case it is *complete*, or only a subset of the possible treatment combinations may appear, in which case the design is *incomplete*. Each occurrence of a treatment combination in a design is a *run*. Designs in which each treatment combination appears exactly once are *single replicate designs*. Incomplete factorial designs are also known as *fractional factorial designs*.

19.5 Remark Fractional factorial designs are sometimes referred to as 1/2 replicates (if half of the possible treatment combinations appear in the experiment), 1/3 replicates, and so on. A one-half replicate of a $2 \times 2 \times 3$ factorial design could have treatment combinations $\{(000), (110), (101), (011), (102), (012)\}$.

A factorial design, whether complete or incomplete, may have one, or more, blocking factors associated with each unit. If each treatment combination appears exactly once in each block, then the blocks are *complete* and each block constitutes a complete replicate of the factorial design. Otherwise, the blocks are *incomplete*.

19.6 Remarks

1. A complete factorial design can be conducted in either complete or incomplete blocks. The same is true for fractional factorial designs.
2. It is possible for blocks to have more units than there are distinct treatment combinations in the experiment. Then the blocks constitute either two, or more, complete replicates of the treatment combinations in the experiment or they can be viewed as a union of one or more complete blocks and an incomplete block.
3. Tables of incomplete blocks and fractional factorial designs for 2^k, 3^k and 2^k3^n designs can be found in §19.8. More extensive tables can be found in [33], [35], [7], and [12].

19.2 Symmetrical Prime-Power Factorial Designs

Let s be a prime or a prime power. Then the treatment combinations in a symmetrical s^k factorial design are k-tuples with each entry coming from $\mathbb{F}_s$. Usually one uses a k-tuple to represent both the treatment combination and the response observed when that treatment combination is applied to an experimental unit.

19.7 Example Suppose that s=3 and k=2. Then the treatment combinations are

$$\{(00), (01), (02), (10), (11), (12), (20), (21), (22)\}.$$

In an s^k factorial design the *main effect* of factor A, say, compares the responses among the s sets $T_\theta = \{(x_1, x_2, \ldots, x_k) | x_a = \theta\}$, $\theta \in \mathbb{F}_s$. The sets T_θ partition the s^k treatment combinations into s sets each of size s^{k-1}. Denote this partition by $\mathcal{P}(A)$.

The *interaction effect* of factors A and B, say, compares the responses among the sets within the $s-1$ partitions given by $\{(x_1, x_2, \ldots, x_k) | x_a + \theta x_b = \gamma\}$, $\theta, \gamma \in \mathbb{F}_s, \theta \neq 0$. For fixed θ, denote these partitions by $\mathcal{P}(AB^\theta)$. Higher order interactions are defined similarly. For instance, the three factor interaction between the factors A, B, and C compares the responses among the sets within the $(s-1)^2$ partitions given by $\mathcal{P}(AB^\theta C^\delta)$, $\theta, \delta \in \mathbb{F}_s, \theta \neq 0, \delta \neq 0$.

19.8 Example Suppose that $s = 3$ and $k = 2$. Then letting the factors be A and B, $\mathcal{P}(A)$ $= \{\{(x_1, x_2) | x_1 = \theta\}, \theta \in \mathbb{F}_3\} = \{\{(00), (01), (02)\}, \{(10), (11), (12)\}, \{(20), (21), (22)\}\}$. The two partitions associated with the interactions between factors A and B are $\mathcal{P}(AB)= \{\{(x_1, x_2) | x_1 + x_2 = \theta\}, \theta \in \mathbb{F}_3\}$ and $\mathcal{P}(AB^2) = \{\{(x_1, x_2) | x_1 + 2x_2 = \theta\}, \theta \in \mathbb{F}_3\}$. Thus $\mathcal{P}(AB) = \{\{(00), (12), (21)\}, \{(01), (10), (22)\}, \{(02), (11), (20)\}\}$ and $\mathcal{P}(AB^2) = \{\{(00), (11), (22)\}, \{(02), (10), (21)\}, \{(01), (12), (20)\}\}$

19.9 Table Partitions for the effects in a $2 \times 2 \times 2$ factorial design.

Effect	Partition	
A	{(000), (001), (010), (011)}	{(100), (101), (110), (111)}
B	{(000), (001), (100), (101)}	{(010), (011), (110), (111)}
C	{(000), (010), (100), (110)}	{(001), (011), (101), (111)}
AB	{(000), (001), (110), (111)}	{(100), (101), (010), (011)}
AC	{(000), (010), (101), (111)}	{(100), (110), (001), (011)}
BC	{(000), (100), (011), (111)}	{(010), (110), (001), (101)}
ABC	{(000), (011), (101), (110)}	{(100), (111), (001), (010)}

19.10 Remarks

1. The sets of the partitions are the pencils of the corresponding affine plane [41].
2. The first nonzero coefficient in each equation can be made to equal 1.

Order the treatment combinations lexicographically and let τ_i represent the expected response of the ith treatment combination. Then a *contrast* is a linear function $\sum_i \lambda_i \tau_i$ with $\sum_i \lambda_i = 0$. If treatment combination i appears n_i times in the design, then two contrasts with coefficients λ_i and μ_i are *orthogonal* if and only if $\sum_i \lambda_i \mu_i / n_i = 0$.

19.11 Example Orthogonal contrasts for a $2 \times 2 \times 2$ factorial design. (These are unique.)

Effect	(000)	(001)	(010)	(011)	(100)	(101)	(110)	(111)
A	−1	−1	−1	−1	1	1	1	1
B	−1	−1	1	1	−1	−1	1	1
C	−1	1	−1	1	−1	1	−1	1
AB	1	1	−1	−1	−1	−1	1	1
AC	1	−1	1	−1	−1	1	−1	1
BC	1	−1	−1	1	1	−1	−1	1
ABC	−1	1	1	−1	1	−1	−1	1

19.12 Example Possible orthogonal contrasts for a 3×3 factorial design. (These are not unique.)

Effect	(00)	(01)	(02)	(10)	(11)	(12)	(20)	(21)	(22)
A	−1	−1	−1	0	0	0	1	1	1
A	−1	−1	−1	2	2	2	−1	−1	−1
B	−1	0	1	−1	0	1	−1	0	1
B	−1	2	−1	−1	2	−1	−1	2	−1
AB	−1	0	1	0	1	−1	1	−1	0
AB	−1	2	−1	2	−1	−1	−1	−1	2
AB^2	−1	1	0	0	−1	1	1	0	−1
AB^2	−1	1	2	2	−1	−1	−1	2	−1

Consider a factorial design with t treatment combinations and with observed responses $y_{ij}, 1 \le i \le t, 1 \le j \le n_i$. If a contrast has coefficients $\lambda_i, 1 \le i \le t$, then the *sum of squares* associated with the contrast is given by $(\sum_i \sum_j \lambda_i y_{ij}/n_i)^2 / (\sum_i \lambda_i^2 / n_i)$.

The set of all treatment contrasts forms a vector space of dimension $t-1$, the *treatment contrast space.* Thus any set of $t-1$ orthogonal contrasts is maximal. Adding the sums of squares associated with each contrast in a maximal set gives a unique sum of squares, the *treatment sum of squares.* The dimension of the space is the number of *degrees of freedom* of the corresponding sum of squares.

Let $T_1, T_2, \ldots, T_m$ be a partition of the treatment combinations. Then a *contrast among the* T_i is one in which all the treatment combinations in each T_i have the same coefficient in the contrast. This is written as $\sum_i \lambda_i T_i$ where necessarily $\sum_i \lambda_i |T_i| = 0$.

19.13 Example The first two contrasts in Example 19.12 are orthogonal contrasts on $\mathcal{P}(A)$. The third and fourth contrasts in Example 19.12 are orthogonal contrasts on $\mathcal{P}(B)$, the next two are orthogonal contrasts on $\mathcal{P}(AB)$, and the final two are orthogonal contrasts on $\mathcal{P}(AB^2)$.

19.14 Remarks

1. The main effects and interaction effects define orthogonal subspaces of the treatment contrast space.
2. Any basis of orthogonal contrasts can be used to calculate the sum of squares associated with any effect.
3. The coefficients of orthogonal polynomials of degree s provide one set of $s-1$ orthogonal contrasts. These can be used as the coefficients for the sets in the corresponding partitions. The various contrasts are the linear effect, the quadratic effect, the linear $\times$ quadratic interaction, and so on. Related matters are discussed in [27, 26, 29].
4. If the aim of an experiment is to model a polynomial response, then response surface designs are more appropriate. See [9].
5. In practice, interactions involving several factors can be regarded as negligible.

Denote the blocks by $B_1, B_2, \ldots, B_b$. These form a partition of the treatment combinations. Contrasts among the B_i are elements of the *block contrast space* which is a $(b-1)$-dimensional subspace of the treatment contrast space. The corresponding sum of squares is the *block sum of squares.*

A main (or interaction) effect is *confounded with blocks* if the inner product of a block contrast and an effect contrast is nonzero. If every contrast in an effect subspace is in the block contrast space, then that effect is said to be *completely confounded with blocks.* If some contrasts in an effect subspace are in the block contrast space and some are not, then that effect is *partly confounded with blocks.*

19.15 Example Consider a $2 \times 2 \times 2$ complete factorial design conducted in two blocks of size four. If the two blocks correspond to any of the partitions in Table 19.9, then the corresponding effect is completely confounded with blocks. For example, if the blocks are $\{(000), (011), (101), (110)\}$ and $\{(100), (111), (001), (010)\}$, then the three factor interaction is completely confounded with blocks. If the two blocks do not correspond to any of these partitions, then two, or more, effects are partly confounded. For instance, suppose that the blocks are $\{(000), (100), (110), (101)\}$ and $\{(010), (001), (011), (111)\}$. Then the block contrast has coefficients (using lexicographic order for the treatment combinations) $(-1,-1,1,-1,1,-1,1,1)$ which are orthogonal to the contrasts corresponding to each of the two factor interactions but not to the other four effects. Thus the three main effects and the three factor interaction are all partly confounded with blocks.

19.16 Remarks

1. It is usual for all blocks to be of the same size, which must then be s^m, $m \le k$.
2. It is usual for all blocks to be obtained from intersections of sets in $\mathcal{P}$'s. The corresponding effects are then confounded with blocks, and the blocks constitute a subspace of dimension m and its cosets.

If the blocks in an incomplete factorial design constitute a subspace and its cosets, then the subspace is the *principal block* (or *intrablock subgroup*). The set of coset representatives form the *interblock subgroup.*

19.17 Theorem [41] Consider an s^k factorial design in s^{k-m} blocks of size s^m where the blocks are given by a principal block and its cosets. Then there are $(s^{k-m}-1)/(s-1)$ effects confounded with blocks.

19.18 Example Let $s=3$, $k=3$, and $m=1$. Hence there are nine blocks of size 3. Suppose one decides to confound $\mathcal{P}(111)$ and $\mathcal{P}(112)$ (which are both part of the three factor interaction). Then the subspace is $\{(x_1, x_2, x_3)|x_1 + x_2 + x_3 = 0, x_1 + x_2 + 2x_3 = 0\}=\{(000), (120), (210)\}$. There are $(3^{3-1} - 1)/(3-1) = 4$ effects, each corresponding to two degrees of freedom, confounded in total. In addition to the two effects that were chosen to confound one has also confounded $\mathcal{P}(001)$ (since $2(x_1+x_2+x_3)+(x_1+x_2+2x_3) = x_3 = 0$) and $\mathcal{P}(110)$ (since $2(x_1 + x_2 + x_3) + 2(x_1 + x_2 + 2x_3) = x_1 + x_2 = 0$). Thus one has confounded part of a main effect and part of a two-factor interaction, so in practice this is not a good allocation.

There are $(k - m)$ independent confounded effects. The remaining confounded effects, that are determined as a linear combination of these, are the *generalized interactions.*

19.19 Remarks

1. The independent confounded effects are usually interactions involving several factors, as these are least likely to be significant. It is necessary to be careful that none of the generalized interactions that are confounded is of interest, but there is no quick and easy way to calculate all the generalized interactions.
2. The blocks can be determined most easily by evaluating the principal block and then obtaining all its cosets.
3. The order of runs in an experiment may be important, as changing some factor levels may be very difficult. See [4, 24] for a discussion of some of the issues involved.
4. Tables of factorial designs in incomplete blocks and the associated confounding schemes for 2^k, 3^k, and 2^k3^n designs, for up to nine factors, are given in §19.8. More extensive tables may be found in found in [33], [7], and [12].

If several replicates of a complete factorial are conducted, and if the same effects are confounded in each replicate, then these effects are *totally confounded*; effects confounded in some replicates but not others are *partially confounded.*

19.3 Fractional Symmetrical Prime-Power Factorial Designs

A $1/(s^m)$ replicate of an s^k factorial design has s^{k-m} distinct treatment combinations appearing in it. Assume that these constitute the principal block of some incomplete block design. The set of confounded effects gives the defining equations for the treatment combinations in the fractional replicate. In this setting the effects are termed the *defining contrasts.* Two effects are *aliased* if, in the fractional replicate, the $\mathcal{P}$ corresponding to the two effects (and hence the estimates) are indistinguishable. An s^{k-m} design is often written to mean a $1/s^m$ replicate of an s^k factorial design.

19.20 Example Consider a 1/2 replicate of a 2^3 factorial design with treatment combinations $\{(000), (110), (101), (011)\}$. These are all the solutions of the equation $x_1+x_2+x_3 = 0$ and so the defining contrast is $\mathcal{P}(111)$, the three factor interaction. For $\mathcal{P}(100)$, for example, the partition in this 1/2 replicate is $\{(000), (011)\}$ and $\{(110), (101)\}$. This is the same as the partition $\mathcal{P}(011)$. Hence the main effect of A and the two factor interaction of B and C are aliased. All the partitions for this case appear in the table next.

Effects	Partition
A, BC	$\{(000), (011)\}$, $\{(110), (101)\}$
B, AC	$\{(000), (101)\}$, $\{(110), (011)\}$
C, AB	$\{(000), (110)\}$, $\{(101), (011)\}$

19.21 Remarks

1. The defining contrasts in a fractional factorial design determine both the effects about which no information will be available and the aliasing structure of the design.
2. The defining contrasts should all correspond to higher order interactions, as should the generalized interactions of these.
3. The aliasing should be such that all main effects can be estimated independently of each other and preferably of two-factor interactions. The higher the order of the interactions aliased with main effects the more desirable is the design. See §19.6.
4. Sometimes irregular fractions are used. Typically these include the principal block and some of the cosets. See [35].

19.22 Example Let $s = 5$, $k = 4$, and $m = 2$. Suppose that the effects corresponding to the partitions $\mathcal{P}(a,b,c,d)$ and $\mathcal{P}(w,x,y,z)$ are the defining contrasts. Then the generalized interactions are calculated as $\alpha(a,b,c,d) + \beta(w,x,y,z)$, $\alpha, \beta \in \mathbb{F}_5$, α, β not both 0. Hence, it is not possible for all of the defining contrasts to involve all four factors. Suppose $a = b = c = 1$, $d = 0$. Then the generalized interactions are of the form $(\alpha + \beta w,\ \alpha + \beta x,\ \alpha + \beta y,\ \beta z)$. Thus one wants z to be nonzero. One does not want w, x, and y all to be zero; otherwise, there is a confounding of the main effect of the fourth factor. So assume that $w = 1$. Then $\alpha + \beta w = \alpha + \beta$. If $\alpha + \beta = 0$, then one would like $\alpha + \beta x$ and $\alpha + \beta y$ to be nonzero (so that the contrast involves three factors). Hence one wants $\beta(x-1) \neq 0$ and $\beta(y-1) \neq 0$. Thus $x, y \neq 1$. Try $z = 1$, $x = y = 2$. Then the defining contrasts are $(1,1,1,0)$, $(1,2,2,1)$, and their generalized interactions $(0,0,1,1) = 4(1,1,1,0) + (1,2,2,1)$ and $(1,0,0,4) = 2(1,1,1,0) + 4(1,2,2,1)$. Both involve confounding part of a two-factor interaction. Can one avoid this? Try $z = 1$, $x = 2$, $y = 3$. The defining contrasts are now $(1,1,1,0)$, $(1,2,3,1)$ and the generalized interactions $(1,0,4,4) = 2(1,1,1,0) + 4(1,2,3,1)$ and $(0,1,2,1) = 4(1,1,1,0) + (1,2,3,1)$. This is a satisfactory arrangement. The treatment combinations to be used in the experiment are then $\{(x_1,x_2,x_3,x_4) | x_1 + x_2 + x_3 = \theta, x_1 + 2x_2 + 3x_3 + x_4 = \eta\}$ for some arbitrary, but fixed, θ, η in F_5. The confounded effects may be written as $I \equiv ABC \equiv ABC^2D^3 \equiv AC^4D^4 \equiv BC^2D$.

19.23 Remarks

1. The designs advocated by Taguchi are in fact fractional factorial designs constructed as described here. For instance, the L_4 is a 2^{3-1}; the L_8 is a 2^{7-4}; the L_9 is a 3^{4-2}; the L_{12} is a Plackett-Burman Design (see §19.7). See [8] and [25] for a more complete discussion.
2. A computer program for determining defining contrasts for two-level designs with $3 \leq k \leq 12$ is given in [42].
3. Another systematic method, suitable for computer implementation, for the determination of defining contrasts and confounded effects is given in [18].
4. It is possible to run a fractional factorial design in incomplete blocks. Then there will be both confounded effects and aliased effects. The tables in §19.8 indicate various blocking possibilities. Tables of blocked fractions for 2-level designs with up to 15 factors are given in [7]. Alias relationships, including some blocking options, for two-level designs with up to 11 factors and at most 64 units are given in [35].

19.4 Single Replicate Generalized Cyclic Designs

In an $s_1 \times s_2 \times \ldots \times s_k$ factorial design, use the integers modulo s_i to denote the s_i levels of factor i, $1 \leq i \leq k$. Let $\mathbb{Z}_{s_i}$ be the integers modulo s_i and let G_t be the abelian group of treatment combinations with addition defined componentwise. A single replicate factorial design in incomplete blocks is said to be a *generalized cyclic design* if one block, B_0, say, is a subgroup of G_t and the other blocks are the cosets of B_0 in G_t. B_0 is the *principal block* of the design. If B_0 can be generated by g treatment combinations then the generalized cyclic design is a *g-generator design.*

19.24 Example In a 2^3 design in which the three-factor interaction is confounded, $B_0 = \{(000), (011), (110), (101)\}$ and any two of the nonzero treatment combinations generate B_0. Hence $g = 2$.

Let $a = \text{lcm}(s_1, s_2, \ldots, s_k)$. Let the abelian group $G_c = \{(c_1, \ldots, c_k) | c_i \in \mathbb{Z}_{s_i}\}$ with addition defined componentwise. Define the bilinear form $[\mathbf{c}, \mathbf{t}] = \sum_i (c_i t_i a / s_i)$ (mod a). Then the *annihilator* of B_0, $(B_0)^0$, is $\{\mathbf{c} | \mathbf{c} \in G_c, [\mathbf{c}, \mathbf{t}] = 0 \ \forall \mathbf{t} \in B_0\}$. The *confounding scheme* of the design constructed from B_0 is completely specified by the elements in $C = (B_0)^0$ in the sense that each element $\mathbf{c} \in C$ corresponds to a single degree of freedom from the interaction of the factors corresponding to the nonzero entries of $\mathbf{c}$. C is the *defining contrasts subgroup.*

19.25 Remark Either B_0 or C can be used to completely specify the design since $B_0 = C^0$.

19.26 Example Let $s = k = 3$, $g = 1$ and let B_0 be generated by (120). Then $B_0 = \{(120), (210), (000)\}$ (see Example 19.18). Thus $a = \text{lcm}(3) = 3$ and so $C = \{\mathbf{c} | c_1 + 2c_2 = 0\} = \{(000), (001), (002), (110), (220), (111), (112), (221), (222)\}$. Hence in this design two degrees of freedom of the main effect of the third factor, two degrees of freedom of the two factor interaction between factors 1 and 2, and four degrees of freedom of the three factor interaction are confounded, as found in Example 19.18.

19.27 Theorem [45] Let B_0 be the principal block of a 1-generator s^k factorial design with generator $\mathbf{t} = (t_1, \ldots, t_k)$. Define three subsets of G_c by

1. $C_1 = \bigcup_{i | t_i = 0} \{\mathbf{c} | c_i = 1, c_j = 0, i \neq j\}$,
2. $C_2 = \bigcup_{i | t_i \neq 0} \{\mathbf{c} | c_i = s / \gcd(t_i, s), c_j = 0, i \neq j\}$,
3. $C_3 = \bigcup_{i,j | t_i \neq 0, t_j \neq 0} \{\mathbf{c} | c_i = s - (t_j / g), c_j = t_i / g, c_m = 0$ otherwise, where $g = \gcd(t_i, t_j)\}$.

Then $C = (B_0)^0$ is generated by $C_1 \cup C_2 \cup C_3$.

19.28 Example For Example 19.26, $C_1 = \{(001)\}$, $C_2 = \{(000)\}$ and $C_3 = \{(110)\}$.

19.29 Theorem [45] Let B_0 be the principal block of a g-generator s^k factorial design with generators $\mathbf{t}_1, \mathbf{t}_2, \ldots, \mathbf{t}_g$. Let $\langle \mathbf{t}_i \rangle$ be the block generated by $\mathbf{t}_i$, $i = 1, \ldots, g$. Let $D_i = \langle \mathbf{t}_i \rangle^0$. Then $C = (B_0)^0 = D_1 \cap \ldots \cap D_g$.

19.30 Example For Example 19.24, let $\mathbf{t}_1 = (011)$ and $\mathbf{t}_2 = (110)$. Then $D_1 = \langle (100), (011) \rangle$, $D_2 = \langle (001), (110) \rangle$, and $C = \langle (111) \rangle$, the three factor interaction, as expected.

19.31 Remark Let B_0 be the principal block of a g-generator $s_1^{n_1} \times s_2^{n_2} \times \ldots \times s_k^{n_k}$ factorial design. If $\mathbf{t}_i$ is one of the generators then let $\mathbf{t}_{i1}$ be the first n_1 positions of $\mathbf{t}_i$, $\mathbf{t}_{i2}$ be the next n_2 positions of $\mathbf{t}_i$, and so on. Then B_0 is also generated by the gk generators $(\mathbf{t}_{j1}, 0, 0, \ldots, 0)$, $(0, \mathbf{t}_{j2}, 0, 0, \ldots, 0)$, $\ldots$, $(0, 0, \ldots, 0, \mathbf{t}_{jk})$, $j = 1, \ldots, g$, provided that $\gcd(s_i, s_j) = 1$ for $i \neq j$.

19.32 Theorem [45] Let B_0 be the principal block of a g-generator $s_1^{n_1} \times s_2^{n_2} \times \ldots \times s_k^{n_k}$ factorial design with $\gcd(s_i, s_j) = 1$ for $i \neq j$ and with generators $\mathbf{t}_1, \mathbf{t}_2, \ldots, \mathbf{t}_g$. Let $B_{0i} = \langle \mathbf{t}_{1i}, \mathbf{t}_{2i}, \ldots, \mathbf{t}_{gi} \rangle$ and let $D_i{=}(B_{0i})^0$, $i = 1, \ldots, k$. Then $C{=}(B_0)^0{=}\{\mathbf{c}|\mathbf{c} = \mathbf{c}_1 \mathbf{c}_2 \ldots \mathbf{c}_g,\ \mathbf{c}_i \in D_i\}$.

19.33 Example Consider a $2^3 \times 3^3$ design with $B_0{=}\langle(011120), (110120)\rangle$. Hence

$$B_0 = \{(000000), (000120), (000210), (011000), (011120), (011210), (110000), (110120), (110210), (101000), (101120), (101210)\}.$$

Thus there are 18 blocks and hence 17 degrees of freedom confounded with blocks. Using the results in Example 19.30, $D_1{=}\langle(111)\rangle$ and using the results in Example 19.28, $D_2{=}\langle(001), (110)\rangle$. Hence $(B_0)^0{=}\langle(111001), (111110)\rangle$ and is of order 18.

A factorial design is *connected* if, given any two treatments in the design, it is possible to construct a chain of treatments, starting with one of the treatments and ending with the other, such that any two adjacent treatments in the chain appear together in at least one block. Any two treatments can be compared in a connected design.

19.34 Theorem [16] Let $B_0 = R \oplus S$, where $S \subseteq G_t$ is a subgroup of order $d \leq k$ and $R \subseteq G_t$ is a set of order k/d. Let S be the largest subgroup which lets B_0 be expressed in this way. Then a necessary and sufficient condition for the design developed from B_0 to be connected is that $\langle R \rangle{+}S{=}G_t$

Consider a factor with ab levels. Associate the levels of this factor with an ordered pair, with the first element from $\mathbb{Z}_a$ and the second from $\mathbb{Z}_b$. The two new factors are *pseudo-factors.*

19.35 Remarks

1. The main effect of the original factor is represented by the main effects and the two-factor interaction of the pseudo-factors.
2. More than two pseudo-factors may be used for each factor.
3. By using pseudo-factors it is possible to apply Theorem 19.32 to construct generalized cyclic designs for factorial designs with any number of levels.

19.36 Example Consider a 3×6^2 design. Replace each of the two factors with six levels by two factors, one with two levels and one with three levels. One then has a $2^2 \times 3^3$ design and looks for a suitable design within this framework. Assume that the first and third factors give the levels for the first original factor with six levels and that the second and fourth factors give the levels of the second original factor with six levels. Then to avoid confounding main effects C must not contain any element of the form $(a, 0, b, 0, 0)$ (part of the main effect of the first factor), $(0, a, 0, b, 0)$ (part of the main effect of the second factor) or $(0, 0, 0, 0, a)$ (part of the main effect of the third factor).

19.5 Other Single Replicate Asymmetrical Factorial Designs

A *bilinear classical design* is a design such that for a subgroup C of G_c, the principal block is C^0 and the other blocks are the cosets of C^0. (See [2]).

19.37 Theorem [44] The classes of single replicate generalized cyclic designs and bilinear classical designs are identical.

Consider a $s_1^{n_1} \times s_2^{n_2} \times \ldots \times s_k^{n_k}$ factorial design and for each $\mathbf{t} \in G_t$, let $\mathbf{t}_j$ be the n_j positions corresponding to the factors with s_j possible levels. For each $\mathbf{c} \in G_c$, define $\mathbf{c}_j$ similarly. Consider k fixed subgroups $C_i \subset G_{c_i}, i = 1, \ldots, k$. Then a *general classical design* has principal block $B_0 = \{\mathbf{t} | \mathbf{t} = \mathbf{t}_1\mathbf{t}_2 \ldots \mathbf{t}_k \in G_t, \sum_j c_{ij}t_{ij} = 0 \pmod{s_i}\ \forall \mathbf{c}_i \in C_i, i = 1, \ldots, k\}$. The other blocks are the cosets of B_0. (See [43]).

19.38 Theorem [44]

1. The class of generalized classical designs using the rings $\mathbb{Z}_{s_i}, i = 1, \ldots, k$, forms a subclass of the generalized cyclic designs.
2. The generalized cyclic designs form a subclass of the general classical designs if $s_1, \ldots, s_k$ are relatively prime.

Consider an s^k single replicate factorial design with s^{k-m} blocks of size s^m, where the principal block is given by $B_0 = \{\mathbf{t} | t_i \in Z_s, 1 \le i \le m, t_j = \sum_i a_{ij}t_i \pmod{s}, a_{ij} \in \mathbb{Z}_s, j = m+1, \ldots, k\}$. If the a_{ij} are chosen so that the elements of B_0, when written as the columns of a $k \times s^m$ array, form an orthogonal array of strength h (see §II.5), the corresponding design is an *orthogonal array design*. The other blocks in the design are obtained from B_0 by adding the elements of G_t with the first m positions equal to 0. (See [15]).

19.39 Example Let $s = k = 3$, $m = 2$, and $t_3 = t_1 + t_2$. Then $B_0 = \{(000), (011), (022), (101), (112), (120), (202), (210), (221)\}$, which is an orthogonal array of strength 2. The other two blocks in the design are $B_0 + (001)$ and $B_0 + (002)$.

19.40 Table a_{ij} for $s = 4, 5$, or 7 and $m = 3, 4, 5$, or 6 (from [15]). (For $s = 4$ the elements of $\mathbb{F}_4$ are α and β.)

s		$m=3$			$m=4$				$m=5$					$m=6$					
		a_{1j}	a_{2j}	a_{3j}	a_{1j}	a_{2j}	a_{3j}	a_{4j}	a_{1j}	a_{2j}	a_{3j}	a_{4j}	a_{5j}	a_{1j}	a_{2j}	a_{3j}	a_{4j}	a_{5j}	a_{6j}
4	t_1	1	1	1	1	1	1	1											
	t_2	1	α	β															
5	t_1	1	1	1	1	1	1	1	1	1	1	1	1						
	t_2	1	3	2	1	2	3	4											
	t_3	1	4	3															
7	t_1	1	1	1	1	1	1	1	1	1	1	1	1	1	1	1	1	1	1
	t_2	1	3	2	1	4	3	2	1	3	4	5	2	1	2	3	4	5	6
	t_3	1	4	3	1	5	4	6	1	2	3	4	6						
	t_4	1	5	4	1	6	5	4											
	t_5	1	2	5															

19.41 Theorem [44] The orthogonal array designs are equivalent to a subclass of the class of generalized cyclic designs.

The *pseudo-factor class of designs* is the class of designs obtained when all the factor levels are replaced by prime-leveled pseudo-factors. (See [44]).

Two classes of designs are *degrees of freedom equivalent* if, for each design in one class, there exists a design in the other class with the same number of degrees of freedom confounded for each effect. (See [44]).

19.42 Theorem [44] Consider an $s_1^{n_1} \times s_2^{n_2} \times \ldots s_k^{n_k}$ factorial design such that each s_i has no prime power divisor, $i = 1, \ldots, k$. Then the classes of generalized cyclic designs and pseudo-factor designs are degrees of freedom equivalent.

19.43 Remark If any s_i has prime power divisors, then Theorem 19.42 does not hold [44].

19.44 Theorem [43] For any single-generator generalized cyclic design there exists a degrees of freedom equivalent pseudo-factor design.

19.45 Remark For a g-generator generalized cyclic design with $g > 1$, the result of Theorem 19.44 does not hold and [43] gives a counterexample.

Associate with each plot in the incomplete block design a pair of plot factors, one denoting the block in which the plot is located and one for the particular plot within that block. The treatment combination assigned to plot $\mathbf{u}$ is $\mathbf{t}=K'\mathbf{u}$. The matrix K' is the *key matrix.* This method of construction is the *DSIGN method.* (See [39]).

19.46 Remarks

1. The DSIGN method works for any of the simple block structures given in [38].
2. Methods to identify the confounded effects have been proposed [2, 5]. For single replicate designs in incomplete blocks the method of [2] gives the bilinear classical designs. See also [41] for a simple account.
3. The Kronecker product designs of [37, 20] must confound some main effect degrees of freedom in any single replicate block design.

19.6 Ways To Compare Factorial Designs

A fractional factorial design is of *resolution* R if no p factor interaction is aliased with another effect containing less than $R-p$ factors. $R_s(k,m)$ denotes the maximum resolution of an s^{k-m} design.

19.47 Remarks

1. The resolution is usually written in roman numerals.
2. In a resolution III design no main effects (1-factor interactions) are aliased with each other.
3. In a resolution V design, no main effects are aliased with 1-, 2-, or 3-factor effects and no 2-factor effects are aliased with each other.
4. An orthogonal array of strength t is a fractional factorial design with resolution $t+1$. See §I.5.

19.48 Example Consider the 2^{3-1} design with treatment combinations $\{(000), (011), (101), (110)\}$. The design is of resolution III since main effects are aliased with two-factor interactions. In the 2^{3-1} design with treatment combinations $\{(000), (001), (110), (111)\}$, the main effects of A and B are aliased so the design is not of resolution III.

19.49 Theorem

1. [1] Suppose that $s_1 < s_2 < \ldots < s_k$. In a $s_1^{n_1} \times \ldots \times s_k^{n_k}$ design of resolution IV, a lower bound on the number of runs required is $s_k(\sum_i (s_i-1)n_i - (s_k-2))$.
2. [13] $R_s(k+(s^m-1)/(s-1), m) \geq R_s(k,m) + s^{m-1}$.

19.50 Remarks

1. The only designs known which meet this bound are 2^k designs, $2^2 \times 2a$ designs, and some two-factor resolution V designs. See [1].
2. An algorithm for partitioning 2^k designs into 2^m mutually exclusive 2^{k-m} fractional factorial designs, each of resolution III, $5 \leq k \leq 20$, is given in [36].

Let $\mathbf{c} \in C$, the defining contrasts subgroup of an s^{k-m} fractional factorial design. The *wordlength* of $\mathbf{c}$ is the number of nonzero entries in $\mathbf{c}$. Let a_i be the number of vectors in C with wordlength i, $1 \le i \le k$. Then $\mathbf{w}=(a_1, \ldots, a_k)$ is the *wordlength pattern* of the design with principal block C^0. The resolution of a design is the smallest i such that a_i is positive. If two designs, d_1 and d_2, are both resolution r and if a_r for d_1 is less than a_r for d_2 then d_1 has *less aberration* than d_2. An s^{k-m} design has *minimum aberration* if no other s^{k-m} design has less aberration. The s^m*-lag* of the vector $\mathbf{w}$ is the juxtaposition of a zero vector of length s^m and the vector $\mathbf{w}$. This is written as $lag(\mathbf{w}, s^m)$. (See [19, 13]).

19.51 Example The wordlength pattern of the design in Example 19.26 is (2, 2, 4).

19.52 Theorem [13] For any s^{k-m} fractional factorial design with wordlength pattern $\mathbf{w}$ there exists an $s^{(k+(s^m-1)/(s-1))-m}$ design with wordlength pattern $lag(\mathbf{w}, s^{m-1})$.

19.53 Theorem [13] Let $F_1, \ldots, F_k$ denote the k factors in the design.

1. The minimum aberration 2^{k-1} designs have defining contrast subgroup $C = \langle(11\ldots1)\rangle$ (equivalently $I \equiv F_1F_2\ldots F_k$).
2. Let $k-2 = 3a+r, 0 \le r < 3$. Then a minimum aberration 2^{k-2} design has defining contrasts
   $I \equiv F_1\ldots F_{2a}F_{k-1} \quad \equiv F_{a+1}\ldots F_{3a}F_k, \quad$ if $r=0$;
   $I \equiv F_1\ldots F_{2a+1}F_{k-1} \equiv F_{a+1}\ldots F_{3a+1}F_k,$ if $r=1$;
   $I \equiv F_1\ldots F_{2a+1}F_{k-1} \equiv F_{a+1}\ldots F_{3a+2}F_k,$ if $r=2$.
3. Let $k = 7a+r, 0 \le r \le 6$. For $i = 1, \ldots, 7$, define
   $B_i = \{(ia-a+1), (ia-a+2), \ldots, (ia), (7a+i)\}$, when $i \le r$ and
   $B_i = \{(ia-a+1), (ia-a+2), \ldots, (ia)\}$, otherwise.
   The elements of B_i are the subscripts of the F's and they subdivide the k letters into seven approximately equal blocks. Then a minimum aberration 2^{k-3} design has defining contrast $I \equiv B_7B_6B_4B_3 \equiv B_7B_5B_4B_2 \equiv B_6B_5B_4B_1$.
4. Let $k = 15a+r, 0 \le r \le 15$. For $i = 1, \ldots, 15$, let
   $B_i = \{(ia-a+1), (ia-a+2), \ldots, (ia), (15a+i)\}$, for $i \le r$ and
   $B_i = \{(ia-a+1), (ia-a+2), \ldots, (ia)\}$ otherwise.
   Then a minimum aberration 2^{k-4} design has defining contrast
   $I \equiv B_{15}B_{14}B_{12}B_9B_8B_7B_6B_1 \ \equiv B_{15}B_{13}B_{11}B_9B_8B_7B_5B_2$
   $\equiv B_{15}B_{14}B_{11}B_{10}B_8B_6B_5B_3 \equiv B_{15}B_{13}B_{12}B_{10}B_7B_6B_5B_4,$ if $r \ne 5$.
   If $r = 5$, then interchange B_{15} and B_5 in the defining contrast.

19.54 Example Let $k = 6$. To construct a 2^{6-3} minimum aberration design, observe that $6 = 7\times 0+6$ so $a = 0$ and $r = 6$. Then $B_1 = \{1\}, B_2 = \{2\}, B_3 = \{3\}, B_4 = \{4\}, B_5 = \{5\}, B_6 = \{6\}, B_7 = \emptyset$. The defining contrast is I $\equiv F_3F_4F_6 \equiv F_2F_4F_5 \equiv F_1F_4F_5F_6$, and hence the principal block is $\{(x_1, x_2, x_3, x_4, x_5, x_6) | x_3+x_4+x_6 = 0, x_2+x_4+x_5 = 0, x_1+x_4+x_5+x_6 = 0\}$ =$\{$(000000), (001110), (010101), (011011), (100111), (101001), (110010), (111100)$\}$. The wordlength pattern for this design is (0, 0, 4, 3, 0, 0) and so the design is resolution III.

19.55 Remark [13] Resolution has a periodic property for fractions with $m \le 4$. This is why the recursive-type constructions work. For other m, one needs to find the least k above which the periodicity holds.

19.56 Theorem [13]

1. For any fixed m, there exists a positive integer K_m such that for $k > K_m$ $R_s(k+(s^m-1)/(s-1), m) = R_s(k, m) + s^{m-1}$.

2. For any fixed m, there exists a positive integer M_m such that for $k > M_m$, the minimum aberration property is periodic (that is, if a minimum aberration s^{k-m} design has the wordlength pattern $\mathbf{w}$ then there exists a minimum aberration $s^{(k+(s^m-1)/(s-1))-m}$ design with wordlength pattern lag($\mathbf{w}$, s^{m-1}).

> Consider an s^{k-m} fractional factorial design, d_1, with wordlength pattern $\mathbf{w} = (a_1, a_2, \ldots, a_k)$. The *ith moment of the design* is defined to be $M_i(d_1) = \sum_j j^i a_j$. Consider two designs d_1 and d_2. Let f be the first i such that $M_i(d_1) \neq M_i(d_2)$ and suppose $M_f(d_1) < M_f(d_2)$. Then d_2 has *better moments* if f is odd and d_1 has *better moments* if f is even. A design has *optimal moments* if no other design has better moments. (See [17]).

19.57 Example The design in Example 19.54 has M_1=24 and M_2=84.

19.58 Remark The optimal moments criterion has a periodic property for any s^{k-m} design. See [13].

19.7 Miscellaneous Topics

> In a 2^{k-m} factorial design a *fold-over design* refers to one in which, if treatment combination $(x_1, x_2, \ldots, x_k)$ is in the fraction, so is the treatment combination $(1+x_1, 1+x_2, \ldots, 1+x_k)$.

19.59 Remark Folding over a design of resolution III gives a design of resolution IV.

19.60 Example Folding over the design in Example 19.54 gives the design with the additional treatment combinations {(111111), (110001), (101010), (100100), (011000), (010110), (001101), (000011)}. Hence the defining contrasts are $I \equiv F_1F_4F_5F_6 \equiv F_2F_3F_5F_6 \equiv F_1F_2F_3F_4$, and the wordlength pattern is (0, 0, 0, 3, 0) so the design is resolution IV.

> *Plackett-Burman designs* are saturated two-level fractional factorial designs that exist when the number of runs is a multiple of 4.

19.61 Example A 12 run Plackett-Burman design is given on the right.

1	0	1	0	0	0	1	1	1	0	1
1	1	0	1	0	0	0	1	1	1	0
0	1	1	0	1	0	0	0	1	1	1
1	0	1	1	0	1	0	0	0	1	1
1	1	0	1	1	0	1	0	0	0	1
1	1	1	0	1	1	0	1	0	0	0
0	1	1	1	0	1	1	0	1	0	0
0	0	1	1	1	0	1	1	0	1	0
0	0	0	1	1	1	0	1	1	0	1
1	0	0	0	1	1	1	0	1	1	0
0	1	0	0	0	1	1	1	0	1	1
0	0	0	0	0	0	0	0	0	0	0

19.62 Remark The Plackett-Burman designs are standard fractional factorial designs if the number of runs is a power of two. Otherwise they are obtained from one initial row which is cyclically shifted to obtain all but one of the runs. The final run is (00...0).

19.63 Remark The alias structure of the Plackett-Burman designs can be very messy and the designs should be used with caution. See [35, 32, 9], but [21] takes a more positive approach.

19.64 Table Some small Plackett-Burman designs.

$k = 11$	11011100010
$k = 19$	1100111101010000110
$k = 23$	11111010110011001010000
$k = 35$	01011100011111011100100001010110010

A *deletion design* is obtained by deleting one or more levels from one or more factors in the original design. *Collapsing levels* is a many-to-one correspondence which maps the s levels of a factor onto $u < s$ levels.

19.65 Example Consider a 3^2 design. The treatment combinations are

$$\{(00), (01), (02), (10), (11), (12), (20), (21), (22)\}.$$

Obtain a 2×3 design by deleting the last level of the first factor, say. This gives treatment combinations

$$\{(00), (01), (02), (10), (11), (12)\}.$$

One can also obtain a 2×3 design by collapsing the last two levels of the first factor. This gives treatment combinations

$$\{(00), (01), (02), (10), (11), (12), (10), (11), (12)\}.$$

The second approach means that the levels are no longer equally replicated but the levels are in proportional frequency. See §IV.32.

In a *search design* the parameters in the model are divided into two sets. One is a set of unknown parameters, all of which are to be estimated. The second is a set of parameters which is known to have at most v nonzero entries. The aim of a search design is to identify and estimate these v parameters. Any design can be used as a search design, but some layouts have better properties than others. See [11].

19.66 Remark Designs may need to have more than one blocking factor, for instance the rows and columns in a latin square may represent two directions in the field. It is then necessary to consider more elaborate confounding schemes. See [23, 31].

19.8 Small Factorial Designs with 2 and 3 Level Factors

19.67 Remark In this section tables of small 2^{k-m}, 3^{k-m} and asymmetrical designs with all factors having two or three levels are given.

19.68 Table Small 2^{k-m} factorial designs. To save space the table is arranged as follows. For a given number of factors (k) and degree of fractionation (m), the treatment combinations and the defining contrast (in full, in which case the generators are the first m effects, or by giving the generators) are listed, and the aliasing for the small order interactions is indicated. The resolution and the word length pattern are also given. Theorem 19.53 was used to generate the designs. Also indicated are a subdivision of the treatment combinations for the fractional design into blocks and (at least the generators of) the corresponding confounded effects. The treatments in the smallest block size, usually 4, are separated by semi-colons; larger blocks are obtained by grouping together an appropriate number of the blocks of size 4, starting from the left-hand side of the first line. The tables can be also used to construct incomplete blocks for a complete factorial. In this context the defining contrast is the set of confounded effects and a set of coset representatives is given to obtain the other blocks of the design (see Remark 19.16). To save space ab denotes $(1, 1, 0, \ldots, 0)$ and AB denotes $\mathrm{P}(1, 1, 0, \ldots, 0)$. No designs requiring more than 64 treatment combinations have been given. Larger designs can be found in [33].

$k = 3$

$m = 1$, $R = 3$, wordlength pattern $= (0,\ 0,\ 1)$

Defining contrast: $I \equiv ABC$

Aliasing: $A \equiv BC$, $B \equiv AC$, $C \equiv AB$

Treatment combinations: $(1), ab, ac, bc$
Coset representatives: a

$k = 4$
$m = 1$, $R = 4$, wordlength pattern = (0, 0, 0, 1)
Defining contrast: $I \equiv ABCD$
Aliasing: $A \equiv BCD$, $B \equiv ACD$, $C \equiv ABD$, $D \equiv ABC$, $AB \equiv CD$, $AC \equiv BD$, $AD \equiv BC$
Treatment combinations: $(1), ab, cd, abcd$; ac, ad, bd, bc
Coset representatives: a
Blocks of size 4: $AB \equiv CD$

$k = 4$
$m = 2$, $R = 2$, wordlength pattern = (0, 1, 2, 0)
Defining contrast: $I \equiv AC \equiv ABD \equiv BCD$
Aliasing: $A \equiv C \equiv BD \equiv ABCD$, $B \equiv ABC \equiv AD \equiv CD$, $D \equiv ACD \equiv AB \equiv BC$
Treatment combinations: $(1), abc, bd, acd$
Coset representatives: a, b, c

$k = 5$
$m = 1$, $R = 5$, wordlength pattern = (0, 0, 0, 0, 1)
Defining contrast: $I \equiv ABCDE$
Aliasing: main effects with 4 factor interactions, 2 factor interactions with 3 factor interactions
Treatment combinations: $(1), abcd, abce, de$; $ab, cd, abde, ce$; $ac, bd, be, acde$; $ad, bc, ae, bcde$
Coset representatives: a
Blocks of size 8: $AB \equiv CDE$
Blocks of size 4: $AB \equiv CDE$, $AC \equiv BDE$, $BC \equiv ADE$

$k = 5$
$m = 2$, $R = 3$, wordlength pattern = (0, 0, 2, 1, 0)
Defining contrast: $I \equiv ABD \equiv BCE \equiv ACDE$
Aliasing: $A \equiv BD \equiv ABCE \equiv CDE$, $B \equiv AD \equiv CE \equiv ABCDE$, $C \equiv ABCD \equiv BE \equiv ADE$, $D \equiv AB \equiv BCDE \equiv ACE$, $E \equiv ABDE \equiv BC \equiv ACD$, $AC \equiv BCD \equiv ABE \equiv DE$, $AE \equiv BDE \equiv ABC \equiv CD$
Treatment combinations: $(1), bcd, acde, abe$; ce, bde, ad, abc
Coset representatives: a, b, c
Blocks of size 4: $AE \equiv BDE \equiv ABC \equiv CD$

$k = 5$
$m = 3$, $R = 2$, wordlength pattern = (0, 2, 4, 1, 0)
Defining contrast: $I \equiv CD \equiv BDE \equiv ADE \equiv BCE \equiv AB \equiv ACE \equiv ABCD$
Aliasing: $A \equiv ACD \equiv ABDE \equiv DE \equiv ABCE \equiv B \equiv CE \equiv BCD$, $C \equiv D \equiv BCDE \equiv ACDE \equiv BE \equiv ABC \equiv AE \equiv ABD$, $E \equiv CDE \equiv BD \equiv AD \equiv BC \equiv ABE \equiv AC \equiv ABCDE$,
Treatment combinations: $(1), cde, abe, abcd$
Coset representatives: a, b, c, d, e, ac, ad

$k = 6$
$m = 1$, $R = 6$, wordlength pattern = (0, 0, 0, 0, 0, 1)
Defining contrast: $I \equiv ABCDEF$
Aliasing: main effects with 5 factor interactions, 2 factor interactions with 4 factor

interactions, 3 factor interactions with 3 factor interactions
Treatment combinations: $(1), bcde, acdf, abef;\ ab, ef, acde, bcdf;\ bc, de, abdf, acef;$ $ac, df, abde, bcef;\ bd, adef, ce, abcf;\ ad, abce, cf, bdef;\ cd, af, be, abcdef;$ $abcd, ae, bf, cdef$
Coset representatives: a
Blocks of size 16: $ABC \equiv DEF$
Blocks of size 8: $ABC \equiv DEF$, $ABD \equiv CEF$, $CD \equiv ABEF$
Blocks of size 4: As for 8 and $ACE \equiv BDF$, $ADE \equiv BCF$, $AF \equiv BCDE$, $BE \equiv ACDF$

$k = 6$
$m = 2$, $R = 4$, wordlength pattern = (0, 0, 0, 3, 0, 0)
Defining contrast: $I \equiv ABCE \equiv BCDF \equiv ADEF$
Aliasing: $A \equiv BCE \equiv ABCDF \equiv DEF$, $B \equiv ACE \equiv CDF \equiv ABDEF$, other main effects similarly. $AB \equiv CE \equiv ACDF \equiv BDEF$, other interactions by multiplication.
Treatment combinations: $(1), bcdf, adef, abce;\ cef, bde, acd, abf;\ abcdef, df, bc, ae;$ cde, bef, acf, abd
Coset representatives: a, b, d
Blocks of size 8: $ABD \equiv CDE \equiv ACF \equiv BEF$
Blocks of size 4: As for 8 and $ABF \equiv CEF \equiv ACD \equiv BDE$, $DF \equiv ABCDEF \equiv BC \equiv AE$

$k = 6$
$m = 3$, $R = 3$, wordlength pattern = (0, 0, 4, 3, 0, 0)
Defining contrast: $I \equiv CDF \equiv BDE \equiv ADEF \equiv BCEF \equiv ABF \equiv ACE \equiv ABCD$
Aliasing: $A \equiv ACDF \equiv ABDE \equiv DEF \equiv ABCEF \equiv BF \equiv CE \equiv BCD$, others by multiplication
Treatment combinations: $(1), bcef, adef, abcd;\ cde, bdf, acf, abe$
Coset representatives: a, b, c, d, e, f, ad
Blocks of size 4: $AD \equiv ACF \equiv ABE \equiv EF \equiv ABCDEF \equiv BDF \equiv CDE \equiv BC$

$k = 6$
$m = 4$, $R = 2$, wordlength pattern = (0, 3, 8, 3, 0, 1)
Defining contrast: $I \equiv AF \equiv BE \equiv CEF \equiv DEF \equiv ABEF \equiv BCF \equiv ACE \equiv ABC \equiv CD \equiv BDF \equiv ADE \equiv ABD \equiv BCDE \equiv ACDF \equiv ABCDEF$
Aliasing: Main effects A and F, B and E and C and D are aliased
Treatment combinations: $(1), bcde, acdf, abef$
Coset representatives: $a, b, c, d, e, f, ab, ac, ad, ae, af, bc, bd, abc, abd$

$k = 7$
$m = 1$, $R = 7$, wordlength pattern = (0, 0, 0, 0, 0, 0, 1)
Defining contrast: $I \equiv ABCDEFG$
Aliasing: Main effects with 6 factor interactions, 2 factor interactions with 5 factor interactions and 3 factor interactions with 4 factor interactions
Treatment combinations: $(1), aceg, bcde, acdf, abdg, defg, bcfg, abef;\ acef, bc, de,$ $abeg, abdf, acdg, fg, bcdefg;\ ac, abde, bcdg, eg, acdefg, bcef, abfg, df;\ acfg, acde,$ $bcdf, dg, bceg, abdefg, ef, ab;\ abcd, ae, adfg, cg, bf, bdeg, abcefg, cdef;\ ad, abce, cf,$ $bg, aefg, cdeg, abcdfg, bdef;\ bd, ce, abcf, ag, befg, abcdeg, cdfg, adef;\ cd, be, adeg,$ $abcg, cefg, af, bdfg, abcdef$
Coset representatives: a
Blocks of size 32: $ABC \equiv DEFG$
Blocks of size 16: As for 32 and $ADE \equiv BCFG$, $AFG \equiv BCDE$

Blocks of size 8: As for 16 and $BDF \equiv ACEG$, $BEG \equiv ACDF$, $CDG \equiv ABEF$, $CEF \equiv ABDG$
Blocks of size 4: Best aliasing from, giving generators only, $ABC \equiv DEFG$, $ADE \equiv BCFG$, $ABE \equiv CDFG$, $ABG \equiv CDEF$. Two factor interactions CE, BD, CG, BF, EG, DF not estimable. Blocks are not given but can be calculated from the aliasing relationship.

$k = 7$
$m = 2$, $R = 4$, wordlength pattern = (0, 0, 0, 1, 2, 0, 0)
Defining contrast: $I \equiv ABCF \equiv BCDEG \equiv ADEFG$
Aliasing: Main effects with 3 and 4 factor interactions; some 2 factor interactions are aliased (eg AB and CF)
Treatment combinations: $(1), abcf, bdefg, acdeg$; $cef, abe, bcdg, adfg$; $eg, abcefg, bdf, acd$; $cfg, abg, bcde, adef$; $de, abcdef, bfg, acg$; $cdf, abd, bceg, aefg$; $dg, abcdfg, bef, ace$; $cdefg, abdeg, bc, af$
Coset representatives: a, b, d
Blocks of size 16: $ABD \equiv CDF \equiv ACEG \equiv BEFG$
Blocks of size 8: As for 16 and $CFG \equiv ABG \equiv BDEF \equiv ACDE$, $ABCDFG \equiv DG \equiv AEF \equiv BCE$
Blocks of size 4: As for 8 and $AC \equiv BF \equiv ABDEG \equiv CDEFG$ and generalized interactions.

$k = 7$
$m = 3$, $R = 4$, wordlength pattern = (0, 0, 0, 7, 0, 0, 0)
Defining contrast: $I \equiv CDFG \equiv BDEG \equiv ADEF$ (generators only)
Aliasing: Main effects with 3 factor interactions, 2 factor interactions with each other.
Treatment combinations: $(1), efg, abcd, abcdefg$; $cdfg, cde, abfg, abe$; $bdf, bdeg, acf, aceg$; $bcg, bcef, adg, adef$
Coset representatives: a, b, c, d, e, f, g
Blocks of size 8: $CD \equiv FG \equiv BCEG \equiv ACEF$
Blocks of size 4: As for 8 and $BD \equiv BCFG \equiv EG \equiv ABEF$, $BC \equiv BDFG \equiv CDEG \equiv ABCDEF$

$k = 7$
$m = 4$, $R = 3$, wordlength pattern = (0, 0, 7, 7, 0, 0, 1)
Defining contrast: $I \equiv AFG \equiv BEG \equiv CEF \equiv DEFG$ (generators only)
Aliasing: Main effects with 2 factor interactions
Treatment combinations: $(1), defg, bcfg, bcde, aceg, acdf, abef, abdg$
Coset representatives: $a, b, c, d, e, f, g, ab, ac, ad, ae, af, ag, bc, abc$

$k = 8$
$m = 2$, $R = 5$, wordlength pattern= (0, 0, 0, 0, 2, 1, 0, 0)
Defining contrast: $I \equiv ABCDG \equiv CDEFH \equiv ABEFGH$
Aliasing: Main effects with 4 and 5 factor interactions, 2 factor interactions with 3 and 4 factor interactions
Treatment combinations: $(1), bcdegh, adefh, abcfg$; $dfg, bcefh, aegh, abcd$; $ceg, bdh, acdfgh, abef$; $abdeg, cdef, bfgh, ach$; $dgh, bce, aefg, abcdfh$; $fh, bcdefg, ade, abcgh$; $cdeh, bg, acf, abdefgh$; $cefgh, bdf, acdg, abeh$; $eh, bcdg, adf, abcefgh$; $abcdeh, defgh, bcf, ag$; $cgh, bde, acdefg, abfh$; $abdgh, cdfh, befg, ace$; $deg, bch, afgh, abcdef$; $ef, bcdfgh, adh, abceg$; $cd, begh, acefh, abdfg$; $acdegh, cfg, bdefh, ab$
Coset representatives: a, c, e
Blocks of size 32: $ACE \equiv BDEG \equiv ADFH \equiv BCFGH$

Blocks of size 16: As for 32 and $BDF \equiv ACFG \equiv BCEH \equiv ADEGH$, $ABCDEF \equiv EFG \equiv ABH \equiv CDGH$
Blocks of size 8: As for 16 and $BC \equiv ADG \equiv BDEFH \equiv ACEFGH$ (and generalized interactions)
Blocks of size 4: As for 8 and $AF \equiv BCDFG \equiv ACDEH \equiv BEGH$ (and generalized interactions)

$k = 8$
$m = 3$, $R = 4$, wordlength pattern = (0, 0, 0, 3, 4, 0, 0, 0)
Defining contrast: $I \equiv CDFG \equiv BDEG \equiv ADEFH$ (generators only)
Aliasing: Main effects with interactions with at least 3 factors, 2 factor interactions with interactions with at least 2 factors
Treatment combinations: $(1), bcef, acdfgh, abdegh$; $abcd, cegh, bfgh, adef$; $abcdefg, cfh, beh, adg$; $acdeh, abdfh, efg, bcg$; $dgh, bcdefgh, acf, abe$; $abcgh, cde, bdf, aefgh$; $abcefh, cdfg, bdeg, ah$; $aceg, abfg, defh, bcdh$
Coset representatives: a, b, c, d, e, f, g
Blocks of size 16: $BCH \equiv BDFGH \equiv CDEGH \equiv ABCDEF$
Blocks of size 8: As for 16 and $ABF \equiv ABCDG \equiv ADEFG \equiv BDEH$ (and generalized interactions)
Blocks of size 4: As for 8 and $DG \equiv CF \equiv BE \equiv AEFGH$ (and generalized interactions)

$k = 8$
$m = 4$, $R = 4$, wordlength pattern = (0, 0, 0, 14, 0, 0, 0, 1)
Defining contrast: $I \equiv AFGH \equiv BEGH \equiv CEFH \equiv DEFG$ (generators only)
Aliasing: Main effects with interactions with at least 3 factors, 2 factor interactions with each other
Treatment combinations: $(1), bcde, afgh, abcdefgh$; $cdgh, begh, acdf, abef$; $cefh, bdfh, aceg, abdg$; $abch, defg, bcfg, adeh$
Coset representatives: $a, b, c, d, e, f, g, h, ab, ac, ad, ae, af, ag, ah$
Blocks of size 8: $AF \equiv GH \equiv ABEFGH \equiv ACEH \equiv ADEG$ (and generalized interactions)
Blocks of size 4: As for 8 and $AG \equiv FH \equiv ABEH \equiv ACEFGH \equiv ADEF$ (and generalized interactions)

$k = 9$
$m = 3$, $R = 4$, wordlength pattern = (0, 0, 0, 1, 4, 2, 0, 0, 0)
Defining contrast: $I \equiv CDFG \equiv BDEGJ \equiv ADEFH$ (generators only)
Aliasing: Main effects with interactions with at least 3 factors, 2 factor interactions with interactions with at least 2 factors
Treatment combinations: $(1), cdfg, befgj, bcdej, adeghj, acefhj, abdfh, abcgh$; $abcfj, fghj, cdhj, beh, bcdefgh, adef, aceg, abdgj$; $dfj, cgj, bdeg, bcef, aefgh, acdeh, abhj, abcdfghj$; $dgh, cfh, bdefhj, bceghj, aej, acdefgj, abfg, abcd$; $efg, cde, bj, bcdfgj, adfhj, acghj, abdegh, abcefh$; $ehj, cdefghj, bfgh, bcdh, adg, acf, abdefj, abcegj$; $degj, cefj, bdf, bcg, ah, acdfgh, abefghj, abcdehj$;
$abcdefg, defh, cegh, bdghj, bcfhj, afgj, acdj, abe$
Coset representatives: a, b, c, d, e, f, g
Blocks of size 32: $ABE \equiv ABCDEFG \equiv ADGJ \equiv BDFH$ (and generalized interactions)
Blocks of size 16: As for 32 and $ACD \equiv AFG \equiv ABCEGJ \equiv CEFH$ (and generalized interactions)
Blocks of size 8: As for 16 and $AH \equiv ACDFGH \equiv ABDEGHJ \equiv DEF$ (and

generalized interactions)

$k = 9$
$m = 4$, $R = 4$, wordlength pattern = (0, 0, 0, 6, 8, 0, 0, 1, 0)
Defining contrast: $I \equiv AFGHJ \equiv BEGHJ \equiv CEFH \equiv DEFG$ (generators only)
Aliasing: Main effects with interactions with at least 3 factors, 2 factor interactions with interactions with at least 2 factors
Treatment combinations: $(1), efj, abcdghj, abcdefgh$; $dgj, defg, abch, abcefhj$; $chj, cefh, abdg, abdefgj$; $cdgh, cdefghj, abj, abef$; $bfghj, begh, acdf, acdej$; $bdfh, bdehj, acfgj, aceg$; $bcfg, bcegj, adfhj, adeh$; $bcdfj, bcde, afgh, aeghj$
Coset representatives: $a, b, c, d, e, f, g, h, j, ac, ad, ae, af, ag, ah$
Blocks of size 16: $AB \equiv BFGHJ \equiv AEGHJ \equiv ABCEFH \equiv ABDEFG$ (and generalized interactions)
Blocks of size 8: As for 16 and
$AC \equiv ACBEGHJ \equiv AEFH \equiv ACDEFG \equiv CFGHJ$ (and generalized interactions)
Blocks of size 4: As for 8 and
$AD \equiv ADBEGHJ \equiv ADCEFH \equiv AEFG \equiv DFGHJ$ (and generalized interactions)

$k = 10$
$m = 4$, $R = 4$, wordlength pattern = (0, 0, 0, 2, 8, 5, 0, 0, 0, 0)
Defining contrast: $I \equiv AFGHJ \equiv BEGHJ \equiv CEFHK \equiv DEFGK$
Aliasing: Main effects with interactions with at least 3 factors, 2 factor interactions with interactions with at least 2 factors
Treatment combinations: $(1), defg, bfghj, bdehj, acfhjk, acdeghjk, abcgk, abcdefk$; $cefh, cdgh, bcegj, bcdfj, aejk, adfgjk, abefghk, abdhk$; $cefgk, cdk, bcehjk, bcdfghjk, aeghj, adfhj, abef, abdg$; $ghk, defhk, bfjk, bdegjk, acfgj, acdej, abch, abcdefgh$; $efghjk, dhjk, bek, bdfgk, aceg, acdf, abcefhj, abcdghj$; $cgjk, cdefjk, bcfhk, bcdeghk, afgh, adeh, abj, abdefgj$; $chj, cdefghj, bcfg, bcde, afk, adegk, abghjk, abdefhjk$; $efj, dgj, begh, bdfh, acehk, acdfghk, abcefgjk, abcdjk$
Coset representatives: $a, b, c, d, e, f, g, h, j, k, ac, ad, ae, ag, ah$
Blocks of size 32: $ABJ \equiv BFGH \equiv AEGH \equiv ABCEFHJK \equiv ABDEFGJK$ (and generalized interactions)
Blocks of size 16: $DFH \equiv ADGJ \equiv BDEFGJ \equiv CDEK \equiv EGHK$ (and generalized interactions)
Blocks of size 8: $BEG \equiv ABEFHJ \equiv HJ \equiv BCFGHK \equiv BDFK$ (and generalized interactions)

19.69 Table Small 3^{k-m} factorial designs. The treatment combinations are written as k-tuples with entries of 0, 1 or 2. The values of k (the number of factors) and m (the degree of fractionation) are given. Blocks are indicated by semi-colons, as in Table 19.68. Designs with 27 and 81 treatment combinations have been given. For the smaller values of k, coset representatives, to generate the other blocks of a completely randomized design, have been given. (No designs with only nine treatment combinations have been given since these designs can estimate at most four main effects.)

Designs for 27 units

$k = 4$, $m = 1$
Defining contrast: $I \equiv ABCD$
Aliasing: $A \equiv AB^2C^2D^2 \equiv BCD$ and similarly, $AB \equiv ABC^2D^2 \equiv CD$ etc.

Treatment combinations: 0000, 0012, 0021, 1200, 2100, 1212, 2121, 1221, 2112; 0102, 0111, 0120, 1002, 2202, 1011, 2220, 1020, 2211; 0201, 0210, 2001, 2010, 0222, 2022, 1101, 1110, 1122.
Coset representatives: 1000, 2000
Blocks of size 9: $AB \equiv ABC^2D^2 \equiv CD$

$k = 5$, $m = 2$
Defining contrast: $I \equiv ABC \equiv CDE$
Aliasing: $A \equiv AB^2C^2 \equiv ACDE$ etc.
Treatment combinations: 00000, 11111, 22222; 01222, 12000, 20111; 02111, 10222, 21000; 00012, 11120, 22201; 01201, 12012, 20120; 02120, 10201, 21012; 00021, 11102, 22210; 01210, 12021, 20102; 02102, 10210, 21021
Coset representatives: 10000, 20000, 00100, 00200, 00010, 00020, 10100, 00110
Blocks of size 9: ABD
Blocks of size 3: ABD, ADE

$k = 6$, $m = 3$
Defining contrast: $I \equiv ABC \equiv ADE \equiv BDF$
Aliasing: $A \equiv AB^2C^2 \equiv AD^2E^2 \equiv ABDF$ etc.
Treatment combinations: 000000, 111111, 222222; 012121, 120202, 201010; 021212, 102020, 210101; 000122, 111200, 222011; 012210, 120021, 201102; 021001, 102112, 210220; 000211, 111022 , 222100; 012002, 120110, 201221; 021120, 102201, 210012
Coset representatives: 100000, 200000, 010000, 020000, 001000, 002000, 000100, 000200, 000010, 000020, 000001, 000002, 110000, 101000, 100100, 100010, 100001, 011000, 010100, 010001, 000110, 000101, 220000, 200200, 200002, 200001
Blocks of size 9: ACD
Blocks of size 3: ACD, AF^2

$k = 7$, $m = 4$
Defining contrast: $I \equiv ABC \equiv ADE \equiv BDF \equiv CEG$
Aliasing: $A \equiv AB^2C^2 \equiv AD^2E^2 \equiv ABDF \equiv ACEG$ etc.
Treatment combinations: 0000000, 1111111, 2222222; 0121212, 1202020, 2010101; 0212121, 1020202, 2101010; 0001221, 1112002, 2220110; 0122100, 1200211, 2011022; 0210012, 1021120, 2102201; 0002112, 1110220, 2221001; 0120021, 1201102, 2012210; 0211200, 1022011, 2100122
Blocks of size 9: ABG
Blocks of size 3: ABG, A^2F

$k = 8$, $m = 5$
Defining contrast: $I \equiv ABC \equiv ADE \equiv BDF \equiv CEG \equiv CDFH$
Aliasing: $A \equiv AB^2C^2 \equiv AD^2E^2 \equiv ABDF \equiv ACEG \equiv ACDFH$ etc.
Treatment combinations: 00000000, 01212122, 02121211, 10202021, 11111110, 12020202, 20101012, 21010101, 22222220; 00012210, 01221002, 02100121, 10211201, 11120020, 12002112, 20110222, 21022011, 22201100; 00021120, 01200212, 02112001, 10220111, 11102200, 12011022, 20122102, 21001221, 22210010
Blocks of size 9: A^2FH

$k = 9$, $m = 6$
Defining contrast: $I \equiv ABC \equiv ADE \equiv BDF \equiv CEG \equiv CDFH \equiv ABEJ$
Aliasing: $A \equiv AB^2C^2 \equiv AD^2E^2 \equiv ABDF \equiv ACEG \equiv ACDFH \equiv AF^2G^2J^2$ etc.
Treatment combinations: 000000000, 000122101, 000211202, 012002122, 012121220, 012210021, 021001211, 021120012, 021212110, 102020210, 102112011, 102201112, 111022002, 111111100, 111200201, 120021121, 120110222, 120202020, 201010120,

201102221, 201221022, 210012212, 210101010, 210220111, 222011001, 222100102, 222222200

Designs for 81 units

$k = 5$, $m = 1$
Defining contrast: $I \equiv ABCDE$
Aliasing: $A \equiv AB^2C^2D^2E^2$ etc.
Treatment combinations: 00000, 10122, 20211; 01110, 11202, 21021; 02220, 12012, 22101; 00012, 10101, 20220; 01122, 11211, 21000; 02202, 12021, 22110; 00021, 10110, 20202; 01101, 11220, 21012; 02211, 12000, 22122; 00102, 10221, 20010; 01212, 11001, 21120; 02022, 12111, 22200; 00111, 10200, 20022; 01221, 11010, 21102; 02001, 12120, 22212; 00120, 10212, 20001; 01200, 11022, 21111; 02010, 12102, 22221; 00201, 10020, 20112; 01011, 11100, 21222; 02121, 12210, 22002; 00210, 10002, 20121; 01020, 11112, 21201; 02100, 12222, 22011; 00222, 10011, 20100; 01002, 11121, 21210; 02112, 12201, 22020
Coset representatives: 10000, 20000
Blocks of size 27:ABC^2
Blocks of size 9: ABC^2, ACD^2
Blocks of size 3: ABC^2, ACD^2, ABD

$k = 6$, $m = 2$
Defining contrast: $I \equiv ABCD \equiv AB^2EF$
Aliasing: $A \equiv AB^2C^2D^2 \equiv ABE^2F^2$ etc.
Treatment combinations: 000000, 112212, 221121; 010201, 122110, 201022; 020102, 102011, 211220; 001212, 110121, 222000; 011110, 120022, 202201; 021011, 100220, 212102; 002121, 111000, 220212; 012022, 121201, 200110; 022220, 101102, 210011; 000012, 112221, 221100; 010210, 122122, 201001; 020111, 102020, 211202; 001221, 110100, 222012; 011122, 120001, 202210; 021020, 100202, 212111; 002100, 111012, 220221; 012001, 121210, 200122; 022202, 101111, 210020; 000021, 112200, 221112; 010222, 122101, 201010; 020120, 102002, 211211; 001212, 110112, 222021; 011101, 120010, 202222; 021002, 100211, 212120; 002100, 111021, 220200; 012010, 121222, 200101; 022211, 101120, 210002
Coset representatives: 100000, 200000, 010000, 020000, 001000, 002000, 000010, 000020
Blocks of size 27: AC^2E
Blocks of size 9: AC^2E, AC
Blocks of size 3: AC^2E, AC, EF

$k = 7$, $m = 3$
Defining contrast: $I \equiv ABCDE \equiv ACDE^2FG^2 \equiv AC^2DE^2G$
Aliasing: $A \equiv AB^2C^2D^2E^2 \equiv AC^2D^2EF^2G \equiv ABC^2DE^2G$ etc.
Treatment combinations: 0000000, 1111212, 2222121; 0122100, 1200012, 2011221; 0211200, 1022112, 2100021; 0010210, 1121122, 2202001; 0102010, 1210222, 2021101; 0221110, 1002022, 2110201; 0020120, 1101002, 2212211; 0112220, 1220102, 2001011; 0201020, 1012202, 2120111; 0001202, 1112111, 2220020; 0120002, 1201211, 2012120; 0212102, 1020011, 2101220; 0011112, 1122021, 2200200; 0100212, 1211121, 2022000; 0222012, 1000221, 2111100; 0021022, 1102201, 2210110; 0110122, 1221001, 2002210; 0202222, 1010101, 2121010; 0002101, 1110010, 2221222; 0121201, 1202110, 2010022; 0210001, 1021210, 2102122; 0012011, 1120220, 2201102; 0101111, 1212020, 2020202; 0220211, 1001120, 2112002; 0022221, 1100100, 2211012; 0111021, 1222200, 2000112; 0200121, 1011000, 2122212
Blocks of size 27: AG

Blocks of size 9: AG, FG
Blocks of size 3: AG, FG, CG

$k = 8$, $m = 4$
Defining contrast: $I \equiv ABCDEFGH \equiv AB^2CE^2G \equiv BCEF^2G \equiv ACD^2E^2FG^2$
Aliasing: $A \equiv AB^2C^2D^2E^2F^2G^2H^2 \equiv ABC^2EG^2 \equiv ABCEF^2G \equiv AC^2DEF^2G$
etc.
Treatment combinations: 00000000, 12212112, 21121221; 00001212, 12210021, 21122100; 00002121, 12211200, 21120012; 01100202, 10012011, 22221120; 01101111, 10010220, 22222002; 01102020, 10011102, 22220211; 02200101, 11112210, 20021022; 02201010, 11110122, 20022201; 02202222, 11111001, 20020110; 00120021, 12002100, 21211212; 00121200, 12000012, 21212121; 00122112, 12001221, 21210000; 01220220, 10102002, 22011111; 01221102, 10100211, 22012020; 01222011, 10101120, 22010202; 02020122, 11202201, 20111010; 02021001, 11200110, 20112222; 02022210, 11201022, 20110101; 00210012, 12122121, 21001200; 00211221, 12120000, 21002112; 00212100, 12121212, 21000021; 01010211, 10222020, 22101102; 01011120, 10220202, 22102011; 01012002, 10221111, 22100220; 02110110, 11022222, 20201001; 02111022, 11020101, 20202210; 02112201, 11021010, 20200122
Blocks of size 27: AB^2CD^2
Blocks of size 9: AB^2CD^2, AB
Blocks of size 3: AB^2CD^2, AB, BH^2

$k = 9$, $m = 5$
Defining contrast:
$I \equiv ABCDEF \equiv ABD^2E^2G^2 \equiv AB^2C^2EG^2 \equiv AC^2DE^2GJ \equiv BC^2D^2EGH$
Aliasing:
$A \equiv AB^2C^2D^2E^2F^2 \equiv AB^2DEG \equiv ABCE^2G \equiv ACD^2EG^2J^2 \equiv ABC^2D^2EGH$
etc.
Treatment combinations: 000000000, 110121221, 220212112; 001212020, 111000211, 221121102; 002121010, 112212201, 222000122; 010122100, 120210021, 200001212; 011001120, 121122011, 201210202; 012210110, 122001001, 202122222; 020211200, 100002121, 210120012; 021120220, 101211111, 211002002; 022002210, 102120101, 212211022; 000111122, 110202010, 220020201; 001020112, 111111000, 221202221; 002202102, 112020020, 222111211; 010200222, 120021110, 200112001; 011112212, 121200100, 201021021; 012021202, 122112120, 202200011; 020022022, 100110210, 210201101; 021201012, 101022200, 211110121; 022110002, 102201220, 212022111; 000222211, 110010102, 220101020; 001101201, 111222122, 221010010; 002010221, 112101112, 222222000; 010011011, 120102202, 200220120; 011220001, 121011222, 201102110; 012102021, 122220212, 202011100; 020100111, 100221002, 210012220; 021012101, 101100022, 211221210; 022221121, 102012012, 212100200
Blocks of size 27: ABG^2
Blocks of size 9: ABG^2, BG
Blocks of size 3: ABG^2, BG, HJ

19.70 Table Small $2^{k_1} \times 3^{k_2}$ designs. For values of k_1 and k_2 at least four, these designs can be constructed from the designs in the previous tables using the method illustrated in Example 19.33. Some smaller designs are given, where $k = k_1 + k_2$.

$k = 2$
2×3
Possible block sizes are 3 (which would confound A) or 2 (which would confound B).

$k = 3$
$2^2 \times 3$
Possible block sizes are 6, so AB is confounded (and hence $A \equiv B$ in a fractional plan); 4, where C is confounded; 3, where all information about the two-level factors is lost.
Blocks of size 6: generator: (111); coset representatives: (100)
2×3^2
Possible block sizes are 9, where A is confounded; 6, so 2 degrees of freedom of the two-factor interaction BC is lost and in a fractional design $B \equiv C$; 3, 3, where A and AB or AC are confounded.
Blocks of size 6: generator: (111); coset representatives: (010), (020).

$k = 4$
$2^3 \times 3$
Possible block sizes are: 12, 8, 6, 4 and 3. For 8 and 4, D is confounded with blocks. For 3, the main effects of A, B, C are all confounded with blocks.
Blocks of size 12: confound $I \equiv ABC$; aliasing: $A \equiv BC$, $D \equiv ABCD$; generators: (1100), (1011); coset representative: (1000).
Blocks of size 6: confound: $I \equiv AB \equiv AC$; aliasing: $A \equiv B \equiv C$ (so a fraction should not be used); generator: (1111); coset representatives: (1000), (1100), (0100).
$2^2 \times 3^2$
Possible block sizes are: 18, 12, 9, 6, 4 and 3. For 9, 4 and 3 some main effects must be confounded with blocks.
Blocks of size 18: confound: $I \equiv AB$; aliasing: $A \equiv B$ (so a fraction should not be used); generators: (1110), (0001); coset representative: (1000).
Blocks of size 12: confound: $I \equiv CD$; aliasing: $C \equiv D$ (so a fraction should not be used); generators: (1012), (0100); coset representatives: (0010), (0020).
Blocks of size 6: confound: $I \equiv AB \equiv CD$; aliasing: $A \equiv B$ (so a fraction should not be used); generator: (1112); coset representatives: (1000), (0010), (0020), (1010), (1020).
2×3^3
Possible block sizes are: 27, 18, 9, 6. For 27 and 9 information on some main effects will be confounded.
Blocks of size 18: confound $I \equiv BCD$; aliasing: $A \equiv ABCD$, $B \equiv BC^2D^2$; generators: (1111), (0120); coset representatives: (0100), (0200).
Blocks of size 6: confound: $I \equiv BC \equiv CD$; aliasing: $B \equiv C \equiv D$ (so a fraction should not be used); generator: (1120); coset representatives: (0100, (0200), (0001), (0002), (0101), (0202), (0201), (0102).

$k = 5$
$2^4 \times 3$
Possible blocks sizes are: 24, 16, 12, 8, 6. For 16 and 8 the main effect of E is confounded with blocks.
Blocks of size 24: confound: $I \equiv ABCD$; aliasing: $A \equiv BCD$, $E \equiv ABCDE$; generators: (11001), (00110), (10100); coset representative: (10000) .
Blocks of size 12: confound: $I \equiv AB \equiv CD \equiv ABCD$; aliasing: $A \equiv B \equiv ACD \equiv BCD$ (so a fraction should not be used); generators: (11001), (00110) ; coset representatives: (10000), (00100), (10100).
Blocks of size 6: confound: $I \equiv AB \equiv AC \equiv AD$ (generators only); aliasing: $A \equiv B \equiv C \equiv D$ (so a fraction should not be used); generator: (11111); coset representatives: (10000), (11000), (01000), (10100), (00100), (10010), (00010).
$2^3 \times 3^2$

Possible blocks sizes are: 36, 24, 18, 12, 9, 8. For 9 and 8 main effects are confounded with blocks.
Blocks of size 36: confound: $I \equiv ABC$; aliasing: $A \equiv BC$, $D \equiv ABCD$; generators: (11010), (10101); coset representative: (10000).
Blocks of size 24: confound: $I \equiv DE$; aliasing: $D \equiv E$ (so a fraction should not be used); generators: (10012), (01000), (00100); coset representatives: (00010), (00020).
Blocks of size 18: confound: $I \equiv AB \equiv AC \equiv BC$; aliasing: $A \equiv B \equiv C \equiv ABC$ (so a fraction should not be used); generators: (11110), (00001); coset representatives: (10000), (01000), (11000).
Blocks of size 12: confound: $I \equiv ABC \equiv DE$; aliasing: $A \equiv BC \equiv ADE$, $D \equiv ABCD \equiv E$ (so a fraction should not be used); generators: (11000), (10112); coset representatives: (00010), (00020), (10000), (10010), (10020).
$2^2 \times 3^3$
Possible blocks sizes are: 54, 36, 27, 18, 12. Only 36 does not alias main effects in a fraction.
Blocks of size 36: confound: $I \equiv CDE$; aliasing: $A \equiv ACDE \equiv AC^2D^2E^2$, $C \equiv CD^2E^2$; generators: (10111), (01120).
2×3^4
Possible blocks sizes are: 81, 54, 27, 18. For all but 54 main effects are confounded or aliased.
Blocks of size 54: confound: $I \equiv BCDE$; aliasing: $A \equiv ABCDE \equiv AB^2C^2D^2E^2$, $B \equiv BC^2D^2E^2 \equiv CDE$; generators: (11200), (00012), (01110).

19.9 See Also

§II.4	Orthogonal arrays of higher index.
§II.5	Orthogonal arrays of higher strength.
§IV.32	Orthogonal main effect plans.
§VI.7	Finite geometry.
[3, 14, 28]	More on factorial designs and groups.
[6]	Looks at confounding through blocking rather than through rings.
[34]	For more details on deletion designs.
[40]	For an example where Taguchi's methods failed.

References

[1] D.A. Anderson and A.M. Thomas, Resolution IV fractional factorial designs for the general asymmetric factorial, *Commun. Stat. Theory and Method* 8 (1979), 931-943.

[2] R. Bailey, Patterns of confounding in factorial designs, *Biometrika* 64 (1977), 597-603.

[3] R. Bailey, Factorial design and abelian groups, *Linear Algebra Applic.* 70 (1985), 349-368.

[4] R.A. Bailey, C.S. Cheng, and P. Kipnis, Construction of trend-resistant factorial designs, *Statistica Sinica* 2 (1992), 393-411.

[5] R.A. Bailey, F.H.L. Gilchrist and H.D. Patterson, Identification of effects and confounding patterns in factorial designs, *Biometrika* 64 (1977), 347-354.

[6] J.H. Beder, The problem of confounding in two-factor experiments, *Commun. Stat. Theory and Methods* 18 (1989), 2165-2188.

[7] S. Bisgaard, Blocking generators for small 2^{k-p} designs, *J. Quality Technology* 26 (1994), 288-296.

[8] G.E.P. Box, S. Bisgaard and C.A. Fung, An explanation and critique of Taguchi's contribution to quality engineering, *Qual. Reliab. Eng. Int.* 4 (1988), 123-131.

[9] G.E.P. Box and N.R. Draper, *Empirical Model Building and Response Surfaces*, John Wiley and Sons, New York, 1987.

[10] J.J. Cannon, An introduction to the group theory language CAYLEY, in: *Computational Group Theory* (M. Atkinson; ed.) 145-183, Academic Press, San Diego, 1984.

[11] K. Chatterjee, Search designs for searching three-factor interaction effects in the general symmetric and asymmetric factorials, *Sankhya B* 53 (1991), 304-326.

[12] J. Chen, D.X. Sun and C.F.J. Wu, A catalogue of two-level and three-level fractional factorial designs with small runs, *Int. Stat. Rev.* 61 (1993), 131-145.

[13] J. Chen and C.F.J. Wu, Some results on s^{n-k} fractional factorial designs with minimum aberration or optimal moments, *Ann. Statist.* 19 (1991), 1028-1041.

[14] B.J. Collings, Generating the intrablock and interblock subgroups for confounding in general factorial experiments, *Ann. Statist.* 12 (1984), 1500-1509.

[15] S.C. Cotter, A general method of confounding for symmetrical factorial experiments, *J. Roy. Statist. Soc. B* 36 (1974), 267-276.

[16] A.M. Dean and S.M. Lewis, A note on the connectivity of generalized cyclic designs, *Commun. Stat. Theory Methods* 15 (1986), 3429-3433.

[17] M.F. Franklin, Constructing tables of minimum aberration p^{n-m} designs, *Technometrics* 26 (1984), 225-232.

[18] M.F. Franklin, Selecting defining contrasts and confounding effects in p^{n-m} factorial experiments, *Technometrics* 27 (1985), 165-172.

[19] A. Fries and W.G. Hunter, Minimum aberration 2^{k-p} designs, *Technometrics* 22 (1980), 601-608.

[20] S.C. Gupta, Some new methods for constructing block designs having orthogonal factorial structure, *J. Roy. Statist. Soc. B* 45 (1983), 297-307.

[21] M. Hamada and C.F.J. Wu, Analysis of designed experiments with complex aliasing, *J. Qual. Technol.* 24 (1992), 130-137.

[22] K. Hinkelmann and O. Kempthorne, *Design and Analysis of Experiments. Vol. I*, John Wiley and Sons, New York, 1994.

[23] J.A. John and S.M. Lewis, Factorial experiments in generalised cyclic row-column designs, *J. Roy. Statist. Soc. B* 45 (1983), 245-251.

[24] P.W.M. John, Time trends and factorial experiments, *Technometrics* 32 (1990), 275-287.

[25] R.N. Kacker, E.S. Lagergren and J.J. Filliben, Taguchi's fixed element arrays are fractional factorials, *J. Qual. Technol.* 23 (1991), 107-116.

[26] J.P.C. Kleijnen, A practical comment on resolution IV designs: Bad properties of quantitative factors, *Commun. Stat. Theory Methods* 16 (1987), 805-812.

[27] A. Koblinsky, Orthogonal factorial designs for quantitative factors, *Stat. Decisions Suppl.* 2 (1985), 275-285.

[28] A. Koblinsky, Confounding in relation to duality of finite abelian groups, *Linear Algebra Applic.* 70 (1985), 321-347.

[29] A.M. Kshirsagar and S.A.N. Stehouwer, Efficiency of Bailey's trigonometric levels in factorial experiments, *Commun. Stat. Theory Methods* 16 (1987), 1655-1673.

[30] R.V. Lenth, Quick and easy analysis of unreplicated factorials, *Technometrics* 31 (1989), 469-473.

[31] S.M. Lewis, Composite generalised cyclic row-column designs, *Sankhya B* 48 (1986), 373-379.

[32] D.K.J. Lin and N.R. Draper, Projection properties of Plackett and Burman designs, *Technometrics* 34 (1992), 423-428.

[33] R.A. McLean and V.L. Anderson, *Applied Fractional Factorial Designs*, Marcel Dekker, New York, 1984.

[34] J.M. Mahoney and J.E. Angus, Estimability and efficiency in nearly orthogonal $2^{m_1} \times 3^{m_2}$ deletion designs, *Statistica Sinica* 3 (1993), 197-207.
[35] D.C. Montgomery, *Design and Analysis of Experiments*, John Wiley and Sons, New York, 1991.
[36] C.A.Mount-Campbell and J.B. Neuhardt, On the number of 2^{n-p} fractional factorials of resolution III, *Commun. Stat. Theory Methods* 10 (1981), 2101-2111.
[37] R. Mukerjee, Construction of effect-wise orthogonal factorial designs, *J. Stat. Plann. Infer.* 5 (1981), 221-229.
[38] J.A. Nelder, The analysis of randomised experiments with orthogonal block structure, *Proc. Royal Society A* 283 (1965), 147-162.
[39] H.D. Patterson and R.A. Bailey, Design keys for factorial experiments, *Appl. Stat.* 37 (1978), 335-343.
[40] R.A. Pease, The Taguchi Method, *Electron. Design* 41 (1993), 85.
[41] A.P. Street and D.J.Street, *Combinatorics of Experimental Design*, Clarendon Press, Oxford, 1987.
[42] T.P. Turiel, A computer program to determine defining contrasts and factor combinations for the two-level fractional factorial designs of resolution III, IV and V, *J. Qual. Technol.* 20 (1988), 267-272.
[43] D.T. Voss, Single-replicate generalised cyclic factorial designs as pseudo-factor designs, *Ann. Statist.* 16 (1988), 1723-1726.
[44] D.T. Voss and A.M. Dean, A comparison of classes of single-replicate factorial designs, *Ann. Statist.* 15 (1987), 376-384.
[45] D.T. Voss and A.M. Dean, On confounding in single replicate factorial experiments, *Utilitas Math.* 33 (1988), 59-64.

20 Frequency Squares

Charles F. Laywine

20.1 Definitions and Enumeration

Let $S = \{s_1, s_2, \ldots, s_m\}$ be a set of distinct symbols. A *frequency square* (F-square) $F(n; \lambda_1, \lambda_2, \ldots, \lambda_m)$ of *order* n on S with *frequency vector* $(\lambda_1, \lambda_2, \ldots, \lambda_m)$ is an $n \times n$ array in which symbol $s_i, i = 1, 2, \ldots, m$ occurs exactly λ_i times in each row and column. The order $n = \lambda_1 + \cdots + \lambda_m$.

An F-square $F(n; \lambda_1, \ldots, \lambda_m)$ is in *standard form* if, in the first row and column, all occurrences of s_i precede those of s_j whenever $i < j$.

20.1 Remark Let L be a latin square on n symbols and let $S_1, S_2, \ldots, S_m$ be a partition of the symbol set with $|S_i| = \lambda_i$. If every symbol in S_i is replaced by a new symbol s_i, then the resulting square is an F-square $F(n; \lambda_1, \ldots, \lambda_m)$.

20.2 Remark If $\lambda_1 = \lambda_2 = \cdots = \lambda_m = \lambda$, then $F(n; \lambda_1, \ldots, \lambda_m)$ is denoted by $F(n; \lambda)$.

20.3 Theorem Let $F(n; \lambda_1, \ldots, \lambda_m)$ and $f(n; \lambda_1, \ldots, \lambda_m)$ be respectively the number of F-squares and the number of standard form F-squares of order n with frequency vector $(\lambda_1, \ldots, \lambda_m)$. Then $F(n; \lambda_1, \ldots, \lambda_m) = \frac{n!(n-1)!f(n;\lambda_1,\ldots,\lambda_m)}{\lambda_1!(\lambda_1-1)!(\lambda_2!)^2\ldots(\lambda_m!)^2}$ [1].

20.2 Orthogonality

Two F-squares $F_1(n; \lambda_1, \ldots, \lambda_m)$ and $F_2(n; \mu_1, \ldots, \mu_k)$ based respectively on symbol sets $S = \{s_1, \ldots, s_m\}$ and $T = \{t_1, \ldots, t_k\}$ are *orthogonal* if each ordered pair (s_i, t_j) occurs exactly $\lambda_i \mu_j$ times when F_1 is superimposed on F_2.

In an *orthogonal set* of F-squares (MOFS) every pair of F-squares is orthogonal.

20.4 Remark If $F_1(n; 1)$ is orthogonal to $F_2(n; 1)$, then F_1 and F_2 are orthogonal latin squares.

20.5 Remark If two F-squares F_1 and F_2 are constructed from a pair of orthogonal latin squares by the method given in Remark 20.1, then F_1 and F_2 are orthogonal F-squares.

20.6 Theorem Let $F_1, F_2, \ldots, F_t$ be a set of MOFS of order n based respectively on $m_1, m_2, \ldots, m_t$ symbols. Then t satisfies the relation $\sum_{i=1}^{t} m_i - t \leq (n-1)^2$ [1].

If equality holds in Theorem 20.6, then the set of MOFS is *complete.*

20.7 Remark If $F_1, F_2, \ldots, F_t$ are all of type $F(n; \lambda)$ and form a complete set, then $t = (n-1)^2/(m-1)$.

20.8 Theorem For m a prime power and $h \geq 1$ there exists a complete set of MOFS of type $F(m^h; m^{h-1})$. These may be constructed

1. from symmetric factorial designs [1];
2. from linear permutation polynomials in $2h$ indeterminates over the field $\mathbb{F}_m$ [1];
3. from a complete set of MOLS of order m^h and an AD(m^{h-1}, m) (an affine resolvable design of order m and block size m^{h-1}) [3];
4. from an AD(m^{2h-1}, m) with at least $2h-2$ independent prime classes;
5. from a complete set of MOLS of order m and an AD(m^{h-1}, m).

20.9 Remark The existence of a complete set of MOFS of type $F(m^h; m^{h-1})$ for m not a prime power is an open question; however, there is one known example of complete sets whose order is not a prime power. These are sets of type $F(4t; 2t)$ which can be constructed from a Hadamard matrix of order $4t$.

20.3 Designs Derivable from Complete Sets of MOFS

20.10 Theorem If there exists a complete set of MOFS of type $F(m^h; m^{h-1})$ and an affine design AD(m^{h-1}, m), then there is an AD(m^{2h-1}, m).

20.11 Theorem [3] Consider some bijection from the symbols of a complete set of desarguesian MOLS of order m^h to the points of the geometry AG(h, m). Suppose that for some parallel class in the geometry the members of each subset of m^{h-1} symbols (identified with the points in a hyperplane) are replaced by a common unique symbol in each of the m^{h-1} MOLS. When done for each of the $(m^h - 1)/(m-1)$ classes, a complete set of $F(m^h; m^{h-1})$ MOFS is obtained, equivalent to the geometry AG$(2h, m)$.

20.4 Nonisomorphic Sets

Two sets of MOFS are *isomorphic* if the sets have the same number of F-squares and the squares of one set can be transformed into the squares of the second set by a series of row and column permutations on all F-squares of the first set.

20.12 Theorem [4] There are exactly three nonisomorphic sets of MOFS of type $F(4;2)$.

20.13 Table Nonisomorphic complete sets of $F(4;2)$ MOFS.

Set									
H_1 :	0 0 1 1	0 1 0 1	0 1 1 0	0 0 1 1	0 0 1 1	0 1 0 1	0 1 0 1	0 1 1 0	0 1 1 0
	0 0 1 1	0 1 0 1	0 1 1 0	1 1 0 0	1 1 0 0	1 0 1 0	1 0 1 0	1 0 0 1	1 0 0 1
	1 1 0 0	1 0 1 0	1 0 0 1	0 0 1 1	1 1 0 0	0 1 0 1	1 0 1 0	0 1 1 0	1 0 0 1
	1 1 0 0	1 0 1 0	1 0 0 1	1 1 0 0	0 0 1 1	1 0 1 0	0 1 0 1	1 0 0 1	0 1 1 0
H_2 :	0 0 1 1	0 1 0 1	0 1 1 0	0 0 1 1	0 0 1 1	0 1 0 1	0 1 0 1	0 1 1 0	0 1 1 0
	0 0 1 1	0 1 0 1	0 1 1 0	1 1 0 0	1 1 0 0	1 0 1 0	1 0 1 0	1 0 0 1	1 0 0 1
	1 1 0 0	1 0 1 0	1 0 0 1	0 0 1 1	1 1 0 0	0 1 1 0	1 0 0 1	0 1 0 1	1 0 1 0
	1 1 0 0	1 0 1 0	1 0 0 1	1 1 0 0	0 0 1 1	1 0 0 1	0 1 1 0	1 0 1 0	0 1 0 1
H_3 :	0 0 1 1	0 1 0 1	0 1 1 0	0 0 1 1	0 0 1 1	0 1 0 1	0 1 0 1	0 1 1 0	0 1 1 0
	0 0 1 1	0 1 0 1	0 1 1 0	1 1 0 0	1 1 0 0	1 0 1 0	1 0 1 0	1 0 0 1	1 0 0 1
	1 1 0 0	1 0 1 0	1 0 0 1	0 1 0 1	1 0 1 0	0 1 1 0	1 0 0 1	0 0 1 1	1 1 0 0
	1 1 0 0	1 0 1 0	1 0 0 1	1 0 1 0	0 1 0 1	1 0 0 1	0 1 1 0	1 1 0 0	0 0 1 1

20.14 Table The number N of nonisomorphic complete sets of MOFS of type $F(q^h; q^{h-1})$ for q a prime power [4].

Type	(4,2)	(8,4)	(9,3)	(16,4)	(16,8)	(25,5)
N	3	≥ 17	≥ 22	≥ 401	≥ 362	≥ 11125

20.5 See Also

§II.1	Frequency squares are analogs of latin squares for higher index.
§II.2	MOFS are generalizations of MOLS.
[1]	Contains a chapter on frequency squares with several of the results along with the original references.
[2]	Examines relationships between frequency squares and codes, and frequency squares and orthogonal arrays.

References

[1] J. Denés and A.D. Keedwell, *Latin Squares: New Developments in the Theory and Applications, Ann. Discrete Math.* 46, North-Holland, Amsterdam, 1991.

[2] W.T. Federer and J.P. Mandeli, Orthogonal F-rectangles, orthogonal arrays and codes, *J. Combin. Theory A* 43 (1986), 149-164.

[3] C.F. Laywine and G.L. Mullen, Generalizations of Bose's equivalence between complete sets of mutually orthogonal latin squares and affine planes, *J. Combin. Theory A* 61 (1992), 13-35.

[4] S.J. Suchower, Nonisomorphic complete sets of F-rectangles with varying numbers of symbols, in: *Finite Fields: Theory, Applications, and Algorithms* (G.L. Mullen and P.J.-S. Shiue; eds.), Contemporary Mathematics, American Mathematical Society, 1994.

21 Generalized Quadrangles

Stanley E. Payne

21.1 Definitions and Examples

A *finite generalized quadrangle* (GQ) is an incidence structure $\mathcal{S} = (P, B, I)$ with pointset P and lineset B satisfying the following:

1. each point is incident with $1+t$ lines ($t \geq 1$) and two distinct points are incident with at most one line;
2. each line is incident with $1+s$ points ($s \geq 1$) and two distinct lines are incident with at most one point;
3. if x is a point and L is a line not incident with x, then there is a unique pair $(y, M) \in P \times B$ for which $xIMIyIL$.

The integers s and t are the *parameters* of the GQ and $\mathcal{S}$ has *order* (s,t); if $s = t$, $\mathcal{S}$ has *order* s.

21.1 Remark There is a point-line duality for GQs of order (s,t) for which in any definition or theorem the words *point* and *line* are interchanged and the parameters s and t are interchanged. Without further notice, the dual of a given theorem or definition is assumed to have also been given. In particular, when listing parameters of known GQs one may assume that $s \leq t$.

21.2 Remark A generalized quadrangle is a partial geometry with $\alpha = 1$. See §IV.34.

21.3 Proposition If $|P| = v$ and $|B| = b$, then $v = (1+s)(1+st)$ and $b = (1+t)(1+st)$. Further, $s+t$ divides $st(s+1)(t+1)$, and for $s \neq 1 \neq t$ it holds that $s \leq t^2$ and $t \leq s^2$.

21.4 Remark For q any prime power, GQs with parameters (s,t), $s \leq t$ are known for $(s,t) \in \{(q-1, q+1), (q,q), (q,q^2), (q^2,q^3)\}$. With $s = 2$, there are unique examples with $t = 2$ and $t = 4$ (Examples 21.5 and 21.6) and none with $t = 3$. With $s = 3$, there is an example with $t = 3$ (not self-dual, but unique up to point-line duality), unique examples with $t = 5$ and $t = 9$, and no GQ with $t = 4, 6, 7, 8$. With $s = 4$, existence would imply that $t \in \{4, 6, 8, 11, 12, 16\}$. There is a unique example with $t = 4$. Nothing is known about $t = 11$ or $t = 12$. In the other cases unique examples are known, but the uniqueness question is settled only in the case $t = 4$.

21.5 Example The GQ with $s = t = 2$. Start with the set $I_6 = \{1, 2, 3, 4, 5, 6\}$. A *duad* is an unordered pair $ab = ba$ of elements of I_6. A *syntheme* is a triple of disjoint duads. Incidence is ordinary containment. There are fifteen duads and fifteen synthemes, and it is easily checked that the syntheme-duad geometry is a model of the unique GQ of order 2 often denoted $W(2)$. See Example VI.3.8 for the point-block incidence graph of $W(2)$.

21.6 Example To the syntheme-duad model, adjoin the 12 symbols $1, 2, \ldots, 6, 1', \ldots, 6'$ as points. Through each duad ab adjoin the lines $\{a, ab, b'\}$ and $\{a', ab, b\}$. The resulting incidence structure is a model of the unique GQ with parameters $(2, 4)$.

21.2 Classical Constructions

21.7 Construction Let $P = \{x_{ij} : i, j = 0, 1, \ldots s\}$, $B = \{L_0, L_1, \ldots, M_0, M_1, \ldots, M_s\}$, $x_{ij} I L_k$ if and only if $i = k$, $x_{ij} I M_k$ if and only if $j = k$. Then $\mathcal{S} = (P, B, I)$ is a GQ of order $(s, 1)$. Such a (trivial) GQ is a *grid*. A dual grid has order $(1, t)$.

21.8 Construction $Q(d, q)$. Consider a nonsingular quadric Q of projective index 1 (i.e., Witt index 2) of the projective space PG(d, q), with $d = 3, 4$, or 5. Then the points of Q together with the lines of Q form a GQ denoted $Q(d, q)$ of order (q, q^{d-3}). The quadric Q has the following canonical equation: $x_0x_1 + x_2x_3 = 0$, when $d = 3$, in which case $Q(3, q)$ is a grid. When $d = 4$ the equation is $x_0{}^2 + x_1x_2 + x_3x_4 = 0$, and when $d = 5$ it is $f(x_0, x_1) + x_2x_3 + x_4x_5 = 0$, where f is an irreducible binary quadratic form.

21.9 Construction $H(d, q^2)$. Let H be a nonsingular hermitian variety of the projective space PG(d, q^2), $d = 3$, or $d = 4$. Then the points of H together with the lines on H form a GQ denoted $H(d, q^2)$ of order (q^2, q) or (q^2, q^3) depending upon whether $d = 3$ or $d = 4$, and H has the canonical equation $x_0{}^{q+1} + x_1{}^{q+1} + \cdots + x_d{}^{q+1} = 0$.

21.10 Construction $W(q)$. The points of PG$(3, q)$, together with the totally isotropic lines with respect to a symplectic polarity, form a GQ denoted $W(q)$ of order (q, q). Specifically, for points $\mathbf{x} = (x_0, x_1, x_2, x_3)$ and $\mathbf{y} = (y_0, y_1, y_2, y_3)$, define $\mathbf{x}$ and $\mathbf{y}$ to be collinear if and only if $x_0y_1 - x_1y_0 + x_2y_3 - x_3y_2 = 0$.

21.11 Remark The *classical* GQs from Construction 21.10 were all discovered by J. Tits and are the only ones embedded as full subgeometries of finite projective space. The point-line dual of $Q(4, q)$ is isomorphic to $W(q)$; $Q(4, q)$ (and hence also $W(q)$) is self-dual if and only if q is even. The point-line dual of $Q(5, q)$ is isomorphic to $H(3, q^2)$.

21.3 GQs from Ovals and Ovoids

21.12 Remark In the constructions of this section, let $\pi = \text{PG}(2, q)$ be embedded as a plane of $\Sigma = \text{PG}(3, q)$, q any prime power. Let $\Omega = \{X_0, \ldots, X_q\}$ be an oval of π, so Ω consists of $q + 1$ points of π, no three on a line. Let t_i be the line of π tangent to Ω at X_i, $0 \le i \le q$. Let P_1 be the pointset of $\Sigma \setminus \pi$.

21.13 Construction $T_2(\Omega)$. There is a GQ $\mathcal{S} = \mathcal{S}(\Omega) = (P, B, I)$ of order q defined as follows. The pointset $P = P_1 \cup P_2$, with P_1 as given above, and with P_2 the set of planes of Σ containing at least one t_i. B is the set of lines of Σ meeting Ω in exactly one point. Incidence I is just that inherited from Σ. For example, the "point" π of $\mathcal{S}$ is incident with the $1 + q$ lines $t_0, \ldots, t_q$. The resulting GQ, denoted $T_2(\Omega)$, has order q.

21.14 Construction $\mathcal{S}(\Omega^-)$. Let $\Omega^- = \{X_1, \ldots, X_q\} = \Omega \setminus \{X_0\}$. Each point $X_j \in \Omega^-$ is on two tangents to Ω^-: the line $l_j = X_0X_j$ and the tangent t_j, $1 \le j \le q$. Put $P_1^- = P_1 = \Sigma \setminus \pi$. Let $P_2^- = \{p : p \ne \pi \text{ is a plane of } \Sigma \text{ with } p \cap \pi = t_j,\ j = 1, \ldots, q\}$; $P_3^- = \{p : p \ne \pi \text{ is a plane of } \Sigma \text{ with } p \cap \pi = l_j,\ j = 1, \ldots, q\}$. Put $P^- = P_1^- \cup P_2^- \cup P_3^-$ and let B^- be the set of lines of Σ not in π and meeting π in a point of Ω^-. With the incidence I^- induced by that of Σ, the structure $\mathcal{S}^- = \mathcal{S}(\Omega^-) = (P^-, B^-, I^-)$ is a GQ of order $(q + 1, q - 1)$.

21.15 Construction $\mathcal{S}(\Omega^+)$. When q is even ($q = 2^e$), there is a unique hyperoval $\Omega^+ = \Omega \cup \{X_\infty\}$ containing Ω. There is a GQ $\mathcal{S}^+ = \mathcal{S}(\Omega^+) = (P^+, B^+, I^+)$ of order $(q - 1, q + 1)$ defined as follows. Let $P^+ = P_1$ and let B^+ be the set of all lines of Σ meeting π at a point of Ω^+. The incidence I^+ is that induced by incidence in Σ.

21.16 Construction $T_3(\mathcal{O})$. Let $\mathcal{O}$ be an *ovoid* of $\Sigma = \mathrm{PG}(3,q)$, i.e., $\mathcal{O}$ is a set of $1+q^2$ points of Σ with no three on a line of Σ. Further, let Σ be embedded as an hyperplane in $\mathrm{PG}(4,q) = \mathcal{P}$. Define points as (i) the points of $\mathcal{P} \setminus \Sigma$, (ii) the hyperplanes Z of $\mathcal{P}$ for which $|Z \bigcap \mathcal{O}| = 1$, and (iii) one new symbol (∞). Lines are defined as (a) the lines of $\mathcal{P}$ that are not contained in Σ and meet $\mathcal{O}$ in a (necessarily unique) point, and (b) the points of $\mathcal{O}$. Incidence is defined as follows: A point of type (i) is incident only with lines of type (a); here the incidence is that of $\mathcal{P}$. A point of type (ii) is incident with all lines of type (a) contained in it and with the unique element of $\mathcal{O}$ in it. The point (∞) is incident with no line of type (a) and all lines of type (b). There are no other incidences.

The incidence structure just defined is a GQ with order $(s,t) = (q,q^2)$. It is usually denoted $T_3(\mathcal{O})$. When $\mathcal{O}$ is an elliptic quadric, $T_3(\mathcal{O})$ is isomorphic to $Q(5,q)$. The only nonclassical example known is for $q = 2^e$ with $e = 2h+1 \geq 3$ and $\mathcal{O}$ the Suzuki-Tits ovoid.

21.4 GQs from q-Clans

A *q-clan* is a set $\mathcal{C} = \{A_t : t \in F\}$ containing q 2×2 matrices over $\mathbb{F}_q$ whose pairwise differences are each *anisotropic*, i.e., the quadratic forms they determine represent zero only trivially.

21.17 Remark In addition to the constructions of GQs in §21.2 and §21.3, many examples of GQs with order $(s,t) = (q^2,q)$, q a prime power, have been constructed as a kind of group coset geometry starting with a q-clan . When q is even, these GQs have subquadrangles of order q associated with ovals in $\mathrm{PG}(2,q)$. Moreover, these families of ovals, known as *herds* of ovals, coexist with the q-clans and hence with certain GQs with parameters (q^2,q).

21.18 Construction Let $G = \{(\alpha,c,\beta) | \alpha,\beta \in F^2, c \in F\}$ be the group of order q^5 with binary operation $(\alpha,c,\beta) \circ (\alpha',c',\beta') = (\alpha+\alpha', c+c'+\beta\circ\alpha', \beta+\beta')$, where $\alpha\circ\beta = \alpha\beta^T$. Define subgroups of G having order q^2: $A(\infty) = \{(\vec{0},0,\beta) \in G | \beta \in F^2\}$, $A(t) = \{(\alpha, \alpha A_t \alpha^T, \alpha K_t) | \alpha \in F^2\}$. The center of G is $Z = \{(\vec{0},c,\vec{0}) | c \in F\}$. For $A \in \mathcal{J} = \{A(t) : t \in F \cup \{\infty\}\}$, put $A^* = AZ$. $\mathcal{J}$ is a 4-gonal family for G (cf. [7]) so provides the following construction of a GQ. Points are (i) the symbol (∞); (ii) right cosets A^*g, $g \in G$, $A \in \mathcal{J}$; (iii) elements g of G. Lines are (a) symbols $[A]$, $A \in \mathcal{J}$; (b) right cosets Ag, $g \in g$, $A \in \mathcal{J}$. Then (∞) is incident with each line of type (a) and none of type (b). The point A^*g is incident with $[A]$ and with each line Ah contained in A^*g. The point g is incident with each line Ag containing it.

21.19 Table Known families of q-clans. The parameter t is to vary over all elements of F. See [5] and [8] for more details and many references to the literature.

1. Classical: $A_t = \begin{pmatrix} t & bt \\ 0 & ct \end{pmatrix}$, x^2+bx+c irreducible over F.
2. $A_t = \begin{pmatrix} t & 3t^2 \\ 0 & 3t^3 \end{pmatrix}$; $q \equiv 2 \pmod 3$.
3. $A_t = \begin{pmatrix} t & 0 \\ 0 & -mt^\sigma \end{pmatrix}$, q odd, $\sigma \in Aut(F)$, m a nonsquare in F.
4. $A_t = \begin{pmatrix} t & 5t^3 \\ 0 & 5t^5 \end{pmatrix}$, $q \equiv \pm 2 \pmod 5$.
5. The Thas-Fisher example for q odd (cf. [5]).

6. $A_t = \begin{pmatrix} t & 3t^2 \\ 3t^2 & k^{-1}t(1+kt^2)^2 \end{pmatrix}$, $q = 5^e$, k a nonsquare in F.

7. $A_t = \begin{pmatrix} t & -t^3 \\ -t^3 & -t(t^8+m^2)/m \end{pmatrix}$, $q = 3^e$, m a nonsquare in F.

8. $A_t = \begin{pmatrix} \frac{d^2t^4+d^3t^3+(d^2+d^4)t^2}{t^4+d^2t^2+1} + (t/d)^{1/2} & t^{1/2} \\ 0 & \frac{(d^2+d^4)t^3+d^3t^2+d^2t}{t^4+d^2t^2+1} + (t/d)^{1/2} \end{pmatrix}$, where $q = 2^e$, x^2+dx+1 irreducible over F.

21.20 Remark Most of what was known about generalized quadrangles up through 1983 is presented in Payne and Thas [7]. The appearance of Kantor [3] in 1980 was seminal for the work that followed, leading to the notion of q-clans. Thas [8] gives an excellent survey of the results published between 1984 and 1992. The so-called Subiaco q-clan of Table 21.19 (8) (originally given in a slightly different form) was found in 1993 by Cherowitzo et al. ([2]; see also [1] and [5]) and has sparked a great deal of interest. For $q = 2^e \geq 16$ it provides a unique nonclassical GQ of order (q^2, q). For $e \not\equiv 2 \pmod 4$ it gives a unique oval in $PG(2,q)$, and for $e \equiv 2 \pmod 4$ it gives two inequivalent ovals in $PG(2,q)$. For $q \geq 32$, the GQs constructed from the different q-clans given above are all distinct. For $q \leq 4096$ a few sporadic examples of q-clans have been found using a computer.

21.5 Other Designs Derived From GQs

21.21 Remark The classical GQs can be used to construct projective planes, inversive planes, Laguerre planes of odd order, and Minkowski planes of even order. The q-clans used to construct GQs of order (q^2, q) can also be used to construct several other types of geometries including translation planes of dimension at most two over their kernel.

21.22 Construction A GQ of order q yields (potentially two) symmetric $2-(q^3+q^2+q+1, q^2+q+1, q+1)$ designs whose incidence matrix contains the identity matrix, one from the point-collinearity graph and the other from the line-concurrency graph of the GQ.

21.23 Construction A GQ with parameters $(q-1, q+1)$ yields a symmetric $2-(q^2(q+2), q(q+1), q)$ design whose incidence matrix has a zero diagonal, from the line-concurrency graph. Here distinct concurrent lines represent an incident point-block pair of the design.

21.24 Construction A GQ with parameters (q, q^2) yields an orthogonal array $OA_q(3, q, 1+q^2)$. Here the $1+q^2$ lines through a fixed point (∞) can have their q other points labeled with q symbols of a set F. Each of the q^4 points not collinear with (∞) determines a "row" of length $1+q^2$ of symbols from F, since it is collinear with exactly one point on each line through (∞). The fact that each triple of pairwise noncollinear points all collinear with (∞) are collinear with exactly q points not collinear with (∞) provides an orthogonal array of strength 3.

21.6 See Also

§I.5	Symmetric designs come from GQs via Constructions 21.22 and 21.23.
§II.5	Certain GQs yield orthogonal arrays via Construction 21.24.
§IV.34	A GQ is a partial geometry with $\alpha = 1$.
§VI.7	The classical GQs are embedded in finite projective space.
[7]	A book that surveys GQs up through 1983.
[8]	A survey of results on GQs from 1983 through 1992.

References

[1] L. Bader, G. Lunardon and S.E. Payne, On q-clan geometry, $q = 2^e$, *Simon Stevin* 1 (1994), 301-328
[2] W. Cherowitzo, T. Penttila, I. Pinneri and G. Royle, Flocks and ovals, to appear.
[3] W.M. Kantor, Generalized quadrangles associated with $G_2(q)$, *J. Combin. Theory A* 29 (1980), 212-219.
[4] S.E. Payne, An essay on skew translation generalized quadrangles, *Geom. Ded.* 32 (1989), 93-118.
[5] S.E. Payne, Collineations of the generalized quadrangles associated with q-clans, *Ann. Discrete Math.* 52 (1992), 449-461.
[6] S.E. Payne, T. Penttila, and I. Pinneri, Isomorphisms between Subiaco q-clan geometries, to appear.
[7] S.E. Payne and J.A. Thas, *Finite Generalized Quadrangles*, Pitman, London, 1984.
[8] J.A. Thas, Generalized polygons, in: *Handbook of Incidence Geometry* (F. Buekenhout; ed.), North-Holland, Amsterdam, 1995.

22 Graph Decompositions and Designs

KATHERINE HEINRICH

22.1 Definitions

Let $G = (V(G), E(G))$ be a graph. A $(\lambda K_n, G)$-*design* is a partition of the edges of λK_n into subgraphs (G-*blocks*) each of which is isomorphic to G.

A $(\lambda K_n, G)$-design is *balanced* if each vertex of λK_n belongs to exactly the same number of G-blocks.

An *(almost) G-factor* is a spanning subgraph of ($\lambda K_n - v$, for some vertex v) λK_n such that each component of the (almost) G-factor is isomorphic to G. A $(\lambda K_n, G)$-design is *(almost) resolvable* if the G-blocks can be partitioned into (almost) G-factors.

A $(\lambda K_n, G)$-design is *cyclic* if the permutation $(12\cdots n)$ on the vertex set of λK_n permutes the G-blocks.

22.1 Remark A $(\lambda K_n, K_k)$-design and a resolvable $(\lambda K_n, K_k)$-design are nothing but a (n, k, λ)-BIBD and a resolvable (n, k, λ)-BIBD, respectively.

22.2 Existence

22.2 Theorem If there exists a $(\lambda K_n, G)$-design, then

1. $\lambda n(n-1) \equiv 0 (\text{mod } 2|E(G)|)$, and
2. $\lambda(n-1) \equiv 0 (\text{mod } d)$, where d is the g.c.d. of the degrees of the vertices of G.

If the design is balanced, then

3. $\lambda |V(G)|(n-1) \equiv 0 (\text{mod } 2|E(G)|)$.

If the design is resolvable, then

4. $n \equiv 0 (\text{mod } |V(G)|)$, and
5. there exist integers $x_1, \ldots, x_k$ such that $\sum_{i=1}^{k} x_i d_i = \lambda(n-1)$ and $\sum_{i=1}^{k} x_i = \frac{\lambda k(n-1)}{2|E(G)|}$, where $k = |V(G)|$ and $(d_1, \cdots, d_k)$ is the degree sequence of G.

22.3 Theorem [20] For a given graph G and integer λ, there exists an integer $N(G, \lambda)$ such that if $n > N(G, \lambda)$, the necessary conditions for the existence of a $(\lambda K_n, G)$-design are sufficient.

22.3 Path Decompositions

P_k is the path on k vertices.

22.4 Theorem [10] There exists a $(\lambda K_n, P_k)$-design if and only if $\lambda n(n-1) \equiv 0 (\text{mod } 2(k-1))$ and $n \geq k$.

22.5 Theorem [10] There exists a balanced $(\lambda K_n, P_k)$-design if and only if $\lambda n(n-1) \equiv 0 (\text{mod } 2(k-1))$, $\lambda(k-2)(n-1) \equiv 0 (\text{mod } 2(k-1))$ and $\lambda(n-1) \equiv 0 (\text{mod } k-1)$.

22.6 Theorem [10] There exists a resolvable $(\lambda K_n, P_k)$-design if and only if $n \equiv 0 (\text{mod } k)$ and $\lambda k(n-1) \equiv 0 (\text{mod } 2(k-1))$.

22.7 Theorem [22] There exists an almost resolvable $(\lambda K_n, P_k)$-design if and only if $n \equiv 1 (\text{mod } k)$ and $\lambda k n \equiv 0 (\text{mod } 2(k-1))$.

22.4 Star Decompositions

22.8 Theorem [8] A $(\lambda K_n, K_{1,k-1})$-design exists if and only if $\lambda n(n-1) \equiv 0 (\text{mod } 2(k-1))$ and $n \geq 2(k-1)$ for $\lambda = 1$; $n \geq k$ for even λ and $n \geq k + \frac{k-1}{\lambda}$ for odd $\lambda \geq 3$.

22.9 Theorem [8] For $k \geq 3$, the following three statements are equivalent.

1. There exists a balanced $(\lambda K_n, K_{1,k-1})$-design.
2. There exists a cyclic $(\lambda K_n, K_{1,k-1})$-design.
3. $\lambda(n-1) \equiv 0 (\text{mod } 2(k-1)), n \geq k-1$.

22.10 Theorem [12] For k even, there is a resolvable $(\lambda K_n, K_{1,k-1})$-design if and only if $n \equiv k (\text{mod } \frac{k(k-1)}{x})$ and $\lambda \equiv 0 (\text{mod } 2x)$, where x is a factor of $k-1$.

22.11 Theorem [12] If k is odd, a resolvable $(\lambda K_n, K_{1,k-1})$-design exists if either $n = k^{2q}$, or λ is even and $n = k^q$, or $n \equiv k (\text{mod } \frac{2k(k-1)}{x})$ and $\lambda \equiv 0 (\text{mod } x)$, where x is even and divides $k-1$.

22.12 Theorem [21] If k is odd, a resolvable $(K_n, K_{1,k-1})$-design exists if and only if $n \equiv 1 (\text{mod } 2(k-1))$ and $n \equiv 0 \ (\text{mod } k)$.

22.5 Cycle Decompositions

22.13 Remark A $(\lambda K_n, C_k)$-design is balanced. A (K_n, C_k)-design is a *k-cycle system.* The necessary conditions for the existence of a $(\lambda K_n, C_k)$-design are $\lambda n(n-1) \equiv 0 (\text{mod } 2k)$ and $\lambda(n-1) \equiv 0 (\text{mod } 2)$.

22.14 Theorem [5, 7] For $k \in \{3, 4, 5, 6, 7, 8, 10, 12, 14, 16\}$, the necessary conditions for the existence of a $(\lambda K_n, C_k)$-design are sufficient.

22.15 Theorem [2, 3, 14] The necessary conditions for the existence of a (K_n, C_k)-design are sufficient for the following k and n:

1. $k = n; 3 \le k \le 50; k = p^r, 2p^r$, where p is a prime;
2. $n \equiv 1 \pmod{2k}$;
3. k is odd and $n \equiv k \pmod{2k}$;
4. $k \equiv 2 (\mathrm{mod}\ 8), n = \frac{5k}{2}$; and
5. $k \equiv 6 (\mathrm{mod}\ 8), n = \frac{3k}{2}$.

22.16 Theorem [16] If there exists a (K_a, C_k)-design for some integer $2\lfloor \frac{k}{2} \rfloor < a \le 6\lfloor \frac{k}{2} \rfloor$, then there exists a (K_{2kx+a}, C_k)-design for all integers $x \ge 0$.

22.17 Theorem [18, 19] For even k or $k = 3, 5$, there exists a cyclic (K_{2kp+1}, C_k)-design for any p. For $k = 3, 5$, there exists a cyclic (K_{2kp+k}, C_k)-design for any p.

22.18 Theorem [14] There exists a resolvable (K_n, C_k)-design if and only if k is odd and $n \equiv k (\mathrm{mod}\ 2k)$. See also §IV.29.

22.19 Remark There is no almost resolvable (K_n, C_k)-design for $n > 1$. However, it is natural to consider the almost resolvable $(2K_n, C_k)$-designs.

22.20 Theorem [14] There exists an almost resolvable $(2K_n, C_k)$-design if and only if $n \equiv 1 (\mathrm{mod}\ k)$.

22.6 k-Cube Decompositions

22.21 Remark Let Q_k be the k-cube. The necessary conditions for the existence of a (K_n, Q_k)-design are $n(n-1) \equiv 0 (\mathrm{mod}\ k2^k)$ and $n \equiv 1 (\mathrm{mod}\ k)$.

22.22 Theorem [13] Let n and k be positive integers. Then

1. If $n \equiv 1 \pmod{k2^k}$, there exists a (K_n, Q_k)-design.
2. If there exists a (K_n, Q_k)-design with n odd (or k even), then $n \equiv 1 \pmod{k2^k}$.
3. If a (K_n, Q_k)-design exists, where n is even and k is odd, then $n \equiv 1 (\mathrm{mod}\ k)$ and $n \equiv 0 \pmod{2^k}$.

22.23 Theorem [15] There exists a (K_n, Q_3)-design if and only if $n \equiv 1$ or $16 \pmod{24}$.

22.24 Remark If n is odd or k is even, statements (1) and (2) in Theorem 22.22 yield a necessary and sufficient condition for the existence of a (K_n, Q_k)-design.

22.7 Graph Decompositions with Special Properties

A $(2K_n, G)$-design is *self-orthogonal* if for any two G-blocks G_1 and G_2, $|E(G_1) \cap E(G_2)| = 1$. (This implies $|E(G)| = n - 1$.)

A $(\lambda K_n, G)$-design is self-complementary if G is self-complementary and the complement of each G-block yields another $(\lambda K_n, G)$-design.

22.25 Theorem [1, 11] If $2K_n$ has a self-orthogonal factorization into hamiltonian paths, then

1. $2K_{rn}$ has a self-orthogonal factorization into hamiltonian paths, where $r = 5^a 9^b 13^c 17^d 29^e$ and a, b, c, d, e are nonnegative integers, $a + b + c + d + e \ge 1$.
2. $2K_{4n}$ has a self-orthogonal factorization into $(n-1)$-cycles.

22.26 Theorem [1] If $n = p^e > 3$ where p is a prime or $4 \leq n \leq 36$, then $2K_n$ has a self-orthogonal factorization into $(n-1)$-cycles.

22.27 Remark Combining Theorems 22.25 and 22.26 and the fact that $2K_n$ has a self-orthogonal factorization into hamiltonian paths for $2 \leq n \leq 20, n \neq 4$, one obtains infinitely many values of n so that $2K_n$ has either a self-orthogonal hamiltonian path factorization or a self-orthogonal $(n-1)$-cycle factorization [1].

22.28 Theorem [1] There is a cyclic, self-orthogonal factorization of $2K_n$ into isomorphic factors whenever $n \equiv 1, 2, 4, 5, 7, 10, 11 (\text{mod } 12)$.

22.29 Theorem [16] A self-complementary (K_n, P_4)-design exists if and only if $n \equiv 1 (\text{mod } 3)$.

22.30 Theorem [14] There exists a self-complementary (K_n, C_5)-design if and only if $n \equiv 1$ or $5 (\text{mod } 10)$ and $n \neq 15$.

22.31 Theorem [17] Let $G = (\{a,b,c,d,e\}, \{ab, bc, cd, de, bd\})$. Then there exists a self-complementary $(\lambda K_n, G)$-design exactly when

1. $\lambda \equiv 1, 3, 7$ or $9 (\text{mod } 10)$ and $n \equiv 1$ or $5 (\text{mod } 10)$, and if $\lambda = 1$, then $n \neq 15$,
2. $\lambda \equiv 5 (\text{mod } 10)$ and n is odd, $n \neq 3$,
3. $\lambda \equiv 2, 4, 6$ or $8 (\text{mod } 10)$ and $n \equiv 0$ or $1 (\text{mod } 5)$,
4. $\lambda \equiv 0 (\text{mod } 10)$ and $n \neq 2, 3, 4$.

22.32 Remark The graph decompositions discussed so far are all for fixed graphs G. One natural question to ask is if for any $t, t|\frac{n(n-1)}{2}$, there exists a graph H with $|E(H)| = t$ and a (K_n, H)-design.

22.33 Theorem [9] For any positive integer t, if $t|\frac{n(n-1)}{2}$, then there exists a graph H with t edges and a (K_n, H)-design.

22.8 Small Graph Decompositions

22.34 Table If G has no isolated vertices, $|V(G)| \leq 5$ and G is not covered by any of the preceding theorems of this section, the known existence of (K_n, G)-designs is given here. In column three, "same" means the sufficient conditions are the same as the necessary conditions, as given in Theorem 22.2 (1) and (2), and the sign "+" means the necessary conditions together with additional conditions as given comprise the sufficient conditions; in other cases only existence results are given. Further details for the constructions are cited in [4, 6, 17].

Graph	Necessity	Sufficiency
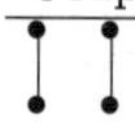	$n \equiv 0, 1 \pmod 4$	same
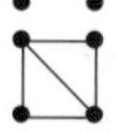	$n \equiv 0, 1 \pmod 5$	$+ n \neq 5$
	$n \equiv 0, 1 \pmod 8$	same
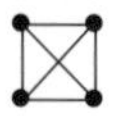	$n \equiv 1, 4 \pmod{12}$	same
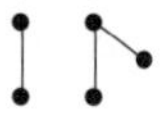	$n \equiv 0, 1 \pmod 3$	same
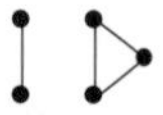	$n \equiv 0, 1 \pmod 8$	same

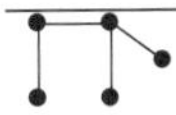	$n \equiv 0, 1 \pmod 8$	same
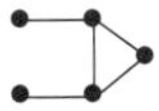	$n \equiv 0, 1 \pmod 5$	same
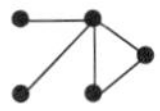	$n \equiv 0, 1 \pmod 5$	$+n > 6$
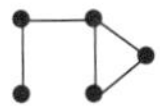	$n \equiv 0, 1 \pmod 5$	$+n > 6$
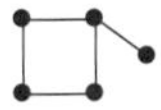	$n \equiv 0, 1 \pmod 5$	$+n > 6$
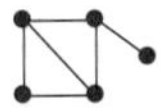	$n \equiv 0, 1, 4, 9 \pmod{12}$	same
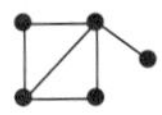	$n \equiv 0, 1, 4, 9 \pmod{12}$	same
	$n \equiv 0, 1, 4, 9 \pmod{12}$	$+n > 13$
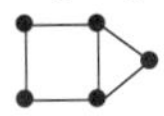	$n \equiv 0, 1, 4, 9 \pmod{12}$	+ possibly $n \neq 24$
	$n \equiv 1, 9 \pmod{12}$	same
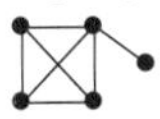	$n \equiv 0, 1 \pmod 7$	$+n > 8$ and possibly $n \neq 119$, 120, 147, 203, 204
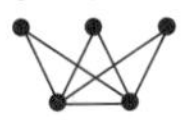	$n \equiv 1, 7 \pmod{14}$	same
	$n \equiv 0, 1 \pmod 7$	$+n \neq 8, 14$ and possibly $n \neq 36$, 56, 92, 98, 120
	$n \equiv 0, 1 \pmod 7$	$+n \neq 8$
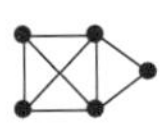	$n \equiv 0, 1 \pmod{16}$	$n \equiv 1 \pmod{16}$ except possibly for $n = 65$; $n \equiv 0 \pmod{16}$ if exists for $n = 65$
	$n \equiv 0, 1 \pmod{16}$	$n \equiv 1 \pmod{16}$; $n \neq 16$, $n = 64$
	$n \equiv 0, 1 \pmod 9$	$n \neq 9, 10, 18$; $n \equiv 1 \pmod{18}$, possibly $n \neq 37$, 55, 73, 109, 397, 415, 469, 487, 505, 541, 613, 685
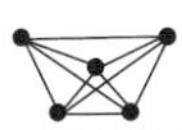	$n \equiv 1, 5 \pmod{20}$	same

22.9 See Also

§I.1	Decompositions into complete graphs are BIBDs.
§IV.9	Decompositions into cycles are cycle systems.

References

[1] B. Alspach, K. Heinrich and G. Liu, Orthogonal factorizations of graphs, in: *Contemporary Design Theory: A Collection of Surveys* (J.H. Dinitz, D.R. Stinson; eds.) John Wiley & Sons, New York, 1992, 13-40.

[2] B. Alspach and B. N. Varma, Decomposing complete graphs into cycles of length $2p^e$, *Ann. Discrete Math.* 9 (1980), 155-162.

[3] E.T. Bell, Decomposition of K_n into cycles of length at most fifty, *Ars Combin.* 40 (1995), 49-58.

[4] J.C. Bermond, C. Huang, A. Rosa and D. Sotteau, Decomposition of complete graphs into isomorphic subgraphs with five vertices, *Ars Combin.* 10 (1980), 211-254.

[5] J.C. Bermond, C. Huang and D. Sotteau, Balanced cycle and circuit designs: Even cases, *Ars Combin.* 5 (1978), 293-318.
[6] J.C. Bermond and J. Schönheim, G-decomposition of K_n, where G has four vertices or less, *Discrete Math.* 19 (1977), 113-120.
[7] J.C. Bermond and D. Sotteau, Balanced cycle and circuit designs: Odd cases, *Proc. Colloq. Oberhof Illmenau* (1978), 11-32.
[8] J. Bosák, *Decompositions of Graphs*, Kluwer Academic Publishers, Boston, 1990.
[9] F. Harary, R.W. Robinson and N.C. Wormald, Isomorphic factorizations I: Complete graphs, *Trans. Amer. Math. Soc.* 242 (1978), 234-260.
[10] K. Heinrich, Path-decompositions, *Le Matematiche (Catania)* XLVII (1992), 241-258.
[11] J.D. Horton and G.M. Nonay, Self-orthogonal Hamilton path decompositions, *Discrete Math.* 97 (1991), 251-264.
[12] C. Huang, Resolvable balanced bipartite designs, *Discrete Math.* 14 (1976), 319-335.
[13] A. Kotzig, Decomposition of complete graphs into isomorphic cubes, *J. Combin. Theory B* 31 (1981), 292-296.
[14] C.C. Lindner and C.A. Rodger, Decomposition into cycles II: Cycle systems, in: *Contemporary Design Theory: A Collection of Surveys* (J.H. Dinitz, D.R. Stinson; eds.) John Wiley & Sons, New York, 1992, 325-369.
[15] M. Maheo, Strongly graceful graphs, *Discrete Math.* 29 (1980), 39-46.
[16] C.A. Rodger, Graph-decompositions, *Le Matematiche (Catania)* XLV (1990), 119-140.
[17] C.A. Rodger, Self-complementary graph decompositions, *J. Austral. Math. Soc. A* 53 (1992), 17-24.
[18] A. Rosa, On cyclic decompositions of the complete graph into polygons with an odd number of edges, *Casopis Pest. Math.* 91 (1966), 53-63.
[19] A. Rosa, On cyclic decompositions of the complete graph into $(4m+2)$-gons, *Mat.-Fyz. Cas.* 16 (1966), 349-352.
[20] R.M. Wilson, Decompositions of complete graphs into subgraphs isomorphic to a given graph, *Congr. Numer.* 15 (1976), 647-659.
[21] M. Yu, On tree factorizations of K_n, *J. Graph Theory* 17 (1993), 713-725.
[22] M. Yu, Almost resolvable path decompositions, *J. Graph Theory* 15 (1991), 523-541.

23 Graphical Designs

Yeow Meng Chee

23.1 Definitions, Example, and Existence

A nontrivial t-(v, K, λ) tBD $(X, \mathcal{A})$ is *graphical* if

1. X is the set of $v = \binom{n}{2}$ edges of the complete graph K_n, and
2. $\mathcal{A}$ is a collection of graphs closed under isomorphism: whenever $A \in \mathcal{A}$ and σ is an automorphism of K_n, then $\sigma(A) \in \mathcal{A}$.

23.1 Remark A graphical t-$(\binom{n}{2}, K, \lambda)$ design has the symmetric group S_n as an automorphism group.

23.2 Table [8] Parameters of the only simple graphical t-$(10, k, \lambda)$ designs, $t \geq 2$ (note $10 = \binom{5}{2}$). $(t, k, \lambda) = (2,3,4)$, $(2,4,2)$, $(2,4,4)$, $(2,4,8)$, $(2,4,10)$, $(2,4,12)$, $(2,5,16)$, $(2,5,20)$, $(3,4,1)$, $(3,5,6)$.

23.3 Table [8] Parameters of the only simple graphical t-$(15,k,\lambda)$ designs, $t \geq 2$.

(t,k,λ)							
$(2,3,1)$	$(2,4,6)$	$(2,4,24)$	$(2,4,30)$	$(2,4,36)$	$(2,7,3)$	$(2,7,24)$	$(2,7,27)$
$(2,7,30)$	$(2,7,33)$	$(2,7,36)$	$(2,7,39)$	$(3,5,30)$	$(3,6,100)$	$(3,7,60)$	$(3,7,75)$
$(3,7,90)$	$(3,7,135)$	$(3,7,150)$	$(3,7,165)$	$(3,7,180)$	$(3,7,225)$	$(3,7,240)$	$(4,7,60)$
$(2,5,\lambda) : 16 \leq \lambda \leq 142,\ \lambda \equiv 0,2,4,6 \pmod{10},\ \lambda \neq 20, 50$							
$(2,6,\lambda) : 10 \leq \lambda \leq 355,\ \lambda \equiv 0,10 \pmod{15}$							
$(2,7,\lambda) : 48 \leq \lambda \leq 642,\ \lambda \equiv 0 \pmod{3}$							

23.4 Example All graphical t-$(v,K,1)$ designs.

(t,v,K)	Graphical Representation
$(1,6,2)$	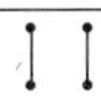
$(2,15,3)$	
$(2,15,\{3,5\})$	
$(3,10,4)$	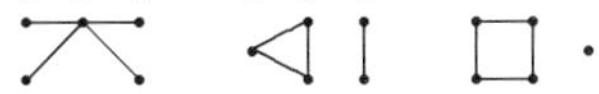
$(4,15,\{5,7\})$	

23.5 Theorem [5] The only graphical t-$(v,K,1)$ designs are given in Example 23.4.

23.6 Theorem [5] The only simple graphical t-$(v,K,2)$ designs with $t > 1$ have the parameters: 2-$(6,\{3,4\},2)$, 2-$(10,\{3,4\},2)$, 2-$(10,4,2)$, 2-$(15,\{3,6\},2)$, 2-$(15,\{5,6\},2)$, and 3-$(10,\{4,5,6\},2)$.

23.7 Table [5] Parameters of the only simple graphical t-$(v,K,3)$ designs with $v \leq 15$ (excluding those presented in Tables 23.2 and 23.3).

(t,v,K)					
$(1,6,\{2,3\})$	$(1,6,\{2,4\})$	$(1,10,2)$	$(1,10,3)$	$(1,15,3)$	$(2,10,\{4,6\})$
$(2,15,\{3,4\})$	$(2,15,\{3,5,6\})$	$(2,15,\{3,10\})$	$(3,10,\{4,6\})$	$(3,15,\{4,5,7\})$	$(3,15,\{7,8\})$

23.8 Theorem [2, 3] A simple graphical 2-$(v,3,\lambda)$ design exists if and only if $(v,\lambda) \in \{(10,4), (15,1), (28,6), (28,10), (55,25)\}$. The only simple graphical 3-$(v,4,\lambda)$ design is the 3-$(10,4,1)$ design. There exist no simple graphical $4-(v,5,\lambda)$ designs for any v and λ.

23.9 Table [2] Parameters of the only simple graphical 2-$(v,4,\lambda)$ designs.

(v,λ)						
$(10,2)$	$(10,4)$	$(10,8)$	$(10,10)$	$(10,12)$	$(15,6)$	$(15,24)$
$(15,30)$	$(15,36)$	$(21,6)$	$(21,12)$	$(21,18)$	$(21,36)$	$(21,42)$
$(21,45)$	$(21,48)$	$(21,51)$	$(21,54)$	$(21,57)$	$(21,60)$	$(21,63)$
$(21,66)$	$(21,69)$	$(21,72)$	$(21,75)$	$(21,78)$	$(21,81)$	$(21,84)$
$(28,5)$	$(28,55)$	$(28,80)$	$(28,85)$	$(28,95)$	$(28,110)$	$(28,120)$
$(28,125)$	$(28,135)$	$(28,150)$	$(36,15)$	$(36,90)$	$(36,111)$	$(36,120)$
$(36,135)$	$(36,165)$	$(36,210)$	$(36,231)$	$(36,240)$	$(36,255)$	$(36,276)$
$(45,63)$	$(45,105)$	$(45,252)$	$(45,357)$	$(45,378)$	$(45,420)$	$(55,168)$
$(55,336)$	$(55,504)$	$(78,630)$	$(78,1080)$	$(78,1350)$	$(91,836)$	$(91,1430)$
$(91,1496)$	$(105,1320)$	$(105,1326)$	$(105,1650)$	$(105,1656)$	$(105,1782)$	$(105,1788)$
$(105,1980)$	$(105,1986)$	$(105,2112)$	$(105,2118)$	$(105,2442)$	$(105,2448)$	$(153,4935)$
$(153,5025)$	$(253,14535)$					

23.10 Table [7, 9, 4] The parameters of *some* known sporadic simple graphical t-(v,k,λ) designs are: 3-$(21,5,3)$, 3-$(21,5,30)$, 3-$(21,5,33)$, 3-$(21,5,39)$, 3-$(21,5,69)$, $3-(21,5,75)$, and 3-$(28,5,30)$.

23.2 Bigraphical Designs

A nontrivial t-(v, K, λ) tBD $(X, \mathcal{A})$ is *bigraphical* if
1. X is the set of $v = mn$ edges of the complete bipartite graph $K_{m,n}$, and
2. whenever $A \in \mathcal{A}$ and σ is an automorphism of $K_{m,n}$ that fixes the two independent sets, then $\sigma(A) \in \mathcal{A}$.

23.11 Remark A bigraphical t-(mn, K, λ) design has $S_m \times S_n$ as an automorphism group.

23.12 Example A bigraphical 3-$(16, \{4, 6\}, 1)$ design.

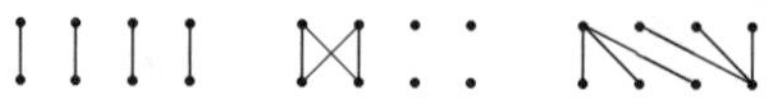

23.13 Theorem [6] The only bigraphical t-$(v, K, 1)$ designs with $t > 1$ have the following parameters: 2-$(9, 3, 1)$, 3-$(8, 4, 1)$, 3-$(16, 4, 1)$, 3-$(16, \{4, 6\}, 1)$, and 5-$(16, \{6, 8\}, 1)$.

23.14 Table [11] Parameters of the only bigraphical t-$(v, K, 2)$ designs.

(t, v, K)					
$(1, mn, m) : 2 \le m \le n$			$(1, mn, n) : 2 \le m \le n$		
$(1, 3n, 2n) : n \ge 3$			$(1, 3n, 2) : n \ge 3$		
$(1, 3m, 2m) : m \in \{2, 3\}$			$(1, 3m, 2) : m \in \{2, 3\}$		
$(1, mn, \{m, n\}) : 2 \le m \le n$			$(1, 4, 2)$	$(1, 6, 2)$	$(1, 9, 3)$
$(2, 9, 3)$	$(2, 9, \{3, 4\})$	$(2, 9, \{3, 6\})$	$(2, 12, 3)$	$(2, 12, \{3, 4\})$	$(2, 12, \{3, 5\})$
$(2, 16, \{3, 4\})$	$(2, 16, 4)$	$(2, 16, 6)$	$(3, 8, 4)$	$(3, 8, \{4, 6\})$	$(3, 10, 4)$
$(3, 10, \{4, 5\})$	$(3, 16, 4)$	$(3, 16, \{4, 6\})$	$(3, 20, 4)$	$(3, 20, \{4, 5\})$	$(3, 25, \{4, 5\})$
$(5, 12, 6)$	$(5, 16, \{6, 8\})$				

23.15 Remark [1] A bigraphical 3-$(77, 20, 3135283200)$ is known.

23.3 Generalization to Hypergraphs

A nontrivial t-(v, K, λ) tBD $(X, \mathcal{A})$ is *r-hypergraphical* if
1. X is the set of $v = \binom{n}{r}$ edges of the complete r-hypergraph $K_n^{(r)}$, and
2. $\mathcal{A}$ is closed under isomorphism of hypergraphs: whenever $A \in \mathcal{A}$ and σ is automorphism of $K_n^{(r)}$, then $\sigma(A) \in \mathcal{A}$.

23.16 Remarks

1. An r-hypergraphical t-$(\binom{n}{r}, K, \lambda)$ design has S_n as an automorphism group.
2. A 2-hypergraphical t-(v, K, λ) design is a graphical t-(v, K, λ) design.

23.17 Example A 3-hypergraphical 2-$(56, 3, 18)$ design.

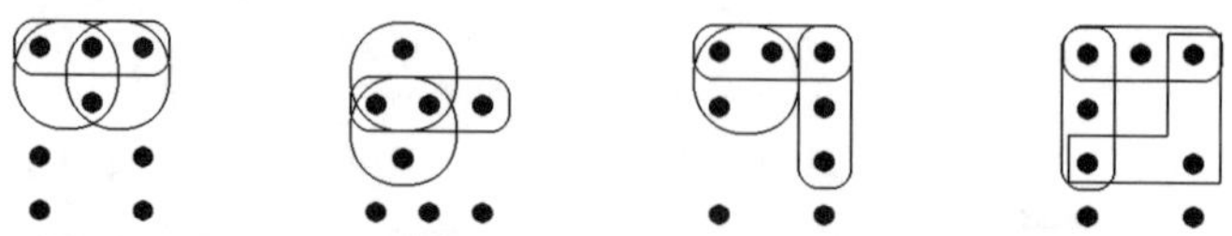

23.18 Theorem A simple 3-hypergraphical 2-$(v, 3, \lambda)$ design exists if and only if $(v, \lambda) \in \{(10, 4), (56, 18), (56, 24), (120, 54), (1540, 722)\}$.

23.19 Table The four distinct 3-hypergraphical 3-$(20,4,1)$ designs. To obtain the blocks of each of the designs, develop the base blocks with the group $G = \langle\alpha,\beta\rangle$, where

$$\alpha = (0\ 10\ 16\ 7\ 2)(1\ 11\ 4\ 13\ 5)(3\ 12\ 17\ 19\ 9)(6\ 14\ 18\ 8\ 15),$$
$$\beta = (2\ 3)(5\ 6)(7\ 8)(11\ 12)(13\ 14)(16\ 17).$$

Design #	Base Blocks
1	$\{0,7,12,19\}$ $\{0,1,7,14\}$ $\{0,3,6,7\}$ $\{0,1,2,16\}$
2	$\{0,7,12,19\}$ $\{0,1,7,14\}$ $\{0,3,7,13\}$ $\{0,1,2,3\}$ $\{0,1,4,10\}$
3	$\{0,7,12,19\}$ $\{0,1,5,7\}$ $\{0,7,14,18\}$ $\{0,3,6,7\}$ $\{0,1,2,16\}$
4	$\{0,7,12,19\}$ $\{0,3,7,13\}$ $\{0,1,2,3\}$ $\{0,1,4,10\}$ $\{0,1,5,7\}$ $\{0,7,14,18\}$

23.20 Theorem For fixed positive integers $r \geq 2$, t, and λ, there exist only finitely many r-hypergraphical t-$(\binom{n}{r}, K, \lambda)$ designs.

23.4 See Also

§I.1	Some graphical designs are block designs with symmetric groups as automorphism groups.
§IV.49	Graphical designs are t-wise balanced designs with symmetric groups as automorphism groups.
[10]	Sections 3.4 and 3.5 present another interpretation of graphical designs, obtained from strongly regular graphs.

References

[1] P.J. Cameron and C. E. Praeger, Block transitive t-designs I, *Discrete Math.*, to appear.

[2] Y.M. Chee, Graphical t-designs with block sizes three and four, *Discrete Math.* 91 (1991), 201-206.

[3] Y.M. Chee, On graphical quintuple systems, *J. Symbolic Computation* 13 (1992), 677-681.

[4] Y.M. Chee, The existence of a simple 3-$(28,5,30)$ design, *Discrete Math.* 118 (1993), 251-252.

[5] L.G. Chouinard II, E.S. Kramer and D.L. Kreher, Graphical t-wise balanced designs, *Discrete Math.* 46 (1983), 227-240.

[6] D.G. Hoffman and D.L. Kreher, The bigraphical t-wise balanced designs of index 1, *J. Combin. Des.* 2 (1994), 41-48.

[7] E.S Kramer, An $S_3(3,5,21)$ using graphs, *Discrete Math.* 81 (1990), 223-224.

[8] E.S. Kramer and D.M. Mesner, t-designs on hypergraphs, *Discrete Math.* 79 (1976), 263-296.

[9] D.L. Kreher, Y.M. Chee, D. de Caen, C.J. Colbourn and E.S. Kramer, Some new simple t-designs, *J. Combin. Math. Combin. Comput.* 7 (1990), 53-90.

[10] V.D. Tonchev, *Combinatorial Configurations: Designs, Codes, Graphs*, Longman, New York, 1988.

[11] L.M. Weiss and D.L. Kreher, The bigraphical t-wise balanced designs of index two, *J. Combin. Des.* 3 (1995), 233-254.

24 Hadamard Matrices and Designs

R. Craigen

24.1 Definitions and Examples

A *Hadamard matrix*, H, of *order* n is an $n \times n$ (± 1)-matrix satisfying $HH^t = nI$ (its rows are pairwise orthogonal). $H(n)$ denotes a Hadamard matrix of order n.

24.1 Examples $H(n)$ for $n = 2, 4, 8$. Here $-$ represents -1.

$$\begin{pmatrix} 1 & 1 \\ 1 & - \end{pmatrix} \qquad \begin{pmatrix} 1&1&1&1 \\ 1&1&-&- \\ 1&-&1&- \\ 1&-&-&1 \end{pmatrix} \qquad \begin{pmatrix} 1&1&1&1&1&1&1&1 \\ 1&1&-&-&1&1&-&- \\ 1&-&1&-&1&-&1&- \\ 1&-&-&1&1&-&-&1 \\ 1&1&1&1&-&-&-&- \\ 1&1&-&-&-&-&1&1 \\ 1&-&1&-&-&1&-&1 \\ 1&-&-&1&-&1&1&- \end{pmatrix}$$

Two Hadamard matrices are *equivalent* if one can be transformed into the other by a series of row or column permutations and negations. A Hadamard matrix is *normalized* if all entries in its first row and first column are equal to 1.

24.2 Example These two Hadamard matrices of order 4 are equivalent. The first is normalized. Negate the second column and third row of the first, and then interchange the third and fourth columns, to obtain the second.

$$\begin{pmatrix} 1&1&1&1 \\ 1&1&-&- \\ 1&-&1&- \\ 1&-&-&1 \end{pmatrix} \qquad \begin{pmatrix} 1&-&1&1 \\ 1&-&-&- \\ -&-&1&- \\ 1&1&1&- \end{pmatrix}$$

A Hadamard matrix H is *skew* if $H = A + I$, where A is skew. H is *regular* if the row and column sums are constant and *graphical* if it is symmetric with all diagonal entries equal to 1.

24.3 Examples $A = H(12)$ is skew; $B = H(16)$ is both regular and graphical:

$$A = \begin{pmatrix}
1&-&-&-&-&-&-&-&-&-&-&- \\
1&1&1&-&1&1&1&-&-&-&1&- \\
1&-&1&1&-&1&1&1&-&-&-&1 \\
1&1&-&1&1&-&1&1&1&-&-&- \\
1&-&1&-&1&1&-&1&1&1&-&- \\
1&-&-&1&-&1&1&-&1&1&1&- \\
1&-&-&-&1&-&1&1&-&1&1&1 \\
1&1&-&-&-&1&-&1&1&-&1&1 \\
1&1&1&-&-&-&1&-&1&1&-&1 \\
1&1&1&1&-&-&-&1&-&1&1&- \\
1&-&1&1&1&-&-&-&1&-&1&1 \\
1&1&-&1&1&1&-&-&-&1&-&1
\end{pmatrix} \quad B = \begin{pmatrix}
1&-&-&-&-&1&1&1&-&1&1&1&-&1&1&1 \\
-&1&-&-&1&-&1&1&1&-&1&1&1&-&1&1 \\
-&-&1&-&1&1&-&1&1&1&-&1&1&1&-&1 \\
-&-&-&1&1&1&1&-&1&1&1&-&1&1&1&- \\
-&1&1&1&1&-&-&-&-&1&1&1&-&1&1&1 \\
1&-&1&1&-&1&-&-&1&-&1&1&1&-&1&1 \\
1&1&-&1&-&-&1&-&1&1&-&1&1&1&-&1 \\
1&1&1&-&-&-&-&1&1&1&1&-&1&1&1&- \\
-&1&1&1&-&1&1&1&1&-&-&-&-&1&1&1 \\
1&-&1&1&1&-&1&1&-&1&-&-&1&-&1&1 \\
1&1&-&1&1&1&-&1&-&-&1&-&1&1&-&1 \\
1&1&1&-&1&1&1&-&-&-&-&1&1&1&1&- \\
-&1&1&1&-&1&1&1&-&1&1&1&1&-&-&- \\
1&-&1&1&1&-&1&1&1&-&1&1&-&1&-&- \\
1&1&-&1&1&1&-&1&1&1&-&1&-&-&1&- \\
1&1&1&-&1&1&1&-&1&1&1&-&-&-&-&1
\end{pmatrix}.$$

24.2 Necessary Conditions and Conjectures

24.4 Theorem If there is an $H(n)$, then $n = 1, 2$ or $4k$, k a positive integer.

24.5 Remark Theorem 24.4 can be proved by considering, when $n \geq 3$, any three rows with entries ± 1 that are pairwise orthogonal.

24.6 Theorem If $H = H(n)$ is regular or graphical, then n is a square.

A *Hadamard design* is a SBIBD$(4n-1, 2n-1, n-1)$; a *Menon design* is a SBIBD$(4u^2, 2u^2 \pm u, u^2 \pm u)$.

24.7 Construction A Hadamard design SBIBD$(4n-1, 2n-1, n-1)$ exists if and only if $H(4n)$ exists. Given $H(4n)$, normalize it by suitable row and column negations. Removing the first row and first column leaves a $(4n-1) \times (4n-1)$ matrix, the *core* of the Hadamard matrix. Replace all -1 entries by 0 entries; the result is an SBIBD$(4n-1, 2n-1, n-1)$. The process can be reversed to recover $H(4n)$ from an SBIBD$(4n-1, 2n-1, n-1)$.

24.8 Construction A 3-$(4n, 2n, 2n-1)$ can be constructed from $H(4n)$. Let $(V, \mathcal{B})$ be an SBIBD$(4n-1, 2n-1, n-1)$ constructed from the Hadamard matrix. The *Hadamard 3-design* has elements $V \cup \{\infty\}$, and blocks $\{V \backslash B : B \in \mathcal{B}\} \cup \{B \cup \{\infty\} : B \in \mathcal{B}\}$. Its derived design with respect to the element ∞ is the chosen SBIBD$(4n-1, 2n-1, n-1)$.

24.9 Construction A Menon design SBIBD$(4u^2, 2u^2-u, u^2-u)$ exists if and only if a regular $H(4u^2)$ exists. Simply replace all -1 entries in the $H(4u^2)$ by 0 entries to get the incidence matrix of a design with parameters $(4u^2, 2u^2-u, u^2-u)$ or $(4u^2, 2u^2+u, u^2+u)$; if the latter occurs, complement the design (equivalently, negate all entries in the regular Hadamard matrix at the outset).

24.10 Conjecture (Hadamard [6]) $H(n)$ exists if and only if $n = 1, 2$ or $4k$.

24.11 Conjecture (Seberry) A skew $H(n)$ exists if and only if $n = 1, 2$ or $4k$.

24.12 Conjecture (Ryser) If $n > 4$, then there is no circulant $H(n)$.

24.13 Conjecture For any order $4k^2$, there exists:

1. a regular Hadamard matrix;
2. a graphical Hadamard matrix;
3. a regular Hadamard matrix equivalent to a graphical Hadamard matrix;
4. a regular graphical Hadamard matrix.

24.14 Conjecture (Wallis, Lin) If $H(n)$ exists, then so do:

1. a symmetric $H(n)$;
2. a skew $H(n)$ equivalent to a symmetric $H(n)$.

24.15 Remark Conjecture 24.10 is confirmed for $n \leq 428$; Conjecture 24.11 for $n < 188$; Conjecture 24.12 for $n < 12996$; Conjecture 24.13 for $k < 45$ (all 4 cases); case (1) of Conjecture 24.14 for $n < 92$, and case (2) for $n < 36$; there is a regular $H(36)$ that is not equivalent to a graphical one; there is an $H(28)$ that is skew but not equivalent to a symmetric one and a $H(28)$ that is symmetric but not equivalent to a skew one.

24.3 Existence

24.16 Remark The earliest Hadamard matrix construction, Construction 24.17, was given by Sylvester [11] in 1867, 26 years before Hadamard's paper [6]. Hadamard [6] used it together with matrices of orders 2, 12, and 20 to produce infinitely many orders.

24.17 Construction [11] If there exists a Hadamard matrix $H(n) = (h_{ij})$ and a Hadamard matrix $H(m) = (\ell_{ij})$, then there is a Hadamard matrix $H(nm) = (k_{ij})$, obtained by setting $k_{na+i,nb+j} = \ell_{ab}h_{ij}$ for $0 \le a, b < m$ and $0 \le i, j < n$.

24.18 Corollary If $H(n)$ exists, then $H(2^t n)$ exists for all $t \ge 0$.

24.19 Theorem [7] Let q be an odd prime power. If $q \equiv 3 \pmod 4$, then $H(q+1)$ exists; if $q \equiv 1 \pmod 4$, then $H(2(q+1))$ exists.

24.20 Theorem If there is a conference matrix of order n, then there is a Hadamard matrix of order $n(n-1)^k$ if $n \equiv 0 \pmod 4$ and order $2n(n-1)^k$ if $n \equiv 2 \pmod 4$.

24.21 Theorem [1, 2] Hadamard matrices of orders $4a$ and $4b$ imply $H(8ab)$; Hadamard matrices of orders $4a, 4b, 4c, 4d$ imply $H(16abcd)$.

Two matrices A and B are *amicable* if $AB^T = BA^T$.

The *Williamson* and *Goethals-Seidel* arrays:

$$\begin{pmatrix} A & -B & -C & -D \\ B & A & -D & C \\ C & D & A & -B \\ D & -C & B & A \end{pmatrix}, \quad \begin{pmatrix} A & -BR & -CR & -DR \\ BR & A & -D^tR & C^tR \\ CR & D^tR & A & -B^tR \\ DR & -C^tR & B^tR & A \end{pmatrix}.$$

Let A, B, C, D be $n \times n$ $(0, \pm 1)$-matrices such that $AA^t + BB^t + CC^t + DD^t = 4nI$. If they are pairwise amicable they are *Williamson type matrices.* If they are symmetric and circulant, they are *Williamson matrices.* If they are circulant (or *type* I [4]), in which case R is a back-circulant (or *type* II) monomial matrix, they are *Goethals-Seidel type matrices.*

24.22 Theorem If A, B, C, D are Williamson matrices of order n, then the Williamson array gives $H(4n)$.

24.23 Remark Williamson matrices have been constructed for all orders $n \le 40$, with the exception of $n = 35$, which was eliminated by Djokovic, by means of an exhaustive computer search.

24.24 Examples Williamson matrices A, B, C, D for small $H(4n)$. The first rows of the circulant matrices are given.

5	7	9	11	13	15
+ − − − −	+ − − − − − −	+ + − − − − − − +	+ − + + − − − − + + −	+ − + − − + + + + − − + −	+ + − + + − + − − + − + + − +
+ − − − −	+ + − − − − +	+ − + − − − − + −	+ + − + + − − + + − +	+ − − − + + + + + + − − −	+ + − + + + + − − + + + + − +
+ + − − +	+ − + − − + −	+ − − + − − + − −	+ + − + − + + − + − +	+ + − + − − + + − − + − +	+ + − + + − − − − − − + + − +
+ − + + −	+ − − + + − −	+ − − − + + − − −	+ + − − − − − − − − +	+ − − − − + − − + − − − −	+ − + − − − + + + + − − − + −

17	19
$+-++-+++--+++-++-$	$++--+-++++++++-+--+$
$+--+-++++++++-+--$	$+-+---++++++++---+-$
$++-+---+--+---+-+$	$++-++--+-++-+--++-+$
$+---+++----+++---$	$++--+++-+--+-+++--+$

An *orthogonal design* $\mathrm{OD}(n; s_1, \ldots, s_m)$ with $n \geq \sum_{i=1}^{m} s_i$ is an $n \times n$ matrix M, with entries from $0, \pm x_1, \ldots, \pm x_m$ where $x_1, \ldots x_m$ are commuting indeterminates, which satisfies $MM^T = \left(\sum_{i=1}^{m} s_i x_i^2\right) I_n$. An $\mathrm{OD}(4n; n, n, n, n)$ is a *Baumert-Hall array.*

24.25 Theorem If there are Williamson type matrices of order m and an orthogonal design $\mathrm{OD}(4n; n, n, n, n)$, then there is a $H(4mn)$.

24.26 Theorem [5] If A, B, C, D are Goethals-Seidel type matrices, then the Goethals-Seidel array gives a skew $H(4n)$.

24.27 Theorem Let n be an odd positive integer whose binary expansion has N nonzero digits. There exists a Hadamard matrix of order $2^t n$ in the following cases:

1. $n \equiv 1 \pmod 4$ and $t \geq 2N$;
2. $n \equiv 3 \pmod 4$ and $t \geq 2N - 1$;
3. $t \geq 6\lfloor \frac{\log_2 \frac{n-1}{2}}{16} \rfloor + 2$.

24.28 Theorem If there is a Hadamard matrix of order n, then there is a regular graphical Hadamard matrix of order n^2.

24.29 Theorem [10] If there is a BIBD$(u(2u-1), 4u^2-1, 2u-1, u, 1)$, then there is a regular graphical Hadamard matrix of order u^2.

24.30 Table Values of $n < 1000$ for which Williamson type matrices of order n are not known (in this range, they are known for all orders $2n$).

47, 59, 65, 67, 77, 103, 105, 107, 111, 119, 133, 143, 151, 155, 151, 155, 161, 163, 167, 171, 179, 183, 185, 191, 203, 207, 209, 215, 219, 221, 227, 237, 245, 247, 249, 251, 253, 259, 267, 273, 275, 283, 287, 291, 299, 303, 305, 319, 323, 333, 341, 343, 345, 347, 357, 369, 377, 391, 395, 403, 407, 417, 425, 437, 445, 447, 455, 463, 467, 475, 483, 485, 487, 493, 495, 513, 515, 519, 523, 527, 533, 539, 545, 551, 553, 561, 563, 567, 571, 575, 581, 583, 585, 587, 589, 595, 603, 605, 621, 623, 629, 631, 633, 635, 637, 643, 655, 657, 665, 667, 671, 679, 689, 693, 695, 697, 699, 707, 711, 713, 723, 725, 731, 733, 751, 765, 771, 779, 783, 785, 791, 793, 795, 803, 813, 817, 823, 825, 831, 833, 837, 845, 851, 859, 865, 869, 873, 875, 883, 891, 897, 899, 903, 905, 909, 913, 915, 919, 921, 923, 927, 935, 943, 947, 949, 959, 963, 969, 973, 979, 981, 985, 989, 993, 995

24.31 Table Odd $n < 500$ for which a skew Hadamard matrix of order $4n$ is not known (t in brackets, when provided, indicates that one of order $2^t n$ is known).

47(4), 59(12), 69(3), 81(3), 89(4), 97(9), 101(10), 103(3), 107(10), 109(9), 119(4), 145(5), 149(4), 151(5), 153(3), 167(4), 169(5), 177(12), 179(8), 191, 193(3), 201(3), 205(3), 209(4), 213(4), 223(3), 225(4), 229(3), 233(4), 235(3), 239(4), 245(4), 247(6), 249(4), 251(6), 253(4), 257(4), 259(5), 261(3), 265(4), 269(8), 275(4), 277(5), 283(11), 285(3), 287(4), 289(3), 295(5), 299(4), 301(3), 303(3), 305(4), 309(3), 311(26), 317(6), 319(3), 325(5), 329(6), 331(3), 335(7), 337(18), 341(4), 343(6), 345(4), 347(18), 349(3), 353(4), 359(4), 361(3), 369(4), 373(7), 377(6), 385(3), 389(15), 391(4), 397(5), 401(10), 403(5), 409(3), 413(4), 419(4), 423(4), 429(3), 433(3), 435(4), 441(3), 443(6), 445(3), 449, 451(3), 455(4), 457(9), 459(3), 461(17), 463(7), 465(3), 469(3), 473(5), 475(4), 479(12), 481(3), 485(4), 487(5), 489(3), 491(46), 493(3)

24.32 Table Odd $n < 200$ for which a symmetric Hadamard matrix of order $4n$ is not known (t in brackets, when provided, indicates that one of order $2^t n$ is known).

23(4), 29(4), 39(3), 43(3), 47(4), 59(12), 65(3), 67(5), 73(7), 81(3), 89(4), 93(3), 101(10), 103(3), 107(10), 109(9), 113(8), 119(3), 127(25), 133(3), 149(4), 151(5), 153(3), 163(3), 167(4), 179(8), 183(3), 189(3), 191, 193(3), 195(3)

24.33 Table Values of t such that a Hadamard matrix of order $2^t n$ is known, for odd $n < 3999$ (n is obtained by adding the three indices of the entry). A period indicates $t = 2$.

	000	100	200	300	400	500	600	700	800	900
00	. . . . .	. . . 3 .	. . . . .	. . . . .	. . . . .	. . . . .	. . . . .	. . . . .	. . . . .	. . . 3 .
10	. . . . .	. . . . .	. . . . .	3	 3	. . 3 . .	. . . . .	 4	. . . . .	. . . 3 3
20	. . . . .	. . . . .	. 3 . . .	. . . . .	. . . . .	. 3 . . .	. . . . .	3	. 3 . . .	. . . . .
30	. . . . .	. . . . .	 4	. . . . .	. . . . .	. . . 3 .	3	 5	 4	. 4 . . .
40	. . . . .	. . . . .	. . . . .	. . . 3 .	. 3 . . .	. . . . .	. 3 . 3 .	 4	. . . . .	. . . 3 .
50	. . . . .	. . . . .	3	 4	. . . . .	. . . . .	. 3 . . 4	3	. 3 . . 3	. . 3 . .
60	. . . . .	. . . 3 .	. . . . .	. . . . .	. 3 . . .	. . . . .	 3	. . . . .	. 3 . . .	. . . . .
70	. . . . .	 3	. . . . .	. . . . .	 4	3 3 . . .	. . . . .	. . . . .	. . . . .	5 . . . .
80	. . . . .	. . . . .	. 3 . . .	. . . . .	. . . 3 .	. . . . .	. . . . .	. . . 3 3	. 3 . . .	. . . . .
90	. . . . .	3	. . . . .	. . . . .	5	 5	. . . . .	. . . . .	. . . . .	3 . . . .
	1000	1100	1200	1300	1400	1500	1600	1700	1800	1900
00	. . . . .	. 3 . . .	. . . . .	. 3 . . .	. . . . .	 3	. . . . .	. 3 . . .	. . . . .	. . . 3 .
10	 3	. . 3 . .	. . . . .	. . 3 . 4	. . . . .	. . . . .	 3	. 3 . . 3	. . . . .	. 7 3 . .
20	. . . . .	. 3 . . .	. 6 . . .	. . . 3 .	. 3 . 3 .	. . . 3 .	. . . . .	. 3 . . .	. 3 . . .	 3
30	6 . . . 3	. 3 . . .	. . . . .	. . . . .	. . . 6 8	. . . . .	. . . . .	. . . 3 .	3	. 3 . . .
40	. . . . .	. . . . .	. . . . .	. . . . .	3 . . 7 .	. 3 . . .	. . . . .	3	3 . . 3 .	 4
50	3	. . . . .	. . 3 3 4	 3	6	 4	. . . . .	. . . . .	. . . . .	. . . 3 .
60	. 3 . . .	 4	. . . . .	. . . 3 .	. . . . .	. . . 6 .	3 3 . 3 3	. . . . .	. . . . .	. 3 . . 4
70	. . . . .	. . . 4 .	. . . . .	. . . . .	3 3 . . .	7 . . . 5	. . . . .	. . . . .	3 . . . 3	 4
80	. . . 3 .	. . . 3 .	. 3 . . .	. . . . .	. 3 . 3 .	. 3 . . 3	 3	. 6 . 3 3	. 3 . . .	4 . . 7 .
90	. 3 . . .	. . . . .	3	. . . 3 .	 7	. . . . .	 3	. 4 5 . .	. 3 . . .	. . . . .
	2000	2100	2200	2300	2400	2500	2600	2700	2800	2900
00	. . . . .	. . . . .	. 3 . 4 .	. . . . .	. . . . .	. 3 . . .	. . . . .	. . . . .	. 3 . . .	. 4 . . .
10	. . . . .	 3	. . 4 . .	. . 4 . .	. 3 . . .	. 4 3 . .	. . . . .	3	. . 3 . 3	. 6 . . .
20	. . . 3 .	. . . . .	. . . 3 .	. . . 4 .	. 4 . . 4	. . . . .	 3	. . . . .	. 3 . . .	3 . . 3 .
30	. 3 . . 7	. . . . .	. . . . .	. . 3 . 4	. . . . .	3	. . . . .	3 3 . . .	. . . . .	 6
40	. . . . .	. 3 . . .	. . . . .	3 . . 3 .	. . . . .	. 6 3 . .	. . . 3 .	. . . . .	3 3 . . .	. . . . .
50	. 3 . . .	. . 3 . 3	5	. . . . .	 3	. . . . .	 3	. . . . .	. . 4 . .	3 . . . .
60	. 7 . . .	. . . . .	 3	 3	3	. . . . .	3	. . . 3 .	. . 3 . .	. 3 . . .
70	. . . . .	4	. . . . .	7	. . . . .	3 . . . 3	7 . . 6 .	. 3 . . .	 7	3 4 . . .
80	. 3 . 4 .	. . . . .	. . . 7 .	. 3 . . .	 3	 4	. 7 . 8 .	. . . . .	. . . 5 .	. . . 3 .
90	. . 3 . 3	. . . . .	. 6 . . .	 8	. . . . .	3	 7	. . . . .	. 3 . . 4	. . 6 . 8
	3000	3100	3200	3300	3400	3500	3600	3700	3800	3900
00	. . . . .	4 4 . 3 .	. 3 . . .	3 . . 6 4	4 . . . 3	. . . . .	. . . . .	. . . . .	. 3 . . .	. . . 6 .
10	. 3 . 3 .	. . . 4 7	. . 4 . .	. . . . .	 3	3	. . . . .	. . 3 . 6	. . . 4 .	 3
20	. 3 . . .	. . . . .	. . . . .	. . . . .	. . . . .	. . . . .	. 3 . . 4	. . . . .	. 3 . . .	. 6 . . .
30	. . . . .	. . . . .	. . . . .	. . . . .	. . . 6 .	. 4 . . 6	5 . . . 3	. . . . .	3	3 . . . .
40	. . . . .	 3	. . . 3 .	. 8 . 3 .	. . . . .	. . . 8 .	. 7 . . .	 3	. . . 3 3	3 3 3 . 5
50	. . . . .	. . . . .	. 3 . . 3	 7	. 3 . . .	 3	 4	. . 3 . .	3 3 . . .	. 3 . 4 4
60	. . . . .	. 3 . 4 .	. 4 . . 3	. . . . .	. 3 . 3 .	. . . . .	. . 3 . 6	. . . 4 .	. . . . .	. . . . .
70	 3	. 4 3 . .	6 3 . . 3	7	. 3 . . .	6	. . . 4 .	 3	. 3 . . .	. . . . .
80	. 3 . . .	 3	. . . . .	. . . . .	. . . . .	. . . . .	. . . . .	. . . . .	. . . . .	. . . . .
90	. 7 . 4 .	. 3 . . .	. . 6 . 4	. . . . .	. . . 3 .	. . 3 . .	3	4 . . . 3	. . . . .	. . . . .
	1 3 5 7 9	1 3 5 7 9	1 3 5 7 9	1 3 5 7 9	1 3 5 7 9	1 3 5 7 9	1 3 5 7 9	1 3 5 7 9	1 3 5 7 9	1 3 5 7 9

	4000	4100	4200	4300	4400	4500	4600	4700	4800	4900
00	8..8.	5...8	8.484	.8.83	568.8	.48.7	88..3	.8..8	4	4834.
10	563.4	..4..	.3...	38.58	.4.8.	.8.4.	8	.4...	6...8	84..8
20	..4.8	83.6.	.4348	...83	7	.4.48	.045.	...83	34.64	33.8.
30	87.47	8	..383	48.6.	.48.4	7.484	.8.6.	..3.8	.648.	8...3
40	.8..4	568..	7	.8..8	304.3	6346.	343.4	76.5.	68...	.4848
50	.34..	.3...	..3.8	36..0	87..4	4..48	.34.5	60385	.43.4	.3.8.
60	835.8	3..68	.348.	.83..	8...4	8.4..	67663	..3.4	.34.8	.3...
70	64588	7...0	8...3	..3.8	..853	83...	...48	3388.	.7..5	...48
80	.4.54	88...	.85.8	.8.83	..363	.45.8	3.48.	87.6.	3..48	338.4
90	..348	.8..5	.488.	68.48	5.4.4	...83	855..	.87.8	..853	.8.88

	5000	5100	5200	5300	5400	5500	5600	5700	5800	5900
00	5.348	..885	87847	8	3.853	..4.3	863.8	..685	.73..	.48.4
10	..48.	...88	34..8	36884	673..	3.3.8	36455	88.8.	..8.4	.33.3
20	..7.3	.4..8	57...	.8.6.	.43.8	..58.	.334.	38..8	3.48.	88583
30	.4..8	336.4	6...4	..8.8	.4.80	.3..7	..8.8	838.4	63.6.	788.8
40	..8..	.3.45	.8364	33.55	8...3	8.8.8	..683	...4.	...88	.3.8.
50	83.43	..8.8	3	88348	...88	486..	6....	..7.4	.765.	8...8
60	363.8	.68..	68..7	88..4	7....	88.45	...84	53.85	...4.	868.8
70	.33.0	673..	688.4	3...5	83.4.	8	8...4	83.48	54864	.6..8
80	.7...	7..34	.8.88	.3688	.85.8	.08..	58.65	6.354	...4.	.7343
90	.63.8	8.8.8	.8...	.4..8	738.3	6748.	5.5.4	384..	.3.45	..058

	6000	6100	6200	6300	6400	6500	6600	6700	6800	6900
00	73.87	63.43	.8..8	.3..4	.7385	.4.58	3488.	..38.	6.8.4	3.484
10	.736.	.6848	.....	.388.	.4..8	834..	63343	.6358	83..4	88..3
20	8	.04.3	637..	.63.8	.38.3	.3..5	.4.48	..48.	.338.	..383
30	348..	6....	.87.4	38..0	68..3	54884	38.50	6....	38.44	84...
40	8836.	...88	...8.	88.6.	4	5.8.3	8748.	38..8	.84..	.8.6.
50	.4..8	..454	845..	8.3.8	.7.5.	..3.5	763.4	8.4.8	88...	.83.8
60	..484	87.87	34348	.0483	.83.8	..7.8	.088.	.....	.4.84	..4.5
70	8..4.	..338	..8.3	8..4.	.44.8	.34.3	.8.65	3.8.4	.335.	.8388
80	.4364	838..	83...	.8.48	4	.7...	..368	.388.	68.85	768.8
90	...5.	.73..	.3.88	.7..8	88443	3.5.8	.748.	8..4.	.8.88	3845.

	7000	7100	7200	7300	7400	7500	7600	7700	7800	7900
00	.8.38	..568	.6484	8834.	3.3.4	8.4..	.83..	.87.8	.3.88	67...
10	4	.8853	..3..	..3.8	.88.0	43..5	383.8	3..83	6...8	.8834
20	38.5.	8334.	86.44	.48..	.8.48	3.8.8	..483	88848	.6..8	.6.88
30	.3463	387.8	.4884	8...3	.4.88	.3...	87.8.	..3.8	.7884	.308.
40	.43.8	3	83.63	.88.8	8468.	837..	88.58	3.6..	.3.33	...88
50	8..80	.3..3	8.8.8	..4..	.7.4.	..854	73845	68...	56544	.48.3
60	63388	.4848	88...	.846.	8...4	.4.8.	.8.85	5.354	833..	83.8.
70	.4..8	.3.85	88.8.	.64.8	865..	67..3	58..8	..888	6334.	4
80	834..	.7.85	..8.8	..688	83...	.43.8	.38..	.8..3	..3.4	836.7
90	686..	3..44	84..3	8....	58884	.48.5	.3.87	..554	3868.	.8.47

	8000	8100	8200	8300	8400	8500	8600	8700	8800	8900
00	.48.8	8.88.	.874.	..7.8	88.53	83.43	.8.54	.3.5.	63..5	.8854
10	.....	.4..3	.83.8	884.0	68.88	.4..4	8.885	.8085	3.8.8	3485.
20	.7.4.	8.3.8	.08.3	8..8.	68358	.04.4	.836.	.4.84	..48.	88388
30	.8..8	5.8.3	8..8.	..5.8	.6.43	.83..	56548	34...	.336.	6.3.8
40	3685.	.8..8	..7.8	874.3	88.48	5.8.8	333.3	.8.6.	.8448	..685
50	..088	..368	54883	..0.7	..8.8	8.48.	.734.	56..4	73..5	.7.63
60	8..58	.84.5	83388	58.68	..68.	88...	..5.8	...85	8..4.	...64
70	.7.84	68..8	..8.8	.4458	83488	.6768	434.0	.3..5	8...8	.4.8.
80	.338.	.4.54	448.4	883..	..838	.7453	.3...	64..8	..484	.834.
90	8	...57	8..45	.47.8	.36.5	33...	8	..88.	8..83	.6.88
	13579	13579	13579	13579	13579	13579	13579	13579	13579	13579

	9000	9100	9200	9300	9400	9500	9600	9700	9800	9900
00	. . 6 . 3	8 3 4 8 .	. 8 7 . 8	8 8 6 8 .	8 3 . 6 8	 8	8 0 8 . 3	. 8 5 8 .	7 6 . . 4	. 0 . . 5
10	6 . 0 . 7	8 . . . 4	5 7 6 . 3	. 8 6 8 .	. 4 7 6 8	7 3 4 8 .	6 8 . 4 .	 8	3 . 8 8 0	8 8 . . 8
20	. 6 3 . 8	8 4 8 . .	. 3 . 8 .	3 6 5 . 8	. . 3 8 3	. 8 6 8 3	5 8 3 8 8	. . 4 8 .	8 3 5 4 5	. 8 3 . 4
30	. . 8 8 .	. 3 3 4 5	. . 8 . 8	. 8 . . 4	8	8 8 . . 4	. 3 4 . 5	. . . 8 .	. 4 5 6 8	. 4 4 5 4
40	8	 4	. 0 . . 8	. . . 4 8	3 . 8 8 8	. 3 6 8 4	8 8 . 4 3	5 . . 3 8	. . 4 5 .	8 8 3 8 7
50	6 4 5 8 8	. 6 6 . 4	. 8 . 8 .	. 8 5 . 8	8 . 4 . .	6 3 4 8 3	. . . 4 4	. 6 . 8 7	8 3 . 8 .	5 8 5 . 8
60	. . 8 . 3	. 3 . 8 3	. 4 5 8 4	3 3 4 5 5	. . 8 4 5	. 8 3 4 8	. . 6 . 3	. 8 8 6 .	. . 3 5 8	3 3 8 8 .
70	8 7 . . 5	. . 5 . 8	. 4 4 . 0	. 8 3 . .	. 8 8 . 8	. 6 . 5 .	8 3 . . .	. 8 5 . 8	7 8 . . 3	4 7 4 4 5
80	3 6 . 4 4	5 0 . 5 7	. . . 6 3	5 8 . . 4	8 8 8 . 5	8 7 . 8 5	. 4 3 . 8	. 8 . 8 .	8 3 7 8 .	. 8 5 . 8
90	. 4 8 5 5	8 8 . . .	8 . 3 . 4	5	. 7 5 . 3	. . 7 5 8	. 3 4 . 3	. 8 4 8 .	. 6 3 . 8	8 . . . .
	1 3 5 7 9	1 3 5 7 9	1 3 5 7 9	1 3 5 7 9	1 3 5 7 9	1 3 5 7 9	1 3 5 7 9	1 3 5 7 9	1 3 5 7 9	1 3 5 7 9

24.34 Theorem There is a unique equivalence class of Hadamard matrices of each order 1,2,4,8,12. The number of classes for orders 16,20,24,28,32, and 36 are $5, 3, 60, 487, \geq 66000$, and ≥ 110, respectively.

24.4 See Also

§I.3	Hadamard matrices are used to construct 3-designs; see Theorem I.3.21.
§I.5	Hadamard matrices are equivalent to Hadamard designs, and regular Hadamard matrices are equivalent to Menon designs.
§II.4	Hadamard matrices are equivalent to certain orthogonal arrays and transversal designs of higher index; see Theorem II.4.6.
§IV.11	Generalized Hadamard matrices.
§IV.12	Some difference sets correspond to Hadamard matrices; see Remark IV.12.17.
§IV.16.1	Hadamard matrices realize the maximum determinant among $(1, -1)$-matrices of order n.
§IV.31	Orthogonal designs provide some of the most important constructions for Hadamard matrices.
§IV.52	A skew Hadamard matrix of order n is equivalent to a skew conference matrix of order n. Hadamard matrices can be constructed from weighing matrices $W(n, n-a)$, $a \leq 4$. A Hadamard matrix of order n is a $W(n, n)$.
§VI.1	A Hadamard matrix of order n gives an optimal error-correcting code of length n.
§VI.5	Regular Hadamard matrices are used to construct strongly regular graphs.
§VI.6	Hadamard matrices are used to construct regular two-graphs (e.g., in Remark VI.6.41.
[3]	A concise survey examining diverse topics connected to Hadamard matrices, less detailed than the other two cited.
[4]	A survey on constructing Hadamard matrices from orthogonal designs.
[9]	This survey contains more detailed information about Williamson and Williamson type matrices, skew Hadamard matrices, etc.
[13]	Important new constructions of Hadamard matrices.

References

[1] S. S. Agaian, *Hadamard Matrices and their Applications* (Lecture Notes in Mathematics 1168), Springer-Verlag, Heidelberg, 1985.

[2] R. Craigen, J. Seberry and X.-M. Zhang, Product of four Hadamard matricess, *J. Combin. Theory A* 55 (1992), 318-320.

[3] R. Craigen and W.D. Wallis, Hadamard matrices: 1893–1993, *Congr. Numer.* 97 (1993), 99-129.

[4] A. Geramita and J. Seberry, *Quadratic Forms, Orthogonal Designs, and Hadamard Matrices*, Marcel Dekker, New York, 1979.

[5] J.M. Goethals and J.J. Seidel, A skew-Hadamard matrix of order 36, *J. Australian Math. Soc.* 11 (1970), 343-344.

[6] J. Hadamard, Résolution d'une question relative aux déterminants, *Bull. Sciences Math.* 17 (1893), 240-246.

[7] R.E.A.C. Paley, On orthogonal matrices, *J. Math. Phys.* 12 (1933), 311-320.

[8] J. Seberry, Tables of Hadamard matrices of order $2^t p$, $p < 4000$, unpublished, 1990.

[9] J. Seberry and M. Yamada, Hadamard matrices, sequences, and block designs, in: *Contemporary Design Theory: A Collection of Surveys* (J.H. Dinitz, D.R. Stinson; eds.) John Wiley & Sons, New York, 1992, 431-560.

[10] S. S. Shrikhande and N. K. Singh, On a method of constructing symmetrical balanced incomplete block designs, *Sankyh-a* 24A (1962), 25-32.

[11] J. J. Sylvester, Thoughts on orthogonal matrices, simultaneous sign-successions, and tesselated pavements in two or more colours, with applications to Newton's rule, ornamental tile-work, and the theory of numbers, *Phil. Mag.* 34 (1867), 461-475.

[12] J. S. Wallis, On the existence of Hadamard matrices, *J. Combin. Theory A* 21 (1976), 188-195.

[13] M.Y. Xia, Some infinite classes of special Williamson matrices and difference sets, *J. Combin. Theory A* 61 (1992), 230-242.

25 Hall Triple Systems

LUCIEN BENETEAU

25.1 Definition and Examples

A *Hall triple system* (HTS) is a pair $(S, \mathcal{L})$ where S is a set of elements (*points*) and $\mathcal{L}$ a set of *lines* satisfying:

1. Every line is a 3-subset of S;
2. Any two distinct points lie in exactly one line;
3. Any two intersecting lines generate a subsystem isomorphic to the affine plane of nine points AG(2,3).

25.1 Example Let S be some $(n+1)$-dimensional vector space over $\mathbb{F}_3$, with $n \geq 3$. Let $\{e_0, e_1, \ldots, e_n\}$ be a basis for S. For any two points $x = \sum \alpha^i e_i$ and $y = \sum \beta^i e_i$ set $z = x \circ y$ as defined by: $x + y + z = (\alpha^1 - \beta^1)(\alpha^2\beta^3 - \alpha^3\beta^2)e_0$. This defines a binary operation on S. One has either $x = y = z$ or the three points x, y, z are pairwise distinct. The 3-subsets of the form $\{x, y, z\}$ such that $z = x \circ y$ provide S with a structure of an HTS. This particular HTS is referred to as $H(n)$.

25.2 Example Any affine space $AG(n, 3)$ over $\mathbb{F}_3$ with the usual lines may be viewed as an HTS. Such an HTS is an *affine* HTS.

25.2 Existence

25.3 Theorem The cardinality of any HTS is 3^m for some integer $m \geq 2$. Nonaffine HTS of order 3^m exist for any $m \geq 4$ and do not exist for $m = 2$ or 3.

25.4 Remark The existence of a nonaffine HTS of order $3^m > 3^3$ is provided by $H(m-1)$ in Example 25.1. For the orders 3^4 and 3^5 there is a unique nonaffine HTS, namely $H(3)$ and $H(4)$, respectively.

25.5 Table The maximum number of nonisomorphic HTSs of order 3^m.

Order	3^1	3^2	3^3	3^4	3^5	3^6	3^7	3^8	3^9	3^m
Affine HTS	1	1	1	1	1	1	1	1	1	1
Nonaffine HTS	0	0	0	1	1	3	12	>42	?	?

25.3 The Rank of a HTS

The rank r of a HTS is the smallest integer r such that there exists a generator set of r points.

25.6 Theorem In a Hall triple system of order 3^m the rank r is at most $m+1$, with equality only in the affine case. Moreover *any* minimal generator subset consists of exactly r distinct points.

25.7 Table The number of nonisomorphic HTSs of order 3^m of rank r.

Order	3^4	3^5	3^6	3^7	3^8
Rank 3 HTS	1	0	0	0	0
Rank 4 HTS	1	1	1	1	4
Rank 5 HTS	0	1	2	6	>16
Rank 6 HTS	0	0	1	5	?
Rank 7 HTS	0	0	0	1	?
Rank 8 HTS	0	0	0	0	1

25.4 Equivalent Quasigroups and Groups

25.8 Remark All the classification theorems concerning the HTSs have been derived from their algebraic description in quasigroup theory, either in terms of distributive Steiner quasigroups or in terms of exponent 3 commutative Moufang loops.

The HTSs are a special case of Steiner triple systems. Any Steiner triple system may be provided with a structure of a *Steiner quasigroup* by setting $z = x \circ y$ when $\{x, y, z\}$ is a block of the design or when $x = y = z$.

A *Steiner quasigroup* is a commutative quasigroup whose binary operation $\circ$ satisfies: $x \circ x = x$ and: $(x \circ y) \circ y = x$.

25.9 Remark Given a Steiner quasigroup one recovers a Steiner triple system by letting the blocks be the sets $\{x, y, z\}$ when $z = x \circ y$ and x and y are two distinct points.

25.10 Theorem The correspondence $(S,\mathcal{L}) \longrightarrow (S,\circ)$ between Steiner triple systems and Steiner quasigroups is one-to-one up to isomorphism.

25.11 Theorem A Hall triple system is a Steiner triple system whose related Steiner quasigroup is distributive, in the sense that the identity $a\circ(x\circ y) = (a\circ x)\circ(a\circ y)$. holds.

An *exponent 3 commutative Moufang loop* is a commutative quasigroup whose binary operation $x+y$ admits an identity element ω and satisfies the two identities $x+x+x=\omega$, and $(x+y)+(x+z)=(x+x)+(y+z)$.

25.12 Proposition An exponent 3 commutative Moufang loop arises from a Hall triple system S if one defines the sum of two points as: $x+y=(x\circ y)\circ\alpha$, where α is an arbitrary fixed point of S. This binary law admits an identity element which is α. Different choices of the point α yield isomorphic loops.

25.13 Remark The kinship between HTSs and exponent 3 commutative Moufang loops recalls the kinship between affine spaces and vector spaces. The correspondence is one-to-one up to isomorphism. For various investigations concerning HTSs, the best setting is loop theory where powerful results due to Bruck can be used.

A group G is a *Fischer group* (in the restrictive sense) if the set $S=S(G)$ of order 2 elements generates G, if it contains at least two elements, and satisfies order$(xy)=3$ for any two distinct elements x and y from S.

25.14 Proposition $S(G)$ may be provided with a structure of a distributive Steiner quasigroup by $x\circ y = xyx$. So the lines of the related HTS are the sets of three involutions from G of the form (x,y,xyx). Any HTS arises in this way from various (nonisomorphic) Fischer groups. Among the different Fischer groups associated with some fixed HTS, say $(S,\mathcal{L})$, there is one and only one centerless group (up to isomorphism); it can be described as the group of permutations of S spanned by all the reflections of the form: $x \longrightarrow \sigma\circ x$ with σ ranging in S.

25.5 Constructing HTSs via Exterior Algebra

25.15 Remark The construction of Example 25.1 may be generalized as follows. Consider some $(n+1)$-dimensional vector space S be over $\mathbb{F}_3$, with $n>2$, and with some chosen basis say $e_0, e_1, e_2, \ldots e_n$. For any two points $x=\sum\alpha^i e_i$ and $y=\sum\beta^i e_i$ set $z=x\circ y$ when $x+y+z=\sum_{i<j<k}\Lambda_{ijk}(\alpha^i-\beta^i)(\alpha^j\beta^k-\alpha^k\beta^j)e_0$ where (Λ_{ijk}) for $i<j<k$ form a nonvanishing sequence of elements from $\mathbb{F}_3$. This defines a binary operation of a distributive Steiner quasigroup on S. The corresponding Hall triple system has rank $n+1$.

25.16 Theorem Any rank $n{+}1$ Hall triple system of order 3^{n+1} arises in the manner described in Remark 25.15.

Let V be the n-dimensional vector space over $\mathbb{F}_3$ spanned by $e_1, e_2\ldots, e_n$. Consider again a nonvanishing sequence (Λ_{ijk}) $i<j<k$ of elements from $\mathbb{F}_3$. The equalities:

$$\phi(e_i,e_j,e_k)=\Lambda_{ijk} \text{ for } i<j<k$$

define a unique symplectic trilinear form ϕ from V^3 onto $\mathbb{F}_3$. Any two symplectic trilinear forms ϕ and ϕ' are *equivalent* when $\phi'(x,y,z)=\phi(f(x),f(y),f(z))$ for some automorphism f of V.

25.17 Proposition Two sequences (Λ_{ijk}') and (Λ_{ijk}) give rise to equivalent symplectic trilinear forms if their related HTSs are isomorphic.

25.18 Proposition The so-defined correspondence between rank $n+1$ HTS of order 3^{n+1} and nonvanishing symplectic trilinear forms of V (n dimensional vector space over $\mathbb{F}_3$) is one-to-one up to isomorphism for Hall triple systems and up to equivalence for symplectic trilinear forms.

25.19 Remark This correspondence may be extended by considering rank $n+1$ HTS or order greater than 3^{n+1} and nonvanishing symplectic trilinear mappings.

25.20 Remark Let V and W be vector spaces over $\mathbb{F}_3$ of dimension n and m respectively. The set $\mathrm{AT}(n,m)$ of symplectic trilinear mappings from V^3 to W whose images have rank m may be provided with a natural notion of equivalence (up to compositions with isomorphisms of V and W). Denote by $\alpha(n,m)$ the number of such equivalence classes. It can be shown that each class is related to an essentially unique HTS (see [2]). As a by-product one derives Theorem 25.21.

25.21 Theorem If $m = 1, 2$ or 3, then $\alpha(n,m)$ coincides with the maximum number $\sigma(n,m)$ of pairwise nonisomorphic rank $n+1$ HTS of order 3^{n+m}. If $m \geq 4$, then $\sigma(n,m) \geq 3 + \alpha(n,m)$.

25.22 Remark The main significance of Theorem 25.21 is that the classification of the Hall triple systems whose 3-order is not too large with respect to their rank r can be dealt with by counting orbits in $\mathrm{AT}(n,m)$. But among the HTS whose 3-order is larger than $3+r$, there are some systems that are not constructible from symplectic trilinear mappings.

25.6 See Also

§I.4.5	Hall triple systems are a type of Steiner triple system.
§IV.36	Related material on quasigroups.
[3, 4, 6]	Additional information about the various quasigroups and groups related to HTS.
[5]	Cubic hypersurfaces might give rise to some HTS. However, in the few known examples of applications of the structure theorem the arising HTSs are affine while the existence of cubic hypersurfaces whose related HTSs might be nonaffine is still an open question.

References

[1] L. Beneteau, Topics about Moufang loops and Hall triple systems, *Simon Stevin* 54 (1980), 107-124.

[2] L. Beneteau and J. Lacaze, Symplectic trilinear forms, related designs and quasigroups, *Comm. Algebra* 16 (1988), 1035-1051.

[3] O. Chein, H.O. Pflugfelder and J.D.H. Smith, *Quasigroups and Loops; Theory and Applications*, Heldermann Verlag, Berlin, 1990.

[4] M. Hall Jr., Automorphism of Steiner triple systems, *IBM J. Res. Develop.* (1960), 406-472.

[5] Y.I. Manin, *Cubic Forms*, North-Holland, Amsterdam, 1974.

[6] H.P. Young, Affine triple systems and matroid designs, *Math. Zeitschr.* 132 (1973), 343-359.

26 Howell Designs

Jeffrey H. Dinitz

26.1 Definitions, Examples, and Existence

Let S be a set of $n+1$ *symbols*. A *Howell design* $H(s, 2n)$ (on symbol set S) is an $s \times s$ array, H, which satisfies the properties:

1. every cell of H either is empty or contains an unordered pair of symbols from S,
2. each symbol of S occurs once in each row and column of H, and
3. every unordered pair of symbols occurs in at most one cell of H.

The pairs of symbols in the cells of an $H(s, 2n)$ can be thought of as the edges of a s regular graph on $2n$ symbols, the *underlying graph* of the Howell design.

An $H^*(s, 2n)$ is an $H(s, 2n)$ whose underlying graph contains a maximal independent set, one of size $2n - s$.

26.1 Examples An $H(4,6)$, $H(6,12)$ and an $H^*(4,8)$ with independent set $\{1, 2, 3, 4\}$.

04		13	25
23	14	05	
	35	24	01
15	02		34

1 7	2 8	3 9	4 a	5 b	6 12
2 3	12 7	4 5	8 9	6 1	a b
b 12	3 4	7 8	5 6	9 a	1 2
5 a	6 b	1 12	2 7	3 8	4 9
6 9	1 a	2 b	3 12	4 7	5 8
4 8	5 9	6 a	1 b	2 12	3 7

15	26	37	48
47	38	25	16
28	17	46	35
36	45	18	27

26.2 Example An $H(6,8)$ is given in Example V.7.32.

26.3 Theorem [1, 10] Let s and n be positive integers, where n is even and $s+1 \leq 2n \leq 2s$. Then there exists an $H(s, 2n)$ if and only if $(s, 2n) \neq (2,4), (3,4), (5,6)$, or $(5,8)$.

26.4 Remark The proof of nonexistence of a $H(2,4)$ is trivial. The ordered pairs $(3,4)$ and $(5,6)$ correspond to (nonexistent) Room squares of sides 3 and 5. The nonexistence of an $H(5,8)$ is more difficult to establish; a reasonably short proof is given in [8].

26.5 Remark The underlying graph of an $H(2n-1, 2n)$ is the complete graph K_{2n}. The underlying graph of an $H(2m, 2m+2)$ is the *cocktail party graph* $K_{2m+2} - f$, where f is a 1-factor. Also, an $H(s, 2s)$ with underlying graph $K_{s,s}$ is equivalent to a pair of mutually orthogonal latin squares of order s. Existence or nonexistence of Howell designs with specified underlying graphs has been determined for all graphs on at most 10 vertices [8].

26.2 Related Combinatorial Objects

26.6 Theorem The existence of a pair of orthogonal latin squares of side n, implies the existence of an $H^*(n, 2n)$.

26.7 Remarks To construct an $H^*(n, 2n)$ from a pair of orthogonal latin squares of side n (on disjoint symbol sets), merely superimpose the two latin squares. The symbols from either of the symbol sets forms a maximal independent set in the underlying graph $K_{n,n}$. The $H^*(4, 8)$ in Example 26.1 was constructed in this manner. It is also interesting that there exists an $H(6, 12)$ (Example 26.1), yet there is no pair of orthogonal latin squares of side 6.

26.8 Theorem An $H(2n-1, 2n)$ is a Room square of side $2n-1$. See §IV.39.

26.9 Theorem The existence of an $H(s, 2n)$ is equivalent to the existence of a pair of orthogonal 1-factorizations on the underlying graph of the $H(s, 2n)$ (an s–regular graph on $2n$ vertices). See §VI.4.

26.10 Remark As with Room squares, Howell designs are often constructed by the use of starters and adders. The starter-adder construction for Howell designs follows.

Suppose G is an additive abelian group of order s, and $s+1 \le 2n \le 2s$. A *Howell n-starter* in G is a set

$$S = \{\{s_i, t_i\} : 1 \le i \le s-n\} \cup \{\{s_i\} : s-n+1 \le i \le n\}$$

which satisfies the two properties:

1. $\{s_i : 1 \le i \le n\} \cup \{t_i : 1 \le i \le s-n\} = G$;
2. $(s_i - t_i) \ne \pm(s_j - t_j)$ if $i \ne j$.

If S is a Howell n-starter, then a set $A = \{\{a_i\} : 1 \le i \le n\}$ is an *adder* for S if the elements in A are distinct, and the set

$$S + A = \{\{s_i + a_i, t_i + a_i\} : 1 \le i \le s-n\} \cup \{\{s_i + a_i\} : s-n+1 \le i \le n\}$$

is again a Howell n-starter.

26.11 Example $S = \{\{5,6\}, \{2,0\}, \{1\}, \{3\}, \{4\}\}$ is a Howell 5-starter in the group $G = \mathbb{Z}_7$. $A = \{4, 5, 0, 1, 2\}$ is an adder for S.

26.12 Theorem If A is an adder for S, then an $H^*(s, 2n)$ exists. (In the case where the group G is cyclic, the resulting Howell design is *cyclic*).

26.13 Construction Let S is a Howell n-starter and let A be an adder for S. An $H^*(s, 2n)$ will be constructed on the symbol set $G \cup \{\infty_1, \infty_2, \ldots, \infty_{2n-s}\}$. The first row of the Howell design H has the pairs $\{s_i, t_i\}$ in cell $H(0, -a_i)$ for $1 \le i \le s-n$ and pairs $\{s_i, \infty_{i-s+n}\}$ in cell $H(0, -a_i)$ for $s-n+1 \le i \le n$. The remainder of the Howell design is filled out by developing the square cyclically, i.e., place pairs $\{s_i + x, t_i + x\}$ in cell $H(x, -a_i + x)$ and place pairs $\{s_i + x, \infty_{i-s+n}\}$ in cell $H(x, -a_i + x)$ where all arithmetic is in the group G. This construction is illustrated in Example 26.14.

26.14 Example An $H(7, 10)$ constructed from the starter and adder in Example 26.11.

$\infty_1 1$		20	56		$4\infty_3$	$3\infty_2$
$\infty_2 4$	$\infty_1 2$		31	60		$\infty_3 5$
$6\infty_3$	$\infty_2 5$	$\infty_1 3$		42	01	
	$0\infty_3$	$\infty_2 6$	$\infty_1 4$		53	12
23		$\infty_3 1$	$0\infty_2$	$\infty_1 5$		64
05	34		$\infty_3 2$	$1\infty_2$	$\infty_1 6$	
	16	45		$3\infty_3$	$2\infty_2$	$\infty_1 0$

26.15 Construction [6] There exist cyclic Howell designs $H(n, 2n)$ and $H(n+1, 2n)$ when $n \ge 3$ is odd. For the $H(n, 2n)$, let $S = \{\{0\}, \{n-1\}, \{n-2\}, \ldots, \{3\}, \{2\}, \{1\}\}$ and $A = \{0, n-1, n-2, \ldots, 3, 2, 1\}$ (arithmetic modulo n). For $H(n+1, 2n)$, let $S = \{\{0\}, \{n\}, \ldots, \{4\}, \{2,3\}, \{1\}\}$ and $A = \{0, n, n-1, \ldots, \frac{n+1}{2}+3, \frac{n+1}{2}+1, \frac{n+1}{2}, \ldots, 3, 2, 1\}$.

26.16 Theorem [6] There do *not* exist cyclic Howell designs $H(n, 2n)$ and $H(n+1, 2n)$ when n is even.

26.17 Table Starters and adders for small (cyclic) Howell designs $H(s, 2n)$ where $s+1 < 2n < 2s$ and $s \leq 11$.

$H(4,6)$	$S=$	{3,2}	4	1						
	$A=$	1	2	0						
$H(5,8)$	No Howell design									
$H(6,8)$	No starter-adder [6]									
$H(6,10)$	$S=$	{2,3}	{0}	{5}	{4}	{1}				
	$A=$	2	0	4	3	1				
$H(7,10)$	$S=$	{5,6}	{2,0}	{1}	{3}	{4}				
	$A=$	4	5	0	1	2				
$H(7,12)$	No starter-adder (Theorem 26.16)									
$H(8,10)$	No starter-adder [6]									
$H(8,12)$	$S=$	{3,2}	{7,4}	{5}	{6}	{0}	{1}			
	$A=$	3	4	5	6	7	0			
$H(8,14)$	$S=$	{2,3}	{0}	{7}	{6}	{5}	{4}	{1}		
	$A=$	2	0	7	5	4	3	1		
$H(9,12)$	$S=$	{5,3}	{0,6}	{2,7}	{8}	{4}	{1}			
	$A=$	4	5	6	7	8	0			
$H(9,14)$	$S=$	{0,7}	{3,8}	{1}	{6}	{5}	{4}	{2}		
	$A=$	8	4	0	7	6	5	3		
$H(9,16)$	No starter-adder (Theorem 26.16)									
$H(10,12)$	$S=$	{4,5}	{9,1}	{3,6}	{8,2}	{0}	{7}			
	$A=$	0	1	3	5	8	4			
$H(10,14)$	$S=$	{4,5}	{9,1}	{3,6}	{0}	{2}	{7}	{8}		
	$A=$	0	1	3	7	6	4	5		
$H(10,16)$	$S=$	{0,8}	{3,9}	{1}	{7}	{6}	{5}	{4}	{2}	
	$A=$	9	3	0	8	7	5	4	2	
$H(10,18)$	$S=$	{2,3}	{0}	{9}	{8}	{7}	{6}	{5}	{4}	{1}
	$A=$	2	0	9	8	6	5	4	3	1
$H(11,14)$	$S=$	{5,8}	{3,7}	{9,10}	{4,6}	{0}	{2}	{1}		
	$A=$	5	8	9	10	6	7	0		
$H(11,16)$	$S=$	{2,6}	{5,7}	{9,8}	{3}	{10}	{11}	{4}	{1}	
	$A=$	4	6	7	5	8	9	10	0	
$H(11,18)$	$S=$	{0,9}	{3,10}	{1}	{8}	{7}	{6}	{5}	{4}	{2}
	$A=$	10	4	0	9	8	7	6	5	3
$H(11,20)$	No starter-adder (Theorem 26.16)									

26.3 Orthogonal 1-Factorizations

A d-dimensional Howell design $H_d(s, 2n)$ is a d-dimensional array in which every cell either is empty or contains an unordered pair of symbols from an s-set and such that each two-dimensional projection is an $H(s, 2n)$. An $H_3(s, 2n)$ is a *Howell cube.*

26.18 Proposition An $H_d(s, 2n)$ is equivalent to d mutually orthogonal 1-factorizations of the underlying graph (an s regular graph on $2n$ vertices).

$\nu(s, 2n)$ denotes the maximum value of d such that an $H_d(s, 2n)$ exists.

26.19 Proposition $\nu(s, 2n) \leq s-1$.

26.20 Conjecture [8] $\nu(s, 2n) \leq n - 1$.

26.21 Theorem Graphs on $2n$ vertices having at least $n - 1$ orthogonal 1-factorizations include:

1. K_{2n}, if $2n - 1 \equiv 3 \pmod 4$ is a prime power, or $2n = 10$ (see §IV.39.6).
2. $K_{n,n}$, if n is a prime power (Theorem II.2.19).
3. K_{2n} minus a 1-factor, if $2n = 2^j + 2$, $j \geq 2$ [7].

26.22 Table Values for $\nu(s, 2n)$, $2n \leq 10$.

$\nu(2,4) = \nu(3,4) = \nu(5,6) = \nu(5,8) = 1$.
$\nu(3,6) = \nu(4,6) = \nu(6,8) = \nu(6,10) = \nu(7,10) = 2$.
$\nu(4,8) = \nu(7,8) = 3$.
$\nu(5,10) = \nu(8,10) = \nu(9,10) = 4$.

26.23 Example [2] A Howell cube $H_3(6, 12)$. The underlying graph G is the icosahedron with antipodal points joined. This is the only known case where $\nu(s, 2s)$ exceeds the maximum number of mutually orthogonal latin squares of order s. (The cube is presented as three orthogonal one factorizations of G).

1,2 3,4 5,6 7,8 9,10 11,12	1,2 4,8 3,6 9,11 7,12 5,10	1,2 5,7 4,9 3,11 6,12 8,10
3,8 4,6 5,7 1,10 9,11 2,12	3,4 5,7 2,8 1,6 9,12 10,11	3,4 1,7 2,5 8,12 9,11 6,10
2,8 3,5 4,9 6,10 1,11 7,12	5,6 4,9 2,3 1,10 7,11 8,12	5,6 3,8 2,9 1,4 7,12 10,11
1,6 2,9 4,7 5,10 3,11 8,12	7,8 5,9 1,4 2,12 6,10 3,11	2,3 4,6 7,8 5,10 1,11 9,12
1,7 2,3 5,9 4,8 6,12 10,11	3,8 2,5 4,7 1,11 9,10 6,12	1,6 3,5 4,8 9,10 7,11 2,12
1,4 2,5 3,6 8,10 7,11 9,12	1,7 4,6 3,5 2,9 8,10 11,12	2,8 4,7 5,9 3,6 1,10 11,12

26.24 Example [9] A Howell cube $H_3(7, 12)$.

1,2 3,4 8,9 6,11 5,10 7,12	1,2 4,9 3,7 6,8 5,11 10,12	1,2 7,9 5,6 3,10 4,11 8,12
1,3 6,7 2,9 5,11 8,12 4,11	1,3 7,9 2,5 6,11 8,10 4,12	1,3 2,8 4,9 6,10 7,11 5,12
1,4 6,8 2,5 7,11 3,10 9,12	1,4 2,8 6,7 3,9 5,10 11,12	1,4 2,7 3,8 5,9 6,11 10,12
1,5 7,9 2,8 4,6 3,11 10,12	1,5 3,8 2,9 4,11 6,10 7,12	1,5 3,4 6,7 8,10 2,11 9,12
1,6 2,7 3,9 5,12 4,11 8,10	1,6 5,9 3,4 2,10 7,11 8,12	1,6 2,5 3,7 8,9 4,10 11,12
1,7 4,9 3,8 5,6 2,10 11,12	1,7 8,9 4,6 3,10 5,12 2,11	1,7 2,9 6,8 5,10 3,11 4,12
1,8 3,7 5,9 6,10 2,11 4,12	1,8 5,6 2,7 4,10 3,11 9,12	1,8 4,6 3,9 2,10 5,11 7,12

26.4 Enumeration of Small Howell Designs

$\text{IR}_d(G)$ denotes the number of inequivalent sets of d pairwise orthogonal 1-factorizations of a graph G. $\text{NH}_d(s, 2n)$ denotes the number of nonisomorphic d-dimensional Howell designs $H_d(s, 2n)$.

26.25 Remark $\text{NH}_d(s, 2n) = \sum_G \text{IR}_d(G)$, where the sum is taken over all s−regular graphs G on $2n$ vertices.

26.26 Table $\text{NH}_d(s, 2n)$, $2n < 10$ [8, 9]. Since $\text{NH}_d(2n-1, 2n) = \nu(2n-1)$, these values are given in Table IV.39.24.

s	2	3	3	4	4	4	4	4	5	5	6	6	5	5	5	5	6	6	7	7	8	8	8	8
$2n$	4	6	6	6	6	8	8	8	8	8	8	8	10	10	10	10	10	10	10	10	10	10	10	10
d	1	1	2	1	2	1	2	3	1	2	1	2	1	2	3	4	1	2	1	2	1	2	3	4
$\text{NH}_d(s, 2n)$	1	2	1	1	1	16	1	1	19	0	13	3	3472	6	1	1	13277	18	14241	901	3192	18220	3	1

26.27 Remark [3] Isomorphism of two Howell designs $H(s, 2n)$ can be decided in time $O(n^{O(\log n)})$.

26.5 Application to Duplicate Bridge

26.28 Remark Cyclic Howell designs are used as *Howell movements* in duplicate bridge. The rows of the design represent the rounds, the columns represent the boards, and the diagonals represent the tables. See [5] and §V.26.5.

A cyclic Howell design (Howell movement) is *compact* if the n filled cells in any row are adjacent (cyclically).

26.29 Theorem If there exists a compact cyclic Room square of side $2n-1$, then n is odd.

26.30 Theorem Suppose $0 \le t \le 3$ and $n \ge 2t+1$. Then there exists a compact cyclic $H(n+t, 2n)$.

26.6 See Also

§IV.39	Howell designs are generalizations of Room squares.
§IV.45	Starters are used to construct Howell designs.
§VI.4	A connection with 1-factorizations is given by Theorem 26.9.
§V.7.10	More on scheduling duplicate bridge tournaments
[4]	A broad survey of Room squares with a section on Howell designs. This survey contains most of the information in this section.
[1, 10]	Contains the existence proof of Howell designs.
[6]	The first paper on Howell designs.
[5]	A book on directing duplicate bridge tournaments. Contains a chapter on Howell movements and many schedules of play.

References

[1] B.A. Anderson, P.J. Schellenberg and D.R. Stinson, The existence of Howell designs of even side, *J. Combin. Theory A* 36 (1984), 23-55.

[2] E.F. Brickell, A few results in message authentication, *Congr. Numer.* 43 (1984), 141-154.

[3] C.J. Colbourn and M.J. Colbourn, Combinatorial isomorphism problems involving 1-factorizations, *Ars Combin.* 9 (1980), 191-200.

[4] J.H. Dinitz and D.R. Stinson, Room squares and related designs, in: *Contemporary Design Theory: A Collection of Surveys* (J.H. Dinitz, D.R. Stinson; eds.) John Wiley & Sons, New York, 1992, 137-204.

[5] A. Groner, *Duplicate Bridge Direction*, Baron Barclay Bridge Supplies, Louisville, 1993.

[6] S.H.Y. Hung and N.S. Mendelsohn, On Howell designs, *J. Combin. Theory A* 16 (1974), 174-198.

[7] E.R. Lamken and S.A. Vanstone, Designs with mutually orthogonal resolutions, *Europ. J. Combin.* 7 (1986), 249-257

[8] A. Rosa and D. R. Stinson, One-factorizations of regular graphs and Howell designs of small order, *Utilitas Math.* 29 (1986), 99-124.

[9] E. Seah and D.R. Stinson, An assortment of new Howell designs, *Utilitas Math.* 31 (1987), 175-188.

[10] D.R. Stinson, The existence of Howell designs of odd side, *J. Combin. Theory A* 32 (1982), 53-65.

27 Maximal Sets of MOLS

Anthony B. Evans

27.1 Definition and Examples

A k Max MOLS(n) is a set of k MOLS of side n that is maximal, i.e., that cannot be extended to a set of $k+1$ MOLS of side n.

27.1 Example 1 Max MOLS(5), 1 Max MOLS(7), and 2 Max MOLS(8).

1	2	5	3	4
4	1	2	5	3
2	3	1	4	5
3	5	4	1	2
5	4	3	2	1

1	2	3	4	5	6	7
2	1	4	5	6	7	3
3	4	5	7	1	2	6
4	5	7	6	3	1	2
5	6	2	3	7	4	1
6	7	1	2	4	3	5
7	3	6	1	2	5	4

1	2	3	4	5	6	7	8
2	3	4	1	8	5	6	7
3	4	1	2	7	8	5	6
4	1	2	3	6	7	8	5
5	6	7	8	1	2	3	4
6	7	8	5	4	1	2	3
7	8	5	6	3	4	1	2
8	5	6	7	2	3	4	1

1	3	6	8	2	4	5	7
2	4	5	7	3	1	8	6
3	1	8	6	4	2	7	5
4	2	7	5	1	3	6	8
5	7	2	4	6	8	1	3
6	8	1	3	7	5	4	2
7	5	4	2	8	6	3	1
8	6	3	1	5	7	2	4

27.2 Example 2 Max MOLS(7) and 3 Max MOLS(8).

1	2	3	4	5	6	7
2	3	4	5	6	7	1
3	4	5	6	7	1	2
4	5	6	7	1	2	3
5	6	7	1	2	3	4
6	7	1	2	3	4	5
7	1	2	3	4	5	6

1	4	7	2	6	5	3
2	5	1	3	7	6	4
3	6	2	4	1	7	5
4	7	3	5	2	1	6
5	1	4	6	3	2	7
6	2	5	7	4	3	1
7	3	6	1	5	4	2

1	5	2	6	3	7	4	8
5	1	6	2	7	3	8	4
2	6	3	7	4	8	1	5
6	2	7	3	8	4	5	1
3	7	4	8	1	5	2	6
7	3	8	4	5	1	6	2
4	8	1	5	2	6	3	7
8	4	5	1	6	2	7	3

1	8	6	3	4	5	7	2
5	4	2	7	8	1	3	6
2	5	7	4	1	6	8	3
6	1	3	8	5	2	4	7
3	6	8	1	2	7	5	4
7	2	4	5	6	3	1	8
4	7	5	2	3	8	6	1
8	3	1	6	7	4	2	5

1	4	7	8	2	3	6	5
5	8	3	4	6	7	2	1
2	1	8	5	3	4	7	6
6	5	4	1	7	8	3	2
3	2	5	6	4	1	8	7
7	6	1	2	8	5	4	3
4	3	6	7	1	2	5	8
8	7	2	3	5	6	1	4

27.2 Constructions, Bounds, and Existence Results

27.3 Theorem If there exists an $(n, r; 1)$-difference matrix D (see §IV.11) for which $mD = (D \dots D)$, i.e., m consecutive copies of D, is maximal and if either $m = 1$ or there exist $r-1$ MOLS of side m, then there exists an $r-1$ Max MOLS(nm).

27.4 Theorem If n does not divide m and there exists a projective plane of order n and $n-1$ MOLS of side m, then there exists an $n-1$ Max MOLS(nm).

27.5 Theorem If $n = mp^r$, p a prime, $\gcd(m, p) = 1$, and either $m = 1$ or there exist $p-1$ MOLS of side m, then there exists a $p-1$ Max MOLS(n).

27.6 Theorem [2, 4] Let $d = n-1-k$, $b(x) = \frac{1}{2}x^4 + x^3 + x^2 + \frac{1}{2}x$, and $m(x) = 8x^3 - 18x^2 + 8x + 4 - 2R(x^2 - x - 1) + 9R(R-1)(x-1)/2$, where $R \in \{0, 1, 2\}$, $R \equiv d+1 \pmod 3$. If there exists a k Max MOLS(n), then either $k = n-1$ (corresponding to a projective plane of order n), or $b(d-1) \geq n$ (Bruck's bound) and $m(d) \geq 3n$ (Metsch's bound).

27.7 Theorem If G is a group of order n with a unique element of order 2, then the Cayley table of G is a 1 Max MOLS(n).

27.8 Theorem Let L be a latin square of side $4t+2$ (alternatively $4t+1$) with a subsquare of side $2t+1$ ($2t$) in which all entries are from a set of $2t+1$ ($2t$) except for t ($\frac{t-1}{2}$) or fewer cells. Then L is a 1 Max MOLS(n).

27.9 Theorem If $n \not\equiv 3 \pmod 4$, then there exists a 1 Max MOLS(n). If $n \equiv 3 \pmod 4$, then there exists a 1 Max MOLS(n) and/or a 2 Max MOLS(n).

27.10 Theorem If $n = q_1 \cdots q_r$ is the factorization of n into prime powers that are relatively prime with $q_1 < \ldots < q_r$, then there exists a $q_1 - 1$ Max MOLS(n).

27.11 Theorem Let $p \geq 7$ be a prime. If $p \equiv 3 \pmod 4$, then there exists a $(p-3)/2$ Max MOLS(p). If $p \equiv 1 \pmod 4$, then there exists a $(p-1)/2$ Max MOLS(p).

27.12 Theorem If $q = p^r$, p a prime, then there exists a k Max MOLS(q^2) for $k = q^2 - q - 1$, and if $p \geq 5$, for $k = q^2 - q - 2$ and $q^2 - q$. [2, 3]

27.13 Table The values of k for which a k Max MOLS(n) is known to exist. B_n is the best known bound for which no k Max MOLS(n), $B_n \leq k < n-1$, exists. This extends the table in [2].

n	B_n	k	n	B_n	k	n	B_n	k
2		1	22	19	1	42	38	1, 2
3	1	2	23	20	(1 or 2), 10, 22	43	39	(1 or 2), 20, 42
4	2	1, 3	24	20	1, 2, 5	44	40	1, 3
5	2	1, 4	25	21	1, 4, 12, 18-20, 24	45	41	1, 2, 4
6	2	1	26	22	1	46	42	1
7	3	1, 2, 6	27	23	2, 12, 26	47	43	(1 or 2), 22, 46
8	5	1-3, 7	28	24	1, 3	48	44	1, 2
9	6	1-3, 5, 8	29	25	1, 14, 28	49	45	1, 6, 24, 40-42, 48
10	7	1, 2	30	26	1, 2	50	46	1
11	8	2-4, 10	31	27	(1 or 2), 14, 30	51	47	2
12	9	1-3, 5	32	28	1, 31	52	48	1, 3
13	10	1, 2, 4, 6, 12	33	29	1, 2	53	49	1, 26, 52
14	11	1	34	30	1	54	50	1
15	12	2, 4	35	31	(1 or 2), 4	55	51	(1 or 2), 4
16	13	1, 11, 15	36	32	1-3	56	52	1, 3, 6
17	14	1, 4, 8, 16	37	33	1, 18, 36	57	53	1, 2
18	15	1	38	34	1	58	54	1
19	16	(1 or 2), 3, 6, 8, 18	39	35	2	59	55	(1 or 2), 28, 58
20	17	1, 3, 4	40	36	1, 4	60	56	1-4
21	18	1, 2, 4	41	37	1, 20, 40	61	57	1, 30, 60

27.14 Remark While many of the sets of MOLS given in §II.2 may indeed be maximal there is often no indication that these sets of MOLS have been checked for maximality.

27.3 See Also

§II.2	Information about latin squares and MOLS.
§IV.11	Maximal difference matrices yield maximal sets of MOLS.
[1]	Contains implicit difference matrix constructions of maximal sets of MOLS.
[2]	A survey of latin squares, including MOLS and maximal sets of MOLS.

References

[1] A.B. Evans, *Orthomorphism Graphs of Groups*, (Lecture Notes in Mathematics 1535), Springer-Verlag, New York, 1992.

[2] D. Jungnickel, Latin squares, their geometries and their groups. A survey, in: *Coding Theory and Design Theory, Part II*, IMA Vol. Math. Appl. 21, Springer-Verlag, New York, 1990, 166-225.

[3] D. Jungnickel, Maximal partial spreads and transversal-free translation nets, *J. Combin. Theory A* 62 (1993), 66-92.

[4] K. Metsch, Improvement of Bruck's completion theorem, *Des. Codes Crypt.* 1 (1991), 99-116.

28 Mendelsohn Designs

Eric Mendelsohn

28.1 Definitions and Examples

The k-tuple $(x_0, x_1, x_2, \ldots, x_{k-1})$ *cyclically contains* the ordered pairs (x_0, x_1), $(x_1, x_2), \ldots, (x_{k-1}, x_0)$, and no others. The notation $(x_0, x_1) \subset_{Z_k} (x_0, x_1, \ldots, x_{k-1})$ is used or, where no confusion arises, simply "$\subset$" is used. Elements x_i, x_{i+t} are t-*apart* in the k-tuple, where $i + t$ is taken modulo k.

A (v, k, λ)-*Mendelsohn design*, or MD(v, k, λ), is a set V together with a collection $\mathcal{B}$ of ordered k-tuples of distinct elements from V (*blocks*), such that each ordered pair (x, y) with $x \neq y$ is cyclically contained in λ blocks. An M$(v, 3, \lambda)$ is a *Mendelsohn triple system*, MTS(v, λ).

28.1 Example Any triple system can be made into a Mendelsohn triple system by replacing $\{a, b, c\}$ by both (a, b, c) and (a, c, b).

28.2 Example An MTS(4, 1). Let $\mathcal{B} = \{(0,1,2), (1,0,3), (2,1,3), (0,2,3)\}$ with $V = \{0, 1, 2, 3\}$.

The *converse* of a Mendelsohn design is the design obtained by reversing the order of its blocks.

A *perfect Mendelsohn design* (PMD) is an MD(v, k, λ), such that for every $i = 1, 2, 3, \ldots, k-1$, each ordered pair (x, y) is i-apart in exactly λ blocks. Every MTS is perfect.

The *underlying design* of a PMD(v, k, λ) is the block design with parameters $(v, k, (k-1)\lambda)$ obtained by "forgetting" the order structure on the blocks.

A Mendelsohn design is *simple* if it has no repeated blocks. It is *pure* if its underlying design has no repeated blocks.

28.3 Example A pure MD(5, 3, 3). Let $\mathcal{B} = [(0,1,2), (0,2,4), (0,3,1), (0,4,3)]$ (mod 5) with $V = \{0, 1, 2, 3, 4\}$.

28.4 Example Small $(v, 4, 2)$-PMDs.

$$\begin{array}{lcl} v=4: & \mathcal{B} = & [(\infty,0,1,2),(\infty,0,2,1)] \pmod 3 \\ v=6: & \mathcal{B} = & \{(1,2,3,4),(1,2,5,3),(1,6,3,2),(1,4,2,6),(1,6,2,5), \\ & & (1,4,5,2),(1,5,6,3),(1,3,6,4),(1,3,4,5),(1,5,4,6), \\ & & (2,3,5,4),(2,4,3,6),(2,6,5,3),(2,4,6,5),(3,5,6,4)\}. \\ v=8: & \mathcal{B} = & [(\infty,0,1,3),(\infty,0,-1,-3),(0,1,-2,2),(0,-1,2,-2)] \pmod 7 \\ v=10: & \mathcal{B} = & [(\infty,0,3,1),(\infty,0,4,8),(0,1,2,4),(0,3,5,1),(0,-3,3,1)] \pmod 9 \\ v=12: & \mathcal{B} = & [(\infty,0,1,5),(\infty,0,-1,-5),(0,1,3,8),(0,-1,-3,-8), \\ & & (0,-4,-2,-5),(0,3,1,5)] \pmod{11} \\ v=14: & \mathcal{B} = & [(\infty,0,-1,-2),(\infty,0,-4,2),(0,1,3,6),(0,4,1,-5), \\ & & (0,6,1,3),(0,1,6,-4),(0,2,-4,4)] \pmod{13} \end{array}$$

28.5 Theorem A necessary condition for a MD(v,k,λ) to exist is that $\lambda v(v-1) \equiv 0 \pmod k$).

A Mendelsohn design is *resolvable* (RMD) if

1. $v \equiv 0 \pmod k$) and its blocks can be partitioned into $\lambda(v-1)$ resolution classes (its underlying design is resolvable) or
2. $v \equiv 1 \pmod k$) and its blocks can be partitioned into λv near-resolution classes (its underlying design is nearly resolvable).

28.6 Example A resolvable $(6,3,6)$-Mendelsohn design. Here $[(0,1,3), (\infty,4,2)]$; $[(0,2,1), (\infty,4,3)]$; $[(0,3,4), (\infty,2,1)]$ give three resolution classes; develop modulo 5 for the remaining 12.

28.7 Example A resolvable MTS(10,1), blocks written as rows.

0	1	2	3	4	5	6	7	8	9
231	480	398	975	702	826	514	649	167	053
794	369	541	106	635	437	873	085	952	246
865	572	607	428	189	901	092	213	340	781

28.2 Relations with Quasigroups and Latin Squares

28.8 Theorem If $(Q,\cdot)$ is a quasigroup satisfying $x\cdot x = x$ and $x\cdot(y\cdot x) = y$ for all $x,y\in Q$ (a *semisymmetric* quasigroup), let $\mathcal{B} = \{(x,y,x\cdot y) : x,y\in Q\}$. Then $(Q,\mathcal{B})$ is a Mendelsohn triple system MTS$(|Q|,1)$. Conversely, if $(Q,\mathcal{B})$ is an MTS, then $(Q,\cdot)$, defined by $x\cdot x = x$ for all $x\in Q$, and $x\cdot y = z$ whenever $(x,y,z)\in\mathcal{B}$, is a semisymmetric quasigroup.

28.9 Theorem Let $w_0(x,y) = x$, $w_1(x,y) = y$, $w_2(x,y) = x\cdot y$, and $w_n(x,y) = w_{n-2}(x,y)\cdot w_{n-1}(x,y)$. Suppose that $(Q,\cdot)$ exists such that

1. $x\cdot x = x$ for all $x\in Q$;
2. $w_k(x,y) = x$ for all $x,y\in Q$; and
3. for any a,b, $w_r(a,x) = b$ has a unique solution for $r = 1,2,\ldots,k-1$.

Let $\mathcal{B} = \{(w_0(x,y),\cdots,w_{k-1}(x,y)) : x,y\in Q, x\neq y\}$. Then $(Q,\mathcal{B})$ is a PMD of order v, blocksize k, and index 1. The converse also holds.

28.10 Theorem Let p be an odd integer and (X,B) be a $(p,k,1)$-PMD. Then if B^* is the matrix obtained by writing out the blocks of B and all their cyclic shifts as columns, and X^* is $[x,x,\cdots,x]^T$, $x\in X$, then $[B^* \mid X^*]$ is an orthogonal array corresponding to $k-2$ idempotent MOLS of side p.

If $(V, \mathcal{B})$ is a resolvable $(v, k, 1)$-PMD (RPMD) with $v \equiv 1 \pmod{k}$), then the quasigroups $(V, *_i)$ for $1 \leq i \leq k-1$ *associated with* $(V, \mathcal{B})$ are defined by $x *_i x = x$ and $x *_i z = y$ if (x, y) are i-apart in the block whose resolution class misses z.

28.11 Remark The $k - 1$ quasigroups so obtained have multiplication tables which are idempotent MOLS. Example 28.7 yields

0	4	7	6	2	1	9	8	3	5
2	1	5	0	8	9	4	3	6	7
3	5	2	8	7	6	0	1	9	4
1	6	9	3	5	7	8	2	4	0
7	8	1	2	4	3	5	9	0	6
8	7	4	9	6	5	1	0	2	3
5	9	0	1	3	8	6	4	7	2
9	2	6	5	0	4	3	7	1	8
6	0	3	4	9	2	7	5	8	1
4	3	8	7	1	0	2	6	5	9

0	8	6	1	7	9	2	5	4	3
3	1	4	6	9	0	5	2	7	8
1	7	2	4	0	8	9	3	5	6
2	9	8	3	6	4	7	1	0	5
9	0	5	8	4	7	1	6	3	2
6	2	1	7	3	5	4	8	9	0
8	3	7	0	5	2	6	9	1	4
4	5	0	9	2	3	8	7	6	1
5	4	9	2	1	6	3	0	8	7
7	6	3	5	8	1	0	4	2	9

As noted by Ganter, Mathon, and Rosa in 1978, these two MOLS have four disjoint common transversals:

(0,0)	(1,1)	(2,2)	(3,3)	(4,4)	(5,5)	(6,6)	(7,7)	(8,8)	(9,9)
(0,1)	(1,0)	(2,4)	(3,9)	(4,6)	(5,3)	(6,5)	(7,8)	(8,2)	(9,7)
(0,4)	(1,6)	(2,5)	(3,0)	(4,7)	(5,9)	(6,8)	(7,3)	(8,1)	(9,2)
(0,9)	(1,8)	(2,0)	(3,2)	(4,1)	(5,6)	(6,7)	(7,4)	(8,5)	(9,3)

Each row specifies the ten cells of a transversal.

28.3 Mendelsohn Designs from BIBDs

28.12 Theorem If p is an odd prime and if a $(v, p, 1)$-BIBD exists, then a $(v, p, 1)$-PMD exists. If in addition $v \equiv 1 \pmod{p}$) and if the BIBD is near resolvable, then the PMD is near resolvable.

A $(v, k, (k-1)\lambda)$-BIBD is *orientable* if and only if a cyclic order can be placed on its blocks to form a MD(v, k, λ).

28.13 Theorem For every $v \equiv 0, 1 \pmod 3$, $v \geq 6$ there is a nonorientable $(v, 3, 2)$-BIBD.

28.14 Theorem There is an algorithm for determining whether a $(v, 3, 2)$-BIBD can be oriented which has $O(v^2)$ running time. It is NP-complete to determine whether a $(v, 3, 2\lambda)$-BIBD can be oriented if $\lambda > 1$.

Let $D = (V, \mathcal{B})$ be a Steiner triple system and let the map $f : V \to V$ be an involution with one fixed point. If f has the properties that the image of a block is never a block, that $(V, \mathcal{B} \cup f(\mathcal{B}))$ is orientable, and that f preserves cyclic order in this orientation, then D is an *antipodal* triple system.

28.15 Example Any cyclic $TS(v, 1)$, with $v \equiv 1 \pmod 6$ and $f : x \to -x \pmod v$, is antipodal.

28.16 Theorem There exists an antipodal $TS(v, 1)$ for all $v \equiv 1, 3 \pmod 6$ [8].

28.4 Existence

28.17 Table The basic existence results. A stronger condition implies the weaker one: e.g., an RPMD implies RMD, PMD, and an MD.

$k = 3$

λ	Type of Design	Existence Result	Reference
1	MTS	$v \equiv 0, 1 \pmod 3$, $v \neq 6$	[9]
	pure MTS	$v \equiv 0, 1 \pmod 3$, $v \neq 3, 6$	[3]
	RMTS	$v \equiv 0, 1 \pmod 3$, $v \neq 6$	[6]
λ	MTS	$\lambda \equiv 0 \pmod{\gcd(v-2, 3)}$, $v \neq 6$	[7]

$k \geq 4$

k	PMD	RMD	RPMD
4	$\lambda v(v-1) \equiv 0 \pmod 4$ except $v = 4, \lambda$ odd and $v = 8, \lambda = 1$ [4]	$\lambda = 1$ $v \equiv 0, 1 \pmod 4$ $v \neq 4, 12$	$\lambda = 1:$ $v \equiv 0, 1 \pmod 4$, $v \neq 4, 8$ and at most 49 exceptions < 340 $\lambda > 1$: all v admissible except $v = 4$, λ odd.
5	$v \equiv 0, 1 \pmod 5$, $v \neq 6, 10$ for $\lambda = 1, 15$ and 20 are 14 possible exceptions [2]		$v \equiv 5 \pmod{20}$ and RBIBD $(v, 5, 1)$ exists (Theorem 28.12)
6	$\lambda = 1, v > 6$ $v \equiv 0, 1 \pmod 6$ 150 exceptions < 2604 [10] $\lambda = 3$, $v \geq 6$ See Table 28.18		$\lambda = 1, v \equiv 1 \pmod 6$ at most 31 exceptions < 1921
7	$\lambda = 1$ $v \equiv 0, 1 \pmod 7$ $v > 421$ See Table 28.19	λ even, $\lambda v(v-1) \equiv 0 \pmod 7$ for all $\lambda \geq 16$, at most 29 exceptions with $\lambda \leq 16$ [5]	$v \equiv 7 \pmod{42}$ and RBIBD $(v, 7, 1)$ (Theorem 28.12)

28.18 Table [11] Values of v for which the existence of a $(v, 6, 3)$-PMD is open.
6, 10, 12, 16, 18, 20, 22, 24, 26, 28, 30, 32, 33, 34, 38, 39, 40, 42, 44, 45, 48, 51, 52, 54, 55, 60, 62

28.19 Table [5] Values of v for which the existence of a $(v, 7, 1)$-PMD is open.
14, 15, 21, 22, 35, 36, 42, 70, 84, 85. 98, 99, 105, 106, 126, 141, 147, 148, 154, 155, 168, 182, 183, 196, 238, 245, 252, 153, 266, 267, 273, 274, 294, 315, 322, 364, 378, 420

28.20 Theorem [11] The necessary conditions are asymptotically sufficient for the existence of a (v, k, λ)-PMD, a (v, k, λ)-RPMD with $v \equiv 0 \pmod k$, and a (v, p, λ)-RMD where p is an odd prime.

28.21 Theorem Let p be an odd prime, then there exists a $(p^2, p, 1)$-RMD. If $p \equiv 1 \pmod k$, then there exists a $(p^r, k, 1)$-RPMD, for all $r > 1$.

28.5 Enumeration

Two MTSs are *equivalent* if they are isomorphic or if one is isomorphic to the converse of the other.

28.22 Table [7] The number $M(v)$ of inequivalent MTS(v,1).

v	3	4	6	7	9	10
$M(v)$	1	1	0	3	18	144

28.23 Example The three inequivalent MTS(7,1).

1. $[(0,1,3)\ (0,3,1)]$ (mod 7);
2. $[(0,1,3)\ (0,3,2)]$ (mod 7);
3. {(0,1,2), (0,2,1), (0,3,4), (0,4,3), (0,5,6), (0,6,5) (1,3,5), (1,6,3), (1,5,4), (1,4,6), (2,5,3), (2,3,6) (2,4,5), (3,6,4)}.

28.24 Table The 18 inequivalent MTS(9). The numbering on the left indicates the (9,3,2)-BIBD from which this MTS was obtained by orienting the blocks (see Table I.1.16 for the list of (9,3,2)-BIBDs).

	MTSs of order 9											
1	012	102	034	304	135	315	238	328	147	417	246	426
	056	506	257	527	458	548	168	618	367	637	078	708
2	012	102	034	304	135	315	237	328	147	418	246	426
	056	506	258	527	457	548	168	617	367	638	078	708
4	012	102	034	304	135	317	237	326	148	415	246	428
	056	506	258	527	538	457	167	618	368	647	078	708
5	012	102	034	304	135	317	238	326	148	415	246	427
	056	506	257	528	537	458	167	618	368	647	078	708
7	012	102	034	304	136	315	237	328	145	418	246	427
	056	507	258	526	357	548	068	167	638	647	708	178
10a	012	102	034	304	135	317	236	328	146	418	245	427
	056	506	157	258	537	548	168	267	638	647	078	708
10b	012	102	034	304	135	317	238	326	148	416	247	425
	056	506	157	528	537	458	618	627	368	467	078	708
11	012	102	034	304	136	315	238	327	147	418	245	426
	056	507	516	257	358	548	068	628	637	467	708	178
12	012	102	034	304	136	315	237	328	147	418	245	426
	056	507	516	258	357	548	068	627	638	467	708	178
20	012	102	034	305	138	314	235	326	046	417	247	428
	507	157	516	258	456	548	068	618	627	367	708	378
21	012	102	034	305	138	314	235	328	046	417	247	426
	507	157	516	256	458	548	068	618	367	637	708	278
23a	012	102	034	305	138	314	236	327	046	417	245	428
	507	157	516	258	356	548	068	618	267	647	708	378
23b	012	102	034	305	138	314	236	327	046	417	248	425
	507	157	516	528	356	458	068	618	267	647	708	378
24	012	103	023	213	045	406	146	415	247	428	348	437
	057	517	258	526	356	538	608	168	627	367	078	718
30	012	103	023	214	135	328	045	406	417	246	347	438
	057	156	258	527	536	548	608	168	267	637	078	718
31	012	103	023	214	135	327	046	405	418	247	348	436
	507	156	258	526	538	457	068	167	628	637	708	178
33	012	103	024	215	304	135	237	328	146	418	426	347
	056	507	258	536	457	548	068	167	627	638	708	178
34	012	103	024	215	304	136	237	326	145	418	427	348
	056	507	258	357	538	546	068	167	628	647	708	178

28.25 Remark For $v > 10$ no exact value of $M(v)$ is known. Theorem 28.26 gives an asymptotic bound.

28.26 Theorem $M(v) = \exp(\frac{v^2 \ln v}{3}(1 + o(1)))$ as $v \to \infty$. This is the same bound as for $TS(v, 2)$'s. It is not known what proportion of $TS(v, 2)$'s are orientable.

28.6 See Also

§I.4	Steiner systems.
§I.6	Resolvable and near-resolvable designs.
§IV.15	Directed designs.
§IV.22	Graph decompositions and designs.
§IV.36	Quasigroups; self-orthogonal Mendelsohn triple systems.
§IV.17	Embedding partial quasigroups.
§IV.53	A directed Whist tournament is a RPMD with block size four.
[1]	A wealth of information on related quasigroups.
[7]	A survey on Mendelsohn and directed triple systems.

References

[1] F.E. Bennett, Quasigroup identities and Mendelsohn designs, *Canad. J. Math.* 51 (1989), 341-368.

[2] F.E. Bennett, C.J. Colbourn and L. Zhu, Existence of certain types of 3 HMOLS, *Discrete Math.*, to appear.

[3] F.E. Bennett, E. Mendelsohn and N.S. Mendelsohn, Resolvable perfect cyclic designs, *J. Combin. Theory A* 29 (1980), 142-150.

[4] F.E. Bennett, X. Zhang and L. Zhu, Perfect Mendelsohn designs with block size 4, *Ars Combin.* 29 (1990), 65-72.

[5] F.E. Bennett, J. Yin and L. Zhu, On the existence of perfect Mendelsohn designs with $k = 7$ and λ even, *Discrete Math.* 113 (1993), 7-25.

[6] J.-C. Bermond, A. Germa and D. Sotteau, Resolvable decomposition of K_n^*, *J. Combin. Theory A* 26 (1979), 179-185.

[7] C.J. Colbourn and A. Rosa, Directed and Mendelsohn triple systems, in: *Contemporary Design Theory: A Collection of Surveys* (J.H. Dinitz, D.R. Stinson; eds.) John Wiley & Sons, New York, 1992, 97-136.

[8] P. Gibbons, E. Mendelsohn and H. Shen, Antipodal triple systems, *Australas. J. Combin.* 9 (1994), 137-151.

[9] N.S. Mendelsohn, A natural generalization of Steiner triple systems, in: *Computations in Number Theory* (A.O. Atkin, B.J. Birch; eds.), Academic Press, New York, 1971, 323-338.

[10] Y. Miao and L. Zhu, Perfect Mendelsohn designs with block size six, *Discrete Math.*, to appear.

[11] X. Zhang, On the existence of perfect Mendelsohn designs and resolvable perfect Mendelsohn designs, preprint.

29 The Oberwolfach Problem

Brian Alspach

29.1 Definition

Let $n \geq 3$ and $\ell_1, \ell_2, \ldots, \ell_t$ be integers satisfying $3 \leq \ell_i$ for $1 \leq i \leq t$ and $\ell_1 + \ell_2 + \cdots + \ell_t = n$. Let $F(\ell_1, \ell_2, \ldots, \ell_t)$ denote the 2-regular graph on n vertices whose components are cycles of lengths $\ell_1, \ell_2, \ldots, \ell_t$, respectively. Let K_n denote the complete graph with n vertices and $K_n - I$ denote the complete graph with a 1-factor removed when n is even. The problem of determining whether or not the edge set of K_n, n odd, or $K_n - I$, n even, can be partitioned into subgraphs isomorphic to $F(\ell_1, \ell_2, \ldots, \ell_t)$ is the *Oberwolfach problem*, denoted $\mathrm{OP}(\ell_1, \ell_2, \ldots, \ell_t)$. In the special case that the 2-factor is made up of m cycles of the same length ℓ, the notation $\mathrm{OP}(m; \ell)$ is used.

29.1 Remark A solution to $\mathrm{OP}(m; 3)$, m odd, is a Kirkman triple system, and a solution to $\mathrm{OP}(m; 3)$, m even, is a nearly Kirkman triple system (§I.6.3). Also, any solution of an Oberwolfach problem is a special case of a resolvable cycle system (§IV.9.5).

29.2 Example Consider the complete graph K_7 with vertex set $\{1, 2, \ldots, 7\}$. The three 2-factors $1471, 23562$ and $2572, 34613$ and $4673, 45124$ form a solution of $\mathrm{OP}(3, 4)$.

29.3 Theorem [1, 2, 4] Other than $\mathrm{OP}(3, 3)$ and $\mathrm{OP}(3, 3, 3, 3)$, neither of which has a solution, $\mathrm{OP}(m; \ell)$ has a solution for all $m \geq 1$ and all $\ell \geq 3$.

29.4 Remark There also are a few cases settled where not all cycles have the same length.

29.5 Theorem [7] Neither $\mathrm{OP}(4, 5)$ nor $\mathrm{OP}(3, 3, 5)$ has a solution. (Note: there is no published proof of the fact that $\mathrm{OP}(3, 3, 5)$ has no solution.)

29.6 Theorem [5, 6, 7] The following Oberwolfach problems all have solutions:

1. $\mathrm{OP}(3, k)$ for all odd $k \geq 5$,
2. $\mathrm{OP}(3, 4k, 4k)$ for all $k \geq 1$,
3. $\mathrm{OP}(3, 8k - 2)$ for all $k \geq 1$,
4. $\mathrm{OP}(4, k)$ for all even $k \geq 4$,
5. $\mathrm{OP}(k, k, 2\lceil k/2 \rceil + 2c)$ for all $k \geq 3$ and each $c = 0, 2, 3, \ldots$,
6. $\mathrm{OP}(2k + 1, 2k + 1, 2k + 2)$ for all $k \geq 1$,
7. $\mathrm{OP}(2k + 1, 4, \ldots, 4)$ for all $k \geq 1$ and arbitrarily many cycles of length 4, and
8. $\mathrm{OP}(2k + 1, 4k, \ldots, 4k)$ for all $k \geq 1$ and arbitrarily many cycles of length $4k$.

In addition, when the number n of vertices is even and satisfies $4 \leq n \leq 16$, every Oberwolfach problem for $K_n - I$ has a solution except $\mathrm{OP}(3, 3)$ and $\mathrm{OP}(3, 3, 3, 3)$ [5].

29.2 Some Generalizations

Let λK_n denote the complete multigraph with n vertices in which there are λ multiple edges joining every pair of distinct vertices. Similarly, when n is even, $\lambda K_n - I$ denotes the multigraph obtained from λK_n by removing a 1-factor I.

Let OP($\lambda; \ell_1, \ell_2, \ldots, \ell_t$) denote the problem of determining whether or not the edge set of λK_n or $\lambda K_n - I$, when n is even and λ is odd, can be partitioned into subgraphs isomorphic to $F(\ell_1, \ell_2, \ldots, \ell_t)$.

29.7 Remark The only result in this direction, other than the obvious result one obtains by replicating solutions for K_n, is Theorem 29.8 for the case in which all cycles have the same length.

29.8 Theorem [3] The problem OP($\lambda; m; \ell$) has a solution if and only if none of the following are satisfied:

1. $\lambda \equiv 2 \pmod 4$, $m = 2$ and $\ell = 3$,
2. λ is odd, $m = 2$ and $\ell = 3$, or
3. $\lambda = 1$, $m = 4$ and $\ell = 3$.

29.9 Remark Another natural graph to consider is the complete bipartite graph $K_{n,n}$. Piotrowski [8] has proved Theorem 29.10.

29.10 Theorem Other than $K_{6,6}$ whose edge set cannot be partitioned into 2-factors $F(6,6)$, the edge set of $K_{n,n}$ can be partitioned into 2-factors $F(\ell_1, \ell_2, \ldots, \ell_t)$ if and only if n is even, $\ell_i \geq 4$ and even for all $1 \leq i \leq t$, and $\ell_1 + \ell_2 + \cdots + \ell_t = 2n$.

29.3 See Also

§IV.9	Resolvable cycle systems are solutions to the Oberwolfach problem with uniform cycle length.
§IV.22	Graph designs are more general graph decompositions.
§V.7	The Oberwolfach problem arises from a scheduling problem.
§VI.4	Solutions of the Oberwolfach problem are 2-factorizations of the complete graph.

References

[1] B. Alspach and R. Häggkvist, Some observations on the Oberwolfach problem, *J. Graph Theory* 9 (1985), 177-187.

[2] B. Alspach, P. Schellenberg, D. Stinson and D. Wagner, The Oberwolfach problem and factors of uniform odd length cycles, *J. Combin. Theory A* 52 (1989), 20-43.

[3] P. Gvozdjak, On the Oberwolfach problem for complete multigraphs, preprint.

[4] D. Hoffman and P. Schellenberg, The existence of C_k-factorizations of $K_{2n} - F$, *Discrete Math.* 97 (1991), 243-250.

[5] C. Huang, A. Kotzig and A. Rosa, On a variation of the Oberwolfach problem, *Discrete Math.* 27 (1979), 261-277.

[6] E. Köhler, Das Oberwolfacher problem, *Mitt. Math. Ges. Hamburg* 10 (1973), 52-57.

[7] E. Köhler, Über das Oberwolfacher problem, in: *Beiträge zur Geometrischen Algebra* (H. Arnold, W. Benz, H. Wefelscheid; eds.), Birkhäuser Verlag, Basel, 1977, 189-201.

[8] W. Piotrowski, The solution of the bipartite analogue of the Oberwolfach problem, *Discrete Math.* 97 (1991), 339-356.

30 Ordered Designs and Perpendicular Arrays

JÜRGEN BIERBRAUER

30.1 Definitions, Examples, and Constructions

An *ordered design* $\text{OD}_\lambda(t,k,v)$ is a $k \times \lambda \cdot \binom{v}{t} \cdot t!$ array with v entries such that

1. each column has k distinct entries, and
2. each tuple of t rows contains each column tuple of t distinct entries precisely λ times.

A *perpendicular array* $\text{PA}_\lambda(t,k,v)$ is a $k \times \lambda \cdot \binom{v}{t}$ array with v entries such that

1. each column has k distinct entries, and
2. each set of t rows contains each set of t distinct entries as a column precisely λ times.

30.1 Remark Often, ordered designs and perpendicular arrays are defined using the transpose of the definitions given.

For $0 \le s \le t$, a $\text{PA}_\lambda(t,k,v)$ is an s-$\text{PA}_\lambda(t,k,v)$ if, for each $w \le t$ and $u \le \min(s,w)$, the following holds: Let E_1, E_2 be disjoint sets of entries, $E_1 \cap E_2 = \emptyset$ with $|E_1| = u$ and $|E_2| = w-u$. Then the number of columns containing $E_1 \cup E_2$ and having E_2 in a given set U of $w-u$ rows is a constant, independent of the choice of E_1, E_2 and U. 1-PA are *authentication perpendicular arrays* (APA).

30.2 Example An $\text{OD}_1(2,3,6)$ (equivalent to an idempotent latin square of order 6).

1	1	1	1	1	2	2	2	2	2	3	3	3	3	3	4	4	4	4	4	5	5	5	5	5	6	6	6	6	6
2	3	4	5	6	1	3	4	5	6	1	2	4	5	6	1	2	3	5	6	1	2	3	4	6	1	2	3	4	5
3	5	2	6	4	3	1	6	4	5	4	6	5	1	2	2	5	6	3	1	6	4	2	1	3	5	1	4	3	2

30.3 Example A 3-$\text{PA}_3(3,4,6)$ (at the same time $\text{APA}_4(4,4,6)$) in compact form.

∞	∞	0	0	0	0	0	0	0	0	0
0	0	∞	∞	4	2	4	2	1	4	3	2
4	3	1	3	∞	∞	1	3	2	3	1	4
2	4	2	1	3	4	∞	∞	3	2	4	1

The 60 columns of the full 3-$\text{PA}_3(3,4,6)$ are the images of the columns to the left under the group generated by the permutation $(\infty)(0\,1\,2\,3\,4)$ on the entries.

30.4 Example An APA(2,3,11) is given in Example V.3.11.

30.5 Remark OD and PA may be viewed as (multi-)sets of injective mappings from a k-set into a v-set by identifying them with the (multi-)set of their columns and interpreting each column as a mapping from the set of rows to the set of entries.

30.6 Theorem An $\text{OD}_\lambda(2,k,v)$, $k \le v$, exists if and only if an idempotent $\text{OA}_\lambda(2,k,v)$ exists. See §II.2 and §II.4.

30.7 Theorem [1]

1. If there is a t-wise balanced design t-(v,K,λ) and for every $\ell \in K$ an s-$\text{PA}_\mu(t,k,\ell)$ exists, then an s-$\text{PA}_{\lambda\cdot\mu}(t,k,v)$ exists.
2. If there is an $\text{OD}_\lambda(t,k,v)$ and an $\text{OD}_\mu(y,\ell,v-k)$, where $y=\min(\ell,t)$, then an $\text{OD}_{\lambda\cdot\mu\cdot\binom{v-k}{y}\cdot y!}(t,k+\ell,v)$ exists.

3. If there is an s-PA$_\lambda(t,k,v)$ and an x-PA$_\mu(y,\ell,v-k)$, where $x=\min(\ell,s), y=\min(\ell,t)$, then an s-PA$_{\lambda\cdot\mu\cdot\binom{v-k}{y}}(t,k+\ell,v)$ exists.
4. If an OD$_\lambda(t,u,k)$, an OD$_\nu(y,\ell,v-k)$, and a t-(v,k,μ) design exist, where $y=\min(\ell,t)$, then an OD$_{\lambda\cdot\mu\cdot\nu\cdot\binom{v-k}{y}\cdot y!}(t,u+\ell,v)$ exists.
5. If an s-PA$_\lambda(t,u,k)$, an x-PA$_\nu(y,\ell,v-k)$ and a t-(v,k,μ) design exist, where $x=\min(\ell,s), y=\min(\ell,t)$, then an s-PA$_{\lambda\cdot\mu\cdot\nu\cdot\binom{v-k}{y}}(t,u+\ell,v)$ exists.

30.2 Existence

30.8 Remark Deleting ℓ rows, $(t\le k-\ell)$, of an OD$_\lambda(t,k,v)$ or PA$_\lambda(t,k,v)$ results in an OD$_\lambda(t,k-\ell,v)$ (PA$_\lambda(t,k-\ell,v)$, respectively). The corresponding statement for s-PA$_\lambda(t,k,v)$ is not true.

30.9 Remark The notions of 0-PA$_\lambda(t,k,k)$, APA$_\lambda(t,k,k),\ldots,t$-PA$_\lambda(t,k,k)$ all coincide.

30.10 Theorem The following exist, and λ is minimal in all cases:

1. OD$_1(t,t,v), v\ge t$ (trivially).
2. APA$_1(1,k,v), k\le v$ [8].
3. APA$_1(1,1,v)$, $APA_2(2,2,v)$, $APA_3(3,3,v)$ for all v [1].
4. APA$_2(2,q,q), q$ power of 2 [6].
5. APA$_1(2,q,q), q$ odd prime power [6].
6. APA$_1(2,3,v), v\ge 7$ odd. There is no APA$_1(2,3,5)$ [8].
7. OD$_1(2,3,v), v\ge 3$ (equivalently: an idempotent latin square of order v).
8. PA$_1(2,4,v), v\ge 5$ odd [5, 4].
9. APA$_2(2,4,v), v\ge 4$ [5].
10. PA$_1(2,5,v), v\ge 5$ odd,$v\ne 39$ [5, 4].
11. APA$_1(2,5,v), v\equiv 1,5 \pmod{10}, v\ge 11, v\ne 15$ [8].
12. APA$_1(2,5,p), p\equiv 3 \pmod 4$ prime, $p\ge 11$ [1].
13. APA$_1(2,k,q), q\equiv 3 \pmod 4$ prime power, k odd, $q>2^{k(k-1)}$ [1].
14. APA$_1(2,k,q), q$ odd prime power, k odd, $k\mid q-1$ [2].
15. APA$_1(2,p^m,p^n), p$ odd prime, $m\le n$ [1].
16. PA$_1(3,4,v), v\equiv 1,2 \pmod 3$ [3].
17. PA$_3(3,4,v), v\equiv 0 \pmod 3$ [3].
18. OD$_1(3,4,v), v\ge 4, v\ne 7$. There is no OD$_1(3,4,7)$ [11].
19. 3-PA$_3(3,4,v), v$ even, $v\ge 4$ (minimal even as APA) [1].
20. 3-PA$_6(3,4,v), v$ odd, $v\ge 5, v\ne 7$ (minimal even as APA) [1].

30.11 Theorem Each t-transitive group G of permutations on k symbols is an OD$_\lambda(t,k,k)$ (and an APA$_{\lambda\cdot t!}(t,k,k)$), where λ is the order of the pointwise stabilizer of t symbols. Each t-homogeneous group G of permutations on k symbols is an APA$_\lambda(t,k,k)$, where λ is the order of the set-stabilizer of t symbols. Special cases are the following:

1. The groups PGL$_2(q), q$ a prime power, form OD$_1(3,q+1,q+1)$; the groups PSL$_2(q), q\equiv 3 \pmod 4$ are APA$_3(3,q+1,q+1)$. The special cases of this last family when the prime power $q\equiv 3,11 \pmod{12}$ form the only known infinite family of APA$_\lambda(t,k,k)$ with $t>2$ and minimal λ.
2. The groups AGL$_1(q)$, (q a prime power), of order $q\cdot(q-1)$ form an OD$_1(2,q,q)$; the groups $ASL_1(q)$, (q a prime power $\equiv 3 \pmod 4$) of order $q\cdot(q-1)/2$ form APA$_1(2,q,q)$.

3. $\mathrm{AGL}_1(8)$ is an $\mathrm{APA}_1(3,8,8)$, $\mathrm{PSL}_2(8)$ is an $\mathrm{APA}_4(4,9,9)$, $A\Gamma L_2(32)$ is an $\mathrm{APA}_1(3,32,32)$, $\mathrm{P\Gamma L}_2(32)$ is an $\mathrm{APA}_4(4,33,33)$, and the alternating group A_7 in its doubly transitive action on 15 points is an $\mathrm{OD}_{12}(2,15,15)$. The Mathieu groups yield $\mathrm{OD}_1(4,11,11)$, $\mathrm{OD}_1(5,12,12)$, $\mathrm{OD}_{48}(3,22,22)$, $\mathrm{OD}_{48}(4,23,23)$, and $\mathrm{OD}_{48}(5,24,24)$.
4. The groups of Lie rank 1 are 2-transitive. The following are some small examples. The Ree groups ${}^2G_2(3) = \mathrm{P\Gamma L}_2(8)$ and ${}^2G_2(27)$ in their action on the Ree unitals are $\mathrm{OD}_2(2,28,28)$ and $\mathrm{OD}_{26}(2,19684,19684)$, respectively. The unitary group $U_3(5) = \mathrm{PSU}_3(5^2)$ is an $\mathrm{OD}_8(2,126,126)$, the Suzuki group $Sz(8) = {}^2B_2(8)$ in its action on the Tits ovoid is an $\mathrm{OD}_7(2,65,65)$.

30.12 Remark See §V.1 for further discussion of the groups mentioned in Theorem 30.11.

30.13 Theorem [1] $\mathrm{APA}_2(2,k,q)$ exist for every prime power q and $2 \le k \le q$. Here λ is minimal if kq is even.

30.14 Theorem [1] $\mathrm{APA}_3(3,k,q+1)$ exist for every prime power $q \equiv 3 \pmod 4$ and $k \le q+1$. Here $\lambda = 3$ is minimal unless $q \equiv 7 \pmod{12}$ and $k \equiv 2$ or $5 \pmod 6$.

30.15 Theorem [1] $\mathrm{APA}_{12}(4,4,v)$ exists for all $v \ge 4$. This is minimal unless $v \equiv 0 \pmod 3$. Restrictions of the group $\mathrm{PGL}_2(8)$ yield $\mathrm{APA}_4(4,k,9), 4 \le k \le 9$.

30.16 Theorem [10] If a (class-)regular $\mathrm{OA}_\lambda(t,t+1,v)$ exists, then an $\mathrm{OD}_\lambda(t,t+1,v+t)$ can be constructed. If $v = \prod_i p_i^{a_i}$ is the expression of v as a product of prime powers and $a_i(p_i - 1) > t$ for all i, then a regular $\mathrm{OA}_1(t,t+1,v)$ exists.

30.17 Example As an illustration of Theorem 30.16, a regular $\mathrm{OA}_1(3,4,5)$ exists, and hence an $\mathrm{OD}_1(3,4,8)$ exists.

30.18 Remark In Tables 30.19 and 30.20, italicized entries are those which are not known to be minimal.

30.19 Table (a) Smallest λ for which $\mathrm{APA}_\lambda(t,k,k)$ are known to exist and (b) smallest λ for which $\mathrm{APA}_\lambda(2,k,v)$ are known to exist.

(a)

$t\backslash k$	3	4	5	6	7	8	9	10	11	12	13
2	1	2	1	2	1	2	1	2	1	2	1
3	3	3	1	3	3	1	3	*6*	*4*	3	*30*
4		12	2	4	3	4	*4*	*24*	*24*	*192*	*1728*
5			10	10	5	5	*4*	*20*	*120*	*120*	*960*

(b)

$k\backslash v$	5	6	7	8	9	10	11	12	13	14
5	1	2	2	2	*2*	2	1	2	*2*	2
6		2	2	2	2	*8*	2	2	2	*12*
7			1	2	2	*8*	*2*	*10*	*2*	*6*

30.20 Table (a) Smallest λ for which $\mathrm{APA}_\lambda(3,k,v)$ are known to exist and (b) smallest λ for which $\mathrm{APA}_\lambda(4,k,v)$ are known to exist.

(a)

$k\backslash v$	5	6	7	8	9	10	11	12	13	14
5	1	3	*12*	*3*	*6*	3	*4*	3	*15*	*5*
6		3	*12*	3	6	*6*	*24*	3	*30*	*6*
7			3	3	*6*	*6*	*48*	3	*30*	*6*
8				1	6	*6*	*48*	3	*30*	*6*
9					3	*6*	*48*	3	*30*	*6*

(b)

$k\backslash v$	4	5	6	7	8	9	10	11	12
4	12	12	4	12	12	4	12	12	*12*
5		2	4	6	*24*	*4*	*24*	2	*16*
6			4	12	*24*	4	*24*	12	*16*
7				3	12	*4*	*24*	*24*	*96*
8					4	4	*24*	*24*	*192*

30.3 Large Sets

A *large set* $\text{LOD}_\lambda(t,k,v)$ or $\text{LPA}_\lambda(t,k,v)$ of $\text{OD}_\lambda(t,k,v)$ and $\text{PA}_\lambda(t,k,v)$, respectively, is a partition of the set of all v-ary k-tuples with distinct entries into $\text{OD}_\lambda(t,k,v)$ (respectively into $\text{PA}_\lambda(t,k,v)$).

30.21 Theorem [11]

1. $\text{LPA}_1(2,3,v)$ exist if and only if $v \geq 3$ is odd.
2. $\text{LOD}_1(2,3,v)$ exist for all $v \geq 3, v \notin \{6,14,62\}$. There is no $\text{LOD}_1(2,3,6)$.

30.4 See Also

[8, 1]	Contain a description of an application of PAs in the cryptographic theory of unconditional secrecy and authentication
[1]	If there is an $\text{APA}_1(2,k+2,v)$ which restricts to an $\text{APA}_1(2,k,v)$, then a *Room k-cube of side v* exists.
[7, 12]	If a $\text{PA}_1(2,k,v)$ admits an abelian automorphism group of order v in its operation on the entries, then $k-1$ MOLS of order v exist. See §IV.11.
[2]	A $\text{PA}_1(2,k,v)$ admitting the cyclic group $\mathbb{Z}_k$ in its operation on the rows is equivalent with a *Steiner k-cycle system of order v.*
[9]	Teirlinck's construction of large sets of t-designs for arbitrary t makes use of $\text{OD}_1(t,t+1,v)$.

References

[1] J. Bierbrauer and Y. Edel, Theory of perpendicular arrays, *J. Combin. Des.* 2 (1994), 375-406.

[2] A. Granville, A. Moisiadis and R. Rees, Nested Steiner n-cycle systems and perpendicular arrays, *J. Combin. Math. Combin. Comput.* 3 (1988), 163-167.

[3] E.S. Kramer, S.S. Magliveras, T.v. Trung and Q. Wu, Some perpendicular arrays for arbitrarily large t, *Discrete Math.* 96 (1991), 101-110.

[4] C.C. Lindner, R.C. Mullin and G.H.J. van Rees, Separable orthogonal arrays, *Utilitas Math.* 31 (1987), 25-32.

[5] R.C. Mullin, P.J. Schellenberg, G.H.J. van Rees and S.A. Vanstone, On the construction of perpendicular arrays, *Utilitas Math.* 18 (1980), 141-160.

[6] C.R. Rao, Combinatorial arrangements analogous to orthogonal arrays, *Sankhya* A23 (1961), 283-286.

[7] P.J. Schellenberg, G.H.J. van Rees and S.A. Vanstone, Four pairwise orthogonal latin squares of order 15, *Ars Combin.* 6 (1978), 141-150.

[8] D.R. Stinson, The combinatorics of authentication and secrecy codes, *J. Cryptology* 2 (1990), 23-49.

[9] L. Teirlinck, Non-trivial t-designs without repeated blocks exist for all t, *Discrete Math.* 65 (1987), 301-311.

[10] L. Teirlinck, On large sets of disjoint ordered designs, *Ars Combin.* 25 (1988), 31-37.

[11] L. Teirlinck, Large sets of disjoint designs and related structures, in: *Contemporary Design Theory: A Collection of Surveys* (J.H. Dinitz, D.R. Stinson; eds.) John Wiley & Sons, New York, 1992, 561-592.

[12] D.T. Todorov, Three mutually orthogonal latin squares of order 14, *Ars Combin.* 20 (1985), 45-48.

31 Orthogonal Designs

JENNIFER SEBERRY
R. CRAIGEN

An *orthogonal design* of *order n* and *type* $(s_1, \ldots, s_u)$, s_i positive integers, is an $n \times n$ matrix X, with entries $\{0, \pm x_1, \ldots, \pm x_u\}$ (the x_i commuting indeterminates) satisfying $XX^T = \left(\sum_{i=1}^{u} s_i x_i^2\right) I_n$. This is written as $\mathrm{OD}(n; s_1, s_2, \ldots, s_u)$.

31.1 Remark Alternatively, X has s_i entries of the type $0, \pm x_i$ and the distinct rows are orthogonal under the euclidean inner product. One may view X as a matrix with entries in the field of fractions of the integral domain $\mathbb{Z}[x_1, \ldots, x_u]$, so if $f = (\Sigma_{i=1}^{u} s_i x_i^2)$, then X is an invertible matrix with inverse $\frac{1}{f}X^T$. Thus $XX^T = fI_n$, and so this alternative definition that the row vectors are orthogonal applies equally well to the column vectors of X.

31.2 Remark An orthogonal design with no zeros and in which each of the entries is replaced by $+1$ or -1 is a *Hadamard matrix*. See §IV.24.

31.3 Examples Some small orthogonal designs. OD(4; 1, 1, 1, 1) is the Williamson array. See §IV.24.3.

$$\begin{bmatrix} x & y \\ y & -x \end{bmatrix} \quad \begin{bmatrix} a & -b & -c & -d \\ b & a & -d & c \\ c & d & a & -b \\ d & -c & b & a \end{bmatrix} \quad \begin{bmatrix} a & b & b & d \\ -b & a & d & -b \\ -b & -d & a & b \\ -d & b & -b & a \end{bmatrix} \quad \begin{bmatrix} a & 0 & -c & 0 \\ 0 & a & 0 & c \\ c & 0 & a & 0 \\ 0 & -c & 0 & a \end{bmatrix}$$

OD(2;1,1) OD(4;1,1,1,1) OD(4;1,1,2) OD(4;1,1)

31.4 Lemma (Equating, Killing and Growing [3]) Let D be an $\mathrm{OD}(n; u_1, u_2, ..., u_t)$ orthogonal design, on the t commuting variables $x_1, x_2, ..., x_t$. Then the following orthogonal designs exist:

1. $\mathrm{OD}(n; u_1, u_2, ..., u_i + u_j, ..., u_t)$ on $t - 1$ variables;
2. $\mathrm{OD}(n; u_1, ..., u_{i-1}, u_{i+1}, ..., u_t)$ on $t - 1$ variables;
3. $\mathrm{OD}(2n; u_1, u_2, ..., u_t)$ on t variables;
4. $\mathrm{OD}(2n; 2u_1, 2u_2, ..., 2u_t)$ on t variables;
5. $\mathrm{OD}(2n; u_1, u_1, u_2, ..., u_t)$ on $t + 1$ variables;
6. $\mathrm{OD}(2n; u_1, u_1, 2u_2, ..., 2u_t)$ on $t + 1$ variables.

31.5 Example Let D_1 be the OD(4; 1, 1, 1, 1) of Example 31.3; letting $b = c$ as in Lemma 31.4 (1) gives the OD(4; 1, 1, 2) design in Example 31.3; letting $d = 0$ as in Lemma 31.4 gives the first OD(4; 1, 1, 1) design in Example 31.6. Let D_2 be the OD(2; 1, 1) design of Example 31.3; replacing variables by 2×2 matrices as in Lemma 31.4 (3), (4), (5) gives the OD(4; 1, 1), OD(4; 2, 2), and second OD(4; 1, 1, 1) in Example 31.6.

31.6 Examples More small orthogonal designs.

$$\begin{bmatrix} a & -b & -c & 0 \\ b & a & 0 & c \\ c & 0 & a & -b \\ 0 & -c & b & a \end{bmatrix} \quad \begin{bmatrix} x & 0 & y & 0 \\ 0 & x & 0 & y \\ y & 0 & -x & 0 \\ 0 & y & 0 & -x \end{bmatrix} \quad \begin{bmatrix} x & x & y & y \\ x & -x & y & -y \\ y & y & -x & -x \\ y & -y & -x & x \end{bmatrix} \quad \begin{bmatrix} z & x & 0 & y \\ -x & z & y & 0 \\ 0 & y & -z & -x \\ y & 0 & x & -z \end{bmatrix}$$

OD(4; 1, 1, 1) OD(4; 1, 1) OD(4; 2, 2) OD(4; 1, 1, 1)

For n a positive integer, the *Radon number* of n is $\rho(n) = 8c + 2^d$ when $n = 2^a \cdot b$, b odd, $a = 4c + d$, and $0 \leq d < 4$.

31.7 **Remark** Any matrix C with entries $\{0, \pm y_i\}$, where y_i are variables can be written as $C = y_1C_1 + y_2C_2 + \ldots y_uC_u$, where, for C to be orthogonal either (i) $C_iC_j^T + C_jC_i^T = 0$, $i \neq j$ and $y_iy_j = y_jy_i$ for any pair C_i, C_j or (ii) $C_kC_\ell^T = C_\ell C_k^T = 0$, $k \neq \ell$ and $y_ky_\ell = -y_\ell y_k$ for any pair C_k, C_ℓ. Both cases allow us to use matrix representation theory to bound the number of variables that can occur in the orthogonal matrix. In case (i), for the orthogonal design $\mathrm{OD}(n; s_1, s_2, \cdots, s_u)$, the number of variables does not exceed $\rho(n)$.

31.1 Necessary Conditions and More Examples

31.8 **Remark** There are both combinatorial and algebraic/number-theoretic conditions restricting the existence of orthogonal designs.

Let p be a prime and let $a = p^\alpha u$ and $b = p^\beta v$ where $p \not| u$ and $p \not| v$. The *Hilbert norm residue symbol* $(a, b)_p$ for the prime p can be computed as

$$(a,b)_p = \begin{cases} (-1^{\alpha\beta(p-1)/2}) \left(\frac{u}{p}\right)^\beta \left(\frac{v}{p}\right)^\alpha & \text{if } p \neq 2 \\ -1^{(u-1)(v-1)/4 + \alpha(v^2-1)/8 + \beta(u^2-1)/8} & \text{if } p = 2 \end{cases}$$

where $\left(\frac{u}{p}\right)$ is the Legendre symbol.

31.9 **Theorems** (Geramita and Seberry, [3])

1. If there is an $\mathrm{OD}(2n; s_1, s_2)$, n odd, then $s_1 + s_2 < n - 1$. Also s_1, s_2 and $s_1 + s_2$ are each the sum of two squares.
2. If there is an $\mathrm{OD}(4n; s_1, s_2)$, n odd, then $(-1, s_1s_2)_p(s_1, s_2)_p = 1$ for all primes p and if $s_1 + s_2 = n - 1$, then an $\mathrm{OD}(4n; 1, s_1, s_2)$ exists.
3. If there is an $OD(4n; s_1, s_2, s_3)$, n odd, then for all primes p
$$(-1, s_1s_2s_3)_p(s_1, s_2)_p(s_1, s_3)_p(s_2, s_3)_p = 1.$$
In addition, if $s_1 + s_2 + s_3 = n - 1$, then an $OD(4n; 1, s_1, s_2, s_3)$ exists;
4. If there is an $\mathrm{OD}(4n; s_1, s_2, s_3, s_4)$, n odd, then $\prod_{1 \leq i < j \leq 4}(s_i, s_j)_p = 1$ for all primes p, $s_1s_2s_3s_4$ is a square, and $s_1 + s_2 + s_3 + s_4 \neq n - 1$.

31.10 **Theorem** (Eades Sum-Fill Theorem) The Goethals-Seidel construction can only be used to construct an $\mathrm{OD}(4n; s_1, s_2, \cdots, s_\ell))$, n odd, if there exists an integer matrix P, with all entries of modulus $\leq n$, which satisfies $PP^T = diag(s_1, s_2, s_3, s_4)$.

31.11 **Remark** Necessary algebraic conditions for an $\mathrm{OD}(4n; s_1, s_2, \cdots, s_\ell)$, n odd, are given in [3, p71] but introduce considerable new notation outside the scope here.

31.12 **Examples** Some orthogonal designs of special interest are:

1. The Williamson array – an OD(4;1,1,1,1). The first is the right representation of the quaternions while the second is the left representation of the quaternions.

$$\begin{bmatrix} A & B & C & D \\ -B & A & -D & C \\ -C & D & A & -B \\ -D & -C & B & A \end{bmatrix} \qquad \begin{bmatrix} A & B & C & D \\ -B & A & D & -C \\ -C & -D & A & B \\ -D & C & -B & A \end{bmatrix}$$

2. An OD(8;1,1,1,1,1,1,1,1).

$$\left[\begin{array}{rrrr|rrrr} A & B & C & D & E & F & G & H \\ -B & A & D & -C & F & -E & -H & G \\ -C & -D & A & B & G & H & -E & -F \\ -D & C & -B & A & H & -G & F & -E \\ \hline -E & -F & -G & -H & A & B & C & D \\ -F & E & -H & G & -B & A & -D & C \\ -G & H & E & -F & -C & D & A & -B \\ -H & -G & F & E & -D & -C & B & A \end{array}\right]$$

3. The Baumert-Hall array, $A(x, y, z, w)$. An OD(12;3,3,3,3).

$$\left[\begin{array}{rrrrrrrrrrrr} y & x & x & x & -z & z & w & y & -w & w & z & -y \\ -x & y & x & -x & w & -w & z & -y & -z & z & -w & -y \\ -x & -x & y & x & w & -y & -y & w & z & z & w & -z \\ -x & x & -x & y & -w & -w & -z & w & -z & -y & -y & -z \\ & & & & & & & & & & & \\ -y & -y & -z & -w & z & x & x & x & -w & -w & z & -y \\ -w & -w & -z & y & -x & z & x & -x & y & y & -z & -w \\ w & -w & w & -y & -x & -x & z & x & y & -z & -y & -z \\ -w & -z & w & -z & -x & x & -x & z & -y & y & -y & w \\ & & & & & & & & & & & \\ -y & y & -z & -w & -z & -z & w & y & w & x & x & x \\ z & -z & -y & -w & -y & -y & -w & -z & -x & w & x & -x \\ -z & -z & y & z & -y & -w & y & -w & -x & -x & w & x \\ z & -w & -w & z & y & -y & y & z & -x & x & -x & w \end{array}\right]$$

4. The Plotkin array – an OD(24;3,3,3,3,3,3,3,3). Let $A(x, y, z, w)$ be as in part (3) and let $B(x, y, z, w)$ be the array

$$\left[\begin{array}{rrrrrrrrrrrr} y & x & x & x & -w & w & z & y & -z & z & w & -y \\ -x & y & x & -x & -z & z & -w & -y & w & -w & z & -y \\ -x & -x & y & x & -y & -w & y & -w & -z & -z & w & z \\ -x & x & -x & y & w & w & -z & -w & -y & z & y & z \\ & & & & & & & & & & & \\ -w & -w & -z & -y & z & x & x & x & -y & -y & z & -w \\ y & y & -z & -w & -x & z & x & -x & -w & -w & -z & y \\ -w & w & -w & -y & -x & -x & z & x & z & y & y & z \\ z & -w & -w & z & -x & x & -x & z & y & -y & y & w \\ & & & & & & & & & & & \\ z & -z & y & -w & y & y & w & -z & w & x & x & x \\ y & -y & -z & -w & -z & -z & -w & -y & -x & w & x & -x \\ z & z & y & -z & w & -y & -y & w & -x & -x & w & x \\ -w & -z & w & -z & -y & y & -y & z & -x & x & -x & w \end{array}\right],$$

then the required design is

$$\left[\begin{array}{ll} A(x_1, x_2, x_3, x_4) & B(x_5, x_6, x_7, x_8) \\ B(-x_5, x_6, x_7, x_8) & -A(-x_1, x_2, x_3, x_4) \end{array}\right].$$

5. The Goethals-Seidel array can be found in §IV.24.3.

6. The Welch array – an OD(20;5,5,5,5) is constructed from sixteen block circulant matrices.

$$\left[\begin{array}{ccccc|ccccc|ccccc|ccccc}
-d & b & -c & -c & -b & c & a & -d & -d & -a & -b & -a & c & -c & -a & a & -b & -d & d & -b \\
-b & -d & b & -c & -c & -a & c & a & -d & -d & -a & -b & -a & c & -c & -b & a & -b & -d & d \\
-c & -b & -d & b & -c & -d & -a & c & a & -d & -c & -a & -b & -a & c & d & -b & a & -b & -d \\
-c & -c & -b & -d & b & -d & -d & -a & c & a & c & -c & -a & -b & -a & -d & d & -b & a & -b \\
b & -c & -c & -b & -d & a & -d & -d & -a & c & -a & c & -c & -a & -b & -b & -d & d & -b & a \\
\hline
-c & a & d & d & -a & -d & -b & -c & -c & b & -a & b & -d & d & b & -b & -a & -c & c & -a \\
-a & -c & a & d & d & b & -d & -b & -c & -c & b & -a & b & -d & d & -a & -b & -a & -c & c \\
d & -a & -c & a & d & -c & b & -d & -b & -c & d & b & -a & b & -d & c & -a & -b & -a & -c \\
d & d & -a & -c & a & -c & -c & b & -d & -b & -d & d & b & -a & b & -c & c & -a & -b & -a \\
a & d & d & -a & -c & -b & -c & -c & b & -d & b & -d & d & b & -a & -a & -c & c & -a & -b \\
\hline
b & -a & -c & c & -a & a & b & -d & d & b & -d & -b & c & c & b & -c & a & -d & -d & -a \\
-a & b & -a & -c & c & b & a & b & -d & d & b & -d & -b & c & c & -a & -c & a & -d & -d \\
c & -a & b & -a & -c & d & b & a & b & -d & c & b & -d & -b & c & -d & -a & -c & a & -d \\
-c & c & -a & b & -a & -d & d & b & a & b & c & c & b & -d & -b & -d & -d & -a & -c & a \\
-a & -c & c & -a & b & b & -d & d & b & a & -b & c & c & b & -d & a & -d & -d & -a & -c \\
\hline
-a & -b & -d & d & -b & b & -a & c & -c & -a & c & a & d & d & -a & -d & b & c & c & -b \\
-b & -a & -b & -d & d & -a & b & -a & c & -c & -a & c & a & d & d & -b & -d & b & c & c \\
d & -b & -a & -b & -d & -c & -a & b & -a & c & d & -a & c & a & d & c & -b & -d & b & c \\
-d & d & -b & -a & -b & c & -c & -a & b & -a & d & d & -a & c & a & c & c & -b & -d & b \\
-b & -d & d & -b & -a & -a & c & -c & -a & b & a & d & d & -a & c & b & c & c & -b & -d
\end{array}\right]$$

31.13 Example The Ono-Sawade-Yamamoto array – an OD(36; 9, 9, 9, 9).

$$\begin{array}{|ccccccccc|ccccccccc|ccccccccc|ccccccccc|}
a & a & a & b & -c & -d & -b & d & c & b & -a & a & b & -c & d & b & -d & c & -c & -a & a & -b & -c & -d & b & d & -c & -d & -a & a & b & c & -d & -b & -d & -c \\
a & a & a & -d & b & -c & c & -b & d & a & b & -a & d & b & -c & c & b & -d & a & -c & -a & -d & -b & -c & -c & b & d & a & -d & -a & -d & b & c & -c & -b & -d \\
a & a & a & -c & -d & b & d & c & -b & -a & a & b & -c & d & b & -d & c & b & -a & a & -c & -c & -d & -b & d & -c & b & -a & a & -d & c & -d & b & -d & -c & -b \\
-b & d & c & a & a & a & b & -c & -d & b & -d & c & b & -a & a & b & -c & d & b & d & -c & -c & -a & a & -b & -c & -d & -b & -d & -c & -d & -a & a & b & c & -d \\
c & -b & d & a & a & a & -d & b & -c & c & b & -d & a & b & -a & d & b & -c & -c & b & d & a & -c & -a & -d & -b & -c & -c & -b & -d & a & -d & -a & -d & b & c \\
d & c & -b & a & a & a & -c & -d & b & -d & c & b & -a & a & b & -c & d & b & d & -c & b & -a & a & -c & -c & -d & -b & -d & -c & -b & -a & a & -d & c & -d & b \\
b & -c & -d & -b & d & c & a & a & a & b & -c & d & b & -d & c & b & -a & a & -b & -c & -d & b & d & -c & -c & -a & a & b & c & -d & -b & -d & -c & -d & -a & a \\
-d & b & -c & c & -b & d & a & a & a & d & b & -c & c & b & -d & a & b & -a & -d & -b & -c & -c & b & d & a & -c & -a & -d & b & c & -c & -b & -d & a & -d & -a \\
-c & -d & b & d & c & -b & a & a & a & -c & d & b & -d & c & b & -a & a & b & -c & -d & -b & d & -c & b & -a & a & -c & c & -d & b & -d & -c & -b & -a & a & -d \\
\hline
-b & a & -a & -b & -c & d & -b & -d & c & a & a & a & b & c & d & -b & -d & -c & d & -a & a & b & -c & d & -b & d & c & -c & a & -a & b & -c & -d & -b & d & -c \\
-a & -b & a & d & -b & -c & c & -b & -d & a & a & a & d & b & c & -c & -b & -d & a & d & -a & d & b & -c & c & -b & d & -a & -c & a & -d & b & -c & -c & -b & d \\
a & -a & -b & -c & d & -b & -d & c & -b & a & a & a & c & d & b & -d & -c & -b & -a & a & d & -c & d & b & d & c & -b & & -a & -c & -c & -d & b & d & -c & -b \\
-b & -d & c & -b & a & -a & -b & -c & d & -b & -d & -c & a & a & a & b & c & d & -b & d & c & d & -a & a & b & -c & d & -b & d & -c & -c & a & -a & b & -c & -d \\
c & -b & -d & -a & -b & a & d & -b & -c & -c & -b & -d & a & a & a & d & b & c & c & -b & d & a & d & -a & d & b & -c & -c & -b & d & -a & -c & a & -d & b & -c \\
-d & c & -b & a & -a & -b & -c & d & -b & -d & -c & -b & a & a & a & c & d & b & d & c & -b & -a & a & d & -c & d & b & d & -c & -b & a & -a & -c & -c & -d & b \\
-b & -c & d & -b & -d & c & -b & a & -a & b & c & d & -b & -d & -c & a & a & a & b & -c & d & -b & d & c & d & -a & a & b & -c & -d & -b & d & -c & -c & a & -a \\
d & -b & -c & c & -b & -d & -a & -b & a & d & b & c & -c & -b & -d & a & a & a & d & b & -c & c & -b & d & a & d & -a & -d & b & -c & -c & -b & d & -a & -c & a \\
-c & d & -b & -d & c & -b & a & -a & -b & c & d & b & -d & -c & -b & a & a & a & -c & d & b & d & c & -b & -a & a & d & -c & -d & b & d & -c & -b & a & -a & -c \\
\hline
c & a & -a & -b & c & -d & b & d & c & -d & a & -a & b & -c & -d & -b & -d & c & a & a & a & -b & -c & d & b & -d & c & -b & -a & a & -b & -c & -d & -b & d & c \\
-a & c & a & -d & -b & c & c & b & d & -a & -d & a & -d & b & -c & c & -b & -d & a & a & a & d & -b & -c & c & b & -d & a & -b & -a & -d & -b & -c & c & -b & d \\
a & -a & c & c & -d & -b & d & c & b & a & -a & -d & -c & -d & b & -d & c & -b & a & a & a & -c & d & -b & -d & c & b & -a & a & -b & -c & -d & -b & d & c & -b \\
b & d & c & c & a & -a & -b & c & -d & -b & -d & c & -d & a & -a & b & -c & -d & b & -d & c & a & a & a & -b & -c & d & -b & d & c & -b & -a & a & -b & -c & -d \\
c & b & d & -a & c & a & -d & -b & c & c & -b & -d & -a & -d & a & -d & b & -c & c & b & -d & a & a & a & d & -b & -c & c & -b & d & a & -b & -a & -d & -b & -c \\
d & c & b & a & -a & c & c & -d & -b & -d & c & -b & a & -a & -d & -c & -d & b & -d & c & b & a & a & a & -c & d & -b & d & c & -b & -a & a & -b & -c & -d & -b \\
-b & c & -d & b & d & c & c & a & -a & b & -c & -d & -b & -d & c & -d & a & -a & -b & -c & d & b & -d & c & a & a & a & -b & -c & -d & -b & d & c & -b & -a & a \\
-d & -b & c & c & b & d & -a & c & a & -d & b & -c & c & -b & -d & -a & -d & a & d & -b & -c & c & b & -d & a & a & a & -d & -b & -c & c & -b & d & a & -b & -a \\
c & -d & -b & d & c & b & a & -a & c & -c & -d & b & -d & c & -b & a & -a & -d & -c & d & -b & -d & c & b & a & a & a & -c & -d & -b & d & c & -b & -a & a & -b \\
\hline
d & a & -a & b & c & d & -b & d & -c & c & -a & a & b & c & -d & -b & d & c & b & a & -a & b & -c & -d & b & d & c & a & a & a & -b & c & -d & b & d & -c \\
-a & d & a & d & b & c & -c & -b & d & a & c & -a & -d & b & c & c & -b & d & -a & b & a & -d & b & -c & c & b & d & a & a & a & -d & -b & c & -c & b & d \\
a & -a & d & c & d & b & d & -c & -b & -a & a & c & c & -d & b & d & c & -b & a & -a & b & -c & -d & b & d & c & b & a & a & a & c & -d & -b & d & -c & b \\
-b & d & -c & d & a & -a & b & c & d & -b & d & c & c & -a & a & b & c & -d & b & d & c & b & a & -a & b & -c & -d & b & d & -c & a & a & a & -b & c & -d \\
-c & -b & d & -a & d & a & d & b & c & c & -b & d & a & c & -a & -d & b & c & c & b & d & -a & b & a & -d & b & -c & -c & b & d & a & a & a & -d & -b & c \\
d & -c & -b & a & -a & d & c & d & b & d & c & -b & -a & a & c & c & -d & b & d & c & b & a & -a & b & -c & -d & b & d & -c & b & a & a & a & c & -d & -b \\
b & c & d & -b & d & -c & d & a & -a & b & c & -d & -b & d & c & c & -a & a & b & -c & -d & b & d & c & b & a & -a & -b & c & -d & b & d & -c & a & a & a \\
d & b & c & -c & -b & d & -a & d & a & -d & b & c & c & -b & d & a & c & -a & -d & b & -c & c & b & d & -a & b & a & -d & -b & c & -c & b & d & a & a & a \\
c & d & b & d & -c & -b & a & -a & d & c & -d & b & d & c & -b & -a & a & c & -c & -d & b & d & c & b & a & -a & b & c & -d & -b & d & -c & b & a & a & a
\end{array}$$

31.2 Asymptotic Existence Results

31.14 Conjecture When $n \equiv 4 \pmod 8$, there exists an $\mathrm{OD}(n; 1, k)$ when $k \leq n-1$, k the sum of three or fewer integer squares and $n-2$ the sum of two or fewer integer squares.

31.15 Conjecture When $n \equiv 0 \pmod 8$, there exists an $\mathrm{OD}(n; 1, k)$ for all $k \leq n-1$.

31.16 Remarks Conjecture 31.14 has been verified for all $n \leq 84$. Conjecture 31.15 is true for $2^t \cdot 3, 2^t \cdot 5, 2^t \cdot 7, 2^t \cdot 9$ and $2^t \cdot 13$ for $t \geq 3$, $2^t \cdot 15$ and $2^t \cdot 21$, for $t \geq 4$. Lemma 31.4 yields Theorem 31.17.

31.17 Theorem Let a, b, c, x, y, z be nonnegative integers such that $a+b+c=n$ and $x+y+z=2n$. Suppose that for all possible choices of a, b, c an $\mathrm{OD}(n; a, b, c)$ exists. Then for all possible choices of x, y, z an $\mathrm{OD}(2n; x, y, z)$ exists. Hence for all possible choices of x, y, z, $x+y+z=2^m$, an $\mathrm{OD}(2^m; x, y, z)$ and an $\mathrm{OD}(2^m; x, y, z)$, $x+y \leq m$, exists.

31.18 Theorem (Robinson [3]) If $n > 40$, there is no $\mathrm{OD}(n : 1, 1, 1, 1, 1, n-5)$.

31.19 Theorem (Robinson [3]) The following orthogonal designs exist:

1. $\mathrm{OD}(2^t; 1, 1, 1, 1, 2, 2, 4, 4, \cdots, 2^{t-2}, 2^{t-2})$,
2. $\mathrm{OD}(2^t; 1, 1, 2, 1, 2, 4, 8, \cdots, 2^{t-3}, 3, 6, 12, \cdots, 3 \cdot 2^{t-3})$,
3. $\mathrm{OD}(2^t; 1, 1, 2, 4, 8, \cdots, 2^{t-3}, 2^{t-3}, 2^{t-2}, 3, 3, 6, \cdots, 3 \cdot 2^{t-4})$,
4. $\mathrm{OD}(2^t; 1, 1, 2, 4, 8, \cdots, 2^{t-3}, 3, 6, 9, 18, \cdots, 9 \cdot 2^{t-5}, 3 \cdot 2^{t-4})$,
5. $\mathrm{OD}(2^t; 1, 2, 3, 2^{t-4}, 3 \cdot 2^{t-3}, 3, 3, 6, 6, 12, 12, \cdots, 3 \cdot 2^{t-5}, 3 \cdot 2^{t-5})$,

A set of 4 *T-matrices*, T_i, $i = 1, \ldots, 4$ of order t are four circulant matrices which have entries $0, +1$ or -1 and which satisfy

1. $T_i * T_j = 0$, $i \neq j$, ($*$ denotes the Hadamard product)
2. $\sum_{i=1}^{4} T_i$ is a $(1, -1)$ matrix,
3. $\sum_{i=1}^{4} T_i T_i^T = tI_t$,
4. $t = t_1^2 + t_2^2 + t_3^2 + t_4^2$ where t_i is the row(column) sum of T_i.

31.20 Theorem (Seberry–Yamada–Turyn [5]) Suppose there are T-matrices of order t. Further suppose there is an $\mathrm{OD}(4s; u_1, \ldots, u_n)$ constructed of sixteen circulant (or type one) $s \times s$ blocks on the variables $x_1, \ldots, x_n$. Then there is an $\mathrm{OD}(4st; tu_1, \ldots, tu_n)$. In particular if there is an $\mathrm{OD}(4s; s, s, s, s)$ constructed of sixteen circulant $s \times s$ blocks, then there is an $\mathrm{OD}(4st; st, st, st, st)$. This means there exist orthogonal designs $\mathrm{OD}(20t; 5t, 5t, 5t, 5t)$ and $\mathrm{OD}(36t; 9t, 9t, 9t, 9t)$.

31.21 Remark T-matrices of order t give Hadamard matrices of order $4t$. T-matrices are known for many orders including (see [5]):

1,...,72,74,...,78,80,...,82,84,...,88,90,..., 96,98,...,102,104,...,106,108,110,...,112, 114,..., 126,128,...,130,132,136,138,140,...,148,150,152,..., 156,158,...,162,164,165, 166,168,...,172,174,..., 178,180,182,184,...,190,192,194,...,196,198,200,..., 210.

31.22 Conjecture There exists an $\mathrm{OD}(4t; t, t, t, t)$ for every positive integer t.

31.3 Existence Results for Small Orders

31.23 Table Orthogonal designs OD($2n : a_1, a_2$), n odd. (E denotes exists, U means existence is unresolved).

$2n$	(a_1, a_2)	Status
2	(1,1)	E
6	(1,1), (1,4), (2,2)	E
10	(1,1), (1,4), (2,2), (4,4)	E
14	(1,1),(1,4),(1,9),(2,2),(2,8),(4,4),(4,9),(5,5)	E
18	(1,1),(1,4),(2,2),(2,8),(4,4),(5,5)	E
	(1,9), (4,9)	U
22	(1,1),(1,4),(1,16),(2,2),(2,8),(4,4),(4,16),(5,5),(8,8)	E
	(1,9),(2,18), (4,9),(9,9)	U
26	(1,1),(1,4),(1,9),(1,16),(2,2),(2,8),(2,18),(4,4),(4,16),(5,5),(8,8),(9,9),(10,10)	E
	(4,9),(5,20)	U
>26	(1,1),(1,4),(1,9),(2,2),(2,8),(4,4),(4,16),(5,5),(8,8),(10,10),(13,13)	E
>38	(1,16)	E

31.24 Table Orthogonal designs OD($4n : a_1, a_2, \ldots, a_u$), n odd.

$4n$	u	Cases	∃	Comment
4	2	4	4	Existence completely settled
4	3	2	2	Existence completely settled
4	4	1	1	Existence completely settled
12	2	36	33	Existence completely settled
12	3	53	31	Existence completely settled
12	4	53	12	Existence completely settled
20	2	100	89	Existence completely settled
20	3	237	97	(3,7,8) unresolved
20	4	359	35	(1,3,6,8),(1,4,4,9), (2,2,5,5) unresolved
28	2	196	169	Existence completely settled
28	3	640	217	(1,5,20) unresolved

31.25 Table Orthogonal designs OD($8n : a_1, a_2, \cdots, a_u$)

$8n$	u	Comment
8	$1, \ldots, 8$	All exist
16	$1, \ldots, 5$	All exist
16	6	(1,1,1,1,4,7) and (1,1,2,2,2,7) unresolved
16	7	94 possible tuples; 13 do not exist; 81 are known
16	8	67 possible tuples; 30 do not exist; 37 are known
16	9	45 possible tuples; 36 do not exist; 9 are known
24	1,2	All exist
24	3	(4,4,15), (7,7,7) and (7,8,8) unresolved
24	4	717 possible tuples
24	6	996 possible tuples
24	8	eight designs known

31.4 See Also

§IV.24	There is a very close connection between Hadamard matrices and ODs.
§IV.52	Weighing matrices are a special class of ODs.
§IV.42	Zero autocorrelation sequences are related to ODs.
[5]	An extensive survey of Hadamard matrices which also includes much of the material in this section and numerous references.
[2, 6]	Contain theorems dealing with the number of variables in an OD.
[3]	The foundational reference work for ODs.
[4]	Contains many new results on existence of ODs.

References

[1] R. Craigen, Construction of Orthogonal Matrices, Ph.D. Thesis, University of Waterloo, 1991.

[2] A.V. Geramita and N.J. Pullman, Radon's function and Hadamard arrays, *Lin. Multilin. Algebra* 2 (1974), 147-150.

[3] A.V. Geramita and J. Seberry, *Orthogonal Designs: Quadratic Forms and Hadamard Matrices*, Marcel Dekker, New York-Basel, 1979.

[4] C. Koukouvinos, N. Platis and J. Seberry, Necessary and sufficient conditions for some two variable orthogonal designs in order 36, *Congr. Numer.*, to appear.

[5] J. Seberry and M. Yamada, Hadamard matrices, sequences, and block designs, in: *Contemporary Design Theory: A Collection of Surveys* (J.H. Dinitz, D.R. Stinson; eds.) John Wiley & Sons, New York, 1992, 431-560.

[6] W.W. Wolfe, Orthogonal Designs–Amicable Orthogonal Designs, Ph.D. Thesis, Queen's University, Kingston, Canada, 1975.

32 Orthogonal Main Effect Plans

Deborah J. Street

32.1 Definitions and Example

A *main effect plan* (MEP) has k *rows* (or *factors*), n *columns* (or *runs*) and s_i *symbols* (or *levels*) in row i, $1 \leq i \leq k$. Let n_{Ix} be the number of times that symbol x appears in row I and let n_{Jy} be the number of times that symbol y appears in row J. Let N_{IJ} be the $s_i \times s_j$ incidence matrix of rows I and J. The (x, y) position in N_{IJ} is the number of columns with x in row I and y in row J. The plan is *orthogonal* if the (x, y) position in N_{IJ} is $(n_{Ix}n_{Jy})/n$, for all pairs of rows I and J. An OMEP with these parameters is represented by $s_1 \times s_2 \times \ldots \times s_k//n$. OMEPs with the smallest possible number n of runs are *minimal*.

32.1 Example A $2 \times 3 \times 4//16$ OMEP.

$$\begin{matrix} 1&1&1&1&1&1&1&1&2&2&2&2&2&2&2&2 \\ 1&1&2&2&3&3&3&3&1&1&2&2&3&3&3&3 \\ 1&3&2&4&1&2&3&4&2&4&1&3&1&2&3&4 \end{matrix}$$

An OMEP is *partially replicated* (PROMEP) if it has at least one repeated run.

32.2 Example A $2 \times 3 \times 4//16$ PROMEP. (Replicated runs in boldface.)

1 1 1 1 **1 1 1 1** 2 2 2 **2 2 2 2**
1 1 2 2 **3 3 3 3** 1 1 2 2 **3 3 3 3**
1 3 1 3 **2 2 4 4** 2 4 2 4 **1 1 3 3**

32.2 Existence

32.3 Theorem [2, 4] An $s_1 \times s_2 \times s_3//n$ OMEP, with $s_1 \leq s_2 \leq s_3$, and with the smallest entry of N_{23} equal to 1, exists if and only if

1. n is of the form $(s_2 + x)(s_3 + y)$ for some positive integers x and y; and
2. $s_1 \leq gcd(s_2 + x, s_3 + y)$.

32.4 Remarks

1. The extension of the necessity to more than three factors is obvious, but there is then no guarantee that the OMEP exists.
2. When the smallest entry in N_{23} exceeds 1, a sufficient condition is given in [4] for an OMEP to exist.

32.5 Theorem [1, 4] Let N_{12} and $N_{k-2,k-1}$ be the incidence matrices of rows 1 and 2 and rows $k-2$ and $k-1$, respectively, of a $s_1 \times s_2 \times \ldots \times s_k//n$ OMEP, with $s_1 \leq s_2 \leq \ldots \leq s_k$.

1. A lower bound for the number of repeated runs is $\sum_{ij}(N_{12}(i,j) - s_k)$, where the summation is over the pairs (i,j) for which $N_{12}(i,j) > s_k$.
2. Let S and T be the sets of levels for factors $k-2$ and $k-1$, respectively, with replication greater than s_k. An upper bound for the number of repeated runs is $$\sum_{i \in S} \sum_{j \in T}(N_{(k-2,k-1)}(i,j) - 1).$$

32.3 Construction and Related Objects

32.6 Construction [4] To construct an $s_1 \times s_2 \times s_3//(s_2 + x)(s_3 + y)$ OMEP with $s_1 \leq gcd(s_2 + x, s_3 + y)$ let r_{ij} be the number of times that level j appears in row i. Let $g = \gcd(s_2 + x, s_3 + y)$, $s_2 + x = gb$ and $s_3 + y = gc$. Choose the r_{3j} so that each $r_{3j} \geq gb$ and $gb \mid r_{3j}$ for all j. The third row of the OMEP consists of r_{31} 1s followed by r_{32} 2s and so on. It can be shown that $r_{2j} = b_j gc$. Let **b** be b_1 1s followed by b_2 2s and so on. The second row of the OMEP consists of gc copies of **b** placed side-by-side. It can be shown that $r_{1i} = a_i gbc$. Let **a** be $a_1 b$ 1s followed by $a_2 b$ 2s and so on. The first row of the OMEP consists of gc copies of **a** placed side-by-side with the entries of **a** being cycled by b each time that **a** is written down.

32.7 Example A $2 \times 4 \times 5//24$ OMEP. Here $x = 0, y = 1, g = 2$, **b**=(1,2,3,4) and **a**=(1,1,2,2).

1 1 2 2 2 2 1 1 1 1 2 2 2 2 1 1 1 1 2 2 2 2 1 1
1 2 3 4 1 2 3 4 1 2 3 4 1 2 3 4 1 2 3 4 1 2 3 4
1 1 1 1 2 2 2 2 3 3 3 3 4 4 4 4 5 5 5 5 5 5 5 5

32.8 Remark A new OMEP can be obtained from an existing OMEP by equating all occurrences of one level of a factor with that of another level of the same factor.

32.9 Table Known, minimal three-factor PROMEPs with at most 25 runs and associated bounds. L is the lower bound for the number of repeated runs, U is the upper bound and KRR is the set of repeated runs for which an OMEP is known. Designs with $s_1 \leq gcd(s_2, s_3)$ have $n = s_2 s_3$ and hence no repeated runs and are not listed.

s_1	s_2	s_3	n	L	U	KRR	s_1	s_2	s_3	n	L	U	KRR	s_1	s_2	s_3	n	L	U	KRR
2	2	3	8	0	2	0, 2	2	2	11	24	0	2	0, 2	3	3	4	16	0	2	0,1,2
2	2	5	12	0	2	0, 2	2	3	4	16	0	4	0,2,4	3	3	5	18	0	3	0,1,3
2	2	7	16	0	2	0, 2	2	3	5	18	0	3	1,3	3	4	5	25	0	4	0,1,2,3
2	2	9	20	0	2	0, 2	2	4	5	24	0	4	0,2,4	4	4	5	25	0	2	0,1,2

32.10 Remarks

1. An $OA_\lambda(k, s)$ is an $s \times \ldots \times s // \lambda s^2$ OMEP in which each level of each factor is replicated equally often.
2. An orthogonal array with a variable number of symbols of strength 2 (OAVS) is an OMEP but not vice versa.
3. A set of $k-2$ mutually orthogonal F-squares of order t can be used to construct a k-factor OMEP with $n = t^2$ runs and two factors each with t levels.
4. OMEPs form a class of fractionally replicated asymmetrical factorial designs.

32.4 See Also

§II.2	Orthogonal arrays give examples of OMEPs, and are used extensively in their construction.
§II.4	Orthogonal arrays of higher index are also used to construct OMEPs.
§IV.20	Orthogonal F-squares.
§IV.19	Factorial designs.
[1]	Algorithms for constructing OMEPs.
[2, 3, 4]	Constructions for OMEPs.
[2, 4]	Existence results for OMEPs.
[5]	A survey of OMEPs and OAVS with extensive bibliography.

References

[1] L. Burgess and D.J. Street, Algorithms for constructing orthogonal main effect plans, *Utilitas Math.* 46 (1994), 33-48.

[2] M. Jacroux, A note on the determination and construction of minimal orthogonal main-effect plans, *Technometrics* 34 (1992), 92-96.

[3] M. Jacroux, On the construction of minimal partially replicated orthogonal main-effect plans, *Technometrics* 35 (1993), 32-36.

[4] D.J. Street, Constructions for orthogonal main effect plans, *Utilitas Math.* 44 (1994), 115-123.

[5] D.J. Street and L. Burgess, A survey of orthogonal main effect plans and related structures, *Congr. Numer.* 99 (1994), 223-239.

33 Packings

Douglas R. Stinson

33.1 Definitions and Examples

Let $v \geq k \geq t$. A t-(v,k,λ) *packing* is a pair $(X,\mathcal{B})$, where X is a v-set of elements (*points*) and $\mathcal{B}$ is a collection of k-subsets of X (*blocks*), such that every t-subset of points occurs in at most λ blocks in $\mathcal{B}$. If $\lambda > 1$, then $\mathcal{B}$ is allowed to contain repeated blocks.

The *packing number* $D_\lambda(v,k,t)$ is the maximum number of blocks in any t-(v,k,λ) packing. A t-(v,k,λ) packing $(X,\mathcal{B})$ is *optimal* if $|\mathcal{B}| = D_\lambda(v,k,t)$. If $\lambda = 1$, often one writes $D(v,k,t)$ for $D_1(v,k,t)$.

33.1 Examples For each of the given parameter sets t-(v,k,λ) one example of an optimal t-(v,k,λ) packing is given.

t-(v,k,λ)	Packing
2-$(6,3,1)$	{1,2,3} {1,4,5} {2,4,6} {3,5,6}
2-$(8,3,1)$	{1,3,5} {1,4,8} {1,6,7} {2,3,7} {2,4,6} {2,5,8} {3,6,8} {4,5,7}
2-$(11,3,1)$	{1,2,3} {1,4,5} {1,6,7} {1,8,11} {1,9,10} {2,6,8} {2,7,9} {2,10,11} {3,6,9} {3,7,11} {3,8,10} {4,6,10} {4,7,8} {4,9,11} {5,6,11} {5,8,9} {5,7,10}
2-$(9,4,1)$	{1,2,3,4} {1,5,6,7} {3,6,8,9}
2-$(10,4,1)$	{1,2,3,4} {1,5,6,7} {2,5,8,9} {3,6,8,10} {4,7,9,10}
2-$(14,5,1)$	{1,2,3,4,5} {1,6,7,8,9} {2,6,10,11,12} {3,7,10,13,14}
3-$(9,5,1)$	{1,2,3,4,5} {1,2,6,7,8} {3,4,6,7,9}
3-$(16,7,1)$	{1,2,3,4,5,6,7} {1,2,8,9,10,11,12} {3,4,8,9,13,14,15} {5,6,9,10,13,14,16}
4-$(10,6,1)$	{1,2,3,4,5,6} {1,2,3,7,8,9} {1,4,5,7,8,10} {2,4,6,7,9,10} {3,5,6,8,9,10}

33.2 Remark The survey paper by Mills and Mullin [4] covers much of the material in this section. For this reason individual references are not given in general.

33.2 Equivalent Combinatorial Objects and Bounds

33.3 Theorem A t-(v,k,λ) packing with $\lambda\binom{v}{t}/\binom{k}{t}$ blocks is equivalent to a t-(v,k,λ) design or an $S_\lambda(t,k,v)$ (possibly containing repeated blocks).

33.4 Theorem A t-$(v,k,1)$ packing with b blocks is equivalent to a binary code of length v, size b, constant weight k, and minimum Hamming distance at least $2(k-t+1)$. (See §V.1).

33.5 Theorem (*First Johnson bound*) $D_\lambda(v,k,t) \leq \left\lfloor \frac{v D_\lambda(v-1,k-1,t-1)}{k} \right\rfloor$. Iterating this bound yields $D_\lambda(v,k,t) \leq U_\lambda(v,k,t)$, where

$$U_\lambda(v,k,t) = \left\lfloor \frac{v}{k} \left\lfloor \frac{v-1}{k-1} \cdots \left\lfloor \frac{\lambda(v-t+1)}{k-t+1} \right\rfloor \right\rfloor \right\rfloor .$$

Further, if $\lambda(v-1) \equiv 0 \pmod{k-1}$ and $\lambda v(v-1) \equiv -1 \pmod{k}$, then
$$D_\lambda(v,k,2) \leq U_\lambda(v,k,2) - 1.$$

33.6 Theorem (*Complementing*) By taking the complement of each block in a packing with $\lambda = 1$, one has that $D(v,k,t) = D(v, v-k, v-2k+t)$.

33.7 Theorem (*Second Johnson bound*) Suppose $d = D_\lambda(v,k,t) = qv + r$, where $0 \leq r \leq v-1$. Then $q(q-1)v + 2qr \leq (t-1)d(d-1)$. From this it follows that
$$D_\lambda(v,k,t) \leq \left\lfloor \frac{v(k+1-t)}{k^2 - v(t-1)} \right\rfloor.$$

33.3 Determination of Packing Numbers

33.8 Theorem If $v \equiv 2 \pmod 6$ and $\lambda \equiv 4 \pmod 6$, or if $v \equiv 5 \pmod 6$ and $\lambda \equiv 1 \pmod 3$, then $D_\lambda(v,3,2) = U_\lambda(v,3,2) - 1$. Otherwise, $D_\lambda(v,3,2) = U_\lambda(v,3,2)$.

33.9 Remark $D_\lambda(v,3,2) = U_\lambda(v,3,2)$ except when Theorem 33.7 applies.

33.10 Theorem $D(v,4,2) = U_1(v,4,2) - \epsilon$, where
$$\epsilon = \begin{cases} 1 & \text{if } v \equiv 7 \text{ or } 10 \pmod{12},\ v \neq 10, 19 \\ 1 & \text{if } v = 9 \text{ or } 17 \\ 2 & \text{if } v = 8, 10 \text{ or } 11 \\ 3 & \text{if } v = 19 \\ 0 & \text{otherwise.} \end{cases}$$

33.11 Theorem Suppose $\lambda > 1$. If $v = 9$ and $\lambda = 2$, or if $v = 3$ and $\lambda = 3$, then $D_\lambda(v,4,2) = U_\lambda(v,4,2) - 1$. Otherwise, $D_\lambda(v,4,2) = U_\lambda(v,4,2)$.

33.12 Theorem

1. If $v \equiv 1, 2, 3$ or $4 \pmod 6$, then $D(v,4,3) = U_1(v,4,3)$.
2. If $v \equiv 0 \pmod 6$, then $D(v,4,3) = \left\lfloor \frac{vD(v-1,3,2)}{4} \right\rfloor$.
3. If $v = 5$ or 11, then $D(v,4,3) = U_1(v,4,3)$.

33.13 Remark For $v \equiv 5 \pmod 6$, $v > 11$, essentially nothing is known about the numbers $D(v,4,3)$.

33.14 Theorem (Mullin and Yin [5])

1. If $v \equiv 3 \pmod{20}$, $v \neq 243$, then $D(v,5,2) = U_1(v,5,2)$.
2. If $v \equiv 9, 17 \pmod{20}$, $v \neq 17, 29, 49$, then $D(v,5,2) = U_1(v,5,2) - 1$.

33.15 Theorem (Yin [6])

1. If $v \equiv 7, 9 \pmod{10}$, $v > 439$, then $D_2(v,5,2) = U_2(v,5,2) - 1$.
2. If $v \not\equiv 7, 9 \pmod{10}$, $v > 526$, then $D_2(v,5,2) = U_2(v,5,2)$.

33.16 Theorem (Yin [7])

1. If $v \equiv 0, 1 \pmod 5$ and $\lambda \equiv 0 \pmod 4$, then $D_\lambda(v,5,2) = U_\lambda(v,5,2)$.
2. Suppose $v \equiv 3 \pmod 5$. If $\lambda \equiv 4 \pmod{20}$, then $D_\lambda(v,5,2) = U_\lambda(v,5,2) - 1$; if $\lambda \equiv 0, 8, 12, 16 \pmod{20}$, then $D_\lambda(v,5,2) = U_\lambda(v,5,2)$.
3. Suppose $v \equiv 2, 4 \pmod 5$. If $\lambda \equiv 12 \pmod{20}$, then $D_\lambda(v,5,2) = U_\lambda(v,5,2) - 1$; if $\lambda \equiv 0, 4, 8, 16 \pmod{20}$, $(v,\lambda) \neq (7,4), (22,16), (22,36)$ or $(27,16)$, then $D_\lambda(v,5,2) = U_\lambda(v,5,2)$.

33.17 Table Known values $D(v,k,t)$ for $v \le 20$, $2 \le t \le 5$ and $1 \le k-t \le 4$.

$t=2$

$v\backslash k$	3	4	5	6
3	1			
4	1	1		
5	2	1	1	
6	4	1	1	1
7	7	2	1	1
8	8	2	1	1
9	12	3	2	1
10	13	5	2	1
11	17	6	2	2
12	20	9	3	2
13	26	13	3	2
14	28	14	4	2
15	35	15	6	3
16	37	20	6	3
17	44	20	7	3
18	48	22	9	4
19	57	25	12	4
20	60	30	16	5

$t=3$

$v\backslash k$	4	5	6	7
4	1			
5	1	1		
6	3	1	1	
7	7	1	1	1
8	14	2	1	1
9	18	3	1	1
10	30	6	2	1
11	35	11	2	1
12	51	12	4	2
13	65	18	4	2
14	91	28	7	2
15	105	42	10	3
16	140	48	16	4
17	156	68	17	5
18	198		21	6
19	228		28	8
20	285		40	10

$t=4$

$v\backslash k$	5	6	7	8
5	1			
6	1	1		
7	3	1	1	
8	8	1	1	1
9	18	3	1	1
10	36	5	1	1
11	66	11	2	1
12		22	3	1
13		26	4	2
14		42	8	2
15		70	15	3
16		112	16	4
17			24	6
18				9
19				12
20			80	17

$t=5$

$v\backslash k$	6	7	8	9
6	1			
7	1	1		
8	4	1	1	
9	12	1	1	1
10	30	3	1	1
11	66	6	1	1
12	132	12	3	1
13		26	3	1
14		42	7	2
15			15	3
16			30	3
17			34	6
18				10
19				19

33.4 Structure of Optimal Packings

Let $(X,\mathcal{B})$ be a 2-$(v,k,1)$ packing. The *leave graph* of $(X,\mathcal{B})$ is the graph (X,E), where E consists of all 2-subsets of X that occur in no blocks of $\mathcal{B}$.

33.18 Table Optimal 2-$(v,3,1)$ packings.

$v \equiv$	$D(v,3,2)$	Leave Graph	Construction
$1,3 \pmod 6$	$\dfrac{v^2-v}{6}$	empty	$(v,3,1)$-BIBD

$v \equiv$	$D(v,3,2)$	Leave Graph	Construction
$0,2 \pmod 6$	$\frac{v^2-2v}{6}$	$\frac{v}{2}K_2$	$(v+1,3,1)$-BIBD with a point deleted
$4 \pmod 6$	$\frac{v^2-2v-2}{6}$	$K_{1,3} \cup \frac{v-4}{2}K_2$	Delete a point from the 4-cycle in an optimal packing on $v+1$ points
$5 \pmod 6$	$\frac{v^2-v-8}{6}$	cycle of length 4	$(v,\{3,5^*\})$-PBD with the block of size 5 replaced by two blocks of size 3.

33.19 Table Optimal 2-$(v,4,1)$ packings, $v \notin \{8,9,10,11,17,19\}$.

$v \equiv$	$D(v,4,2)$	Leave Graph	Construction
$1,4 \pmod{12}$	$\frac{v^2-v}{12}$	empty	$(v,4,1)$-BIBD
$0,3 \pmod{12}$	$\frac{v^2-3v}{12}$	$\frac{v}{3}K_3$	$(v+1,4,1)$-BIBD with a point deleted
$2,8 \pmod{12}$	$\frac{v^2-2v}{12}$	$\frac{v}{2}K_2$	$\{4\}$-GDD of type $2^{v/2}$
$5,11 \pmod{12}$	$\frac{v^2-2v-3}{12}$	$K_{1,4} \cup \frac{v-5}{2}K_2$	$\{4\}$-GDD of type $2^{(v-5)/2}5^1$ with the group of size 5 replaced by a block of size 4
$7,10 \pmod{12}$	$\frac{v^2-v-18}{12}$	$K_{3,3}$	$(v,\{4,7^*\})$-PBD with the block of size 7 replaced by two blocks of size 4
$6,9 \pmod{12}$	$\frac{v^2-3v-6}{12}$	$(K_6 \backslash K_4) \cup \frac{v-6}{3}K_3$	$\{4\}$-GDD of type $3^{(v-6)/3}6^1$ with the group of size 6 replaced by a block of size 4

33.20 Table Optimal 3-$(v,4,1)$ packings, $v \not\equiv 5 \pmod 6$.

$v \equiv$	$D(v,4,3)$	Construction
$2,4 \pmod 6$	$\frac{v^3-3v^2+2v}{24}$	3-$(v,4,1)$ design
$1,3 \pmod 6$	$\frac{v^3-4v^2+3v}{24}$	3-$(v,4,1)$ design with a point deleted
$0 \pmod 6$	$\frac{v^3-3v^2-6v}{24}$	$G(m,6,4,3)$ with every group replaced by three blocks of size 4 (a $G(m,6,4,3)$ is a $\{4,6\}$-3BD in which the blocks of size 6 form a parallel class). (See §IV.49)

33.5 Resolvable Packings with $\lambda = 1$

A t-(v,k,λ) packing $(X,\mathcal{B})$ is *resolvable* if $\mathcal{B}$ can be partitioned into *parallel classes*, each of which consists of v/k disjoint blocks.

33.21 Theorem Suppose $v \equiv 0 \pmod k$. If $v-1 \equiv 0 \pmod{(k-1)}$, then a resolvable 2-$(v,k,1)$ packing with $U_1(v,k,2)$ blocks is equivalent to a resolvable $(v,k,1)$-BIBD. If

$v-1 \not\equiv 0 \pmod{(k-1)}$, then a resolvable 2-$(v,k,1)$ packing with $U_1(v,k,2)$ blocks is equivalent to a resolvable $\{k\}$-GDD of type $g^{v/g}$, where $g = v - \lfloor\frac{v-1}{k-1}\rfloor(k-1)$.

33.22 Table Resolvable optimal 2-$(v,k,1)$ packings for $k=3,4$.

k	v	Equivalent Design	Existence Results
3	$\equiv 0 \pmod 6$	$\{3\}$-RGDD of type $2^{v/2}$	All admissible $v \neq 6, 12$
3	$\equiv 3 \pmod 6$	Resolvable $(v,3,1)$-BIBD	All admissible v
4	$\equiv 0 \pmod{12}$	$\{4\}$-RGDD of type $3^{v/3}$	All admissible $v \geq 24$, except possibly for $v = 84$, 132, 264, 372, 456, 552, 660, 804 or 852
4	$\equiv 4 \pmod{12}$	Resolvable $(v,4,1)$-BIBD	All admissible v
4	$\equiv 8 \pmod{12}$	$\{4\}$-RGDD of type $2^{v/2}$	No example of these designs is known at present

33.6 See Also

§I.1	BIBDs are packings with void leave graph.
§II.3	Incomplete transversal designs are used extensively in various constructions for packings.
§III.1-6	Pairwise balanced designs.
§IV.50	Resolvable group divisible designs.
§IV.49	t-wise balanced designs.
[1]	3-$(v,4,1)$ packings.
[2]	2-$(v,4,1)$ packings.
[3]	Contains a comprehensive discussion of packings, including tables.
[4]	A general survey of coverings and packings with an extensive bibliography.

References

[1] A.E. Brouwer, On the packing of quadruples without common triples, *Ars Combin.* 5 (1978), 3-6.

[2] A.E. Brouwer, Optimal packings of K_4's into a K_n, *J. Combin. Theory A* 26 (1979), 278-297.

[3] A.E. Brouwer, J.B. Shearer, N.J.A. Sloane and W.D. Smith, A new table of constant weight codes, *IEEE Trans. Inform. Theory* 36 (1990), 1334-1380.

[4] W.H. Mills and R.C. Mullin, Coverings and packings, in: *Contemporary Design Theory: A Collection of Surveys* (J.H. Dinitz, D.R. Stinson; eds.) John Wiley & Sons, New York, 1992, 371-399.

[5] R. C. Mullin and J. Yin, On packings of pairs by quintuples: $v \equiv 3, 9$ or $17 \pmod{20}$, *Ars Combin.* 35 (1993), 161-171.

[6] J. Yin, On the packing of pairs by quintuples with index 2, *Ars Combin.* 31 (1991), 287-301.

[7] J. Yin, On λ-packings of pairs by quintuples: $\lambda \equiv 0 \pmod 4$, *Ars Combin.* 37 (1994), 288-300.

34 Partial Geometries

JOSEPH A. THAS

34.1 Definitions and Examples

A (finite) *partial geometry* is an incidence structure $\mathcal{S} = (P, B, I)$ in which P and B are disjoint (nonempty) sets of objects called *points* and *lines*, and for which I is a symmetric point-line incidence relation, i.e., $I \subseteq (P \times B) \cup (B \times P)$, satisfying the following axioms:

1. each point is incident with $1 + t$ lines ($t \geq 1$) and two distinct points are incident with at most one line;
2. each line is incident with $1+s$ points ($s \geq 1$) and two distinct lines are incident with at most one point;
3. if x is a point and L is a line not incident with x, then there are exactly α ($\alpha \geq 1$) points $y_1, y_2, \cdots, y_\alpha$ and α lines $M_1, M_2, \cdots, M_\alpha$ such that xIM_iIy_iIL, with $i = 1, 2, \cdots, \alpha$.

The integers s, t and α are the *parameters* of the partial geometry. Put $|P| = v$ and $|B| = b$.

34.1 Remarks

1. There is a point-line duality for partial geometries for which in any definition or theorem the words "point" and "line" are interchanged and the parameters s and t are interchanged.
2. R.C. Bose used r, k, t for the parameters $t + 1, s + 1, \alpha$ given above.

34.2 Examples

1. A *generalized quadrangle* is a partial geometry with $\alpha = 1$. See §IV.21 for many constructions and examples.
2. Let P be the set of all points of the n-dimensional projective space PG(n, q) over the Galois field $\mathbb{F}_q$, which are not contained in a fixed subspace PG$(n-2, q)$, with $n \geq 2$. Let B be the set of all lines of PG(n, q) having no point in common with PG$(n-2, q)$. Finally, let I be the natural incidence. Then (P, B, I) is a partial geometry with $s = q, t = q^{n-1} - 1, \alpha = q$. This geometry is denoted by H_q^n [10].

34.2 Existence and Pseudo-Geometric Graphs

34.3 Theorem If $\mathcal{S} = (P, B, I)$ is a partial geometry with parameters s, t, α, then $v = (s+1)(st+1)/\alpha$ and $b = (t+1)(st+1)/\alpha$ [5].

34.4 Theorem The point graph of a partial geometry with parameters s, t, α is a strongly regular graph with parameters $v, k = s(t+1), \lambda = s - 1 + t(\alpha - 1), \mu = \alpha(t+1)$. See §VI.5.

Each strongly regular graph Γ having the parameters of Theorem 34.4 with $t \geq 1, s \geq 1, 1 \leq \alpha \leq s+1$, and $1 \leq \alpha \leq t+1$ is a *pseudo-geometric* (t, s, α)-graph. If the pseudo-geometric graph Γ is the point graph of at least one partial geometry, then it is *geometric.*

34.5 Theorem Any pseudo-geometric $(s^2, s, 1)$-graph is geometric [2].

34.6 Theorem Let $\mathcal{S} = (P, B, I)$ be a partial geometry with parameters s, t, α. Then s, t and α satisfy:

1. the integer $\alpha(s+t+1-\alpha)$ divides $st(s+1)(t+1)$ [5];
2. the Krein inequalities [2], namely
   $(s+1-2\alpha)t \leq (s-1)(s+1-\alpha)^2$ and $(t+1-2\alpha)s \leq (t-1)(t+1-\alpha)^2$.

Partial geometries $\mathcal{S}$ can be divided into four (nondisjoint) classes.

1. $\mathcal{S}$ has $\alpha = s+1$ or, dually, $\alpha = t+1$; when $\alpha = s+1$, then $\mathcal{S}$ is a 2-$(v, s+1, 1)$ design.
2. $\mathcal{S}$ has $\alpha = s$ or, dually, $\alpha = t$; when $\alpha = t$, then $\mathcal{S}$ is a *net* of *order* $s+1$ and *degree* $t+1$.
3. When $\alpha = 1$, $\mathcal{S}$ is a generalized quadrangle. See §IV.21.
4. When $1 < \alpha < \min(s, t)$, then $\mathcal{S}$ is *proper.*

34.7 Remark Example 34.2(2) is a dual net.

34.8 Remark The dual of a net of order $s+1$ and degree $t+1$ is a transversal design $TD(t+1, s+1)$.

Axiom of Pasch, also called the *Axiom of Veblen*: if $L_1 I x I L_2$, $L_1 \neq L_2$, $M_1 \not{I} x \not{I} M_2$, and if L_i has a point in common with M_j for all $i, j \in \{1, 2\}$, then M_1 and M_2 have a point in common.

34.9 Theorem Let $\mathcal{S}$ be a dual net of order $s+1$ and degree $t+1$, with $s < t+1$. Then $\mathcal{S}$ satisfies the axiom of Pasch if and only if it is isomorphic to a dual net H_q^n ($s = q$ and $t = q^{n-1} - 1$) [10].

34.3 Maximal Arcs

In a projective plane $\mathcal{P}$ of order q any nonempty set of k points may be described as a $\{k; d\}$-*arc*, where d ($d \neq 0$) is the greatest number of collinear points in the set. For given q and d ($d \neq 0$), k can never exceed $dq - q + d$, and a $\{dq - q + d; d\}$-arc is a *maximal arc.* Equivalently, a maximal arc may be defined as a nonempty set of points in $\mathcal{P}$ meeting every line in just d points or in none at all.

34.10 Examples

1. The point set of $\mathcal{P}$ is a maximal arc with $d = q+1$.
2. The point set of the affine plane $\mathcal{A}$ obtained by deleting a line from $\mathcal{P}$ is a maximal arc with $d = q$.
3. A single point is a maximal arc with $d = 1$.
4. A *hyperoval* of $\mathcal{P}$, that is, a set of $q+2$ points no three of which are collinear, is a maximal arc with $d = 2$.

34.11 Theorem If K is a $\{dq - q + d; d\}$-arc (i.e., a maximal arc) of the projective plane $\mathcal{P}$, where $d \leq q$, then the set K'=\{lines L of $\mathcal{P}$ | $L \cap K = \phi$\} is a $\{q(q-d+1)/d; q/d\}$-arc (i.e., a maximal arc) of the dual plane.

34.12 Corollary A necessary condition for the existence of a maximal arc, with $d \leq q$, is that d is a factor of q.

34.13 Remark Theorem 34.14 contains the two most important results on the existence of maximal arcs; Part 1 is due to Blokhuis, Ball and Mazzocca (private communication; the case $p = 3$ was settled by Thas), Part 2 to Denniston [4].

34.14 Theorem

1. In $\mathrm{PG}(2,q), q = p^h$ with p an odd prime, and $h > 1$, there are no $\{p^{h+1} - p^h + p; p\}$-arcs and dually no $\{p^{2h-1} - p^h + p^{h-1}; p^{h-1}\}$-arcs.
2. For any factor d of q, $q = 2^h$, there exists a $\{dq - q + d; d\}$-arc in the plane $\mathrm{PG}(2,q)$.
3. There is no (105;5)-arc in $\mathrm{PG}(2,25)$.

34.15 Conjecture In $\mathrm{PG}(2,q)$, q odd, the only maximal arcs are the trivial ones of Examples 34.10 (1), (2), and (3).

34.16 Construction Let K be a maximal $\{qd-q+d; d\}$-arc of a projective plane $\mathcal{P}$ of order q, $2 \leq d < q$. Let P be the set of all points of $\mathcal{P}$ not in K, let B be the set of all lines of $\mathcal{P}$ having a nonempty intersection with K, and let I be the natural incidence. Then $\mathcal{S}(K) = (P, B, I)$ is a partial geometry with parameters

$$s = q - d, \quad t = q(d-1)/d, \quad \alpha = (q-d)(d-1)/d.$$

34.17 Construction Let $\mathcal{P}$ =$\mathrm{PG}(2,q)$ and embed $\mathcal{P}$ in the projective space $\mathrm{PG}(3,q)$. Further, let P' consist of all points of $\mathrm{PG}(3,q)$ not in $\mathcal{P}$, let B' consist of all lines of $\mathrm{PG}(3,q)$ having a unique point in common with the maximal $\{qd-q+d; d\}$-arc K of $\mathcal{P}$, and let I' be the natural incidence. Then $\mathcal{T}_2^*(K) = (P', B', I')$ is a partial geometry with parameters

$$s = q - 1, \quad t = (q+1)(d-1), \quad \alpha = d - 1.$$

34.18 Remark Construction 34.16 is due independently to Thas and Wallis, Construction 34.17 to Thas.

34.19 Corollary By Theorem 34.14 there exist partial geometries with parameters as follows:

Type 1: $s = 2^h - 2^m$, $t = 2^h - 2^{h-m}$, $\alpha = (2^{h-m} - 1)(2^m - 1)$ with $1 \leq m < h$;
Type 2: $s = 2^h - 1$, $t = (2^h + 1)(2^m - 1)$, $\alpha = 2^m - 1$ with $1 \leq m < h$.

Such a partial geometry has $\alpha = 1$ or is proper. A partial geometry of Type 1 is a generalized quadrangle if and only if $h = 2$ and $m = 1$; then $s = t = 2$ and $v = b = 15$. A partial geometry of Type 2 is a generalized quadrangle if and only if $m = 1$; then K is a hyperoval in $\mathcal{P}$, $s = 2^h - 1, t = 2^h + 1, v = 2^{3h}, b = 2^{2h}(2^h + 2)$ with $h \geq 2$.

34.4 Partial Geometries in Projective and Affine Spaces

A *projective partial geometry* $\mathcal{S} = (P, B, I)$ is a partial geometry with parameters $s = q, t, \alpha$ for which the point set P is a subset of the point set of some projective space $\mathrm{PG}(n,q)$, for which the line set B is a set of lines of $\mathrm{PG}(n,q)$, and for which I is the natural incidence. In such a case the partial geometry $\mathcal{S}$ is *embedded* in $\mathrm{PG}(n,q)$. If $\mathrm{PG}(n',q)$ is the subspace of $\mathrm{PG}(n,q)$ generated by all points of P, then $\mathrm{PG}(n',q)$ is the *ambient space* of $\mathcal{S}$.

Similarly one may define an *affine partial geometry* by replacing the projective space $\mathrm{PG}(n,q)$ by the affine space $AG(n,q)$.

34.20 Theorem If $\mathcal{S} = (P, B, I)$ is a partial geometry with parameters s, t, α which is projective with ambient space $\mathrm{PG}(n, s), n \geq 2$, then one of the following holds:

1. $\alpha = s + 1$ and $\mathcal{S}$ is the 2-design formed by all points and all lines of $\mathrm{PG}(n, s)$;
2. $\alpha = 1$ and $\mathcal{S}$ is a classical generalized quadrangle;
3. $\alpha = t + 1, n = 2$, P is the complement of a maximal $\{sd - s + d; d\}$-arc K of $\mathrm{PG}(2, s)$ with $d = s/\alpha$ and $2 \leq d < s$, and B consists of all lines of $\mathrm{PG}(2, s)$ having an empty intersection with K;
4. $\alpha = s, n \geq 2$ and $\mathcal{S}$ is the dual net H_s^n [5].

34.21 Remark Part 2 of Theorem 34.20 is due to Buekenhout, Parts 1, 3, and 4 to De Clerck and Thas.

34.22 Remark All partial geometries embedded in the affine space $AG(n, q), n \geq 2$, were determined by Thas; the three-dimensional case with $\alpha = 1$ was independently settled by Bichara [8]. For $\alpha = 1$ and $n \geq 3$, five interesting sporadic cases occur [8].

34.23 Theorem If $\mathcal{S}$ is a proper partial geometry with parameters s, t, α which is affine with ambient space $AG(n, s + 1), n \geq 2$, then $n = 3$ and $\mathcal{S}$ is a partial geometry $\mathcal{T}_2^*(K)$ with K a maximal arc in the plane at infinity of $AG(3, s + 1)$ [8].

34.5 Enumeration of Partial Geometries

34.24 Theorem Up to duality, the parameters of the known proper partial geometries are the following:

Type 1: $s = 2^h - 2^m, t = 2^h - 2^{h-m}, \alpha = (2^m - 1)(2^{h-m} - 1)$, with $h \neq 2$ and $1 \leq m < h$;
Type 2: $s = 2^h - 1, t = (2^h + 1)(2^m - 1), \alpha = 2^m - 1$, with $1 < m < h$;
Type 3: $s = 2^{2h-1} - 1, t = 2^{2h-1}, \alpha = 2^{2h-2}$, with $h > 1$;
Type 4: $s = 26, t = 27, \alpha = 18$;
Type 5: $s = t = 5, \alpha = 2$;
Type 6: $s = 4, t = 17, \alpha = 2$.

34.25 Remarks

1. Mathon proved by computer that there exist exactly two partial geometries with parameters $s = 6, t = 4, \alpha = 3$ [3]. For $(s, t, \alpha) \neq (6, 4, 3)$ all known partial geometries of Type 1 and 2 are given by the constructions of §34.4.
2. With each partition by $(2h - 1)$-dimensional spaces (i.e., maximal singular subspaces) of the hyperbolic quadric $Q^+(4h - 1, 2)$ of $\mathrm{PG}(4h - 1, 2), h > 1$, there corresponds a partial geometry of Type 3; no other partial geometries with these parameters are known. The geometries of Type 3 are due to De Clerck, Dye, and Thas; for $h = 2$ the geometry was also constructed by Cohen, and by Haemers and van Lint (the three constructions are completely different) [3].
3. With each partition by $(2h - 1)$-dimensional spaces of the hyperbolic quadric $Q^+(4h - 1, 3)$ of $\mathrm{PG}(4h - 1, 3), h > 1$, there corresponds a partial geometry with parameters $s = 3^{2h-1} - 1, t = 3^{2h-1}, \alpha = 2.3^{2h-2}$. Up to now only $Q^+(7, 3)$ is known to have such a partition; this yields the geometry of Type 4. This construction is due to Thas [3].
4. The geometry of Type 5 is due to van Lint and Schrijver, the geometry of Type 6 to Haemers; one geometry of each type is known [3].

34.26 Table The parameters of possible partial geometries (PG) with $v < 100$, excluding designs and nets and their duals; of a dual pair, the one with $s \leq t$ is given. All conditions in §34.2 are taken into account.

v	s	t	α	b	Comment
15	2	2	1	15	Unique PG exists [7]
27	2	4	1	45	Unique PG exists [7]
28	3	4	2	35	Does not exist [3]
40	3	3	1	40	Exactly two PG exist [7]
45	4	6	3	63	Exactly two PG exist; Remark 34.25 (1)
64	3	5	1	96	Unique PG exists [7]
66	5	8	4	99	Does not exist, Lam et al. [3]
70	6	6	4	70	Unknown
75	4	7	2	120	Unknown
76	3	6	1	133	Does not exist, Dixmier and Zara [7]
81	5	5	2	81	One PG known; Remark 34.25 (4)
85	4	4	1	85	Unique PG exists [7]
91	6	10	5	143	Unknown
95	4	9	2	190	Unknown
96	5	6	2	112	Unknown
96	5	9	3	160	Unknown

34.27 Remark For $v = 70$ and $v = 91$ pseudo-geometric graphs are known; for $v = 75$ and $v = 96$ (both cases) no pseudo-geometric graph is known.

34.6 See Also

§I.4	Partial geometries with $\alpha = s + 1$ are Steiner systems.
§II.2	Partial geometries with $\alpha = s$ are transversal designs.
§IV.1	Pseudo-geometric graphs are particular association schemes.
§IV.21	Partial geometries with $\alpha = 1$ are generalized quadrangles.
§VI.5	Pseudo-geometric graphs are particular strongly regular graphs.
§VI.7	Classical geometries and nondesarguesian planes are used to construct partial geometries.
[1]	Excellent survey, without proofs, with extensive bibliography of strongly regular graphs and partial geometries.
[3]	A survey, without proofs, with extensive bibliography. This survey contains most of the information in this section.
[4]	Contains extensive information on maximal arcs.
[5]	Contains a section on projective partial geometries.
[6]	Excellent survey on translation nets.
[7]	Standard reference on generalized quadrangles.
[8]	Determination of all affine partial geometries.
[9]	Contains a survey of maximal arcs.

References

[1] A.E. Brouwer and J.H. van Lint, Strongly regular graphs and partial geometries, in: *Enumeration and Design*, Academic Press, Toronto, 1984, 85-122.

[2] P.J. Cameron, J.-M. Goethals and J.J. Seidel, Strongly regular graphs having strongly regular subconstituents, *J. Algebra* 55 (1978), 257-280.

[3] F. De Clerck and H. Van Maldeghem, Some classes of rank 2 geometries, in: *Handbook of Incidence Geometry* (F. Buekenhout; ed.), Elsevier, Amsterdam, 1995.
[4] J.W.P. Hirschfeld, *Projective Geometries over Finite Fields*, Oxford University Press, Oxford, 1979.
[5] J.W.P. Hirschfeld and J.A. Thas, *General Galois Geometries*, Oxford University Press, Oxford, 1991.
[6] D. Jungnickel, Maximal partial spreads and transversal-free translation nets, *J. Combin. Theory A* 62 (1993), 66-92.
[7] S.E. Payne and J.A. Thas, *Finite Generalized Quadrangles*, Pitman, London, 1984.
[8] J.A. Thas, Partial geometries in finite affine spaces, *Math. Zeitschr.* 158 (1978), 1-13.
[9] J.A. Thas, Projective geometry over a finite field, in: *Handbook of Incidence Geometry* (F. Buekenhout; ed.), Elsevier, Amsterdam, 1995.
[10] J.A. Thas and F. De Clerck, Partial geometries satisfying the axiom of Pasch, *Simon Stevin* 51 (1977), 123-137.

35 Partially Balanced Incomplete Block Designs

Deborah J. Street
Anne Penfold Street

35.1 Definitions and Examples

Let X be a v-set with a symmetric association scheme defined on it. A *partially balanced incomplete block design with m associate classes* (PBIBD(m)) is a design based on the v-set X with b blocks, each of size k, and with each treatment appearing in r blocks. Any two treatments that are i^{th} associates appear together in λ_i blocks of the PBIBD(m). The numbers $v, b, r, k, \lambda_i (1 \leq i \leq m)$ are the *parameters* of the PBIBD(m). One writes PBIBD(v, k, λ_i) and lets N be the $v \times b$ (0,1) incidence matrix of the PBIBD(m).

35.1 Example Let $v = 6, b = 8, r = 4, k = 3, \lambda_1 = \lambda_2 = 2, \lambda_3 = 1$. Associate classes are shown as adjacencies in the following diagram.

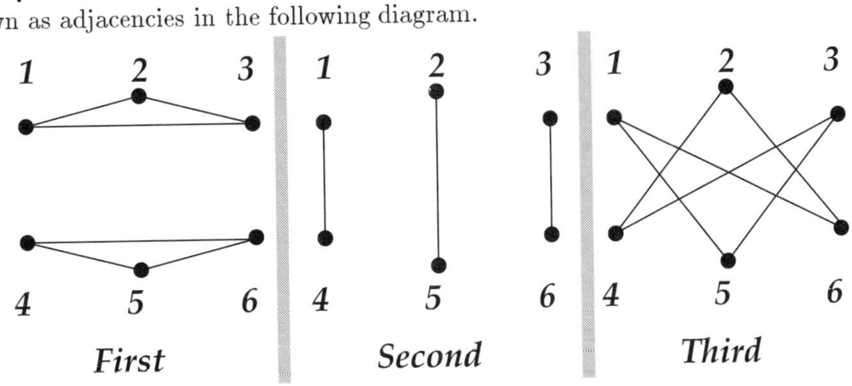

Consider the design which contains the blocks $\{1, 2, 4\}, \{1, 2, 5\}, \{1, 3, 4\}, \{1, 3, 6\}, \{2, 3, 5\}, \{2, 3, 6\}, \{4, 5, 6\}, \{4, 5, 6\}$. Even though $\lambda_1 = \lambda_2$, the first and second associate classes cannot be combined since if they could, then 1 would have first associates 2, 3, and 4; 2 would have first associates 1, 3, and 5; 4 would have first associates 1, 5, and 6. Then $p^1_{11} = 1$ for first associates 1 and 2, and is 0 for first associates 1 and 4. But this number must be a constant, from the definition of an association scheme. See §IV.1 for definitions pertaining to association schemes.

35.2 Remarks

1. PBIBDs were introduced as generalizations of BIBDs because of the easy analysis. BIBDs are universally optimal designs. The optimality of some PBIBD(2)s is established in [3].
2. The structure that the association scheme imposes on the treatments is not necessarily an inherent attribute of the treatments, as is the case in a factorial design; see §IV.19.
3. A PBIBD(m) determines an association scheme, but the converse is not true.
4. A PBIBD(1) is a BIBD, and a PBIBD(2) in which $\lambda_1 = \lambda_2$ is a BIBD.
5. [7] gives conditions for a PBIBD(m), with $\lambda_i = \lambda_j$, to become a PBIBD($m-1$).
6. PBIBDs are used in plant breeding work (see [8]) and in group testing (see [12] and §V.6).

35.3 Theorem [11] Let $A_0, \ldots, A_m$ be the matrices of an association scheme corresponding to a PBIBD(m). Then $NN^T = rI + \sum_i \lambda_i A_i$ and $JN = kJ$. Conversely if N is a (0,1) matrix which satisfies these conditions and the A_i are the matrices of an association scheme, then N is the incidence matrix of a PBIBD(m). See §IV.1.

35.4 Theorem [11] Let each treatment in a PBIBD(m) have n_i i^{th} associates, $1 \le i \le m$. Then the parameters of a PBIBD(m) satisfy $vr = bk$ and $\sum_i n_i \lambda_i = r(k-1)$, while the parameters of an association scheme satisfy $\sum_i n_i = v - 1$, $\sum_u p^h_{ju} = n_j$, and $n_i p^i_{jh} = n_j p^j_{ih}$.

35.2 Two-Associate-Class PBIBDs

There are six types of PBIBD(2)s, based on the underlying association scheme. These are (see [4]) 1. group divisible; 2. triangular; 3. latin-square-type; 4. cyclic; 5. partial geometry type; and 6. miscellaneous.

35.5 Remarks

1. The GDDs of §III.1 are PBIBD(2)s with $\lambda_1 = 0$ and $\lambda_2 = \lambda$.
2. The duals of triangular PBIBD(2)s are the residual designs of symmetric BIBDs with $\lambda = 2$; see Theorem 35.12 and §I.5.

35.6 Theorem [11] Let N be the incidence matrix of a PBIBD($v, k, \lambda_1, \lambda_2$). Let $\gamma = p^2_{12} - p^1_{12}$, $\beta = p^2_{12} + p^1_{12}$, $\delta = \gamma^2 + 2\beta + 1$. Then the eigenvalues of NN^T are rk and $\theta_i = r + ((\lambda_1 - \lambda_2)(\gamma + (-1)^i\sqrt{\delta}) - (\lambda_1 + \lambda_2))/2$, $i = 1, 2$, with multiplicities 1 and $\alpha_i = (n_1 + n_2)/2 + (-1)^i((n_1 - n_2) - \gamma(n_1 + n_2))/(2\sqrt{\delta})$, $i = 1, 2$ respectively.

▷ Group Divisible PBIBDs

Let the v-set X be partitioned into m groups each of size n. Then in a *group divisible association scheme* the first associates are the treatments in the same group and the second associates are all the other treatments. The eigenvalues of NN^T are rk, $r - \lambda_1$ and $rk - v\lambda_2$ with multiplicities 1, $m(n-1)$, and $m - 1$, respectively. A group divisible PBIBD is *singular* if $r = \lambda_1$; *semi-regular* if $r > \lambda_1$, $rk - v\lambda_2 = 0$; and *regular* if $r > \lambda_1$ and $rk > v\lambda_2$.

35.7 Remark In group divisible PBIBDs neither λ need be 0.

35.8 Theorems [10, 11]

1. The parameters of a group divisible association scheme uniquely determine the scheme.
2. The existence of a singular group divisible PBIBD$(mn, kn, \lambda_1 = r, \lambda_2 = \lambda)$ is equivalent to the existence of a BIBD(m, k, λ).
3. In a semi-regular group divisible PBIBD, every block has k/m treatments from each group.
4. The dual of an affine resolvable BIBD$(ak, a(c(a-1)+1), ac+1)$ is a semi-regular group divisible PBIBD with $v = a(ca^2+a+1), b = a^2(c(a-1)+1), r = a(c(a-1)+1), k = a^2c+a+1, m = a^2c+a+1, n, \lambda_1 = 0, \lambda_2 = c(a-1)+1$.
5. Given a BIBD$(v, k, 1)$, then there exists a group divisible PBIBD$(v-1, k, 0, 1)$.

35.9 Examples A BIBD(7,3,1), a singular group divisible PBIBD(21,9,3,1) obtained by replacing each element of the BIBD by a set of three elements, and a semi-regular group divisible PBIBD(6,3,0,1) with groups obtained from the blocks of the BIBD containing 7.

BIBD	PBIBD(21,9,3,1)	PBIBD(6,3,0,1)
$\{1,2,4\}$	$\{1, 8, 15, 2, 9, 16, 4, 11, 18\}$	$\{1,2,4\}$
$\{2,3,5\}$	$\{2, 9, 16, 3, 10, 17, 5, 12, 19\}$	$\{2,3,5\}$
$\{3,4,6\}$	$\{3, 10, 17, 4, 11, 18, 6, 13, 20\}$	$\{3,4,6\}$
$\{4,5,7\}$	$\{4, 11, 18, 5, 12, 19, 7, 14, 21\}$	
$\{5,6,1\}$	$\{5, 12, 19, 6, 13, 20, 1, 8, 15\}$	$\{5,6,1\}$
$\{6,7,2\}$	$\{6, 13, 20, 7, 14, 21, 2, 9, 16\}$	
$\{7,1,3\}$	$\{7, 14, 21, 1, 8, 15, 3, 10, 17\}$	

35.10 Remarks

1. Difference methods can be used to construct group divisible PBIBDs; see §IV.10.
2. The rectangular association scheme [10] and the extended group divisible association scheme [10, 2] are generalizations of the group divisible scheme. Both are examples of the factorial association scheme; see [1, 5].

▷ Triangular PBIBDs

Let $v = n(n-1)/2, n \geq 5$, and arrange the v elements of X in a symmetrical $n \times n$ array with the diagonal entries blank. In the *triangular association scheme* the first associates of a treatment are those in the same row or column of the array; all other treatments are second associates.

35.11 Example The triangular scheme with $n = 6$ and a triangular PBIBD with $v = b = 15, r = k = 3, \lambda_1 = 0, \lambda_2 = 1$. This is design T16 in [4].

Array					
*	1	2	3	4	5
1	*	6	7	8	9
2	6	*	10	11	12
3	7	10	*	13	14
4	8	11	13	*	15
5	9	12	14	15	*

PBIBD		
$\{1,10,15\}$	$\{15,3,6\}$	$\{5,6,13\}$
$\{11,1,14\}$	$\{8,12,3\}$	$\{7,5,11\}$
$\{12,13,1\}$	$\{3,11,9\}$	$\{10,8,5\}$
$\{2,15,7\}$	$\{6,14,4\}$	$\{13,9,2\}$
$\{14,2,8\}$	$\{4,7,12\}$	$\{9,4,10\}$

35.12 Theorems [11] (1) The parameters of the triangular association scheme uniquely determine the scheme for $n \neq 8$. (2) If a symmetric BIBD$((n^2+3n+4)/2, n+2, 2)$ exists, then so does a triangular PBIBD with parameters $v = (n+1)(n+2)/2, b = n(n+1)/2, r = n, k = n+2, \lambda_1 = 1, \lambda_2 = 2$.

35.13 Remarks

1. There are three pseudo-triangular association schemes when $n = 8$; see [10, 11].
2. Triangular schemes and generalized triangular schemes are also known as Johnson schemes; see §IV.1.
3. The triangular scheme arises in diallel cross experiments in which the treatments are the crosses of n genotypes, there are no self-crosses, and the crosses (AB) and (BA) are not distinguished. First associates are then those which have one parental genotype in common [1].

▷ Latin-Square-Type PBIBDs

Let $v = n^2$ and arrange the v treatments in an $n \times n$ array. Superimpose on this array a set of $i - 2$ mutually orthogonal latin squares of order n. Let the first associates of any treatment be those in the same row or column of the array or associated with the same symbols in one of the latin squares. This is an *L_i-type association scheme.* If $i = n$, then the scheme is group divisible; if $i = n + 1$, then all the treatments are first associates of each other.

35.14 Example Let $n = 4$. Then $v = 16$ and an L_3-type PBIBD(2) with $b = 16$, $r = k = 3$, $\lambda_1 = 0$ and $\lambda_2 = 1$ is given below. It is design LS18 of [4].

Square			
1	2	3	4
4	1	2	3
3	4	1	2
2	3	4	1

PBIBD			
$\{10, 7, 16\}$	$\{7, 9, 4\}$	$\{13, 4, 6\}$	$\{2, 16, 9\}$
$\{11, 8, 13\}$	$\{8, 10, 1\}$	$\{14, 1, 7\}$	$\{3, 13, 10\}$
$\{12, 5, 14\}$	$\{5, 11, 2\}$	$\{15, 2, 8\}$	$\{4, 14, 11\}$
$\{9, 6, 15\}$	$\{6, 12, 3\}$	$\{16, 3, 5\}$	$\{1, 15, 12\}$

35.15 Remark $L_2(n)$ is also the 2-dimensional Hamming scheme; see §IV.1.

▷ Cyclic PBIBDs

Let $\{1, 2, \ldots, v-1\} = D \cup E$. A non-group divisible association scheme defined on $\mathbb{Z}_v$ is *cyclic* if $D = -D$, and if the set of $n_1(n_1 - 1)$ differences of distinct elements of D has each element of D p^1_{11} times and each element of E p^2_{11} times. The first associates of i are $i + D$.

35.16 Example Let $v = 17$, $D = \{3, 5, 6, 7, 10, 11, 12, 14\}$. Then the initial blocks {0,1,4,5} and {1,8,10,16} developed modulo 17 give a PBIBD(17,4,1,2).

▷ Partial-Geometry-Type PBIBDs

In a *partial-geometry-type association scheme*, two treatments are first associates if they are incident with a line of the geometry and second associates if they are not incident with a line of the geometry.

35.17 Remarks

1. Using the notation of §IV.34.1, if a partial geometry with parameters s, t, and α exists, then so does a PBIBD(2) with parameters $v = (s+1)(ts+\alpha)/\alpha$, $b = (t+1)(ts+\alpha)/\alpha$, $r = t+1$, $k = s+1$, $\lambda_1 = 1$ and $\lambda_2 = 1$; see [4].
2. A partial geometry with $\alpha = t$ is a PBIBD(2) with $k = r$ and $t + 1$ replicates based on a L_{t+1} type association scheme, or a lattice design with $t+1$ replicates; see [4].

35.3 See Also

§I.1	BIBDs are PBIBD(1)s.
§II.4	More on affine resolvable designs.
§III.5	GDDs form a class of PBIBDs.
§IV.1	Association schemes underlie PBIBDs.
§IV.34	Partial geometries underlie some PBIBDs.
§V.6	PBIBDs are used in group testing.
[4]	Extensive tables of PBIBD(2)s.
[1]	Most recent overview of PBIBDs.
[6, 9]	Extend nested row-column designs to PBIBDs.
[11]	Chapter 11 gives a description of two class PBIBDs and contains proofs of most of the material in this section (and more).

References

[1] R.A. Bailey, Partially balanced designs, in: *Encyclopedia of Statistical Science* Vol. 6 (N.L. Johnson, S. Kotz and C. Read; eds.), John Wiley & Sons, New York, 1985, 593-610.

[2] C.-T. Chang and K. Hinkelmann, A new series of EGD-PBIB designs, *J. Stat. Plann. Infer.* 16 (1987), 1-13.

[3] C.-S. Cheng and R.A. Bailey, Optimality of some two-associate-class partially balanced incomplete block designs, *Ann. Statist.* 19 (1991), 1667-1671.

[4] W.H. Clatworthy, *Tables of Two-Associate-Class Partially Balanced Designs*, National Bureau of Standards (U.S.) Applied Mathematics Series No. 63, 1973.

[5] S.C. Gupta, Generating generalised cyclic designs with factorial balance, *Comm. Statist.* A 16 (1987), 1885-1900.

[6] S.C. Gupta and M. Singh, Partially balanced incomplete block designs with nested rows and columns, *Utilitas Math.* 40 (1991), 291-302.

[7] S. Kageyama, Reduction of associate classes for block designs and related combinatorial arrangements, *Hiroshima Math. J.* 4 (1974), 527-618.

[8] O. Mayo, *The Theory of Plant Breeding*, Clarendon Press, Oxford, 1987.

[9] J.P. Morgan and N. Uddin, Some constructions for rectangular, latin square and pseudo-cyclic nested row-column designs, *Utilitas Math.* 38 (1990), 43-51.

[10] D. Raghavarao, *Constructions and Combinatorial Problems in Design of Experiments*, John Wiley & Sons, New York, 1971.

[11] A.P Street and D.J. Street, *Combinatorics of Experimental Design*, Clarendon Press, Oxford, 1987.

[12] F. Vakil and M. Parnes, On the structure of a class of sets useful in nonadaptive group testing, *J. Stat. Plann. Infer.* 39 (1994), 57-69.

36 Quasigroups

Frank E. Bennett

36.1 Definitions and Examples

A *quasigroup* is an ordered pair $(Q, \oplus)$, where Q is a set and $\oplus$ is a binary operation on Q such that the equations $a \oplus x = b$ and $y \oplus a = b$ are uniquely solvable for every pair of elements a, b in Q. For a finite set Q, the *order* of the quasigroup $(Q, \oplus)$ is $|Q|$.

36.1 Example A quasigroup and its associated latin square.

$\oplus$	0	1	2	3
0	0	1	2	3
1	1	0	3	2
2	2	3	1	0
3	3	2	0	1

0	1	2	3
1	0	3	2
2	3	1	0
3	2	0	1

36.2 Remark The multiplication table of a quasigroup defines a *latin square*; that is, a latin square can be viewed as the multiplication table of a quasigroup with the headline and sideline removed.

A quasigroup $(Q, \cdot)$ is *idempotent* if the identity $x \cdot x = x$ (briefly $x^2 = x$) holds for all x in Q.

If $(Q, \otimes)$ is a quasigroup, one may define on the set Q six binary operations $\otimes_{(1,2,3)}$, $\otimes_{(1,3,2)}$, $\otimes_{(2,1,3)}$, $\otimes_{(2,3,1)}$, $\otimes_{(3,1,2)}$, and $\otimes_{(3,2,1)}$ as follows: $a \otimes b = c$ if and only if:

$$a \otimes_{(1,2,3)} b = c, \qquad a \otimes_{(1,3,2)} c = b, \qquad b \otimes_{(2,1,3)} a = c,$$
$$b \otimes_{(2,3,1)} c = a, \qquad c \otimes_{(3,1,2)} a = b, \qquad c \otimes_{(3,2,1)} b = a.$$

These six (not necessarily distinct) quasigroups $(Q, \otimes_{(i,j,k)})$, where $\{i, j, k\} = \{1, 2, 3\}$, are the *conjugates* of $(Q, \otimes)$.

36.3 Example Examples of six conjugate quasigroups are given in §II.1.

Two quasigroups $(Q, \cdot)$ and $(Q, *)$ defined on the same set Q are *orthogonal* if the equations $x \cdot y = z \cdot t$ and $x * y = z * t$ together imply $x = z$ and $y = t$.

36.4 Remark When two quasigroups $(Q, \cdot)$ and $(Q, *)$ are orthogonal, their corresponding latin squares are also orthogonal in the usual sense.

A quasigroup (latin square) is *self-orthogonal* if it is orthogonal to its $(2, 1, 3)$-conjugate, its *transpose*. More generally, a quasigroup which is orthogonal to its (i, j, k)-conjugate is (i, j, k)*-conjugate orthogonal.*

Adopting the functional notation $a(x, y)$ in place of the infix notation $x * y$ for the operation, two binary operations $a(x, y)$ and $b(x, y)$ defined on the same set Q are *orthogonal operations*, written $a \perp b$, if $|\{(x, y)\colon a(x, y) = i,\ b(x, y) = j\}| = 1$ for every ordered pair i, j in Q.

The *spectrum* of the two-variable quasigroup identity $u(x, y) = v(x, y)$ is the set of all integers n such that there exists a quasigroup of order n satisfying the identity $u(x, y) = v(x, y)$.

36.5 Remark Two-variable quasigroup identities are quite often instrumental in the construction or algebraic description of combinatorial designs. For example, an *idempotent totally symmetric quasigroup* $(Q, \cdot)$, or *Steiner quasigroup*, satisfying the identities $\{x^2 = x, x(xy) = y, (xy)y = x\}$, corresponds to a *Steiner triple system* (STS), where $\{x, y, z\}$ is a triple if and only if $x \cdot y = z$ when x, y, z are distinct, and $x^2 = x$ for all $x \in Q$. Similarly, an *idempotent semisymmetric quasigroup* $(Q, \cdot)$, which satisfies the identities $\{x^2 = x, x(yx) = y\}$, corresponds to a *Mendelsohn triple system* (MTS). Here (x, y, z) is a cyclically ordered triple if and only if $x \cdot y = z$ when x, y, z are distinct, and $x^2 = x$ for all $x \in Q$.

36.6 Remark A Hall triple system (see §IV.25) is a Steiner triple system whose related Steiner quasigroup is distributive, in the sense that the following identity holds: $a \circ (x \circ y) = (a \circ x) \circ (a \circ y)$.

36.7 Remark The most direct method of constructing finite models of a quasigroup $(Q, \cdot)$ satisfying the two-variable identity $u(x, y) = v(x, y)$ is to look for a model of the identity of the form $x \cdot y = \lambda x + \mu y$, where the elements lie in some finite field. In particular, if one is interested in idempotent models, then look for models of the identity of the form $x \cdot y = \lambda x + (1 - \lambda)y$ in $\mathbb{F}_q$, where q is a prime power and $\lambda \neq 0$ or 1. This requires finding a solution to some polynomial equation $f(\lambda) = 0$ in $\mathbb{F}_q$ depending on the identity to be satisfied. More general information on varieties of quasigroups and their related structures can be found in [3, 4, 9, 11] and in their extensive bibliographies.

36.8 Example Let $Q = \mathbb{Z}_7$ and define the operation $x * y = 3x + 5y \pmod 7$ for all x, y in Q. This operation produces an idempotent semisymmetric quasigroup $(Q, *)$ of order 7, which is self-orthogonal. This quasigroup corresponds to a self-orthogonal MTS of order 7 (SOMTS(7)), which is generated by developing the two base blocks $(0, 1, 5)$ and $(0, 3, 1) \pmod 7$. It is also evident that this MTS(7) is separable into a pair of orthogonal STS(7). The quasigroup $(Q, *)$ is readily checked to be both $(1, 3, 2)$-and $(3, 2, 1)$-conjugate orthogonal.

36.9 Example Let the set Q be based on the elements of $\mathbb{F}_4$ and take λ to be a primitive element in $\mathbb{F}_4$. Then the operation $x \cdot y = \lambda x + (1 - \lambda)y$ in $\mathbb{F}_4$ defines an idempotent semisymmetric quasigroup $(Q, \cdot)$, which is self-orthogonal and corresponds to a SOMTS(4). The quasigroup $(Q, \cdot)$ is $(1, 3, 2)$-, and $(3, 2, 1)$-conjugate orthogonal.

36.10 Example Consider the quasigroup identity $((yx)y)y = x$. If one wants idempotent models of this identity, look for models of the form $x \cdot y = \lambda x + (1 - \lambda)y$ in $\mathbb{F}_q$, where q is a prime power and $\lambda \neq 0$ or 1 and where the polynomial equation $f(\lambda) = \lambda^3 - \lambda^2 + 1 = 0$ is satisfied in $\mathbb{F}_q$. Idempotent models of $((yx)y)y = x$ for orders $q = 5, 7, 8$ are readily obtained. A quasigroup satisfying the identity $((yx)y)y = x$ is $(2, 3, 1)$-, $(3, 1, 2)$- and $(3, 2, 1)$-conjugate orthogonal.

36.11 Example A quasigroup satisfying the identity $(xy)y = x(xy)$ is necessarily idempotent. If one looks for models of this identity of the form $x \cdot y = \lambda x + (1 - \lambda)y$ in $\mathbb{F}_q$, one finds such models for all orders $q = 2^k$ where $k \geq 2$. A quasigroup satisfying the given identity is $(1, 3, 2)$- and $(3, 2, 1)$-conjugate orthogonal.

36.2 Equivalent Objects

36.12 Theorem The spectrum of Steiner quasigroups (equivalently, Steiner triple systems) is precisely the set of all positive integers $n \equiv 1$ or 3 (mod 6).

36.13 Theorem [11] An idempotent semisymmetric quasigroup of order n (equivalently, a Mendelsohn triple system of order n) exists for precisely all positive integers $n \equiv 0$ or 1 (mod 3) except for $n = 6$.

36.14 Theorem [6] A self-orthogonal semisymmetric quasigroup of order n (equivalently, a SOMTS(n)) exists for precisely all positive integers $n \equiv 0$ or 1 (mod 3) except for $n = 3, 6, 9, 10, 12$ and except possibly for $n = 18$.

36.15 Remark Let Q be an n-set and let K_n be based on Q. A quasigroup $(Q, \cdot)$ satisfying the three identities $\{x^2 = x, x(yx) = y(xy), (yx)x = y\}$ is a *Steiner pentagon quasigroup*; it corresponds to a *Steiner pentagon system* (K_n, B), where a pentagon (x, y, z, u, v) belongs to B if and only if $xy = z$ and $yx = v$ for $x \neq y$ and $x^2 = x$ for all x in Q. Steiner pentagon quasigroups are also useful in the construction of other types of combinatorial structures, including perfect Mendelsohn designs with block size five and holey SOLSSOMs.

36.16 Theorem [12] The spectrum of Steiner pentagon quasigroups (equivalently, Steiner pentagon systems) is precisely the set of all positive integers $n \equiv 1$ or 5 (mod 10), except for $n = 15$.

36.3 Short Conjugate-Orthogonal Identities

36.17 Theorem [9] Let $a(x, y)$ and $b(x, y)$ be conjugate operations on Q. Then $a \perp b$ if and only if there is a quasigroup word $w(x, y)$ such that $w(a(x, y), b(x, y)) = x$ holds identically.

An identity of the type described in Theorem 36.17 where $w(x, y)$ is a word of length two is a *short conjugate-orthogonal identity*.

Two quasigroup identities $u_1(x, y) = v_1(x, y)$ and $u_2(x, y) = v_2(x, y)$ are *conjugate-equivalent* if, when $(Q, \cdot)$ is a quasigroup satisfying one of them, at least one conjugate of $(Q, \cdot)$ satisfies the other.

36.18 Example The identity $yx \cdot xy = x$ is conjugate-equivalent to the identity $(x \cdot yx)y = x$, since it can readily be verified that the $(1, 3, 2)$-conjugate of a quasigroup satisfying the first identity satisfies the second identity.

36.19 Remark Most of what follows is confined to short conjugate-orthogonal identities and their associated combinatorial structures. Here is provided a set of quasigroup identities that imply that at least two conjugates are orthogonal. See [3, 9] for more details. A name of the identity and/or the associated quasigroup is given when one is in common usage.

36.20 Theorem [3, 9] Any nontrivial short conjugate-orthogonal identity is conjugate-equivalent to one of the following:

1. $xy \cdot yx = x$ *Schröder quasigroup*
2. $yx \cdot xy = x$ *Stein's third law*
3. $(xy \cdot y)y = x$ *C_3-quasigroup*
4. $x \cdot xy = yx$ *Stein's first law; Stein quasigroup*
5. $(yx \cdot y)y = x$
6. $yx \cdot y = x \cdot yx$
7. $xy \cdot y = x \cdot xy$ *Schröder's first law*

The notation xy is shorthand for $(x \cdot y)$. So, for example, the first quasigroup identity is actually $(x \cdot y) \cdot (y \cdot x) = x$

36.4 Basic Constructions

36.21 Remark The standard construction techniques for finite models of the short conjugate-orthogonal identities combine the direct constructions from finite fields with recursive constructions, including the familiar direct product construction and its numerous generalizations in the form of the singular direct product construction. Pairwise balanced designs (PBDs) are of fundamental importance in the recursive construction of two-variable quasigroup models and the following is used quite extensively.

36.22 Theorem Let V be a variety of algebras that is idempotent and based on two-variable identities. Suppose that there is a $(v, K, 1)$-PBD such that for each block size $k \in K$ there is a model of V of order k. Then there is a model of V of order v.

36.23 Example The idempotent quasigroup of order 4 constructed in Example 36.9 represents a model for each of the identities (1), (3), (4), and (7) in Theorem 36.20. Since there is a $(v, \{4\}, 1)$-PBD for all integers $v \equiv 1$ or 4 (mod 12), Theorem 36.22 guarantees the existence of idempotent quasigroups satisfying each of the identities (1), (3), (4), and (7) in Theorem 36.20 for all orders $v \equiv 1$ or 4 (mod 12).

36.24 Example In Example 36.10, finite fields produced idempotent models of the identity (5) in Theorem 36.20 for orders 5, 7, and 8. Together with the closure $B(5, 7, 8)$ from Table III.3.17, Theorem 36.22 guarantees the existence of idempotent quasigroups satisfying the identity $(yx \cdot y)y = x$ for all but a relatively small number of orders.

36.25 Example For each of the identities (2) and (6) in Theorem 36.20, there are idempotent models in $\mathbb{F}_q$ for all prime powers $q \equiv 1$ (mod 4). In particular, there are idempotent quasigroups satisfying each of the two identities for order $k \in \{5, 9, 13, 17, 29\}$. Together with $B(5, 9, 13)$ from Table III.3.17, Theorem 36.22 therefore guarantees the existence of idempotent quasigroups satisfying each of the identities (2) and (6) in Theorem 36.20 for all orders $v \equiv 1$ (mod 4), except for $v = 33$.

36.5 Existence and Equivalences

36.26 Table Existence results for quasigroups in all cases of Theorem 36.20.

Case	Spectrum	Reference
1. Schröder quasigroup	$n \equiv 0, 1 \pmod 4$, $n \neq 5$	[5, 7, 11]
2. Stein's third law	$n \equiv 0, 1 \pmod 4$	[3, 4, 10, 11, 13]
3. C_3-quasigroup	$n \equiv 0, 1 \pmod 3$, $n \neq 6$	[3]
4. Stein quasigroup	$n \geq 1$, $n \notin \{2, 3, 6, 7, 8, 10, 12, 14, 15, 18\}$, except possibly for $n \in \{22, 23, 26, 27, 30, 34, 38, 42, 43, 46, 50, 54, 62, 66, 74, 78, 90, 98, 102, 114, 126\}$	[3]

Case	Spectrum	Reference
5. $(yx \cdot y)y = x$	$n \geq 1, n \notin \{2, 6, 10\}$, except possibly for $n \in \{14, 18, 26, 30, 38, 42, 158\}$	[2, 10, 13]
6. $yx \cdot y = x \cdot yx$	does not contain 2, 3, 4, 6, 7, 8, 10, 11, 12, 14, 15; contains all positive integers $n \equiv 1 \pmod 4$, except possibly $n = 33$; other cases open.	[3, 4, 10, 13]
7. Schröder's first law	does not contain 2, 3, 5, 6, 7, 10, 11, 12, 14, 15; contains all other positive integers $n \equiv 0, 1 \pmod 4$ except possibly for $n \in \{20, 21, 24, 41, 44, 48, 53, 60, 69, 77, 101, 164\}$; other cases open.	[3, 4, 10, 13]

36.27 Table Existence results for idempotent quasigroups in cases of Theorem 36.20. In cases (4), (6), and (7), the identity forces the quasigroup to be idempotent, so the result of Table 36.26 is not repeated here.

Case	Spectrum	Reference
1. Schröder quasigroup	$n \equiv 0, 1 \pmod 4$, $n \neq 5, 9$	[5, 7]
2. Stein's third law	$n \equiv 0, 1 \pmod 4$, $n \neq 4, 8$	[3, 4, 10, 11, 13]
3. C_3-quasigroup	$n \equiv 1 \pmod 3$	[3]
5. $(yx \cdot y)y = x$	$n \geq 1$, $n \notin \{2, 3, 4, 6, 9, 10, 12, 13, 14, 15, 16\}$, except possibly for $n \in \{18, 20, 22, 24, 26, 28, 30, 34, 38, 39, 42, 44, 46, 51, 52, 58, 60, 62, 66, 68, 70, 72, 74, 75, 76, 86, 87, 90, 94, 96, 98, 99, 100, 102, 106, 108, 110, 114, 116, 118, 122, 132, 142, 146, 154, 158, 164, 170, 174\}$	[2, 10, 13]

36.28 Remark Tables 36.26 and 36.27 include some recent results of exhaustive searches, as yet unpublished, as follows:

Case	Result	Author
4	There is no solution of order 15	J. Slaney
5, idempotent	There is no solution of order 16	H. Zhang
6	There is no solution of order 15	H. Zhang
6	There is no solution of order 16	J. Zhang
7	There is no solution of order 14 or 15	H. Zhang
7	There is a solution of order 17	W. McCune

36.29 Table Conjugate orthogonality implied in each case of Theorem 36.20.

Case	(1,3,2)	(2,1,3)	(2,3,1)	(3,1,2)	(3,2,1)
1. Schröder quasigroup		X			
2. Stein's third law		X			
3. C_3-quasigroup					X
4. Stein quasigroup	X	X			
5. $(yx \cdot y)y = x$			X	X	X
6. $yx \cdot y = x \cdot yx$			X	X	
7. Schröder's first law	X				X

36.30 Remark The Schröder identity is conjugate invariant in the sense that every conjugate of a Schröder quasigroup is also a Schröder quasigroup. Consequently, Schröder quasigroups provide examples of latin squares with self-orthogonal conjugates. More importantly, Schröder quasigroups have been used in the construction and description of a variety of other combinatorial designs, including a class of edge-colored graph designs which correspond to triple tournaments (see, for example, [1, 3, 5, 7, 11]).

36.31 Proposition A Schröder quasigroup of order n is equivalent to each of the following:

1. a SOLS(n) with the Weisner property (§IV.40.6), and
2. an OA($4, n$) which has the Klein 4-group as a conjugate invariant subgroup.

36.32 Theorem [5] An idempotent Schröder quasigroup of order n missing a sub-quasigroup of order 2 exists for precisely all positive integers $n \equiv 2$ or 3 (mod 4), $n \geq 7$, except for $n = 10$.

36.33 Remark Stein's third law has been quite useful in the construction of combinatorial designs (see, for example, [1, 3, 4, 11]).

36.34 Proposition A quasigroup of order n satisfying Stein's third law is equivalent to an OA($4, n$) which has the cyclic group of order 4 as a conjugate invariant subgroup.

36.35 Proposition An idempotent quasigroup of order n satisfying Stein's third law is equivalent to each of the following:

1. An $(n, 4, 1)$-perfect Mendelsohn design.
2. A directed triple tournament with n players.

36.36 Proposition A C_3-quasigroup of order n is equivalent to an OA($4, n$) which has the cyclic group of order 3 as a conjugate invariant subgroup. When idempotent, it is equivalent to an almost resolvable MTS(n).

36.37 Remark Stein quasigroups are perhaps the most extensively studied of the short conjugate-orthogonal identities. In 1957, Stein [14] first investigated these quasigroups with the hope of finding a counterexample to the Euler conjecture concerning orthogonal latin squares. However, investigations of the quasigroups continued long after the disproof of the Euler conjecture, because their broad spectrum provides a source for SOLS among other things.

36.38 Remark Idempotent models of quasigroups satisfying the identity $(yx \cdot y)y = x$ produce a class of 2-fold perfect resolvable Mendelsohn designs [2].

36.39 Remark Schröder's first law is conjugate invariant. Quasigroups satisfying Schröder's first law produce a class of 2-fold perfect resolvable Mendelsohn designs.

36.6 See Also

§I.4.5	Steiner triple systems are equivalent to Steiner quasigroups.
§II.1	Latin squares correspond to multiplication tables of quasigroups.
§II.2	Quasigroup identities lead to orthogonal arrays with various conjugate orthogonal subgroups.
§III.3	PBD closures are useful in the determination of spectra of quasigroup identities.
§IV.9	Steiner pentagon systems are equivalent to Steiner pentagon quasigroups.
§IV.17	A section on embedding partial quasigroups.
§IV.28	Mendelsohn triple systems are equivalent to idempotent semisymmetric quasigroups.

§IV.40	Self-orthogonal latin squares arise from (2,1,3)-conjugate orthogonal quasigroups.
[4]	An extensive survey on quasigroup identities and their use in the construction of combinatorial designs.
[8]	A textbook on latin squares and quasigroups.

References

[1] R.D. Baker, Quasigroups and tactical systems, *Aequat. Math.* 18 (1978), 296-303.

[2] F.E. Bennett, Quasigroup identities and Mendelsohn designs, *Canad. J. Math.* 41 (1989), 341-368.

[3] F.E. Bennett, The spectra of a variety of quasigroups and related combinatorial designs, *Discrete Math.* 77 (1989), 29-50.

[4] F.E. Bennett and L. Zhu, Conjugate-orthogonal latin squares and related structures, in: *Contemporary Design Theory: A Collection of Surveys* (J.H. Dinitz, D.R. Stinson; eds.) John Wiley & Sons, New York, 1992, 41-96.

[5] F.E. Bennett, R. Wei and L. Zhu, Incomplete idempotent Schröder quasigroups and related packing designs, *Aequat. Math.*, to appear.

[6] F.E. Bennett, H. Zhang and L. Zhu, Self-orthogonal Mendelsohn triple systems, *J. Combin. Theory A*, to appear.

[7] C.J. Colbourn and D.R. Stinson, Edge-coloured designs with block size four, *Aequat. Math.* 36 (1988), 230-245.

[8] J. Dénes and A. D. Keedwell, *Latin Squares and Their Applications*, Academic Press, New York, 1974.

[9] T. Evans, Algebraic structures associated with latin squares and orthogonal arrays, *Congr. Numer.* 13 (1975), 31-52.

[10] M. Fujita, J. Slaney and F. Bennett, Automatic generation of some results in finite algebra, in: *Proc. 13th International Joint Conference on Artificial Intelligence* (Ruzena Bajcsy; ed.), Morgan Kaufmann, 1993, 52-57.

[11] C.C. Lindner, Quasigroup identities and orthogonal arrays, in: *Surveys in Combinatorics* (E.K. Lloyd; ed.), Cambridge University Press, London, 1983, 77-105.

[12] C.C. Lindner and D.R. Stinson, Steiner pentagon systems, *Discrete Math.* 52 (1984), 67-74.

[13] J. Slaney, M. Fujita and M. Stickel, Automated reasoning and exhaustive search: Quasigroup existence problems, *Comput. Math. Applic.*, to appear.

[14] S.K. Stein, On the foundations of quasigroups, *Trans. Amer. Math. Soc.* 85 (1957), 228-256.

37 Quasi-Symmetric Designs

Mohan S. Shrikhande

37.1 Definition and Examples

A t-(v, k, λ) design D is *quasi-symmetric* (q.s.) with intersection numbers x and y ($x < y$) if any two blocks of D intersect in either x or y points.

37.1 Remark This concept originated in statistics in the investigation of duals of designs with $\lambda = 1$.

37.2 Example Any 2-$(v, k, 1)$ design with $b > v$ is q.s. with $x = 0$ and $y = 1$.

37.3 Example Let D be a multiple of a symmetric 2-(v, k, λ) design. Then D is q.s. with $x = \lambda$ and $y = k$.

37.4 Example The unique 3-$(8, 4, 1)$ design (Example I.4.27) is q.s. with $x = 0$ and $y = 2$.

37.5 Remark This is the smallest case of an infinite class of q.s. designs, known as Hadamard 3-designs (see Example I.5.22). More generally, affine designs provide further examples of q.s. designs.

37.6 Example Let D be a quasi-residual 2-(v, k, λ) design (i.e., $r = k + \lambda$) with $\lambda = 2$. Then D is quasi-symmetric with $x = 1$ and $y = 2$.

37.2 The Connection to Strongly Regular Graphs

The *block graph* Γ of q.s. 2-(v, k, λ) design D is the graph with vertex set being the blocks of D, and where blocks B_i and B_j are adjacent if $|B_i \cap B_j| = y$.

37.7 Theorem Let D be a q.s. 2-(v, b, r, k, λ) design with intersection numbers x and y. Then Γ is a strongly regular graph with parameters (n, a, c, d) where $n = b$, $a = [k(r-1) - x(b-1)]/(y-x)$, $c = a + \theta_1\theta_2 + \theta_1 + \theta_2$, $d = a + \theta_1\theta_2$, $\theta_1 = (r - \lambda - k + x)/(y-x)$, $\theta_2 = (x-k)/(y-x)$.

37.8 Remark Theorem 37.9 implies that not every strongly regular graph is a block graph of q.s. design.

37.9 Theorem There is no q.s. 2-design whose block graph is the lattice graph $L_2(n)$ or its complement.

37.10 Theorem Let D be a q.s. 2-(v, k, λ) design with intersection numbers x and y. Then

1. $y - x$ divides $k - x$ and $r - \lambda$;
2. $k(r-1)(x+y-1) + xy(1-b) = k(k-1)(\lambda-1)$;
3. if $x = 0$, then $(r-1)(y-1) = (k-1)(\lambda-1)$;
4. if D has no repeated blocks, then $b \leq \binom{v}{2}$, with equality if and only if D is a 4-design;
5. if $x = 0$, then $b \leq v(v-1)/k$, with equality if and only if D is a 3-design.

37.3 Quasi-Symmetric t-Designs ($t \geq 3$)

37.11 Remark If a symmetric 2-(v, k, λ) design is extendible, then its extension is a q.s. 3-$(v+1, k+1, \lambda)$ design with $x = 0$ and $y = \lambda + 1$. Any q.s. 3-design with $x = 0$ is an extension of a symmetric 2-design.

37.12 Theorem If D is a q.s. 3-(v, k, λ) design with $x = 0$, then $y = \lambda + 1$ and one of the following holds:

1. $v = 4(\lambda + 1)$, $k = 2(\lambda + 1)$;
2. $v = (\lambda + 1)(\lambda^2 + 5\lambda + 5)$, $k = (\lambda + 1)(\lambda + 2)$; or
3. $v = 496$, $k = 40$, $\lambda = 3$.

37.13 Remark Designs in case 1 of Theorem 37.12 are the Hadamard 3-designs. In case 2, for $\lambda = 1$, it is the unique extension of PG(2, 4) and nothing is known for $\lambda \geq 2$. The existence in case 3 is unknown.

37.14 Theorem If D is a q.s. 3-(v, k, λ) design with $x = 1$, then D is the 4-(23, 7, 1) design or its residual, the 3-(22, 7, 4) design.

37.15 Theorem If D is a q.s. 4-(v, k, λ) design, then D is the 4-(23, 7, 1) design or its complement.

37.16 Theorem Quasi-symmetric t-designs do not exist for $t \geq 5$.

37.17 Conjecture Quasi-symmetric 3-designs up to complements are given by Theorems 37.12 and 37.14.

37.4 Quasi-Symmetric Designs and Codes

37.18 Remark The tools of coding theory are important in the study of q.s. designs. One of the connections is given by Lemma 37.19 which is used to prove Theorem 37.20. See §V.1 for definitions.

37.19 Lemma Let N be a $b \times v$ incidence matrix of a q.s. 2-(v, k, λ) design with $k \equiv x \equiv y \pmod 2$.

1. If k is even, then the binary code of length v with generator matrix N is self-orthogonal.
2. Let B be the bordered $b \times (v + 1)$ matrix obtained by adding an extra column of 1s to N. If k is odd, then B generates a binary self-orthogonal code of length $v + 1$.

37.20 Theorem The quasi-symmetric designs 2-(21, 6, 4), $x = 0$, $y = 2$, 2-(21, 7, 12), $x = 1$, $y = 3$, and 2-(22, 7, 16), $x = 1$, $y = 3$ are unique.

37.5 Classifying Quasi-Symmetric 2-Designs

37.21 Remark The classification of q.s. 2-designs is a difficult open problem. In [7], four classes of q.s. designs were distinguished: multiples of symmetric 2-designs, Steiner systems with $v > k^2$, strongly resolvable designs (i.e., those q.s. designs having a complete multipartite block graph), and residuals of biplanes. If a q.s. design or its complement does not belong to one of these classes it is referred to as *exceptional.*

37.22 Table An updated table of exceptional q.s. designs with $2k \leq v \leq 70$. References not given here can be found in [10].

No.	v	k	λ	r	b	x	y	Existence	Ref.
1	19	7	7	21	57	1	3	No	[7]
2	19	9	16	36	76	3	5	No	
3	20	10	18	38	76	4	6	No	
4	20	8	14	38	95	2	4	No	
5	21	9	12	30	70	3	5	No	[1]
6	21	8	14	40	105	2	4	No	[1]
7	21	6	4	16	56	0	2	Yes(1)	[9]
8	21	7	12	40	120	1	3	Yes(1)	[9]
9	22	8	12	36	99	2	4	No	
10	22	6	5	21	77	0	2	Yes(1)	

No.	v	k	λ	r	b	x	y	Existence	Ref.
11	22	7	16	56	176	1	3	Yes(1)	[9]
12	23	7	21	77	253	1	3	Yes(1)	
13	24	8	7	23	69	2	4	No	
14	28	7	16	72	288	1	3	No	[9]
15	28	12	11	27	63	4	6	Yes($\geq$ 22)	[4], [6], [9]
16	29	7	12	56	232	1	3	No	[9]
17	31	7	7	35	155	1	3	Yes(5)	
18	33	15	35	80	176	6	9	?	
19	33	9	6	24	88	1	3	No	
20	35	7	3	17	85	1	3	No	[1]
21	35	14	13	34	85	5	8	?	
22	36	16	12	28	63	6	8	Yes($\geq$ 32)	[5], [6]
23	37	9	8	36	148	1	3	?	
24	39	12	22	76	247	3	6	?	
25	41	9	9	45	205	1	3	?	
26	41	20	57	120	246	8	11	?	
27	41	17	34	85	205	5	8	No	[1]
28	42	21	60	123	246	9	12	?	
29	42	18	51	123	287	6	9	?	
30	43	18	51	126	301	6	9	No	[1]
31	43	16	40	112	301	4	7	?	
32	45	21	70	154	330	9	13	?	
33	45	9	8	44	220	1	3	Yes	[9]
34	45	18	34	88	220	6	9	?	
35	45	15	42	132	396	3	6	?	
36	46	16	72	216	621	4	7	?	
37	46	16	8	24	69	4	6	?	
38	49	9	6	36	196	1	3	Yes	[3]
39	49	16	45	144	441	4	7	?	
40	49	13	13	52	196	1	4	?	
41	51	21	14	35	85	6	9	No	[1]
42	51	15	7	25	85	3	5	No	[1]
43	52	16	20	68	221	4	7	?	
44	55	16	40	144	495	4	8	?	
45	55	15	63	243	891	3	6	?	
46	55	15	7	27	99	3	5	?	
47	56	16	18	66	231	4	8	?	
48	56	15	42	165	616	3	6	?	
49	56	12	9	45	210	0	3	?	
50	56	21	24	66	176	6	9	?	
51	56	20	19	55	154	5	8	?	
52	56	16	6	22	77	4	6	Yes	
53	57	27	117	252	532	12	17	No	[1]
54	57	9	3	21	133	1	3	?	
55	57	15	30	120	456	3	6	?	
56	57	12	11	56	266	0	3	?	
57	57	24	23	56	133	9	12	No	[1]
58	57	21	25	70	190	6	9	?	
59	57	21	10	28	76	7	9	No	[2]
60	60	15	14	59	236	3	6	?	
61	60	30	58	118	236	14	18	No	
62	61	25	160	400	976	9	13	?	
63	61	21	21	63	183	6	9	?	
64	63	15	35	155	651	3	7	Yes	
65	63	24	92	248	651	8	12	?	
66	63	18	17	62	217	3	6	?	
67	64	24	46	126	336	8	12	?	
68	65	20	19	64	208	5	8	?	
69	66	30	29	65	143	12	15	?	
70	69	33	176	374	782	15	21	No	[1]
71	69	18	30	120	460	3	6	?	
72	70	10	6	46	322	0	2	?	
73	70	30	58	138	322	10	14	?	

37.6 See Also

§V.1	The tools of coding theory are used in the study of q.s. designs.
§VI.5	The connection to strongly regular graphs is given in Theorem 37.7.
[8]	A monograph on q.s. designs and some related topics.
[4], [5]	Quasi-symmetric designs with the symmetric difference property.
[9]	Links between q.s. designs and self-dual codes.
[10]	For references in Table 37.22.

References

[1] A.R. Calderbank, Geometric invariants for quasi-symmetric designs, *J. Combin. Theory A* 47 (1988), 101-110.

[2] W.H. Haemers, There exists no (76, 21, 2, 7) strongly regular graph, in: *Finite Geometry and Combinatorics*, (F. De Clerck et al.; eds.) Cambridge University Press, London, 1993, 175-176.

[3] W.C. Huffman and V.D. Tonchev, The existence of extremal self-dual [50, 25, 10] codes and quasi-symmetric 2-(49, 9, 6) designs, *Des. Codes Crypt.* 6 (1995), 97-106.

[4] D. Jungnickel and V.D. Tonchev, On symmetric and quasi-symmetric designs with symmetric difference property and their codes, *J. Combin. Theory A* 59 (1992), 40-50.

[5] D. Jungnickel and V.D. Tonchev, Exponential number of quasi-symmetric SDP designs and codes meeting the Grey-Rankin bound, *Des. Codes Crypt.* 1 (1991), 247-253.

[6] C. Lam, L. Thiel and V. Tonchev, On quasisymmetric 2-(28, 12, 11) and 2-(36, 16, 12) designs, *Des. Codes Crypt.* 5 (1995), 43-55.

[7] A. Neumaier, Regular sets and quasi-symmetric 2-designs, in: *Combinatorial Theory* (D. Jungnickel, K. Vedder; eds.), *Lecture Notes in Math.* 969, Springer-Verlag, Berlin, 1982, pp. 258-275.

[8] M.S. Shrikhande and S.S. Sane, *Quasi-Symmetric Designs*, Cambridge University Press, London, 1991.

[9] V.D. Tonchev, Quasi-symmetric designs and self-dual codes, *Europ. J. Combin.* 7 (1986), 67-73.

[10] V.D. Tonchev, Self-orthogonal designs, *Contemp. Math.* 111 (1990), 203-219.

38 (r, λ)-designs

G.H. John van Rees

38.1 Definitions and Example

An (r, λ)-*design* is a pair $(V, \mathcal{B})$ where V is a set of v elements and $\mathcal{B}$ is a collection of b subsets (*blocks*) of V such that every distinct pair of elements occurs in precisely λ blocks and every variety occurs in precisely r blocks. Let $r - \lambda = n$. Let k_i be the size of the ith block in the (r, λ)-design.

38.1 Example A (6,3)-design on 5 elements.

{12345}, {123}, {124}, {135}, {145}, {235}, {245}, {34}, {34}, {1}, {2}, {5}

A block containing all the elements is a *complete* block. A set of v blocks containing just one element (*singletons* or *singleton blocks*) whose union is V is a *complete set of singletons.*

An (r, λ)-design which is composed of λ complete blocks and consequently n complete sets of singletons is *trivial.* Non-trivial (r, λ)-designs are sometimes called *(r, λ)-systems.* Further, if $k_i \geq 2$, the designs are *regular pairwise balanced designs.*

Designs that only contain blocks of cardinality 1, $v-1$, or v are *near-trivial.* An (r, λ)-design which contains a complete block or a complete set of singletons is *reducible.*

38.2 Connections and General Results

38.2 Remarks

1. Any (r, λ)-design can be changed into an $(r+1, \lambda)$-design or an $(r+1, \lambda+1)$-design by the addition of a complete set of singletons or the addition of a complete block, respectively. Quite often the non-near-trivial or irreducible (r, λ)-designs are the interesting designs.
2. Any pairwise balanced design (PBD) can be made into an (r, λ)-design by the addition of singleton blocks. Therefore many theorems on PBDs can be transformed into theorems about (r, λ)-designs (see §III.1-6).
3. If all the blocks of a nontrivial (r, λ)-design have the same cardinality k, then the design is a (v, b, r, k, λ) BIBD. If v and b of an (r, λ)-design are such that v, b, r, λ could be the parameters of a BIBD, then all blocks have cardinality k and the (r, λ)-design is a BIBD (see §I).
4. The incidence matrix of an (r, λ)-design is a binary balanced equidistant $(v, 2n, b)$-code.
5. All blocking sets of BIBDs are (r, λ)-designs, where a *blocking set* is a set S of points of the design such that any block contains a point of S and a point outside S.
6. Many schedules of sporting events use (r, λ)-designs directly, use their dual, or incorporate an (r, λ)-design as part of their structure.

38.3 Theorem In any (r, λ)-design, $b \geq vr^2/(r+\lambda(v-1))$ and $k_i(n+\lambda v-\lambda k_i) \leq n(n+\lambda v)$.

38.4 Theorem If an (r, λ)-design is nontrivial, then $v \leq r(r-1)+1$ with equality if n is the order of a finite projective plane.

38.5 Theorem For a resolvable (r, λ)-design, $b \geq v+r-1$. If $b = v+r-1$, then it is an affine resolvable BIBD with 0, 1 or more complete blocks adjoined.

38.3 Doehlert-Klee Designs (DK designs)

An (r, λ)-design is *elliptic, parabolic, or hyperbolic* depending whether $\lambda(v-1) - r(r-1)$ is less than, equal to, or greater than 0.

38.6 Theorem All irreducible (r, λ)-designs for $\lambda = 1$ or 2 are elliptic or parabolic. This is *not* true for $\lambda \geq 3$.

A *Doehlert-Klee* (DK) design is an (r, λ)-design with $r^2 = \lambda b$.

38.7 Remarks The duals of DK designs are used as experimental designs. For a given b, $K(b)$ denotes the maximum v such that a DK design exists.

38.8 Table Lower and upper bounds for undetermined $K(b)$, $b \leq 100$.

b	45	50	54	63	90	98	99
$K(b)$	16-40	26-46	25-51	31-60	28-87	50-95	31-97

38.4 Embedding Results

38.9 Remark Given any (r, λ)-design, one can obtain a $(2n, n)$-design by complementing certain blocks. If the (r, λ)-design has $v > (n+2)^2/2$, then an $(n+1, 1)$-design can be obtained. For this reason most results are proved using $(2n, n)$-designs or $(r, 1)$-designs.

$v_0(r, \lambda)$ is the least integer such that if D is an (r, λ)-design on v elements with $v > v_0(r, \lambda)$, then D is trivial.

$v_1(r, \lambda)$ is the least integer such that if D is an (r, λ)-design on v elements with $v > v_1(r, \lambda)$, then D is reducible.

$v_p(r, 1)$ is the largest integer such that there exists a nontrivial $(r, 1)$-design D on $v_p(r, 1)$ elements in which the number of blocks is less than or equal to $n^2 + n + 1$.

38.10 Theorem $v_1(r, \lambda) \leq v_0(r, \lambda)$ and $v_p(r, 1) \leq v_1(r, 1)$.

38.11 Theorem $v_0(r, \lambda) \leq \max\{n^2 + n + 1, \lambda + 2\}$ with equality if and only if there exists a finite projective plane of order n or $\lambda \geq n^2 + n + 1$.

38.12 Theorem If n is not the order of a projective plane, then $v_0(r, \lambda) \leq \max\{\lambda + 2, n^2 - a\}$ if $r > 2n$ and $v_0(r, \lambda) \leq \max\{\lambda + 2, n^2 - a - 1\}$ if $r \leq 2n$, where $n > 2a^2 + 3a + 2$ and a is a positive integer.

38.13 Theorem $v_1(r, \lambda) = n^2 + n + 1$ if and only if n $(n \geq 3)$ is the order of a finite projective plane and $r = n + 1$, $2n$ or n^2 or if $\lambda = n^3 + n^2 - n$.

38.14 Theorem $v_0(7, 1) = 31$, and $25 \leq v_p(7, 1) \leq 28$.

38.15 Table v_1 for small values of (r, λ).

(r, λ)	(3,1)	(4,2)	(6,3)	(7,4)	(8,5)	(9,6)	(10,7)
$v_1(r, \lambda)$	7	7	13	12	9	13	7

38.5 See Also

§III.1	PBDs which are regular (have constant replication number) are (r, λ)-designs.
§IV. 18	Equidistant permutation arrays are closely related to (r, λ)-designs.

References

[1] M. Deza, R.C. Mullin and S.A. Vanstone, Recent results on (r, λ)-designs and their related configurations, *Rev. Tec. Ing. Univ. Zulia* 4 (1981), 139-158.

[2] D. McCarthy and S.A. Vanstone, On the structure of regular pairwise balanced designs, *Discrete Math.* 25 (1979), 237-244.

[3] R.G. Stanton and S.A. Vanstone, Some theorems on DK designs, *Ars Combin.* 8 (1979), 117-129.

39 Room Squares

Jeffrey H. Dinitz

39.1 Definitions, Example, and Existence

Let S be a set of $n+1$ elements (*symbols*). A *Room square* of side n (on symbol set S) is an $n \times n$ array, F, which satisfies the following properties:

1. every cell of F either is empty or contains an unordered pair of symbols from S,
2. each symbol of S occurs once in each row and column of F,
3. every unordered pair of symbols occurs in precisely one cell of F.

A Room square of side n is *standardized* (with respect to the symbol ∞) if the cell (i, i) contains the pair $\{\infty, i\}$.

A standardized Room square of side n is *skew* if for every pair of cells (i, j) and (j, i) (with $i \neq j$) exactly one is filled.

39.1 Example Skew Room squares of side 7 and 9.

∞0			15		46	23
34	∞1			26		50
61	45	∞2			30	
	02	56	∞3			41
52		13	60	∞4		
	63		24	01	∞5	
		04		35	12	∞6

∞1		49	37	28		56		
89	∞2				57	34		16
	58	∞3		69	24		17	
	36	78	∞4		19		25	
	79		12	∞5	38		46	
45					∞6	18	39	27
		26	59	13		∞7		48
67	14					29	∞8	35
23		15	68	47				∞9

39.2 Theorem A (skew) Room square of side n exists if and only if n is odd and $n \neq 3$ or 5.

39.2 Equivalent Combinatorial Objects

39.3 Theorem The existence of the following are equivalent:

1. a Room square of side n;
2. a Room frame of type 1^n (§39.7);
3. two pairwise orthogonal-symmetric latin squares of order n (§II.2.7);
4. a Howell design $H(n, n+1)$ (§IV.26);
5. two pairwise orthogonal 1-factorizations of K_{n+1} (§VI.4); and
6. two orthogonal resolutions of a $(n+1, 2, 1)$-BIBD (§I.6).

39.4 Remark A Room square of side n can be used to schedule a round robin tournament with $n+1$ teams which has the following properties:

1. Every team plays every other team exactly once during the tournament.
2. Every team plays in exactly one game in each round.
3. Every team plays at every location exactly once.

39.3 Construction of Room Squares

39.5 Theorem If there exists two orthogonal starters in a group of order n, then there exists a Room square of side n (see §IV.45).

39.6 Remark The construction of a Room square from a pair of orthogonal starters is given explicitly in Construction 45.6.

39.7 Remark A very effective method for constructing Room squares is by use of a *hill-climbing algorithm.* A description of this type of algorithm (for Steiner triple systems) is given in §VI.9.3. A description of the specific algorithm for Room squares can be found in [3].

39.4 Enumeration of Room Squares

Let F_1 and F_2 be Room squares of side n on a symbol set S.

1. F_1 and F_2 are *isomorphic* if F_2 can be obtained from F_1 by any sequence of permuting the rows, permuting the columns, and permuting the symbols in S.
2. F_1 and F_2 are *equivalent* if they are isomorphic, or if F_1 is isomorphic to the transpose of F_2.

39.8 Table [8] The six inequivalent Room squares of order 7.

R_1

01		45	67			23
57	02				13	46
	56	03	12		47	
	37		04	26		15
36	14	27		05		
24			35	17	06	
		16		34	25	07

R_2

01	67				45	23
57	02	46		13		
		03		47	12	56
	15		04	26	37	
36			27	05		14
24		17	35		06	
	34	25	16			07

R_3

01		45	67		23	
57	02		13			46
	56	03			47	12
	37		04	26	15	
36	14	27		05		
24				17	06	35
		16	25	34		07

R_4

01	67			23		45
	02	46			57	13
56		03		47	12	
	35	27	04	16		
			17	05	34	26
37	14		25		06	
24		15	36			07

R_5

01			25	67		34
46	02				37	15
27	56	03		14		
	13	57	04			26
	47		36	05	12	
35		24	17		06	
		16		23	45	07

R_6

01	37		25	46		
	02	15	36		47	
26		03	17			45
			04	27	35	16
	14	67		05		23
57		24		13	06	
34	56				12	07

NR(n) is the number of nonisomorphic Room squares of side n.

IR(n) is the number of inequivalent Room squares of side n.

39.9 Theorem $\text{NR}(n) \leq 2\text{IR}(n)$.

39.10 Table Nonisomorphic and inequivalent Room squares of side $n \leq 9$.

n	3	5	7	9
Nonisomorphic	0	0	10	511562
Inequivalent	0	0	6	257630

39.11 Remark For $n > 10$, no exact values of $\mathrm{IR}(n)$ and $\mathrm{NR}(n)$ are known. Theorem 39.12 is a lower bound for these values when n gets large.

39.12 Theorem For odd $n \geq 153$, $\mathrm{NR}(n) > .19e^{.04n^2}$.

39.13 Remark There are three inequivalent skew Room squares of order 7; they are R_1, R_5, and R_6 in Table 39.8. The unique skew room square of side 9 is given in Example 39.1.

39.5 Incomplete Room Squares

An (n, s)-*incomplete Room square* F is an $n \times n$ array for which
1. every cell of F is empty or contains an unordered pair of symbols from an $(n+1)$-set N;
2. there is an empty $s \times s$ subarray G of F;
3. every symbol of $N \setminus S$ occurs once in each row and column of F, where $S \subset N$ is an $(s+1)$-set;
4. each symbol of S occurs once in each row and each column not meeting G, but in no row or column meeting G;
5. the pairs in F are precisely those $\{x, y\} \in \binom{N}{2} \setminus \binom{S}{2}$.

39.14 Remark If an (n, s)-incomplete Room square exists and a Room square of side s exists, then a Room square of side n exists.

39.15 Example An (11,3)-incomplete Room square. (Note $S = \{0, 1, 2, Y\}$.)

			48			37	6X			59
			69			5X	38			47
				39	4X			57	68	
67	8X		3Y			04	15	29		
58	79			4Y	03				2X	16
9X		78		06	5Y		24		13	
			05	7X	89	6Y		14		23
	46	3X		25		19	7Y			08
	35	49	1X		26			8Y	07	
34		56			17	28		0X	9Y	
			27	18			09	36	45	XY

39.16 Theorem Let s and t be odd positive integers. The necessary condition for the existence of an (s, t) incomplete Room square is that $s \geq 3t + 2$.

39.17 Theorem For every pair of odd positive integers (s, t) with $s \geq 3t + 2$ there exists an (s, t) incomplete Room square, with the possible exception of $(s, t) = (67, 21)$ [6, 4].

39.6 Higher Dimensions

A Room d-cube of side n is a d-dimensional cube of side n with the property that every 2-dimensional projection is a Room square of side n.

$\nu(n)$ denotes the maximum d for which there exists a Room d-cube of side n.

39.18 Remark The existence of a Room d-cube of side n is equivalent to the existence of d pairwise orthogonal 1-factorizations of K_{n+1}. The construction goes as follows: Edge $\{x, y\}$ is in 1-factor f_{a_i} in 1-factorization F_i for $1 \leq i \leq d$, if and only if cell $(a_1, a_2, \ldots, a_d)$ contains the symbols x and y.

39.19 Table The unique set of three pairwise orthogonal 1-factorizations of K_8 (Room 3-cube of side 7).

F_1				F_2				F_3			
∞0	16	25	34	∞0	13	26	45	∞0	15	23	46
∞1	20	36	45	∞1	24	30	56	∞1	26	34	50
∞2	31	40	56	∞2	35	41	60	∞2	30	45	61
∞3	42	51	60	∞3	46	52	01	∞3	41	56	02
∞4	53	62	01	∞4	50	63	12	∞4	52	60	13
∞5	64	03	12	∞5	61	04	23	∞5	63	01	24
∞6	05	14	23	∞6	02	15	34	∞6	04	12	35

39.20 Table [5] The unique set of four pairwise orthogonal 1-factorizations of K_{10} (Room 4-cube of side 9).

F_1					F_2					F_3					F_4				
01	23	45	67	89	01	29	36	48	57	01	26	39	47	58	01	25	34	68	79
02	13	46	58	79	02	15	34	69	78	02	14	37	56	89	02	18	35	49	67
03	12	47	59	68	03	16	28	45	79	03	17	25	48	69	03	15	27	46	89
04	16	25	39	78	04	17	26	35	89	04	18	27	36	59	04	13	28	57	69
05	18	24	37	69	05	14	27	39	68	05	19	28	34	67	05	16	29	38	47
06	19	27	35	48	06	12	37	39	58	06	15	24	38	79	06	14	23	59	78
07	15	28	36	49	07	19	25	38	46	07	13	29	45	68	07	12	39	48	56
08	17	29	34	56	08	13	24	59	67	08	16	23	49	67	08	19	26	37	45
09	14	26	38	57	09	18	23	47	56	09	12	35	46	78	09	17	24	36	58

39.21 Theorem $\nu(2n-1) \to \infty$ for $n \to \infty$.

39.22 Theorem For every odd $n \geq 3$, $\nu(n) \leq n-2$, and if $n \geq 17$, then $\nu(n) \geq 5$.

39.23 Theorem If $q = 2^k t + 1$ is a prime power with t odd, then $\nu(q) \geq t$.

39.24 Table Lower bounds for $\nu(n)$.

n	$\nu(n) \geq$	n	$\nu(n) \geq$	n	$\nu(n) \geq$	n	$\nu(n) \geq$
1	$=\infty$	27	13	53	17	79	39
3	$=1$	29	13	55	5	81	5
5	$=1$	31	15	57	5	83	41
7	$=3$	33	5	59	29	85	5
9	$=4$	35	5	61	21	87	5
11	5	37	15	63	5	89	11
13	5	39	5	65	5	91	5
15	4	41	9	67	33	93	5
17	5	43	21	69	5	95	5
19	9	45	5	71	35	97	5
21	5	47	23	73	9	99	5
23	11	49	5	75	5	101	31
25	7	51	5	77	5	103	51

39.7 Room Frames

If $\{S_1, \ldots, S_n\}$ is a partition of a set S, an $\{S_1, \ldots, S_n\}$-*Room frame* is an $|S| \times |S|$ array, F, indexed by S, satisfying:

1. every cell of F either is empty or contains an unordered pair of symbols of S,
2. the subarrays $S_i \times S_i$ are empty, for $1 \leq i \leq n$ (these subarrays are *holes*),
3. each symbol $x \notin S_i$ occurs once in row (or column) s for any $s \in S_i$, and
4. the pairs occurring in F are those $\{s, t\}$, where $(s, t) \in (S \times S) \setminus \bigcup_{i=1}^{n} (S_i \times S_i)$.

If F is a Room frame as above, then

1. The *type* of a Room frame F is the multiset $\{|S_i| : 1 \leq i \leq n\}$. An "exponential" notation is used to describe types; a Room frame has type $t_1{}^{u_1} t_2{}^{u_2} \ldots t_k{}^{u_k}$ if there are u_i S_j's of cardinality t_i, $1 \leq i \leq k$.
2. The *order* of the frame is $|S|$.
3. A Room frame of type t^u (one hole size) is *uniform.*

39.25 Example A Room frame of type $1^1 3^4$.

	$9c$		$5b$	$2d$				36		$7a$	48	
				$8b$	ad			$5c$			79	16
				$1c$	$9b$		$5d$				$6a$	78
59						ac	$1b$		$7d$	68		
$2a$	18	$9d$						$4b$	$3c$			
		ab	$8d$				$4c$				13	29
$8c$								$1d$	$2b$	49		$3a$
	$6d$		$7c$			$3b$				12		45
$6b$		17			$2c$	$4d$				35		
$3d$	$7b$	$6c$			14						25	
		58	$1a$	39				27	46			
			69	$4a$		28	37		15			
47	$5a$				38	19	26					

39.26 Theorem [2, 7, 4] Suppose t and u are positive integers, $u \geq 4$, and $(t, u) \neq (1, 5), (2, 4)$. Then there exists a uniform Room frame of type t^u if and only if $t(u-1)$ is even, except possibly when $u = 4$ and $t = 14$.

39.8 See Also

§IV.26	Howell designs are generalizations of Room squares.
§IV.45	Starters are used to construct Room squares.
§VI.4	A connection with 1-factorizations is given by Theorem 39.3.
§V.7	Room squares give a schedule of play for a round robin tournament.
[3]	A broad survey of Room squares with an extensive bibliography. This survey contains most of the information in this section.
[8]	A textbook dealing with Room squares.
[1]	A survey of orthogonal factorizations.
[4]	The proofs of Theorems 39.17 and 39.26.

References

[1] B. Alspach, K. Heinrich and G. Liu, Orthogonal factorizations of graphs, in: *Contemporary Design Theory: A Collection of Surveys* (J.H. Dinitz, D.R. Stinson; eds.) John Wiley & Sons, New York, 1992, 13-40.

[2] J.H. Dinitz and E.R. Lamken, Uniform Room frames with five holes, *J. Combin. Des.* 1 (1993), 323-328.

[3] J.H. Dinitz and D.R. Stinson, Room squares and related designs, in: *Contemporary Design Theory: A Collection of Surveys* (J.H. Dinitz, D.R. Stinson; eds.) John Wiley & Sons, New York, 1992, 137-204.

[4] J.H. Dinitz, D.R. Stinson and L. Zhu, On the spectra of certain classes of Room frames, *Elect. J. Combin.* 1 (1994), #R7.

[5] J.H. Dinitz and W.D. Wallis, Four orthogonal one-factorizations on ten points, *Ann. Discrete Math.* 26 (1985), 143-150.

[6] B. Du and L. Zhu, The existence of incomplete Room squares, *J. Combin. Math. Combin. Comput.* 14 (1993), 183-192.

[7] G. Ge and L. Zhu, On the existence of Room frames of type t^u, *J. Combin. Des.* 1 (1993), 183-191.

[8] W.D. Wallis, A.P. Street and J.S. Wallis, *Combinatorics: Room Squares, Sum-Free Sets, Hadamard Matrices*, Springer-Verlag, Berlin, 1972.

40 Self-Orthogonal Latin Squares (SOLS)

L. Zhu

40.1 Definitions and Examples

A *self-orthogonal* latin square (SOLS) is a latin square that is orthogonal to its transpose.

40.1 Examples Self-orthogonal latin squares of orders 4, 5, 7, 8, 9.

1	3	4	2
4	2	1	3
2	4	3	1
3	1	2	4

1	5	4	3	2
3	2	1	5	4
5	4	3	2	1
2	1	5	4	3
4	3	2	1	5

1	7	6	5	4	3	2
3	2	1	7	6	5	4
5	4	3	2	1	7	6
7	6	5	4	3	2	1
2	1	7	6	5	4	3
4	3	2	1	7	6	5
6	5	4	3	2	1	7

1	4	7	6	3	5	8	2
7	2	1	3	6	8	5	4
6	8	3	1	7	2	4	5
8	6	5	4	2	7	1	3
4	1	2	8	5	3	6	7
2	7	4	5	8	6	3	1
5	3	8	2	4	1	7	6
3	5	6	7	1	4	2	8

1	5	9	6	2	4	8	7	3
8	2	6	9	7	3	5	1	4
2	1	3	7	9	8	4	6	5
7	3	2	4	8	9	1	5	6
6	8	4	3	5	1	9	2	7
3	7	1	5	4	6	2	9	8
9	4	8	2	6	5	7	3	1
4	9	5	1	3	7	6	8	2
5	6	7	8	1	2	3	4	9

40.2 Remark A self-orthogonal latin square has at least one transversal, its main diagonal. So, any self-orthogonal latin square can be changed to an idempotent one by renaming the symbols.

40.3 Theorem [4] Self-orthogonal latin squares exist for all orders $n \neq 2, 3, 6$.

40.2 Equivalent Objects

40.4 Theorem A self-orthogonal latin square of order n (SOLS(n)) is equivalent to

1. An orthogonal array of strength two and index one, OA(4,n), which is invariant under the row permutation (12)(34) (see §II.2).
2. A $(2,1,3)$-conjugate orthogonal latin square (or quasigroup) of order n, i.e., a latin square which is orthogonal to its $(2,1,3)$-conjugate (see §IV.36).
3. A spouse avoiding mixed doubles round robin tournament (SAMDRR) with n couples (see §V.7.11 and Remark 40.5).

40.5 Remark [1] A spouse avoiding mixed doubles round robin tournament is a schedule of matches for n couples in tennis, such that (1) husband and wife never appear in the same match either as partners or opponents; (2) each pair of players of the same sex oppose each other exactly once; (3) each pair of players of the opposite sex, not husband and wife, play in exactly one match as partners and in exactly one match as opponents. An idempotent SOLS (a_{ij}) of order n can be used to schedule the $n(n-1)/2$ matches in the spouse avoiding mixed doubles round robin tournament as follows: In the unique match in which Mr. i opposes Mr. j ($i \neq j$), the partner of Mr. i is Mrs. a_{ij} (and the partner of Mr. j is Mrs. a_{ji}).

40.3 Enumeration and Basic Facts

Two self-orthogonal latin squares L and L' of order n are *isotopic* if there are bijections θ, θ, α from the rows, columns, and symbols of L to the rows, columns, and symbols, respectively, of L' that map L or its transpose to L'.

40.6 Remark Any self-orthogonal latin square of order 4 or 5 is isotopic to the one presented in Example 40.1.

40.7 Theorem If $\mu \in \mathbb{F}_q \setminus \{0,1\}$, then $L_\mu = (\mu i + (1-\mu) j)$ is a self-orthogonal latin square of order q (see Theorem IV.41.8(1) and Constructions IV.41.9 and IV.41.10).

40.8 Example A SOLS(11) from Theorem 40.7, taking $q = 11$ and $\mu = 2$ (with a denoting 10), is shown on the right.

0	a	9	8	7	6	5	4	3	2	1
2	1	0	a	9	8	7	6	5	4	3
4	3	2	1	0	a	9	8	7	6	5
6	5	4	3	2	1	0	a	9	8	7
8	7	6	5	4	3	2	1	0	a	9
a	9	8	7	6	5	4	3	2	1	0
1	0	a	9	8	7	6	5	4	3	2
3	2	1	0	a	9	8	7	6	5	4
5	4	3	2	1	0	a	9	8	7	6
7	6	5	4	3	2	1	0	a	9	8
9	8	7	6	5	4	3	2	1	0	a

40.9 Theorem If a SOLS(m) and a SOLS(n) exist, then a SOLS(mn) also exists.

40.10 Theorem Suppose that a $(v, K, 1)$-PBD exists and that, for each $k \in K$, a SOLS(k) exists. Then a SOLS(v) exists.

40.11 Remark Since a $(v, \{4,5,7,8,9,11\}, 1)$-PBD exists for all $v \geq 4$, where $v \neq 6$, 10, 12, 14, 15, 18, 19, 23, 26, 27, 30, (see Table III.3.17) the proof of Theorem 40.3 can be reduced to orders $v = 10, 12, 14, 15, 18, 26, 30$. SOLS($v$) for $v = 12, 15, 18, 26$, and 30 are presented in §IV.41.3. SOLS(10) and SOLS(14) are given in Example 40.13.

40.12 Theorem Let G be an additive abelian group of order m and let $X = \{x_1, x_2, \ldots, x_n\}$ be an n-set disjoint from G. Consider the elements of X as infinite elements, i.e., $x + g = x$ for any $x \in X$ and $g \in G$. Suppose vector $\mathbf{e} = (e_1, e_2, \ldots, e_m)$ based on $G \bigcup X$ and vectors $\mathbf{f} = (f_1, f_2, \ldots, f_n)$ and $\mathbf{g} = (g_1, g_2, \ldots, g_n)$ based on G, satisfy

1. $\{e_1, e_2, \ldots, e_m\} \bigcup \{f_1, f_2, \ldots, f_n\} = G \bigcup X$,
2. $\{e_1 - 1, e_2 - 2, \ldots, e_m - m\} \bigcup \{g_1, g_2, \ldots, g_n\} = G \bigcup X$, and
3. $\{e_{m-i} - e_i + i : 1 \leq i \leq m, e_i, e_{m-i} \in G\} \bigcup \{\pm(f_i - g_i) : 1 \leq i \leq n\} = G$.

Then there is a SOLS($m + n$) missing a sub-SOLS(n), denoted by ISOLS($m + n, n$). Further, if a SOLS(n) exists, then a SOLS($m + n$) also exists.

If $G = \mathbb{Z}_m$ and $X = \emptyset$ in Theorem 40.12, then the latin square is *cyclic* or *cyclically generated.* If $G = \mathbb{Z}_m$ and $|X| = 1$, the latin square is *pseudo-cyclic.*

40.13 Examples SOLS(10) and SOLS(14) generated pseudo-cyclically. In the SOLS(14), 10, 11, and 12 are replaced by a, b, and c, respectively.

0	2	8	6	x	7	1	5	4	3
5	1	3	0	7	x	8	2	6	4
7	6	2	4	1	8	x	0	3	5
4	8	7	3	5	2	0	x	1	6
2	5	0	8	4	6	3	1	x	7
x	3	6	1	0	5	7	4	2	8
3	x	4	7	2	1	6	8	5	0
6	4	x	5	8	3	2	7	0	1
1	7	5	x	6	0	4	3	8	2
8	0	1	2	3	4	5	6	7	x

0	8	3	c	9	2	5	a	6	b	1	4	x	7
x	1	9	4	0	a	3	6	b	7	c	2	5	8
6	x	2	a	5	1	b	4	7	c	8	0	3	9
4	7	x	3	b	6	2	c	5	8	0	9	1	a
2	5	8	x	4	c	7	3	0	6	9	1	a	b
b	3	6	9	x	5	0	8	4	1	7	a	2	c
3	c	4	7	a	x	6	1	9	5	2	8	b	0
c	4	0	5	8	b	x	7	2	a	6	3	9	1
a	0	5	1	6	9	c	x	8	3	b	7	4	2
5	b	1	6	2	7	a	0	x	9	4	c	8	3
9	6	c	2	7	3	8	b	1	x	a	5	0	4
1	a	7	0	3	8	4	9	c	2	x	b	6	5
7	2	b	8	1	4	9	5	a	0	3	x	c	6
8	9	a	b	c	0	1	2	3	4	5	6	7	x

40.14 Remark For odd prime power q, there are $(q-3)/2$ SOLS(q) which together with their transposes form $q - 3$ MOLS. For even prime power q, there are $(q - 2)/2$ SOLS(q) which together with their transposes form $q - 2$ MOLS [6].

40.4 Frame SOLS

A *frame* SOLS (or *holey* SOLS) is a self-orthogonal latin square of order n with n_i missing sub-SOLS (*holes*) of order h_i ($1 \leq i \leq k$), which are disjoint and spanning (i.e., $\sum_{1 \leq i \leq k} n_i h_i = n$). It is denoted by FSOLS($h_1^{n_1} \ldots h_k^{n_k}$), where $h_1^{n_1} \ldots h_k^{n_k}$ is the *type* of the FSOLS.

40.15 Example FSOLS(2^n) for $n = 4$ and 5.

		7	6	8	4	3	5
		5	7	3	8	4	6
6	8			7	1	5	2
8	5			2	7	6	1
3	7	1	8			2	4
7	4	8	2			1	3
5	6	2	1	4	3		
4	3	6	5	1	2		

		5	9	8	6	4	3	2	7
		8	4	7	9	2	5	6	3
9	5			6	1	0	8	7	4
4	8			0	7	9	1	5	6
7	9	1	6			8	2	3	0
8	6	7	0			3	9	1	2
3	4	9	1	2	8			0	5
5	2	0	8	9	3			4	1
6	3	4	7	1	2	5	0		
2	7	6	5	3	0	1	4		

40.16 Example FSOLS($2^n 1^1$) for $n = 4$ and 5.

		6	8	9	7	3	5	4
		7	5	8	9	6	4	3
8	6			7	1	9	2	5
5	7			2	8	1	9	6
4	9	1	7			2	3	8
9	3	8	2			4	1	7
6	4	5	9	3	2			1
3	5	9	6	1	4			2
7	8	2	1	4	3	5	6	

		5	6	b	9	a	3	7	8	4
		6	5	a	7	3	4	8	b	9
b	a			8	2	5	9	6	1	7
8	9			2	b	1	6	5	7	a
9	7	b	8			4	a	2	3	1
3	b	7	a			9	1	4	2	8
4	6	a	b	9	1			3	5	2
a	5	2	9	1	4			b	6	3
6	3	1	7	4	8	b	2			5
5	4	8	1	7	3	2	b			6
7	8	9	2	3	a	6	5	1	4	

40.17 Example FSOLS($1^6 3^1 2^1$) [9] and a FSOLS($2^4 3^1$). The FSOLS($2^4 3^1$) is based on $\mathbb{Z}_8 \bigcup \{a, b, c\}$ constructed cyclically modulo 8. The holes are based on $\{0, 4\}$, $\{1, 5\}$, $\{2, 6\}$, $\{3, 7\}$, and $\{a, b, c\}$.

	7	5	2	a	9	4	b	3	6	8
6		8	7	3	a	1	5	b	4	9
9	4		a	8	1	b	2	6	5	7
5	b	9		7	2	6	a	1	8	3
7	6	b	3		8	2	4	a	9	1
b	8	4	9	1		a	3	5	7	2
3	a	6	1	b	5				2	4
4	1	a	6	2	b				3	5
a	5	2	b	4	3				1	6
8	9	7	5	6	4	3	1	2		
2	3	1	8	9	7	5	6	4		

	7	3	c		2	b	a	1	6	5
a		0	4	c		3	b	2	7	6
b	a		1	5	c		4	3	0	7
5	b	a		2	6	c		4	1	0
	6	b	a		3	7	c	5	2	1
c		7	b	a		4	0	6	3	2
1	c		0	b	a		5	7	4	3
6	2	c		1	b	a		0	5	4
2	3	4	5	6	7	0	1			
3	4	5	6	7	0	1	2			
7	0	1	2	3	4	5	6			

40.18 Example FSOLS($2^5 u^1$) for $u = 3$ or 4. The square is based on $\mathbb{Z}_{10} \bigcup X$, where $X = \{a, b, c\}$ or $\{a, b, c, d\}$, constructed cyclically modulo 10. The holes are based on $\{0, 5\}$, $\{1, 6\}$, $\{2, 7\}$, $\{3, 8\}$, $\{4, 9\}$, and X.

	3	8	7	1		9	c	b	a	2	4	6
a		4	9	8	2		0	c	b	3	5	7
b	a		5	0	9	3		1	c	4	6	8
c	b	a		6	1	0	4		2	5	7	9
3	c	b	a		7	2	1	5		6	8	0
	4	c	b	a		8	3	2	6	7	9	1
7		5	c	b	a		9	4	3	8	0	2
4	8		6	c	b	a		0	5	9	1	3
6	5	9		7	c	b	a		1	0	2	4
2	7	6	0		8	c	b	a		1	3	5
1	2	3	4	5	6	7	8	9	0			
8	9	0	1	2	3	4	5	6	7			
9	0	1	2	3	4	5	6	7	8			

	7	a	b	3		c	8	6	d	2	9	1	4
d		8	a	b	4		c	9	7	3	0	2	5
8	d		9	a	b	5		c	0	4	1	3	6
1	9	d		0	a	b	6		c	5	2	4	7
c	2	0	d		1	a	b	7		6	3	5	8
	c	3	1	d		2	a	b	8	7	4	6	9
9		c	4	2	d		3	a	b	8	5	7	0
b	0		c	5	3	d		4	a	9	6	8	1
a	b	1		c	6	4	d		5	0	7	9	2
6	a	b	2		c	7	5	d		1	8	0	3
4	5	6	7	8	9	0	1	2	3				
2	3	4	5	6	7	8	9	0	1				
7	8	9	0	1	2	3	4	5	6				
3	4	5	6	7	8	9	0	1	2				

40.19 Theorems

1. If there exists a FSOLS($a^n b^1$), then $n \geq 1 + 2b/a$.
2. A FSOLS(2^n) exists if and only if $n \geq 4$ [7].
3. A FSOLS($2^n 1^1$) exists if and only if $n \geq 4$ [8].
4. For any $u \geq 3$, a FSOLS($2^n u^1$) exists if and only if $n \geq u + 1$ [8].

40.5 Incomplete SOLS

An *incomplete* SOLS is a self-orthogonal latin square of order n missing a sub-SOLS of order k, denoted by ISOLS(n, k).

40.20 Example An ISOLS(7,2) is shown on the right.

0	2	4	x	y	1	3
y	1	3	0	x	2	4
x	y	2	4	1	3	0
2	x	y	3	0	4	1
1	3	x	y	4	0	2
3	4	0	1	2		
4	0	1	2	3		

40.21 Remark An ISOLS(n, k) is equivalent to a FSOLS$(1^{n-k}k^1)$. An ISOLS$(n,\ 1)$ or an ISOLS$(n,\ 0)$ is equivalent to a SOLS(n).

40.22 Remark If there exists an ISOLS(n, k), then $n \geq 3k + 1$.

40.23 Theorem There exists an ISOLS(n, k) for all values of n and k satisfying $n \geq 3k + 1$, except for $n = 6$ and $(n, k) = (8, 2)$ and possibly excepting $n = 3k + 2$ and $k \in \{4, 6, 8, 10, 14, 16, 18, 20, 22, 26, 28, 32, 34, 46\}$ [9].

40.6 SOLS with Additional Properties

If a SOLS is superimposed with its transpose such that a pair (i, j) is in position (x, y) whenever the pair (x, y) is in position (i, j), then the SOLS has the *Weisner property*.

40.24 Example A SOLS(4) with the Weisner property. The second square is a superimposed SOLS(4) in which the Weisner property is easily seen.

1	3	4	2
4	2	1	3
2	4	3	1
3	1	2	4

11	34	42	23
43	22	14	31
24	41	33	12
32	13	21	44

40.25 Theorem The existence of a SOLS(n) with the Weisner property is equivalent to the existence of a Schröder quasigroup of order n (a quasigroup satisfying the identity $xy \cdot yx = x$) (see Proposition IV.36.31).

40.26 Theorem [2, 5] A SOLS(n) with the Weisner property exists for precisely all positive integers $n \equiv 0$ or 1 (mod 4) except for $n = 5$.

40.27 Theorem [2, 5] An idempotent SOLS(n) with the Weisner property exists for precisely all positive integers $n \equiv 0$ or 1 (mod 4) except for $n = 5$ and 9.

40.28 Theorem [2] An idempotent ISOLS$(n, 2)$ with the Weisner property exists for precisely all positive integers $n \equiv 2$ or 3 (mod 4), $n \geq 7$, except for $n = 10$.

A SOLS is *semisymmetric* if the quasigroup satisfies the identity $x(yx) = y$.

40.29 Example SOLS(n) with the semisymmetric property for $n = 4$ and 7.

1	3	4	2
4	2	1	3
2	4	3	1
3	1	2	4

0	3	6	2	5	1	4
5	1	4	0	3	6	2
3	6	2	5	1	4	0
1	4	0	3	6	2	5
6	2	5	1	4	0	3
4	0	3	6	2	5	1
2	5	1	4	0	3	6

40.30 Remark The existence of a SOLS(n) with the semisymmetric property is equivalent to the existence of a self-orthogonal Mendelsohn triple system of order n.

40.31 Theorem [3] A SOLS(n) with the semisymmetric property exists for precisely all positive integers $n \equiv 0$ or 1 (mod 3) except for $n = 3, 6, 9, 10, 12$, and except possibly for $n = 18$.

40.7 See Also

§II.1	SOLS are special types of latin squares.
§II.2	SOLS are examples of MOLS.
§IV.41	SOLSSOMs are SOLS with a symmetric orthogonal mate.
[1]	Basic facts on SOLS.

References

[1] I. Anderson, *Combinatorial Designs: Construction Methods*, Ellis Horwood Limited, England, 1990.

[2] F.E. Bennett, R. Wei and L. Zhu, Incomplete idempotent Schröder quasigroups and related packing designs, *Aequat. Math.*, to appear.

[3] F.E. Bennett, H. Zhang and L. Zhu, Self-orthogonal Mendelsohn triple systems, *J. Combin. Theory A*, to appear.

[4] R.K. Brayton, D. Coppersmith and A.J. Hoffman, Self-orthogonal latin squares, Coll. Int. Th. Comb., Rome, 1973, *Atti del Convegni Lincei* No. 17, 1976, 509-517.

[5] C.J. Colbourn and D.R. Stinson, Edge-coloured designs with block size four, *Aequat. Math.* 36 (1988), 230-245.

[6] K. Heinrich and L. Zhu, Incomplete self-orthogonal latin squares, *J. Austral. Math. Soc. Ser. A* 42 (1987), 365-384.

[7] D.R. Stinson and L. Zhu, On the existence of MOLS with equal-sized holes, *Aequat. Math.* 33 (1987), 96-105.

[8] Y. Xu and L. Zhu, Existence of frame SOLS of type $2^n u^1$, *J. Combin. Des.* 3 (1995), 115-133.

[9] L. Zhu, Existence of self-orthogonal latin squares ISOLS($6m+2$, 2m), *Ars Combin.* 39 (1995), 65-74.

41 SOLS with a Symmetric Orthogonal Mate (SOLSSOM)

Norman J. Finizio

41.1 Definitions and Examples

A *SOLSSOM*(n) is a pair (S, M) such that S is a $SOLS(n)$ (self orthogonal latin square) and M is a symmetric latin square of order n that is orthogonal to both S and the transpose of S.

A SOLSSOM(n) (S, M) is *regular* if M is (1) *idempotent* if n is odd and (2) *unipotent* with $m_{ii} = n$ if n is even.

A latin square $A = (a_{ij}), a_{ij} \in \mathbb{Z}_n$, of order n is *cyclic* if $a_{i+1,j+1} = a_{ij} + 1$ for all $i, j \in \mathbb{Z}_n$ (all arithmetic is modulo n). A latin square $A = (a_{ij}), a_{ij} \in \mathbb{Z}_{n-1} \cup \{\infty\}$, of order n is *pseudo-cyclic* if the rows and columns are each indexed by $\mathbb{Z}_{n-1} \cup \{\infty\}$ and

1. $a_{i+1,j+1} = a_{ij} + 1, 0 \le i+1, j+1 \le n-1$,
2. $a_{n,j+1} = a_{nj} + 1, 0 \le j+1 \le n-1$,
3. $a_{i+1,n} = a_{in} + 1, 0 \le i+1 \le n-1$, and
4. $a_{nn} = \infty$.

All arithmetic is modulo $n-1$ except $\infty + 1 = \infty$.

A *transversal* SOLSSOM(n) is such that both S and M are cyclic if n is odd and pseudo-cyclic if n is even.

41.1 Example A regular, transversal SOLSSOM(4).

$$S = \begin{array}{|cccc|}\hline 0 & \infty & 1 & 2 \\ 2 & 1 & \infty & 0 \\ \infty & 0 & 2 & 1 \\ 1 & 2 & 0 & \infty \\ \hline\end{array} \qquad M = \begin{array}{|cccc|}\hline \infty & 2 & 1 & 0 \\ 2 & \infty & 0 & 1 \\ 1 & 0 & \infty & 2 \\ 0 & 1 & 2 & \infty \\ \hline\end{array}$$

41.2 Example A regular, transversal SOLSSOM(5).

$$S = \begin{array}{|ccccc|}\hline 0 & 4 & 3 & 2 & 1 \\ 2 & 1 & 0 & 4 & 3 \\ 4 & 3 & 2 & 1 & 0 \\ 1 & 0 & 4 & 3 & 2 \\ 3 & 2 & 1 & 0 & 4 \\ \hline\end{array} \qquad M = \begin{array}{|ccccc|}\hline 0 & 3 & 1 & 4 & 2 \\ 3 & 1 & 4 & 2 & 0 \\ 1 & 4 & 2 & 0 & 3 \\ 4 & 2 & 0 & 3 & 1 \\ 2 & 0 & 3 & 1 & 4 \\ \hline\end{array}$$

41.3 Example A regular SOLSSOM(9).

$$S = \begin{array}{|ccccccccc|}\hline 1 & 9 & 5 & 8 & 4 & 3 & 6 & 2 & 7 \\ 6 & 2 & 7 & 1 & 9 & 5 & 8 & 4 & 3 \\ 8 & 4 & 3 & 6 & 2 & 7 & 1 & 9 & 5 \\ 9 & 5 & 1 & 4 & 3 & 8 & 2 & 7 & 6 \\ 2 & 7 & 6 & 9 & 5 & 1 & 4 & 3 & 8 \\ 4 & 3 & 8 & 2 & 7 & 6 & 9 & 5 & 1 \\ 5 & 1 & 9 & 3 & 8 & 4 & 7 & 6 & 2 \\ 7 & 6 & 2 & 5 & 1 & 9 & 3 & 8 & 4 \\ 3 & 8 & 4 & 7 & 6 & 2 & 5 & 1 & 9 \\ \hline\end{array} \qquad M = \begin{array}{|ccccccccc|}\hline 1 & 3 & 2 & 7 & 9 & 8 & 4 & 6 & 5 \\ 3 & 2 & 1 & 9 & 8 & 7 & 6 & 5 & 4 \\ 2 & 1 & 3 & 8 & 7 & 9 & 5 & 4 & 6 \\ 7 & 9 & 8 & 4 & 6 & 5 & 1 & 3 & 2 \\ 9 & 8 & 7 & 6 & 5 & 4 & 3 & 2 & 1 \\ 8 & 7 & 9 & 5 & 4 & 6 & 2 & 1 & 3 \\ 4 & 6 & 5 & 1 & 3 & 2 & 7 & 9 & 8 \\ 6 & 5 & 4 & 3 & 2 & 1 & 9 & 8 & 7 \\ 5 & 4 & 6 & 2 & 1 & 3 & 8 & 7 & 9 \\ \hline\end{array}$$

41.4 Example SOLSSOM(12).

$$S = \begin{array}{|cccccccccccc|}\hline 1 & 3 & 8 & 10 & 7 & 12 & 11 & 4 & 6 & 9 & 2 & 5 \\ 4 & 2 & 9 & 7 & 11 & 8 & 3 & 12 & 10 & 5 & 6 & 1 \\ 6 & 12 & 3 & 1 & 9 & 2 & 5 & 10 & 4 & 7 & 8 & 11 \\ 11 & 5 & 2 & 4 & 1 & 10 & 9 & 6 & 8 & 3 & 12 & 7 \\ 10 & 1 & 6 & 9 & 5 & 7 & 12 & 2 & 11 & 4 & 3 & 8 \\ 2 & 9 & 10 & 5 & 8 & 6 & 1 & 11 & 3 & 12 & 7 & 4 \\ 8 & 11 & 12 & 3 & 10 & 4 & 7 & 5 & 1 & 6 & 9 & 2 \\ 12 & 7 & 4 & 11 & 3 & 9 & 6 & 8 & 5 & 2 & 1 & 10 \\ 3 & 8 & 7 & 12 & 2 & 5 & 10 & 1 & 9 & 11 & 4 & 6 \\ 7 & 4 & 11 & 8 & 6 & 1 & 2 & 9 & 12 & 10 & 5 & 3 \\ 5 & 10 & 1 & 6 & 12 & 3 & 4 & 7 & 2 & 8 & 11 & 9 \\ 9 & 6 & 5 & 2 & 4 & 11 & 8 & 3 & 7 & 1 & 10 & 12 \\ \hline\end{array} \qquad M = \begin{array}{|cccccccccccc|}\hline 12 & 11 & 8 & 7 & 6 & 2 & 3 & 5 & 10 & 1 & 4 & 9 \\ 11 & 12 & 7 & 8 & 1 & 6 & 5 & 4 & 2 & 10 & 9 & 3 \\ 8 & 7 & 12 & 11 & 4 & 5 & 6 & 1 & 3 & 9 & 10 & 2 \\ 7 & 8 & 11 & 12 & 5 & 3 & 2 & 6 & 9 & 4 & 1 & 10 \\ 6 & 1 & 4 & 5 & 12 & 11 & 10 & 9 & 8 & 2 & 3 & 7 \\ 2 & 6 & 5 & 3 & 11 & 12 & 9 & 10 & 1 & 8 & 7 & 4 \\ 3 & 5 & 6 & 2 & 10 & 9 & 12 & 11 & 4 & 7 & 8 & 1 \\ 5 & 4 & 1 & 6 & 9 & 10 & 11 & 12 & 7 & 3 & 2 & 8 \\ 10 & 2 & 3 & 9 & 8 & 1 & 4 & 7 & 12 & 11 & 6 & 5 \\ 1 & 10 & 9 & 4 & 2 & 8 & 7 & 3 & 11 & 12 & 5 & 6 \\ 4 & 9 & 10 & 1 & 3 & 7 & 8 & 2 & 6 & 5 & 12 & 11 \\ 9 & 3 & 2 & 10 & 7 & 4 & 1 & 8 & 5 & 6 & 11 & 12 \\ \hline\end{array}$$

41.2 Existence

41.5 Theorem [3] Regular SOLSSOM(n) do not exist for $n \in \{2,3,6\}$ but do exist for all positive integers n with the possible exception of $n \in \{10,14,66,70\}$.

41.6 Remarks Whenever a SOLSSOM(n) is known, a regular SOLSSOM(n) is also known. Hence Theorem 41.5 gives the current existence result for SOLSSOM(n) as well. Authorities for the existence of SOLSSOM(n) for $n \le 100$ are [1, 2, 4, 7, 8, 9].

41.7 Theorem If there exists a SOLSSOM(m) and a SOLSSOM(n) with n odd, then there exists a SOLSSOM(mn).

41.8 Theorem Let p and p_i denote primes of the form $4t+1$. There exist transversal SOLSSOM(n) for the following cases:

1. $n = 6k \pm 1, k \ge 1$. Indeed $S = (s_{ij}) = (2i-j) \pmod n$ and $M = (m_{ij}) = (\frac{i}{2}+\frac{j}{2}) \pmod n$;
2. $n = \prod_{i=1}^{k} p_i^{m_i},\ m_i \ge 1,\ k \ge 1$;
3. $n = 3p^m + 1,\ m \ge 1$;
4. $n = 3\prod_{i=1}^{k} p_i^{m_i} + 1,\ p_i = 2^d u_i + 1,\ u_i$ odd, $d \ge 2,\ m_i \ge 1$;
5. $n = 3qp^m,\ q \in \{7,11\},\ \gcd(p^{m-1}(p-1), q-1) = 2$;
6. $n = q\prod_{i=1}^{k} p_i^{m_i} + 1,\ q \in \{7,19,23,31\}, p_i \ge 29,\ m_i \ge 1$;
7. $n \in \{4,8,16,20,21,24,28,32,33,36\}$.

41.9 Construction There exist idempotent SOLSSOM(q) for any odd prime power q. Let $S = (s_{ij}) = (ui + (1-u)j)$ and $M = (m_{ij}) = (\frac{i}{2} + \frac{j}{2})$ where $i,j \in \mathbb{F}_q$ and $u \in \mathbb{F}_q \setminus \{0,1,\frac{1}{2}\}$.

41.10 Construction There exist SOLSSOM(q) for any even prime power $q \ge 4$. Let $S = (s_{ij}) = (ui + (1-u)j)$ and $M = (m_{ij}) = (i+j)$ where $i,j \in \mathbb{F}_q$ and $u \in \mathbb{F}_q \setminus \{0,1\}$.

41.3 SOLSSOMs of Orders up to 30

41.11 Remark SOLSSOM(n) for $n = 4,5,9$ and 12 are given in §41.1. A direct construction for SOLSSOM(n) when $n = 6k \pm 1, k \ge 1$ is given in Theorem 41.8(1) above. A direct construction for a SOLSSOM(7) and a SOLSSOM(27) is given in Construction 41.9, and a direct construction for a SOLSSOM(8) and a SOLSSOM(16) is given in Construction 41.10. Direct constructions for the remaining orders of $n \le 30$ for which SOLSSOMs are known to exist are given here.

41.12 Remark If (S, M) is a transversal SOLSSOM(n), then if n is odd both S and M can be determined by merely presenting their respective first rows. If n is even, then the entry in the cell s_{n1} may also be given. Let $S_1 = (s_{11}s_{12}s_{13}\cdots s_{1n})$ and $M_1 = (m_{11}m_{12}m_{13}\cdots m_{1n})$ be the first rows of S and M, respectively.

41.13 Table Transversal SOLSSOM(n) for $n = 15,20,21,24,28$.

n	
15	$S_1 = (1\ 7\ 0\ 5\ 12\ 10\ 6\ 2\ 4\ \infty\ 13\ 8\ 11\ 3\ 9)$ $s_{n1} = 7$ (generate pseudo-cyclically)
	$M_1 = (1\ 11\ 8\ 2\ 4\ 12\ 9\ \infty\ 3\ 7\ 0\ 13\ 6\ 10\ 5)$
20	$S_1 = (1\ 10\ 14\ 8\ \infty\ 5\ 2\ 9\ 0\ 6\ 13\ 17\ 16\ 4\ 12\ 3\ 15\ 11\ 7\ 18)$ $s_{n1} = 14$
	$M_1 = (\infty\ 1\ 13\ 12\ 8\ 3\ 16\ 6\ 15\ 14\ 5\ 7\ 18\ 10\ 17\ 4\ 9\ 11\ 0\ 2)$
21	$S_1 = (1\ 12\ 16\ 11\ 17\ 8\ 5\ 4\ 13\ 7\ 20\ 6\ 0\ 19\ 10\ 15\ 18\ 3\ 9\ 2\ 14)$
	$M_1 = (1\ 0\ 19\ 10\ 9\ 13\ 12\ 18\ 2\ 4\ 3\ 14\ 16\ 15\ 11\ 6\ 8\ 5\ 7\ 17\ 20)$

24 $S_1 = (1\ 21\ 9\ 22\ 17\ 10\ 16\ 6\ 19\ 3\ 8\ \infty\ 14\ 2\ 20\ 18\ 7\ 12\ 4\ 0\ 5\ 13\ 15\ 11)$ $s_{n1} = 23$
$M_1 = (\infty\ 1\ 22\ 19\ 10\ 9\ 14\ 18\ 13\ 21\ 17\ 3\ 15\ 7\ 12\ 5\ 11\ 8\ 4\ 6\ 16\ 20\ 0\ 2)$
28 $S_1 = (1\ 7\ 17\ 19\ 4\ 3\ 23\ 21\ 13\ 11\ 14\ 10\ 24\ 9\ 22\ 6\ 2\ 26\ 25\ \infty\ 12\ 5\ 15\ 18\ 0\ 8\ 20\ 16)$ $s_{n1} = 24$
$M_1 = (\infty\ 23\ 21\ 11\ 1\ 18\ 10\ 12\ 14\ 7\ 26\ 20\ 0\ 3\ 17\ 15\ 9\ 16\ 25\ 6\ 5\ 4\ 13\ 24\ 8\ 19\ 22\ 2)$

41.14 Remark Some SOLSSOM(n) for orders n not covered in Remark 41.11 or Table 41.13 above can be constructed using the following generalization of the pseudo-cyclic construction. Define $A = (a_{ij}), a_{ij} \in \mathbb{Z}_{n-4} \cup \{\infty_1, \infty_2, \infty_3, \infty_4\}$, of order n by indexing the rows and columns by $\mathbb{Z}_{n-4} \cup \{\infty_1, \infty_2, \infty_3, \infty_4\}$ and letting

1. $a_{i+1,j+1} = a_{ij} + 1, 0 \le i+1, j+1 \le n-5$,
2. $a_{\infty_k, j+1} = a_{\infty_k j} + 1, 0 \le j+1 \le n-5, k = 1,2,3,4$,
3. $a_{i+1,\infty_k} = a_{i,\infty_k} + 1, 0 \le i+1 \le n-5, k = 1,2,3,4$, and
4. $a_{\infty_k,\infty_s} \doteq A_\infty$ ($1 \le k, s \le 4$) be latin squares on the symbols ∞_i.

All arithmetic is modulo $n-4$ except $\infty_i + 1 = \infty_i$.

If a SOLSSOM(n) (S, M) has this structure, then it is only necessary to give the first rows S_1 and M_1 of each; cells $s_{\infty_1}, s_{\infty_2}, s_{\infty_3}, s_{\infty_4}$; and the corners S_∞ and M_∞.

41.15 Table Regular SOLSSOM(n) for $n = 18, 22, 26$, and 30. Each of these squares are of the type described in Remark 41.14 above. In addition, to construct the matrix M for $n = 18, 26$ and 30, one further rule must be applied: if $m_{ij} = \infty_2$, then $m_{i+1,j+1} = \infty_3$, and if $m_{ij} = \infty_3$, then $m_{i+1,j+1} = \infty_2$ $(0 \le i, j \le n-5)$. For the matrix M when $n = 22$ replace ∞_3 with ∞_1 in this latter rule. All arithmetic on these subscripts is modulo $n-4$.

For each of these SOLSSOMs let

$S_\infty =$

∞_1	∞_4	∞_2	∞_3
∞_3	∞_2	∞_4	∞_1
∞_4	∞_1	∞_3	∞_2
∞_2	∞_3	∞_1	∞_4

$M_\infty =$

∞_4	∞_1	∞_2	∞_3
∞_1	∞_4	∞_3	∞_2
∞_2	∞_3	∞_4	∞_1
∞_3	∞_2	∞_1	∞_4

n = 18
$S_1 = (1\ 8\ 0\ 13\ 12\ 11\ 3\ 2\ 10\ \infty_4\ \infty_3\ \infty_2\ \infty_1\ 4\ 9\ 6\ 5\ 7)$,
$(s_{\infty_1}, s_{\infty_2}, s_{\infty_3}, s_{\infty_4}) = (0, 4, 13, 3)$,
$M_1 = (\infty_4\ \infty_2\ 13\ 6\ 9\ 7\ 4\ \infty_1\ 12\ 2\ 5\ 3\ 11\ \infty_3\ 0\ 10\ 8\ 1)$.

n = 22
$S_1 = (1\ 10\ 0\ 17\ 16\ 15\ 9\ 7\ 12\ 2\ 5\ 8\ 14\ \infty_4\ \infty_3\ \infty_2\ \infty_1\ 6\ 3\ 4\ 11\ 13)$,
$(s_{\infty_1}, s_{\infty_2}, s_{\infty_3}, s_{\infty_4}) = (8, 6, 5, 17)$,
$M_1 = (\infty_4\ \infty_2\ 0\ 11\ 3\ 14\ 10\ 13\ 5\ \infty_3\ 15\ 6\ 4\ 9\ 17\ 8\ 16\ \infty_1\ 12\ 1\ 7\ 2)$.

n = 26
$S_1 = (1\ 10\ 0\ 21\ 20\ 19\ 16\ 13\ 11\ 14\ 17\ 2\ 12\ 15\ 9\ 8\ 6\ \infty_4\ \infty_3\ \infty_2\ \infty_1\ 18\ 5\ 4\ 7\ 3)$,
$(s_{\infty_1}, s_{\infty_2}, s_{\infty_3}, s_{\infty_4}) = (8, 11, 21, 4)$,
$M_1 = (\infty_4\ \infty_2\ 0\ 10\ 17\ 14\ 8\ 18\ 1\ 12\ 4\ \infty_1\ 16\ 3\ 15\ 11\ 2\ 9\ 13\ 7\ 20\ \infty_3\ 21\ 19\ 6\ 5)$.

n = 30
$S_1 = (1\ 8\ 0\ 25\ 24\ 23\ 5\ 12\ 16\ 19\ 22\ 13\ 15\ 2\ 9\ 21\ 20\ 14\ 18\ 6\ 11\ \infty_4\ \infty_3\ \infty_2\ \infty_1\ 10\ 3\ 4\ 7\ 17)$,
$(s_{\infty_1}, s_{\infty_2}, s_{\infty_3}, s_{\infty_4}) = (14, 9, 16, 19)$,
$M_1 = (\infty_4\ \infty_2\ 0\ 10\ 3\ 21\ 18\ 20\ 19\ 14\ 6\ 2\ 1\ \infty_1\ 15\ 17\ 22\ 5\ 11\ 13\ 12\ 16\ 25\ 7\ 24\ \infty_3\ 23\ 8\ 4\ 9)$.

41.4 Holey SOLSSOMs

A *holey* SOLSSOM (or *frame* SOLSSOM) is a self-orthogonal latin square S of order n and type $h_1^{n_1}...h_k^{n_k}$, together with a symmetric partitioned latin square M of order n and type $h_1^{n_1}...h_k^{n_k}$, satisfying the property that when superimposed, the ordered pairs are exactly those pairs of symbols which are from different holes. A holey SOLSSOM with this structure is denoted by HSOLSSOM $(h_1^{n_1}...h_k^{n_k})$, where $h_1^{n_1}...h_k^{n_k}$ is the *type* of the HSOLSSOM.

41.16 Theorem [2, 3]

1. HSOLSSOM(2^n) exist for all odd $n \geq 5$ except possibly for $n \in E = \{8, 12, 14, 18, 22, 24, 28, 32\}$.
2. If $h \equiv 2 \pmod 4$, $h \geq 10$, then HSOLSSOM(h^n) exists for all $n \geq 5$ except possibly for $n \in E$. HSOLSSOM(6^n) exists for all $n \geq 5$ except possibly for $n \in \{6, 7, 10, 11, 12, 16, 18, 19, 20, 22, 23, 24, 27, 32, 38, 39\}$.
3. If $h \equiv 0 \pmod 4$, then there exists HSOLSSOM(h^n) if and only if $n \geq 5$.
4. If h is an odd integer, then HSOLSSOM(h^n) exists if and only if $n \geq 5$ is odd except possibly for $h = 3$ and $n \in \{19, 23, 27\}$.

41.17 Example HSOLSSOM(2^5).

$$S = \begin{array}{|cc|cc|cc|cc|cc|}
\hline
 & & 10 & 5 & 9 & 8 & 3 & 6 & 4 & 7 \\
 & & 6 & 9 & 7 & 10 & 5 & 4 & 8 & 3 \\
\hline
6 & 9 & & & 2 & 7 & 1 & 10 & 5 & 8 \\
10 & 5 & & & 8 & 1 & 9 & 2 & 7 & 6 \\
\hline
7 & 10 & 8 & 1 & & & 4 & 9 & 3 & 2 \\
9 & 8 & 2 & 7 & & & 10 & 3 & 1 & 4 \\
\hline
5 & 4 & 9 & 2 & 10 & 3 & & & 6 & 1 \\
3 & 6 & 1 & 10 & 4 & 9 & & & 2 & 5 \\
\hline
8 & 3 & 7 & 6 & 1 & 4 & 2 & 5 & & \\
4 & 7 & 5 & 8 & 3 & 2 & 6 & 1 & & \\
\hline
\end{array}
\qquad
M = \begin{array}{|cc|cc|cc|cc|cc|}
\hline
 & & 7 & 8 & 3 & 4 & 9 & 10 & 5 & 6 \\
 & & 8 & 7 & 4 & 3 & 10 & 9 & 6 & 5 \\
\hline
7 & 8 & & & 9 & 10 & 5 & 6 & 1 & 2 \\
8 & 7 & & & 10 & 9 & 6 & 5 & 2 & 1 \\
\hline
3 & 4 & 9 & 10 & & & 1 & 2 & 7 & 8 \\
4 & 3 & 10 & 9 & & & 2 & 1 & 8 & 7 \\
\hline
9 & 10 & 5 & 6 & 1 & 2 & & & 3 & 4 \\
10 & 9 & 6 & 5 & 2 & 1 & & & 4 & 3 \\
\hline
5 & 6 & 1 & 2 & 7 & 8 & 3 & 4 & & \\
6 & 5 & 2 & 1 & 8 & 7 & 4 & 3 & & \\
\hline
\end{array}$$

41.5 Related Designs

41.18 Theorem Regular SOLSSOM(n) exist if and only if there exists a resolvable spouse avoiding mixed doubles round robin tournament for n couples (see Remark IV.40.5 and §V.7.11).

41.19 Theorem If there exists a regular SOLSSOM(n), then there exists a triplewhist tournament for $4n$ players (see §IV.53.3).

41.20 Theorem If there exists a ($\mathbb{Z}$-cyclic) TWh(n) or if there exists a ($\mathbb{Z}$-cyclic) directed whist tournament DWh(n) on n players, then there exists a transversal SOLSSOM(n).

A k-SOLSSOM(n) is a set $\{(S_1, S_2, \ldots, S_k\}$ of self-orthogonal latin squares, together a symmetric latin square M, for which $\{S_i, S_i^T : 1 \leq i \leq k\} \cup \{M\}$ is a set of $2k+1$ MOLS(n).

41.21 Theorem For any odd prime power q, there exist $((q-3)/2)$-SOLSSOM(q). In addition, for $n \geq 1$ there is a $(2^{n-1}-1)$-SOLSSOM(2^n).

41.22 Theorem [6] There exist 2-SOLSSOM(n) for all $n > 4308$.

41.23 Remark Lee [6] shows that when $n > 135$ and n is odd, a 2-SOLSSOM(n) exists. In addition, when $n \equiv 0 \pmod 8$ and $n \notin \{24, 40, 48\}$, a 2-SOLSSOM(n) exists.

41.6 See Also

§I.6	Resolvable and near-resolvable designs.
§II.2	MOLS, transversal designs and orthogonal arrays.
§IV.40	Self orthogonal latin squares (SOLS).
§IV.53	Whist tournaments.
[5]	Explicit constructions for some small SOLSSOMs.

References

[1] R.D. Baker, Factorization of graphs, Ph.D. dissertation, Ohio State University, 1975.

[2] F.E. Bennett and L. Zhu, The spectrum of HSOLSSOM(h^n) where h is even, *Discrete Math.*, to appear.

[3] F.E. Bennett and L. Zhu, Further results on the existence of HSOLSSOM(h^n), preprint.

[4] B. Du, A few more resolvable spouse avoiding mixed doubles round robin tournaments, *Ars Combin.* 36 (1993), 309-314.

[5] N.J. Finizio, Some quick and easy SOLSSOMs, *Congr. Numer.*, to appear.

[6] T.C.Y. Lee, Tools for constructing RBIBDs, Frames, NRBs and SOLSSOMs, Ph.D. thesis, University of Waterloo, 1995.

[7] C.C. Lindner, R.C. Mullin and D.R. Stinson, On the spectrum of resolvable orthogonal arrays invariant under the Klein group K_4, *Aequat. Math.* 26 (1983), 176-183.

[8] S.P. Wang, On self orthogonal Latin squares and partial transversals of Latin squares, Ph.D. dissertation, Ohio State University, 1978.

[9] L. Zhu, A few more self-orthogonal Latin squares with symmetric orthogonal mates, *Congr. Numer.* 42 (1984), 313-320.

42 Sequences with Zero Autocorrelation

Christos Koukouvinos

42.1 Basic Definitions

Given the sequence $A = (a_1, \ldots, a_n)$ of length n, the *nonperiodic autocorrelation function* $N_A(s)$ is defined as

$$N_A(s) = \sum_{i=1}^{n-s} a_i a_{i+s} \text{ for } s = 0, 1, \ldots, n-1, \quad \text{and } N_A(s) = 0 \text{ for } s \geq n.$$

42.1 Proposition If $A(z) = a_1 + a_2 z + \cdots + a_n z^{n-1}$ is the associated polynomial of the sequence A, then

$$A(z)A(z^{-1}) = \sum_{i=1}^{n}\sum_{j=1}^{n} a_i a_j z^{i-j} = N_A(0) + \sum_{s=1}^{n-1} N_A(s)(z^s + z^{-s}), \quad z \neq 0.$$

Given A as above of length n, the *periodic autocorrelation function* $P_A(s)$ is defined, reducing $i+s$ modulo n, as

$$P_A(s) = \sum_{i=1}^{n} a_i a_{i+s}, \quad s = 0, 1, \ldots, n-1$$

42.2 Remark If $A^* = (a_n, \ldots, a_1)$ is the reversed sequence, then $A^*(z) = z^{n-1}A(z^{-1})$.

The *weight* of the sequence A is the number of nonzero entries in A. m sequences $A_1, A_2, \ldots, A_m$ of length n with entries $0, 1, -1$ are *m-complementary sequences of weight w*, written m-CS(n, w), if

$$N_{A_1}(s) + \cdots + N_{A_m}(s) = \begin{cases} 0, & s = 1, \ldots, n-1 \\ w, & s = 0 \end{cases}$$

or equivalently, if $A_1(z)A_1(z^{-1}) + \cdots + A_m(z)A_m(z^{-1}) = w, \quad z \neq 0$. Such sequences have *zero autocorrelation*.

42.3 Lemma If there exist 4-CS(m, w) A, B, C, and D having lengths m_1, m_2, m_1, and m_2 and weight w, then (A, B), $(A, -B)$, (C, D), and $(C, -D)$ are 4-CS$(m_1 + m_2, 2w)$.

42.4 Example $(1, 0, -1)$, $(1, 1, -1)$, $(1, 1, 1)$, $(1, -1, 1)$ are 4-CS(3,11).

42.5 Remark Golay sequences (§42.2), base and Turyn sequences (§42.3), and normal and near normal sequences (§42.4) are sequences with zero autocorrelation. All are useful in constructing orthogonal designs and Hadamard matrices (§IV.24), as well as in communications engineering, optics, and signal transmission problems [3, 7].

42.2 Golay Sequences

If $A = (a_1, \ldots, a_n)$ and B$=(b_1, \ldots, b_n)$ are two $(1, -1)$ sequences of length n and

$$N_A(s) + N_B(s) = 0 \quad \text{for} \quad s = 1, \ldots, n-1,$$

then A and B are *Golay sequences* of length n.

42.6 Example Golay sequences of lengths $n = 2, 10$, and 26 (write $-$ for -1).

$n = 2$ $\quad 1\,1$ and $1-$

$n = 10$ $\quad 1 - - 1 - 1 - - - 1$ and $1 - - - - - - - 1\,1 -$

$n = 26$ $\quad 1\,1\,1 - - 1\,1\,1 - 1 - - - - - - 1 - 1\,1 - - 1 - - - -$

$\qquad\quad - - - 1\,1 - - - 1 - 1\,1 - 1 - 1 - 1\,1 - - 1 - - - -$

42.7 Remark All known Golay sequences can be constructed from those in Example 42.6.

42.8 Lemma Let $A = (a_1, \ldots, a_n)$ and $B = (b_1, \ldots, b_n)$ be Golay sequences of length n. Suppose k_1 of the a_i are positive and k_2 of the b_i are positive. Then n is even and $n = (k_1 + k_2 - n)^2 + (k_1 - k_2)^2$.

42.9 Proposition Summary of Golay properties.

1. The length n is even and the sum of two squares or $n = 1$.
2. $a_i b_i a_{n-i+1} b_{n-i+1} = -1$.
3. They exist for orders $2^a 10^b 26^c$, a, b, c nonnegative integers [4, 7]. These are *Golay numbers*.
4. They do not exist for length 34, 50, 58, or 68.
5. They do not exist for length divisible by a prime $p \equiv 3 \pmod 4$ [2].
6. The unresolved cases with $n < 200$ are: $n = 74$, 82, 106, 116, 122, 130, 136, 146, 148, 164, 170, 178, 194.

42.3 Base and Turyn Sequences

Four $(1,-1)$ sequences A, B, C, D of lengths $m+p$, $m+p$, m, m are *base sequences*, if:

$$N_A(s)+N_B(s)+N_C(s)+N_D(s)=\begin{cases}0, & s=1,\ldots,m-1\\ 4m+2p, & s=0\end{cases}$$
$$N_A(s)+N_B(s)=0,\quad s=m,\ldots,m+p-1$$

or equivalently, if

$$A(z)A(z^{-1})+B(z)B(z^{-1})+C(z)C(z^{-1})+D(z)D(z^{-1})=4m+2p,\quad z\neq 0.$$

42.10 Theorem If X and Y are Golay sequences of length m, then the sequences $(1,X)$, $(1,-X)$, (Y), (Y) are base sequences of lengths $m+1$, $m+1$, m, m.

42.11 Proposition Base sequences exist for all $m=2^a10^b26^c$, a, b, c nonnegative integers with $p=1$. They also exist for $p=1$ and $m\in\{1,2,\ldots,30\}$ [5, 6] and when $m=g$ and $p=h-g$ where both g and h are Golay numbers.

42.12 Example Base sequences for $m=1,2,3,4,5,6$ and $p=1$ (write $-$ for -1).

$m=1$	$(1\,1)$, $(1\,-)$, (1), (1)
$m=2$	$(1\,1\,1)$, $(1\,1\,-)$, $(1\,-)$, $(1\,-)$
$m=3$	$(1\,1\,-\,1)$, $(1\,1\,-\,-)$, $(1\,1\,1)$, $(1\,-\,1)$
$m=4$	$(1\,1\,-\,1\,1)$, $(1\,1\,1\,1\,-)$, $(1\,1\,-\,-)$, $(1\,-\,1\,-)$
$m=5$	$(1\,1\,-\,1\,-\,1)$, $(1\,1\,1\,-\,-\,-)$, $(1\,1\,-\,1\,1)$, $(1\,1\,-\,1\,1)$
$m=6$	$(1\,1\,-\,1\,-\,1\,1)$, $(1\,1\,-\,-\,-\,1\,-)$, $(-\,1\,1\,1\,1\,1)$, $(-\,-\,1\,1\,-\,1)$

If sequences A, B, C, D of lengths $m+1$, $m+1$, m, m are base sequences that also satisfy $N_{AC}(s-1)+N_{CA}(s)=0$ for $s=1,\ldots,m$, then they are *Turyn sequences*, where the *cross-correlations* are

$$N_{AC}(s)=\sum_{i=1}^{m-s}a_ic_{i+s}\ ,\quad N_{CA}(s)=\sum_{i=1}^{m+1-s}c_ia_{i+s}\ ,\quad s=0,1,\ldots,m.$$

42.13 Example Turyn sequences for $m=6$ and $m=7$ (write $-$ for -1).

$m=6$ $(1\,1\,1\,-\,1\,1\,1)$, $(1\,1\,-\,-\,-\,1\,-)$, $(1\,1\,-\,1\,-\,-)$, $(1\,1\,-\,1\,-\,-)$
$m=7$ $(1\,1\,-\,1\,-\,1\,-\,-)$, $(1\,1\,1\,1\,-\,-\,-\,1)$, $(1\,1\,1\,-\,1\,1\,1)$, $(1\,-\,-\,1\,-\,-\,1)$

42.14 Remark Turyn sequences exist for $m\leq 7$ and $m=12,14$, and they do not exist for any other length ≤ 42. The first open case is 43 [1].

42.15 Theorem If there are Turyn sequences of lengths $m+1$, $m+1$, m, m, then there are base sequences of lengths $2m+2$, $2m+2$, $2m+1$, $2m+1$.

42.4 Normal and Near Normal Sequences

A triple $(F\,;G,H)$ of sequences is a set of *normal sequences* for length n if the following conditions are satisfied:

1. $F=(f_k)$ is a $(1,-1)$ sequence of length n.
2. $G=(g_k)$ and $H=(h_k)$ are $(0,\pm1)$ sequences of length n, such that $G+H=(g_k+h_k)$ is a $(1,-1)$ sequence of length n.
3. $N_F(s)+N_G(s)+N_H(s)=0\ ,\quad s=1,\ldots,n-1.$

42.16 Theorem If F, H are Golay sequences of length n, then the two sets $(F\,;\,H, O_n)$ and $(F\,;\,L, M)$ are normal sequences of length n, where O_n is the sequence of n zeros, and L and M are the symmetric and skew parts of H, respectively.

42.17 Theorem If A, B, C, D are Turyn sequences of lengths $m+1$, $m+1$, m, m, then the sequences $F = A\,/\,C$, $G = B\,/\,O_m$, $H = O_{m+1}\,/\,D$ are normal sequences of length $2m+1$, where $A\,/\,C = (a_1, c_1, a_2, c_2, \ldots, a_m, c_m, a_{m+1})$ and O_m is the zero sequence.

42.18 Theorem If $(F\,;\,G, H)$ are normal sequences of length n, then the following $(1,-1)$ sequences $A = (F, 1)$, $B = (F, -1)$, $C = G + H$, $D = G - H$ of lengths $n+1$, $n+1$, n, n respectively are base sequences and vice versa.

42.19 Example Normal sequences for $n = 7, 9$ (write $-$ for -1).

$n = 7$ $(1\ -\ 1\,1\,1\ -\ -)$, $(1\,1\,0\,1\,0\ -\ 1)$, $(0\,0\,1\,0\,1\,0\,0)$
$n = 9$ $(1\ -\ 1\,1\,1\,1\,1\ -\ -)$, $(1\,1\ -\ 0\,1\,0\,1\ -\ 1)$, $(0\,0\,0\,1\,0\ -\ 0\,0\,0)$

42.20 Remark Normal sequences of length n do not exist if $2n$ cannot be written as the sum of three squares. Normal sequences

1. Exist for $n \in \{1, 2, 3, 4, 5, 7, \ldots, 13, 15, 16, 18, 19, 20, 25, 26, 29, 32\}$ and all $g > 32$, where g is a Golay number.
2. Do not exist for $n \in \{6, 14, 17, 21, 22, 23, 24, 30, 46, 56, 62, 78, 94\}$ or any $n > 94$ where $2n$ is not the sum of three squares.
3. The first three open cases are $n = 27, 28, 31$ [5, 8].

A quadruple $(E, F\,;\,G, H)$ of $(0, \pm 1)$ sequences is a set of *near normal sequences* for length $n = 4m + 1$ if the following conditions are satisfied.

1. $E = (X\,/\,O_{m-1}, 1)$, $F = (Y\,/\,O_{m-1})$, where X and Y are $(1,-1)$ sequences of length m and O_{m-1} the sequence of zeros of length $m-1$, i.e., E and F are respectively of lengths $2m$ and $2m-1$; G and H are $(0, \pm 1)$ sequences of length $2m$, such that $G + H$ is a $(1,-1)$ sequence of length $2m$.
2. $N_E(s) + N_F(s) + N_G(s) + N_H(s) = 0\ ,\quad s = 1, \ldots, 2m-1$

42.21 Remark The sequences G and H of this definition are quasi-symmetric, i.e., if $g_k = 0$, then $g_{2m+1-k} = 0$ and also if $h_k = 0$, then $h_{2m+1-k} = 0$.

42.22 Theorem Let $n = 4m + 1$. Then $(E, F\,;\,G, H)$, with $E = (X\,/\,O_{m-1}, 1)$ and $F = (Y\,/\,O_{m-1})$, are near normal sequences, if and only if the $(1,-1)$ sequences $A = (Y\,/\,X, 1)$, $B = (Y\,/\,(-X), -1)$, $C = G + H$, $D = G - H$ of lengths $2m+1$, $2m+1$, $2m$, $2m$, respectively, are base sequences.

42.23 Example Near normal sequences for $n = 5, 9, 13$ (write $-$ for -1).

$n = 5$ $(1\,1)$, (1), $(1\,-)$, $(0\,0)$
$n = 9$ $(-\,0\,1\,1)$, $(1\,0\,1)$, $(1\,0\,0\,1)$, $(0\,1\ -\ 0)$
$n = 13$ $(1\,0\ -\ 0\,1\,1)$, $(1\,0\,1\,0\,1)$, $(1\,1\ -\ 1\ -\ -)$, $(0\,0\,0\,0\,0\,0)$

42.24 Proposition Near normal sequences for length $n = 4m + 1$ exist for $n \leq 61$ [6, 8].

Four circulant matrices T_1, T_2, T_3, T_4 of order t with entries $0, 1, -1$ are *T-matrices* if:

1. $T_i * T_j = 0$, $i \neq j$ ($*$ denotes the Hadamard product).
2. $T_1 T_1^T + T_2 T_2^T + T_3 T_3^T + T_4 T_4^T = t I_t$.

Four sequences X, Y, Z, W of length n with entries $0, 1, -1$ are *T-sequences* if:

1. $|x_i| + |y_i| + |z_i| + |w_i| = 1\ ,\quad i = 1, \ldots, n.$
2. $N_X(s) + N_Y(s) + N_Z(s) + N_W(s) = \begin{cases} 0\ , & s = 1, \ldots, n-1 \\ n\ , & s = 0. \end{cases}$

If A, B, C, D are base sequences of lengths $m+1$, $m+1$, m, m, then the sequences $(\frac{1}{2}(A+B))$, $(\frac{1}{2}(A-B))$, $(\frac{1}{2}(C+D))$, $(\frac{1}{2}(C-D))$ are *suitable sequences.*

42.25 Remark Suitable sequences are crucial to Yang's [8] constructions for finding longer T-sequences of odd length. The most prolific method for constructing orthogonal designs and Hadamard matrices uses T-matrices or T-sequences. T-sequences of length n imply the existence of T-matrices, but not conversely. There are several important small orders (for example, 67) for which T-matrices are known, but T-sequences are not known.

If there is a multiplication method which uses suitable sequences of lengths $m+p, m+p, m, m$ to produce T-sequences of length $y(2m+p)$, then y is a *Yang* number.

42.26 Proposition Yang numbers are known for $y \in \{3, 5, 7, \ldots, 33, 37, 39, 41, 45, 49, 51, 53, 57, 59, 61, 65, 81\}$ and all $y = 2g + 1 > 81$, when g is a Golay number [5, 6, 8].

42.27 Example The multiplication method for $y = 3$. If A, B, C, D are suitable sequences of lengths $m+p, m+p, m, m$, then the sequences $X = (A, C; 0_{m+p}, 0_m; -B^*, 0_m)$, $Y = (B, D; 0_{m+p}, 0_m; A^*, 0_m)$, $Z = (0_{m+p}, 0_m; A, -C; 0_{m+p}, -D^*)$, $W = (0_{m+p}, 0_m; B, -D; 0_{m+p}, C^*)$ are T-sequences of length $3(2m+p)$. If the suitable sequences are $A = (110), B = (001), C = (1-), D = (00)$ (so $m = 2$, $p = 1$), then the following T-sequences of length 15 are obtained: $X = (1101-00000-0000)$, $Y = (001000000001100)$, $Z = (00000110-100000)$, $W = (0000000100000-1)$.

42.5 See Also

§IV.24	Hadamard matrices and designs.

References

[1] G.M. Edmondson, J. Seberry and M.R. Anderson, On the existence of Turyn sequences of length less than 43, *Math. Comput.* 62 (1994), 351-362.

[2] S. Eliahou, M. Kervaire and B. Saffari, A new restriction on the lengths of Golay complementary sequences, *J. Combin. Theory A* 55 (1990), 49-59.

[3] A.V. Geramita and J. Seberry, *Orthogonal designs: Quadratic Forms and Hadamard Matrices*, Marcel Dekker, New York, 1979.

[4] M.J.E. Golay, Complementary series, *IRE Trans. Information Theory IT-7* (1961), 82-87.

[5] C. Koukouvinos, S. Kounias, J. Seberry, C.H. Yang and J. Yang, On sequences with zero autocorrelation, *Des. Codes Crypt.* 4 (1994), 327-340.

[6] J. Seberry and M. Yamada, Hadamard matrices, sequences and block designs, in: *Contemporary Design Theory: A Collection of Surveys* (J.H. Dinitz, D.R. Stinson; eds.) John Wiley & Sons, New York, 1992, 431-560.

[7] R.J. Turyn, Hadamard matrices, Baumert-Hall units, four symbol sequences, pulse compressions and surface wave encodings, *J. Combin. Theory A* 16 (1974), 313-333.

[8] C.H. Yang, On composition of four-symbol δ-codes and Hadamard matrices, *Proc. Amer. Math. Soc.* 107 (1989), 763-776.

43 Skolem Sequences

Nabil Shalaby

43.1 Definitions and Examples

A *Skolem sequence* of order n is a sequence $S = (s_1, s_2, \ldots, s_{2n})$ of $2n$ integers satisfying the conditions

1. for every $k \in \{1, 2, \ldots, n\}$ there exist exactly two elements $s_i, s_j \in S$ such that $s_i = s_j = k$,
2. if $s_i = s_j = k$ with $i < j$, then $j - i = k$.

Skolem sequences are also written as collections of ordered pairs $\{(a_i, b_i) : 1 \le i \le n, b_i - a_i = i\}$ with $\bigcup_{i=1}^{n} \{a_i, b_i\} = \{1, 2, \ldots, 2n\}$.

43.1 **Example** A Skolem sequence of order 5: $S = (1, 1, 3, 4, 5, 3, 2, 4, 2, 5)$, or, equivalently, the collection $\{(1, 2), (7, 9), (3, 6), (4, 8), (5, 10)\}$.

An *extended Skolem sequence* of order n is a sequence ES= $(s_1, s_2, \ldots s_{2n+1})$ of $2n + 1$ integers satisfying conditions 1 and 2 of the definition and

3. there is exactly one $s_i \in ES$ such that $s_i = 0$.

A *hooked Skolem sequence* HS of order n is an extended Skolem sequence of order n with $s_{2n} = 0$.

43.2 **Remark** The element $s_i = 0$ is the *hook* or *zero* of the sequence.

43.3 **Example** A hooked Skolem sequence of order 6 which is also an extended Skolem sequence of order 6: $(1, 1, 2, 5, 2, 4, 6, 3, 5, 4, 3, 0, 6)$.

A *Langford sequence* of order n and defect $d, n > d$, is a sequence $L = (l_1, l_2, \ldots, l_{2n})$ of $2n$ integers satisfying the conditions

1. for every $k \in \{d, d+1, \ldots, d+n-1\}$ there exist exactly two elements $l_i, l_j \in L$ such that $l_i = l_j = k$,
2. if $l_i = l_j = k$ with $i < j$, then $j - i = k$.

A *near-Skolem sequence* of order n and defect $m, n \ge m$, is a sequence NS= $(s_1, s_2, \ldots, s_{2n-2})$ of $2n - 2$ integers satisfying the conditions

1. for every $k \in \{1, 2, \ldots, m - 1, m + 1, \ldots n\}$ there exist exactly two elements $s_i, s_j \in$NS, such that $s_i = s_j = k$,
2. if $s_i = s_j = k$ with $i < j$ then $j - i = k$.

43.4 **Remark** The *hooked Langford* and *hooked near-Skolem* sequences are defined in a manner similar to that of the hooked Skolem sequence.

43.5 **Example** A Langford sequence of order 5 and defect 3 and a near-Skolem sequence of order 7 and defect 4, respectively.

$$L = (7, 5, 3, 6, 4, 3, 5, 7, 4, 6) \qquad \text{NS} = (1, 1, 6, 3, 7, 5, 3, 2, 6, 2, 5, 7)$$

43.6 **Example** A hooked Langford sequence of order 4 and defect 2 and a hooked near-Skolem sequence of order 7 and defect 3, respectively.

$$\text{HL}= (4, 5, 6, 2, 4, 2, 5, 0, 6) \qquad \text{HNS} = (2, 5, 2, 4, 6, 7, 5, 4, 1, 1, 6, 0, 7)$$

43.2 Existence

43.7 Theorem (see [2, 3])

1. (Skolem) A Skolem sequence of order n exists if and only if $n \equiv 0, 1 \pmod 4$. When $n = 1$, take (1,1). When $n = 4$, take (1,1,3,4,2,3,2,4). When $n = 5$, take (2,4,2,3,5,4,3,1,1,5). When $n > 5$, use the construction (as ordered pairs):

$$n = 4s: \begin{cases} (4s+r+1, 8s-r+1) & r = 1, \dots, 2s \\ (r, 4s-r-1) & r = 1, \dots, s-2 \\ (s+r+1, 3s-r) & r = 1, \dots, s-2 \\ (s-1, 3s),\ (s, s+1),\ (2s, 4s-1),\ (2s+1, 6s) & \end{cases}$$

$$n = 4s+1: \begin{cases} (4s+r-1, 8s-r+3) & r = 1, \dots, 2s \\ (r, 4s-r+1) & r = 1, \dots, s \\ (s+r+2, 3s-r+1) & r = 1, \dots, s-2 \\ (s+1, s+2),\ (2s+1, 6s+2),\ (2s+2, 4s+1) & \end{cases}$$

2. (O'Keefe) A hooked Skolem sequence of order n exists if and only if $n \equiv 2, 3 \pmod 4$. When $n = 2$, take (1,1,2,0,2). When $n = 3$, take (3,1,1,3,2,0,2). For $n \geq 6$, use the construction (as ordered pairs):

$$n = 4s+2: \begin{cases} (r, 4s-r+2) & r = 1, \dots, 2s \\ (4s+r+3, 8s-r+4) & r = 1, \dots, s-1 \\ (5s+r+2, 7s-r+3) & r = 1, \dots, s-1 \\ (2s+1, 6s+2),\ (4s+2, 6s+3), & \\ (4s+3, 8s+5),\ (7s+3, 7s+4) & \end{cases}$$

$$n = 4s-1: \begin{cases} (4s+r, 8s-r-2) & r = 1, \dots, 2s-2 \\ (r, 4s-r-1) & r = 1, \dots, s-2 \\ (s+r+1, 3s-r) & r = 1, \dots, s-2 \\ (s-1, 3s),\ (s, s+1),\ (2s, 4s-1),\ (2s+1, 6s-1),\ (4s, 8s-1) & \end{cases}$$

3a. (Abrham and Kotzig) An extended Skolem sequence of order n exists for all n.

3b. (Baker) An extended Skolem sequence of order n exists for all positions i of the zero, if and only if i is odd and $n \equiv 0, 1 \pmod 4$ or i is even and $n \equiv 2, 3 \pmod 4$.

43.8 Theorem [14] A Langford sequence of order n and defect d exists if and only if

1. $n \geq 2d - 1$.
2. $n \equiv 0, 1 \pmod 4$ and d is odd, or $n \equiv 2, 3 \pmod 4$ and d is even.

A hooked Langford sequence of order n and defect d exists if and only if

1. $n(n - 2d + 1) + 2 \geq 0$.
2. $n \equiv 2, 3 \pmod 4$ and d is odd, or $n \equiv 1, 2 \pmod 4$ and d is even.

43.9 Theorem [13] A near-Skolem sequence of order n and defect m exists if and only if $n \equiv 0, 1 \pmod 4$ and m is odd, or $n \equiv 2, 3 \pmod 4$ and m is even. A hooked near-Skolem sequence of order n and defect m exists if and only if $n \equiv 0, 1 \pmod 4$ and m is even, or $n \equiv 2, 3 \pmod 4$ and m is odd.

43.10 Remark There are numerous open problems of the existence of related generalizations of Skolem-Langford sequences (see [11, 13]).

43.3 Connections with Other Designs

43.11 Construction A cyclic Steiner triple system STS$(6n+1)$. Suppose that $\{1,\ldots,3n\}$ can be partitioned into m triples $\{a,b,c\}$ such that $a+b=c$ or $a+b+c\equiv 0 \pmod{6n+1}$. (This is the *first Heffter difference problem.*) Then the set of all triples $\{0,a,a+b\}$ is a $(6n+1,3,1)$ cyclic difference family, the base blocks for a STS$(6n+1)$. From a Skolem sequence or hooked Skolem sequence of order n (Theorem 43.7), construct the pairs (b_r,a_r) such that $b_r-a_r=r$, for $1\le r\le n$. The set of all triples (r,a_r+n,b_r+n), for $1\le r\le n$ is a solution to the first Heffter difference problem.

43.12 Example The Skolem sequence of order 4, $S=(1,1,4,2,3,2,4,3)$ yields the pairs $\{(2,1),(6,4),(8,5),(7,3)\}$. These pairs in turn give the set of triples $\{(1,5,6),(2,8,10),(3,9,12),(4,7,11)\}$ forming a solution of the first Heffter problem. These triples yield base blocks for a STS(25): $\{0,1,6\},\{0,2,10\},\{0,3,12\},\{0,4,11\}$.

43.13 Construction A cyclic Steiner triple system STS$(6n+3)$ [12]. Suppose $\{1,\ldots,3n+1\}\setminus\{2n+1\}$ can be partitioned into m triples $\{a,b,c\}$ such that $a+b=c$ or $a+b+c\equiv 0 \pmod{6n+3}$. (This is the *second Heffter difference problem.*) Then the set of all triples $\{0,a,a+b\}$, together with the 'short' block $\{0,2n+1,4n+2\}$, is a $(6n+3,3,1)$ cyclic partial difference family, the base blocks for a STS$(6n+3)$. Heffter's second difference problem is solved using extended Skolem sequences of order n with a hook in the nth position; from such a sequence, construct the pairs (b_r,a_r) such that $b_r-a_r=r$, for $1\le r\le n$. Then take the set of all triples (r,a_r+n,b_r+n), for $1\le r\le n$.

An explicit construction for the required extended Skolem sequences (as ordered pairs):

$$n=4s:\quad \begin{cases} (r,4s-r+1) & r=1,\ldots,s-1\\ (s+r-1,3s-r & r=1,\ldots,s-1\\ (4s+r+1,8s-r+1) & r=1,\ldots,s-1\\ (5s+r+1,7s-r+1) & r=1,\ldots,s-1\\ (2s-1,2s),\ (3s,5s+1),\ (3s+1,7s+1),\ (6s+1,8s+1) \end{cases}$$

$$\begin{array}{l} n=4s+1:\\ (n>5)\end{array}\quad \begin{cases} (r,4s-r+2) & r=1,\ldots,2s\\ (5s+r,7s-r+3) & r=1,\ldots,s\\ (4s+r+2,8s-r+3) & r=1,\ldots,s-2\\ (2s+1,6s+2),\ (6s+1,8s+4),\ (7s+3,7s+4) \end{cases}$$

$$\begin{array}{l} n=4s+2:\\ (n>2)\end{array}\quad \begin{cases} (r,4s-r+3) & r=1,\ldots,2s\\ (4s+r+4,8s-r+4) & r=1,\ldots,s-1\\ (5s+r+3,7s-r+3) & r=1,\ldots,s-2\\ (2s+1,6s+3),\ (2s+2,6s+2),\ (4s+4,6s+4),\\ (7s+3,7s+4),\ (8s+4,8s+6) \end{cases}$$

$$n=4s-1:\quad \begin{cases} (r,4s-r) & r=1,\ldots,2s-1\\ (4s+r+1,8s-r) & r=1,\ldots,s-2\\ (5s+r,7s-r-1) & r=1,\ldots,s-2\\ (2s,6s-1),\ (5s,7s+1),\ (4s+1,6s),\ (7s-1,7s) \end{cases}$$

When $n=2$, take the sequence $\{(1,2),(4,6\}$; when $n=5$, take

$$\{(1,5),(2,7),(3,4),(8,10),(9,12)\}.$$

When $n=1$, the sequence does not exist; indeed there is no cyclic STS(9).

43.14 Remark A Skolem sequence of order n is equivalent to a starter (coming from Skolem) in $\mathbb{Z}_{2n+1}$ by taking the set of pairs of subscripts of the numbers $1,\ldots,n$. Thus special types of Skolem and extended Skolem sequences produce special types of starters.

43.15 Example $S = (5, 6, 7, 8, 2, 5, 2, 6, 4, 7, 3, 8, 4, 3, 1, 1)$ is a Skolem sequence of order 8. It gives a strong starter in $\mathbb{Z}_{17}$:

$$\{\{15, 16\}, \{5, 7\}, \{11, 14\}, \{9, 13\}, \{1, 6\}, \{3, 10\}, \{4, 12\}\}$$

43.16 Theorem [2] The existence of an extended Skolem sequence implies the existence of a unique additive permutation.

43.17 Example $S = (3, 1, 1, 3, 4, 2, 0, 2, 4)$ produces a unique additive permutation $y = (3, 1, -1, -3, 4, 2, 0, -2, -4)$ of $f = (-4, -3, -2, -1, 0, 1, 2, 3, 4)$, such that $y + f = z = (-1, -2, -3, -4, 4, 3, 2, 1, 0)$ is another permutation of f.

43.18 Remark A recursive theorem of Rosa (1975) uses the existence of Skolem sequences: If there is a large set of Steiner triple systems on v points, then there is a large set of Steiner triple systems on $2v + 1$ points.

43.4 Enumeration of Sequences

43.19 Theorem [1]

1. The number of distinct Skolem sequences of order $n, \sigma_n \geq 2^{\lfloor n/3 \rfloor}$.
2. The number of distinct extended Skolem sequences of order $n, \epsilon_n \geq 2^{\lfloor n/3 \rfloor}$.
3. The number of extremal Langford sequences with defect d, $l_n \geq 4^{\lfloor (d-1)/3 \rfloor}$, i.e., the Langford sequences with the smallest order n that satisfy the necessary conditions for Theorem 43.8.

43.20 Table The number N_S of distinct Skolem sequences of order $n \leq 13$.

n	1	4	5	8	9	12	13
N_S	1	6	10	504	2656	455936	3040560

43.21 Remark In Table 43.20 the sequence and its reverse are considered distinct.

43.22 Table The number N_H of distinct hooked Skolem sequences of order $n \leq 14$.

n	2	3	6	7	10	11	14
N_H	1	1	38	124	12808	72648	21878816

43.23 Table The number N_E of distinct extended Skolem sequences of order $n \leq 14$.

n	1	2	3	4	5	6	7	8	9
N_E	2	2	6	22	48	160	636	3556	19488

n	10	11	12	13	14
N_E	95872	594320	4459888	32131648	227072544

43.5 See Also

§I.4.5	Skolem sequences generate cyclic Steiner triple systems.
§IV.2	Generalizations of Skolem sequences generate balanced ternary designs.
§IV.10	Skolem sequences generate difference families.
§IV.45	Starters are generalizations of Skolem sequences.

[4]	Disjoint Skolem sequences generate disjoint cyclic Steiner and Mendelsohn systems and 1-covering designs.
[5]	Generalizations of Skolem sequences are used to generate designs with special automorphism types.
[6]	Langford sequences are used to construct 2-factorizations of $2K_n$.
[8]	An application of Heffter's difference problem to the construction of codes resistant to random interference.
[9]	An application of Langford sequences to the construction of binary sequences with controllable complexity.
[10]	Skolem sequences are used to construct k-rotational Steiner triple systems.
[11]	Contains generalizations of Skolem-Langford sequences and their relation to other mathematical topics such as golden section, Wythoff pairs, and partition of sets of numbers.
[13]	Contains generalizations of Skolem sequences and their relations to designs and graph labelings. Also contains a detailed bibliography.

References

[1] J. Abrham, Exponential lower bounds for the number of Skolem and extremal Langford sequences, *Ars Combin.* 22 (1986), 187-198.

[2] J. Abrham and A. Kotzig, Skolem sequences and additive permutations, *Discrete Math.* 37 (1981), 143-146.

[3] C. Baker, Extended Skolem sequences, *J. Combin. Des.* 3 (1995), 363-379.

[4] C. Baker and N. Shalaby, Disjoint Skolem sequences and related disjoint structures, *J. Combin. Des.* 1 (1993), 329-345.

[5] C.J. Cho, Combinatorial Designs with Prescribed Automorphism Types, Ph.D. thesis, McMaster University, Canada, 1983.

[6] C.J. Colbourn and A. Rosa, Repeated edges in 2-factorizations, *J. Graph Theory* 14 (1990), 1-24.

[7] C.J. Colbourn and A. Rosa, *Triple Systems*, Oxford University Press, New York, to be published.

[8] A.R. Eckler, Construction of missile guidance codes resistant to random interference, *Bell Syst. Tech. J.* (1960), 973-994.

[9] E.J. Gorth, Generation of binary sequences with controllable complexity, *IEEE Trans. Inform. Theory* IT-17 (1971), 288-296.

[10] Z. Jiang, Rotational Steiner Triple Systems, Ph.D. thesis, University of Waterloo, Canada, 1995.

[11] R.J. Nowakowski, Generalizations of the Langford-Skolem Problem, M.Sc. thesis, University of Calgary, Canada, 1975.

[12] A. Rosa, Poznámka o cyklikých Steinerových systémoch trojíc, *Mat. Fyz. Časopis* 16 (1966), 285-290.

[13] N. Shalaby, Skolem Sequences: Generalizations and Applications, Ph.D. thesis, McMaster University, Canada, 1992.

[14] J.E. Simpson, Langford sequences: perfect and hooked, *Discrete Math.* 44 (1983), 97-104.

44 Spherical t-Designs

Stuart G. Hoggar

44.1 Definitions and Examples

Let $\Omega = \Omega_d = \{x \in \mathbb{R}^d : |x| = 1\}$. A subset X is *antipodal* if $x \in X$ implies $-x \in X$.

Let $\mathbb{F}$ denote the real numbers $\mathbb{R}$, complex numbers $\mathbf{C}$, quaternions $\mathbf{H}$, or octonions (Cayley numbers) $\mathbf{O}$. The *projective space* $\mathbb{F}\mathbf{P}^{d-1}$, or just $\mathbf{P}$, is the set of idempotents x of trace 1 in the $\mathbb{R}$-linear space of d by d hermitean matrices over $\mathbb{F}(d = 2, 3$ if $\mathbb{F} = \mathbf{O})$, with inner product $\langle x, y\rangle =$ real part of trace(xy) [7].

$X \subseteq \Omega$, $\mathbf{P}$ has *angle set* $A = \{\langle x, y\rangle : x, y \in X, x \neq y\}$. Let $s = |A|$, $v = |X|$. For $X \subseteq \Omega$ not antipodal, assume that $O \notin A$. (It is conjectured that for $t \geq s$ this is always true.) (See Theorem 44.3(d).)

44.1 Remarks For $\mathbb{F} \neq \mathbf{O}$, the map: (unit column vector u in line) $\to uu^*$, where * denotes conjugate transpose, identifies lines through the origin with the idempotents above. $|\langle u, v\rangle| = |u^*v| = \cos\theta$ in $\mathbb{F}^d$ becomes $\langle x, y\rangle = \cos^2\theta$. For all $\mathbb{F}$, $0 \leq \langle x, y\rangle \leq 1$.

Let $m = \frac{1}{2}, 1, 2,$ or 4 according as $\mathbb{F} = \mathbb{R}, \mathbf{C}, \mathbf{H},$ or $\mathbf{O}$; let $N = md$. $(a)_0 = 1 = {}_0(a)$ and $(a)_n = a(a+1)\ldots(a+n-1)$, ${}_n(a) = a(a-1)\ldots(a-n+1)(a \in \mathbb{R}, n \in \mathbf{N})$. Let ε denote 0 or 1. The *special polynomials* $Q_k(x) = Q_k^{\Omega}(x)$, $Q_k^{\varepsilon}(x)$, of degree $k = 0, 1, 2, \ldots$ associated with $E = \Omega$, $\mathbf{P}$ are given by $Q_{2k+\varepsilon}^{\Omega}(x) = x^{\varepsilon} Q_k^{\varepsilon}(x^2)$ and

$$Q_k^{\varepsilon}(x) = \textstyle\sum_{r=0}^{k}(-1)^{k-r}\binom{k}{r}(N + 2k + \varepsilon - 1)\{(N)_{k+r+\varepsilon-1}/(m)_{r+\varepsilon}k!\}x^r.$$

A *spherical t-design* in E is a finite nonempty subset X satisfying

$\sum_{x\in X} Q_k(\langle x, y\rangle) = 0, \forall y \in X, \quad k = 1, 2, \ldots, t.$ (Other definitions, [4, 5, 6].)

In many tables, $t = t(X) = \max\{t : X$ is a t-design$\}$ while * denotes a *tight* design (§44.5) and *Aut* denotes the automorphism group.

44.2 Remark The most useful definition of a spherical t-design is: X is a spherical t-design in E if and only if one is able to determine *exactly* the average value on E of any polynomial f of total degree at most t by sampling f at the points of X, i.e.,

$$\frac{1}{volume\ E}\int_E f(\xi)d\xi = \frac{1}{|X|}\sum_{x\in X} f(x).$$

44.3 Theorem Spherical t-designs exist in every E for all t and d (the proof is nonconstructive but [1] gives the construction in $E = \Omega$ for all t and d), and

1. Any $X \subseteq \Omega$ is a 1-design if and only if its centre of mass is the origin.
2. Any $X \subseteq \Omega$ is a 2-design if and only if it is also the orthogonal projection into (a copy of) $\mathbb{R}^d$ of v orthogonal vectors in $\mathbb{R}^v$.
3. In Ω, $t + 1 \leq v \leq 2t + 1$ implies a regular $(t+1)$-gon; if $v = 2t + 2$, X is the union of two regular $(t+1)$-gons.
4. A t-design in $\mathbb{R}\mathbf{P}^{d-1}$ with $A = \{\alpha^2, \beta^2, \ldots\}$ is equivalent to an antipodal $(2t+1)$-design of size $2v$ with $A = \{-1, \pm\alpha, \pm\beta, \ldots\}$.
5. $v \geq R_k(1)$ (see §44.5), where k is the integral part of $(t - \epsilon)/2$.

44.4 Table The main sources of spherical t-designs. Ex. n, or just n, refers to Example n in [7, 8] where they are listed together and studied as t-designs.

Lattices L [4] ($Aut = W(L)$): their minimal vectors. D_4, E_6, E_7, E_8, Coxeter-Todd lattice K_{12}, Barnes-Wall lattice Λ_{16}, Leech lattice Λ_{24} (= Ex. 11, complex form 28).

Polytopes: their vertices. 1–8, 16, 17, 22–24, the Gosset polytopes $2_{21}, 3_{21}, 4_{21}$ (orbits of $W(E_i)$, $i = 6, 7, 8$). 4 is $(10)_0$ (see Table 44.10). Ex. 8 is 22 [2] made complex. 23: multiply 22 by $1, i, j, k$ and make real.

Reflection groups: normal vectors to their reflecting hyperplanes. 9, 12, 13, 18, 21, 25–27. For 25: multiply 13 by 24 units ($2.A_4$) of $\mathbf{H}$ and make real, or use Λ_{16}.

Other: 10: via group ${}^3D_4(2)$. 14, 15: via matrices and difference sets (see Construction 44.15). 19: via complex cube roots of unity. 20: from Rudvalis group. Using groups, orbits and codes generally [3]. $K_{12}, 4_{21}$ are the real forms of 7, 24.

Knowledge of lattices, polytopes, and reflection groups owes a large debt to Coxeter; see [3, 4, 5] and references therein.

44.5 Examples In the notation below, v^P, v^S, v^{PS} denote the orbits of a vector v whose coordinates are subject, respectively, to permutations, sign changes, or both. Thus for example $(1,2)^P = \{(1,2),(2,1)\}$ and $(1,i)^S = \{(1,i),(-1,i),(1,-i),(-1,-i)\}$.

1. E_8. A 3-design of 120 points in $\mathbf{P}$ (7-design in Ω_8) from vectors $(1,1,0,\ldots,0)^{PS}$, $(1,\ldots,1)^{PS}$ (even numbers of -1's).
2. *Ex.* 9. Cyclic permutations of coordinates and of (ijk) in
$$(1,0,\ldots,0),\ (0,1,w,w,1)^S,\ (0,1,iwi,kwk,i)^S,$$
where w is a cube root of unity in $\mathbf{H}$, $w = (-1+i+j+k)/2$.
3. *Ex.* 12, 26, 10 form a nested sequence of generalized hexagons in $\mathbf{CP}^2$, $\mathbf{HP}^2$, $\mathbf{OP}^2$ under $x \approx y$ if and only if $\langle x, y\rangle = 0$. Vectors for Ex 12: $(1,0,0)^P$, $(1,1,0)^{PS}$, $(a,1,1)^{PS}$, where a is a fixed complex root of $z^2 - z + 2 = 0$.

44.2 Special Bounds and the Value of t

For a finite set B, let $B(x) = \prod_{b\in B}(x-b) = \sum \sigma_r(B)x^{w-r}$ $(0 \le r \le w = |B|)$.

Let $\varepsilon = 1$ if $0 \in A$ and $X \subseteq \mathbf{P}$; otherwise, let $\varepsilon = 0$. Set $e = s - \varepsilon$, $\Lambda = v/A(1)$, and $\sigma_r = \sigma_r(A)$. The *indicator coefficients* $a_i = a_i^{\Omega}, a_i^{\varepsilon}$, for $X \subseteq \Omega, \mathbf{P}$, defined by $\Lambda x^{-\epsilon}A(x) = \sum a_i Q_i(x)$ $(0 \le i \le e)$, are, respectively,
$\Lambda \sum_{r=0}^{(s-i)/2}(i+2r)!\sigma_{s-i-2r}/2^{i+2r}(N)_{i+r}r!$, and $\Lambda\sum_{k=i}^{e}\sigma_{e-k}(m)_{k+\epsilon\ i}(k)/(N)_{k+i+\varepsilon}$.

44.6 Theorem If all $a_i \ge 0$, then all $a_i \le 1$. Let $a_0,\ldots,a_e > 0$. Then (1) X is an $(s+i)$–design if and only if $a_i = 1$ (for any $0 \le i \le e$), (2) X is an s-design if the following *Special Bound* (equivalent to $a_0 \le 1$) is attained, in respective cases $\Omega, \mathbf{P}$, where $u = \lfloor s/2\rfloor$:
$$(N)_u A(1)/\textstyle\sum_{r=0}^{u}(m)_r(N+r)_{u-r}\sigma_{s-2r}, \qquad (N)_e A(1)/\sum_{i-0}^{e}(m)_i(N)_{e-i}\sigma_{e-i}.$$

44.7 Table Unusual cases in $\mathbf{P}$ meeting the Special Bound of Theorem 44.6.

Ex	$\mathbb{F}^d$	v	A	$s\ t$	What Is Special
13	$\mathbf{H}^4$	180	$\{0, \frac{1}{4}, \frac{1}{2}\}$	3 3	Theorem 44.6(b) valid with $\varepsilon - 0$, not 1.
22	$\mathbf{H}^4$	64	$\{\frac{1}{9}, \frac{1}{3}\}$	2 2	$0 \notin A$ the only such case in $\mathbf{P}$?
27	$\mathbf{H}^3$	315	$\{0, \frac{1}{4}, \frac{1}{2}, \frac{(3\pm\sqrt{5})}{8}\}$	5 5	Large $A : s > 3$.
28	$\mathbf{C}^{12}$	32760	$\{0, \frac{1}{12}, \frac{1}{4}, \frac{1}{3}\}$	4 5	Large $A : s > 3$.

44.8 Example Ex. 11d (Table 44.16) is 47104 points in Ω_{23} with $s = 5$, indicator coefficients $a_i = 1, 1, 1, \frac{55}{87}, \frac{25}{29}, \frac{375}{899}$. By Theorem 44.6 it is a 7-design. Indeed, $t(X) = 7$. The respective subdegrees (§44.3) are 275, 7128, 22275, 15400, 2025, by Theorem 44.11(a).

44.3 Strongly Regular Graphs from Spherical t-Designs

X is *regular* if each *subdegree* $d_\alpha(x) = |\{y \in X : \langle x, y\rangle = \alpha\}|$ is a constant, d_α.

X has *subconstituents* $X_\alpha = X_\alpha(x) = \{z \in X : \langle x, z\rangle = \alpha\}$ $(\alpha \in A,\ x \in X)$.

The *graph* Γ of X (with respect to inner product α) has x, y adjacent if and only if $\langle x, y\rangle = \alpha$.

44.9 Theorem

1. The graph Γ, with respect to 0, of a 2-design X in $\mathbf{P}$ with $A = \{0, \alpha\}$ is strongly regular (§VI.5, hence regular), connected, and not a conference or complete graph.
2. Γ has integral eigenvalues: $d_\alpha, r, -q$, where $q \geq 2$, $0 \leq r \leq q(q-2)$, and $\alpha = 1/q$.
3. X is a 3-design if and only if $r = q(q-2)$, when the subconstituents are 2-designs.

44.10 Table Strongly regular graphs from 2-designs in $\mathbf{P}$. Unknown cases for $2 \leq q \leq 4$ are $(q, r) = (3, 1), (3, 3), \mathbb{F} = \mathbb{R}$. Parameters: $d = q + r(q-1)/m(q+r)$, $v = d(1 + mq + (m+1)r)$, $k = \alpha^{-1}(v/d - 1)$, $\mu = mq(q+r)$, $\lambda = \mu - q + r$. Also, note that $n+$ denotes the conjectured extension of Ex. n.

q	r	$\mathbb{F}^d$	v	k	λ	μ	Ex.	q	r	$\mathbb{F}^d$	v	k	λ	μ	Ex.
2*	0	$\mathbb{R}^2$	4	2	0	2	1	4	2	$\mathbb{C}^5$	45	32	22	24	18
4	0	$\mathbb{R}^4$	12	8	4	8	D_4	4*	8	$\mathbb{C}^6$	126	80	52	48	7
4	2	$\mathbb{R}^6$	36	20	10	12	E_6	9	0	$\mathbb{C}^9$	90	81	72	81	19
4	4	$\mathbb{R}^7$	63	32	16	16	E_7	16	64	$\mathbb{C}^{28}$	4060	2304	1328	1280	20
4*	8	$\mathbb{R}^8$	120	56	28	24	E_8	2*	0	$\mathbf{H}^2$	10	8	6	8	3
9	7	$\mathbb{R}^{16}$	256	135	70	72	23	3	0	$\mathbf{H}^3$	21	18	15	18	16+
9	39	$\mathbb{R}^{22}$	1408	567	246	216	$(11b)_0$	4	0	$\mathbf{H}^4$	36	32	28	32	21
9*	63	$\mathbb{R}^{23}$	2300	891	378	324	11b	4*	8	$\mathbf{H}^5$	165	128	100	96	9
2*	0	$\mathbb{C}^2$	6	4	2	4	2	9	0	$\mathbf{H}^9$	171	162	153	162	19+
3	0	$\mathbb{C}^3$	12	9	6	9	16	2*	0	$\mathbf{O}^2$	18	16	14	16	4
3*	3	$\mathbb{C}^4$	40	27	18	18	6	3	0	$\mathbf{O}^3$	39	36	33	36	16+
4	0	$\mathbb{C}^4$	20	16	12	16	17								

44.4 t-Designs, Subdegrees, and Association Schemes

Parameters: $c_i = v(m)_i/(N)_i$, $C^\varepsilon_{ik} = v(m+\varepsilon)_i(m)_{k+\varepsilon}/(N)_{i+k+\varepsilon}$, $C_{ij} = C^0_{ij}$.

X is an *association scheme* (§IV.1, hence regular) if, for all $\alpha, \beta \in A$, the *intersection number* $p_{\alpha\beta}(x, y) = |\{z \in X : \langle x, z\rangle = \alpha, \langle y, z\rangle = \beta\}|$ has the same value $p^\gamma_{\alpha\beta}$ for all $x, y \in X$ with $\langle x, y\rangle = \gamma$. The *intersection matrix* P^γ has α, β entry $p^\gamma_{\alpha\beta}$. Case $s = 2$ is a strongly regular graph with respect to either 'angle'.

Intersection polynomials $g_{ik}(x) = g^\Omega_{ik}(x)$, $g^\varepsilon_{ik}(x)$ for Ω, $\mathbf{P}$ are given by $g^\Omega_{pq}(x) = 0$ if $p-q$ is odd, $g^\Omega_{2i+\varepsilon,2k+\varepsilon}(x) = x^\varepsilon g^\varepsilon_{ik}(x^2)$, and $g^\varepsilon_{ik}(x) = C^\varepsilon_{ik}\sum_{r=0}\binom{k}{r}\{{}_r(i)/(m+\varepsilon)_r\}x^r$.

44.11 Theorem Let $B = A\backslash\{\alpha\}$, $g_a = B(\alpha)$, $\sigma_r^\lambda = \sigma_r(A\backslash\{\lambda\})$.

1. If $t \geq s-1$, then X is regular, the sub-degrees d_α for cases Ω, $\mathbf{P}$ being $[c_0\sigma_{s-1}+c_1\sigma_{s-3}+\ldots-B(1)]/B(\alpha)$, $[c_0\sigma_{s-1}+c_1\sigma_{s-2}+\ldots-B(1)]/B(\alpha)$ $(\sigma_r = \sigma_r^\alpha)$. When $A = \{\alpha, \beta\}$ we have $d_\alpha = [1+\beta(v-1)-cv/d]/(\beta-\alpha)$, where $c = 0$ for Ω and 1 for $\mathbf{P}$.
2. If $t \geq 2s-2$, then X is a (Q-polynomial) association scheme with intersection numbers $P_{\alpha\beta}^\gamma = \frac{1}{g_\alpha g_\beta}\sum_{i,j=0}^{s-1}(g_{ij}(\gamma)-\gamma^i-\gamma^j)\sigma_{s-i-1}^\alpha\sigma_{s-j-1}^\beta$.

44.12 Table Some t-designs in $\mathbf{P}$ with $A = \{0, \alpha, \beta\}$ and $t \geq 2s-3$.

$\mathbb{F}^d$	A	v	s	t	d_0	d_α	d_β	Ex	scheme
$\mathbb{R}^{16}$	$\{0, \frac{1}{16}, \frac{1}{4}\}$	2160	3	3	855	1024	280	25	
$\mathbb{R}^{24}$	$\{0, \frac{1}{16}, \frac{1}{4}\}$	98280	3	5	46575	47104	4600	11*	Yes
$\mathbb{R}^{23}$	$\{0, \frac{1}{16}, \frac{1}{4}\}$	46575	3	3	21582	22528	2464	$(11)_0$	Yes
$\mathbf{C}^3$	$\{0, \frac{1}{4}, \frac{1}{2}\}$	21	3	3	4	8	8	12	Yes
$\mathbf{C}^4$	$\{0, \frac{1}{4}, \frac{1}{2}\}$	60	3	3	15	32	12	24	No
$\mathbf{H}^3$	$\{0, \frac{1}{4}, \frac{1}{2}\}$	63	3	3	6	32	24	26	Yes
$\mathbf{H}^4$	$\{0, \frac{1}{4}, \frac{1}{2}\}$	180	3	3	27	128	24	13	No
$\mathbf{O}^3$	$\{0, \frac{1}{4}, \frac{1}{2}\}$	819	3	5	18	512	288	10*	Yes

44.13 Table Symmetric intersection matrices written as vectors $P = [p_{11}\ldots, p_{22}\ldots, p_{33}\ldots, \ldots]$. Ordering $0 < \alpha < \beta < \ldots$. For Ex. 27, $0 < \frac{(3-\sqrt{5})}{2} < \frac{1}{4} < \frac{1}{2} < \frac{(3+\sqrt{5})}{2}$.

(**10**) [1 0 16, 256 256, 16], [0 9 9, 304 198, 81], [1 16 1, 352 144, 142].
(**11**) [21582 22528 2464, 22528 2048, 88], [22275 22275 2025, 22528 2300, 275], [24948 20736 891, 23552 2816, 892].
(**12**) [1 0 2, 4 4, 2], [0 2 2, 3 2, 4], [1 2 1, 2 4, 2].
(**26**) [1 0 4, 16 16, 4], [0 3 3, 16 12, 9], [1 4 1, 16 12, 10].
(**27**) [1 0 0 8 0, 0 16 0 16, 64 64 16, 8 0, 0], [0 0 5 0 5, 5 5 20 1, 85 40 25, 20 0, 1] [0 1 4 4 1, 1 17 8 5, 77 44 17, 16 8, 1], [1 0 8 1 0, 8 16 8 0, 88 32 16, 30 8, 8], [0 5 5 0 0, 1 25 0 1, 85 40 5, 20 20, 5].

44.5 Tight t-Designs

Write $\varepsilon = 1$ if $0 \in A$, otherwise $\varepsilon = 0$. Let $R_e(x) = \sum_{i=0}^e Q_i(X)$ $(e = s - \varepsilon)$.

A t-design X is *tight* if any of the (equivalent) following holds: (1) $v = R_e(1)$, (2) $a_0 = a_1 = \ldots = a_e = 1$, (3) $t = 2s - \varepsilon$. The elements of A are then the roots of $x^\varepsilon R_e(x)$. Every t-design satisfies $t \leq 2s - \varepsilon$, and the *Absolute Bound* $v \leq R_e(1)$.

44.14 Table Feasible $t, A, \mathbb{F}$ for tight t-designs, and some examples. Here $d \geq 2$ except for the $\mathbb{F} = \mathbb{R}$ cases, when $d \geq 3$ (a tight t-design in Ω_2 is a regular $(t+1)$-gon, $t \geq 2$). The diagonals of an icosahedron form the unique tight 2-design in $\mathbb{R}\mathbf{P}^2$.

t	A	$\mathbb{R}\mathbf{P}^{d-1}$	$\mathbf{C}\mathbf{P}^{d-1}$	$\mathbf{H}\mathbf{P}^{d-1}$	$\mathbf{O}\mathbf{P}^{d-1}$	t	A	Ω (Non-Antipodal)
2	$\{\alpha\}$	$3_{21}, S3$	5, 8	?	?	2	$\{\alpha\}$	Regular simplex
3	$\{0, \alpha\}$	E_8, 11b	2, 6, 7	3, 9	4			only
5	$\{0, \alpha, \beta\}$	11 only	None	None	10 only	4	$\{\alpha\beta\}$	2_{21}, $Mc1$

44.15 Construction in Ω involving (mostly) tight t-designs [5].

1. *From a Graph.* Let M be the ± 1 adjacency matrix of a graph on v vertices with least eigenvalue λ, of multiplicity f. Then $M - \lambda I$ is the Gram matrix of vectors

in equiangular lines in $\mathbb{R}^{v-f}$ with $\cos^2\theta = 1/\lambda^2$. If $t \geq 1$, one can append $-x$ for every vector x to get an antipodal $2t+1$ design of $2v$ vectors, $A = \{-1, \pm\cos\theta\}$; $0, 1$ adjacency gives $\{-1, 0, \pm\cos\theta\}$.

2. *Derived designs.* For a t-design X in Ω_{d+1}, write $A^* = A\backslash\{-1\}, s^* = |A^*|$. The *derived design* $D_\alpha(X)$ of X with respect to $\alpha \in A^*$ (and $\mathbf{n} \in \mathbf{X}$) is the orthogonal projection of the subconstituent X_α onto the hyperplane $\langle x, n\rangle = 0$, scaled to lie in Ω_d. It is a $(t+1-s^*)$-design with angle set $\{\beta - \alpha^2)/(1-\alpha^2) : \beta \in A^*\}$.

44.16 Table Designs in Ω_d by Constructions 44.15. Note $D_0(X) = X_0$.

Ex	v	d	A	t	Source	Graph/Aut
$S3^*$	552	23	$\{-1, \pm\frac{1}{5}\}$	5	Graph	Co_3
$Mc1^*$	275	22	$\{-\frac{1}{4}, \frac{1}{6}\}$	4	$D_{1/5}$	McLauchlan
$Mc2$	162	21	$\{-\frac{2}{7}, \frac{1}{7}\}$	3	$D_{1/6}$	
$Mc3$	105	20	$\{-\frac{5}{16}, \frac{1}{8}\}$	2	$D_{1/7}$	
E_8^*	240	8	$\{-1, 0, \pm\frac{1}{2}\}$	7	Lattice	$W(E_8)$
3_{21}^*	56	7	$\{-1, \pm\frac{1}{3}\}$	5	$D_{1/2}$	Chang graphs
2_{21}^*	27	6	$\{-\frac{1}{2}, \frac{1}{4}\}$	4	$D_{1/3}$	Schläfli graph
	16	5	$\{-\frac{3}{5}, \frac{1}{5}\}$	3	$D_{1/4}$	Clebsch graph
	10	4	$\{-\frac{2}{3}, \frac{1}{6}\}$	2	$D_{1/5}$	Petersen graph
$11a^*$	196560	24	$\{-1, 0, \pm\frac{1}{4}, \pm\frac{1}{2}\}$	11	Λ_{24}	$2.Co_1$
$11b^*$	4600	23	$\{-1, 0, \pm\frac{1}{3}\}$	7	$D_{1/2}$	Co_2
$11c$	891	22	$\{-\frac{1}{2}, -\frac{1}{8}, \frac{1}{4}\}$	5	$D_{1/3}$	
$11d$	47104	23	$\{-\frac{3}{5}, -\frac{1}{3}, -\frac{1}{15}, \frac{1}{5}, \frac{7}{15}\}$	7	$D_{1/4}(\Lambda_{24})$	
$H1$	100	22	$\{-\frac{4}{11}, \frac{1}{11}\}$	3	Graph	Higman-Sims
G	2048	23	$\{-\frac{9}{23}, -\frac{1}{23}, \frac{7}{23}\}$	3	Golay code	M_{24}

44.6 See Also

§IV.1	Association schemes.
§VI.5	Strongly regular graphs.
[3]	Cubature, orbits, harmonics, polytopes, root systems.
[5]	The foundation paper for spherical designs.
[6]	General formulation in terms of spherical polynomial spaces.

References

[1] B. Bajnok, Construction of spherical t-designs, *Geom. Ded.* 43 (1992), 167-179.

[2] E. Bannai and S.G. Hoggar, Tight t-designs and squarefree integers, *Europ. J. Combin.* 10 (1989), 113-135.

[3] J. Chandler et al. , *The Geometric Vein*, Springer-Verlag, Berlin 1981, pp. 203-230.

[4] J.H. Conway and N.J.A. Sloane, *Sphere Packings, Lattices, and Groups*, Springer-Verlag, Berlin, 1988. See also *Discrete Math.* 106/107 (1992), 255-264.

[5] P. Delsarte, J-M. Goethals and J.J. Seidel, Spherical codes and designs, *Geom. Ded.* 6 (1977), 363-388.

[6] C.D. Godsil, *Algebraic Combinatorics*, Chapman and Hall, 1993.

[7] S.G. Hoggar, t-designs in projective spaces, *Europ. J. Combin.* 3 (1982), 233-254.

[8] S.G. Hoggar, t-designs with general angle set, *Europ. J. Combin.* 13 (1992), 257-271.

45 Starters

Jeffrey H. Dinitz

45.1 Definitions and Examples

A *starter* in the odd order abelian group G (written additively), where $|G| = g$ is a set of unordered pairs $S = \{\{s_i, t_i\} : 1 \leq i \leq (g-1)/2\}$ which satisfies:

1. $\{s_i : 1 \leq i \leq (g-1)/2\} \cup \{t_i : 1 \leq i \leq (g-1)/2\} = G\backslash\{0\}$, and
2. $\{\pm(s_i - t_i) : 1 \leq i \leq (g-1)/2\} = G\backslash\{0\}$.

A *strong starter* is a starter $S = \{\{s_i, t_i\}\}$ in the abelian group G with the additional property that $s_i + t_i = s_j + t_j$ implies $i = j$, and for any i, $s_i + t_i \neq 0$.

A *skew starter* is a starter $S = \{\{s_i, t_i\}\}$ in the abelian group G with the additional property that $s_i + t_i = \pm(s_j + t_j)$ implies $i = j$, and for any i, $s_i + t_i \neq 0$.

45.1 Remarks

1. These definitions extend to nonabelian groups.
2. A skew starter is a strong starter.
3. Much of the material of this section is covered in the survey paper [1].

45.2 Example A strong starter in $\mathbb{Z}_{17}$.

$$\{9,10\}, \{3,5\}, \{13,16\}, \{11,15\}, \{1,6\}, \{2,8\}, \{7,14\}, \{4,12\}$$

Let $S = \{\{s_i, t_i\} : 1 \leq i \leq (g-1)/2\}$ and $T = \{\{u_i, v_i\} : 1 \leq i \leq (g-1)/2\}$ be two starters in G. Without loss of generality, assume that $s_i - t_i = u_i - v_i$, for all i. Then S and T are *orthogonal* starters if $u_i - s_i = u_j - s_j$ implies $i = j$, and if $u_i \neq s_i$ for all i.

In any abelian group G of odd order, the set of pairs $P = \{\{x, -x\} : x \in G\}$ is a starter, the *patterned* starter. See Example VI.4.33.

45.3 Example Two orthogonal starters in $\mathbb{Z}_7$ (P is the patterned starter).

$$S = \{\{2,3\}, \{4,6\}, \{1,5\}\}, \qquad P = \{\{3,4\}, \{6,1\}, \{5,2\}\}$$

An *adder* for the starter $S = \{\{s_i, t_i\} : 1 \leq i \leq (g-1)/2\}$ is an ordered set $A_S = \{a_1, a_2, \ldots, a_{(g-1)/2}\}$ of $(g-1)/2$ distinct nonzero elements from G such that the set $T = \{\{s_i + a_i, t_i + a_i\} : 1 \leq i \leq (g-1)/2\}$ is also a starter in the group G.

45.4 Remarks The starters S and T (in the definition) are orthogonal starters. Conversely, given any two orthogonal starters, S and T the adder A_S is easily found as the "differences" between the pairs in the two starters. In Example 45.3 the adder is $A_S = \{1, 2, 4\}$.

45.2 The Connection Between Starters and Other Designs

45.5 Theorem The existence of d pairwise orthogonal starters in an abelian group of order n implies the existence of a Room d-cube of side n. (See §IV.39)

45.6 Construction Let $S = \{\{s_i, t_i\} : 1 \le i \le (n-1)/2\}$ and $T = \{\{u_i, v_i\} : 1 \le i \le (n-1)/2\}$ be two orthogonal starters in G (usually $\mathbb{Z}_n$), (n odd) with $A_S = \{a_i : 1 \le i \le (n-1)/2\}$ the associated adder. Let R be an $n \times n$ array indexed by the elements of G. Each $\{s_i, t_i\} \in S$ is placed in the first row in cell $R(0, -a_i)$. This is now cycled so that the pair $\{s_i + x, t_i + x\}$ is in the cell $R(x, -a_i + x)$, where all arithmetic is performed in G. Finally, for each $x \in G$ place the pair $\{\infty, x\}$ in cell $R(x, x)$. Then R is a Room square of side n.

45.7 Example The Room square of side 7 in Example IV.39.1 was constructed from the orthogonal starters given in Example 45.3 using Construction 45.6.

45.8 Remark If $S = \{\{s_i, t_i\} : 1 \le i \le (g-1)/2\}$ is a starter, then $-S = \{\{-s_i, -t_i\} : 1 \le i \le (g-1)/2\}$ is also a starter.

45.9 Theorem If there exists a strong starter S in an abelian group of odd order n, then the starters S, $-S$ and P are pairwise orthogonal.

45.10 Theorem If there exists a starter in an abelian group of odd order $2n+1$, then there exists an odd balanced tournament design OBTD(n) (see §IV.3 and §V.7).

45.11 Theorem Let G be a finite group of order $2n-1$. A starter in G is equivalent to

1. A (2,1)-orthomorphism in G (see §IV.5.7).
2. A 1-rotational 1-factorization of K_{2n} if $G = \mathbb{Z}_{2n-1}$ (see §VI.4.5).

45.12 Remark The connection between starters and Skolem sequences is given in Remark IV.43.14. The construction of a round robin tournament from a starter is given in Remark V.7.6.

45.3 Constructions for Orthogonal Starters

Let q be a prime power that can be written in the form $q = 2^k t + 1$, where $t > 1$ is odd and let ω be a primitive element in the field $\mathbb{F}_q$. Then define

1. C_0 to be the multiplicative subgroup of $\mathbb{F}_q \backslash \{0\}$ of order t.
2. $C_i = \omega^i C_0$, $0 \le i \le 2^k - 1$ to be the cosets of C_0 (*cyclotomic classes*).
3. $\Delta = 2^{k-1}$, $H = \cup_{i=0}^{\Delta-1} C_i$ and $C_i^a = (1/(a-1))C_i$.

45.13 Theorem Let $T = \{\{x, \omega^\Delta x\} : x \in H\}$. Then T is a skew starter in the additive subgroup of $\mathbb{F}_q$.

45.14 Theorem For each $a \in C_\Delta$, let $S_a = \{\{x, ax\} : x \in \cup_{i=0}^{\Delta-1} C_i^a\}$. Then for any $a \in C_\Delta$, S_a is a strong starter in the additive group of $\mathbb{F}_q$. Further, S_a and S_b are orthogonal if $a, b \in C_\Delta$ with $a \ne b$. Hence, the set $\{S_a | a \in C_\Delta\}$ is a set of t pairwise orthogonal starters of order q.

45.15 Theorem Let $p > 5$ be a Fermat prime. There exists a strong starter in the additive group of $\mathbb{F}_p$.

45.16 Remark No strong starter in the additive group of $\mathbb{F}_3$, $\mathbb{F}_5$, or $\mathbb{F}_9$ exists.

45.17 Remark A very effective method for constructing strong starters uses a hill-climbing algorithm. A description of hill-climbing (for Steiner triple systems) is given in §VI.9.3. Hill-climbing for strong starters is described in [1].

45.4 Existence and Enumeration of Starters

45.18 Table [2] The number of distinct starters in the $\mathbb{Z}_n$, for $n \leq 27$.

n	3	5	7	9	11	13	15	17	19	21	23	25	27
# Starters	1	1	3	9	25	133	631	3857	25905	188181	1515283	13376125	128102625

45.19 Remark It remains a conjecture that there exists a strong starter in $\mathbb{Z}_n$ for all $n > 9$. It has been shown by computer to hold when $7 \leq n \leq 999$, $n \neq 9$. More generally, Horton [2] conjectured that any abelian group of odd order has a strong starter, with the exceptions of the groups $\mathbb{Z}_3$, $\mathbb{Z}_5$, $\mathbb{Z}_9$, and $\mathbb{Z}_3 \times \mathbb{Z}_3$. The strongest general existence result for strong starters is Theorem 45.20.

45.20 Theorem Let G be an abelian group of order n where $n > 5$ and n is relatively prime to 6. Then there is a strong starter in G.

Two strong starters S_1 and S_2 in an abelian group G are *equivalent* if there exists $\theta \in Aut(G)$ such that for some $f \in Aut(G)$, $\theta(\{x,y\}) = \{f(x), f(y)\} \in S_2$ for all $\{x,y\} \in S_1$.

45.21 Table The number of distinct and inequivalent strong starters in $\mathbb{Z}_n$, for $n \leq 27$ [3].

n	3	5	7	9	11	13	15	17	19	21	23	25	27
Distinct	0	0	1	0	4	4	32	224	800	6600	27554	158680	1249650
Inequivalent	0	0	1	0	2	2	4	14	52	555	1257	7934	69425

45.22 Table [3] The inequivalent strong starters in $\mathbb{Z}_n$, for $n \leq 19$.

$n = 7$								
1.	2,3	4,6	1,5					
$n = 11$								
1.	1,2	7,9	3,6	4,8	5,10			
2.	2,3	5,7	6,9	8,1	10,4			
$n = 13$								
1.	1,2	8,10	3,6	5,9	7,12	11,4		
2.	3,4	5,7	9,12	10,1	6,11	2,8		
$n = 15$								
1.	2,3	6,8	7,10	9,13	11,1	14,5	12,4	
2.	2,3	8,10	4,7	12,1	9,14	5,11	6,13	
3.	2,3	12,14	8,11	6,10	4,9	1,7	13,5	
4.	3,4	7,9	11,14	2,6	8,13	10,1	5,12	
$n = 17$								
1.	1,2	3,5	10,13	11,15	4,9	6,12	7,14	8,16
2.	1,2	3,5	12,15	7,11	9,14	4,10	6,13	8,16
3.	1,2	4,6	8,11	12,16	10,15	3,9	7,14	5,13
4.	1,2	5,7	8,11	12,16	9,14	4,10	13,3	15,6
5.	1,2	6,8	12,15	7,11	4,9	10,16	13,3	14,5
6.	1,2	7,9	3,6	12,16	10,15	8,14	4,11	5,13
7.	1,2	8,10	4,7	11,15	9,14	16,5	6,13	12,3
8.	1,2	12,14	4,7	6,10	11,16	3,9	8,15	5,13
9.	2,3	4,6	9,12	7,11	13,1	10,16	8,15	14,5
10.	2,3	7,9	12,15	1,5	8,13	10,16	4,11	6,14
11.	2,3	11,13	9,12	6,10	16,4	1,7	8,15	14,5
12.	3,4	5,7	12,15	6,10	13,1	8,14	9,16	11,2
13.	3,4	7,9	12,15	2,6	8,13	10,16	11,1	14,5
14.	4,5	10,12	15,1	7,11	3,8	13,2	9,16	6,14

$n = 19$

1.	1,2	3,5	7,10	11,15	8,13	12,18	9,16	17,6	14,4
2.	1,2	3,5	7,10	12,16	8,13	9,15	11,18	17,6	14,4
3.	1,2	3,5	7,10	12,16	13,18	9,15	4,11	6,14	8,17
4.	1,2	3,5	9,12	13,17	6,11	10,16	8,15	18,7	14,4
5.	1,2	3,5	13,16	11,15	4,9	6,12	7,14	10,18	8,17
6.	1,2	3,5	15,18	10,14	8,13	6,12	4,11	9,17	7,16
7.	1,2	4,6	5,8	13,17	11,16	9,15	7,14	10,18	3,12
8.	1,2	4,6	14,17	5,9	11,16	7,13	8,15	10,18	3,12
9.	1,2	5,7	9,12	13,17	11,16	4,10	15,3	6,14	18,8
10.	1,2	5,7	10,13	14,18	3,8	11,17	9,16	4,12	6,15
11.	1,2	5,7	11,14	12,16	8,13	4,10	15,3	17,6	9,18
12.	1,2	6,8	7,10	14,18	12,17	9,15	4,11	16,5	13,3
13.	1,2	6,8	12,15	18,3	9,14	10,16	4,11	5,13	17,7
14.	1,2	6,8	13,16	7,11	4,9	12,18	10,17	14,3	15,5
15.	1,2	7,9	5,8	10,14	11,16	12,18	15,3	17,6	4,13
16.	1,2	7,9	12,15	13,17	6,11	18,5	3,10	8,13	14,4
17.	1,2	7,9	14,17	12,16	3,8	4,10	11,18	5,13	6,15
18.	1,2	8,10	4,7	14,18	11,16	9,15	5,12	17,6	13.3
19.	1,2	8,10	12,15	13,17	4,9	5,11	18,6	14,3	7,16
20.	1,2	8,10	14,17	11,15	4,9	7,13	18,6	16,5	3,12
21.	1,2	11,13	3,6	5,9	12,17	10,16	8,15	18,7	14,4
22.	1,2	11,13	7,10	12,16	4,9	18,5	15,3	6,14	8,17
23.	1,2	11,13	12,15	3,7	5,10	17,4	9,16	6,14	18,8
24.	1,2	11,13	15,8	3,7	4,9	10,16	15,12	6,14	8,17
25.	1,2	12,14	6,9	7,11	13,18	17,4	3,10	8,16	15,5
26.	1,2	12,14	6,9	18,3	11,16	4,10	8,15	5,13	17,7
27.	1,2	12,14	10,13	5,9	3,8	11,17	16,4	18,7	6,15
28.	1,2	14,16	6,9	11,15	8,13	18,5	3,10	4,12	17,7
29.	1,2	15,17	13,16	4,8	5,10	6,12	7,14	3,11	9,18
30.	1,2	16,18	5,8	6,10	12,17	9,15	7,14	3,11	4,13
31.	2,3	5,7	12,15	9,13	6,11	14,1	16,4	10,18	8,17
32.	2,3	5,7	15,18	6,10	8,13	11,17	9,16	12,1	14,4
33.	2,3	6,8	9,12	16,1	13,18	5,11	10,17	7,15	14,4
34.	2,3	7,9	5,8	12,16	10,15	14,1	11,18	17,6	4,13
35.	2,3	7,9	12,15	1,5	8,13	10,16	11,18	17,6	14,4
36.	2,3	8,10	14,17	12,16	4,9	5,11	13,1	18,7	6,15
37.	2,3	9,11	4,7	14,18	12,17	10,16	1,8	5,13	6,15
38.	2,3	10,12	5,8	7,11	13,18	14,1	16,4	9,17	6,15
39.	2,3	12,14	6,9	13,17	11,16	4,10	1,8	18,7	15,5
40.	3,4	5,7	15,18	10,14	8,13	11,17	2,9	12,1	16,6
41.	3,4	8,10	12,15	1,5	13,18	11,17	2,9	6,14	7,16
42.	3,4	8,10	12,15	16,1	13,18	5,11	2,9	6,14	17,7
43.	3,4	9,11	5,8	10,14	12,17	15,2	13,1	18,7	16,6
44.	3,4	10,12	15,18	1,5	8,13	11,17	2,9	6,14	7,16
45.	3,4	10,12	15,18	16,1	8,13	5,11	2,9	6,14	17,7
46.	3,4	11,13	7,10	12,16	15,1	18,5	2,9	6,14	8,17
47.	3,4	14,16	7,10	1,5	8,13	9,15	11,18	17,6	12,2
48.	4,5	8,10	11,14	9,13	17,3	15,2	18,6	12,1	7,16
49.	4,5	12,14	17,1	6,10	3,8	9,15	11,18	13,2	7,16
50.	7,8	3,5	14,17	6,10	16,2	9,15	11,18	12,1	4,13
51.	7,8	3,5	14,17	9,13	16,2	4,10	11,18	12,1	6,15
52.	8.9	1,3	11,14	13,17	2,7	10,16	18,6	4,12	15,5

45.23 Example [2] A strong starter in $\mathbb{Z}_3 \times \mathbb{Z}_9$ ((a, b) written as ab).

$$\{18,04\}\ \{05,03\}\ \{15,27\}\ \{28,17\}\ \{07,21\}\ \{12,22\}\ \{10,06\}$$
$$\{01,23\}\ \{08,02\}\ \{14,13\}\ \{11,26\}\ \{20,24\}\ \{16,25\}$$

45.24 Remark There are few general nonexistence results on starters. Probably the most interesting is Theorem 45.25, due to Mullin and Wallis.

45.25 Theorem Suppose G is an abelian group of order $n \equiv 3 \pmod 6$ in which the 3-Sylow subgroup is cyclic. Then there is no skew starter in G.

45.26 Table A strong starter in $\mathbb{Z}_n$ for $21 \leq n \leq 81$.

n	Starter
21	{4,5}, {13,15}, {8,11}, {20,3}, {18,2}, {6,12}, {9,16}, {14,1}, {10,19}, {7,17}
23	{7,8}, {1,3}, {17,20}, {10,14}, {6,11}, {15,21}, {18,2}, {4,12}, {13,22}, {9,19}, {5,16}
25	{3,4}, {11,13}, {18,21}, {2,6}, {7,12}, {17,23}, {15,22}, {1,9}, {10,19}, {14,24}, {5,16}, {8,20}
27	{15,16}, {20,22}, {2,5}, {4,8}, {14,19}, {6,12}, {23,3}, {18,26}, {1,10}, {7,17}, {13,24}, {9,21}, {25,11}
29	{25,26}, {14,16}, {2,5}, {17,21}, {15,20}, {6,12}, {4,11}, {22,1}, {23,3}, {9,19}, {7,18}, {27,10}, {24,8}, {28,13}
31	{17,18}, {24,26}, {11,14}, {2,6}, {5,10}, {21,27}, {28,4}, {22,30}, {7,16}, {13,23}, {29,9}, {8,20}, {12,25}, {1,15}, {19,3}
33	{20,21}, {27,29}, {4,7}, {32,3}, {6,11}, {10,16}, {24,31}, {5,13}, {17,26}, {9,19}, {14,25}, {18,30}, {22,2}, {1,15}, {8,23}, {12,28}
35	{23,24}, {9,11}, {4,7}, {16,20}, {27,32}, {22,28}, {1,8}, {13,21}, {5,14}, {19,29}, {26,2}, {3,15}, {18,31}, {33,12}, {30,10}, {25,6}, {17,34}
37	{1,2}, {24,26}, {14,17}, {16,20}, {6,11}, {30,36}, {21,28}, {25,33}, {3,12}, {5,15}, {18,29}, {34,9}, {19,32}, {8,22}, {35,13}, {31,10}, {27,7}, {23,4}
39	{9,10}, {27,29}, {4,7}, {36,1}, {20,25}, {17,23}, {35,3}, {33,2}, {12,21}, {6,16}, {26,37}, {18,30}, {15,28}, {38,13}, {19,34}, {8,24}, {5,22}, {14,32}, {31,11}
41	{12,13}, {14,16}, {21,24}, {4,8}, {26,31}, {11,17}, {29,36}, {1,9}, {38,6}, {18,28}, {32,2}, {22,34}, {33,5}, {30,3}, {20,35}, {23,39}, {10,27}, {7,25}, {37,15}, {40,19}
43	{16,17}, {25,27}, {29,32}, {10,14}, {4,9}, {39,2}, {15,22}, {28,36}, {42,8}, {40,7}, {20,31}, {23,35}, {13,26}, {41,12}, {34,6}, {3,19}, {21,38}, {30,5}, {18,37}, {24,1}, {33,11}
45	{31,32}, {41,43}, {39,42}, {18,22}, {4,9}, {30,36}, {13,20}, {6,14}, {8,17}, {38,3}, {33,44}, {25,37}, {11,24}, {15,29}, {19,34}, {5,21}, {40,12}, {10,28}, {16,35}, {7,27}, {26,2}, {1,23}
47	{25,26}, {27,29}, {40,43}, {14,18}, {8,13}, {45,4}, {37,44}, {20,28}, {23,32}, {2,12}, {6,17}, {30,42}, {22,35}, {1,15}, {31,46}, {3,19}, {41,11}, {21,39}, {5,24}, {34,7}, {36,10}, {16,38}, {33,9}
49	{13,14}, {16,18}, {4,7}, {30,34}, {21,26}, {40,46}, {32,39}, {35,43}, {29,38}, {23,33}, {48,10}, {8,20}, {2,15}, {17,31}, {37,3}, {45,12}, {24,41}, {42,11}, {6,25}, {27,47}, {1,22}, {36,9}, {5,28}, {44,19}
51	{42,43}, {30,32}, {2,5}, {3,7}, {44,49}, {23,29}, {41,48}, {6,14}, {15,24}, {8,18}, {16,27}, {40,1}, {25,38}, {36,50}, {11,26}, {19,35}, {4,21}, {28,46}, {20,39}, {13,33}, {47,17}, {9,31}, {22,45}, {10,34}, {12,37}
53	{51,52}, {29,31}, {11,14}, {6,10}, {34,39}, {44,50}, {21,28}, {12,20}, {24,33}, {38,48}, {43,1}, {7,19}, {27,40}, {8,22}, {3,18}, {42,5}, {32,49}, {23,41}, {26,45}, {16,36}, {47,15}, {13,35}, {2,25}, {46,17}, {37,9}, {4,30}
55	{2,3}, {5,7}, {27,30}, {12,16}, {45,50}, {9,15}, {14,21}, {40,48}, {17,26}, {37,47}, {54,10}, {41,53}, {31,44}, {22,36}, {23,38}, {18,34}, {46,8}, {24,42}, {20,39}, {29,49}, {35,1}, {11,33}, {51,19}, {4,28}, {43,13}, {6,32}, {25,52}

n	Starter
57	{18,19}, {37,39}, {10,13}, {51,55}, {17,22}, {41,47}, {46,53}, {8,16}, {3,12}, {38,48}, {21,32}, {33,45}, {23,36}, {42,56}, {49,7}, {14,30}, {9,26}, {2,20}, {24,43}, {15,35}, {31,52}, {5,27}, {40,6}, {44,11}, {25,50}, {28,54}, {34,4}, {1,29}
59	{40,41}, {47,49}, {32,35}, {29,33}, {4,9}, {58,5}, {12,19}, {14,22}, {44,53}, {1,11}, {39,50}, {24,36}, {2,15}, {20,34}, {16,31}, {27,43}, {13,30}, {51,10}, {37,56}, {6,26}, {45,7}, {55,18}, {23,46}, {38,3}, {42,8}, {54,21}, {25,52}, {48,17}, {28,57}
61	{11,12}, {50,52}, {42,45}, {2,6}, {38,43}, {41,47}, {1,8}, {16,24}, {25,34}, {10,20}, {26,37}, {46,58}, {27,40}, {14,28}, {53,7}, {19,35}, {48,4}, {39,57}, {17,36}, {13,33}, {9,30}, {44,5}, {56,18}, {31,55}, {59,23}, {3,29}, {49,15}, {32,60}, {54,22}, {21,51}
63	{43,44}, {55,57}, {45,48}, {18,22}, {31,36}, {8,14}, {59,3}, {42,50}, {17,26}, {39,49}, {4,15}, {28,40}, {61,11}, {33,47}, {19,34}, {56,9}, {7,24}, {51,6}, {2,21}, {12,32}, {58,16}, {38,60}, {30,53}, {1,25}, {10,35}, {20,46}, {27,54}, {13,41}, {23,52}, {62,29}, {37,5}
65	{20,21}, {15,17}, {54,57}, {31,35}, {11,16}, {63,4}, {7,14}, {37,45}, {59,3}, {43,53}, {29,40}, {62,9}, {10,23}, {13,27}, {33,48}, {44,60}, {38,55}, {46,64}, {52,6}, {30,50}, {18,39}, {25,47}, {28,51}, {42,1}, {24,49}, {58,19}, {5,32}, {8,36}, {12,41}, {61,26}, {56,22}, {2,34}
67	{60,61}, {21,23}, {14,17}, {54,58}, {41,46}, {63,2}, {13,20}, {64,5}, {1,10}, {42,52}, {15,26}, {32,44}, {25,38}, {19,33}, {51,66}, {8,24}, {30,47}, {27,45}, {31,50}, {53,6}, {18,39}, {37,59}, {55,11}, {16,40}, {9,34}, {48,7}, {43,3}, {28,56}, {36,65}, {49,12}, {4,35}, {57,22}, {29,62}
69	{1,2}, {21,23}, {31,34}, {9,13}, {46,51}, {37,43}, {41,48}, {50,58}, {16,25}, {54,64}, {52,63}, {68,11}, {49,62}, {26,40}, {14,29}, {56,3}, {27,44}, {4,22}, {57,7}, {19,39}, {12,33}, {65,18}, {15,38}, {36,60}, {5,30}, {53,10}, {66,24}, {17,45}, {32,61}, {47,8}, {28,59}, {35,67}, {42,6}, {55,20}
71	{64,65}, {2,4}, {69,1}, {42,46}, {8,13}, {19,25}, {38,45}, {6,14}, {44,53}, {21,31}, {39,50}, {49,61}, {15,28}, {54,68}, {17,32}, {62,7}, {9,26}, {34,52}, {22,41}, {40,60}, {30,51}, {67,18}, {47,70}, {3,27}, {10,35}, {29,55}, {36,63}, {5,33}, {37,66}, {57,16}, {12,43}, {24,56}, {58,20}, {48,11}, {59,23}
73	{17,18}, {43,45}, {25,28}, {46,50}, {39,44}, {27,33}, {49,56}, {55,63}, {11,20}, {19,29}, {24,35}, {64,3}, {72,12}, {54,68}, {42,57}, {67,10}, {13,30}, {62,7}, {60,6}, {1,21}, {38,59}, {53,2}, {65,15}, {16,40}, {23,48}, {69,22}, {31,58}, {8,36}, {41,70}, {52,9}, {47,5}, {34,66}, {4,37}, {71,32}, {26,61}, {51,14}
75	{46,47}, {4,6}, {22,25}, {30,34}, {28,33}, {67,73}, {55,62}, {15,23}, {56,65}, {7,17}, {49,60}, {41,53}, {1,14}, {37,51}, {12,27}, {29,45}, {18,35}, {54,72}, {42,61}, {16,36}, {57,3}, {66,13}, {9,32}, {50,74}, {71,21}, {43,69}, {58,10}, {24,52}, {48,2}, {38,68}, {64,20}, {8,40}, {11,44}, {5,39}, {59,19}, {70,31}, {26,63}
77	{14,15}, {66,68}, {67,70}, {30,34}, {1,6}, {54,60}, {4,11}, {23,31}, {56,65}, {8,18}, {28,39}, {45,57}, {36,49}, {24,38}, {43,58}, {63,2}, {47,64}, {7,25}, {55,74}, {9,29}, {73,17}, {26,48}, {75,21}, {22,46}, {27,52}, {50,76}, {32,59}, {41,69}, {13,42}, {5,35}, {62,16}, {12,44}, {20,53}, {37,71}, {61,19}, {51,10}, {3,40}, {72,33}
79	{31,32}, {50,52}, {13,16}, {57,61}, {55,60}, {21,27}, {66,73}, {18,26}, {71,1}, {76,7}, {74,6}, {37,49}, {56,69}, {9,23}, {44,59}, {38,54}, {36,53}, {29,47}, {24,43}, {5,25}, {14,35}, {65,8}, {75,19}, {77,22}, {33,58}, {64,11}, {51,78}, {68,17}, {12,41}, {40,70}, {15,46}, {10,42}, {30,63}, {48,3}, {4,39}, {45,2}, {62,20}, {34,72}, {28,67}
81	{45,46}, {58,60}, {28,31}, {17,21}, {24,29}, {48,54}, {40,47}, {64,72}, {57,66}, {68,78}, {11,22}, {44,56}, {38,51}, {27,41}, {4,19}, {74,9}, {65,1}, {62,80}, {42,61}, {39,59}, {5,26}, {67,8}, {73,15}, {75,18}, {10,35}, {23,49}, {6,33}, {2,30}, {3,32}, {76,25}, {70,20}, {63,14}, {36,69}, {37,71}, {53,7}, {43,81}, {13,50}, {77,34}, {16,55}, {12,52}

45.5 Other Types of Starters

Let G be an additive abelian group of order g, and let H be a subgroup of order h of G, where $g-h$ is even (i.e., g and h are both even or both odd). A *frame starter* in $G\backslash H$ is a set of unordered pairs $S=\{\{s_i,t_i\}:1\le i\le (g-h)/2\}$ such that

1. $\{s_i:1\le i\le (g-h)/2\}\cup\{t_i:1\le i\le (g-h)/2\}=G\backslash H$; and
2. $\{\pm(s_i-t_i):1\le i\le (g-h)/2\}=G\backslash H$

45.27 Remark A starter is the special case of a frame starter when $H=\{0\}$. The concepts of strong, skew, and orthogonal for frame starters are as for starters (§45.1), replacing $\{0\}$ by H and $(g-1)/2$ by $(g-h)/2$.

45.28 Example A skew frame starter in $\mathbb{Z}_{10}\backslash\{0,5\}$: $S=\{\{3,4\},\{7,9\},\{1,8\},\{2,6\}\}$.

45.29 Theorem Suppose a pair of orthogonal frame starters in $G\backslash H$ exist, where $|G|=g$ and $|H|=h$. Then there exists a Room frame of type $h^{g/h}$ (see §IV.39.7).

45.30 Remark The construction for a Room frame from a pair of orthogonal frame starters is an easy generalization of Construction 45.6.

45.31 Theorem Let $q\equiv 1 \pmod 4$ be a prime power such that $q=2^k t+1$, where $t>1$ is odd. Then there exist t orthogonal frame starters in $(\mathbb{F}_q\times(\mathbb{Z}_2)^n)\backslash(\{0\}\times(\mathbb{Z}_2)^n)$ for all $n\ge 1$.

45.32 Remark Howell starters are discussed in §IV.26.2.

45.6 See Also

§IV.3	The connection to tournament designs is given in Theorem 45.10.
§IV.39	Starters are used to construct Room squares.
§IV.43	Starters are related to Skolem sequences.
§IV.53	Starters are related to $\mathbb{Z}$-cyclic Whist tournaments.
§VI.4.5	The existence of a starter in the cyclic group implies the existence of a 1-rotational 1-factorization of the complete graph.
§V.7	Starters are particularly useful in the construction of round robin tournaments.
[1]	A comprehensive survey on Room squares including starters; contains much of the information given in this section.
[4]	A textbook dealing with Room squares and starters.

References

[1] J.H. Dinitz and D.R. Stinson, Room squares and related designs, in: *Contemporary Design Theory: A Collection of Surveys* (J.H. Dinitz, D.R. Stinson; eds.) John Wiley & Sons, New York, 1992, pp. 137-204.

[2] J.D. Horton, Orthogonal starters in finite abelian groups, *Discrete Math.* 79 (1989/90), 265-278.

[3] W.L. Kocay, D.R. Stinson and S.A. Vanstone, On strong starters in cyclic groups, *Discrete Math.* 56 (1985), 45-60.

[4] W.D. Wallis, A.P. Street and J.S. Wallis, *Combinatorics: Room Squares, Sum-Free Sets, Hadamard Matrices*, Springer-Verlag, Berlin, 1972.

46 Trades and Defining Sets

Anne Penfold Street

46.1 Trades

Let $1 \le t < k < v$, and let V be a v-set. Its k-subsets are *blocks*. A pair of distinct collections of m blocks each, (T_1, T_2), is *mutually t-balanced* if each t-subset of V is covered by precisely the same number of blocks of T_1 as of T_2. If T_1 and T_2 are disjoint, then (T_1, T_2) is a (v, k, t) *trade of volume* m. If $m = 0$, the trade is *void.* The *foundation* of the trade, *found*(T), is the set of elements of V covered by T_1 and T_2. A trade may contain repeated blocks.

46.1 Remarks

1. The single collection T_1 is often referred to as a trade.
2. A (v, k, t) trade is a (v', k, t) trade for every $v' \ge v$.

46.2 Theorem [8] For a nonvoid (v, k, t) trade T to exist, $|found(T)| \ge k + t + 1$; hence, $v \ge k+t+1$, and $m \ge 2^t$. There always exist (v, k, t) trades with $|found(T)| = k+t+1$ and $m = 2^t$. They have unique structures.

A (v, k, t) trade with $|found(T)| = k + t + 1$ and $m = 2^t$ is a *smallest* trade. (This has also been called a minimal trade.)

46.3 Example A smallest $(6, 3, 2)$ trade:

$$T_1 = \{124,\ 156,\ 235,\ 346\}, \quad T_2 = \{125,\ 146,\ 234,\ 356\}.$$

46.4 Theorem [14] Let $v \ge k + t + 1$. Then:

1. There exist trades of volumes $2^t + 2^{t-1}$ for $t \ge 1$, $2^t + 2^{t-1} + 2^{t-2}$ for $t \ge 2$, $2^t + 2^{t-1} + 2^{t-2} + 2^{t-3}$ for $t \ge 3$, and 2^{t+1} for all t.
2. There does not exist a trade of volume m for $2^t < m < 2^t + 2^{t-1}$.

46.5 Theorem [2, 7] Let $P_{v,k,t} = [p_{X,Y}]$ be the $\binom{v}{t} \times \binom{v}{k}$ *inclusion matrix* where, for X a t-subset of V and Y a block,

$$p_{X,Y} = \begin{cases} 1 & \text{if } X \subseteq Y, \\ 0 & \text{otherwise.} \end{cases}$$

If a nonvoid (v, k, t) trade exists, then $P_{v,k,t}$ has full rank $\binom{v}{t}$ and hence its kernel is a $\mathbb{Z}$-module of dimension $\binom{v}{k} - \binom{v}{t}$. The trade (T_1, T_2) corresponds to the $\binom{v}{k}$-dimensional column vector $F_1 - F_2$, where $P_{v,k,t}F_1 = P_{v,k,t}F_2$. That is, the set of all (v, k, t) trades is the kernel of $P_{v,k,t}$.

A t-(v, k, λ_t) design is defined by a $\binom{v}{k}$-dimensional column vector B if and only if each component of B is a nonnegative integer, and $P_{v,k,t}B = \lambda J$, where J is the column vector of all ones. In the design, the total number of blocks, b, is the sum of the components of B, and the number of distinct blocks (or *support size*), b^*, is the number of nonzero components of B. The vector $B + F_1 - F_2$ also corresponds to a t-(v, k, λ_t) design, where the value of b^* (but not of b) may be different.

46.6 Theorem [7] Every smallest (v, k, t) trade is associated with a polynomial

$$\phi(x_1, \cdots, x_v) = (x_{b_1} - x_{c_1}) \cdots (x_{b_{t+1}} - x_{c_{t+1}}) x_{b_{t+2}} x_{b_{t+3}} \cdots x_{b_k},$$

where $1 \leq b_i, c_i \leq v$ and $b_1, \cdots, b_k, c_1, \cdots, c_{t+1}$ are all distinct. Multiplying out, and identifying x_i with i, for $i \in V$, gives the smallest trade (T_1, T_2), where T_1 and T_2 consist of the positive and negative terms, respectively.

46.7 Example Let $(v, k, t) = (8, 4, 2)$. Then $(x_2 - x_3)(x_4 - x_5)(x_6 - x_7)x_8$ gives the trade

$$T_1 = \{2468, 2578, 3478, 3568\},\ T_2 = \{2478, 2568, 3468, 3578\}.$$

If $w = (w^1, \cdots, w^n)$ is a nonzero vector in $\mathbb{Z}_n$ and if $1 \leq i \leq n$ is the smallest positive integer for which $w^i \neq 0$, then i is the *starting index* of w. An ordered set of nonzero vectors $\{w_1, \cdots, w_k\}$ such that x_i is the starting index of w_i is *semitriangular* if $1 \leq x_1 < \cdots < x_k \leq n$. The *starting block* of a trade is well-defined by the lexicographic ordering of the set of blocks, since T and $-T$ have the same starting block.

46.8 Proposition For b_i, c_i as in Theorem 46.6, $b_1 \cdots b_k$ is a starting block if:

$$b_1 < \cdots < b_{t+1}; \quad b_i < c_i \text{ for } 1 \leq i \leq t+1; \quad b_{t+1} < \cdots < b_k.$$

46.9 Theorem [3, 12] Let T be a smallest trade associated with the polynomial of Theorem 46.6 and satisfying the inequalities of Proposition 46.8. Then:

1. the block $B_s = j_1 \cdots j_k$ is a starting block if and only if $j_h \leq v - k - t + 2h - 2$ for $1 \leq h \leq t+1$ and $j_h \leq v - k + h$ for $t + 2 \leq h \leq k$;
2. there are $\binom{v}{k} - \binom{v}{t}$ starting blocks for given (v, k, t);
3. from a starting block $B_s = b_1 \cdots b_k$, a trade can be constructed with
$$c_i = \min(\{b_i + 1, \cdots, v\} \setminus \{c_{i+1}, \cdots, c_{t+1}, b_{i+1}, \cdots, b_{t+1}\});$$
4. the trades satisfying (3) form a semitriangular basis for the $\mathbb{Z}$-module of all (v, k, t) trades.

46.10 Example Let $(v, k, t) = (8, 4, 2)$. By Theorem 46.9 (1), $B_s = j_1 j_2 j_3 j_4$ is a starting block if and only if $j_1 \leq 2$, $j_2 \leq 4$, $j_3 \leq 6$, $j_4 \leq 8$. By Theorem 46.9(2) there are $\binom{8}{4} - \binom{8}{2} = 42$ starting blocks; these run from 1234 to 2468. For 2468, by Theorem 46.9(3), $c_3 = \min(\{b_3 + 1, \cdots, 8\}) = 7$, $c_2 = \min(\{b_2 + 1, \cdots, 8\} \setminus \{c_3, b_3\}) = 5$, and $c_1 = \min(\{b_1 + 1, \cdots, 8\} \setminus \{c_2, c_3, b_2, b_3\}) = 3$, giving the polynomial of Example 46.7. The 42 trades constructed from these starting blocks form the basis referred to in Theorem 46.9(4).

A trade T is *minimal* if it does not properly contain a trade; that is, if for every proper subcollection R, $\emptyset \subset R \subset T$, there does not exist a disjoint collection Q such that (R, Q) is mutually t-balanced.

46.11 Examples

1. A minimal $(8, 3, 2)$ trade:
$$T_1 = \{123, 147, 258, 267, 348, 456\},\ T_2 = \{127, 134, 238, 256, 458, 467\}.$$
2. A nonminimal $(6, 3, 2)$ trade:
$$T_1 = \{137, 156, 235, 267, 346, 457\},\ T_2 = \{136, 157, 237, 256, 345, 467\}.$$
Although no subcollection of T_1 forms a trade with a subcollection of T_2, $R = \{137, 156, 235, 267\} \subseteq T_1$ forms a trade with $Q = \{135, 167, 237, 256\}$.

46.2 Defining Sets

A set of blocks which is a subset of a unique t-(v,k,λ_t) design $D = (V, \mathcal{B})$ is a *defining set* of that design, denoted by $d(t\text{-}(v,k,\lambda_t))$ or dD. A defining set is *minimal* ($d_m D$) if it does not properly contain a defining set of D, and *smallest* ($d_s D$) if no defining set of D has fewer blocks; $|d_s D|$ is denoted by s.

46.12 Example Defining sets of the 2-$(9,3,1)$ design of Example I.1.15.

1. A smallest defining set: $S = \{456, 789, 258, 369\}$.
2. A minimal defining set: $S = \{123, 456, 147, 249, 159\}$.

46.13 Theorem [9, 16] Let D be a t-(v,k,λ_t) design, with minimal trades $T_1, T_2, \ldots, T_n \subseteq \mathcal{B}$ and minimal defining sets $S_1, S_2, \ldots, S_m \subseteq \mathcal{B}$. Then:

1. $S \subseteq \mathcal{B}$ is a defining set of D if and only if $S \cap T_i \neq \emptyset$ for each $i = 1, \ldots, n$;
2. if $T \subseteq \mathcal{B}$ and if $T \cap S_i \neq \emptyset$ for each $i = 1, \ldots, m$, then $T \supseteq T_i$ for some i.

46.14 Theorem [9, 16] Let S be a defining set of a design D, and let $\rho \in Aut(D)$. Then $\rho(S)$ is a defining set of $\rho(D) = D$; $Aut(S) \leq Aut(D)$; and $Aut(D) = \{\rho \mid \rho(S) \subseteq D\}$.

A group is *single-transposition-free* (STF) if none of its elements is a single transposition (ij). A design is STF if its automorphism group is STF.

46.15 Remark All t-$(v,k,1)$ designs and all symmetric 2-(v,k,λ) designs are STF.

46.16 Theorem [9, 16] Let S be a defining set of an STF design D. Then at least $v-1$ elements occur in the blocks of S; if the elements i and j appear only once each in the blocks of S, they appear in different blocks; and if D is simple, it contains $|Aut(D)|/|Aut(S)|$ isomorphic copies of S.

46.17 Theorem [9, 16] Let D be a simple STF t-design and let $S \subseteq D$, where S contains at least $v-1$ distinct elements, and there are $|Aut(D)|/|Aut(S)|$ configurations in D isomorphic to S. If S is not contained in any design with the same parameters as D, but not isomorphic to D, then S is a defining set of D.

46.18 Table The sizes of smallest defining sets for some small t-designs. For references to proofs, see [16]. Here (q) indicates the designs based on quadratic residues, and x^y means y nonisomorphic designs with the given parameters have $s = x$. When $y = 1$, it is omitted.

t	v	b	r	k	λ	s	t	v	b	r	k	λ	s
2	7	7	3	3	1	3	3	8	14	7	4	1	3
2	9	12	4	3	1	4	3	12	22	11	6	2	5
2	13	13	4	4	1	6	3	10	30	12	4	1	4
2	6	10	5	3	2	3	3	8	28	14	4	2	6^4
2	16	20	5	4	1	7	3	16	30	15	8	3	7^3, 8, 9
2	21	21	5	5	1	8	3	10	36	18	5	3	8^4, 6^2
2	11	11	5	5	2	5	3	20	38	19	10	4	8^6
2	13	26	6	3	1	9, 8	3	8	42	21	4	3	9^9, 7
2	7	14	6	3	2	6^4	3	22	77	21	6	1	8
2	10	15	6	4	2	8, 6, 5	3	24	46	23	12	5	8 (q)
2	16	16	6	6	2	9, 7, 7	3	28	54	27	14	6	≤ 11 (q)
2	15	35	7	3	1	Remark 46.19	3	32	62	31	16	7	≤ 10 (q)
2	8	14	7	4	3	6^4							

t	v	b	r	k	λ	s	t	v	b	r	k	λ	s
2	15	15	7	7	3	7^3, 8, 9							
2	9	24	8	3	2	10^5, 9^{21}, 8^{10}	4	11	66	30	5	1	5
2	9	18	8	4	3	8^9, 6^2	4	11	66	36	6	3	5
2	7	21	9	3	3	9^9, 7	4	23	253	77	7	1	8
2	19	19	9	9	4	8^6							
2	23	23	11	11	5	8 (q)							
2	27	27	13	13	6	≤ 11 (q)	5	12	132	66	6	1	5
2	31	31	15	15	7	≤ 10 (q)	5	24	759	253	8	1	8

46.19 Remark There are exactly 80 nonisomorphic STS(15) (see Table I.1.21). Moran has recently found that 25 of the STS(15) have smallest defining sets of at most 11 blocks, 43 others (including #79, #80) of at most 12 blocks, five others (including #6, #8) of at most 13 blocks, three others (#4, #5, #7) of precisely 13 blocks, three others (#2, #3, #16) of precisely 14 blocks, and the one from PG(3,2) (#1) of precisely 16 blocks.

46.20 Theorem [6] In the projective geometry $\mathrm{PG}(d,2)$, with $d \geq 2$, there exists a set of $d+1$ hyperplanes such that no point of $\mathrm{PG}(d,2)$ is incident with all of them. The hyperplanes of such a set cover a total of $(\frac{2}{3})4^d - (\frac{1}{2})3^d - 2^d + (\frac{5}{6})$ lines which form a minimal defining set of the point-line design of $\mathrm{PG}(d,2)$, that is, a $\mathrm{STS}(2^{d+1}-1)$. In the affine geometry $\mathrm{AG}(d,3)$, there is a set of $2d$ hyperplanes, two from each of d parallel classes, such that no line of $\mathrm{AG}(d,3)$ is incident with any d of them, and there is a point of $AG(d,3)$ incident with none of them. The hyperplanes of such a set cover a total of $[3^d(3^d-1) - 7^d + 1]/6$ lines which form a minimal defining set of the point-line design of $\mathrm{AG}(d,3)$, that is, a $\mathrm{STS}(3^d)$ (see §VI.7 for information on these geometries).

46.3 See Also

[1]	Construction of designs with a given number of common blocks and, hence, incidentally of trades.
[4]	Smallest trades between STS(15).
[5]	A treatment of trades (null designs) based on the approach of Graver and Jurkat and of Frankl.
[10]	Algorithm for finding smallest defining sets of designs, based on Theorems 46.13 to 46.16.
[11]	A survey of trades and their applications to constructing t-designs with varying support sizes.
[13]	Minimal defining sets for Hadamard designs.
[15]	Relations between defining sets in designs and their extensions.
[16]	A survey of defining sets with bibliography.

References

[1] C.J. Colbourn, D.G. Hoffman and C.C. Lindner, Intersections of $S(2,4,v)$ designs, *Ars Combin.* 33 (1992), 97-111.

[2] W. Foody and A.S. Hedayat, On theory and applications of BIB designs with repeated blocks, *Ann. Statist.* 5 (1977), 932-945; corrigendum, *Ann. Statist.* 7 (1979), 925.

[3] P. Frankl, Intersection theorems and mod p rank of inclusion matrices, *J. Combin. Theory A* 54 (1990), 85-94.

[4] P.B. Gibbons, Computing techniques for the construction and analysis of block designs, Computer Science, University of Toronto, Technical Report 92/76 (1976).
[5] C.D. Godsil, *Linear Algebra and Designs*, manuscript, University of Waterloo, 1994.
[6] R.A.H. Gower, Minimal defining sets for another family of Steiner triple systems, *J. Combin. Des.*, to appear.
[7] R.L. Graham, S.-Y.R. Li and W.-C.W. Li, On the structure of t-designs, *SIAM J. Alg. Disc. Meth.* 1 (1980), 8-14.
[8] J.E. Graver and W.B. Jurkat, The module structure of integral designs, *J. Combin. Theory A* 15 (1973), 75-90.
[9] K. Gray, Defining sets of single-transposition-free designs, *Utilitas Math.* 38 (1990), 97-103.
[10] C.S. Greenhill, An algorithm for finding smallest defining sets of t-designs, *J. Combin. Math. Combin. Comput.* 14 (1993), 39-60.
[11] A.S. Hedayat, The theory of trade-off for t-designs, in: *Coding Theory and Design Theory Part II: Design Theory* (D. Ray-Chaudhuri; ed.), Springer-Verlag, Berlin, 1990, 101-126.
[12] G.B. Khosrovshahi and S. Ajoodani-Namini, A new basis for trades, *SIAM J. Disc. Math.* 3 (1990), 364-372.
[13] T. Kunkle and J. Seberry, A few more small defining sets for $SBIBD(4t-1, 2t-1, t-1)$, *Bull. ICA* 12 (1994), 61-64.
[14] E.S. Mahmoodian and N. Soltankhah, On the existence of (v, k, t) trades, *Australas. J. Combin.* 6 (1992), 279-291.
[15] T. Moran, Smallest defining sets for 2-$(9, 4, 3)$ and 3-$(10, 5, 3)$ designs, *Australas. J. Combin.* 10 (1994), 279-291; corrigendum, to appear.
[16] A.P. Street, Defining sets for block designs: An update, in: *Combinatorics Advances* (C.J. Colbourn, E.S. Mahmoodian; eds.), Kluwer Academic Press, Norwell MA, to appear.

47 (t,m,s)-Nets

Charles J. Colbourn

47.1 Definitions and Basics

Let $\mathcal{I}_s$ be the unit cube of dimension s. Let $b \geq 2$ and $m \geq 1$ be integers, let $d_1, \ldots, d_s$ be nonnegative integers, and let $a_1, \ldots, a_s$ be integers satisfying $0 \leq a_i < b^{d_i}$. An *elementary interval in base* b in $\mathcal{I}_s$ is the parallelepiped of points $(x_1, \ldots, x_s)$ satisfying $a_i \leq x_i b^{d_i} < a_i + 1$.

Let $s \geq 1$, $b \geq 2$ and $m \geq t \geq 0$ be integers. A (t, m, s)*-net in base* b is a multiset M of b^m points in $\mathcal{I}_s$ so that every elementary interval in base b in which $\sum_{i=1}^{s} d_i = m - t$ (i.e., the parallelepiped has volume b^{t-m}) contains precisely b^t points in M.

47.1 Remarks For a collection of σ points in the unit s-cube, the average number of points in a parallelepiped of volume v is σv. The *local discrepancy* of a particular parallelepiped is the difference between this average and the actual number of points found. The *discrepancy* is the maximum of the local discrepancies over all parallelepipeds. The main motivation for (t, m, s)-nets is that they provide point collections with “low” discrepancy (for certain elementary intervals, in fact, the discrepancy is guaranteed to be zero) [4].

Sets of low discrepancy are used in numerical integration, and (t,m,s)-nets are used directly in the generation of pseudorandom numbers [4].

47.2 Theorem A (t,m,s)-net in base b gives a (u,n,r)-net in base b whenever $t \le u \le n \le m$ and $1 \le r \le s$.

47.3 Remark The goal is to construct (t,m,s)-nets with $m-t$ "large", since they have stronger uniformity properties. When $m-t=0$, any set of b^m points can be chosen. When $m-t=1$, take the points $(i/b,\ldots,i/b) \in \mathcal{I}_s$ for $i=0,1,\ldots,b-1$, each b^{m-1} times. Hence the cases of interest are for $m \ge t+2$.

47.2 The Connection to Orthogonal Arrays

47.4 Construction [4] A $(0,2,s)$-net in base b is equivalent to $s-2$ MOLS of order b, as follows. Let $P_1,\ldots,P_{b^2}$ be the b^2 points of a $(0,2,s)$-net in base $b \ge 2$. Let p_{ij} be the coordinate of P_j in dimension i. For $1 \le i \le s$, define the $b \times b$ square $S_i = (s^{(i)}_{jk})$ with entries in $\{0,1,\ldots,b-1\}$ by $s^{(i)}_{jk} = \lfloor b p_{i,(h-1)b+k} \rfloor$ for $1 \le h,k \le b$. Then $S_1,\ldots,S_b$ forms a set of orthogonal squares on elements $\{0,1,\ldots,b-1\}$. Apply row and column permutations to each square so that S_1 has constant rows and S_2 has constant columns; then $S_3,\ldots,S_s$ is a set of $s-2$ MOLS of order b. This can be reversed to give a $(0,2,s)$-net in base b from $s-2$ MOLS of order b.

A *cubical orthogonal array* $\mathrm{COA}_\lambda(t,k,n)$ (also called a generalized orthogonal array) is a $t \times k \times \lambda n^t$ array $C=(c_{ijk})$ with elements from a set of size n, meeting a condition on certain subarrays as follows. For any set $S = \{(x_i,y_i) : 1 \le i \le t,\ 1 \le x_i \le t,\ 1 \le y_i \le k\}$, for which $(x_i,y_i) \in S$ and $x_i > 1$ implies that $(x_i-1,y_i) \in S$, define a $t \times \lambda n^t$ array $A=(a_{ik})$ by setting $a_{ik} = c_{x_i,y_i,k}$; then, A contains every $t \times 1$ column vector the same number λ of times.

47.5 Remark In a $\mathrm{COA}_\lambda(t,k,n)$ with $t \le k$, the $k \times \lambda n^t$ array obtained by taking first coordinate 1 is an $\mathrm{OA}_\lambda(t,k,n)$.

47.6 Theorem [1] Let $s \ge 1$, $b \ge 2$, $t \ge 0$ and $m \ge t+1$. A (t,m,s)-net in base b exists if and only if a $\mathrm{COA}_{b^t}(m-t,s,b)$ exists.

47.7 Remark A $\mathrm{COA}_\lambda(2,k,n)$ exists if and only if an $\mathrm{OA}_\lambda(2,k,n)$ exists, and hence a $(t,t+2,s)$-net in base b is equivalent to an $\mathrm{OA}_{b^t}(2,s,b)$. However, the existence of a $\mathrm{COA}_\lambda(t,k,n)$ for $t \ge 3$ is not equivalent to an $\mathrm{OA}_\lambda(t,k,n)$ in general. Mullen et al. have also shown equivalences with various classes of hypercubes.

47.8 Theorem If there is a q-linear code $[n,l,d]$, $d \ge 3$, then there is a $(n-l-d+1,n-l,s)$-net, where $s = \begin{cases} \lfloor (n-1)/h \rfloor & \text{if } d = 2h+2 \\ \lfloor n/h \rfloor & \text{if } d = 2h+1 \end{cases}$.

47.3 See Also

§II.5	Orthogonal arrays of higher strength.
§V.1	Linear codes.
[4]	Applications of (t,m,s)-nets

References

[1] K.M. Lawrence, Combinatorial Bounds and Constructions in the Theory of Uniform Point Distributions in Unit Cubes, Connections with Orthogonal Arrays and a Poset Generalization of a Related Problem in Coding Theory, Ph.D. thesis, University of Wisconsin, 1995.

[2] K.M. Lawrence, A. Mahalanabis, G.L.Mullen and W.C. Schmid, Construction of digital (t,m,s)-nets from linear codes, preprint.

[3] G.L. Mullen and W.C. Schmid, An equivalence between (t, m, s)-nets and strongly orthogonal hypercubes, preprint.

[4] H. Niederreiter, Point sets and sequences with small discrepancy, *Monats. Math.* 104 (1987), 273-337.

48 Tuscan Squares

Hong-Yeop Song
Jeffrey H. Dinitz

48.1 Definitions, Examples and Enumeration

An $r \times n$ *tuscan-k rectangle* has r rows and n columns such that (1) each row is a permutation of the n symbols and (2) for any two distinct symbols a and b and for each m from 1 to k, there is *at most* one row in which b is m steps to the right of a.

A *tuscan rectangle* is a tuscan-1 rectangle. When $r = n$, it is a *tuscan square* of order n. A tuscan square is in *standard form* when the top row and the left-most column contain the symbols in natural order. A *roman* square is both tuscan and latin and was originally called a *row complete latin square.* A tuscan-$(n-1)$ rectangle is a *florentine* rectangle and is a *vatican* rectangle when it is also latin.

48.1 Remark If there exists an $r \times n$ tuscan rectangle, then $r \leq n$.

48.2 Table [9, 8] The number of standard form tuscan-k squares of order n.

$k\backslash n$	2	3	4	5	6	7	8	9	10	11	12	13
1	1	0	1	0	736	466144	$\geq 3 \cdot 10^7$	$\geq 3 \cdot 10^7$	≥ 72	≥ 1	≥ 964	≥ 1
2		0	1	0	1	0	6	?	≥ 2	?	≥ 2	?
3			1	0	1	0	0	0	≥ 1	?	≥ 1	?
4				0	1	0	0	0	≥ 1	?	≥ 1	?
5					1	0	0	0	≥ 1	?	≥ 1	?
6						0	0	0	≥ 1	?	≥ 1	?
7							0	0	≥ 1	?	≥ 1	?
8								0	1	?	≥ 1	?
9									1	0	≥ 1	?
10										0	≥ 1	?
11											≥ 1	?
12												?

48.3 **Example** Florentine squares of orders 2, 4, and 6. A tuscan square of order 7 that is not tuscan-2. A tuscan-2 square of order 8 that is not tuscan-3.

1	2
2	1

1	2	3	4
2	4	1	3
3	1	4	2
4	3	2	1

1	2	3	4	5	6
2	4	6	1	3	5
3	6	2	5	1	4
4	1	5	2	6	3
5	3	1	6	4	2
6	5	4	3	2	1

6	1	5	2	4	3	7
2	6	3	5	4	7	1
5	7	2	3	1	4	6
4	2	5	1	6	7	3
3	6	2	1	7	4	5
1	3	2	7	5	6	4
7	6	5	3	4	1	2

1	2	3	4	5	6	7	8
2	1	6	8	7	3	5	4
3	2	7	1	8	4	6	5
4	1	7	5	3	8	6	2
5	8	1	4	7	2	6	3
6	1	5	2	4	8	3	7
7	4	2	8	5	1	3	6
8	2	5	7	6	4	3	1

A *circular tuscan-k array* of size $r \times n$ satisfies the same condition as a tuscan-k rectangle with r *circular* rows containing n symbols.

48.4 **Remark** Adjoining a column of asterisks to the left of an $n \times n$ tuscan square makes it an $n \times (n+1)$ circular tuscan array (since the left-most and right-most columns must be permutations). Rotating the rows so that another symbol fills the left-most column, and then deleting it, gives a new $n \times n$ tuscan square. This is the *add-asterisk transformation.*

48.5 **Remark** A *polygonal-path* (PP) tuscan-k square has columns which are cyclic shifts of each other. If it becomes a circular tuscan-k array when a column of asterisks is adjoined at the far-left, it is a *polygonal-path* circular tuscan-k array.

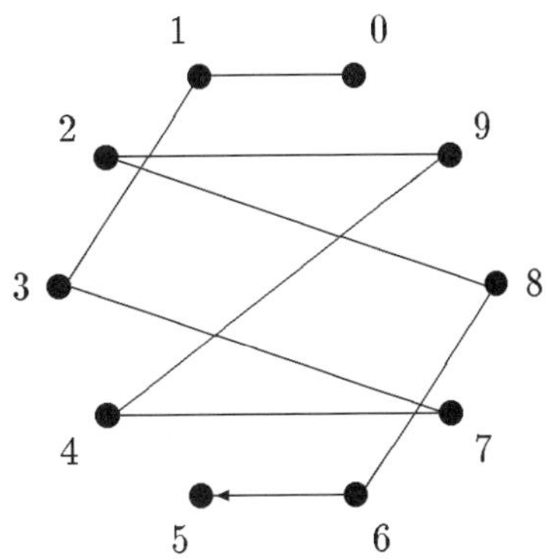

48.6 **Example** Successive rotation of the symmetric polygonal-path, 0 1 3 7 4 9 2 8 6 5, (shown) produces a partition of K_{10}^* into 10 hamiltonian directed paths. When written as 10 rows, this becomes a polygonal-path tuscan-2 square which is not tuscan-3 [8]. For the general case, see Theorem 48.15(2).

48.2 Existence and Related Combinatorial Objects

48.7 **Theorem** If n is even, there exists an $n \times n$ roman square (row complete latin square).

48.8 **Construction** To construct a roman square of order $n = 2m$, use the "zig-zag" polygonal-path, namely, $0, 1, 2m-1, 2, 2m-2, \ldots, m+1, m$, and its successive rotations.

48.9 **Theorems** (*existence*)

1. If $n > 0$, $n \neq 3, 5$, then there exists a tuscan square of order n [12].
2. For every integer $n > 5$, there exists a nonlatin tuscan square of order n all of whose add-asterisk transforms are nonlatin [6].
3. If p is a prime and either $p \equiv 7 \pmod{12}$ or $p \equiv 5 \pmod{24}$, then there exists a tuscan-2 square of order $2p$ whose transpose is also a tuscan-2 square [1].
4. If p is the smallest prime factor of n, then $C(x, y) \equiv xy \pmod{n}$ for $x = 1, 2, \ldots, p-1$ and $y = 0, 1, \ldots, n-1$ will construct a $(p-1) \times n$ circular florentine array (CFA).
5. If n is even, there does not exist a $2 \times n$ CFA [8].
6. A PP CFA of size $n \times (n+1)$ exists if and only if $n+1$ is an odd prime [7].

48.10 Remark There exists a roman square of order pq whenever $p < q$ are odd primes and 2 is a primitive root mod p (see Theorem 48.15(1) and Theorem IV.5.29). Roman squares of odd order n are known for $n \in \{9, 15, 21, 27, 39, 55, 57\}$ [2]. Roman squares of orders 3, 5 and 7 do not exist.

48.11 Remark It is not known whether any nonlatin florentine squares exist.

48.12 Remark Two tuscan-k rectangles are *isomorphic* if their standard forms differ by (1) a substitution of symbols, (2) permuting the rows, and/or (3) taking the reverse order of symbols in every row. Every known florentine square is isomorphic to the one from Theorem 48.9(4) with n prime.

48.13 Remark A symmetric PP circular tuscan-2 array of size $n \times (n+1)$ is known for every even n such that $8 \leq n \leq 50$ [8].

48.14 Remark For $n \leq 20$, every PP tuscan-3 square of order n is vatican. For $n \leq 28$ except for $n = 22$, every symmetric PP tuscan-3 square is vatican. For $n = 22$, there exists a *unique* symmetric PP tuscan-3 square which is not tuscan-4. Three *inequivalent* symmetric PP tuscan-3 squares (not tuscan-4) are known for $n = 30$ [8].

48.15 Theorems (*connections to other combinatorial objects*)

1. If a group of order n is sequenceable, there exists a tuscan square of order n whose transpose is also tuscan (a *complete latin square*). (See §IV.5.3.)
2. A tuscan square of order n is equivalent to a decomposition of the complete directed graph with n vertices into n hamiltonian directed paths [6, 12].
3. There exists a PP vatican square of order n if and only if there exists a singly periodic Costas array of order n [7] (see §IV.7.3).
4. An $n \times (n+1)$ circular tuscan-k array induces k MOLS of order $n+1$ [7].
5. An $r \times n$ circular florentine array induces r MOLS, $L_1, L_2, \ldots, L_r$, of order n. To construct these, first rotate the rows of the CFA so that the leftmost symbol in every row is the same. Then let the top row of L_x be row x of the CFA for each x from 1 to r. The remaining rows of each latin square are obtained by using cyclic shifts of its top row such that (1) L_1 has a constant diagonal and (2) for $x \neq 1$, L_x has the top row of L_1 as its diagonal [10].
6. An $r \times n$ circular florentine array is equivalent to an $(n, r+1; 1)$ difference matrix over the cyclic group of order n [10] (see §IV.11).
7. A pair of orthogonal latin squares cannot both be vatican [10].

48.3 Florentine Rectangles and Circular Florentine Arrays

48.16 Examples [10] The two nonisomorphic 6×7 florentine rectangles having a constant middle column which are not from Theorem 48.9(4). The two nonisomorphic 8×9 florentine rectangles having a constant middle column.

1	2	3	4	5	6	7
2	5	7	4	3	1	6
3	7	1	4	6	5	2
5	3	6	4	7	2	1
6	1	5	4	2	7	3
7	6	2	4	1	3	5

1	2	3	4	5	6	7
2	6	1	4	7	5	3
3	1	7	4	6	2	5
5	7	6	4	1	3	2
6	5	2	4	3	7	1
7	3	5	4	2	1	6

1	2	3	4	5	6	7	8	9
2	1	7	9	5	8	3	6	4
3	7	2	8	5	4	1	9	6
4	9	8	1	5	7	6	3	2
6	8	4	7	5	2	9	1	3
7	3	1	6	5	9	2	4	8
8	6	9	3	5	1	4	2	7
9	4	6	2	5	3	8	7	1

1	2	3	4	5	6	7	8	9
2	9	6	1	5	8	3	7	4
3	6	9	7	5	4	2	1	8
4	1	7	9	5	3	8	6	2
6	8	4	3	5	1	9	2	7
7	3	2	8	5	9	1	4	6
8	7	1	6	5	2	4	9	3
9	4	8	2	5	7	6	3	1

$F(n)$ = the maximum such that an $F(n) \times n$ florentine rectangle exists.
$F_c(n)$ = the maximum such that an $F_c(n) \times n$ circular florentine array exists.

48.17 Table Possible values of $F_c(n)$ and $F(n)$ [10, 11].

n	$F_c(n)$	n	$F_c(n)$
1	1	21	5,...,19
3	2	23	22
5	4	25	4,...,24
7	6	27	4,...,26
9	2	29	28
11	10	31	30
13	12	33	3,...,30
15	4	35	4,...,33
17	16	37	36
19	18	39	3,...,38

n	$F(n)$	n	$F(n)$	n	$F(n)$
1	1	11	10	21	7,...,21
2	2	12	12	22	22
3	2	13	12,13	23	22,23
4	4	14	7,...,14	24	6,...,24
5	4	15	7,...,15	25	6,...,25
6	6	16	16	26	6,...,26
7	6	17	16,17	27	6,...,27
8	7	18	18	28	28
9	8	19	18,19	29	28,29
10	10	20	6,...,20	30	30

48.18 Example A 4×27 circular florentine array [11].

0	1	2	3	4	5	6	7	8	9	10	11	12	13	14	15	16	17	18	19	20	21	22	23	24	25	26
0	4	10	1	19	21	11	14	2	22	20	12	3	9	18	24	15	7	5	25	13	16	6	8	26	17	23
0	15	1	13	2	16	3	19	4	18	5	17	20	6	21	7	10	22	9	23	8	24	11	25	14	26	12
0	20	16	4	14	25	8	12	26	24	21	5	9	17	10	18	22	6	3	1	15	19	2	13	23	11	7

48.4 Applications of Tuscan Squares

48.19 Remark [3] In a sequence of visual impressions one flash card may have some effect on the impression given by the next. This bias can be cancelled by using n sequences corresponding to the rows of an $n \times n$ tuscan-1 square.

48.20 Remark [10, 11] In frequency-hopped, multiple-access communication systems any two users should avoid hopping simultaneously to the same carrier frequency, an event termed a *hit*. Assigning a row of an $r \times n$ circular florentine array to each user as a frequency hopping sequence minimizes the number of hits between any two users.

48.5 See Also

§IV.7	Proposition IV.7.19 gives a connection between vatican squares and Costas arrays.
[4]	A discussion of "horizontally complete" (i.e., roman) squares can be found on pages 299-302.
[5]	Chapter 3 deals with row complete latin squares and sequencings of groups

References

[1] B.A. Anderson, A family of $N \times N$ Tuscan-2 squares with $N + 1$ composite, *Ars Combin.* 32 (1991), 33-55.

[2] D.S. Archdeacon, J.H. Dinitz, D.R. Stinson and T.W. Tillson, Some new row-complete Latin squares, *J. Combin. Theory A* 29 (1980), 395-398.

[3] B.R. Bugelski, A note on Grants's discussion of the Latin square principle in the design of experiments, *Psychological Bull.* 46 (1949), 49-50.

[4] J. Dénes and A.D. Keedwell, *Latin Squares and their Applications*, English University Press Ltd., London, 1974.

[5] J. Dénes and A.D. Keedwell, *Latin Squares: New Developments in the Theory and Application*, North-Holland, Amsterdam, 1991.

[6] T. Etzion, Hamiltonian decomposition of K_n^*, patterns with distinct differences, and Tuscan squares, *Discrete Math.* 91 (1991), 259-276.

[7] T. Etzion, S.W. Golomb and H. Taylor, Tuscan-k squares, *Adv. Appl. Math.* 10 (1989), 164-174.

[8] S.W. Golomb, T. Etzion and H. Taylor, Polygonal path constructions for Tuscan-k squares, *Ars Combin.* 30 (1990), 97-140.

[9] S.W. Golomb and H. Taylor, Tuscan squares - A new family of combinatorial designs, *Ars Combin.* 20B (1985), 115-132.

[10] H.Y. Song, On Aspects of Tuscan Squares, Ph.D. thesis, University of Southern California, 1991.

[11] H.Y. Song and S.W. Golomb, Generalized Welch-Costas sequences and their application to vatican arrays, *Contemp. Math.* 168 (1994), 341-351.

[12] T.W. Tillson, A hamiltonian decomposition of K_{2m}^*, $2m \geq 8$, *J. Combin. Theory B* 29 (1980), 68-74.

49 t-Wise Balanced Designs

EARL S. KRAMER

49.1 Definitions

A *t-wise balanced design* (tBD) of type t-(v, K, λ) is a pair $(\mathbf{X}, \mathcal{B})$ where $\mathbf{X}$ is a v element set of points and $\mathcal{B}$ is a collection of subsets of $\mathbf{X}$ (*blocks*) with the property that the size of every block is in the set K and every t-element subset of $\mathbf{X}$ is contained in exactly λ blocks.

If K is a set of positive integers strictly between t and v, then the tBD is *proper*. If $\mathcal{B}$ contains all k-subsets of $\mathbf{X}$ for some k, then $(\mathbf{X}, \mathcal{B})$ is a *trivial design*.

49.1 Remark While emphasis is generally placed on proper and nontrivial tBDs, improper tBDs are often required for minimal exact coverings and also trivial tBDs are convenient to include when listing general theorems.

A t-(v, K, λ) design is also denoted by $S_\lambda(t, K, v)$ or by $B_t[K, \lambda; v]$, when then one may write $v \in B_t(K, \lambda) = \{v : \text{there is a } t\text{BD of type } t\text{-}(v, K, \lambda)\}$. If $K = \{k\}$, so $|K| = 1$, then the tBD is a t-design. If $\lambda = 1$, the notation $S(t, K, v)$ is often used and the design is a *Steiner system*.

49.2 Example Let $\mathbf{X} = \{1, 2, \ldots, 12\}$ with base blocks $\{1\ 2\ 3\ 4\ 5\ 6\}$, $\{7\ 8\ 9\ 10\ 11\ 12\}$, $\{1\ 2\ 7\ 8\}$, $\{1\ 2\ 9\ 11\}$, $\{1\ 2\ 10\ 12\}$, $\{3\ 5\ 7\ 8\}$, $\{3\ 5\ 9\ 11\}$, $\{3\ 5\ 10\ 12\}$, $\{4\ 6\ 7\ 8\}$, $\{4\ 6\ 9\ 11\}$ and $\{4\ 6\ 10\ 12\}$. Applying the automorphism $(1\ 2\ 3\ 4\ 5)(7\ 8\ 9\ 10\ 11)$ yields 2+45=47 blocks which form a 3-$(12, \{4, 6\}, 1)$ design. This tBD is not balanced for t=2 since some pairs appear four times while others appear five times.

A *graphical* (*bigraphical*) tBD is a tBD whose point set $\mathbf{X}$ is the edges of the complete graph $K_p(K_{m,n})$ and whose automorphism group is the symmetric group $S_p(S_m \times S_n)$ (see §IV.23).

49.3 Remark Although there are only finitely many graphical (bigraphical) Steiner tBDs for $t \geq 2$, these do provide nice presentations for such tBDs and t-designs. The Steiner cases for $t \geq 2$ are (1) graphical: $S(2,3,15)$, $S(2,\{3,5\},15)$, $S(3,4,10)$, $S(4,\{5,7\},15)$; (2) bigraphical: $S(2,3,9)$, $S(3,4,8)$, $S(3,4,16)$, $S(3,\{4,6\},16)$, $S(5,\{6,8\},16)$ (see [1], [8] and §IV.23 for details).

49.2 Elementary Properties

49.4 Remark By taking the derived incidence structure, Theorem 49.5 is immediate.

49.5 Theorem A t-(v,K,λ) design implies there exists a $(t-1)$-$(v-1,K-1,\lambda)$ design where $K-1 = \{k_i - 1 : k_i \in K\}$.

49.6 Remark A t-design is always an s-design for $0 \leq s \leq t$, but a t-(v,K,λ) design is not necessarily a $(t-1)$-(v,K,λ) design. For example, there is a 13-(24,{14,15},11) (see [9]) which is balanced for t=1, 2, 3, 4, 5, and 13, but is not balanced for $6 \leq t \leq 12$.

49.7 Theorem If $t \geq 2$, then there is a tBD which is not s-balanced for any $0 < s < t$.

49.8 Example Let $N_n = \{1,2,\ldots,n\}$ and $\mathbf{X} = N_n \cup \{\infty\}$ for $t < n$. Let $\mathcal{B} = N_n \cup \{S \cup \{\infty\} : S \in \mathcal{P}_{t-1}(N_n)\}$, where $\mathcal{P}_{t-1}(N_n)$ is the set of all subsets of N_n of size $(t-1)$. Then $(\mathbf{X},\mathcal{B})$ is a tBD which is not s-balanced for any $0 < s < t$.

49.9 Remark Under certain conditions a tBD does possess balance for other values of $0 < s < t$ (see [7]).

49.10 Theorem If $(\mathbf{X}, \mathcal{B}_1 \cup \mathcal{B}_2)$ is a t-wise and a $(t-1)$-wise balanced design where the blocks in $\mathcal{B}_i$ have size k_i, $(i = 1,2; k_1 \neq k_2)$, then each $(\mathbf{X},\mathcal{B}_i)$, $i = 1,2$, is a $(t-1)$-design.

For $0 \leq s \leq t$, and with $t \leq k$ for all $k \in K$ define
$$\alpha(K,s) = \gcd\{(k-s)(k-s-1)\cdots(k-t+1) : k \in K\}.$$

49.11 Theorem If $v \in B_t(K,\lambda)$, then $\lambda(v-s)(v-s-1)\cdots(v-t+1) \equiv 0 \pmod{\alpha(K,s)}$, for $0 \leq s < t$.

49.12 Remarks The necessary conditions for tBDs given by Theorem 49.11 are not sufficient since such conditions are satisfied for the parameters 2-(45,{6,7},1), but such a tBD cannot exist. However, Wilson [12] has shown that the necessary conditions for PBDs (pairwise balanced designs) are sufficient for large enough v. Hartman [6] mentions that one would like to prove that the necessary conditions for a 3-wise balanced design are sufficient if v is large enough. Hanani (see [3, 4, 5]) was a pioneer in the use of tBDs and first gave several properties which are listed in Theorem 49.13.

49.13 Theorem

1. $K \subset B_t(K,1)$.
2. If $K' \subset K$, then $B_t(K',\lambda) \subset B_t(K,\lambda)$.
3. If λ' divides λ, then $B_t(K,\lambda') \subset B_t(K,\lambda)$.
4. $B_t(K,\lambda) \cap B_t(K,\lambda') \subset B_t(K, n\lambda + n'\lambda')$ where λ and λ' are any positive integers, and n and n' are any nonnegative integers.
5. If $v \in B_t(K',\lambda')$ and $K' \subset B_t(K,\lambda)$, then $v \in B_t(K,\lambda\lambda')$.

49.14 Remark There are several recursive results when $t = 3$ (see [6] for references and discussion). One such result is Theorem 49.15.

49.15 Theorem If there exists an $S(3, K, g + u)$ with a subsystem $S(3, K, u)$, where u is even and $g \equiv u + 2 \pmod 6$, then there exists an $S(3, K \cup \{4, 6\}, 3g + u)$.

49.16 Theorem [9] A nontrivial tBD with $t \geq 2$ and $\lambda = 1$ has $k \leq (v + t - 3)/2$ for all $k \in K$.

49.17 Theorem Let $(\mathbf{X}, \mathcal{B})$ be a nontrivial tBD with $t \geq 2$ and $\lambda = 1$, and let $k \in K$.

1. If $t = 2$ or 4, then $k \leq (v - 1)/2$ with equality occurring infinitely often.
2. If $t = 3$ or 5, then $k \leq v/2$ with equality occurring infinitely often.

49.18 Remark Theorem 49.17 makes use of an infinite family mentioned in Remark 49.21. Theorem 49.19 is an asymptotic result (see [2]) for $S(t, K, v)$s which allow blocks of size t.

49.19 Theorem The number of nonisomorphic $S(t, K, v)$s is $v^{(\binom{v}{t}/(t+1))(1+o(1))}$ for fixed $t \geq 2$.

49.3 Some Families of tBDs for t $\geq$ 3

49.20 Remark The previous theorems, plus others for $|K| = 1$, are powerful in building other designs from known tBDs.

49.21 Remark Although at present there are no known infinite families of Steiner systems for $t \geq 4$, Wilson has constructed the family of designs with parameters $S(5, \{6, 8\}, 2^n)$ for all $n \geq 4$. From this family it is easy to produce $S(5, \{6, 8, 2^{n-1}\}, 2^n)$, $S(4, \{5, 7\}, 2^n - 1)$ and $S(4, \{5, 7, 2^{n-1} - 1\}, 2^n - 1)$ for all $n \geq 4$ (see [9] for a discussion).

49.22 Remark Tonchev [11] has constructed the infinite family $S(4, \{5, 6\}, 4^n + 1)$ for $n \geq 2$.

49.23 Remark Some of the following results may use trivial tBDs.

49.24 Theorem [4] If v is even, then $v \in B_3(\{4, 6\}, 1)$.

49.25 Theorem [4] If $v \geq 4$, then $v \in B_3(\{4, 5, 6, 7, 9, 11, 13, 15, 19, 23, 27, 29, 31\}, 1)$.

49.26 Theorem [13] If $v \geq 5$, then $v \in B_3(\{5, 6, \ldots, 40\} \setminus \{17, 21, 22, 25, 26\}, 1)$.

49.4 All Proper tBDs for t $\geq$ 3 with λ = 1 and v $\leq$ 16

49.27 Remark In [10] the proper t-wise balanced designs t-$(v, K, 1)$ for $t \geq 3, \lambda = 1$, and $v \leq 16$ with at least two block sizes are characterized. While extensions of $S(3, 4, 16)$s are not examined there, all other possible extensions of $S(3, K, v)$s for $v \leq 16$ are determined. A particularly interesting extension is the $S(4, \{5, 6\}, 17)$ design. Table 49.28 summarizes the results in [10].

49.28 Table The proper $S(t, K, v)$s for $t \geq 3$ and $v \leq 16$. For a given t and v, the number of blocks of each size k is listed. Also listed is the number of nonisomorphic proper $S(t, K, v)$ designs with the prescribed number of blocks of each size. For each entry, a reference number for a construction is given.

Con. #	t	v	$k=4$	$k=5$	$k=6$	$k=7$	$k=8$	Number
49.30	3	8	14					1
49.31	3	10	30					1
49.32	3	12	45		2			1
49.33	3	14	56		7			4
49.34	3	14	91					4
49.35	3	16	112				2	≥ 31021
49.35	3	16	126				1	≥ 31021
49.36	3	16	60		16			1
49.37	3	16	100		8			1
49.38	3	16	20	48				1
49.39	3	16	35	42				1
49.40	3	16	60	32				6
49.41	3	16	85	22				2
49.42	3	16	100	16				506
49.43	3	16	105	14				22
49.44	3	16	110	12				8
49.35	3	16	140					≥ 31021
49.31	4	11		66				1
49.45	4	15		168		15		1
49.46	4	17		140	112			1
	4	17		476				?
49.31	5	12			132			1
49.45	5	16			448		30	1

49.29 Remark In the constructions to follow, a design of type $a^x b^y$ means there are x blocks of size a and y blocks of size b. A given subgroup H (G is the full automorphism group of the design) acts on the listed base blocks to generate the design. Any additional generators to produce G are mentioned, along with $|G|$.

49.30 Construction The $S(3,4,8)$ of type 4^{14}. $H = \langle(1\ 2\ 3\ 4\ 5\ 6\ 7)\rangle$. Base blocks: $\{1\ 2\ 4\ 8\}$, $\{3\ 5\ 6\ 7\}$. $G = \langle H,\ (2\ 3\ 6)(5\ 7\ 8)\rangle$ and $|G| = 168$.

49.31 Construction The $S(3,4,10)$ of type 4^{30}; the $S(4,5,11)$ of type 5^{66}; the $S(5,6,12)$ of type 6^{132}. Let $X = \{1, 2, 3, ..., 9, a, b, c\}$ and $H = \langle(1\ 2\ 3\ 4\ 5\ 6\ 7\ 8\ 9\ \mathrm{a}\ \mathrm{b})\rangle$. To generate the $S(4,5,11)$ let H act on the base blocks: $\{1\ 2\ 3\ 4\ 6\}$, $\{1\ 2\ 3\ 7\ \mathrm{a}\}$, $\{1\ 2\ 3\ 8\ 9\}$, $\{1\ 2\ 4\ 5\ 8\}$, $\{1\ 2\ 4\ 7\ 9\}$, $\{1\ 2\ 6\ 8\ \mathrm{a}\}$. Taking the derived design through the point b produces the $S(3,4,10)$. To get the $S(5,6,12)$ add point c to each block of the $S(4,5,11)$ and then take the complement of each such block in X to get the needed $66 + 66 = 132$ blocks of size 6. The group for the $S(3,4,10)$ is $G = \langle(1\ 2\ 3\ 6\ 9\ 7\ 5\ 10\ 4\ 8),\ (2\ 7\ 6\ 5)(3\ 8\ 10\ 4)\rangle$ where $|G| = 1440$. The group for the $S(4,5,11)$ is $M_{11} = \langle H,\ (1\ 3\ 9\ 5\ 4)(2\ 6\ 7\ 10\ 8),\ (3\ 4\ 6)(5\ 9\ 10)(\ 7\ 8\ 11)\rangle$ where $|M_{11}| = 7920$. The group for the $S(5,6,12)$ is $M_{12} = \langle M_{11},\ (1\ 10)(2\ 5)(3\ 7)(4\ 8)(6\ 9)(11\ 12)\rangle$ where $|M_{12}| = 95040$.

49.32 Construction An $S(3,\{4,6\},12)$ of type $4^{45}6^2$. $H = \langle(1\ 2\ 3\ 4\ 5)(7{,}8{,}9{,}10{,}11)\rangle$. Base blocks: $\{1{,}2{,}3{,}4{,}5{,}6\}$, $\{7{,}8{,}9{,}10{,}11{,}12\}$, $\{1{,}2{,}7{,}8\}$, $\{1{,}2{,}9{,}11\}$, $\{1{,}2{,}10{,}12\}$, $\{3{,}5{,}7{,}8\}$, $\{3{,}5,\ 9{,}11\}$, $\{3{,}5{,}10{,}12\}$, $\{4{,}6{,}7{,}8\}$, $\{4{,}6{,}9{,}11\}$, $\{4{,}6{,}10{,}12\}$. $G = \langle H{,}(1\ 7)(2\ 8)(3\ 9)(4\ 10)(5\ 11)(6\ 12),\ (3\ 4\ 5\ 6)(9\ 10\ 11\ 12)\rangle$ and $|G| = 240$.

49.33 Construction An $S(3,\{4,6\},14)$ of type $4^{56}6^7$. $H = \langle(1\ 2\ 3\ 4\ 5\ 6\ 7\ 8\ 9\ 10\ 11\ 12\ 13\ 14)\rangle$. Base blocks: $\{1\ 2\ 5\ 7\}$, $\{1\ 2\ 6\ 10\}$, $\{1\ 2\ 12\ 14\}$, $\{1\ 3\ 6\ 9\}$, $\{1\ 2\ 4\ 8\ 9\ 11\}$. $G = \langle H,\ (1\ 2)(3\ 6)(8\ 9)(10\ 13)\rangle$ and $|G| = 2688$.

49.34 Construction An $S(3,4,14)$ of type 4^{91}. $H = \langle(1\ 2\ 3\ 4\ 5\ 6\ 7)(8\ 9\ 10\ 11\ 12\ 13\ 14)\rangle$. Base blocks: $\{1\ 2\ 3\ 6\}$, $\{1\ 2\ 4\ 14\}$, $\{1\ 2\ 8\ 9\}$, $\{1\ 2\ 10\ 11\}$, $\{1\ 2\ 12\ 13\}$, $\{1\ 3\ 8\ 10\}$, $\{1$

3 9 11}, {1 4 8 11}, {1 4 10 13}, {1 5 9 13}, {1 6 10 12}, {1 11 13 14}, {8 9 10 12}. $G = \langle H,$ (2 3 5)(7 6 4)(11 14 13)(9 10 12)⟩ and $|G|= 42$.

49.35 Construction An $S(3,4,16)$ of type 4^{140}; an $S(3,\{4,8\},16)$ of type $4^{126}8^1$; and an $S(3,\{4,8\},16)$ of type $4^{112}8^2$. To generate an $S(3,4,16)$ of type 4^{140} (here it is the geometry $AG(4,2)$), use $H = \langle$(1 2 5 3 11 6 13 4 12 8 15 7 16 14) (9 10)⟩, along with base blocks: {1 2 3 9}, {1 2 4 12}, {1 2 5 6}, {1 2 7 13}, {1 2 8 16}, {1 2 10 15}, {1 3 4 15}, {1 3 10 12}, {1 4 5 8}, {1 4 9 10}, {1 5 9 13}, {1 5 11 15}. $G = \langle H,$ (4 5 16)(6 8 12)(7 14 15)(10 13 11), (5 7)(6 13)(8 11)(14 16)⟩. Then $|G|$=322,560 and G is triply transitive on {1,2,...,16}. To get an $S(3,\{4,8\},16)$ of type $4^{126}8^1$ replace the 14 blocks (from the $S(3,4,16)$) of an $S(3,4,8)$ whose points lie entirely in the set $X = \{1,2,3,5,6,7,9,13\}$, by the single 8-block whose points are precisely those of X. For this design $G = \langle$(1 2 6 7 9 5 3) (8 10 16 14 11 15 12), (1 2 3 7 13 5)(4 12 15 11 14 8)(6 9)(10 16)⟩ and $|G|$=10752. To get an $S(3,\{4,8\},16)$ of type $4^{112}8^2$ replace the 14 blocks (from the $S(3,\{4,8\},16)$) of an $S(3,4,8)$ whose points lie entirely in the set $Y = \{4,8,10,11,12,14,15,16\}$ by the single 8-block whose points are those of Y. For this design $G = \langle$(1 2 6 7 9 5 3) (8 10 16 14 11 15 12), (1 4 2 8 3 15 9 11) (5 10 13 16 7 12 6 14)⟩ and $|G|$=21504.

49.36 Construction The $S(3,\{4,6\},16)$ of type $4^{60}6^{16}$. $H = \langle$(1 2 3 4 5 6 7 8) (9 10 11 12 13 14 15 16)⟩. Base blocks: {1 2 3 6 9 11}, {1 3 10 13 14 15}, {1 2 4 13}, {1 2 7 14}, {1 2 12 15}, {1 3 5 7}, {1 4 10 11}, {1 5 9 13}, {1 5 11 15}, {1 9 10 12}, {1 9 14 16}, {9 11 13 15}. $G = \langle H,$ (2 3 11)(5 15 12 8 14 7)(6 9)(10 13 16)⟩. $|G|$=11520 and G is doubly transitive on points.

49.37 Construction The $S(3,\{4,6\},16)$ of type $4^{100}6^8$. $H = G = \langle$(1 2 3 4)(5 6 7 8) (9 10 11 12)(13 14 15 16), (3 4)(5 16)(7 15)(8 9)(10 14)(12 13)⟩. Base blocks: {1 2 3 5 9 13}, {5 6 10 11 13 15}, {1 2 6 11}, {1 2 7 14}, {1 5 6 14}, {1 5 7 12}, {1 5 8 15}, {1 5 11 16}, {1 6 7 8}, {5 6 9 12}, {5 7 9 11}. $|G| = 24$ and G has two orbits on points 1 to 4 and 5 to 16.

49.38 Construction The $S(3,\{4,5\},16)$ of type $4^{20}5^{48}$. $H = \langle$(1 2 3 4 5 6 7 8 9 10 11 12 13 14 15)⟩. Base blocks: {1 2 3 6 13}, {1 2 4 8 16}, {2 4 6 11 12}, {3 6 9 12 15}, {1 2 10 14}, {1 6 11 16}. $G = \langle H,$ (2 4 5 13)(3 8 9 14)(6 16)(7 15 10 12)⟩. $|G|$=960 and G is 2-transitive on points.

49.39 Construction The $S(3,\{4,5\},16)$ of type $4^{35}5^{42}$. $H = \langle$(1 4 7 11 12 5 14)(2 13 6 3 15 9 10)⟩. Base blocks: {1 2 3 6 8}, {1 2 4 12 13}, {1 2 5 7 15}, {1 3 4 7 9}, {1 4 5 8 10}, {1 9 10 13 15}, {1 2 9 16}, {1 3 10 16}, {1 4 11 16}, {1 6 13 16}, {1 8 15 16}. $G = \langle H,$ (1 3 12 15 11)(2 6 9 4 10)(5 7 14 8 13)⟩. $|G|= 2520$ and G fixes 16 and is doubly transitive on 1,...,15. This design uniquely extends to an $S(4,\{5,6\},17)$.

49.40 Construction An $S(3,\{4,5\},16)$ of type $4^{60}5^{32}$. $|H| = 32$ and $H = \langle$(1 7 3 12) (2 8 15 5)(4 14 13 6)(9 10 16 11), (1 9 3 16)(2 13 14 5)(4 6 8 15) (7 12)⟩. Base blocks: {1 2 3 7 14}, {1 2 4 15 16}, {1 2 12 13}, {1 3 9 16}, {1 7 9 10}, {1 7 13 15}, {1 8 13 16}, {2 4 5 14}, {2 6 14 15}. $G = \langle H,$ (3 5 4 6)(7 9 15 11) (8 16 10 14)(12 13)⟩. $|G|$=320 and G is point-transitive.

49.41 Construction An $S(3,\{4,5\},16)$ of type $4^{85}5^{22}$. $|H| = 24$ where $H = \langle$(1 2 3 6) (7 14 8 10)(9 16)(11 13 15 12), (1 4 6)(2 3 5)(7 8 12)(11 14 15)⟩. Base blocks:{1 2 3 11 14}, {1 2 4 15 16}, {7 8 9 12 13}, {1 2 10 13}, {1 3 9 16}, {1 7 9 11}, {1 7 13 14}, {1 8 13 16}, {7 8 11 15}, {7 9 15 16}. $G = \langle H,$ (1 2 3 5 4) (7 8 9 12 13)(10 15 11 16 14)⟩. $|G|$=120 and G has orbits on point sets of sizes 6 and 10.

49.42 Construction An $S(3,\{4,5\},16)$ of type $4^{100}5^{16}$. $H = \langle(1\ 2\ 9\ 10\ 13\ 11\ 5\ 16\ 4\ 14\ 15\ 6)(3\ 12\ 8\ 7)\rangle$. Base blocks: {1 2 3 4 16}, {1 4 7 12 13}, {1 2 6 8}, {1 2 10 15}, {1 2 13 14}, {1 3 5 8}, {1 3 7 10}, {1 3 9 15}, {1 4 5 9}, {1 5 10 14}, {1 7 8 14}, {1 7 11 15}, {3 7 8 12}. $G = \langle H, (1\ 2\ 9\ 10\ 6)(3\ 5\ 4\ 8\ 7)\ (11\ 12\ 14\ 15\ 13)\rangle$. $|G|= 1920$ and G is transitive on points.

49.43 Construction An $S(3,\{4,5\},16)$ of type $4^{105}5^{14}$. $|G| = 42$ and $H = G = \langle(1\ 2\ 11\ 13\ 3\ 12\ 4)(5\ 9\ 6\ 8\ 10\ 7\ 16),\ (1\ 5\ 12\ 10\ 3\ 16)(2\ 7\ 11\ 8\ 4\ 6)(9\ 13)(14\ 15)\rangle$. Base blocks: {1 2 3 4 16}, {1 2 6 9}, {1 2 7 10}, {1 2 13 14}, {1 7 14 15}.

49.44 Construction An $S(3,\{4,5\},16)$ of type $4^{110}5^{12}$. $|G| = 12$ and $H = G = \langle(1\ 6\ 2\ 5\ 9\ 11\ 8\ 4\ 7\ 10\ 3\ 16)(12\ 14\ 15\ 13)\rangle$. Base blocks: {1 2 3 4 16}, {1 2 5 15}, {1 2 7 14}, {1 2 8 12}, {1 2 11 13}, {1 4 6 8}, {1 4 7 12}, {1 4 13 14}, {1 5 8 10}, {1 5 12 13}, {1 6 13 15}, {1 7 9 13}, {12 13 14 15}.

49.45 Construction The $S(4,\{5,7\},15)$ of type $5^{168}7^{15}$. $H = \langle(1\ 2\ 5\ 8\ 15\ 11\ 13\ 7\ 12\ 10\ 6\ 3\ 14\ 4\ 9)\rangle$. Base blocks:{1 2 3 4 13 14 15}, {1 2 3 5 9}, {1 2 3 6 10}, {1 2 3 7 11}, {1 2 3 8 12}, {1 2 4 5 10}, {1 2 4 7 12}, {1 2 4 8 11}, {1 2 6 7 14}, {1 2 6 8 13}, {1 2 10 11 14}, {1 3 4 5 11}, {1 8 10 13 14}. $G = \langle H,(1\ 6\ 9\ 4\ 7\ 12)\ (2\ 8\ 10)(3\ 5\ 11)(14\ 15)\rangle$ and $|G|$=20,160. **For the $S(5,\{6,8\},16)$ of type $6^{448}8^{30}$** add 16 to each of the blocks of the above $S(4,\{5,7\},15)$ to get 183 blocks. For the remaining 295 blocks use the above H and the additional base blocks: {1 2 3 4 5 6 7 8}, {1 2 3 5 10 15}, {1 2 3 5 11 14}, {1 2 3 5 12 13}, {1 2 3 6 11 13}, {1 2 3 6 12 14}, {1 2 3 7 9 14}, {1 2 3 7 10 13}, {1 2 3 7 12 15}, {1 2 3 8 9 13}, {1 2 3 8 10 14}, {1 2 4 5 9 15}, {1 2 4 6 10 15}, {1 2 4 6 11 14}, {1 2 4 6 12 13}, {1 2 4 7 9 13}, {1 2 4 7 10 14}, {1 2 4 8 10 13}, {1 2 5 7 10 11}, {1 3 4 5 10 13}, {1 4 6 8 11 12}. $G = \langle H,\ (1\ 2\ 5\ 15\ 10)(4\ 8\ 9\ 6\ 12)(7\ 14\ 16\ 13\ 11)\rangle$ and $|G|$=322,560 where $G = AGL(4,2)$.

49.46 Construction The $S(4,\{5,6\},17)$ of type $5^{140}6^{112}$. $H = \langle(1\ 2\ 3\ 4\ 5\ 6\ 7\ 8)\ (9\ 10\ 11\ 12\ 13\ 14\ 15\ 16)\ (17)\rangle$. Base blocks:{1 2 3 4 6 11}, {1 2 5 9 12 14}, {1 2 6 9 15 16}, {1 10 12 13 14 15}, {1 2 3 7 14 15}, {1 2 9 10 11 13}, {1 2 3 12 13 16}, {1 3 6 9 10 14}, {1 2 4 5 13 15}, {1 7 9 13 14 16}, {1 2 4 10 14 16}, {1 3 5 11 13 14}, {1 2 5 7 11 16}, {1 4 9 11 14 15}, {1 2 3 9 17}, {1 2 4 12 17}, {1 2 5 6 17}, {1 2 10 15 17}, {1 3 4 15 17}, {1 3 5 7 17}, {1 3 6 13 17}, {1 3 10 12 17}, {1 3 14 16 17}, {1 4 9 10 17}, {1 4 13 14 17}, {1 5 9 13 17}, {1 5 10 14 17}, {1 5 11 15 17}, {1 5 12 16 17}, {1 8 10 13 17}, {1 9 11 16 17}, {1 9 12 15 17}, {1 11 12 13 17}, {1 13 15 16 17}, {9 10 13 14 17}, {9 11 13 15 17}. $G = \langle H,\ (4\ 5\ 16)(6\ 8\ 12)(7\ 14\ 15)(10\ 13\ 11)\rangle$. $|G|$=40320, where G fixes 17 and is triply transitive on {1,2,...,16}.

49.5 See Also

§III.1	PBDs are tBDs with $t = 2$.

References

[1] L.G. Chouinard II, E.S. Kramer and D.L. Kreher, Graphical t-wise balanced designs, *Discrete Math.* 46 (1983), 227-240.

[2] C.J. Colbourn, D.G. Hoffman, K.T. Phelps, V. Rödl and P.M. Winkler, The number of t-wise balanced designs, *Combinatorica* 11 (3) (1991), 207-218.

[3] H. Hanani, On some tactical configurations, *Canad. J. Math.* 15 (1963), 702-722.

[4] H. Hanani, Truncated finite planes, *Proc. Symp. Pure Math. (A.M.S.)* 19 (1971), 115-120.

[5] H. Hanani, A class of three-designs, *J. Combin. Theory A* 26 (1979), 1-19.
[6] A. Hartman, The fundamental construction for 3-designs, *Discrete Math.* 124 (1994), 107-132.
[7] S. Kageyama, A property of t-wise balanced designs, *Ars Combin.* 31 (1991), 237-238.
[8] D.G. Hoffman and D.L. Kreher, The bigraphical t-wise balanced designs of index 1, *J. Combin. Des.* 2 (1994), 41-48.
[9] E.S. Kramer, Some results on t-wise balanced designs, *Ars Combin.* 15 (1983), 179-192.
[10] E.S. Kramer and R.A. Mathon, Proper $S(t, K, v)$'s for $t \geq 3$, $v \leq 16$, $|K| > 1$ and their extensions, *J. Combin. Des.* 3 (1995), 411-425.
[11] V.D. Tonchev, A class of Steiner 4-wise balanced designs derived from the Preparata codes, *J. Combin. Des.*, to appear.
[12] R.M. Wilson, An existence theory for pairwise balanced designs, Parts I, II, and III, *J. Combin. Theory A* 13 (1972), 222-245; 13 (1972), 246-273; and 18 (1975), 71-79.
[13] Qiu-rong Wu, On Properties and Constructions of t-Designs, λ-Designs and PerpendiculAr arrays, Ph.D. thesis, University of Nebraska, 1991.

50 Uniformly Resolvable Designs

PETER DANZIGER
PETER RODNEY

50.1 Definitions and Example

A parallel class in a pairwise balanced design is *uniform* if every block in the parallel class is of the same size.

A *uniformly resolvable design*, URD(v, K, λ, R), is a resolvable PBD(v, K, λ) such that all of the parallel classes are uniform. R is a multiset, where $|R| = |K|$ and for each $k \in K$ there corresponds a positive $r_k \in R$ such that there are exactly r_k parallel classes of size k.

50.1 Remark While this definition requires that $r_k > 0$, one could allow $r_k = 0$. In this case there are no blocks of size k and the design is k-*free*; otherwise, it is k-*essential*.

50.2 Remark If $K = \{g, k\}$ and $R = \{r_g, r_k\}$ write k-URD(v, g, λ, r), where $r = r_g$ and hence $r_k = \frac{\lambda(v-1) - r(g-1)}{k-1}$.

50.3 Example URD(12, $\{2, 3\}$, 1, $\{5, 3\}$) (also 2-URD(12,3,1,3) or 3-URD(12,2,1,5)) on point set $\mathbb{Z}_3 \times \mathbb{Z}_4$. The parallel classes are shown.

$[\{(0,0),(0,1)\},\ \{(0,2),(0,3)\}] \bmod (3,-)$
$[\{(0,0),(0,2)\},\ \{(0,1),(0,3)\}] \bmod (3,-)$
$[\{(0,0),(0,3)\},\ \{(0,1),(0,2)\}] \bmod (3,-)$
$[\{(0,j),(1,j+2)\},\ \{(0,j+1),(1,j+3)\},$
$\{(0,j+2),(2,j+3)\},\ \{(0,j+3),(2,j)\},$
$\{(1,j),(2,j+1)\},\ \{(1,j+1),(2,j+2)\}]\ j = 0, 2$

$[\{(0,0),(1,0),(2,0)\}] \bmod (-,4)$
$[\{(0,0),(1,1),(2,3)\}] \bmod (-,4)$
$[\{(0,0),(1,3),(2,2)\}] \bmod (-,4)$

50.2 Existence and Related Combinatorial Designs

50.4 Theorem The necessary conditions for the existence of a URD(v, K, λ, R) are

1. $v \equiv 0$ (mod k) for each $k \in K$; and
2. $\lambda(v-1) = \sum_{k \in K} r_k(k-1)$.

In addition, if $\lambda = 1$ and $r_k > 1$, then $v \geq k^2$.

50.5 Theorem A 3-URD($v, 2, 1, r$) exists if and only if $v \equiv 0 \pmod 6$, $1 \leq r \leq v-3$ and $(v, r) \notin \{(6,1), (12,1)\}$ [4].

50.6 Theorem [2] Suppose that $k = 2$, or that $k = 3$ and g is odd. Let $\epsilon_{k,g} = 1$ if $g \equiv 0 \pmod k$ and k otherwise. Then

1. If n and $u \equiv 0 \bmod (k-1)$ and there exists a resolvable TD($\epsilon_{k,g}g, u$), then there exists a k-URD($\epsilon_{k,g}gnu, g, \lambda, r$) for all $1 \leq r \leq \lambda n$.
2. There is a k-URD($\epsilon_g qg, g, \lambda, r$) for all odd prime powers $q \geq g$ and all $1 \leq r \leq \frac{\lambda(q-1)}{k-1}$ except possibly when $k = 3$ and $q \equiv 5 \pmod 6$.

50.7 Proposition

1. If $K = \{k\}$ then a URD(v,K,λ,r_k) is equivalent to a resolvable (v,k,λ) BIBD, where $r_k = \frac{\lambda(v-1)}{k-1}$. See §I.6.
2. A k-URD(v,g,1,1) is equivalent to a k-RGDD of type $g^{v/g}$.
3. A k-URD(kn, n, 1, 1) is equivalent to a resolvable TD(k,n). See §II.2.
4. A 3-URD($v, 2, 1, 1$) is equivalent to a nearly Kirkman triple system on v points.

50.3 Resolvable Group Divisible Designs

50.8 Example [5] A 3-RGDD of type 6^4 with point set $\mathbb{Z}_{10} \times \mathbb{Z}_2 \cup \{\infty_i \mid i = 1,2,3,4\}$ and groups $\{(i,0), (i+5,0), (i,1), (i+5,1)\}$, $i = 1,2,3,4$; $\{\infty_i \mid i = 1,2,3,4\}$. Parallel classes are obtained by developing the following mod(10, −).

$$\begin{array}{llll}[\{(1,0),(7,0),(8,1)\}, & \{(3,0),(4,0),(6,0)\} & \{(0,0),(2,1),\infty_1\}, & \{(2,0),(0,1),\infty_2\} \\ \ \{(1,1),(7,1),(8,0)\}, & \{(3,1),(4,1),(6,1)\} & \{(5,0),(9,1),\infty_3\}, & \{(9,0),(5,1),\infty_4\}]\end{array}$$

50.9 Theorem The necessary conditions for the existence of a (k, λ)-RGDD of type $g^{v/g}$ are $v \equiv 0 \pmod k$, $v \equiv 0 \pmod g$ and $\lambda(v-g) \equiv 0 \pmod{k-1}$.

50.10 Theorem [6] A $(2, \lambda)$-RGDD of type $g^{v/g}$ exists if and only if v is even and $v \equiv 0 \pmod g$.

50.11 Theorem [1, 5] A $(3, \lambda)$-RGDD of type $g^{v/g}$ exists if and only if $\lambda(v-g)$ is even, $v \equiv 0 \pmod 3$, and $(\lambda, g, v) \notin \{(1,2,12), (1,6,18)\} \cup \{(2j+1, 2, 6), (4j+2, 1, 6) \mid j \geq 0\}$.

50.12 Theorem [3, 6]

1. If $g \geq 5$, $(g, 6) = 1$, then a 4-RGDD of type $g^{v/g}$ exists if and only if $v \equiv 0 \pmod{12}$.
2. If $g = 1$ or 4, then a 4-RGDD of type $g^{v/g}$ exists if and only if $v \equiv 4 \pmod{12}$ (equivalent to a resolvable (v, 4, 1) BIBD; see §I.6).
3. If $g = 3$, then a 4-RGDD of type $g^{v/g}$ exists if and only if $v \equiv 0 \pmod{12}$ and $v \geq 24$, except possibly for $v \in \{84, 132, 264, 372, 456, 552, 660, 804, 852\}$.

50.4 See Also

§I.6	Resolvable BIBDs are a class of URDs.
§I.6	Kirkman frames are used to construct RGDDs.
§II.2	Resolvable TDs are a class of URDs.
§III.2	Resolvable PBDs are the underlying block set of URDs.

References

[1] A.M. Assaf and A. Hartman, Resolvable group divisible designs with block size 3, *Discrete Math.* 77 (1989), 5-20.

[2] P. Danziger and E. Mendelsohn, Uniformly resolvable designs, *J. Combin. Math. Combin. Comput.*, to appear.

[3] H. Shen, Resolvable group divisible designs with block size 4, *J. Combin. Math. Combin. Comput.* 1 (1987), 125-130.

[4] R. Rees, Uniformly resolvable pairwise balanced designs with block sizes two and three, *J. Combin. Theory A* 45 (1987), 207-225.

[5] R. Rees and D.R. Stinson, On resolvable group-divisible designs with block size 3, *Ars Combin.* 23 (1987), 107-120.

[6] R. Rees and D.R. Stinson, Frames with block size four, *Canad. J. Math.* 44 (1992), 107-120.

51 Vector Space Designs

Dijen K. Ray-Chaudhuri

51.1 Preliminaries

Let P be a finite partially ordered set with a rank function r, $P_\ell = \{p : p \in P, r(p) = \ell\}$. Let t be a nonnegative integer, K a set of nonnegative integers, and $F : P_t \to \mathbb{Z}$. Let $P_K = \cup_{k \in K} P_k$. A *signed* t-(P, K, F) *design* is a function $f : P_K \to \mathbb{Z}$ such that for any $T \in P_t$, $\sum_{B \geq T} f(B) = F(T)$.

If $f(B) \neq 0$, B is a *block*. If for all blocks $f(B) > 0$, it is a t-(P, K, F) *design*. If $f(K) \in \{0, 1\}$, it is a *design without repeated blocks*. If $F(T) = \lambda$, for all $T \in P_t$, then it is a t-(P, K, λ) *design*. If K is a singleton set $\{k\}$, then a t-(P, K, λ) design is a t-(P, k, λ) *design*.

51.1 Example A t-(v, k, λ) design (§I.3) on a v-set X is a t-(P, k, λ) design where $P = (P(X), \subseteq)$, the subsets of X ordered by inclusion.

51.2 Example An orthogonal array $OA_\lambda(t, k, v)$ (§II.5) is a t-(P, k, λ) design where V is a set of v symbols, I is a set of size k. Elements of P are $(J, \underline{X})$, $J \subseteq I$, $\underline{X} : J \to V$. $(J, \underline{X}) \leq (L, \underline{Y})$ if and only if $J \subseteq L$, $\underline{Y}|_J = \underline{X}$.

51.3 Example Projective space: V is the vector space of dimension v over a finite field of order q, elements of P are the subspaces of V ordered by inclusion. Here the design is a t-$[v, K, \lambda; q]$-*design*.

51.4 Example Affine space: V is the vector space of dimension v, H is a dimension $(v-1)$ subspace, and elements of P are the subspaces not contained in H ordered by inclusion.

51.5 Example Ordered design: P is similar to Example 51.2 except that the mappings $\underline{X}$, $\underline{Y}$ are required to be injective.

51.6 Example Perpendicular array: elements of P are the same as in Example 51.5; however, $(J, \underline{X}) \leq (L, \underline{Y})$ if and only if $J \subseteq L$ and $\underline{Y}(J) = \underline{X}(J)$.

51.7 Example Orthogonal arrays on projective spaces: start with a vector space V of dimension v over a finite field F, a vector space I of dimension k over F. As elements of P, take ordered pairs $(J, \underline{X})$ where J is a subspace of I, $\underline{X}$ a linear mapping from J to V. Also $(J, \underline{X}) \leq (L, \underline{Y})$ if and only if $J \subseteq L$ and $\underline{X}|_J = \underline{Y}$.

51.8 Remark Analogs of ordered designs and perpendicular arrays can be defined naturally for projective spaces and affine spaces.

51.9 Remark Necessary and sufficient conditions for existence of signed designs when $K = \{k\}$:

1. Example 51.1: $(v \geq k \geq t)$ for $0 \leq i \leq t$, $\lambda\binom{v-i}{t-i} \equiv 0 \bmod \binom{k-i}{t-i}$ [4, 12].
2. Example 51.2: $k \geq t \geq 1$ [5, 6].
3. Example 51.3: $(v \geq k \geq t)$ for $0 \leq i \leq t$, $\lambda\left[{v-i \atop t-i}\right]_q \equiv 0 \bmod \left[{k-i \atop t-i}\right]_q$, where $\left[{m \atop n}\right]_q$ denotes the number of n dimensional subspaces of an m dimensional vector space over a field of order q [5, 6] (see Proposition VI.7.37).
4. Example 51.4: $(v \geq k \geq t)$ same conditions as in Example 51.3, for i in the range $1 \leq i \leq t$ [5, 6].

51.10 Conjecture (*v-large existence*) In Examples 51.1, 51.2, 51.3, and 51.4, there exists a constant v_0 depending on k, t, and possibly q such that if the necessary conditions of Remark 51.9 hold and $v \geq v_0$, then a t-(P, k, λ) design exists in the corresponding poset P.

51.11 Remark Conjecture 51.10 is proved for t-designs with $t = 2$ by Wilson [13]. For orthogonal arrays with $t = 2$, $\lambda = 1$, Conjecture 51.10 is proved by Chowla, Erdös, and Strauss [1]. Recently, Blanchard [2] proved Conjecture 51.10 for orthogonal arrays with general t and $\lambda = 1$.

51.12 Theorem (*λ-large existence*) In Examples 51.1, 51.2, 51.3, and 51.4, there exists a constant λ_0 depending on k, t, and possibly q such that if $\lambda \geq \lambda_0$ and the necessary conditions of Remark 51.9 hold, then a t-(P, k, λ) design exists in the corresponding partially ordered set P.

51.13 Remark Theorem 51.12 for Example 51.1 is proved by Wilson [12], and is proved by Ray-Chaudhuri and Singhi [5] for Examples 51.2, 51.3 and 51.4. Blanchard [3] proved a similar λ-large theorem for perpendicular arrays.

51.2 Constructions of Projective Space Designs

51.14 Remark The first nontrivial example of projective space designs was given by Thomas in [11] for $q = 2$, $t = 2$, $k = 3$, $(v, 6) = 1$, and $\lambda = 7$.

51.15 Remark Ray-Chaudhuri and Schram [7] used quadratic forms to construct projective space designs for q prime power, $t = 2$, $k = 3$, v odd, $\lambda = q^2 + q + 1$, and $t = 3$, $k = 4$, v even, $\lambda = q^6 + q^5 + q^4$. The family for $t = 2$ was also constructed by Suzuki [10].

51.16 Construction $F = \mathbb{F}_q$, $V = \mathbb{F}_{q^v}$, v odd, $k = t+1$ and $Q : V^k \to V$, $Q(x_1, x_2, \cdots, x_k) = \sum \alpha_{ij} x_i x_j$, $\alpha_{ij} \in F$. If $(x_1, x_2, \cdots, x_k)$ satisfies $\sum \alpha_{ij} x_i x_j = y$ for a nondegenerate quadratic form, $\sum_{i,j} \alpha_{ij} x_i x_j$, $\alpha_{ij} \in F$, declare the subspace $\langle x_1, x_2, \cdots, x_k \rangle$ over F to be a block. After some refinement, this method of generating blocks works.

Fix $y \in V$. A duplet is an ordered pair $(Q, \underline{a})$, where $\underline{a} = (a_1, a_2, \cdots, a_k) \in V^k$, $k = t+1$, Q is a quadratic form of rank k with k variables and coefficients from F, and $Q(\overline{a}) = y$. Define an equivalence relation on duplets generated by the related pairs $(Q, \underline{a}) \cong (QL^{-1}, L\underline{a})$ if L is a nonsingular $k \times k$ matrix over F, and $(Q, \underline{a}) \cong (R, \underline{a})$ if $Q(\underline{x}) = \sum \alpha_{ij} x_i x_j$, $R(\underline{x}) = \sum \beta_{ij} x_i x_j$, $\alpha_{ij} = \beta_{ij}$ whenever $a_i a_j \neq 0$.

For a duplet $(Q, \underline{a})$, the subspace $\langle a_1, a_2, \cdots, a_k \rangle$ over F is the *associated block*. Let $\mathcal{D}(i, y)$ be the set of equivalence classes of duplets whose associated blocks have dimension i and value y (i.e., $Q(\underline{a}) = y$).

Define $\mathcal{D}(i, y)(B)$ to be the number of equivalence classes of duplets whose associated block is B, with dimension i, value y, and $\mathcal{B}(t, y)(B) = \mathcal{D}(t+1, y)(B)$ if dim $B = t+1$, $\mathcal{B}(t, y)(B) = q^t \mathcal{D}(t, y)(B)$ if dim $B = t$, and 0 otherwise.

51.17 Theorem For v odd and any $y \in V$, $\mathcal{B}(t, y)$ is a t-$[v, \{t, t+1\}, \lambda; q]$ design, where λ is the number of nondegenerate quadratic forms in $t+1$ variables with coefficients from $F = \mathbb{F}_q$ divided by $(q^{t+1} - q^t)$.

51.18 Remark In Theorem 51.17 if $t = 1$, then $\lambda = q$; if $t = 2$, then $\lambda = q^3 - 1$; if $t = 3$, then $\lambda = q^3(q^3 - 1)$; if $t = 4$, then $\lambda = q^2(q^3 - 1)(q^5 - 1)$. When $y = 0$, one has a special case which results in bigger equivalence classes of duplets $(Q, \underline{a})$, $Q(\underline{a}) = 0$. For $y = 0$, further define $(Q, \underline{a}) \cong (cQ, \underline{a})$, for $c \in F$, $c \neq 0$. Theorem 51.19 results.

51.19 Theorem Let $\mathcal{B}'(t, 0)$ be equal to $\mathcal{B}(t, 0)$ divided by $(q-1)$. If v is odd, then $\mathcal{B}'(t, 0)$ is a t-$[v, \{t, t+1\}, \lambda'; q]$ design where λ' is the number of rank $t+1$ quadratic forms in $t+1$ variables with coefficients from $F = \mathbb{F}_q$ divided by $(q-1)(q^{t+1} - q^t)$.

51.20 Remark In Theorem 51.19, for $t = 2$, $\mathcal{D}(2, 0)$ is empty, i.e., for all duplets $(Q, \underline{a})$, the associated blocks have dimension 3. So one gets a $2 - [v, 3, \lambda; q]$ design with $\lambda = q^2 + q + 1$. If $(v, 6) = 1$, this design has no repeated blocks.

51.21 Construction A 3-design with $k = 4$: For v odd, $\mathcal{B}'(3, 0)$ is a 3-design on $V = \mathbb{F}_{q^v}$ with blocks of dimension 3 and dimension 4. Define $V^* = V \oplus F_\infty$, so V^* is a $(v+1)$-dimensional vector space over F. Start with the design $\mathcal{B}'(3, 0)$ on V. If B is a block of dimension 3, adjoin ∞ to get a block of dim 4. If B is a block of dim 4, keep all the 4-dim subspaces of $\langle B, \infty \rangle$ not containing ∞. This gives a $t = 3$, $k = 4$, $\lambda = q^6 + q^5 + q^4$ design for the space V^*.

51.22 Construction Method of inflating blocks: Given a t-$[v, k, \lambda; q]$ design, replace each k-dim block by all $k+1$ subspaces which contain it; the result is a t-design with blocks of dim $k+1$. This method produces vector space 2-designs with a single block size k, $k \geq 3$, v odd.

51.3 A Generalization of the Concept of Large Set of Designs

Let P, K and P_k be as in §51.1. Let $\underline{P_k}$ denote the function $f : P_k \to \{0, 1\}$ which takes the value 1 for each element of P_k and 0 elsewhere. For $i = 1, 2, \ldots c$, let f_i be a t-(P, K, λ) design for a fixed λ. If there exists λ_k, $k \in K$, positive integers such that $\sum_{i=1}^{c} f_i = \sum_{k \in K} \lambda_k \underline{P_k}$, then $(f_1, f_2, \ldots, f_c)$ is a *generalized large set of t-(P, K, λ) designs.*

51.23 Remark The following theorems of Ray-Chaudhuri and Schram [8] give large sets of designs in this sense.

51.24 Theorem $\sum_{y\in V}\mathcal{B}(t,y)=\lambda_1\underline{P_{t+1}}+\lambda_2\underline{P_t}$, where λ_1 is the number of rank $t+1$ quadratic forms in $t+1$ variables with coefficients from F and $\lambda_2=\lambda_1/(q-1)$, each $\mathcal{B}(t,y)$ is a vector space t-design with parameters described in Theorem 51.19.

51.25 Remark For $t=1$, a simpler situation occurs. For $y,z\neq 0$, $\mathcal{B}(1,y)$ and $\mathcal{B}(1,z)$ are isomorphic if $y=cz$ for some $c\in F^*$.

51.26 Theorem $\sum\mathcal{B}(1,F^*y)=q^2\underline{P_2}+q\underline{P_1}$, the sum is over $F^*y\in V^*/F^*$. $\mathcal{B}(1,y)$ has only one block of dimension 1, all other blocks are of dimension 2.

51.4 The Number of Designs

51.27 Theorem [9] Let $n^v_{k,t}(\lambda)$ be the number of distinct t-(v,k,λ) designs. For fixed v,k,t, $n^v_{k,t}(\lambda)$ is a quasi-polynomial in λ, i.e., there exists a period N and polynomials $\phi_i(\lambda)$, $i=0,1,2,\cdots,(N-1)$ such that for $\lambda\equiv i \pmod N$, $n^v_{k,t}(\lambda)=\phi_i(\lambda)$.

51.28 Remark Singhi and Shrinkhande also proved the reciprocity relations $n^v_{k,t}(-1)=n^v_{k,t}(-2)=\cdots=n^v_{k,t}(-\rho+1)=0$ and $n^v_{k,t}(\lambda)=(-1)^d n^v_{k,t}(-\rho-\lambda)$, $d=\binom{v}{k}-\binom{v}{t}$, $\rho=\binom{v-t}{k-t}$, $k\geq t, v\geq k+t$. Similar results for Examples 51.2, 51.3, and 51.4 are proved by Ray-Chaudhuri and Singhi [5, 6].

51.29 Remark Consider the linear equations $Y_B=f(B)$, $\sum_{T\subset B}Y_B-Y_\infty=0$, one equation for each $T\in P_t$. If f is a t-(v,k,λ) design, $Y_B=f(B), B\in P_k$, $Y_\infty=\lambda$ is a solution. If $\underline{Y}$ and $\underline{Y}'$ are solutions corresponding to t-(v,k,λ) and t-(v,k,λ') designs, then $\underline{Y}+\underline{Y}'$ is a solution corresponding to a t-$(v,k,\lambda+\lambda')$ design. The solutions form a monoid M. Form the monoid algebra. Take one indeterminate X_i for each Y_i, corresponding to $\underline{Y}$; take the monomial $X_1^{Y_1}X_2^{Y_2}\cdots$, $R_M=$ the ring generated by these monomials over a coefficient field F. R_M is a Cohen-Macaulay ring. Then the Hilbert series $F(R_M,\theta_1,\theta_2,\cdots)=\sum_{\underline{Y}\in M}\theta_1^{Y_1}\theta_2^{yY2}\cdots$ is a rational function, $F(R_M,\theta_1=\theta_2=\cdots=\theta)$ is a rational function in θ and the coefficient of θ^l with $l=\lambda\binom{v}{t}/\binom{k}{t}+\lambda$ is $n^v_{k,t}(\lambda)$. Since $F(R_M,\theta)$ is a rational function, $n^v_{k,t}(\lambda)$ is a quasi-polynomial in λ.

51.5 See Also

§I.3	t-Designs.
§II.5	Orthogonal arrays of higher strength.

References

[1] S. Chowla, P. Erdös, E.G. Straus, On the maximal number of pairwise orthogonal Latin squares of a given order, *Canad. J. Math.* 12 (1960), 204-208.

[2] J.L. Blanchard, The existence of orthogonal arrays of any strength with large order, preprint.

[3] J.L. Blanchard, private communication.

[4] J.E. Graver and W.B. Jurkat, The module structure of integral designs, *J. Combin. Theory A* 15 (1973), 75-90.

[5] D.K. Ray-Chaudhuri and N.M. Singhi, On existence and number of orthogonal arrays, *J. Combin. Theory A* 47 (1988), 28-36.

[6] D.K. Ray-Chaudhuri and N.M. Singhi, q-analogues of t-designs and their existence, *Linear Alg. Applic.* 114-115 (1989), 57-68.
[7] D.K. Ray-Chaudhuri and E.J. Schram, Designs on vector spaces constructed using quadratic forms, *Geom. Ded.* 42 (1992), 1-42.
[8] D.K. Ray-Chaudhuri and E.J. Schram, A large set of designs on vector spaces, *J. Number Theory* 47 (1994), 247-272.
[9] S.S. Shrikhande and N.M. Singhi, A reciprocity relation for t-designs, *Europ. J. Combin.* 8 (1987), 59-68.
[10] H. Suzuki, 2-designs over GF(2^m), *Graphs Combin.* 6 (1990), 293-296.
[11] S. Thomas, Designs over finite fields, *Geom. Ded.* 24 (1987), 237-242.
[12] R.M. Wilson, The necessary conditions for t-designs are sufficient for something, *Utilitas Math.* 4 (1973), 207-215.
[13] R.M. Wilson, An existence theory for pairwise balanced designs, III. Proof of the existence conjecture, *J. Combin. Theory A* 18 (1975), 71-79.

52 Weighing Matrices and Conference Matrices

R. Craigen

52.1 Definitions and Examples

Let n and w be positive integers. A *weighing matrix* W, of *weight* w and *order* n, is an $n \times n$ $(0, \pm 1)$-matrix satisfying $WW^t = wI$. Equivalently, one can write $W = W(n, w)$. One generally does not allow $w = 0$.

A $W(n, n-1)$ is a *conference matrix* of order n.

Two weighing matrices are *equivalent* if one can be transformed into the other by a series of permutations and/or negations of its rows and columns.

Two matrices, A and B, are *disjoint* if nonzero entries of A occur only in positions where B is zero. They are *amicable* if $AB^t = BA^t$ and *antiamicable* if $AB^t = -BA^t$. A *circulant matrix*, $A = \text{circ}(a_1, \ldots, a_n)$, is a square $n \times n$ matrix whose (i, j) entry is a_{j-i+1} (index reduced modulo n). A is *block-circulant* if it is $mn \times mn$ and $a_1, \ldots, a_n$ are $m \times m$ submatrices. A matrix is *regular* if its row and column sums are equal, and *quasisymmetric* if its pattern of zeros is symmetric.

52.1 Remark In displays, $-$ represents -1.

52.2 Example Some classical weighing matrices: $A = W(4, 3)$ (a skew conference matrix), $B = W(7, 4)$ (B is circulant), $C = W(10, 9)$ (a regular conference matrix), and $D = W(12, 10)$ (quasisymmetric).

$$A = \begin{pmatrix} 0 & - & - & - \\ 1 & 0 & 1 & - \\ 1 & - & 0 & 1 \\ 1 & 1 & - & 0 \end{pmatrix}, \qquad B = \begin{pmatrix} - & 1 & 1 & 0 & 1 & 0 & 0 \\ 0 & - & 1 & 1 & 0 & 1 & 0 \\ 0 & 0 & - & 1 & 1 & 0 & 1 \\ 1 & 0 & 0 & - & 1 & 1 & 0 \\ 0 & 1 & 0 & 0 & - & 1 & 1 \\ 1 & 0 & 1 & 0 & 0 & - & 1 \\ 1 & 1 & 0 & 1 & 0 & 0 & - \end{pmatrix},$$

$$C = \begin{pmatrix} 0&1&1&1&1&1&1&-&-&- \\ 1&0&1&1&1&-&-&-&1&1 \\ 1&1&0&1&-&1&-&1&-&1 \\ 1&1&1&0&-&-&1&1&1&- \\ 1&1&-&-&0&1&1&-&1&1 \\ 1&-&1&-&1&0&1&1&-&1 \\ 1&-&-&1&1&1&0&1&1&- \\ -&-&1&1&-&1&1&0&1&1 \\ -&1&-&1&1&-&1&1&0&1 \\ -&1&1&-&1&1&-&1&1&0 \end{pmatrix}, \qquad D = \begin{pmatrix} 0&1&1&1&-&1&0&-&1&1&-&- \\ 1&0&1&1&1&-&-&0&-&1&1&- \\ -&1&0&1&1&1&-&-&0&-&1&1 \\ 1&-&1&0&1&1&1&-&-&0&-&1 \\ 1&1&-&1&0&1&1&1&-&-&0&- \\ 1&1&1&-&1&0&-&1&1&-&-&0 \\ 0&-&-&1&1&-&0&-&1&-&-&- \\ -&0&-&-&1&1&-&0&-&1&-&- \\ 1&-&0&-&-&1&-&-&0&-&1&- \\ 1&1&-&0&-&-&-&-&-&0&-&1 \\ -&1&1&-&0&-&1&-&-&-&0&- \\ -&-&1&1&-&0&-&1&-&-&-&0 \end{pmatrix}.$$

52.2 Necessary Conditions

52.3 Theorem Suppose a $W(n, w)$ exists.

1. If $n > 1$ is odd, then w is a perfect square and $n \geq w + \sqrt{w} + 1$. Equivalent inequalities: (i) $w \leq n - \frac{1+\sqrt{4n-3}}{2}$; (ii) $(n-w)^2 - (n-w) + 1 \geq n$. In the case of equality, a projective plane of order $\sqrt{w}$ must exist. See §VI.7.2.
2. If $n \equiv 2 \pmod 4$, then w is a sum of two squares; if $n > 2$, then $w < n$.

52.4 Theorem Suppose there is a skew $W(n, w)$. Then:

1. $n \equiv 0 \pmod 2$;
2. if $n \equiv 2 \pmod 4$, then w is a square, and $w < n - 1$ if $n \neq 2$;
3. if $n \equiv 4 \pmod 8$, then w is a sum of 3 squares; if $w = n - 2$, then w is a sum of 2 squares.

52.5 Theorem If $W = W(n, w)$ is regular with row sum k, then $w = k^2$.

52.3 Structure, Conjectures, Miscellanea

52.6 Theorem If $W = W(n, n-1)$, then either $n \equiv 2 \pmod 4$ and W is equivalent to a symmetric matrix, or $n \equiv 0 \pmod 4$ and W is equivalent to a skew matrix.

52.7 Conjecture (*weighing matrix conjecture* [5]) If $n \equiv 0 \pmod 4$ and $w \leq n$, then $W(n, w)$ exists.

52.8 Remark Conjecture 52.7 has been confirmed for $n = 4k \leq 88$. One might also conjecture that a $W(n, w)$ exists when n is even and w is less than n and a sum of two squares (confirmed for $n \leq 32$). There is no $W(9, 4)$, so the corresponding result for square weights is certainly not true.

52.9 Conjecture (*skew weighing matrix conjecture* [5]) If $n \equiv 4 \pmod 8$, $w = a^2 + b^2 + c^2 < n$, then a skew $W(n, w)$ exists unless $w = n - 2$ is not a sum of squares. If $n \equiv 0 \pmod 8$ and $w < n$, then a skew $W(n, w)$ exists.

52.10 Remark Conjecture 52.9 has been confirmed for $n = 4k \leq 88$ and $n = 8k \leq 96$. It also holds for $2^t \cdot 3$, $2^t \cdot 5$, $2^t \cdot 7$, $2^t \cdot 9$ and $2^t \cdot 13$ for $t \geq 3$, $2^t \cdot 15$ and $2^t \cdot 21$, for $t \geq 4$ [6]. As for square weights in even orders, there is no skew $W(18, 16)$.

52.11 Conjecture If there is a $W(n, w)$, n odd, then there is a symmetric $W(n, w)$

52.12 Remark Conjecture 52.11 is confirmed up to order 13; there exists a $W(35, 16)$ that is not equivalent to a symmetric matrix. One might also conjecture in this case that there exists a regular $W(n, w)$. This can also be confirmed up to order 13. Conjecture 52.11 does not extend to even orders: There is a skew $W(16, 15)$ but no symmetric $W(16, 15)$ [5].

52.13 Conjecture (*half-full weighing matrix conjecture*) If w is even, then there exists a $W(2w, w)$.

52.14 Remark Conjecture 52.13 has been confirmed for $w = 2k \leq 212$. It is probably also true that a $W(2w, w)$ exists whenever w is an odd sum of two squares (confirmed for $w \leq 25$).

52.15 Conjecture (*conference matrix conjecture*) A conference matrix $W(n, n-1)$ exists for all $n \equiv 2 \pmod 4$ as long as $n - 1$ is a sum of two squares.

52.16 Remark Conjecture 52.15 is confirmed for $n = 2k \leq 61$ with k odd. For the case $n \equiv 0 \pmod 4$, see §IV.24.

52.17 Remark $W(n, w)$ have been classified up to equivalence for $w \leq 5$ [1] and for $n \leq 13$ [7]. Similarly, the block structure of $W(n, n-a)$ has been largely determined for $a \leq 4$ [4]; for instance, if $n > 3$ and $a = 2$, then the matrix is equivalent to one partitioned into 2×2 blocks, whose diagonal blocks are zero and whose off-diagonal blocks are rank 1 if $n \equiv 2 \pmod 4$ and rank 2 otherwise.

52.4 Constructions and Existence

52.18 Theorem If w is a square and n is large enough, then $W(n, w)$ exists. If w is a sum of two squares and $n \equiv 0 \pmod 2$ is large enough, then $W(n, w)$ exists. For any w, if $n \equiv 0 \pmod 4$ is large enough, then $W(n, w)$ exists. See [5].

52.19 Theorem If $W = W(n, w)$ is skew, then

1. $W + I = W(n, w + 1)$, and
2. $\begin{pmatrix} W & -W-I \\ W+I & W \end{pmatrix} = W(2n, 2w + 1)$ is skew.

52.20 Theorem If $W = W(n, w)$ has diagonal 0, then

$$\begin{pmatrix} W+I & W-I \\ W^t-I & -W^t-I \end{pmatrix} = W(2n, 2w + 2).$$

52.21 Theorem If there exist amicable $m \times m$ $(0, \pm 1)$-matrices $A_1, \ldots, A_k$ with $\sum_{i=1}^{k} A_i {A_i}^t = wI$ and an orthogonal design $\text{OD}(n; a_1, \ldots, a_k)$, then there is a $W(mn, w)$. See [5].

52.22 Theorem If p is an odd prime power, then there is a $W(p^k + \cdots + p^2 + p + 1, p^k)$, $k \geq 0$, a $W(p+1, p)$, a symmetric $W(2p^2, p^2)$, and a symmetric $W(p^2 + h(p+1), p^2)$; $h \geq p$; if p is any prime power, then there is a circulant $W(p^2 + p + 1, p^2)$. See [2, 5].

52.23 Theorem If $w \leq \min_i(n_i)$ and $W(n_1 + 1, w), \ldots, W(n_k + 1, w)$ exist, then so does $W(n_1 \cdots n_k + 1, w^k)$. If an even number of the given matrices are skew and the remainder are symmetric, then the resulting matrix is symmetric. If an odd number of them are skew and the remainder are symmetric, then the resulting matrix is skew.

52.24 Theorem [3] Let M be a $(0,1)$-matrix having row sums $r_1, \ldots, r_m$ and column sums $c_1, \ldots, c_n$. If, for fixed integers a, b, there are weighing matrices $A_i = W(r_i, a)$, $i = 1, ..., m$ and $B_j = W(c_j, b)$, $j = 1, ..., n$, then there is a $W(r_1 + \cdots + r_m, ab)$. If M is symmetric and $a = b$, then the resulting matrix may be taken symmetric; if in addition M has zero diagonal, then the resulting matrix may be taken skew.

52.25 Remark Theorem 52.24 is based on the method of *weaving*, which generalizes both direct sum and Kronecker (direct) product. Further: if $W(m, w)$ and $W(n, w)$ exist, both symmetric (respectively skew), then the direct sum gives a symmetric (respectively skew) $W(m + n, w)$; if $W(m, a)$ and $W(n, b)$ exist, then the Kronecker product gives $W(mn, w)$ that is symmetric if the weighing matrices are both symmetric, and skew if one is symmetric and the other is skew; the direct sum of $W(n, w)$ with itself is equivalent to a $W(2n, w)$ that can be made skew or symmetric.

52.26 Theorem Suppose there are k $(0, \pm 1)$-sequences of length n having zero periodic autocorrelation (§IV.42) and weight w.

1. If $k = 1$, then there is a circulant $W(n, w)$.
2. If $k = 2$, then there is a block-circulant $W(2n, w)$ (2×2 blocks).
3. If $k = 4$, then there is a $W(4n, w)$.
4. If $k > 2$, then there is a block-circulant $W(2^{2k}n, 2w + 1)$ ($2^{2k} \times 2^{2k}$ blocks).
5. If the sequences are all disjoint and quasisymmetric, then there is a block-circulant $W(2^{2k-1}n, w)$ ($2^{2k-1} \times 2^{2k-1}$ blocks).

52.27 Remark How to read the Tables 52.28 - 52.34. If n appears without parenthesis in the record corresponding to w, this indicates that $W(n, w)$ is known to exist; if n appears in parentheses, then its existence has not been resolved; if n does not appear, then it has been shown that a $W(n, w)$ does not exist. Consecutive admissible n that exist are indicated by dots. For example, in Table 52.28 when $w = 49$, the entry 104...110 indicates that $W(n, 49)$ exists for $n =$ 104,105,106,107,108, 109 and 110, while 10... (with $w = 4$) indicates that $W(n, 4)$ exists for all $n \geq 10$. Consecutive unresolved orders are indicated similarly, within parentheses.

52.28 Table Weighing matrices $W(n, w)$, $w = a^2 \leq 256$ (any n admissible).

w	n
1	1...
4	4, 6, 7, 8, 10...
9	10, 12...16, (17), 18...
16	16, 18, 20, 21, 22, (23), 24, (25), 26, (27), 28, (29), 30...
25	26, 28, 30, 31, 32, (33, 34, 35), 36, (37, 38, 39), 40, (41), 42, (43), 44, (45), 46, (47), 48, (49), 50, (51), 52, (53), 54...
36	36, (38), 40, 42, 44, (45), 46, (47), 48, (49), 50, (51), 52, (53), 54, (55), 56, (57, 58, 59), 60, (61), 62, 63, 64, (65), 66, (67), 68, (69), 70, (71), 72...
49	50, 52, (54), 56, 57, (58, 59), 60, (61, 62, 63), 64, (65, 66, 67), 68, (69, 70, 71), 72, (73, 74, 75), 76, (77), 78, (79), 80, (81, 82, 83), 84, (85, 86, 87), 88, (89, 90, 91), 92, (93, 94, 95), 96, (97), 98, (99), 100, (101), 102, (103), 104...110, (111), 112, 113, 114, (115), 116, 117, (118, 119), 120, 121, 122, (123), 124, 125, 126, (127), 128, 129, 130, (131), 132, 133, 134, (135), 136, 137, 138, (139), 140, 141, (142, 143), 144...150, (151), 152...
64	64, 66, 68, 70, 72, 73, 74, (75), 76, (77), 78, (79), 80, (81), 82, (83), 84, (85), 86, (87), 88, (89), 90, (91), 92, (93), 94, (95), 96, (97), 98, (99), 100, (101), 102, (103), 104, (105), 106, (107), 108, (109), 110, (111), 112, (113), 114, (115), 116...
81	82, 84, (86), 88, 90, 91, 92, (93, 94, 95), 96, (97, 98, 99), 100, (101, 102, 103), 104, (105, 106, 107), 108, 109, 110, (111), 112, (113...116), 117, 118, (119), 120, 121, 122, (123), 124, (125), 126, 127, 128, (129), 130...

w	n
100	100, (102), 104, (106), 108, 110, 112, (113), 114, (115), 116, (117), 118, (119), 120, (121), 122, (123), 124, (125), 126, (127), 128, (129), 130, (131), 132, (133), 134, (135), 136, (137), 138, (139), 140, (141), 142, (143), 144, (145), 146, (147), 148, (149), 150, (151), 152, (153), 154, (155), 156, (157), 158, (159), 160, (161), 162, (163), 164, (165), 166, (167), 168, (169), 170, (171), 172, (173), 174, (175), 176, (177), 178, (179), 180, (181), 182, (183), 184, (185), 186...
121	122, (124, 126), 128, (130), 132, 133, (134...143), 144, 145, (146...155), 156, (157, 158, 159), 160, (161...165), 166, (167), 168, (169...175), 176, (177, 178, 179), 180, (181...191), 192, (193, 194, 195), 196, 197, 198, (199), 200, (201, 202, 203), 204, (205, 206, 207), 208, (209), 210, (211), 212, (213, 214, 215), 216, (217, 218, 219), 220, (221, 222, 223), 224, (225), 226, (227), 228, (229, 230, 231), 232, (233, 234, 235), 236, (237, 238, 239), 240, (241), 242, (243), 244, (245, 246, 247), 248, (249), 250, (251), 252...256, (257), 258, (259), 260, 261, 262, (263), 264, 265, 266, (267), 268, (269, 270, 271), 272, 273, (274, 275), 276, 277, 278, (279), 280, (281), 282, (283), 284, (285), 286, (287), 288, 289, 290, (291), 292, 293, 294, (295), 296, (297), 298...302, (303), 304, 305, (306, 307), 308, 309, 310, (311), 312, 313, 314, (315), 316, 317, 318, (319), 320, 321, 322, (323), 324, 325, 326, (327), 328...334, (335), 336, 337, 338, (339), 340...346, (347), 348, 349, 350, (351), 352...366, (367), 368, 369, 370, (371), 372...
144	144, (146, 148, 150, 152, 154, 156, 158, 159), 160, (161...167), 168, (169...179), 180, (181, 182, 183), 184, (185, 186, 187), 188, (189), 190, (191), 192, (193), 194, (195), 196, (197), 198, (199), 200, (201), 202, (203), 204, (205), 206, (207), 208, (209), 210, (211), 212...
169	170, (172, 174, 176, 178, 180), 182, 183, (184...195), 196, (197...207), 208, (209), 210, (211...221), 222, (223), 224, (225), 226, (227), 228, (229), 230, (231), 232, (233), 234, (235), 236, (237), 238, (239), 240, (241), 242, (243), 244, (245), 246, (247), 248, (249), 250, (251), 252, (253), 254, (255), 256, (257), 258, (259), 260, (261), 262, (263), 264, (265), 266, (267), 268, (269), 270, (271), 272, (273), 274, (275), 276, (277), 278, (279), 280, (281), 282, (283), 284, (285), 286, (287), 288, (289), 290, (291), 292, (293), 294, (295), 296, (297), 298, (299), 300, (301), 302, (303), 304, (305), 306, (307), 308, (309), 310, (311), 312, (313), 314, (315), 316, (317), 318, (319), 320, (321), 322, (323), 324, (325), 326, (327), 328, (329), 330, (331), 332, (333), 334, (335), 336, (337), 338, (339), 340, (341), 342, (343), 344, (345), 346, (347), 348, (349), 350...354, (355), 356, (357), 358, (359), 360, (361), 362, (363), 364, 365, 366, (367), 368, (369), 370, (371), 372, (373), 374, (375), 376, (377), 378...
196	196, (198), 200, (202, 204, 206), 208, (210...225), 226, (227), 228, (229...239), 240, (241...255), 256, (257...271), 272, (273...285), 286, (287), 288, (289...299), 300, (301, 302, 303), 304, (305...311), 312, (313...319), 320, (321, 322, 323), 324, (325, 326, 327), 328, (329, 330, 331), 332, (333, 334, 335), 336, (337, 338, 339), 340, (341), 342, (343), 344, (345), 346, (347), 348, (349), 350, (351), 352, (353, 354, 355), 356, 357...
225	226, (228, 230, 232, 234, 236, 238), 240, (241...255), 256, 260, (261...271), 272, (273...285), 286, (287), 288, (289...299), 300, (301, 302, 303), 304, (305...313), 314, (315), 316, (317), 318, (319), 320, (321), 322, (323), 324, (325), 326, (327), 328, (329), 330, (331), 332, (333), 334, (335), 336, (337), 338, (339), 340, (341), 342...
256	256, 258, 260, 262, 264, 266, 268, 270, 272, 273, 274, (275) ... in this range, even orders all exist, odd are all unresolved ... (313), 314, 315, 316, (317), 318, (319), 320, (321), 322, (323), 324, (325), 326, (327), 328, (329), 330, (331), 332, (333), 334, (335), 336, (337), 338, (339), 340, 341, 342, (343), 344, (345), 346, (347), 348, (349), 350, (351), 352, (353), 354, (355), 356, 357, 358, (359) ... even orders exist, odd are unresolved ... (379), 380, 381, 382, (383), 384, (385), 386, (387), 388, (389), 390, (391), 392, (393), 394, (395), 396, (397), 398, 399, 400, (401), 402, (403), 404, (405), 406, (407), 408, (409), 410, (411), 412, (413), 414, (415), 416, (417), 418, (419), 420...

52.29 Table Weighing matrices $W(n,w)$, $a^2 \neq w = b^2 + c^2 \leq 50$, $w = 125, 128$ (even n only).

w	n
2	2...
5	6...
8	8...
10	12...
13	14...
17	18...
18	20...32, (34), 36...
20	20...
26	28...
29	30, 32, (34), 36, (38), 40, (42), 44, (46), 48, (50), 52, (54), 56, (58), 60...
32	32...
34	36, (38), 40, 42, 44, 46, 48, (50), 52, (54), 56, (58), 60, (62), 64, (66), 68, (70), 72, (74), 76...
37	38, 40, (42), 44, 46, 48, (50), 52, (54), 56, (58), 60, (62), 64, (66), 68, (70), 72...
40	40...
41	42...48, (50), 52, (54), 56, (58), 60, (62), 64, (66), 68, (70), 72, (74), 76, (78), 80, (82), 84...
45	46, 48, (50), 52, 54, 56, (58), 60, (62), 64, (66), 68, (70), 72, (74), 76, 78, 80, (82), 84, (86), 88...
50	52, (54), 56, (58), 60, 62, 64, (66), 68, (70), 72, (74), 76, (78), 80, (82), 84, (86), 88, (90), 92, (94), 96, (98), 100, (102), 104, (106), 108...
125	126, (128...154), 156, (158...166), 168, (170, 172, 174), 176, (178), 180, (182, 184), 186, (188, 190), 192, (194...203), 204, (206), 208...
128	128...

52.30 Table Weighing matrices $W(n,w)$, $a^2 + b^2 \neq w \leq 99$, ($n \equiv 0 \pmod 4$ only).

w	n
for w = 3, 6, 7, 11, 12, 14, 15, 19, 21, 22, 23, 24, 27, 28, 30, 31, 33, 35, 38, 39, 42, 43, 44, 46, 47, 48, 51, 54, 55, 56, 57, 59, 60, 62, 63, 66, 67, 69, 70, 71, 75, 76, 77, 78, 86, 88, 92, 94, 96: $W(4k,w)$ exists whenever $w \leq 4k$.	
79	80, 84, 88, (92), 96, 100, 104, 108, 112, (116), 120, (124), 128, (132), 136, (140), 144, (148), 152, (156), 160...
83	84, 88, 92, 96, 100, 104, (108), 112, (116), 120, (124), 128, (132), 136, (140), 144, (148), 152, (156), 160, (164), 168...
87	88, 92, 96, (100), 104, (108), 112, (116), 120, (124), 128, (132), 136, (140), 144, (148), 152, (156), 160, (164), 168, (172), 176...
91	92, 96, (100), 104, 108, 112, (116), 120, (124), 128, (132), 136, (140), 144, (148), 152, 156, 160, (164), 168, (172), 176...
93	96, (100), 104, (108), 112, (116), 120, (124), 128, (132), 136, (140), 144, (148), 152, (156), 160, (164), 168, (172), 176, (180), 184, (188), 192, (196), 200, (204), 208, (212), 216, (220), 224, (228), 232, (236), 240, (244), 248, (252), 256...
95	96, (100), 104, 108, 112, (116), 120, (124), 128, (132), 136, 140, 144, (148), 152, (156), 160...
99	100, 104, (108), 112, (116), 120, (124), 128, (132), 136, (140), 144, (148), 152, 156, 160, (164), 168, (172), 176, (180), 184, (188), 192, (196), 200, (204), 208, 212, 216, (220), 224, 228, 232, (236), 240, 244, 248, (252), 256...

52.31 Table Symmetric weighing matrices $W(n, w)$, $w = a^2 \leq 81$ (all n admissible).

w	n
1	1...
4	4, 6, 7, 8, 10...
9	10, 12, 13, (14), 15, 16, (17), 18, (19), 20, (21), 22...
16	16, 18, 20, 21, (22, 23), 24, (25), 26, (27), 28, (29), 30...
25	26, (28, 30), 31, (32...35), 36, (37...41), 42, (43), 44, (45, 46, 47), 48, (49), 50, (51), 52, (53), 54, (55), 56, 57, 58, (59), 60, (61), 62, (63), 64, (65), 66, (67), 68, 69, 70, (71), 72, 73, 74, (75), 76, (77), 78...
36	36, (38), 40, (42), (44...47), 48, (49), 50, (51), 52, (53, 54, 55), 56, (57, 58, 59), 60, (61), 62, (63), 64, (65), 66, (67, 68, 69), 70, (71), 72, (73, 74, 75), 76, (77, 78, 79), 80, (81, 82, 83), 84, (85), 86, (87), 88, (89), 90, 91, 92, (93, 94, 95), 96, (97), 98, (99), 100, (101), 102, (103), 104, 105, 106, (107), 108, (109), 110, (111), 112, (113), 114, (115), 116, (117), 118, (119), 120, (121), 122, (123), 124, (125), 126, 127, 128, (129), 130...134, (135), 136, (137), 138...148, (149), 150, 151, 152, 153, 154, 155, 156, 157, 158, (159), 160...
49	50, (52, 54, 56), 57, (58, 59, 60, 61, 62, 63), 64, (65, 66, 67, 68, 69, 70, 71), 72, (73, 74, 75, 76, 77), 78, (79), 80, (81, 82, 83, 84, 85, 86, 87), 88, (89, 90, 91, 92, 93, 94, 95, 96, 97), 98, (99), 100, (101, 102, 103), 104, 105, 106, 107, 108, (109, 110, 111), 112, 113, 114, (115), 116, (117, 118, 119), 120, 121, 122, (123), 124, (125, 126, 127), 128, 129, 130, (131), 132, (133), 134, 135, 136, 137, 138, (139), 140, (141), 142, (143), 144, 145, (146, 147), 148, (149), 150, (151), 152, (153), 154, 155, 156, 157, 158, (159), 160...
64	64, (66, 68, 70), 72, 73, (74, 75, 76, 77, 78, 79), 80, (81), 82, (83), 84, (85), 86, (87), 88, (89), 90, (91), 92, (93), 94, (95), 96, (97), 98, (99), 100, (101), 102, (103), 104, (105), 106, (107), 108, (109), 110, (111), 112, (113), 114, (115), 116, (117), 118, (119), 120, (121), 122, (123), 124, (125), 126, (127), 128, (129), 130, (131), 132, (133), 134, (135), 136, 137, 138, (139), 140, (141), 142, (143), 144, 145, 146, 147, 148, (149), 150, (151), 152...
81	82, 84, 86, 88, 90), 91, 92, (93, 94, 95), 96, (97, 98, 99), 100, (101...108), 109, (110, 111), 112, (113...116), 117, 118, (119), 120, 121, 122, (123), 124, (125), 126, 127, 128, (129), 130...

52.32 Table Skew weighing matrices $W(n, w)$, $w = a^2 \leq 81$ ($n \equiv 0 \pmod 2$ only).

w	n
1	2...
4	6...
9	12, 14, 16, (18), 20, (22), 24...
16	20...
25	28, (30), 32, (34), 36, (38), 40...
36	(38), 40, (42), 44, (46), 48, 50, 52, (54), 56, (58), 60, (62), 64, (66), 68, (70), 72, (74), 76, 78, 80, (82), 84, (86), 88...
49	52, (54), 56, (58), 60, (62), 64, (66), 68, (70), 72, (74), 76, 78, 80, (82), 84, (86), 88, (90), 92, (94), 96, (98), 100, (102), 104, 106, 108, (110), 112, 114, 116, (118), 120, 122, 124, (126), 128...
64	(66), 68, (70), 72, (74), 76, (78), 80, (82), 84, (86), 88, (90), 92, (94), 96, (98), 100, (102), 104, (106), 108, (110), 112, (114), 116, (118), 120, 122, 124, 126, 128, (130), 132, (134), 136, (138), 140, (142), 144, 146, 148, (150), 152, 154, 156, (158), 160, (162), 164, 166, 168, (170), 172, (174), 176, (178), 180, (182), 184...
81	84, (86), 88, (90), 92, (94), 96, (98), 100, (102), 104, (106), 108, 110, 112, (114, 116), 118, 120, 122, (124, 126), 128, 130, (132), 134...

52.33 Table Skew weighing matrices $W(n, w)$, $a^2 \neq w = b^2 + c^2 + d^2 < 100$ (except $w = 93, 94$, $n \equiv 0 \pmod 4$ only).

w	n
for w = 2, 3, 5, 6, 8, 10, 11, 12, 13, 14, 17, 18, 19, 20, 21, 22, 24, 26, 27, 29, 30, 32, 33, 34, 35, 37, 38, 40, 41, 43, 44, 45, 46, 48, 50, 51, 52, 53, 54, 56, 57, 58, 59, 62, 65, 67, 68, 69, 70, 72, 73, 74, 75, 76, 80, 83, 84, 88, 96, skew $W(4k, w)$ exists whenever $4k > w$.	
42	48...
61	64...112, (116), 120, (124), 128...
66	72...
77	80, 84, 88, (92), 96, (100), 104, 108, 112, (116), 120, (124), 128, (132), 136, (140), 144, (148), 152, (156), 160...
78	80, 84, 88, (92), 96, (100), 104, 108, 112, (116), 120, (124), 128, (132), 136, (140), 144, (148), 152, (156), 160...
82	84, 88, (92), 96, (100), 104, 108, 112, (116), 120, (124), 128, (132), 136, (140), 144, (148), 152, (156), 160, (164), 168...
85	88, (92), 96, (100), 104, 108, 112, (116), 120, (124), 128, (132), 136, (140), 144, (148), 152, (156), 160, (164), 168, (172), 176...
86	88, (92), 96, (100), 104, 108, 112, (116), 120, (124), 128, (132), 136, (140), 144, (148), 152, (156), 160, (164), 168, (172), 176...
89	(92), 96, (100), 104, 108, 112, (116), 120, (124), 128, (132), 136, (140), 144, (148), 152, (156), 160, (164), 168, (172), 176, (180), 184, (188), 192, (196), 200...
90	92, 96, (100), 104, 108, 112, (116), 120, (124), 128, (132), 136, (140), 144, (148), 152, (156), 160, (164), 168, (172), 176, (180), 184...
91	92, 96, (100), 104, (108), 112, (116), 120, (124), 128, (132), 136, (140), 144, (148), 152, (156), 160, (164), 168, (172), 176, (180), 184...
97	(100), 104, 108, 112, (116), 120, (124), 128, (132), 136, (140), 144, (148), 152, (156), 160, (164), 168, (172), 176, (180), 184, (188), 192, (196), 200, (204), 208...
98	(100), 104, (108), 112, (116), 120, (124), 128, (132), 136, (140), 144, (148), 152, 156, 160, (164), 168, (172), 176, (180), 184, (188), 192, (196), 200, (204), 208, 212, 216, (220), 224, 228, 232, (236), 240, 244, 248, (252), 256...
99	100, 104, (108), 112, (116), 120, (124), 128, (132), 136, (140), 144, (148), 152, 156, 160, (164), 168, (172), 176, (180), 184, (188), 192, (196), 200...

52.34 Table Skew weighing matrices $W(n, w)$, $a^2 + b^2 + c^2 \neq w \leq 100$ ($n \equiv 0 \pmod 8$ only).

w	n
for w = 7, 15, 23, 28, 31, 39, 47, 55, 60, 63, 65, 87, 92, skew $W(8k, w)$ exists whenever $w \leq 8k$.	
79	80, 88, 96, 104, 112, 120, 128, (136), 144, (152), 160...
95	96, (104), 112, 120, 128, (136), 144, (152), 160, 168, 176, (184), 192...

52.5 See Also

§VI.8	A $W(k^2 + k + 1, k^2)$ is always complementary to a projective plane of order k. If there is a conference matrix of order $k+1$, the reverse implication also holds.
§IV.24	A $W(n, n)$ is also a Hadamard matrix of order n.
§IV.31	Weighing matrices are one-parameter orthogonal designs.
§IV.42	Sequences with zero autocorrelation are used to construct weighing matrices.
[2]	Some recent constructions involving odd prime powers.
[3]	A synopsis of weaving and some variations used to obtain weighing matrices; cites references giving various related methods for constructing them.
[5]	A survey on orthogonal designs (including weighing matrices).

References

[1] H.C. Chan, C.A. Rodger and J. Seberry, On inequivalent weighing matrices, *Ars Combin.* 21 (1986), 299-333.

[2] R. Craigen, A new class of weighing matrices with square weights, *Bull. ICA* 3 (1991), 33-42.

[3] R. Craigen, The craft of weaving matrices, *Congr. Numer.* 92 (1993), 9-28.

[4] R. Craigen, The structure of weighing matrices having large weights, *Des. Codes Crypt.* 5 (1995), 199-216.

[5] A. Geramita and J. Seberry, *Quadratic Forms, Orthogonal Designs, and Hadamard Matrices*, Marcel Dekker, New York, 1979.

[6] C. Koukouvinos and J. Seberry, Some new weighing matrices using sequences with zero autocorrelation function, *Australas. J. Combin.* 8 (1993), 143-152.

[7] H. Ohmori, Classification of weighing matrices of order 13 and weight 9, *Discrete Math.* 116 (1993), 55-78.

53 Whist Tournaments

Ian Anderson

53.1 Definitions and Examples

A *whist tournament* Wh($4n$) for $4n$ players is a schedule of games each involving two players opposing two others, such that

1. the games are arranged into $4n-1$ rounds, each of n games;
2. each player plays in exactly one game in each round;
3. each player partners every other player exactly once;
4. each player opposes every other player exactly twice.

Each game is denoted by an ordered 4-tuple (a,b,c,d) in which the pairs $\{a,c\}$, $\{b,d\}$ are *partner pairs.* The other pairs are *opponent pairs*; in particular $\{a,b\}$, $\{c,d\}$ are *opponent pairs of the first kind*, and $\{a,d\}$, $\{b,c\}$ are *opponent pairs of the second kind.*

53.1 Example A Wh(8) with players $\infty,0,1,\ldots,6$.

Round			Round		
1	$(\infty,4,0,5)$	$(1,2,3,6)$	5	$(\infty,1,4,2)$	$(5,6,0,3)$
2	$(\infty,5,1,6)$	$(2,3,4,0)$	6	$(\infty,2,5,3)$	$(6,0,1,4)$
3	$(\infty,6,2,0)$	$(3,4,5,1)$	7	$(\infty,3,6,4)$	$(0,1,2,5)$
4	$(\infty,0,3,1)$	$(4,5,6,2)$			

A *whist tournament* Wh($4n+1$) for $4n+1$ players is defined as above, except that Conditions 1, 2 are replaced by:

1′. the games are arranged into $4n+1$ rounds each of n games;

2′. each player plays in one game in each of $4n$ rounds, but does not play in the remaining round.

53.2 Remarks

1. Think of (a, b, c, d) as giving the order in which the players sit around a table.
2. A Wh$(4n)$ is a resolvable $(4n, 4, 3)$ BIBD in which the blocks are ordered so that each pair of elements occurs as a partner (opponent) pair once (twice).

53.3 Theorem A Wh$(4n)$ and a Wh$(4n+1)$ exist for all $n \geq 1$ (see [1]).

53.2 Z-Cyclic Whist Tournaments

A Wh$(4n)$ is $\mathbb{Z}$-cyclic if the players are $\infty, 0, 1, \ldots, 4n-2$ and each round is obtained from the previous one by adding 1 modulo $4n-1$ to each non-∞ entry. A Wh$(4n+1)$ is $\mathbb{Z}$-cyclic if the players are $0, 1, \ldots, 4n$ and the rounds are similarly developed modulo $4n+1$.

53.4 Example A $\mathbb{Z}$-cyclic Wh(5) with players $0, 1, 2, 3, 4$.

Round	1	2	3	4	5
Games	(1,2,4,3)	(2,3,0,4)	(3,4,1,0)	(4,0,2,1)	(0,1,3,2)

53.5 Remarks

1. By convention ∞, 0 are partners in the initial round of a $\mathbb{Z}$-cyclic Wh$(4n)$, and 0 is absent from the initial round of a $\mathbb{Z}$-cyclic Wh$(4n+1)$.
2. Example 53.1 is $\mathbb{Z}$-cyclic.
3. The partner pairs in the initial round of a $\mathbb{Z}$-cyclic Wh$(4n+1)$ form a starter in $\mathbb{Z}_{4n+1}$. See §IV.45.
4. $\mathbb{Z}$-cyclic whist tournaments are given by presenting only the initial round.

53.6 Theorem If $p = 4n+1$ is prime and w is a primitive root of p, then the games $(w^i, w^{i+n}, w^{i+2n}, w^{i+3n})$, $0 \leq i \leq n-1$, form the initial round of $\mathbb{Z}$-cyclic Wh$(4n+1)$.

53.7 Remark Let P denote any product of primes p with each $p \equiv 1 \pmod 4$.

53.8 Theorem A $\mathbb{Z}$-cyclic Wh$(4n)$ is known to exist in the following cases:

1. $4n \leq 64$.
2. $4n = 4^\alpha (\alpha \geq 1)$.
3. $4n = qP + 1$, $q \in \{7, 11, 19, 23, 31, 43, 47, 59\}$.
4. $4n = 3P + 1$, where all $p - 1$ are divisible by the same power of 2.
5. $4n = qr^2P + 1$, q and r distinct primes, $q, r \equiv 3 \pmod 4$, $q < 60$, $r < 500$.

53.9 Theorem A $\mathbb{Z}$-cyclic Wh$(4n+1)$ is known to exist in the following cases:

1. $4n + 1 = P$ or $49P$.
2. $4n + 1 = 21, 33, 45$.
3. $4n + 1 = 3pq$, $q \in \{7, 11\}$, p prime, $p \equiv 1 \pmod 4$.
4. $4n + 1 = r^2$, r prime, $r \equiv 3 \pmod 4$, $r < 500$ [6].

53.10 Examples Initial rounds of $\mathbb{Z}$-cyclic Wh$(4n)$, $n \leq 12$.

$4n =$ 4 $(\infty, 1, 0, 2)$.
8 $(\infty, 4, 0, 5), (1, 2, 3, 6)$.
12 $(\infty, 4, 0, 5), (1, 2, 10, 8), (3, 6, 7, 9)$.
16 $(\infty, 5, 0, 10), (1, 4, 2, 8), (6, 13, 9, 7), (11, 12, 3, 14)$.
20 $(\infty, 13, 0, 17), (8, 2, 10, 1), (3, 18, 15, 4), (7, 12, 16, 9), (11, 14, 5, 6)$.
24 $(\infty, 22, 0, 10), (14, 20, 7, 12), (19, 18, 15, 8), (4, 6, 13, 9), (11, 3, 17, 5), (21, 1, 16, 2)$.

28 $(\infty, 23, 0, 15), (18, 9, 21, 20), (14, 19, 2, 25), (4, 1, 22, 5),$
$(6, 12, 7, 26), (8, 24, 10, 17), (13, 11, 3, 16).$
32 $(\infty, 18, 0, 25), (12, 4, 2, 5), (6, 28, 9, 20), (23, 11, 8, 24),$
$(15, 13, 29, 19), (1, 14, 3, 10), (17, 16, 22, 27), (21, 7, 30, 26).$
36 $(\infty, 16, 0, 32), (28, 14, 3, 27), (31, 13, 4, 2), (20, 24, 34, 33),$
$(26, 1, 19, 6), (7, 12, 8, 15), (17, 23, 21, 29), (18, 9, 30, 11),$
$(22, 10, 5, 25).$
40 $(\infty, 0, 13, 26), (1, 8, 25, 5), (2, 16, 11, 10), (4, 32, 22, 20),$
$(14, 38, 18, 21), (15, 24, 23, 29), (17, 35, 33, 6), (27, 31, 34, 12),$
$(28, 36, 3, 37), (30, 7, 19, 9).$
44 $(\infty, 20, 0, 35), (13, 30, 29, 19), (12, 17, 31, 38), (25, 1, 32, 2),$
$(3, 14, 5, 24), (33, 28, 16, 8), (21, 18, 39, 23), (27, 42, 36, 34),$
$(7, 41, 37, 4), (10, 6, 22, 9), (11, 40, 15, 26).$ [5]
48 $(\infty, 45, 0, 17), (36, 9, 20, 15), (18, 1, 33, 5), (6, 19, 7, 29),$
$(43, 46, 21, 35), (32, 39, 27, 22), (38, 40, 12, 28), (37, 16, 24, 23),$
$(25, 41, 34, 2), (42, 31, 13, 4), (44, 3, 30, 26), (8, 11, 10, 14).$ [5]

53.11 Examples Initial rounds of $\mathbb{Z}$-cyclic Wh$(4n+1)$, $3 \le n \le 11$.

$4n+1 = 13$ $(1, 8, 12, 5), (2, 3, 11, 10), (4, 6, 9, 7).$
17 $(1, 13, 16, 4), (3, 5, 14, 12), (9, 15, 8, 2), (10, 11, 7, 6).$
21 $(1, 12, 2, 15), (5, 6, 3, 18), (4, 7, 11, 20), (8, 13, 19, 17), (14, 10, 9, 16).$
25 $(5, 3, 1, 14), (9, 17, 16, 20), (15, 6, 21, 7), (10, 4, 8, 13),$
$(2, 12, 19, 22), (11, 18, 24, 23).$
29 $(1, 2, 9, 27), (16, 3, 28, 26), (24, 19, 13, 10), (7, 14, 5, 15),$
$(25, 21, 22, 8), (23, 17, 4, 12), (20, 11, 6, 18).$
33 $(1, 3, 5, 20), (4, 17, 28, 29), (31, 25, 26, 2), (9, 12, 22, 30),$
$(13, 27, 24, 19), (6, 16, 7, 23), (32, 11, 18, 14), (15, 8, 21, 10).$
37 $(1, 2, 17, 4), (16, 32, 13, 27), (34, 31, 23, 25), (26, 15, 35, 30), (9, 18, 5, 36),$
$(33, 29, 6, 21), (10, 20, 22, 3), (12, 24, 19, 11), (7, 14, 8, 28).$
41 $(1, 17, 4, 15), (36, 12, 21, 13), (10, 6, 40, 27), (32, 38, 5, 7),$
$(18, 19, 31, 24), (33, 11, 9, 29), (16, 26, 23, 35), (2, 28, 8, 3),$
$(37, 14, 25, 22), (20, 34, 39, 30).$
45 $(1, 12, 2, 40), (32, 3, 25, 5), (27, 10, 36, 22), (44, 34, 31, 19),$
$(35, 20, 29, 43), (14, 42, 24, 13), (37, 33, 26, 7), (6, 11, 9, 15),$
$(18, 21, 23, 39), (16, 17, 30, 38), (41, 28, 4, 8).$ [5]

53.3 Directed Whist and Triplewhist Tournaments

A *directed whist tournament* DWh(v) is a Wh(v) with Condition 4 replaced by: each player has every other player once as an opponent on the right, and once on the left.

53.12 Theorem A DWh$(4n+1)$ exists for all $n \ge 1$, and a DWh$(4n)$ exists at least for all $n \ge 85$.

53.13 Theorem A $\mathbb{Z}$-cyclic DWh$(4n+1)$ exists when $4n+1$ is a product of primes $p \equiv 1 \pmod 4$; also when $4n+1 = 21, 33$.

53.14 Examples $\mathbb{Z}$-cyclic DWh(m)s for $m = 21$ and 33.

$m = 21$ (1,8,2,14), (3,7,6,11), (4,10,12,15), (5,19,16,17), (9,20,18,13)
$m = 33$ (25,27,14,32), (26,6,10,29), (23,21,4,13), (16,12,7,15),
(3,2,24,31), (5,17,20,11), (30,22,28,9), (18,8,19,1)

53.15 Remarks

1. A DWh(v) is a resolvable perfect Mendelsohn design with block size 4. See §IV.28.
2. The Wh(p) of Theorem 53.6 are all DWh(p).

A *triplewhist tournament* TWh(v) is a Wh(v) with Condition 4 replaced by: each player has every other player once as an opponent of the first kind and once as an opponent of the second kind.

53.16 Theorem A TWh($4n+1$) exists for all sufficiently large $n \geq 1$. A TWh($4n$) exists for all $n \geq 1$ except possibly $n = 14, 54, 62, 70$.

53.17 Theorem There is no $\mathbb{Z}$-cyclic TWh(p) where p is a prime, $p < 29$.

53.18 Theorem A $\mathbb{Z}$-cyclic TWh($4n + 1$) exists whenever $4n + 1$ is a prime $p \equiv 5, 9$, or 13 (mod 16), $p \geq 29$; for $p \equiv 1$ (mod 16), $97 \leq p \leq 15000$ (except possibly 257); and for all products of such primes [7].

53.19 Theorem A $\mathbb{Z}$-cyclic TWh($4n$) exists in the following cases:

1. $4n = qP + 1$, P as in the Remark 53.7, $q \in \{7, 19, 23, 31\}$.
2. $4n = 3P + 1$, P as in Theorem 53.8(4).

53.20 Remark The Wh($4n$) in Example 53.10 are TWh($4n$) in the cases $4n \leq 40$, $4n \neq 12$. The Wh($4n + 1$) in Example 53.11 are TWh($4n + 1$) in the cases $21 \leq 4n + 1 \leq 41$.

53.21 Table Initial rounds of TWh(p), p prime, $p \equiv 1$ (mod 4), $50 < p < 100$. Elements in $\mathbb{Z}_p$ [7].

$$\begin{aligned}
p = 53: &\quad (1, 4, 12, 2) \times 1, 2^4, 2^8, \ldots, 2^{48} \\
61: &\quad (1, 2, 35, 4) \times 1, 2^4, 2^8, \ldots, 2^{56} \\
73: &\quad (1, 31, 42, 65) \times 31^{8i+2j} \quad (0 \leq i \leq 8,\ 0 \leq j \leq 1) \\
89: &\quad (1, 6, 83, 38) \times 6^{8i+2j} \quad (0 \leq i \leq 10,\ 0 \leq j \leq 1) \\
97: &\quad (1, 7, 90, 62) \times 7^{32i+2j} \quad (0 \leq i \leq 2,\ 0 \leq j \leq 7).
\end{aligned}$$

53.4 Connections with Other Combinatorial Structures

53.22 Theorem If TWh(v) exists, then a resolvable spouse avoiding mixed doubles round robin tournament for v couples (abbreviated SAMDRR(v)) exists (see §V.7.11).

53.23 Theorem If a resolvable SAMDRR(n) exists, then a TWh($4n$) exists.

53.24 Theorem If a $\mathbb{Z}$-cyclic TWh($4n + 1$) exists, then there are 4 MOLS of order $4n + 1$.

53.25 Remark The tournaments of Theorem 53.19(2) yield resolvable 1-rotational designs [3].

53.5 See Also

§IV.28	Resolvable Mendelsohn designs are generalizations of DWh(v).
§IV.41	TWh(v) are related to SOLSSOMs and to SAMDRRs, as in Theorems 53.22 and 53.23.
[1]	Contains a chapter on whist tournaments.
[2]	Surveys the development of constructions of whist tournaments and contains most of the information in this section.

References

[1] I. Anderson, *Combinatorial Designs*, Ellis Horwood, 1990.

[2] I. Anderson, A hundred years of whist tournaments, *J. Combin. Math. Combin. Comput.* 19 (1995), 129-150.

[3] I. Anderson and N.J. Finizio, Cyclically resolvable designs and triplewhist tournaments, *J. Combin. Des.* 1 (1993), 347-358.

[4] R.D. Baker, Whist tournaments, *Congr. Numer.* 14 (1975), 89-100.

[5] N.J. Finizio, A few more $\mathbb{Z}$-cyclic whist tournaments, *J. Combin. Math. Combin. Comput.* 19 (1995), 93-95.

[6] P.A. Leonard, Some new $\mathbb{Z}$-cyclic whist tournaments, *Utilitas Math.*, to appear.

[7] Y.S. Liaw, $\mathbb{Z}$-cyclic triplewhist tournaments, Mathematics Preprint Series, Glasgow, 1994.

54 Youden Designs, Generalized

Charles J. Colbourn

54.1 Definition and Basics

A *generalized Youden design* GYD(v, b_1, b_2) is a $b_1 \times b_2$ array B with $b_1 = m_1v + c_1$ and $b_2 = m_2v + c_2$, $0 \leq c_1, c_2 < v$, such that

1. each cell contains one element from a set V of v elements;
2. every element occurs either m_1 or $m_1 + 1$ times in each row and either m_2 or $m_2 + 1$ times in each column;
3. every pair of distinct elements occur together in a row the same number of times, and together in a column the same number of times.

54.1 Examples A GYD(4,6,6), GYD(6,10,15), and a GYD(9,12,12).

GYD(4,6,6)

```
abcdbd
badccb
cdabda
dcbaac
cadbab
dbaccd
```

GYD(6,10,15)

```
abcdefabcdefcdb
bcdefabcdefabca
cdefabcdefabdae
defabcdefabcafd
efabcdefabcdfec
fabcdefabcdeebf
ceafedcbafdbaec
fcbabeadedfcedb
baedfadcfcbefba
abdbacfedecfcfd
```

GYD(9,12,12)

```
abcdefghiabc
bcdefghiadef
cdefghiabghi
defghiabciae
efghiabcdcdh
fghiabcdefgb
ghiabcdefhfa
hiabcdefgbid
iabcdefghecg
abcihgfdeadg
defacbhigbeh
ghiedfabccfi
```

A *balanced block design* BBD(v, b_1, b_2) is a collection $\mathcal{B}$ of b_1 multisets (*blocks*) of elements from a v-set V, each of size b_2, so that

1. $b_2 = m_2v + c_2$, $0 \leq c_2 < v$, and every $x \in V$ appears in every $B \in \mathcal{B}$ either m_2 or $m_2 + 1$ times;
2. every pair of distinct elements occur together in the same block of $\mathcal{B}$ the same number of times.

54.2 Remark A GYD(v, b_1, b_2) is equivalent to an array in which the multisets defined by the rows form a BBD(v, b_1, b_2), and the multisets defined by the columns form a BBD(v, b_2, b_1). When $b_1 < v$ and $b_2 = v$, a GYD is a Youden square (§IV.55), and the BBD(v, b_2, b_1) is a symmetric BIBD.

54.3 Remarks If a BBD(v, b_1, b_2) $(V, \mathcal{B})$ exists with $b_2 = m_2 v + c_2$, $0 \le c_2 < v$, then by removing m_2 copies of each element from each $B \in \mathcal{B}$, one obtains a collection $\widehat{\mathcal{B}}$ for which $(V, \widehat{\mathcal{B}})$ is a BIBD. Thus the existence of a BBD(v, b_1, b_2) is equivalent to that of a BIBD with parameters $(v, b = b_1, r = \frac{b_1 c_2}{v}, k = c_2, \lambda = \frac{b_1 c_2 (c_2 - 1)}{v(v-1)})$. Hence one obtains a necessary condition for the existence of a GYD(v, b_1, b_2).

54.4 Theorem (*necessary conditions*) If a GYD(v, b_1, b_2) exists with $b_1 = m_1 v + c_1$ and $b_2 = m_2 v + c_2$, $0 \le c_1, c_2 < v$, then

1. $v \mid c_1 c_2$;
2. $v - 1 \mid b_1 c_2 (c_2 - 1)$; and
3. $v - 1 \mid b_2 c_1 (c_1 - 1)$.

54.2 Regular GYDs

A GYD(v, b_1, b_2) is *doubly regular* if $v \mid b_1$ and $v \mid b_2$.

54.5 Construction A doubly regular GYD. Let $b_1 = m_1 v$ and $b_2 = m_2 v$. Let (ℓ_{ij}), $0 \le i, j < v$ be a $v \times v$ latin square. Then define a $b_1 \times b_2$ array $R = (r_{ij})$ by setting $r_{av+i, bv+j} = \ell_{ij}$, for $0 \le a < m_1$, $0 \le i < v$, $0 \le b < m_2$, and $0 \le j < v$. In other words, R can be blocked into a $m_1 \times m_2$ pattern in which each block is a copy of L. Then R is a GYD(v, b_1, b_2).

A GYD(v, b_1, b_2) is *(simply) regular* if $v \mid b_1$ or $v \mid b_2$.

54.6 Construction A regular GYD(v, b_1, b_2). If $v \mid b_1$ and $v \mid b_2$, apply Construction 54.5. Otherwise, if $v \mid b_2$, use the following to form a GYD(v, b_2, b_1), and transpose the result. So suppose that $b_1 = m_1 v$ and $b_2 = m_2 v + c_2$, with $0 < c_2 < v$. If no BBD(v, b_1, b_2) exists, there is no GYD(v, b_1, b_2). Otherwise, let $(V, \mathcal{B})$ be a BBD(v, b_1, b_2). Let $V = \{p_1, \ldots, p_v\}$, and let $\mathcal{B} = \{B_1, \ldots, B_{b_1}\}$. Form a bipartite multigraph G on the vertices $\{p_1, \ldots, p_v\} \cup \{\ell_1, \ldots, \ell_{b_1}\}$; vertices p_i and ℓ_j are connected by σ edges if p_i appears exactly σ times in B_j. Now using Algorithm 54.7, assign colors to the edges of G. The edge colors are used to form a $b_1 \times b_2$ array R; each edge color determines the column in which the corresponding element of the block is to be placed. When edge $e = \{p_i, \ell_j\}$ gets color c, place p_i in cell (i, c) of R. The resulting array R is a GYD provided that

1. the number of edge colors is b_2;
2. for each vertex ℓ_j and for each color c, there is exactly one edge colored c incident at ℓ_j; and
3. for each vertex p_i and for each color c, there are exactly m_1 edges colored c incident at p_i.

Algorithm 54.7 guarantees that these conditions hold.

54.7 Algorithm for balanced edge-coloring.

1. Initialize by arbitrarily assigning different colors to each of the b_2 edges incident at each vertex p_i (using a total of b_2 colors).

2. While there is a vertex p_i and a color c for which fewer than m_1 edges incident at p_i are colored c (a *deficient vertex*), one finds an 'alternating path' as follows. Choose a second color c' that is the color of more than m_1 edges incident at p_i. Starting at $p_{s_0} = p_i$, form a sequence of distinct edges $e_0, \ldots, e_{2n-1}$, where $e_{2i} = \{p_{s_i}, \ell_{t_i}\}$ is an edge of color c' for $0 \le i < n$, and $e_{2i+1} = \{\ell_{t_i}, p_{s_{i+1}}\}$ is an edge of color c for $0 \le i < n$. This can be formed in a greedy manner, stopping only when color c appears on more edges incident at p_n than does color c'. When such a vertex s_n is encountered, interchange the colors c and c' on all of the edges $e_0, \ldots, e_{2n-1}$. If a deficient vertex still exists, repeat this step; otherwise, stop.

54.8 Remark Let γ_{ij} denote the number of edges colored j that are incident at p_i. Considering $M = \sum_{i=1}^{v} \sum_{j=1}^{b_2} |\gamma_{ij} - m_1|$ as the total deficiency, Algorithm 54.7 reduces the total deficiency by at least two whenever an alternating path is switched. Thus the method is efficient, taking at most $vb_2m_1/2$ alternating paths.

54.9 Theorem A GYD(v, b_1, b_2) with $v|b_1$ exists if and only if the 'row' BBD(v, b_1, b_2) exists.

54.3 Nonregular GYDs

54.10 Lemma If a GYD(v, m_1v, b_2) and a GYD(v, b_1, b_2) both exist, then a GYD$(v, m_1v + b_1, b_2)$ exists.

54.11 Theorem [2] Let v, b_1 and b_2 be positive integers, and write $b_i = m_iv + c_i$ with $0 \le c_i < v$ for $i = 1, 2$. Let $c_1c_2 = tv$. If there exists a BBD(v, b_1, c_2) $(V, \mathcal{B})$ and a BBD(v, b_2, c_1) $(V, \mathcal{D})$, so that $\mathcal{B}$ contains a collection $\widehat{\mathcal{B}}$ of c_1 blocks in every element occurs exactly t times, and the blocks of $\widehat{\mathcal{B}}$ can be arranged as the rows of a $c_1 \times c_2$ array whose columns appear as blocks of $\mathcal{D}$, then a GYD(v, b_1, b_2) exists.

54.12 Corollary If $c_1c_2 = v$, and a BBD(v, b_1, c_2) and a BBD(v, b_2, c_1) both exist having a parallel class of blocks, then a GYD(v, b_1, b_2) exists.

54.13 Lemma If a GYD(v, b_1, a_2v) contains a GYD(v, b_1, b_2), complementing the rows of the latter with respect to those of the former gives a GYD$(v, b_1, a_2v - b_2)$.

54.14 Theorem [3] Let q be a prime power, and $v = q^m$, $m \ge 2$. Then, for all integers $t_1, t_2 \ge 0$, there exists a GYD$(v, (t_1q+1)q^{m-1}\frac{q^m-1}{q-1}, (t_2q+1)q^{m-1}\frac{q^m-1}{q-1})$.

54.15 Remark Ash [1] gives explicit examples to show that, for $v \le 25$ and $b_1, b_2 \le 50$, the necessary conditions of Theorem 54.4 are sufficient except when $v = 15$ and $\{b_1, b_2\} = \{21, 35\}$ (since no BIBD with parameters (15,21,7,5,2) exists), and possibly when $v = 25$ and $b_1 = b_2 = 40$. In particular, she lists nonregular solutions for $(v, b_1, b_2) \in \{(4,6,6)$, (6,10,15), (6,15,20), (8,14,28), (8,28,28), (8,28,42), (9,12,12), (9,12,24), (9,24,24), (10,15,36), (10,18,45), (12,33,44), (16,20,20), (21,30,35), $(25,30,30)\}$.

54.4 See Also

§I.1	Existence of BIBDs.
§IV.55	Youden squares are special types of GYDs.
[1]	A table of explicit solutions for $v \le 25$, $b_1, b_2 \le 50$.

References

[1] A. Ash, Generalized Youden designs: constructions and tables, *J. Stat. Plann. Infer.* 5 (1981), 1-25.
[2] J. Kiefer, Balanced block designs and generalized Youden designs I. Construction (patchwork), *Ann. Statist.* 3 (1975), 109-118.
[3] E. Seiden and C.Y. Wu, A geometric construction of generalized Youden designs for v a power of a prime, *Ann. Statist.* 6 (1978), 452-460.

55 Youden Squares

DONALD A. PREECE

55.1 Definition and Example

Let S be a set of v *symbols* (or *treatments*). A *Youden square* of size $k \times v$ (in short, a $k \times v$ Youden square) is a $k \times v$ rectangular array of symbols, $k < v$, satisfying

1. each symbol of S occurs exactly once in each row, and
2. the symbols in each column constitute a k-subset of S, and the v such subsets constitute the blocks of a symmetric design.

55.1 Example A 3×7 Youden square.

4	7	2	3	1	5	6
6	5	1	7	4	2	3
7	1	6	2	3	4	5

55.2 Remark A $k \times v$ Youden square Y with $S = \{1, 2, ..., v\}$ is equivalent to a $v \times v$ array A such that

1. $A(i, j) = p$ for $i, j = 1, 2, \ldots, v$, if $Y(p, j) = i$ for $p = 1, 2, \ldots, k$,
2. $A(i, j)$ is empty, otherwise.

The squareness of A is the historical reason for the rectangular Y being described as a Youden square, not a Youden rectangle.

55.3 Remark If the entry in each nonempty cell of A is replaced by the entry 1, and the entry 0 is put into each empty cell of A, then A becomes the incidence matrix of the symmetric design on which Y is based.

55.2 Existence

55.4 Theorem A Youden square can be constructed from any symmetric design by suitably ordering the elements of each block of the symmetric design. (See §I.5 and Construction IV.54.6.)

55.5 Remark A $k \times v$ Youden square with $k = v - 1$ can be obtained by deleting any row of any $v \times v$ latin square.

55.6 Remark Any $k \times v$ Youden square can be completed to a $v \times v$ latin square by the addition of $v-k$ rows (Theorem II.1.38). If $k < v-1$, the added rows constitute a $(v-k) \times v$ Youden square.

55.3 Enumeration of Youden Squares

Let Y_1 and Y_2 be $k \times v$ Youden squares with $S = \{1, 2, \ldots, v\}$.

1. Y_1 and Y_2 belong to the same *transformation set (isotopy class)* if Y_2 can be obtained from Y_1 by any combination of permuting the rows, permuting the columns and permuting the symbols.
2. Y_2 is the *dual* of Y_1 if, given that the $(i,j)th$ element of Y_2 is q ($i = 1, 2, \ldots, k; j = 1, 2, \ldots, v$), then the $(i,q)th$ element of Y_1 is j ($i = 1, 2, \ldots, k; q = 1, 2, \ldots, v$).
3. Y_1 and Y_2 belong to the same *species (main class)* if Y_1 and either Y_2 or the dual of Y_2 belong to the same transformation set.

55.7 Table The number of transformation sets and species of small Youden squares.

$k \times v$	2×3	3×4	3×7	4×5	4×7	5×6	6×7
Transformation Sets	1	2	1	3	6	40	3479
Species	1	2	1	3	6	33	1986

55.8 Table The six species of 4×7 Youden squares.

1	2	3	4	5	6	7
2	3	4	5	6	7	1
3	4	5	6	7	1	2
5	6	7	1	2	3	4

1	2	5	6	7	3	4
2	4	3	1	5	6	7
3	6	4	5	2	7	1
5	3	7	4	6	1	2

1	6	5	4	2	3	7
2	4	3	5	7	6	1
3	2	7	6	5	1	4
5	3	4	1	6	7	2

1	6	5	4	2	3	7
2	3	7	6	5	1	4
3	2	4	5	7	6	1
5	4	3	1	6	7	2

1	6	5	4	2	3	7
2	3	4	5	6	7	1
3	2	7	6	5	1	4
5	4	3	1	7	6	2

1	6	5	4	2	3	7
2	3	4	5	7	6	1
3	2	7	6	5	1	4
5	4	3	1	6	7	2

55.4 Double Youden Rectangles

Let S be a set of v elements and T be a distinct set of k elements. A *double Youden rectangle* of size $k \times v$ (in short, a $k \times v$ double Youden rectangle) is a $k \times v$ rectangular array, $k < v$, such that

1. Each entry is an ordered pair $(x, y) \in S \times T$.
2. If the elements from T are disregarded, the array becomes a $k \times v$ Youden square whose symbols are taken from S.
3. Each element from T is paired exactly once with each element from S.
4. Each element from T occurs exactly once in each column.
5. Each element from T occurs exactly n or $n+1$ times in each row, where n is the integral part of v/k, and if n occurrences of each element from T are removed from each row, leaving $v - nk$ elements from T in each row, then either (i) $v - nk = 1$ or (ii) the k sets of $v - nk$ remaining elements of T in the k rows constitute the blocks of a symmetric design with k blocks.

55.9 Example A 4×13 double Youden rectangle that can be constructed from a standard deck of 52 playing cards, with $S = \{A, 2, 3, ..., 10, J, Q, K\}$ and $T = \{\spadesuit, \diamondsuit, \clubsuit, \heartsuit\}$.

A♠	3♣	4♡	7♡	8♣	2♣	10♢	J♠	5♠	6♢	Q♢	K♠	9♡
2♢	5♡	3♢	4♠	6♠	7♢	8♠	9♣	10♣	K♡	J♡	Q♣	A♢
4♣	J♢	6♣	K♢	5♢	9♠	7♣	8♡	Q♡	10♠	A♣	2♡	3♠
10♡	2♠	Q♠	5♣	A♡	6♡	3♡	4♢	9♢	J♣	7♠	8♢	K♣

55.10 Table Existence results for double Youden rectangles, $k \leq 12$.

$k \times v$	3×7	4×7	4×13	5×11	5×21	6×11	6×31
Existence	No	Yes	Yes	Yes	Yes	Yes	?

$k \times v$	7×15	8×15	8×57	9×19	9×37	9×73	10×19
Existence	Yes	Yes	?	Yes	?	?	?

$k \times v$	10×31	10×91	11×23	11×56	12×23	12×133	
Existence	?	?	Yes	?	?	?	

55.11 Remark For all known double Youden rectangles, $v - nk = 1$ or $v - nk = k - 1$.

55.12 Theorem If $k \equiv 3 \pmod 4$ is a prime power, $k \geq 7$, then there exists a $k \times (2k+1)$ double Youden rectangle.

55.5 Balanced Superimpositions of Two Youden Squares

Let S_1 and S_2 each be v-sets. A *balanced superimposition* of two $k \times v$ Youden squares is a $k \times v$ rectangular array D such that

1. Each entry is an ordered pair $(x, y) \in S_1 \times S_2$.
2. If the elements from S_1 (or S_2) are disregarded, the array becomes a $k \times v$ Youden square whose symbols are taken from S_2 (or S_1).
3. If n_{21} is the $v \times v$ $(0,1)$–matrix whose (i,j)th element $(i,j = 1, 2, \ldots, v)$ is the number of times that the ith element of S_2 is paired with the jth element of S_1, then n_{21} is the incidence matrix of a symmetric design.
4. If n_1 is the $v \times v$ $(0,1)$–matrix whose (i,j)th element $(i,j = 1, 2, \ldots, v)$ is the number of times that the ith element of S_1 occurs in the jth column of D, and n_2 is defined likewise for S_2, then
$$n_{21}^t n_2 n_1^t + n_1 n_2^t n_{21} = n_{21} n_1 n_2^t + n_2 n_1^t n_{21}^t = dI + eJ$$
where d and e are integers, I is the $v \times v$ identity matrix and J is the $v \times v$ matrix whose elements are all 1.

55.13 Example Two balanced superimpositions of two 3×7 Youden squares with $S_1 = S_2 = \{1, 2, \ldots, 7\}$.

23	34	45	56	67	71	12
35	46	57	61	72	13	24
52	63	74	15	26	37	41

$(d = -2,\ e = 8)$

24	35	46	57	61	72	13
37	41	52	63	74	15	26
56	67	71	12	23	34	45

$(d = 5,\ e = 7)$

55.14 Remark A balanced superimposition of two $k \times v$ Youden squares with $k = v - 1$ can be obtained by deleting any row of any $v \times v$ Graeco-Latin square.

55.15 Theorem If q is any positive integer such that $4q + 3$ is a prime power, then balanced superimpositions of two Youden squares exist for the sizes $(2q + 1) \times (4q + 3)$ and $(2q + 2) \times (4q + 3)$.

55.16 Remark Balanced superimpositions of two Youden squares have also been found for sizes 7×15, 8×15, 6×16, and 10×16.

55.6 See Also

§I.5	Existence of symmetric designs.
§IV.54	Contains a generalization, GYDs.
[1]	A historical, mathematical and statistical survey of Youden squares, with full bibliography. This survey contains most of the information in this section.
[2, 3]	These two papers, taken together, give an almost up-to-date account of knowledge about double Youden rectangles.

References

[1] D.A. Preece, Fifty years of Youden squares: A review, *Bull. Inst. Math. Applic.* 26 (1990), 65-75.

[2] D.A. Preece, Double Youden rectangles of size 6×11, *Math. Scientist* 16 (1991), 41-45.

[3] D.A. Preece, Double Youden rectangles — an update with examples of size 5×11, *Discrete Math.* 125 (1994), 309-317.

PART V

APPLICATIONS

1 Codes

VLADIMIR D. TONCHEV

1.1 A Brief Introduction to Coding Theory

A *code* C of *length* n over an alphabet F_q of size q (or a *q-ary* code), is a subset $C \subseteq F_q^n$ of the set of all n-tuples with components from F_q. A code is *binary* if $q = 2$, *ternary* if $q = 3$, and so on. If q is a prime power and F_q is the finite field $\mathbb{F}_q$ of order q, then *linear codes* are linear subspaces of the n-dimensional vector space $\mathbb{F}_q^n$. A k-dimensional subspace of $\mathbb{F}_q^n$ is a (linear) $[n, k]$ code over $\mathbb{F}_q$. The elements of a code are the *words* or the *codewords*.

A *generator matrix* of a linear $[n, k]$ code C is any matrix of rank k (over $\mathbb{F}_q$) with rows from C.

The *dual* (or *orthogonal*) code $C^\perp$ of a code C is its orthogonal space with respect to the usual dot product in $\mathbb{F}_q^n$:

$$C^\perp = \{\, y \in \mathbb{F}_q^n \mid yx = 0 \text{ for all } x \in C\}.$$

The dual code of an $[n, k]$ code is an $[n, n-k]$ code.

A *parity check matrix* of C is any generator matrix of $C^\perp$.

1.1 **Remark** If $G = (I_k, A)$ is a generator matrix of C, then $H = (-A^T, I_{n-k})$ is a generator matrix of $C^\perp$, or, equivalently, a parity check matrix of C.

The set F_q^n becomes a metric space when equipped with the *Hamming distance*:

$$d(x, y) = |\{\, i \mid x_i \neq y_i \,\}| \text{ for } x = (x_1, \ldots, x_n),\ y = (y_1, \ldots, y_n),\ x, y \in F_q^n.$$

The norm induced by the Hamming distance is the (Hamming) *weight*:

$$w(x) = d(x, \overline{0}) = |\{\, i \mid x_i \neq 0\}| \text{ for } x = (x_1, \ldots, x_n) \in \mathbb{F}_q^n,$$

where $\overline{0} = (0, \ldots, 0)$ is the *zero* vector.

1.2 **Proposition** The Hamming distance is invariant under translation, i.e.,

$$d(x + z, y + z) = d(x, y) \text{ for } x, y, z \in \mathbb{F}_q^n.$$

An important parameter of a code C is its *minimum distance* d where $d = \min_{x,y \in C,\ x \neq y} d(x, y)$, or, equivalently, in the case of a linear code, its *minimum weight*: $d = \min_{x \in C,\ x \neq \overline{0}} w(x)$. An $[n, k, d]$ *code* is an $[n, k]$ code with minimum distance (or weight) d.

1.3 **Theorem** The minimum distance of a linear code with a parity check matrix H is d if and only if every $d - 1$ columns in H are linearly independent, and there is a set of d linearly dependent columns in H.

The *weight distribution* of a code C is the sequence $\{A_i\}_{i=0}^n$, where A_i is the number of codewords of weight i. The polynomial $A(x) = \sum_{i=0}^n A_i x^i$ is the *weight enumerator* of C.

1.4 **Theorem** (*MacWilliams identity*) Let C be an $[n, k]$ code over $\mathbb{F}_q$ with weight enumerator $A(x)$, and let $B(x)$ be the weight enumerator of $C^\perp$. Then

$$q^k B(x) = (1 + (q-1)x)^n A\left(\tfrac{1-x}{1+(q-1)x}\right).$$

Two codes are *equivalent* if one of the codes can be obtained from the other by permuting the coordinates (in all codewords), and permuting the symbols within one or more coordinate positions.

1.5 **Remark** Permutations of the symbols within a coordinate position that are realized by multiplication of a coordinate with a nonzero field element preserve the linearity of a code.

Two codes are *isomorphic* if they differ by a permutation of the coordinates only.

An *automorphism* of a code is any equivalence of the code with itself.

1.6 **Example** The rows of the matrices

$$C_1 = \begin{pmatrix} 0 & 0 & 0 & 0 \\ 0 & 1 & 1 & 1 \\ 0 & 2 & 2 & 2 \end{pmatrix}, \quad C_2 = \begin{pmatrix} 0 & 0 & 1 & 0 \\ 1 & 0 & 2 & 1 \\ 2 & 0 & 0 & 2 \end{pmatrix},$$

form equivalent (but not isomorphic) ternary codes of length $n = 4$ and minimum distance $d = 3$. The code C_2 is obtained from C_1 by interchanging the first two coordinate positions, and applying the permutation $(0, 1, 2)$ on the symbols in the third position. Any permutation that fixes the first coordinate and permutes the coordinates $2, 3, 4$ arbitrarily is an automorphism of C_1. The code C_1 is linear (of dimension $k = 1$), while C_2 is nonlinear. The minimum weight of C_2 is 1. The weight enumerator of C_1 is $1 + 2x^3$.

A *sphere* of radius r about a vector $x \in F_q^n$ is the set of all n-tuples whose distance is at most r from x.

1.7 **Proposition** The spheres of radius $e = \lfloor \frac{d-1}{2} \rfloor$ about the codewords of a code with minimum distance d are disjoint.

1.8 **Remarks** The minimum distance determines the *error-correcting ability* of the code. More precisely, a code C with minimum distance d can correct up to $e = \lfloor \frac{d-1}{2} \rfloor$ errors, or C is an *e-error-correcting* code. Suppose that the codewords of C are used as communication messages that are transmitted via a noisy channel. The noise may result in random changes of the values of some coordinates in the transmitted words. Assume that the number of errors (the number of coordinates being changed) in any single codeword x does not exceed $e = \lfloor \frac{d-1}{2} \rfloor$, where d is the minimum distance. The errors in the received vector can be corrected by using the *maximum likelihood principle*: At the receiving end, consider any received word y as the image of the codeword from C that is closest to y in terms of the Hamming distance. By Proposition 1.7, the codeword x being sent is properly determined by this procedure provided that $d(x, y) \le e$.

The length n, the minimum distance d, and the total number of codewords (or *size*) $M = |C|$ are the basic parameters of a code C. Increasing d is usually possible at the expense of decreasing M or increasing n. If d and M are fixed, the most interesting codes for the given d, M are those of shortest length n. If d and n are fixed, one looks for a code of largest possible size M.

In general, there are three fundamental optimization problems imposed by fixing two of the parameters n, d, M and optimizing with respect to the third. Explicit solutions of any of these optimization problems are rarely known in general. However, there are estimates (*bounds*) for the optimal values in terms of inequalities. On many occasions, the codes that are optimal with respect to a given bound are closely related to combinatorial designs. This chapter is a brief survey of such relations between codes and designs.

1.9 **Remark** The codes defined in this chapter are *block* codes, or fixed-length codes. There are other types of codes that allow codewords of variable length, or even streams of symbols rather than words of a fixed length. Such codes are not discussed here since the relationships between codes and combinatorial designs are best studied and understood in the case of block codes.

1.2 Designs in Codes

The *support* of a nonzero vector $x = (x_1, \ldots, x_n)$, $x_i \in F_q = \{0, 1, \ldots, q-1\}$ is the set of indices of its nonzero coordinates: $supp(x) = \{\, i \mid x_i \neq 0\}$.

1.10 **Example** The support of any nonzero word in C_1 from Example 1.6 is $\{2, 3, 4\}$, while the support of the first codeword in C_2 is $\{3\}$.

The (*support*) *design* of a code of length n for a given nonzero weight w is the design with points the n coordinate indices, and blocks the supports of all codewords of weight w.

1.11 **Example** The matrix

$$G = \begin{pmatrix} 1 & 1 & 0 & 1 & 0 & 0 & 0 \\ 0 & 1 & 1 & 0 & 1 & 0 & 0 \\ 0 & 0 & 1 & 1 & 0 & 1 & 0 \\ 0 & 0 & 0 & 1 & 1 & 0 & 1 \end{pmatrix}$$

is a generator matrix of a binary $[7, 4, 3]$ code (the *Hamming code* of length 7) with weight distribution $A_0 = A_7 = 1$, $A_3 = A_4 = 7$. The seven codewords of weight 3 form the incidence matrix of a 2-(7,3,1) design (the projective plane of order 2).

1	1	0	1	0	0	0
0	1	1	0	1	0	0
0	0	1	1	0	1	0
0	0	0	1	1	0	1
1	0	0	0	1	1	0
0	1	0	0	0	1	1
1	0	1	0	0	0	1

The seven codewords of weight 4 support the complementary 2-(7,4,2) design.

1.12 **Remark** To avoid repeated blocks in the nonbinary case ($q > 2$), only one support is taken for all words with the same support (e.g., one support for all $q - 1$ nonzero multiples of a word in a linear q-ary code). The regularity of a support design depends on the structure of the underlying code. For example, if the code is *cyclic*, that is, having the cyclic permutation $(1\ 2\ \ldots\ n)$ as an automorphism, the supports of any weight $w \neq 0$ form a 1-(n, w, r) design for some r.

1.13 **Theorem** If a code is invariant under a permutation group that acts t-homogeneously (in particular, t-transitively) on the coordinates, the supports of the codewords of any nonzero weight form a t-design.

1.14 **Remark** The Hamming code of Example 1.11 is cyclic. The full automorphism group of the code ($PSL_3(2)$) is of order 168 and acts 2-transitively on the set of coordinates.

1.15 **Example** The *Reed-Muller* (RM) code RM(r, m) of order r ($0 \le r \le m$) is a binary $[2^m, \sum_{i=0}^{r} \binom{m}{i}, 2^{m-r}]$ code consisting of all binary vectors of length 2^m that are value-vectors of polynomials in binary variables $v_1, \ldots, v_m$ of degree at most r. The dual code of $RM(r, m)$ is $RM(m - r - 1, m)$. Any RM code is invariant under the 3-transitive group of all affine transformations of the 2^m-dimensional binary vector space. Consequently, the codewords of any fixed weight support a 3-design.

1.3 Perfect Codes

1.16 Theorem (*sphere packing or Hamming bound*) A q-ary code C of length n and minimum distance $d = 2e + 1$ satisfies:

$$|C| \leq \frac{q^n}{\sum_{i=0}^{e} \binom{n}{i}(q-1)^i}.$$

A code that attains the sphere packing bound is *perfect*. Equivalently, a code is perfect if the spheres of radius e around all codewords cover the whole space.

1.17 Examples

- (The *trivial* perfect codes):
  - ∗ The whole space F_q^n is a perfect code with $e = 0$.
  - ∗ The zero vector and the all-one vector of odd length n form a binary perfect code with $e = \frac{n-1}{2}$.
  - ∗ The zero vector itself is a perfect code for $e = n$.
- *The binary Golay code* G_{23} is a linear $[23, 12, 7]$ code with a generator matrix being the circulant with first row the incidence vector of the (nonzero) quadratic residues modulo 23.
- *The ternary Golay code* G_{11} is a linear $[11, 6, 5]$ code with a generator matrix $A = \left(I_6 \, {}^{1\,1\,1\,1\,1}_{\quad C}\right)$, where C is the circulant with first row $(0, 1, 2, 2, 1)$.
- *The Hamming code* over $\mathbb{F}_q$ is a linear $[\frac{q^m-1}{q-1}, \frac{q^m-1}{q-1} - m, 3]$ code with a parity check matrix having as columns a set of representatives of all 1-dimensional subspaces of $\mathbb{F}_q^m$. In particular, the parity check matrix of the binary Hamming code of length $2^m - 1$ consists of all nonzero binary columns of length m.

If C is a code of length n over $\mathbb{F}_q$, the *extended* code $\overline{C}$ is

$$\overline{C} = \{(c_1, \ldots, c_n, c_{n+1}) \mid (c_1, \ldots, c_n) \in C,\ c_{n+1} = \textstyle\sum_{i=1}^{n} c_i\}.$$

Conversely, C is the *shortened* code of $\overline{C}$.

1.18 Theorem (Assmus and Mattson [4])

1. A linear $[n, k, d = 2e+1]$ code over $\mathbb{F}_q$ is perfect if and only if the support design of the minimum weight codewords is an $(e+1)$-$(n, 2e+1, (q-1)^e)$ design.
2. A binary (linear or nonlinear) code of length n and minimum distance $d = 2e+1$ that contains the zero vector is perfect if and only if the support design of the minimum weight codewords is a Steiner system $S(e+1, 2e+1, n)$. Moreover, the supports of the minimum weight codewords in the extended code form a Steiner system $S(e+2, 2e+2, n+1)$.

1.19 Table Perfect codes and their designs. The designs obtained from trivial perfect codes are trivial. The parameters of all nontrivial perfect codes and related designs are

e	q	n	Code	Design
1	Any prime power	$\frac{q^m-1}{q-1}$	Hamming code	2-$(\frac{q^m-1}{q-1}, 3, q-1)$
2	3	11	Ternary Golay code G_{11}	4-(11,5,1)
3	2	23	Binary Golay code G_{23}	4-(23,7,1)

1.20 Remark Theorem 1.18(1) guarantees only a 3-(11,5,4) design in the ternary Golay code G_{11}. Since G_{11} is invariant under the 4-transitive Mathieu group M_{11}, this 3-design is a 4-design by Theorem 1.13.

1.21 Remark The minimum weight vectors in the extended Golay codes $\overline{G_{11}}$ and $\overline{G_{23}}$ support a 5-(12,6,1) and a 5-(24,8,1) design, respectively. These codes and the related designs are invariant under the 5-transitive Mathieu groups M_{12} and M_{24} respectively.

1.22 Remark The Hamming codes and the Golay codes are (up to equivalence) the only nontrivial linear perfect codes. Any (nontrivial) nonlinear perfect code over a finite field is either a translate of a Hamming or a Golay code, or is a single-error-correcting code ($e = 1$) with parameters of a Hamming code. There are many examples of nonlinear perfect codes with $e = 1$ (binary and nonbinary) that are not translates of the Hamming code [54, 83, 26, 57, 58].

1.23 Remark The extended code of the binary Hamming code of length $2^m - 1$ is the Reed-Muller code RM$(m-2, m)$ of Example 1.15.

1.4 The Assmus-Mattson Theorem

1.24 Remark Theorem 1.25, proved by Assmus and Mattson in 1969, provides an important method for the construction of t-designs as support designs in linear codes with relatively few nonzero weights.

1.25 Theorem (*Assmus-Mattson theorem*) Let C be an $[n, k, d]$ linear code over $\mathbb{F}_q$ and $C^\perp$ be the dual $[n, n-k, d^\perp]$ code. Denote by n_0 the largest integer $\le n$ such that $n_0 - \frac{n_0+q-2}{q-1} < d$, and define $n_0^\perp$ similarly for the dual code $C^\perp$. Suppose that for some integer t, $0 < t < d$, there are at most $d - t$ nonzero weights w in $C^\perp$ such that $w \le n - t$. Then:

1. the support design for any weight u, $d \le u \le n_0$ in C is a t-design;
2. the support design for any weight w, $d^\perp \le w \le min\{n - t, n_0^\perp\}$ in $C^\perp$ is a t-design.

1.26 Remark The largest value of t for any known t-design derived from a code via the Assmus-Mattson theorem is $t = 5$. All such 5-designs come from self-dual codes (Table 1.41).

An $[n, k]$ code C is *self-orthogonal* if $C \subset C^\perp$ and *self-dual* if $C = C^\perp$. The length of a self-dual code is even and $n = 2k$. An $[n, \frac{n}{2}]$ code C is *formally* self-dual if C and $C^\perp$ have the same weight distribution.

1.27 Remark Self-orthogonal, and in particular, self-dual codes over $\mathbb{F}_2$ and $\mathbb{F}_3$, as well as some formally self-dual codes over $\mathbb{F}_4$ have regular gaps in their weight distribution: all weights are divisible by 2 if $q = 2$; by 3 if $q = 3$; and by 2 if $q = 4$ and $C^\perp$ coincides with the hermitian conjugate of C.

A code is *even* if all weights are even, and *doubly-even* if all weights are divisible by 4. A code is *singly-even* if it is even but not doubly-even.

1.28 Remark Binary doubly-even self-dual codes exist for any $n \equiv 0 \pmod 8$.

A self-dual code whose distance achieves equality in Theorem 1.29 is *extremal*.

1.29 Theorem The minimum distance d of a self-dual $[n, \frac{n}{2}]$ code C satisfies

$$d \leq \begin{cases} 2[\frac{n}{8}]+2 & \text{if} \quad q=2 \quad \text{and } C \text{ is singly-even,} \\ 4[\frac{n}{24}]+4 & \text{if} \quad q=2 \quad \text{and } C \text{ is doubly-even,} \\ 3[\frac{n}{12}]+3 & \text{if} \quad q=3 \\ 2[\frac{n}{6}]+2 & \text{if} \quad q=4 \quad \text{and } C \text{ is even.} \end{cases}$$

1.30 Example The matrix $(I_4, J_4 - I_4)$ generates an extremal binary $[8,4,4]$ doubly-even self-dual code with weight distribution $A_0 = A_8 = 1$, $A_4 = 14$. Since $d = 4$ and there is only one nonzero weight smaller than $n = 8$, the support design for weight $w = 4$ is a 3-design ($d - t = 4 - 3 = 1$), hence a 3-(8,4,1) design (the unique SQS(8)). This code is equivalent to the extended Hamming code of Example 1.11, or the Reed-Muller code RM(1,3) of Example 1.15, and the related 3-design can also be explained by Theorems 1.13 or 1.18(2).

1.31 Theorem

1. An extremal binary singly-even $[n, \frac{n}{2}]$ code yields 3-, 2-, or 1-designs provided that $n \equiv 0, 2$, or 4 (mod 8).
2. An extremal binary doubly-even $[n, \frac{n}{2}]$ code yields 5-, 3-, or 1-designs provided that $n \equiv 0, 8$, or 16 (mod 24).
3. An extremal ternary code with $n \equiv 0$ (mod 12) yields 5-designs.
4. An extremal even quaternary code with $n \equiv 0$ (mod 6) yields 5-designs.

1.32 Theorem [42] If C is an extremal binary even formally self-dual code of length $n \equiv 2$ (mod 8), then the words of any fixed weight in $C \cup C^{\perp}$ form a 3-design.

1.33 Remark Extremal self-dual codes can exist only for finitely many values of n, but the spectrum is unknown.

1.34 Construction (*quadratic residue (QR) codes*) Let $n = p$ be a prime and q be a quadratic residue modulo p. The quadratic residue (QR) code of length p over $\mathbb{F}_q$ is an $[p, \frac{1}{2}(p+1), d \geq \sqrt{p}]$ code with a generator matrix the zero-one circulant of order p with first row the incidence vector of the (nonzero) quadratic residues modulo p. If $p \equiv 3$ (mod 4) the extended QR code is self-dual.

1.35 Examples

- The binary QR code for $p = 7$ is the Hamming code of Example 1.11.
- The binary QR code for $p = 23$ is the $[23, 12, 7]$ binary Golay code.
- The ternary QR code for $p = 11$ is the $[11, 6, 5]$ ternary Golay code.
- The binary extended QR code of length 18 ($p = 17$) is an extremal even $[18, 9, 6]$ formally self-dual code with weight distribution

$$A_0 = A_{18} = 1, \; A_6 = A_{12} = 102, \; A_8 = A_{10} = 153.$$

By Theorem 1.31(1), the words of any fixed weight form a 2-design. By Theorem 1.32, the union of the 102 words of weight 6 in the code and the 102 words of weight 6 in the dual code form a 3-design, hence a 3-(18,6,5) design.

1.36 Construction (*Pless symmetry codes* [59, 60]) Let p be a prime and $Q = (q_{ij})_{p \times p} = (\chi(j-i))$ where χ is the Legendre symbol modulo p. Let S_p be the matrix displayed on the right. Then the *symmetry code* $C(p)$ is the $[2p+2, p+1]$ ternary code with generator matrix (I, S_p). The code $C(p)$ is self-dual if $p \equiv 2$ (mod 3).

$$\begin{pmatrix} 0 & 1 & \cdots & 1 \\ \chi(-1) & & & \\ \cdot & & & \\ \cdot & & Q & \\ \cdot & & & \\ \chi(-1) & & & \end{pmatrix}$$

1.37 Remark The symmetry code $C(11)$ is equivalent to the extended ternary Golay code.

1.38 Construction (*self-dual codes from Hadamard matrices* [84, 34]) Let H be a Hadamard matrix of order $n = 8t + 4$ such that the number of $+1$'s in each row is congruent to s (mod 4), s odd. Then $(I, (H+J)/2)$ is a generator matrix of a binary self-dual $[2n, n]$ code. The code is singly-even if $s = 1$, and doubly-even if $s = 3$.

1.39 Examples

1. The Hadamard matrix of order 4 produces the extended Hamming $[8, 4, 4]$ code of Example 1.30.
2. The Hadamard matrix of order 12 yields the extended Golay $[24, 12, 8]$ code [84].
3. The three inequivalent Hadamard matrices of order 20 produce 118 inequivalent doubly-even extremal $[40, 20, 8]$ codes [84, 11], and 6 singly-even $[40, 20, 8]$ codes [34].
4. The Paley matrix of order 28 produces 5 inequivalent extremal doubly-even $[56, 28, 12]$ codes [10, 45].

1.40 Remark For other constructions of extremal self-dual codes from combinatorial designs, see [7, 62, 73, 82, 85].

1.41 Table 5-Designs from Self-Dual Codes.

Design	Code	Comments
5-(12,6,1)	$[12, 6, 6]$, $q = 3$	Extended ternary Golay code
5-(18,8,6) 5-(18,10,180)	$[18, 9, 8], q = 4$	Extended cyclic code [53]
5-(24,8,1) 5-(24,12,48)	$[24, 12, 8]$, $q = 2$	Extended binary Golay code
5-(24,9,6) 5-(24,12,576) 5-(24,15,8580)	$[24, 12, 9]$, $q = 3$	Extended QR code; Pless symmetry code $C(11)$
5-(30,12,220) 5-(30,14,5390) 5-(30,16,123000)	$[30, 15, 12]$, $q = 4$	Extended QR code
5-(36,12,45) 5-(36,15,5577) 5-(36,18,209685) 5-(36,21,2438973)	$[36, 18, 12]$, $q = 3$	Pless symmetry code $C(17)$
5-(48,12,8) 5-(48,16,1365) 5-(48,20,36176) 5-(48,24,190680)	$[48, 24, 12]$, $q = 2$	Extended QR code
5-(48,15,364) 5-(48,18,50456) 5-(48,21,2957388) 5-(48,24,71307600) 5-(48,27,749999640)	$[48, 24, 15]$, $q = 3$	Extended QR code; Pless symmetry code $C(23)$
5-(60,18,3060) 5-(60,21,449820) 5-(60,24,34337160) 5-(60,27,1271766600) 5-(60,30,24140500956) 5-(60,33,239329029060)	$[60, 30, 18]$, $q = 3$	Pless symmetry code $C(29)$; Extended QR code

1.5 Nonlinear Codes and Designs

An (n, M, d) *code* is a binary code C of length n, minimum distance d, and size $M = |C|$.

The *distance distribution* of an (n, M, d) code C is the sequence $\{B_i\}_{i=0}^n$ where B_i is the number of ordered pairs of codewords at distance i apart, divided by M.

1.42 Remark The distance and weight distribution of a linear code coincide.

The *MacWilliams transform* $\{B'_i\}_{i=0}^n$ of the distance distribution $\{B_i\}_{i=0}^n$ of an (n, M, d) code C is defined by the identity

$$2^n \sum_{j=0}^n B_j x^j = M \sum_{i=0}^n B'_i (1+x)^{n-i}(1-x)^i.$$

Let $\sigma_1 < \sigma_2 < \ldots < \sigma_{s'}$ be the nonzero subscripts i for which $B'_i \neq 0$. Then $d' = \sigma_1$ is the *dual distance* of C, and s' is the *external distance* of C.

1.43 Remark If $\{B_i\}_{i=0}^n$ is the distance distribution of a linear code C, then $\{B'_i\}_{i=0}^n$ is the weight distribution of the dual code $C^\perp$ (cf. Theorem 1.4).

1.44 Theorem [22] Let s be the number of distinct nonzero distances in an (n, M, d) code C with distance distribution$\{B_i\}_{i=0}^n$ such that $\overline{0} \in C$. Let $\overline{s} = s$ if $B_n = 0$ and $\overline{s} = s - 1$ if $B_n = 1$.

1. If $\overline{s} < d'$, then the codewords of any weight $w \geq d' - \overline{s}$ in C form a $(d' - \overline{s})$-design.
2. If $d - s' \leq s' < d$, the codewords of any fixed weight form a $(d - s')$-design.

1.45 Construction (*Nordstrom-Robinson or Semakov-Zinov'ev code* [54]) The codewords in the extended binary [24, 12, 8] Golay code that have at most one nonzero coordinate among the 8 nonzero positions of a given minimum weight codeword, after deleting those 8 coordinates, form a nonlinear (16,256,6) code with weight and distance distribution $B_0 = B_{16} = 1$, $B_6 = B_{10} = 112$, $B_8 = 30$, and MacWilliams transform $\{B'_i = B_i\}$. Hence, $d = d' = 6$, $s = s' = 4$, and by Theorem 1.44(1) the nonzero codewords support 3-designs (with parameters 3-(16,6,4), 3-(16,8,3), and 3-(16,10,24)).

1.46 Theorem For any binary $(n, M, d = 2t + 1)$ code

$$M \leq \frac{2^n}{\sum_{i=0}^t \binom{n}{i} + \frac{1}{\lfloor n/(t+1) \rfloor}\binom{n}{t}\left(\frac{n-t}{t+1} - \lfloor \frac{n-t}{t+1} \rfloor\right)}.$$

A binary code for which equality holds in Theorem 1.46 is *nearly perfect.*

1.47 Remark If $t + 1$ divides $n + 1$, then the bound of Theorem 1.46 coincides with the sphere packing bound of Theorem 1.16. Thus any perfect code is also nearly perfect.

1.48 Theorem

1. The minimum weight codewords in a nearly perfect $(n, M, d = 2t + 1)$ code C that contains the zero vector form a t-$(n, 2t+1, [(n-t)/(t+1)])$ design.
2. The minimum weight codewords in the extended code $\overline{C}$ form a $(t+1)$-$(n+1, 2t+2, [(n-t)/(t+1)])$ design.

1.49 Remark The code of Construction 1.45 is the first member of an infinite class of nonlinear binary codes, *Preparata codes.* The shortened Preparata codes are nearly perfect, thus the minimum weight vectors yield designs via Theorem 1.48. By Theorem 1.44, the codewords of any other weight also yield designs.

1.50 Remark The nearly perfect codes are a subclass of a more general class of *uniformly packed codes* (cf. [71, 12, 29, 76] and [17], Ch. 16) that also yield designs via Theorem

1.44. For example, the Goethals code C (Table 1.51) is a nonlinear uniformly packed code.

1.51 Table Some nonlinear binary codes that yield 3-designs

(n, M, d)	Weight w	Comment
$(2^m, 2^{2^m-m-1}, 4)$	even $w \geq 4$	Extended perfect code
$(4^m, 2^{4^m-4m}, 6)$	even $w \geq 6$	Preparata code [54]
$(4^m, 2^{4m}, 2^{2m-1} - 2^{m-1})$	$2^{2m-1} \pm 2^{m-1}$ 2^{2m-1}	Kerdock code [54]
$(4^m, 2^{4^m-6m+1}, 8)$	even $w \geq 8$	Goethals code C [27]
$(4^m, 2^{6m-1}, 2^{2m-1} - 2^m)$	$2^{2m-1} \pm 2^m$ $2^{2m-1} \pm 2^{m-1}$ 2^{2m-1}	Goethals code D [27]

1.52 Remark The Preparata and Kerdock code for $m = 2$ coincide with the code of Construction 1.45.

1.53 Remark The Preparata, Kerdock, and Goethals codes were found in the 1970s as nonlinear binary codes that contained more codewords than any linear binary code of the same length and minimum distance. In addition, the distance distribution of the Preparata and Kerdock code, as well as those of the Goethals codes C and D, of Table 1.51 are related by the MacWilliams transform as they are dual linear codes (see Theorem 1.4). Some Preparata and Kerdock codes, as well as the Goethals codes, can be viewed as binary images of linear codes over the ring $\mathbb{Z}_4$, and the corresponding linear codes over $\mathbb{Z}_4$ are indeed dual codes [33].

1.54 Theorem [5] Let D be the 3-$(4^m, 6, \frac{1}{3}(4^m - 4))$ design formed by the minimum weight words in the Preparata code.

1. The 4-subsets of coordinate indices that are not covered by any block of D form a Steiner system $S = S(3, 4, 4^m)$.
2. The 3-$(4^m, 4, 2)$ design S^* consisting of two identical copies of S is extendible to a 4-$(4^m + 1, 5, 2)$ design F by adding one new point to all blocks of S^*, and adding as further new blocks all 5-subsets that are contained in blocks of D.

1.55 Remark The class of 4-designs of Theorem 1.54(2) is the only known infinite class of 4-designs derived from codes. Although these designs contain repeated blocks, they provide an infinite class of 4-designs with the smallest known fixed λ $(= 2)$.

1.6 Codes with Majority Decoding

A linear code whose dual code supports the blocks of a t-design (see §1.2) admits one of the simplest decoding algorithms, *majority decoding.* Essentially, for each symbol y_i of the received codeword $y = (y_1, \ldots, y_n)$ a set of values $y_i^{(j)}$, $j = 1, \ldots, r$ of certain linear functions defined by the blocks of the design are computed, and the "true" value of y_i is decided to be the one that appears most frequently among $y_i^{(1)}, \ldots, y_i^{(r)}$.

1.56 Theorem [65] Assume that the dual code of a linear code C supports a t-(n, w, λ) design $\mathcal{D}$ $(t \geq 2)$. Let $A_l = \sum_{j=0}^{l-1}(-1)^j\binom{l-1}{j}\lambda_{j+2}$, where $\lambda_i = \lambda\binom{n-i}{t-i}/\binom{w-i}{t-i}$, $(0 \leq i \leq t)$ is the number of blocks of $\mathcal{D}$ containing a set of i fixed points. Define further

$$A'_l = \begin{cases} A_l & \text{if } \ l \leq t-1, \\ A_{t-1} & \text{if } \ l > t-1. \end{cases}$$

Then C can correct up to l errors by majority decoding, where l is the largest integer that satisfies the inequality $\sum_{i=1}^{l} A'_i \leq \lfloor(\lambda_1 + A'_l - 1)/2\rfloor$. In particular, if $t = 2$, the code can correct up to $l \leq \lfloor(r + \lambda - 1)/(2\lambda)\rfloor$ errors by majority decoding, where $r = \lambda(n-1)/(w-1)$ [67].

1.57 Remark One way to construct a code over $\mathbb{F}_p$ whose dual code contains designs is to start with a given t-(v, k, λ) design and consider the code having the block-by-point incidence matrix A of that design as a parity check matrix. The number of errors correctable by majority decoding is determined by the design parameters, while the dimension of the code is $v - rank_p A$, where $rank_p A$ (the p-*rank* of A) is the rank of A over $\mathbb{F}_p$.

1.58 Theorem [31]. Let A be the incidence matrix of a 2-design with parameters v, k, λ, r, b, and let p be a prime.

1. If p does not divide $r(r - \lambda)$, then $rank_p A = v$.
2. If p divides r but does not divide $r - \lambda$, then $rank_p A \geq v - 1$.
3. If $rank_p A < v - 1$, then p divides $r - \lambda$.

1.59 Table Small 2-(v, k, λ) designs and their p-ranks. N_d denotes the number of nonisomorphic designs for the given parameters, and l is the number of correctable errors (Theorem 1.56). The designs 1, 3, 6, 12 are finite desarguesian projective planes, while 2, 5, 11 are finite affine planes (see §VI.7).

No.	v	k	λ	r	N_d	# Errors l	$r-\lambda$	p	$\rho = rank_p A$	$dim C = v - \rho$
1	7	3	1	3	1	1	2	2	4	3
2	9	3	1	4	1	2	3	3	6	3
3	13	4	1	4	1	2	3	3	7	6
4	6	3	2	5	1	1	3	3	5	1
5	16	4	1	5	1	2	4	2	9	7
6	21	5	1	5	1	2	4	2	10	11
7	11	5	2	5	1	1	3	3	6	5
8	13	3	1	6	2	3	5	5	13	0
9	7	3	2	6	4	1	4	2	4,5,6,7	3,2,1,0
10	10	4	2	6	3	1	4	2	5,6,7	5,4,3
11	25	5	1	6	1	3	5	5	15	10
12	31	6	1	6	1	3	5	5	16	15
13	16	6	2	6	3	1	4	2	6,7,8	10,9,8
14	15	3	1	7	80	3	6	2	$11 \leq \rho \leq 15$	$0 \leq dim \leq 4$
15	8	4	3	7	4	1	4	2	4,5,6,7	4,3,2,1

1.60 Remark The p-ranks of the geometric designs defined by the points and subspaces of a given dimension in a finite projective or affine space have been determined [28, 31, 52, 72]. Theorem 1.61 gives the ranks of the designs formed by the hyperplanes in a finite projective or affine space.

1.61 Theorem Let $q = p^m$, p prime.

1. The p-rank of the 2-$(\frac{q^{n+1}-1}{q-1}, \frac{q^n-1}{q-1}, \frac{q^{n-1}-1}{q-1})$ design formed by the hyperplanes in the n-dimensional projective space PG(n, q) over $\mathbb{F}_q$ is $\binom{p+n-1}{n}^m + 1$.
2. The p-rank of the 2-$(q^n, q^{n-1}, \frac{q^{n-1}-1}{q-1})$ design formed by the hyperplanes in the n-dimensional affine space AG(n, q) over $\mathbb{F}_q$ is $\binom{p+n-1}{n}^m$.

1.62 Remark Among the 80 (15,3,1) designs (or STS(15)s) (Table 1.59, No. 14) there is a unique design with minimal 2-rank 11, being the geometric design formed by the lines in $\mathrm{PG}(3,2)$. The binary code generated by this design is equivalent to the Hamming code of length 15 (see §1.3). Similarly, the 2-(8,4,3) design with minimal 2-rank 4 (Table 1.59, No. 15) is a geometric design formed by the planes in $\mathrm{AG}(3,2)$.

1.63 Conjecture (Hamada [31]). The p-rank of any design $\mathcal{D}$ with the parameters of a geometric design $\mathcal{G}$ in $\mathrm{PG}(n,q)$ or $\mathrm{AG}(n,q)$ ($q=p^m$) is greater or equal to the p-rank of $\mathcal{G}$, with equality if and only if $\mathcal{D}$ is isomorphic to $\mathcal{G}$.

1.64 Remark Hamada's conjecture has been proved for the hyperplane design in a binary affine or projective space [32], the design of the lines in a binary projective or a ternary affine space ($\mathrm{STS}(2^n-1)$ or $\mathrm{STS}(3^n)$) [25], and the design of the planes in a binary affine space ($\mathrm{SQS}(2^m)$), [79]. The only known parameters, for which there is more than one design with the same rank as a geometric design (hence showing that the "only if" part of the Hamada's conjecture is not true in general), are 2-(31,7,7) ($\mathrm{PG}(4,2)$) and 2-(32,8,35) (or 3-(32,8,7), $\mathrm{AG}(5,2)$) [78]. These designs can be constructed from codes; see Example 1.67.

1.7 Designs, Intersection Numbers, and Codes

1.65 Remark Linear codes provide a powerful tool for the study of designs having constant block intersection numbers modulo some prime p, in particular, symmetric and quasi-symmetric designs (see §I.5 and §IV.37), both for their construction and for deriving necessary conditions for their existence.

1.66 Theorem [78, 79] Assume that $\mathcal{D}$ is a 2-(v,k,λ) design with block intersection numbers $s_1,s_2,\ldots,s_m$. Denote by C the binary code spanned by the block-by-point incidence matrix of $\mathcal{D}$, and by $\overline{C}$, the extended binary code.

1. If $k,s_1,\ldots,s_m$ are all even, then C is self-orthogonal.
2. If $v,k,s_1,\ldots,s_m$ are all odd, then $\overline{C}$ is self-orthogonal.
3. If $v,k,s_1,\ldots,s_m$ are all even, then C is contained in a length v self-dual code.
4. If $v\equiv 0 \pmod 8$, $k\equiv 0 \pmod 4$ and $s_1,\ldots,s_m$ are all even, then C is contained in a doubly-even self-dual code of length v.
5. If $v\equiv 7 \pmod 8$, $k\equiv 3 \pmod 4$ and $s_1,\ldots,s_m$ are all odd, then $\overline{C}$ is contained in a doubly-even self-dual code of length $v+1$.
6. The dual code $C^\perp$ has minimum distance $d^\perp \geq (r+\lambda)/\lambda$.
7. The dual of the extended code $\overline{C}^\perp$ has minimum distance
$$\overline{d}^\perp \geq \min\{(b+r)/r,(r+\lambda)/\lambda\}.$$

1.67 Example [78] Assume that $\mathcal{D}$ is a quasi-symmetric 2-(31,7,7) design with intersection numbers $x=1, y=3$ (an example being the design of the planes in the binary projective 4-space, $\mathrm{PG}(4,2)$). By Theorem 1.66, 5 and 7, $\overline{C}^\perp$ is contained in a *doubly-even* self-dual $[32,16,D]$ code with minimum distance $D\geq \overline{d}^\perp \geq 5$, hence $D\geq 8$. On the other hand, $D\leq 8$ for any self-dual doubly-even [32,16] code by Theorem 1.29. Thus $\overline{C}^\perp$ is contained in an extremal doubly-even [32,16,8] code. Up to equivalence, there are precisely five such codes [18], one of these five codes being the Reed-Muller code $\mathrm{RM}(3,5)$ (see Example 1.15). Any doubly-even self-dual [32,16,8] code contains exactly 620 codewords of minimum weight 8 which support a 3-(32,8,7) design (with intersection numbers 0, 2, 4) by Theorem 1.25. Therefore, any quasi-symmetric 2-(31,7,7) design is a derived design of a 3-(32,8,7) design supported by a doubly-even extremal [32,16,8] code. Consequently, since the five doubly-even [32,16,8] codes all have transitive automorphism groups [18] and are generated by their minimum weight codewords, there are exactly five nonisomorphic quasi-symmetric 2-(31,7,7) designs with intersection numbers $x = 1, y = 3$. Similarly, there are exactly 5 nonisomorphic 3-(32,8,7) designs with intersection numbers 0, 2, 4.

1.68 Example [78] The 5-(48,12,8) design supported by the minimum weight codewords in the extended binary QR code, being a doubly-even self-dual [48,24,12] code (see Table 1.41), has intersection numbers 0, 2, 4, 6. The derived design consisting of all blocks through a given triple of points, after having that triple deleted, is a 2-(45,9,8) design with intersection numbers 1, 3, that is, a quasi-symmetric design with $x = 1, y = 3$.

1.69 Example [36] Table 1.70 contains two generating permutations and the base blocks of a quasi-symmetric 2-(49,9,6) design with $x = 1, y = 3$. The extended code $\overline{C}$ is a self-dual [50,25,10] code which is extremal with respect to an improved bound for self-dual codes given in [20]. The minimum weight codewords correspond to the incidence vectors of the blocks extended by an overall parity check. The weight enumerator of the code is $A(x) = 1 + 196x^{10} + 11368x^{12} + 31752x^{14} + \cdots$.

1.70 Table A quasi-symmetric 2-(49,9,6) design.

Generating permutations:
$\varphi = (1\ 2\ \ldots\ 7)(8\ 9\ \ldots\ 14)\cdots(43\ 44\ \ldots\ 49)$;
$\psi = (1\ 8\ 22)(2\ 12\ 24)(3\ 9\ 26)(4\ 13\ 28)\ (5\ 10\ 23)\ (6\ 14\ 25)(7\ 11\ 27)$
$(15)(16\ 19\ 17)(18\ 20\ 21)(29\ 46\ 37)(30\ 43\ 39)(31\ 47\ 41)(32\ 44\ 36)$
$(33\ 48\ 38)(34\ 45\ 40)(35\ 49\ 42)$
Automorphism group order: 21
Base blocks:
1 3 7 16 33 34 39 45 46; 4 5 6 12 14 22 30 32 33; 4 13 15 24 29 36 44 45 49;
5 6 9 19 21 25 30 34 36; 14 15 23 27 31 34 35 40 42; 1 9 11 16 17 19 33 35 36;
3 4 8 9 10 15 29 37 41; 1 2 4 11 16 25 31 39 43; 3 8 15 17 21 28 32 38 43;
32 34 35 39 41 42 46 48 49; 4 7 12 14 25 26 35 40 43

1.71 Remark By further exploration of the relationship between quasi-symmetric designs and self-dual codes (Theorem 1.66(4,5)), and an appropriate generalization for codes over other fields, one obtains numerical necessary conditions for the existence of 2-designs with constant block intersection numbers modulo some prime, in particular, quasi-symmetric designs.

1.72 Theorem (Calderbank, [14]). Let $\mathcal{D}$ be a 2-(v, k, λ) design with intersection numbers $s_1 \equiv s_2 \equiv \ldots \equiv s_m \equiv s \pmod 2$. Then

1. $r \equiv \lambda \pmod 4$,
2. $s \equiv 0 \pmod 2$, $k \equiv 0 \pmod 4$, $v \equiv \pm 1 \pmod 8$, or
3. $s \equiv 1 \pmod 2$, $k \equiv v \pmod 4$, $v \equiv \pm 1 \pmod 8$.

1.73 Theorem (Calderbank, [15]). Let p be an odd prime and let $\mathcal{D}$ be a 2-(v, k, λ) design with intersection numbers $s_1 \equiv s_2 \equiv \ldots \equiv s_m \equiv s \pmod p$. Then either

1. $r \equiv \lambda \pmod{p^2}$,
2. $v \equiv 0 \pmod 2$, $v \equiv k \equiv s \equiv 0 \pmod p$, $(-1)^{\frac{v}{2}}$ is a square in $\mathbb{F}_p$,
3. $v \equiv 1 \pmod 2$, $v \equiv k \equiv s \not\equiv 0 \pmod p$, $(-1)^{\frac{v-1}{2}}$ is a square in $\mathbb{F}_p$,
4. $r \equiv \lambda \equiv 0 \pmod p$, and

(a) $v \equiv 0 \pmod 2$, $v \equiv k \equiv s \not\equiv 0 \pmod p$,
(b) $v \equiv 0 \pmod 2$, $k \equiv s \not\equiv 0 \pmod p$, v/s is a nonsquare in $\mathbb{F}_p$,
(c) $v \equiv 1 \pmod{2p}$, $r \equiv 0 \pmod{p^2}$, $k \equiv s \not\equiv 0 \pmod p$,
(d) $v \equiv p \pmod{2p}$, $k \equiv s \equiv 0 \pmod p$,
(e) $v \equiv 1 \pmod 2$, $k \equiv s \equiv 0 \pmod p$, v is a nonsquare in $\mathbb{F}_p$, or
(f) $v \equiv 1 \pmod 2$, $k \equiv s \equiv 0 \pmod p$, v and $(-1)^{\frac{v-1}{2}}$ are squares in $\mathbb{F}_p$.

1.74 Table Some nonexistent quasi-symmetric designs ruled out by Theorems 1.72 and 1.73.

Number	v	k	λ	r	s_1	s_2	Theorem
1	21	9	12	30	3	5	1.72
2	21	8	14	40	2	4	1.72
3	35	7	3	17	1	3	1.72
4	41	17	34	85	5	8	1.73
5	43	18	51	126	6	9	1.73
6	51	21	14	35	6	9	1.73
7	51	15	7	25	3	5	1.72
8	57	27	117	252	12	17	1.73
9	57	24	23	56	9	12	1.73
10	69	33	176	374	15	21	1.72

1.75 Remark Theorems 1.72 and 1.73 apply for any symmetric 2-(v, k, λ) design, but are no more restrictive than the Bruck-Ryser-Chowla Theorem (see §I.5.2).

The *orbit matrix* $M = (m_{ij})$ of a 2-(v, k, λ) design $\mathcal{D}=(X, \mathcal{B})$ with respect to an automorphism group G is an integral matrix with rows indexed by the orbits of G on the point set X and columns indexed by the orbits of G on the block set $\mathcal{B}$, where m_{ij} is the number of points from the ith point orbit contained in a fixed block from the jth block orbit.

1.76 Theorem If G is a cyclic group of a prime order p which does not fix any point or block and $p|r-\lambda$, then the rows of the orbit matrix M generate a self-orthogonal code over $\mathbb{F}_p$.

1.77 Remark Theorem 1.76 can be used for finding orbit matrices of designs by means of codes [77, 61]. Theorem 1.78, due to Anstee, Hall, and Thompson [1] and Bridges, Hall, and Hayden [13] (see also [30], Ch. 17), combined with Theorem 1.4, is especially useful in the study of automorphisms of symmetric designs and finite planes, as well as for the construction of self-dual codes with given automorphism of a prime order [19], [35].

1.78 Theorem Let V be an n-dimensional vector space over a field F. Assume that G is a permutation group on the coordinates of V and that $|G|$ has an inverse in F. Suppose that $C \subseteq V$ is a G module and $\Theta = \frac{1}{|G|}\sum_{g\in G} g$, $H = C\Theta$, $W = V\Theta$. Then with an appropriate orthonormal basis for W, $(C\Theta)^{\perp}_{W} = H^{\perp}_{W} = (C^{\perp})\Theta$.

1.8 Steiner Triple Systems and Linear Codes

Given a design $\mathcal{D}$ with v points and b blocks and a point by block $(v \times b)$ incidence matrix A, the *code of points* (or *point code*) of $\mathcal{D}$ over $\mathbb{F}_p$ is the linear span of the rows of A over $\mathbb{F}_p$, while the *code of blocks* is the linear span of the columns of A. The point code is of length b, while the code of blocks has length v. The dimension of either code is $rank_p A$.

1.79 Remark Most known results about codes of designs, including Theorem 1.25, are about codes spanned by the blocks. The notion of a code of blocks or points is essentially the same if the design is symmetric ($v = b$). However, if the design is not symmetric, the information provided by these two codes may differ significantly.

An important step in the study of the code of a design is the computation of the dimension (see Theorem 1.58). The p-rank of a Steiner triple system is determined by Theorem 1.80 due to Doyen, Hubaut and Vandensavel [25] using the projective and affine dimension as defined by Teirlinck [74].

1.80 Theorem The p-rank of the incidence matrix of an $S = \text{STS}(v)$ is:

1. v if p is a prime greater than 3;
2. $v - d_P - 1$ if $p = 2$ and d_P is the projective dimension of S;
3. $v - d_A - 1$ if $p = 3$ and d_A is the affine dimension of S.

1.81 Remark The computation of the binary codes of Steiner triple systems on 15 or fewer points [88] showed that nonisomorphic STS(v)s generate inequivalent point codes (see Tables 1.84, 1.86). At the same time, the point code of a given Steiner triple system on v points may and does sometimes contain several nonisomorphic systems with incidence matrices formed by codewords of weight $r = (v-1)/2$ (Table 1.84). In contrast, the binary codes generated by the blocks of any two systems with the same parameters and 2-rank are equivalent (see Table 1.85). The latter is a general property of the binary codes of blocks of Steiner triple systems.

1.82 Theorem (Assmus [2]) Any two Steiner triple systems on v points with the same 2-rank have equivalent binary codes of blocks.

1.83 Remark The binary code of blocks of a given $S = \text{STS}(v)$ can be used for the construction of all STS(v)s having the same 2-rank as S [2]. The data from Tables 1.84 and 1.86 and some further results from [6] suggest an open problem, to prove or disprove: The binary point codes of nonisomorphic Steiner triple systems are inequivalent.

1.84 Table The point codes of the 80 STS(15)s (see Table I.1.21).

No.	2-Rank	STS(15)s in C	A_4	A_7	A_8	A_{11}	A_{12}	A_{15}
1	11	1	0	15	0	0	105	448
2	12	1,2	1	21	3	33	199	851
3	13	1,2,3	3	33	9	99	387	1657
4	13	1,2,4	2	29	10	112	389	1644
5	13	1,2,5	2	28	11	117	384	1634
6	13	6	0	22	15	137	382	1612
7	13	1,7	0	18	15	153	378	1588
8	14	1,2,3,4,5,8	4	44	25	275	764	3220
9	14	1,2,4,9	3	39	23	289	777	3213
10	14	1,2,4,5,10	3	38	24	294	772	3203
11	14	6,11	1	30	26	320	776	3177
12	14	1,2,4,12	3	40	24	286	774	3215
13	14	1,2,4,5,13	3	40	26	288	766	3207
14	14	1,2,3,4,14	4	45	24	270	769	3230
15	14	6,15	1	31	29	319	765	3171
16	14	1,2,3,16	7	57	21	231	763	3269
17	14	1,2,7,17	1	30	30	324	760	3161
18	14	1,2,6,18	1	32	28	314	770	3181
19	14	1,19	0	22	27	349	774	3140

No.	2-Rank	STS(15)s in C	A_4	A_7	A_8	A_{11}	A_{12}	A_{15}
20	14	1,20	0	28	27	325	780	3176
21	14	1,21	0	28	27	325	780	3176
22	14	1,22	0	25	24	334	789	3170
23	15	1,2,4,5,9,10,12,23	4	51	46	668	1567	6330
24	15	1,2,4,9,12,24	4	53	48	662	1561	6334
25	15	1,2,4,5,9,10,12,13,25	4	54	51	661	1550	6328
26	15	1,2,3,4,5,8,10,12,14,26	5	60	52	642	1542	6345
27	15	6,11,15,27	2	42	49	699	1566	6294
28	15	1,2,4,6,11,18,28	2	44	51	693	1560	6298
29	15	1,2,4,9,12,29	4	53	48	662	1561	6334
30	15	6,11,30	2	42	49	699	1566	6294
31	15	1,2,4,5,10,31	4	49	48	678	1557	6310
32	15	6,11,32	2	41	46	700	1577	6300
33	15	1,2,20,33	1	39	49	707	1573	6291
34	15	1,2,20,34	1	39	49	707	1573	6291
35	15	1,2,21	1	41	51	701	1567	6295
36	15	36	0	35	50	720	1575	6278
37	15	37	0	27	42	744	1599	6262
38	15	38	0	32	49	731	1576	6264
39	15	1,2,21,39	1	39	49	707	1573	6291
40	15	1,2,21,40	1	41	51	701	1567	6295
41	15	1,2,21,41	1	39	49	707	1573	6291
42	15	42	0	30	47	737	1582	6260
43	15	6,43	0	34	51	725	1570	6268
44	15	44	0	32	45	727	1592	6280
45	15	1,45	0	33	48	726	1581	6274
46	15	46	0	30	43	733	1598	6276
47	15	1,2,21,47	1	36	44	714	1590	6293
48	15	48	0	32	45	727	1592	6280
49	15	49	0	30	43	733	1598	6276
50	15	50	0	26	43	749	1594	6252
51	15	1,51	0	33	48	726	1581	6274
52	15	1,52	0	34	47	721	1586	6284
53	15	1,2,22,53	1	36	44	714	1590	6293
54	15	1,2,22,54	1	37	47	713	1579	6287
55	15	55	0	33	48	726	1581	6274
56	15	1,56	0	32	45	727	1592	6280
57	15	57	0	26	39	745	1610	6268
58	15	1,2,22,58	1	32	40	726	1602	6285
59	15	1,2,20,59	1	41	51	701	1567	6295
60	15	60	0	28	45	743	1588	6256
61	14	1,61	0	22	21	343	798	3164
62	15	1,2,22,62	1	32	36	722	1618	6301
63	15	1,2,19,63	1	29	39	737	1603	6271
64	15	1,2,61,64	1	35	45	719	1585	6283
65	15	65	0	30	43	733	1598	6276
66	15	66	0	28	41	739	1604	6272
67	15	67	0	26	39	745	1610	6268
68	15	68	0	29	40	734	1609	6282
69	15	69	0	26	39	745	1610	6268
70	15	1,70	0	33	48	726	1581	6274

No.	2-Rank	STS(15)s in C	A_4	A_7	A_8	A_{11}	A_{12}	A_{15}
71	15	71	0	27	38	740	1615	6278
72	15	72	0	26	39	745	1610	6268
73	15	73	0	25	44	754	1589	6242
74	15	74	0	31	46	732	1587	6270
75	15	75	0	28	45	743	1588	6256
76	15	1,76	0	35	50	720	1575	6278
77	15	77	0	23	30	748	1643	6286
78	15	1,78	0	25	44	754	1589	6242
79	15	1,7,79	0	21	48	774	1569	6202
80	15	80	0	15	30	780	1635	6238

1.85 Table The codes of blocks of the 80 STS(15)s.

2-Rank	A_1	A_2	A_3	A_4	A_5	A_6	A_7	Remark
11	0	0	35	105	168	280	435	Hamming code
12	1	7	63	189	357	595	835	
13	3	21	119	357	735	1225	1635	
14	7	49	231	693	1491	2485	3235	
15	15	105	455	1365	3003	5005	6435	Complete space

1.86 Table The point codes of STS(v)s with $v < 15$.

v	b	2-Rank	A_2	A_3	A_4	A_5	A_6	A_7	A_8	A_9	A_{10}	A_{11}	A_{12}
7	7	4	0	7	7	0	0	1					
9	12	9	12	0	135	0	216	0	135	0	12	0	1
13_1	26	13	0	0	13	0	52	0	403	0	1144	0	2483
13_2	26	13	0	0	8	0	47	0	423	0	1164	0	2453

1.9 Designs with the Symmetric Difference Property

A symmetric 2-design has the *symmetric difference property*, or is an *SDP design* [40], if the symmetric difference of any *three* blocks is either a block or the complement of a block. The parameters of a symmetric 2-(v, k, λ) SDP design with $k < v/2$ are of the form

(*) $$v = 2^{2m},\ k = 2^{2m-1} - 2^{m-1},\ \lambda = 2^{2m-2} - 2^{m-1}.$$

1.87 Remark The designs with the symmetric difference property provide a number of interesting links between two-weight codes, uniformly-packed codes, Reed-Muller codes, and codes achieving the Grey-Rankin bound. They are also closely related to quasi-symmetric designs, difference sets, bent functions, strongly regular graphs, and point sets in a projective space having prescribed intersection numbers with the hyperplanes.

A nonsymmetric 2-(v, k, λ) design with parameters of the form

(**) $$v = 2^{2m-1} - 2^{m-1},\ k = 2^{2m-2} - 2^{m-1},\ \lambda = 2^{2m-2} - 2^{m-1} - 1,$$

or

(* * *) $$v = 2^{2m-1} + 2^{m-1}A,\ k = 2^{2m-2},\ \lambda = 2^{2m-2} - 2^{m-1}$$

has the *symmetric difference property*, or is an *SDP design*, if the symmetric difference of any *two* blocks is either a block or the complement of a block.

1.88 Remark Any nonsymmetric SDP design is quasi-symmetric. However, there are quasi-symmetric designs with parameters (**) or (* * *) which are not SDP designs [38, 47].

1.89 Theorem

1. [39]. A derived (residual) design of a symmetric SDP design is a quasi-symmetric SDP design with parameters (**) ((* * *), respectively).
2. [86]. Any quasi-symmetric SDP design is embeddable as a derived or residual design into a unique symmetric SDP design. Moreover, two derived or residual designs of a symmetric SDP design $\mathcal{D}$ with respect to blocks B', B'' are isomorphic if and only if B' and B'' are in the same orbit under the automorphism group of $\mathcal{D}$.

1.90 Theorem [23] The rows of the incidence matrix of any symmetric SDP design are the minimum weight codewords in a binary linear code spanned by the first order Reed-Muller code $\mathrm{RM}(1, 2m)$ and the characteristic function of an elementary Abelian difference set in $\mathrm{AG}(2m, 2)$ (or equivalently, the vector of values of a *bent function* on $2m$ variables [66]).

An (n, k, h_1, h_2) *set* S in the projective space $\mathrm{PG}(k-1, q)$ is a set of n points with the property that every hyperplane meets S in either h_1 or h_2 points.

A linear code C is *projective* if the minimum distance of $C^{\perp}$ is at least 3 and is a *two-weight code* if there are only two distinct nonzero weights in C.

1.91 Theorem [21] The existence of a projective $[n, k]$ two-weight code over $\mathbb{F}_q$ with weights w_1, w_2 is equivalent to the existence of an $(n, k, n-w_1, n-w_2)$ set in $\mathrm{PG}(k-1, q)$.

1.92 Theorem [21]. The graph whose vertices are the codewords in a projective 2-weight code, in which two words are adjacent if their mutual distance is w_1, is strongly regular (see §VI.5).

1.93 Theorem [51, 76] A single-error-correcting code C is uniformly packed if and only if $C^{\perp}$ is a two-weight code.

A set of *elliptic* type is an (n, k, h_1, h_2) set in $\mathrm{PG}(2m-1, 2)$ with parameters
$$n = 2^{2m-1} - 2^{m-1} - 1,\ k = 2m,\ h_1 = 2^{2m-2} - 2^{m-1} - 1, h_2 = 2^{2m-2} - 1.$$

A set of *hyperbolic* type is an (n, k, h_1, h_2) set in $\mathrm{PG}(2m-1, 2)$ with parameters
$$n = 2^{2m-1} + 2^{m-1} - 1,\ k = 2m,\ h_1 = 2^{2m-2} + 2^{m-1} - 1,\ h_2 = 2^{2m-2} - 1.$$

1.94 Theorem [86]

1. The rows of the point by block incidence matrix of any quasi-symmetric SDP design are the minimum weight codewords in a binary linear code spanned by the dual code of the Hamming code of length $2^{2m} - 1$ and the incidence vector of a set of elliptic or hyperbolic type in $\mathrm{PG}(2m-1, 2)$.
2. The point orbits under the full automorphism group of an SDP quasi-symmetric design correspond to equivalence classes of sets of elliptic or hyperbolic type in $\mathrm{PG}(2m-1, 2)$.

A binary linear code is *self-complementary* if it contains the all-one vector.

1.95 Theorem (Grey-Rankin bound) If C is a self-complementary $[n, k, d]$ code, then $|C| \leq 8d(n-d)/(n-(n-2d)^2)$, provided the denominator is positive.

1.96 Theorem [38] The linear span of the block-by-point incidence matrix of a quasi-symmetric SDP design is a self-complementary code optimal with respect to the Grey-Rankin bound.

1.97 Theorem The 2-rank of any symmetric 2-design with parameters (*) is greater or equal to $2m+2$. The 2-rank is exactly $2m+2$ if the design is an SDP design.

1.98 Conjecture (*Dillon's conjecture*) The 2-rank of a symmetric 2-design with parameters $(*)$ is $2m+2$ if and only if the design is an SDP design.

1.99 Conjecture (*all-one vector conjecture*) The all-one vector is contained in the binary code of any symmetric design with parameters $(*)$.

1.100 Remark Conjecture 1.99 (also due to Dillon) implies Conjecture 1.98. Conjecture 1.99, and consequently Conjecture 1.98, are true for designs on 64 or fewer points [39].

1.101 Example There are exactly four affine equivalence classes of bent functions on six variables [66] which give rise to four nonisomorphic symmetric 2-(64,28,12) SDP designs via Theorem 1.90. These designs all have transitive automorphism groups [55], hence by Theorem 1.89 there are four nonisomorphic quasi-symmetric 2-(28,12,11) (2-(36,16,12)) SDP designs obtained as derived (residual, respectively) designs. By Theorem 1.94(2), the point orbits of these quasi-symmetric SDP designs give rise to five projective equivalence classes of sets of elliptic type, and seven classes of sets of hyperbolic type in $PG(5,2)$ [86]. By Theorems 1.91 and 1.93, the classification of these sets implies that up to equivalence, there are exactly five two-weight binary $[27,6,12]$ codes (uniformly packed $[27,21,3]$ codes), and exactly seven binary two-weight $[35,6,16]$ codes (uniformly packed $[35,29,3]$ codes, respectively). Tables 1.102 and 1.103 contain generator matrices of these 2-weight codes [87].

1.102 Table Generator matrices for the binary projective two-weight [27,6,12] codes.

Code C	Generator Matrix	$\|AutC\|$
1	0 0 0 0 0 0 0 0 0 0 0 1 1 1 1 1 1 1 1 1 1 1 1 1 1 1 1 0 0 0 0 0 0 0 1 1 1 1 0 0 0 0 0 0 0 0 1 1 1 1 1 1 1 1 0 0 0 0 0 1 1 0 0 1 1 0 0 0 0 1 1 1 1 0 0 0 0 1 1 1 1 0 0 0 1 1 1 1 1 1 1 1 0 0 1 1 0 0 1 1 0 0 1 1 0 0 1 1 0 1 1 0 1 0 1 0 1 0 1 0 1 1 1 0 1 1 1 0 1 1 1 0 1 0 0 1 0 1 0 1 0 1 0 1 1 0 0 0 0 1 1 1 0 1 1 1 0 1 1 1 0 1	51840
2	0 0 0 0 0 0 0 0 0 0 0 0 0 0 0 1 1 1 1 1 1 1 1 1 1 1 1 0 0 0 0 0 0 0 0 0 1 1 1 1 1 1 0 0 0 0 0 0 1 1 1 1 1 1 0 0 0 1 1 1 1 1 1 0 0 0 0 1 1 0 0 0 0 1 1 0 0 0 0 1 1 0 0 1 0 0 0 1 1 1 0 1 1 1 0 1 0 0 0 1 0 1 0 0 0 1 0 1 0 1 0 0 0 1 0 1 1 1 0 1 1 1 0 0 1 1 1 0 1 0 1 1 1 0 1 1 0 0 0 1 0 1 0 1 0 0 0 1 0 1 0 0 1 1 1 1 1 0 1 0 0 0	384
3	0 0 0 0 0 0 0 0 0 0 0 0 0 0 0 1 1 1 1 1 1 1 1 1 1 1 1 0 0 0 0 0 0 0 1 1 1 1 1 1 1 1 0 0 0 0 1 1 1 1 1 1 1 1 0 0 0 0 0 1 1 0 0 0 0 1 1 1 1 0 0 1 1 0 0 0 0 1 1 1 1 0 0 1 1 1 0 1 0 0 0 1 0 0 0 1 0 1 0 1 0 1 1 1 0 0 0 1 0 1 0 1 1 0 1 0 1 1 1 0 1 1 1 1 1 0 0 0 0 1 1 0 1 1 0 1 0 0 0 1 1 1 0 0 1 1 0 0 1 1 1 1 1 1 1 0 0 1 0 0 1 1	160
4	0 0 0 0 0 0 0 0 0 0 0 0 0 0 0 1 1 1 1 1 1 1 1 1 1 1 1 0 0 0 0 0 0 0 1 1 1 1 1 1 1 1 0 0 0 0 0 0 0 0 1 1 1 1 0 0 0 0 0 1 1 0 0 0 0 1 1 1 1 0 0 0 0 1 1 1 1 0 0 1 1 0 0 1 1 1 0 1 0 0 0 1 0 1 1 1 0 1 1 1 0 1 1 1 0 1 0 1 0 1 0 1 1 0 1 0 1 1 1 0 0 0 1 1 0 0 1 0 0 1 1 0 0 0 0 1 1 1 0 1 0 0 1 0 1 0 0 0 1 1 0 0 1 1 0 1 0 1 0 0 0 0	120
5	0 0 0 0 0 0 0 0 0 0 0 0 0 0 0 1 1 1 1 1 1 1 1 1 1 1 1 0 0 0 0 0 0 0 0 0 1 1 1 1 1 1 0 0 0 0 0 0 1 1 1 1 1 1 0 0 0 0 0 0 1 1 1 0 0 0 1 1 1 0 0 0 1 1 1 0 0 0 1 1 1 0 0 0 1 1 1 0 0 1 0 0 1 0 0 1 0 1 1 0 1 1 0 0 0 0 1 1 0 1 1 0 0 1 0 1 1 0 1 0 1 1 0 0 0 1 1 0 1 0 0 1 0 0 0 1 0 1 0 1 0 1 1 0 0 1 1 0 1 0 0 0 0 0 1 0 0 1 0 1 0 1	24

1.103 Table Generator matrices for the binary projective two-weight [35,6,16] codes.

Code C	Generator Matrix	$\|AutC\|$
1	0 0 0 0 0 0 0 0 0 0 0 0 0 0 0 0 0 0 0 1 1 1 1 1 1 1 1 1 1 1 1 1 1 1 1	40320
	0 0 0 0 0 0 0 0 0 0 0 1 1 1 1 1 1 1 1 0 0 0 0 0 0 0 0 1 1 1 1 1 1 1 1	
	0 0 0 0 0 1 1 1 1 1 1 0 0 0 0 0 0 1 1 0 0 0 0 1 1 1 1 0 0 0 0 1 1 1 1	
	0 0 0 1 1 0 0 0 0 1 1 0 0 0 0 1 1 1 1 0 0 1 1 0 0 1 1 0 0 1 1 0 0 1 1	
	0 1 1 0 1 0 0 1 1 0 1 0 0 1 1 0 1 0 1 0 1 1 1 0 1 0 0 0 1 0 0 0 1 0 0	
	1 0 1 0 1 0 1 0 1 1 0 0 1 0 1 1 0 1 0 0 0 0 1 0 0 0 1 0 0 0 1 1 1 0 1	
2	0 0 0 0 0 0 0 0 0 0 0 0 0 0 0 0 0 0 0 1 1 1 1 1 1 1 1 1 1 1 1 1 1 1 1	1344
	0 0 0 0 0 0 0 0 0 1 1 1 1 1 1 1 1 1 1 0 0 0 0 0 0 1 1 1 1 1 1 1 1 1 1	
	0 0 0 1 1 1 1 1 1 0 0 0 0 1 1 1 1 1 1 0 0 0 0 1 1 0 0 0 0 1 1 1 1 1 1	
	0 0 1 0 0 0 1 1 1 0 0 0 1 0 0 0 1 1 1 0 0 0 1 0 1 0 1 1 1 0 0 0 1 1 1	
	0 1 0 0 0 1 0 1 1 0 0 1 0 0 0 1 0 1 1 0 1 1 1 0 1 0 0 0 1 0 1 1 0 0 1	
	1 0 0 0 1 0 1 0 1 0 1 1 1 0 1 1 0 0 1 0 0 1 1 1 1 0 0 1 1 1 0 1 0 1 1	
3	0 0 0 0 0 0 0 0 0 0 0 0 0 0 0 1 1 1 1 1 1 1 1 1 1 1 1 1 1 1 1 1 1 1 1	384
	0 0 0 0 0 0 0 0 0 1 1 1 1 1 1 0 0 0 0 0 0 0 0 0 0 1 1 1 1 1 1 1 1 1 1	
	0 0 0 1 1 1 1 1 1 0 0 0 0 1 1 0 0 0 0 1 1 1 1 1 1 0 0 0 0 1 1 1 1 1 1	
	0 0 1 0 0 0 1 1 1 0 1 1 1 0 1 0 1 1 1 0 0 0 1 1 1 0 1 1 1 0 0 0 1 1 1	
	0 1 0 0 0 1 0 1 1 1 0 1 1 1 0 0 0 0 1 0 1 1 0 0 1 0 0 0 1 0 1 1 0 0 1	
	1 0 0 0 1 0 1 0 1 0 0 0 1 0 1 1 0 1 0 0 0 1 0 1 0 0 0 1 1 1 0 1 0 1 1	
4	0 0 0 0 0 0 0 0 0 0 0 0 0 0 0 0 0 0 0 1 1 1 1 1 1 1 1 1 1 1 1 1 1 1 1	96
	0 0 0 0 0 0 0 0 0 0 0 1 1 1 1 1 1 1 1 0 0 0 0 0 0 0 0 1 1 1 1 1 1 1 1	
	0 0 0 0 0 1 1 1 1 1 1 0 0 0 0 1 1 1 1 0 0 0 0 0 0 1 1 0 0 0 0 1 1 1 1	
	0 0 1 1 1 0 0 0 1 1 1 0 0 0 1 0 1 1 1 0 0 0 1 1 1 0 1 0 0 0 1 0 0 0 1	
	0 1 0 1 1 0 1 1 0 0 1 0 1 1 1 0 0 0 1 0 0 1 0 0 1 0 0 0 1 1 0 0 1 1 0	
	1 0 0 0 1 0 0 1 0 1 0 0 0 1 1 1 0 1 0 0 1 0 0 1 0 1 1 0 0 1 1 0 0 1 1	
5	0 0 0 0 0 0 0 0 0 0 0 0 0 0 0 0 0 0 0 1 1 1 1 1 1 1 1 1 1 1 1 1 1 1 1	120
	0 0 0 0 0 0 0 0 0 0 0 1 1 1 1 1 1 1 1 0 0 0 0 0 0 0 0 1 1 1 1 1 1 1 1	
	0 0 0 0 0 1 1 1 1 1 1 0 0 0 0 1 1 1 1 0 0 0 0 1 1 1 1 0 0 0 0 0 0 1 1	
	0 0 1 1 1 0 0 0 1 1 1 0 1 1 1 0 1 1 1 0 1 1 1 0 0 0 1 0 0 0 1 1 1 0 1	
	0 1 0 1 1 0 1 1 0 0 1 0 0 0 1 0 0 0 1 1 0 0 1 0 1 1 0 0 1 1 0 1 1 0 0	
	1 1 1 0 1 1 0 1 0 1 1 0 0 1 1 0 0 1 1 0 0 1 1 1 0 1 0 1 0 1 1 0 1 0 0	
6	0 0 0 0 0 0 0 0 0 0 0 0 0 0 0 0 0 0 0 1 1 1 1 1 1 1 1 1 1 1 1 1 1 1 1	24
	0 0 0 0 0 0 0 0 0 0 0 1 1 1 1 1 1 1 1 0 0 0 0 0 0 0 0 1 1 1 1 1 1 1 1	
	0 0 0 0 0 0 1 1 1 1 1 0 0 0 1 1 1 1 1 0 0 0 0 0 1 1 1 0 0 0 0 0 1 1 1	
	0 0 0 1 1 1 0 0 1 1 1 0 0 1 0 0 1 1 1 0 0 0 1 1 0 1 1 0 1 1 1 1 0 1 1	
	0 1 1 0 0 1 0 1 0 0 1 0 1 0 0 0 0 1 1 0 1 1 0 1 1 0 1 1 0 0 1 1 0 0 0	
	1 0 1 0 1 0 0 0 0 1 1 0 1 1 0 1 1 0 1 1 0 1 1 1 0 1 0 1 0 1 0 1 1 0 1	
7	0 0 0 0 0 0 0 0 0 0 0 0 0 0 0 0 0 0 0 1 1 1 1 1 1 1 1 1 1 1 1 1 1 1 1	84
	0 0 0 0 0 0 0 0 0 0 0 1 1 1 1 1 1 1 1 0 0 0 0 0 0 0 0 1 1 1 1 1 1 1 1	
	0 0 0 0 0 0 1 1 1 1 1 0 0 0 1 1 1 1 1 0 0 0 1 1 1 1 1 0 0 0 1 1 1 1 1	
	0 0 0 1 1 1 0 0 1 1 1 0 0 1 0 0 1 1 1 0 1 1 0 0 0 1 1 0 0 0 0 0 0 1 1	
	0 1 1 0 0 1 0 1 0 0 1 0 1 0 0 0 0 1 1 0 0 1 0 0 1 0 1 0 0 1 0 1 1 1 1	
	1 0 1 0 1 0 0 0 0 1 1 0 1 1 0 1 1 0 1 0 0 0 0 1 1 0 1 0 1 0 0 0 1 0 1	

1.10 Maximal Codes and Hadamard Matrices

Given n, d ($d \leq n$), $A(n,d)$ denotes the maximum size of a binary code of length n and minimum distance d.

1.104 Theorem (*Plotkin bound* [63]) If d is even and $d > n/2$, then

1. $A(n,d) \leq 2\left\lfloor \frac{d}{2d-n} \right\rfloor$, and $A(2d,d) \leq 4d$.

If d is odd and $d > (n-1)/2$, then

2. $A(n,d) \leq 2\left\lfloor \frac{d+1}{2d+1-n} \right\rfloor$, and $A(2d+1,d) \leq 4d+4$.

An (n, M, d) code of size M that achieves equality in the Plotkin bound (Theorem 1.104) for the given n and d is *maximal*.

1.105 Examples Let $n = 4\lambda$ and A be the incidence matrix of a symmetric Hadamard 2-$(4\lambda - 1, 2\lambda, \lambda)$ design. Then:

1. The rows of A and the zero vector form a maximal $(n-1, n, n/2)$ code $\mathcal{C}_n$ that achieves equality in Theorem 1.104(1).
2. The words of $\mathcal{C}_n$ together with their complements, obtained by replacing 0 by 1 and vice-versa, form a maximal $(n-1, 2n, n/2-1)$ code $\mathcal{B}_n$ that achieves equality in Theorem 1.104(2).
3. The words of $\mathcal{C}_n$ beginning with zero after deleting the first coordinate form a maximal $(n-2, n/2, n/2)$ code $\mathcal{C}'_n$ that achieves equality in Theorem 1.104(1).

1.106 Theorem (Levenshtein [49]) Maximal codes exist for all admissible n and d provided that Hadamard matrices exist of any order divisible by 4.

Given two codes C_1, C_2 with parameters (n_1, M_1, d_1) and (n_2, M_2, d_2) respectively, and nonnegative integers a, b, the code $C = aC_1 \oplus bC_2$ is an (n, M, d) code with $n = an_1 + bn_2$, $M = \min(M_1, M_2)$, and $d \geq ad_1 + bd_2$, defined as follows: glue together a copies of C_1 followed by b copies of C_2; then delete the last $M_2 - M_1$ rows if $M_2 > M_1$, or the last $M_1 - M_2$ rows if $M_1 \geq M_2$.

1.107 Construction (Levenshtein [49]) A maximal (n, M, d) code attaining the bound in Theorem 1.104(1), with n and d given such that d is even and $2d > n \geq d$. Let

$$k = \lfloor \tfrac{d}{2d-n} \rfloor,\ a = d(2k+1) - n(k+1),\ b = kn - d(2k-1).$$

Then a, b are nonnegative integers, $n = (2k-1)a + (2k+1)b$, $d = ka + (k+1)b$, and the code

$$C = \begin{cases} \frac{a}{2}\mathcal{C}'_{4k} \oplus \frac{b}{2}\mathcal{C}'_{4k+4} & \text{if } n \text{ is even;} \\ a\mathcal{C}'_{2k} \oplus \frac{b}{2}\mathcal{C}'_{4k+4} & \text{if } n \text{ is odd and } k \text{ is even;} \\ \frac{a}{2}\mathcal{C}'_{4k} \oplus b\mathcal{C}_{2k+2} & \text{if } n \text{ and } k \text{ are odd,} \end{cases}$$

is a code of length n, minimum distance d and size $M = 2k = 2\lfloor d/(2d-n) \rfloor$.

1.11 Equidistant Codes, Designs and Orthogonal Arrays

An $(n, M, d; q)$ *code* is a q-ary code of length n, size M, and minimum (Hamming) distance d.

An $(n, M, d; q)$ code is *equidistant* if the distance between any two codewords is d.

1.108 Theorem The distance of any equidistant $(n, M, d; q)$ code satisfies $d \leq \frac{nM(q-1)}{(M-1)q}$.

An equidistant code that achieves equality in Theorem 1.108 is *optimal.*

1.109 Construction [69] An optimal equidistant $(n, M, d; q)$ code exists if and only if there exists a resolvable 2-design with parameters $v = M$, $k = \frac{M}{q}$, $\lambda = n - d$, $r = n$, $b = nq$. Label the rows of the identity matrix of order q, I_q by the symbols $0, 1, \ldots, q-1$. Let $\mathcal{D}$ be a resolvable 2-design with parameters v, k, λ, r, b and $v \times b$ incidence matrix $A = [A_1 \ldots A_i \ldots A_r]$, where A_i is the incidence matrix of the $q = \frac{v}{k}$ blocks from the ith parallel class. Replace each row of A_i $(1 \leq i \leq r)$ by the corresponding symbol from $\{0, 1, \ldots, q-1\}$ according to the labeling of the rows of the identity matrix. The resulting array is an optimal equidistant $(n, M, d; q)$-code. In the other direction, given an optimal equidistant $(n, M, d; q)$ code C, replace each symbol i in C by the row of I_q with label i. The resulting (0,1)-matrix is the incidence matrix of a resolvable 2-design.

1.110 Theorem

1. The columns of an $M \times n$ array, being an optimal equidistant $(n, M, d; q)$ code, form an orthogonal array $\mathrm{OA}(M, n, q, 1)$.
2. The columns of an optimal equidistant $(n, M, d; q)$ code corresponding to an affine resolvable 2-design, form an orthogonal array $\mathrm{OA}(M, n, q, 2)$ meeting the Bose-Bush bound (see §II.4).

1.111 Remark Any affine resolvable 2-design has a unique resolution. Therefore, there is a one-to-one correspondence between the isomorphism classes of affine resolvable designs with given parameters and the equivalence classes of corresponding optimal equidistant codes or orthogonal arrays.

1.112 Example Up to isomorphism, there are exactly 68 affine resolvable 2-designs with parameters $v = 27, k = 9, \lambda = 4, r = 13, b = 39$ [48], hence 68 optimal equidistant $(13, 27, 9; 3)$ codes, and 68 $\mathrm{OA}(27, 13, 3, 2)$. Table 1.113 contains the six $\mathrm{OA}(27, 13, 3, 2)$ corresponding to the six designs with point transitive automorphism groups.

1.113 Table $\mathrm{OA}(27, 13, 3, 2)$ and $(13, 27, 9; 3)$ codes derived from transitive affine 2-(27,9,4) designs

1:	2:
111111112002200220202020021	111111112002200220202020021
112002021111110220200202201	112002021111112002022020021
112002020220021111112020021	112002022002201111110202201
112002022002202002021111111	112002020220020220201111111
201010221102021120021120201	201010221120201102201102021
201010220211202011200211021	201010222011020211022011201
201010222020110202112002111	201010220202112020110220111
200122011102022011202002111	200122011120200211020220111
200122012020111120020211021	200122010202111102202011201
200122010211200202111120201	200122012011022020111102021
202201101102020202110211021	202201101010221022102210101
202201100211201120022002111	202201100122102210011001221
202201102020112011201120201	202201102201010101220122011

3:	4:
111111112000220220200202201	111111112002202020020220021
110020221111110220202020021	112002021111110202200220021
110020220222001111110202201	112002022002201111112002201
110020222000222002021111111	112002020220022020021111111
201210021102021120021120201	201010221102021120201120201
200112021210022011200211021	201010220211200211022011021
202011021021020202112002111	201010222020112002110202111
201200212210101200120012121	200122011122000011222200111
200102212021102121002100211	200122010201212102010121201
202001212102100012211221001	200122012010121220101012021
201220100021211010222201011	202201101100220022112211001
200122100102212201101022101	202201100212102110200102121
202021100210210122010110221	202201102021011201021020211

5:	6:
1111111120022020020202022201	1111111122020000202222020021
1100202211111102220200202201	1102020211111122020020200021
1100202220022011111112020021	1102020222020011111110202201
1100202220220022002021111111	1010202212020212222001111001
201210021102021102201102021	2100120201200201200211222111
2012100202112002110220112 01	2021100210122002110202101 21
2001120212120020201102201 11	2022110002011210120220012 11
2012002120201110221022101 01	2010022121102002211020012 11
2020012122101001012201220 11	2201002100021100121011222 21
2020012110220122100110012 21	2000221110220110022122001 11
2001221021001212120000212 11	2012012002210121201002101 21
2001221000212121002121120 01	2001201202112021002110212 01
2020211001012200211212001 21	2002211021001202012101120 21

1.12 Constant Weight Codes and Designs

> A *constant weight code* consists of words of a given fixed weight.

> Given n, d, w ($d \leq n,\ w \leq n$), $A(n, d, w)$ denotes the maximum size of a binary constant weight code of length n, minimum distance d, and weight w.

1.114 Examples

1. The incidence vectors of the blocks of a t-(v, k, λ) design with maximum block intersection number s form a constant weight code of weight $w = k$, length $n = v$, and minimum distance $d = 2(k - s)$. In particular, the incidence vectors of a t-$(v, k, 1)$ packing form a constant weight code of weight $w = k$, length $n = v$, and minimum distance $d = 2(k - t + 1)$. See §IV.33.
2. The rows of the point by block incidence matrix of a 2-design with parameters v, k, λ, r, b form a constant weight code with $n = b,\ w = r,\ d = 2(r - \lambda)$.

1.115 Theorem (*Johnson bound* [37]).

1. If $w \geq \delta$, then $A(n, 2\delta, w) \leq \left\lfloor \frac{n}{w} \left\lfloor \frac{n-1}{w-1} \left\lfloor \cdots \left\lfloor \frac{n-w+\delta}{\delta} \right\rfloor \cdots \right\rfloor \right\rfloor \right\rfloor$.
2. If $w^2 - nw + n\delta > 0$, then $A(n, 2\delta, w) \leq \left\lfloor \frac{n\delta}{w^2 - nw + n\delta} \right\rfloor$.

1.116 Remark The inequalities of Theorem 1.115 have the relaxations

1. $A(n, 2\delta, w) \leq \frac{n(n-1)\ldots(n-w+\delta)}{w(w-1)\ldots\delta}$, and
2. $A(n, 2\delta, w) \leq \frac{n\delta}{w^2 - nw + n\delta}$.

These weaker bounds are equivalent to the bounds of Theorem 1.115 if the numbers encountered happen to be integer.

1.117 Theorem (Semakov and Zinov'ev [70])

1. An $(n, M, 2\delta)$ code C of a constant weight w achieves equality in Remark 1.116(1) if and only if C is the block by point incidence matrix of a Steiner system $S(t, k, v)$ with $v = n,\ k = w,\ t = w - \delta + 1$.
2. An $(n, M, 2\delta)$ code C of a constant weight w achieves equality in Remark 1.116(2) if and only if C is the point by block incidence matrix of a 2-design with $b = n,\ r = w,\ \lambda = w - \delta$.

1.118 Remark By Theorem 1.117, the existence of a Steiner system or a 2-design implies the existence of an optimal constant weight code meeting either of the bounds in Remark 1.116. Therefore, any table for such designs gives a table of values for $A(n, 2\delta, w)$. For example, the existence of an $S(5, 6, 12)$ implies that $A(12, 4, 6) = 132$.

1.13 See Also

§I.5	Symmetric designs.
§II.4, II.5	Orthogonal arrays.
§IV.24	Hadamard matrices.
§IV.33	Packings with constant block size are constant weight codes.
§IV.37	Quasi-symmetric designs.
§V.1	Finite Groups and Designs.
§VI.5	Strongly regular graphs.
§VI.7	Projective and affine geometries.
[9]	A table of binary linear codes with largest minimum distance for given length and dimension. Brouwer's tables for the minimum distance of linear codes are also available online on the World Wide Web. The address (URL) is `http://www.cwi.nl/htbin/aeb/lincodbd/`q`/`n`/`k where $q = 2, 3$ or 4, n is the code length, and k is the code dimension.
[8]	An extensive table of constant weight codes of length $n \leq 28$.
[54]	An extensive general introduction.
[50, 75, 89]	More on perfect codes (§1.2).
[56, 64, 68]	Tables of cyclic codes.
[17, 51, 54]	More on nonlinear codes (§1.5).
[3, 83]	More on codes with majority decoding (§1.6).
[30, 46, 47], [80, 81, 85]	More on intersection numbers and codes (§1.7).
[2, 3, 43], [44, 6, 88]	More on STS and codes (§1.8).
[16, 24, 41]	More on codes and designs with the symmetric difference property (§1.9).
[83]§2.3, [48]	More on orthogonal arrays, resolvable designs and optimal equidistant codes (§1.11).
[83]§2.4	More on constant weight codes (§1.12).

References

[1] R. Anstee, M. Hall Jr. and J. Thompson, Planes of order 10 do not have a collineation of order 5, *J. Combin. Theory A* 29 (1980), 39-58.

[2] E. F. Assmus Jr., On 2-ranks of Steiner triple systems, *Electron. J. Combin.* 2 (1995), Research Paper 9.

[3] E.F. Assmus Jr. and J.D. Key, *Designs and their Codes*, Cambridge University Press, London, 1992.

[4] E.F. Assmus Jr. and H.F. Mattson Jr., Coding and Combinatorics, *SIAM Review* 16 (1974), 349-388.

[5] A.H. Baartmans, I. Bluskov and V.D. Tonchev, The Preparata codes and a class of 4-designs, *J. Combin. Des.* 2 (1994), 167-170.

[6] A. Baartmans, I. Landjev and V.D. Tonchev, On the binary codes of Steiner triple systems, *Des. Codes Crypt.*, to appear.

[7] V.K. Bhargava and J.M. Stein, (v, k, λ) configurations and self-dual codes, *Inform. Control* 28 (1975), 352-355.

[8] A.E. Brouwer, J.B. Shearer, N.J.A. Sloane and W.D. Smith, A new table of constant weight codes, *IEEE Trans. Inform. Theory* 36 (1990), 1334-1380.

[9] A.E. Brouwer and T. Verhoeff, An updated table of minimum-distance bounds for binary linear codes, *IEEE Trans. Inform. Theory* 39 (1993), 662-677.

[10] F.C. Bussemaker and V.D. Tonchev, New extremal doubly-even codes of length 56 derived from Hadamard matrices of order 28, *Discrete Math.* 76 (1989), 45-49.

[11] F.C. Bussemaker and V.D. Tonchev, Extremal doubly-even codes of length 40 derived from Hadamard matrices of order 20, *Discrete Math.* 82 (1990), 317-321.

[12] L.A. Bassalygo, G.V. Zaitsev and V.A. Zinov'ev, Uniformly packed codes, *Problemi Peredachi Informatsii* 10 (1974) 9-14.

[13] W.G. Bridges, M. Hall Jr. and J.L. Hayden, Codes and designs, *J. Combin. Theory A* 31 (1981), 155-174.

[14] A.R. Calderbank, The application of invariant theory to the existence of quasi-symmetric designs, *J. Combin. Theory A* 44 (1987), 94-109.

[15] A.R. Calderbank, Geometric invariants for quasi-symmetric designs, *J. Combin. Theory A* 47 (1988), 101-110.

[16] R. Calderbank and W.M. Kantor, The geometry of two-weight codes, *Bull. London Math. Soc.* 18 (1986), 97-122.

[17] P.J. Cameron and J.H. van Lint, *Graphs, Codes and Designs*, Cambridge University Press, London, 1980.

[18] J.H. Conway and V. Pless, On the enumeration of self-dual codes, *J. Combin. Theory A* 28 (1980), 26-53.

[19] J.H. Conway and V. Pless, On primes dividing the group order of a doubly-even (72,36,16) code and the group order of a quaternary (24,12,10) code, *Discrete Math.* 38 (1982), 143-156.

[20] J.H. Conway and N.J.A. Sloane, A new bound on the minimal distance of self-dual codes, *IEEE Trans. Inform. Theory* 36 (1990), 1319-1333.

[21] P. Delsarte, Weights of linear codes and strongly regular normed spaces, *Discrete Math.* 3 (1972), 47-64.

[22] P. Delsarte, Four fundamental parameters of a code and their combinatorial significance, *Inform. Control* 23 (1973), 407-438.

[23] J.F. Dillon and J.R. Schatz, Block designs with the symmetric difference property, in: *Proceedings NSA Mathematical Sciences Meetings* (R.L. Ward, ed.), U.S. Government Printing Office, Washington, D.C., 1987, 159-164.

[24] S.M. Dodunekov, S.B. Encheva and S.N. Kapralov, On the $[28, 7, 12]$ binary self-complementary codes and their residuals, *Des. Codes Crypt.* 4 (1994), 57-68.

[25] J. Doyen, X. Hubaut and M. Vandensavel, Ranks of incidence matrices of Steiner triple systems, *Math. Zeitschr.* 163 (1978), 251-259.

[26] T. Etzion and A. Vardy, Perfect binary codes: constructions, properties and enumeration, *IEEE Trans. Inform. Theory* 40 (1994), 754-763.

[27] J.M. Goethals, Two dual families of nonlinear binary codes, *Electron. Lett.* 10 (1974), 471-472.

[28] J.M. Goethals and P. Delsarte, On a class of majority logic decodable cyclic codes, *IEEE Trans. Inform. Theory* 14 (1968), 182-188.

[29] J.M. Goethals and H.C.A. van Tilborg, Uniformly packed codes, *Philips Res. Repts* 30 (1975), 9-36.

[30] M. Hall Jr., *Combinatorial Theory*, 2nd Ed., John Wiley & Sons, New York, 1986.

[31] N. Hamada, On the p-rank of the incidence matrix of a balanced or partially balanced incomplete block design and its applications to error correcting codes, *Hiroshima Math. J.* 3 (1973), 153-226.

[32] N. Hamada and H. Ohmori, On the BIB design having the minimum p-rank, *J. Combin. Theory A* 18 (1975), 131-140.

[33] A.R. Hammons Jr., P.V. Kumar, A.R. Calderbank, N.J.A. Sloane and P. Solé, The $\mathbb{Z}_4$-linearity of Kerdock, Preparata, Goethals and related codes, *IEEE Trans. Inform. Theory* 40 (1994), 301-319.

[34] M. Harada and V.D. Tonchev, Singly-even self-dual codes and Hadamard matrices, *Lect. Notes Comput. Sci.* 948 (1995), 279-284.

[35] W.C. Huffman, Automorphisms of codes with applications on extremal doubly even codes of length 48, *IEEE Trans. Inform. Theory* 28 (1982), 511-521.

[36] W.C. Huffman and V.D. Tonchev, The existence of extremal self-dual [50,25,10] codes and quasi-symmetric 2-(49,9,6) designs, *Des. Codes Crypt.* 6 (1995), 97-106.

[37] S.M. Johnson, A new upper bound for error-correcting codes, *IEEE Trans. Inform. Theory* 8 (1962), 203-207.

[38] D. Jungnickel and V.D. Tonchev, Exponential number of quasi-symmetric SDP designs and codes meeting the Grey-Rankin bound, *Des. Codes Crypt.* 1 (1991), 247-253.

[39] D. Jungnickel and V.D. Tonchev, On symmetric and quasi-symmetric designs with the symmetric difference property and their codes, *J. Combin. Theory A* 59 (1992), 40-50.

[40] W.M. Kantor, Symplectic groups, symmetric designs and line ovals, *J. Algebra* 33 (1975), 43-58.

[41] W.M. Kantor, Exponential numbers of two-weight codes, difference sets and symmetric designs, *Discrete Math.* 46 (1983), 95-98.

[42] G.T. Kennedy and V. Pless, On designs and formally self-dual codes, *Des. Codes Crypt.* 4 (1994), 43-55.

[43] J.D. Key, Ternary codes of Steiner triple systems, *J. Combin. Des.* 2 (1994), 25-30.

[44] J.D. Key and F.E. Sullivan, Codes of Steiner triple and quadruple systems, *Des. Codes Crypt.* 2 (1993), 117-126.

[45] H. Kimura, Extremal doubly even (56,28,12) codes and Hadamard matrices of order 28, *Australas. J. Combin.* 10 (1994), 153-161.

[46] C.W.H. Lam, The search for a finite projective plane of order 10, *Amer. Math. Monthly* 98 (1991), 305-318.

[47] C. Lam, L. Thiel and V.D. Tonchev, On quasi-symmetric 2-(28,12,11) and 2-(36,16,12) designs, *Des. Codes Crypt.* 5 (1995), 43-55.

[48] C. Lam and V.D. Tonchev, Classification of affine resolvable 2-(27,9,4) designs, *J. Stat. Plann. Infer.*, to appear.

[49] V.I. Levenshtein, The application of Hadamard matrices to a problem in coding, *Problemy Kibernetiki* 5 (1961), 123-136.

[50] J.H. van Lint, Recent results on perfect codes and related topics, in: *Combinatorics, Part I* (M. Hall Jr. and J.H. van Lint; eds.), *Mathematical Centre Tracts* 55 (1974), 158-178.

[51] J.H. van Lint, *Introduction to Coding Theory*, Springer-Verlag, Berlin, 1982.

[52] F.J. MacWilliams and H.B. Mann, On the p-rank of the design matrix of a difference set, *Inform. Control* 12 (1968), 474-488.

[53] F.J. MacWilliams, A.M. Odlyzko, N.J.A. Sloane and H.N. Ward, Self-dual codes over $GF(4)$, *J. Combin. Theory A* 25 (1978), 288-318.

[54] F.J. MacWilliams and N.J.A. Sloane, *The Theory of Error-Correcting Codes*, North-Holland, Amsterdam 1977.

[55] C. Parker, E. Spence and V.D. Tonchev, Designs with the symmetric difference property on 64 points and their groups, *J. Combin. Theory A* 67 (1994), 23-43.

[56] W.W. Peterson and E.J. Weldon Jr., *Error-Correcting Codes*, Massachusetts Inst. Technology Press, Cambridge MA, 1972.

[57] K.T. Phelps, A combinatorial construction of perfect codes, *SIAM J. Alg. Discr. Math.* 4 (1983), 398-403.

[58] K.T. Phelps, A general construction for error correcting codes, *SIAM J. Alg. Discr. Math.* 5 (1984), 224-228.

[59] V. Pless, Symmetry codes over $GF(3)$ and new five-designs, *J. Combin. Theory* 5 (1968), 215-228.

[60] V. Pless, *Introduction to the Theory of Error-Correcting Codes*, John Wiley & Sons, New York, 1989.

[61] V.S. Pless and V.D. Tonchev, Self-dual codes over $GF(7)$, *IEEE Trans. Inform. Theory* 33 (1987), 723-727.

[62] V. Pless, V. Tonchev and J. Leon, On the existence of a certain (64,32,12) extremal code, *IEEE Trans. Inform. Theory* 39 (1993), 214-215.

[63] M. Plotkin, Binary codes with specified minimum distances, *IEEE Trans. Inform. Theory* 6 (1960), 445-450.

[64] G. Promhouse and S. Tavares, The minimum distance of all binary cyclic codes of odd lengths from 69 to 99, *IEEE Trans. Inform. Theory* 24 (1978), 438-442.

[65] M. Rahman and I.F. Blake, Majority logic decoding using combinatorial designs, *IEEE Trans. Inform. Theory* 21 (1975), 585-587.

[66] O.S. Rothaus, On "bent" functions, *J. Combin. Theory A* 20 (1976), 300-305.

[67] L.D. Rudolph, A class of majority-logic decodable codes, *IEEE Trans. Inform. Theory* 13 (1967), 305-307.

[68] D. Schomaker and M. Wirtz, On binary cyclic codes of length from 101 to 127, *IEEE Trans. Inform. Theory* 38 (1992), 516-518.

[69] N.V. Semakov and V.A. Zinov'ev, Equidistant maximal q-ary codes and resolvable balanced incomplete block designs, *Problemy Peredachi Informatsii* 4 (1968), 1-7.

[70] N.V. Semakov and V.A. Zinov'ev, Balanced codes and tactical configurations, *Problemy Peredachi Informatsii* 5 (1969), 28-36.

[71] N.V. Semakov, V.A. Zinov'ev and G.V. Zaitsev, Uniformly packed codes, *Problemi Peredachi Informatsii* 7 (1971), 38-50.

[72] K.J.C. Smith, On the p-rank of the incidence matrix of points and hyperplanes in a finite projective geometry, *J. Combin. Theory* 7 (1969), 122-129.

[73] E. Spence and V.D. Tonchev, Extremal self-dual codes from symmetric designs, *Discrete Math.* 110 (1992), 265-268.

[74] L. Teirlinck, On projective and affine hyperplanes, *J. Combin. Theory A* 28 (1980), 290-306.

[75] A. Tietäväinen, On the nonexistence of perfect codes over finite fields, *SIAM J. Appl. Math.* 24 (1973), 88-96.

[76] H.C.A. van Tilborg, *Uniformly Packed Codes*, Thesis, Technical University Eindhoven, 1976.

[77] V.D. Tonchev, Hadamard matrices of order 28 with automorphisms of order 7, *J. Combin. Theory A* 40 (1985), 62-81.

[78] V.D. Tonchev, Quasi-symmetric 2-(31,7,7) designs and a revision of Hamada's conjecture, *J. Combin. Theory A* 42 (1986), 104-110.

[79] V.D. Tonchev, Quasi-symmetric designs and self-dual codes, *Europ. J. Combin.* 7 (1986), 67-73.

[80] V.D. Tonchev, A characterization of designs related to the Witt system $S(5,8,24)$, *Math. Zeitschr.* 191 (1986), 225-230.

[81] V.D. Tonchev, A characterization of designs related to dodecads in the Witt system $S(5,8,24)$, *J. Combin. Theory A* 43 (1986), 219-227.

[82] V.D. Tonchev, Symmetric designs without ovals and extremal self-dual codes, *Ann. Discrete Math.* 37 (1988), 451-458.
[83] V.D. Tonchev, *Combinatorial Configurations*, Longman, Wiley, New York, 1988.
[84] V.D. Tonchev, Self-orthogonal designs and extremal doubly-even codes, *J. Combin. Theory A* 52 (1989), 197-205.
[85] V.D. Tonchev, Self-orthogonal designs, *Contemp. Math.* 111 (1990), 219-235.
[86] V.D. Tonchev, Quasi-symmetric designs, codes, quadrics, and hyperplane sections, *Geom. Ded.* 48 (1993), 295-308.
[87] V.D. Tonchev, The uniformly packed binary [27,21,3] and [35,29,3] codes, *Discrete Math.*, to appear.
[88] V.D. Tonchev and R.S. Weishaar, Steiner triple systems of order 15 and their codes, *J. Stat. Plann. Infer.*, to appear.
[89] V.A. Zinov'ev and V.K. Leontjev, The nonexistence of perfect codes over Galois fields, *Problems Control Information Theory* 2 (1973), 123-132.

2 Computer Science: Selected Applications

CHARLES J. COLBOURN

2.1 Sorting in Rounds

2.1 **Remarks** Given n distinct elements in some order, the linear order of the elements is to be established through binary comparisons. Each comparison (question) establishes the relative order (answer) for a pair of elements, and having determined the relative ordering of several pairs of elements, additional orderings may be deduced by implication. For example, if $x > y$, $y > z$ and $z > w$, then it can be deduced that $x > w$. This is an example of a 3-step implication; a 2-step implication is a *direct implication.* Given n elements and a positive integer k, the problem of *sorting in rounds* involves determining the complete order of the elements in k rounds. In each round, a number of questions is answered simultaneously, and the questions asked within one round can be formulated on the basis of answers from all previous rounds; the object is to ask as few questions as necessary. One application of the problem is in situations where comparisons are made via correspondence, for example, in testing consumer preferences [7].

2.2 **Theorem** Sorting n items in one round requires all $\binom{n}{2}$ comparisons. Sorting with one question per round requires $O(n\log n)$ rounds.

2.3 **Remarks** The cases of interest are hence for $1 < k < n\log n$. Consider sorting in two rounds. Let V be the set to be sorted. A question (comparison) is an unordered pair of elements from V. Let E be the set of questions asked in the first round, and observe that $G = (V, E)$ is an undirected graph. Now an answer for the question $\{x, y\}$ is either (x, y) or (y, x), according to whether $x > y$ or not. Replacing each $e = \{x, y\} \in E$ by the answer (x, y) or (y, x) transforms G into an *oriented* graph. In fact, the sets of possible answers correspond precisely to the *acyclic* orientations of G. The *d-step transitive closure* of an oriented graph G is the oriented graph with the vertex set of G and an arc from vertex v to vertex w if and only if there is a path (of length at most d) from v to w in G. A 2-round sorting algorithm using r processors to sort n elements corresponds to a graph G with n vertices and at most r edges, such that the transitive closure G^c of any acyclic orientation of G contains at least $\binom{n}{2} - r$ arcs; the edges of G correspond to the questions asked in the first round, and the (at most r) edges not in G^c are the questions which remain to be asked in the second round.

2.4 **Theorem** [1] Sorting n items in two rounds requires at least $\Omega(n^{5/3})$ and at most $O(n^{7/4})$ comparisons.

2.5 **Remarks** The upper bound employs projective geometries as follows. As noted above, a graph G corresponding to an efficient 2-round sorting algorithm using direct implications only is a sparse graph G for which the 2-step transitive closure of every acyclic orientation of G is dense. Now the points of $\mathrm{PG}(m, q)$ can be put into one-to-one correspondence with the hyperplanes (point x_H corresponding to hyperplane H) such that the following graph G is well-defined. G has one vertex v_H for each point x_H of $\mathrm{PG}(m, q)$, with v_{H_1} joined to v_{H_2} in G if and only if $x_{H_1} \subset H_2$. Then for $m = 4$, G has $(1 + o(1))n^{7/4}$ edges, for $n \to \infty$ with m fixed, and the 2-step transitive closure of any acyclic orientation of G has at most $O(n^{7/4})$ missing edges [1]. This gives an algorithm for sorting $n = (q^5 - 1)/(q - 1)$ elements, for any prime or prime power q, in two rounds, using only direct implications. If the number n of elements to be sorted does not have the form $(q^5 - 1)/(q - 1)$, then "virtual elements" can be added and handled appropriately, without increasing the complexity of the sorting scheme.

2.2 Bus Networks

2.6 **Remarks** The generic *interconnection problem* is to establish communication paths between *input nodes* V_I and *output nodes* V_O; one allows the case that the input nodes and output nodes are the same. *Switch nodes* V_S may be used; these simply relay a message from one communications channel to another. Our task is to connect the nodes in $V = V_I \cup V_O \cup V_S$ using some communication medium. Two main media are bidirectional point-to-point *links*, and (multipoint) *buses*. Ignoring implementation issues, a link is just a 2-subset of nodes, while a bus is a k-subset of nodes, for $k \geq 2$. A network can therefore be viewed as a set system $(V, \mathcal{B})$, where $\mathcal{B}$ is the collection of links or buses; for simplicity, treat links as a special case of buses. The physical realization of a network imposes some basic constraints:

1. The number of buses should be small.
2. Each node should connect to a small number of buses.
3. The number of switch nodes should be small.
4. Each bus should involve a small number of nodes.

The first three constraints address the construction cost, while the fourth is to ensure that each bus carries an acceptably small amount of traffic. The network must also be *connecting*, in that there is a communication path from each input node to each output node. The *distance* from an input node to an output node is the number of buses in the shortest communication path connecting them. The *diameter* or *depth* is the maximum distance from an input to an output. A fifth requirement is then:

5. The diameter should be small.

Naturally, these five goals are conflicting, and any network design is a compromise solution. Let us consider first a network $(V, \mathcal{B})$ with diameter one, and $V = V_I = V_O$.

2.7 **Theorem** [12] For the diameter to be one, every 2-subset of V must appear in some block of V; hence, $(V, \mathcal{B})$ is a 2-$(|V|, K, 1)$ covering (See §IV.8). To minimize the number of buses, and also the number of buses meeting each node, take $(V, \mathcal{B})$ to be a projective plane.

2.8 **Remarks** If there is no plane on $|V|$ elements, Mickunas [12] suggests simply omitting points from a larger plane to obtain a pairwise balanced design which meets the criteria. Bermond, Bond, and Saclé [3] and Bermond and Bond [2] consider the problem of building networks with specified maximum bus size k, in order to meet the requirement for small buses. Such a network is a covering with small block sizes; the minimum number of blocks is achieved by a 2-$(v, k, 1)$ design when there is such a design on the required number of elements.

2.3 Shifting Networks

2.9 **Remarks** Now require that n input nodes can simultaneously communicate with n output nodes, given a specified mapping of inputs to outputs. This requires *disjoint* communication paths, which share no common bus or intermediate node. A good example of this situation arises in the design of shifting networks. A *barrel shifter* is a network whose nodes are 0,1,...,$n-1$, the integers modulo n. Given a shift distance s, $1 \le s < n$, every node must transfer a value to the node whose label is s larger; more precisely, for each $0 \le i < n$, node i must establish a connection to node $i+s$ (modulo n), and all n communication paths are to be disjoint. Kilian, Kipnis, and Leiserson [10] developed a barrel shifter which has diameter one; as they remark, when implemented in VLSI, this means that the shift is accomplished in a single clock tick.

A $(v, k; n)$-*difference cover* is a set $\mathbf{X} = \{X_1, \ldots, X_n\}$, where $X_i = \{a_{i1}, \ldots, a_{ik}\}$ for $1 \le i \le n$, and for every difference d with $1 \le d < v$, there is at least one choice of i, j, and ℓ with $a_{i\ell} - a_{ij} \equiv d \pmod v$.

2.10 **Remarks** Developing the blocks of a $(v, k; n)$-difference cover modulo v gives a covering of pairs by k-sets on v points containing nv blocks. In this covering, the n pairs $(i, i+s \bmod v)$ for $0 \le i < v$ occur in n different blocks. Any difference cover can then be used to design a barrel shifter, taking the nv blocks as buses. Each node finds the first orbit in which $\{0, s\}$ appears, say in block B. Node x now writes its value to the bus $B + x \pmod n$, and reads its value from the bus $B + x - s \pmod n$. In this way each node x reads the value node $x - s \bmod n$ wrote, and each communication path corresponds to a unique block in the orbit.

Kilian, Kipnis, and Leiserson [10] observe that to minimize the total number of buses and the number of buses incident at a node, a good choice for the difference cover is a cyclic difference set. They also use difference covers to design 'permutation architectures', which realize permutations other than just cyclic shifts.

2.4 Superconcentrators and Expanders

An n-*superconcentrator* is a network with n inputs and n outputs in which disjoint communication paths can be established from the inputs to the outputs in any of the $n!$ possible orderings.

A network $(V_I \cup V_O, E)$ with $V_I \cap V_O = \emptyset$ is a (n, α, β)-*expander* if every set of α inputs is directly connected to at least β output nodes.

2.11 Remarks Restrict superconcentrators to have only links and no larger buses. A superconcentrator of depth one requires all n^2 connections (i.e., each input connected to each output); hence, superconcentrators of depth greater than one are of interest. Nevertheless, superconcentrators are typically constructed using expanders (see, for example, Chung [5]).

Any depth-one network with $V_I \cap V_O = \emptyset$ can be equivalently written as a set system $(V_I, \mathcal{B})$, where $\mathcal{B} = \{\{v_i : \{v_i, v_o\} \in E\} : v_o \in V_O\}$. In this setting, an (n, α, β)-expander is a set system with n elements and n blocks, so that every set of α elements intersects at least β of the blocks. Intuitively, β is largest when the blocks intersect each other as little as possible. At the same time, however, for β to be large, each element must appear in a large number of blocks. As one would expect, to maximize the expansion, one should choose to balance the block sizes and balance the sizes of block intersections. Hence, symmetric designs are considered.

2.12 Theorem [1] In the design of points and hyperplanes of the projective geometries PG(d, q), every set of α elements intersects $\beta \geq (\alpha n)/(\alpha + q - 1)$ blocks. Hence, for all $\alpha = \mathrm{o}(n)$, $\alpha = \mathrm{o}(\beta)$; such a network is termed *highly expanding.*

2.13 Remarks Moreover, Alon remarks that these expanders have essentially the smallest number of links of any network with equivalent expansion properties. Using projective geometries for expanders, Alon establishes the existence of n-superconcentrators of depth three with $\mathrm{O}(n^{4/3})$ links.

2.14 Remark Steiner systems in general have also been used to construct expander graphs for applications in shared memory access [14] and in simulation of PRAMs on bounded-degree networks [15].

2.5 Distributed Consensus

2.15 Remarks Consider a database distributed over n sites (e.g., computers) which are to be linked in a decentralized communication network. A function which depends on the contents of the database is to be computed, and the value is to be made known to all sites. The evaluation is to be symmetric, and each site acts identically to all others; there is no site hierarchy. The objective is to accomplish this consensus in a small number of rounds of message passing.

2.16 Theorem [11] Consensus can be achieved in two rounds of message passing using $\mathrm{O}(n\sqrt{n})$ messages, and $\Omega(n\sqrt{n})$ messages are required.

2.17 Remarks The upper bound can be established with projective planes. Assume first that $n = m^2 + m + 1$ and that a projective plane on n points exists; how to modify the scheme to handle general n is discussed in Remark 2.19. Let the sites correspond to the points of the plane. The blocks of the plane can be put into one-to-one correspondence with the points, such that each point is associated with a unique block containing it. (This "system of distinct representatives" can be found using a bipartite graph matching algorithm.) The point i associated with each block B_i is the *leader* of block B_i. Each point i sends messages to all other points in block B_i and to the leaders of all other blocks in which point i occurs. Then, at each round, each point i receives exactly $2m$ messages, one from each of the $2m$ (distinct) points it sends messages to. Since the index in the corresponding design is $\lambda = 1$, the communication paths outlined above provide a "backbone" in this topology such that two successive rounds of communication suffice for the propagation of data from each site to every other.

2.18 **Example** As an application, consider the use of this set-up and these communication paths to solve the *decentralized extrema finding* problem. Each of the n sites contains a value; the goal is to determine the maximum of the values and to make this maximum known to all sites. Let $v(i)$ be the value at site i, and let each site i in the network carry out the following steps:

1. Send $v(i)$ to the sites on its communication paths
2. Await receipt of $2m$ values (messages)
3. Compute the maximum M_i of these received values and $v(i)$
4. Send M_i to the sites on its communication paths
5. Await receipt of $2m$ values
6. Compute the maximum M of these values and M_i.

The value M computed by each site in this last step is the desired maximum.

2.19 **Remarks** As mentioned in Remark 2.17, if $n = m^2 + m + 1$ for some $m = p^k$, then a projective plane on n points exists and can be used. Otherwise, a projective plane with point set of smallest cardinality exceeding n can be employed; "virtual sites" are then used in addition to the original n sites, and the communication scheme can be modified appropriately, affecting its complexity by only a constant factor.

Decentralized extrema finding is of use, for example, in the coordination of distributed checkpoints. Other applications of this communication protocol include the computation of associative functions which have inverses and commit protocols in distributed database systems.

2.6 Storage in Redundant Disk Arrays

2.20 **Remarks** In online transaction processing systems, such as those for airline reservations, various data records are to be stored in a collection of disks for fast data retrieval and update. By partitioning each data record D_i into k equal-sized *data blocks* $D_{i,1}, \ldots, D_{i,k}$, and storing each on a different disk, one can retrieve the k data blocks in parallel, thereby accelerating the retrieval of data. Remaining "on the air" is critical for many online systems; thus disk failures, although perhaps infrequent, should result in minimal disruption to the system. In particular, a single disk failure ought not interrupt the operation of the system. For this reason, with each data record D_i, one associates a *parity block* P_i, calculated as the bitwise modulo 2 sum of the data blocks $D_{i,j}$, $1 \le j \le k$. Then if one disk containing the data block $D_{i,j}$ fails, its contents can be recalculated with little delay as the modulo 2 sum of the remaining $k-1$ data blocks of D_i and its parity block. A complete hardware failure of a single disk need not result in any loss of data, since the same recovery technique can be used to rebuild the contents of the failed disk on a standby. In practice, reconstruction of individual data blocks, reconstruction of the whole disk contents, and ongoing user transactions continue in parallel until the standby disk can replace the failed disk. Naturally, a disk failure increases system load; one main objective is to balance the increase across the remaining disks.

A *disk mapping scheme* for placing data blocks $D_{i,j}$ $(1 \le i \le b,\ 1 \le j \le k)$ and parity blocks P_i $(1 \le i \le b)$ onto v disks is a collection $\{(B_i, x_i) : 1 \le i \le b\}$, where $|B_i| = k+1$, $x_i \in B_i$, and $B_i \subseteq \{1, \ldots, v\}$.

2.21 **Remark** If $B_i = \{b_{i,1}, \ldots, b_{i,k}, x_i\}$, one places data block $D_{i,j}$ on disk $b_{i,j}$, and parity block P_i on disk x_i. The definition ensures that all data and parity blocks for a given data record are assigned to different disks.

2.22 Remarks Two *load balancing* requirements during data access are (1) every disk is assigned the same number of data blocks, and (2) every disk is assigned the same number of parity blocks. (Parity blocks are more frequently accessed than are data blocks, since a change to a single data block necessitates a change to the parity block.) A load balancing requirement during disk reconstruction is (3) the number of data records with one data or parity block on disk i and another on disk j is a constant independent of the choice of i and j; this ensures that the surviving disks participate equally in the reconstruction of a failed disk.

Now $\mathcal{B} = \{B_1, \ldots, B_b\}$ is a collection of sets, each of size $k + 1$, with elements from $V = \{1, \ldots, v\}$. The load balance requirements ensure that $(V, \mathcal{B})$ has constant replication number and constant index (frequency of occurrence of unordered pairs); hence it is a BIBD. Condition (2) adds a further requirement that every block of the BIBD have a distinguished element, so that every element occurs as the distinguished element equally often. This condition can be satisfied whenever $b \equiv 0 \pmod v$ (using a technique such as Algorithm IV.54.7).

2.23 Remark Muntz and Lui [13] first suggested the use of BIBDs for storage mapping in redundant disk arrays. Holland, Gibson, and Sieworuk [9] explore the use of BIBDs in detail.

2.7 See Also

§II.1.9	Conflict-free latin squares are used in assigning array data to parallel memory devices.
§III.1	Pairwise balanced designs can be used as bus network architectures.
§IV.12	Difference sets are used to construct barrel shifters.
§VI.7	Projective geometries are used to construct expander graphs; also, for construction of PG(m, q)
§VI.8	Projective planes are used to determine message routing.
[4]	Describes the use of projective planes for almost sorting.
[6]	A discussion of a number of other applications in computer science.

References

[1] N. Alon, Expanders, sorting in rounds and superconcentrators of limited depth, in: *Proc. Seventeenth ACM Symposium on the Theory of Computing*, 1985, 98-102.

[2] J.C. Bermond and J. Bond, Combinatorial designs and hypergraphs of diameter one, LRI Rapport de Recherche No. 329, Centre d'Orsay, Universit'e de Paris-Sud, 1986.

[3] J.C. Bermond, J. Bond and J.F. Saclé, Large hypergraphs of diameter 1, *Graph Theory and Combinatorics, Proc. Coll. Cambridge,*, 1984, 19-28.

[4] B. Bollobás and M. Rosenfeld, Sorting in one round, *Israel J. Math.* 38 (1981), 154-160.

[5] F.R.K. Chung, On concentrators, superconcentrators, generalizers and nonblocking networks, *Bell Syst. Tech. J.* 58 (1979), 1765-1777.

[6] C.J. Colbourn and P.C. van Oorschot, Applications of combinatorial designs in computer science, *ACM Computing Surveys* 21 (1989) 223-250.

[7] R. Häggkvist and P. Hell, Parallel sorting with constant time for comparisons, *SIAM J. Computing* 10 (1981), 465-472.

[8] L. Hellerstein, G.A. Gibson, R.M. Karp, R.H. Katz and D.A. Patterson, Coding techniques for handling failures in large disk arrays, *Algorithmica* 12 (1994), 182-208.

[9] M. Holland, G.A. Gibson and D.P. Sieworuk, Architectures and algorithms for on-line failure recovery in redundant disk arrays, *J. Parallel Distributed Databases* 2 (1994).

[10] J. Kilian, S. Kipnis and C.E. Leiserson, The organization of permutation architectures with bussed interconnections, in: *Proc. 28th Symposium on Foundations of Computer Science*, 1987, 305-315.

[11] T.V. Lakshman and V.K. Agrawala, Efficient decentralized consensus protocols, *IEEE Trans. Software Engineering* SE-12 (1986), 600-607.

[12] M.D. Mickunas, Using projective geometry to design bus connection networks, in: *Proc. Workshop on Interconnection Networks for Parallel and Distributed Processing*, 1980, 47-55.

[13] R. Muntz and J. Lui, Performance analysis of disk arrays under failure, in: *Proc. Conf. Very Large Data Bases*, 1990, 162-173.

[14] A. Pietracaprina and F.P. Preparata, Practical constructive schemes for deterministic shared-memory access, preprint.

[15] A. Pietracaprina and G. Pucci, Improved deterministic PRAM simulation on the mesh, *Lect. Notes Comput. Sci.* 944 (1995), 372-383.

3 Applications of Designs to Cryptography

K. Gopalakrishnan
Douglas R. Stinson

3.1 Secrecy Codes

One fundamental goal of cryptography is to allow two people, Alice and Bob, to communicate over an insecure channel in such a way that an observer, Oscar, cannot understand what is being said. Alice encrypts the plaintext message, using a predetermined key, and sends the resulting ciphertext over the channel. Even if Oscar intercepts the ciphertext, he cannot determine what the plaintext was; but Bob, who knows the key, can decrypt the ciphertext and obtain the plaintext. This can be described more formally as follows.

A *cryptosystem* or *secrecy code* is a five-tuple $(\mathcal{P}, \mathcal{C}, \mathcal{K}, \mathcal{E}, \mathcal{D})$, where the following conditions are satisfied:

1. $\mathcal{P}$ is a finite set of possible *plaintexts*.
2. $\mathcal{C}$ is a finite set of possible *ciphertexts*
3. $\mathcal{K}$, the *keyspace*, is a finite set of possible *keys*.
4. For each $K \in \mathcal{K}$, there is an injective *encryption rule* $e_K \in \mathcal{E}$ and a corresponding *decryption rule* $d_K \in \mathcal{D}$. Each $e_K : \mathcal{P} \to \mathcal{C}$ and $d_K : \mathcal{C} \to \mathcal{P}$ are such that $d_K(e_K(x)) = x$ for every plaintext $x \in \mathcal{P}$.

Alice and Bob jointly choose a secret random key $K \in \mathcal{K}$. At a later time, when Alice wants to send $x \in \mathcal{P}$ to Bob, she encrypts x using the encryption rule e_K. That is, she computes $y = e_K(x)$, and sends y to Bob over the channel. When Bob receives y, he decrypts it using the decryption function d_K, obtaining x. Informally, *perfect secrecy* means that observation of a ciphertext gives no information about the corresponding plaintext. This idea can be stated more precisely using probability distributions. Suppose there is a probability distribution $p_{\mathcal{P}}$ on $\mathcal{P}$. A cryptosystem provides *perfect secrecy* whenever $p_{\mathcal{P}}(x|y) = p_{\mathcal{P}}(x)$ for every $x \in \mathcal{P}$ and $y \in \mathcal{C}$.

Shannon proved Theorem 3.1 in 1949; it is a fundamental result and is probably the first linking design theory and cryptography.

The *encryption matrix* of a cryptosystem is the $|\mathcal{K}| \times |\mathcal{P}|$ matrix in which the entry in row K ($K \in \mathcal{K}$) and column x ($x \in \mathcal{P}$) is $e_K(x)$.

3.1 **Theorem** In any cryptosystem that provides perfect secrecy, $|\mathcal{K}| \geq |\mathcal{P}|$. Further, equality occurs if and only if the encryption matrix is a latin square of order $|\mathcal{P}|$ and the encryption rules are used with equal probability.

3.2 **Example** The *Vernam one-time pad.* $\mathcal{P} = \mathcal{C} = \mathcal{K} = (\mathbb{Z}_2)^n$ for some integer $n \geq 1$. For any $x = (x_1, \ldots, x_n) \in (\mathbb{Z}_2)^n$ and $K = (K_1, \ldots, K_n) \in (\mathbb{Z}_2)^n$, define

$$e_K(x) = (x_1 + K_1 \pmod 2, \ldots, x_n + K_n \pmod 2).$$

The encryption matrix is a latin square of order 2^n, and thus perfect secrecy is achieved.

Example 3.2 assumes that each key is used for only one encryption. The theory can be extended in a natural way. Suppose each key is used to encrypt up to t plaintexts, where $t \geq 1$ is an integer. *Perfect t-fold secrecy* means that observation of a set of at most t ciphertexts gives no information about the corresponding set of t plaintexts. As before, one uses probability distributions to define the notion precisely. For any $s \leq t$, suppose there is a probability distribution $p_{\binom{\mathcal{P}}{s}}$ on the set of s-subsets of $\mathcal{P}$. Let $P(s)$ denote the property:

$$p_{\binom{\mathcal{P}}{s}}(X|Y) = p_{\binom{\mathcal{P}}{s}}(X) \text{ for every } X \subseteq \mathcal{P} \text{ and } Y \subseteq \mathcal{C} \text{ such that } |X| = |Y| = s.$$

A cryptosystem provides *perfect t-fold secrecy* whenever $P(s)$ is satisfied for $1 \leq s \leq t$.

3.3 **Theorem** Suppose a perpendicular array $PA_\lambda(t, k, v)$ with $k \geq 2t - 1$ exists. Then a cryptosystem (with k plaintexts, v ciphertexts, and $\lambda\binom{v}{t}$ encryption rules) that provides perfect t-fold secrecy exists.

Here is an outline of a proof of Theorem 3.3. If the transpose of a $PA_\lambda(t, k, v)$ is used as the encryption matrix of a cryptosystem and each of the $\lambda\binom{v}{t}$ encryption rules is chosen with equal probability, then it is not difficult to see that property $P(t)$ is satisfied. However, property $P(s)$ must be satisfied for all $1 \leq s \leq t$. In other words, the $PA_\lambda(t, k, v)$ must also be a $PA_{\lambda(s)}(s, k, v)$ for $1 \leq s \leq t$. Such a perpendicular array is *inductive.* Any $PA_\lambda(t, k, v)$ is inductive provided $k \geq 2t - 1$. Hence the result follows. Theorem 3.4 gives a partial converse.

3.4 **Theorem** Suppose a cryptosystem provides perfect t-fold secrecy, and $|\mathcal{P}| \geq 2t - 1$. Then $|\mathcal{K}| \geq \binom{\mathcal{P}}{t}$ with equality if and only if the encryption matrix is the transpose of a perpendicular array $PA(t, |\mathcal{P}|, |\mathcal{P}|)$ and the encryption rules are used with equal probability.

3.2 Authentication Codes

3.5 **Remark** Authentication codes were invented in 1974 by Gilbert, MacWilliams, and Sloane and have been extensively studied in recent years. The general theory of unconditional authentication was developed by Simmons. Recent research in authentication codes has involved considerable use of various combinatorial designs. This section presents three different types of authentication codes (viz., authentication codes without secrecy, general authentication codes, and authentication codes with secrecy) and discusses their relation to combinatorial designs.

▷ Authentication without Secrecy

The three participants are again Alice, Bob, and Oscar. Suppose that Oscar has the ability to introduce his own messages into the channel and/or to modify existing messages. The purpose of an authentication code is to protect the *integrity* of the message. That is, when Bob receives a message from Alice, he wants to be sure that it was really sent by Alice and was not tampered with by Oscar.

Two types of attacks are considered. When Oscar places a (new) message m' into the channel, this is *impersonation.* When Oscar sees a message m and changes it to a message $m' \neq m$, this is *substitution.*

An *authentication code without secrecy* is a four-tuple $(\mathcal{S}, \mathcal{A}, \mathcal{K}, \mathcal{E})$, where the following conditions are satisfied:

1. $\mathcal{S}$ is a finite set of plaintexts called *source states.*
2. $\mathcal{A}$ is a finite set of *authenticators.*
3. $\mathcal{K}$ is a finite set of *keys.*
4. For each $K \in \mathcal{K}$, there is an *authentication rule* $e_K \in \mathcal{E}$. Each $e_K : \mathcal{S} \to \mathcal{A}$.

As before, Alice and Bob jointly choose a secret key $K \in \mathcal{K}$. At a later time, when Alice wants to communicate a source state $s \in \mathcal{S}$ to Bob, she uses the authentication rule e_K to produce the authenticator $a = e_K(s)$. The *message* $m = (s, a)$ is then sent over the channel. When Bob receives m, he verifies that $a = e_K(s)$ to authenticate the source state s.

When Oscar performs impersonation or substitution, his goal is to have his bogus message $m' = (s', a')$ accepted as authentic by Bob, thus misleading Bob as to the state of the source. That is, if e is the authentication rule being used (which is not known to Oscar), then he is hoping that $a' = e(s')$.

Assume that there is some probability distribution on $\mathcal{S}$, which is known to all the participants. Given this probability distribution, Alice and Bob choose a probability distribution for $\mathcal{E}$, an *authentication strategy.* Once Alice and Bob have chosen their authentication strategy, it is possible to compute, for $i = 0, 1$, a *deception probability* denoted by P_{d_i}, which is the probability that Oscar can deceive Bob by impersonation and substitution, respectively. In computing the deception probabilities, it is assumed that Oscar is using an optimal strategy.

It is not difficult to show that both $P_{d_0} \geq 1/|\mathcal{A}|$ and $P_{d_1} \geq 1/|\mathcal{A}|$. One is not only interested in minimizing the deception probabilities but also in minimizing the number of authentication rules in the code. This is because the key, K, must be exchanged secretly between Alice and Bob, and thus the amount of information that must be kept secret is minimized.

An authentication code is represented by the $|\mathcal{E}| \times |\mathcal{S}|$ *authentication matrix*, in which the rows are indexed by authentication rules, the columns are indexed by source states, and the entry in row e and column s is $e(s)$. Authentication codes (without secrecy) with minimum possible deception probabilities and the minimum number of encoding rules have been shown to be equivalent to orthogonal arrays.

3.6 Theorem Suppose there exists an authentication code without secrecy for k source states and having l authenticators, in which $P_{d_0} = P_{d_1} = 1/l$. Then

1. $|\mathcal{E}| \geq l^2$, and equality occurs if and only if the authentication matrix is the transpose of an orthogonal array $OA(2, k, l)$ (with $\lambda = 1$) and the authentication rules are used with equal probability;
2. $|\mathcal{E}| \geq k(l-1) + 1$, and equality occurs if and only if the authentication matrix is the transpose of an orthogonal array $OA_\lambda(2, k, l)$ where $\lambda = (k(l-1)+1)/l^2$, and the authentication rules are used with equal probability.

Here is an extension of the impersonation/substitution model. Suppose the same encoding rule is used for the transmission of i messages, which Oscar observes in the channel. Then, suppose that Oscar inserts a new message of his own choosing, hoping to have it accepted by Bob as authentic. This is termed *spoofing* of *order i*. The deception probability P_{d_i} is the probability that Oscar succeeds in deceiving Bob with an order i spoofing attack. An authentication code is *t-fold secure against spoofing* if $P_{d_i} = 1/l$ for $0 \le i \le t$. Such codes are related to orthogonal arrays of strength t as indicated by Theorem 3.7.

3.7 **Theorem** Given an authentication code without secrecy that is t-fold secure against spoofing. Then $|\mathcal{E}| \ge |\mathcal{A}|^t$, and equality occurs if and only if the authentication matrix is the transpose of an orthogonal array $OA(t, |\mathcal{S}|, |\mathcal{E}|)$ and the authentication rules are used with equal probability.

This is actually an easy generalization of the first part of Theorem 3.6. The generalization of the second part to the case of t-fold secure authentication codes remains an open problem.

▷ General Authentication Codes

In the case of general authentication codes, one has an arbitrary message space $\mathcal{M}$; i.e., a message does not consist of a source state with an appended authenticator. In this scenario, it is not a requirement that the message reveals the source state. The following bounds on deception probabilities hold for general codes:

$$P_{d_0} \ge \frac{|\mathcal{S}|}{|\mathcal{M}|} \quad \text{and} \quad P_{d_1} \ge \frac{|\mathcal{S}|-1}{|\mathcal{M}|-1}.$$

Codes satisfying these bounds with equality and having the minimum number of authentication rules have been characterized in terms of Steiner 2-designs $S(2, k, v)$.

3.8 **Theorem** Suppose $P_{d_0} = |\mathcal{S}|/|\mathcal{M}|$ and $P_{d_1} = (|\mathcal{S}| - 1)/(|\mathcal{M}| - 1)$. Then

$$|\mathcal{E}| \ge \frac{|\mathcal{M}|(|\mathcal{M}|-1)}{|\mathcal{S}|(|\mathcal{S}|-1)},$$

and equality occurs if and only if the authentication matrix is an $S(2, |\mathcal{S}|, |\mathcal{M}|)$, the authentication rules are used equiprobably, and the source states occur equiprobably.

t-fold secure general authentication codes have been studied as well. In the general case, the deception probabilities are bounded below according to the formula

$$P_{d_i} \ge \frac{|\mathcal{S}|-i}{|\mathcal{M}|-i}.$$

A general authentication code is *t-fold secure against spoofing* if for $0 \le i \le t$,

$$P_{d_i} = \frac{|\mathcal{S}|-i}{|\mathcal{M}|-i}.$$

3.9 **Theorem** Suppose a general authentication code is t-fold secure against spoofing. Then

$$|\mathcal{E}| \ge \frac{\binom{|\mathcal{M}|}{t+1}}{\binom{|\mathcal{S}|}{t+1}},$$

and equality occurs only if the authentication matrix is an $S(t+1, |\mathcal{S}|, |\mathcal{M}|)$. Conversely, if there exists an $S(t+1, k, v)$, then there exists an authentication code for k equiprobable source states, having v possible messages and $\binom{v}{t+1}/\binom{k}{t+1}$ equiprobable authentication rules, that is t-fold secure against spoofing.

▷ Authentication with Secrecy

(t, t')-codes are authentication codes that are t-fold secure against spoofing and simultaneously provide perfect t'-fold secrecy. First consider the simple case of $t = t' = 1$.

Such codes can be constructed from Steiner 2-designs that satisfy an additional numerical condition.

3.10 Theorem Suppose there is an $S(2,k,v)$, where $v-1 \equiv 0 \pmod{k(k-1)}$. Then there is a $(1,1)$-code for k equiprobable source states, having v messages and $v(v-1)/(k(k-1))$ equiprobable encoding rules.

The authentication capability of the code is obvious as it is analogous to the situation of general authentication codes. To provide for perfect secrecy, every block of the $S(2,k,v)$ must be ordered in such a way that every point occurs in each possible position in exactly $(v-1)/(k(k-1))$ blocks. A necessary condition for this to be possible is $v-1 \equiv 0 \pmod{k}(k-1)$, since every point occurs in exactly $(v-1)/(k-1)$ blocks. That this condition is also sufficient can be demonstrated by finding suitable systems of distinct representatives.

Finally, consider $(t-1,t)$ codes. It was shown in §3.1 that (inductive) perpendicular arrays $\mathrm{PA}_\lambda(t,k,v)$ yield codes having t-fold secrecy. If one employs a PA that enjoys an extra property, the code constructed from a $\mathrm{PA}_\lambda(t,k,v)$ is also $(t-1)$-fold secure against spoofing.

An *authentication perpendicular array* $\mathrm{APA}_\lambda(t,k,v)$ is a $\mathrm{PA}_\lambda(t,k,v)$ such that, for any $s \le t-1$ and for any $s+1$ distinct symbols x_i $(1 \le i \le s+1)$, it holds that among the columns of the PA that contain all the symbols x_i $(1 \le i \le s+1)$, the s symbols x_i $(1 \le i \le s)$ occur in all possible subsets of s columns equally often.

This property is precisely what is needed for a $\mathrm{PA}_\lambda(t,k,v)$ to be $(t-1)$-fold secure against spoofing.

3.11 Example An APA$(2,3,11)$. Construct a 3×55 array A by developing the following five columns modulo 11:

$$\begin{array}{ccccc} 0 & 0 & 0 & 0 & 0 \\ 1 & 9 & 3 & 4 & 5 \\ 2 & 7 & 6 & 8 & 10 \end{array}$$

Any unordered pair $\{x,y\}$ occurs in three columns of A. Within these three columns, x occurs once in each of the three rows, as does y.

3.12 Theorem If there exists an $\mathrm{APA}_\lambda(t,k,v)$, then there exists an authentication code that is $(t-1)$-fold secure against spoofing and provides t-fold secrecy for k equiprobable source states, having v messages and $\lambda\binom{v}{t}$ encoding rules. The code is optimal if and only if $\lambda = 1$.

3.3 Threshold Schemes

3.13 Example Suppose that a bank has a vault that must be opened every day. The bank employs three senior tellers, but they do not want to trust any individual with the combination. Hence, they would like a system whereby any two of the three senior tellers can gain access to the vault. This problem can be solved by means of threshold schemes.

Threshold schemes, which are a special case of secret sharing schemes, were first described by Shamir and Blakley in 1979. Blakley's solution uses finite geometries, while Shamir's scheme is based on polynomial interpolation (actually orthogonal arrays in disguise).

A *perfect* (t,w)*-threshold scheme* is a method of sharing a secret key K among a finite set $\mathcal{P}$ of w participants, in such a way that any t participants can compute the value of K, but no group of $t-1$ (or fewer) participants can compute any information about the value of k (in an information-theoretic sense) from the shares they hold.

The value of K is chosen by a special participant, the *dealer*. The dealer is denoted by D and it is assumed that $D \notin \mathcal{P}$. When D wants to share the key K among the participants in $\mathcal{P}$, he gives each participant some partial information, a *share*. The shares should be distributed secretly, so no participant knows the share given to another participant.

At a later time, a subset of participants $B \subseteq \mathcal{P}$ pool their shares in an attempt to compute the secret key K. If $|B| \geq t$, then they should be able to compute the value of K as a function of the shares they collectively hold; if $|B| < t$, then they should not be able to compute K. In Example 3.13, a $(2,3)$-threshold scheme is wanted.

Each share is chosen from a specified *share set* $\mathcal{S}$. It is well known that in any perfect threshold scheme $|\mathcal{K}| \leq |\mathcal{S}|$. A scheme in which equality is attained is an *ideal scheme*.

The first result in this section is a combinatorial characterization of ideal threshold schemes.

3.14 Theorem An ideal (t,w)-threshold scheme with $|\mathcal{K}| = v$ is equivalent to an $OA(t, w+1, v)$.

3.15 Construction It is easy to obtain a threshold scheme from an orthogonal array. Associate the first w rows of the array with the w participants, and the last row with the key. If the dealer wants to share key K, then he chooses a random column of the array such that K occurs in the last row and gives out the remaining w elements in that column as the shares. Now, any t shares uniquely determine a column of the array, and the key is revealed as the element in the last row. But $t-1$ shares leave the key completely undetermined, since for any hypothetical value of the key, say K, there is a unique column containing the $t-1$ given shares and having K in the last row.

A perfect threshold scheme is *anonymous* if the following two properties are satisfied:

1. The w participants receive w distinct shares.
2. The key can be computed solely as a function of t shares, without the knowledge of which participant holds which share. Thus, the key computation can be performed by a black box that is given t shares and does not know the identities of the participants holding those shares.

3.16 Theorem [16] In an anonymous (t,w)-threshold scheme, $|\mathcal{S}| \geq |\mathcal{K}|(w-t+1)+t-1$. Further, equality occurs if and only if there exists a Steiner system $S(t,w,v)$ that can be partitioned into $S(t-1,w,v)$.

3.17 Example As an illustration of Theorem 3.16, let us see how a large set of $S(2,3,9)$ can be used to construct an anonymous $(3,3)$ threshold scheme with seven possible keys. The large set S consists of seven disjoint $S(2,3,9)$. Denote them by $(X, \mathcal{B}_i)$ $(0 \leq i \leq 6)$. Now, the seven keys correspond to the seven designs in the large set; the shares are the points in X. If the dealer's secret is K, then he chooses a random block B from $\mathcal{B}_K$ and gives the three points in B to the three participants as their shares. Now, three shares uniquely determine the block B, which occurs in exactly one of the seven designs (since S is a large set). Hence, three participants can compute K. But from two shares, say x and y, nothing can be determined, since for each possible secret K, there is a (unique) block in $\mathcal{B}_K$ that contains x and y.

3.4 Correlation-Immune and Resilient Functions

Let S be a set, $|S| = q$. Let $n \geq m \geq 1$ be integers and suppose $f : S^n \longrightarrow S^m$ is a function.

1. f has *balanced input* if every possible input vector occurs with equal probability $1/q^n$.
2. f is an (n, m, t, q)-*correlation-immune* function if the probability of obtaining an output m-tuple $(y_1, \ldots, y_m)$ is exactly the same when the input is balanced as it is when the values of t inputs are fixed and the remaining $n - t$ inputs are chosen independently at random.
3. f has *balanced output* if every possible output m-tuple occurs with equal probability $1/q^m$.
4. f is an (n, m, t, q)-*resilient function* if it is t th-order correlation-immune and has balanced output.

3.18 Example Some resilient functions (all arithmetic performed in $\mathbb{F}_q$.)

1. $m = 1$, $t = n - 1$. Define $f(x_1, \ldots, x_n) = x_1 + \ldots + x_n$.
2. $m = n - 1$, $t = 1$. Define $f(x_1, \ldots, x_n) = (x_1 + x_2, x_2 + x_3, \ldots, x_{n-1} + x_n)$.
3. $m = 2$, $n = 3h$, $t = 2h - 1$, and $q = 2$. Define

$$f(x_1, \ldots, x_{3h}) = (x_1 + \ldots + x_{2h}, x_{h+1} + \ldots + x_{3h}).$$

The following theorems describe the connection between correlation-immune functions, resilient functions, and orthogonal arrays.

3.19 Theorem An (n, m, t, q)-resilient function is equivalent to a large set of orthogonal arrays $\mathrm{LOA}_{q^{n-m-t}}(t, n, q)$.

The following is an outline of the proof. Start with a (n, m, t, q)-resilient function. For any $y \in \mathbb{F}_q^m$, the inverse image $f^{-1}(y)$ yields an $\mathrm{OA}_{q^{n-m-t}}(t, n, q)$ and the q^m orthogonal arrays thus obtained form a large set of orthogonal arrays $\mathrm{LOA}_{q^{n-m-t}}(t, n, q)$. The converse is also straightforward.

3.20 Example As an illustration, consider the following binary resilient function:

$$f(x_1, x_2, x_3, x_4, x_5, x_6) = (x_1 + x_2 + x_3 + x_4, x_3 + x_4 + x_5 + x_6),$$

where addition is modulo 2. This is a $(6, 2, 3, 2)$-resilient function, and by Theorem 3.19, it is equivalent to a large set of $\mathrm{OA}_2(3, 6, 2)$. There are four orthogonal arrays in the large set, one of which is obtained from $f^{-1}(0, 0)$:

$$\begin{array}{cccccccccccccccc}
0 & 0 & 1 & 1 & 0 & 0 & 1 & 1 & 0 & 0 & 1 & 1 & 0 & 0 & 1 & 1 \\
0 & 0 & 1 & 1 & 0 & 0 & 1 & 1 & 1 & 1 & 0 & 0 & 1 & 1 & 0 & 0 \\
0 & 0 & 0 & 0 & 1 & 1 & 1 & 1 & 0 & 0 & 0 & 0 & 1 & 1 & 1 & 1 \\
0 & 0 & 0 & 0 & 1 & 1 & 1 & 1 & 1 & 1 & 1 & 1 & 0 & 0 & 0 & 0 \\
0 & 1 & 0 & 1 & 0 & 1 & 0 & 1 & 0 & 1 & 0 & 1 & 0 & 1 & 0 & 1 \\
0 & 1 & 0 & 1 & 0 & 1 & 0 & 1 & 1 & 0 & 1 & 0 & 1 & 0 & 1 & 0
\end{array}$$

A *generalized large set of orthogonal arrays* $\mathrm{GLOA}[q^m]_{\lambda_1,\ldots,\lambda_{q^m}}(t, n, q)$ consists of q^m simple arrays $\mathrm{OA}_{\lambda_i}(t, n, q)$ whose union contains all possible n-tuples among its columns, so

$$\textstyle\sum_{i=1}^{q^m} \lambda_i = q^{n-t}.$$

3.21 Theorem An (n, m, t, q)-correlation-immune function is equivalent to a generalized large set of orthogonal arrays $\mathrm{GLOA}[q^m]_{\lambda_1,\ldots,\lambda_{q^m}}(t, n, q)$.

3.22 Remark Theorem 3.23 gives a construction for resilient functions from linear codes.

3.23 Theorem If there is a linear k-dimensional code of length n over $\mathbb{F}_q$ having minimum Hamming distance d, then there is a $(n, m, d-1, q)$-resilient function.

3.5 Key Distribution Patterns

Suppose there are n users in a large network, say $P_1, P_2, \ldots, P_n$, where every pair of users might wish to communicate in a secure way. Assume that a conventional (i.e., symmetric or private key) cryptosystem is being used for secret communication. Then the problem can be solved by letting a trusted authority generate and distribute a secret key for each pair of users. The drawback to this simpleminded solution is that it involves $\binom{n}{2}$ keys and requires each user to store $n-1$ keys. In a large network, this could give rise to considerable key storage problems.

The number of keys that must be stored by each user can be considerably reduced under some weaker security conditions. An approach follows which is based on combinatorial designs. It was proposed by Mitchell and Piper [6] .

A *key distribution pattern* (KDP) with parameters (n, v, m) is a pair $(X, \mathcal{B})$, where X is a set of v keys, say $\{K_1, K_2, \ldots, K_v\}$, and $\mathcal{B}$ is a set of n m-subsets of X, say $\{B_1, B_2, \ldots, B_n\}$.

KDPs are used as follows. Each one of the n users is given a set of m keys. To be precise, the user P_i is given the set of keys B_i. The key to be used by users P_i and P_j to communicate with one another is computed by applying a special function (e.g., a one-way function or a resilient function) to the set of common keys held by the two users (i.e., to $B_i \cap B_j$).

A KDP is *w-collusion resistant* if the key used for communication between any two users, say P_i and P_j, cannot be computed by a coalition of w other malicious users who pool their sets of keys. More formally, the property can be stated as follows.

A KDP is *w-collusion resistant* if for every pair of users P_i, P_j and for every set of w users C such that $C \cap \{P_i, P_j\} = \emptyset$, it holds that $(B_i \bigcap B_j) \setminus \left(\bigcup_{P_\ell \in C} B_\ell\right) \neq \emptyset$.

w-collusion resistant key distribution patterns can be constructed from certain 3-designs.

3.24 Theorem Suppose a 3-(n, k, λ_3) design exists such that $\lambda_2 > w\lambda_3$. Then, there exists a (n, b, λ_1) w-collusion resistant key distribution pattern.

The proof is simple: Use the *dual* of the 3-(n, k, λ_3) design as the key distribution pattern. The number of users in the KDP is n (the number of points in the 3-design). The number of keys in the KDP is b (the number of blocks in the 3-design). In the 3-design, each point occurs in exactly λ_1 blocks, and so each user gets λ_1 keys in the KDP.

By definition, for any $1 \le i < j \le n$ it holds that $|B_i \cap B_j| = \lambda_2$, and further that $|B_i \cap B_j \cap B_\ell| = \lambda_3$ for every $P_\ell \in C$. Since $\lambda_2 > w\lambda_3$, it follows immediately that the key distribution pattern is w-collusion resistant.

3.6 See Also

§I.2, §I.1	Balanced incomplete block designs.
§II.5	Orthogonal arrays of strength $t > 2$.
§II.2	Orthogonal arrays of strength 2.
§IV.30.1	Perpendicular arrays.
§V.1	Codes and orthogonal arrays.

[8]	Shannon's classic paper on secrecy systems.
[15]	A survey on combinatorial aspects of cryptography.
[10]	A survey on authentication codes.
[11, 14]	Surveys on secret sharing schemes.
[5]	The paper that introduced authentication codes.
[12, 13]	Gives connections between authentication codes and designs.
[7, 2]	The papers that introduced threshold schemes.
[16]	Gives connection between anonymous threshold schemes and designs.
[9]	The paper that introduced correlation-immune functions.
[3, 4]	The papers that introduced resilient functions.
[1]	Recent results on resilient functions.
[6]	The paper that introduced key distribution patterns.

References

[1] J. Bierbrauer, K. Gopalakrishnan and D.R. Stinson, Orthogonal arrays, resilient functions, error correcting codes and linear programming bounds, *SIAM J. Discrete Math.*, to appear.

[2] G.R. Blakley, Safeguarding cryptographic keys, in: *AFIPS Conf. Proc.* 48 (1979), 313-317.

[3] C.H. Bennett, G. Brassard and J.M. Robert, How to reduce your enemy's information, *Lect. Notes Comput. Sci.* 218 (1986), 468-476.

[4] B. Chor, O. Goldreich, J. Håstad, J. Friedman, S. Rudich and R. Smolensky, The bit extraction problem or t-resilient functions, *IEEE Symp. Foundations Comp. Sci.* 26 (1985), 396-407.

[5] E.N. Gilbert, F.J. MacWilliams and N.J.A. Sloane, Codes which detect deception, *Bell Syst. Tech. J.* 53 (1974), 405-424.

[6] C.J. Mitchell and F.C. Piper, Key storage in secure networks, *Disc. Appl. Math.* 21 (1988), 215-228.

[7] A. Shamir, How to share a secret, *Commun. ACM* 22 (1979), 612-613.

[8] C.E. Shannon, Communication theory of secrecy systems, *Bell Syst. Tech. J.* 28 (1949), 656-715.

[9] T. Siegenthaler, Correlation-immunity of nonlinear combining functions for cryptographic applications, *IEEE Trans. Inform. Theory* 30 (1984), 776-780.

[10] G.J. Simmons, A survey of information authentication, in: *Contemporary Cryptology, The Science of Information Integrity*, IEEE Press, 1992, 379-420.

[11] G.J. Simmons, An introduction to shared secret and/or shared control schemes and their application, in: *Contemporary Cryptology, The Science of Information Integrity*, IEEE Press, 1992, 441-500.

[12] D.R. Stinson, The combinatorics of authentication and secrecy codes, *J. Cryptology* 2 (1990), 23-49.

[13] D.R. Stinson, Combinatorial characterizations of authentication codes, *Des. Codes Crypt.* 2 (1992), 175-187.

[14] D.R. Stinson, An explication of secret sharing schemes, *Des. Codes Crypt.* 2 (1992), 357-390.

[15] D.R. Stinson, Combinatorial designs and cryptography, in: *Surveys in Combinatorics, 1993*, Cambridge Univ. Press, London, 257-287.

[16] D.R. Stinson and S.A. Vanstone, A combinatorial approach to threshold schemes, *SIAM J. Discrete Math.* 1 (1988), 230-237.

4 Derandomization

K. Gopalakrishnan
Douglas R. Stinson

4.1 Monte Carlo Algorithms

A *decision problem* is a problem which has a "yes" or "no" answer for every problem instance I.

A *yes-biased Monte Carlo Algorithm* having error probability ϵ is a randomized algorithm to solve a decision problem such that the following conditions are satisfied:

1. If the instance I is a no-instance, then the algorithm answers "no."
2. If the instance I is a yes-instance, then the probability that the algorithm answers "yes" is at least $1 - \epsilon$.

A *no-biased* Monte Carlo Algorithm is defined analogously.

4.1 **Algorithm** Solovay-Strassen Algorithm for "Composites"

> Input: an odd integer n;
> choose a random integer a, $1 \le a \le n-1$;
> **if** $\left(\frac{a}{n}\right) = a^{(n-1)/2} \pmod{n}$
> **then** answer "no: n is prime"
> **else** answer "yes: n is composite".

In this algorithm $\left(\frac{a}{n}\right)$ is a Jacobi symbol (See §VI.2). The Solovay-Strassen algorithm for "Composites" is a yes-biased Monte Carlo algorithm having error probability $\epsilon \le 1/2$.

4.2 **Remark** Despite the fact that Monte Carlo algorithms can give wrong answers, the error probability can be made as small as desired by repeated application of the algorithm. Suppose one has a yes-biased algorithm, and applies the algorithm k times. If at least one "yes" answer is obtained, then the algorithm is finished, since the instance I must be a yes-instance. On the other hand, if I is a yes-instance, then the probability of getting k "no" answers is at most ϵ^k.

4.2 Two-Point Based Sampling

4.3 **Remark** The analysis in Remark 4.2 is based on the assumption that the random numbers used in successive trials are independent. In the implementation of randomized algorithms in practice, one typically uses a pseudo-random number generator, which invalidates this assumption.

Two-point sampling, an alternative technique recently developed by Chor and Goldreich [3], is amenable to more realistic implementation and analysis. This technique essentially uses orthogonal arrays. Suppose $\log n$ random bits are needed for one trial of the algorithm and k trials of the algorithm are to be conducted. Instead of generating $k \log n$ random bits, which would ensure independence, one can use orthogonal arrays to generate pairwise independent random numbers as follows. Consider any $OA(2, k, n)$. This array has n^2 rows. Generate $2 \log n$ random bits and use them to index a specific row of the OA. Now apply the Monte Carlo algorithm using the k points which are elements of the row selected. By elementary combinatorial arguments [4], the probability of getting k "no" answers for a yes-instance is bounded above by $\frac{\epsilon}{1+(k-1)(1-\epsilon)}$.

4.4 **Example** An application of the two-point sampling technique, to the problem of random pattern VLSI chip testing, deserves special mention due to its practical importance. Consider a combinational circuit with n inputs. One can test the circuit by applying all 2^n possible test vectors to the circuit and checking for the response of the circuit. As the number of test patterns is too high for exhaustive testing, usually the circuit is tested with some smaller number of random input patterns. The probability of not detecting a specific fault in the circuit can then be analyzed. But the analysis is invalid if successive test vectors applied are not independent.

In view of this problem, Spenser [7] developed a pseudo-random pattern generator based on the idea of two-point sampling. This generator can be efficiently implemented as a modification of the classical linear feedback shift register (LFSR), and it is amenable to a realistic analysis. This generator is, in fact, an orthogonal array $OA(2, t, 2^n)$.

4.3 Total Derandomization

4.5 **Remark** In some randomized algorithms for search problems, it is possible to eliminate completely the random bits needed, thus converting the algorithm into a deterministic algorithm. This process is *(total) derandomization.* Using orthogonal arrays of strength t, derandomization can be accomplished for randomized algorithms whose analysis only depends on t-wise rather than fully independent random choices. This method was first developed by Luby [6] for the case $t = 2$ in the context of the maximal independent set problem and later generalized to arbitrary strength t by Alon, Babai and Itai [1].

4.6 **Example** An illustration of the idea using the "Monochromatic K_4's" problem [2].

For any integer $n \geq 4$, there is an edge-coloring of K_n with two colors such that the number of monochromatic induced K_4's is at most $\binom{n}{4}/32$. This can be seen easily using a probabilistic argument as follows. For a 2-edge-coloring c of K_n, let $M(c)$ denote the number of monochromatic induced K_4's. Let $\overline{M}$ denote the average value of $M(c)$ over all colorings c. Then

$$\begin{aligned} \overline{M} \times 2^{\binom{n}{2}} &= \textstyle\sum_c M(c) &&= \textstyle\sum_{A \subseteq V, |A|=4} |\{c : |c(A)| = 1\}| \\ &= \binom{n}{4} \times 2 \times 2^{\binom{n}{2}-6} &&= \binom{n}{4} \times 2^{\binom{n}{2}-5} \end{aligned}$$

Hence $\overline{M} = \binom{n}{4}/32$. So there must exist a coloring with at most $\binom{n}{4}/32$ monochromatic induced K_4's.

How would one actually find such a coloring? A randomized algorithm would simply generate a random coloring and check whether it has the desired property. Since the total number of possible colorings, $2^{\binom{n}{2}}$, is exponential in n, it is not possible to test every coloring in polynomial time. Hence, there is no guarantee that a suitable coloring will be found in time $poly(n)$.

However, it is not necessary that the color of every edge be chosen independently at random. It suffices to consider a sample space S where every subset of six edges has colors chosen independently at random. Since only a 6-wise independent sample space is needed, the rows of an $\text{OA}_\lambda\left(6, \binom{n}{2}, 2\right)$ can be used as the sample space S. A deterministic algorithm checks the colorings determined by all the rows of the OA, until one is found in which the number of monochromatic induced K_4's is at most $\binom{n}{4}/32$.

The proof that this approach works is as follows: Within the sample space $\mathbf{S}$,

$$\begin{aligned}\overline{M} \times \lambda 2^6 &= \textstyle\sum_{c \in \mathbf{S}} M(c) \\ &= \textstyle\sum_{A \subseteq V, |A|=4} |\{c \in \mathbf{S} : |c(A)| = 1\}| \\ &= \binom{n}{4} \times 2\lambda\end{aligned}$$

and so $\overline{M} = \binom{n}{4}/32$, as before.

Now, it is possible to check deterministically any particular coloring in polynomial time. Thus, one has a polynomial-time deterministic algorithm provided that $\lambda = poly(n)$. This can be accomplished as follows. Let m be such that $\binom{n}{2} \leq 2^m - 1$. Then construct an orthogonal array $\mathrm{OA}_{2^{3(m-2)}}\left(6, \binom{n}{2}, 2\right)$ as the dual code of a BCH code with designed distance 7 (See §VI.1). The size of the sample space is $O(n^6)$ and so it is a $poly(n)$-sized sample space.

4.7 **Remark** More generally, suppose that the analysis of a randomized algorithm does not require the complete independence of n random bits, but only t-wise independence. Then, one can use any $OA_\lambda(t, n, 2)$ as the sample space instead of the sample space with 2^n points. As in Example 4.6, appropriate sample spaces can be obtained from BCH codes with designed distance $t + 1$, as follows.

4.8 **Theorem** Suppose $n = 2^m - 1$ and $t = 2e$. Then there exists an orthogonal array $OA_{2^{e(m-2)}}(2e, n, 2)$, which yields a t-wise independent sample space of size $(n + 1)^e$.

4.9 **Remark** This approach can also be parallelized. Thus, it yields both a $poly(n)$-time sequential algorithm, and a $polylog(n)$-time parallel algorithm with $poly(n)$ processors.

4.4 See Also

§II.2	for information on orthogonal arrays of strength 2
§II.5	for information on orthogonal arrays of strength $t > 2$
[5]	for a general survey of randomized algorithms
[8]	for a collection of references on pairwise independence and derandomization techniques

References

[1] N. Alon, L. Babai and A. Itai, A fast and simple randomized parallel algorithm for the maximal independent set problem, *J. Algorithms* 7 (1986), 567-583.

[2] N. Alon and J. H. Spencer, *The Probabilistic Method*, Wiley, 1992.

[3] B. Chor and O. Goldreich, On the power of two-point based sampling, *J. Complexity* 5 (1989), 96-106.

[4] K. Gopalakrishnan and D. R. Stinson, A simple analysis of the error probability of two-point based sampling, preprint.

[5] R. Gupta, S. A. Smolka and S. Bhaskar, On randomization in sequential and distributed algorithms, *ACM Computing Surveys* 26 (1994), 7-86.

[6] M. Luby, A simple parallel algorithm for the maximal independent set problem, *SIAM J. Comput.* 15 (1986), 1036-1053.

[7] T. H. Spenser, Provably good pattern generators for a random pattern test, *Algorithmica* 11 (1994), 429-442.

[8] A. Wigderson, The amazing power of pairwise independence, *Proc. 26th ACM Symposium on the Theory of Computing*, 1994, 645-647.

5 Optimality and Efficiency: Comparing Block Designs

Deborah J. Street

5.1 **Remark** Often several essentially different designs of the same size can be found. Is there any sensible way to decide which design is to be preferred for various applications? Resolvability is often a desirable feature, as it means that one or more resolution classes may be processed or harvested at a time and allowance made for this in the analysis. Another useful criterion is how well treatment differences are estimated. One reasonable measure of good estimation is small variance. Ways to compare designs based on this idea include the pairwise efficiency factors of the design and the optimality of the design.

Let Y_{ij} be the response observed in block j from treatment i (if treatment i is applied in block j). Write $E(Y_{ij}) = \mu + \tau_i + \beta_j$, where μ is the overall mean, τ_i is the effect of treatment i, and β_j is the effect of block j. This is the *linear model* for the design. One assumes that the variance of each response is a constant, $Var(Y_{ij}) = \sigma^2$, independent of every other response. The aim of the experiment is to estimate the treatment effects. This estimate is denoted by $\widehat{\tau}$. To do this write the linear model as $E(\mathbf{Y}) = \mathbf{j}\mu + T\tau + B\beta$. Then $T^TB = N$, the $v \times b$ treatment-block incidence matrix. Solving for $\widehat{\tau}$ gives the *reduced normal equations* $A\widehat{\tau} = \mathbf{q}$, where $A = rI - NN^T/k$ is the *information matrix*, $\mathbf{T}$ is the vector of treatment totals, $\mathbf{B}$ is the vector of block totals, and $\mathbf{q} = \mathbf{T} - N\mathbf{B}/k$.

5.2 **Remarks**

1. The expressions above are correct for any incomplete block design in which all treatments are replicated equally often and all blocks have the same size. Otherwise let R be a diagonal matrix with the treatment replication numbers on the diagonal and let K be a diagonal matrix with the block sizes on the diagonal. Then, in general, $A = R - NK^{-1}N^T$ and $\mathbf{q} = \mathbf{T} - NK^{-1}\mathbf{B}$.
2. A is not of full rank; one requires a generalized inverse of A to get an expression for $\widehat{\tau}$. A *generalized inverse* of A is a matrix G such that $AGA = A$. To calculate G write A in *canonical form* as $A = \sum_i \lambda_i \mathbf{x}_i\mathbf{x}_i^T$, where λ_i is an eigenvalue of A with corresponding eigenvector $\mathbf{x}_i$ and the eigenvectors have been normalized. Then $G = \sum_i \lambda_i^{-1}\mathbf{x}_i\mathbf{x}_i^T$, where this summation is over the nonzero eigenvalues only. Then $\widehat{\tau} = G\mathbf{q}$ and $Var(\widehat{\tau}) = \sigma^2 G$.
3. For a BIBD it is possible to evaluate G in general. Since $NN^T = (r - \lambda)I + \lambda J$, by substituting, one obtains $A = \lambda v/k(I - J/v)$. Now $(I - J/v)^2 = (I - J/v)$ so $G = k/\lambda v(I - J/v)$.

5.3 **Example** Consider a (7,3,1) BIBD with blocks {1,2,4}, {2,3,5}, {3,4,6}, {4,5,7}, {5,6,1}, {6,7,2}, {7,1,3}.

Then $E(Y_{11}) = \mu + \tau_1 + \beta_1$, $E(Y_{21}) = \mu + \tau_2 + \beta_1$, and Y_{31} does not exist as treatment 3 is not applied in block 1.

The matrix T is 21×7 and records which treatment generates each response. The matrix B is 21×7 and records in which block each response is observed. $A = 3I - (2I + J)/3$, $T_1 = Y_{11} + Y_{15} + Y_{17}$, $T_2 = Y_{21} + Y_{22} + Y_{26}$ and so on. $G = 3/7(I - J/7)$ and $\widehat{\tau} = 3/7(I - J/7)(\mathbf{T} - N\mathbf{B}/3)$.

> A design is *connected* if, for any two treatments, it is possible to find a list of treatments, starting with one and ending with the other, such that adjacent treatments in the list are in the same block.

5.4 Proposition A design is connected if and only if the rank of A is $v-1$ (see [6]).

5.5 Remark BIBDs are connected. Group divisible PBIBDs with $\lambda_1 = 0$ (§IV.35.2) are disconnected.

> Blocks are a grouping of the plots into homogeneous subsets. If there is no need to block the plots, then the design is a *completely randomized design.* In a completely randomized design, with each treatment replicated r times, $G = I/r$ so the variance of the estimate of any treatment difference, denoted $Var(\widehat{\tau}_i - \widehat{\tau}_j)$, is $2\sigma^2/r$. The *pairwise efficiency factor* of a connected block design is the ratio of $2\sigma^2/r$ to $Var(\widehat{\tau}_i - \widehat{\tau}_j)$ in the block design. The ratio of $2\sigma^2/r$ to the average variance of the treatment differences in a connected design is the *average efficiency factor.*

5.6 Remark In a BIBD, all estimates of treatment differences have the same variance, which is $2\sigma^2 k/\lambda v$. Thus each pairwise efficiency factor is $\lambda v/rk$ and hence so is the average efficiency factor. This expression depends only on the parameters of the BIBD and not on the particular BIBD.

> Since $Var(\mathbf{x}_i^T\widehat{\tau}) = \mathbf{x}_i^T G\mathbf{x}_i\sigma^2 = \sigma^2\lambda_i$, the efficiency factor of the contrast $\mathbf{x}_i^T\tau$ is λ_i/r. The values λ_i/r are the *canonical efficiency factors* of the design with information matrix A. The *efficiency factor* of the design is the harmonic mean of the canonical efficiency factors.

5.7 Remarks

1. For a BIBD, the canonical efficiency factors are $\lambda v/kr$.
2. See [5] for bounds on the efficiency factors of various designs.
3. The tables by Clatworthy [2] give the efficiency factors for the various PBIBDs that are tabulated.

> Consider a set of possible block designs for v treatments all with block size k. A design in the set is *A-optimal* if it has minimum trace $\mathrm{tr}(G)$; *D-optimal* if it has minimum determinant $\det(G)$; and *E-optimal* if the maximum eigenvalue of G is a minimum. *(M,S)-optimality* is a two-stage criterion based on the idea that the canonical efficiency factors should be as equal as possible. At the first stage a class of designs which maximizes $\sum_i \lambda_i/r$ is formed. At the second stage the design, or designs, within this class which minimize $\sum_i(\lambda_i/r)^2$ are obtained. These designs are the (M,S)-optimal designs.

5.8 Remarks

1. When the canonical efficiency factors are all equal, then the design is optimal under all four criteria. Hence BIBDs are A-, D-, E- and (M,S)-optimal.
2. Other alphabetic optimality criteria have been discussed and the idea of universal optimality, which encompasses many of the alphabetic criteria, was introduced by Kiefer [8].
3. Row-column designs, which are extensions of latin squares and include generalized Youden designs (§IV.54), have optimality values associated with them. See [6] for a discussion. Bounds for the values are given in [7]. Some of the PBIBDs tabulated in [2] can be written as Youden designs and are so written in the tabulation when this is possible.
4. On occasion some row-column combinations may not be possible. Then the rows and columns have a *nonorthogonal structure* and the question arises as to the

optimal structure for such designs. Shah and Sinha [13] have developed some designs which balance the occurrence of each of the treatments as far as possible in both rows and columns; this is often referred to as *combinatorial balance*. Kunert [9] has shown that these conditions are more restrictive than necessary and develops some conditions for optimal designs based on the A matrix.

5.1 See Also

§IV.35	PBIBDs.
§IV.54	Generalized Youden designs.
[1]	Discusses the optimality of some PBIBDs.
[3]	A discussion of approximations to BIBDs, with particular reference to parameters (22,8,4), for which a BIBD is unknown.
[4]	Gives an algorithm for optimal response surface designs.
[10, 15]	Gives designs optimal for general dependence structures, including change-over designs
[11]	Discusses exchange algorithms for constructing D-optimal designs.
[12]	A good general introduction.
[14]	Optimal designs when the responses are binary.
[16]	An interchange algorithm to find optimal block designs.
[17]	Contains a simulated annealing algorithm for optimal block designs.

References

[1] C.-S. Cheng and R.A. Bailey, Optimality of some two-associate-class partially balanced incomplete-block designs, *Ann. Statist.* 19 (1991), 1667-1671.

[2] W.H. Clatworthy, *Tables of Two-Associate-Class Partially Balanced Designs*, National Bureau of Standards (U.S.) Applied Mathematics Series No. 63.

[3] A.S. Hedayat, J. Stufken and W.G. Zhang, Virtually balanced incomplete block designs for $v = 22$, $k = 8$, and $\lambda = 4$, *J. Combin. Des.* 3 (1995), 195-201.

[4] R.H. Hardin and N.J.A. Sloane, A new approach to the construction of optimal designs, *J. Stat. Plann. Infer.* 37 (1993), 339-369.

[5] R. Jarrett, A review of bounds for the efficiency factor of block designs, *Austral. J. Statist.* 31 (1989), 118-129.

[6] J.A. John, *Cyclic Designs*, Chapman-Hall, London, 1987.

[7] J.A. John and D.J. Street, Bounds for the efficiency factor of row-column designs, *Biometrika* 79 (1992), 658-661.

[8] J. Kiefer, Construction and optimality of generalized Youden designs, in: *A Survey of Statistical Design and Linear Models* (J.N. Srivastava, ed.), North-Holland, Amsterdam, 1975, 333-353.

[9] J. Kunert, A note on optimal designs with a non-orthogonal row-column-structure, *J. Stat. Plann. Infer.* 37 (1993), 265-270.

[10] R.J. Martin and J.A. Eccleston, Optimal incomplete block designs for general dependence structures, *J. Stat. Plann. Infer.* 28 (1991), 67-81.

[11] N.K. Nguyen and A.J. Miller, A review of some exchange algorithms for constructing discrete D-optimal designs, *Comput. Stat. Data Anal.* 14 (1992), 489-498.

[12] K.R. Shah and B.K. Sinha, *Theory of Optimal Designs*, Springer-Verlag, Berlin, 1989.

[13] K.R. Shah and B.K. Sinha, Optimality aspects of row-column designs with non-orthogonal structure, *J. Stat. Plann. Infer.* 36 (1993), 331-346.

[14] R.R. Sitter and C.F.J. Wu, Optimal designs for binary response experiments: Fieller, *D*, and *A* criteria, *Scand. J. Statist.* 20 (1993), 329-341.
[15] D.J. Street, A note on strongly equi-neighboured designs, *J. Stat. Plann. Infer.* 30 (1992), 99-105.
[16] T. Tjur, An algorithm for the optimization of block designs, *J. Stat. Plann. Infer.* 36 (1993), 277-282.
[17] W.N. Venables and J.A. Eccleston, Randomized search strategies for finding optimal or near optimal block and row-column designs, *Austral. J. Statist.* 35 (1993), 371-382.

6 Group Testing

Charles J. Colbourn

6.1 Nonadaptive Group Testing

6.1 Remarks A population $\mathcal{P}$ of b items contains a number d of *defective* items, and the remaining $b-d$ items are *good*. Items can be pooled together for testing: for a subset $X \subseteq \mathcal{P}$, the *group test* reports "yes" if X contains one or more defective elements, and reports "no" otherwise. The objective is to determine, using a number of group tests performed in parallel, precisely which items are defective.

6.2 Theorem Let $\mathcal{P}$ be a set of b items, and let $\mathbf{X}$ be a collection of subsets of $\mathcal{P}$ corresponding to the group tests performed. Then $(\mathcal{P}, \mathbf{X})$ is a solution to the nonadaptive group testing problem if and only if, for any possible sets D_1 and D_2 of defective items, $\{X : D_1 \cap X \neq \emptyset,\ X \in \mathbf{X}\} = \{X : D_2 \cap X \neq \emptyset,\ X \in \mathbf{X}\}$ only if $D_1 = D_2$.

6.3 Remarks The dual of a solution $(\mathcal{P}, \mathbf{X})$ is a pair $(V, \mathcal{B})$, where the v group tests of $\mathbf{X}$ are in one-to-one correspondence with the points of V, and the b items are in correspondence with the blocks of $\mathcal{B}$ (for each item, the corresponding block contains the elements corresponding to the group tests containing the item). Typically $(V, \mathcal{B})$ is referred to as a solution to the group testing problem; the goal is to maximize the number of blocks (items tested) as a function of the number of points (group tests performed).

Often one knows with high probability that the number of defectives d does not exceed some threshold value p. In the *hypergeometric* problem, one assumes that the number of defectives never exceeds p, and hence one requires that $(V, \mathcal{B})$ has the union of any two distinct sets, each containing at most p blocks, themselves distinct. In the *strict* problem, one is required to correctly identify the set of defective items when $d \leq p$ and is also required to report when $d > p$. In the latter case, the specific set of defective items need not be determined, however.

In the *error correction* version of the problem, some group tests are permitted to report "false positives"; an *a priori* bound q on the number of such false positives is assumed.

6.4 Theorem [1] $(V, \mathcal{B})$ is a solution to the strict group testing problem with threshold p and error correction for q false positives if and only if, for every union of p or fewer blocks, every other block contains at least $q+1$ points not in this union.

6.5 **Corollary** Any packing $(V, \mathcal{B})$ of t-sets into k-sets having $k \geq p(t-1)+q+1$ is a solution to the strict group testing problem with threshold p and error correction for q false positives.

6.6 **Theorem** [1] A Steiner system $S(t, 2t-1, v)$ is a solution to the strict group testing problem with $p = 2$ and $q = 0$ that has the maximum number of blocks of any solution.

6.2 See Also

§I.4	Existence results on Steiner systems.
[3]	A comprehensive monograph on the combinatorics of group testing.
[2, 4]	Contains related results on nonadaptive group testing.
[5]	Discusses testing with two defectives.
[6]	Applications of group testing in communications.

References

[1] D.J. Balding and D.C. Torney, Optimal pooling designs with error detection, *J. Combin. Theory A*, to appear.

[2] K.A. Bush, W.T. Federer, H. Pesotan and D. Raghavarao, New combinatorial designs and their applications to group testing, *J. Stat. Plann. Infer.* 10 (1984), 335-343.

[3] D.Z. Du and F.K. Hwang, *Combinatorial Group Testing and Its Applications*, World Scientific, 1993.

[4] F.K. Hwang and V.T. Sós, Non-adaptive hypergeometric group testing, *Stud. Sci. Math. Hung.* 22 (1987), 257-263.

[5] F. Vakil and M. Parnes, On the structure of a class of sets useful in nonadaptive group testing, *J. Stat. Plann. Infer.* 39 (1994), 57-69.

[6] J.K. Wolf, Born again group testing: Multiaccess communications, *IEEE Trans. Inform. Theory* IT-31 (1985), 185-191.

7 Scheduling a Tournament

Jeffrey H. Dinitz
Esther R. Lamken
Walter D. Wallis

7.1 Round Robin Tournaments

A *round robin tournament* is a tournament with the property that every team plays every other team precisely once during the tournament.

7.1 **Remarks** The underlying graph for a round robin tournament with n teams is K_n, the complete graph on n vertices. The number of games is $\binom{n}{2}$. If n is even, then it is possible to play the tournament in $n-1$ *rounds* each consisting of $n/2$ games involving all n teams. The underlying design of a round robin tournament played in rounds of $n/2$ games is a 1-factorization of K_n (see §VI.4). In applications to games and sports, often the schedule of play must satisfy a number of additional requirements. Many of these are addressed in this section.

A *tournament design* defined on a $2n$ set V is an arrangement of the $\binom{2n}{2}$ distinct unordered pairs of the elements of V into an $n \times (2n-1)$ array such that every element of V is contained in precisely one cell of each column.

7.2 **Example** A tournament design on eight teams, $\infty, 0, 1, \ldots, 6$ (the pairing $\{a, b\}$ is represented by ab).

0∞	1∞	2∞	3∞	4∞	5∞	6∞
16	20	31	42	53	64	05
25	36	40	51	62	03	14
34	45	56	60	01	12	23

7.3 **Remarks** A tournament design provides a schedule of play for a round robin tournament with $2n$ teams played in $2n-1$ rounds with the property that every team plays precisely once in each round (column).

An *odd tournament design*, OTD(n), defined on a $2n+1$ set V is an arrangement of the $\binom{2n+1}{2}$ distinct unordered pairs of the elements of V into an $n \times (2n+1)$ array such that every element of V is contained in at most one cell of each column.

7.4 **Example** An OTD(3), with teams named $0, 1, \ldots, 6$.

16	20	31	42	53	64	05
25	36	40	51	62	03	14
34	45	56	60	01	12	23

7.5 **Remarks** An OTD(n) is equivalent to a near 1-factorization of K_{2n+1}. An OTD(n) provides a schedule of play for a round robin tournament with the property that every team sits out precisely one round of the tournament.

7.6 **Remark** It is always possible to construct both a tournament design for $2n$ teams and an OTD(n). The easiest way is to use a starter in the cyclic group (§IV.45) as the first column and develop each subsequent column cyclically from the first. Examples 7.2 and 7.4 were constructed from the patterned starter $P = \{\{1,6\},\{2,5\},\{3,4\}\}$ in $\mathbb{Z}_7$.

7.7 **Remark** A *Swiss tournament* is an incomplete simple round robin tournament played in rounds. In the first round, matches are chosen at random or according to some prearranged seeding plan. After each subsequent round, players' scores are calculated (the method may be simply 1 point for each win, 1/2 for each tie, or it may involve the scores in the matches). In the next round, the two highest scoring players form the first match, then the next two, and so on, with the proviso that no repeated pairs are allowed. A random draw is used to decide between equal scores.

At any stage, let G be the underlying graph of the union of the rounds already held. In order for the tournament to continue, the complement of G must contain a 1-factor; in other words, the set of factors chosen so far must not be *premature* (see [19]). In practice a limit of approximately $\log_2 n$ rounds is used for n teams.

7.2 Oriented 1-Factorizations

7.8 **Remark** Oriented 1-factorizations can be used to satisfy several requirements for scheduling round robin tournaments in sports.

7.9 Example Home-Away schedules. In many sports, a team regularly plays at a specified location, its home stadium or field. If a game is played between team i and team j at team j's home location, then it is a *home* game for team j and an *away* game for team i, and the edge ij in the underlying graph is oriented by using the directed edge (i, j). A home-away schedule is often preferred which alternates, as much as possible, home and away games. A *break* in a schedule is a pair of successive rounds in which a given team is at home, or away, in both rounds.

7.10 Example A home-away schedule for 6 teams with 4 breaks (for teams 2,3,4,5) [27].

Day	Matches			Teams 1	2	3	4	5	6
1	(1,6)	(2,5)	(4,3)	A	A	H	A	H	H
2	(6,2)	(3,1)	(5,4)	H	H	A	H	A	A
3	(3,6)	(4,2)	(1,5)	A	H	A	A	H	H
4	(6,4)	(5,3)	(2,1)	H	A	H	H	A	A
5	(5,6)	(1,4)	(3,2)	A	H	A	H	A	H

7.11 Theorem [27]

1. There is a home-away schedule for a tournament design on $2n$ teams with exactly $2n - 2$ breaks.
2. There is a home-away schedule for an OTD(n) without breaks.

7.12 Remarks If two teams share the same home ground, then the two teams cannot both play at home at the same time, except when they play each other. A home-away schedule with the property that the teams can be paired so that only one member of a pair is at home in each round has *Property P*.

7.13 Theorem [25] Given any 1-factorization of K_{2n}, there exists a way of pairing the vertices and a way of orienting the edges so that the resulting home-away schedule has Property P.

7.14 Remark There are a number of other schedule requirements that can be satisfied by using oriented 1-factorizations; see [25].

7.3 Balanced Tournament Designs

7.15 Remarks If the locations (courts, fields) are of unequal attractiveness, then a schedule is required that balances the effect of the locations. A tournament design or an OTD(n) with the additional property that no element of V is contained in more than two cells of any row is a *balanced tournament design*, BTD(n), or an *odd balanced tournament design*, OBTD(n), respectively. These designs provide schedules of play for the tournament with the additional property that no one plays more than twice at any location. Balanced tournament designs can be constructed with additional structure. Existence results and examples of several different types of balanced tournament designs can be found in §IV.3.

7.4 Softball Balanced Tournament Designs

7.16 Remark In many applications, the tournament schedules have several different requirements which must be satisfied simultaneously. Scheduling softball league tournaments motivated the following definition; see [7].

Let $B_1, B_2, \ldots, B_m$ denote a collection of bias categories for a tournament design for $2n$ teams. Each B_k is an n_k-dimensional vector $(b_{k1}, b_{k2}, \ldots, b_{kn_k})$, $n_k \geq 1$. The b_{kj} are the *attributes*. Each team $i \in V$ has a frequency vector with respect to each B_k, $(\lambda^i_{k1}, \lambda^i_{k2}, \ldots, \lambda^i_{kn_k})$ where λ^i_{kj} denotes the number of times in the tournament design that team i has attribute b_{kj}. Team i has *balance with respect to bias category* B_k if $\max_{j,\ell} |\lambda^i_{kj} - \lambda^i_{k\ell}| \leq 1$. A tournament design is *balanced with respect to bias category* B_k if (1) each team $i \in V$ has balance with respect to B_k, and (2) for each j $\max_{i,p} |\lambda^i_{kj} - \lambda^p_{kj}| \leq 1$.

A *softball balanced tournament design for* $2n$ *teams*, SBTD(n), is a tournament design that is balanced with respect to each of its bias categories.

7.17 Example A SBTD(6) [7]. This design is balanced with respect to B_1=(M,T,W,Th) (day of the week bias), B_2=(6:20PM,7:25PM,8:30PM) (time of day bias), B_3=(H,A) (home team bias), B_4=(number of games any given team plays in one week), and B_5=(number of games any given team plays in one day).

		M	T	W	TH
Week 1	6:20 PM	(4,11)	(10,6)	(8,10)	(2,3)
	7:25 PM	(3,12)	(9,7)	(11,5)	(7,1)
	8:30 PM	(1,2)	(5,8)	(12,4)	(6,9)
Week 2	6:20 PM	(8,2)	(12,5)	(5,4)	(7,11)
	7:25 PM	(9,1)	(4,3)	(3,9)	6,12)
	8:30 PM	(11,6)	(10,7)	(2,10)	(1,8)
Week 3	6:20 PM	(7,12)	(3,1)	(11,9)	(1,5)
	7:25 PM	(5,6)	(8,11)	(12,8)	(2,4)
	8:30 PM	(4,10)	(9,2)	(6,7)	(10,3)
Week 4	6:20 PM	(5,3)	(7,8)	(3,7)	(12,10)
	7:25 PM	(10,11)	(6,2)	(4,6)	(8,9)
	8:30 PM	(9,12)	(1.4)	(11,1)	(2,5)
Week 5	6:20 PM	(10,9)	(11,2)	(2,12)	(8,6)
	7:25 PM	(4,8)	(12,1)	(1,10)	(3,11)
	8:30 PM	(5,7)	(6,3)	(9,5)	(7,4)
Week 6	6:20 PM	(6,1)	(9,4)		
	7:25 PM	(7,2)	(5,10)		
	8:30 PM	(8,3)	(12,11)		

7.18 Remarks Some infinite classes of SBTDs are described in [7]. Finizio also considers cases where interdivisional play is separated from divisional play and league schedules where the teams may play against each other more than once. Not all of the designs in [7] have the property that the games can be partitioned into rounds.

7.5 Tournament Arrays

7.19 Remarks If there are fewer than n locations for a tournament for $2n$ or $2n+1$ teams, then not every team can play in each round of the tournament. However, tournaments can still be scheduled for round robin tournaments with certain balance categories.

A *tournament array* TA(m, c) is a $c \times t$ array of the $\binom{m}{2}$ distinct unordered pairs of the elements of an m set V such that every element of V occurs at most once in each column of the array.

7.20 Remark Necessary conditions for the existence of a $TA(m,c)$ are $ct = \binom{m}{2}$ and $1 \le c \le \lfloor \frac{m}{2} \rfloor$.

A *court balanced tournament array*, CBTA(m,c), is a tournament array with the additional property that each element occurs exactly α times in each row. This requires that $c\alpha = m-1$.

7.21 Example A CBTA$(10,3)$ [14].

10	15	26	27	38	39	23	50	64	75	86	79	80	14	49
36	37	18	19	24	20	58	69	13	84	95	60	47	25	70
28	29	34	35	16	17	67	78	89	90	40	45	12	30	56

7.22 Theorem [14] The necessary conditions for the existence of a CBTA(m,c) are also sufficient.

7.23 Remarks Tournament arrays can also be balanced with respect to the number of rounds that teams rest between matches. These designs are *interval-balanced*; see [18].

7.24 Example An interval-balanced TA$(9,2)$ [18]. The number of rounds between two matches for team i is either 1 or 2.

12	56	91	45	89	52	96	35	49	15	92	61	39	46	95	62	79	36
34	78	23	67	13	74	81	27	68	73	84	57	28	17	83	41	58	24

7.25 Remark A particularly interesting instance of interval balance is when there is only one available court. In this case, a schedule of play is required that maximizes the minimum number of pairs occurring between any two appearances of any one team. For m teams, this minimum is $\lfloor \frac{m-3}{2} \rfloor$. When this is achieved in a schedule of m teams, the tournament (design) is a *pair design*, PD(m).

7.26 Theorem [21] A PD(m) exists for all positive integers $m \ge 2$.

7.27 Construction A construction of pair designs for all m. First assume that m is odd. The sequence for the first m pairs is

$$Q_m = (\{1,2\},\{3,4\},\ldots,\{m-2,m-1\},\{m,1\},\{2,3\},\ldots,\{m-1,m\})$$

Let P be the permutation $(1)(3\ 5\ 7\ \cdots\ m-4\ m-2\ m\ m-1\ m-3\cdots 6\ 4\ 2)$ and $L = \frac{m-3}{2}$. The pair design PD(m) is $Q_m, PQ_m, P^2Q_m, \ldots, P^LQ_m$. When m is even, let $Q_m = (\{1,2\},\{3,4\},\ldots,\{m-1,m\})$, $L = n-2$, and $P = (1)(3\ 5\ 7\ \cdots\ m-3\ m-1\ m\ m-2\ m-4\cdots 6\ 4\ 2)$; again, the pair design is $Q_m, PQ_m, P^2Q_m, \ldots, P^LQ_m$.

7.28 Remark The PD(m) from Construction 7.27 can be used to construct an interval-balanced TA(m,c) by placing the pairs of the pair designs down a column of the TA until filled, then moving to the next column. The interval balanced TA$(9,2)$ of Example 7.24 was constructed from a PD(9) by this method (see [18]).

7.6 Room Squares and Howell Designs

7.29 Example Room squares. If there are $n+1$ teams and n locations, a round robin tournament can be scheduled with the properties that every team plays in exactly one game in each round, and every team plays exactly once at each location. This tournament corresponds to a Room square of side n. See §IV.39.

7.30 Remarks Howell designs, $H(s,2n)$, can be used to schedule tournaments with the properties that every team plays in each round and every team plays at each of s

locations precisely once. However, when $n \le s \le 2n-2$, the schedule no longer has the property that every team plays every other team once during the tournament. See §IV.26.

7.31 Example Spouse avoiding tournament. Suppose there are $2n$ players and n couples, and suppose every person is to play every person except their spouse precisely once in the tournament. An $H(2n-2,2n)$ can be used to construct a schedule of play where every person plays in each round; every person plays at each of $2n-2$ locations precisely once; and every person plays each other person except their spouse precisely once. The underlying graph is the complete graph minus a 1-factor, the *cocktail party graph.*

7.32 Example An $H(6,8)$. The columns correspond to the rounds, the rows to the locations and the 4 couples are $\{1,4\},\{2,3\},\{5,6\},\{7,8\}$.

67	12	58		34	
24	57	13		68	
38		27	45		16
15		46	37		28
	48		26	17	35
	36		18	25	47

7.33 Example Orthogonal latin squares (§II.2). Suppose there are n people on each of two teams of a tournament. A pair of orthogonal latin squares of side n (equivalently, an $H^*(n,2n)$) can be used to schedule a tournament with the properties that every person from team 1 plays each person on team 2; there are no games between two people from the same team; every person plays in each round; and every person plays at each of n locations precisely once during the tournament. To construct this schedule, let L_1 and L_2 be two orthogonal latin squares of side n with L_i on the symbols T_i (representing the players on team i) for $i=1,2$. The schedule is constructed by superimposing the two latin squares. The pairs in the cells represent the games, the rows are the rounds and the columns are the locations.

7.34 Remark Many of the earliest examples of Howell designs (dating back more than 100 years) were constructed in order to schedule bridge tournaments. See [22] and §7.10.

7.7 Golf Designs

7.35 Remarks There are often occasions when the teams or players may compete more than once against each other.

Let V be a set of $2n+1$ elements. A *golf design of order* $2n+1$ is a set of $2n-1$ symmetric idempotent latin squares, $L_1, L_2, \ldots, L_{2n-1}$, defined on V such that L_i and L_j are disjoint off the main diagonal; that is, for $x,y \in V$, $L_i(x,y) \ne L_j(x,y)$ for $i,j = 1,2,\ldots,2n-1$.

7.36 Remarks A golf design provides a schedule of play with the following properties: $2n+1$ teams play a round robin tournament in $2n+1$ rounds; in each round one team does not play, and the others meet in $2n-1$ matches all played at the home ground of the non-playing team. In addition, $2n-1$ round robin tournaments are to be played so that each pair of teams meets exactly once at each of the $2n-1$ neutral grounds. (Such tournaments are scheduled in golf.)

7.37 Theorem [25, 3] Let $v = 2n + 1$. There exists a golf design of order v for n a positive integer except for $n = 2$ and possibly for $n = 20$.

7.38 Example A golf design of order 7 [24].

L_1

0	2	4	5	1	6	3
2	1	6	4	5	3	0
4	6	2	0	3	1	5
5	4	0	3	6	2	1
1	5	3	6	4	0	2
6	3	1	2	0	5	4
3	0	5	1	2	4	6

L_2

0	3	6	1	5	4	2
3	1	5	0	6	2	4
6	5	2	4	1	0	3
1	0	4	3	2	6	5
5	6	1	2	4	3	0
4	2	0	6	3	5	1
2	4	3	5	0	1	6

L_3

0	4	5	6	2	3	1
4	1	3	2	0	6	5
5	3	2	1	6	4	0
6	2	1	3	5	0	4
2	0	6	5	4	1	3
3	6	4	0	1	5	2
1	5	0	4	3	2	6

L_4

0	5	3	2	6	1	4
5	1	4	6	3	0	2
3	4	2	5	0	6	1
2	6	5	3	1	4	0
6	3	0	1	4	2	5
1	0	6	4	2	5	3
4	2	1	0	5	3	6

L_5

0	6	1	4	3	2	5
6	1	0	5	2	4	3
1	0	2	6	5	3	4
4	5	6	3	0	1	2
3	2	5	0	4	6	1
2	4	3	1	6	5	0
5	3	4	2	1	0	6

7.39 Example The golf design of Example 7.38 is used to construct five tournaments. The unordered cells which contain i in L_j provide the matches hosted by i in T_j.

Host	T_1	T_2	T_3	T_4	T_5
0	23 16 45	13 25 46	14 26 35	15 24 36	12 34 56
1	04 25 36	03 24 56	06 23 45	05 26 34	02 35 46
2	01 35 46	06 15 34	04 13 56	03 16 45	05 14 36
3	24 15 06	01 26 45	05 12 46	02 14 56	04 16 25
4	02 13 56	05 16 23	01 25 36	06 12 35	03 15 26
5	03 14 26	04 12 36	02 16 34	01 23 46	06 13 24
6	12 05 34	02 14 35	03 15 24	04 13 25	01 23 45

7.8 Elimination Tournaments

An *elimination tournament* on n teams is a tournament played in rounds in which only the winner of a match at each round can proceed to a match at the next round. In other words, the loser is eliminated at each round.

7.40 Proposition An elimination tournament on n teams can always be constructed in $\lceil \log_2 n \rceil$ rounds and this is best possible.

7.41 Construction An elimination tournament on n teams can be constructed in $r = \lceil \log_2 n \rceil$ rounds by first adding s phantom teams where $s = 2^r - n$. Next construct a complete binary rooted tree with 2^r leaves and associate each leaf with a team so that no two phantom teams are paired. The teams paired with phantom teams automatically win their first match and essentially have a *bye* in the first round. The winner of a match in round k moves on to the next round and plays the winner of the match in round k that has minimum distance on the tree.

7.42 Remark Elimination tournaments with exactly 2^k teams for some k have the nicest schedules. Major tournaments such as the NCAA basketball tournament (64 teams) and the U.S. Open tennis tournament (128 men) use this format. In an elimination tournament with n teams there are always $n-1$ games, no matter how the tournament is scheduled. (Each game has a loser and the tournament has $n-1$ losers.)

7.43 Remark A common variant of the (single) elimination tournament is the *double elimination* tournament. In this tournament the first time a team loses it is placed in the losers bracket. The team continues to play until its second loss, at which time it is out of the tournament. The winners of the two brackets meet in the final championship with the team from the losers bracket required to win two games in order to be champion.

7.9 Whist Tournaments

7.44 Remarks A whist tournament, Wh($4n$), for $4n$ players is a schedule of games each consisting of two players opposing two others such that the games are arranged in $4n-1$ rounds with the following properties: each round consists of n games; each person plays in exactly one game in each round; each person partners every other person exactly once; and each person opposes every other person exactly twice. Whist tournaments can also be played with $4n+1$ players; in this case there are $4n+1$ rounds and each person sits out during exactly one round of play. Additional structure is required for directed whist tournaments and triplewhist tournaments. These designs date back to the 1800s. Whist tournaments are now used in bridge for *individual* competitions with the property that each player has a variety of partners instead of fixed partnerships. See §IV.53 for a more extensive discussion of whist tournaments including many constructions.

7.10 Bridge Tournaments

7.45 Remark In bridge tournaments, the most common form of competition is where the competitors are fixed partnerships or *pairs*. The competition is a simple round robin consisting of spanning rounds. Each deal of the cards is called a *board* (because the deals were originally preserved using wooden containers) and is played in several rounds. The two pairs who play against each other in a match are allocated a (numbered) table, and the two pairs are designated by the directions north-south and east-west.

7.46 Example A bridge tournament for eight pairs on four tables $\{1, 2, 3, 4\}$ playing boards $\{a, b, c, d, e, f, g\}$ in seven rounds. For example, in Round 1, pair 8 (NS) plays board a against pair 7 (EW) at Table 1. (In practice $a, b, \ldots$ would each represent a set of three or four boards.)

Round	1	2	3	4	5	6	7
Table	1234	1234	1234	1234	1234	1234	1234
NS	8241	8352	8463	8574	8615	8726	8137
EW	7653	1764	2175	3216	4327	5431	6542
Boards	$abce$	$bcdf$	$cdeg$	$defa$	$efgb$	$fgac$	$gabd$

7.47 Remarks There are two combinatorial aspects to these competitions. The *movement* is the arrangement which assigns boards to each table, assigns directions to each pair, and directs pairs to tables in each round. (For practical purposes, the number of boards in each set must be equal.) Example 7.46 is a movement for eight pairs.

The second aspect is the *scoring*. The scores of all east-west pairs on the same set of boards are compared, as are the scores of all north-souths. In Example 7.46, the boards in set a lead to comparisons between the sets $\{8,4,2,1\}$ and $\{7,6,3,5\}$. The whole tournament gives rise to 14 sets:

$$\{8421\},\{2853\},\{4386\},\{5487\},\{1658\},\{2768\},\{3178\},$$
$$\{7635\},\{6174\},\{5721\},\{6132\},\{3724\},\{4135\},\{5246\}.$$

These sets constitute the *underlying block design* of the tournament. The tournament is *balanced* if the underlying design is (pairwise) balanced. For example, the above tournament is balanced because its underlying design is an (8,4,3) BIBD.

7.48 Example The most common bridge tournament is the *Mitchell movement* for $2n$ pairs (n odd). The pairs are split into two groups of size n, NS and EW. Pairs in NS always play north-south. The tournament consists of n rounds, played at n tables, using n sets of boards. Tables, board-sets, rounds and both sets of players are labeled with the integers modulo n. In round i, pair j NS and pair $j+i$ EW meet at table j and play board-set $j-i$. The comparisons are balanced within sets NS and EW, but members of different sets are never compared, so this tournament produces two winners. Another "balance" feature is that all pairs meet the same set of opponents; this produces a fairer tournament.

7.49 Remark The Mitchell movement is equivalent to a pair of orthogonal latin squares: a circulant and a back-circulant square of side n. If n is even, these squares are not an orthogonal pair, and the movement does not work, but minor adjustments can be made which result in playable tournaments which are close to balanced (see [9]).

7.50 Example Another common movement used in bridge tournaments is the *Howell movement.* This movement on $2n$ pairs satisfies the following properties:

1. every board is played by at most one pair of teams (one in each direction) in each round;
2. every team plays every other team exactly once during the tournament;
3. every team plays exactly one set of boards in each round;
4. every team plays each board exactly once;
5. every team competes equally often against any other team.

The movement in Example 7.46 is an eight-pair, seven-table Howell movement.

Let R be a Room square (§IV.39) of order $2n$ (side $2n-1$) with each entry considered as an ordered pair. From each row of R construct a pair of complementary blocks — one block consisting of the elements that occur as the first coordinate of an ordered pair in that row, and one block consisting of the elements that occur as the second coordinate. If the resulting set of $2(2n-1)$ blocks is the set of blocks of a $(2n,n,n-1)$ BIBD, then the (ordered) Room square is a *balanced* Room square and is denoted BRS$(2n)$.

7.51 Construction Let R be a balanced Room square of order $2n$ (side $2n-1$). The rows, columns, and symbols of R correspond respectively to the boards, rounds, and teams of a Howell movement on $2n$ teams.

7.52 Theorems (see [5, 15])

1. If a BRS$(2n)$ exists, then $2n$ is a multiple of 8.
2. A BRS$(q+1)$ exists for all prime powers $q \equiv 3 \pmod 4$.
3. A BRS$(2q+2)$ exists for all prime powers $q \equiv 3 \pmod 4$.
4. There exists a BRS(2^n) if n is odd, or if n is even and $4 \le n \le 18$.

5. BRS(n) exist for all $n \equiv 0 \pmod 4$ such that $8 \le n \le 100$, with the two possible exceptions $n = 36$ and $n = 92$.

7.53 Remark If the number of boards in each set is at least k, then it is possible for k tables to play the same set of boards in the same round ("sharing boards"), allowing a relaxation of condition 1 in Example 7.50.

7.54 Example Since there does not exist a Room square of side 5, no Howell movement for six pairs is possible. In this case, the following generalized Howell design can be used. There are five sets of boards; set e is shared between all three tables in round 5.

Round	1	2	3	4	5
Table	123	123	123	123	123
NS	623	634	645	651	612
EW	154	215	321	432	543
Boards	*adb*	*bdc*	*cab*	*dac*	*eee*

7.55 Remark Construction 7.51 can be generalized by using a Howell design (§IV.26). Such a movement is a *three-quarter Howell movement.* Because of the imbalance between the sets of opponents, these designs are used primarily in club games when the number of pairs is small, and seldom in serious competition.

7.56 Remark For extensive discussions of bridge tournaments including explicit master sheets for Mitchell and Howell movements, see [9, 13].

7.11 Spouse Avoiding Doubles Tournaments

A *spouse avoiding mixed doubles round robin tournament for n couples*, a SAMDRR(n), is a tournament with the following properties: (1) a match consists of two players of different genders playing against two players (of different genders), (2) every two players of the same gender play against each other exactly once, (3) every player plays with each player of the opposite gender (except their spouse) exactly once as a partner and exactly once as an opponent.

7.57 Remark Self orthogonal latin squares (SOLS) can be used to construct a SAMDRR.

7.58 Example The array A is a SOLS(4), and the array B is a symmetric orthogonal mate for A. A and B are used to construct the matches listed below.

$$A = \begin{array}{|cccc|} \hline 1 & 4 & 2 & 3 \\ 3 & 2 & 4 & 1 \\ 4 & 1 & 3 & 2 \\ 2 & 3 & 1 & 4 \\ \hline \end{array} \qquad B = \begin{array}{|cccc|} \hline 4 & 1 & 2 & 3 \\ 1 & 4 & 3 & 2 \\ 2 & 3 & 4 & 1 \\ 3 & 2 & 1 & 4 \\ \hline \end{array}$$

Cell	Match	Cell	Match
(1,2)	H_1W_4 vs H_2W_3	(3,4)	H_3W_2 vs H_4W_1
(1,3)	H_1W_2 vs H_3W_4	(2,4)	H_2W_1 vs H_4W_3
(1,4)	H_1W_3 vs H_4W_2	(2,3)	H_2W_4 vs H_3W_1

7.59 Remarks This schedule has the additional property that it can be played in three rounds where each person plays in one game in each round. This can be guaranteed by using a SOLS with a symmetric orthogonal mate, a SOLSSOM. The symmetric orthogonal mate B is used to construct the rounds. The unordered cells which contain the 1s, $\{1,2\}$ and $\{3,4\}$, provide the matches in round 1; the cells which contain the 2s provide the matches in round 2 and the cells which contain the 3s provide the matches in round 3. See §IV.40 and §IV.41.

7.12 Resolvable Coverings of Pairs

7.60 **Example** A schedule of play can be constructed for 24 people with the following properties: the game is played at tables of size 4; every person plays at a table with every other person at least once; and each person plays in every round (a round is six tables of size four). The minimum number of rounds required for this tournament is 8. A solution is constructed using a resolvable minimum cover with block size 4 [12] and is displayed as Example IV.8.28.

7.61 **Remark** Example 7.60 is a special case of a more general tournament structure with the following properties: the game is played at tables of size k; there are nk players; every person plays at a table with every other person at least (most) once; and each person plays in every round (a round is n tables of size k). In addition, the tournament must be completed in the smallest (largest) number of rounds possible. The existence of such a tournament is equivalent to a resolvable minimum cover (maximum packing) of pairs by k-sets. When $nk \equiv 1 \pmod{k(k-1)}$ both the minimum cover and maximum packing are equivalent to a resolvable $(nk, k, 1)$ BIBD. For results on resolvable BIBDs see §I.6, for results on resolvable packings see §IV.33.5, and for results on resolvable coverings see [12] and §IV.8.6.

7.62 **Example** [6] A tournament for 12 competitors can be constructed with the following properties: the game is played in groups of size 4; there are eight rounds (a round is three groups of size 4); every competitor plays in a group with every other competitor either two or three times; and each competitor plays in every round. There are exactly two nonisomorphic solutions [20]. One is:

0123	017T	0359	4791	8E15	0678	47E0	8234
4567	245E	148T	5802	9046	134E	5783	9E07
89TE	3689	267E	6TE3	T237	259T	6912	T156

7.63 **Remark** Example 7.62 is a tournament where the condition on the number of times pairs of competitors play has been relaxed. To generalize the example, consider a game played in groups of size k with nk players and r rounds, each round being a partition of the players into n disjoint groups and every person plays at a table with every other person either λ or $\lambda+1$ times ("approximate balance"). This design is a *resolvable regular graph design*, RRGD$(n, k : r)$, with $\lambda = \lfloor \frac{r(k-1)}{n-1} \rfloor$. There is an RRGD$(12, 4 : r)$ unless $2 \le r \le 7$ (see [11] and §IV.22).

7.13 Tournaments for Trios

7.64 **Example** Suppose $3n$ golfers wish to play for n weeks. There are n starting times available and the golfers play in groups of three, so that (1) each golfer plays exactly once per week, (2) no golfer has the same starting time twice, and (3) no two golfers play in the same trio twice.

If condition 2 is ignored, then a resolvable $(3n, 3, 1)$ BIBD could be used and the tournament could extend for $\frac{3n-1}{2}$ weeks with every pair of golfers in a trio exactly once. However, a solution to the problem as given is even easier to construct using

MOLS. Let L_1, L_2, and L_3 be three MOLS of side n where L_i is on the symbol set P_i for $i = 1, 2, 3$ (the $3n$ players). Now superimpose the three squares. The triples in each cell are the trios, the columns correspond to the weeks, and the rows correspond to the n starting times.

If $n = 2, 3, 6$, or perhaps 10 this construction cannot be used. There is no solution for 2 or 3, and 10 is undecided. There are precisely six solutions for $n = 6$, up to isomorphism [16]; one is

	Week 1	Week 2	Week 3	Week 4	Week 5	Week 6
Time 1	0 1 2	3 4 5	6 7 8	9 10 11	12 13 14	15 16 17
Time 2	9 12 15	0 6 13	1 3 10	4 14 16	7 11 17	2 5 8
Time 3	3 7 14	8 10 12	2 11 16	1 13 17	5 6 15	0 4 9
Time 4	4 8 17	1 11 15	5 9 14	2 6 12	0 3 16	7 10 13
Time 5	5 11 13	7 9 16	0 12 17	3 8 15	2 4 10	1 6 14
Time 6	6 10 16	2 14 17	4 13 15	0 5 7	1 8 9	3 11 12

7.65 Remark Many times it is not the most important aspect of the tournament that every player plays against every other player. For example, if a tournament is played during the weeks of the summer it may be more important that the number of rounds be fewer than 12. An example is Joyce Cook's golf league; thirty-nine golfers wish to play for ten weeks, under the conditions in 7.64. A solution exists using the first ten columns of three MOLS of side 13.

7.14 Miscellaneous Examples

7.66 Example World Speedway Championships (motorcycle). The final is 20 heats for 16 riders (4 in a race) and the schedule is precisely the affine plane AG(2, 4), divided into parallel classes. So there are five rounds where each round consisting of 4 heats is a parallel class in the affine plane. See [8, 2].

7.67 Example World Cup Soccer. Contested over every four years; 24 teams qualify for the tournament after two years of preliminary matches. The teams are then divided into six groups (of four teams each) in the first round and play round robin (three games). The top 16 teams then advance to a single elimination tournament.

7.68 Example National Football League. In the NFL there are two conferences (the NFC and AFC) each composed of three divisions (West, Central, and East) containing five teams each. Each team plays a 16 game season consisting of the following: a home-and-away round robin within the division (eight games); in the interconference games, each team plays four games against teams in the other conference (in 1996, the NFC East plays the AFC East, etc.); within the conference, each team plays four other teams. The teams that are played within the same conference are determined by the team's last years finish; for example, the first-place team in a division plays the other two first-place finishers plus a second- and a third-place team while the last-place team plays the other fifth-place teams plus a fourth- and third-place team. This is *parity* scheduling. Also, each team has one bye week per season. The actual week-by-week schedule is constructed on a big board at NFL headquarters with input from the league, the networks, and the sponsors (but no mathematicians).

7.69 Example Jai-alai. In the game of jai-alai, several players compete in each match but only two can compete at any one time. At any time there is an ordered list of competitors. The first two in the list play; the winner stays at the top of the list and the loser goes to the bottom. Thus the tournament structure is related to a pop-up stack. The winner scores one or more points (there are different methods for different types of events, and in different countries), and the first to amass a certain number of points wins. See [4] for a discussion of the advantages of different starting positions.

7.15 See Also

§IV.3	Balanced tournament designs are tournament designs with an additional balance on the courts.
§IV.39	Room squares provide a possible arrangement for a round robin tournament.
§IV.26	Howell designs give incomplete tournaments played in rounds.
§IV.45	Starters are useful in constructing (cyclic) round robin tournaments
[9, 13]	Contain many schedules for bridge tournaments.
[22]	Contains a number of interesting schedules for duplicate bridge.
[17]	A reference for ranking the teams in a tournament after the results of the games are known.
[23]	Studies schedules of league play where there is more than one conference and the number of meetings between a pair of teams in the same conference may be different than the number of meetings between teams from different conferences.
[10]	The International Yacht Racing Union provides schedules for round robin tournaments for three to ten boats. These have additional properties dealing with port and starboard starts and some interval balance.
[1, 26]	Discusses tournaments where there is an advantage to certain starting positions in each heat of a race. The schedules constructed in these papers balance for the starting positions.

References

[1] E. Barbut and A. Bialostocki, On regular r-tournaments, *Ars Combin.* 34 (1992), 97–106.

[2] F. Budden, Speedway tournaments in the classroom, *Math. Gaz.* 61 (1977), 266–272.

[3] C.J. Colbourn and G. Nonay, A golf design of order 11, *J. Stat. Plann. Infer.*, to appear.

[4] C. S. Dibley and W. D. Wallis, The effect of starting position in jai-alai, *Congressus Num.* **32** (1981), 253–259.

[5] J.H. Dinitz and D.R. Stinson, Room squares and related designs, in: *Contemporary Design Theory: A Collection of Surveys* (J.H. Dinitz, D.R. Stinson; eds.) John Wiley & Sons, New York, 1992, 173-176.

[6] B. Elenbogen and B. Maxim, Scheduling a bridge club (a case study in discrete optimization), *Math. Mag.* 65 (1992), 18-26.

[7] N.J. Finizio, Tournament designs balanced with respect to several bias categories, *Bull. ICA* 9 (1993), 69-95.

[8] T.J. Fletcher, Speedway tournaments, *Math. Gaz.* 60 (1976), 256-264.

[9] A. Groner, *Duplicate Bridge Direction*, Baron Barclay Bridge Supplies, Louisville, 1993.

[10] IYRU Umpires and Match Racing Manual, Edition 3/93.

[11] D.L. Kreher, G.F. Royle and W.D. Wallis, A family of resolvable regular graph designs, *Discrete Math.*, to appear.

[12] E.R. Lamken, W.H. Mills and R. Rees, Resolvable minimum covers with quadruples, in preparation.
[13] I. McKinnon, *Bridge Directing Complete*, McKinnon, Sydney, 1979.
[14] E. Mendelsohn and P. Rodney, The existence of court balanced tournament designs, *Discrete Math.* 133 (1994), 207-216.
[15] E.T. Parker and A.N. Mood, Some balanced Howell rotations for duplicate bridge movements, *Am. Math. Monthly* 62 (1955), 714-716.
[16] N.C.K. Phillips and W.D. Wallis, All solutions to a tournament problem, *Congr. Numer.*, to appear.
[17] K.B. Reid and L.W. Beineke, Tournaments, in: *Selected Topics in Graph Theory* (L.W. Beineke, R.J. Wilson; eds.) Academic Press, New York, 1978, 169-204.
[18] P. Rodney, The existence of interval-balanced tournament designs, preprint.
[19] A. Rosa and W.D. Wallis, Premature sets of one-factors, or, how not to schedule a round-robin tournament, *Discrete Applied Math.* 4 (1982), 291-297.
[20] G.F. Royle and W.D. Wallis, Constructing bridge club designs, *Bull. ICA* 11 (1994), 122-125.
[21] G.J. Simmons and J.A. Davis, Pair designs, *Commun. Stat.* 4 (1975), 255-272.
[22] P.H. Smith, Resolvable and Doubly Resolvable Designs with Special Reference to the Scheduling of Duplicate Bridge Tournaments, Ph.D. Thesis, University of Montana, 1977.
[23] T.H. Straley, Scheduling designs for a league tournament, *Ars Combin.* 15 (1983), 193-200.
[24] W.D. Wallis, The problem of the hospitable golfers, *Ars Combin.* 15 (1983), 149-152.
[25] W.D. Wallis, One-factorizations of complete graphs, in: *Contemporary Design Theory: A Collection of Surveys* (J.H. Dinitz, D.R. Stinson; eds.) John Wiley & Sons, New York, 1992, 593–631.
[26] W.D. Wallis, Tournaments for triads, in preparation.
[27] D. de Werra, Scheduling in sports, in: *Studies on Graphs and Discrete Programming* (P. Hansen; ed.) North-Holland, Amsterdam, 1981, 381-395.

8 Winning the Lottery

Charles J. Colbourn

8.1 Keno and Lotto

8.1 **Remarks** Lotteries are a common form of revenue generation for governments, charities, sports clubs, and the like. They serve to redistribute wealth randomly among the wagering populace. In a lottery, wagers are placed on the outcome of a random, or apparently random, event. Payoffs are based upon how "close" the outcome predicted by the bettor comes to the actual outcome. The organization running the lottery, the *house*, announces the structure of payoffs for winning bets, specifying payment in a fixed amount for smaller winnings and payment on a pari-mutuel basis for the larger prizes (i.e., a share of the collected money minus the house's take and the amount paid in fixed amount prizes).

In the *numbers game*, the objective is to guess a k-digit integer chosen by an apparently random method (such as the last k digits of the Canadian national debt at the end of the business day); winnings are paid if one gets the first ℓ or more of the k digits correct.

8.2 **Remarks** Most state and government lotteries operate on the same principle. The house specifies a set V of n integers and announces that a subset S of p integers in V is to be chosen. The bettor chooses a set T of k integers from V, and pays an amount c to the house to register T as a wager. Once all bets are placed, the house arranges for the set S to be chosen, typically by some random process such as airborne numbered ping-pong balls landing in clear plastic chutes.

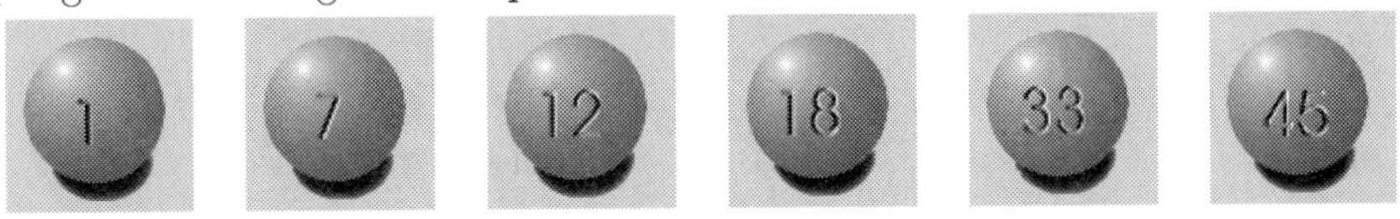

The bettor is rewarded based upon the size of the intersection $S \cap T$. Typically, the larger the size of the intersection, the larger is the reward. When $|S \cap T|$ is small, no money is paid on that bet. For intermediate values of $|S \cap T|$, payment in a predetermined fixed amount is made. The money remaining after small awards are paid and the house takes its percentage is divided among all bets with large values of $|S \cap T|$. In government lotteries, it is reasonable to assume that there is no lollapalooza (change in the rules after play has begun) or crooked play.

8.3 **Remarks** *Keno* is a common casino version of this kind of lottery, in which the house chooses the integers n and p (often $n = 80$ and $p = 20$); the house then allows the bettor to select from a set of possible values for k, usually with $k < p$. For each possible selection of k, a schedule of winnings is posted in advance.

Lotto is the version used in most government run lotteries. With Lotto, usually the values of n, p, and k are all fixed; in addition, $p \leq k$, and typically $p = k$.

8.2 Actual Lotteries

8.4 **Table** A sample of lotteries active in 1995. Format p/n means that p numbers are chosen from n; in each case, $k = p$ numbers are also selected by the bettor on each ticket. Some of these lotteries also have 'quick-pick' and 'bonus number' options.

State	Name	Format	State	Name	Format
Arizona	Fantasy 5	5/35	Massachusetts	Cash 5	5/35
	Lotto	6/42		Megabucks	6/42
Australia	Lotto	6/45		Millions	6/49
Austria	Lotto	6/45	Michigan	Cash 5	5/39
Belgium	Lotto	6/42		Lotto	6/49
California	Fantasy 5	5/39	Minnesota	Gopher 5	5/39
	Super Lotto	6/51	Missouri	SuperCash 5	5/42
Canada	Lotto	6/49		Show Me 5	5/30
	BC Lotto	6/49		Lotto	6/48
Colorado	Lotto	6/42	Montana	Cash 5	5/39
Connecticut	Cash 5	5/35	Netherlands	Lotto	6/41
	Lotto	6/44	New Hampshire	Lotto	6/36
DC	Quick Cash	6/39	New Jersey	Cash 5	5/38
Delaware	Lotto	6/36		Pick 6	6/46
England	Lotto	6/49	New York	Take 5	5/39
Finland	Lotto	7/39		Lotto 54	6/54

State	Name	Format	State	Name	Format
Florida	Fantasy 5	5/26	Nigeria	Lotto	5/90
	Lotto	6/49	Ohio	Kicker	6/10
France	Loto	6/49		Buckeye 5	5/37
Georgia	Fantasy 5	5/35		Lotto	6/47
	Lotto	6/46	Oneida Nation	Big Green	6/36
Germany	Lotto	6/49	Oregon	Megabucks	6/44
Ghana	Lotto	5/90	Pennsylvania	Cash 5	5/39
Hong Kong	Mark Six	6/45		Wild Card	6/48
Hungary	Lotto	5/90	PowerBall		5/45
Idaho	Fantastic 5	5/32	Puerto Rico	Lotto	6/38
Illinois	Little Lotto	5/30	Rhode Island	Lot-O-Bucks	5/40
	Lotto	6/54		Rhody Cash	5/30
Indiana	Lucky 5	5/36	Singapore	Toto	6/42
	Hoosier Lotto	6/48	Sweden	Lotto	7/35
Iowa	Cash Game	5/35	Texas	Lotto	6/50
	SuperCash	6/42	Tri-State	Cash 5	5/35
Ireland	Lotto	6/42		Megabucks	6/40
Israel	Lotto	6/45	Tri-West	Lotto	6/33
Kansas	Cash Lotto	6/33	Venezuela	Lotto	6/42
Kentucky	Cash 5	5/35	Vermont	Lotto	6/30
	Lotto	6/49	Virginia	Cash 5	5/34
Louisiana	Lotto	6/44		Lotto	6/44
Maryland	Match 5	5/39	Washington	Lotto	6/49
	Lotto	6/49	West Virginia	Cash 25	6/25

8.3 The Odds and Lottery Schemes

8.5 **Remarks** Suppose that n, p, and k are fixed. There are $\binom{n}{p}$ possible tickets. To calculate the probability that exactly i of your k choices occur among the chosen p numbers from the n-set, compute the number $\binom{k}{i}\binom{n-k}{p-i}$ of tickets containing exactly i numbers on your ticket. The ratio r_i of $\binom{k}{i}\binom{n-k}{p-i}$ to $\binom{n}{p}$ is the probability of an i-way match. The *expected return* is the value $\sum_{i=0}^{p} r_i w_i$, where w_i is the payoff for an i-way match. Normally, this cannot be calculated with certainty, because p-way matches typically do not have a fixed payment amount. Nevertheless, in lotteries currently active, the expected return over all bettors ranges from 45 to 60% of the amount wagered (in other words, the house keeps 40 to 55% of the total wagered).

Unless one believes in luck, psychic phenomena, or divine intervention, this probability depends in no way upon which numbers are chosen. However, the expected value of the winnings depends upon what numbers other bettors choose; if a pari-mutuel prize is shared, winnings are accordingly lower than if it is won outright. Betting numbers less frequently chosen by others increases the expected return, but leaves unchanged the probability of winning. Orkin [9] reports a study of Lotto 6/49 in which the least frequently chosen ticket is {20,30,39,40,41,48}, and the most frequently chosen is {3,7,9,11,25,27}. In general, choosing multiples of 10 and numbers greater than 31 appears to be sensible; until, of course, everyone else starts doing so as well!

8.6 **Remarks** Each i-subset has the same probability $\binom{p}{i}/\binom{n}{i}$ of appearing in the p-subset chosen. If one bets once in a draw, the chances of getting an i-match are independent of the k numbers you choose. However, if more than one set of numbers are bet, the probability of an i-match depends on the sets selected. As a trivial illustration, in Lotto 6/49, of the 13983816 possible tickets, one can in principle buy 13723192 distinct sets of numbers and yet win absolutely nothing. This would require buying every combination except for those 260624 which win some prize (and is as unlikely as buying a single ticket and getting a 6-match — but much less lucrative).

While one can buy 98.14% of the possible tickets and win nothing, it is possible to buy as few as 174 tickets (0.00124% of the possible combinations) and guarantee at least a 3-match. This appears to suggest that there is a system for improving your odds. However, the expected return on each ticket remains unchanged, except for a small effect caused by repetitions on your tickets which may cause you to 'share' a pari-mutuel win between two or more of your tickets — this reduces your expected return. What a lottery system can do is improve your odds of winning something, as opposed to suffering a total loss. It does not in any way improve your expected return, just your minimum return.

8.4 The Lottery Problem

An (n,k,p,t)-*lottery scheme* is an n-set V of *elements* and a set $\mathcal{B}$ of k-element subsets of V (*blocks*), so that for each p-subset P of V, there is a block $B \in \mathcal{B}$ for which $|P \cap B| \geq t$. $L(n,k,p,t)$ denotes the smallest number of blocks in an (n,k,p,t)-lottery scheme; when $n \geq t > p$, $L(n,k,p,t) = \infty$.

8.7 **Proposition** An (n,k,t,t)-lottery scheme is a t-$(n,k,1)$ covering; hence $L(n,k,t,t) = C(n,k,t)$. In general, write $p = \tau(t-1)+1+\nu$ for $0 \leq \nu < t-1$. Then $L(n,k,p,t) \leq \min_{n-\nu=\sum_{i=1}^{\tau}\sigma_i} \sum_{i=1}^{\tau} C(\sigma_i,k,t)$. See §IV.8.

8.8 **Proposition** If $n = n_1+n_2$ and $p = p_1+p_2-1$, then $L(n,k,p,t) \leq L(n_1,k,p_1,t) + L(n_2,k,p_2,t)$.

8.9 **Theorem** [7] $L(n,k,p,2) \geq \dfrac{n(n-p+1)}{k(k-1)(p-1)}$.

8.10 **Proposition** If $n = s(p-1)$, s is an integer, and a Steiner system $S(2,k,s)$ exists, then equality holds in Theorem 8.9.

8.11 **Theorem** [1]

$$\begin{aligned} L(2m+1,3,3,2) &= C(m,3,2) + C(m+1,3,2), \\ L(4m+2,3,3,2) &= C(2m+1,3,2) + C(2m+1,3,2), \text{ and} \\ L(4m,3,3,2) &= C(2m-1,3,2) + C(2m+1,3,2) \end{aligned}$$

8.12 **Theorem** [5] $L(n,k,p,2) \geq \frac{1}{k}\min_{n=\sum_{i=1}^{p-1}\alpha_i}\left(\sum_{i=1}^{p-1}\alpha_i\left\lceil\frac{\alpha_i-1}{k-1}\right\rceil\right)$.

8.5 Turán Designs and Constant Weight Covering Codes

An (n,p,t)-*Turán design* $(V,\mathcal{D})$ is an n-set V and a set $\mathcal{D}$ of t-element subsets of V (*blocks*) for which every p-subset of V contains at least one of the blocks of $\mathcal{D}$. The *Turán number* $T(n,p,t)$ is the smallest number of blocks in an (n,p,t)-Turán design.

8.13 **Proposition** Complementing each block of an (n,p,t)-Turán design gives a $(v-p)$-$(v,v-t,1)$ covering, and vice versa. Hence $T(n,p,t) = C(n,n-t,n-p)$.

8.14 **Theorem** [3] $T(n,p,t) \geq \dfrac{\binom{n}{t}}{\binom{p-1}{t-1}} \cdot \dfrac{n-p+1}{n-t+1}$.

8.15 Theorem [2] $L(n,k,p,t) \geq \dfrac{T(n,p,t)}{\binom{k}{t}} \geq \dfrac{\binom{n}{t}}{\binom{p-1}{t-1}\binom{k}{t}} \cdot \dfrac{n-p+1}{n-t+1}$.

Let $\mathcal{V}$ be a nonempty set of binary vectors (*codewords*), each of length n and each having weight k (i.e., k 1s and $n-k$ 0s). $\mathcal{V}$ is an (n,k,p,d)-*constant weight covering code (CWCC)* if every binary vector of weight p and length n has Hamming distance at most d from at least one codeword.
$K(n,k,p,d)$ denotes the minimum number of codewords in a (n,k,p,d)-CWCC.

8.16 Remark The Hamming distance h between any binary vector of weight p and any codeword satisfies $|k-p| \leq h \leq k+p$ and $h \equiv k+p \pmod 2$. Hence $K(n,k,p,k+p-2\varepsilon) = K(n,k,p,k+p-2\varepsilon+1)$, and thus one usually assumes that $d \equiv k+p \pmod 2$.

8.17 Proposition $K(n,k,p,d) = L(n,k,p,(k+p-d)/2)$.

8.6 Some Lottery Wheels

8.18 Remarks A *lottery wheel* is a system for buying multiple tickets, usually for a group or syndicate. There are numerous vendors of software packages, books, newsletters, and so on. Many make preposterous claims about improving your odds; they are known to the losers (and many bystanders) as *snake oil salesmen*. Most lottery wheels sold are predicated on the assumption that some numbers are *hot* (more likely to be chosen), and some are *cold* (less likely to be chosen). In such cases, certain numbers may be chosen for inclusion, or eliminated from the selection. Of interest here, however, is the construction of sets of tickets that guarantee an i-match at least. Combinatorial designs are of considerable use here, as the following examples show.

8.19 Examples Lottery wheels for Lotto $6/n$, $44 \leq n \leq 51$, that guarantee at least a 3-match. Serotic [11] sells solutions for for Lotto $6/n$ with 355 tickets for $n = 44$, 380 tickets for $n = 45$, 411 tickets for $n = 46$ and 496 tickets for $n = 49$; wheels with fewer tickets are given here, free. Partition the set of n numbers into three sets A, B and C, where $A = \{1,\ldots,22\}$, $B = \{23,\ldots,44\}$, and $C = \{45,\ldots,n\}$. On each of A and B, purchase 77 tickets corresponding to the blocks of the Steiner system $S(3,6,22)$ (§I.4). Holding these 154 tickets, one must obtain a 3-match at least, unless the six winning numbers contain at most two from A, and at most two from B. Evidently, this cannot happen for $n \in \{44,45\}$. Hence $L(n,6,6,3) \leq 154$ for $n \in \{44,45\}$.

If $n = 46$, purchase the additional tickets $\{45,46,1,2,3,4\}$, $\{45,46,5,6,7,8\}$, $\{45,46,9,10,11,12\}$, $\{45,46,13,14,15,16\}$, $\{45,46,17,18,19,20\}$. With the 159 tickets now held, one can fail to get a 3-match only if the chosen numbers are 45, 46, 21, 22, and two numbers from B. Any additional ticket containing $\{21,22,45\}$ completes the wheel, so $L(46,6,6,3) \leq 160$.

If $n = 47$, start with the 154 tickets, and purchase in addition

$$\{45,46,47,1,2,3\}, \{45,46,47,4,5,6\}, \{45,46,47,7,8,9\}, \{45,46,47,10,11,12\},$$
$$\{45,46,47,13,14,15\}, \{45,46,47,16,17,18\}, \{45,46,47,19,20,21\}.$$

This shows $L(47,6,6,3) \leq 161$.

If $n = 48$, start with the 154 tickets, and purchase in addition $\{45,46,47,48\} \cup \{2i-1,2i\}$ for $1 \leq i \leq 11$. This shows $L(48,6,6,3) \leq 165$.

If $n = 49$, start with the 154 tickets. If one purchases tickets $\{45,46,47,48,49\} \cup \{i\}$ for $1 \leq i \leq 22$, one obtains $L(49,6,6,3) \leq 176$. But one can do better. Purchase only those tickets with $6 \leq i \leq 22$ (17 additional tickets). With the 171 tickets in hand, one can only fail to 3-match if the six chosen numbers have two in C, two in B

and two in $\{1, 2, 3, 4, 5\}$. Purchase tickets $\{1, 2, 45, 46, 47, 48\}$ and $\{3, 4, 45, 46, 47, 48\}$. Any failure to match must now select 49, two from $\{1, 2, 3, 4, 5\}$ except $\{1,2\}$ or $\{3,4\}$, one from $\{45, 46, 47, 48\}$, and two from B. One additional ticket $\{1,2,3,4,5,49\}$ suffices now to show $L(49, 6, 6, 3) \leq 174$.

If $n = 50$, start with the 154 tickets. Let $\{B_1, B_2, B_3\}$ be a 2-(6,4,1) covering of the points in C. For example, $B_1 = \{45, 46, 47, 48\}$, $B_2 = \{45, 46, 49, 50\}$, and $B_3 = \{47, 48, 49, 50\}$ is such a covering. For $i \in \{1, 2, 3\}$ and $1 \leq j \leq 11$, purchase the ticket $B_i \cup \{2j - 1, 2j\}$. Purchase the final ticket $\{45, 46, 47, 48, 49, 50\}$. This establishes that $L(50, 6, 6, 3) \leq 154 + 33 + 1 = 188$. In fact, the need for the last ticket can be removed by using (carefully chosen) different 2-(6,4,1) coverings for various values of j; hence, $L(50, 6, 6, 3) \leq 187$.

If $n = 51$, start with the 154 tickets. Let $\{B_i : 1 \leq i \leq 7\}$ be the blocks of a Steiner triple system (minimum 2-(7,3,1) covering) on $\{45, \ldots, 51\}$. For $1 \leq i \leq 7$ and $1 \leq j \leq 7$, one could purchase the 49 tickets $B_i \cup \{3j - 2, 3j - 1, 3j\}$. If no 3-match occurs in the 203 tickets held, the only possibility is that three or more of the elements of C are chosen, and the elements chosen do not include a block of the Steiner triple system. One could remedy this by purchasing the four tickets specified by a minimum 3-(7,6,1) covering on C. However, a more careful selection of the 49 tickets removes the need for this. Let $\mathcal{B}_1, \ldots, \mathcal{B}_7$ be seven STS(7)s on the set $\{45, \ldots, 51\}$, so that every possible triple on the 7-set occurs in at least one of the first six STS(7)s (use Construction 8.20). Label the seven blocks of $\mathcal{B}_j$ as $B_{1j}, \ldots, B_{7j}$. Now to form the 49 tickets, for $1 \leq i \leq 7$ and $1 \leq j \leq 7$, purchase tickets $B_{ij} \cup \{3j - 2, 3j - 1, 3j\}$. Hence $L(51, 6, 6, 3) \leq 203$.

8.20 Construction [6] Six Steiner triple systems of order 7, among which each of the 35 possible triples appears at least once.

123	145	167	247	256	346	357
123	147	156	246	257	345	367
124	135	167	236	257	347	456
125	137	146	234	267	356	457
126	134	157	237	245	356	467
127	136	145	235	246	347	567

Each row is an STS(7). An analogous structure with only five rows would be a large set of STS(7)s; however, none exists. See §I.3.4.

8.21 Construction A 3-(18,6,1) covering with 48 blocks. Let

$$P_1 = \{\{0, 1, 2\}, \{3, 4, 5\}, \{6, 7, 8\}\},\ P_2 = \{\{0, 3, 6\}, \{1, 4, 7\}, \{2, 5, 8\}\},$$
$$P_3 = \{\{0, 4, 8\}, \{1, 5, 6\}, \{2, 3, 7\}\},\ P_4 = \{\{0, 5, 7\}, \{1, 3, 8\}, \{2, 4, 6\}\}$$

be the parallel classes of a Kirkman triple system of order 9. On $\{0, \ldots, 8\} \times \{0, 1\}$, for $1 \leq i \leq 4$ and $B, C \in P_i$, include the 6-set $(B \times \{0\}) \cup (C \times \{1\})$, to obtain the first 36 blocks. Next for $i = 1, 2$, $\{B, C\} \subset P_i$, and $\ell = 0, 1$, include the 6-set $(B \cup C) \times \{\ell\}$ to get the remaining 12 sets of the covering.

8.22 Examples Lottery wheels for Lotto $6/n$, $36 \leq n \leq 43$, guaranteeing a 3-match. The technique is similar to Examples 8.19. Here one partitions $\{1, \ldots, n\}$ into three sets A, B and C, with $A = \{1, \ldots, 18\}$, $B = \{19, \ldots, 36\}$, and $C = \{37, \ldots, n\}$. Use Construction 8.21 to purchase 48 tickets on A, and a further 48 tickets on B. Continuing as in Examples 8.19, one easily obtains the following bounds on $L(n, 6, 6, 3)$:

n	36	37	38	39	40	41	42	43
$L(n, 6, 6, 3) \leq$	96	96	101	102	105	112	123	138

8.23 **Example** A lottery wheel for Lotto 5/34. Use Proposition 8.8 to write $L(34,5,5,3) \leq L(17,5,3,3)+L(17,5,3,3)$. There is a Steiner system $S(3,5,17)$, so $L(17,5,3,3) = 68$. Hence, $L(34,5,5,3) \leq 136$.

8.24 **Remark** In [11, 12], numerous wheels are presented for the case when the bettor believes that certain numbers are lucky, or that certain pairs are more likely to appear together. For example, if twelve numbers in Lotto $6/n$ are thought to be lucky, system SD9 from [12] guarantees a 4-match when 6 of the twelve numbers are chosen, and only six tickets are needed (if the twelve numbers are represented by $\{0,1,2,3\} \times \{0,1,2\}$, one purchases tickets $\{i,j\} \times \{0,1,2\}$ for $0 \leq i < j \leq 3$).

8.25 **Example** (System 25 in [11]) A 42 ticket wheel ensuring a 4-match if six among 18 numbers are chosen. Take $\{0,1,2,3,4,5\} \times \mathbb{Z}_3$ as the 18 numbers. Purchase tickets $\{(0,i),(1,j),(2,i+k),(3,j+k),(4,i+j+k),(5,i+j+2k)\}$ for $i,j,k \in \mathbb{Z}_3$ (a total of 27 tickets). Then purchase tickets $\{i,j\} \times \mathbb{Z}_3$ for $0 \leq i < j \leq 5$, making a total of 42 tickets.

8.7 See Also

§I.4	Steiner 2-designs give optimal lottery schemes.
§IV.8	Coverings are used to construct lottery schemes and are complementary to Turán designs.
[10]	A history of lottery schemes from biblical times until 1974.
[11]	A compilation of lottery wheels, many based on designs.
[13]	A much more detailed survey of Turán numbers and designs.
[4]	Constructions of constant weight covering codes, and applications to quantizers.

References

[1] A.E. Brouwer, Some Lotto numbers from an extension of Turán's theorem, Math. Centre (Amsterdam) Report, 1978.

[2] A.E. Brouwer and M. Voorhoeve, Turán theory and the Lotto problem, *Math. Centre Tracts* 106 (1979), 99-105.

[3] D. de Caen, Extension of a theorem of Moon and Moser on complete subgraphs, *Ars Combin.* 16 (1983), 5-10.

[4] T. Etzion, V. Wei and Z. Zhang, Bounds on the sizes of constant weight covering codes, *Des. Codes Crypt.* 5 (1995), 217-239.

[5] Z. Füredi, G.J. Székely and Z. Zubor, On the lottery problem, *J. Combin. Des.*, to appear.

[6] T.S. Griggs and A. Rosa, Large sets of Steiner triple systems, in: *Surveys in Combinatorics* (P. Rowlinson; ed.) Cambridge University Press, London, 1995, 25-39.

[7] H. Hanani, D. Ornstein and V.T. Sós, On the lottery problem, *Magyar Tud. Acad. Mat. Kutató* 9 (1964), 155-158.

[8] M. Morley and G.H.J. van Rees, Lottery schemes and covers, *Utilitas Math.* 37 (1990), 159-166.

[9] M. Orkin, *Can You Win?*, Freeman, New York, 1991.

[10] J. Scarne, *Scarne's New Complete Guide to Gambling*, Simon & Schuster, New York, 1986.

[11] R. Serotic, *The Only Way to Win at Lotto*, LSI Publishing, Menlo Park CA, 1985.

[12] R. Serotic, *Lotto: How to Play and Win, for Experts*, LSI Publishing, Menlo Park CA, 1987.

[13] A. Sidorenko, What we know and what we do not know about Turán numbers, *Graphs Combin.* 11 (1995), 179-199.

PART VI

RELATED MATHEMATICS AND COMPUTATIONAL METHODS

1 Finite Groups and Designs

Leo G. Chouinard II
Robert Jajcay
Spyros S. Magliveras

1.1 Introduction

A *group* $(P, \cdot)$ is a set P, together with a binary operation $\cdot$ on P, for which
1. if $x, y \in P$, then $x \cdot y \in P$,
2. an *identity* element $e \in P$ exists, i.e., $x \cdot e = e \cdot x = x$ for all $x \in P$;
3. $\cdot$ is *associative*, i.e. $x \cdot (y \cdot z) = (x \cdot y) \cdot z$ for all $x, y, z \in P$;
4. every element $x \in P$ has an *inverse*, an element x^{-1} for which $x \cdot x^{-1} = x^{-1} \cdot x = e$.

If $G = (P, \cdot)$ is a group, $Q \subseteq P$, and $H = (Q, \cdot)$ is also a group, then H is a *subgroup* of G, written $H \leq G$. It is *proper* if H is not the trivial group $\{e\}$, nor the whole group G. The *index* $[G : H]$ of H in G is $|G|/|H|$.

If G is a group and $H \leq G$, a *right coset* of H in G is any subset of the form $Hx = \{hx : h \in H\}$ for a fixed element $x \in G$. A *left coset* of H in G is similarly defined to be any subset of G of the form $xH = \{xh : h \in H\}$ for $x \in G$ fixed.

1.1 **Remarks** All groups and sets are assumed here to be finite. As is customary, the symbol 1 denotes the rational integer 1, the identity element of a group, as well as the trivial subgroup $\{1\}$. The meaning intended is always clear from the context. If G is a group and $H \leq G$, two right cosets Hx, Hy are identical if and only if $xy^{-1} \in H$; otherwise, they are disjoint. Hence, the number of distinct right cosets of H in G is $[G : H]$. Similarly, the number of distinct left cosets of H in G is $[G : H]$.

A *permutation* π on the set X is a one to one and onto mapping $\pi : X \to X$. The set of all permutations on X, with composition of functions as a binary operation, forms a group, the *symmetric group* on X. It is denoted by $Sym(X)$ or if $X = \{1, 2, \ldots, n\}$, by S_n. If G is a subgroup of $Sym(X)$, then G is a *permutation group*.

1.2 **Examples** of permutation groups of particular importance in design theory arise from considering a graph, design, or geometry. A permutation of the vertices, elements, or points which "preserves the structure" is an *automorphism*. For example, a permutation on the elements of a simple BIBD is an automorphism when it maps blocks to blocks, and sets that are not blocks to sets that are also not blocks. The collection of all automorphisms forms a permutation group, the *automorphism group* of the structure.

1.3 **Remarks** Automorphism groups of combinatorial structures form a rich source of permutation groups. Although most combinatorial objects admit only the trivial automorphism or have very small automorphism groups, objects with large automorphism groups are elegant and sought after. Designs, graphs, and geometries with large groups are often the easiest to construct, and the first to be discovered among members of specified families. Moreover, such objects can be presented efficiently, frequently in terms of generators of their automorphism groups and certain base blocks.

A subgroup N of a group G is a *normal* subgroup, written $N \lhd G$, if $a^{-1}Na \subseteq N$ for every $a \in G$.

A nontrivial group is *simple* if it has no proper, normal subgroups.

If $N \lhd G$, then, for each $x \in G$, the right coset Nx is identical with the left coset xN, and Nx is a *coset* of N in G. If $N \lhd G$, the collection G/N of all distinct cosets of N forms a group under the operation $Nx \cdot Ny = Nxy$, the *quotient* or *factor* group of G modulo N.

A *normal series* of a group G is a chain of subgroups

$$1 = G_n \lhd G_{n-1} \lhd \cdots \lhd G_1 \lhd G_0 = G$$

whose *factor groups* are G_i/G_{i+1} for $0 \leq i < n$.

A finite group is *solvable* when it has a normal series in which each factor group is cyclic of prime order.

1.4 **Remarks** A number of references and computational tools were of considerable help in deriving structural information and completing Tables 1.37 and 1.41. The *Atlas* [5] was of particular utility, especially for the information on the primitive spectrum of those simple groups in the list which are included in the *Atlas*. Of utility also were the libraries of groups in the algebraic computational systems *Cayley* [2], and *Gap* [12], as well as standard references [1], [6], [8], [10], [13]. To obtain and check the results, extensive computation on a SPARC II Sun workstation was used, with the computational package *Pythia*, developed by Chouinard and Magliveras, as well as the computational algebra systems *Gap* and *Cayley*.

1.2 Permutation Groups

A *group action* is a triple (X, G, exp), where X is a set, G a group, and $exp : X \times G \to X$ a mapping written as exponentiation, $((x, g) \mapsto x^g)$, such that:

1. $x^1 = x$ for any $x \in X$, where 1 is the identity of G, and
2. $(x^g)^h = x^{(gh)}$ for any $x \in X$, and any $g, h \in G$.

The notation (X, G, exp) is abbreviated to $G|X$. If $G|X$ is a group action, then $|X|$ is the *degree* of $G|X$.

1.5 **Example** (right regular action) Let G be any group, and let $X = G$. Define an action $G|X$ by $(x, g) \mapsto xg$. Now, (1) and (2) are satisfied since $x \cdot 1 = x$, and $(xg)h = x(gh)$.

1.6 **Example** (permutation groups) Let G be a permutation group, i.e. a subgroup of the symmetric group $\mathcal{S}_n$. If $x \in X$, and $g \in G$, let $x^g = g(x)$.

1.7 **Example** (right action on cosets) Let G be a group, $H \leq G$, and $\Omega = \{Hx_i : x_i \in G\}$ be the collection of all distinct right cosets of H in G. It is easy to check that the mapping $(Hx, g) \mapsto Hxg$ defines an action $G|\Omega$.

1.8 **Remark** The action in Example 1.5 is a special case of the action in Example 1.7, where H is the identity subgroup of G.

1.9 **Example** Let $G|X$ be a group action. Then G acts on the collection of all k-subsets of X in a natural way. For $S \subseteq X$ and $g \in G$, define S^g by $S^g = \{x^g : x \in S\}$. Some beautiful designs constructed in this way are given in §IV.23, where the symmetric group is acting on the set of all 2-subsets of a set of size n.

1.10 **Example** (G acts on G by conjugation) Let G be any group, and let $X = G$. Define

$exp : X \times G \to X$ by $x^g = g^{-1}xg$.

Let $G|X$ be a group action and let $x \in X$. Then the *orbit of x under G* is $x^G = \{x^g : g \in G\}$. The orbits under G partition X. The *stabilizer of x in G* is $G_x = \{g \in G : x^g = x\}$.

1.11 Proposition G_x is a subgroup of G, and when G is finite, $|G| = |x^G| \cdot |G_x|$.

Let $G|X$ be a group action and let $g \in G$. The set of all elements of X *fixed* by g, $\{x \in X : x^g = x\}$, is denoted by $Fix(g)$. The cardinality of $Fix(g)$ is the *character* of g, denoted by $\chi_{G|X}(g)$ or simply $\chi(g)$, and the mapping $\chi : g \mapsto \chi(g)$ is the *character* of the group action.

The *kernel* of a given group action $G|X$, denoted by $ker(G|X)$, is the set $\{g \in G : Fix(g) = X\}$. Then $ker(G|X)$ is a normal subgroup of G. $G|X$ is *faithful* if and only if $|ker(G|X)| = 1$.

1.12 Example A permutation group is a faithful group action.

In a group G two elements x, y are *conjugate* if and only if $x = y^g = g^{-1}yg$ for some $g \in G$. Thus, two elements are conjugate if and only if they are in the same G-orbit in the action of Example 1.10. The G-orbits of G in the action of Example 1.10 are the *conjugacy classes* of G. G is the disjoint union of its conjugacy classes: $G = K_1 + K_2 + \cdots + K_c$ and c is the *class number* of G. Similarly, two subgroups $H_1, H_2 \leq G$ are *conjugate* in G (denoted by $H_1 \sim H_2$) if and only if $H_1 = H_2^g$ for some $g \in G$.

A group action $G|X$ is *transitive* if and only if X is a single G-orbit. Thus, $G|X$ is transitive if and only if for any two elements $x, y \in X$ there exists a group element $g \in G$ such that $x = y^g$.

1.13 Remarks If G is a group and $H \leq G$, the action described in Example 1.7 is transitive. Moreover, any transitive action $G|X$ is isomorphic to one of the type described in Example 1.7. If G is a group and $H_1, H_2 \leq G$, then the two actions of G on the two collections of right cosets of H_1, and H_2 respectively are isomorphic if and only if the two subgroups H_1 and H_2 are conjugate. These observations are fundamental for the determination of all inequivalent transitive permutation representations of a given group.

If $G|X$ is a permutation group, G is *$\frac{1}{2}$-transitive* or *semi-regular* if all G-orbits of X are of length $|G|$. G is *regular* or *sharply transitive* on X if G is transitive and semi-regular.

G is *k-transitive* (or *k-fold transitive*) on X if the induced action of G on the collection of all ordered k-subsets of X is transitive. $G|X$ is *k-homogeneous*, or *k^*-transitive*, if G is transitive on the k-subsets of X.

G is *sharply k-transitive* (*sharply k-homogeneous*) on X if G is *k-transitive* (*k-homogeneous*) on X and the stabilizer in G of an ordered (unordered) k-subset of X is the identity subgroup.

1.14 Remarks $G|X$ is k-transitive if and only if $G_x|(X - \{x\})$ is $(k-1)$-transitive for each $x \in X$. If $G|X$ is k-transitive, then it is k-homogeneous. Surprisingly, the converse is also true when $k \geq 5$ [11]. See Theorem 1.29.

A permutation group $G|X$ is of *type k^** if it is k^*-transitive, but not k-transitive. If $G|X$ is a transitive permutation group, the *rank* of G is the number of orbits of G_x on X.

Let $G|X$ be a transitive group action. A subset Δ of X is a *block of imprimitivity* if for any $g \in G$, $\Delta \cap \Delta^g = \Delta$ or $\emptyset$. The subsets $\emptyset, X$ and the singleton sets $\{x\}, x \in X$ are always blocks of imprimitivity for $G|X$, the *trivial* blocks. If the only blocks of imprimitivity for $G|X$ are the trivial ones, then $G|X$ is a *primitive* action.

1.15 Remarks A combinatorial design having an automorphism group that is transitive on its points necessarily has isomorphic derived structures with respect to any two points. Thus transitive permutation groups are important to design theory. A transitive permutation group is either primitive or imprimitive. In Tables 1.37 and 1.38, the primitive groups of degree up to 50 are presented. Transitive imprimitive groups have primitive homomorphic images of degree dividing the degree of the given group; Proposition 1.16 shows how imprimitive groups can be reduced to primitive groups of smaller degree. Intransitive groups can be formed as subdirect products of transitive groups ([10], p.63).

1.16 Proposition Let Δ be a nontrivial block of imprimitivity of the imprimitive group action $G|X$ and set $\Omega = \{\Delta_1, \ldots, \Delta_m\} = \Delta^G$, the orbit of Δ under G. Then

1. Ω is a partition of X, a *system of imprimitivity.*
2. If for each $g \in G$ one defines $\bar{g} : \Omega \mapsto \Omega$ by
$$\bar{g} = \begin{pmatrix} \Delta_1 & \Delta_2 & \cdots & \Delta_m \\ \Delta_1^g & \Delta_2^g & \cdots & \Delta_m^g \end{pmatrix},$$
then $\bar{G} = \{\bar{g} : g \in G\}$ forms a permutation group on Ω and $g \mapsto \bar{g}$ is a homomorphism of G onto $\bar{G}$,
3. $\bar{G}$ is transitive on Ω and is isomorphic to G/N where N is the set of permutations in G that fix each of $\Delta_1, \Delta_2, \ldots, \Delta_m$ as sets, i.e., the kernel of the homomorphism $g \mapsto \bar{g}$. N is the *kernel of imprimitivity* for this system of imprimitivity.

1.17 Proposition If $G|X$ is a transitive group action and $x, y \in X$, then the respective stabilizer subgroups G_x and G_y are conjugate in G. Moreover, $G|X$ is primitive if and only if G_x is a maximal subgroup of G.

For k a positive integer, G is $(k+\frac{1}{2})$-*transitive* if the stabilizer of each $x \in X$ is $(k-\frac{1}{2})$-transitive on $X - \{x\}$. G is $(k+1)$-*primitive* if G_x is k-primitive on $X - \{x\}$.

1.18 Theorem (Zassenhaus, 1934; Tits, 1952; Huppert, 1962) Let p be an odd prime, and m any odd positive integer. Then there is a unique sharply 3-transitive group of degree p^m+1, namely $\mathrm{PGL}_2(p^m)$. If m is even, then there are precisely two sharply 3-transitive groups of degree p^m+1. These groups are $\mathrm{PGL}_2(p^m)$ and $\mathrm{P\Delta L}_2(p^m)$. If $p = 2$, and m is a positive integer, then $L_2(2^m)$ is the unique sharply 3-transitive group of degree $p^m+1 = 2^m+1$.

1.19 Remark (Feit, 1960; Ito, 1962; Suzuki, 1962) All doubly transitive groups which contain an element fixing exactly two points but no element except for the identity fixing more than two points have been determined. These all have degree p^m+1, or 2^p, where p is a prime.

1.20 Theorem (Huppert) Every solvable, primitive group of degree $1+p$, p a prime, is 2-transitive. Solvability is a necessary condition.

1.21 Theorem (Jordan) If G is a primitive solvable permutation group, then its degree is a prime power p^a, and it contains a normal, elementary abelian subgroup N of order p^a. If A is a transitive abelian subgroup of G, then $p > 3$ implies that $A = N$. When $p = 3$, A is elementary abelian.

1.22 Theorem (Jordan) If $G|X$ is k-transitive ($k \geq 2$) and N is a proper normal subgroup of G, then N is $(k-1)$-transitive, except when G is the symmetric group on X, or if $|X|$ is a power of 2 and $k = 3$.

1.23 Theorem (Ito) Let G be $(k + \frac{1}{2})$-transitive on X, and let N be a nonregular normal subgroup of G, then N is k-transitive, except possibly when $k = 1$ or 2.

1.24 Theorem (Wielandt and Huppert) If G is k-primitive, then a nonregular normal subgroup is k-transitive.

1.25 Theorem (Ito) For $k > 2$, a nonregular normal subgroup of a k-transitive permutation group $G|X$ is $(k-1)$-primitive.

An *isomorphism* from group $(P, \cdot)$ to group $(Q, \circ)$ is a one-to-one correspondence $\phi : P \to Q$ for which $\phi(x \cdot y) = \phi(x) \circ \phi(y)$ for all $x, y \in P$. An *automorphism* is an isomorphism from a group onto itself. The automorphism is *inner* if it is conjugation ($x \mapsto a^{-1}xa$) by a fixed group element $a \in G$; otherwise, it is an *outer automorphism*. The collection of all automorphisms of a group G is a subgroup of $Sym(G)$, denoted by $Aut(G)$. The inner automorphisms of G form a normal subgroup $I(G)$ of $Aut(G)$. The quotient group $Aut(G)/I(G)$ is the *outer automorphism* group of G.

1.26 Theorem (Nagao) Let G be a 6-transitive permutation group of degree n. If the outer automorphism group of every simple subgroup of G is solvable, then G is $\mathcal{S}_n$ or $\mathcal{A}_n$.

1.27 Theorem The outer automorphism group of every finite simple group is solvable. (This was known as the Schreier Conjecture.)

1.28 Remark Theorem 1.27 has been proved using the classification theorem of finite simple groups. No proof has yet been found which does not use the classification theorem.

1.29 Theorem (O'Nan) If G is a nontrivial t-transitive group for which the composition factors of all its proper subgroups have solvable outer automorphism groups, then $t \leq 5$. Hence, there do not exist 6-transitive groups, except for $\mathcal{S}_n$ and $\mathcal{A}_n$.

1.30 Theorem (Wagner) If $G|X$ is 3-transitive and $|X| > 3$ is odd, then any nontrivial normal subgroup of G is also 3-transitive on X.

A permutation group $G|X$ is *set-transitive* if it is k-homogeneous for each $k \leq |X|$.

1.31 Theorem (Beaumont and Peterson) Apart from $\mathcal{S}_n$ and $\mathcal{A}_n$, there are exactly four set-transitive groups

1. a metacyclic group of degree 5 and order 20,
2. a group $G \cong \mathcal{S}_5$ represented on six points,
3. $L_2(8)$ of order 504, represented on nine points, and
4. $\mathrm{P\Gamma L}_2(8) = Aut(L_2(8))$ of order 1512, represented on nine points.

1.32 Theorem (Livingstone and Wagner) If $G|X$ is a permutation group of degree n, and $2 < k \leq n/2$, then the number of G-orbits on unordered k-sets is at least the number of G-orbits on ordered $(k-1)$-sets. Thus if G is k^*-transitive, then it must be $(k-1)$-transitive. When $k \geq 5$, G must also be k-transitive.

1.33 Remark (Livingstone and Wagner) Permutation groups $G|X$ of type k^* exist only for $k = 2, 3$, and 4.

1.34 Theorem (Kantor) If $G|X$ is of type 4^*, and G_x is odd, then $G \cong L_2(5)$, $\mathrm{PGL}_2(5)$, $L_2(8)$, $\mathrm{P\Gamma L}_2(8)$, or $\mathrm{P\Gamma L}_2(32)$, in their usual permutation representations.

1.3 Groups and Group Operations

1.35 Table Notation for group operations and names for groups.

Symbol	Description
Operations on Groups:	
$A \times B$	The *direct product* of groups A and B.
A^n	The group $A^{n-1} \times A$, where A is a group, and n a positive integer.
$A \cdot B$	A *semidirect* product (*split extension*) of A by B (the action of B by conjugation on the normal subgroup A is not specified here).
$\mathbb{Z}_m \cdot_j \mathbb{Z}_n$	The semidirect product where the action of a generator of $\mathbb{Z}_n$ on $\mathbb{Z}_m$ is multiplication by $j \in \mathbb{Z}_m$.
$A \backslash B$	A general extension of A whose factor group by the normal subgroup A is B, but which is not a semidirect product of A by B.
$A \wr B$	The wreath product of A by B (which is isomorphic to some $A^n \cdot B$, $B \leq \mathcal{S}_n$).
$\mathcal{A}_n \cap H$	When H is a subgroup of $\mathcal{S}_n$ and contains odd permutations, the subgroup of index 2 in H consisting of the even permutations present.
Basic Groups:	
1	The trivial group.
$\mathbb{Z}_n$	The cyclic group of order n.
D_n	The dihedral group of order $2n$.
$\mathcal{S}_n$	The symmetric group on a set of n elements.
$\mathcal{A}_n$	The alternating group on a set of n elements.
V_q	The elementary abelian group of order q, q a prime power.
Q_8	The quaternion group of order 8.
Linear Groups (q a prime power):	
$\mathrm{GL}_n(q)$	The general linear group of all invertible matrices of dimension n over $\mathbb{F}_q$. This is equivalent to the group of $\mathbb{F}_q$-automorphisms of vector space $V = \mathbb{F}_q^n = \mathbb{F}_q \oplus \cdots \oplus \mathbb{F}_q$.
$\mathrm{AGL}_n(q)$	The group of *affine* transformations on V of the form $x \mapsto Ax + b$, where $A \in \mathrm{GL}_n(q)$ and $b \in \mathbb{F}_q^n$. Thus $\mathrm{AGL}_n(q) \cong \mathbb{F}_q^n \cdot \mathrm{GL}_n(q)$.
$F_{q,v}$	The subgroup of $\mathrm{AGL}_1(q)$ of order $q \cdot v$ of transformations of the form $x \mapsto Ax+b$ where $A^v = 1 \in \mathbb{F}_q$ and $b \in \mathbb{F}_q$. This is also known as the *Frobenius* group of order qv.
$\mathrm{SL}_n(q)$	The *special linear* group of dimension n over $\mathbb{F}_q$, the group of transformations of determinant 1 in $\mathrm{GL}_n(q)$.
$\mathrm{ASL}_n(q)$	The extension $\mathbb{F}_q^n \cdot \mathrm{SL}_n(q) < \mathrm{AGL}_n(q)$.
$L_n(q)$	(or $\mathrm{PSL}(n, \mathbb{F}_q)$ or $\mathrm{PSL}_n(q)$ or $\mathrm{LF}_n(q)$) The projective special linear group formed by taking $\mathrm{SL}_n(q)$ modulo its center.
$\mathrm{PGL}_n(q)$	$\mathrm{GL}_n(q)$ modulo its center.
$\Gamma\mathrm{L}_n(q)$	For $q = p^a$, p a prime, $a > 1$, the extension of $\mathrm{GL}_n(q)$ by the Galois group of $\mathbb{F}_q$ over $\mathbb{F}_p$.

Symbol	Description
$\Sigma L_n(q)$	For $q = p^a$, p a prime, $a > 1$, the extension of $SL_n(q)$ by the Galois group of $\mathbb{F}_q$ over $\mathbb{F}_p$. $A\Gamma L_n(q)$, $A\Sigma L_n(q)$, $P\Gamma L_n(q)$ and $P\Sigma L_n(q)$ are similarly defined from these.
$P\Delta L_2(q)$	If $p \neq 2$ and $q = p^2$, then $P\Gamma L_2(q) \cong PSL_2(q) \cdot V_4$ contains three extensions of the form $PSL_2(q) \cdot \mathbb{Z}_2$. Two of these are $PGL_2(q)$ and $P\Sigma L_2(q)$. J.H. Conway suggested denoting the third extension of this sort by $P\Delta L_2(q)$ (the *diagonal* of the other two).
$U_n(q)$	(or $PSU_n(q)$, or $PSU_n(q^2)$) The *projective special unitary* group, a subgroup of $L_n(q^2)$.
$S_{2n}(q)$	(or $PSp_{2n}(q)$) The *projective symplectic* group, a subgroup of $L_{2n}(q)$.
Lie Type and Sporadic Simple Groups:	
$Sz(8)$, $Sz(32)$	The *Suzuki* groups of orders 29,120 and 32,537,600, respectively.
$G_2(3)$	The Chevalley group of order 4,245,696.
$R(2)'$	($\cong {}^2F_4(2)'$) The *Ree-Tits* group of order 17,971,200.
M_n	for $n = 11, 12, 22, 23, 24$, The five well-known Mathieu simple groups.
J_a	The smallest Janko group, of order 175,600 with smallest transitive representation of degree 266.
HaJ	The *Janko-Hall* group of order 604,800 with minimal transitive representation on 100 points.
J_3	The Janko-Higman-McKay sporadic simple group of order 50,232,960.
HS	The *Higman-Sims* simple group of order 44,352,000.

1.4 Primitive Groups

1.36 Remarks Table 1.37 contains information about the *primitive* groups of degree at most 50. Table 1.38 lists permutations which generate the groups in Table 1.37.

Table 1.37 has eight columns. The first column is an index of the form $d.i$, where d is the degree of the primitive permutation group and i is a serial index distinguishing primitive groups of the same degree. Column 2 contains the *name* of the group, which often also provides structural information about the group. Column 3 contains the *order* of the group. Column 4 provides information about the *depth* of transitivity of the group. For example, 3p! indicates that the group is *sharply 3-primitive*. The "p" notation indicates primitivity, the numeral 3 indicates the depth, (i.e., that the stabilizer of a point is 2-primitive on $d-1$ points), and the exclamation mark indicates that the stabilizer of two points is sharply primitive (hence also regular) on $d-2$ points. A notation such as 3t! means that the group is sharply 3-transitive. Similarly, 3t indicates that the group is 3-transitive, 3* that the group is 3-homogeneous, while 3*! indicates that the group is sharply 3-homogeneous. If the group is 1-primitive but not 2-primitive, then the stabilizer of a point is not primitive on $d-1$ points, and in these cases the orbit lengths for the action of the point-stabilizer are provided. For example, group 15.1 ($\mathcal{A}_6$ acting on 15 points) is just primitive but not 2-primitive. The stabilizer of a point, which happens to be isomorphic to $\mathcal{S}_4$, has 3 orbits on points, of lengths 1, 6, and 8 respectively. A superscript over an orbit length indicates the number of times that orbit length is encountered, so, for example, for group 16.1 column 4 contains $[1, 5^3]$, which means that the stabilizer of a point has 3 orbits of length 5, besides the fixed point.

Column 5 specifies generators of the group in the current primitive representation. When a generator is labeled in the form a_k, it is the *canonical* element of order k and degree d. This means that one starts writing cycles of length k, encompassing serially the points $1, 2, \cdots, d$ for as long as one can. If k does not divide d, then the last res points are left fixed, where res is the remainder of the division of d by k. For example, for group 10.8, $a_5 = (1,2,3,4,5)(6,7,8,9,10)$, while for group 15.5, $a_7 = (1,2,3,4,5,6,7)(8,9,10,11,12,13,14)(15)$. Other generators are listed in Table 1.38 under the relevant degree. The subscript of a generator indicates its order. An exponent over a specified generator, such as in a_{14}^2 for group 14.1, has the usual meaning. A generator listed in the form $ax+b$, simply indicates the element $x \mapsto ax+b$ over the Galois field $\mathbb{F}_p$ where $d = p$. Every primitive group listed has been given as a two-generator group with the exception of group 25.11 which is not a 2-generator group.

Column 6 describes the structure of the stabilizer of a point for the specific group. If group G is at least 2-primitive on X, the point stabilizer G_x, $x \in X$, is primitive on $X - \{x\}$ and can be described by referring to an entry in Table 1.37. This is the case, for example, for all primitive groups of degree 12. When the stabilizer of a point G_x is not primitive on the remaining points $X - \{x\}$, it is sometimes possible to describe G_x as being isomorphic (as an abstract group) to an earlier group in the table. For example, the group $G = M_{21} \cong L_3(4)$, of index 21.4, is doubly transitive, with point stabilizer G_x which is imprimitive on 20 points. Here, a nontrivial system of imprimitivity consists of a set of 5 blocks, each of size 4, and G_x is a split extension of the kernel of imprimitivity V_{16}, by the image of imprimitivity $\mathcal{A}_5$. If G_x is represented on the cosets of (maximal) $\mathcal{A}_5$, then G_x is a primitive group of degree 16, and is isomorphic to the group 16.11. Similar is the meaning of symbol $\approx$ in column 6 for groups 21.5, 21.6, and 21.7.

Column 7 describes the lattice of the groups appearing as primitive groups for a specific degree, under *conjugate subgroup inclusion*. In most cases groups are represented so that any conjugate subgroup inclusion is in fact a *subgroup inclusion*. The indices that appear in this column refer to groups under the same degree. For example, group 37.1 is contained in groups 37.2 and 37.3. Group 37.2 in groups 37.4 and 37.5, and so on. In a few cases, a specific conjugate inclusion is given, as for group 16.12.

In the eighth column, examples of combinatorial structures are given whose full automorphism group is the group listed in the second column of the same row. A *strongly regular graph* with parameters v, k, λ, μ is denoted by $\mathrm{srg}(v, k, \lambda, \mu)$. A t-(v, k, λ) design is denoted by $S_\lambda(t, k, v)$. All t-(v, k, λ) designs listed are simple. Note that $PG_n(q)$ denotes the *projective space* of (projective) dimension n over the field of q elements. This list of combinatorial structures is by no means complete!

Table 1.38 displays generators for the groups of Table 1.37. The amount of data needed to specify group generators is minimized, by listing as few permutations as possible in Table 1.38 for each specific degree, and by generating various groups by powers of elements listed in Table 1.38, or powers of canonical elements. This provides economy in representation of the data. For example, although there are 22 primitive groups of degree 16, only six permutations of degree 16 need be listed in Table 1.38.

1.37 Table Primitive groups of degree at most 50.

Degree $n.i$	Group G	Order $\|G\|$	Trans/Rank	Generators	G_α	min H, $G < H \le \mathcal{S}_n$	Combinatorial Structures
5.1	$\mathbb{Z}_5$	5	$[1^5]$	$a_5 : x+1$	1	2	
.2	D_5	10	$[1,2^2]$	$x+1, -x$	$\mathbb{Z}_2$	3, 4	
.3	$\mathrm{AGL}_1(5)$	20	2t!	$x+1, 2x$	$\mathbb{Z}_4$	5	
.4	$\mathcal{A}_5$	60	3p!	a_3, a_5	$\mathcal{A}_4$	5	
.5	$\mathcal{S}_5$	120	5p!	a_4, a_5	$\mathcal{S}_4$		
6.1	$L_2(5) \cong \mathcal{A}_5$	60	2p	a_5, b_3	5.2	2, 3	
.2	$\mathrm{PGL}_2(5) \cong \mathcal{S}_5$	120	3t!	a_5, b_6	5.3	4	
.3	$\mathcal{A}_6$	360	4p!	a_3, a_5	5.4	4	
.4	$\mathcal{S}_6$	720	6p!	a_5, a_6	5.5		
7.1	$\mathbb{Z}_7$	7	$[1^7]$	$a_7 : x+1$	1	2, 3	
.2	D_7	14	$[1,2^3]$	$x+1, -x$	$\mathbb{Z}_2$	4	
.3	$F_{7,3}$	21	2*!	$x+1, 2x$	$\mathbb{Z}_3$	4, 5	
.4	$\mathrm{AGL}_1(7)$	42	2t!	$x+1, 3x$	$\mathbb{Z}_6$	7	
.5	$L_3(2)$	168	2t	a_7, b_7	$\mathcal{S}_4$	6	$S_1(2,3,7)$
.6	$\mathcal{A}_7$	2520	5p!	a_3, a_7	6.3	7	
.7	$\mathcal{S}_7$	5040	7p!	a_6, a_7	6.4		
8.1	$\mathrm{AGL}_1(8)$	56	2p!	a_7, b_7	7.1	2	
.2	$V_8 \cdot F_{7,3}$	168	3*	a_7, b_6	7.3	5	
.3	$L_2(7) \cong L_3(2)$	168	3*	a_7, c_7	7.3	4, 6	
.4	$\mathrm{PGL}_2(7)$	336	3t!	a_7, b_8	7.4	7	
.5	$\mathrm{AGL}_3(2)$	1344	3t	a_7, d_7	7.5	7	$S_1(3,4,8)$
.6	$\mathcal{A}_8$	20160	6p!	a_4, a_7	7.6	7	
.7	$\mathcal{S}_8$	40320	8p!	a_7, a_8	7.7		
9.1	$F_{9,4}$	36	$[1,4^2]$	d_4, e_4	$\mathbb{Z}_4$	2,3,4	
.2	$V_9 \cdot D_4$	72	$[1,4^2]$	f_6, g_6	D_4	5	
.3	$\mathrm{AGL}_1(9)$	72	2t!	b_3, c_8	$\mathbb{Z}_8$	5	
.4	$V_9 \cdot Q_8$	72	2t!	b_4, c_4	Q_8	5, 6	
.5	$V_9 \cdot (\mathbb{Z}_8 \cdot_3 \mathbb{Z}_2)$	144	2t	c_8, e_6	$\mathbb{Z}_8 \cdot_3 \mathbb{Z}_2$	7	
.6	$V_9 \cdot (Q_8 \cdot \mathbb{Z}_3)$	216	2t	c_6, d_6	$Q_8 \cdot \mathbb{Z}_3$	7, 10	
.7	$\mathrm{AGL}_2(3)$	432	2t	a_8, b_8	$\mathrm{GL}_2(3)$	11	$S_1(2,3,9)$
.8	$L_2(8)$	504	4*	a_9, b_9	8.1	9	
.9	$\mathrm{P\Gamma L}_2(8)$	1512	4*	a_9, b_6	8.2	10	
.10	$\mathcal{A}_9$	9!/2	7p!	a_4, a_9	8.6	11	
.11	$\mathcal{S}_9$	9!	9p!	a_8, a_9	8.7		

Degree $n.i$	Group G	Order $\|G\|$	Trans/Rank	Generators	G_α	min H, $G < H \le \mathcal{S}_n$	Combinatorial Structures
10.1	$\mathcal{A}_5$	60	[1,3,6]	b_3, d_5	$\mathcal{S}_3$	2, 3	$S_2(2,3,10)$
.2	$\mathcal{S}_5$	120	[1,3,6]	b_6, d_5	D_6	4	srg$(10,3,0,1)$
.3	$L_2(9) \cong \mathcal{A}_6$	360	2p	b_5, c_5	9.1	4, 5, 6	
.4	$\mathrm{P\Sigma L}_2(9) \cong \mathcal{S}_6$	720	2p	b_6, c_6	9.2	7	$S_2(2,4,10)$
.5	$\mathrm{P\Delta L}_2(9) \cong M_{10}$	720	3t!	c_8, d_8	9.4	7	$S_3(3,5,10)$
.6	$\mathrm{PGL}_2(9)$	720	3t!	a_{10}, b_{10}	9.3	7	
.7	$\mathrm{P\Gamma L}_2(9)$	1440	3t	a_{10}, b_8	9.5	9	$S_1(3,4,10)$
.8	$\mathcal{A}_{10}$	10!/2	8p!	a_5, a_9	9.10	9	
.9	$\mathcal{S}_{10}$	10!	10p!	a_9, a_{10}	9.11		
11.1	$\mathbb{Z}_{11}$	11	$[1^{11}]$	$a_{11} : x+1$	1	2, 3	
.2	$F_{11,2}$	22	$[1,2^5]$	$x+1,\ -x$	$\mathbb{Z}_2$	4	
.3	$F_{11,5}$	55	2*!	$x+1,\ 4x$	$\mathbb{Z}_5$	4, 5	
.4	$\mathrm{AGL}_1(11)$	110	2t	$x+1,\ 2x$	$\mathbb{Z}_{10}$	8	
.5	$L_2(11)$	660	2p	a_{11}, b_5	10.1	6	$S_2(2,5,11)$
.6	M_{11}	7920	4t!	a_{11}, b_{11}	10.5	7	$S_1(4,5,11)$
.7	$\mathcal{A}_{11}$	11!/2	9p!	a_5, a_{11}	10.8	8	
.8	$\mathcal{S}_{11}$	11!	11p!	a_{10}, a_{11}	10.9		
12.1	$L_2(11)$	660	3*	a_{11}, d_{11}	11.3	2,3	
.2	$\mathrm{PGL}_2(11)$	1320	3t!	a_{12}, b_{12}	11.4	6	
.3	M_{11}	7920	3p	a_{11}, c_{11}	11.5	4	
.4	M_{12}	95040	5t!	a_{11}, b_{11}	11.6	5	$S_1(5,6,12)$
.5	$\mathcal{A}_{12}$	12!/2	10p!	a_6, a_{11}	11.7	6	
.6	$\mathcal{S}_{12}$	12!	12p!	a_{11}, a_{12}	11.8		
13.1	$\mathbb{Z}_{13}$	13	$[1^{13}]$	$a_{13} : x+1$	1	2, 3	$S_5(2,6,13)$
.2	$F_{13,2}$	26	$[1,2^6]$	$x+1, -x$	$\mathbb{Z}_2$	4, 5	$S_2(3,4,13)$
.3	$F_{13,3}$	39	$[1,3^4]$	$x+1, 3x$	$\mathbb{Z}_3$	5, 7	$S_1(2,3,13)$
.4	$F_{13,4}$	52	$[1,4^3]$	$x+1, 8x$	$\mathbb{Z}_4$	6	$S_2(3,4,13)$
.5	$F_{13,6}$	78	$[1,6^2]$	$x+1, 4x$	$\mathbb{Z}_6$	6, 8	
.6	$\mathrm{AGL}_1(13)$	156	2t!	$x+1, 2x$	$\mathbb{Z}_{12}$	9	$S_2(2,3,13)$,$S_5(2,5,13)$
.7	$L_3(3)$	5616	2t	a_{13}, b_{13}	≈ 9.7	8	$\mathrm{PG}_2(3)$
.8	$\mathcal{A}_{13}$	13!/2	11p!	a_6, a_{13}	12.5	9	
.9	$\mathcal{S}_{13}$	13!	13p!	a_{12}, a_{13}	12.6		
14.1	$L_2(13)$	1092	2p	a_{14}^2, b_{13}	13.5	2, 3	
.2	$\mathrm{PGL}_2(13)$	2184	3t!	a_{14}, b_{13}	13.6	4	$S_2(3,4,14)$

Degree $n.i$	Group G	Order $\|G\|$	Trans/Rank	Generators	G_α	min H, $G < H \leq \mathcal{S}_n$	Combinatorial Structures
14.3	$\mathcal{A}_{14}$	14!/2	12p!	a_7, a_{13}	13.8	4	
.4	$\mathcal{S}_{14}$	14!	14p!	a_{13}, a_{14}	13.9		
15.1	$\mathcal{A}_6$	360	[1,6,8]	a_{15}^3, b_2	$\mathcal{S}_4$	2, 3	
.2	$\mathcal{S}_6$	720	[1,6,8]	a_{15}^3, c_2	$\mathcal{S}_4 \times \mathbb{Z}_2$	4	
.3	$\mathcal{A}_7$	2520	2t	a_{15}^3, b_7	$L_3(2)$	4	
.4	$L_4(2) \cong \mathcal{A}_8$	20160	2t	a_{15}, c_2	$\mathrm{AGL}_3(2)$	5	$\mathrm{PG}_3(2)$
.5	$\mathcal{A}_{15}$	15!/2	13p!	a_7, a_{15}	14.3	6	
.6	$\mathcal{S}_{15}$	15!	15p!	a_{14}, a_{15}	14.4		
16.1	$V_{16} \cdot \mathbb{Z}_5$	80	$[1,5^3]$	a_{16}^8, b_{15}^3	$\mathbb{Z}_5$	2, 3	$S_1(3,4,16)$
.2	$V_{16} \cdot D_5$	160	$[1,5^3]$	a_{16}^4, b_{15}^3	D_5	5, 6, 9, 11	
.3	$\mathrm{AGL}_1(16)$	240	2t!	a_{16}^8, b_{15}	$\mathbb{Z}_{15}$	6	
.4	$V_{16} \cdot (S_3 \times \mathbb{Z}_3)$	288	[1,6,9]	b_6^2, c_6	$S_3 \times \mathbb{Z}_3$	8, 15	
.5	$V_{16} \cdot F_{5,4}$	320	[1,5,10]	a_{16}^2, b_{15}^3	$F_{5,4}$	10, 13, 14	
.6	$V_{16} \cdot (D_5 \times \mathbb{Z}_3)$	480	2t	a_{16}^4, b_{15}	$D_5 \times \mathbb{Z}_3$	10, 15	
.7	$V_{16} \cdot (F_{9,4})$	576	[1,6,9]	b_4, b_6^2	9.1	12, 17	
.8	$V_{16} \cdot (\mathcal{S}_3 \times \mathcal{S}_3)$	576	[1,6,9]	b_4^2, c_6	$S_3 \times S_3$	12, 16	
.9	$V_{16} \cdot \mathcal{A}_5$	960	[1,5,10]	a_{16}^4, b_3	10.1	13, 17	
.10	$V_{16} \cdot G_\alpha$	960	2t	a_{16}^2, b_{15}	$A_8 \cap (F_{5,4} \times \mathcal{S}_3)$	16	
.11	$V_{16} \cdot \mathcal{A}_5$	960	2t	a_{16}^4, b_6^2	$\mathcal{A}_5$	14, 15, 17	
.12	$S_4 \wr S_2$	1152	[1,6,9]	b_4, c_6	9.2	$18^{b_4 b_3}$	
.13	$V_{16} \cdot \mathcal{S}_5$	1920	[1,5,10]	a_{16}^2, b_3	$\mathcal{S}_5$	18	$\mathrm{srg}(16,5,0,2)$
.14	$V_{16} \cdot \mathcal{S}_5$	1920	2t	a_{16}^2, b_6	10.2	16, 18, 19	
.15	$\mathrm{AGL}_2(4)$	2880	2t	b_6^2, b_{15}	$\mathcal{A}_5 \times \mathbb{Z}_3$	16	
.16	$\mathrm{A\Gamma L}_2(4)$	5760	2t	b_6, b_{15}	$A_8 \cap (\mathcal{S}_5 \times \mathcal{S}_3)$	20	$S_1(2,4,16)$
.17	$V_{16} \cdot \mathcal{A}_6$	5760	2p	b_3, b_6^2	15.1	18, $19^{b_{15}}$	
.18	$V_{16} \cdot \mathcal{S}_6$	11520	2p	b_3, b_6	15.2	20	
.19	$V_{16} \cdot \mathcal{A}_7$	40320	3t	a_{16}^2, b_7	15.3	20	
.20	$\mathrm{AGL}_4(2)$	322560	3t	b_3, b_{15}	15.4	21	$S_1(3,4,16)$
.21	$\mathcal{A}_{16}$	16!/2	14p!	a_8, a_{15}	15.5	22	
.22	$\mathcal{S}_{16}$	16!	16p!	a_{15}, a_{16}	15.6		
17.1	$\mathbb{Z}_{17}$	17	$[1^{17}]$	$a_{17} : x+1$	1	2	
.2	D_{17}	34	$[1,2^8]$	$x+1, -x$	$\mathbb{Z}_2$	3, 6	
.3	$F_{17,4}$	68	$[1,4^4]$	$x+1, 13x$	$\mathbb{Z}_4$	4, 7	
.4	$F_{17,8}$	136	$[1,8^2]$	$x+1, 9x$	$\mathbb{Z}_8$	5, 8	

Degree $n.i$	Group G	Order $\|G\|$	Trans/Rank	Generators	G_α	min H, $G < H \leq \mathcal{S}_n$	Combinatorial Structures
17.5	$\mathrm{AGL}_1(17)$	272	2t!	$x+1, 3x$	$\mathbb{Z}_{16}$	10	$S_4(4,5,17)$
.6	$L_2(16)$	4080	3t!	a_{17}, b_8^4	16.3	7	$S_6(4,7,17)$
.7	$\mathrm{PZL}_2(16)$	8160	3t	a_{17}, b_8^2	16.6	8	
.8	$\mathrm{P\Gamma L}_2(16)$	16320	3t	a_{17}, b_8	16.10	9	$S_1(3,5,17)$
.9	$\mathcal{A}_{17}$	17!/2	15p!	a_8, a_{17}	16.21	10	
.10	$\mathcal{S}_{17}$	17!	17p!	a_{16}, a_{17}	16.22		
18.1	$L_2(17)$	2448	2p	b_{17}, a_{18}^2	17.4	2, 3	
.2	$\mathrm{PGL}_2(17)$	4896	3t!	a_{18}, b_{17}	17.5	4	
.3	$\mathcal{A}_{18}$	18!/2	16p!	a_9, a_{17}	17.9	4	
.4	$\mathcal{S}_{18}$	18!	18p!	a_{17}, a_{18}	17.10		
19.1	$\mathbb{Z}_{19}$	19	$[1^{19}]$	$a_{19} : x+1$	1	2, 3	$S_1(2,3,19)$
.2	D_{19}	38	$[1,2^9]$	$x+1, -x$	$\mathbb{Z}_2$	4	
.3	$F_{19,3}$	57	$[1,3^6]$	$x+1, 7x$	$\mathbb{Z}_3$	4, 5	$S_1(2,3,19)$
.4	$F_{19,6}$	114	$[1,6^3]$	$x+1, 8x$	$\mathbb{Z}_6$	6	
.5	$F_{19,9}$	171	2*!	$x+1, 4x$	$\mathbb{Z}_9$	6, 7	$S_{112}(5,8,19)$
.6	$\mathrm{AGL}_1(19)$	342	2t!	$x+1, 2x$	$\mathbb{Z}_{18}$	8	
.7	$\mathcal{A}_{19}$	19!/2	17p!	a_9, a_{19}	18.3	8	
.8	$\mathcal{S}_{19}$	19!	19p!	a_{18}, a_{19}	18.4		
20.1	$L_2(19)$	3420	3*	a_{19}, b_{20}^2	19.5	2, 3	$S_{112}(6,9,20)$
.2	$\mathrm{PGL}_2(19)$	6840	3t!	a_{19}, b_{20}	19.6	4	
.3	$\mathcal{A}_{20}$	20!/2	18p!	a_{10}, a_{19}	19.7	4	
.4	$\mathcal{S}_{20}$	20!	20p!	a_{19}, a_{20}	19.8		
21.1	$\mathrm{PGL}_2(7)$	336	$[1,4,8^2]$	a_7, b_2	D_8	9	
.2	$\mathcal{A}_7$	2520	$[1,10^2]$	a_7, c_4^2	10.2	3, 8	
.3	$\mathcal{S}_7$	5040	$[1,10^2]$	a_7, c_4	$\mathcal{S}_5 \times \mathcal{S}_2$	9	
.4	$M_{21} \cong L_3(4)$	20160	2t	a_{21}^3, b_4^2	≈ 16.11	5, 6	
.5	$\mathrm{P\Sigma L}_3(4)$	40320	2t	a_{21}^3, b_4	≈ 16.14	7	
.6	$\mathrm{PGL}_3(4)$	60480	2t	a_{21}, b_4^2	≈ 16.15	7, 8	
.7	$\mathrm{P\Gamma L}_3(4)$	120960	2t	a_{21}, b_4	≈ 16.16	9	$\mathrm{PG}_2(4)$
.8	$\mathcal{A}_{21}$	21!/2	19p!	a_{10}, a_{21}	20.3	9	
.9	$\mathcal{S}_{21}$	21!	21p!	a_{20}, a_{21}	20.4		
22.1	M_{22}	443520	3t	a_{11}, b_{10}^2	21.4	2, 3	$S_1(3,6,22)$
.2	$M_{22} \cdot \mathbb{Z}_2$	887040	3t	a_{11}, b_{10}	21.5	4	
.3	$\mathcal{A}_{22}$	22!/2	20p!	a_{11}, a_{21}	21.8	4	

Degree $n.i$	Group G	Order $\lvert G\rvert$	Trans/Rank	Generators	G_α	min H, $G < H \leq \mathcal{S}_n$	Combinatorial Structures
22.4	$\mathcal{S}_{22}$	22!	22p!	a_{21}, a_{22}	21.9		
23.1	$\mathbb{Z}_{23}$	23	$[1^{23}]$	$a_{23} : x+1$	1	2, 3	
.2	D_{23}	46	$[1,2^{11}]$	$x+1, -x$	$\mathbb{Z}_2$	4	
.3	$F_{23,11}$	253	2*	$x+1, 2x$	$\mathbb{Z}_{11}$	4, 5	$S_1(4,5,23)$
.4	$\mathrm{AGL}_1(23)$	506	2t!	$x+1, 5x$	$\mathbb{Z}_{22}$	7	
.5	M_{23}	10200960	4t	a_{23}, b_2	22.1	6	$S_1(4,7,23)$
.6	$\mathcal{A}_{23}$	23!/2	21p!	a_{11}, a_{23}	22.3	7	
.7	$\mathcal{S}_{23}$	23!	23p!	a_{22}, a_{23}	22.4		
24.1	$L_2(23)$	6072	3*	a_{23}, b_{24}^2	23.3	2, 3	$S_1(5,6,24)$
.2	$\mathrm{PGL}_2(23)$	12144	3t!	a_{23}, b_{24}	23.4	5	
.3	M_{24}	244823040	5t	a_{23}, b_2	23.5	4	$S_1(5,8,24)$
.4	$\mathcal{A}_{24}$	24!/2	22p!	a_{12}, a_{23}	23.6	5	
.5	$\mathcal{S}_{24}$	24!	24p!	a_{23}, a_{24}	23.7		
25.1	$F_{25,3}$	75	$[1,3^8]$	a_5, b_{24}^8	$\mathbb{Z}_3$	2, 3	
.2	$F_{25,6}$	150	$[1,6^4]$	a_5, b_{24}^4	$\mathbb{Z}_6$	7,8,9,13	
.3	$V_{25} \cdot \mathcal{S}_3$	150	$[1,3^4,6^2]$	b_{10}, b_{24}^8	$\mathcal{S}_3$	9	
.4	$D_5 \wr \mathcal{S}_2$	200	$[1,4^4,8]$	b_2, b_{20}^2	D_4	11, 22	
.5	$F_{25,8}$	200	$[1,8^3]$	b_{24}^6, c_8	$\mathbb{Z}_8$	10, $14^{c_{10}}$, 15	
.6	$V_{25} \cdot Q_8$	200	$[1,8^3]$	b_4, d_4	Q_8	11, 13	
.7	$F_{25,12}$	300	$[1,12^2]$	a_5, b_{24}^2	$\mathbb{Z}_{12}$	12, 14, 15, 17	
.8	$V_{25} \cdot (\mathbb{Z}_3 \cdot \mathbb{Z}_4)$	300	$[1,12^2]$	b_{24}^4, c_4	$\mathbb{Z}_3 \cdot \mathbb{Z}_4$	12, 20	
.9	$V_{25} \cdot D_6$	300	$[1,6^4]$	b_{24}^4, b_{10}	D_6	12	
.10	$V_{25} \cdot (\mathbb{Z}_8 \cdot_5 \mathbb{Z}_2)$	400	[1,8,16]	b_{20}^2, c_8	$\mathbb{Z}_8 \cdot_5 \mathbb{Z}_2$	16, $18^{c_{20}}$, 25	
.11	$V_{25} \cdot (Q_8 \cdot \mathbb{Z}_2)$	400	$[1,8^3]$	b_{24}^6, b_4, b_2	$Q_8 \cdot \mathbb{Z}_2$	16, 17, 24	
.12	$V_{25} \cdot (\mathcal{S}_3 \times \mathbb{Z}_4)$	600	$[1,12^2]$	b_{24}^2, b_{10}	$\mathcal{S}_3 \times \mathbb{Z}_4$	18, 21	
.13	$V_{25} \cdot (Q_8 \cdot \mathbb{Z}_3)$	600	2t!	b_{24}^4, b_4	$Q_8 \cdot \mathbb{Z}_3$	17, 20	
.14	$\mathrm{AGL}_1(25)$	600	2t!	b_{24}, a_5	$\mathbb{Z}_{24}$	18	
.15	$V_{25} \cdot (\mathbb{Z}_3 \cdot \mathbb{Z}_8)$	600	2t!	b_{24}^2, c_8	$\mathbb{Z}_3 \cdot \mathbb{Z}_8$	18, 19	
.16	$\mathrm{AGL}_1(5) \wr \mathcal{S}_2$	800	[1,8,16]	b_{20}, b_8	$\mathbb{Z}_4 \wr \mathbb{Z}_2$	19, 26	
.17	$V_{25} \cdot G_\alpha$	1200	2t	b_{24}^2, b_{20}^2	$(Q_8 \cdot \mathbb{Z}_2) \cdot \mathbb{Z}_3$	19, 21	
.18	$V_{25} \cdot (\mathbb{Z}_{24} \cdot_5 \mathbb{Z}_2)$	1200	2t	b_{24}, b_{10}	$\mathbb{Z}_{24} \cdot_5 \mathbb{Z}_2$	23	
.19	$V_{25} \cdot G_\alpha$	2400	2t	b_{24}^2, b_{20}	$(Q_8 \cdot \mathbb{Z}_3) \cdot \mathbb{Z}_4$	23	
.20	$\mathrm{ASL}_2(5)$	3000	2t	b_{24}^4, c_{20}^2	$\approx \mathbb{Z}_2 \backslash 6.1$	21	
.21	$\mathcal{A}_{25} \cap \mathrm{AGL}_2(5)$	6000	2t	b_{24}^2, c_{20}	$\approx \mathbb{Z}_4 \backslash 6.1$	23, 27	

Degree $n.i$	Group G	Order $\|G\|$	Trans/Rank	Generators	G_α	min H, $G < H \leq \mathcal{S}_n$	Combinatorial Structures
25.22	$\mathcal{A}_5 \wr \mathcal{S}_2$	7200	[1,8,16]	b_6, b_{12}^4	16.4	24	
.23	$\mathrm{AGL}_2(5)$	12000	2t	b_{24}, b_{20}	$\mathrm{GL}_2(5)$	28	
.24	$\mathcal{A}_5^2 \cdot V_4$	14400	[1,8,16]	b_6, b_{12}^2	16.8	26, 27	
.25	$\mathcal{A}_5^2 \cdot \mathbb{Z}_4$	14400	[1,8,16]	b_6^2, b_{12}	16.7	26	
.26	$\mathcal{S}_5 \wr \mathcal{S}_2$	28800	[1,8,16]	b_6, b_{12}	16.12	28	
.27	$\mathcal{A}_{25}$	25!/2	23p!	a_{12}, a_{25}	24.4	28	
.28	$\mathcal{S}_{25}$	25!	25p!	a_{24}, a_{25}	24.5		
26.1	$L_2(25)$	7800	2p	a_{26}^2, b_{10}^2	25.7	2,3,4	
.2	$\mathrm{P\Sigma L}_2(25)$	15600	2p	a_{26}^2, b_{10}	25.12	5,6	
.3	$\mathrm{PGL}_2(25)$	15600	3t!	a_{26}, b_{10}^2	25.14	5	
.4	$\mathrm{P\Delta L}_2(25)$	15600	3t!	a_{26}^2, b_{12}	25.15	5	
.5	$\mathrm{P\Gamma L}_2(25)$	31200	3t	a_{26}, b_{10}	25.18	7	
.6	$\mathcal{A}_{26}$	26!/2	24p!	a_{13}, a_{25}	25.27	7	
.7	$\mathcal{S}_{26}$	26!	26p!	a_{25}, a_{26}	25.28		
27.1	$V_{27} \cdot \mathcal{A}_4$	324	[1,4^2,6,12]	a_9, b_6^2	$\mathcal{A}_4$	3,4,5	
.2	$F_{27,13}$	351	[1,13^2]	a_9^3, b_{26}^2	$\mathbb{Z}_{13}$	6,7	$S_1(4,6,27)$
.3	$V_{27} \cdot \mathcal{S}_4$	648	[1,6,8,12]	a_9, b_4	$\mathcal{S}_4$	8,12	
.4	$V_{27} \cdot \mathcal{S}_4$	648	[1,4^2,6,12]	a_9, c_4	$\mathcal{S}_4$	8	
.5	$V_{27} \cdot (V8 \cdot \mathbb{Z}_3)$	648	[1,6,8,12]	a_9, b_6	$V_8 \cdot \mathbb{Z}_3$	8	
.6	$\mathrm{AGL}_1(27)$	702	2t!	a_9^3, b_{26}	$\mathbb{Z}_{26}$	9	
.7	$V_{27} \cdot F_{13,3}$	1053	[1,13^2]	a_9, d_6^2	13.3	9,12	
.8	$V_{27} \cdot (\mathcal{S}_4 \times \mathbb{Z}_2)$	1296	[1,6,8,12]	b_4, c_6	$\mathcal{S}_4 \times \mathbb{Z}_2$	13	
.9	$\mathrm{A\Gamma L}_1(27)$	2106	2t	a_9, d_6	$\mathbb{Z}_{26} \cdot \mathbb{Z}_3$	13	
.10	$U_4(2) \cong S_4(3)$	25920	[1,10,16]	a_9, b_{12}^2	16.9	11	
.11	$U_4(2) \cdot \mathbb{Z}_2$	51840	[1,10,16]	a_9, b_{12}	16.13	14	srg(27, 10, 1, 5)
.12	$\mathrm{ASL}_3(3)$	151632	2t	a_9, b_{24}^2	≈ 13.7	13,14	
.13	$\mathrm{AGL}(3,3)$	303264	2t	a_9, b_{24}	$\approx \mathbb{Z}_2 \backslash 13.7$	15	
.14	$\mathcal{A}_{27}$	27!/2	25p!	a_{13}, a_{27}	26.6	15	
.15	$\mathcal{S}_{27}$	27!	27p!	a_{26}, a_{27}	26.7		
28.1	$\mathrm{PGL}_2(7)$	336	[1,3,6^2,12]	a_7, b_6	D_6	6, 10	
.2	$L_2(8)$	504	[1,9^3]	a_7, b_9^3	D_9	3	
.3	$\mathrm{P\Gamma L}_2(8)$	1512	2t	a_7, b_9	$D_9 \cdot \mathbb{Z}_3$	12	
.4	$U_3(3)$	6048	2t	a_7, b_{12}^2	$T_{27} \cdot \mathbb{Z}_8$	6	

Degree $n.i$	Group G	Order $\|G\|$	Trans/Rank	Generators	G_α	min H, $G < H \leq \mathcal{S}_n$	Combinatorial Structures
28.5	$L_2(27)$	9828	2p	a_7, b_{26}^2	27.2	7,9	$S_1(5,7,28)$
.6	$\mathrm{P\Sigma U}_3(3)$	12096	2t	a_7, b_{12}	$T_{27} \cdot (\mathbb{Z}_8 \cdot_3 \mathbb{Z}_2)$	12	
.7	$\mathrm{PGL}_2(27)$	19656	3t!	a_7, b_{26}	27.6	11	
.8	$\mathcal{A}_8$	20160	[1,12,15]	a_7, c_6	$\mathcal{S}_6$	10	
.9	$\mathrm{P\Sigma L}_2(27)$	29484	2p	a_7, c_{12}^2	27.7	11,13	
.10	$\mathcal{S}_8$	40320	[1,12,15]	a_7, b_8	$\mathcal{S}_6 \times \mathbb{Z}_2$	12	
.11	$\mathrm{P\Gamma L}_2(27)$	58968	3t	a_7, c_{12}	27.9	14	
.12	$S_6(2)$	1451520	2p	a_7, d_{12}	27.11	13	
.13	$\mathcal{A}_{28}$	28!/2	26p!	a_{14}, a_{27}	27.14	14	
.14	$\mathcal{S}_{28}$	28!	28p!	a_{27}, a_{28}	27.15		
29.1	$\mathbb{Z}_{29}$	29	$[1^{29}]$	$a_{29} : x+1$	1	2,4	
.2	$F_{29,2}$	58	$[1,2^{14}]$	$x+1, 28x$	$\mathbb{Z}_2$	3,5	
.3	$F_{29,4}$	116	$[1,4^7]$	$x+1, 12x$	$\mathbb{Z}_4$	6	
.4	$F_{29,7}$	203	$[1,7^4]$	$x+1, 16x$	$\mathbb{Z}_7$	5	
.5	$F_{29,14}$	406	$[1,14^2]$	$x+1, 4x$	$\mathbb{Z}_{14}$	6,7	
.6	$\mathrm{AGL}_1(29)$	812	2t!	$x+1, 2x$	$\mathbb{Z}_{28}$	8	
.7	$\mathcal{A}_{29}$	29!/2	27p!	a_{14}, a_{29}	28.13	8	
.8	$\mathcal{S}_{29}$	29!	29p!	a_{28}, a_{29}	28.14		
30.1	$L_2(29)$	12180	2p	a_{29}, b_{30}^2	29.5	2,3	
.2	$\mathrm{PGL}_2(29)$	24360	3t!	a_{29}, b_{30}	29.6	4	
.3	$\mathcal{A}_{30}$	30!/2	28p!	a_{15}, a_{29}	29.7	4	
.4	$\mathcal{S}_{30}$	30!	30p!	a_{29}, a_{30}	29.8		
31.1	$\mathbb{Z}_{31}$	31	$[1^{31}]$	$a_{31} : x+1$	1	2, 3, 4	
.2	$F_{31,2}$	62	$[1,2^{15}]$	$x+1,\ -x$	$\mathbb{Z}_2$	5, 6	
.3	$F_{31,3}$	93	$[1,3^{10}]$	$x+1,\ 5x$	$\mathbb{Z}_3$	5, 7, 9	
.4	$F_{31,5}$	155	$[1,5^6]$	$x+1,\ 2x$	$\mathbb{Z}_5$	6, 7, 10	$S_{36}(4,6,31)$, $S_{10}(5,6,31)$
.5	$F_{31,6}$	186	$[1,6^5]$	$x+1,\ 6x$	$\mathbb{Z}_6$	8	
.6	$F_{31,10}$	310	$[1,10^3]$	$x+1,\ 15x$	$\mathbb{Z}_{10}$	8	
.7	$F_{31,15}$	465	$[1,15^2]$	$x+1,\ 7x$	$\mathbb{Z}_{15}$	8, 11	
.8	$\mathrm{AGL}_1(31)$	930	2t!	$x+1,\ 3x$	$\mathbb{Z}_{30}$	12	
.9	$L_3(5)$	372000	2t	$a_{31},\ b_{31}$	$\mathrm{AGL}_2(5)$	11	
.10	$L_5(2)$	9999360	2t	$a_{31},\ c_{31}$	$\mathrm{AGL}_4(2)$	11	
.11	$\mathcal{A}_{31}$	31!/2	29p!	$a_{15},\ a_{31}$	30.3	12	

Degree $n.i$	Group G	Order $\lvert G\rvert$	Trans/Rank	Generators	G_α	min H, $G < H \le \mathcal{S}_n$	Combinatorial Structures
31.12	$\mathcal{S}_{31}$	31!	31p!	a_{30}, a_{31}	30.4		
32.1	$\mathrm{AGL}_1(32)$	992	2p!	a_{31}, b_{10}^5	31.1	2	
.2	$\mathrm{A\Gamma L}_1(32)$	4960	3*!	a_{31}, b_{10}	31.4	5	$S_{36}(5,7,32)$,$S_{10}(6,7,32)$
.3	$L_2(31)$	14880	3*	a_{31}, b_{32}^2	31.7	4, 6	
.4	$\mathrm{PGL}_2(31)$	29760	3t!	a_{31}, b_{32}	31.8	7	
.5	$\mathrm{AGL}_5(2)$	319979520	3t	a_{31}, b_{31}	31.10	6	
.6	$\mathcal{A}_{32}$	32!/2	30p!	a_{16}, a_{31}	31.11	7	
.7	$\mathcal{S}_{32}$	32!	32p!	a_{31}, a_{32}	31.12		
33.1	$L_2(32)$	32736	3p!	a_{33}, b_{15}^5	32.1	2	
.2	$\mathrm{P\Gamma L}_2(32)$	163680	4*	a_{33}, b_{15}	32.2	3	$S_{36}(6,8,33)$,$S_{10}(7,8,33)$
.3	$\mathcal{A}_{33}$	33!/2	31p!	a_{16}, a_{33}	32.6	4	
.4	$\mathcal{S}_{33}$	33!	33p!	a_{32}, a_{33}	32.7		
34.1	$\mathcal{A}_{34}$	34!/2	32p!	a_{17}, a_{33}	33.3	2	
.2	$\mathcal{S}_{34}$	34!	34p!	a_{33}, a_{34}	33.4		
35.1	$\mathcal{A}_7$	2520	[1,4,12,18]	$a_{15,5}^3, b_{12}^2$	$\approx \mathbb{Z}_2\backslash 9.1$	2, 3	
.2	$\mathcal{S}_7$	5040	[1,4,12,18]	$a_{15,5}^3, b_{12}$	≈ 9.5	4	
.3	$\mathcal{A}_8$	20160	[1,16,18]	$a_{15,5}, b_{12}^2$	16.8	4	
.4	$\mathcal{S}_8$	40320	[1,16,18]	$a_{15,5}, b_{12}$	16.12	5	srg(35, 16, 6, 8)
.5	$\mathcal{A}_{35}$	35!/2	33p!	a_{17}, a_{35}	34.1	6	
.6	$\mathcal{S}_{35}$	35!	35p!	a_{34}, a_{35}	34.2		
36.1	$L_2(8)$	504	$[1,7^3,14]$	a_{36}^{12}, b_8^4	D_7	5	
.2	$\mathrm{PGL}_2(9)$	720	$[1,5,10^3]$	b_{10}, c_8^2	D_{10}	4,15	
.3	M_{10}	720	[1,5,10,20]	b_{10}^2, c_8	$F_{5,4}$	$4{,}15^{b_{12}}$	
.4	$\mathrm{P\Gamma L}_2(9)$	1440	[1,5,10,20]	b_{10}, c_8	$F_{5,4} \times \mathbb{Z}_2$	17	
.5	$\mathrm{P\Gamma L}_2(8)$	1512	[1,14,21]	a_{36}^4, b_8^4	≈ 7.4	14,20	
.6	$U_3(3)$	6048	$[1,7^2,21]$	a_{36}^{12}, c_{12}^2	7.5	8	
.7	$\mathcal{A}_5 \wr \mathcal{S}_2$	7200	[1,10,25]	b_{10}, b_{12}^4	25.4	9,15	
.8	$\mathrm{P\Sigma U}_3(3)$	12096	[1,14,21]	a_{36}^{12}, c_{12}	21.1	20	srg(36, 14, 4, 6)
.9	$\mathcal{A}_5^2 \cdot V_4$	14400	[1,10,25]	b_{10}, b_{12}^2	25.11	12,17	
.10	$\mathcal{A}_5^2 \cdot \mathbb{Z}_4$	14400	[1,10,25]	b_{10}^2, b_{12}	25.10	12,18	
.11	$S_4(3) \cong U_4(2)$	25920	[1,15,20]	a_{36}^4, b_6^2	15.2	13	
.12	$\mathcal{S}_5 \wr \mathcal{S}_2$	28800	[1,10,25]	b_{10}, b_{12}	25.16	19	
.13	$U_4(2) \cdot \mathbb{Z}_2$	51840	[1,15,20]	a_{36}^4, b_6	$\mathcal{S}_6 \times \mathcal{S}_2$	20	srg(36, 15, 6, 6)

Degree $n.i$	Group G	Order $\|G\|$	Trans/Rank	Generators	G_α	min H, $G < H \le \mathcal{S}_n$	Combinatorial Structures
36.14	$\mathcal{A}_9$	181440	[1,14,21]	a_{36}^4, b_8^2	21.3	16, 21	
.15	$\mathcal{A}_6 \wr \mathcal{S}_2$	259200	[1,10,25]	a_{36}^{12}, b_{10}	25.22	17	
.16	$\mathcal{S}_9$	362880	[1,14,21]	a_{36}^4, b_8	$\mathcal{S}_7 \times \mathcal{S}_2$	22	
.17	$\mathcal{A}_6^2 \cdot V_4$	518400	[1,10,25]	a_{36}^6, b_{10}	25.24	19	
.18	$\mathcal{A}_6^2 \cdot \mathbb{Z}_4$	518400	[1,10,25]	a_{36}^3, b_{10}^2	25.25	19	
.19	$\mathcal{S}_6 \wr \mathcal{S}_2$	1036800	[1,10,25]	a_{36}^3, b_{10}	25.26	22	
.20	$S_6(2)$	1451520	2p	a_{36}^4, c_{12}	35.4	21	
.21	$\mathcal{A}_{36}$	36!/2	34p!	a_{18}, a_{35}	35.5	22	
.22	$\mathcal{S}_{36}$	36!	36p!	a_{35}, a_{36}	35.6		
37.1	$\mathbb{Z}_{37}$	37	$[1^{37}]$	$a_{37} : x+1$	1	2,3	
.2	$F_{37,2}$	74	$[1, 2^{18}]$	$x+1, -x$	$\mathbb{Z}_2$	4,5	
.3	$F_{37,3}$	111	$[1, 3^{12}]$	$x+1, 26x$	$\mathbb{Z}_3$	5,6	
.4	$F_{37,4}$	148	$[1, 4^9]$	$x+1, 31x$	$\mathbb{Z}_4$	7	
.5	$F_{37,6}$	222	$[1, 6^6]$	$x+1, 27x$	$\mathbb{Z}_6$	7,8	
.6	$F_{37,9}$	333	$[1, 9^4]$	$x+1, 16x$	$\mathbb{Z}_9$	8,11	
.7	$F_{37,12}$	444	$[1, 12^3]$	$x+1, 8x$	$\mathbb{Z}_{12}$	9	
.8	$F_{37,18}$	666	$[1, 18^2]$	$x+1, 4x$	$\mathbb{Z}_{18}$	9,10	
.9	$\mathrm{AGL}_1(37)$	1332	2t!	$x+1, 2x$	$\mathbb{Z}_{36}$	11	
.10	$\mathcal{A}_{37}$	37!/2	35p!	a_{18}, a_{37}	36.21	11	
.11	$\mathcal{S}_{37}$	37!	37p!	a_{36}, a_{37}	36.22		
38.1	$L_2(37)$	25308	2p	a_{37}, b_{38}^2	37.8	2,3	
.2	$\mathrm{PGL}_2(37)$	50616	3t!	a_{37}, b_{38}	37.9	4	
.3	$\mathcal{A}_{38}$	38!/2	36p!	a_{19}, a_{37}	37.10	4	
.4	$\mathcal{S}_{38}$	38!	38p!	a_{36}, a_{37}	37.11		
39.1	$\mathcal{A}_{39}$	39!/2	37p!	a_{19}, a_{39}	38.3	2	
.2	$\mathcal{S}_{39}$	39!	39p!	a_{38}, a_{39}	38.4		
40.1	$S_4(3) \cong U_4(2)$	25920	[1,12,27]	a_{40}^8, b_5	$T_{27} \cdot (Q_8 \cdot \mathbb{Z}_3)$	3	
.2	$S_4(3) \cong U_4(2)$	25920	[1,12,27]	a_{40}^8, b_{10}^2	27.4	4	
.3	$U_4(2) \cdot \mathbb{Z}_2$	51840	[1,12,27]	a_{40}^4, b_5	$T_{27} \cdot (Q_8 \cdot \mathcal{S}_3)$	5	
.4	$U_4(2) \cdot \mathbb{Z}_2$	51840	[1,12,27]	a_{40}^8, b_{10}	27.8	8	
.5	$L_4(3)$	6065280	2t	a_{40}^2, b_5	$\mathrm{ASL}_3(3)$	6,7	
.6	$\mathrm{PGL}_4(3)$	12130560	2t	a_{40}, b_5	$\mathrm{AGL}_3(3)$	8	
.7	$\mathcal{A}_{40}$	40!/2	38p!	a_{20}, a_{39}	39.1	8	

Degree $n.i$	Group G	Order $\lvert G \rvert$	Trans/Rank	Generators	G_α	min H, $G < H \le \mathcal{S}_n$	Combinatorial Structures
40.8	$\mathcal{S}_{40}$	40!	40p!	a_{39}, a_{40}	39.2		
41.1	$\mathbb{Z}_{41}$	41	$[1^{41}]$	$a_{41} : x+1$	1	2,4	
.2	$F_{41,2}$	82	$[1,2^{20}]$	$x+1, 40x$	$\mathbb{Z}_2$	3,6	
.3	$F_{41,4}$	164	$[1,4^{10}]$	$x+1, 32x$	$\mathbb{Z}_4$	5,7	
.4	$F_{41,5}$	205	$[1,5^{8}]$	$x+1, 10x$	$\mathbb{Z}_5$	6	
.5	$F_{41,8}$	328	$[1,8^{5}]$	$x+1, 27x$	$\mathbb{Z}_8$	8	
.6	$F_{41,10}$	410	$[1,10^{4}]$	$x+1, 25x$	$\mathbb{Z}_{10}$	7	
.7	$F_{41,20}$	820	$[1,20^{2}]$	$x+1, 36x$	$\mathbb{Z}_{20}$	8,9	
.8	$\mathrm{AGL}_1(41)$	1640	2t!	$x+1, 6x$	$\mathbb{Z}_{40}$	10	
.9	$\mathcal{A}_{41}$	41!/2	39p!	a_{20}, a_{41}	40.7	10	
.10	$\mathcal{S}_{41}$	41!	41p!	a_{40}, a_{41}	40.8		
42.1	$L_2(41)$	34440	2p	a_{41}, b_{42}^2	41.7	2,3	
.2	$\mathrm{PGL}_2(41)$	68880	3t!	a_{41}, b_{42}	41.8	4	
.3	$\mathcal{A}_{42}$	42!/2	40p!	a_{21}, a_{41}	41.9	4	
.4	$\mathcal{S}_{42}$	42!	42p!	a_{41}, a_{42}	41.10		
43.1	$\mathbb{Z}_{43}$	43	$[1^{43}]$	$a_{43} : x+1$	1	2,3,5	
.2	$F_{43,2}$	86	$[1,2^{21}]$	$x+1, 42x$	$\mathbb{Z}_2$	4,6	
.3	$F_{43,3}$	129	$[1,3^{14}]$	$x+1, 36x$	$\mathbb{Z}_3$	4,7	
.4	$F_{43,6}$	258	$[1,6^{7}]$	$x+1, 37x$	$\mathbb{Z}_6$	8	
.5	$F_{43,7}$	301	$[1,7^{6}]$	$x+1, 41x$	$\mathbb{Z}_7$	6,7	
.6	$F_{43,14}$	602	$[1,14^{3}]$	$x+1, 27x$	$\mathbb{Z}_{14}$	8	
.7	$F_{43,21}$	903	$[1,21^{2}]$	$x+1, 9x$	$\mathbb{Z}_{21}$	8,9	
.8	$\mathrm{AGL}_1(43)$	1806	2t!	$x+1, 3x$	$\mathbb{Z}_{42}$	10	
.9	$\mathcal{A}_{43}$	43!/2	41p!	a_{21}, a_{43}	42.3	10	
.10	$\mathcal{S}_{43}$	43!	43p!	a_{42}, a_{43}	42.4		
44.1	$L_2(43)$	39732	2p	a_{43}, b_{44}^2	43.7	2,3	
.2	$\mathrm{PGL}_2(43)$	79464	3t!	a_{43}, b_{44}	43.8	4	
.3	$\mathcal{A}_{44}$	44!/2	42p!	a_{22}, a_{43}	43.9	4	
.4	$\mathcal{S}_{44}$	44!	44p!	a_{43}, a_{44}	43.10		
45.1	$\mathrm{PGL}_2(9)$	720	$[1,4,8^3,16]$	b_8^2, b_{10}	D_8	3	
.2	M_{10}	720	$[1,4,8,16^2]$	b_8, b_{10}^2	$\mathbb{Z}_8 \cdot_3 \mathbb{Z}_2$	3, 6	
.3	$\mathrm{P\Gamma L}_2(9)$	1440	$[1,4,8,16^2]$	b_8, b_{10}	$\mathbb{Z}_8 \cdot Aut\ \mathbb{Z}_8$	7	
.4	$S_4(3) \cong U_4(2)$	25920	$[1,12,32]$	b_{10}^2, d_8^2	$\mathbb{Z}_2 \backslash 16.4$	5, 8	

Degree $n.i$	Group G	Order $\|G\|$	Trans/Rank	Generators	G_α	min H, $G < H \leq \mathcal{S}_n$	Combinatorial Structures
45.5	$U_4(2) \cdot \mathbb{Z}_2$	51840	[1,12,32]	b_{10}^2, d_8	$\mathbb{Z}_2 \backslash 16.8$	9	srg(45, 12, 3, 3)
.6	$\mathcal{A}_{10}$	1814400	[1,16,28]	b_{10}^2, c_8^2	28.10	7	
.7	$\mathcal{S}_{10}$	3628800	[1,16,28]	b_{10}, c_8	$\mathcal{S}_8 \times \mathcal{S}_2$	8	
.8	$\mathcal{A}_{45}$	45!/2	43p!	a_{22}, a_{45}	44.3	9	
.9	$\mathcal{S}_{45}$	45!	45p!	a_{44}, a_{45}	44.4		
46.1	$\mathcal{A}_{46}$	46!/2	44p!	a_{23}, a_{45}	45.8	2	
.2	$\mathcal{S}_{46}$	46!	46p!	a_{45}, a_{46}	45.9		
47.1	$\mathbb{Z}_{47}$	47	$[1^{47}]$	$a_{47} : x+1$	1	2, 3	
.2	$F_{47,2}$	94	$[1,2^{23}]$	$x+1, 46x$	$\mathbb{Z}_2$	4	
.3	$F_{47,23}$	1081	$[1,23^2]$	$x+1, 25x$	$\mathbb{Z}_{23}$	4, 5	$S_1(4,5,47)$
.4	$\mathrm{AGL}_1 47$	2162	2t!	$x+1, 5x$	$\mathbb{Z}_{46}$	6	
.5	$\mathcal{A}_{47}$	47!/2	45p!	a_{23}, a_{47}	46.1	6	
.6	$\mathcal{S}_{47}$	47!	47p!	a_{46}, a_{47}	46.2		
48.1	$L_2(47)$	51888	2p	a_{48}^2, b_{47}	47.3	2,3	$S_1(5,6,48)$
.2	$\mathrm{PGL}_2(47)$	103776	3t!	a_{48}, b_{47}	47.4	4	
.3	$\mathcal{A}_{48}$	48!/2	46p!	a_{24}, a_{47}	47.5	4	
.4	$\mathcal{S}_{48}$	48!	48p!	a_{47}, a_{48}	47.6		
49.1	$F_{49,4}$	196	$[1,4^{12}]$	a_7, b_{48}^{12}	$\mathbb{Z}_4$	3,4,5,6,7	
.2	$V_{49} \cdot \mathcal{S}_3$	294	$[1,3^6,6^5]$	b_{42}^3, b_{14}	$\mathcal{S}_3$	8, 12	
.3	$F_{49,8}$	392	$[1,8^6]$	a_7, b_{48}^6	$\mathbb{Z}_8$	9,10,11,13	
.4	$V_{49} \cdot Q_8$	392	$[1,8^6]$	b_{48}^{12}, b_{12}^3	Q_8	9,14,15,16	
.5	$D_7 \wr \mathcal{S}_2$	392	$[1,4^6,8^3]$	b_{14}, b_{48}^{12}	D_4	11,17,18	
.6	$V_{49} \cdot (\mathbb{Z}_3 \cdot \mathbb{Z}_4)$	588	$[1,12^4]$	b_{48}^{12}, d_6	$\mathbb{Z}_3 \cdot \mathbb{Z}_4$	$17, 20, 22^{b_{42}}, 30$	
.7	$F_{49,12}$	588	$[1,12^4]$	a_7, b_{48}^4	$\mathbb{Z}_{12}$	13,14,18,20	
.8	$V_{49} \cdot D_6$	588	$[1,6^6,12]$	b_{42}^3, d_6	D_6	17,21	
.9	$V_{49} \cdot (\mathbb{Z}_2 \backslash D_4)$	784	$[1,16^3]$	b_{48}^6, b_{12}^3	$\mathbb{Z}_2 \backslash D_4$	19,22,23	
.10	$F_{49,16}$	784	$[1,16^3]$	a_7, b_{48}^3	$\mathbb{Z}_{16}$	19,24	
.11	$V_{49} \cdot D_8$	784	$[1,8^6]$	b_{14}, b_{48}^6	D_8	19, 25	
.12	$V_{49} \cdot G_\alpha \cong 7.3 \wr \mathcal{S}_2$	882	$[1,6^2,9^2,18]$	b_{14}, c_6^2	$\mathbb{Z}_3 \wr \mathcal{S}_2$	21, 33	
.13	$F_{49,24}$	1176	$[1,24^2]$	a_7, b_{48}^2	$\mathbb{Z}_{24}$	23, 24, 25	
.14	$V_{49} \cdot (Q_8 \times \mathbb{Z}_3)$	1176	$[1,24^2]$	b_{48}^4, b_{12}	$Q_8 \times \mathbb{Z}_3$	23, 26	
.15	$V_{49} \cdot (Q_8 \cdot \mathbb{Z}_3)$	1176	$[1,24^2]$	b_{48}^{12}, b_3	$Q_8 \cdot \mathbb{Z}_3$	22,26	
.16	$V_{49} \cdot (Q_8 \cdot \mathbb{Z}_3)$	1176	$[1,8^3,24]$	b_{48}^{12}, b_6	$Q_8 \cdot \mathbb{Z}_3$	26	

Degree $n.i$	Group G	Order $\|G\|$	Trans/Rank	Generators	G_α	min H, $G < H \leq \mathcal{S}_n$	Combinatorial Structures
49.17	$V_{49} \cdot (\mathbb{Z}_3 \cdot D_4)$	1176	$[1,12^4]$	b_{48}^{12}, b_{42}^3	$\mathbb{Z}_3 \cdot D_4$	27,31	
.18	$V_{49} \cdot (D_4 \times \mathbb{Z}_3)$	1176	$[1,12^2,24]$	b_{14}, b_{48}^4	$D_4 \times \mathbb{Z}_3$	25,27	
.19	$V_{49} \cdot (\mathbb{Z}_{16} \cdot_7 \mathbb{Z}_2)$	1568	$[1,16^3]$	b_{14}, b_{48}^3	$\mathbb{Z}_{16} \cdot_7 \mathbb{Z}_2$	28,31	
.20	$V_{49} \cdot G_\alpha$	1764	[1,12,36]	b_{48}^4, c_6	$\mathbb{Z}_3 \times (\mathbb{Z}_3 \cdot \mathbb{Z}_4)$	$27,29^{b_{42}}$, 32, 37	
.21	$V_{49} \cdot (\mathcal{S}_3 \times \mathbb{Z}_6)$	1764	$[1,12,18^2]$	b_{14}, c_6	$\mathcal{S}_3 \times \mathbb{Z}_6$	27,36	
.22	$V_{49} \cdot (\mathbb{Z}_2 \backslash \mathcal{S}_4)$	2352	2t!	b_{48}^6, b_8^2	$\mathbb{Z}_2 \backslash \mathcal{S}_4$	29,30	
.23	$V_{49} \cdot G_\alpha$	2352	2t!	b_{48}^2, b_{12}	$\mathbb{Z}_3 \times (\mathbb{Z}_2 \backslash D_4)$	28,29	
.24	$\text{AGL}_1(49)$	2352	2t!	a_7, b_{48}	$\mathbb{Z}_{48}$	28	
.25	$\text{A}\Sigma\text{L}_1(49)$	2352	$[1,24^2]$	b_{48}^2, b_{14}	$\mathbb{Z}_{24} \cdot_7 \mathbb{Z}_2$	28	
.26	$V_{49} \cdot G_\alpha$	3528	$[1,24^2]$	b_{48}^4, b_6	$(Q_8 \cdot \mathbb{Z}_3) \times \mathbb{Z}_3$	29	
.27	$7.4 \wr \mathcal{S}_2$	3528	[1,12,36]	b_{48}^4, b_{42}^3	$\mathbb{Z}_6 \wr \mathcal{S}_2$	34,38	
.28	$\text{A}\Gamma\text{L}_1(49)$	4704	2t	b_{48}, b_{14}	$\mathbb{Z}_{48} \cdot_7 \mathbb{Z}_2$	34	
.29	$V_{49} \cdot G_\alpha$	7056	2t	b_{48}^2, b_8^2	$(\mathbb{Z}_2 \backslash \mathcal{S}_4) \times \mathbb{Z}_3$	32	
.30	$\text{ASL}_2(7)$	16464	2t	b_{48}^6, b_8	$\text{SL}_2(7)$	31,32	
.31	$\text{AS}^{\pm}L_2(7)$	32928	2t	b_{48}^3, b_{42}^3	$(\pm 1) \cdot SL_2(7)$	34	
.32	$\text{AS}^2 L_2(7)$	49392	2t	b_{48}^2, b_{42}^2	$\langle 2 \rangle \times SL_2(7)$	34,39	
.33	$L_3(2) \wr \mathcal{S}_2 \cong 7.5 \wr \mathcal{S}_2$	56448	[1,12,36]	b_{24}^{12}, b_{42}^3	$\mathcal{S}_4 \wr \mathcal{S}_2$	35	
.34	$\text{AGL}_2(7)$	98784	2t	b_{48}, b_{42}	$\text{GL}_2(7)$	40	
.35	$\mathcal{A}_7 \wr \mathcal{S}_2$	12700800	[1,12,36]	b_{24}^4, b_{42}^3	36.15	36	
.36	$(\mathcal{A}_7 \times \mathcal{A}_7) \cdot V_4$	25401600	[1,12,36]	b_{24}^2, b_{42}^3	36.17	38	
.37	$(\mathcal{A}_7 \times \mathcal{A}_7) \cdot \mathbb{Z}_4$	25401600	[1,12,36]	b_{24}, b_{42}^6	36.18	38,39	
.38	$\mathcal{S}_7 \wr \mathcal{S}_2$	50803200	[1,12,36]	b_{24}, b_{42}^3	36.19	40	
.39	$\mathcal{A}_{49}$	49!/2	47p!	a_{24}, a_{49}	48.3	40	
.40	$\mathcal{S}_{49}$	49!	49p!	a_{48}, a_{49}	48.4		
50.1	$L_2(49)$	58800	2p	a_7, b_{50}^2	49.13	2,3,4	
.2	$\text{P}\Delta\text{L}_2(49)$	117600	3t!	a_7, b_{16}	49.23	6,8	
.3	$\text{PGL}_2(49)$	117600	3t!	a_7, b_{50}	49.24	6	
.4	$\text{P}\Sigma\text{L}_2(49)$	117600	2p	b_{14}, b_{50}^2	49.25	6	
.5	$\text{PSU}_3(5^2)$	126000	[1,7,42]	a_7, b_8^2	7.6	7	
.6	$\text{P}\Gamma\text{L}_2(49)$	235200	3t	b_{14}, b_{50}	49.28	9	
.7	$\text{P}\Sigma\text{U}_3(5^2)$	252000	[1,7,42]	a_7, b_8	7.7	8	srg(50, 7, 0, 1)
.8	$\mathcal{A}_{50}$	50!/2	48p!	a_{25}, a_{49}	49.39	9	
.9	$\mathcal{S}_{50}$	50!	50p!	a_{49}, a_{50}	49.40		

1.38 Table Generators for primitive groups of degree at most 50.

Deg	Elt	Permutation	Elt	Permutation	Elt	Permutation	Elt	Permutation
6	b_3	(1 3 4)(2 5 6)	b_6	(1 5 2 4 3 6)				
7	b_7	(1 4 6 2 3 7 5)						
8	b_6	(1 2 8 5 7 6)(3 4)	b_7	(1 5 6 8 2 4 3)	b_8	(1 2 8 3 4 7 6 5)	c_7	(1 8 7 2 5 3 6)
	d_7	(1 8 2 7 5 4 6)						
9	b_3	(1 5 9)(2 3 8)(4 7 6)	b_4	(1 6 4 5)(2 9 8 7)	b_6	(1 9 3 4 7 5)(2 8)	b_8	(1 7 5 2 8 4 3 9)
	b_9	(1 8 5 2 9 4 7 3 6)	c_4	(1 9 8 2)(3 7 5 4)	c_6	(1 2 4 6 3 9)(5 7)	c_8	(2 9 3 6 7 5 4 8)
	d_4	(1 6 4 5)(2 9 8 7)	d_6	(1 5 6 2 8 4)(3 9)	e_4	(1 3 6 2)(4 5 9 7)	e_6	(1 7 5 6 9 4)(2 8 3)
	f_6	(1 4 3)(2 8 6 7 9 5)	g_6	(1 7 8 9 6 3)(2 5 4)				
10	b_3	(1 6 8)(2 7 10)(3 9 5)	b_5	(1 10 4 7 5)(2 8 6 9 3)	b_6	(1 8 6)(2 3 7 9 10 5)	b_8	(1 8 6 2 9 5 3 10)(4 7)
	b_{10}	(1 2 5 8 9 7 4 10 6 3)	c_5	(1 3 4 5 7)(2 10 9 8 6)	c_6	(1 8 9 3 5 6)(2 7 4)	c_8	(1 6 10 9 3 8 4 5)(2 7)
	d_5	(1 10 9 6 4)(2 8 3 5 7)	d_8	(1 7 2 6 5 9 4 10)(3 8)				

Deg	Elt	Permutation	Elt	Permutation
11	b_5	(1 11 10 3 6)(2 5 4 9 7)	b_{11}	(1 2 7 9 4 3 6 11 5 8 10)
12	b_{11}	(1 12 10 3 7 5 8 4 2 11 9)	c_{11}	(2 3 7 8 10 4 5 9 6 11 12)
	d_{11}	(1 2 10 11 9 5 4 12 8 7 3)	b_{12}	(1 7 2 5 12 6 3 11 9 10 8 4)
13	b_{13}	(1 3 4 6 8 10 7 13 2 5 12 11 9)		
14	b_{13}	(1 9 4 3 8 10 13 5 11 14 2 7 6)		
15	b_2	(1 9)(2 5)(4 15)(8 13)(10 11)(12 14)	b_7	(2 6 3 8 12 4 9)(5 7 13 11 10 14 15)
	c_2	(1 5)(6 13)(7 8)(10 12)		
16	b_3	(2 16 4)(3 14 9)(5 6 11)(7 12 8)	b_4	(2 5)(3 9 8 13)(4 6 11 16)(7 14 10 12)
	b_6	(1 8 10 15 7 3)(2 11 4)(6 9 13 16 14 12)	b_7	(1 2 11 12 7 16 13)(3 15 10 9 8 6 5)
	b_{15}	(1 2 10 14 11 8 5 12 3 13 9 6 15 4 7)	c_6	(1 3 13)(2 14 16 7 5 15)(4 11 10 8 9 12)
17	b_8	(1 15 13 10 16 12 17 6)(2 8 7 4 9 11 5 14)	b_{10}	(1 16 8 4 2 14 9 12 5 7)(3 6 15 11 13)(10 17)
	b_{17}	(1 11 6 12 17 4 7 10 14 2 8 3 13 5 16 15 9)		
18	b_{17}	(1 12 13 8 2 15 16 9 7 11 10 6 14 4 18 17 3)		
20	b_{20}	(1 3 5 17 18 10 16 19 13 11 2 20 4 14 12 6 9 15 7 8)		
21	b_2	(2 17)(3 12)(4 19)(5 16)(6 13)(7 18)(9 21)(11 14)(15 20)	b_4	(1 14 8 12)(3 21 7 18)(4 20 16 19)(5 11)(9 17 10 15)
	c_4	(1 21 4 10)(2 9 11 3)(5 12 15 14)(7 13 20 8)(17 18)		
22	b_{10}	(1 15 17 2 21 14 7 3 22 10)(4 12 11 6 16 13 9 19 20 5)(8 18)		
23	b_2	(1 4)(2 6)(8 23)(9 10)(11 16)(12 19)(13 21)(14 20)		
24	b_2	(6 7)(8 19)(9 14)(10 22)(11 12)(13 24)(15 17)(20 23)		
	b_{24}	(1 2 24 3 4 15 19 22 18 16 8 7 17 23 14 5 11 21 20 12 10 6 9 13)		
25	b_2	(1 18)(2 23)(4 8)(5 13)(6 19)(7 24)(10 14)(11 20)(12 25)(17 21)		

Deg	Elt	Permutation
25	b_4	(2 3 5 4)(6 16 21 11)(7 18 25 14)(8 20 24 12)(9 17 23 15)(10 19 22 13)
	b_6	(1 9 25)(2 14 22 11 7 15)(3 19 23 16 8 20)(4 24 21 6 10 5)(13 17)
	b_8	(1 2 7 9 19 18 13 11)(3 12 6 4 17 8 14 16)(5 22 10 24 20 23 15 21)
	b_{10}	(1 3 5 2 4)(6 22 10 21 9 25 8 24 7 23)(11 16 15 20 14 19 13 18 12 17)
	b_{12}	(1 10 17)(2 5 20 16 6 7)(3 15 19 21 8 12 4 25 18 11 9 22)(13 14 24 23)
	b_{20}	(1 9 17 15 3 6 19 12 5 8 16 14 2 10 18 11 4 7 20 13)(21 24 22 25 23)
	b_{24}	(1 4 17 9 20 22 7 8 14 18 15 24 19 16 3 11 5 23 13 12 6 2 10 21)
25	c_4	(2 3 5 4)(6 19 21 13)(7 16 25 11)(8 18 24 14)(9 20 23 12)(10 17 22 15)
	c_8	(2 6 3 11 5 21 4 16)(7 8 13 15 25 24 19 17)(9 18 12 10 23 14 20 22)
	c_{20}	(2 7 21 12 15 3 13 16 23 24 5 25 6 20 17 4 19 11 9 8)(10 14 22 18)
	d_4	(1 7 3 22)(4 17 5 12)(6 8 23 21)(9 18 25 11)(10 13 24 16)(14 19 20 15)
26	b_{10}	(1 17 16 9 21 15 12 18 14 6)(2 13 19 26 24)(3 25 10 5 4 23 11 8 20 22)
	b_{12}	(1 22 14 25 11 23 17 7 18 5 10 8)(2 16 9 3 4 20 15 12 26 13 21 19)(6 24)
27	b_4	(1 18 5 3)(6 17 26 12)(7 11 21 20)(8 19 27 16)(9 15 22 10)(13 25 23 14)
	b_6	(1 5 17 22 15 6)(2 10 14 18 13 9)(3 27 12 11 26 8)(4 21 25 19 23 7)(20 24)
	b_{12}	(1 25 27 5 13 21)(2 15 18 16)(3 23 17 12)(4 19 8 9 20 14 6 26 11 10 7 22)
	b_{24}	(1 27 21 13 4 10 18 19 9 14 17 15 3 12 2 24 26 22 25 8 20 5 16 11)(6 7 23)
	b_{26}	(1 10 18 5 20 2 27 23 25 3 4 24 21 16 7 26 6 15 9 14 12 19 8 13 17 11)
	c_4	(1 2 18 15)(3 22 13 7)(4 19 26 8)(5 11 23 9)(10 14)(12 17 27 16)(20 25 21 24)
27	c_6	(1 11 26 19 9 24)(2 14 21 15 6 20)(3 16 27 17 13 23)(4 18 22 25 8 7)(5 12)
	d_6	(1 27 19 14 9 20)(2 17 21 16 7 6)(3 11 15 24 4 12)(5 8 26 13 10 22)(23 25)
28	b_6	(1 23 24 4 18 19)(2 8 7 20 21 22)(3 25 16 17 5 14)(6 9 28 15 26 13)(10 12 27)
	b_8	(1 13 26 20 16 23 7 17)(2 25 21 5 12 19 4 28)(3 9 18 6 10 24 22 11)(8 14 15 27)
	b_9	(1 27 22 26 10 23 13 12 3)(2 21 7 24 15 14 25 16 19)(4 5 17 9 18 6 28 8 20)
	b_{12}	(1 11 5 15 14 19 28 10 17 24 2 6)(3 18 13 22 16 7 25 23 4 21 20 8)(9 26 27)
	b_{26}	(1 22 4 27 23 24 5 3 7 8 21 9 25 2 26 16 20 18 10 11 14 19 12 15 17 6)
	c_6	(1 3 14)(2 27)(4 26)(5 20 9 11 19 21)(6 12 17)(8 18)(10 23 16 28 15 25)(13 22)
	c_{12}	(1 24 13 8 10 15 16 7 19 14 4 27)(2 18 11 17 23 21 6 3 9 20 22 12)(5 28 25 26)
	d_{12}	(1 14 21 20 23 4 19 22 7 25 27 9)(3 28 5 11)(6 17 26 10 16 24)(12 13 15 18)
30	b_{30}	(1 20 19 6 28 8 3 15 23 12 13 11 16 27 4 10 21 26 24 25 14 22 5 29 9 2 18 17 7 30)
31	b_{31}	(1 13 15 7 30 10 19 23 3 18 29 6 9 14 25 2 27 28 26 4 21 24 16 5 17 22 12 20 31 11 8)
	c_{31}	(1 26 4 24 19 13 10 20 23 6 17 16 15 7 22 11 12 21 9 8 14 18 5 25 3 2 27 31 30 29 28)
32	b_{10}	(1 21 25 13 6 14 10 26 32 8)(2 12 19 29 30 5 31 4 7 3)(9 16 11 27 15 23 24 18 17 20)(22 28)
	b_{31}	(1 5 19 9 23 3 24 15 2 21 12 25 6 20 28 8 11 17 14 22 7 26 29 32 4 30 31 13 10 16 18)

Deg	Elt	Permutation
32	b_{32}	(1 29 23 21 13 25 27 11 12 31 30 6 14 9 32 7 2 10 17 16 4 5 20 22 3 26 24 18 15 19 8 28)
33	b_{15}	(1 20 29 24 7 32 17 16 8 18 30 6 10 13 23)(2 22 14 33 4 21 31 11 9 28 15 19 25 12 5)(3 26 27)
35	b_{12}	(1 21 31 28 9 12 29 15 3 25 33 27)(2 5 30 10 26 34 16 22 35 17 7 18)(4 14 6 8 13 32)(11 20 19 23)
	$a_{15,5}$	(a_{15} for degree 30) $\cdot$ (31 32 33 34 35)
36	b_6	(1 10 16 20 28 29)(2 15 35 13 12 3)(4 19 27)(6 26 34)(7 31 25)(8 14)(9 17 24 21 22 23)(11 32 30)
	b_8	(1 17 33 30 2 28 25 24)(3 22 31 10 4 6 36 27)(5 35 11 34 7 23 20 13)(8 29 15 9 18 26 32 19)
	b_{10}	(1 11 33 28 22 15 23 14 20 13)(2 25 31 30 9)(3 8 6 26 18 19 36 16 21 32)(4 27 17 34 35 12 10 29 7 5)
	b_{12}	(1 36 6 21 14 22 26 2 34 31 8 28)(3 35 32 10 11 24 23 4 19 17 18 7)(5 30 16 9 29 12 13 33 25 15 20 27)
	c_8	(1 14 30 5 35 27 21 6)(2 16 17 19)(3 18 34 4 12 31 20 15)(7 9 36 25 10 33 24 28)(8 32 23 11 29 22 13 26)
	c_{12}	(1 35 2 8 7 26)(3 10 28 13 23 34 17 27 30 21 5 12)(4 18 11 20 29 24 36 15 16 14 19 31)(6 9 32 22 25 33)
38	b_{38}	(1 20 4 18 29 38 12 23 37 21 3 32 36 17 6 33 30 10 16 15 28 14 22 7 2 34 19 27 13 26 25 31 11 8 35 24 5 9)
40	b_5	(1 29 26 4 13)(2 28 38 30 22)(3 8 39 16 32)(5 7 27 14 15)(6 25 23 37 9)(10 17 11 40 35)(12 34 18 24 31)(19 36 21 33 20)
	b_{10}	(1 2 3 4 27 20 12 5 6 25)(7 26 17 9 8)(10 13 11 24 33 38 22 31 18 30)(14 34 21 36 37 2315 35 16 29)(19 40 39 32 28)
42	b_{42}	(1 38 15 6 25 14 16 18 7 26 17 35 31 24 12 19 41 4 39 9 40 37 11 3 27 22 30 42 2 10 5 29 21 36 33 23 34 28 32 13 20 8)
44	b_{44}	(1 41 29 9 42 43 8 22 15 26 31 13 2 34 16 21 32 25 39 4 5 38 18 6 3 12 36 10 28 27 14 30 40 24 44 23 7 17 33 20 19 37 11 35)
45	b_8	(1 9 20 37 41 45 34 12)(2 11 8 31 26 5 27 40)(3 39 32 7)(4 6 25 42 36 35 44 28)(13 33 38 19 23 22 21 30)(14 29 43 24 18 17 16 15)
	b_{10}	(1 2 3 4 5 6 7 8 9 10)(11 12 13 14 15 16 17 18 19 20)(21 22 23 24 25 26 27 28 29 30)(31 32 33 34 35 36 37 38 39 40)(41 42 43 44 45)
	c_8	(1 5 16 11 33 13 32 19)(2 15 3 36 12 23 25 40)(4 29 22 41)(6 24 18 37 21 45 34 14)(7 35 28 31 8 30 20 17)(9 10 38 27 42 39 26 44)
	d_8	(1 45 18 31)(2 21 12 34 13 39 22 4)(3 41 35 40)(5 38 10 37 32 16 14 25)(6 9 8 29 26 28 17 27)(7 43 11 19 15 44 36 24)(20 23 42 30)
48	b_{47}	(1 32 48 12 4 43 30 45 20 27 34 9 24 11 2 42 6 22 5 37 23 13 33 38 46 39 19 25 7 18 44 14 28 26 40 10 36 47 29 35 15 8 16 21 41 31 17)
49	b_3	(2 3 5)(4 7 6)(8 35 16)(9 30 20)(10 32 17)(11 34 21)(12 29 18)(13 31 15)(14 33 19)(22 40 46)(23 42 43)(24 37 47)(25 39 44)(26 41 48)(27 36 45)(28 38 49)
	b_6	(2 4 3 7 5 6)(8 45 12 43 13 46)(9 48 14 49 10 44)(11 47)(15 40 16 36 18 42)(17 39 20 41 19 38)(21 37)(22 35 27 29 23 31)(24 34)(25 30 26 33 28 32)
	b_8	(2 16 11 19 7 42 47 39)(3 31 21 30 6 27 37 28)(4 46 24 48 5 12 34 10)(8 23 33 44 43 35 25 14)(9 38 36 13 49 20 15 45)(17 26 22 18 41 32 29 40)
	b_{12}	(2 12 4 27 3 16 7 46 5 31 6 42)(8 23 22 18 15 45 43 35 29 40 36 13)(9 34 25 37 17 11 49 24 33 21 41 47)(10 38 28 14 19 26 48 20 30 44 39 32)
	b_{14}	(1 7 49 48 41 40 33 32 25 24 17 16 9 8)(2 14 43 6 42 47 34 39 26 31 18 23 10 15)(3 21 44 13 36 5 35 46 27 38 19 30 11 22)(4 28 45 20 37 12 29)
	b_{24}	(1 33 14 23 38 48 15 32 7 26 10 44 36 34 21 25 3 47 8 30 42 27 17 46)(2 40 13 16 39 6 19 11)(4 5 12 9 37 41 20 18)(22 31 49)(24 45 43 29 35 28)
	b_{42}	(1 6 14 2 23 41 16 21 22 17 38 7 31 29 37 32 4 15 46 44 3 47 19 30 12 10 18 13 34 45 27 25 33 28 49 11 42 40 48 36 8 26)(5 39 24 9 43 35 20)
	b_{48}	(1 3 42 34 24 31 46 15 33 29 14 23 36 22 48 5 25 26 21 38 12 40 44 32 41 39 7 8 18 11 45 27 9 13 35 19 6 20 43 37 17 16 28 4 30 2 47 10)
	c_6	(2 4 3 7 5 6)(8 43)(9 46 10 49 12 48)(11 45 14 47 13 44)(15 36)(16 39 17 42 19 41)(18 38 21 40 20 37)(22 29)(23 32 24 35 26 34)(25 31 28 33 27 30)
	d_6	(2 4 3 7 5 6)(8 36 29 43 15 22)(9 39 31 49 19 27)(10 42 33 48 16 25)(11 38 35 47 20 23)(12 41 30 46 17 28)(13 37 32 45 21 26)(14 40 34 44 18 24)
50	b_8	(1 27 15 19)(2 37 40 43 34 33 32 14)(3 38)(4 7 16 25 28 18 35 49)(5 22 6 23 45 36 30 47)(8 48 17 9)(10 41 11 42 24 46 31 26)(12 21 13 50 20 44 29 39)
	b_{14}	(1 26 2 27 3 28 4 22 5 23 6 24 7 25)(8 12 9 13 10 14 11)(15 33 16 34 17 35 18 29 19 30 20 31 21 32)(36 43 37 44 38 45 39 46 40 47 41 48 42 49)
	b_{16}	(1 6 11 4 7 8 38 39 10 44 24 37 36 45 26 35)(2 21 19 47 18 20 22 28 49 32 9 23 25 48 3 30)(5 17 46 15 27 40 12 13 16 31 34 14 50 29 41 43)(33 42)
	b_{50}	(1 34 30 17 7 10 44 37 41 18 39 25 26 28 40 20 43 8 32 31 16 47 29 48 14 49 24 12 11 35 46 27 3 19 21 15 4 22 2 6 45 33 36 23 13 9 42 5 50 38)

1.5 Simple Groups

1.39 Remark The relatively recent classification of all finite simple groups has been a major accomplishment of modern group theory. As a result of a remarkable effort by mathematicians from all continents, spanning about two and a half decades (1955-1980), all finite simple groups have been classified. See [9] for an introduction to the proof.

Apart from the cyclic groups $\mathbb{Z}_p$ of prime order, and the alternating groups $\mathcal{A}_n$, there are 16 infinite classes of simple groups of *Lie* type, and an additional 26 *sporadic simple groups* that do not fall in any of the other classes. The simple groups of Lie type consist of 6 classes of so called *classical* groups, and another 10 classes of so called *exceptional simple groups.* The classical simple groups consist of *linear, symplectic, orthogonal* and *unitary* groups. The smallest sporadic simple group is the Mathieu group M_{11} of order 7920, and the largest is a group denoted by F_1, of order $2^{46}.3^{20}.5^9.7^6.11^2.13^3.17.19.23.29.31.41.47.59.71$. F_1 is called the *Fischer-Griess* group or the (friendly) *giant.* Its existence was established by Griess, who based his construction on results by several other people notably on work of S. Norton, but also of Conway, Fischer, Livingstone, Thorne, and Smith, and his own earlier work. The uniqueness of a simple group of "type F_1" [7] was established by Thompson and Norton. This remarkable giant is constructed as a group of complex matrices of degree 196, 883. It is worth noting that every finite simple group has a representation as a primitive permutation group of rank at most 8, where the stabilizer of a point is a (smaller) simple group.

1.40 Remark Data related to the simple groups of order less than 100 million are displayed in seven columns. (The 106 groups $L_2(q)$ for $q > 5^3$ are omitted; their structure is known and described fully by Dickson [6].) The first column gives the name of the simple group using the conventions in §1.3. Column 2 displays the order of the group. The third column, labeled $\Omega.\{\sigma, \mu\}$ gives information in the form $n.x$. Here, n is a positive integer indicating the order of the outer automorphism group of G, and x is one of the two Greek letters σ, μ. If a μ appears then the group is *minimal simple*, i.e., the group is simple with all of its subgroups solvable. If a σ appears, then the simple group has nonsolvable subgroups, some of which are listed in column 6. In a few instances, as in the case of $L_3(7)$ and $U_4(3)$, in position n of column 3 the isomorphism type of the outer-automorphism group is listed instead of its order. The next column gives information about the conjugacy classes of the group. Only a *signature* which describes the number of conjugacy classes of nonidentity elements of a given order is given. As an example, for group $U_3(3)$ the signature $2^1 3^2 4^3 6^1 7^2 8^2 12^2$ says that there is a single conjugacy class of involutions, two classes of elements of order 3, three classes of elements of order 4, and so on.

Column 5 displays the *primitive spectrum* of the group. This is the list of degrees of the various (inequivalent) primitive permutation representations for the group. An exponent here denotes that there is more than one inequivalent primitive representation of a particular degree for the given group. For example, $\mathcal{A}_6$ has two inequivalent primitive representations of degree 6 (corresponding to two conjugacy classes of $\mathcal{A}_5$'s in $\mathcal{A}_6$), one primitive representation of degree 10, and two inequivalent primitive representations of degree 15. Column 6 displays some nonsolvable subgroups that are contained in the given group. Here the existence of several conjugacy classes of a certain nonsolvable group is indicated by a *multiplicity index* positioned in front of the subgroup specified. For example, the simple group $U_3(5) \cong PSU_3(5^2)$ contains three conjugacy classes of $\mathcal{A}_7$'s, three classes of M_{10}'s, and a class of $\mathcal{S}_5$'s. If a group G is generated as $G = \langle x, y\rangle$, where $|x| = \ell$, $|y| = m$, and $|xy| = n$, G is an (ℓ, m, n) group. The seventh column provides several ways in which G is an (ℓ, m, n) group.

1.41 Table Simple groups of order less than 100,000,000.

Group G	$\lvert G\rvert$	$\Omega.\{\sigma,\mu\}$	Conjugacy Classes	Primitive Spec.	n.s. $H < G$	Generated as (ℓ, m, n)
$\mathcal{A}_5 \cong L_2(4) \cong L_2(5)$	60	$2.\mu$	$2^1 3^1 5^2$	$5, 6, 10$	-	(2,3,5)
$L_3(2) \cong L_2(7)$	168	$2.\mu$	$2^1 3^1 4^1 7^2$	$7^2, 8$	-	(2,3,7),(2,4,7),(3,3,4),(2,7,7)
$\mathcal{A}_6 \cong L_2(9)$	360	$4.\sigma$	$2^1 3^2 4^1 5^2$	$6^2, 10, 15^2$	${}_2\mathcal{A}_5$	(2,4,5),(3,3,4),(3,4,5),(2,5,5)
$L_2(8)$	504	$3.\mu$	$2^1 3^1 7^3 9^3$	$9, 28, 36$	-	(2,3,7),(2,3,9),(2,7,9),(2,7,7)
$L_2(11)$	660	$2.\sigma$	$2^1 3^1 5^2 6^1 11^2$	$11^2, 12, 55$	${}_2\mathcal{A}_5$	(2,3,11),(2,5,6),(2,5,11),(2,5,5)
$L_2(13)$	1092	$2.\mu$	$2^1 3^1 6^1 7^3 13^2$	$14, 78, 91^2$	-	(2,3,7),(2,3,13),(2,6,7),(2,7,13)
$L_2(17)$	2448	$2.\mu$	$2^1 3^1 4^1 8^2 9^3 17^2$	$18, 136, 153, 102^2$	-	(2,3,8),(2,3,9),(2,3,17)
$\mathcal{A}_7$	2520	$2.\sigma$	$2^1 3^2 4^1 5^1 6^1 7^2$	$7, 15^2, 21, 35$	$\mathcal{A}_6, \mathcal{A}_5$	(2,4,7),(2,5,7),(2,6,7),(2,7,7)
$L_2(19)$	3420	$2.\sigma$	$2^1 3^1 5^2 9^3 10^2 19^2$	$20, 57^2, 171, 190$	${}_2\mathcal{A}_5$	(2,3,9),(2,3,19),(2,5,9),(2,5,5)
$L_2(16)$	4080	$4.\sigma$	$2^1 3^1 5^2 15^4 17^8$	$17, 68, 120, 136$	$\mathcal{A}_5$	(2,3,17),(2,5,17),(2,3,15),(2,5,15)
$L_3(3)$	5616	$2.\mu$	$2^1 3^2 4^1 6^1 8^2 13^4$	$13^2, 144, 234$	-	(2,3,13),(2,4,13),(2,8,8),(3,4,13)
$U_3(3)$	6048	$2.\sigma$	$2^1 3^2 4^3 6^1 7^2 8^2 12^2$	$28, 36, 63^2$	$L_2(7)$	(2,7,17),(2,6,7),(3,4,7),(3,6,7)
$L_2(23)$	6072	$2.\mu$	$2^1 3^1 4^1 6^1 12^2 11^5 23^2$	$24, 253^3, 276$	-	(2,3,11),(2,3,12),(2,4,6),(2,6,11)
$L_2(25)$	7800	$4.\sigma$	$2^1 3^1 4^1 5^2 6^1 12^2 13^6$	$26, 65^2, 300, 325$	$\mathcal{S}_5$	(2,3,13),(2,3,12),(2,4,13),(2,5,13)
M_{11}	7920	$1.\sigma$	$2^1 3^1 4^1 5^1 6^1 8^2 11^2$	$11, 12, 55, 66, 165$	$L_2(11), \mathcal{S}_5, M_{10}$	(2,5,11),(2,5,8),(2,4,11),(3,4,5)
$L_2(27)$	9828	$6.\mu$	$2^1 3^2 7^3 14^3 13^6$	$28, 351, 378, 819$	-	(2,3,7),(2,3,13),(2,7,13),(2,7,7)
$L_2(29)$	12180	$2.\sigma$	$2^1 3^1 5^2 7^3 14^3 15^4 29^2$	$30, 203^2, 406, 435$	${}_2\mathcal{A}_5$	(2,3,7),(2,5,7),(2,3,15),(2,5,15)
$L_2(31)$	14880	$2.\sigma$	$2^1 3^1 4^1 5^2 8^2 15^4 16^4 31^2$	$32, 248^2, 465, 496$	${}_2\mathcal{A}_5$	(2,3,15),(2,3,8),(2,4,5),(2,5,16)
$\mathcal{A}_8 \cong L_4(2)$	20160	$2.\sigma$	$2^2 3^2 4^2 5^1 6^2 7^2 15^2$	$8, 15^2, 28, 35, 36$	$\mathcal{A}_7, L_2(7), \mathcal{S}_5$	(2,5,7),(2,7,15),(2,4,9),(3,4,15)
$L_3(4) \cong M_{21}$	20160	$12.\sigma$	$2^2 3^1 4^3 5^2 7^2$	$21^2, 56^3, 120^3, 280$	$L_2(7), \mathcal{A}_6, \mathcal{A}_5$	(2,5,7),(2,5,5),(2,4,7),(3,4,7)
$L_2(37)$	25308	$2.\mu$	$2^1 3^1 9^3 6^1 18^3 19^9 37^2$	$38, 666, 703, 2109$	-	(2,3,37),(2,3,9),(2,9,19),(2,18,19)
$U_4(2) \cong S_4(3)$	25920	$2.\sigma$	$2^2 3^4 4^2 5^1 6^6 9^2 12^2$	$27, 36, 40^2, 45$	$\mathcal{S}_5, \mathcal{S}_6$	(2,4,9),(2,5,9),(3,4,9),(3,5,6)
$Sz(8)$	29120	$3.\mu$	$2^1 4^2 5^1 7^3 13^3$	$65, 560, 1456, 2080$	-	(2,5,7),(2,5,13),(2,7,13),(2,4,5)
$L_2(32)$	32736	$5.\mu$	$2^1 3^1 11^5 33^{10} 31^{15}$	$33, 496, 528$	-	(2,3,11),(2,3,31),(2,11,31),(2,31,33)
$L_2(41)$	34440	$2.\mu$	$2^1 3^1 4^1 5^2 7^3 10^2$ $20^4 21^6 41^2$	$42, 820, 861, 1435^2$	${}_2\mathcal{A}_5$	(2,3,7),(2,3,10),(2,5,20),(2,7,20)
$L_2(43)$	39732	$2.\mu$	$2^1 3^1 7^3 11^5 21^6$ $22^5 43^2$	$44, 903, 946, 3311$	-	(2,3,7),(2,3,11),(2,7,11),(3,7,11)
$L_2(47)$	51888	$2.\mu$	$2^1 3^1 4^1 6^1 8^2 12^2$ $23^{11} 24^4 47^2$	$48, 1081, 1128, 2162^2$	-	(2,3,8),(2,3,23),(2,3,47),(3,4,47)

Group G	$\|G\|$	$\Omega.\{\sigma,\mu\}$	Conjugacy Classes	Primitive Spec.	n.s. $H < G$	Generated as (ℓ,m,n)
$L_2(49)$	58800	$4.\sigma$	$2^1 3^1 4^1 5^2 6^1 7^2$ $8^2 12^2 24^4 25^{10}$	$50, 175^2, 400,$ $1176, 1225$	${}_2PGL_2(7)$ ${}_2\mathcal{A}_5$	(2,3,12),(2,7,25), (3,4,5),(3,4,6)
$U_3(4)$	62400	$4.\sigma$	$2^1 3^1 4^1 5^6 10^4 13^4 15^4$	$65, 208, 416, 1600$	$\mathcal{A}_5$	(2,3,13),(2,3,15),(2,10,13),(2,13,15)
$L_2(53)$	74412	$2.\mu$	$2^1 3^1 9^3 13^6 26^6 27^9 53^2$	$54, 1378, 1431, 6201$	-	(2,3,13),(2,3,9),(2,9,26),(2,13,27)
M_{12}	95040	$2.\sigma$	$2^2 3^2 4^2 5^1 6^2 8^2$ $10^1 11^2$	$12^2, 66^2, 144, 220^2,$ $396, 495^2, 1320$	${}_2M_{11},\ {}_2\mathcal{A}_6,$ ${}_3L_2(11),\ \mathcal{S}_5$	(2,3,11),(2,3,10) (2,5,6),(2,5,11)
$L_2(59)$	102660	$2.\sigma$	$2^1 3^1 5^2 6^1 10^2$ $15^4 29^{14} 30^4 59^2$	$60, 1711^3, 1770$	${}_2\mathcal{A}_5$	(2,3,29),(2,3,10), (2,3,15),(3,5,10)
$L_2(61)$	113460	$2.\sigma$	$2^1 3^1 5^2 6^1 10^2$ $15^4 30^4 31^{15} 61^2$	$62, 1891^3, 1830$	${}_2\mathcal{A}_5$	(2,3,31),(2,3,10) (2,5,31),(2,15,31)
$U_3(5)$	126000	$5.\sigma$	$2^1 3^1 4^1 5^4 6^1 7^2 8^2 10^1$	$50^3, 126, 175^3, 525$	${}_3\mathcal{A}_{7,3}\, M_{10}, \mathcal{S}_5$	(2,5,7),(2,4,10),(3,5,7),(3,7,10)
$L_2(67)$	150348	$2.\mu$	$2^1 3^1 11^5 17^8 33^{10} 34^8 67^2$	$68, 2211, 2278, 12529$	-	(2,3,11),(2,3,17),(2,11,67),(2,17,33)
Ja_1	175560	$1.\sigma$	$2^1 3^1 5^2 6^1 7^1$ $10^2 11^1 15^2 19^3$	$266,\ 1045,\ 1463,$ $1540, 1596, 2926, 4180$	$L_2(11),\ \mathcal{A}_5$	(2,3,7),(2,3,11), (2,3,19),(5,7,19)
$L_2(71)$	178920	$2.\sigma$	$2^1 3^1 4^1 5^2 6^1 7^3 9^3$ $12^2 18^3 35^{12} 36^6 71^2$	$72,\ 2485,\ 2556,\ 2982^2$	${}_2\mathcal{A}_5$	(2,3,7),(2,3,71), (2,5,6), (2,6,36),(2,9,35),(5,12,71)
$\mathcal{A}_9$	181440	$2.\sigma$	$2^2 3^3 4^2 5^1 6^2 7^1$ $9^2 10^1 12^1 15^2$	$9, 36, 84, 120^2$ $126,\ 280,\ 840$	$\mathcal{A}_8, \mathcal{S}_7$	(2,3,15),(2,4,7), (3,7,9),(3,7,15)
$L_2(73)$	194472	$2.\mu$	$2^1 3^1 4^1 6^1 9^3$ $12^2 18^3 36^6 37^{18} 73^2$	$74, 2628, 2701,$ 8103^2	-	(2,3,37),(2,3,9), (2,3,73),(2,12,37),(3,6,37)
$L_2(79)$	246480	$2.\sigma$	$2^1 3^1 4^1 5^2 8^2$ $10^2 13^6 20^4 39^{12} 40^8 79^2$	$80, 3081, 3160,$ 4108^2	${}_2\mathcal{A}_5$	(2,3,8),(2,3,13), (2,3,79),(2,13,39),(3,8,13)
$L_2(64)$	262080	$6.\sigma$	$2^1 3^1 5^2 7^3 9^3 13^6 21^6 63^{18} 65^{24}$	$65, 520^2, 2016, 2080$	$\mathcal{A}_5,\ L_2(8)$	(2,3,13),(2,3,63),(2,5,63),(5,13,63)
$L_2(81)$	265680	$4.\sigma$	$2^1 3^2 4^1 5^2 8^2$ $10^2 20^4 40^8 41^{20}$	$82, 369^2, 3240,$ 3321^2	${}_2\mathcal{A}_5$ ${}_2\mathcal{A}_6$	(2,3,41),(2,5,41), (2,10,41),(3,5,8)
$L_2(83)$	285852	$2.\mu$	$2^1 3^1 6^1 7^3 14^3$ $21^6 41^{20} 42^5 83^2$	$84, 3403^2, 3486$ 22821	-	(2,3,7),(2,3,41), (2,41,83),(3,7,41)
$L_2(89)$	352440	$2.\sigma$	$2^1 3^1 4^1 5^2 9^3 11^5$ $15^4 22^5 44^{10} 45^{12} 89^2$	$90, 3916, 4005^2,$ $14685^2, 11748^2$	${}_2\mathcal{A}_5$	(2,3,11),(2,3,9), (2,3,89), (2,11,89),(3,5,45),(3,5,89)
$L_3(5)$	372000	$2.\sigma$	$2^1 3^1 4^3 5^2 6^1 8^2$ $10^1 12^2 20^2 24^4 31^{10}$	$31^2, 3100, 3875, 4000$	$\mathcal{S}_5$	(2,3,31),(2,5,31),(3,8,31), (3,10,24),(3,12,31),(3,20,24)
M_{22}	443520	$2.\sigma$	$2^1 3^1 4^2 5^1 6^1 7^2$ $8^1 11^2$	$22, 77, 176^2, 231,$ $330, 616, 672$	$M_{21},\ {}_2\mathcal{A}_7,$ $L_3(2), \mathcal{A}_6, L_2(11)$	(2,5,7),(2,5,11),(3,4,7), (3,4,11),(3,5,7),(3,5,8)

Group G	$\lvert G\rvert$	$\Omega.\{\sigma,\mu\}$	Conjugacy Classes	Primitive Spec.	n.s. $H < G$	Generated as (ℓ, m, n)
$L_2(97)$	456288	$2.\mu$	$2^1 3^1 4^1 6^1 7^3 8^2 12^2 16^4$ $24^4 48^8 49^{20} 97^2$	$98, 4656, 7453^2,$ 19012^2	-	(2,3,7),(2,3,8),(2,3,49), (3,12,49),(3,7,24),(3,12,49)
$L_2(101)$	515100	$2.\sigma$	$2^1 3^1 5^2 10^2 17^8$ $25^{10} 50^{10} 51^{16} 101^2$	$102, 1717^2, 5050,$ 5151^2	${}_2\mathcal{A}_5$	(2,3,17),(2,3,10), (2,17,25), (3,17,50),(5,17,51),(5,17,101)
$L_2(103)$	546312	$2.\mu$	$2^1 3^1 4^1 13^6 17^8$ $26^6 51^{16} 52^{12} 103^2$	$104, 5253^2,$ $5356, 277823^2$	-	(2,3,13),(2,3,17),(2,51,52), (3,4,52),(4,13,51)
HaJ	604800	$2.\sigma$	$2^2 3^2 4^1 5^4 6^2 7^1$ $8^1 10^4 12^1 15^2$	100,280,315, 525, 840, 1008, 1800, 2016, 10080	$U_3(3), PGL_2(9)$ ${}_4\mathcal{A}_5, L_3(2)$	(2,3,7),(2,3,10),(2,5,7), (2,5,15),(2,6,7),(3,7,10)
$L_2(107)$	612468	$2.\mu$	$2^1 3^1 6^1 9^3 18^3$ $27^9 53^{26} 54^9 107^2$	$108, 5671^2,$ 5778, 51039	-	(2,3,53),(2,3,9), (2,9,54),(3,9,27),(3,9,53)
$L_2(109)$	647460	$2.\mu$	$2^1 3^1 5^2 6^1 9^3 11^5 18^3$ $27^9 54^9 55^{20} 109^2$	110, 5886, $5995^2, 10791^2$	${}_2\mathcal{A}_5$	(2,3,11),(2,3,9),(2,11,55), (2,54,55),(5,6,27),(5,11,18)
$L_2(113)$	721392	$2.\mu$	$2^1 3^1 4^1 7^3 8^2 14^3 19^9$ $28^6 56^{12} 57^{18} 113^2$	114, 6328, $6441^2, 30058^2$	-	(2,3,7),(2,3,19),(2,7,7), (3,7,57),(2,57,57),(3,19,28)
$L_2(11^2)$	885720	$4.\sigma$	$2^1 3^1 4^1 5^2 6^1 10^2 11^2 12^2$ $15^4 20^4 30^4 60^8 61^{30}$	$122, 671^3, 7260,$ $7381^2, 36905^2$	${}_2\mathcal{A}_5$ ${}_2L_2(11)$	(2,3,15),(2,3,61), (2,10,61),(3,5,61),(3,11,61),
$L_2(5^3)$	976500	$6.\sigma$	$2^1 3^1 5^2 7^3 9^3 21^5$ $31^{15} 62^{15} 63^{18}$	126, 7750, $7875^2, 16275$	$\mathcal{A}_5$	(2,3,7),(2,3,31), (2,31,63),(5,7,21)
$S_4(4)$	979200	$4.\sigma$	$2^3 3^2 4^2 5^5 6^2 10^4$ $15^4 17^4$	$85^2, 120^2, 136^2,$ 1360	${}_2\mathcal{A}_5, \mathcal{S}_6,$ ${}_2\mathcal{A}_5 \times \mathcal{A}_5, {}_2L_2(16)$	(2,5,17),(2,4,5), (2,10,15),(3,5,6),(3,10,17)
$S_6(2)$	1451520	$1.\sigma$	$2^4 3^3 4^5 5^1 6^7 7^1$ $8^2 9^1 10^1 12^3 15^1$	28, 36, 63, 120, 135, 315, 336, 960	$U_4(2), U_3(3), \mathcal{S}_8,$ ${}_2\mathcal{S}_6, L_3(2), L_2(8)$	(2,5,7),(2,3,15), (4,5,12),(5,6,7),(6,7,9)
$\mathcal{A}_{10}$	1814400	$2.\sigma$	$2^2 3^3 4^3 5^2 6^3 7^1$ $8^1 9^2 10^1 12^2 15^1$ 21^2	10,45,120,126, 210, 945, 2520	$\mathcal{S}_8, \mathcal{A}_9, \mathcal{A}_7 \times \mathbb{Z}_3,$ $M_{10}, \mathcal{A}_5 \times \mathcal{A}_5,$ $\mathcal{S}_5,\ \mathcal{A}_6 \times \mathcal{A}_4$	(2,3,8),(2,3,9) (3,4,9),(3,7,8) (4,5,9),(6,7,8)
$L_3(7)$	1876896	$\mathcal{S}_3.\sigma$	$2^1 3^1 4^1 6^1 7^4 8^2$ $14^1 16^4 19^6$	$57^2, 5586^3,$ $26068^2, 32928$	$L_2(7)$	(2,3,19),(3,7,8), (3,7,19),(3,8,16)
$U_4(3)$	3265920	$D_8.\sigma$	$2^1 3^4 4^2 5^1 6^3 7^2$ $8^1 9^4 12^1$	$112, 126^2, 162^2, 280, 540,$ $567^2, 1296^4,$ $2835, 4536^2$	${}_3\mathcal{A}_6,\ {}_2U_4(2),$ ${}_2L_3(4),\ U_3(3)$ ${}_4\mathcal{A}_7,\ {}_2M_{10}$	(2,5,7),(3,7,7),(3,7,9), (4,5,6),(4,5,7),(4,5,9), (5,6,7),(6,7,8)
$G_2(3)$	4245696	$2.\sigma$	$2^1 3^5 4^2 6^4 7^1 8^2$ $9^3 12^2 13^2$	$351^2, 364^2, 378^2,$ 2808, 3159, 3888, 7371	${}_2U_3(3),\ {}_2L_3(3),$ $L_2(8),\ L_3(2), L_2(13)$	
$S_4(5)$	4680000	$2.\sigma$	$2^2 3^2 4^2 5^6 6^3 10^6$ $12^2 13^3 15^3 20^2 30^2$	$156^2, 300, 325, 4875,$ 6500, 9750, 13000	${}_4\mathcal{A}_5,\ \mathcal{A}_5 \times \mathcal{A}_5,\ L_2(25)$ $\mathcal{A}_6,\ \mathcal{S}_3 \times \mathcal{S}_5$	(2,5,13),(3,10,13),(4,13,15) (3,10,15),(4,5,13),

Group G	$\lvert G\rvert$	$\Omega.\{\sigma,\mu\}$	Conjugacy Classes	Primitive Spec.	n.s. $H < G$	Generated as (ℓ, m, n)
$U_3(8)$	5515776	$12.\sigma$	$2^1 3^3 4^3 6^2 7^3$ $9^3 19^6 21^6$	$513, 3648, 25536^3,$ $34048, 96768$	$L_2(8)$	(2,7,19),(2,7,7),(6,19,19), (19,19,19), (19,19,21),(19,21,21)
$U_3(7)$	5663616	$2.\sigma$	$2^1 3^1 4^3 6^1 7^2 8^{10} 12^2 14^1$ $16^4 24^4 28^2 43^{14} 48^8 56^4$	$344, 2107, 14749,$ $16856, 43904$	${}_2L_2(7)$	(2,3,43),(3,8,43), (3,14,28),(6,8,16),(3,3,43)
$L_4(3)$	6065280	$4.\sigma$	$2^2 3^4 4^3 5^1 6^5 8^1 9^2$ $10^1 12^3 13^4 20^2$	$40^2, 117^2, 130,$ $2106, 8424, 10530$	${}_2L_3(3), {}_2U_4(2)$ $\mathcal{A}_6,\ \mathcal{S}_6$	(2,3,13),(2,13,13),(3,6,13), (3,8,13),(3,13,20),(4,12,13)
$L_5(2)$	9999360	$2.\sigma$	$2^2 3^2 3^3 5^1 6^2 7^2$ $8^1 12^1 14^2 15^2 21^2 31^6$	$31^2, 155^2,$ 64512	${}_2L_4(2),$ ${}_2\mathcal{S}_3 \times L_3(2)$	(2,3,31),(3,14,21),(4,5,6), (4,6,14),(4,7,14),(4,7,31)
M_{23}	10200960	$1.\sigma$	$2^1 3^1 4^1 5^1 6^1 7^2$ $8^1 11^2 14^2 15^2 23^2$	$23, 253^2, 506,$ $1288, 1771, 40320$	$M_{22}, L_3(4), \mathcal{A}_7,$ $\mathcal{A}_8, M_{11}, \mathcal{A}_5$	(2,5,11),(2,5,23),(4,8,14), (4,11,15),(5,6,23),(5,7,11)
$U_5(2)$	13685760	$2.\sigma$	$2^2 3^6 4^3 5^1 6^{14} 8^1$ $9^4 11^2 12^9 15^2 18^2$	$165, 176, 297,$ $1408, 3520, 20736$	$U_4(2),$ ${}_2\mathcal{A}_5, L_2(11)$	(3,9,15),(4,8,9),(5,6,12), (5,6,15),(5,8,11),(6,9,11)
$L_3(8)$	16482816	$6.\sigma$	$2^1 3^1 4^1 7^{11} 9^3$ $14^6 21^6 63^{18} 73^{24}$	$73^2, 5604,$ $75264, 98112$	${}_2L_2(8),$ $L_3(2)$	(3,4,63),(3,7,73) (4,7,63),(4,9,63),(7,9,14)
${}^2F_4(2)'$	17971200	$2.\sigma$	$2^2 3^1 4^3 5^1 6^1 8^4$ $10^1 12^2 13^2 16^4$	$1600^2, 1755, 2304,$ $2925, 12480^2, 14976$	${}_2L_3(3),$ $L_2(25), {}_2\mathcal{A}_6$	
$\mathcal{A}_{11}$	19958400	$2.\sigma$	$2^2 3^3 4^3 5^2 6^5 7^1 8^1 9^1 10^1$ $11^2 12^3 14^1 15^2 20^1 21^2$	$11, 55, 165, 330,$ $462, 2520^2$	$\mathcal{A}_{10},\ \mathcal{A}_7 \times \mathcal{A}_4, \mathcal{A}_8,$ $\mathcal{S}_9,\ \mathcal{A}_6 \times \mathcal{A}_5, {}_2M_{11}$	(3,4,11),(3,8,11),(4,8,11), (5,8,20),(6,7,9),(6,8,9)
$Sz(32)$	32537600	$5.\mu$	$2^1 4^2 5^1 25^5$ $31^{15} 41^{10}$	$1025, 198400,$ $325376, 524800$	-	(2,4,25),(2,5,31) (2,5,41),(2,31,41)
$L_3(9)$	42456960	$4.\sigma$	$2^1 3^2 4^3 5^2 6^1 7^2$ $8^{10} 10^2 12^2 13^4$ $16^4 20^4 24^4 40^8 80^{16} 91^{24}$	$91^2, 7020, 7560,$ $58968, 110565,$ 155520	$U_3(3),$ $L_3(3)$ $\mathrm{PGL}_2(9),\ {}_2GL_2(9)$	(4,8,24),(4,16,91) (7,8,16),(8,10,12)
$U_3(9)$	42573600	$4.\sigma$	$2^1 3^2 4^1 5^6 6^1 8^2 10^{14} 15^4 16^4$ $20^4 30^4 40^8 73^{24} 80^{16}$	$730, 5913, 59130,$ $70956, 194400$	${}_2\mathcal{A}_6$	(2,5,73),(2,15,10), (2,4,20),(2,40,40)
HS	44352000	$2.\sigma$	$2^2 3^1 4^3 5^3 6^2 7^1$ $8^3 10^2 11^2 12^1$ $15^1 20^2$	$100, 176^2, 1100^2,$ $3850, 4125, 5600^2,$ $5775, 15400, 36960$	$M_{22}, L_3(2), \mathcal{S}_5, \mathcal{A}_6,$ ${}_2U_3(5), L_3(4), \mathcal{A}_5$ $\mathcal{S}_6, \mathcal{S}_8,\ {}_2M_{11}$	(4,5,6),(4,7,11) (5,7,8),(5,7,11) (5,8,10),(6,7,8)
J_3	50232960	$2.\sigma$		$6156, 14688^2, 17442,$ $20520, 23256, 25840,$ $26163, 43605$	$L_2(16),\ {}_2L_2(19)$ $L_2(17), \mathcal{A}_6$ ${}_2\mathcal{A}_5$	
$U_3(11)$	70915680	$\mathcal{S}_3.\sigma$	$2^1 3^1 4^3 5^2 6^1 8^2$ $10^2 11^4 12^4 20^4$ $22^1 37^1 240^8 44^2$	$1332, 13431, 53724^3,$ $196988^3, 246235,$ $638880, 984940$	${}_4L_2(11),$ ${}_3\mathcal{A}_6$	(4,12,12),(8,37,44) (10,10,37),(37,37,37)

1.6 See Also

§I.4	Gives some Steiner systems with large automorphism groups.
§I.3	t-designs with large automorphism groups.
§IV.36	Quasigroups are generalizations of groups.
§VI.9	Groups are used extensively to reduce the computation time in algorithms to produce designs.
[1]	A nice summary treatment of multiply transitive groups.
[8, 10, 13]	The fundamental standard references for permutation groups and group actions.

References

[1] T. Beth, D. Jungnickel and H. Lenz, *Design Theory*, Cambridge University Press, London, 1986.
[2] J. Cannon, *Computational Group Theory System Cayley*, Sydney, Australia, 1994.
[3] J. Cannon, *Computational Group Theory System Magma*, Sydney, Australia, 1994.
[4] R.W. Carter, *Simple Groups of Lie Type*, John Wiley & Sons, New York, 1972.
[5] J.H. Conway, R.T. Curtis, S.P. Norton, R.A. Parker and R.A. Wilson, *Atlas of Finite Groups*, Clarendon Press, Oxford, 1985.
[6] L.E. Dickson, *Linear Groups*, Dover Publications, 1958.
[7] D. Gorenstein, *Finite Simple Groups - An Introduction to their Classification*, Plenum Press, New York and London, 1982.
[8] D. Gorenstein, *Finite Groups*, Harper and Row, 1968.
[9] D. Gorenstein, R. Lyons and R. Solomon, *The Classification of the Finite Simple Groups*, American Math. Society, 1994.
[10] M. Hall Jr., *The Theory of Groups*, Macmillan, New York, 1961.
[11] D. Livingstone and A. Wagner, Transitivity of finite permutation groups on unordered sets, *Math. Zeitschr.* 90 (1965), 393-403.
[12] A. Niemeyer, N. Nickel et al., *Computational Group Theory System Gap*, Lehrstuhl D für Mathematik, RWTH Aachen, Germany, 1993.
[13] H. Wielandt, *Finite Permutation Groups*, Academic Press, New York, 1964.

2 Number Theory and Finite Fields

Hugh Williams

2.1 Table of Primes and Primitive Roots

A number ω is a *primitive root* for p if $\omega^x \not\equiv 1 \pmod{p}$ for any $0 < x < p-1$. (Equivalently: if ω is a generator of $\mathbb{Z}_p^*$, the multiplicative group of $\mathbb{Z}_p$.)

2.1 **Table** For each prime $p < 10000$ the least primitive root ω is given.

p	ω	p	ω	p	ω	p	ω	p	ω	p	ω	p	ω	p	ω	p	ω
3	2	5	2	7	3	11	2	13	2	17	3	19	2	23	5	29	2
31	3	37	2	41	6	43	3	47	5	53	2	59	2	61	2	67	2

p	ω	p	ω	p	ω	p	ω	p	ω	p	ω	p	ω	p	ω	p	ω
71	7	73	5	79	3	83	2	89	3	97	5	101	2	103	5	107	2
109	6	113	3	127	3	131	2	137	3	139	2	149	2	151	6	157	5
163	2	167	5	173	2	179	2	181	2	191	19	193	5	197	2	199	3
211	2	223	3	227	2	229	6	233	3	239	7	241	7	251	6	257	3
263	5	269	2	271	6	277	5	281	3	283	3	293	2	307	5	311	17
313	10	317	2	331	3	337	10	347	2	349	2	353	3	359	7	367	6
373	2	379	2	383	5	389	2	397	5	401	3	409	21	419	2	421	2
431	7	433	5	439	15	443	2	449	3	457	13	461	2	463	3	467	2
479	13	487	3	491	2	499	7	503	5	509	2	521	3	523	2	541	2
547	2	557	2	563	2	569	3	571	3	577	5	587	2	593	3	599	7
601	7	607	3	613	2	617	3	619	2	631	3	641	3	643	11	647	5
653	2	659	2	661	2	673	5	677	2	683	5	691	3	701	2	709	2
719	11	727	5	733	6	739	3	743	5	751	3	757	2	761	6	769	11
773	2	787	2	797	2	809	3	811	3	821	2	823	3	827	2	829	2
839	11	853	2	857	3	859	2	863	5	877	2	881	3	883	2	887	5
907	2	911	17	919	7	929	3	937	5	941	2	947	2	953	3	967	5
971	6	977	3	983	5	991	6	997	7	1009	11	1013	3	1019	2	1021	10
1031	14	1033	5	1039	3	1049	3	1051	7	1061	2	1063	3	1069	6	1087	3
1091	2	1093	5	1097	3	1103	5	1109	2	1117	2	1123	2	1129	11	1151	17
1153	5	1163	5	1171	2	1181	7	1187	2	1193	3	1201	11	1213	2	1217	3
1223	5	1229	2	1231	3	1237	2	1249	7	1259	2	1277	2	1279	3	1283	2
1289	6	1291	2	1297	10	1301	2	1303	6	1307	2	1319	13	1321	13	1327	3
1361	3	1367	5	1373	2	1381	2	1399	13	1409	3	1423	3	1427	2	1429	6
1433	3	1439	7	1447	3	1451	2	1453	2	1459	3	1471	6	1481	3	1483	2
1487	5	1489	14	1493	2	1499	2	1511	11	1523	2	1531	2	1543	5	1549	2
1553	3	1559	19	1567	3	1571	2	1579	3	1583	5	1597	11	1601	3	1607	5
1609	7	1613	3	1619	2	1621	2	1627	3	1637	2	1657	11	1663	3	1667	2
1669	2	1693	2	1697	3	1699	3	1709	3	1721	3	1723	3	1733	2	1741	2
1747	2	1753	7	1759	6	1777	5	1783	10	1787	2	1789	6	1801	11	1811	6
1823	5	1831	3	1847	5	1861	2	1867	2	1871	14	1873	10	1877	2	1879	6
1889	3	1901	2	1907	2	1913	3	1931	2	1933	5	1949	2	1951	3	1973	2
1979	2	1987	2	1993	5	1997	2	1999	3	2003	5	2011	3	2017	5	2027	2
2029	2	2039	7	2053	2	2063	5	2069	2	2081	3	2083	2	2087	5	2089	7
2099	2	2111	7	2113	5	2129	3	2131	2	2137	10	2141	2	2143	3	2153	3
2161	23	2179	7	2203	5	2207	5	2213	2	2221	2	2237	2	2239	3	2243	2
2251	7	2267	2	2269	2	2273	3	2281	7	2287	19	2293	2	2297	5	2309	2
2311	3	2333	2	2339	2	2341	7	2347	3	2351	13	2357	2	2371	2	2377	5
2381	3	2383	5	2389	2	2393	3	2399	11	2411	6	2417	3	2423	5	2437	2
2441	6	2447	5	2459	2	2467	2	2473	5	2477	2	2503	3	2521	17	2531	2
2539	2	2543	5	2549	2	2551	6	2557	2	2579	2	2591	7	2593	7	2609	3
2617	5	2621	2	2633	3	2647	3	2657	3	2659	2	2663	5	2671	7	2677	2
2683	2	2687	5	2689	19	2693	2	2699	2	2707	2	2711	7	2713	5	2719	3
2729	3	2731	3	2741	2	2749	6	2753	3	2767	3	2777	3	2789	2	2791	6
2797	2	2801	3	2803	2	2819	2	2833	5	2837	2	2843	2	2851	2	2857	11
2861	2	2879	7	2887	5	2897	3	2903	5	2909	2	2917	5	2927	5	2939	2
2953	13	2957	2	2963	2	2969	3	2971	10	2999	17	3001	14	3011	2	3019	2
3023	5	3037	2	3041	3	3049	11	3061	6	3067	2	3079	6	3083	2	3089	3
3109	6	3119	7	3121	7	3137	3	3163	3	3167	5	3169	7	3181	7	3187	2
3191	11	3203	2	3209	3	3217	5	3221	10	3229	6	3251	6	3253	2	3257	3

p	ω	p	ω	p	ω	p	ω	p	ω	p	ω	p	ω	p	ω	p	ω
3259	3	3271	3	3299	2	3301	6	3307	2	3313	10	3319	6	3323	2	3329	3
3331	3	3343	5	3347	2	3359	11	3361	22	3371	2	3373	5	3389	3	3391	3
3407	5	3413	2	3433	5	3449	3	3457	7	3461	2	3463	3	3467	2	3469	2
3491	2	3499	2	3511	7	3517	2	3527	5	3529	17	3533	2	3539	2	3541	7
3547	2	3557	2	3559	3	3571	2	3581	2	3583	3	3593	3	3607	5	3613	2
3617	3	3623	5	3631	15	3637	2	3643	2	3659	2	3671	13	3673	5	3677	2
3691	2	3697	5	3701	2	3709	2	3719	7	3727	3	3733	2	3739	7	3761	3
3767	5	3769	7	3779	2	3793	5	3797	2	3803	2	3821	3	3823	3	3833	3
3847	5	3851	2	3853	2	3863	5	3877	2	3881	13	3889	11	3907	2	3911	13
3917	2	3919	3	3923	2	3929	3	3931	2	3943	3	3947	2	3967	6	3989	2
4001	3	4003	2	4007	5	4013	2	4019	2	4021	2	4027	3	4049	3	4051	10
4057	5	4073	3	4079	11	4091	2	4093	2	4099	2	4111	12	4127	5	4129	13
4133	2	4139	2	4153	5	4157	2	4159	3	4177	5	4201	11	4211	6	4217	3
4219	2	4229	2	4231	3	4241	3	4243	2	4253	2	4259	2	4261	2	4271	7
4273	5	4283	2	4289	3	4297	5	4327	3	4337	3	4339	10	4349	2	4357	2
4363	2	4373	2	4391	14	4397	2	4409	3	4421	3	4423	3	4441	21	4447	3
4451	2	4457	3	4463	5	4481	3	4483	2	4493	2	4507	2	4513	7	4517	2
4519	3	4523	5	4547	2	4549	6	4561	11	4567	3	4583	5	4591	11	4597	5
4603	2	4621	2	4637	2	4639	3	4643	5	4649	3	4651	3	4657	15	4663	3
4673	3	4679	11	4691	2	4703	5	4721	6	4723	2	4729	17	4733	5	4751	19
4759	3	4783	6	4787	2	4789	2	4793	3	4799	7	4801	7	4813	2	4817	3
4831	3	4861	11	4871	11	4877	2	4889	3	4903	3	4909	6	4919	13	4931	6
4933	2	4937	3	4943	7	4951	6	4957	2	4967	5	4969	11	4973	2	4987	2
4993	5	4999	3	5003	2	5009	3	5011	2	5021	3	5023	3	5039	11	5051	2
5059	2	5077	2	5081	3	5087	5	5099	2	5101	6	5107	2	5113	19	5119	3
5147	2	5153	5	5167	6	5171	2	5179	2	5189	2	5197	7	5209	17	5227	2
5231	7	5233	10	5237	3	5261	2	5273	3	5279	7	5281	7	5297	3	5303	5
5309	2	5323	5	5333	2	5347	3	5351	11	5381	3	5387	2	5393	3	5399	7
5407	3	5413	5	5417	3	5419	3	5431	3	5437	5	5441	3	5443	2	5449	7
5471	7	5477	2	5479	3	5483	2	5501	2	5503	3	5507	2	5519	13	5521	11
5527	5	5531	10	5557	2	5563	2	5569	13	5573	2	5581	6	5591	11	5623	5
5639	7	5641	14	5647	3	5651	2	5653	5	5657	3	5659	2	5669	3	5683	2
5689	11	5693	2	5701	2	5711	19	5717	2	5737	5	5741	2	5743	10	5749	2
5779	2	5783	7	5791	6	5801	3	5807	5	5813	2	5821	6	5827	2	5839	6
5843	2	5849	3	5851	2	5857	7	5861	3	5867	5	5869	2	5879	11	5881	31
5897	3	5903	5	5923	2	5927	5	5939	2	5953	7	5981	3	5987	2	6007	3
6011	2	6029	2	6037	5	6043	5	6047	5	6053	2	6067	2	6073	10	6079	17
6089	3	6091	7	6101	2	6113	3	6121	7	6131	2	6133	5	6143	5	6151	3
6163	3	6173	2	6197	2	6199	3	6203	2	6211	2	6217	5	6221	3	6229	2
6247	5	6257	3	6263	5	6269	2	6271	11	6277	2	6287	7	6299	2	6301	10
6311	7	6317	2	6323	2	6329	3	6337	10	6343	3	6353	3	6359	13	6361	19
6367	3	6373	2	6379	2	6389	2	6397	2	6421	6	6427	3	6449	3	6451	3
6469	2	6473	3	6481	7	6491	2	6521	6	6529	7	6547	2	6551	17	6553	10
6563	5	6569	3	6571	3	6577	5	6581	14	6599	13	6607	3	6619	2	6637	2
6653	2	6659	2	6661	6	6673	5	6679	7	6689	3	6691	2	6701	2	6703	5
6709	2	6719	11	6733	2	6737	3	6761	3	6763	2	6779	2	6781	2	6791	7
6793	10	6803	2	6823	3	6827	2	6829	2	6833	3	6841	22	6857	3	6863	5
6869	2	6871	3	6883	2	6899	2	6907	2	6911	7	6917	2	6947	2	6949	2
6959	7	6961	13	6967	5	6971	2	6977	3	6983	5	6991	6	6997	5	7001	3

p	ω	p	ω	p	ω	p	ω	p	ω	p	ω	p	ω	p	ω	p	ω
7013	2	7019	2	7027	2	7039	3	7043	2	7057	5	7069	2	7079	7	7103	5
7109	2	7121	3	7127	5	7129	7	7151	7	7159	3	7177	10	7187	2	7193	3
7207	3	7211	2	7213	5	7219	2	7229	2	7237	2	7243	2	7247	5	7253	2
7283	2	7297	5	7307	2	7309	6	7321	7	7331	2	7333	6	7349	2	7351	6
7369	7	7393	5	7411	2	7417	5	7433	3	7451	2	7457	3	7459	2	7477	2
7481	6	7487	5	7489	7	7499	2	7507	2	7517	2	7523	2	7529	3	7537	7
7541	2	7547	2	7549	2	7559	13	7561	13	7573	2	7577	3	7583	5	7589	2
7591	6	7603	2	7607	5	7621	2	7639	7	7643	2	7649	3	7669	2	7673	3
7681	17	7687	6	7691	2	7699	3	7703	5	7717	2	7723	3	7727	5	7741	7
7753	10	7757	2	7759	3	7789	2	7793	3	7817	3	7823	5	7829	2	7841	12
7853	2	7867	3	7873	5	7877	2	7879	3	7883	2	7901	2	7907	2	7919	7
7927	3	7933	2	7937	3	7949	2	7951	6	7963	5	7993	5	8009	3	8011	14
8017	5	8039	11	8053	2	8059	3	8069	2	8081	3	8087	5	8089	17	8093	2
8101	6	8111	11	8117	2	8123	2	8147	2	8161	7	8167	3	8171	2	8179	2
8191	17	8209	7	8219	2	8221	2	8231	11	8233	10	8237	2	8243	2	8263	3
8269	2	8273	3	8287	3	8291	2	8293	2	8297	3	8311	3	8317	6	8329	7
8353	5	8363	2	8369	3	8377	5	8387	2	8389	6	8419	3	8423	5	8429	2
8431	3	8443	2	8447	5	8461	6	8467	2	8501	7	8513	5	8521	13	8527	5
8537	3	8539	2	8543	5	8563	2	8573	2	8581	6	8597	2	8599	3	8609	3
8623	3	8627	2	8629	6	8641	17	8647	3	8663	5	8669	2	8677	2	8681	15
8689	13	8693	2	8699	2	8707	5	8713	5	8719	3	8731	2	8737	5	8741	2
8747	2	8753	3	8761	23	8779	11	8783	5	8803	2	8807	5	8819	2	8821	2
8831	7	8837	2	8839	3	8849	3	8861	2	8863	3	8867	2	8887	3	8893	5
8923	2	8929	11	8933	2	8941	6	8951	13	8963	2	8969	3	8971	2	8999	7
9001	7	9007	3	9011	2	9013	5	9029	2	9041	3	9043	3	9049	7	9059	2
9067	3	9091	3	9103	6	9109	10	9127	3	9133	6	9137	3	9151	3	9157	6
9161	3	9173	2	9181	2	9187	3	9199	3	9203	2	9209	3	9221	2	9227	2
9239	19	9241	13	9257	3	9277	5	9281	3	9283	2	9293	2	9311	7	9319	3
9323	2	9337	5	9341	2	9343	5	9349	2	9371	2	9377	3	9391	3	9397	2
9403	3	9413	3	9419	2	9421	2	9431	7	9433	5	9437	2	9439	22	9461	3
9463	3	9467	2	9473	3	9479	7	9491	2	9497	3	9511	3	9521	3	9533	2
9539	2	9547	2	9551	11	9587	2	9601	13	9613	2	9619	2	9623	5	9629	2
9631	3	9643	2	9649	7	9661	2	9677	2	9679	3	9689	3	9697	10	9719	17
9721	7	9733	2	9739	3	9743	5	9749	2	9767	5	9769	13	9781	6	9787	3
9791	11	9803	2	9811	3	9817	5	9829	10	9833	3	9839	7	9851	2	9857	5
9859	2	9871	3	9883	2	9887	5	9901	2	9907	2	9923	2	9929	3	9931	10
9941	2	9949	2	9967	3	9973	11										

2.2 **Conjecture** (Artin) 2 is an primitive root for infinitely many primes.

2.3 **Theorem** [1] For any three distinct primes p, q, and r, at least one of p, q, or r is a primitive root for infinitely many primes.

Let S denote a finite set of primes. If N represents the product of the elements of S, then the integer x, $1 < x \leq N-1$ is a *common primitive root* of the elements of S if and only if for each $p \in S$, x is congruent to ω (mod p) where ω is a primitive root of p.

2.4 **Remark** The number of common primitive roots equals $\prod_{p \in S} \phi(p-1)$ where ϕ is the Euler ϕ-function (§2.2).

2.5 **Table** Table of (all) common primitive roots for $S = \{p, q\}$ with $1 < p, q \leq 19$.

p	q	Common Primitive Roots
3	5	2 8
3	7	5 17
3	11	2 8 17 29
3	13	2 11 20 32
3	17	5 11 14 20 23 29 41 44
3	19	2 14 29 32 41 53
5	7	3 12 17 33
5	11	2 7 8 13 17 18 28 52
5	13	2 7 28 32 33 37 58 63
5	17	3 7 12 22 23 27 28 37 48 57 58 62 63 73 78 82
5	19	2 3 13 22 32 33 48 52 53 67 72 78
7	11	17 19 24 40 52 61 68 73
7	13	19 24 33 45 54 59 80 89
7	17	3 5 10 12 24 31 40 45 54 61 73 75 80 82 96 108
7	19	3 10 33 40 52 59 89 108 110 117 124 129
11	13	2 6 7 19 24 28 41 46 50 63 72 84 85 106 123 128
11	17	6 7 24 28 29 39 40 41 46 57 61 62 63 73 74 79 90 95 96 105 107 112 116 129 139 150 156 160 167 173 182 184
11	19	2 13 29 40 41 51 52 72 79 90 105 116 117 127 128 129 162 167 173 184 193 200 204 205
13	17	6 7 11 20 24 28 37 41 45 46 54 58 63 71 80 97 124 141 150 158 163 167 175 176 180 184 193 197 201 210 214 215
13	19	2 15 32 33 41 59 67 71 72 89 97 98 110 124 128 136 154 162 167 184 193 219 223 241
17	19	3 10 14 22 29 40 41 48 71 78 79 90 91 97 105 108 109 116 124 129 143 146 147 148 165 167 173 181 184 192 193 211 224 231 241 243 249 250 260 261 262 269 279 295 299 300 317 318

2.6 **Table** The least common primitive root ω for pairs of primes $p, q < 50$, with $q > 19$.

$q \backslash p$	3	5	7	11	13	17	19	23	29	31	37	41	43
23	5	7	5	7	7	5	10						
29	2	2	3	2	2	3	2	10					
31	11	3	3	13	11	3	3	11	3				
37	2	2	5	2	2	5	2	5	2	3			
41	11	7	12	6	6	6	13	7	11	11	13		
43	5	3	3	18	9	3	3	5	3	3	5	12	
47	5	13	5	13	11	5	10	5	10	11	5	11	5

2.2 Number Theoretic Functions

The Euler $\phi(n)$ function denotes the number of positive integers less than n that are relatively prime to n.

2.7 **Theorems**

1. $\phi(p) = p - 1$ for any prime p.
2. $\phi(ab) = \phi(a) \times \phi(b)$ if a and b are relatively prime.
3. If $n > 1$, then $\phi(n) = n \prod_{p|n}(1 - 1/p)$.
4. For $n \geq 1$, $\sum_{d|n} \phi(d) = n$.
5. If a and n are relatively prime, then $a^{\phi(n)} \equiv 1 \pmod{n}$.

$\sigma_k(n)$ is the sum of the kth powers of the positive divisors of n.
$\sigma(n) = \sigma_1(n)$ is the sum of the positive divisors of n.
$\tau(n) = \sigma_0(n)$ is the number of positive divisors of n.

2.8 Theorems

1. If p is prime, then $\sigma(p) = p + 1$ and $\tau(p) = 2$.
2. If a and b are relatively prime, then $\sigma(ab) = \sigma(a) \times \sigma(b)$ and $\tau(ab) = \tau(a) \times \tau(b)$.
3. If $n = p_1^{e_1} p_2^{e_2} \cdots p_r^{e_r}$, then
   (a) $\tau(n) = (e_1 + 1)(e_2 + 1) \cdots (e_r + 1)$, and
   (b) $\sigma(n) = \prod_i^r \frac{p_i^{e_i+1} - 1}{p_i - 1}$.

The *Möbius function* $\mu(n)$ is

$$\mu(n) = \begin{cases} 1 & \text{if } n = 1 \\ 0 & \text{if } a^2 | n \text{ for some } a > 1 \\ (-1)^r & \text{if } n = p_1 p_2 \cdots p_r, \text{ distinct prime factors} \end{cases}$$

2.9 Theorems

1. $\mu(p) = -1$ for any prime p.
2. $\mu(ab) = \mu(a) \times \mu(b)$ if a and b are relatively prime.
3. If $n > 1$, then $\sum_{d|n} \mu(d) = 0$.
4. (*Möbius inversion formula*) If $F(n) = \sum_{d|n} f(d)$ for every positive integer n, then $f(n) = \sum_{d|n} \mu(d) F(n/d)$.

2.10 Table Values of $\phi(n), \sigma(n), \tau(n)$, and $\mu(n)$ for $1 \le n \le 1000$. Using `Maple V`, the `numtheory` package can be used with the commands `phi(n)`, `sigma(n)`, `tau(n)` and `mobius(n)` to calculate these functions.

n	ϕ	σ	τ	μ	n	ϕ	σ	τ	μ	n	ϕ	σ	τ	μ	n	ϕ	σ	τ	μ	n	ϕ	σ	τ	μ
1	1	1	1	1	2	1	3	2	-1	3	2	4	2	-1	4	2	7	3	0	5	4	6	2	-1
6	2	12	4	1	7	6	8	2	-1	8	4	15	4	0	9	6	13	3	0	10	4	18	4	1
11	10	12	2	-1	12	4	28	6	0	13	12	14	2	-1	14	6	24	4	1	15	8	24	4	1
16	8	31	5	0	17	16	18	2	-1	18	6	39	6	0	19	18	20	2	-1	20	8	42	6	0
21	12	32	4	1	22	10	36	4	1	23	22	24	2	-1	24	8	60	8	0	25	20	31	3	0
26	12	42	4	1	27	18	40	4	0	28	12	56	6	0	29	28	30	2	-1	30	8	72	8	-1
31	30	32	2	-1	32	16	63	6	0	33	20	48	4	1	34	16	54	4	1	35	24	48	4	1
36	12	91	9	0	37	36	38	2	-1	38	18	60	4	1	39	24	56	4	1	40	16	90	8	0
41	40	42	2	-1	42	12	96	8	-1	43	42	44	2	-1	44	20	84	6	0	45	24	78	6	0
46	22	72	4	1	47	46	48	2	-1	48	16	124	10	0	49	42	57	3	0	50	20	93	6	0
51	32	72	4	1	52	24	98	6	0	53	52	54	2	-1	54	18	120	8	0	55	40	72	4	1
56	24	120	8	0	57	36	80	4	1	58	28	90	4	1	59	58	60	2	-1	60	16	168	12	0
61	60	62	2	-1	62	30	96	4	1	63	36	104	6	0	64	32	127	7	0	65	48	84	4	1
66	20	144	8	-1	67	66	68	2	-1	68	32	126	6	0	69	44	96	4	1	70	24	144	8	-1
71	70	72	2	-1	72	24	195	12	0	73	72	74	2	-1	74	36	114	4	1	75	40	124	6	0
76	36	140	6	0	77	60	96	4	1	78	24	168	8	-1	79	78	80	2	-1	80	32	186	10	0
81	54	121	5	0	82	40	126	4	1	83	82	84	2	-1	84	24	224	12	0	85	64	108	4	1
86	42	132	4	1	87	56	120	4	1	88	40	180	8	0	89	88	90	2	-1	90	24	234	12	0
91	72	112	4	1	92	44	168	6	0	93	60	128	4	1	94	46	144	4	1	95	72	120	4	1
96	32	252	12	0	97	96	98	2	-1	98	42	171	6	0	99	60	156	6	0	100	40	217	9	0
101	100	102	2	-1	102	32	216	8	-1	103	102	104	2	-1	104	48	210	8	0	105	48	192	8	-1
106	52	162	4	1	107	106	108	2	-1	108	36	280	12	0	109	108	110	2	-1	110	40	216	8	-1
111	72	152	4	1	112	48	248	10	0	113	112	114	2	-1	114	36	240	8	-1	115	88	144	4	1
116	56	210	6	0	117	72	182	6	0	118	58	180	4	1	119	96	144	4	1	120	32	360	16	0
121	110	133	3	0	122	60	186	4	1	123	80	168	4	1	124	60	224	6	0	125	100	156	4	0

n	ϕ	σ	τ	μ	n	ϕ	σ	τ	μ	n	ϕ	σ	τ	μ	n	ϕ	σ	τ	μ	n	ϕ	σ	τ	μ
126	36	312	12	0	127	126	128	2	-1	128	64	255	8	0	129	84	176	4	1	130	48	252	8	-1
131	130	132	2	-1	132	40	336	12	0	133	108	160	4	1	134	66	204	4	1	135	72	240	8	0
136	64	270	8	0	137	136	138	2	-1	138	44	288	8	-1	139	138	140	2	-1	140	48	336	12	0
141	92	192	4	1	142	70	216	4	1	143	120	168	4	1	144	48	403	15	0	145	112	180	4	1
146	72	222	4	1	147	84	228	6	0	148	72	266	6	0	149	148	150	2	-1	150	40	372	12	0
151	150	152	2	-1	152	72	300	8	0	153	96	234	6	0	154	60	288	8	-1	155	120	192	4	1
156	48	392	12	0	157	156	158	2	-1	158	78	240	4	1	159	104	216	4	1	160	64	378	12	0
161	132	192	4	1	162	54	363	10	0	163	162	164	2	-1	164	80	294	6	0	165	80	288	8	-1
166	82	252	4	1	167	166	168	2	-1	168	48	480	16	0	169	156	183	3	0	170	64	324	8	-1
171	108	260	6	0	172	84	308	6	0	173	172	174	2	-1	174	56	360	8	-1	175	120	248	6	0
176	80	372	10	0	177	116	240	4	1	178	88	270	4	1	179	178	180	2	-1	180	48	546	18	0
181	180	182	2	-1	182	72	336	8	-1	183	120	248	4	1	184	88	360	8	0	185	144	228	4	1
186	60	384	8	-1	187	160	216	4	1	188	92	336	6	0	189	108	320	8	0	190	72	360	8	-1
191	190	192	2	-1	192	64	508	14	0	193	192	194	2	-1	194	96	294	4	1	195	96	336	8	-1
196	84	399	9	0	197	196	198	2	-1	198	60	468	12	0	199	198	200	2	-1	200	80	465	12	0
201	132	272	4	1	202	100	306	4	1	203	168	240	4	1	204	64	504	12	0	205	160	252	4	1
206	102	312	4	1	207	132	312	6	0	208	96	434	10	0	209	180	240	4	1	210	48	576	16	1
211	210	212	2	-1	212	104	378	6	0	213	140	288	4	1	214	106	324	4	1	215	168	264	4	1
216	72	600	16	0	217	180	256	4	1	218	108	330	4	1	219	144	296	4	1	220	80	504	12	0
221	192	252	4	1	222	72	456	8	-1	223	222	224	2	-1	224	96	504	12	0	225	120	403	9	0
226	112	342	4	1	227	226	228	2	-1	228	72	560	12	0	229	228	230	2	-1	230	88	432	8	-1
231	120	384	8	-1	232	112	450	8	0	233	232	234	2	-1	234	72	546	12	0	235	184	288	4	1
236	116	420	6	0	237	156	320	4	1	238	96	432	8	-1	239	238	240	2	-1	240	64	744	20	0
241	240	242	2	-1	242	110	399	6	0	243	162	364	6	0	244	120	434	6	0	245	168	342	6	0
246	80	504	8	-1	247	216	280	4	1	248	120	480	8	0	249	164	336	4	1	250	100	468	8	0
251	250	252	2	-1	252	72	728	18	0	253	220	288	4	1	254	126	384	4	1	255	128	432	8	-1
256	128	511	9	0	257	256	258	2	-1	258	84	528	8	-1	259	216	304	4	1	260	96	588	12	0
261	168	390	6	0	262	130	396	4	1	263	262	264	2	-1	264	80	720	16	0	265	208	324	4	1
266	108	480	8	-1	267	176	360	4	1	268	132	476	6	0	269	268	270	2	-1	270	72	720	16	0
271	270	272	2	-1	272	128	558	10	0	273	144	448	8	-1	274	136	414	4	1	275	200	372	6	0
276	88	672	12	0	277	276	278	2	-1	278	138	420	4	1	279	180	416	6	0	280	96	720	16	0
281	280	282	2	-1	282	92	576	8	-1	283	282	284	2	-1	284	140	504	6	0	285	144	480	8	-1
286	120	504	8	-1	287	240	336	4	1	288	96	819	18	0	289	272	307	3	0	290	112	540	8	-1
291	192	392	4	1	292	144	518	6	0	293	292	294	2	-1	294	84	684	12	0	295	232	360	4	1
296	144	570	8	0	297	180	480	8	0	298	148	450	4	1	299	264	336	4	1	300	80	868	18	0
301	252	352	4	1	302	150	456	4	1	303	200	408	4	1	304	144	620	10	0	305	240	372	4	1
306	96	702	12	0	307	306	308	2	-1	308	120	672	12	0	309	204	416	4	1	310	120	576	8	-1
311	310	312	2	-1	312	96	840	16	0	313	312	314	2	-1	314	156	474	4	1	315	144	624	12	0
316	156	560	6	0	317	316	318	2	-1	318	104	648	8	-1	319	280	360	4	1	320	128	762	14	0
321	212	432	4	1	322	132	576	8	-1	323	288	360	4	1	324	108	847	15	0	325	240	434	6	0
326	162	492	4	1	327	216	440	4	1	328	160	630	8	0	329	276	384	4	1	330	80	864	16	1
331	330	332	2	-1	332	164	588	6	0	333	216	494	6	0	334	166	504	4	1	335	264	408	4	1
336	96	992	20	0	337	336	338	2	-1	338	156	549	6	0	339	224	456	4	1	340	128	756	12	0
341	300	384	4	1	342	108	780	12	0	343	294	400	4	0	344	168	660	8	0	345	176	576	8	-1
346	172	522	4	1	347	346	348	2	-1	348	112	840	12	0	349	348	350	2	-1	350	120	744	12	0
351	216	560	8	0	352	160	756	12	0	353	352	354	2	-1	354	116	720	8	-1	355	280	432	4	1
356	176	630	6	0	357	192	576	8	-1	358	178	540	4	1	359	358	360	2	-1	360	96	1170	24	0
361	342	381	3	0	362	180	546	4	1	363	220	532	6	0	364	144	784	12	0	365	288	444	4	1
366	120	744	8	-1	367	366	368	2	-1	368	176	744	10	0	369	240	546	6	0	370	144	684	8	-1
371	312	432	4	1	372	120	896	12	0	373	372	374	2	-1	374	160	648	8	-1	375	200	624	8	0
376	184	720	8	0	377	336	420	4	1	378	108	960	16	0	379	378	380	2	-1	380	144	840	12	0
381	252	512	4	1	382	190	576	4	1	383	382	384	2	-1	384	128	1020	16	0	385	240	576	8	-1
386	192	582	4	1	387	252	572	6	0	388	192	686	6	0	389	388	390	2	-1	390	96	1008	16	1

n	ϕ	σ	τ	μ	n	ϕ	σ	τ	μ	n	ϕ	σ	τ	μ	n	ϕ	σ	τ	μ	n	ϕ	σ	τ	μ
391	352	432	4	1	392	168	855	12	0	393	260	528	4	1	394	196	594	4	1	395	312	480	4	1
396	120	1092	18	0	397	396	398	2	-1	398	198	600	4	1	399	216	640	8	-1	400	160	961	15	0
401	400	402	2	-1	402	132	816	8	-1	403	360	448	4	1	404	200	714	6	0	405	216	726	10	0
406	168	720	8	-1	407	360	456	4	1	408	128	1080	16	0	409	408	410	2	-1	410	160	756	8	-1
411	272	552	4	1	412	204	728	6	0	413	348	480	4	1	414	132	936	12	0	415	328	504	4	1
416	192	882	12	0	417	276	560	4	1	418	180	720	8	-1	419	418	420	2	-1	420	96	1344	24	0
421	420	422	2	-1	422	210	636	4	1	423	276	624	6	0	424	208	810	8	0	425	320	558	6	0
426	140	864	8	-1	427	360	496	4	1	428	212	756	6	0	429	240	672	8	-1	430	168	792	8	-1
431	430	432	2	-1	432	144	1240	20	0	433	432	434	2	-1	434	180	768	8	-1	435	224	720	8	-1
436	216	770	6	0	437	396	480	4	1	438	144	888	8	-1	439	438	440	2	-1	440	160	1080	16	0
441	252	741	9	0	442	192	756	8	-1	443	442	444	2	-1	444	144	1064	12	0	445	352	540	4	1
446	222	672	4	1	447	296	600	4	1	448	192	1016	14	0	449	448	450	2	-1	450	120	1209	18	0
451	400	504	4	1	452	224	798	6	0	453	300	608	4	1	454	226	684	4	1	455	288	672	8	-1
456	144	1200	16	0	457	456	458	2	-1	458	228	690	4	1	459	288	720	8	0	460	176	1008	12	0
461	460	462	2	-1	462	120	1152	16	1	463	462	464	2	-1	464	224	930	10	0	465	240	768	8	-1
466	232	702	4	1	467	466	468	2	-1	468	144	1274	18	0	469	396	544	4	1	470	184	864	8	-1
471	312	632	4	1	472	232	900	8	0	473	420	528	4	1	474	156	960	8	-1	475	360	620	6	0
476	192	1008	12	0	477	312	702	6	0	478	238	720	4	1	479	478	480	2	-1	480	128	1512	24	0
481	432	532	4	1	482	240	726	4	1	483	264	768	8	-1	484	220	931	9	0	485	384	588	4	1
486	162	1092	12	0	487	486	488	2	-1	488	240	930	8	0	489	324	656	4	1	490	168	1026	12	0
491	490	492	2	-1	492	160	1176	12	0	493	448	540	4	1	494	216	840	8	-1	495	240	936	12	0
496	240	992	10	0	497	420	576	4	1	498	164	1008	8	-1	499	498	500	2	-1	500	200	1092	12	0

2.11 Table Inverse of the Euler ϕ-function.

x	Integers n for which $\phi(n) = x$	x	Integers n for which $\phi(n) = x$
1	1, 2	2	3, 4, 6
4	5, 8, 10, 12	6	7, 9, 14, 18
8	15, 16, 20, 24, 30	10	11, 22
12	13, 21, 26, 28, 36, 42	16	17, 32, 34, 40, 48, 60
18	19, 27, 38, 54	20	25, 33, 44, 50, 66
22	23, 46	24	35, 39, 45, 52, 56, 70, 72, 78, 84, 90
28	29, 58	30	31, 62
32	51, 64, 68, 80, 96, 102, 120	36	37, 57, 63, 74, 76, 108, 114, 126
40	41, 55, 75, 82, 88, 100, 110, 132, 150	42	43, 49, 86, 98
44	69, 92, 138	46	47, 94
48	65, 104, 105, 112, 130, 140, 144, 156, 168, 180, 210	52	53, 106
54	81, 162	56	87, 116, 174
58	59, 118	60	61, 77, 93, 99, 122, 124, 154, 186, 198
64	85, 128, 136, 160, 170, 192, 204, 240	66	67, 134
70	71, 142	72	73, 91, 95, 111, 117, 135, 146, 148, 152, 182, 190, 216, 222, 228, 234, 252, 270
78	79, 158	80	123, 164, 165, 176, 200, 220, 246, 264, 300, 330
82	83, 166	84	129, 147, 172, 196, 258, 294
88	89, 115, 178, 184, 230, 276	92	141, 188, 282
96	97, 119, 153, 194, 195, 208, 224, 238, 260, 280, 288, 306, 312, 336, 360, 390, 420	100	101, 125, 202, 250

$\pi(n)$ is the number of primes $p \leq n$.
p_n denotes the nth prime. $(p_1 = 2, p_2 = 3, p_3 = 5 \ldots)$

2.12 Theorem (*prime number theorem*)

$$\lim_{n\to\infty} \frac{\pi(n)}{n/\log n} = 1.$$

2.13 Theorems Bounds on $\pi(x)$ [2].

1. $\frac{x}{\log x}(1+\frac{1}{2\log x}) < \pi(x) < \frac{x}{\log x}(1+\frac{3}{2\log x})$, if $x \geq 59$.
2. $x/(\log x - \frac{1}{2}) < \pi(x) < x/(\log x - \frac{3}{2})$, if $x \geq 67$.
3. $x/\log x < \pi(x) < 1.25x/\log x$, if $x \geq 17$, $x \neq 113$.

2.14 Theorems Other bounds. [2]

1. $n(\log n + \log\log n - \frac{3}{2}) < p_n < n(\log n + \log\log n - \frac{1}{2})$, if $n \geq 20$.
2. $\log\log x + B - 1/(2\log^2 x) < \sum_{p\leq x} 1/p < \log\log x + B + 1/(2\log^2 x)$, if $x \geq 286$ where $B \approx 0.261497212847643$.
3. $\log x + E - 1/(2\log x) < \sum_{p\leq x}(\log p)/p < \log x + E + 1/(2\log x)$, if $x > 319$ where $E \approx -1.332582275733221$.

If p denotes an odd prime and $\gcd(a,p)=1$, the *Legendre symbol* $\left(\frac{a}{p}\right)$ is 1 if a is a quadratic residue ($x^2 \equiv a \pmod p$), and -1 if a is a quadratic nonresidue modulo p.

2.15 Theorems

1. $\left(\frac{a}{p}\right) \equiv a^{(p-1)/2} \pmod p$.
2. $\left(\frac{a}{p}\right)\left(\frac{b}{p}\right) = \left(\frac{ab}{p}\right)$.
3. If $a \equiv b \pmod p$, then $\left(\frac{a}{p}\right) = \left(\frac{b}{p}\right)$.
4. (*quadratic reciprocity*) If p and q are distinct odd primes, then

$$\left(\frac{p}{q}\right)\left(\frac{q}{p}\right) = (-1)^{[(p-1)/2][(q-1)/2]}.$$

2.3 Table of Indices for Primes

2.16 Table For each prime $p \leq 113$ two tables are given. Let g be least primitive element of the group $\mathbb{F}_p^*$ and assume $g^x = y$. The table on the left has a y in position x, while the one on the right has an x in position y.

3

N	0	1	2	3	4	5	6	7	8	9
0		2	1							

I	0	1	2	3	4	5	6	7	8	9
0	1	2	1							

5

N	0	1	2	3	4	5	6	7	8	9
0		4	1	3	2					

I	0	1	2	3	4	5	6	7	8	9
0	1	2	4	3	1					

7

N	0	1	2	3	4	5	6	7	8	9
0		6	2	1	4	5	3			

I	0	1	2	3	4	5	6	7	8	9
0	1	3	2	6	4	5	1			

11

N	0	1	2	3	4	5	6	7	8	9
0		10	1	8	2	4	9	7	3	6
1	5									

I	0	1	2	3	4	5	6	7	8	9
0	1	2	4	8	5	10	9	7	3	6
1	1									

13

N	0	1	2	3	4	5	6	7	8	9
0		12	1	4	2	9	5	11	3	8
1	10	7	6							

I	0	1	2	3	4	5	6	7	8	9
0	1	2	4	8	3	6	12	11	9	5
1	10	7	1							

17

N	0	1	2	3	4	5	6	7	8	9
0		16	14	1	12	5	15	11	10	2
1	3	7	13	4	9	6	8			

I	0	1	2	3	4	5	6	7	8	9
0	1	3	9	10	13	5	15	11	16	14
1	8	7	4	12	2	6	1			

19

N	0	1	2	3	4	5	6	7	8	9
0		18	1	13	2	16	14	6	3	8
1	17	12	15	5	7	11	4	10	9	

I	0	1	2	3	4	5	6	7	8	9
0	1	2	4	8	16	13	7	14	9	18
1	17	15	11	3	6	12	5	10	1	

23

N	0	1	2	3	4	5	6	7	8	9
0		22	2	16	4	1	18	19	6	10
1	3	9	20	14	21	17	8	7	12	15
2	5	13	11							

I	0	1	2	3	4	5	6	7	8	9
0	1	5	2	10	4	20	8	17	16	11
1	9	22	18	21	13	19	3	15	6	7
2	12	14	1							

29

N	0	1	2	3	4	5	6	7	8	9
0		28	1	5	2	22	6	12	3	10
1	23	25	7	18	13	27	4	21	11	9
2	24	17	26	20	8	16	19	15	14	

I	0	1	2	3	4	5	6	7	8	9
0	1	2	4	8	16	3	6	12	24	19
1	9	18	7	14	28	27	25	21	13	26
2	23	17	5	10	20	11	22	15	1	

31

N	0	1	2	3	4	5	6	7	8	9
0		30	24	1	18	20	25	28	12	2
1	14	23	19	11	22	21	6	7	26	4
2	8	29	17	27	13	10	5	3	16	9
3	15									

I	0	1	2	3	4	5	6	7	8	9
0	1	3	9	27	19	26	16	17	20	29
1	25	13	8	24	10	30	28	22	4	12
2	5	15	14	11	2	6	18	23	7	21
3	1									

37

N	0	1	2	3	4	5	6	7	8	9
0		36	1	26	2	23	27	32	3	16
1	24	30	28	11	33	13	4	7	17	35
2	25	22	31	15	29	10	12	6	34	21
3	14	9	5	20	8	19	18			

I	0	1	2	3	4	5	6	7	8	9
0	1	2	4	8	16	32	27	17	34	31
1	25	13	26	15	30	23	9	18	36	35
2	33	29	21	5	10	20	3	6	12	24
3	11	22	7	14	28	19	1			

41

N	0	1	2	3	4	5	6	7	8	9
0		40	26	15	12	22	1	39	38	30
1	8	3	27	31	25	37	24	33	16	9
2	34	14	29	36	13	4	17	5	11	7
3	23	28	10	18	19	21	2	32	35	6
4	20									

I	0	1	2	3	4	5	6	7	8	9
0	1	6	36	11	25	27	39	29	10	19
1	32	28	4	24	21	3	18	26	33	34
2	40	35	5	30	16	14	2	12	31	22
3	9	13	37	17	20	38	23	15	8	7
4	1									

43

N	0	1	2	3	4	5	6	7	8	9
0		42	27	1	12	25	28	35	39	2
1	10	30	13	32	20	26	24	38	29	19
2	37	36	15	16	40	8	17	3	5	41
3	11	34	9	31	23	18	14	7	4	33
4	22	6	21							

I	0	1	2	3	4	5	6	7	8	9
0	1	3	9	27	38	28	41	37	25	32
1	10	30	4	12	36	22	23	26	35	19
2	14	42	40	34	16	5	15	2	6	18
3	11	33	13	39	31	7	21	20	17	8
4	24	29	1							

47

N	0	1	2	3	4	5	6	7	8	9
0		46	18	20	36	1	38	32	8	40
1	19	7	10	11	4	21	26	16	12	45
2	37	6	25	5	28	2	29	14	22	35
3	39	3	44	27	34	33	30	42	17	31
4	9	15	24	13	43	41	23			

I	0	1	2	3	4	5	6	7	8	9
0	1	5	25	31	14	23	21	11	8	40
1	12	13	18	43	27	41	17	38	2	10
2	3	15	28	46	42	22	16	33	24	26
3	36	39	7	35	34	29	4	20	6	30
4	9	45	37	44	32	19	1			

53

N	0	1	2	3	4	5	6	7	8	9
0		52	1	17	2	47	18	14	3	34
1	48	6	19	24	15	12	4	10	35	37
2	49	31	7	39	20	42	25	51	16	46
3	13	33	5	23	11	9	36	30	38	41
4	50	45	32	22	8	29	40	44	21	28
5	43	27	26							

I	0	1	2	3	4	5	6	7	8	9
0	1	2	4	8	16	32	11	22	44	35
1	17	34	15	30	7	14	28	3	6	12
2	24	48	43	33	13	26	52	51	49	45
3	37	21	42	31	9	18	36	19	38	23
4	46	39	25	50	47	41	29	5	10	20
5	40	27	1							

59

N	0	1	2	3	4	5	6	7	8	9
0		58	1	50	2	6	51	18	3	42
1	7	25	52	45	19	56	4	40	43	38
2	8	10	26	15	53	12	46	34	20	28
3	57	49	5	17	41	24	44	55	39	37
4	9	14	11	33	27	48	16	23	54	36
5	13	32	47	22	35	31	21	30	29	

I	0	1	2	3	4	5	6	7	8	9
0	1	2	4	8	16	32	5	10	20	40
1	21	42	25	50	41	23	46	33	7	14
2	28	56	53	47	35	11	22	44	29	58
3	57	55	51	43	27	54	49	39	19	38
4	17	34	9	18	36	13	26	52	45	31
5	3	6	12	24	48	37	15	30	1	

61

N	0	1	2	3	4	5	6	7	8	9
0		60	1	6	2	22	7	49	3	12
1	23	15	8	40	50	28	4	47	13	26
2	24	55	16	57	9	44	41	18	51	35
3	29	59	5	21	48	11	14	39	27	46
4	25	54	56	43	17	34	58	20	10	38
5	45	53	42	33	19	37	52	32	36	31
6	30									

I	0	1	2	3	4	5	6	7	8	9
0	1	2	4	8	16	32	3	6	12	24
1	48	35	9	18	36	11	22	44	27	54
2	47	33	5	10	20	40	19	38	15	30
3	60	59	57	53	45	29	58	55	49	37
4	13	26	52	43	25	50	39	17	34	7
5	14	28	56	51	41	21	42	23	46	31
6	1									

67

N	0	1	2	3	4	5	6	7	8	9
0		66	1	39	2	15	40	23	3	12
1	16	59	41	19	24	54	4	64	13	10
2	17	62	60	28	42	30	20	51	25	44
3	55	47	5	32	65	38	14	22	11	58
4	18	53	63	9	61	27	29	50	43	46
5	31	37	21	57	52	8	26	49	45	36
6	56	7	48	35	6	34	33			

I	0	1	2	3	4	5	6	7	8	9
0	1	2	4	8	16	32	64	61	55	43
1	19	38	9	18	36	5	10	20	40	13
2	26	52	37	7	14	28	56	45	23	46
3	25	50	33	66	65	63	59	51	35	3
4	6	12	24	48	29	58	49	31	62	57
5	47	27	54	41	15	30	60	53	39	11
6	22	44	21	42	17	34	1			

71

N	0	1	2	3	4	5	6	7	8	9
0		70	6	26	12	28	32	1	18	52
1	34	31	38	39	7	54	24	49	58	16
2	40	27	37	15	44	56	45	8	13	68
3	60	11	30	57	55	29	64	20	22	65
4	46	25	33	48	43	10	21	9	50	2
5	62	5	51	23	14	59	19	42	4	3
6	66	69	17	53	36	67	63	47	61	41
7	35									

I	0	1	2	3	4	5	6	7	8	9
0	1	7	49	59	58	51	2	14	27	47
1	45	31	4	28	54	23	19	62	8	56
2	37	46	38	53	16	41	3	21	5	35
3	32	11	6	42	10	70	64	22	12	13
4	20	69	57	44	24	26	40	67	43	17
5	48	52	9	63	15	34	25	33	18	55
6	30	68	50	66	36	39	60	65	29	61
7	1									

73

N	0	1	2	3	4	5	6	7	8	9
0		72	8	6	16	1	14	33	24	12
1	9	55	22	59	41	7	32	21	20	62
2	17	39	63	46	30	2	67	18	49	35
3	15	11	40	61	29	34	28	64	70	65
4	25	4	47	51	71	13	54	31	38	66
5	10	27	3	53	26	56	57	68	43	5
6	23	58	19	45	48	60	69	50	37	52
7	42	44	36							

I	0	1	2	3	4	5	6	7	8	9
0	1	5	25	52	41	59	3	15	2	10
1	50	31	9	45	6	30	4	20	27	62
2	18	17	12	60	8	40	54	51	36	34
3	24	47	16	7	35	29	72	68	48	21
4	32	14	70	58	71	63	23	42	64	28
5	67	43	69	53	46	11	55	56	61	13
6	65	33	19	22	37	39	49	26	57	66
7	38	44	1							

79

N	0	1	2	3	4	5	6	7	8	9
0		78	4	1	8	62	5	53	12	2
1	66	68	9	34	57	63	16	21	6	32
2	70	54	72	26	13	46	38	3	61	11
3	67	56	20	69	25	37	10	19	36	35
4	74	75	58	49	76	64	30	59	17	28
5	50	22	42	77	7	52	65	33	15	31
6	71	45	60	55	24	18	73	48	29	27
7	41	51	14	44	23	47	40	43	39	

I	0	1	2	3	4	5	6	7	8	9
0	1	3	9	27	2	6	18	54	4	12
1	36	29	8	24	72	58	16	48	65	37
2	32	17	51	74	64	34	23	69	49	68
3	46	59	19	57	13	39	38	35	26	78
4	76	70	52	77	73	61	25	75	67	43
5	50	71	55	7	21	63	31	14	42	47
6	62	28	5	15	45	56	10	30	11	33
7	20	60	22	66	40	41	44	53	1	

83

N	0	1	2	3	4	5	6	7	8	9
0		82	1	72	2	27	73	8	3	62
1	28	24	74	77	9	17	4	56	63	47
2	29	80	25	60	75	54	78	52	10	12
3	18	38	5	14	57	35	64	20	48	67
4	30	40	81	71	26	7	61	23	76	16
5	55	46	79	59	53	51	11	37	13	34
6	19	66	39	70	6	22	15	45	58	50
7	36	33	65	69	21	44	49	32	68	43
8	31	42	41							

I	0	1	2	3	4	5	6	7	8	9
0	1	2	4	8	16	32	64	45	7	14
1	28	56	29	58	33	66	49	15	30	60
2	37	74	65	47	11	22	44	5	10	20
3	40	80	77	71	59	35	70	57	31	62
4	41	82	81	79	75	67	51	19	38	76
5	69	55	27	54	25	50	17	34	68	53
6	23	46	9	18	36	72	61	39	78	73
7	63	43	3	6	12	24	48	13	26	52
8	21	42	1							

89

N	0	1	2	3	4	5	6	7	8	9
0		88	16	1	32	70	17	81	48	2
1	86	84	33	23	9	71	64	6	18	35
2	14	82	12	57	49	52	39	3	25	59
3	87	31	80	85	22	63	34	11	51	24
4	30	21	10	29	28	72	73	54	65	74
5	68	7	55	78	19	66	41	36	75	43
6	15	69	47	83	8	5	13	56	38	58
7	79	62	50	20	27	53	67	77	40	42
8	46	4	37	61	26	76	45	60	44	

I	0	1	2	3	4	5	6	7	8	9
0	1	3	9	27	81	65	17	51	64	14
1	42	37	22	66	20	60	2	6	18	54
2	73	41	34	13	39	28	84	74	44	43
3	40	31	4	12	36	19	57	82	68	26
4	78	56	79	59	88	86	80	62	8	24
5	72	38	25	75	47	52	67	23	69	29
6	87	83	71	35	16	48	55	76	50	61
7	5	15	45	46	49	58	85	77	53	70
8	32	7	21	63	11	33	10	30	1	

97

N	0	1	2	3	4	5	6	7	8	9
0		96	34	70	68	1	8	31	6	44
1	35	86	42	25	65	71	40	89	78	81
2	69	5	24	77	76	2	59	18	3	13
3	9	46	74	60	27	32	16	91	19	95
4	7	85	39	4	58	45	15	84	14	62
5	36	63	93	10	52	87	37	55	47	67
6	43	64	80	75	12	26	94	57	61	51
7	66	11	50	28	29	72	53	21	33	30
8	41	88	23	17	73	90	38	83	92	54
9	79	56	49	20	22	82	48			

I	0	1	2	3	4	5	6	7	8	9
0	1	5	25	28	43	21	8	40	6	30
1	53	71	64	29	48	46	36	83	27	38
2	93	77	94	82	22	13	65	34	73	74
3	79	7	35	78	2	10	50	56	86	42
4	16	80	12	60	9	45	31	58	96	92
5	72	69	54	76	89	57	91	67	44	26
6	33	68	49	51	61	14	70	59	4	20
7	3	15	75	84	32	63	24	23	18	90
8	62	19	95	87	47	41	11	55	81	17
9	85	37	88	52	66	39	1			

101

N	0	1	2	3	4	5	6	7	8	9
0		100	1	69	2	24	70	9	3	38
1	25	13	71	66	10	93	4	30	39	96
2	26	78	14	86	72	48	67	7	11	91
3	94	84	5	82	31	33	40	56	97	35
4	27	45	79	42	15	62	87	58	73	18
5	49	99	68	23	8	37	12	65	92	29
6	95	77	85	47	6	90	83	81	32	55
7	34	44	41	61	57	17	98	22	36	64
8	28	76	46	89	80	54	43	60	16	21
9	63	75	88	53	59	20	74	52	19	51
10	50									

I	0	1	2	3	4	5	6	7	8	9
0	1	2	4	8	16	32	64	27	54	7
1	14	28	56	11	22	44	88	75	49	98
2	95	89	77	53	5	10	20	40	80	59
3	17	34	68	35	70	39	78	55	9	18
4	36	72	43	86	71	41	82	63	25	50
5	100	99	97	93	85	69	37	74	47	94
6	87	73	45	90	79	57	13	26	52	3
7	6	12	24	48	96	91	81	61	21	42
8	84	67	33	66	31	62	23	46	92	83
9	65	29	58	15	30	60	19	38	76	51
10	1									

103

N	0	1	2	3	4	5	6	7	8	9
0		102	44	39	88	1	83	4	30	78
1	45	61	25	72	48	40	74	70	20	80
2	89	43	3	24	69	2	14	15	92	86
3	84	57	16	100	12	5	64	93	22	9
4	31	50	87	77	47	79	68	85	11	8
5	46	7	58	97	59	62	34	17	28	98
6	26	36	101	82	60	73	42	13	56	63
7	49	67	6	33	35	41	66	65	53	18
8	75	54	94	38	29	71	19	23	91	99
9	21	76	10	96	27	81	55	32	52	37
10	90	95	51							

I	0	1	2	3	4	5	6	7	8	9
0	1	5	25	22	7	35	72	51	49	39
1	92	48	34	67	26	27	32	57	79	86
2	18	90	38	87	23	12	60	94	58	84
3	8	40	97	73	56	74	61	99	83	3
4	15	75	66	21	2	10	50	44	14	70
5	41	102	98	78	81	96	68	31	52	54
6	64	11	55	69	36	77	76	71	46	24
7	17	85	13	65	16	80	91	43	9	45
8	19	95	63	6	30	47	29	42	4	20
9	100	88	28	37	82	101	93	53	59	89
10	33	62	1							

107

N	0	1	2	3	4	5	6	7	8	9
0		106	1	70	2	47	71	43	3	34
1	48	22	72	14	44	11	4	29	35	78
2	49	7	23	62	73	94	15	104	45	32
3	12	27	5	92	30	90	36	38	79	84
4	50	40	8	59	24	81	63	66	74	86
5	95	99	16	52	105	69	46	42	33	21
6	13	10	28	77	6	61	93	103	31	26
7	91	89	37	83	39	58	80	65	85	98
8	51	68	41	20	9	76	60	102	25	88
9	82	57	64	97	67	19	75	101	87	56
10	96	18	100	55	17	54	53			

I	0	1	2	3	4	5	6	7	8	9
0	1	2	4	8	16	32	64	21	42	84
1	61	15	30	60	13	26	52	104	101	95
2	83	59	11	22	44	88	69	31	62	17
3	34	68	29	58	9	18	36	72	37	74
4	41	82	57	7	14	28	56	5	10	20
5	40	80	53	106	105	103	99	91	75	43
6	86	65	23	46	92	77	47	94	81	55
7	3	6	12	24	48	96	85	63	19	38
8	76	45	90	73	39	78	49	98	89	71
9	35	70	33	66	25	50	100	93	79	51
10	102	97	87	67	27	54	1			

109

N	0	1	2	3	4	5	6	7	8	9
0		108	57	52	6	76	1	40	63	104
1	25	83	58	67	97	20	12	93	53	75
2	82	92	32	33	7	44	16	48	46	34
3	77	14	69	27	42	8	2	5	24	11
4	31	45	41	30	89	72	90	17	64	80
5	101	37	73	49	105	51	103	19	91	47
6	26	10	71	36	18	35	84	95	99	85
7	65	78	59	56	62	96	81	15	68	23
8	88	100	102	70	98	61	87	86	38	28
9	21	107	39	66	74	43	13	4	29	79
10	50	9	94	55	22	60	106	3	54	

I	0	1	2	3	4	5	6	7	8	9
0	1	6	36	107	97	37	4	24	35	101
1	61	39	16	96	31	77	26	47	64	57
2	15	90	104	79	38	10	60	33	89	98
3	43	40	22	23	29	65	63	51	88	92
4	7	42	34	95	25	41	28	59	27	53
5	100	55	3	18	108	103	73	2	12	72
6	105	85	74	8	48	70	93	13	78	32
7	83	62	45	52	94	19	5	30	71	99
8	49	76	20	11	66	69	87	86	80	44
9	46	58	21	17	102	67	75	14	84	68
10	81	50	82	56	9	54	106	91	1	

113

N	0	1	2	3	4	5	6	7	8	9
0		112	12	1	24	83	13	8	36	2
1	95	74	25	22	20	84	48	5	14	99
2	107	9	86	41	37	54	34	3	32	89
3	96	50	60	75	17	91	26	67	111	23
4	7	94	21	47	98	85	53	31	49	16
5	66	6	46	52	15	45	44	100	101	71
6	108	102	62	10	72	105	87	109	29	42
7	103	77	38	63	79	55	11	82	35	73
8	19	4	106	40	33	88	59	90	110	93
9	97	30	65	51	43	70	61	104	28	76
10	78	81	18	39	58	92	64	69	27	80
11	57	68	56							

I	0	1	2	3	4	5	6	7	8	9
0	1	3	9	27	81	17	51	40	7	21
1	63	76	2	6	18	54	49	34	102	80
2	14	42	13	39	4	12	36	108	98	68
3	91	47	28	84	26	78	8	24	72	103
4	83	23	69	94	56	55	52	43	16	48
5	31	93	53	46	25	75	112	110	104	86
6	32	96	62	73	106	92	50	37	111	107
7	95	59	64	79	11	33	99	71	100	74
8	109	101	77	5	15	45	22	66	85	29
9	87	35	105	89	41	10	30	90	44	19
10	57	58	61	70	97	65	82	20	60	67
11	88	38	1							

2.4 Prime Powers

2.17 **Table** Prime powers $q = p^\alpha$, $q \leq 10000$, $\alpha \geq 2$.

q	p	q	p	q	p	q	p	q	p	q	p	q	p	q	p	q	p
4	2	8	2	9	3	16	2	25	5	27	3	32	2	49	7	64	2
81	3	121	11	125	5	128	2	169	13	243	3	256	2	289	17	343	7
361	19	512	2	529	23	625	5	729	3	841	29	961	31	1024	2	1331	11
1369	37	1681	41	1849	43	2048	2	2187	3	2197	13	2209	47	2401	7	2809	53
3125	5	3481	59	3721	61	4096	2	4489	67	4913	17	5041	71	5329	73	6241	79
6561	3	6859	19	6889	83	7921	89	8192	2	9409	97						

2.5 Indices and Power Residues for Prime Power Fields

2.18 **Remark** In Tables 2.19, 2.20, 2.21, and 2.22, if g is a generator of $\mathbb{F}_{p^n}$ and if $g^u = v$, then u is in cell v of the array where the row and column indices are added to get the value of v.

2.19 **Table** Indices and power residues for $\mathbb{F}_{2^n}$, $n \leq 7$.

$\mathbb{F}_{2^2}$
$g = x$

	000	001	010	011	100	101	110	111
0		3	1	2				

$\mathbb{F}_{2^3}$
$g = x$

	000	001	010	011	100	101	110	111
0		7	1	3	2	6	4	5

$\mathbb{F}_{2^4}$
$g = x + 1$

	000	001	010	011	100	101	110	111
0		15	12	1	9	2	13	7
1000	6	8	14	11	10	5	4	3

$\mathbb{F}_{2^5}$
$g = x$

	000	001	010	011	100	101	110	111
0		31	1	14	2	28	15	22
1000	3	5	29	26	16	7	23	11
10000	4	25	6	10	30	13	27	21
11000	17	18	8	19	24	9	12	20

$\mathbb{F}_{2^6}$
$g = x + 1$

	000	001	010	011	100	101	110	111
0		63	56	1	49	2	57	20
1000	42	21	58	25	50	53	13	3
10000	35	4	14	16	51	40	18	54
11000	43	37	46	22	6	26	59	31
100000	28	32	60	10	7	8	9	27
101000	44	29	33	38	11	23	47	61
110000	36	45	30	5	39	17	15	34
111000	62	55	19	48	52	12	24	41

$\mathbb{F}_{2^7}$
$g = x$

	000	001	010	011	100	101	110	111
0		127	1	89	2	51	90	59
1000	3	21	52	113	91	85	60	13
10000	4	102	22	95	53	118	114	47
11000	92	107	86	110	61	75	14	66
100000	5	28	103	17	23	122	96	37
101000	54	32	119	69	115	72	48	82
110000	93	45	108	64	87	57	111	11
111000	62	9	76	78	15	35	67	80
1000000	6	42	29	25	104	99	18	124
1001000	24	41	123	98	97	40	38	39
1010000	55	43	33	7	120	26	70	30
1011000	116	100	73	105	49	125	83	19
1100000	94	101	46	117	109	106	65	74
1101000	88	126	58	50	112	20	12	84
1110000	63	44	10	56	77	8	79	34
1111000	16	27	36	121	68	31	81	71

2.20 Table Indices and power residues for $\mathbb{F}_{3^n}$, $n \leq 4$.

$\mathbb{F}_{3^2}$
$g = x$

	00	01	02	10	11	12	20	21	22
0		8	4	1	2	7	5	3	6

$\mathbb{F}_{3^3}$
$g = x$

	00	01	02	10	11	12	20	21	22
0		26	13	1	19	4	14	17	6
100	2	22	23	20	8	24	5	12	3
200	15	10	9	18	16	25	7	11	21

$\mathbb{F}_{3^4}$
$g = x$

	00	01	02	10	11	12	20	21	22
0		80	40	1	15	49	41	9	55
100	2	37	64	16	18	13	50	30	60
200	42	24	77	10	20	70	56	53	58
1000	3	45	67	38	62	75	65	34	26
1100	17	29	36	19	52	23	14	8	79
1200	51	7	28	31	32	6	61	33	44
2000	43	27	5	25	66	74	78	35	22
2100	11	68	47	21	4	73	71	46	72
2200	57	76	69	54	39	48	59	63	12

2.21 Table Indices and power residues for $\mathbb{F}_{5^2}$.

$\mathbb{F}_{5^2}$, $g = x$

	00	01	02	03	04	10	11	12	13	14
0		24	6	18	12	1	17	14	15	10
20	7	21	23	16	20	19	8	4	11	9
40	13	22	3	2	5					

2.22 Table Indices and power residues for $\mathbb{F}_{7^2}$.

$\mathbb{F}_{7^2}$, $g = x$

	0	1	2	3	4	5	6
0		48	16	8	32	40	24
10	1	31	11	26	12	14	5
20	17	28	47	30	27	21	42
30	9	22	34	39	13	20	19
40	33	43	44	37	15	10	46
50	41	18	45	3	6	23	4
60	25	29	38	36	2	35	7

2.6 Zech Logarithms for Prime Power Fields

Let α be a root of $p(x)$, a generator of the multiplicative group of elements of $\mathbb{F}_q$. The *Zech logarithm* is the function $Z : \{1, \ldots, p^n - 1\} \to \mathbb{Z}$ such that

$$\alpha^{Z(k)} = 1 + \alpha^k$$

$Z(k) = \emptyset$ denotes $1 + \alpha^k = 0$.

2.23 Remark The Zech logarithm provides a direct method for adding elements α^i and α^j $(i > j)$ in $\mathbb{F}_q$, since $\alpha^i + \alpha^j = \alpha^j(\alpha^{i-j} + 1) = \alpha^j \cdot \alpha^{Z(i-j)}$. The values of the Zech logarithm function depend on the polynomial $p(x)$ used to obtain the field.

2.24 Table Zech logarithms for $\mathbb{F}_{2^3}$, $\mathbb{F}_{2^4}$, $\mathbb{F}_{2^5}$, $\mathbb{F}_{3^2}$, $\mathbb{F}_{3^3}$, $\mathbb{F}_{5^2}$, and $\mathbb{F}_{7^2}$ (generator polynomials are the same as given in Tables 2.19-2.22).

$\mathbb{F}_{2^3}$

k	1	2	3	4	5	6	7
$Z(k)$	3	6	1	5	4	2	$\emptyset$

$\mathbb{F}_{2^4}$

k	1	2	3	4	5	6	7	8	9	10	11	12	13	14	15
$Z(k)$	12	9	4	3	10	8	13	6	2	5	14	1	7	11	$\emptyset$

$\mathbb{F}_{2^5}$

k	1	2	3	4	5	6	7	8	9	10	11	12	13	14	15	16	17	18	19	20
$Z(k)$	14	28	5	25	3	10	16	19	24	6	23	20	30	1	22	7	18	17	8	12
k	21	22	23	24	25	26	27	28	29	30	31									
$Z(k)$	27	15	11	9	4	29	21	2	26	13	$\emptyset$									

$\mathbb{F}_{3^2}$

k	1	2	3	4	5	6	7	8
$Z(k)$	2	7	6	$\emptyset$	3	5	1	4

$\mathbb{F}_{3^3}$

k	1	2	3	4	5	6	7	8	9	10	11	12	13	14	15	16	17	18	19	20
$Z(k)$	19	22	5	1	12	14	11	24	15	9	21	3	$\emptyset$	17	10	25	6	16	4	8
k	21	22	23	24	25	26														
$Z(k)$	7	23	2	20	18	13														

$\mathbb{F}_{5^2}$

k	1	2	3	4	5	6	7	8	9	10	11	12	13	14	15	16	17	18	19	20
$Z(k)$	17	5	2	11	13	18	21	4	19	1	9	$\emptyset$	22	15	10	20	14	12	8	7
k	21	22	23	24																
$Z(k)$	23	3	16	6																

$\mathbb{F}_{72}$

k	1	2	3	4	5	6	7	8	9	10	11	12	13	14	15	16	17	18	19	20
$Z(k)$	31	35	6	41	1	23	25	32	22	46	26	14	20	5	10	8	28	45	9	19
k	21	22	23	24	25	26	27	28	29	30	31	32	33	34	35	36	37	38	39	40
$Z(k)$	42	34	4	$\emptyset$	29	12	21	47	38	27	11	40	43	39	7	2	15	36	13	24
k	41	42	43	44	45	46	47	48												
$Z(k)$	18	17	44	37	3	33	30	16												

2.7 Prime Power Decompositions up to 2500

2.25 Table Integer factorizations of integers up to 2500 as products of prime powers. Prime numbers are shown in bold.

	0	1	2	3	4	5	6	7	8	9
0	0	1	**2**	**3**	2^2	**5**	2·3	**7**	2^3	3^2
1	2·5	**11**	$2^2\cdot 3$	**13**	2·7	3·5	2^4	**17**	$2\cdot 3^2$	**19**
2	$2^2\cdot 5$	3·7	2·11	**23**	$2^3\cdot 3$	5^2	2·13	3^3	$2^2\cdot 7$	**29**
3	2·3·5	**31**	2^5	3·11	2·17	5·7	$2^2\cdot 3^2$	**37**	2·19	3·13
4	$2^3\cdot 5$	**41**	2·3·7	**43**	$2^2\cdot 11$	$3^2\cdot 5$	2·23	**47**	$2^4\cdot 3$	7^2
5	$2\cdot 5^2$	3·17	$2^2\cdot 13$	**53**	$2\cdot 3^3$	5·11	$2^3\cdot 7$	3·19	2·29	**59**
6	$2^2\cdot 3\cdot 5$	**61**	2·31	$3^2\cdot 7$	2^6	5·13	2·3·11	**67**	$2^2\cdot 17$	3·23
7	2·5·7	**71**	$2^3\cdot 3^2$	**73**	2·37	$3\cdot 5^2$	$2^2\cdot 19$	7·11	2·3·13	**79**
8	$2^4\cdot 5$	3^4	2·41	**83**	$2^2\cdot 3\cdot 7$	5·17	2·43	3·29	$2^3\cdot 11$	**89**
9	$2\cdot 3^2\cdot 5$	7·13	$2^2\cdot 23$	3·31	2·47	5·19	$2^5\cdot 3$	**97**	$2\cdot 7^2$	$3^2\cdot 11$
10	$2^2\cdot 5^2$	**101**	2·3·17	**103**	$2^3\cdot 13$	3·5·7	2·53	**107**	$2^2\cdot 3^3$	**109**
11	2·5·11	3·37	$2^4\cdot 7$	**113**	2·3·19	5·23	$2^2\cdot 29$	$3^2\cdot 13$	2·59	7·17
12	$2^3\cdot 3\cdot 5$	11^2	2·61	3·41	$2^2\cdot 31$	5^3	$2\cdot 3^2\cdot 7$	**127**	2^7	3·43
13	2·5·13	**131**	$2^2\cdot 3\cdot 11$	7·19	2·67	$3^3\cdot 5$	$2^3\cdot 17$	**137**	2·3·23	**139**
14	$2^2\cdot 5\cdot 7$	3·47	2·71	11·13	$2^4\cdot 3^2$	5·29	2·73	$3\cdot 7^2$	$2^2\cdot 37$	**149**
15	$2\cdot 3\cdot 5^2$	**151**	$2^3\cdot 19$	$3^2\cdot 17$	2·7·11	5·31	$2^2\cdot 3\cdot 13$	**157**	2·79	3·53
16	$2^5\cdot 5$	7·23	$2\cdot 3^4$	**163**	$2^2\cdot 41$	3·5·11	2·83	**167**	$2^3\cdot 3\cdot 7$	13^2
17	2·5·17	$3^2\cdot 19$	$2^2\cdot 43$	**173**	2·3·29	$5^2\cdot 7$	$2^4\cdot 11$	3·59	2·89	**179**
18	$2^2\cdot 3^2\cdot 5$	**181**	2·7·13	3·61	$2^3\cdot 23$	5·37	2·3·31	11·17	$2^2\cdot 47$	$3^3\cdot 7$
19	2·5·19	**191**	$2^6\cdot 3$	**193**	2·97	3·5·13	$2^2\cdot 7^2$	**197**	$2\cdot 3^2\cdot 11$	**199**
20	$2^3\cdot 5^2$	3·67	2·101	7·29	$2^2\cdot 3\cdot 17$	5·41	2·103	$3^2\cdot 23$	$2^4\cdot 13$	11·19
21	2·3·5·7	**211**	$2^2\cdot 53$	3·71	2·107	5·43	$2^3\cdot 3^3$	7·31	2·109	3·73
22	$2^2\cdot 5\cdot 11$	13·17	2·3·37	**223**	$2^5\cdot 7$	$3^2\cdot 5^2$	2·113	**227**	$2^2\cdot 3\cdot 19$	**229**
23	2·5·23	3·7·11	$2^3\cdot 29$	**233**	$2\cdot 3^2\cdot 13$	5·47	$2^2\cdot 59$	3·79	2·7·17	**239**
24	$2^4\cdot 3\cdot 5$	**241**	$2\cdot 11^2$	3^5	$2^2\cdot 61$	$5\cdot 7^2$	2·3·41	13·19	$2^3\cdot 31$	3·83
25	$2\cdot 5^3$	**251**	$2^2\cdot 3^2\cdot 7$	11·23	2·127	3·5·17	2^8	**257**	2·3·43	7·37
26	$2^2\cdot 5\cdot 13$	$3^2\cdot 29$	2·131	**263**	$2^3\cdot 3\cdot 11$	5·53	2·7·19	3·89	$2^2\cdot 67$	**269**
27	$2\cdot 3^3\cdot 5$	**271**	$2^4\cdot 17$	3·7·13	2·137	$5^2\cdot 11$	$2^2\cdot 3\cdot 23$	**277**	2·139	$3^2\cdot 31$
28	$2^3\cdot 5\cdot 7$	**281**	2·3·47	**283**	$2^2\cdot 71$	3·5·19	2·11·13	7·41	$2^5\cdot 3^2$	17^2
29	2·5·29	3·97	$2^2\cdot 73$	**293**	$2\cdot 3\cdot 7^2$	5·59	$2^3\cdot 37$	$3^3\cdot 11$	2·149	13·23
30	$2^2\cdot 3\cdot 5^2$	7·43	2·151	3·101	$2^4\cdot 19$	5·61	$2\cdot 3^2\cdot 17$	**307**	$2^2\cdot 7\cdot 11$	3·103
31	2·5·31	**311**	$2^3\cdot 3\cdot 13$	**313**	2·157	$3^2\cdot 5\cdot 7$	$2^2\cdot 79$	**317**	2·3·53	11·29
32	$2^6\cdot 5$	3·107	2·7·23	17·19	$2^2\cdot 3^4$	$5^2\cdot 13$	2·163	3·109	$2^3\cdot 41$	7·47
33	2·3·5·11	**331**	$2^2\cdot 83$	$3^2\cdot 37$	2·167	5·67	$2^4\cdot 3\cdot 7$	**337**	$2\cdot 13^2$	3·113
34	$2^2\cdot 5\cdot 17$	11·31	$2\cdot 3^2\cdot 19$	7^3	$2^3\cdot 43$	3·5·23	2·173	**347**	$2^2\cdot 3\cdot 29$	**349**
35	$2\cdot 5^2\cdot 7$	$3^3\cdot 13$	$2^5\cdot 11$	**353**	2·3·59	5·71	$2^2\cdot 89$	3·7·17	2·179	**359**
36	$2^3\cdot 3^2\cdot 5$	19^2	2·181	$3\cdot 11^2$	$2^2\cdot 7\cdot 13$	5·73	2·3·61	**367**	$2^4\cdot 23$	$3^2\cdot 41$
37	2·5·37	7·53	$2^2\cdot 3\cdot 31$	**373**	2·11·17	$3\cdot 5^3$	$2^3\cdot 47$	13·29	$2\cdot 3^3\cdot 7$	**379**
38	$2^2\cdot 5\cdot 19$	3·127	2·191	**383**	$2^7\cdot 3$	5·7·11	2·193	$3^2\cdot 43$	$2^2\cdot 97$	**389**
39	2·3·5·13	17·23	$2^3\cdot 7^2$	3·131	2·197	5·79	$2^2\cdot 3^2\cdot 11$	**397**	2·199	3·7·19
40	$2^4\cdot 5^2$	**401**	2·3·67	13·31	$2^2\cdot 101$	$3^4\cdot 5$	2·7·29	11·37	$2^3\cdot 3\cdot 17$	**409**
41	2·5·41	3·137	$2^2\cdot 103$	7·59	$2\cdot 3^2\cdot 23$	5·83	$2^5\cdot 13$	3·139	2·11·19	**419**
42	$2^2\cdot 3\cdot 5\cdot 7$	**421**	2·211	$3^2\cdot 47$	$2^3\cdot 53$	$5^2\cdot 17$	2·3·71	7·61	$2^2\cdot 107$	3·11·13
43	2·5·43	**431**	$2^4\cdot 3^3$	**433**	2·7·31	3·5·29	$2^2\cdot 109$	19·23	2·3·73	**439**
44	$2^3\cdot 5\cdot 11$	$3^2\cdot 7^2$	2·13·17	**443**	$2^2\cdot 3\cdot 37$	5·89	2·223	3·149	$2^6\cdot 7$	**449**
45	$2\cdot 3^2\cdot 5^2$	11·41	$2^2\cdot 113$	3·151	2·227	5·7·13	$2^3\cdot 3\cdot 19$	**457**	2·229	$3^3\cdot 17$
46	$2^2\cdot 5\cdot 23$	**461**	2·3·7·11	**463**	$2^4\cdot 29$	3·5·31	2·233	**467**	$2^2\cdot 3^2\cdot 13$	7·67
47	2·5·47	3·157	$2^3\cdot 59$	11·43	2·3·79	$5^2\cdot 19$	$2^2\cdot 7\cdot 17$	$3^2\cdot 53$	2·239	**479**

	0	1	2	3	4	5	6	7	8	9
48	$2^5\cdot3\cdot5$	$13\cdot37$	$2\cdot241$	$3\cdot7\cdot23$	$2^2\cdot11^2$	$5\cdot97$	$2\cdot3^5$	**487**	$2^3\cdot61$	$3\cdot163$
49	$2\cdot5\cdot7^2$	**491**	$2^2\cdot3\cdot41$	$17\cdot29$	$2\cdot13\cdot19$	$3^2\cdot5\cdot11$	$2^4\cdot31$	$7\cdot71$	$2\cdot3\cdot83$	**499**
50	$2^2\cdot5^3$	$3\cdot167$	$2\cdot251$	**503**	$2^3\cdot3^2\cdot7$	$5\cdot101$	$2\cdot11\cdot23$	$3\cdot13^2$	$2^2\cdot127$	**509**
51	$2\cdot3\cdot5\cdot17$	$7\cdot73$	2^9	$3^3\cdot19$	$2\cdot257$	$5\cdot103$	$2^2\cdot3\cdot43$	$11\cdot47$	$2\cdot7\cdot37$	$3\cdot173$
52	$2^3\cdot5\cdot13$	**521**	$2\cdot3^2\cdot29$	**523**	$2^2\cdot131$	$3\cdot5^2\cdot7$	$2\cdot263$	$17\cdot31$	$2^4\cdot3\cdot11$	23^2
53	$2\cdot5\cdot53$	$3^2\cdot59$	$2^2\cdot7\cdot19$	$13\cdot41$	$2\cdot3\cdot89$	$5\cdot107$	$2^3\cdot67$	$3\cdot179$	$2\cdot269$	$7^2\cdot11$
54	$2^2\cdot3^3\cdot5$	**541**	$2\cdot271$	$3\cdot181$	$2^5\cdot17$	$5\cdot109$	$2\cdot3\cdot7\cdot13$	**547**	$2^2\cdot137$	$3^2\cdot61$
55	$2\cdot5^2\cdot11$	$19\cdot29$	$2^3\cdot3\cdot23$	$7\cdot79$	$2\cdot277$	$3\cdot5\cdot37$	$2^2\cdot139$	**557**	$2\cdot3^2\cdot31$	$13\cdot43$
56	$2^4\cdot5\cdot7$	$3\cdot11\cdot17$	$2\cdot281$	**563**	$2^2\cdot3\cdot47$	$5\cdot113$	$2\cdot283$	$3^4\cdot7$	$2^3\cdot71$	**569**
57	$2\cdot3\cdot5\cdot19$	**571**	$2^2\cdot11\cdot13$	$3\cdot191$	$2\cdot7\cdot41$	$5^2\cdot23$	$2^6\cdot3^2$	**577**	$2\cdot17^2$	$3\cdot193$
58	$2^2\cdot5\cdot29$	$7\cdot83$	$2\cdot3\cdot97$	$11\cdot53$	$2^3\cdot73$	$3^2\cdot5\cdot13$	$2\cdot293$	**587**	$2^2\cdot3\cdot7^2$	$19\cdot31$
59	$2\cdot5\cdot59$	$3\cdot197$	$2^4\cdot37$	**593**	$2\cdot3^3\cdot11$	$5\cdot7\cdot17$	$2^2\cdot149$	$3\cdot199$	$2\cdot13\cdot23$	**599**
60	$2^3\cdot3\cdot5^2$	**601**	$2\cdot7\cdot43$	$3^2\cdot67$	$2^2\cdot151$	$5\cdot11^2$	$2\cdot3\cdot101$	**607**	$2^5\cdot19$	$3\cdot7\cdot29$
61	$2\cdot5\cdot61$	$13\cdot47$	$2^2\cdot3^2\cdot17$	**613**	$2\cdot307$	$3\cdot5\cdot41$	$2^3\cdot7\cdot11$	**617**	$2\cdot3\cdot103$	**619**
62	$2^2\cdot5\cdot31$	$3^3\cdot23$	$2\cdot311$	$7\cdot89$	$2^4\cdot3\cdot13$	5^4	$2\cdot313$	$3\cdot11\cdot19$	$2^2\cdot157$	$17\cdot37$
63	$2\cdot3^2\cdot5\cdot7$	**631**	$2^3\cdot79$	$3\cdot211$	$2\cdot317$	$5\cdot127$	$2^2\cdot3\cdot53$	$7^2\cdot13$	$2\cdot11\cdot29$	$3^2\cdot71$
64	$2^7\cdot5$	**641**	$2\cdot3\cdot107$	**643**	$2^2\cdot7\cdot23$	$3\cdot5\cdot43$	$2\cdot17\cdot19$	**647**	$2^3\cdot3^4$	$11\cdot59$
65	$2\cdot5^2\cdot13$	$3\cdot7\cdot31$	$2^2\cdot163$	**653**	$2\cdot3\cdot109$	$5\cdot131$	$2^4\cdot41$	$3^2\cdot73$	$2\cdot7\cdot47$	**659**
66	$2^2\cdot3\cdot5\cdot11$	**661**	$2\cdot331$	$3\cdot13\cdot17$	$2^3\cdot83$	$5\cdot7\cdot19$	$2\cdot3^2\cdot37$	$23\cdot29$	$2^2\cdot167$	$3\cdot223$
67	$2\cdot5\cdot67$	$11\cdot61$	$2^5\cdot3\cdot7$	**673**	$2\cdot337$	$3^3\cdot5^2$	$2^2\cdot13^2$	**677**	$2\cdot3\cdot113$	$7\cdot97$
68	$2^3\cdot5\cdot17$	$3\cdot227$	$2\cdot11\cdot31$	**683**	$2^2\cdot3^2\cdot19$	$5\cdot137$	$2\cdot7^3$	$3\cdot229$	$2^4\cdot43$	$13\cdot53$
69	$2\cdot3\cdot5\cdot23$	**691**	$2^2\cdot173$	$3^2\cdot7\cdot11$	$2\cdot347$	$5\cdot139$	$2^3\cdot3\cdot29$	$17\cdot41$	$2\cdot349$	$3\cdot233$
70	$2^2\cdot5^2\cdot7$	**701**	$2\cdot3^3\cdot13$	$19\cdot37$	$2^6\cdot11$	$3\cdot5\cdot47$	$2\cdot353$	$7\cdot101$	$2^2\cdot3\cdot59$	**709**
71	$2\cdot5\cdot71$	$3^2\cdot79$	$2^3\cdot89$	$23\cdot31$	$2\cdot3\cdot7\cdot17$	$5\cdot11\cdot13$	$2^2\cdot179$	$3\cdot239$	$2\cdot359$	**719**
72	$2^4\cdot3^2\cdot5$	$7\cdot103$	$2\cdot19^2$	$3\cdot241$	$2^2\cdot181$	$5^2\cdot29$	$2\cdot3\cdot11^2$	**727**	$2^3\cdot7\cdot13$	3^6
73	$2\cdot5\cdot73$	$17\cdot43$	$2^2\cdot3\cdot61$	**733**	$2\cdot367$	$3\cdot5\cdot7^2$	$2^5\cdot23$	$11\cdot67$	$2\cdot3^2\cdot41$	**739**
74	$2^2\cdot5\cdot37$	$3\cdot13\cdot19$	$2\cdot7\cdot53$	**743**	$2^3\cdot3\cdot31$	$5\cdot149$	$2\cdot373$	$3^2\cdot83$	$2^2\cdot11\cdot17$	$7\cdot107$
75	$2\cdot3\cdot5^3$	**751**	$2^4\cdot47$	$3\cdot251$	$2\cdot13\cdot29$	$5\cdot151$	$2^2\cdot3^3\cdot7$	**757**	$2\cdot379$	$3\cdot11\cdot23$
76	$2^3\cdot5\cdot19$	**761**	$2\cdot3\cdot127$	$7\cdot109$	$2^2\cdot191$	$3^2\cdot5\cdot17$	$2\cdot383$	$13\cdot59$	$2^8\cdot3$	**769**
77	$2\cdot5\cdot7\cdot11$	$3\cdot257$	$2^2\cdot193$	**773**	$2\cdot3^2\cdot43$	$5^2\cdot31$	$2^3\cdot97$	$3\cdot7\cdot37$	$2\cdot389$	$19\cdot41$
78	$2^2\cdot3\cdot5\cdot13$	$11\cdot71$	$2\cdot17\cdot23$	$3^3\cdot29$	$2^4\cdot7^2$	$5\cdot157$	$2\cdot3\cdot131$	**787**	$2^2\cdot197$	$3\cdot263$
79	$2\cdot5\cdot79$	$7\cdot113$	$2^3\cdot3^2\cdot11$	$13\cdot61$	$2\cdot397$	$3\cdot5\cdot53$	$2^2\cdot199$	**797**	$2\cdot3\cdot7\cdot19$	$17\cdot47$
80	$2^5\cdot5^2$	$3^2\cdot89$	$2\cdot401$	$11\cdot73$	$2^2\cdot3\cdot67$	$5\cdot7\cdot23$	$2\cdot13\cdot31$	$3\cdot269$	$2^3\cdot101$	**809**
81	$2\cdot3^4\cdot5$	**811**	$2^2\cdot7\cdot29$	$3\cdot271$	$2\cdot11\cdot37$	$5\cdot163$	$2^4\cdot3\cdot17$	$19\cdot43$	$2\cdot409$	$3^2\cdot7\cdot13$
82	$2^2\cdot5\cdot41$	**821**	$2\cdot3\cdot137$	**823**	$2^3\cdot103$	$3\cdot5^2\cdot11$	$2\cdot7\cdot59$	**827**	$2^2\cdot3^2\cdot23$	**829**
83	$2\cdot5\cdot83$	$3\cdot277$	$2^6\cdot13$	$7^2\cdot17$	$2\cdot3\cdot139$	$5\cdot167$	$2^2\cdot11\cdot19$	$3^3\cdot31$	$2\cdot419$	**839**
84	$2^3\cdot3\cdot5\cdot7$	29^2	$2\cdot421$	$3\cdot281$	$2^2\cdot211$	$5\cdot13^2$	$2\cdot3^2\cdot47$	$7\cdot11^2$	$2^4\cdot53$	$3\cdot283$
85	$2\cdot5^2\cdot17$	$23\cdot37$	$2^2\cdot3\cdot71$	**853**	$2\cdot7\cdot61$	$3^2\cdot5\cdot19$	$2^3\cdot107$	**857**	$2\cdot3\cdot11\cdot13$	**859**
86	$2^2\cdot5\cdot43$	$3\cdot7\cdot41$	$2\cdot431$	**863**	$2^5\cdot3^3$	$5\cdot173$	$2\cdot433$	$3\cdot17^2$	$2^2\cdot7\cdot31$	$11\cdot79$
87	$2\cdot3\cdot5\cdot29$	$13\cdot67$	$2^3\cdot109$	$3^2\cdot97$	$2\cdot19\cdot23$	$5^3\cdot7$	$2^2\cdot3\cdot73$	**877**	$2\cdot439$	$3\cdot293$
88	$2^4\cdot5\cdot11$	**881**	$2\cdot3^2\cdot7^2$	**883**	$2^2\cdot13\cdot17$	$3\cdot5\cdot59$	$2\cdot443$	**887**	$2^3\cdot3\cdot37$	$7\cdot127$
89	$2\cdot5\cdot89$	$3^4\cdot11$	$2^2\cdot223$	$19\cdot47$	$2\cdot3\cdot149$	$5\cdot179$	$2^7\cdot7$	$3\cdot13\cdot23$	$2\cdot449$	$29\cdot31$
90	$2^2\cdot3^2\cdot5^2$	$17\cdot53$	$2\cdot11\cdot41$	$3\cdot7\cdot43$	$2^3\cdot113$	$5\cdot181$	$2\cdot3\cdot151$	**907**	$2^2\cdot227$	$3^2\cdot101$
91	$2\cdot5\cdot7\cdot13$	**911**	$2^4\cdot3\cdot19$	$11\cdot83$	$2\cdot457$	$3\cdot5\cdot61$	$2^2\cdot229$	$7\cdot131$	$2\cdot3^3\cdot17$	**919**
92	$2^3\cdot5\cdot23$	$3\cdot307$	$2\cdot461$	$13\cdot71$	$2^2\cdot3\cdot7\cdot11$	$5^2\cdot37$	$2\cdot463$	$3^2\cdot103$	$2^5\cdot29$	**929**
93	$2\cdot3\cdot5\cdot31$	$7^2\cdot19$	$2^2\cdot233$	$3\cdot311$	$2\cdot467$	$5\cdot11\cdot17$	$2^3\cdot3^2\cdot13$	**937**	$2\cdot7\cdot67$	$3\cdot313$
94	$2^2\cdot5\cdot47$	**941**	$2\cdot3\cdot157$	$23\cdot41$	$2^4\cdot59$	$3^3\cdot5\cdot7$	$2\cdot11\cdot43$	**947**	$2^2\cdot3\cdot79$	$13\cdot73$
95	$2\cdot5^2\cdot19$	$3\cdot317$	$2^3\cdot7\cdot17$	**953**	$2\cdot3^2\cdot53$	$5\cdot191$	$2^2\cdot239$	$3\cdot11\cdot29$	$2\cdot479$	$7\cdot137$
96	$2^6\cdot3\cdot5$	31^2	$2\cdot13\cdot37$	$3^2\cdot107$	$2^2\cdot241$	$5\cdot193$	$2\cdot3\cdot7\cdot23$	**967**	$2^3\cdot11^2$	$3\cdot17\cdot19$
97	$2\cdot5\cdot97$	**971**	$2^2\cdot3^5$	$7\cdot139$	$2\cdot487$	$3\cdot5^2\cdot13$	$2^4\cdot61$	**977**	$2\cdot3\cdot163$	$11\cdot89$
98	$2^2\cdot5\cdot7^2$	$3^2\cdot109$	$2\cdot491$	**983**	$2^3\cdot3\cdot41$	$5\cdot197$	$2\cdot17\cdot29$	$3\cdot7\cdot47$	$2^2\cdot13\cdot19$	$23\cdot43$
99	$2\cdot3^2\cdot5\cdot11$	**991**	$2^5\cdot31$	$3\cdot331$	$2\cdot7\cdot71$	$5\cdot199$	$2^2\cdot3\cdot83$	**997**	$2\cdot499$	$3^3\cdot37$
100	$2^3\cdot5^3$	$7\cdot11\cdot13$	$2\cdot3\cdot167$	$17\cdot59$	$2^2\cdot251$	$3\cdot5\cdot67$	$2\cdot503$	$19\cdot53$	$2^4\cdot3^2\cdot7$	**1009**
101	$2\cdot5\cdot101$	$3\cdot337$	$2^2\cdot11\cdot23$	**1013**	$2\cdot3\cdot13^2$	$5\cdot7\cdot29$	$2^3\cdot127$	$3^2\cdot113$	$2\cdot509$	**1019**
102	$2^2\cdot3\cdot5\cdot17$	**1021**	$2\cdot7\cdot73$	$3\cdot11\cdot31$	2^{10}	$5^2\cdot41$	$2\cdot3^3\cdot19$	$13\cdot79$	$2^2\cdot257$	$3\cdot7^3$
103	$2\cdot5\cdot103$	**1031**	$2^3\cdot3\cdot43$	**1033**	$2\cdot11\cdot47$	$3^2\cdot5\cdot23$	$2^2\cdot7\cdot37$	$17\cdot61$	$2\cdot3\cdot173$	**1039**
104	$2^4\cdot5\cdot13$	$3\cdot347$	$2\cdot521$	$7\cdot149$	$2^2\cdot3^2\cdot29$	$5\cdot11\cdot19$	$2\cdot523$	$3\cdot349$	$2^3\cdot131$	**1049**
105	$2\cdot3\cdot5^2\cdot7$	**1051**	$2^2\cdot263$	$3^4\cdot13$	$2\cdot17\cdot31$	$5\cdot211$	$2^5\cdot3\cdot11$	$7\cdot151$	$2\cdot23^2$	$3\cdot353$
106	$2^2\cdot5\cdot53$	**1061**	$2\cdot3^2\cdot59$	**1063**	$2^3\cdot7\cdot19$	$3\cdot5\cdot71$	$2\cdot13\cdot41$	$11\cdot97$	$2^2\cdot3\cdot89$	**1069**
107	$2\cdot5\cdot107$	$3^2\cdot7\cdot17$	$2^4\cdot67$	$29\cdot37$	$2\cdot3\cdot179$	$5^2\cdot43$	$2^2\cdot269$	$3\cdot359$	$2\cdot7^2\cdot11$	$13\cdot83$
108	$2^3\cdot3^3\cdot5$	$23\cdot47$	$2\cdot541$	$3\cdot19^2$	$2^2\cdot271$	$5\cdot7\cdot31$	$2\cdot3\cdot181$	**1087**	$2^6\cdot17$	$3^2\cdot11^2$
109	$2\cdot5\cdot109$	**1091**	$2^2\cdot3\cdot7\cdot13$	**1093**	$2\cdot547$	$3\cdot5\cdot73$	$2^3\cdot137$	**1097**	$2\cdot3^2\cdot61$	$7\cdot157$
110	$2^2\cdot5^2\cdot11$	$3\cdot367$	$2\cdot19\cdot29$	**1103**	$2^4\cdot3\cdot23$	$5\cdot13\cdot17$	$2\cdot7\cdot79$	$3^3\cdot41$	$2^2\cdot277$	**1109**

	0	1	2	3	4	5	6	7	8	9
111	2·3·5·37	11·101	$2^3\cdot 139$	3·7·53	2·557	5·223	$2^2\cdot 3^2\cdot 31$	**1117**	2·13·43	3·373
112	$2^5\cdot 5\cdot 7$	19·59	2·3·11·17	**1123**	$2^2\cdot 281$	$3^2\cdot 5^3$	2·563	$7^2\cdot 23$	$2^3\cdot 3\cdot 47$	**1129**
113	2·5·113	3·13·29	$2^2\cdot 283$	11·103	$2\cdot 3^4\cdot 7$	5·227	$2^4\cdot 71$	3·379	2·569	17·67
114	$2^2\cdot 3\cdot 5\cdot 19$	7·163	2·571	$3^2\cdot 127$	$2^3\cdot 11\cdot 13$	5·229	2·3·191	31·37	$2^2\cdot 7\cdot 41$	3·383
115	$2\cdot 5^2\cdot 23$	**1151**	$2^7\cdot 3^2$	**1153**	2·577	3·5·7·11	$2^2\cdot 17^2$	13·89	2·3·193	19·61
116	$2^3\cdot 5\cdot 29$	$3^3\cdot 43$	2·7·83	**1163**	$2^2\cdot 3\cdot 97$	5·233	2·11·53	3·389	$2^4\cdot 73$	7·167
117	$2\cdot 3^2\cdot 5\cdot 13$	**1171**	$2^2\cdot 293$	3·17·23	2·587	$5^2\cdot 47$	$2^3\cdot 3\cdot 7^2$	11·107	2·19·31	$3^2\cdot 131$
118	$2^2\cdot 5\cdot 59$	**1181**	2·3·197	$7\cdot 13^2$	$2^5\cdot 37$	3·5·79	2·593	**1187**	$2^2\cdot 3^3\cdot 11$	29·41
119	2·5·7·17	3·397	$2^3\cdot 149$	**1193**	2·3·199	5·239	$2^2\cdot 13\cdot 23$	$3^2\cdot 7\cdot 19$	2·599	11·109
120	$2^4\cdot 3\cdot 5^2$	**1201**	2·601	3·401	$2^2\cdot 7\cdot 43$	5·241	$2\cdot 3^2\cdot 67$	17·71	$2^3\cdot 151$	3·13·31
121	$2\cdot 5\cdot 11^2$	7·173	$2^2\cdot 3\cdot 101$	**1213**	2·607	$3^5\cdot 5$	$2^6\cdot 19$	**1217**	2·3·7·29	23·53
122	$2^2\cdot 5\cdot 61$	3·11·37	2·13·47	**1223**	$2^3\cdot 3^2\cdot 17$	$5^2\cdot 7^2$	2·613	3·409	$2^2\cdot 307$	**1229**
123	2·3·5·41	**1231**	$2^4\cdot 7\cdot 11$	$3^2\cdot 137$	2·617	5·13·19	$2^2\cdot 3\cdot 103$	**1237**	2·619	3·7·59
124	$2^3\cdot 5\cdot 31$	17·73	$2\cdot 3^3\cdot 23$	11·113	$2^2\cdot 311$	3·5·83	2·7·89	29·43	$2^5\cdot 3\cdot 13$	**1249**
125	$2\cdot 5^4$	$3^2\cdot 139$	$2^2\cdot 313$	7·179	2·3·11·19	5·251	$2^3\cdot 157$	3·419	2·17·37	**1259**
126	$2^2\cdot 3^2\cdot 5\cdot 7$	13·97	2·631	3·421	$2^4\cdot 79$	5·11·23	2·3·211	7·181	$2^2\cdot 317$	$3^3\cdot 47$
127	2·5·127	31·41	$2^3\cdot 3\cdot 53$	19·67	$2\cdot 7^2\cdot 13$	$3\cdot 5^2\cdot 17$	$2^2\cdot 11\cdot 29$	**1277**	$2\cdot 3^2\cdot 71$	**1279**
128	$2^8\cdot 5$	3·7·61	2·641	**1283**	$2^2\cdot 3\cdot 107$	5·257	2·643	$3^2\cdot 11\cdot 13$	$2^3\cdot 7\cdot 23$	**1289**
129	2·3·5·43	**1291**	$2^2\cdot 17\cdot 19$	3·431	2·647	5·7·37	$2^4\cdot 3^4$	**1297**	2·11·59	3·433
130	$2^2\cdot 5^2\cdot 13$	**1301**	2·3·7·31	**1303**	$2^3\cdot 163$	$3^2\cdot 5\cdot 29$	2·653	**1307**	$2^2\cdot 3\cdot 109$	7·11·17
131	2·5·131	3·19·23	$2^5\cdot 41$	13·101	$2\cdot 3^2\cdot 73$	5·263	$2^2\cdot 7\cdot 47$	3·439	2·659	**1319**
132	$2^3\cdot 3\cdot 5\cdot 11$	**1321**	2·661	$3^3\cdot 7^2$	$2^2\cdot 331$	$5^2\cdot 53$	2·3·13·17	**1327**	$2^4\cdot 83$	3·443
133	2·5·7·19	11^3	$2^2\cdot 3^2\cdot 37$	31·43	2·23·29	3·5·89	$2^3\cdot 167$	7·191	2·3·223	13·103
134	$2^2\cdot 5\cdot 67$	$3^2\cdot 149$	2·11·61	17·79	$2^6\cdot 3\cdot 7$	5·269	2·673	3·449	$2^2\cdot 337$	19·71
135	$2\cdot 3^3\cdot 5^2$	7·193	$2^3\cdot 13^2$	3·11·41	2·677	5·271	$2^2\cdot 3\cdot 113$	23·59	2·7·97	$3^2\cdot 151$
136	$2^4\cdot 5\cdot 17$	**1361**	2·3·227	29·47	$2^2\cdot 11\cdot 31$	3·5·7·13	2·683	**1367**	$2^3\cdot 3^2\cdot 19$	37^2
137	2·5·137	3·457	$2^2\cdot 7^3$	**1373**	2·3·229	$5^3\cdot 11$	$2^5\cdot 43$	$3^4\cdot 17$	2·13·53	7·197
138	$2^2\cdot 3\cdot 5\cdot 23$	**1381**	2·691	3·461	$2^3\cdot 173$	5·277	$2\cdot 3^2\cdot 7\cdot 11$	19·73	$2^2\cdot 347$	3·463
139	2·5·139	13·107	$2^4\cdot 3\cdot 29$	7·199	2·17·41	$3^2\cdot 5\cdot 31$	$2^2\cdot 349$	11·127	2·3·233	**1399**
140	$2^3\cdot 5^2\cdot 7$	3·467	2·701	23·61	$2^2\cdot 3^3\cdot 13$	5·281	2·19·37	3·7·67	$2^7\cdot 11$	**1409**
141	2·3·5·47	17·83	$2^2\cdot 353$	$3^2\cdot 157$	2·7·101	5·283	$2^3\cdot 3\cdot 59$	13·109	2·709	3·11·43
142	$2^2\cdot 5\cdot 71$	$7^2\cdot 29$	$2\cdot 3^2\cdot 79$	**1423**	$2^4\cdot 89$	$3\cdot 5^2\cdot 19$	2·23·31	**1427**	$2^2\cdot 3\cdot 7\cdot 17$	**1429**
143	2·5·11·13	$3^3\cdot 53$	$2^3\cdot 179$	**1433**	2·3·239	5·7·41	$2^2\cdot 359$	3·479	2·719	**1439**
144	$2^5\cdot 3^2\cdot 5$	11·131	2·7·103	3·13·37	$2^2\cdot 19^2$	$5\cdot 17^2$	2·3·241	**1447**	$2^3\cdot 181$	$3^2\cdot 7\cdot 23$
145	$2\cdot 5^2\cdot 29$	**1451**	$2^2\cdot 3\cdot 11^2$	**1453**	2·727	3·5·97	$2^4\cdot 7\cdot 13$	31·47	$2\cdot 3^6$	**1459**
146	$2^2\cdot 5\cdot 73$	3·487	2·17·43	7·11·19	$2^3\cdot 3\cdot 61$	5·293	2·733	$3^2\cdot 163$	$2^2\cdot 367$	13·113
147	$2\cdot 3\cdot 5\cdot 7^2$	**1471**	$2^6\cdot 23$	3·491	2·11·67	$5^2\cdot 59$	$2^2\cdot 3^2\cdot 41$	7·211	2·739	3·17·29
148	$2^3\cdot 5\cdot 37$	**1481**	2·3·13·19	**1483**	$2^2\cdot 7\cdot 53$	$3^3\cdot 5\cdot 11$	2·743	**1487**	$2^4\cdot 3\cdot 31$	**1489**
149	2·5·149	3·7·71	$2^2\cdot 373$	**1493**	$2\cdot 3^2\cdot 83$	5·13·23	$2^3\cdot 11\cdot 17$	3·499	2·7·107	**1499**
150	$2^2\cdot 3\cdot 5^3$	19·79	2·751	$3^2\cdot 167$	$2^5\cdot 47$	5·7·43	2·3·251	11·137	$2^2\cdot 13\cdot 29$	3·503
151	2·5·151	**1511**	$2^3\cdot 3^3\cdot 7$	17·89	2·757	3·5·101	$2^2\cdot 379$	37·41	2·3·11·23	$7^2\cdot 31$
152	$2^4\cdot 5\cdot 19$	$3^2\cdot 13^2$	2·761	**1523**	$2^2\cdot 3\cdot 127$	$5^2\cdot 61$	2·7·109	3·509	$2^3\cdot 191$	11·139
153	$2\cdot 3^2\cdot 5\cdot 17$	**1531**	$2^2\cdot 383$	3·7·73	2·13·59	5·307	$2^9\cdot 3$	29·53	2·769	$3^4\cdot 19$
154	$2^2\cdot 5\cdot 7\cdot 11$	23·67	2·3·257	**1543**	$2^3\cdot 193$	3·5·103	2·773	7·13·17	$2^2\cdot 3^2\cdot 43$	**1549**
155	$2\cdot 5^2\cdot 31$	3·11·47	$2^4\cdot 97$	**1553**	2·3·7·37	5·311	$2^2\cdot 389$	$3^2\cdot 173$	2·19·41	**1559**
156	$2^3\cdot 3\cdot 5\cdot 13$	7·223	2·11·71	3·521	$2^2\cdot 17\cdot 23$	5·313	$2\cdot 3^3\cdot 29$	**1567**	$2^5\cdot 7^2$	3·523
157	2·5·157	**1571**	$2^2\cdot 3\cdot 131$	$11^2\cdot 13$	2·787	$3^2\cdot 5^2\cdot 7$	$2^3\cdot 197$	19·83	2·3·263	**1579**
158	$2^2\cdot 5\cdot 79$	3·17·31	2·7·113	**1583**	$2^4\cdot 3^2\cdot 11$	5·317	2·13·61	$3\cdot 23^2$	$2^2\cdot 397$	7·227
159	2·3·5·53	37·43	$2^3\cdot 199$	$3^3\cdot 59$	2·797	5·11·29	$2^2\cdot 3\cdot 7\cdot 19$	**1597**	2·17·47	3·13·41
160	$2^6\cdot 5^2$	**1601**	$2\cdot 3^2\cdot 89$	7·229	$2^2\cdot 401$	3·5·107	2·11·73	**1607**	$2^3\cdot 3\cdot 67$	**1609**
161	2·5·7·23	$3^2\cdot 179$	$2^2\cdot 13\cdot 31$	**1613**	2·3·269	5·17·19	$2^4\cdot 101$	$3\cdot 7^2\cdot 11$	2·809	**1619**
162	$2^2\cdot 3^4\cdot 5$	**1621**	2·811	3·541	$2^3\cdot 7\cdot 29$	$5^3\cdot 13$	2·3·271	**1627**	$2^2\cdot 11\cdot 37$	$3^2\cdot 181$
163	2·5·163	7·233	$2^5\cdot 3\cdot 17$	23·71	2·19·43	3·5·109	$2^2\cdot 409$	**1637**	$2\cdot 3^2\cdot 7\cdot 13$	11·149
164	$2^3\cdot 5\cdot 41$	3·547	2·821	31·53	$2^2\cdot 3\cdot 137$	5·7·47	2·823	$3^3\cdot 61$	$2^4\cdot 103$	17·97
165	$2\cdot 3\cdot 5^2\cdot 11$	13·127	$2^2\cdot 7\cdot 59$	3·19·29	2·827	5·331	$2^3\cdot 3^2\cdot 23$	**1657**	2·829	3·7·79
166	$2^2\cdot 5\cdot 83$	11·151	2·3·277	**1663**	$2^7\cdot 13$	$3^2\cdot 5\cdot 37$	$2\cdot 7^2\cdot 17$	**1667**	$2^2\cdot 3\cdot 139$	**1669**
167	2·5·167	3·557	$2^3\cdot 11\cdot 19$	7·239	$2\cdot 3^3\cdot 31$	$5^2\cdot 67$	$2^2\cdot 419$	3·13·43	2·839	23·73
168	$2^4\cdot 3\cdot 5\cdot 7$	41^2	$2\cdot 29^2$	$3^2\cdot 11\cdot 17$	$2^2\cdot 421$	5·337	2·3·281	7·241	$2^3\cdot 211$	3·563
169	$2\cdot 5\cdot 13^2$	19·89	$2^2\cdot 3^2\cdot 47$	**1693**	$2\cdot 7\cdot 11^2$	3·5·113	$2^5\cdot 53$	**1697**	2·3·283	**1699**
170	$2^2\cdot 5^2\cdot 17$	$3^5\cdot 7$	2·23·37	13·131	$2^3\cdot 3\cdot 71$	5·11·31	2·853	3·569	$2^2\cdot 7\cdot 61$	**1709**
171	$2\cdot 3^2\cdot 5\cdot 19$	29·59	$2^4\cdot 107$	3·571	2·857	$5\cdot 7^3$	$2^2\cdot 3\cdot 11\cdot 13$	17·101	2·859	$3^2\cdot 191$
172	$2^3\cdot 5\cdot 43$	**1721**	2·3·7·41	**1723**	$2^2\cdot 431$	$3\cdot 5^2\cdot 23$	2·863	11·157	$2^6\cdot 3^3$	7·13·19
173	2·5·173	3·577	$2^2\cdot 433$	**1733**	$2\cdot 3\cdot 17^2$	5·347	$2^3\cdot 7\cdot 31$	$3^2\cdot 193$	2·11·79	37·47

	0	1	2	3	4	5	6	7	8	9
174	$2^2\cdot3\cdot5\cdot29$	**1741**	$2\cdot13\cdot67$	$3\cdot7\cdot83$	$2^4\cdot109$	$5\cdot349$	$2\cdot3^2\cdot97$	**1747**	$2^2\cdot19\cdot23$	$3\cdot11\cdot53$
175	$2\cdot5^3\cdot7$	$17\cdot103$	$2^3\cdot3\cdot73$	**1753**	$2\cdot877$	$3^3\cdot5\cdot13$	$2^2\cdot439$	$7\cdot251$	$2\cdot3\cdot293$	**1759**
176	$2^5\cdot5\cdot11$	$3\cdot587$	$2\cdot881$	$41\cdot43$	$2^2\cdot3^2\cdot7^2$	$5\cdot353$	$2\cdot883$	$3\cdot19\cdot31$	$2^3\cdot13\cdot17$	$29\cdot61$
177	$2\cdot3\cdot5\cdot59$	$7\cdot11\cdot23$	$2^2\cdot443$	$3^2\cdot197$	$2\cdot887$	$5^2\cdot71$	$2^4\cdot3\cdot37$	**1777**	$2\cdot7\cdot127$	$3\cdot593$
178	$2^2\cdot5\cdot89$	$13\cdot137$	$2\cdot3^4\cdot11$	**1783**	$2^3\cdot223$	$3\cdot5\cdot7\cdot17$	$2\cdot19\cdot47$	**1787**	$2^2\cdot3\cdot149$	**1789**
179	$2\cdot5\cdot179$	$3^2\cdot199$	$2^8\cdot7$	$11\cdot163$	$2\cdot3\cdot13\cdot23$	$5\cdot359$	$2^2\cdot449$	$3\cdot599$	$2\cdot29\cdot31$	$7\cdot257$
180	$2^3\cdot3^2\cdot5^2$	**1801**	$2\cdot17\cdot53$	$3\cdot601$	$2^2\cdot11\cdot41$	$5\cdot19^2$	$2\cdot3\cdot7\cdot43$	$13\cdot139$	$2^4\cdot113$	$3^3\cdot67$
181	$2\cdot5\cdot181$	**1811**	$2^2\cdot3\cdot151$	$7^2\cdot37$	$2\cdot907$	$3\cdot5\cdot11^2$	$2^3\cdot227$	$23\cdot79$	$2\cdot3^2\cdot101$	$17\cdot107$
182	$2^2\cdot5\cdot7\cdot13$	$3\cdot607$	$2\cdot911$	**1823**	$2^5\cdot3\cdot19$	$5^2\cdot73$	$2\cdot11\cdot83$	$3^2\cdot7\cdot29$	$2^2\cdot457$	$31\cdot59$
183	$2\cdot3\cdot5\cdot61$	**1831**	$2^3\cdot229$	$3\cdot13\cdot47$	$2\cdot7\cdot131$	$5\cdot367$	$2^2\cdot3^3\cdot17$	$11\cdot167$	$2\cdot919$	$3\cdot613$
184	$2^4\cdot5\cdot23$	$7\cdot263$	$2\cdot3\cdot307$	$19\cdot97$	$2^2\cdot461$	$3^2\cdot5\cdot41$	$2\cdot13\cdot71$	**1847**	$2^3\cdot3\cdot7\cdot11$	43^2
185	$2\cdot5^2\cdot37$	$3\cdot617$	$2^2\cdot463$	$17\cdot109$	$2\cdot3^2\cdot103$	$5\cdot7\cdot53$	$2^6\cdot29$	$3\cdot619$	$2\cdot929$	$11\cdot13^2$
186	$2^2\cdot3\cdot5\cdot31$	**1861**	$2\cdot7^2\cdot19$	$3^4\cdot23$	$2^3\cdot233$	$5\cdot373$	$2\cdot3\cdot311$	**1867**	$2^2\cdot467$	$3\cdot7\cdot89$
187	$2\cdot5\cdot11\cdot17$	**1871**	$2^4\cdot3^2\cdot13$	**1873**	$2\cdot937$	$3\cdot5^4$	$2^2\cdot7\cdot67$	**1877**	$2\cdot3\cdot313$	**1879**
188	$2^3\cdot5\cdot47$	$3^2\cdot11\cdot19$	$2\cdot941$	$7\cdot269$	$2^2\cdot3\cdot157$	$5\cdot13\cdot29$	$2\cdot23\cdot41$	$3\cdot17\cdot37$	$2^5\cdot59$	**1889**
189	$2\cdot3^3\cdot5\cdot7$	$31\cdot61$	$2^2\cdot11\cdot43$	$3\cdot631$	$2\cdot947$	$5\cdot379$	$2^3\cdot3\cdot79$	$7\cdot271$	$2\cdot13\cdot73$	$3^2\cdot211$
190	$2^2\cdot5^2\cdot19$	**1901**	$2\cdot3\cdot317$	$11\cdot173$	$2^4\cdot7\cdot17$	$3\cdot5\cdot127$	$2\cdot953$	**1907**	$2^2\cdot3^2\cdot53$	$23\cdot83$
191	$2\cdot5\cdot191$	$3\cdot7^2\cdot13$	$2^3\cdot239$	**1913**	$2\cdot3\cdot11\cdot29$	$5\cdot383$	$2^2\cdot479$	$3^3\cdot71$	$2\cdot7\cdot137$	$19\cdot101$
192	$2^7\cdot3\cdot5$	$17\cdot113$	$2\cdot31^2$	$3\cdot641$	$2^2\cdot13\cdot37$	$5^2\cdot7\cdot11$	$2\cdot3^2\cdot107$	$41\cdot47$	$2^3\cdot241$	$3\cdot643$
193	$2\cdot5\cdot193$	**1931**	$2^2\cdot3\cdot7\cdot23$	**1933**	$2\cdot967$	$3^2\cdot5\cdot43$	$2^4\cdot11^2$	$13\cdot149$	$2\cdot3\cdot17\cdot19$	$7\cdot277$
194	$2^2\cdot5\cdot97$	$3\cdot647$	$2\cdot971$	$29\cdot67$	$2^3\cdot3^5$	$5\cdot389$	$2\cdot7\cdot139$	$3\cdot11\cdot59$	$2^2\cdot487$	**1949**
195	$2\cdot3\cdot5^2\cdot13$	**1951**	$2^5\cdot61$	$3^2\cdot7\cdot31$	$2\cdot977$	$5\cdot17\cdot23$	$2^2\cdot3\cdot163$	$19\cdot103$	$2\cdot11\cdot89$	$3\cdot653$
196	$2^3\cdot5\cdot7^2$	$37\cdot53$	$2\cdot3^2\cdot109$	$13\cdot151$	$2^2\cdot491$	$3\cdot5\cdot131$	$2\cdot983$	$7\cdot281$	$2^4\cdot3\cdot41$	$11\cdot179$
197	$2\cdot5\cdot197$	$3^3\cdot73$	$2^2\cdot17\cdot29$	**1973**	$2\cdot3\cdot7\cdot47$	$5^2\cdot79$	$2^3\cdot13\cdot19$	$3\cdot659$	$2\cdot23\cdot43$	**1979**
198	$2^2\cdot3^2\cdot5\cdot11$	$7\cdot283$	$2\cdot991$	$3\cdot661$	$2^6\cdot31$	$5\cdot397$	$2\cdot3\cdot331$	**1987**	$2^2\cdot7\cdot71$	$3^2\cdot13\cdot17$
199	$2\cdot5\cdot199$	$11\cdot181$	$2^3\cdot3\cdot83$	**1993**	$2\cdot997$	$3\cdot5\cdot7\cdot19$	$2^2\cdot499$	**1997**	$2\cdot3^3\cdot37$	**1999**
200	$2^4\cdot5^3$	$3\cdot23\cdot29$	$2\cdot7\cdot11\cdot13$	**2003**	$2^2\cdot3\cdot167$	$5\cdot401$	$2\cdot17\cdot59$	$3^2\cdot223$	$2^3\cdot251$	$7^2\cdot41$
201	$2\cdot3\cdot5\cdot67$	**2011**	$2^2\cdot503$	$3\cdot11\cdot61$	$2\cdot19\cdot53$	$5\cdot13\cdot31$	$2^5\cdot3^2\cdot7$	**2017**	$2\cdot1009$	$3\cdot673$
202	$2^2\cdot5\cdot101$	$43\cdot47$	$2\cdot3\cdot337$	$7\cdot17^2$	$2^3\cdot11\cdot23$	$3^4\cdot5^2$	$2\cdot1013$	**2027**	$2^2\cdot3\cdot13^2$	**2029**
203	$2\cdot5\cdot7\cdot29$	$3\cdot677$	$2^4\cdot127$	$19\cdot107$	$2\cdot3^2\cdot113$	$5\cdot11\cdot37$	$2^2\cdot509$	$3\cdot7\cdot97$	$2\cdot1019$	**2039**
204	$2^3\cdot3\cdot5\cdot17$	$13\cdot157$	$2\cdot1021$	$3^2\cdot227$	$2^2\cdot7\cdot73$	$5\cdot409$	$2\cdot3\cdot11\cdot31$	$23\cdot89$	2^{11}	$3\cdot683$
205	$2\cdot5^2\cdot41$	$7\cdot293$	$2^2\cdot3^3\cdot19$	**2053**	$2\cdot13\cdot79$	$3\cdot5\cdot137$	$2^3\cdot257$	$11^2\cdot17$	$2\cdot3\cdot7^3$	$29\cdot71$
206	$2^2\cdot5\cdot103$	$3^2\cdot229$	$2\cdot1031$	**2063**	$2^4\cdot3\cdot43$	$5\cdot7\cdot59$	$2\cdot1033$	$3\cdot13\cdot53$	$2^2\cdot11\cdot47$	**2069**
207	$2\cdot3^2\cdot5\cdot23$	$19\cdot109$	$2^3\cdot7\cdot37$	$3\cdot691$	$2\cdot17\cdot61$	$5^2\cdot83$	$2^2\cdot3\cdot173$	$31\cdot67$	$2\cdot1039$	$3^3\cdot7\cdot11$
208	$2^5\cdot5\cdot13$	**2081**	$2\cdot3\cdot347$	**2083**	$2^2\cdot521$	$3\cdot5\cdot139$	$2\cdot7\cdot149$	**2087**	$2^3\cdot3^2\cdot29$	**2089**
209	$2\cdot5\cdot11\cdot19$	$3\cdot17\cdot41$	$2^2\cdot523$	$7\cdot13\cdot23$	$2\cdot3\cdot349$	$5\cdot419$	$2^4\cdot131$	$3^2\cdot233$	$2\cdot1049$	**2099**
210	$2^2\cdot3\cdot5^2\cdot7$	$11\cdot191$	$2\cdot1051$	$3\cdot701$	$2^3\cdot263$	$5\cdot421$	$2\cdot3^4\cdot13$	$7^2\cdot43$	$2^2\cdot17\cdot31$	$3\cdot19\cdot37$
211	$2\cdot5\cdot211$	**2111**	$2^6\cdot3\cdot11$	**2113**	$2\cdot7\cdot151$	$3^2\cdot5\cdot47$	$2^2\cdot23^2$	$29\cdot73$	$2\cdot3\cdot353$	$13\cdot163$
212	$2^3\cdot5\cdot53$	$3\cdot7\cdot101$	$2\cdot1061$	$11\cdot193$	$2^2\cdot3^2\cdot59$	$5^3\cdot17$	$2\cdot1063$	$3\cdot709$	$2^4\cdot7\cdot19$	**2129**
213	$2\cdot3\cdot5\cdot71$	**2131**	$2^2\cdot13\cdot41$	$3^3\cdot79$	$2\cdot11\cdot97$	$5\cdot7\cdot61$	$2^3\cdot3\cdot89$	**2137**	$2\cdot1069$	$3\cdot23\cdot31$
214	$2^2\cdot5\cdot107$	**2141**	$2\cdot3^2\cdot7\cdot17$	**2143**	$2^5\cdot67$	$3\cdot5\cdot11\cdot13$	$2\cdot29\cdot37$	$19\cdot113$	$2^2\cdot3\cdot179$	$7\cdot307$
215	$2\cdot5^2\cdot43$	$3^2\cdot239$	$2^3\cdot269$	**2153**	$2\cdot3\cdot359$	$5\cdot431$	$2^2\cdot7^2\cdot11$	$3\cdot719$	$2\cdot13\cdot83$	$17\cdot127$
216	$2^4\cdot3^3\cdot5$	**2161**	$2\cdot23\cdot47$	$3\cdot7\cdot103$	$2^2\cdot541$	$5\cdot433$	$2\cdot3\cdot19^2$	$11\cdot197$	$2^3\cdot271$	$3^2\cdot241$
217	$2\cdot5\cdot7\cdot31$	$13\cdot167$	$2^2\cdot3\cdot181$	$41\cdot53$	$2\cdot1087$	$3\cdot5^2\cdot29$	$2^7\cdot17$	$7\cdot311$	$2\cdot3^2\cdot11^2$	**2179**
218	$2^2\cdot5\cdot109$	$3\cdot727$	$2\cdot1091$	$37\cdot59$	$2^3\cdot3\cdot7\cdot13$	$5\cdot19\cdot23$	$2\cdot1093$	3^7	$2^2\cdot547$	$11\cdot199$
219	$2\cdot3\cdot5\cdot73$	$7\cdot313$	$2^4\cdot137$	$3\cdot17\cdot43$	$2\cdot1097$	$5\cdot439$	$2^2\cdot3^2\cdot61$	13^3	$2\cdot7\cdot157$	$3\cdot733$
220	$2^3\cdot5^2\cdot11$	$31\cdot71$	$2\cdot3\cdot367$	**2203**	$2^2\cdot19\cdot29$	$3^2\cdot5\cdot7^2$	$2\cdot1103$	**2207**	$2^5\cdot3\cdot23$	47^2
221	$2\cdot5\cdot13\cdot17$	$3\cdot11\cdot67$	$2^2\cdot7\cdot79$	**2213**	$2\cdot3^3\cdot41$	$5\cdot443$	$2^3\cdot277$	$3\cdot739$	$2\cdot1109$	$7\cdot317$
222	$2^2\cdot3\cdot5\cdot37$	**2221**	$2\cdot11\cdot101$	$3^2\cdot13\cdot19$	$2^4\cdot139$	$5^2\cdot89$	$2\cdot3\cdot7\cdot53$	$17\cdot131$	$2^2\cdot557$	$3\cdot743$
223	$2\cdot5\cdot223$	$23\cdot97$	$2^3\cdot3^2\cdot31$	$7\cdot11\cdot29$	$2\cdot1117$	$3\cdot5\cdot149$	$2^2\cdot13\cdot43$	**2237**	$2\cdot3\cdot373$	**2239**
224	$2^6\cdot5\cdot7$	$3^3\cdot83$	$2\cdot19\cdot59$	**2243**	$2^2\cdot3\cdot11\cdot17$	$5\cdot449$	$2\cdot1123$	$3\cdot7\cdot107$	$2^3\cdot281$	$13\cdot173$
225	$2\cdot3^2\cdot5^3$	**2251**	$2^2\cdot563$	$3\cdot751$	$2\cdot7^2\cdot23$	$5\cdot11\cdot41$	$2^4\cdot3\cdot47$	$37\cdot61$	$2\cdot1129$	$3^2\cdot251$
226	$2^2\cdot5\cdot113$	$7\cdot17\cdot19$	$2\cdot3\cdot13\cdot29$	$31\cdot73$	$2^3\cdot283$	$3\cdot5\cdot151$	$2\cdot11\cdot103$	**2267**	$2^2\cdot3^4\cdot7$	**2269**
227	$2\cdot5\cdot227$	$3\cdot757$	$2^5\cdot71$	**2273**	$2\cdot3\cdot379$	$5^2\cdot7\cdot13$	$2^2\cdot569$	$3^2\cdot11\cdot23$	$2\cdot17\cdot67$	$43\cdot53$
228	$2^3\cdot3\cdot5\cdot19$	**2281**	$2\cdot7\cdot163$	$3\cdot761$	$2^2\cdot571$	$5\cdot457$	$2\cdot3^2\cdot127$	**2287**	$2^4\cdot11\cdot13$	$3\cdot7\cdot109$
229	$2\cdot5\cdot229$	$29\cdot79$	$2^2\cdot3\cdot191$	**2293**	$2\cdot31\cdot37$	$3^3\cdot5\cdot17$	$2^3\cdot7\cdot41$	**2297**	$2\cdot3\cdot383$	$11^2\cdot19$
230	$2^2\cdot5^2\cdot23$	$3\cdot13\cdot59$	$2\cdot1151$	$7^2\cdot47$	$2^8\cdot3^2$	$5\cdot461$	$2\cdot1153$	$3\cdot769$	$2^2\cdot577$	**2309**
231	$2\cdot3\cdot5\cdot7\cdot11$	**2311**	$2^3\cdot17^2$	$3^2\cdot257$	$2\cdot13\cdot89$	$5\cdot463$	$2^2\cdot3\cdot193$	$7\cdot331$	$2\cdot19\cdot61$	$3\cdot773$
232	$2^4\cdot5\cdot29$	$11\cdot211$	$2\cdot3^3\cdot43$	$23\cdot101$	$2^2\cdot7\cdot83$	$3\cdot5^2\cdot31$	$2\cdot1163$	$13\cdot179$	$2^3\cdot3\cdot97$	$17\cdot137$
233	$2\cdot5\cdot233$	$3^2\cdot7\cdot37$	$2^2\cdot11\cdot53$	**2333**	$2\cdot3\cdot389$	$5\cdot467$	$2^5\cdot73$	$3\cdot19\cdot41$	$2\cdot7\cdot167$	**2339**
234	$2^2\cdot3^2\cdot5\cdot13$	**2341**	$2\cdot1171$	$3\cdot11\cdot71$	$2^3\cdot293$	$5\cdot7\cdot67$	$2\cdot3\cdot17\cdot23$	**2347**	$2^2\cdot587$	$3^4\cdot29$
235	$2\cdot5^2\cdot47$	**2351**	$2^4\cdot3\cdot7^2$	$13\cdot181$	$2\cdot11\cdot107$	$3\cdot5\cdot157$	$2^2\cdot19\cdot31$	**2357**	$2\cdot3^2\cdot131$	$7\cdot337$
236	$2^3\cdot5\cdot59$	$3\cdot787$	$2\cdot1181$	$17\cdot139$	$2^2\cdot3\cdot197$	$5\cdot11\cdot43$	$2\cdot7\cdot13^2$	$3^2\cdot263$	$2^6\cdot37$	$23\cdot103$

	0	1	2	3	4	5	6	7	8	9
237	$2\cdot3\cdot5\cdot79$	**2371**	$2^2\cdot593$	$3\cdot7\cdot113$	$2\cdot1187$	$5^3\cdot19$	$2^3\cdot3^3\cdot11$	**2377**	$2\cdot29\cdot41$	$3\cdot13\cdot61$
238	$2^2\cdot5\cdot7\cdot17$	**2381**	$2\cdot3\cdot397$	**2383**	$2^4\cdot149$	$3^2\cdot5\cdot53$	$2\cdot1193$	$7\cdot11\cdot31$	$2^2\cdot3\cdot199$	**2389**
239	$2\cdot5\cdot239$	$3\cdot797$	$2^3\cdot13\cdot23$	**2393**	$2\cdot3^2\cdot7\cdot19$	$5\cdot479$	$2^2\cdot599$	$3\cdot17\cdot47$	$2\cdot11\cdot109$	**2399**
240	$2^5\cdot3\cdot5^2$	7^4	$2\cdot1201$	$3^3\cdot89$	$2^2\cdot601$	$5\cdot13\cdot37$	$2\cdot3\cdot401$	$29\cdot83$	$2^3\cdot7\cdot43$	$3\cdot11\cdot73$
241	$2\cdot5\cdot241$	**2411**	$2^2\cdot3^2\cdot67$	$19\cdot127$	$2\cdot17\cdot71$	$3\cdot5\cdot7\cdot23$	$2^4\cdot151$	**2417**	$2\cdot3\cdot13\cdot31$	$41\cdot59$
242	$2^2\cdot5\cdot11^2$	$3^2\cdot269$	$2\cdot7\cdot173$	**2423**	$2^3\cdot3\cdot101$	$5^2\cdot97$	$2\cdot1213$	$3\cdot809$	$2^2\cdot607$	$7\cdot347$
243	$2\cdot3^5\cdot5$	$11\cdot13\cdot17$	$2^7\cdot19$	$3\cdot811$	$2\cdot1217$	$5\cdot487$	$2^2\cdot3\cdot7\cdot29$	**2437**	$2\cdot23\cdot53$	$3^2\cdot271$
244	$2^3\cdot5\cdot61$	**2441**	$2\cdot3\cdot11\cdot37$	$7\cdot349$	$2^2\cdot13\cdot47$	$3\cdot5\cdot163$	$2\cdot1223$	**2447**	$2^4\cdot3^2\cdot17$	$31\cdot79$
245	$2\cdot5^2\cdot7^2$	$3\cdot19\cdot43$	$2^2\cdot613$	$11\cdot223$	$2\cdot3\cdot409$	$5\cdot491$	$2^3\cdot307$	$3^3\cdot7\cdot13$	$2\cdot1229$	**2459**
246	$2^2\cdot3\cdot5\cdot41$	$23\cdot107$	$2\cdot1231$	$3\cdot821$	$2^5\cdot7\cdot11$	$5\cdot17\cdot29$	$2\cdot3^2\cdot137$	**2467**	$2^2\cdot617$	$3\cdot823$
247	$2\cdot5\cdot13\cdot19$	$7\cdot353$	$2^3\cdot3\cdot103$	**2473**	$2\cdot1237$	$3^2\cdot5^2\cdot11$	$2^2\cdot619$	**2477**	$2\cdot3\cdot7\cdot59$	$37\cdot67$
248	$2^4\cdot5\cdot31$	$3\cdot827$	$2\cdot17\cdot73$	$13\cdot191$	$2^2\cdot3^3\cdot23$	$5\cdot7\cdot71$	$2\cdot11\cdot113$	$3\cdot829$	$2^3\cdot311$	$19\cdot131$
249	$2\cdot3\cdot5\cdot83$	$47\cdot53$	$2^2\cdot7\cdot89$	$3^2\cdot277$	$2\cdot29\cdot43$	$5\cdot499$	$2^6\cdot3\cdot13$	$11\cdot227$	$2\cdot1249$	$3\cdot7^2\cdot17$

2.8 Irreducible Polynomials of Degree n

An *irreducible polynomial* of *degree* n over $\mathbb{F}_p$ is a polynomial $\sum_{i=0}^{n} a_i x^i$ that cannot be expressed as the product of two polynomials each of smaller degree, with computations of coefficients in $\mathbb{F}_p$.

2.26 Remark The polynomial $\sum_{i=0}^{n} a_i x^i$ is written here by specifying the vector $a_n \cdots a_1 a_0$.

2.27 Table Irreducible polynomials for $\mathbb{F}_2$.

n	Polynomials				
1	10	11			
2	111				
3	1011	1101			
4	10011	11001	11111		
5	100101	101001	101111	110111	111011
	111101				
6	1000011	1001001	1010111	1011011	1100001
	1100111	1101101	1110011	1110101	
7	10000011	10001001	10001111	10010001	10011101
	10100111	10101011	10111001	10111111	11000001
	11001011	11010011	11010101	11100101	11101111
	11110001	11110111	11111101		
8	100011011	100011101	100101011	100101101	100111001
	100111111	101001101	101011111	101100011	101100101
	101101001	101110001	101110111	101111011	110000111
	110001011	110001101	110011111	110100011	110101001
	110110001	110111101	111000011	111001111	111010111
	111011101	111100111	111110011	111110101	111111001
9	1000000011	1000010001	1000010111	1000011011	1000100001
	1000101101	1000110011	1001001011	1001011001	1001011111
	1001100101	1001101001	1001101111	1001110111	1001111101
	1010000111	1010010101	1010011001	1010100011	1010100101
	1010101111	1010110111	1010111101	1011001111	1011010001
	1011011011	1011110101	1011111001	1100000001	1100010011
	1100010101	1100011111	1100100011	1100110001	1100111011
	1101001001	1101001111	1101011011	1101100001	1101101011
	1101101101	1101110011	1101111111	1110000101	1110001111
	1110100001	1110110101	1110111001	1111000111	1111001011
	1111001101	1111010101	1111011001	1111100011	1111101001
	1111111011				

n	Polynomials				
10	10000001001	10000001111	10000011011	10000011101	10000100111
	10000101101	10000110101	10001000111	10001010011	10001100011
	10001100101	10001101111	10010000001	10010001011	10010011001
	10010101001	10010101111	10011000101	10011001001	10011010111
	10011100111	10011101101	10011110011	10011111111	10100001011
	10100001101	10100011001	10100011111	10100100011	10100110001
	10100111101	10101000011	10101010111	10101100001	10101100111
	10101101011	10110000101	10110001111	10110010111	10110011011
	10110100001	10110101011	10110111001	10111000001	10111000111
	10111100101	10111110111	10111111011	11000010011	11000010101
	11000100011	11000100101	11000110001	11000110111	11001000011
	11001001111	11001010001	11001011011	11001111001	11001111111
	11010000101	11010001001	11010100111	11010101101	11010110101
	11010111111	11011000001	11011001101	11011010011	11011011111
	11011110111	11011111101	11100001111	11100010001	11100010111
	11100011101	11100100001	11100101011	11100110101	11100111001
	11101000111	11101001101	11101010101	11101011001	11101100011
	11101111011	11101111101	11110000001	11110000111	11110001101
	11110010011	11110101001	11110110001	11111000101	11111011011

2.28 Table Irreducible polynomials for $\mathbb{F}_3$.

n	Polynomials							
1	10	11	12					
2	101	112	122					
3	1021	1022	1102	1112	1121	1201	1211	1222
4	10012	10022	10102	10111	10121	10202	11002	11021
	11101	11111	11122	11222	12002	12011	12101	12112
	12121	12212						
5	100021	100022	100112	100211	101011	101012	101102	101122
	101201	101221	102101	102112	102122	102202	102211	102221
	110002	110012	110021	110101	110111	110122	111011	111121
	111211	111212	112001	112022	112102	112111	112201	112202
	120001	120011	120022	120202	120212	120221	121012	121111
	121112	121222	122002	122021	122101	122102	122201	122212
6	1000012	1000022	1000111	1000121	1000201	1001012	1001021	1001101
	1001122	1001221	1002011	1002022	1002101	1002112	1002211	1010201
	1010212	1010222	1011001	1011011	1011022	1011122	1012001	1012012
	1012021	1012112	1020001	1020101	1020112	1020122	1021021	1021102
	1021112	1021121	1022011	1022102	1022111	1022122	1100002	1100012
	1100111	1101002	1101011	1101101	1101112	1101212	1102001	1102111
	1102121	1102201	1102202	1110001	1110011	1110122	1110202	1110221
	1111012	1111021	1111111	1111112	1111222	1112011	1112201	1112222
	1120102	1120121	1120222	1121012	1121102	1121122	1121212	1121221
	1122001	1122002	1122122	1122202	1122221	1200002	1200022	1200121
	1201001	1201111	1201121	1201201	1201202	1202002	1202021	1202101
	1202122	1202222	1210001	1210021	1210112	1210202	1210211	1211021
	1211201	1211212	1212011	1212022	1212121	1212122	1212212	1220102
	1220111	1220212	1221001	1221002	1221112	1221202	1221211	1222022
	1222102	1222112	1222211	1222222				
7	10000102	10000121	10000201	10000222	10001011	10001012	10001102	10001111
	10001201	10001212	10002112	10002122	10002211	10002221	10010122	10010222
	10011002	10011101	10011211	10012001	10012022	10012111	10012202	10020121
	10020221	10021001	10021112	10021202	10022002	10022021	10022101	10022212
	10100011	10100012	10100102	10100122	10100201	10100221	10101101	10101112
	10101202	10101211	10102102	10102201	10110022	10110101	10110211	10111001
	10111102	10111121	10111201	10112002	10112012	10112021	10112111	10112122

n	Polynomials							
	10120021	10120112	10120202	10121002	10121102	10121201	10121222	10122001
	10122011	10122022	10122212	10122221	10200001	10200002	10200101	10200112
	10200202	10200211	10201021	10201022	10201121	10201222	10202011	10202012
	10210001	10210121	10210202	10211101	10211111	10211122	10211221	10212011
	10212022	10212101	10212112	10212212	10220002	10220101	10220222	10221122
	10221202	10221212	10221221	10222012	10222021	10222111	10222202	10222211
	11000101	11000222	11001022	11001112	11001211	11002012	11002022	11002121
	11002202	11010001	11010022	11010121	11010221	11011111	11011202	11012002
	11012102	11012212	11020021	11020022	11020102	11020112	11020201	11020222
	11021111	11021122	11021201	11021212	11022101	11022122	11022211	11022221
	11100002	11100022	11100121	11100212	11101012	11101022	11101102	11101111
	11101121	11102002	11102111	11102222	11110001	11110012	11110111	11110112
	11110211	11110222	11111011	11111021	11111201	11111222	11112011	11112221
	11120102	11120111	11120122	11120212	11120221	11121001	11121101	11121202
	11122021	11122112	11122201	11122222	11200201	11200202	11201012	11201021
	11201101	11201111	11201221	11201222	11202002	11202121	11202211	11202212
	11210002	11210011	11210021	11210101	11211001	11211022	11211122	11211212
	11211221	11212012	11212112	11212202	11220001	11220112	11220211	11221022
	11221102	11221112	11221121	11222011	11222102	11222122	11222201	11222221
	12000121	12000202	12001021	12001112	12001211	12002011	12002021	12002101
	12002222	12010021	12010022	12010102	12010121	12010201	12010211	12011102
	12011111	12011212	12011221	12012112	12012122	12012202	12012221	12020002
	12020021	12020122	12020222	12021101	12021212	12022001	12022111	12022201
	12100001	12100021	12100111	12100222	12101011	12101021	12101201	12101212
	12101222	12102001	12102121	12102212	12110111	12110122	12110201	12110212
	12110221	12111002	12111101	12111202	12112022	12112102	12112121	12112211
	12120002	12120011	12120112	12120121	12120211	12120212	12121012	12121022
	12121102	12121121	12122012	12122122	12200101	12200102	12201011	12201022
	12201121	12201122	12201202	12201212	12202001	12202111	12202112	12202222
	12210002	12210112	12210211	12211021	12211201	12211211	12211222	12212012
	12212102	12212122	12212201	12212221	12220001	12220012	12220022	12220202
	12221002	12221021	12221111	12221122	12221221	12222011	12222101	12222211

2.29 Table Irreducible polynomials for $\mathbb{F}_5$.

n	Polynomials									
1	10	11	12	13	14					
2	102	103	111	112	123	124	133	134	141	142
3	1011	1014	1021	1024	1032	1033	1042	1043	1101	1102
	1113	1114	1131	1134	1141	1143	1201	1203	1213	1214
	1222	1223	1242	1244	1302	1304	1311	1312	1322	1323
	1341	1343	1403	1404	1411	1412	1431	1434	1442	1444
4	10002	10003	10014	10024	10034	10044	10102	10111	10122	10123
	10132	10133	10141	10203	10221	10223	10231	10233	10303	10311
	10313	10341	10343	10402	10412	10413	10421	10431	10442	10443
	11004	11013	11023	11024	11032	11041	11042	11101	11113	11114
	11124	11133	11142	11202	11212	11213	11221	11222	11234	11244
	11301	11303	11321	11342	11344	11402	11411	11414	11441	11443
	12004	12013	12014	12021	12022	12033	12042	12102	12121	12123
	12131	12134	12201	12203	12211	12222	12224	12302	12311	12312
	12324	12332	12333	12344	12401	12414	12422	12433	12434	12443
	13004	13012	13023	13031	13032	13043	13044	13102	13121	13124
	13131	13133	13201	13203	13232	13234	13241	13302	13314	13322
	13323	13334	13341	13342	13401	13413	13423	13424	13432	13444
	14004	14011	14012	14022	14033	14034	14043	14101	14112	14123
	14134	14143	14144	14202	14214	14224	14231	14232	14242	14243
	14301	14303	14312	14314	14331	14402	14411	14413	14441	14444

2.30 Table Irreducible polynomials for $\mathbb{F}_7$.

n	Polynomials									
1	10	11	12	13	14	15	16			
2	101	102	104	113	114	116	122	123	125	131
	135	136	141	145	146	152	153	155	163	164
	166									
3	1002	1003	1004	1005	1011	1016	1021	1026	1032	1035
	1041	1046	1052	1055	1062	1065	1101	1103	1112	1115
	1124	1126	1131	1135	1143	1146	1151	1152	1153	1154
	1163	1165	1201	1203	1214	1216	1223	1226	1233	1235
	1242	1245	1251	1255	1261	1262	1263	1264	1304	1306
	1311	1314	1322	1325	1333	1334	1335	1336	1341	1343
	1352	1354	1362	1366	1401	1403	1413	1416	1422	1425
	1431	1432	1433	1434	1444	1446	1453	1455	1461	1465
	1504	1506	1511	1513	1521	1524	1532	1534	1542	1545
	1552	1556	1563	1564	1565	1566	1604	1606	1612	1615
	1621	1623	1632	1636	1641	1644	1653	1654	1655	1656
	1662	1664								

2.9 Primitive Polynomials of Degree n

A polynomial $p(x)$ is a *primitive polynomial* of degree n over $\mathbb{F}_p$ if it is the minimal polynomial over $\mathbb{F}_p$ of a primitive element of $\mathbb{F}_{p^n}$

2.31 Remark Each polynomial $\sum_{i=0}^{n} a_i x^i$ is written as a vector $a_n \cdots a_1 a_0$.

2.32 Table Primitive polynomials over $\mathbb{F}_2$.

n	Polynomials				
1	11				
2	111				
3	1011	1101			
4	10011	11001			
5	100101	101001	101111	110111	111011
	111101				
6	1000011	1011011	1100001	1100111	1101101
	1110011				
7	10000011	10001001	10001111	10010001	10011101
	10100111	10101011	10111001	10111111	11000001
	11001011	11010011	11010101	11100101	11101111
	11110001	11110111	11111101		
8	100011101	100101011	100101101	101001101	101011111
	101100011	101100101	101101001	101110001	110000111
	110001101	110101001	111000011	111001111	111100111
	111110101				
9	1000010001	1000011011	1000100001	1000101101	1000110011
	1001011001	1001011111	1001101001	1001101111	1001110111
	1001111101	1010000111	1010010101	1010100011	1010100101
	1010101111	1010110111	1010111101	1011001111	1011010001
	1011011011	1011110101	1011111001	1100010011	1100010101
	1100011111	1100100011	1100110001	1100111011	1101001111
	1101011011	1101100001	1101101011	1101101101	1101110011
	1101111111	1110000101	1110001111	1110110101	1110111001
	1111000111	1111001011	1111001101	1111010101	1111011001
	1111100011	1111101001	1111111011		

n	Polynomials				
10	10000001001	10000011011	10000100111	10000101101	10001100101
	10001101111	10010000001	10010001011	10011000101	10011010111
	10011100111	10011110011	10011111111	10100001101	10100011001
	10100100011	10100110001	10100111101	10101000011	10101010111
	10101101011	10110000101	10110001111	10110010111	10110100001
	10111000111	10111100101	10111110111	10111111011	11000010011
	11000010101	11000100101	11000110111	11001000011	11001001111
	11001011011	11001111001	11001111111	11010001001	11010110101
	11011000001	11011010011	11011011111	11011111101	11100010111
	11100011101	11100100001	11100111001	11101000111	11101001101
	11101010101	11101011001	11101100011	11101111101	11110001101
	11110010011	11110110001	11111011011	11111110011	11111111001

2.33 Table Primitive polynomials over $\mathbb{F}_3$.

n	Polynomials							
1	11							
2	112	122						
3	1021	1121	1201	1211				
4	10012	10022	11002	11122	11222	12002	12112	12212
5	100021	100211	101011	101201	101221	102101	102211	110021
	110101	110111	111011	111121	111211	112001	112111	112201
	120001	120011	120221	121111	122021	122101		
6	1000012	1000022	1001012	1002022	1010212	1010222	1011022	1011122
	1012012	1012112	1020112	1020122	1021112	1022122	1100002	1101002
	1101112	1101212	1102202	1110122	1110202	1111012	1111112	1111222
	1112222	1120102	1120222	1121012	1121102	1121212	1122202	1200002
	1201202	1202002	1202122	1202222	1210112	1210202	1211212	1212022
	1212122	1212212	1220102	1220212	1221202	1222022	1222102	1222222

2.34 Table Primitive polynomials over $\mathbb{F}_5$.

n	Polynomials									
1	12	13								
2	112	123	133	142						
3	1032	1033	1042	1043	1102	1113	1143	1203	1213	1222
	1223	1242	1302	1312	1322	1323	1343	1403	1412	1442
4	10122	10123	10132	10133	10412	10413	10442	10443	11013	11023
	11032	11042	11113	11202	11212	11303	11342	11443	12013	12022
	12033	12042	12123	12203	12222	12302	12332	12433	13012	13023
	13032	13043	13133	13203	13232	13302	13322	13423	14012	14022
	14033	14043	14143	14202	14242	14303	14312	14413		

2.35 Table Primitive polynomials over $\mathbb{F}_7$.

n	Polynomials									
1	12	14								
2	113	123	125	135	145	153	155	163		
3	1032	1052	1062	1112	1124	1152	1154	1214	1242	1262
	1264	1304	1314	1322	1334	1352	1354	1362	1422	1432
	1434	1444	1504	1524	1532	1534	1542	1552	1564	1604
	1612	1632	1644	1654	1662	1664				
4	10135	10145	10333	10335	10343	10345	10433	10443	10515	10525
	10533	10543	10555	10565	10613	10623	10635	10645	10653	10663
	11013	11063	11103	11105	11153	11213	11223	11225	11233	11245

n	Polynomials									
	11323	11355	11365	11405	11423	11443	11455	11463	11523	11533
	11545	11605	11625	11643	11653	11665	12025	12055	12123	12135
	12143	12203	12205	12253	12303	12323	12325	12363	12365	12403
	12435	12465	12553	12563	12643	12655	13015	13023	13053	13065
	13103	13115	13135	13155	13205	13213	13225	13243	13323	13345
	13355	13443	13445	13455	13465	13513	13525	13535	13553	13655
	14015	14023	14053	14065	14103	14125	14145	14165	14205	14233
	14255	14263	14325	14335	14353	14415	14425	14433	14435	14523
	14545	14555	14563	14625	15025	15055	15133	15145	15153	15203
	15205	15223	15303	15313	15315	15353	15355	15403	15415	15445
	15513	15523	15625	15633	16013	16063	16103	16105	16123	16235
	16243	16253	16255	16263	16315	16325	16353	16405	16413	16425
	16433	16453	16535	16543	16553	16605	16615	16623	16633	16655

2.36 Remark See also Table IV.10.34.

2.10 Factorizations of $x^n - 1$

2.37 Table Factorizations of $x^n - 1$ over $\mathbb{F}_2$

n	Factorizations
1	$x+1$
2	$(x+1)^2$
3	$(x^2+x+1)(x+1)$
4	$(x+1)^4$
5	$(x^4+x^3+x^2+x+1)(x+1)$
6	$(x^2+x+1)^2(x+1)^2$
7	$(x^3+x^2+1)(x+1)(x^3+x+1)$
8	$(x+1)^8$
9	$(x^6+x^3+1)(x^2+x+1)(x+1)$
10	$(x^4+x^3+x^2+x+1)^2(x+1)^2$
11	$(x+1)(x^{10}+x^9+x^8+x^7+x^6+x^5+x^4+x^3+x^2+x+1)$
12	$(x^2+x+1)^4(x+1)^4$
13	$(x^{12}+x^{11}+x^{10}+x^9+x^8+x^7+x^6+x^5+x^4+x^3+x^2+x+1)(x+1)$
14	$(x^3+x^2+1)^2(x+1)^2(x^3+x+1)^2$
15	$(x^4+x+1)(x^4+x^3+x^2+x+1)(x^2+x+1)(x^4+x^3+1)(x+1)$
16	$(x+1)^{16}$
17	$(x^8+x^5+x^4+x^3+1)(x+1)(x^8+x^7+x^6+x^4+x^2+x+1)$
18	$(x^6+x^3+1)^2(x^2+x+1)^2(x+1)^2$
19	$(x^{18}+x^{17}+x^{16}+x^{15}+x^{14}+x^{13}+x^{12}+x^{11}+x^{10}+x^9+x^8+x^7+x^6+x^5+x^4+x^3+x^2+x+1)(x+1)$
20	$(x^4+x^3+x^2+x+1)^4(x+1)^4$
21	$(x^3+x^2+1)(x^6+x^4+x^2+x+1)(x^2+x+1)(x^6+x^5+x^4+x^2+1)(x+1)(x^3+x+1)$
22	$(x+1)^2(x^{10}+x^9+x^8+x^7+x^6+x^5+x^4+x^3+x^2+x+1)^2$
23	$(x^{11}+x^9+x^7+x^6+x^5+x+1)(x+1)(x^{11}+x^{10}+x^6+x^5+x^4+x^2+1)$
24	$(x^2+x+1)^8(x+1)^8$
25	$(x^4+x^3+x^2+x+1)(x+1)(x^{20}+x^{15}+x^{10}+x^5+1)$
26	$(x^{12}+x^{11}+x^{10}+x^9+x^8+x^7+x^6+x^5+x^4+x^3+x^2+x+1)^2(x+1)^2$
27	$(x^6+x^3+1)(x^2+x+1)(x+1)(x^{18}+x^9+1)$
28	$(x^3+x^2+1)^4(x+1)^4(x^3+x+1)^4$
29	$(x+1)(1+x+x^2+x^3+x^4+x^5+x^6+x^7+x^8+x^9+x^{10}+x^{11}+x^{12}+x^{13}+x^{14}+x^{15}+x^{16}+x^{17}+x^{18}+x^{19}+x^{20}+x^{21}+x^{22}+x^{23}+x^{24}+x^{25}+x^{26}+x^{27}+x^{28})$
30	$(x^4+x+1)^2(x^4+x^3+x^2+x+1)^2(x^2+x+1)^2(x^4+x^3+1)^2(x+1)^2$

n	Factorizations
31	$(x^5+x^4+x^3+x^2+1)(x^5+x^3+x^2+x+1)(x^5+x^3+1)(x^5+x^4+x^3+x+1)(x+1)(x^5+x^2+1)(x^5+x^4+x^2+x+1)$
32	$(x+1)^{32}$
33	$(x^{10}+x^7+x^5+x^3+1)(x^2+x+1)(x^{10}+x^9+x^5+x+1)(x+1)(x^{10}+x^9+x^8+x^7+x^6+x^5+x^4+x^3+x^2+x+1)$
34	$(x^8+x^5+x^4+x^3+1)^2(x+1)^2(x^8+x^7+x^6+x^4+x^2+x+1)^2$
35	$(x^3+x^2+1)(x^4+x^3+x^2+x+1)(x^{12}+x^{10}+x^9+x^8+x^7+x^4+x^2+x+1)(x+1)(x^3+x+1)(x^{12}+x^{11}+x^{10}+x^8+x^5+x^4+x^3+x^2+1)$
36	$(x^6+x^3+1)^4(x^2+x+1)^4(x+1)^4$
37	$(1+x+x^2+x^3+x^4+x^5+x^6+x^7+x^8+x^9+x^{10}+x^{11}+x^{12}+x^{13}+x^{14}+x^{15}+x^{16}+x^{17}+x^{18}+x^{19}+x^{20}+x^{21}+x^{22}+x^{23}+x^{24}+x^{25}+x^{26}+x^{27}+x^{28}+x^{29}+x^{30}+x^{31}+x^{32}+x^{33}+x^{34}+x^{35}+x^{36})(x+1)$
38	$(x^{18}+x^{17}+x^{16}+x^{15}+x^{14}+x^{13}+x^{12}+x^{11}+x^{10}+x^9+x^8+x^7+x^6+x^5+x^4+x^3+x^2+x+1)^2(x+1)^2$
39	$(x^{12}+x^{11}+x^{10}+x^9+x^5+x^4+x^3+x^2+1)(x^{12}+x^{11}+x^{10}+x^9+x^8+x^7+x^6+x^5+x^4+x^3+x^2+x+1)(x^{12}+x^{10}+x^9+x^8+x^7+x^3+x^2+x+1)(x^2+x+1)(x+1)$
40	$(x^4+x^3+x^2+x+1)^8(x+1)^8$
41	$(x^{20}+x^{19}+x^{17}+x^{16}+x^{14}+x^{11}+x^{10}+x^9+x^6+x^4+x^3+x+1)(x^{20}+x^{18}+x^{17}+x^{16}+x^{15}+x^{14}+x^{11}+x^{10}+x^9+x^6+x^5+x^4+x^3+x^2+1)(x+1)$
42	$(x^3+x+1)^2(x^2+x+1)^2(x^6+x^4+x^2+x+1)^2(x^6+x^5+x^4+x^2+1)^2(x+1)^2(x^3+x^2+1)^2$
43	$(x^{14}+x^{11}+x^{10}+x^9+x^8+x^7+x^6+x^5+x^4+x^3+1)(x^{14}+x^{12}+x^{10}+x^7+x^4+x^2+1)(x+1)(x^{14}+x^{13}+x^{11}+x^7+x^3+x+1)$
44	$(x^{10}+x^9+x^8+x^7+x^6+x^5+x^4+x^3+x^2+x+1)^4(x+1)^4$
45	$(x^6+x^3+1)(x^4+x^3+1)(x^2+x+1)(x^4+x+1)(x^{12}+x^3+1)(x^4+x^3+x^2+x+1)(x^{12}+x^9+1)(x+1)$
46	$(x^{11}+x^{10}+x^6+x^5+x^4+x^2+1)^2(x+1)^2(x^{11}+x^9+x^7+x^6+x^5+x+1)^2$
47	$(x^{23}+x^{19}+x^{18}+x^{14}+x^{13}+x^{12}+x^{10}+x^9+x^7+x^6+x^5+x^3+x^2+x+1)(x+1)(x^{23}+x^{22}+x^{21}+x^{20}+x^{18}+x^{17}+x^{16}+x^{14}+x^{13}+x^{11}+x^{10}+x^9+x^5+x^4+1)$
48	$(x^2+x+1)^{16}(x+1)^{16}$
49	$(x^3+x+1)(x^{21}+x^{14}+1)(x^{21}+x^7+1)(x+1)(x^3+x^2+1)$
50	$(x^{20}+x^{15}+x^{10}+x^5+1)^2(x^4+x^3+x^2+x+1)^2(x+1)^2$

2.11 Cyclotomic Numbers for e = 2, 3, 5 and 7

Let $q=ef+1$ be a prime power and let H be the subgroup of $\mathbb{F}_q^*$ of order f with $\{H=C_0,C_1,C_2,\ldots,C_{e-1}\}$ the set of (multiplicative) cosets of H in $\mathbb{F}_q^*$ (i.e. $C_i=g^iC_0$ where g is the least primitive element of $\mathbb{F}_q$). The *cyclotomic number* (i,j) is $|\{x\in C_i : x+1\in C_j\}|$.

2.38 Theorem [3] Some basic facts about cyclotomic numbers ($q=ef+1$).

1. For any integers m,n $(i+me,j+me)=(i,j)$.
2. $(i,j)=\begin{cases}(j,i) & \text{if } f \text{ is even}\\ (j+e/2,i+e/2) & \text{if } f \text{ is odd.}\end{cases}$
3. $\sum_{j=0}^{e-1}(i,j)=f-\theta_i$ where $\theta_i=\begin{cases}1 & \text{if } f \text{ is even and } i=0\\ 1 & \text{if } f \text{ is odd and } i=e/2\\ 0 & \text{otherwise.}\end{cases}$
4. $\sum_{j=0}^{e-1}(i,j)=f-\alpha_j$ where $\alpha_j=\begin{cases}1 & \text{if } j=0\\ 0 & \text{otherwise}\end{cases}$.
5. The cyclotomic numbers $(0,h)$ are odd or even according as $2\in C_h$ or not. In particular, exactly one of the numbers $(0,j)$ is odd.

2.39 Theorem (See [3]) Some connections between cyclotomic numbers and difference sets. (See §IV.12).

1. C_0 is a difference set in $G = \mathbb{F}_q$ if and only if $\lambda = (i, 0) = (f-1)/e$ for all $i = 0, 1 \ldots e-1$.
2. Fix e so that $q = ef + 1$, and let C_0 correspond to e. If C_0 is a difference set in G, then e is even and f is odd.
3. $C_0 + \{0\}$ is a difference set in G if and only if $1 + (0,0) = (i,0) = (f+1)/e$ for all $i = 1, 2, \ldots (e-2)/2$.
4. Fix e so that $q = ef + 1$, and let C_0 correspond to e. If $C_0 + \{0\}$ is a difference set in G, then e is even and f is odd.
5. If $q = 2f + 1 \equiv 3 \pmod 4$, then C_0 is a difference set in G.
6. If $q \equiv 1 \pmod 4$, then C_0 is a difference set in G if and only if $q = 1 + 4t^2$ (i.e., if and only if $k = (q-1)/4$ is an odd square).

2.40 Table Cyclotomic numbers for $e = 2$ ($p = 2f + 1$).

	0	1
0	$(f-2)/2$	$f/2$
1	$f/2$	$f/2$

f even

	0	1
0	$(f-1)/2$	$(f+1)/2$
1	$(f-1)/2$	$(f-1)/2$

f odd

2.41 Table Cyclotomic numbers for $e = 3$ ($p = 3f + 1$).

	0	1	2
0	0	0	1
1	0	1	1
2	1	1	0

$p = 7$

	0	1	2
0	0	1	2
1	1	2	1
2	2	1	1

$p = 13$

	0	1	2
0	2	1	2
1	1	2	3
2	2	3	1

$p = 19$

	0	1	2
0	3	4	2
1	4	2	4
2	2	4	4

$p = 31$

	0	1	2
0	2	5	4
1	5	4	3
2	4	3	5

$p = 37$

	0	1	2
0	3	6	4
1	6	4	4
2	4	4	6

$p = 43$

	0	1	2
0	6	5	8
1	5	8	7
2	8	7	5

$p = 61$

	0	1	2
0	6	9	6
1	9	6	7
2	6	7	9

$p = 67$

	0	1	2
0	8	6	9
1	6	9	9
2	9	9	6

$p = 73$

	0	1	2
0	6	9	10
1	9	10	7
2	10	7	9

$p = 79$

	0	1	2
0	12	9	10
1	9	10	13
2	10	13	9

$p = 97$

	0	1	2
0	12	12	9
1	12	9	13
2	9	13	12

$p = 103$

2.42 Table Cyclotomic numbers for $e = 5$, $(p = 5f + 1)$.

$p = 11$

	0	1	2	3	4
0	0	1	0	0	0
1	1	0	0	1	0
2	0	0	0	1	1
3	0	1	1	0	0
4	0	0	1	0	1

$p = 31$

	0	1	2	3	4
0	2	0	0	2	1
1	0	1	2	1	2
2	0	2	2	1	1
3	2	1	1	0	2
4	1	2	1	2	0

$p = 41$

	0	1	2	3	4
0	0	3	2	2	0
1	3	0	2	1	2
2	2	2	2	1	1
3	2	1	1	2	2
4	0	2	1	2	3

$p = 61$

	0	1	2	3	4
0	2	3	0	4	2
1	3	2	2	3	2
2	0	2	4	3	3
3	4	3	3	0	2
4	2	2	3	2	3

$p = 71$

	0	1	2	3	4
0	0	5	4	2	2
1	5	2	2	3	2
2	4	2	2	3	3
3	2	3	3	4	2
4	2	2	3	2	5

$p = 101$

	0	1	2	3	4
0	0	3	6	4	6
1	3	6	4	3	4
2	6	4	4	3	3
3	4	3	3	6	4
4	6	4	3	4	3

$p = 131$

	0	1	2	3	4
0	6	5	2	8	4
1	5	4	6	5	6
2	2	6	8	5	5
3	8	5	5	2	6
4	4	6	5	6	5

$p = 151$

	0	1	2	3	4
0	5	6	6	8	4
1	6	4	8	4	8
2	6	8	8	4	4
3	8	4	4	6	8
4	4	8	4	8	6

$p = 181$

	0	1	2	3	4
0	8	3	6	8	10
1	3	10	8	7	8
2	6	8	8	7	7
3	8	7	7	6	8
4	10	8	7	8	3

$p = 191$

	0	1	2	3	4
0	12	8	4	8	5
1	8	5	8	9	8
2	4	8	8	9	9
3	8	9	9	4	8
4	5	8	9	8	8

2.43 Table Cyclotomic numbers for $e = 7$, $(p = 7f + 1)$.

$p = 29$

	0	1	2	3	4	5	6
0	0	1	0	0	2	0	0
1	1	0	1	0	0	1	1
2	0	1	0	1	1	1	0
3	0	0	1	2	0	1	0
4	2	0	1	0	0	0	1
5	0	1	1	1	0	0	1
6	0	1	0	0	1	1	1

$p = 43$

	0	1	2	3	4	5	6
0	2	0	0	0	2	0	1
1	0	1	1	1	0	2	1
2	0	1	0	2	1	1	1
3	0	1	2	2	0	1	0
4	2	0	1	0	0	1	2
5	0	2	1	1	1	0	1
6	1	1	1	0	2	1	0

$p = 71$

	0	1	2	3	4	5	6
0	0	0	2	2	2	2	1
1	0	1	3	1	2	0	3
2	2	3	2	0	1	1	1
3	2	1	0	2	2	1	2
4	2	2	1	2	2	1	0
5	2	0	1	1	1	2	3
6	1	3	1	2	0	3	0

	0	1	2	3	4	5	6
0	0	2	2	4	0	5	2
1	2	2	2	3	3	2	2
2	2	2	5	2	1	1	3
3	4	3	2	0	3	1	3
4	0	3	1	3	4	3	2
5	5	2	1	1	3	2	2
6	2	2	3	3	2	2	2

$p = 113$

	0	1	2	3	4	5	6
0	2	2	5	2	4	0	2
1	2	2	4	2	3	1	4
2	5	4	0	1	3	3	2
3	2	2	1	4	3	3	3
4	4	3	3	3	2	2	1
5	0	1	3	3	2	5	4
6	2	4	2	3	1	4	2

$p = 127$

	0	1	2	3	4	5	6
0	6	1	2	4	6	4	4
1	1	4	5	6	2	5	5
2	2	5	4	5	3	3	6
3	4	6	5	6	2	3	2
4	6	2	3	2	4	6	5
5	4	5	3	3	6	2	5
6	4	5	6	2	5	5	1

$p = 197$

	0	1	2	3	4	5	6
0	2	5	8	2	6	2	4
1	5	4	3	6	4	5	3
2	8	3	2	5	3	3	6
3	2	6	5	6	4	3	4
4	6	4	3	4	2	6	5
5	2	5	3	3	6	8	3
6	4	3	6	4	5	3	5

$p = 211$

	0	1	2	3	4	5	6
0	6	2	4	5	2	6	8
1	2	8	5	3	7	4	5
2	4	5	6	4	6	6	3
3	5	3	4	2	7	6	7
4	2	7	6	7	5	3	4
5	6	4	6	6	3	4	5
6	8	5	3	7	4	5	2

$p = 239$

	0	1	2	3	4	5	6
0	12	5	8	6	4	2	2
1	5	2	7	6	6	7	7
2	8	7	2	7	5	5	6
3	6	6	7	4	6	5	6
4	4	6	5	6	6	6	7
5	2	7	5	5	6	8	7
6	2	7	6	6	7	7	5

$p = 281$

References

[1] R. Gupta and M.R. Murty, A remark on Artin's conjecture, *Inventiones Math.* 78 (1984), 127-130.

[2] J.B. Rosser and L. Schoenfeld, Approximate formulas for some functions of prime numbers, *Illinois J. Math.* 6 (1962), 64-94.

[3] T. Storer, *Cyclotomy and Difference Sets*, Markham, Chicago, 1967.

3 Graphs and Multigraphs

GORDON F. ROYLE

3.1 All Graphs

A graph G is a pair $(V(G), E(G))$ where $V(G)$ is a finite set, the *vertices* of G, and $E(G)$ is a set of unordered pairs of distinct vertices of G (*edges*).

Graphs G and H are *isomorphic* if there is a bijection $\phi : V(G) \to V(H)$ such that $\{v, w\} \in E(G)$ if and only if $\{\phi(v), \phi(w)\} \in E(H)$. An *automorphism* of a graph G is an isomorphism from G to itself. The set of all automorphisms of G forms a group under composition and is denoted $\mathrm{Aut}(G)$.

If $\{v, w\} \in E(G)$, then v and w are *adjacent*, written $v \sim w$. A graph is *connected* if for any two vertices v and w there is a sequence of vertices $v = v_0 \sim v_1 \sim \cdots \sim v_{n-1} \sim v_n = w$.

3.1 Table The connected graphs on up to five vertices.

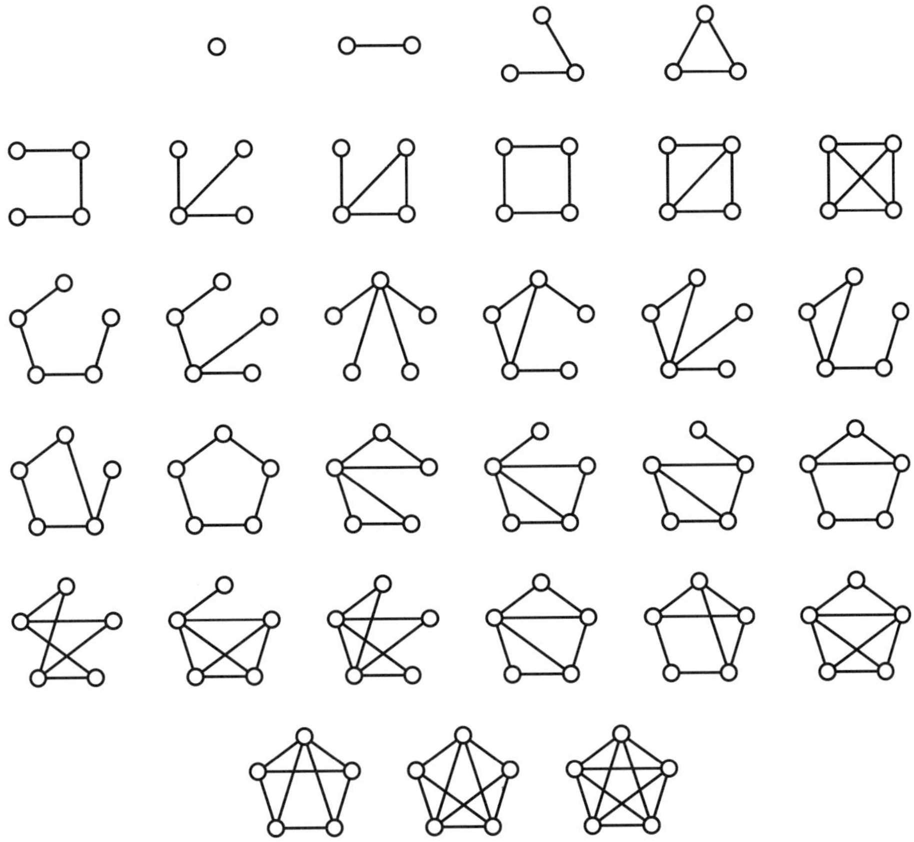

3.2 Remark The numbers of pairwise nonisomorphic graphs can be computed exactly by using Polya's enumeration theory (see [5]). All numbers given in this chapter refer to numbers of pairwise nonisomorphic graphs.

3.3 Table Numbers of graphs on up to 16 vertices.

n	All Graphs	Connected Graphs
1	1	1
2	2	1
3	4	2
4	11	6
5	34	21
6	156	112
7	1044	853
8	12346	11117
9	274668	261080
10	12005168	11716571
11	1018997864	1006700565
12	165091172592	164059830476
13	50502031367952	50335907869219
14	29054155657235488	29003487462848061
15	31426485969804308768	31397381142761241960
16	64001015704527557894928	63969560113225176176277

3.4 Theorem Every finite abstract group occurs as the automorphism group of some graph (see [4]).

3.5 Theorem Almost all graphs have trivial automorphism group.

3.6 Example The smallest graphs with trivial automorphism group have six vertices and seven edges; there are exactly three of them.

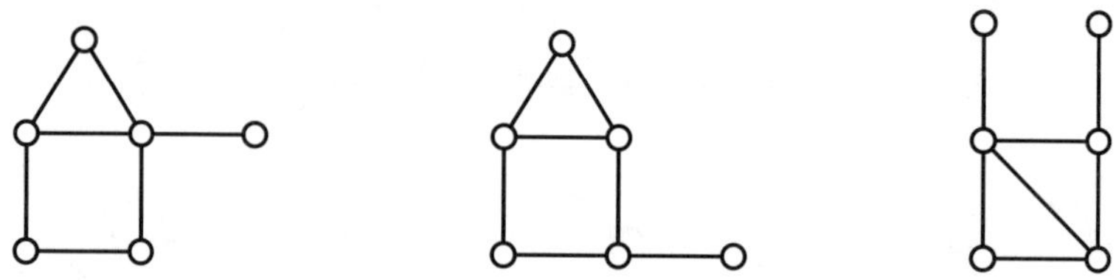

3.2 Regular Graphs

A graph is *regular* with *degree k* if each vertex is adjacent to exactly k other vertices. A regular graph of degree 3 is a *cubic graph.*

A *path* of length n is a sequence $v_0 \sim v_1 \sim \cdots \sim v_n$ of distinct vertices. The *distance* between any two vertices is the length of the shortest path between them, and the *diameter* of the graph is the longest of these distances. A *cycle* of length n is a sequence $v_0 \sim v_1 \sim \cdots \sim v_{n-1} \sim v_0$ of distinct vertices (except that v_0 occurs twice), and the *girth* of a graph is the length of the shortest cycle.

A graph is *bipartite* if the vertex set can be partitioned into two sets $V(G) = X \cup Y$ such that every edge has one vertex in X and the other in Y.

3.7 Remark The smallest regular graphs with trivial automorphism group are 4-regular graphs on 10 vertices, or 3-regular graphs on 12 vertices.

3.8 Example The point/block incidence graph of an incidence structure is a bipartite graph. The projective plane $PG(2,2)$ yields a 14-vertex cubic bipartite graph with girth 6, *Heawood's graph*, and the generalized quadrangle $W(2)$ (Example IV.21.5) yields a 30-vertex cubic bipartite graph with girth 8, *Tutte's 8-cage.*

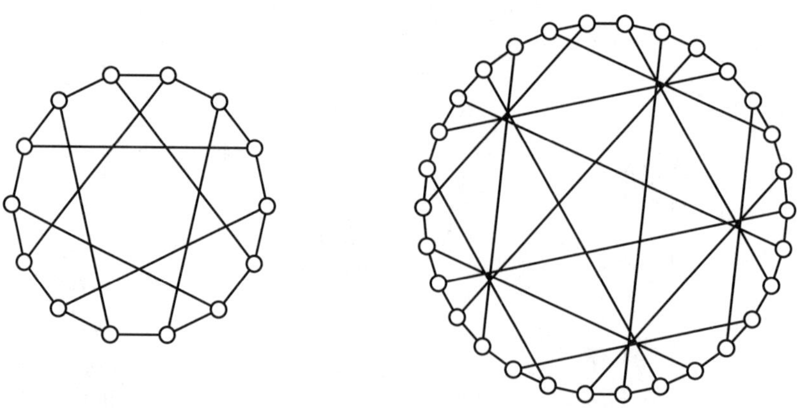

3.9 Remarks The exact number of cubic graphs can be computed analytically by results of Robinson and Wormald [8]. However the results in Table 3.10 for cubic graphs of a given girth were found by constructing all the graphs using the algorithm in [3]. An explicit listing of all cubic graphs on $2n$ vertices for $2n \leq 20$ is available via anonymous ftp at the address `ftp.cs.uwa.edu.au` in the subdirectory `pub/graphs/cubics`. Instructions are in the `README` file.

3.10 Table Numbers of connected cubic graphs tabulated according to girth g. Missing entries in the top right of the table are zeros, whereas missing entries at the bottom are unknown values.

n	$g \geq 3$	$g \geq 4$	$g \geq 5$	$g \geq 6$	$g \geq 7$	$g \geq 8$
4	1					
6	2	1				
8	5	2				
10	19	6	1			
12	85	22	2			
14	509	110	9	1		
16	4060	792	49	1		
18	41301	7805	455	5		
20	510489	97546	5783	32		
22	7319447	1435720	90938	385		
24	117940535	23780814	1620479	7574	1	
26	2094480864	432757568	31478584	181227	3	
28	40497138011		656783890	4624501	21	
30	845480228069			122090544	546	1
32	18941522184590				30368	0
34	453090162062723					1
36	11523392072541432					3
38	310467244165539782					13
40	8832736318937756165					155

3.11 Remark The smallest cubic graphs with girth 9, 10, and 12 have 58, 70, and 126 vertices, respectively. The smallest cubic graph of girth 11 is unknown.

3.12 Table Connected cubic graphs on up to 10 vertices.

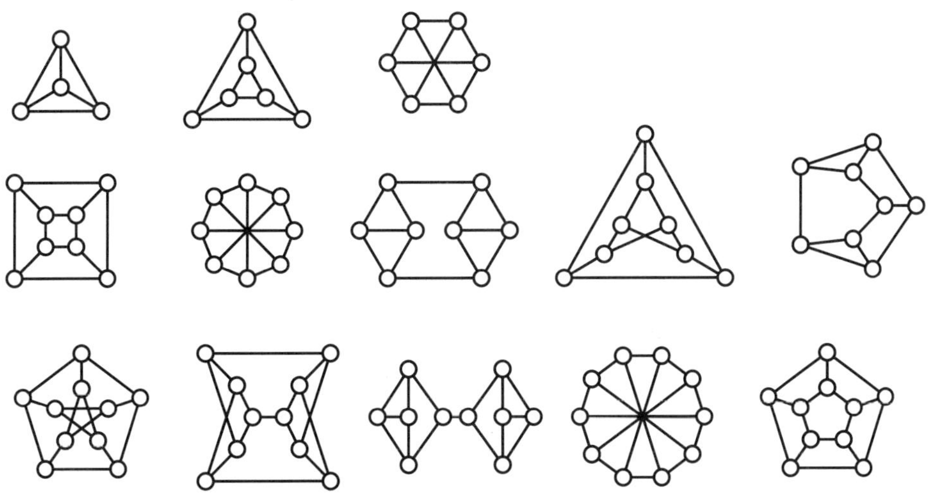

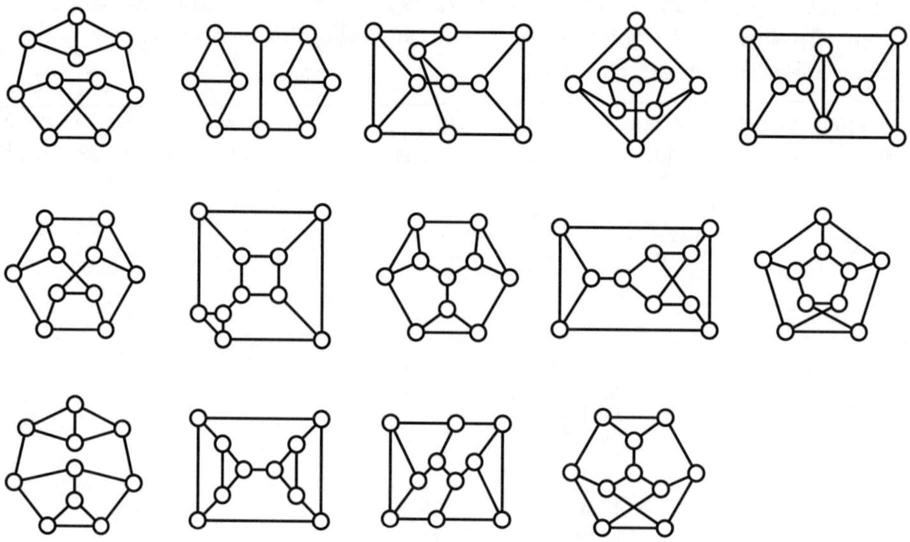

3.13 Table Numbers of connected k-regular graphs on up to 16 vertices.

$n\backslash k$	4	5	6	7	8	9	10	11
5	1							
6	1	1						
7	2		1					
8	6	3	1	1				
9	16		4		1			
10	59	60	21	5	1	1		
11	265		266		6		1	
12	1544	7848	7849	1547	94	9	1	1
13	10778		367860		10786		10	
14	88168	3459383	?	?	3459386	88193	540	13
15	805491		?		?		805579	
16	8037418	?	?	?	?	?	?	8037796

3.14 Table Numbers of connected k-regular bipartite graphs on up to 20 vertices.

$n\backslash k$	2	3	4	5	6	7	8	9	10
6	1	1							
8	1	1	1						
10	1	2	1	1					
12	1	5	4	1	1				
14	1	13	14	4	1	1			
16	1	38	129	41	7	1	1		
18	1	149	1980	1981	157	8	1	1	
20	1	703	62611	304495	62616	725	12	1	1

3.3 Transitive Graphs

A graph G is *vertex-transitive* if for any two vertices v and w there is an automorphism $\sigma \in \mathrm{Aut}(G)$ such that $\sigma(v) = w$. (Equivalently, $\mathrm{Aut}(G)$ is transitive on $V(G)$.) A vertex-transitive graph is necessarily regular.

A graph G is a *Cayley graph* if there is a subgroup $A \leq \mathrm{Aut}(G)$ such that for any two vertices v and w there is *exactly one* automorphism $\sigma \in A$ such that $\sigma(v) = w$. (Equivalently, A acts regularly on $V(G)$.)

3.15 Table Numbers of vertex-transitive graphs on up to 26 vertices tabulated by degree k [6]. Totals include the degrees not tabulated.

$n\backslash k$	0	1	2	3	4	5	6	7	8	9	10	11	12	Total
1	1													1
2	1	1												2
3	1		1											2
4	1	1	1	1										4
5	1		1		1									3
6	1	1	2	2	1	1								8
7	1		1		1		1							4
8	1	1	2	3	3	2	1	1						14
9	1		2		3		2		1					9
10	1	1	2	3	4	4	3	2	1	1				22
11	1		1		2		2		1		1			8
12	1	1	4	7	11	13	13	11	7	4	1	1		74
13	1		1		3		4		3		1		1	14
14	1	1	2	3	6	6	9	9	6	6	3	2	1	56
15	1		3		8		12		12		8		3	48
16	1	1	3	7	16	27	40	48	48	40	27	16	7	286
17	1		1		4		7		10		7		4	36
18	1	1	4	7	16	24	38	45	54	54	45	38	24	380
19	1		1		4		10		14		14		10	60
20	1	1	4	11	28	47	83	115	149	168	168	149	115	1214
21	1		3		11		29		48		56		48	240
22	1	1	2	3	11	18	38	52	79	94	109	109	94	816
23	1		1		5		15		30		42		42	188
24	1	1	6	20	74	167	373	652	1064	1473	1858	2064	2064	15506
25	1		2		9		25		57		86		104	464
26	1	1	2	5	16	29	71	117	204	286	397	466	523	4236

3.16 Remark The vast majority of the small vertex-transitive graphs are Cayley graphs, but it is not known if this situation continues for higher orders.

3.17 Table Numbers of vertex-transitive graphs which are not Cayley graphs.

$n\backslash k$	3	4	5	6	7	8	9	10	11	12	Total
10	1			1							2
15		1		1		1		1			4
16		1	1	1	1	1	1	1	1		8
18			1			1	1			1	4
20	3	4	4	7	8	7	8	8	7	8	82
24			1	5	7	9	11	11	12	12	112
26	1		2	3	7	6	8	13	14	12	132

3.4 Multigraphs and Digraphs

A *multigraph* G is a pair $(V(G), E(G))$ where $V(G)$ is a finite set, the *vertices* of G, and $E(G)$ is a multiset of unordered pairs of distinct vertices (*edges*).

A *directed graph* or *digraph* D is a pair $(V(D), A(D))$ where $V(D)$ is a finite set, the *vertices* of D, and $A(D)$ is a set of ordered pairs of distinct vertices (*arcs*).

3.18 Remark Both multigraphs with a given number of edges and digraphs with a given number of arcs can be counted by applications of Polya's enumeration theory.

3.19 Table Numbers of multigraphs on up to eight vertices, tabulated by number of edges e.

e	$n=3$	$n=4$	$n=5$	$n=6$	$n=7$	$n=8$
0	1	1	1	1	1	1
1	1	1	1	1	1	1
2	2	3	3	3	3	3
3	3	6	7	8	8	8
4	4	11	17	21	22	23
5	5	18	35	52	60	64
6	7	32	76	132	173	197
7	8	48	149	313	471	588
8	10	75	291	741	1303	1806
9	12	111	539	1684	3510	5509
10	14	160	974	3711	9234	16677
11	16	224	1691	7895	23574	49505
12	19	313	2874	16310	58464	143761
13	21	420	4730	32604	140340	406091
14	24	562	7620	63363	326792	1114890
15	27	738	11986	119745	738090	2970964
16	30	956	18485	220546	1619321	7685972
17	33	1221	27944	396428	3455129	19311709
18	37	1550	41550	696750	7180856	47170674
19	40	1936	60744	1198812	14555856	112123118

3.20 Table Connected cubic multigraphs on up to eight vertices (excluding the cubic graphs in Table 3.12).

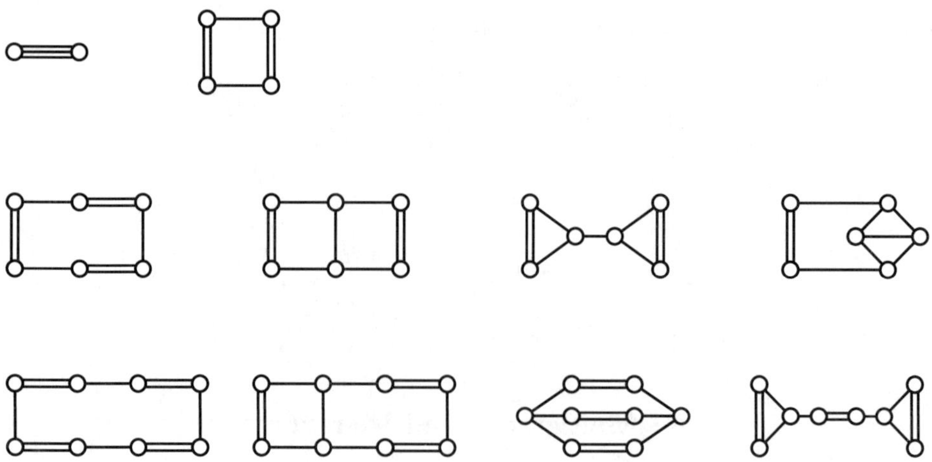

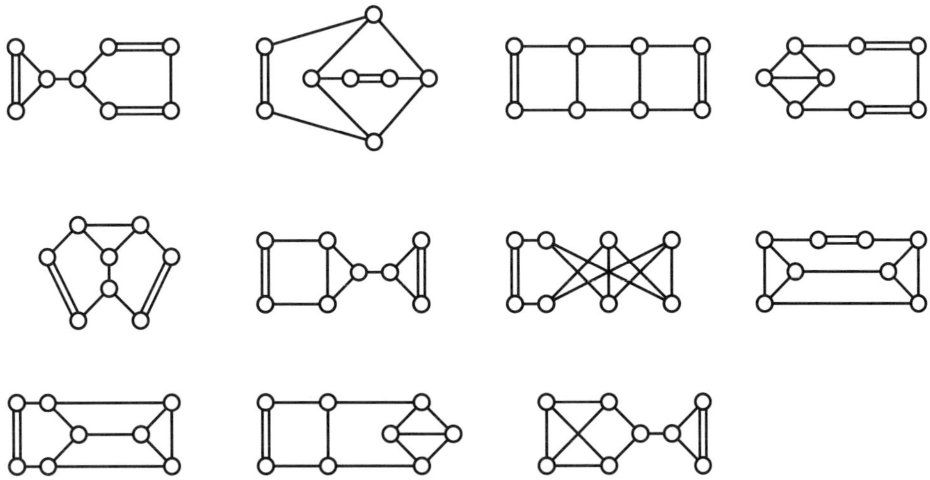

3.21 **Table** The number N_c of connected cubic multigraphs on up to 16 vertices (including the cubic graphs).

n	2	4	6	8	10	12	14	16
N_c	1	2	6	20	91	509	3608	31856

3.22 **Table** Numbers of digraphs on up to 10 vertices.

n	Number
1	1
2	3
3	16
4	218
5	9608
6	1540944
7	882033440
8	1793359192848
9	13027956824399552
10	341260431952972580352

3.5 Some Graphs from Designs

3.23 **Remarks** Graphs are constructed from designs in a variety of different ways. In particular many examples of strongly regular or distance-regular graphs come from designs.

3.24 **Example** The point/block incidence graph of a design often leads to interesting graphs, as already seen with Heawood's graph. The point/block incidence graph of the unique 2-(11,5,2) design (the biplane of order 3, Example I.1.19) is a bipartite distance-regular graph of diameter 3. In fact, symmetric designs are equivalent to bipartite distance-regular graphs of diameter 3.

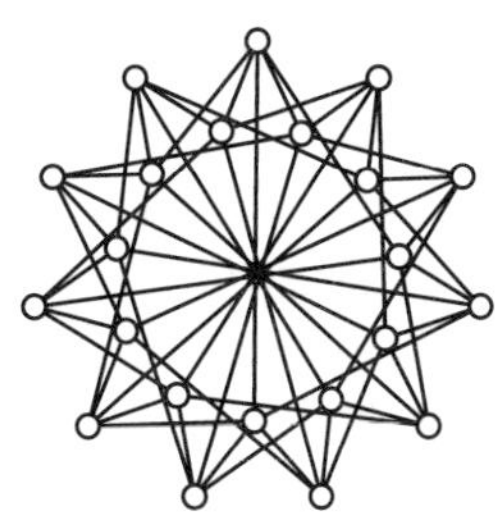

3.25 **Example** There are several ways to define a graph on the blocks of a design. The following graph is defined on the blocks of the unique 2-(6,3,2) design (Example I.1.11)

where two vertices are adjacent if the corresponding blocks meet in exactly two points. This graph is a strongly regular graph, *Petersen's graph*; two of the many possible drawings are given.

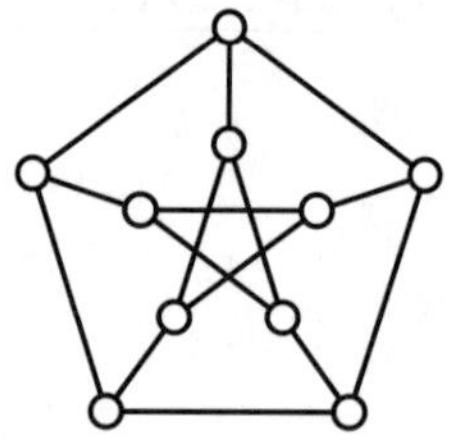

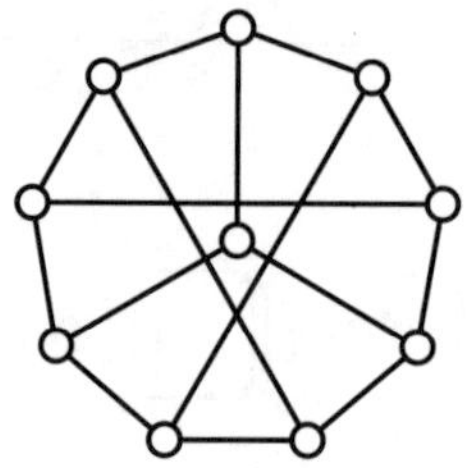

3.26 Remark This construction always results in a strongly regular graph if the design is quasi-symmetric.

3.27 Example A (16,5,0,2) strongly regular graph, the *Clebsch graph.* If A is the adjacency matrix of this graph, then $A + I$ is a point/block incidence matrix for a 2-(16,6,2) design, namely the biplane of order 4 with a doubly transitive automorphism group. (See Example I.1.25).

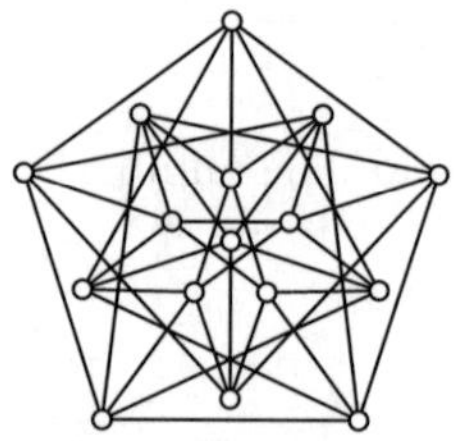

3.28 Remark If a symmetric design has a polarity, then its point/block incidence matrix A can be chosen to be symmetric. If the polarity has the property that no point lies on its polar block, then the diagonal of the matrix is 0 and the matrix is the adjacency matrix of a strongly regular graph. If, as in Example 3.27, every point lies on its polar block then $A - I$ is the adjacency matrix of a strongly regular graph.

3.29 Remark The 2-(16,6,2) design mentioned in Example 3.27 has two further polarities with no absolute points; the corresponding graphs are (16,6,2,2) strongly regular graphs: the *Shrikhande graph* and the lattice graph $L_2(4)$.

3.6 See Also

[5]	A reference for many enumeration techniques related to graphs and other combinatorial structures.
[7]	Very comprehensive listings of data relating to graphs, including thousands of pictures of graphs and hundreds of tables.
[1, 4]	The standard references in graph theory.

References

[1] J.A. Bondy and U.S.R. Murty, *Graph Theory with Applications*, North-Holland, Amsterdam, 1976.

[2] P.J. Cameron and J.H. van Lint, *Designs, Graphs, Codes and their Links*, Cambridge University Press, London, 1991.

[3] G. Brinkmann, Fast generation of cubic graphs, *J. Graph Theory*, to appear.

[4] F. Harary, *Graph Theory*, Addison-Wesley, Reading MA, 1969.

[5] F. Harary and E. Palmer, *Graphical Enumeration*, Academic Press, New York, 1973.

[6] B.D. McKay and G.F. Royle, The transitive graphs with at most 26 vertices, *Ars Combin.* 30 (1990), 161-176.
[7] R.C. Read and R.J Wilson, *An Atlas of Graphs*, Oxford University Press, to appear.
[8] R.W. Robinson and N.C. Wormald, Numbers of cubic graphs, *J. Graph Theory* 7 (1983), 463-467.

4 Factorizations of Graphs

LARS D. ANDERSEN

4.1 Definitions and Examples

A *factor* of a multigraph G is a submultigraph of G including all vertices of G (a *spanning* submultigraph). For a nonnegative integer r, an *r-factor* is a factor which is regular of degree r.

A *factorization* of a multigraph G is a set $\mathcal{F} = \{F_1, \ldots, F_k\}$ of edge-disjoint factors of G whose edge-sets partition the edge-set of G. A factorization into r-factors for some nonnegative integer r is an *r-factorization.*

4.1 **Remark** As the factors in a factorization of a multigraph are permitted to have isolated vertices, almost any graph decomposition problem is a factorization problem. The focus here is on r-factorizations. Only regular multigraphs can have r-factorizations.

Two factorizations $\mathcal{F} = \{F_1, \ldots, F_k\}$ and $\mathcal{G} = \{G_1, \ldots, G_\ell\}$ of a multigraph are *orthogonal* if each factor of $\mathcal{F}$ has at most one edge in common with each factor of $\mathcal{G}$.

4.2 **Remark** In [2] two factorizations satisfying the definition above are termed *suborthogonal.* If each factor of $\mathcal{F}$ has *exactly* one edge in common with each factor of $\mathcal{G}$, then $\mathcal{F}$ and $\mathcal{G}$ are *orthogonal.* This notation, although quite useful, is not adopted here.

4.3 **Remark** Two r-factorizations of a (simple) graph G of order n cannot be orthogonal unless $r = 1$, and then only if G is regular of degree at least $\frac{n}{2}$.

4.4 **Example** A 2-factorization and three pairwise orthogonal 1-factorizations of $K_{4,4}$. The vertices $\{1, 2, 3, 4\}$ are completely joined to $\{A, B, C, D\}$. Each factorization is a line, factors are separated by semicolons, and only edge-sets are shown.

1A A2 2B B1 3D D4 4C C3; 1C C2 2D D1 3A A4 4B B3

1A 2B 3C 4D; 1B 2A 3D 4C; 1C 2D 3A 4B; 1D 2C 3B 4A
1A 2D 3B 4C; 1B 2C 3A 4D; 1C 2B 3D 4A; 1D 2A 3C 4B
1A 2C 3D 4B; 1B 2D 3C 4A; 1C 2A 3B 4D; 1D 2B 3A 4C

4.2 Existence Results for Factorizations

4.5 **Remarks** Deciding whether a cubic graph has a 1-factorization is NP-complete [18]. It follows that the 1-factorization problem is NP-complete in general, and that the problem of deciding whether a graph has two disjoint 1-factors is also NP-complete. See §VI.9 for a discussion of NP-completeness in design theory; see [15] for a general discussion. Deciding whether a graph has a 1-factor is polynomial, and there is a good characterization of graphs having a 1-factor [4].

4.6 Theorem The following multigraphs have 1-factorizations; therefore, they have r-factorizations for any r dividing their degree:

1. Regular bipartite multigraphs — in particular $K_{n,n}$ and the hypercube Q^n.
2. Regular graphs of order $2n$ and degree at least $(\sqrt{7}-1)n$ — in particular, K_{2n}.
3. Planar cubic multigraphs without bridges [3].

4.7 Remark Many other classes of multigraphs are known to have 1-factorizations. Petersen's graph is a notable example of a cubic graph *not* having a 1-factorization. Cubic graphs which do not have a 1-factorization are *snarks* (see [7]).

4.8 Conjecture Any regular graph of order $2n$ and degree at least n has a 1-factorization.

4.9 Theorem A multigraph has a 2-factorization if and only if it is regular of even degree.

4.10 Remark It follows from Theorem 4.9 that a regular multigraph has r-factorizations for all even r dividing the degree. Theorem 4.11, about graphs only, allows r to be odd.

4.11 Theorem [16] Let G be a graph of order $2n$ which is regular of degree at least n, and let $r \geq 2$. Then G has an r-factorization if and only if r divides the degree of G.

4.3 Equivalent Combinatorial Objects

4.12 Theorem The existence of the following are equivalent:

1. A 1-factorization of K_n.
2. A symmetric latin square of order $n-1$.
3. A unipotent (constant diagonal) symmetric latin square of order n.
4. A resolution of an $(n, 2, 1)$-BIBD.
5. A round robin tournament with n players played in $n-1$ rounds (see §V.7.1).

4.13 Theorem The existence of the following are equivalent:

1. Two orthogonal 1-factorizations of K_n.
2. Two orthogonal-symmetric latin squares of order $n-1$ (see [14]).
3. Two orthogonal resolutions of an $(n, 2, 1)$-BIBD.
4. A Room square of side $n-1$ (see §IV.39).
5. A Room frame of type 1^{n-1} (see §IV.39).
6. A Howell design $H(n-1, n)$ (see §IV.26).

4.14 Theorem For any integer $d \geq 2$, the existence of the following are equivalent:

1. d pairwise orthogonal 1-factorizations of K_n.
2. d pairwise orthogonal-symmetric latin squares of order $n-1$.
3. d pairwise orthogonal resolutions of an $(n, 2, 1)$-BIBD.
4. A Room d-cube of side $n-1$ (see §IV.39).

4.15 Theorem For any positive integer d, the existence of the following are equivalent:

1. d pairwise orthogonal 1-factorizations of $K_{n,n}$.
2. d pairwise orthogonal latin squares of order n.
3. A transversal design $\mathrm{TD}(d+2, n)$ (see §II.2).
4. An orthogonal array $\mathrm{OA}(d+2, n)$ (see §II.4).

4.16 Remark For $d = 1$, Theorem 4.15 states that the existence of a 1-factorization of $K_{n,n}$ is equivalent the existence of a latin square of order n (see Theorem II.1.5 for other equivalences).

4.17 Remark 1-factorizations of K_{2n} and $K_{n,n}$ are simple models of round robin tournaments and tournaments where each player from one team plays each player from another team, respectively (see §V.7).

4.4 Enumeration and Tables

LF($2n$) is the number of 1-factorizations of a labeled K_{2n}.

Two 1-factorizations $F = \{f_1, f_2, \ldots, f_k\}$ and $H = \{h_1, h_2, \ldots, h_k\}$ of G are *isomorphic* if there exists a map ϕ from the vertex-set of G onto itself such that $\{f_1\phi, f_2\phi, \ldots, f_k\phi\} = \{h_1, h_2, \ldots, h_k\}$. Here, $f_i\phi$ is the set of all the edges $\{x\phi, y\phi\}$ where $\{x, y\}$ is an edge in f_i.

NF($2n$) is the number of pairwise nonisomorphic 1-factorizations of K_{2n}.

4.18 Table The known values for the number of distinct and the number of nonisomorphic 1-factorizations of K_{2n} [13, 26].

$2n$	2	4	6	8	10	12
LF($2n$)	1	1	6	6,240	1,225,566,720	252,282,619,805,368,320
NF($2n$)	1	1	1	6	396	526,915,620

4.19 Remark A discussion of the computer algorithm used to obtain the results on K_{12} in Table 4.18 can be found in Example VI.9.7.

4.20 Theorem

1. $\text{NF}(2n) \leq ((2n-1)!)^{n-1}$;
2. $\lim_{n\to\infty} \log(\text{NF}(2n))/(2n^2 \log(2n)) = 1$ [6].

4.21 Table The six nonisomorphic factorizations of K_8. Each line is a factorization, with the number at the end. Each block of digits is a 1-factor, with 0 omitted, so that, for example, 1234567 denotes the 1-factor with edges 01, 23, 45, and 67.

```
1234567 2134657 3124756 4152637 5142736 6172435 7162534  1
1234567 2134657 3124756 4152637 5142736 6172534 7162435  2
1234567 2134657 3124756 4162537 5172634 6142735 7152436  3
1234567 2134657 3124756 4162735 5172634 6142537 7152436  4
1234567 2134657 3142756 4162537 5172634 6123547 7152436  5
1234567 2143657 3162547 4172635 5123746 6152734 7132456  6
```

4.22 Table The 396 nonisomorphic factorizations of K_{10}. Notation as in Table 4.21.

```
123456789 213465879 312495768 415263978 514283769 617243859 716293548 819253647 918273456  1
123456789 213465879 312495768 415263978 514283769 617243859 716293548 819273456 918253647  2
123456789 213465879 312475968 415263978 514283769 617253849 716293548 819243657 918273456  3
123456789 213465879 312475968 415263978 514283769 617253849 716293548 819273456 918243657  4
123456789 213465879 312495768 415263978 514283769 617293548 716243859 819273456 918253647  5
123456789 213465879 312475968 415263978 514283769 619273548 716253849 817293456 918243657  6
123456789 213465879 312475968 415263978 514283769 617253849 718293456 819243657 916273548  7
123456789 213465879 312495768 415263978 516293847 619253748 714283659 817243569 918273456  8
123456789 213465879 312495768 415263978 516293748 617243859 714283569 819273456 918253647  9
123456789 213465879 312475968 416283957 518263749 617293548 715243869 819273456 914253678 10
123456789 213465879 312495768 416273859 518293647 614253978 719283456 815243769 917263548 11
123456789 213465879 312495768 415283769 514263978 618293547 719243856 816273459 917253648 12
123456789 213465879 312495768 415283769 516293478 614273859 719263548 817243956 918253647 13
123456789 213465879 312475968 415263978 516283749 614293857 719253648 817243569 918273456 14
123456789 213465879 312475968 415263978 516283749 617293548 714253869 819243657 918273456 15
```

123456789	213465879	312475968	415263978	516283749	619273548	714293856	817253469	918243657	16
123456789	213465879	312475968	415263978	518273469	617293548	716253849	819243756	914283657	17
123456789	213465879	312495768	415263978	516293847	614283759	719253648	817243569	918273456	18
123456789	213465879	312495768	415263978	514283769	617293548	718243659	819273456	916253847	19
123456789	213465879	312475968	415263978	516293748	617253849	714283569	819243657	918273456	20
123456789	213465879	312475968	415263978	514283769	618293457	719253648	816273549	917243856	21
123456789	214365879	315264978	412375968	513274869	618293457	719243856	817253946	916283547	22
123456789	213465879	312475968	415263978	514283769	619273548	718253649	817293456	916243857	23
123456789	213465879	312475968	416253978	518243769	617293548	715263849	819273456	914283657	24
123456789	214365879	315264978	412375968	513284769	619243857	718293456	817253946	916273548	25
123456789	213465879	312475968	415263978	514273869	617283549	718293456	819243657	916253748	26
123456789	213465879	312475968	415263978	518273649	617293548	714253869	819243756	916283457	27
123456789	213465879	314256978	412395768	518273649	615293847	716283459	819243756	917263548	28
123456789	213465879	312475968	415263978	516283749	617293548	714253869	819273456	918243657	29
123456789	213465879	312475968	416253978	518273469	617283549	714293856	819243657	915263748	30
123456789	213465879	314256978	412395768	516293847	617283459	718263549	819243756	915273648	31
123456789	213465879	314265978	412385769	516283947	619253748	718293456	815273649	917243568	32
123456789	214365879	315264978	412375968	513274869	618243957	719253846	816293547	917283456	33
123456789	213465879	312495768	415263978	517243869	618293547	719283456	814273659	916253748	34
123456789	213465879	312475968	416253978	518273469	617293548	715263849	819243756	914283657	35
123456789	213465879	312495768	416253978	517283469	618243759	719263548	815293647	914273856	36
123456789	213465879	312495768	415263978	516293748	619283547	718253469	817243659	914273856	37
123456789	213465879	312495768	415263978	517283469	614273859	719253648	816293547	918243756	38
123456789	213465879	314256978	412375968	516283947	619243857	718293456	815273649	917263548	39
123456789	213465879	312495768	415263978	516293847	614283759	718243569	819273456	917253648	40
123456789	213465879	312475968	415263978	516283749	614293857	718243569	819273456	917253648	41
123456789	213465879	312475968	415263978	516293748	617283549	718253469	819243657	914273856	42
123456789	214365879	315264978	412375968	513284769	618293457	719243856	817253946	916273548	43
123456789	213465879	312495768	415263978	517243869	614283759	716293548	819273456	918253647	44
123456789	213465879	314256978	412375968	516283947	617293548	719243856	815273649	918263457	45
123456789	213465879	314256978	415273968	516283749	617293548	718263459	819243657	912384756	46
123456789	213465879	312475968	415263978	518273469	617283549	714293856	819243657	916253748	47
123456789	213465879	312475968	415263978	518243769	617293548	716253849	819273456	914283657	48
123456789	214365879	315264978	412375968	513274869	618243957	719283456	817293546	916253847	49
123456789	213465879	314256978	412375968	516293847	618243957	719283456	817263549	915273648	50
123456789	213465879	312475968	416253978	518243769	617293548	715283649	819273456	914263857	51
123456789	213465879	312475968	415263978	514283769	618273549	719253648	817293456	916243857	52
123456789	213465879	314265978	415273968	516293748	617283549	719243856	812345769	918253647	53
123456789	213465879	314256978	415273968	516293748	619283457	712384956	817243659	918263547	54
123456789	213465879	312475968	416293857	519273648	615243978	718263549	814253769	917283456	55
123456789	213465879	314256978	412375968	516283947	619273548	715263849	817293456	918243657	56
123456789	213465879	314256978	412375968	517263948	615293847	719283456	816273549	918243657	57
123456789	213465879	312475968	415263978	514283769	618293457	719243856	816273549	917253648	58
123456789	213465879	314256978	412375968	518273649	615293847	719263548	817243956	916283457	59
123456789	213465879	314256978	412395768	516283749	617243859	719263548	815293647	918273456	60
123456789	213465879	312475968	415263978	519273648	614293857	716283549	817253469	918243756	61
123456789	213465879	312475968	415263978	516293748	617253849	718243569	819273456	914283657	62
123456789	213465879	312475968	415263978	516273849	617293548	718253469	819243756	914283657	63
123456789	213465879	312475968	415263978	517283649	614293857	718243569	819273456	916253748	64
123456789	214365879	315264978	412395768	517283469	619253847	718293546	816243759	913274856	65
123456789	214365879	315264978	412395768	518243769	619283547	713294856	816273459	917253846	66
123456789	213465879	312475968	415263978	519273648	617253849	714283569	816293457	918243756	67
123456789	213465879	314256978	415273968	516293748	617283549	718243659	819263457	912384756	68
123456789	213465879	312475968	415263978	516273849	617293548	718253469	819243657	914283756	69
123456789	213465879	314256978	415273968	516283749	617293548	718243659	819263457	912384756	70
123456789	213465879	312495768	415263978	516293748	619283547	718243659	817253469	914273856	71
123456789	214365879	315264978	412385769	519243768	618273459	713294856	816253947	917283546	72
123456789	213465879	312495768	415263978	517293648	619283547	714253869	816273459	918243756	73
123456789	213465879	314256978	412375968	516283947	618293457	719243856	815273649	917263548	74
123456789	213465879	314256978	412375968	518273649	615293847	719263548	816243957	917283456	75
123456789	213465879	314256978	412375968	516273849	615283947	718293456	819243657	917263548	76
123456789	213465879	314256978	417283956	518263749	619243857	715293468	812364759	916273548	77
123456789	213465879	314256978	412395768	516293847	618243759	719283456	815273649	917263548	78
123456789	214365879	315264978	412375968	513284769	618243957	719253846	817293456	916273548	79
123456789	213465879	314256978	412375968	518263947	619243857	716293548	815273649	917283456	80
123456789	213465879	312475968	415263978	516273849	618293457	719253648	817243569	914283756	81
123456789	213465879	314256978	412395768	518263749	619273548	716243859	817293456	915283647	82
123456789	213465879	314256978	412395768	518293647	617243859	719283456	816273549	915263748	83
123456789	213465879	312495768	415283769	516243978	614273859	718293456	819263547	917253648	84
123456789	213465879	314256978	412395768	516293847	618243759	719263548	815273649	917283456	85
123456789	213465879	314256978	412395768	519283647	617243859	718293456	816273549	915263748	86
123456789	213465879	312475968	415263978	516283749	619273548	714253869	817293456	918243657	87
123456789	213465879	312475968	416253978	518273469	617293548	715283649	819243756	914263857	88
123456789	213465879	314256978	412395768	519273648	618243759	716283549	817293456	915263847	89
123456789	213465879	314256978	416273859	518293647	617283549	715243968	819263457	912374856	90
123456789	214365879	315264978	412395768	513284769	618243759	719253846	817293456	916273548	91
123456789	213465879	314256978	415273968	519263748	612384759	716283549	817293456	918243657	92
123456789	213465879	314256978	415273968	516293847	617283459	719263548	812364957	918243756	93
123456789	214365879	315264978	412395768	517243869	619253748	718293456	813274659	916283547	94
123456789	213465879	312475968	416253978	518243769	617283549	714293856	819263457	915273648	95
123456789	213465879	312475968	415263978	519273648	617283549	714253869	816293457	918243756	96
123456789	213465879	314256978	415273968	516293847	617283459	719263548	812374956	918243657	97
123456789	213465879	312475968	415263978	514283769	617293548	718253649	819273456	916243857	98
123456789	213465879	312495768	415263978	516293847	617283459	719253648	814273569	918243756	99
123456789	213465879	312475968	415263978	516293748	617283549	714253869	819273456	918243657	100
123456789	213465879	314256978	415273968	516293748	617283549	718263459	819243657	912384756	101
123456789	213465879	314295678	417253869	518273649	619283547	715263948	816243759	912345768	102
123456789	214365879	315284769	419253768	516243978	618293457	712384659	817263549	913274856	103

123456789	214365879	315284769	417293568	519263478	618243957	712384956	813274659	916253748	104
123456789	213465879	314256978	412395768	519263748	618273549	716283459	815293647	917243856	105
123456789	213465879	312475968	415263978	518243769	617283549	719253648	816293457	914273856	106
123456789	213465879	314256978	412375968	516283947	618293457	719243856	817263549	915273648	107
123456789	213465879	314256978	412395768	518263749	617243859	719283456	815293647	916273548	108
123456789	214365879	315284769	417293568	518273946	612345978	719243856	816253749	913264857	109
123456789	214365879	315264978	412395768	518273469	619283547	713294856	816243759	917253846	110
123456789	214365879	315284769	416293578	519243768	612384957	718263459	817253946	913274856	111
123456789	214365879	315274968	412385769	516293748	619283547	718243956	817263459	913254678	112
123456789	213465879	312475968	416253978	518273469	617293548	719243856	815263749	914283657	113
123456789	213465879	314256978	415293768	518263947	617283459	719243856	816273549	912364857	114
123456789	213465879	312475968	416283957	518273649	615293478	719263548	814253769	917243856	115
123456789	213465879	314256978	412395768	517283649	618243759	716293548	819273456	915263847	116
123456789	213465879	312475968	416283957	518273649	615293478	719243856	814253769	917263548	117
123456789	213465879	314275968	415283769	516243978	617253849	718293456	819263547	912364857	118
123456789	213465879	314256978	412395768	518273649	617293548	719283456	816243759	915263847	119
123456789	213465879	314265978	412385769	517243968	615293748	718253649	819273456	916283547	120
123456789	214365879	315264978	412375968	513284769	619273548	718293456	816243957	917253846	121
123456789	213465879	312475968	415263978	519273648	618293457	714283569	816253749	917243856	122
123456789	213465879	312475968	415263978	518273469	617283549	719243856	814293657	916253748	123
123456789	214365879	315264978	412395768	517283469	618243759	719253846	816293547	913274856	124
123456789	213465879	314256978	415273968	517293648	618243759	712384956	819263547	916283457	125
123456789	214365879	315284769	418293756	516273948	612345978	719253846	817263549	913245768	126
123456789	214365879	315284769	418293756	516273948	619243857	712354968	817263459	913254678	127
123456789	214365879	315264978	412395768	517283469	618293547	719253846	816243759	913274856	128
123456789	213465879	314256978	412395768	518293647	617283459	719243856	816273549	915263748	129
123456789	213465879	312475968	415263978	518243769	619273548	716253849	817293456	914283657	130
123456789	213465879	314256978	416273859	517293648	615283947	718263549	819243756	912345768	131
123456789	213465879	312475968	416293857	518273649	615243978	719283456	814253769	917263548	132
123456789	214365879	315284769	417293568	518273946	612345978	719243856	813264957	916253748	133
123456789	214365879	315264978	412395768	518273469	619253847	713294856	816243759	917283546	134
123456789	213465879	312495768	415283769	514263978	618273459	719243856	816293547	917253648	135
123456789	213465879	312495768	415283769	516243978	614273859	719263548	817293456	918253647	136
123456789	213465879	312475968	416253978	518273469	617293548	715263849	819243657	914283756	137
123456789	213465879	314256978	412395768	518263749	619273548	716243859	815293647	917283456	138
123456789	213465879	314256978	412395768	517263849	618243759	716293548	819273456	915283647	139
123456789	213465879	312475968	416253978	518243769	617283549	715293648	819273456	914263857	140
123456789	213465879	314256978	412395768	518293647	617283549	719243856	816273459	915263748	141
123456789	213465879	314256978	412395768	518273649	615293847	719283456	816243759	917263548	142
123456789	213465879	314256978	412375968	518273649	615283947	719263548	816293457	917243856	143
123456789	213465879	314256978	412395768	519263847	617293548	715283649	816273459	918243756	144
123456789	213465879	314256978	412395768	518273649	615293847	719263548	816243759	917283456	145
123456789	213465879	314256978	415293768	518263947	617243859	719283456	812364957	916273548	146
123456789	213465879	314265978	415273968	512374869	617283549	718293456	819243657	916253847	147
123456789	213465879	312495768	415283769	516243978	619273548	714263859	817293456	918253647	148
123456789	213465879	314256978	412395768	519283647	615293748	716243859	817263549	918273456	149
123456789	214365879	315274968	412385769	513294678	618243759	719263548	816253947	917283456	150
123456789	213465879	314256978	415273968	516293847	612374859	718263549	819243657	917283456	151
123456789	213465879	314256978	412395768	516293847	617283459	719263548	815273649	918243756	152
123456789	214365879	315264978	412395768	518243769	619253847	713294856	816273459	917283546	153
123456789	213465879	314256978	415273968	516293847	619283457	712364859	817263549	918243756	154
123456789	214365879	315284769	418293756	517263948	619243857	712354968	816273459	913254678	155
123456789	214365879	315264978	412375968	513284769	619273548	718253946	816293457	917243856	156
123456789	214365879	315264978	412385769	519243768	613284759	718293456	817253946	916273548	157
123456789	213465879	314256978	412395768	517283649	615293847	718263459	819243756	916273548	158
123456789	213465879	314265978	412395768	519273648	615293847	718243569	816253749	917283456	159
123456789	214365879	315264978	412385769	517243968	619283547	718293456	813274659	916253748	160
123456789	213465879	314265978	415273869	512374968	617293548	718243956	819253647	916283457	161
123456789	213465879	312495768	416253978	517243869	618273459	719263548	814293756	915283647	162
123456789	213465879	314256978	412375968	516273948	619243857	718293456	817263549	915283647	163
123456789	214365879	315264978	413295768	517243869	618273459	719283546	816253947	912374856	164
123456789	214365879	315294768	416273859	518243769	619283457	712394856	817263549	913254678	165
123456789	214365879	315264978	412395768	518243769	613284759	719253846	817293456	916273548	166
123456789	214365879	315264978	412375968	513284769	619243857	718253946	817293456	916273548	167
123456789	213465879	314256978	415273968	516293748	619283547	712384956	817243659	918263457	168
123456789	214365879	315264978	412385769	519243768	613284759	718253946	817293456	916273548	169
123456789	213465879	314256978	412375968	516283947	615273849	719263548	817293456	918243657	170
123456789	213465879	314256978	415273968	516293748	619283547	712384956	817263459	918243657	171
123456789	214365879	315264978	412375968	513274869	618243957	719283456	816293547	917253846	172
123456789	214365879	315264978	412375968	513284769	618293457	719253846	817243956	916273548	173
123456789	213465879	314256978	412395768	519283647	618273549	716243859	817293456	915263748	174
123456789	213465879	314256978	417293856	516283749	612394857	718243659	819263547	915273468	175
123456789	213465879	314256978	412375968	518273649	615283947	719263548	817293456	916243857	176
123456789	213465879	314265978	415273968	512374869	617283549	719243856	816293457	918253647	177
123456789	213465879	314256978	412395768	519263847	617283549	715293648	816273459	918243756	178
123456789	213465879	314256978	412395768	518263749	617293548	719243856	816273459	915283647	179
123456789	213465879	314256978	412395768	516293847	618243759	715283649	819273456	917263548	180
123456789	213465879	312475968	416253978	518273469	617283549	719243856	814293657	915263748	181
123456789	214365879	315294768	416273859	518243769	612394857	719283456	817263549	913254678	182
123456789	213465879	314256978	412375968	518263947	619283457	716293548	815273649	917243856	183
123456789	213465879	314256978	412395768	519273648	617283549	718293456	816243759	915263847	184
123456789	213465879	312475968	416253978	517283649	614293857	719263548	815273469	918243756	185
123456789	213465879	314256978	415273968	516293748	612384759	718263549	819243657	917283456	186
123456789	213465879	314256978	415273968	516293748	612384957	719283456	817243659	918263547	187
123456789	213465879	314256978	415293768	516273948	619283457	718263549	817243659	912384756	188
123456789	213465879	314275968	416293578	519263748	615283947	718253469	812364957	917243856	189
123456789	213465879	314265978	415293768	516273948	619283457	718253649	812354769	917243856	190
123456789	213465879	314256978	412395768	516293847	617283549	718243659	819273456	915263748	191

123456789	213465879	314256978	415293768	516273948	617283549	718243659	819263457	912384756	192
123456789	214365879	315264978	412395768	518243769	619253847	713284659	817293456	916273548	193
123456789	213465879	314256978	415273968	518293647	612374859	716283549	819263457	917243856	194
123456789	214365879	315264978	412395768	517283469	618293547	719243856	813274659	916253748	195
123456789	214365879	315294768	418253769	517263849	612394857	719283546	816273459	913245678	196
123456789	214365879	315274968	412385769	516293748	619283547	718263459	817253946	913245678	197
123456789	214365879	315264978	412375968	516273948	618293547	719253846	813245769	917283456	198
123456789	214365879	315274968	416253978	518293746	613284759	719263548	812345769	917243856	199
123456789	214365879	315284769	416253978	519243768	618293457	712384956	813274659	917263548	200
123456789	214365879	315264978	417253869	519283746	618243957	713294856	816273459	912354768	201
123456789	213465879	314265978	418293657	512384769	615283749	719253468	817243956	916273548	202
123456789	213465879	314265978	415293768	512384769	619283457	718253649	817243956	916273548	203
123456789	213465879	314265978	415293768	512384769	619273548	718243956	817253649	916283457	204
123456789	214365879	315264978	413285769	517243968	619253847	718293456	812374659	916273548	205
123456789	214365879	315294768	416253978	518243769	612384957	719263548	813274659	917283456	206
123456789	213465879	314256978	417283956	516293748	612384957	718243659	819263547	915273468	207
123456789	214365879	315264978	412385769	517293468	619283547	718243956	813274659	916253748	208
123456789	214365879	315264978	412385769	517243968	619253748	718293456	813274659	916283547	209
123456789	214365879	315264978	412375968	516273948	618293457	719243856	813254769	917283546	210
123456789	214365879	315264978	412385769	519243768	618273459	713294856	817253946	916283547	211
123456789	213465879	314256978	412395768	516293847	615283749	718243659	819273456	917263548	212
123456789	213465879	314256978	412395768	516293748	617243859	718263549	819273456	915283647	213
123456789	214365879	315274968	416253978	517283469	619243857	718293546	813264759	912374856	214
123456789	213465879	314256978	412395768	519283647	615273849	716293548	817263459	918243756	215
123456789	213465879	314256978	412375968	516273948	615293847	719283456	817263549	918243657	216
123456789	213465879	314256978	415273968	516293847	619283457	718243659	812374956	917263548	217
123456789	214365879	315264978	413295768	517283469	619273548	718243956	812374659	916253847	218
123456789	214365879	315284769	417293568	519263478	618243957	712384659	816253749	913274856	219
123456789	214365879	315294768	418253769	517263948	612384957	719283546	816273459	913245678	220
123456789	213465879	314256978	412375968	519263847	617293548	718243956	815273649	916283457	221
123456789	213465879	314256978	415273968	517283649	612384759	719263548	816293457	918243756	222
123456789	213465879	314256978	415293768	516273948	612384759	718263549	819243657	917283456	223
123456789	213465879	314256978	416273859	512374968	617293548	719283456	815263947	918243657	224
123456789	214365879	315274968	413285769	516243978	619253847	718293456	812374659	917263548	225
123456789	213465879	314256978	412375968	516293847	618243957	719263548	815273649	917283456	226
123456789	213465879	314256978	416293857	517243968	618273459	715283649	819263547	912374856	227
123456789	213465879	314256978	417293856	516273948	615283749	718243659	819263547	912345768	228
123456789	213465879	312495768	415283769	516243978	618293547	714263859	819273456	917253648	229
123456789	213465879	314256978	415293768	516273849	619283547	712394856	817243659	918263457	230
123456789	213465879	314256978	415273968	519263748	612384957	718243659	817293456	916283547	231
123456789	213465879	314256978	412395768	518263749	617283459	719243856	815293647	916273548	232
123456789	213465879	314256978	415273968	516293748	619283457	718243659	817263549	912384756	233
123456789	213465879	312495768	415283769	514263978	618273459	719253648	816293547	917243856	234
123456789	214365879	315284769	417293568	519263478	618253749	712384659	816243957	913274856	235
123456789	213465879	314265978	417283569	516273948	618293457	712384956	819253647	915243768	236
123456789	213465879	314275968	416253978	519263748	618293547	715283649	812345769	917243856	237
123456789	213465879	314265978	412395768	518273469	619253748	716283549	815293647	917243856	238
123456789	213465879	314256978	412395768	516283749	619273548	718243659	817293456	915263847	239
123456789	213465879	314265978	415293768	518273469	612394857	719243856	817253649	916283547	240
123456789	213465879	314265978	412385769	518293647	615273948	719243568	816253749	917283456	241
123456789	213465879	314256978	412375968	517263948	618273549	719283456	815293647	916243857	242
123456789	214365879	315284769	419263857	517243968	612345978	718293546	816253749	913274856	243
123456789	213465879	314256978	412375968	517293648	615283947	719243856	816273549	918263457	244
123456789	213465879	314256978	416283957	517293648	618273459	712384956	819263547	915243768	245
123456789	214365879	315264978	412395768	518293746	619283547	713254869	816273459	917243856	246
123456789	214365879	315284769	416253978	519243768	612384957	718293546	817263459	913274856	247
123456789	214365879	315264978	412395768	518273469	613284759	719243856	817293546	916253748	248
123456789	214365879	315264978	412385769	519243768	618253947	713294856	816273459	917283546	249
123456789	214365879	315264978	412385769	519243768	618253947	713284659	817293456	916273548	250
123456789	214365879	315274968	412385769	516293748	619283547	718263459	817243956	913254678	251
123456789	214365879	315264978	412395768	519273846	618293547	713254869	816243759	917283456	252
123456789	214365879	315284769	419263857	516293478	618243759	712354968	817253946	913274856	253
123456789	214365879	315264978	417253968	518293746	612384759	719283456	813245769	916273548	254
123456789	214365879	315284769	418263957	516293478	619253748	712354968	813274659	917243856	255
123456789	214365879	315264978	417253968	519273846	618243759	713294856	812354769	916283457	256
123456789	214365879	315274968	412385769	519263478	618293547	713284659	817243956	916253748	257
123456789	214365879	315264978	413285769	517293468	619273548	718243956	812374659	916253847	258
123456789	214365879	315284769	417293568	519273846	612345978	718243956	816253749	913264857	259
123456789	214365879	315274968	416253978	519283746	618293457	712354869	813264759	917243856	260
123456789	214365879	315284769	418293756	517263948	612345978	719253846	816273549	913245768	261
123456789	214365879	315264978	417253968	518293746	613284759	719243856	812345769	916273548	262
123456789	213465879	314256978	412395768	516293847	619273548	718243659	815263749	917283456	263
123456789	213465879	314256978	416273859	512374968	617293548	718243956	819263457	915283647	264
123456789	213465879	314256978	412375968	516273948	617283549	718293456	819243657	915263847	265
123456789	214365879	315264978	412375968	513284769	619273548	718243956	816293457	917253846	266
123456789	214365879	315264978	412385769	519273468	613284759	718243956	817293546	916253748	267
123456789	214365879	315264978	412395768	519283746	618273459	713254869	816293547	917243856	268
123456789	214365879	315284769	417293568	519273846	612345978	718243956	813264957	916253748	269
123456789	214365879	315264978	412395768	519273846	618243759	713254869	817293456	916283547	270
123456789	213465879	314256978	412395768	519283647	618273459	716293548	815263749	917243856	271
123456789	213465879	314256978	415273968	517293648	618243759	712384956	819263457	916283547	272
123456789	213465879	314256978	415273968	516293847	618243759	719283456	812364957	917263548	273
123456789	214365879	315264978	417253869	518273946	619283547	713294856	816243759	912345768	274
123456789	214365879	315264978	417283569	518273946	619253847	713294856	816243759	912345768	275
123456789	213465879	314275968	412395678	516283749	619243857	718253469	815293647	917263548	276
123456789	213465879	314265978	415293768	517283649	619273548	718243956	812345769	916253847	277
123456789	213465879	314256978	412395768	517293648	618243759	719283456	816273549	915263847	278
123456789	214365879	315264978	413295768	517243869	619283547	718253946	816273459	912374856	279

123456789	214365879	315284769	419253768	516293478	618243957	712384659	817263549	913274856	280
123456789	214365879	315274968	416253978	519283746	612384759	718293456	813245769	917263548	281
123456789	213465879	314265978	412395768	518273649	619283547	715243869	817293456	916253748	282
123456789	213465879	314275968	412385769	518293647	619253748	716283549	817243956	915263478	283
123456789	213465879	314256978	412395768	516273849	618243759	719283456	815293647	917263548	284
123456789	214365879	315264978	413295768	518273469	619253847	712394856	816243759	917283546	285
123456789	213465879	314256978	416273859	512374968	618293457	719263548	817243956	915283647	286
123456789	213465879	314256978	416273859	519243768	618293457	712394856	817263549	915283647	287
123456789	213465879	314256978	416273859	512374968	618243957	719263548	815293647	917283456	288
123456789	213465879	314256978	412375968	517293648	618273549	719283456	815263947	916243857	289
123456789	213465879	312495768	416273859	517293648	618253947	719283456	815243769	914263578	290
123456789	213465879	314295678	415273869	518263947	619283457	712354968	817243659	916253748	291
123456789	214365879	315264978	412375968	518273946	619283547	713254869	816293457	917243856	292
123456789	214365879	315264978	412395768	519283746	618293547	713254869	816273459	917243856	293
123456789	214365879	315284769	418293756	516273948	612354978	719253846	817263459	913245768	294
123456789	214365879	315274968	412385769	516293478	619253748	718243956	813264759	917283546	295
123456789	214365879	315294768	418253769	516273849	612394857	719283546	817263459	913245678	296
123456789	214365879	315264978	413285769	517243968	619253847	718293546	816273459	912374856	297
123456789	214365879	315264978	413295768	519273846	618243759	712354869	816253947	917283456	298
123456789	214365879	315274968	417253869	516293748	618243957	719283546	813264759	912345678	299
123456789	214365879	315264978	412395768	516293748	618273459	719243856	813254769	917283546	300
123456789	214365879	315264978	413285769	517293468	618253947	719243856	812374659	916273548	301
123456789	214365879	315264978	416273859	519283746	618253947	712354869	817293456	913245768	302
123456789	214365879	315264978	416273859	518293746	619283457	712354869	817243956	913254768	303
123456789	214365879	315284769	419263857	517293468	612354978	718253946	816243759	913274856	304
123456789	213465879	314256978	415293768	517283649	612384759	718243956	819263457	916273548	305
123456789	214365879	315264978	416283759	517243968	618293457	719253846	812354769	913274856	306
123456789	214365879	315274968	412385769	519263478	618253947	713294856	816243759	917283546	307
123456789	214365879	315274968	416293578	519283746	612394857	718263459	813254769	917243856	308
123456789	214365879	315264978	412395768	516293748	618273459	719283546	813254769	917243856	309
123456789	214365879	315264978	413295768	519283746	618253947	712354869	816273459	917243856	310
123456789	214365879	315284769	417293568	519263478	612384957	718243956	813274659	916253748	311
123456789	214365879	315264978	412375968	516273948	618293457	719283546	813254769	917243856	312
123456789	214365879	315294768	416253978	518243769	612384957	719283456	813274659	917263548	313
123456789	213465879	314256978	416283759	519263847	617293548	718243956	815273649	912345768	314
123456789	213465879	314275968	415283769	518293647	612354978	716253948	819263457	917243856	315
123456789	213465879	314256978	415293768	519263847	617283459	718243956	812364957	916273548	316
123456789	214365879	315264978	417253869	518273946	619283457	713294856	816243759	912354768	317
123456789	214365879	315264978	412385769	519273468	618253947	713294856	816243759	917283546	318
123456789	214365879	315284769	417293568	519263478	618243957	712384659	813274956	916253748	319
123456789	214365879	315264978	413285769	518293746	619253847	712345968	817243956	916273548	320
123456789	214365879	315264978	412385769	518273946	619283547	713245968	817293456	916253748	321
123456789	214365879	315284769	416293578	519243768	612384957	718253946	817263459	913274856	322
123456789	214365879	315264978	417283569	518293746	619243857	712394856	816273459	913254768	323
123456789	214365879	315274968	416253978	517283469	618293547	719243856	812374659	913264857	324
123456789	214365879	315284769	416253978	519243768	612384957	718263459	817293546	913274856	325
123456789	214365879	315274968	412385769	513294678	618243759	719283456	816253947	917263548	326
123456789	214365879	315284769	419263857	516293748	618273459	712354968	817253946	913245678	327
123456789	214365879	315264978	412375968	513274869	618253947	719283546	817293456	916243857	328
123456789	214365879	315274968	412385769	513294678	618253947	719283456	816243759	917263548	329
123456789	214365879	315274968	413285769	512394678	619253847	718293456	816243759	917263548	330
123456789	214365879	315264978	413285769	516293748	618273459	719243856	817253946	912354768	331
123456789	214365879	315284769	419263857	516243978	618253749	712345968	817293546	913274856	332
123456789	214365879	315264978	412385769	513294768	618243759	719283456	817253946	916273548	333
123456789	214365879	315264978	412385769	517293468	619283547	718253946	816243759	913274856	334
123456789	214365879	315274968	417253869	518263947	613294857	719283456	816243759	912354678	335
123456789	214365879	315264978	412385769	518293746	619273548	713245968	816253947	917283456	336
123456789	214365879	315264978	417253869	519283746	613294857	718243956	816273459	912354768	337
123456789	214365879	315264978	416273859	518293746	619283457	713254869	817243956	912354768	338
123456789	214365879	315264978	413285769	512394768	618273459	719243856	817293546	916253748	339
123456789	214365879	315284769	418263957	517293468	612354978	719243856	813274659	916253748	340
123456789	214365879	315284769	418293756	516273849	612345978	719263548	817253946	913245768	341
123456789	214365879	315274968	417253869	518263947	619283457	713294856	816243759	912354678	342
123456789	214365879	315284769	417253968	519263478	612384957	718293546	816243759	913274856	343
123456789	214365879	315264978	417253968	519283746	618273459	713294856	812354769	916243857	344
123456789	214365879	315284769	416253978	519243768	612384957	718293456	813274659	917263548	345
123456789	214365879	315284769	418293756	517243968	612345978	719253846	816273549	913264857	346
123456789	213465879	312495768	416253978	518293647	619273548	715243869	817263459	914283756	347
123456789	213465879	314275968	412385769	516243978	619283547	718253649	817293456	915263748	348
123456789	214365879	315264978	413285769	517243968	619273548	718293456	812374659	916253847	349
123456789	214365879	315274968	412385769	518293746	613245978	719283456	816253947	917263548	350
123456789	214365879	315264978	412375968	516273948	618293547	719283456	813245769	917253846	351
123456789	214365879	315284769	416273859	519243768	612394857	718263549	817293456	913254678	352
123456789	214365879	315274968	416283759	519263478	618243957	713294856	812354769	917253846	353
123456789	214365879	315264978	412385769	517243968	618293547	719283456	813274659	916253748	354
123456789	214365879	315264978	413295768	518273469	619283547	712394856	816243759	917253846	355
123456789	214365879	315284769	418273956	516293748	619243857	712345968	817263549	913254678	356
123456789	213465879	314275968	412385769	519243678	615283947	718263549	817293456	916253748	357
123456789	213465879	314256978	412395768	519263847	615283749	718243659	817293456	916273548	358
123456789	214365879	315284769	416253978	519243768	618293457	712384659	817263549	913274856	359
123456789	214365879	315264978	412385769	519283746	618253947	713245968	817293456	916273548	360
123456789	214365879	315264978	413295768	518273469	612384759	719283546	817243956	916253748	361
123456789	214365879	315284769	416253978	519243768	618293457	712384659	813274956	917263548	362
123456789	214365879	315264978	413295768	516273948	618243759	719253846	812354769	917283456	363
123456789	214365879	315264978	413295768	518273469	619283547	712384659	817243956	916253748	364
123456789	214365879	315264978	412385769	518273946	619253748	713245968	816293547	917283456	365
123456789	214365879	315264978	413285769	516293748	618273459	719253846	817243956	912354768	366
123456789	214365879	315264978	413295768	517243869	618253947	719283456	812374659	916273548	367

123456789	214365879	315274968	416283759	518263947	619243857	713254869	817293546	912345678	368
123456789	214365879	315264978	412385769	517293468	618253947	719283546	816243759	913274856	369
123456789	214365879	315264978	412385769	517293468	619253748	718243956	813274659	916283547	370
123456789	214365879	315274968	413285769	516243978	619253847	718263459	817293546	912374856	371
123456789	214365879	315264978	413285769	519243768	618253947	712384659	817293456	916273548	372
123456789	214365879	315284769	416273859	517243968	618293457	719263548	812374956	913254678	373
123456789	213465879	314275968	416293857	518243769	612354978	719283456	815263947	917253648	374
123456789	214365879	315264978	417253968	519273846	618243759	713294856	812345769	916283547	375
123456789	214365879	315274968	416253978	519263847	618243759	713294856	812345769	917283546	376
123456789	214365879	315264978	417253869	518273946	613294857	719283456	816243759	912354768	377
123456789	214365879	315284769	412395678	513274968	618243759	719253846	816293457	917263548	378
123456789	214365879	315274968	416253978	519283746	613294857	718263459	812354769	917243856	379
123456789	213465879	312475968	416253978	518243769	619273548	715283649	817293456	914263857	380
123456789	213465879	314275968	416253978	519263847	615293748	718243569	812364957	917283456	381
123456789	213465879	314275968	412395678	518293647	615283749	719263548	817253469	916243857	382
123456789	214365879	315264978	412385769	516273948	618243759	719283546	817293456	913254768	383
123456789	214365879	315264978	412385769	516293748	618273459	719283546	817243956	913254768	384
123456789	214365879	315264978	416283759	518273946	619243857	712354869	817293456	913254768	385
123456789	214365879	315264978	412385769	516273948	618243759	719283456	817293546	913254768	386
123456789	214365879	315274968	417263859	518243769	613294857	719283456	816253947	912354678	387
123456789	214365879	315284769	416253978	519243768	618273549	712384659	817293456	913264857	388
123456789	214365879	315264978	416283759	518273946	619253847	712354869	817293456	913245768	389
123456789	214365879	315274968	413285769	516293478	618253947	719263548	812374659	917243856	390
123456789	214365879	315274968	416253978	518293746	613284759	719243856	812345769	917263548	391
123456789	214365879	315264978	413285769	519273846	618293547	712345968	817243956	916253748	392
123456789	214365879	315264978	413285769	518273946	619253748	712345968	816293547	917243856	393
123456789	214365879	315274968	416283759	517263948	618293457	719253846	812354769	913245678	394
123456789	214365879	315274968	413285769	512394678	618293547	719243856	817263459	916253748	395
123456789	214365879	315274968	413285769	512394678	618243759	719263548	817293456	916253847	396

4.23 Remark The existence of k pairwise orthogonal 1-factorizations of K_{2n} is equivalent to the existence of a Room k-cube of side $2n-1$ (see §IV.39). In particular, $\nu(2n-1)$ denotes the maximum number of pairwise orthogonal 1-factorizations of K_{2n}.

LB(n) denotes the number of 1-factorizations of a labeled $K_{n,n}$.

NB(n) denotes the number of pairwise nonisomorphic 1-factorizations of $K_{n,n}$.

4.24 Remark The number of latin squares of order n is $n!\,\text{LB}(n)$, and NB(n) is greater than the number of non main class isotopic latin squares of order n and smaller than the number of nonisotopic latin squares of order n. It is in fact greater than half the latter number. Compare with Table II.1.6.

4.25 Table Some known values for the number of distinct and the number of nonisomorphic 1-factorizations of $K_{n,n}$. LB(8) and LB(9) are also known [11].

n	1	2	3	4	5	6	7
LB(n)	1	1	2	24	1344	1,128,960	12,198,297,600
NB(n)	1	1	1	2	2	17	≥ 282

4.26 Remark In principle, formulae exist for the calculation of LB(n), but they are impractical for large n [24, 12].

4.27 Theorem

1. $\text{LB}(n) \geq (n!)^{2n-1}/n^{n^2}$;
2. $\text{NB}(n) \geq (n!)^{2n-3}/(2n^{n^2})$;
3. $\lim_{n\to\infty}(\text{LB}(n))^{1/n^2}/(e^{-2}n) = \lim_{n\to\infty}(\text{NB}(n))^{1/n^2}/(e^{-2}n) = 1$. [25]

4.28 Remark From Theorem 4.27(3), the following weaker statement is easily deduced:

$$\lim_{n\to\infty}(\log(\text{LB}(n))/(n^2\log n) = \lim_{n\to\infty}(\log \text{NB}(n))/(n^2\log n) = 1.$$

4.29 Remarks MOLS correspond to pairwise orthogonal 1-factorizations of $K_{n,n}$, and hence $N(n)$ (defined in §II.2) denotes the maximum number of pairwise orthogonal 1-factorizations of $K_{n,n}$. Bounds on $N(n)$ can be found in §II.2.4 and §II.2.6.

4.30 Table All nonisomorphic 2-factorizations of all regular graphs of even degree at least 4 and order at most 8. Each line is a 2-factorization, and 2-factors are given by their cycles. The complements of the first 5 graphs of degree 4 on 8 vertices are, in order, the graphs in the second line of Table VI.3.12.

Graph	2-Factorizations
K_5	01234; 02413
$K_6 - \{03, 14, 25\}$	012 345; 042315 012354; 024315
$K_7 - \{01, 12, 23, 34, 45, 56, 60\}$	024 1536; 0314625 0241635; 0315264 0246135; 0362514
$K_7 - \{01, 12, 20, 34, 45, 56, 63\}$	035 1426; 046 1325 0314625 0423516
K_7	012 3456; 035 1426; 046 1325 012 3456; 0314625 0423516 0123456; 0241635; 0315264 0123456; 0246135; 0362514
$K_8 - \{01, 03, 04, 12, 15, 23, 26, 37, 45, 47, 56, 67\}$	0246 1357; 0527 1436 0257 1364; 05342716 02416357; 05271346 02461357; 05271436
$K_8 - \{01, 04, 07, 12, 15, 23, 26, 34, 37, 45, 56, 67\}$	025 13647; 03572416 0246 1357; 03614725 0275 1364; 03524716 02461753; 05274136
$K_8 - \{01, 02, 04, 12, 13, 23, 37, 45, 46, 56, 57, 67\}$	035 14726; 06342517 0347 1526; 0536 1427 0527 1436; 03517426 03471526; 05361427 03514726; 05243617
$K_8 - \{01, 02, 03, 12, 15, 26, 34, 37, 45, 46, 57, 67\}$	047 13256; 0536 1427 056 13247; 04163527 04136527; 05324716
$K_8 - \{01, 02, 05, 13, 16, 23, 24, 34, 47, 56, 57, 67\}$	037 14625; 127 04536 036 14527; 04621537 037 12645; 04172536 0364 1527; 06214537 0417 2536; 03721546 03514627; 04521736 03527146; 04512637 03546217; 04152736 03627154; 06412537 03712546; 04153627
$K_8 - \{01, 02, 03, 12, 13, 23, 45, 46, 47, 56, 57, 67\}$ $(= K_{4,4})$	0415 2637; 0617 2435 04152637 05342716
$K_8 - \{01, 23, 45, 67\}$	024 13657; 146 03527; 347 05126 024 13657; 035 14726; 125 06437 024 13657; 125 03746; 05341627 024 13657; 126 05347; 03725046 024 13657; 0347 1526; 05372147 024 13657; 0627 1435; 03746125 0213 4657; 0415 2637; 0617 2435 0213 4657; 0435 1627; 0637 1425 0213 4657; 0437 1526; 0536 1427 0246 1357; 0374 1265; 0527 1436 0214 3657; 0527 1346; 03517426 0246 1375; 0527 1436; 03562174

Graph	2-Factorizations
$K_8 - \{01, 23, 45, 67\}$ (cont.)	0246 1375; 0365 1274; 04352617
	0213 4657; 04152637; 05342716
	0213 4657; 04251637; 05341726
	0213 4657; 04261537; 05271436
	0214 3657; 03164725; 06243517
	0214 3657; 03461527; 05317426
	0246 1357; 03415627; 04736125
	0246 1357; 03472165; 04152637
	0246 1357; 03615274; 05621437
	0246 1357; 03651274; 05261437
	0246 1375; 03562147; 04361725
	0246 1537; 03652147; 04316275
	02134756; 03641725; 04261537
	02137564; 03472516; 05362417
	02137564; 03472615; 06352417
	02137564; 03524716; 05143627
	02143657; 03517426; 04613725
	02146357; 03165274; 05173426
	02146375; 03517426; 04316527
	02415637; 03527164; 05743126
	02416537; 03462175; 04725136
	02431657; 03746215; 04172536
	02435617; 03752146; 04726315
	02436517; 03752614; 05312746
	02461375; 03415627; 04712536
	02461573; 05314726; 04365217
	02465137; 03417526; 05361274
	02465173; 05341627; 04752136

4.31 Remark [8] There are precisely 122 nonisomorphic 2-factorizations of K_9 with the property that each factor is a Hamilton cycle.

4.5 Some 1-Factorizations of Complete Graphs

4.32 Construction If $S = \{\{s_i, t_i\} : 1 \leq i \leq (g-1)/2\}$ is a starter in an abelian group G of order $2n-1$, then $\{\{\infty, g\} \cup (S+g) : g \in G\}$ are the edge-sets of a 1-factorization of K_{2n} (with vertex set $G \cup \{\infty\}$) (see §IV.45)

4.33 Example If S is the patterned starter $\{\{g, -g\} : g \in G\}$ in the cyclic group of order $2n-1$, the resulting 1-factorization of K_{2n} is known as GK_{2n}. It can be obtained by rotating the diagram to the right (shown here for GK_8). In Table 4.21, factorization number 6 is isomorphic to GK_8, and in Table 4.22, factorization number 290 is isomorphic to GK_{10} (see §IV.45).

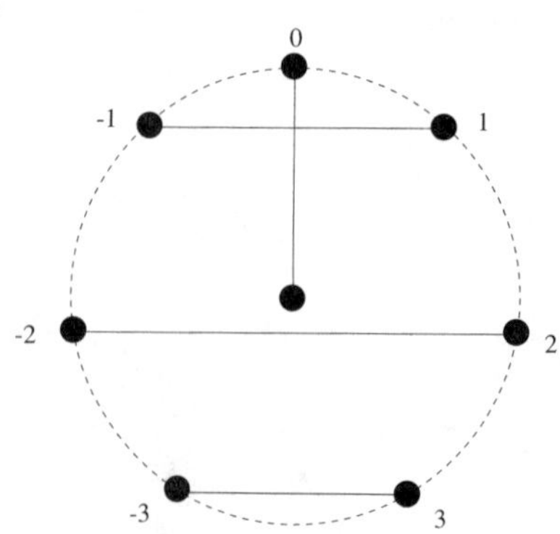

4.34 Construction

1. For even n, a 1-factorization $\mathcal{F}$ of K_{2n} can be obtained by letting $\{F_1, \ldots, F_{n-1}\}$ and $\{G_1, \ldots, G_{n-1}\}$ be 1-factorizations of two disjoint K_ns, letting $\{H_1, \ldots, H_n\}$ be a 1-factorization of the $K_{n,n}$ with all edges joining a vertex of one K_n to a vertex of the other, and putting $\mathcal{F} = \{F_1 \cup G_1, \ldots, F_{n-1} \cup G_{n-1}, H_1, \ldots, H_n\}$.
2. For odd n there is a similar construction; a specific example is given in Construction 4.35.

4.35 Construction A 1-factorization of K_{2n} known as GA_{2n} (best known when n is odd):

n even. Let $\{F_1, \ldots, F_{n-1}\}$ be a GK_n on the vertex set $C_1 \cup \{\infty_1\}$, where C_1 is the group with elements $1_1, \ldots, (n-1)_1$ and addition modulo $n-1$. Let $n_1 = \infty_1$ and choose notation such that the edge $i_1\infty_1$ is in F_i. Similarly, let $\{G_1, \ldots, G_{n-1}\}$ be a GK_n on $C_2 \cup \{\infty_2\}$, where $n_2 = \infty_2$ and the elements of C_2 are $1_2, \ldots, (n-1)_2$, and where the edge $i_2\infty_2$ is in G_i. Further, let $\{H_1, \ldots, H_n\}$ be the 1-factorization of $K_{n,n}$ where H_i has edge-set $\{j_1(j+i)_2 : 1 \le j \le n\}$, numbers reduced modulo n. Then GA_{2n} is the 1-factorization $\{F_1 \cup G_1, \ldots, F_{n-1} \cup G_{n-1}, H_1, \ldots, H_n\}$ of K_{2n}. In Table 4.21, factorization number 2 is isomorphic to GA_8.

n odd. Let $\{F_1, \ldots, F_n\}$ be a GK_{n+1} defined on the vertex set $\{1_1, \ldots, n_1\} \cup \{\infty_1\}$, addition modulo n, with the edge $i_1\infty_1$ in F_i, and let $\{G_1, \ldots, G_n\}$ be a GK_{n+1} on the vertex set $\{1_2, \ldots, n_2, \infty_2\}$ defined in the same way. Let $\{H_1, \ldots, H_n\}$ be as above. Then GA_{2n} is the 1-factorization on the vertex set $\{1_1, \ldots, n_1, 1_2, \ldots n_2\}$ with factors $(F_1 - 1_1\infty_1) \cup (G_1 - 1_2\infty_2) + 1_11_2, \ldots, (F_n - n_1\infty_1) \cup (G_n - n_2\infty_2) + n_1n_2, H_1, \ldots, H_{n-1}$. In Table 4.22, factorization number 396 is isomorphic to GA_{10}.

An *automorphism* of a factorization of a graph G is a mapping of the vertices of G to themselves which also maps factors to factors. The *automorphism group* of a factorization F is the group of all automorphisms of F.

A 1-factorization of K_{2n} is

1. *1-rotational* (or *vertex-1-rotational*, or *factor-cyclic*) if it has an automorphism which fixes one vertex and is cyclic on the remaining vertices;
2. *cyclic* if it has an automorphism which is cyclic on the vertices; and
3. *rigid* if the identity is the only automorphism.

4.36 Remark GK_{2n} is 1-rotational.

4.37 Theorem The order of automorphism group for the 1-factorizations of K_8 (see Table 4.21).

Factorization Number	1	2	3	4	5	6
Group Order	1,344	64	16	96	24	42

4.38 Table The number $f_n(t)$ of nonisomorphic 1-factorizations of K_n for $n = 10$ and 12 with automorphism group of order t. [13]

t	1	2	3	4	5	6	8	9	10	11
$f_{10}(t)$	298	69	5	7	0	6	3	1	0	0
$f_{12}(t)$	526,461,499	437,436	669	14,801	92	245	610	0	10	2

t	12	16	18	20	24	32	40	48	54	55	110	240	432	660
$f_{10}(t)$	2	1	1	0	0	0	1	0	1	0	0	0	1	0
$f_{12}(t)$	138	76	0	2	25	4	0	6	0	1	1	2	0	1

4.39 Theorems

1. If n is prime, GA_{2n} is cyclic.
2. A cyclic 1-factorization of K_{2n} exists if and only if $2n \neq 2^t, t \geq 2$.
3. A rigid 1-factorization of K_{2n} exists if and only if $2n \geq 10$.
4. The number of nonisomorphic rigid 1-factorizations of K_{2n} tends to infinity with n.
5. If $2n-1 \geq 7$ is a prime, the automorphism group of GK_{2n} is the Frobenius group F_{2n-1} of order $(2n-1)(2n-2)$.
6. The order of the automorphism group of GK_{2n}, $2n \geq 8$, is $(2n-1)\phi(2n-1)$.
7. If $n \geq 5$ is a prime, the automorphism group of GA_{2n} is the group $F_n \times \mathbb{Z}_2$ of order $2n(n-1)$.

4.40 Construction Let S be a Steiner triple system of order v on the vertices of K_v, and for each vertex x take the 1-factor F_x of K_{v+1} with edge-set $\{\infty x\} \cup \{yz : \{x,y,z\} \in S\}$. This is a 1-factorization of K_{v+1} (a *Steiner 1-factorization*).

A *near 1-factorization* of a graph is a factorization where each factor has one vertex of degree zero, all other vertices having degree 1.

4.41 Remark Near 1-factorizations of K_{2n-1} correspond to 1-factorizations of K_{2n}; thus they always exist.

4.6 Perfect 1-Factorizations

A 1-factorization $\mathcal{F} = \{F_1, \ldots, F_k\}$ of a graph G is *perfect* if $F_i \cup F_j$ is a Hamilton cycle of G for all $i \neq j$.

4.42 Theorem Perfect 1-factorizations of K_{2n} exist when $2n-1$ or n is a prime, and for the following values of $2n$: 16, 28, 36, 40, 50, 126, 170, 244, 344, 730, 1332, 1370, 1850, 2198, 3126, 6860, 12168, 16808, and 29792 [27].

4.43 Remarks

1. It is conjectured that a perfect 1-factorization of K_{2n} exists for all n.
2. When $2n-1$ is prime, GK_{2n} is perfect.
3. When n is a prime, GA_{2n} is perfect.
4. The smallest number $2n$ for which the existence of a perfect 1-factorization of K_{2n} is in doubt is 52.
5. The existence of a perfect 1-factorization of K_{2n} implies the existence of a symmetric latin square of order $2n-1$ with no proper subsquare (see Remark II.1.33).
6. The study of general graphs which possess a perfect 1-factorization was first done in [19]. A graph is *strongly hamiltonian* if it possesses a perfect 1-factorization.

4.44 Theorem There is a cyclic perfect 1-factorization of K_{2n} if and only if n is prime.

$NP(2n)$ denotes the number of pairwise nonisomorphic perfect 1-factorizations of K_{2n}.

4.45 Theorem $NP(4) = NP(6) = NP(8) = NP(10) = 1; NP(12) = 5; NP(14) = 23;$ $NP(20) \geq 3; NP(24) \geq 5$.

$P(2n)$ denotes the maximum number of pairwise orthogonal perfect 1-factorizations of K_{2n}.

4.46 Theorem $P(4) = P(6) = P(8) = P(10) = 1$.

4.47 Table Some known lower bounds for $P(2n)$.

$2n$	12	20	24	30	38	44	48	54	60	68	72	80	84	126
Lower Bound for $P(2n)$	3	5	9	5	7	3	5	8	5	7	7	9	17	9

4.7 Maximal Sets of 1-Factors

Two s-factors are *compatible* (or *edge-disjoint*) if they contain no edge in common. If S is a set of pairwise compatible s-factors which cannot be extended to an s-factorization, then S is a *premature set* of s-factors. If there is no s-factor compatible with all the members of S, then S is *maximal.*

4.48 Theorem [10] If n is odd, then a maximal set of 1-factors in K_{2n} contains at least n 1-factors. If n is even, then a maximal set in K_{2n} contains at least $n+1$ 1-factors.

4.49 Theorem [10] If d is even and $d < n$ then K_{2n} contains a maximal set of $2n-d-1$ 1-factors.

4.50 Theorems

1. If there is a maximal set of k 1-factors in K_{2n}, and k is even, then $k \geq (4n+4)/3$.
2. [21] There is a maximal set of k 1-factors in K_{2n} whenever $(4n+4)/3 \leq k \leq 2n-3$.

4.51 Theorem There is a premature set of k 1-factors in K_{2n} whenever k is even and $n < k < 2n-4$, or when $k = 2n-4$, $n \geq 5$ and n is odd.

4.52 Theorem For every order $2n$ greater than 8 there is a premature set of $2n-4$ 1-factors in K_{2n} which is not maximal.

4.53 Proposition The question of deciding whether a set S of 1-factors of a graph G is premature is NP-complete [9].

4.8 Some 2-Factorizations of Complete Graphs

4.54 Theorem For any $m \geq 3$, K_n has a 2-factorization where each factor consists of disjoint m-cycles if and only if m is odd and $n \equiv m \pmod{2m}$ (see Table IV.9.18).

4.55 Remarks

1. The case $n = m$ of Theorem 4.54 states that for odd n, K_n has a *hamiltonian decomposition.*
2. The general problem of finding a factorization of K_{2n-1} into *given, isomorphic* 2-factors is known as the *Oberwolfach problem* (see §IV.29).

4.56 Theorem [17] When n is even there exists a maximal set of k pairwise compatible 2-factors of K_n if and only if $k \in \{\lceil \frac{n-\sqrt{n}}{2} \rceil, \lceil \frac{n-\sqrt{n}}{2} + 1 \rceil, \ldots, \frac{n-2}{2}\}$. When n is odd, there exists a maximal set of k pairwise compatible 2-factors of K_n if and only if $k = (n-1)/2$.

4.57 Theorem [17] There exists a maximal set of k pairwise compatible hamiltonian cycles in K_n if and only if $k \in \{\lfloor \frac{n+3}{4} \rfloor, \lfloor \frac{n+3}{4} + 1 \rfloor, \ldots, \lfloor \frac{n-1}{2} \rfloor\}$.

A *Kotzig factorization* of K_{2n-1} is a hamiltonian decomposition together with an orthogonal near 1-factorization.

4.58 Theorem For all $n \geq 2$, K_{2n-1} has a Kotzig factorization.

4.9 See Also

§I.6	A resolvable (v, k, λ) BIBD is a $(k-1)$-factorization of λK_v.
§II.1	Latin squares are 1-factorizations of $K_{n,n}$.
§IV.9	Resolvable cycle systems are 2-factorizations of complete graphs.
§IV.22	Graph decompositions.
§IV.26	Howell Designs are equivalent to orthogonal 1-factorizations of regular graphs.
§IV.29	Oberwolfach designs are 2-factorizations of complete graphs.
§IV.39	Room squares (and Room d-cubes) are related to orthogonal 1-factorizations of K_{2n}.
§IV.45	Starters are used to construct 1-rotational 1-factorizations of K_{2n}.
[26]	A survey of 1-factorizations of complete graphs including most of the material of this section.
[2]	A survey of orthogonal factorizations.
[14]	Contains, among many other subjects related to factorization, a fast algorithm for finding 1-factorizations of K_{2n}.
[5]	A monograph on general factorizations of graphs.
[22]	A survey on perfect 1-factorizations of K_{2n}.
[20, 23]	Together enumerate all nonisomorphic 1-factorizations of simple graphs of order at most 10.
[1]	Hamiltonian decompositions in general.

References

[1] B. Alspach, J.-C. Bermond and D. Sotteau, Decomposition into cycles I. Hamilton decompositions, in: *Cycles and Rays* (G. Hahn, G. Sabidussi and R.E. Woodrow; eds.), Kluwer Academic Publishers, Dordrecht, 1990, 9-18.

[2] B. Alspach, K. Heinrich and G. Liu, Orthogonal factorizations of graphs, in: *Contemporary Design Theory: A Collection of Surveys* (J.H. Dinitz, D.R. Stinson; eds.) John Wiley & Sons, New York, 1992, 13-40.

[3] K. Appel and W. Haken, *Every Planar Map is Four Colorable*, American Math. Society, Providence RI, 1989.

[4] J.A. Bondy and U.S.R. Murty, *Graph Theory with Applications*, Macmillan, London, 1976.

[5] J. Bosák, *Decompositions of Graphs*, Kluwer Academic Publishers, Dordrecht, 1990.

[6] P.J. Cameron, *Parallelisms of Complete Designs*, London Math. Soc. Lecture Notes 23, Cambridge University Press, London, 1976.

[7] A.G. Chetwynd and R.J. Wilson, Snarks and supersnarks, in: *The Theory and Applications of Graphs*, John Wiley & Sons, New York, 1981, 215-241.

[8] C.J. Colbourn, Hamiltonian decompositions of complete graphs, *Ars Combin.* 14 (1982), 261-269.

[9] C.J. Colbourn, Embedding partial Steiner triple systems is NP-complete, *J. Combin. Theory A* 35 (1983), 100-105.

[10] E. Cousins and W.D. Wallis, Maximal sets of one-factors, *Lecture Notes in Mathematics* 452 (1975), 90-94.

[11] J. Dénes and A.D. Keedwell, *Latin Squares and their Applications*, Akadémiai Kiadó, Budapest, English Universities Press Ltd., London, and Academic Press, New York, 1974.

[12] J. Dénes and G.L. Mullen, Enumeration formulas for latin and frequency squares, *Discrete Math.* 111 (1993), 157-163.

[13] J.H. Dinitz, D.K. Garnick and B.D. McKay, There are 526,915,620 nonisomorphic one-factorizations of K_{12}, *J. Combin. Des.* 2 (1994), 273-285.

[14] J.H. Dinitz and D.R. Stinson, Room squares and related designs, in: *Contemporary Design Theory: A Collection of Surveys* (J.H. Dinitz, D.R. Stinson; eds.) John Wiley & Sons, New York, 1992, pp. 137-204.

[15] M.R. Garey and D.S. Johnson, *Computers and Intractability*, Freeman, San Francisco, 1979.

[16] A.J.W. Hilton, Factorizations of regular graphs of high degree, *J. Graph Theory* 9 (1985), 193-196.

[17] D.G. Hoffman, C.A. Rodger and A. Rosa, Maximal sets of 2-factors and Hamiltonian cycles, *J. Combin. Theory B* 57 (1993), 69-76.

[18] I. Holyer, The NP-completeness of edge-coloring, *SIAM J. Computing* 10 (1981), 718-720.

[19] A. Kotzig and J. Labelle, Strongly hamiltonian graphs, *Utilitas Math.* 14 (1978), 99-116.

[20] A. Rosa and D.R. Stinson, One-factorizations of regular graphs and Howell designs of small order, *Utilitas Math.* 29 (1986), 99-124.

[21] R. Rees and W.D. Wallis, The spectrum of maximal sets of one-factors, *Discrete Math.* 97 (1991), 357-369.

[22] E. Seah, Perfect one-factorizations of the complete graph — a survey, *Bull. ICA* 1 (1991), 59-70.

[23] E. Seah and D.R. Stinson, An enumeration of non-isomorphic one-factorizations and Howell designs for the graph K_{10} minus a one-factor, *Ars Combin.* 21 (1986), 145-161.

[24] J. Shao and W. Wei, A formula for the number of latin squares, *Discrete Math.* 110 (1992), 293-296.

[25] J.H. van Lint and R.M. Wilson, *A Course in Combinatorics*, Cambridge University Press, London, 1992.

[26] W.D. Wallis, One-factorizations of complete graphs, in: *Contemporary Design Theory: A Collection of Surveys* (J.H. Dinitz, D.R. Stinson; eds.) John Wiley & Sons, New York, 1992, 593-631.

[27] J.Z. Zhang, Some results on perfect 1-factorizations of K_{2n}, *Dianzi Keji Daxue Xuebao* 21 (1992), 434-436 [in Chinese].

5 Strongly Regular Graphs

ANDRIES E. BROUWER

5.1 Definition and Parameters

A *strongly regular graph* with parameters (v, k, λ, μ) is a finite graph on v vertices, without loops or multiple edges, regular of degree k (with $0 < k < v - 1$, so that there are both edges and nonedges), and such that any two distinct vertices have λ common neighbors when they are adjacent, and μ common neighbors when they are nonadjacent.

5.1 **Remark** The complement of a strongly regular graph with parameters (v, k, λ, μ) is again strongly regular, with parameters $(v, v-k-1, v-2k+\mu-2, v-2k+\lambda)$.

5.2 **Proposition** The 0-1 adjacency matrix A of a strongly regular graph satisfies the equation $A^2 = kI + \lambda A + \mu(J - I - A)$, where I is the identity matrix and J the all-1 matrix, both of size v.

5.3 **Corollary** A has eigenvalues k, r, s with multiplicities 1, f, g, where r, s are found as solutions of $x^2 + (\mu - \lambda)x + (\mu - k) = 0$ (assuming $r \geq 0$, $s \leq -1$) and f, g are found as solutions of $1 + f + g = v$ and $k + fr + gs = 0$.

A parameter set (v, k, λ, μ) is *feasible* when v, k, λ, μ are nonnegative integers satisfying $\mu(v-k-1) = k(k-1-\lambda)$, $0 < k < v-1$, $v-2k+\mu-2 \geq 0$, $v-2k+\lambda \geq 0$, and f, g are nonnegative integers.

5.4 **Proposition** If (v, k, λ, μ) is feasible, then either r, s are integral, or there is a t such that $v = 4t+1$, $k = f = g = 2t+1$, $\lambda = t-1$, $\mu = t$, $r, s = (-1 \pm \sqrt{v})/2$ (the *half case*).

A strongly regular graph Γ is *imprimitive* when either it or its complement is disconnected.

5.5 **Remark** If a strongly regular graph Γ with parameters (v, k, λ, μ) is disconnected, then it is the disjoint union of complete graphs, so $v = a(k+1)$, $\lambda = k-1$, $\mu = 0$, $r = k$, $s = -1$, $f = a-1$, $g = ak$ (so that the eigenvalue k has multiplicity a). These graphs exist and are uniquely determined given a and k; they are aK_{k+1}. If the complement of Γ is disconnected, then Γ is the complement of aK_m, that is, it is the complete multipartite graph $K_{a \times m}$ with parameters $v = am$, $k = \mu = (a-1)m$, $\lambda = (a-2)m$, $r = 0$, $s = -m$, $f = a(m-1)$, $g = a-1$.

5.6 **Remark** Henceforth Γ is primitive, so that $k > \mu > 0$ and $k > r > 0$.

5.7 **Remarks** Table 5.9 gives feasible parameter sets for strongly regular graphs on at most 280 vertices. The spectrum, the known constructions, and the known nonexistence results are given. Space limitations have forced the omission of theory and explanations. All information about (semi)partial geometries, gamma and delta spaces, and so on, has been omitted.

5.2 Table of Strongly Regular Graphs

5.8 **Remarks** The first column of Table 5.9 lists information about existence and enumeration; a minus sign '–' means that there is no such graph; a question mark '?' that no such graph is known; a plus sign '+' that there is at least one such graph; an exclamation mark '!' that there is precisely one such graph, a number n, that at least n such graphs are known; a number n followed by an exclamation mark, that precisely n such graphs exist. Multiplicities are written as exponents. In the comments column, the (or at least some) known constructions are given. It may well happen that several constructions in fact produce the same graph.

Finite fields

Paley(q): For prime powers $q = 4t+1$, the graph with vertex set $\mathbb{F}_q$ where two vertices are adjacent when they differ by a square.

vanLint-Schrijver(u): a graph constructed by the cyclotomic construction in [39], taking the union of u classes.

Combinatorial

$\binom{n}{2}$ or $T(n)$: the graph on the pairs of an n-set, two pairs being adjacent when they have an element in common. It is the *triangular graph.*

n^2: the cartesian product of two complete graphs of size n.

OA(t,n): $(t \geq 3)$ the block graph of an orthogonal array OA(t,n) ($t-2$ MOLS of order n; see §II.2).

$S(2,k,v)$: the block graph of a Steiner system with these parameters (see §I.4). More generally, given a quasi-symmetric design (§IV.37), the graph on the blocks, adjacent when they meet in one of the two possible cardinalities, is strongly regular. Much is known, but there is no room for details. An old survey is Neumaier [41]. From the Witt design $S(5,8,24)$ one finds five quasi-symmetric designs: 2-(23,7,21), 2-(22,7,16), 2-(21,7,12), 2-(22,6,5), 2-(21,6,4). Tonchev constructed a quasi-symmetric 2-(45,9,8) design with intersection numbers 1, 3 from the extended binary quadratic residue code of length 48. See also [35].

Goethals-Seidel(k,r): a graph constructed from a Steiner system $S(2,k,v)$ (with $r = (v-1)/(k-1)$) and a Hadamard matrix of order $r+1$ as in [27]. See also §IV.24.

Pasechnik(h): graphs obtained from the tensor product of two nonsymmetric association schemes on h points obtained from a skew Hadamard matrix of order $h+1$; see [44].

Regular two-graphs

two-graph: the graph belongs to the switching class of a regular two-graph (see §VI.6).

two-graph$-*$: the graph with an isolated point added belongs to the switching class of a regular two-graph. (These are usually comments only, but the existence of a two-graph on $v+1$ points implies the existence of a strongly regular graph on v points.)

RSHCD$\pm$: regular two-graphs from regular symmetric Hadamard matrices with constant diagonal (see [9, 27]).

Taylor two-graph for $U_3(q)$: derived from Taylor's regular two-graphs (see [9, 56, 57]).

skewhad^2: derived from the tensor product of the cores of two skew Hadamard matrices of the same order (see [25]).

Projective graphs

$A_{n-1,2}(q)$ or $\left[{n \atop 2}\right]_q$: the graph on the lines in PG$(n-1,q)$, adjacent when they have a point in common.

$Bilin_{2\times d}(q)$: the graph on the $2\times d$ matrices over $\mathbb{F}_q$, adjacent when their difference has rank 1.

Polar graphs

$O^{\varepsilon}_{2d}(q)$: the graph on the isotropic points on a nondegenerate quadric in PG$(2d-1,q)$, where two points are joined when the connecting line is totally singular.

$O_{2d+1}(q)$: such a graph in PG$(2d,q)$.

$Sp_{2d}(q)$: the graph on the points of PG$(2d-1,q)$ provided with a nondegenerate symplectic form, where two points are joined when the connecting line is totally isotropic.

$U_d(q)$: the graph on the isotropic points of PG$(d-1,q^2)$ provided with a nondegenerate hermitian form, where two points are joined when the connecting line is totally isotropic.

$NO^{\varepsilon}_{2d}(2)$: the graph on the nonisotropic points of PG$(2d-1,2)$ provided with a nondegenerate quadratic form, where two points are joined when they are orthogonal, i.e., when the connecting line is a tangent.

$NO^{\varepsilon}_{2d}(3)$: the graph on one class of nonisotropic points of PG$(2d-1,3)$ provided with a nondegenerate quadratic form, where two points are joined when they are orthogonal, i.e., when the connecting line is elliptic.

$NO^{\varepsilon}_{2d+1}(q)$: the graph on one class of nondegenerate hyperplanes of PG$(2d, q)$ provided with a nondegenerate quadratic form, where two hyperplanes are joined when their intersection is degenerate.

$NO^{\varepsilon\perp}_{2d+1}(5)$: the graph on one class of nonisotropic points of PG$(2d, 5)$ provided with a nondegenerate quadratic form, where two points are joined when they are orthogonal.

$NU_n(q)$: the graph on the nonisotropic points of PG$(n-1, q)$ provided with a nondegenerate hermitian form, where two points are joined when the connecting line is a tangent.

Affine polar graphs

$VO^{\varepsilon}_{2d}(q)$: the graph on the vectors of a vector space of dimension $2d$ over $\mathbb{F}_q$ provided with a nondegenerate quadratic form Q, where two vectors u and v are joined when $Q(v-u) = 0$.

$VNO^{\varepsilon}_{2d}(q)$: ($q$ odd) the graph on the vectors of a vector space of dimension $2d$ over $\mathbb{F}_q$ provided with a nondegenerate quadratic form Q, where two vectors u and v are joined when $Q(v-u)$ is a nonzero square.

Affine difference sets Let V be an n-dimensional vector space over $\mathbb{F}_q$ and let X be a set of directions (i.e., a subset of the projective space PV). Two vectors are *adjacent* when the line joining them has a direction in X. Then $v = q^n$ and $k = (q-1)|X|$. This graph is strongly regular if and only if there are two integers w_1, w_2 such that all hyperplanes of PV miss either w_1 or w_2 points of X. If this is the case, then $r = k - qw_1$, $s = k - qw_2$ (assuming $w_1 < w_2$), and hence $\mu = k + (k-qw_1)(k-qw_2)$, $\lambda = k - 1 + (k - qw_1 + 1)(k - qw_2 + 1)$. On the other hand, one can view X as the set of columns of the generator matrix of a code, and then the requirement that hyperplanes meet X in one of two cardinalities translates into the requirement that one has a 2-weight code. Thus, projective 2-weight codes, projective sets that have two intersection sizes with hyperplanes, and strongly regular graphs defined by a multiplication invariant difference set on a vector space are equivalent objects. These observations are due to Delsarte [19]. For an extensive discussion, see [11]. One more infinite family is given in Brouwer [4]. There are many sporadic examples — a full discussion is omitted for reasons of space. All constructions 'from a Baer subplane', 'from a hyperoval', 'from a partial spread' etc., refer to this situation, and specify the set X.

Geometric

GQ(s, t): (the collinearity graph of) a generalized quadrangle with $s + 1$ points on each line, and $t + 1$ lines on each point. See §IV.21 and [46].

Haemers(q): the graph constructed from a GQ$(q + 1, q - 1)$ as described in [9], §8A.

Nonexistence: The Half-Case and Conference Matrices

Conf: In the Half Case (parameters $(v, k, \lambda, \mu) = (4t + 1, 2t, t - 1, t)$ for some integer t), a strongly regular graph exists if and only if a conference matrix C of order $v + 1$ exists. Thus v must be a sum of two squares. See §IV.52.

Nonexistence: The Krein Conditions A strongly regular graph carries a two-class association scheme, and the standard theory of association schemes applies. In particular, the Krein parameters q^k_{ij} are nonnegative (see Krein [37], Scott [49], Higman [29], Delsarte [20]). See also §IV.1. Expressed in the parameters, this condition becomes

Krein$_1$: $(r + 1)(k + r + 2rs) \le (k + r)(s + 1)^2$.

Krein$_2$: $(s + 1)(k + s + 2rs) \le (k + s)(r + 1)^2$.

Nonexistence: The Absolute Bound One has $v \le \frac{1}{2}f(f+3)$ and even $v \le \frac{1}{2}f(f+1)$ if $q^1_{11} \ne 0$ (see Koornwinder [36]; Delsarte, Goethals and Seidel [21]; Seidel [52]; Neumaier [42]). This condition is noted by 'Abs'.

$\mu = 1$: If $\mu = 1$, then the set of neighbors of a point is a union of $(\lambda+1)$-cliques, so that $(\lambda+1)|k$. The total number of $(\lambda+2)$-cliques equals $vk/(\lambda+1)(\lambda+2)$ which must be an integer.

5.9 Table Strongly Regular Graphs on at most 280 vertices.

$\exists$	v	k	λ	μ	r^f	s^g	Comments
!	5	2	0	1	0.618^2	-1.618^2	pentagon; Paley(5); Seidel two-graph−∗
!	9	4	1	2	1^4	-2^4	Paley(9); $q^1_{11}=0$; $q^2_{22}=0$; 3^2; two-graph−∗
!	10	3	0	1	1^5	-2^4	Petersen graph [47]; $NO_4^-(2)$; $NO_3^{-\perp}(5)$; switch OA(2,3) + ∗; two-graph
		6	3	4	1^4	-2^5	$\binom{5}{2}$; two-graph
!	13	6	2	3	1.303^6	-2.303^6	Paley(13); two-graph−∗
!	15	6	1	3	1^9	-3^5	$O_5(2)$ polar graph; $Sp_4(2)$ polar graph; $NO_4^-(3)$; two-graph−∗
		8	4	4	2^5	-2^9	$\binom{6}{2}$; two-graph−∗
!	16	5	0	2	1^{10}	-3^5	$q^2_{22}=0$; vanLint-Schrijver(1); $VO_4^-(2)$ affine polar graph; projective binary [5,4] code with weights 2, 4; $RSHCD-$; two-graph
		10	6	6	2^5	-2^{10}	Clebsch graph [16, 18, 15, 51]; $q^1_{11}=0$; vanLint-Schrijver(2); two-graph
2!	16	6	2	2	2^6	-2^9	Shrikhande graph [53]; 4^2; from a partial spread: projective binary [6,4] code with weights 2, 4; $RSHCD+$; two-graph
		9	4	6	1^9	-3^6	OA(3,4); $\mathrm{Bilin}_{2\times2}(2)$; Goethals-Seidel(2,3); $VO_4^+(2)$ affine polar graph; two-graph
!	17	8	3	4	1.562^8	-2.562^8	Paley(17); two-graph−∗
!	21	10	3	6	1^{14}	-4^6	
		10	5	4	3^6	-2^{14}	$\binom{7}{2}$
−	21	10	4	5	1.791^{10}	-2.791^{10}	Conf
!	25	8	3	2	3^8	-2^{16}	5^2
		16	9	12	1^{16}	-4^8	OA(4,5)
15!	25	12	5	6	2^{12}	-3^{12}	complete enumeration by Paulus [45]; Paley(25); OA(3,5); two-graph−∗
10!	26	10	3	4	2^{13}	-3^{12}	complete enumeration by Paulus [45]; switch OA(3,5) + ∗; two-graph
		15	8	9	2^{12}	-3^{13}	S(2,3,13); two-graph
!	27	10	1	5	1^{20}	-5^6	$q^2_{22}=0$; $O_6^-(2)$ polar graph; GQ(2,4); two-graph−∗
		16	10	8	4^6	-2^{20}	Schläfli graph; unique by Seidel [51]; $q^1_{11}=0$; two-graph−∗
−	28	9	0	4	1^{21}	-5^6	Krein_2; Abs
		18	12	10	4^6	-2^{21}	Krein_1; Abs
4!	28	12	6	4	4^7	-2^{20}	Chang graphs [14]; $\binom{8}{2}$; two-graph
		15	6	10	1^{20}	-5^7	$NO_6^+(2)$; Goethals-Seidel(3,3); Taylor two-graph for $U_3(3)$
41!	29	14	6	7	2.193^{14}	-3.193^{14}	complete enumeration by Bussemaker and Spence [pers.comm.]; Paley(29); two-graph−∗
−	33	16	7	8	2.372^{16}	-3.372^{16}	Conf
+	35	16	6	8	2^{20}	-4^{14}	two-graph−∗
		18	9	9	3^{14}	-3^{20}	S(2,3,15); $A_{3,2}(2)$; $O_6^+(2)$ polar graph; two-graph−∗
!	36	10	4	2	4^{10}	-2^{25}	6^2
		25	16	20	1^{25}	-5^{10}	
+	36	14	4	6	2^{21}	-4^{14}	$U_3(3).2/L_2(7).2$ - subconstituent of Hall-Janko graph; $RSHCD-$; two-graph
		21	12	12	3^{14}	-3^{21}	two-graph
!	36	14	7	4	5^8	-2^{27}	$\binom{9}{2}$
		21	10	15	1^{27}	-6^8	
+	36	15	6	6	3^{15}	-3^{20}	OA(3,6); $NO_6^-(2)$; $RSHCD+$; two-graph
		20	10	12	2^{20}	-4^{15}	$NO_5^-(3)$; two-graph
+	37	18	8	9	2.541^{18}	-3.541^{18}	Paley(37); two-graph−∗
+	40	12	2	4	2^{24}	-4^{15}	$O_5(3)$ polar graph; $Sp_4(3)$ polar graph
		27	18	18	3^{15}	-3^{24}	$NU(4,2)$

∃	v	k	λ	μ	r^f	s^g	Comments
+	41	20	9	10	2.702^{20}	-3.702^{20}	Paley(41); two-graph−∗
+	45	12	3	3	3^{20}	-3^{24}	$U_4(2)$ polar graph
		32	22	24	2^{24}	-4^{20}	$NO_5^+(3)$
!	45	16	8	4	6^9	-2^{35}	$\binom{10}{2}$
		28	15	21	1^{35}	-7^9	
+	45	22	10	11	2.854^{22}	-3.854^{22}	Mathon [40]; two-graph−∗
!	49	12	5	2	5^{12}	-2^{36}	7^2
		36	25	30	1^{36}	-6^{12}	OA(6,7)
−	49	16	3	6	2^{32}	-5^{16}	Bussemaker-Haemers-Mathon-Wilbrink [10]
		32	21	20	4^{16}	-3^{32}	
+	49	18	7	6	4^{18}	-3^{30}	OA(3,7); Pasechnik(7)
		30	17	20	2^{30}	-5^{18}	OA(5,7)
+	49	24	11	12	3^{24}	-4^{24}	Paley(49); OA(4,7); two-graph−∗
!	50	7	0	1	2^{28}	-3^{21}	Hoffman-Singleton graph [33]; $U_3(5^2).2/\mathrm{Sym}(7)$
		42	35	36	2^{21}	-3^{28}	
−	50	21	4	12	1^{42}	-9^7	Abs
		28	18	12	8^7	-2^{42}	Abs
+	50	21	8	9	3^{25}	-4^{24}	switch OA(4, 7) + ∗; switch skewhad^2 + ∗; two-graph
		28	15	16	3^{24}	-4^{25}	S(2,4,25); two-graph
+	53	26	12	13	3.140^{26}	-4.140^{26}	Paley(53); two-graph−∗
!	55	18	9	4	7^{10}	-2^{44}	$\binom{11}{2}$
		36	21	28	1^{44}	-8^{10}	
!	56	10	0	2	2^{35}	-4^{20}	Sims-Gewirtz graph [23, 24, 8]; $L_3(4).2^2/\mathrm{Alt}(6).2^2$
		45	36	36	3^{20}	-3^{35}	Witt: intersection 2 graph of a 2-(21,6,4) design with block intersections 0, 2
−	56	22	3	12	1^{48}	-10^7	Krein$_2$; Abs
		33	22	15	9^7	-2^{48}	Krein$_1$; Abs
−	57	14	1	4	2^{38}	-5^{18}	Wilbrink-Brouwer [60]
		42	31	30	4^{18}	-3^{38}	
+	57	24	11	9	5^{18}	-3^{38}	S(2,3,19)
		32	16	20	2^{38}	-6^{18}	
−	57	28	13	14	3.275^{28}	-4.275^{28}	Conf
+	61	30	14	15	3.405^{30}	-4.405^{30}	Paley(61); two-graph−∗
−	63	22	1	11	1^{55}	-11^7	Krein$_2$; Abs
		40	28	20	10^7	-2^{55}	Krein$_1$; Abs
+	63	30	13	15	3^{35}	-5^{27}	$O_7(2)$ polar graph; $Sp_6(2)$ polar graph; two-graph−∗
		32	16	16	4^{27}	-4^{35}	S(2,4,28); $NU(3,3)$; two-graph−∗
!	64	14	6	2	6^{14}	-2^{49}	8^2; from a partial spread of 3-spaces: projective binary [14,6] code with weights 4, 8
		49	36	42	1^{49}	-7^{14}	OA(7,8)
+	64	18	2	6	2^{45}	-6^{18}	GQ(3, 5); from a hyperoval: projective 4-ary [6,3] code with weights 4, 6
		45	32	30	5^{18}	-3^{45}	
−	64	21	0	10	1^{56}	-11^7	Krein$_2$; Abs
		42	30	22	10^7	-2^{56}	Krein$_1$; Abs
+	64	21	8	6	5^{21}	-3^{42}	OA(3,8); Bilin$_{2\times3}(2)$; from a Baer subplane: projective 4-ary [7,3] code with weights 4, 6; Brouwer [4]; from a partial spread of 3-spaces: projective binary [21,6] code with weights 8, 12
		42	26	30	2^{42}	-6^{21}	OA(6,8)
+	64	27	10	12	3^{36}	-5^{27}	from a unital: projective 4-ary [9,3] code with weights 6, 8; $VO_6^-(2)$ affine polar graph; $RSHCD-$; two-graph
		36	20	20	4^{27}	-4^{36}	two-graph
+	64	28	12	12	4^{28}	-4^{35}	OA(4,8); from a partial spread of 3-spaces: projective binary [28,6] code with weights 12, 16; $RSHCD+$; two-graph
		35	18	20	3^{35}	-5^{28}	OA(5,8); Goethals-Seidel(2,7); $VO_6^+(2)$ affine polar graph; two-graph
−	64	30	18	10	10^8	-2^{55}	Abs
		33	12	22	1^{55}	-11^8	Abs
?	65	32	15	16	3.531^{32}	-4.531^{32}	two-graph−∗?
!	66	20	10	4	8^{11}	-2^{54}	$\binom{12}{2}$
		45	28	36	1^{54}	-9^{11}	

∃	v	k	λ	μ	r^f	s^g	Comments
?	69	20	7	5	5^{23}	-3^{45}	
		48	32	36	2^{45}	-6^{23}	S(2,6,46)?
–	69	34	16	17	3.653^{34}	-4.653^{34}	Conf
+	70	27	12	9	6^{20}	-3^{49}	S(2,3,21)
		42	23	28	2^{49}	-7^{20}	
+	73	36	17	18	3.772^{36}	-4.772^{36}	Paley(73); two-graph−∗
?	75	32	10	16	2^{56}	-8^{18}	two-graph−∗?
		42	25	21	7^{18}	-3^{56}	two-graph−∗?
–	76	21	2	7	2^{56}	-7^{19}	Haemers
		54	39	36	6^{19}	-3^{56}	
?	76	30	8	14	2^{57}	-8^{18}	two-graph?
		45	28	24	7^{18}	-3^{57}	two-graph?
?	76	35	18	14	7^{19}	-3^{56}	two-graph?
		40	18	24	2^{56}	-8^{19}	two-graph?
!	77	16	0	4	2^{55}	-6^{21}	S(3,6,22); $M_{22}.2/2^4$: Sym(6); unique by Brouwer [2]; subconstituent of Higman-Sims graph
		60	47	45	5^{21}	-3^{55}	Witt 3-(22,6,1): intersection 2 graph of a 2-(22,6,5) design with block intersections 0, 2
–	77	38	18	19	3.887^{38}	-4.887^{38}	Conf
!	78	22	11	4	9^{12}	-2^{65}	$\binom{13}{2}$
		55	36	45	1^{65}	-10^{12}	
!	81	16	7	2	7^{16}	-2^{64}	9^2; Brouwer [4]; from a partial spread: projective ternary [8,4] code with weights 3, 6
		64	49	56	1^{64}	-8^{16}	OA(8,9)
!	81	20	1	6	2^{60}	-7^{20}	unique by Brouwer and Haemers [7]; $VO_4^-(3)$ affine polar graph; projective ternary [10,4] code with weights 6, 9
		60	45	42	6^{20}	-3^{60}	
+	81	24	9	6	6^{24}	-3^{56}	OA(3,9); $VNO_4^+(3)$ affine polar graph; from a partial spread: projective ternary [12,4] code with weights 6, 9
		56	37	42	2^{56}	-7^{24}	OA(7,9)
+	81	30	9	12	3^{50}	-6^{30}	$VNO_4^-(3)$ affine polar graph; Hamada-Helleseth [28]: projective ternary [15,4] code with weights 9, 12
		50	31	30	5^{30}	-4^{50}	
+	81	32	13	12	5^{32}	-4^{48}	OA(4,9); $\text{Bilin}_{2\times 2}(3)$; $VO_4^+(3)$ affine polar graph; from a partial spread: projective ternary [16,4] code with weights 9, 12
		48	27	30	3^{48}	-6^{32}	OA(6,9)
–	81	40	13	26	1^{72}	-14^{8}	Abs
		40	25	14	13^{8}	-2^{72}	Abs
+	81	40	19	20	4^{40}	-5^{40}	Paley(81); OA(5,9); projective ternary [20,4] code with weights 12, 15; two-graph−∗
+	82	36	15	16	4^{41}	-5^{40}	switch OA(5, 9) + ∗; two-graph
		45	24	25	4^{40}	-5^{41}	S(2,5,41); two-graph
?	85	14	3	2	4^{34}	-3^{50}	
		70	57	60	2^{50}	-5^{34}	
+	85	20	3	5	3^{50}	-5^{34}	$O_5(4)$ polar graph; $Sp_4(4)$ polar graph
		64	48	48	4^{34}	-4^{50}	
?	85	30	11	10	5^{34}	-4^{50}	
		54	33	36	3^{50}	-6^{34}	S(2,6,51)?
?	85	42	20	21	4.110^{42}	-5.110^{42}	two-graph−∗?
?	88	27	6	9	3^{55}	-6^{32}	
		60	41	40	5^{32}	-4^{55}	
+	89	44	21	22	4.217^{44}	-5.217^{44}	Paley(89); two-graph−∗
!	91	24	12	4	10^{13}	-2^{77}	$\binom{14}{2}$
		66	45	55	1^{77}	-11^{13}	
–	93	46	22	23	4.322^{46}	-5.322^{46}	Conf
?	95	40	12	20	2^{75}	-10^{19}	two-graph−∗?
		54	33	27	9^{19}	-3^{75}	two-graph−∗?
+	96	19	2	4	3^{57}	-5^{38}	Haemers(4)
		76	60	60	4^{38}	-4^{57}	
+	96	20	4	4	4^{45}	-4^{50}	GQ(5,3)
		75	58	60	3^{50}	-5^{45}	

∃	v	k	λ	μ	r^f	s^g	Comments
?	96	35	10	14	3^{63}	-7^{32}	
		60	38	36	6^{32}	-4^{63}	
?	96	38	10	18	2^{76}	-10^{19}	two-graph?
		57	36	30	9^{19}	-3^{76}	two-graph?
?	96	45	24	18	9^{20}	-3^{75}	two-graph?
		50	22	30	2^{75}	-10^{20}	two-graph?
+	97	48	23	24	4.424^{48}	-5.424^{48}	Paley(97); two-graph−∗
?	99	14	1	2	3^{54}	-4^{44}	
		84	71	72	3^{44}	-4^{54}	
?	99	42	21	15	9^{21}	-3^{77}	
		56	28	36	2^{77}	-10^{21}	
+	99	48	22	24	4^{54}	-6^{44}	two-graph−∗
		50	25	25	5^{44}	-5^{54}	S(2,5,45); two-graph−∗
!	100	18	8	2	8^{18}	-2^{81}	10^2
		81	64	72	1^{81}	-9^{18}	
!	100	22	0	6	2^{77}	-8^{22}	Higman-Sims graph [30]; $HS.2/M_{22}.2$; unique by Gewirtz [23]; $q_{22}^2 = 0$
		77	60	56	7^{22}	-3^{77}	$q_{11}^1 = 0$
+	100	27	10	6	7^{27}	-3^{72}	OA(3,10)
		72	50	56	2^{72}	-8^{27}	OA(8,10)?
?	100	33	8	12	3^{66}	-7^{33}	
		66	44	42	6^{33}	-4^{66}	
+	100	33	14	9	8^{24}	-3^{75}	S(2,3,25)
		66	41	48	2^{75}	-9^{24}	
−	100	33	18	7	13^{11}	-2^{88}	Abs
		66	39	52	1^{88}	-14^{11}	Abs
+	100	36	14	12	6^{36}	-4^{63}	Hall-Janko graph; $J_2.2/U_3(3).2$; subconstituent of $G_2(4)$ graph; OA(4,10)
		63	38	42	3^{63}	-7^{36}	OA(7,10)?
?	100	44	18	20	4^{55}	-6^{44}	$RSHCD-$?; two-graph?
		55	30	30	5^{44}	-5^{55}	two-graph?
+	100	45	20	20	5^{45}	-5^{54}	OA(5,10)?; $RSHCD+$; two-graph
		54	28	30	4^{54}	-6^{45}	OA(6,10)?; two-graph
+	101	50	24	25	4.525^{50}	-5.525^{50}	Paley(101); two-graph−∗
!	105	26	13	4	11^{14}	-2^{90}	$\binom{15}{2}$
		78	55	66	1^{90}	-12^{14}	
+	105	32	4	12	2^{84}	-10^{20}	Aut$L_3(4)$ on flags (rk 4) [Goethals and Seidel]
		72	51	45	9^{20}	-3^{84}	
?	105	40	15	15	5^{48}	-5^{56}	
		64	38	40	4^{56}	-6^{48}	
?	105	52	21	30	2^{84}	-11^{20}	
		52	29	22	10^{20}	-3^{84}	
−	105	52	25	26	4.623^{52}	-5.623^{52}	Conf
+	109	54	26	27	4.720^{54}	-5.720^{54}	Paley(109); two-graph−∗
?	111	30	5	9	3^{74}	-7^{36}	
		80	58	56	6^{36}	-4^{74}	
+	111	44	19	16	7^{36}	-4^{74}	S(2,4,37)
		66	37	42	3^{74}	-8^{36}	
!	112	30	2	10	2^{90}	-10^{21}	unique by Cameron, Goethals and Seidel [12]; subconstituent of McLaughlin graph; $q_{22}^2 = 0$; $O_6^-(3)$ polar graph
		81	60	54	9^{21}	-3^{90}	$q_{11}^1 = 0$
?	112	36	10	12	4^{63}	-6^{48}	
		75	50	50	5^{48}	-5^{63}	
+	113	56	27	28	4.815^{56}	-5.815^{56}	Paley(113); two-graph−∗
?	115	18	1	3	3^{69}	-5^{45}	
		96	80	80	4^{45}	-4^{69}	
+	117	36	15	9	9^{26}	-3^{90}	Wallis; S(2,3,27); $NO_3^+(3)$; Lines in $AG(3,3)$ (rk 4)
		80	52	60	2^{90}	-10^{26}	
?	117	58	28	29	4.908^{58}	-5.908^{58}	two-graph−∗?
+	119	54	21	27	3^{84}	-9^{34}	$O_8^-(2)$ polar graph; two-graph−∗
		64	36	32	8^{34}	-4^{84}	two-graph−∗
!	120	28	14	4	12^{15}	-2^{104}	$\binom{16}{2}$
		91	66	78	1^{104}	-13^{15}	

∃	v	k	λ	μ	r^f	s^g	Comments
?	120	34	8	10	4^{68}	-6^{51}	
		85	60	60	5^{51}	-5^{68}	
?	120	35	10	10	5^{56}	-5^{63}	
		84	58	60	4^{63}	-6^{56}	
+	120	42	8	18	2^{99}	-12^{20}	$L_3(4)$ on Baer subplanes (rk 5)
		77	52	44	11^{20}	-3^{99}	Witt: intersection 3 graph of a 2-(21,7,12) design with block intersections 1, 3
+	120	51	18	24	3^{85}	-9^{34}	$NO_5^-(4)$; two-graph
		68	40	36	8^{34}	-4^{85}	two-graph
+	120	56	28	24	8^{35}	-4^{84}	two-graph
		63	30	36	3^{84}	-9^{35}	dist. 2 in J(10,3) - Mathon; $NO_8^+(2)$; Goethals-Seidel(3,7); two-graph
!	121	20	9	2	9^{20}	-2^{100}	11^2
		100	81	90	1^{100}	-10^{20}	OA(10,11)
+	121	30	11	6	8^{30}	-3^{90}	OA(3,11)
		90	65	72	2^{90}	-9^{30}	OA(9,11)
?	121	36	7	12	3^{84}	-8^{36}	
		84	59	56	7^{36}	-4^{84}	
+	121	40	15	12	7^{40}	-4^{80}	OA(4,11)
		80	51	56	3^{80}	-8^{40}	OA(8,11)
?	121	48	17	20	4^{72}	-7^{48}	
		72	43	42	6^{48}	-5^{72}	
+	121	50	21	20	6^{50}	-5^{70}	OA(5,11); Pasechnik(11)
		70	39	42	4^{70}	-7^{50}	OA(7,11)
−	121	56	15	35	1^{112}	-21^{8}	Abs
		64	42	24	20^{8}	-2^{112}	Abs
+	121	60	29	30	5^{60}	-6^{60}	Paley(121); $q_{11}^1 = 0$; OA(6,11); two-graph−*
+	122	55	24	25	5^{61}	-6^{60}	switch $\mathrm{OA}(6,11) + *$; switch skewhad$^2 + *$; two-graph
		66	35	36	5^{60}	-6^{61}	S(2,6,61)?; two-graph
+	125	28	3	7	3^{84}	-7^{40}	GQ(4,6)
		96	74	72	6^{40}	-4^{84}	
−	125	48	28	12	18^{10}	-2^{114}	Abs
		76	39	57	1^{114}	-19^{10}	Abs
+	125	52	15	26	2^{104}	-13^{20}	two-graph−*
		72	45	36	12^{20}	-3^{104}	two-graph−*
+	125	62	30	31	5.090^{62}	-6.090^{62}	Paley(125); two-graph−*
+	126	25	8	4	7^{35}	-3^{90}	dist. 1 or 4 in J(9,4) - Mathon, Buekenhout and Hubaut
		100	78	84	2^{90}	-8^{35}	
+	126	45	12	18	3^{90}	-9^{35}	$NO_6^-(3)$
		80	52	48	8^{35}	-4^{90}	
+	126	50	13	24	2^{105}	-13^{20}	Goethals; two-graph
		75	48	39	12^{20}	-3^{105}	two-graph
+	126	60	33	24	12^{21}	-3^{104}	two-graph
		65	28	39	2^{104}	-13^{21}	Taylor two-graph for $U_3(5)$
−	129	64	31	32	5.179^{64}	-6.179^{64}	Conf
+	130	48	20	16	8^{39}	-4^{90}	S(2,4,40); $A_{3,2}(3)$; $O_6^+(3)$ polar graph
		81	48	54	3^{90}	-9^{39}	
?	133	24	5	4	5^{56}	-4^{76}	
		108	87	90	3^{76}	-6^{56}	
?	133	32	6	8	4^{76}	-6^{56}	
		100	75	75	5^{56}	-5^{76}	
?	133	44	15	14	6^{56}	-5^{76}	
		88	57	60	4^{76}	-7^{56}	
−	133	66	32	33	5.266^{66}	-6.266^{66}	Conf
+	135	64	28	32	4^{84}	-8^{50}	two-graph−*
		70	37	35	7^{50}	-5^{84}	$O_8^+(2)$ polar graph; two-graph−*
?	136	30	8	6	6^{51}	-4^{84}	
		105	80	84	3^{84}	-7^{51}	
!	136	30	15	4	13^{16}	-2^{119}	$\binom{17}{2}$
		105	78	91	1^{119}	-14^{16}	
+	136	60	24	28	4^{85}	-8^{50}	two-graph
		75	42	40	7^{50}	-5^{85}	$NO_5^+(4)$; two-graph

∃	v	k	λ	μ	r^f	s^g	Comments
+	136	63	30	28	7^{51}	-5^{84}	$NO_8^-(2)$; two-graph
		72	36	40	4^{84}	-8^{51}	two-graph
+	137	68	33	34	5.352^{68}	-6.352^{68}	Paley(137); two-graph−∗
−	141	70	34	35	5.437^{70}	-6.437^{70}	Conf
+	143	70	33	35	5^{77}	-7^{65}	two-graph−∗
		72	36	36	6^{65}	-6^{77}	S(2,6,66); two-graph−∗
!	144	22	10	2	10^{22}	-2^{121}	12^2
		121	100	110	1^{121}	-11^{22}	OA(11,12)?
+	144	33	12	6	9^{33}	-3^{110}	OA(3,12)
		110	82	90	2^{110}	-10^{33}	OA(10,12)?
+	144	39	6	12	3^{104}	-9^{39}	$L_3(3)$ (rk 8)
		104	76	72	8^{39}	-4^{104}	
+	144	44	16	12	8^{44}	-4^{99}	OA(4,12)
		99	66	72	3^{99}	-9^{44}	OA(9,12)?
?	144	52	16	20	4^{91}	-8^{52}	
		91	58	56	7^{52}	-5^{91}	
+	144	55	22	20	7^{55}	-5^{88}	OA(5,12)
		88	52	56	4^{88}	-8^{55}	OA(8,12)?
−	144	65	16	40	1^{135}	-25^{8}	Krein_2; Abs
		78	52	30	24^{8}	-2^{135}	Krein_1; Abs
+	144	65	28	30	5^{78}	-7^{65}	*RSHCD*−; two-graph
		78	42	42	6^{65}	-6^{78}	two-graph
+	144	66	30	30	6^{66}	-6^{77}	OA(6,12); *RSHCD*+; two-graph
		77	40	42	5^{77}	-7^{66}	OA(7,12); Goethals-Seidel(2,11); two-graph
?	145	72	35	36	5.521^{72}	-6.521^{72}	two-graph−∗?
?	147	66	25	33	3^{110}	-11^{36}	two-graph−∗?
		80	46	40	10^{36}	-4^{110}	two-graph−∗?
?	148	63	22	30	3^{111}	-11^{36}	two-graph?
		84	50	44	10^{36}	-4^{111}	two-graph?
?	148	70	36	30	10^{37}	-4^{110}	two-graph?
		77	36	44	3^{110}	-11^{37}	two-graph?
+	149	74	36	37	5.603^{74}	-6.603^{74}	Paley(149); two-graph−∗
!	153	32	16	4	14^{17}	-2^{135}	$\binom{18}{2}$
		120	91	105	1^{135}	-15^{17}	
?	153	56	19	21	5^{84}	-7^{68}	
		96	60	60	6^{68}	-6^{84}	
?	153	76	37	38	5.685^{76}	-6.685^{76}	two-graph−∗?
?	154	48	12	16	4^{98}	-8^{55}	
		105	72	70	7^{55}	-5^{98}	
−	154	51	8	21	2^{132}	-15^{21}	Krein_2
		102	71	60	14^{21}	-3^{132}	Krein_1
?	154	72	26	40	2^{132}	-16^{21}	
		81	48	36	15^{21}	-3^{132}	
+	155	42	17	9	11^{30}	-3^{124}	S(2,3,31); $A_{4,2}(2)$
		112	78	88	2^{124}	-12^{30}	
+	156	30	4	6	4^{90}	-6^{65}	$O_5(5)$ polar graph; $Sp_4(5)$ polar graph
		125	100	100	5^{65}	-5^{90}	
+	157	78	38	39	5.765^{78}	-6.765^{78}	Paley(157); two-graph−∗
?	160	54	18	18	6^{75}	-6^{84}	
		105	68	70	5^{84}	-7^{75}	
−	161	80	39	40	5.844^{80}	-6.844^{80}	Conf
?	162	21	0	3	3^{105}	-6^{56}	
		140	121	120	5^{56}	-4^{105}	
?	162	23	4	3	5^{69}	-4^{92}	
		138	117	120	3^{92}	-6^{69}	
?	162	49	16	14	7^{63}	-5^{98}	
		112	76	80	4^{98}	-8^{63}	
!	162	56	10	24	2^{140}	-16^{21}	$q_{22}^2 = 0$
		105	72	60	15^{21}	-3^{140}	$U_4(3)/L_3(4)$; unique by Cameron, Goethals and Seidel [12]; subconstituent of McLaughlin graph; $q_{11}^1 = 0$
?	162	69	36	24	15^{23}	-3^{138}	
		92	46	60	2^{138}	-16^{23}	
+	165	36	3	9	3^{120}	-9^{44}	$U_5(2)$ polar graph
		128	100	96	8^{44}	-4^{120}	

∃	v	k	λ	μ	r^f	s^g	Comments
–	165	82	40	41	5.923^{82}	-6.923^{82}	Conf
!	169	24	11	2	11^{24}	-2^{144}	13^2
		144	121	132	1^{144}	-12^{24}	OA(12,13)
+	169	36	13	6	10^{36}	-3^{132}	OA(3,13)
		132	101	110	2^{132}	-11^{36}	OA(11,13)
?	169	42	5	12	3^{126}	-10^{42}	
		126	95	90	9^{42}	-4^{126}	
+	169	48	17	12	9^{48}	-4^{120}	OA(4,13)
		120	83	90	3^{120}	-10^{48}	OA(10,13)
?	169	56	15	20	4^{112}	-9^{56}	
		112	75	72	8^{56}	-5^{112}	
+	169	60	23	20	8^{60}	-5^{108}	OA(5,13)
		108	67	72	4^{108}	-9^{60}	OA(9,13)
?	169	70	27	30	5^{98}	-8^{70}	
		98	57	56	7^{70}	-6^{98}	
+	169	72	31	30	7^{72}	-6^{96}	OA(6,13)
		96	53	56	5^{96}	-8^{72}	OA(8,13)
+	169	84	41	42	6^{84}	-7^{84}	Paley(169); OA(7,13); two-graph$-*$
+	170	78	35	36	6^{85}	-7^{84}	switch $\mathrm{OA}(7,13)+*$; two-graph
		91	48	49	6^{84}	-7^{85}	S(2,7,85)?; two-graph
!	171	34	17	4	15^{18}	-2^{152}	$\binom{19}{2}$
		136	105	120	1^{152}	-16^{18}	
?	171	50	13	15	5^{95}	-7^{75}	
		120	84	84	6^{75}	-6^{95}	
?	171	60	15	24	3^{132}	-12^{38}	
		110	73	66	11^{38}	-4^{132}	
+	173	86	42	43	6.076^{86}	-7.076^{86}	Paley(173); two-graph$-*$
+	175	30	5	5	5^{84}	-5^{90}	GQ(6,4)
		144	118	120	4^{90}	-6^{84}	
?	175	66	29	22	11^{42}	-4^{132}	
		108	63	72	3^{132}	-12^{42}	
+	175	72	20	36	2^{153}	-18^{21}	edges of Hoffman-Singleton graph; two-graph$-*$
		102	65	51	17^{21}	-3^{153}	two-graph$-*$
?	176	25	0	4	3^{120}	-7^{55}	
		150	128	126	6^{55}	-4^{120}	
+	176	40	12	8	8^{55}	-4^{120}	
		135	102	108	3^{120}	-9^{55}	$NU(5,2)$
+	176	45	18	9	12^{32}	-3^{143}	S(2,3,33)
		130	93	104	2^{143}	-13^{32}	
+	176	49	12	14	5^{98}	-7^{77}	Higman symmetric 2-design [31, 55, 54, 3]
		126	90	90	6^{77}	-6^{98}	
+	176	70	18	34	2^{154}	-18^{21}	$S(4,7,23)\setminus S(3,6,22)$; $M_{22}/\mathrm{Alt}(7)$; two-graph
		105	68	54	17^{21}	-3^{154}	Witt 3-(22,7,4): intersection 3 graph of a 2-(22,7,16) design with block intersections 1, 3; two-graph
?	176	70	24	30	4^{120}	-10^{55}	
		105	64	60	9^{55}	-5^{120}	
–	176	70	42	18	26^{10}	-2^{165}	Abs
		105	52	78	1^{165}	-27^{10}	Abs
+	176	85	48	34	17^{22}	-3^{153}	Haemers; two-graph
		90	38	54	2^{153}	-18^{22}	two-graph
–	177	88	43	44	6.152^{88}	-7.152^{88}	Conf
+	181	90	44	45	6.227^{90}	-7.227^{90}	Paley(181); two-graph$-*$
?	183	52	11	16	4^{122}	-9^{60}	
		130	93	90	8^{60}	-5^{122}	
+	183	70	29	25	9^{60}	-5^{122}	S(2,5,61)
		112	66	72	4^{122}	-10^{60}	
–	184	48	2	16	2^{160}	-16^{23}	Krein_2
		135	102	90	15^{23}	-3^{160}	Krein_1
?	185	92	45	46	6.301^{92}	-7.301^{92}	two-graph$-*$?
?	189	48	12	12	6^{90}	-6^{98}	
		140	103	105	5^{98}	-7^{90}	
?	189	60	27	15	15^{28}	-3^{160}	
		128	82	96	2^{160}	-16^{28}	

∃	v	k	λ	μ	r^f	s^g	Comments
?	189	88	37	44	4^{132}	-11^{56}	two-graph−∗?
		100	55	50	10^{56}	-5^{132}	two-graph−∗?
−	189	94	46	47	6.374^{94}	-7.374^{94}	Conf
!	190	36	18	4	16^{19}	-2^{170}	$\binom{20}{2}$
		153	120	136	1^{170}	-17^{19}	
?	190	45	12	10	7^{75}	-5^{114}	
		144	108	112	4^{114}	-8^{75}	
?	190	84	33	40	4^{133}	-11^{56}	two-graph?
		105	60	55	10^{56}	-5^{133}	two-graph?
+	190	84	38	36	8^{75}	-6^{114}	S(2,6,76)
		105	56	60	5^{114}	-9^{75}	
?	190	90	45	40	10^{57}	-5^{132}	two-graph?
		99	48	55	4^{132}	-11^{57}	two-graph?
+	193	96	47	48	6.446^{96}	-7.446^{96}	Paley(193); two-graph−∗
+	195	96	46	48	6^{104}	-8^{90}	two-graph−∗
		98	49	49	7^{90}	-7^{104}	S(2,7,91); two-graph−∗
!	196	26	12	2	12^{26}	-2^{169}	14^2
		169	144	156	1^{169}	-13^{26}	OA(13,14)?
?	196	39	2	9	3^{147}	-10^{48}	
		156	125	120	9^{48}	-4^{147}	
+	196	39	14	6	11^{39}	-3^{156}	OA(3,14)
		156	122	132	2^{156}	-12^{39}	OA(12,14)?
?	196	45	4	12	3^{150}	-11^{45}	
		150	116	110	10^{45}	-4^{150}	
+	196	52	18	12	10^{52}	-4^{143}	OA(4,14)
		143	102	110	3^{143}	-11^{52}	OA(11,14)?
?	196	60	14	20	4^{135}	-10^{60}	
		135	94	90	9^{60}	-5^{135}	
+	196	60	23	16	11^{48}	-4^{147}	S(2,4,49); Huffman-Tonchev [35]: intersection 3 graph of a 2-(49,9,6) design with block intersections 1, 3
		135	90	99	3^{147}	-12^{48}	
+	196	65	24	20	9^{65}	-5^{130}	OA(5,14)
		130	84	90	4^{130}	-10^{65}	OA(10,14)?
?	196	75	26	30	5^{120}	-9^{75}	
		120	74	72	8^{75}	-6^{120}	
?	196	78	32	30	8^{78}	-6^{117}	OA(6,14)?
		117	68	72	5^{117}	-9^{78}	OA(9,14)?
?	196	81	42	27	18^{24}	-3^{171}	
		114	59	76	2^{171}	-19^{24}	
−	196	85	18	51	1^{187}	-34^{8}	Krein_2; Abs
		110	75	44	33^{8}	-2^{187}	Krein_1; Abs
?	196	90	40	42	6^{105}	-8^{90}	*RSHCD*−?; two-graph?
		105	56	56	7^{90}	-7^{105}	two-graph?
+	196	91	42	42	7^{91}	-7^{104}	OA(7,14)?; *RSHCD*+; two-graph
		104	54	56	6^{104}	-8^{91}	OA(8,14)?; two-graph
+	197	98	48	49	6.518^{98}	-7.518^{98}	Paley(197); two-graph−∗
−	201	100	49	50	6.589^{100}	-7.589^{100}	Conf
?	204	28	2	4	4^{119}	-6^{84}	
		175	150	150	5^{84}	-5^{119}	
?	204	63	22	18	9^{68}	-5^{135}	
		140	94	100	4^{135}	-10^{68}	S(2,10,136)?
?	205	68	15	26	3^{164}	-14^{40}	
		136	93	84	13^{40}	-4^{164}	
?	205	96	50	40	14^{40}	-4^{164}	
		108	51	63	3^{164}	-15^{40}	
?	205	102	50	51	6.659^{102}	-7.659^{102}	two-graph−∗?
?	208	45	8	10	5^{117}	-7^{90}	
		162	126	126	6^{90}	-6^{117}	
+	208	75	30	25	10^{64}	-5^{143}	S(2,5,65); $NU(3,4)$
		132	81	88	4^{143}	-11^{64}	
?	208	81	24	36	3^{168}	-15^{39}	
		126	80	70	14^{39}	-4^{168}	
−	209	16	3	1	5^{76}	-3^{132}	$\mu = 1$
		192	176	180	2^{132}	-6^{76}	

∃	v	k	λ	μ	r^f	s^g	Comments
?	209	52	15	12	8^{76}	-5^{132}	
		156	115	120	4^{132}	-9^{76}	
+	209	100	45	50	5^{132}	-10^{76}	two-graph−∗
		108	57	54	9^{76}	-6^{132}	two-graph−∗
−	209	104	51	52	6.728^{104}	-7.728^{104}	Conf
?	210	33	0	6	3^{154}	-9^{55}	
		176	148	144	8^{55}	-4^{154}	
!	210	38	19	4	17^{20}	-2^{189}	$\binom{21}{2}$
		171	136	153	1^{189}	-18^{20}	
?	210	76	26	28	6^{114}	-8^{95}	
		133	84	84	7^{95}	-7^{114}	
?	210	77	28	28	7^{99}	-7^{110}	
		132	82	84	6^{110}	-8^{99}	
?	210	95	40	45	5^{133}	-10^{76}	two-graph?
		114	63	60	9^{76}	-6^{133}	two-graph?
+	210	99	48	45	9^{77}	-6^{132}	Sym(7); Klin; two-graph
		110	55	60	5^{132}	-10^{77}	two-graph
−	213	106	52	53	6.797^{106}	-7.797^{106}	Conf
?	216	40	4	8	4^{140}	-8^{75}	
		175	142	140	7^{75}	-5^{140}	
?	216	43	10	8	7^{86}	-5^{129}	
		172	136	140	4^{129}	-8^{86}	
−	216	70	40	14	28^{12}	-2^{203}	Abs
		145	88	116	1^{203}	-29^{12}	Abs
?	216	75	18	30	3^{175}	-15^{40}	
		140	94	84	14^{40}	-4^{175}	
?	216	86	40	30	14^{43}	-4^{172}	
		129	72	84	3^{172}	-15^{43}	
?	216	90	39	36	9^{80}	-6^{135}	S(2,6,81)?
		125	70	75	5^{135}	-10^{80}	
?	217	66	15	22	4^{154}	-11^{62}	
		150	105	100	10^{62}	-5^{154}	
?	217	88	39	33	11^{62}	-5^{154}	
		128	72	80	4^{154}	-12^{62}	
−	217	108	53	54	6.865^{108}	-7.865^{108}	Conf
?	220	72	22	24	6^{120}	-8^{99}	
		147	98	98	7^{99}	-7^{120}	
+	220	84	38	28	14^{44}	-4^{175}	Tonchev: intersection 3 graph of a 2-(45,9,8) design with block intersections 1, 3
		135	78	90	3^{175}	-15^{44}	
+	221	64	24	16	12^{51}	-4^{169}	S(2,4,52)
		156	107	117	3^{169}	-13^{51}	
?	221	110	54	55	6.933^{110}	-7.933^{110}	two-graph−∗?
+	222	51	20	9	14^{36}	-3^{185}	S(2,3,37)
		170	127	140	2^{185}	-15^{36}	
!	225	28	13	2	13^{28}	-2^{196}	15^2
		196	169	182	1^{196}	-14^{28}	OA(14,15)?
+	225	42	15	6	12^{42}	-3^{182}	OA(3,15)
		182	145	156	2^{182}	-13^{42}	OA(13,15)?
?	225	48	3	12	3^{176}	-12^{48}	
		176	139	132	11^{48}	-4^{176}	
−	225	56	1	18	2^{200}	-19^{24}	Krein_2
		168	129	114	18^{24}	-3^{200}	Krein_1
+	225	56	19	12	11^{56}	-4^{168}	OA(4,15)
		168	123	132	3^{168}	-12^{56}	OA(12,15)?
?	225	64	13	20	4^{160}	-11^{64}	
		160	115	110	10^{64}	-5^{160}	
+	225	70	25	20	10^{70}	-5^{154}	OA(5,15)
		154	103	110	4^{154}	-11^{70}	OA(11,15)?
?	225	80	25	30	5^{144}	-10^{80}	
		144	93	90	9^{80}	-6^{144}	
+	225	84	33	30	9^{84}	-6^{140}	OA(6,15)
		140	85	90	5^{140}	-10^{84}	OA(10,15)?

∃	v	k	λ	μ	r^f	s^g	Comments
–	225	96	19	57	1^{216}	-39^{8}	Krein$_2$; Abs
		128	88	52	38^{8}	-2^{216}	Krein$_1$; Abs
?	225	96	39	42	6^{128}	-9^{96}	
		128	73	72	8^{96}	-7^{128}	
?	225	96	51	33	21^{24}	-3^{200}	
		128	64	84	2^{200}	-22^{24}	
+	225	98	43	42	8^{98}	-7^{126}	OA(7,15)?; Pasechnik(15)
		126	69	72	6^{126}	-9^{98}	OA(9,15)?
+	225	112	55	56	7^{112}	-8^{112}	skewhad^2; OA(8,15)?; two-graph$-*$
+	226	105	48	49	7^{113}	-8^{112}	switch skewhad^2 + $*$; two-graph
		120	63	64	7^{112}	-8^{113}	S(2,8,113)?; two-graph
+	229	114	56	57	7.066^{114}	-8.066^{114}	Paley(229); two-graph$-*$
+	231	30	9	3	9^{55}	-3^{175}	M_{22}; Cameron
		200	172	180	2^{175}	-10^{55}	
!	231	40	20	4	18^{21}	-2^{209}	$\binom{22}{2}$
		190	153	171	1^{209}	-19^{21}	
?	231	70	21	21	7^{110}	-7^{120}	
		160	110	112	6^{120}	-8^{110}	
?	231	90	33	36	6^{132}	-9^{98}	
		140	85	84	8^{98}	-7^{132}	
?	232	33	2	5	4^{144}	-7^{87}	
		198	169	168	6^{87}	-5^{144}	
?	232	63	14	18	5^{144}	-9^{87}	
		168	122	120	8^{87}	-6^{144}	
?	232	77	36	20	19^{28}	-3^{203}	
		154	96	114	2^{203}	-20^{28}	
?	232	81	30	27	9^{87}	-6^{144}	
		150	95	100	5^{144}	-10^{87}	S(2,10,145)?
+	233	116	57	58	7.132^{116}	-8.132^{116}	Paley(233); two-graph$-*$
?	235	42	9	7	7^{94}	-5^{140}	
		192	156	160	4^{140}	-8^{94}	
?	235	52	9	12	5^{140}	-8^{94}	
		182	141	140	7^{94}	-6^{140}	
?	236	55	18	11	11^{59}	-4^{176}	
		180	135	144	3^{176}	-12^{59}	S(2,12,177)?
–	237	118	58	59	7.197^{118}	-8.197^{118}	Conf
?	238	75	20	25	5^{153}	-10^{84}	
		162	111	108	9^{84}	-6^{153}	
+	241	120	59	60	7.262^{120}	-8.262^{120}	Paley(241); two-graph$-*$
+	243	22	1	2	4^{132}	-5^{110}	$3^5.2.M_{11}$ (rk 3); Berlekamp-vanLint-Seidel [1]; Golay code: projective ternary [11,5] code with weights 6, 9
		220	199	200	4^{110}	-5^{132}	
?	243	66	9	21	3^{198}	-15^{44}	
		176	130	120	14^{44}	-4^{198}	
–	243	88	52	20	34^{11}	-2^{231}	Abs
		154	85	119	1^{231}	-35^{11}	Abs
+	243	110	37	60	2^{220}	-25^{22}	$3^5.2.M_{11}$ (rk 3); Delsarte; projective ternary [55,5] code with weights 36, 45
		132	81	60	24^{22}	-3^{220}	
?	243	112	46	56	4^{182}	-14^{60}	two-graph$-*$?
		130	73	65	13^{60}	-5^{182}	two-graph$-*$?
?	244	108	42	52	4^{183}	-14^{60}	two-graph?
		135	78	70	13^{60}	-5^{183}	two-graph?
?	244	117	60	52	13^{61}	-5^{182}	two-graph?
		126	60	70	4^{182}	-14^{61}	two-graph?
?	245	52	3	13	3^{195}	-13^{49}	
		192	152	144	12^{49}	-4^{195}	
?	245	64	18	16	8^{100}	-6^{144}	
		180	131	135	5^{144}	-9^{100}	
?	245	108	39	54	3^{204}	-18^{40}	two-graph$-*$?
		136	81	68	17^{40}	-4^{204}	two-graph$-*$?
?	245	122	60	61	7.326^{122}	-8.326^{122}	two-graph$-*$?
?	246	85	20	34	3^{204}	-17^{41}	
		160	108	96	16^{41}	-4^{204}	

∃	v	k	λ	μ	r^f	s^g	Comments
?	246	105	36	51	3^{205}	-18^{40}	two-graph?
		140	85	72	17^{40}	-4^{205}	two-graph?
?	246	119	64	51	17^{41}	-4^{204}	two-graph?
		126	57	72	3^{204}	-18^{41}	two-graph?
+	247	54	21	9	15^{38}	-3^{208}	S(2,3,39)
		192	146	160	2^{208}	-16^{38}	
?	249	88	27	33	5^{165}	-11^{83}	
		160	104	100	10^{83}	-6^{165}	
–	249	124	61	62	7.390^{124}	-8.390^{124}	Conf
?	250	81	24	27	6^{144}	-9^{105}	
		168	113	112	8^{105}	-7^{144}	
?	250	96	44	32	16^{45}	-4^{204}	
		153	88	102	3^{204}	-17^{45}	
!	253	42	21	4	19^{22}	-2^{230}	$\binom{23}{2}$
		210	171	190	1^{230}	-20^{22}	
–	253	90	17	40	2^{230}	-25^{22}	Krein_2
		162	111	90	24^{22}	-3^{230}	Krein_1
+	253	112	36	60	2^{230}	-26^{22}	$S(4,7,23)$; M_{23}
		140	87	65	25^{22}	-3^{230}	Witt 4-(23,7,1): intersection 3 graph of a 2-(23,7,21) design with block intersections 1, 3
–	253	126	62	63	7.453^{126}	-8.453^{126}	Conf
+	255	126	61	63	7^{135}	-9^{119}	$O_9(2)$ polar graph; $Sp_8(2)$ polar graph; two-graph−∗
		128	64	64	8^{119}	-8^{135}	S(2,8,120); two-graph−∗
!	256	30	14	2	14^{30}	-2^{225}	16^2; from a partial spread: projective 4-ary [10,4] code with weights 4, 8; from a partial spread of 4-spaces: projective binary [30,8] code with weights 8, 16
		225	196	210	1^{225}	-15^{30}	OA(15,16)
+	256	45	16	6	13^{45}	-3^{210}	OA(3,16); $\text{Bilin}_{2\times 4}(2)$; Brouwer [4]; from a partial spread: projective 4-ary [15,4] code with weights 8, 12; from a partial spread of 4-spaces: projective binary [45,8] code with weights 16, 24
		210	170	182	2^{210}	-14^{45}	OA(14,16)
+	256	51	2	12	3^{204}	-13^{51}	vanLint-Schrijver(1); $VO_4^-(4)$ affine polar graph; projective 4-ary [17,4] code with weights 12, 16
		204	164	156	12^{51}	-4^{204}	vanLint-Schrijver(4)
+	256	60	20	12	12^{60}	-4^{195}	OA(4,16); from a partial spread: projective 4-ary [20,4] code with weights 12, 16; Brouwer [4]; from a partial spread of 4-spaces: projective binary [60,8] code with weights 24, 32
		195	146	156	3^{195}	-13^{60}	OA(13,16)
–	256	66	2	22	2^{231}	-22^{24}	Krein_2
		189	144	126	21^{24}	-3^{231}	Krein_1
+	256	68	12	20	4^{187}	-12^{68}	Brouwer [4]; projective binary [68,8] code with weights 32, 40
		187	138	132	11^{68}	-5^{187}	
+	256	75	26	20	11^{75}	-5^{180}	OA(5,16); $\text{Bilin}_{2\times 2}(4)$; $VO_4^+(4)$ affine polar graph; from a partial spread: projective 4-ary [25,4] code with weights 16, 20; from a partial spread of 4-spaces: projective binary [75,8] code with weights 32, 40
		180	124	132	4^{180}	-12^{75}	OA(12,16)
+	256	85	24	30	5^{170}	-11^{85}	vanLint-Schrijver(1); possibly projective binary [85,8] code with weights 40, 48
		170	114	110	10^{85}	-6^{170}	vanLint-Schrijver(2)
+	256	90	34	30	10^{90}	-6^{165}	OA(6,16); from a partial spread: projective 4-ary [30,4] code with weights 20, 24; from a partial spread of 4-spaces: projective binary [90,8] code with weights 40, 48
		165	104	110	5^{165}	-11^{90}	OA(11,16)
+	256	102	38	42	6^{153}	-10^{102}	$2^8.L_2(17)$ (rk 3); Liebeck; vanLint-Schrijver(2); possibly projective 4-ary [34,4] code with weights 24, 28
		153	92	90	9^{102}	-7^{153}	vanLint-Schrijver(3)
+	256	105	44	42	9^{105}	-7^{150}	OA(7,16); from a partial spread: projective 4-ary [35,4] code with weights 24, 28; Brouwer [4]; from a partial spread of 4-spaces: projective binary [105,8] code with weights 48, 56

∃	v	k	λ	μ	r^f	s^g	Comments
		150	86	90	6^{150}	-10^{105}	OA(10,16)
+	256	119	54	56	7^{136}	-9^{119}	$VO_8^-(2)$ affine polar graph; projective binary [119,8] code with weights 56, 64; $RSHCD-$; two-graph
		136	72	72	8^{119}	-8^{136}	two-graph
+	256	120	56	56	8^{120}	-8^{135}	OA(8,16); from a partial spread: projective 4-ary [40,4] code with weights 28, 32; from a partial spread of 4-spaces: projective binary [120,8] code with weights 56, 64; $RSHCD+$; two-graph
		135	70	72	7^{135}	-9^{120}	OA(9,16); Goethals-Seidel(2,15); $VO_8^+(2)$ affine polar graph; two-graph
+	257	128	63	64	7.516^{128}	-8.516^{128}	Paley(257); two-graph−∗
?	259	42	5	7	5^{147}	-7^{111}	
		216	180	180	6^{111}	-6^{147}	
?	260	70	15	20	5^{168}	-10^{91}	
		189	138	135	9^{91}	-6^{168}	
?	261	52	11	10	7^{116}	-6^{144}	
		208	165	168	5^{144}	-8^{116}	
?	261	64	14	16	6^{144}	-8^{116}	
		196	147	147	7^{116}	-7^{144}	
?	261	80	25	24	8^{116}	-7^{144}	
		180	123	126	6^{144}	-9^{116}	
?	261	84	39	21	21^{29}	-3^{231}	
		176	112	132	2^{231}	-22^{29}	
?	261	130	64	65	7.578^{130}	-8.578^{130}	two-graph−∗?
?	265	96	32	36	6^{159}	-10^{105}	
		168	107	105	9^{105}	-7^{159}	
?	265	132	65	66	7.639^{132}	-8.639^{132}	two-graph−∗?
?	266	45	0	9	3^{209}	-12^{56}	
		220	183	176	11^{56}	-4^{209}	
+	269	134	66	67	7.701^{134}	-8.701^{134}	Paley(269); two-graph−∗
?	273	72	21	18	9^{104}	-6^{168}	
		200	145	150	5^{168}	-10^{104}	
?	273	80	19	25	5^{182}	-11^{90}	
		192	136	132	10^{90}	-6^{182}	
+	273	102	41	36	11^{90}	-6^{182}	S(2,6,91)
		170	103	110	5^{182}	-12^{90}	
?	273	136	65	70	6^{168}	-11^{104}	
		136	69	66	10^{104}	-7^{168}	
−	273	136	67	68	7.761^{136}	-8.761^{136}	Conf
!	275	112	30	56	2^{252}	-28^{22}	$q_{22}^2 = 0$; two-graph−∗
		162	105	81	27^{22}	-3^{252}	$McL.2/U_4(3).2$; McLaughlin graph; unique by Goethals and Seidel [26]; $q_{11}^1 = 0$; two-graph−∗
!	276	44	22	4	20^{23}	-2^{252}	$\binom{24}{2}$
		231	190	210	1^{252}	-21^{23}	
?	276	75	10	24	3^{230}	-17^{45}	
		200	148	136	16^{45}	-4^{230}	
?	276	75	18	21	6^{160}	-9^{115}	
		200	145	144	8^{115}	-7^{160}	
−	276	110	28	54	2^{253}	-28^{22}	Krein_2; Abs
		165	108	84	27^{22}	-3^{253}	Krein_1; Abs
?	276	110	52	38	18^{45}	-4^{230}	
		165	92	108	3^{230}	-19^{45}	
+	276	135	78	54	27^{23}	-3^{252}	Conway-GandS; two-graph
		140	58	84	2^{252}	-28^{23}	two-graph
+	277	138	68	69	7.822^{138}	-8.822^{138}	Paley(277); two-graph−∗
+	279	128	52	64	4^{216}	-16^{62}	two-graph−∗
		150	85	75	15^{62}	-5^{216}	two-graph−∗
+	280	36	8	4	8^{90}	-4^{189}	$J_2/3.PGL_2(9)$ (rk 4); $U_4(3)$ polar graph
		243	210	216	3^{189}	-9^{90}	
?	280	62	12	14	6^{155}	-8^{124}	
		217	168	168	7^{124}	-7^{155}	
?	280	63	14	14	7^{135}	-7^{144}	
		216	166	168	6^{144}	-8^{135}	

∃	v	k	λ	μ	r^f	s^g	Comments
+	280	117	44	52	5^{195}	-13^{84}	
		162	96	90	12^{84}	-6^{195}	Sym(9) (rk 5); Mathon and Rosa
?	280	124	48	60	4^{217}	-16^{62}	two-graph?
		155	90	80	15^{62}	-5^{217}	two-graph?
+	280	135	70	60	15^{63}	-5^{216}	$J_2/3.PGL_2(9)$ (rk 4); two-graph
		144	68	80	4^{216}	-16^{63}	two-graph

5.3 See Also

§I.4	Steiner systems.
§II.2	Mutually orthogonal latin squares.
§IV.1	Association schemes.
§IV.21	Generalized quadrangles.
§IV.24	Hadamard matrices.
§IV.52	Conference matrices.
§IV.37	Quasi-symmetric designs.
§VI.6	Two-graphs.
[9, 34]	Early surveys of the construction of strongly regular graphs.
[6]	Parameters of infinite series, the isomorphisms, existence and nonexistence results.

References

[1] E.R. Berlekamp, J.H. van Lint and J.J. Seidel, A strongly regular graph derived from the perfect ternary Golay code, in: *A Survey of Combinatorial Theory*, (J.N. Srivastava et al.; eds.), North-Holland, Amsterdam, 1973, 25-30.

[2] A.E. Brouwer, The uniqueness of the strongly regular graph on 77 points, *J. Graph Theory* 7 (1983), 455-461.

[3] A.E. Brouwer, Polarities of G. Higman's symmetric design and a strongly regular graph on 176 vertices, *Aequat. Math.* 25 (1982), 77-82.

[4] A.E. Brouwer, Some new two-weight codes and strongly regular graphs, *Discrete Applied Math.* 10 (1985), 111-114.

[5] A.E. Brouwer, Distance regular graphs of diameter 3 and strongly regular graphs, *Discrete Math.* 49 (1984), 101-103.

[6] A.E. Brouwer, A.M. Cohen and A. Neumaier, *Distance-Regular Graphs*, Springer-Verlag, Heidelberg, 1989.

[7] A.E. Brouwer and W.H. Haemers, Structure and uniqueness of the (81,20,1,6) strongly regular graph, *Discrete Math.* 106/107 (1992), 77-82.

[8] A.E. Brouwer and W.H. Haemers, The Gewirtz graph — an exercise in the theory of graph spectra, *Europ. J. Combin.* 14 (1993), 397-407.

[9] A.E. Brouwer and J.H van Lint, Strongly regular graphs and partial geometries, in: *Enumeration and Design* (D.M. Jackson, S.A. Vanstone; eds.), Academic Press, Toronto, 1984, 85-122.

[10] F.C. Bussemaker, W.H. Haemers, R. Mathon and H.A. Wilbrink, A (49,16,3,6) strongly regular graph does not exist, *Europ. J. Combin.* 10 (1989), 413-418.

[11] A.R. Calderbank and W.M. Kantor, The geometry of two-weight codes, *Bull. London Math. Soc.* 18 (1986), 97-122.

[12] P.J. Cameron, J.-M. Goethals and J.J. Seidel, Strongly regular graphs having strongly regular subconstituents, *J. Algebra* 55 (1978), 257-280.

[13] L.C. Chang, The uniqueness and nonuniqueness of triangular association schemes, *Sci. Record* 3 (1959), 604-613.

[14] L.C. Chang, Association schemes of partially balanced block designs with parameters $v = 28$, $n_1 = 12$, $n_2 = 15$ and $p_{11}^2 = 4$, *Sci. Record* 4 (1960), 12-18.

[15] W.H. Clatworthy, Partially balanced incomplete block designs with two associate classes and two treatments per block, *J. Res. Nat. Bur. Standards* 54 (1955), 177-190.

[16] A. Clebsch, Über die Flächen vierter Ordnung, welche eine Doppelcurve zweiten Grades besitzen, *J für Math.* 69 (1868), 142-184.

[17] W.S. Connor, The uniqueness of the triangular association scheme, *Ann. Math. Stat.* 29 (1958), 262-266.

[18] H.S.M. Coxeter, Self-dual configurations and regular graphs, *Bull. Amer. Math. Soc.* 56 (1950), 413-455.

[19] Ph. Delsarte, Weights of linear codes and strongly regular normed spaces, *Discrete Math.* 3 (1972), 47-64.

[20] Ph. Delsarte, An algebraic approach to the association schemes of coding theory, *Philips Res. Rep. Suppl.* 10, 1973.

[21] Ph. Delsarte, J.-M. Goethals and J.J. Seidel, Bounds for systems of lines and Jacobi polynomials, *Philips Res. Rep.* 30 (1975), 91-105.

[22] M. van Eupen and R. Hill, An optimal ternary $[69, 5, 45]$ code and related codes, *Des. Codes Crypt.* 4 (1994), 271-282.

[23] A. Gewirtz, Graphs with maximal even girth, *Canad. J. Math.* 21 (1969), 915-934.

[24] A. Gewirtz, The uniqueness of $g(2, 2, 10, 56)$, *Trans. New York Acad. Sci.* 31 (1969), 656-675.

[25] J.-M. Goethals and J.J. Seidel, Orthogonal matrices with zero diagonal, *Canad. J. Math.* 19 (1967), 1001-1010.

[26] J.-M. Goethals and J.J. Seidel, The regular two-graph on 276 vertices, *Discrete Math.* 12 (1975), 143-158.

[27] J.-M. Goethals and J.J. Seidel, Strongly regular graphs derived from combinatorial designs, *Canad. J. Math.* 22 (1970), 597-614.

[28] N. Hamada and T. Helleseth, A characterization of some $\{2v_2 + v_3, 3v_1 + v_2; 3, 3\}$-minihypers and some $[15, 4, 9; 3]$-codes with $B_2 = 0$, *J. Stat. Plann. Infer.*, to appear.

[29] D.G. Higman, Invariant relations, coherent configurations and generalized polygons, in: *Combinatorics* (M. Hall, J.H. van Lint; eds.), Math. Centre Tracts 57, Amsterdam, 1974, 27-43.

[30] D.G. Higman and C. Sims, A simple group of order 44,352,000, *Math. Zeitschr.* 105 (1968), 110-113.

[31] G. Higman, On the simple group of D.G. Higman and C.C. Sims, *Illinois J. Math.* 13 (1969), 74-80.

[32] A.J. Hoffman, On the uniqueness of the triangular association scheme, *Ann. Math. Stat.* 31 (1960), 492-497.

[33] A.J. Hoffman and R.R. Singleton, On Moore graphs with diameters 2 and 3, *IBM J. Res. Develop.* 4 (1960), 497-504.

[34] X. Hubaut, Strongly regular graphs, *Discrete Math.* 13 (1975), 357-381.

[35] W.C. Huffman and V.D. Tonchev, The existence of extremal self-dual $[50, 25, 10]$ codes and quasi-symmetric 2-$(49, 9, 6)$ designs, *Des. Codes Crypt.* 6 (1995), 97-106.

[36] T.H. Koornwinder, A note on the absolute bound for systems of lines, *Proc. KNAW A* 79 = *Indag. Math.* 38 (1976), 152-153.

[37] M.G. Kreĭn, Hermitian-positive kernels II, *Amer. Math. Soc. Transl. Ser.* 2 34 (1963), 109-164 (in particular, p. 139). *Ukr. Mat. Zh.* 2 (1950), 10-59.

[38] J.H. van Lint and J.J. Seidel, Equilateral point sets in elliptic geometry, *Proc. Kon. Ned. Acad. Wet. A* 69 (1966), 335-348. (=*Indag. Math.* 28)

[39] J.H. van Lint and A. Schrijver, Construction of strongly regular graphs, two-weight codes and partial geometries by finite fields, *Combinatorica* 1 (1981), 63-73.

[40] R.A. Mathon, Symmetric conference matrices of order $pq^2 + 1$, *Canad. J. Math.* 30 (1978), 321-331.

[41] A. Neumaier, Regular sets and quasi-symmetric 2-designs, in: *Combinatorial Theory* (D. Jungnickel, K. Vedder; eds.), Lecture Notes in Math. 969, Springer-Verlag, Berlin, 1982, 258-275.

[42] A. Neumaier, New inequalities for the parameters of an association scheme, in: *Combinatorics and Graph Theory*, Proc. Calcutta 1980 (S.B. Rao, ed.), Lecture Notes in Math. 885, Springer, Berlin, 1981, 365-367.

[43] R.E.A.C. Paley, On orthogonal matrices, *J. Math. Phys.* 12 (1933), 311-320.

[44] D.V. Pasechnik, Skew-symmetric association schemes with two classes and strongly regular graphs of type $L(2n-1, 4n-1)$, *Acta Appl. Math.* 29 (1992), 129-138.

[45] A.J.L. Paulus, Conference matrices and graphs of order 26, Technische Hogeschool Eindhoven, report WSK 73/06, Eindhoven, 1983, 89 pp.

[46] S. Payne and J.A. Thas, *Finite generalized quadrangles*, Pitman, New York, 1985.

[47] J. Petersen, Sur le théorème de Tait, *L'Intermédiaire des Mathématiciens* 5 (1898), 225-227.

[48] D. Raghavarao, Some aspects of weighing designs, *Ann. Math. Statist.* 31 (1960), 878-884.

[49] L.L. Scott Jr., A condition on Higman's parameters, *Notices Amer. Math. Soc.* 20 (1973), A-97 (701-20-45).

[50] J. Seberry and A.L. Whiteman, New Hadamard matrices and conference matrices obtained via Mathon's construction, preprint, 1985.

[51] J.J. Seidel, Strongly regular graphs with $(-1, 1, 0)$ adjacency matrix having eigenvalue 3, *Lin. Alg. Appl.* 1 (1968), 281-298.

[52] J.J. Seidel, Strongly regular graphs, in: *Surveys in Combinatorics*, (B. Bollobás; ed.), Cambridge University Press, London, 1979, 157-180.

[53] S.S. Shrikhande, The uniqueness of the L_2 association scheme, *Ann. Math. Statist.* 30 (1959), 781-798.

[54] C.C. Sims, On the isomorphism of two groups of order 44,352,000, in: *The Theory of Finite Groups* (R. Brauer, C.-H. Sah; eds.), Benjamin, 1969, 101-108.

[55] M.S. Smith, On the isomorphism of two simple groups of order 44,352,000, *J. Algebra* 41 (1976), 172-174.

[56] D.E. Taylor, Graphs and block designs associated with the three-dimensional unitary groups, in: *Combinatorial Math.*, (D.A. Holton; ed.) Springer-Verlag, Berlin, 1974, 128-131.

[57] D.E. Taylor, Regular two-graphs, *Proc. London Math. Soc.* 35 (1977), 257-274.

[58] J.A. Thas, Some results concerning $\{(q+1)(n-1); n\}$-arcs and $\{(q+1)(n-1)+1; n\}$-arcs in finite projective planes of order q, *J. Combin. Theory A* 19 228-232.

[59] W.D. Wallis, A.P. Street and J.S. Wallis, *Combinatorics: Room Squares, Sum-Free Sets, Hadamard Matrices*, Lecture Notes in Math. 292, Springer-Verlag, Berlin, 1972.

[60] H.A. Wilbrink and A.E. Brouwer, A $(57, 14, 1)$ strongly regular graph does not exist, *Indag. Math.* 45 (1983) 117-121.

6 Two-Graphs

EDWARD SPENCE

6.1 Definitions and Examples

1. A *two-graph* (V, Δ) consists of a set V whose elements are the vertices, and a collection Δ of unordered triples of the vertices (the *odd triples*), such that each 4-tuple of V contains, as subsets, an even number of elements of Δ. If (V, Δ) is a two-graph, then so also is the pair $(V, V^3 \setminus \Delta)$; the *complement* of (V, Δ).
2. A *clique* of a two-graph (V, Δ) is a subset C of V such that every triple from C is in Δ.
3. The two-graphs (V, Δ) and (V', Δ') are *isomorphic* if there is a bijection $V \to V'$ that induces a bijection $\Delta \to \Delta'$.
4. The automorphism group $\mathrm{Aut}(V, \Delta)$ of a two-graph (V, Δ) is the group of permutations of V that preserve Δ.

6.1 **Remark** Given a (simple) graph $\Gamma = (V, E)$, the set Δ of triples of V whose induced subgraph in Γ contains an odd number of edges gives rise to a two-graph (V, Δ). In fact, every two-graph can be represented in this way.

6.2 **Example** The ladder graph on four vertices and a triangle extended by an isolated vertex give rise to the same two-graph. The odd triples are $\{123, 124, 134, 234\}$.

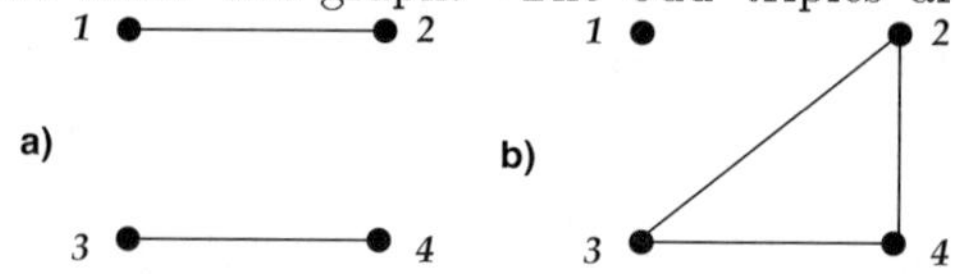

6.3 **Remarks**

1. In Example 6.2, (b) is obtained from (a) by *switching* the adjacencies of vertex 2. In general, switching with respect to a subset $X \subset V$ consists of interchanging adjacency and nonadjacency between X and its complement $V \setminus X$.
2. Graphs $\Gamma_1 = (V, E_1)$ and $\Gamma_2 = (V, E_2)$ represent the same two-graph if they are related by (the equivalence relation of) switching with respect to some $X \subset V$. Thus a two-graph can be identified with the switching class of a graph.

The *switching class* of a graph Γ comprises those graphs with ∓ 1 adjacency matrices DAD, where D is a diagonal matrix with entries ± 1, and A is the ∓ 1 adjacency matrix of Γ (-1 for adjacency).

6.4 **Remark** Since A and DAD (D as above) have the same eigenvalues, the *eigenvalues of a two-graph* are the eigenvalues of the ∓ 1 adjacency matrix of any graph in its switching class.

6.5 **Theorem** [13] Two-graphs, switching classes of graphs and the isomorphism classes of eulerian graphs (every vertex has even valency) on n vertices are equal in number.

6.6 **Theorem** [18] Let $\Gamma = (V, E)$ be a graph in the switching class of a two-graph (V, Δ). Then $\mathrm{Aut}(\Gamma)$ is a subgroup of $\mathrm{Aut}(V, \Delta)$.

6.7 **Theorem** [18] Let $\Gamma_i = (V, E_i)$, $i = 1, 2, \ldots, s$ denote the nonisomorphic graphs in the switching class of a two-graph (V, Δ), and let $|V| = n$, $|\mathrm{Aut}(V, \Delta)| = m$, $|\mathrm{Aut}\Gamma_i| = m_i$. Then $\sum_{i=1}^{s} \frac{m}{m_i} = 2^{n-1}$.

6.8 **Remark** Mallows and Sloane [13] prove that every automorphism of a two-graph is also an automorphism of some graph in the switching class of the two-graph. However, there are two-graphs (V, Δ) each of whose switching classes contain graphs, all of which have automorphism group a *proper* subgroup of $\mathrm{Aut}(V, \Delta)$. All such two-graphs on fewer than $n \leq 10$ vertices are enumerated in [3].

6.9 **Table** A list of the numbers $N_T(n)$ of pairwise nonisomorphic two-graphs on n vertices ($n \leq 10$) as found in [3]. There all two-graphs on $n \leq 9$ vertices are given, together with their eigenvalues.

n	3	4	5	6	7	8	9	10
$N_T(n)$	2	3	7	16	54	243	2038	33120

6.2 Two-Graphs from Trees

6.10 **Construction** (Cameron [4])

1. Let T be a tree with edge set E and let Δ be the set of 3-subsets of E not contained in paths in T. Then (E, Δ) is a two-graph.
2. Let X be the end vertices of a reduced vertex-2-colored (black and white) tree T (no divalent vertices) and let Δ comprise the triples of X whose vertices are pairwise joined by paths that meet in a black vertex. Then (X, Δ) is a two-graph. Use of the color white yields the complementary two-graph.

A two-graph Ω contains a graph Γ as an *induced substructure* if Γ is a subgraph of some graph in the switching class of Ω.

6.11 **Theorem** [4] A two-graph arises from a tree by Construction 6.10(1) if and only if the two-graph contains neither the pentagon nor the hexagon as an induced substructure. Only isomorphic trees yield isomorphic two-graphs.

6.12 **Theorem** [4] A two-graph arises from a tree by Construction 6.10(2) if and only if the two-graph does not contain the pentagon as an induced substructure. Only isomorphic colored reduced trees yield isomorphic two-graphs.

6.13 **Remark** Cameron [5] produces formulae for the numbers of labeled and labeled reduced two-graphs obtained by Construction 6.10. His proofs use various enumeration results for trees.

6.3 Groups Represented by Two-Graphs

Let $\Gamma = (V, E)$ be a graph with vertex set V. *The Coxeter group* of Γ, $\mathrm{Cox}(\Gamma)$ has a generator s_v for each $v \in V$, and defining relations $s_v^2 = 1, (s_v s_w)^3 = 1$ if $v \sim w$, and $(s_v s_w)^2 = 1$ if $v \not\sim w$. The set of all products of generators of even length forms a normal subgroup of index 2, denoted by $\mathrm{Cox}^+(\Gamma)$ [6]. Analogously [21], the *Tsaranov group* of Γ has generators $t_v, v \in V$ and is given by

$$\mathrm{Ts}(\Gamma) := \langle t_v, v \in V \,:\, t_v^3 = 1,\ (t_v t_w^{-1})^2 = 1 \text{ if } v \sim w,\ (t_v t_w)^2 = 1 \text{ if } v \not\sim w \rangle.$$

6.14 Remark Switching Γ with respect to a set of vertices corresponds to replacing Tsaranov's generators for these vertices by their inverses. It follows that the Tsaranov group is an invariant of the two-graph (V, Δ) corresponding to the switching class of Γ. Thus isomorphic two-graphs have isomorphic Tsaranov groups.

6.15 Theorem [21] The group $\mathrm{Ts}(\Delta)$ of the two-graph (V, Δ) is finite if and only if its smallest eigenvalue is greater than -3. If this is the case, then $\mathrm{Ts}(\Delta) \cong \mathrm{Cox}^+(\Pi)$, where Π is a "Coxeter-Dynkin graph" of type A, D, or E.

6.16 Remark All two-graphs with smallest eigenvalue greater than -3 are identified in [21].

6.17 Theorem [21] Let $\Delta(T)$ be the two-graph arising from a tree T by means of Construction 6.10(1). Then $\mathrm{Ts}(\Delta(T)) \cong \mathrm{Cox}^+(T)$.

6.4 Regular Two-Graphs

6.18 Theorem Suppose the graph $\Gamma = (V, E)$ on n vertices has ∓ 1 adjacency matrix A with eigenvalues $\lambda_1, \lambda_2, \ldots, \lambda_d$ and $-\rho$ of multiplicity $n - d$. Then using the values of $\mathrm{trace}(A)$ and $\mathrm{trace}(A^2)$,

$$\rho^2(n-d) \le d(n-1) \text{ with equality if and only if } \lambda_1 = \lambda_2 = \ldots = \lambda_d.$$

Equality in Theorem 6.18 is of particular interest for the two-graph with Γ in its switching class. The following two definitions are equivalent [18]:

1. A two-graph (V, Δ) is *regular* if each pair of vertices in V is contained in the same number of triples of Δ.
2. A two-graph (V, Δ) is *regular* if and only if it has two distinct eigenvalues $\rho_1 > 0 > \rho_2$, say, where $\rho_1\rho_2 = 1 - |V|$. The set $\{\rho_1, \rho_2\}$ is the *spectrum* of the two-graph.

6.19 Remark The switching class of a graph Γ, and hence any two-graph, can be represented geometrically as a set of *equiangular lines* [12]. Let $-\rho = \lambda_{\mathrm{min}}$ be the smallest eigenvalue of the ∓ 1 adjacency matrix A of Γ (on n vertices) and suppose that, as above, $-\rho$ has multiplicity $n - d$. Then $\rho I + A$ is positive semi-definite of rank d and so can be represented as the Gram matrix of the inner products of n vectors in euclidean d-space. Since these vectors have the same norm $\sqrt{\rho}$ and mutual inner products ± 1, any pair of the n lines spanned by them have the same angle ϕ with $\cos\phi = 1/\rho$. Conversely, given a set of n nonorthogonal equiangular lines in euclidean d-space there exists a two-graph from which it can be constructed by the above method [18]. Thus, from Theorem 6.18, the maximum cardinality n of such a set satisfies $n \le d(\rho^2 - 1)/(\rho^2 - d)$ and this bound is achieved if and only if the corresponding two-graph is regular.

6.20 Remark In a regular two-graph with n vertices, the eigenvalues ρ_1 and ρ_2 are odd integers except possibly when $\rho_1 + \rho_2 = 0$. In this latter case the ∓ 1 adjacency matrix A of any graph in the switching class of the regular two-graph satisfies $A^2 = (n-1)I$ (the conference two-graph [18]).

6.21 Theorem [18] Suppose that (V, Δ) is a regular two-graph with spectrum $\{\rho_1, \rho_2\}$ and let $\Gamma = (V, E)$ be any graph in its switching class with an isolated vertex $v \in V$. Then $(V \backslash \{v\}, E)$ is a strongly regular graph (§VI.5) with Seidel spectrum (the eigenvalues of the ∓ 1 adjacency matrix) $\{\rho_0, \rho_1, \rho_2\}$, where $\rho_0 = \rho_1 + \rho_2$. This graph is a *descendant* of the two-graph.

6.22 Remark In the opposite direction, if Γ is a strongly regular graph (see §VI.5) with parameters (v, k, λ, μ), then the graph obtained from Γ by adjoining an isolated vertex belongs to the switching class of a regular two-graph if and only if $k = 2\mu$ [18].

6.23 Theorem [18] Given ρ_2 there are only finitely many possible ρ_1 such that there exists a regular two-graph with spectrum $\{\rho_1, \rho_2\}$ [18].

6.24 Table The only possible spectra for regular two-graphs with $\rho_2 = -3$ or -5 are listed.

$-\rho_2$	3	3	3	3	3	5	5	5	5	5	5	5	5	5	5
ρ_1	1	3	5	9	21	1	3	5	7	15	19	25	35	55	115
n	4	10	16	28	64	6	16	26	36	76	96	126	176	276	576

6.25 Remark Of the pairs (ρ_1, ρ_2) listed it is known that $(21, -3)$ and $(115, -5)$ are impossible [18] while the only ones remaining open are $(15, -5)$ and $(19, -5)$.

▷ Conference Two-Graphs

A *C-matrix* of order n is a square matrix C with zero diagonal and off-diagonal entries ± 1, such that $C^2 = (n-1)I$. C-matrices are essentially symmetric or skew according as $n \equiv 2$ or $0 \pmod 4$ [7]. Symmetric C-matrices (the ones of interest here) are *conference* matrices.

6.26 Remark Conference matrices are discussed in §IV.52.

6.27 Theorem A conference matrix of order n is in the switching class of a regular two-graph with eigenvalues $\rho_1 = \sqrt{n-1}, \rho_2 = -\sqrt{n-1}$. The two-graph is a *conference* two-graph.

6.28 Theorem [18] Necessary conditions for the existence of a conference two-graph with eigenvalues ρ_1, ρ_2 such that $\rho_1 + \rho_2 = 0$ are that $\rho_1^2 \equiv 1 \pmod 4$, and ρ_1^2 is a sum of squares of integers.

6.29 Theorem [17] If $q \equiv 1 \pmod 4$ is a prime power, then there exists a conference two-graph of order $q + 1$ (the Paley two-graph).

6.30 Theorem [26] If a conference two-graph of order $m + 1$ exists, then there also exists a conference two-graph of order $m^d + 1$ for all $d \geq 1$.

6.31 Theorem [14] If $q = 4t - 1$ is a prime power and $p + 1 = 4t + 2$ is the order of a conference two-graph, then there exists a conference two-graph on $pq^2 + 1$ vertices.

6.32 Remark There are two further infinite families of conference two-graphs, known as the Möbius and Minkowski two-graphs, which are constructed in the setting of Möbius and Minkowski geometry (see [19, 27]). The Möbius two-graphs are isomorphic to the Paley two-graphs.

6.33 Construction (*Minkowski two-graphs* [18]) Let $V = V(2, q)$ be the vector space of dimension 2 over $\mathbb{F}_q$, where q is an odd prime power. Suppose that the one-dimensional subspaces of V are partitioned into two disjoint sets P and N of equal size $(q+1)/2$. Then $(V \cup \{\infty\}, \Delta)$, where Δ comprises the set of triples $\{\infty, x, y\}$, with $x, y \in V$, such that $\langle x - y\rangle \in P$, is a self-complementary conference two-graph on $q^2 + 1$ vertices.

6.34 Remark The switching class of a two-graph from Construction 6.33 contains strongly regular graphs of degrees $q(q-1)/2$ and $q(q+1)/2$.

6.35 Theorem [18] If there exist s mutually orthogonal latin squares of order $2s-1$ or there exists a 2-$(2s^2-2s+1, s, 1)$ design, then there exists a conference two-graph on $(2s-1)^2+1$ vertices.

▷ Symplectic and Orthogonal Two-Graphs

1. [18] Let $V = V(2m, 2)$ denote the vector space of dimension $2m$ over $\mathbb{F}_2$ and let $B : V \times V \to \mathbb{F}_2$ be a nondegenerate, alternating bilinear form. The *symplectic* two-graph $\Sigma(2m, 2) = (V, \Delta)$ consists of the set V and the set $\Delta = \{\{u, v, w\} \subseteq V \;:\; u, v, w \text{ are distinct and } B(u, v) + B(v, w) + B(w, u) = 0\}$. Its eigenvalues are $\rho_1 = 1 + 2^m, \rho_2 = 1 - 2^m$.
2. [18] There are essentially two quadratic forms on $V(2m, 2)$ defined by $Q^+(x) = x_1x_2 + x_3x_4 + \cdots + x_{2m-1}x_{2m}$ and $Q^-(x) = x_1^2 + x_2^2 + x_1x_2 + \cdots + x_{2m-1}x_{2m}$. The *orthogonal* two-graph $\Omega^\epsilon(2m, 2)$, $\epsilon = \pm$, consists of the set of points $\Omega^\epsilon := \{x \in V \mid Q^\epsilon(x) = 0\}$, and the set of triples of distinct $u, v, w \in \Omega^\epsilon$ such that $B(u, v) + B(v, w) + B(w, u) = 0$. $\Omega^+(2m, 2)$ has eigenvalues $\rho_1 = 1 + 2^{m-1}, \rho_2 = 1 - 2^m$ and $\Omega^-(2m, 2)$ has eigenvalues $\rho_1 = 1 + 2^m, \rho_2 = 1 - 2^{m-1}$.

6.36 Remark Taylor [25] proved that the automorphism group of $\Omega^\epsilon(2m, 2)$ is the symplectic group, while $\Sigma(2m, 2)$ has automorphism group the symplectic group extended by the translations of $V(2m, 2)$. See §VI.1.

▷ Unitary Two-Graphs

Let H be a nondegenerate hermitian form on the projective plane $\mathrm{PG}(2, q^2)$, where q is an odd prime power, and let $U = \{x \mid H(x, x) = 0\}$ be the corresponding unital. Then $|U| = q^3 + 1$. The *unitary* two-graph $U(q) = (U, \Delta)$ consists of the set Δ of triples $\{x, y, z\}$, with $x, y, z \in U$, for which $H(x, y)H(y, z)H(z, x)$ is a square or nonsquare according as $q \equiv \mp 1 \pmod 4$. The Unitary group $\mathrm{P\Gamma U}(3, q^2)$ acts 2-transitively on Ω.

6.37 Proposition The eigenvalues of $U(q)$ are $\rho_1 = q^2, \rho_2 = -q$ [25]. Its switching class contains a strongly regular graph of degree $q(q^2+1)/2$.

6.38 Remark Taylor [24] proved that the automorphism group of $U(5)$ is $P\Gamma U(3, 5^2)$.

▷ Two Sporadic Regular Two-Graphs

6.39 Construction There is a unique 4-design with parameters 4-$(23, 7, 1)$ (it is a Witt design). Its residual design, a 3-$(22, 7, 4)$ design, is also unique. In each of these two designs distinct blocks meet in one or three points. The residual design has 176 blocks and its point-block incidence matrix M satisfies $MM^t = 40I + 16J$, $MJ = 56J$, $JM = 7J$. The eigenvalues of MM^t are clearly 392 and 40 and hence M^tM has eigenvalues $392, 40, 0$. Thus $A := M^tM - 5I - 2J$, a symmetric matrix of order 176 with zero diagonal and elements ± 1 elsewhere, has eigenvalues 35 and -5. It is therefore the adjacency matrix of a graph in the switching class of a regular two-graph on 176 vertices.

6.40 Construction If N is the point-block incidence matrix of the 4-$(23, 7, 1)$ design, then $NN^t = 56I + 21J$, $NJ = 77J$, $JN = 7J$. N^tN has eigenvalues $539, 56, 0$ and those of $B := N^tN - 5I - 2J$ are $28, 51, -5$. Hence B satisfies $(B - 51I)(B + 5I) = -3J$, $BJ = 28J$, $NB = 51N - 7J$. A straightforward calculation shows that the symmetric matrix C of order 276, with zero diagonal and ± 1 elsewhere, given by $C = \begin{bmatrix} J-I & J-2N \\ J-2N^t & B \end{bmatrix}$ satisfies $(C-55I)(C+5I) = 0$, and hence is in the switching class of a regular two-graph on 276 vertices. This two-graph is unique [9].

▷ Symmetric Hadamard Matrices of Order 36

6.41 Remark Let E be the adjacency matrix of any graph in the switching class of a regular two-graph on 36 vertices. By complementation, if necessary, it may be assumed that the two-graph, and hence E, has eigenvalues 5 and -7, so that $(E-5I)(E+7I) = 0$. If $H := E + I$, then H is a symmetric Hadamard matrix of order 36 with constant diagonal (see §IV.24). In [3] there were constructed 91 such regular two-graphs, 11 from latin squares of order 6 and 80 from Steiner triple systems of order 15 (see also [8]). A further 136 were found using a computer and a backtracking algorithm [22]. Many of these new two-graphs, and some of the older ones, have a particular shape. The corresponding Hadamard matrices contain, as a principal sub-matrix, the matrix A of order 20 defined by $A = \begin{bmatrix} I+P & I-P \\ I-P & I+P \end{bmatrix}$, where P is the ∓ 1 adjacency matrix of the Petersen graph. Assume this to be the case. Then for $\begin{bmatrix} A & N \\ N^t & B \end{bmatrix}$ to be a Hadamard matrix of order 36 it is necessary that $N = \begin{bmatrix} M \\ M \end{bmatrix}$, where M is of size 10×16 and is determined by Theorem 6.42.

6.42 Theorem [22] There exist diagonal ± 1 matrices D_1, D_2 such that $M = D_1 [\,C \quad \mathbf{j}\,] D_2$, where C is the ± 1 incidence matrix of a 2-(10, 4, 2) design (Table I.1.18). Conversely, let C be the ± 1 incidence matrix of a 2-(10, 4, 2) design and suppose that D_1 and D_2 are diagonal matrices of order 10, 16 respectively. If $M = D_1 [\,C \quad \mathbf{j}\,] D_2$, and $B = 6I - \frac{1}{2} M^t M$, then $B - I$ is in the switching class of a regular two-graph of size 16,

$$H = \begin{bmatrix} I+P & I-P & M \\ I-P & I+P & M \\ M^t & M^t & B \end{bmatrix},$$

a symmetric Hadamard matrix of order 36 and $H - I$ is in the switching class of a regular two-graph on 36 vertices.

6.43 Remark There are three nonisomorphic 2-(10, 4, 2) designs and each of them yields regular two-graphs on 36 vertices. In total 100 nonisomorphic regular two-graphs of order 36 are characterized by this method. See [22] for the details.

▷ Completely Regular Two-Graphs

A regular two-graph is *completely regular* if, for all k, every k-clique is in a constant number α_k of $(k+1)$-cliques.

6.44 Theorem There are unique completely regular two-graphs on 10, 16, 28, 36, and 276 vertices. No others exist except possibly on 1128 or 3160 vertices.

6.45 Remark Neumaier [16] allowed the further possibilities of 96, 288, and 640 vertices. These were shown subsequently not to exist [1, 2].

▷ Two-Graph Geometries

6.46 Theorem In a regular two-graph with eigenvalues $\rho_1 > \rho_2$, the maximum size of a clique c is given in terms of the smaller eigenvalue ρ_2 by $|c| \le 1 - \rho_2$ [20].

A *maximal* clique is one in which this upper bound of Theorem 6.46 is achieved.

A *two-graph geometry* is a set C of maximal cliques in a regular two-graph (V, Δ), such that each triple from Δ is in a unique clique from C.

6.47 Examples [10]

1. In a regular two-graph, every odd triple is contained in $1 - \frac{1}{4}(\rho_1 + 3)(\rho_2 + 3)$ 4-cliques. When $\rho_2 = -3$, this number equals 1, and these cliques are maximal. Thus by taking C to be the set of all 4-cliques, three two-graph geometries are obtained corresponding to the three regular two-graphs on 10, 16, and 28 vertices.
2. The known regular two-graph on 176 vertices has the Higman-Sims group acting as a 2-transitive group of automorphisms [25] (see §VI.1). The subgroup M_{22} has an orbit C of size 18480 acting on the maximal 6-cliques of the two-graph and every odd triple is contained in exactly one 6-clique.

▷ Numerical Data

6.48 Table Data concerning the known regular two-graphs with eigenvalues ρ_1 and $\rho_2 = -3, -5$ and the conference two-graphs on ≤ 50 vertices. N is the number of known nonisomorphic two-graphs on n vertices and N^* indicates that no further two-graphs on n vertices can exist. The main reference is [3]. For $n = 30, 36$ see also [22]. In the cases $n = 46, 50$ additional work by Mathon [15] improved $N(46)$ to 97 and $N(50)$ to 19. A further 24 regular two-graphs on 50 vertices were found as a result of the discovery of 12 new Steiner systems $S(2, 4, 25)$ [11] (see Theorem 6.35), and seven more were recently found [23], thus increasing $N(50)$ to 50. The case $n = 126$ corresponds to $U(5)$ and $n = 176, 276$ are the two sporadic two-graphs described earlier in Constructions 6.39 and 6.40.

n	6	10	14	16	18	26	28	30
ρ_1, ρ_2	$\pm\sqrt{5}$	± 3	$\pm\sqrt{13}$	$5, -3$	$\pm\sqrt{17}$	± 5	$9, -3$	$\pm\sqrt{29}$
N	1*	1*	1*	1*	1*	4*	1*	6*
n	36	38	42	46	50	126	176	276
ρ_1, ρ_2	$7, -5$	$\pm\sqrt{37}$	$\pm\sqrt{41}$	$\pm\sqrt{45}$	± 7	$25, -5$	$35, -5$	$55, -5$
N	227	11	18	97	50	1	1	1*

6.49 Remark It can happen that the switching class of a regular two-graph itself contains strongly regular graphs. There are five values of $n \le 50$ for which this is possible, namely $n = 10, 16, 26, 36$, and 50. The numbers of such graphs are listed, together with the numbers of nonisomorphic descendants. Here $|(n, \rho_0, \rho_1, \rho_2)|$ denotes the number of strongly regular graphs on n vertices with Seidel spectrum ρ_0, ρ_1, ρ_2.

n	10	16	26	36	50
ρ_1, ρ_2	$(3, -3)$	$(5, -3)$	$(5, -5)$	$(7, -5)$	$(7, -7)$
$\|(n-1, \rho_1+\rho_2, \rho_1, \rho_2)\|$	1	1	15	3854	312
$\|(n, \rho_1, \rho_1, \rho_2)\|$	1	1	10	180	1482
$\|(n, \rho_2, \rho_1, \rho_2)\|$	1	2	10	32548	1482

6.5 See Also

§I.3	t-designs are used in Constructions 6.39 and 6.40.
§II.2	MOLS are used in Theorem 6.35.
§IV.24	Hadamard matrices are used to construct regular two-graphs (e.g., in Remark 6.41.
§IV.52	Conference matrices are used to construct conference two-graphs (Theorem 6.27).
§VI.5	Strongly regular graphs arise in the switching classes of some two-graphs (e.g., Remark 6.34).

References

[1] A. Blokhuis and A.E. Brouwer, Uniqueness of the Zara graph on 126 points and nonexistence of a completely regular graph on 288 points, in: *Papers Dedicated to J.J. Seidel* (P.J. de Doelder, J. de Graaf, J.H. van Lint; eds.), *Report 84-WSK-03, Techn. Univ. Eindhoven*, 1984, 6-19.

[2] A. Blokhuis and H. Wilbrink, Characterisation theorems for Zara graphs, *Europ. J. Combin.* 10 (1989), 57-68.

[3] F.C. Bussemaker, R.A. Mathon and J.J. Seidel, Tables of two-graphs, in: *Combinatorics and Graph Theory*, (S.B. Rao; ed.), Springer-Verlag, Berlin, 1981, 70-112; Report 79-WSK-05, Techn. Univ. Eindhoven, 1979.

[4] P.J. Cameron, Two-graphs and trees, *Discrete Math.* 127 (1994), 63-74.

[5] P.J. Cameron, Counting two-graphs related to trees, *Elect. J. Comb.* 2 (1995), #R4.

[6] H.S.M. Coxeter and W.O.J. Moser, *Generators and Relations for Discrete Groups*, Springer-Verlag, Berlin, 1965.

[7] P. Delsarte, J.M. Goethals and J.J. Seidel, Orthogonal matrices with zero diagonal II, *Canad. J. Math.* 23 (1971), 816-832.

[8] J.M. Goethals and J.J. Seidel, Strongly regular graphs derived from combinatorial designs, *Canad. J. Math.* 22 (1970), 597-614.

[9] J.M. Goethals and J.J. Seidel, The regular two-graph on 276 vertices, *Discrete Math.* 12 (1975), 143-158.

[10] W.H. Haemers, Regular 2-graphs and extensions of partial geometries, *Europ. J. Combin.* 12 (1991), 115-123.

[11] E.S. Kramer, S.S. Magliveras and R. Mathon, The Steiner systems $S(2,4,25)$ with nontrivial automorphism group, *Discrete Math.* 77 (1989), 137-157.

[12] J.H. van Lint and J.J. Seidel, Equilateral point sets in elliptic geometry, *Proc KNAW A 69; Indag. Math.* 28 (1966), 335-348.

[13] C.L. Mallows and N.J.A. Sloane, Two-graphs, switching classes and Euler graphs are equal in number, *SIAM J. Appl. Math.* 28 (1975), 876-880.

[14] R. Mathon, Symmetric conference matrices of order pq^2+1, *Canad. J. Math.* 30 (1978), 321-331.

[15] R. Mathon, On self-complementary strongly regular graphs, *Discrete Math.* 69 (1988), 263-281.

[16] A. Neumaier, Completely regular two-graphs, *Arch. Math.* 38 (1982), 378-384.

[17] R.E.A.C. Paley, On orthogonal matrices, *J. Math. Phys.* 12 (1933), 311-320.

[18] J.J. Seidel, A survey of two-graphs, *Coll. Intern. Teorie Combin., Atti dei convegni Lincei 17, Roma,* 1976, 481-511.

[19] J.J. Seidel and D.E. Taylor, Two-graphs, a second survey, in: *Algebraic Methods in Graph Theory* (L. Lovász, V.T. Sós; eds.), *Coll. Math. Soc. J. Bolyai* 25, 1981, 689-711.

[20] J.J. Seidel, More about two-graphs, in: *Fourth Czechoslovakian Symposium on Combinatorics, Graphs and Complexity* (J. Nešetřil and M. Fiedler; eds.), 1992, 297-309.
[21] J.J. Seidel and S.V. Tsaranov, Two-graphs, related groups and root systems, *Bull. Soc. Math. Belgique,* XLII (1990), 695-711.
[22] E. Spence, Regular two-graphs on 36 vertices, *Lin. Alg. Appl.,* to appear.
[23] E. Spence, unpublished, 1995.
[24] D.E. Taylor, Graphs and block designs associated with the three-dimensional unitary groups, in: *Lecture Notes in Mathematics* 403, Springer-Verlag, Berlin, 1974.
[25] D.E. Taylor, Regular two-graphs, *Proc. London. Math. Soc.* 35 (1977), 257-274.
[26] R.J. Turyn, On *C*-matrices of arbitrary powers, *Canad. J. Math.* 23 (1971), 531-535.
[27] H. Wilbrink, Two-graphs and geometries, unpublished.

7 Classical Geometries

Albrecht Beutelspacher

7.1 Geometry and Incidence

A *finite incidence structure*, or *finite geometry*, is a finite set of *points*, a finite set of *lines*, and a relation, *incidence*, between them.

7.1 Remark This basic definition of finite geometry is much too weak by itself to define interesting geometries. In what follows, axioms for the incidence relation are added in order to specify "interesting" types of geometries. In this section, all geometries are finite.

7.2 Projective Planes

A *projective plane* is an incidence structure $(\mathcal{P}, \mathcal{L})$ of points and lines such that

1. any two (distinct) points are incident with exactly one line;
2. any two (distinct) lines are incident with exactly one point;
3. there exists a quadrangle, i.e., four points no three collinear.

A geometry that fulfils the axioms (1) and (2), but not (3), is a *near pencil.*

For any two lines ℓ and m of a projective plane there is a one-to-one mapping from ℓ onto m. Therefore, for a projective plane P with a finite number of points there is a positive integer n such that any line of P has exactly $n+1$ points. This number n is the *order* of P.

7.2 Proposition A finite projective plane of order n has n^2+n+1 points, n^2+n+1 lines, and $n+1$ lines through each point.

7.3 Remark One of the most difficult questions in finite geometry is "What positive integers are orders of finite projective planes?" Very little is known: All known examples have prime power order. (This does not imply that any two projective planes of the same prime power order are isomorphic! See §VI.8.) The Bruck-Ryser-Chowla theorem (§I.5.2) excludes an infinite number of integers. Only one further number has been excluded: there is no projective plane of order 10. See Remark VI.9.12 for a discussion of the computer search for a projective plane of order 10.

7.4 Table An overview of the state of knowledge concerning small projective planes.

n	2	3	4	5	6	7	8	9	10	11	12	13	14	15	16
Existence	y	y	y	y	n	y	y	y	n	y	?	y	n	?	y
Number	1	1	1	1	0	1	1	4	0	≥ 1	?	≥ 1	0	?	≥ 22

7.5 Theorem Let P be a geometry consisting of a finite number of points and a finite number of at least two lines for which any two distinct points are on exactly one line, and there is an integer $n \geq 2$ such that any line has exactly $n+1$ points. Then the following assertions are equivalent:

1. P is a projective plane.
2. Any point of P is on at most $n+1$ lines.
3. Any point of P is on exactly $n+1$ lines.
4. There are exactly n^2+n+1 points of P.
5. P is a 2-$(n^2+n+1, n+1, 1)$ design (or, in other words an $S(2, n+1, n^2+n+1)$ Steiner system).

7.6 Construction The most important class of projective planes can be obtained as follows. Let F be a division ring. (So, the construction works in particular if F is a field.) Let (a, b, c) be a triple with $a, b, c \in F$, where at least one of the elements a, b, c is different from 0. By $(a : b : c)$ denote the set of all nonzero multiples of (a, b, c). So $(a : b : c) = \{(fa, fb, fc) | f \in F\backslash\{0\}\}$. Similarly define $[x : y : z] = \{[fx, fy, fz] | f \in F\backslash\{0\}\}$ for any three elements $x, y, z \in F$ with at least one of the elements x, y, z different from 0. Now consider the symbols $(a : b : c)$ as points and the symbols $[x : y : z]$ as lines and say that a point $(a : b : c)$ lies on the line $[x : y : z]$ if and only if $ax + by + cz = 0$. This gives a projective plane.

$(a : b : c)$ and $[x : y : z]$ are the *homogeneous coordinates* of the corresponding point and line.

7.7 Remark Another way of describing these projective planes is as follows. Let V be a 3-dimensional vector space over the division ring F. Then the points of P are the 1-dimensional subspaces of V and the lines are the 2-dimensional subspaces of V. These projective planes are denoted by PG$(2, F)$. In addition, if $F = \mathbb{F}_q$ is a finite field with q elements, then the projective plane obtained in this way has order q. In this case PG$(2, F)$ is also denoted by PG$(2, q)$. So PG$(2, q)$ is a symmetric $(q^2+q+1, q+1, 1)$ BIBD.

7.8 Theorem Since for any prime power q there is a finite field with q elements, then for any prime power q there is a projective plane of order q.

7.9 Remark A consequence of the representation theorem of projective geometry is that the projective planes of the form PG$(2, F)$ are exactly the desarguesian projective planes, those projective planes in which the theorem of Desargues is valid. Equivalently, the projective planes PG$(2, F)$ are precisely those projective planes that can be embedded as a plane in a higher-dimensional projective space (see §VI.8 for a discussion of nondesarguesian projective planes).

7.3 Affine Planes

An *affine plane* is a geometry consisting of a set of points and a set of lines, where each line is a subset of the point set, such that the following axioms hold

1. Any two points are on exactly one line.
2. For any point P outside a line ℓ there is exactly one line through P that has no point in common with ℓ. (This is the *euclidean parallel axiom.*)
3. There exist three points not on a common line.

7.10 Remark If one defines a *parallelism* among the lines by saying that two lines are parallel if they are equal or have no point in common, then the parallelism of an affine plane is an equivalence relation such that any two nonparallel lines meet. The equivalence classes belonging to the parallelism are *parallel classes*. By axiom (2), any parallel class of an affine plane is a partition of the point set by lines.

7.11 Construction Given an affine plane A one can define a new geometry P as follows: the points of P are the points of A and the parallel classes of A; the lines of P are the lines of A and one new line $g^\star$. The incidence is defined as follows: line $g^\star$ is incident with all parallel classes, and each line ℓ of A is now incident with its points in A together with the parallel class to which it belongs. Then P is a projective plane. Conversely, if one removes a line g together with all its points from a projective plane P one gets an affine plane P_g. Two lines of the affine plane P_g are parallel if and only if the projective lines containing them meet the line g in the same point. Sometimes one calls g the *infinite line* of A and the points of g the *infinite points* of A.

For any two lines ℓ and m of an affine plane there is a one-to-one mapping from ℓ onto m. Therefore, for an affine plane A with a finite number of points there is a positive integer n such that any line of A has exactly n points. This number n is the *order* of A.

7.12 Proposition The order of a finite affine plane is the same as the order of the corresponding projective plane. So an integer is the order of an affine plane if and only if it is the order of a projective plane.

7.13 Proposition A finite affine plane of order n has n^2 points, n^2+n lines, and $n+1$ lines through each point.

7.14 Theorem Let A be a geometry consisting of a finite number of points and a finite number of at least two lines for which any two distinct points are on exactly one line, and there is an integer $n \geq 2$ such that any line has exactly n points. Then the following assertions are equivalent:

1. A is an affine plane.
2. Any point of A is on exactly $n+1$ lines.
3. There are exactly n^2 points of A.
4. A is a 2-$(n^2, n, 1)$ design (or, in other words an $S(2, n, n^2)$ Steiner system).

7.15 Construction The usual way to construct affine planes is as follows. Let F be a division ring. Then the geometry $\mathrm{AG}(2, F)$ has as points the pairs (a, b) with $a, b \in F$, and its lines are the sets of points (x, y) satisfying an equation $y = mx + b$ $(m, b \in F)$ or an equation $x = c$ with $c \in F$. Then $\mathrm{AG}(2, F)$ is an affine plane.

7.16 Remark Another way of describing this affine plane is the following. Let V be a 2-dimensional vector space over F. The points of $\mathrm{AG}(2, F)$ are all vectors of V and the lines are all cosets $v + \langle u \rangle$, where $u \in V \setminus \{0\}$. If F is a finite field with q elements, then AG$(2, F)$ is an affine plane of order q. In this case AG$(2, F)$ is also denoted by AG$(2, q)$.

7.17 Remark The representation theorem of affine geometry says that the affine planes AG$(2, F)$ are precisely the desarguesian affine planes. Another way of saying this is that the planes AG$(2, F)$ are precisely those affine planes that can be embedded in a higher-dimensional affine space. See §7.6.

7.4 Linear Spaces

A *linear space* is a geometry (incidence structure) consisting of a set of points and a set of lines, where each line is a subset of the point set, such that any two points are contained in a unique line, and any line contains at least two points.

7.18 Remark A linear space is a pairwise balanced design (PBD) with $\lambda = 1$ (see §III.4).

7.19 Remark These axioms are very weak, so there are lots of nonisomorphic linear spaces. (Two linear spaces S and S' are *isomorphic* if there is a bijection from the point set of S onto the point set of S' such that three points of S are on a common line if and only if the image points are in a common line of S'.)

7.20 Theorem (Doyen) For any integer $v \geq 10$ there are at least 2^v nonisomorphic linear spaces on v points.

7.21 Table The exact numbers of linear spaces with $v \leq 11$ points.

v	0	1	2	3	4	5	6	7	8	9	10	11
Number	1	1	1	2	3	5	10	24	69	384	5250	232929

7.22 Remark Table III.4.20 explicitly lists all linear spaces of order $v \leq 15$ in which all blocks have size at least 3.

7.23 Theorem (de Bruijn and Erdös) If a finite linear space S has at least two lines, then its number b of lines is as least as big as its number v of points; also, $b = v$ holds if and only if S is a near-pencil or a projective plane.

A linear space is *semiaffine* if through any point outside a line ℓ there is at most one line disjoint to ℓ. This is a generalization of projective planes (where "at most one" is "no") and affine planes (where "at most one" is "exactly one").

7.24 Theorem (Dembowski) A finite semiaffine plane is one of the following: a projective plane, a near-pencil, a projective plane minus one of its points, an affine plane with one of its points at infinity added, or an affine plane.

7.25 Remark In the proof of Theorem 7.24, *parameters* of the linear space play a crucial role: apart from b and v, the number r_P of lines through the point P and k_ℓ, the number of points on the line ℓ. Using these parameters the definition of a semiaffine linear space can be expressed as $r_P - k_\ell = \{0, 1\}$ for any nonincident point line pair (P, ℓ). One shows that r_P and k_ℓ can take only very few values and characterizes these cases.

7.5 Projective Spaces

A *projective space* is a geometry consisting of a set P of points and a set L of lines, where each line is a subset of the point set, such that the following axioms hold

1. Any two points are on exactly one line.
2. (*Veblen-Young axiom*) Let A, B, C, D be four distinct points of which no three are collinear. If the lines AB and CD intersect each other, then the lines AD and BC also intersect each other.
3. Any line has at least three points.

7.26 **Examples** Projective spaces can be obtained as follows. Let V be a vector space of dimension $d+1$ over the division ring F. The geometry $P(V)$ that has as its *points* the 1-dimensional subspaces of V and as its *lines* the 2-dimensional subspaces of V is a projective space.

7.27 **Remark** Another way of describing the same geometry is as follows. Let $d+1$ be a positive integer and F a division ring. The points are represented by $(d+1)$-tuples $(a_0, a_1, \ldots, a_d) \neq (0, 0, \ldots, 0)$ with $a_i \in F$, where two such tuples $(a_0, a_1, \ldots, a_d)$ and $(a'_0, a'_1, \ldots, a'_d)$ represent the same point if there is a $c \in F$ with $(a_0, a_1, \ldots, a_d) = c \cdot (a'_0, a'_1, \ldots, a'_d)$. Denote by $(a_0 : a_1 : \ldots : a_d)$ the point corresponding to $(a_0, a_1, \ldots, a_d)$; $(a_0 : a_1 : \ldots : a_d)$ are *homogeneous coordinates.* The line through $(a_0 : a_1 : \ldots : a_d)$ and $(b_0 : b_1 : \ldots : b_d)$ is the set of points

$$\{(b_0 : b_1 : \ldots : b_d)\} \cup \{(a_0 + cb_0 : a_1 + cb_1 : \ldots : a_d + cb_d) \mid c \in F\}.$$

The projective space $P(V)$ is also denoted by $\mathrm{PG}(d, F)$. If F is a finite field with q elements, then write $\mathrm{PG}(d, q)$ instead of $\mathrm{PG}(d, F)$.

7.28 **Theorem** (*Representation theorem of projective geometry*) Any projective space that is not a projective plane is isomorphic to some $P(V)$.

Let P be an arbitrary projective space. A *(linear) subspace* of P is a set U of points of P such that with any two points P, Q, any point on the line PQ is contained in U. Any subspace is, together with the lines contained in it, a projective space.

7.29 **Remark** Any point, any line, and the whole point set are subspaces of P.

If S is an arbitrary set of points of P, then the *span* $\langle S\rangle$ of S is the intersection of all subspaces containing S. The span $\langle S\rangle$ is again a subspace.

A set S of points of P is *independent*, if for any $x \in S$, then $x \notin \langle S \setminus \{x\}\rangle$. Loosely speaking, S is independent if each point of S contributes to $\langle S\rangle$. An independent set that spans a projective space P is a *basis* of P.

7.30 **Theorem** Let U be a subspace of P, and let Q be a point of P outside U. Then the subspace W spanned by U and Q consists precisely of the points on the lines QX with $X \in U$.

7.31 **Remark** A set of points is a basis if and only if it is a maximal independent set if and only if it is a minimal spanning set.

7.32 **Proposition** All bases have the same number of elements.

A projective space P has *dimension* d if a basis has $d+1$ elements. A subspace of dimension 2 is a *plane*; in a d-dimensional projective space, the *hyperplanes* are the subspaces of dimension $d-1$.

7.33 Proposition The dimensions of the span and the intersection of two subspaces U_1 and U_2 are connected by the dimension formula: $dim(\langle U_1, U_2\rangle) = dim(U_1) + dim(U_2) - dim(U_1 \cap U_2)$. In particular, any t-dimensional subspace not contained in a hyperplane H intersects H in a subspace of dimension $t-1$.

The *quotient geometry* (or *residual geometry*) P/U of a projective space P modulo a t-dimensional subspace U has as points the $(t+1)$-dimensional subspaces through U, and as lines the $(t+2)$-dimensional subspaces through U.

7.34 Proposition If P has dimension d, then P/U is again a projective space, whose dimension is $d-t-1$.

Any two lines of a projective space P can bijectively be mapped onto each other. Thus, if P is finite, there is an integer $n \geq 2$ such that any line has exactly $n+1$ points. This number n is the *order* of P.

7.35 Theorem (Tallini, Bose and Burton; see [15]) Let U be a set of points in a d-dimensional finite projective space of order n. If any subspace of dimension $d-t$ has at least one point in common with U, then $|U| \geq n^t + \cdots + n + 1$ with equality if and only if U is a t-dimensional subspace. In particular, a set of points that is hit by any line has at least as many points as a hyperplane, with equality if and only if it is a hyperplane.

7.36 Remark The dimension d and the order n are the essential parameters of a finite projective space in the sense that one can express nearly any other parameter as a function of the dimension and the order. Let P be a finite d-dimensional projective space of order n. There are $n^t + \cdots + n + 1$ points in a t-dimensional subspace of P. The number of points and the number of hyperplanes of P are both equal to $n^d + \cdots + n + 1$. Any subspace and any quotient geometry of P also have order n.

The *gaussian coefficients* $\begin{bmatrix} d \\ i \end{bmatrix}_q$ denote the number of i-dimensional vector subspaces of a d-dimensional vector space over $\mathbb{F}_q$.

7.37 Proposition Let q be a prime power and d and i positive integers with $i \leq d$, then

$$\begin{bmatrix} d \\ i \end{bmatrix}_q = \frac{(q^d-1)(q^{d-1}-1)\cdots(q^{d-i+1}-1)}{(q^i-1)(q^{i-1}-1)\cdots(q-1)}.$$

The points of $\mathrm{PG}(d,q)$ together with the i-dimensional flats of $\mathrm{PG}(d,q)$ as blocks and incidence by natural containment form an incidence structure denoted by $\mathrm{PG}_i(d,q)$.

7.38 Proposition $\mathrm{PG}_i(d,q)$ is a block design with parameters

$$v = \begin{bmatrix} d+1 \\ 1 \end{bmatrix}_q = (q^{d+1}-1)/(q-1),\ k = \begin{bmatrix} i+1 \\ 1 \end{bmatrix}_q = (q^{i+1}-1)/(q-1),\ \lambda = \begin{bmatrix} d-1 \\ i-1 \end{bmatrix}_q$$

(also $b = \begin{bmatrix} d+1 \\ i+1 \end{bmatrix}_q$ and $r = \begin{bmatrix} d \\ i \end{bmatrix}_q$).

7.6 Affine Spaces

A *parallelism* of a geometry consisting of a set of points and a set B of blocks is an equivalence relation ("parallel") on B that satisfies the *euclidean parallel axiom*: For any point P and any block B there is exactly one block C through P that is parallel to B. The equivalence classes of a parallelism are *parallel classes*.

An *affine space* is a geometry A consisting of points and lines such that the following axioms hold:

1. Any two points are on exactly one line.
2. A has a parallelism.
3. (*Triangle axiom*) If A, B, C are three noncollinear points and A', B' are two distinct points such that the line $A'B'$ is parallel to AB, then the line on A' parallel to AC and the line on B' parallel to BC intersect in some point C'.
4. There exist three points not on a common line.

7.39 Remark Most properties of affine spaces follow from the fact that they have a *projective closure*: For any affine space A there exists a projective space P and a hyperplane H of P such that A consists precisely of the points and lines not contained in H. Also, two lines of A are parallel if and only if they meet in the same point of H. The hyperplane H is the *hyperplane at infinity* of A.

The *order* of an affine space is the number of points on one of its lines. The *dimension* of an affine space is the dimension of the corresponding projective space.

7.40 Theorem Let A be a finite affine space of dimension d and order n.

1. Any line of A has exactly n points.
2. The number of points of A is n^d.
3. If U is a t-dimensional subspace of A, then the quotient geometry A/U is a projective space of dimension $d-t+1$ and order n.

$\mathrm{AG}(d,q)$ denotes the affine space of dimension d and order q with the property that its projective closure is desarguesian.

7.41 Remark When one wants to prove that a given geometry is an affine space, it is often very difficult to show that there is a parallelism. In this situation, Theorem 7.42 due to Buekenhout [8] is a great help.

7.42 Theorem Let A be a linear space that contains at least one line with at least four points. If any three noncollinear points generate an affine plane, then A is an affine space.

7.43 Remark Theorem 7.42 is false if one drops the hypothesis that there is a line with at least four points. Hall triple systems (§IV.25) give examples in which all lines have size three.

In $\mathrm{AG}(d,q)$ the cosets of $\{0\}$ are *points*, those of 1-dimensional spaces *lines*, those of $(n-1)$-dimensional spaces *hyperplanes*; in general, the cosets of i-dimensional subspaces are *i-flats*. The points of $\mathrm{AG}(d,q)$ together with the i-dimensional flats of $\mathrm{AG}(d,q)$ as blocks and incidence by natural containment form an incidence structure denoted by $\mathrm{AG}_i(d,q)$.

7.44 Proposition $\mathrm{AG}_i(d,q)$ is a block design with parameters $v=q^d$, $k=q^i$ and $\lambda=\begin{bmatrix} d-1 \\ i-1 \end{bmatrix}_q$ (also $b=q^{d-i}\begin{bmatrix} d \\ i \end{bmatrix}_q$ and $r=\begin{bmatrix} d \\ i \end{bmatrix}_q$).

7.7 Subplanes

Let P be a projective plane with point set $\mathcal{P}$ and line set $\mathcal{L}$. A projective plane P' with point set $\mathcal{P}'$ and line set $\mathcal{L}'$ is a *subplane* of P if $\mathcal{P}' \subseteq \mathcal{P}$ and $\mathcal{L}' \subseteq \mathcal{L}$. In other words, a substructure of P is a subplane if it is a projective plane.

7.45 Theorem (Bruck) Let P be a finite projective plane of order n, and let P' be a subplane of order $m < n$ of P. Then $n = m^2$ or $m^2 + m \leq n$.

7.46 Remark It is unknown whether equality can hold; this would imply that the order $n = m^2 + m$ of P is not a prime power. The case $n = m^2$ is of particular interest.

A subplane of order m of a projective plane of order $n = m^2$ is a *Baer subplane.* These are the "largest" subplanes.

7.47 Theorem Let P be a finite projective plane of order n, and let P' be a subplane of order $m < n$ of P. Then the following assertions are equivalent:

1. P' is a Baer subplane of P.
2. any point outside P' is on (exactly) one line of P'.
3. any line not in P' has exactly one point in common with P'.

7.48 Theorem Let P be a projective plane whose points and lines are represented as homogeneous coordinates over some field F. Let K be a subfield of F and consider the geometry P' consisting of the homogeneous coordinates for points and lines whose entries are in K. Then P' is a subplane of P.

7.49 Remark The subfields of a finite field $\mathbb{F}_{p^a}$ where p is a prime are precisely those fields $\mathbb{F}_{p^b}$ where b divides a. Therefore, the subplanes of a projective plane $\mathrm{PG}(2, p^a)$ with p a prime are exactly the projective planes $\mathrm{PG}(2, p^b)$ where b divides a.

7.50 Proposition A projective plane $\mathrm{PG}(2, q)$ has a Baer subplane if and only if q is a square.

7.8 Baer Subspaces

7.51 Proposition (Bruck) Let P be a finite projective space of dimension d and order q. Let B be a subspace of P of order q'. If $q' < q$, then $q' \leq \sqrt{q}$.

A d-dimensional subspace of P whose order is $\sqrt{q}$ is a *Baer subspace* (sometimes termed a *Baer subgeometry*).

7.52 Theorem If $P = P(V)$ is a finite desarguesian projective space of order $q = p^n$, p prime, then P has a subspace of order q' if and only if $q' = p^m$ with $m \mid n$. In particular, P has a Baer subspace if and only if q is a square.

7.53 Theorem Let B be a Baer subspace of the d-dimensional projective space P.

1. If U is an i-dimensional subspace of P that intersects B in a j-dimensional subspace, then there is precisely one subspace of dimension $2i-j$ of B containing U. In particular, if U has no point in common with B then U is contained in a unique $(2i+1)$-dimensional subspace of B. So, any point of $P - B$ lies on a unique line of B.
2. Any subspace of dimension $d-i$ of P contains at least one subspace of dimension $d - 2i$ of B; this bound is sharp.

7.54 Theorem (Beutelspacher and Ueberberg; see [17]) Let B and B' be two Baer subspaces of a projective space. Then the number of points in B and B' equals the number of hyperplanes that B and B' have in common. In particular, if B and B' are disjoint, then they also have no hyperplane in common.

7.9 Collineations

Let P be a projective space. A *collineation* (*automorphism*) of P is a bijective map α on the point set of P that preserves collinearity; that is, P, Q, R collinear implies $\alpha(P)$, $\alpha(Q)$, $\alpha(R)$ collinear.

7.55 Remark The collineations of P form a group with respect to composition of maps. Also, any collineation maps a t-dimensional subspace of P onto a t-dimensional subspace.

Let $P = P(V)$ be a projective space defined over the F-vector space V. A *semilinear map* of V is a map $f : V \to V$ together with an automorphism σ of F such that $f(v+w) = f(v) + f(w)$ for all $v, w \in V$, and $f(a \cdot v) = \sigma(a) \cdot f(v)$ for all $a \in F$, $v \in V$.

7.56 Theorem (*Fundamental theorem of projective geometry*) Any invertible semilinear map of V induces a collineation of $P(V)$. Conversely, any collineation of $P(V)$ is induced by an invertible semilinear map.

The group of all collineations of the d-dimensional projective space $P(V)$ is denoted by $\mathrm{P\Gamma L}(d+1, F)$. The set $\mathrm{PGL}(d+1, F)$ of collineations that are induced by invertible linear maps of V is a normal subgroup of $\mathrm{P\Gamma L}(d+1, F)$. The set of collineations induced by linear maps with determinant 1, $\mathrm{PSL}(d+1, F)$, is a group of collineations, the *little projective group* of $P(V)$.

7.57 Proposition The group $\mathrm{PSL}(d+1, F)$ acts doubly transitively on the points of $P(V)$.

Let P be an arbitrary projective space. A collineation α of P is a *central collineation* if there exists a point C (the *center*) and a hyperplane H (the *axis*) such that α fixes any line on C and any point on H. If α is a central collineation different from the identity, its center and axis are uniquely determined. A central collineation different from the identity with center C and axis H is an *elation* if $C \in H$ and a *homology* otherwise.

7.58 Theorem The group $\mathrm{PGL}(d+1, F)$ is generated by the central collineations of $P(V)$; the group $\mathrm{PSL}(d+1, F)$ is generated by the elations of $P(V)$.

Let (C, H) be a point-hyperplane pair of a projective space P. P is (C, H)-*transitive* if for any line ℓ on C not contained in H, for any two points P, Q on ℓ and $P, Q \neq C, \ell \cap H$ there is a (unique) central collineation with center C, axis H such that $\alpha(P) = Q$.

7.59 Theorem (Baer) A projective space P is (C, H)-transitive for any point-hyperplane pair (C, H) if and only if P is desarguesian.

A *Baer collineation* is a collineation that fixes a Baer subspace pointwise.

7.60 Theorem (Baer) A collineation of order 2 of a projective space is either a central collineation or a Baer collineation.

7.61 Remark A collineation of a projective space P may have fixed points and fixed hyperplanes.

7.62 Theorem The number of fixed points of a collineation of P equals the number of fixed hyperplanes of P.

7.63 Theorem Let G be an arbitrary automorphism group of a finite projective space P.

1. The number of point orbits of P equals the number of hyperplane orbits.

2. The group G is point-transitive if and only if it is transitive on the hyperplanes.
3. The group G is 2-fold transitive on the points of P if and only if it is 2-fold transitive on the hyperplanes of P.

7.64 Theorem (Wagner) If a group G of collineations of a projective space P acts transitively on the points of P and contains a central collineation different from the identity, then P is desarguesian and G contains the little projective group of P. See [10].

7.10 Spreads

Let t and d be positive integers with $t < d$. A *t-spread* in a d-dimensional projective space P is a set S of t-dimensional subspaces such that any point of P is on exactly one element of S.

7.65 Theorem A finite projective space $P = \mathrm{PG}(d, q)$ contains a t-spread if and only if $t+1$ is a divisor of $d+1$. In particular, $\mathrm{PG}(d, q)$ contains a spread of lines if and only if d is odd.

7.66 Construction One construction for t-spreads is as follows. Let t and d be positive integers such that $t+1$ divides $d+1$. Let q be a prime power. Then the field $F^\star = \mathbb{F}_{q^{d+1}}$ of order q^{d+1} contains $F = \mathbb{F}_{q^{t+1}}$ as a subfield. Consider $F^\star$ as an F-vector space. Then the set of 1-dimensional subspaces of $F^\star$ forms a t-spread in the d-dimensional projective space associated to the $\mathbb{F}_q$-vector space $\mathbb{F}_{q^{d+1}}$. The t-spreads obtained in this way are regular (see below).

A *t-regulus* of $P = \mathrm{PG}(2t+1, q)$ is a set R of $q+1$ mutually skew t-dimensional subspaces such that any line intersecting three elements of R intersects all elements of R. These lines are *transversals* of R.

7.67 Proposition Any three mutually skew t-dimensional subspaces of $\mathrm{PG}(2t+1, q)$ determine uniquely a t-regulus containing them.

7.68 Remark The points covered by a 1-regulus in $\mathrm{PG}(3, q)$ are the points of a hyperbolic quadric.

A t-spread S in $\mathrm{PG}(2t+1, q)$ is *regular* if for any three elements of S, the regulus determined by them is also contained in S. In general, a t-spread S of $P = \mathrm{PG}(d, q)$ is *regular*, if for any two elements U, W of P, S induces a regular t-spread in $\langle U, W\rangle$.

7.69 Theorem A finite projective space $P = \mathrm{PG}(d, q)$ contains a regular t-spread if and only if $t+1$ divides $d+1$.

7.70 Remark The importance of t-spreads lies, among other things, in the fact that, using them, one can construct affine planes. This construction is due to André. Let $P = \mathrm{PG}(2t+2, q)$ be a projective space of dimension $2t+2$, and let H be a hyperplane of P. Let S be a t-spread of H. Then define the geometry $A(S)$ with points those of $P \backslash H$, and lines the $(t+1)$-dimensional subspaces of P that intersect H precisely in an element of S.

7.71 Theorem (André) $A(S)$ is an affine plane of order q^{t+1}. More precisely, it is a *translation plane*, in that the group of elations whose axis is the line at infinity (group of "translations") acts transitively on the points of the affine plane. Conversely, for any translation plane A, there is an integer t, so that a t-spread S in $H = \mathrm{PG}(2t+1, q)$ exists for which $A = A(S)$. In addition, $A(S)$ is desarguesian if and only if S is regular.

7.72 Remark There are lots of spreads that are not regular. One class of examples is obtained as follows. Let $t = 1$. First, the set of transversals of a 1-regulus R is again a 1-regulus, the *opposite* regulus of R. Start with a regular 1-spread S_0. Replace one regulus of S_0 by its opposite regulus to obtain a 1-spread S_1. If S_1 contains a regulus, replace this regulus by its opposite regulus to obtain S_2. Continue in this manner. Any 1-spread so obtained is *subregular.* If one replaces only one regulus of a 1-spread in PG$(3, q)$, one obtains a 1-spread that is not regular. The corresponding affine planes are the Hall planes. See §VI.8.3.

7.11 Unitals

A *unital* is a 2-$(n^3 + 1, n + 1, 1)$ BIBD, i.e., a geometry with $n^3 + 1$ points, $n + 1$ points on each line such that any two distinct points are on exactly one line.

7.73 Remark Here the interest is not in abstract unitals, but in unitals embedded in a projective plane of order $q = n^2$. The classical examples of unitals are obtained as follows.

7.74 Construction Let P be the desarguesian projective plane of order q^2 represented using homogeneous coordinates. Then the points $(x : y : z)$ defined by $xx^q + yy^q + zz^q = 0$ form a unital; this unital is a *hermitian curve.* Another way of describing this unital is as follows. It is the set of absolute points of a unitary polarity of P. (A point P is *absolute* with respect to a polarity Π if P lies on the line $\Pi(P)$. A polarity is *unitary* if it has the maximum number $q^3 + 1$ of absolute points.)

7.75 Remark Another very important class of unitals are the *Buekenhout-Metz unitals.* First, a description of the plane: Let $P =$ PG$(4, q)$ be the 4-dimensional projective space of order q, and let H be a hyperplane of P. Let S be a spread of lines of H, i.e., a set S of lines such that any point of H is on exactly one line of S. Then the geometry $A(S)$ has as points those in $P\backslash H$ and as lines the planes of P that intersect H just in an element of S. Then $A(S)$ is an affine plane of order q^2 (see Theorem 7.71). If the spread is regular, then $A(S)$ is desarguesian. Now fix a line $s \in S$ and consider a point $Q \in s$. Then there exists an ovoid O in the quotient geometry P/Q (s a line in S) such that H is a tangent plane to O. Let U be the set of points of $O\backslash H$ together with the point at infinity of $A(S)$ corresponding to s. Then U is a unital in the projective closure of $A(S)$. If $A(S)$ is desarguesian one can choose O in such a way that U is not a hermitian unital.

7.12 Arcs, Ovals, and Hyperovals

A *k-arc* in a projective plane is a set of k points, no three of which are collinear.

7.76 Theorem If a projective plane of order q has a k-arc then $k \le q+2$. If $k = q+2$, then q is even.

A k-arc is *complete* if it is not properly contained in a $(k + 1)$-arc.

7.77 Theorem Let $P =$ PG$(2, q)$ and let K be a complete k-arc in P with $k < q + 1$.

1. (Segre) If q is even, then $k \le q - \sqrt{q} + 1$.
2. (Segre) If q is odd, then $k \le q - \frac{1}{4}\sqrt{q} + \frac{7}{4}$.
3. (Thas) If q is odd, then $k \le q - \frac{1}{4}\sqrt{q} + \frac{25}{16}$.
4. (Voloch) If q is prime and $q > 5$, then $k \le \frac{44}{45}q + \frac{8}{9}$.
5. (Voloch) If $q = 2^{2e+1}$, $e \ge 1$, then $k \le q - \sqrt{2q} + 2$.
6. (Hirschfeld and Korchmáros) If $q = p^h$, $p \ge 5$, then $k \le q - \frac{1}{2}\sqrt{q} + 5$.

7.78 Remark The bound in Theorem 7.77(1) is sharp when q is an even square. In any projective plane $PG(2, q^2)$ there exists a complete $(q^2 - q + 1)$-arc.

7.79 Theorem Any complete k-arc in a projective plane of order q satisfies $\binom{k-1}{q} \geq q$.

7.80 Remark Szönyi has constructed a family of complete k-arcs for which asymptotically k is of the order of $q^{0.9}$; that is, k is not of the order of q.

> Let P be a projective plane. An *oval* of P is a nonempty set O of points for which no three points of O are collinear (and hence any line intersects O in 0, 1, or 2 points), and through any point P on O there is exactly one line intersecting O just in P (the tangent at P).

7.81 Proposition In a finite projective plane of order q, a k-arc is a oval if and only if $k = q + 1$.

7.82 Examples of ovals in desarguesian projective planes $PG(2, q)$ are the conics. Let $P = PG(2, q)$ be described with homogeneous coordinates. Then a conic is the set of points $(x : y : z)$ satisfying the quadratic equation $x^2 = yz$.

7.83 Theorem (Segre) In a finite desarguesian projective plane of odd order, every oval is a conic.

> A *hyperoval* in a projective plane of order q is a set of $q + 2$ points no three of which are collinear.

7.84 Proposition If a finite projective plane contains a hyperoval, its order must be even. In addition, any $(q + 1)$-arc in a plane of even order q can uniquely be extended to a hyperoval. (The reason for this is that for q even all tangents to a $(q + 1)$-arc pass through a common point, the *nucleus* of the $(q + 1)$-arc.)

> A $\{k; n\}$-*arc* in a projective plane of order q is a nonempty set K of k points such that n is the maximum number of points in K that are collinear. A $\{k; 2\}$-arc is a k-*arc*.

7.85 Theorem (Barlotti) Let K be a $\{k; n\}$-arc in a projective plane of order q. Then $k \leq (q + 1)(n - 1) + 1$.

> The $\{k; n\}$-arcs satisfying $k = (q + 1)(n - 1) + 1$ are *maximal* $\{k; n\}$-*arcs*.

7.86 Examples Trivial maximal $\{k; n\}$-arcs are the singletons ($k = n = 1$) and the whole plane ($n = q + 1$); other examples are the hyperovals.

7.87 Lemma (Barlotti) If a projective plane of order q contains a maximal $\{k; n\}$-arc, then n must be a divisor of q.

7.88 Remark There is one other class of maximal arcs known, namely the Denniston arcs described in Theorem 7.89.

7.89 Theorem (Denniston) Let $PG(2, q)$ be a desarguesian projective plane, i.e., a plane over $\mathbb{F}_q$. If q is even, $q = 2^h$, then for any $n = 2^{h'}$ with $0 \leq h' \leq h$ there exist a maximal $\{k; n\}$-arc in $PG(2, q)$.

7.90 Remark Thas proved that certain translation planes of order 2^{2m} contain maximal $\{k; 2m\}$-arcs. He also stated the conjecture that a desarguesian projective plane of odd order has only trivial maximal arcs.

Maximal arcs are used to construct partial geometries (see §IV.34.3).

7.13 Quadratic Sets

Let P be a projective space and S be a set of points of P. A *tangent* of S is a line with the property that either it intersects S in just one point or is totally contained in S. A *quadratic set* is a set S of points in P for which any line of P that contains at least three points of S is totally contained in S, and for any point Q of S, Q together with the set of points X of P such that XQ is a tangent form a hyperplane or the whole space. This *tangent space* is denoted by S_Q.

A quadratic set is *nondegenerate* if for any point of S, its tangent space is a hyperplane.

7.91 Examples of quadratic sets are the *quadrics*, i.e., point sets in a projective space $P(V)$ whose homogeneous coordinates satisfy an equation of the second degree.

7.92 Theorem (Buekenhout, see [15]) Any quadratic set in a finite projective space is either a quadric or an *ovoid* (a quadratic set that contains no lines).

The *index* of a nondegenerate quadratic set is the positive integer $t+1$, where t is the maximum dimension of a subspace totally contained in the quadratic set.

7.93 Remark An ovoid has index 1, since it contains points but no lines.

7.94 Theorem (Witt) Let S be a nondegenerate quadratic set with index $t+1$ in PG(d, q).

1. If d is even then $t+1 = d/2$. In this case the quadratic set is *parabolic*.
2. If d is odd then either $t+1 = (d-1)/2$, in which case the quadratic set is *elliptic*, or $t+1 = (d+1)/2$, and in this case the quadratic set is *hyperbolic*.

7.95 Remark Among other things, Theorem 7.94 says that there is no ovoid in a finite projective space whose dimension is bigger than 3.

7.14 Blocking Sets

Let P be a projective plane. A *blocking set* is a set S of points of P such that any line contains a point of S and a point outside S.

7.96 Proposition Any projective plane P of order $q \neq 2$ contains a blocking set. For instance, let m_1 and m_2 be two lines of P intersecting in a point Q. Let $Q_i \neq Q$ be a point on m_i, and let Q_3 be a point other than Q_1 or Q_2 on Q_1Q_2. Then the set $S = (m_1 \cup m_2 \backslash \{Q_1, Q_2\}) \cup \{Q_3\}$ is a blocking set having $2q$ points.

7.97 Examples of blocking sets are Baer subplanes and unitals in projective planes of square order.

7.98 Theorem (Bruen) Let P be a finite projective plane of order q, and let S be a blocking set of P. Then $q + \sqrt{q} + 1 \leq |\, S \,| \leq q^2 - \sqrt{q}$.

7.99 Theorem (Blokhuis). Let S be a blocking set in a desarguesian projective plane of order p^{2e+1}, p a prime. Then

1. If $e = 0$, then $|\, S \,| \geq p + (p+3)/2$.
2. If $e \geq 1$, then $|\, S \,| \geq p^{2e+1} + p^{e+1} + 1$.

These bounds are sharp when $e = 0$ or $e = 1$.

A blocking set S is *irreducible* (or *minimal*) if through any point Q of S there is a line intersecting S just in Q. In other words, a blocking set is irreducible if every point is "needed".

7.100 Theorem (Bruen and Thas) Let S be an irreducible blocking set in a projective plane of order q. Then $|S| \leq q\sqrt{q}+1$, with equality if and only if S is a unital.

7.101 Remark Blocking sets in projective spaces of dimension at least 3 have also been examined. There are basically two ways of generalizing the definition. First, one could ask for sets S of points in PG(d, q) such that any t-dimensional subspace contains a point of S and a point outside S ("Italian definition"). Such "t-blocking sets" exist for large q, but there are no "nice" examples known. Using Ramsey theory, Mazzocca and Tallini prove Theorem 7.102. The other possibility to generalize blocking sets is to ask for sets S of points in PG(d, q) such that any $(d-t)$-dimensional subspace contains a point of S and no t-dimensional subspace is contained in S; this is a blocking set "in the German sense".

7.102 Theorem (Mazzocca and Tallini; see [17]) There exists a constant $b(q)$ depending only on q such that PG(d, q) contains a t-blocking set (in the Italian sense) if and only if $d \leq b(q)$.

7.103 Theorem (Beutelspacher) Let S be a t-blocking set (in the German sense) in PG(d, q). Then $|S| \geq q^t + \cdots + 1 + q^{t-1}\sqrt{q}$, with equality if and only if S is the point set of a cone over a Baer subplane.

7.104 Remark For a generalization see [17, 14], dealing with the case where q is not a square.

7.15 Parallelisms

Let P be a d-dimensional projective space. A *parallelism (of lines)* (or *packing*) is an equivalence relation among the lines of P such that through any point of P there is exactly one line of each equivalence class. In other words, a parallelism is a set of 1-spreads of P such that any line is contained in exactly one of those spreads.

7.105 Remark Since a finite projective space has a 1-spread if and only if d is odd, a necessary condition for the existence of a parallelism is that d is odd.

7.106 Theorem

1. [5] If $d = 2^i - 1$, then PG(d, q) has a parallelism.
2. [3] PG$(d, 2)$ has a parallelism if and only if d is odd.

7.107 Remark One could also generalize parallelisms to t-dimensional subspaces instead of lines, but nothing is known in this case.

7.16 See Also

§I.5	Projective planes are symmetric designs.
§I.4	Affine and projective spaces give Steiner systems.
§II.2	Projective planes are equivalent to certain transversal designs, orthogonal arrays, and MOLS.
§III.1, §III.2	Finite geometries provide constructions for PBDs.
§III.4	A view of linear spaces as PBDs.
§IV.25	Hall triple systems generalize affine planes.
§VI.8	Nondesarguesian planes.
[9]	A useful collection of detailed survey articles in finite geometry.
[6]	If not stated otherwise, all references can be found here.

References

[1] L.M. Batten, *Combinatorics of Finite Geometries*, Cambridge University Press, London, 1986.

[2] L.M. Batten and A. Beutelspacher, *The Theory of Linear Spaces: Combinatorics of Points and Lines*, Cambridge University Press, London, 1993.

[3] R. Baker, Partitioning the planes of $AG_{2m}(2)$ into 2-designs, *Discrete Math.* 15 (1976), 205-211.

[4] T. Beth, D. Jungnickel and H. Lenz, *Design Theory*, Cambridge University Press, London, 1986.

[5] A. Beutelspacher, On parallelisms in finite projective spaces, *Geom. Ded.* 3 (1974), 35-40.

[6] A. Beutelspacher, Projective Planes, in: [9], pp. 107-136.

[7] A. Beutelspacher and U. Rosenbaum, *Projektive Geometrie*, Verlag Vieweg, 1992 (English edition: Cambridge University Press, to appear).

[8] F. Buekenhout, Une caractérisation des espaces affines basée sur la notion de droite, *Math. Zeitschr.* 111 (1969), 367-371.

[9] F. Buekenhout (ed.), *Handbook of Incidence Geometry*, Elsevier, New York, 1995.

[10] P.J. Cameron and W. Kantor, 2-transitive and antiflag transitive collineation groups of finite projective spaces, *J. Algebra* 60 (1979), 384-422.

[11] P.J. Cameron, *Projective and Polar Spaces*, QMS Maths Notes 13.

[12] H.S.M. Coxeter, *Projective Geometry*, University of Toronto Press, 1974.

[13] P. Dembowski, *Finite Geometries*, Springer-Verlag, Berlin, 1968.

[14] U. Heim, On t-blocking sets in projective spaces, to appear.

[15] J.W.P. Hirschfeld, *Projective Geometries over Finite Fields*, Oxford University Press, Oxford, 1979.

[16] J.W.P. Hirschfeld, *Finite Projective Spaces of Three Dimensions*, Oxford University Press, Oxford, 1985.

[17] J.W.P. Hirschfeld and J.A. Thas, *General Galois Geometries*, Oxford University Press, Oxford, 1991.

[18] D.R. Hughes and F.C. Piper, *Projective Planes*, Springer-Verlag, Berlin, 1973.

8 Projective Planes, Nondesarguesian

Marialuisa J. de Resmini

8.1 Basic Definitions and Properties

A *projective plane* is an incidence structure $(\mathcal{P}, \mathcal{L})$ of points and lines such that:

1. any two (distinct) points are incident with exactly one line;
2. any two (distinct) lines are incident with exactly one point;
3. there exists a quadrangle, i.e. four points no three collinear.

8.1 **Remark** All lines have the same cardinality equal to that of the set of lines on a point. Therefore, if the projective plane is finite ($|\mathcal{P}|, |\mathcal{L}| \in \mathbb{N}$), then there is an integer $n \geq 2$ such that all lines have $n+1$ points and on each point there are $n+1$ lines. Consequently, the plane contains n^2+n+1 points and as many lines and so a plane of order n is an $(n^2+n+1, n+1, 1)$ block design. The integer n is the *order* of the plane. For the remainder of this section, let π denote a finite projective plane of order n.

8.2 **Theorem** Whenever n is a prime power, $n = q = p^m$, p a prime, $m \geq 1$ an integer, then at least one projective plane (denoted PG$(2, n)$) exists, namely the one coordinatized over $\mathbb{F}_q$, the field with q elements.

8.3 **Construction** To construct PG$(2, q)$ from $\mathbb{F}_q$ define the points to be the homogeneous triples $(x, y, z) \neq (0, 0, 0)$, the lines to be the homogeneous triples $[u, v, w] \neq [0, 0, 0]$, $(x, y, z, u, v, w \in \mathbb{F}_q)$, with incidence defined by:

$$(x, y, z) \text{ is on } [u, v, w] \text{ if and only if } ux + vy + wz = 0.$$

For any π, its *dual plane* π^* is defined as the plane whose points are the lines of π and whose lines are the points of π.

A *subplane* π_0 of π is a subset of the elements of π which form a projective plane having the same incidence relation as π.

8.4 **Theorem** (Bruck) If π is a projective plane of order n and π_0 is a proper subplane $(\pi_0 \neq \pi)$ of order m, then either $n = m^2$, or $n \geq m^2 + m$.

When $n = m^2$, then π_0 is a *Baer subplane* of π and has the property that each line of π meets it in either $m + 1$ points or in a unique point (i.e. is a tangent). For all other possible subplanes there exist exterior lines.

A finite projective plane is *desarguesian* if it is coordinatized by a field. Any plane for which this is not true is *nondesarguesian.*

A projective plane is *pappian* if the Pappus configuration occurs for any pair of distinct lines l, l'. That is, given any distinct A, B, C on l and any distinct A', B', C' on l', the points $AB' \cap A'B$, $AC' \cap A'C$, $BC' \cap B'C$ are collinear. See Example IV.6.2

8.5 **Theorem** Every pappian plane (finite or infinite) is desarguesian. Conversely, in the finite case (by Wedderburn's theorem that every finite skew field is a field) every desarguesian plane is pappian.

A *collineation* of π is an automorphism of π, i.e. a bijection of the point set which preserves the incidences. All collineations of π form the *full collineation group* of π, *Aut*π, under the usual product of mappings. A *collineation group* of π is a subgroup of *Aut*π.

8.6 **Remark** If $\alpha \neq 1$ is a collineation of π, then $Fix(\alpha)$, the set of elements fixed by α, when not empty, is either a subplane or consists of a point P and a line ℓ. In the former case, α is a *planar collineation*; it is a *Baer collineation* if $Fix(\alpha)$ is a Baer subplane. In the latter case, α is a (P, ℓ)-*perspectivity*; in particular, a (P, ℓ)-*homology* if $P \notin \ell$ and a (P, ℓ)-*elation* if P is on ℓ. P is the *center* and ℓ is the *axis* of the perspectivity. All (P, ℓ)- perspectivities of π form a subgroup of Aut π. An involution is either planar or a perspectivity.

π is (P, ℓ)-*transitive* if for any distinct points A, B not on ℓ and collinear with $P\,(A \neq P \neq B)$ there is a (P, ℓ)-perspectivity α in Aut π such that $A^\alpha = B$. $\Sigma = Aut\pi$, $\Sigma_{(P,\ell)}$ denotes the group of all perspectivities with center P and axis ℓ.

8.7 **Proposition** If π has order n, then

1. π is (P, ℓ)-transitive for P on ℓ if and only if $|\Sigma_{(P,\ell)}| = n$.
2. π is (P, ℓ)-transitive for P not on ℓ if and only if $|\Sigma_{(P,\ell)}| = n - 1$.

π is *(m, ℓ)-transitive* if it is (P, ℓ)-transitive for all points P on m; dually, π is *(P, Q)-transitive* if it is (P, ℓ)-transitive for all lines ℓ on Q. If π is (ℓ, ℓ)-transitive for some line ℓ, then ℓ is a *translation line* of π and π is a *translation plane* with respect to ℓ. A *translation point* for π is a point P such that π is (P, P)-transitive; then π is a *dual translation plane* with respect to P.

8.8 Remark If π is a translation plane with respect to a line ℓ, then usually $\pi^\ell = \pi \setminus \ell$ is called a translation plane; the translation plane is considered as an affine plane, its line at infinity being the translation line. Also, the translation group of such an affine plane is the group of all (ℓ, ℓ)-elations.

A projective plane in which every line is a translation line is a *Moufang plane.*

8.9 Proposition Finite Moufang planes are desarguesian. (See [8] for details.)

8.10 Remark Let π be a projective plane of order $n \geq 5$ and G any collineation group of π. Define $T(G) = \{(P, \ell) : G \text{ is } (P, \ell)\text{-transitive}\}$. The projective planes can be classified by looking at all possibilities for $T(G)$. This provides the *Lenz-Barlotti classification* of projective planes (see [2]). However, projective planes can be classified according to the coordinatizing algebra as the properties of such an algebra translate the geometric properties of the plane as well as those of its collineation group.

The *Desargues configuration* is a symmetric (10_3) configuration consisting of two triangles with vertices A_i, B_i, C_i and opposite sides a_i, b_i, c_i, $i = 1, 2$, such that the lines A_1A_2, B_1B_2, C_1C_2 are concurrent at a point P (the triangles are *in perspective* from P) and the points a_1a_2, b_1b_2, c_1c_2 are collinear on a line ℓ (the triangles are *in perspective* from ℓ). (If P is on ℓ, then the configuration is the *little Desargues configuration* and is not a (10_3).) (See §IV.6 for more on configurations.)

8.11 Example The Desargues configuration.

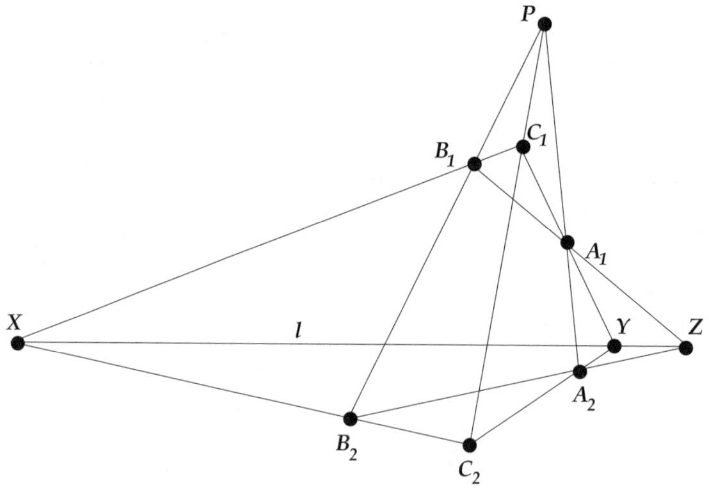

8.12 Proposition Let α be a (P, ℓ)-perspectivity of π. If Δ is a triangle having no vertex or side fixed by α, then Δ and Δ^α are in perspective from both P and ℓ.

8.13 Theorem (Ostrom) Any finite projective plane of order $n \geq 5$ contains a pair of triangles which are in perspective from a point and a line.

8.14 Remark It is impossible to embed the Desargues (10_3) in a plane of order $n < 5$.

A projective plane π is (P,ℓ)-*desarguesian* if for each pair of nondegenerate triangles Δ_i ($i = 1, 2$) with vertices A_i, B_i, C_i and opposite sides a_i, b_i, c_i, such that (1) Δ_1 and Δ_2 are in perspective from P, and (2) $X = a_1a_2$ and $Y = b_1b_2$ are on the line t, and the point $Z = c_1c_2$ is on t. (See Example 8.11.)

8.15 Theorem (Baer) A projective plane π is (P,ℓ)-transitive if and only if is (P,ℓ)-desarguesian.

8.2 Coordinatization of Projective Planes

8.16 Construction Let π be a projective plane of order n and R a set of n symbols with two distinguished symbols 0 and 1, $0 \neq 1$. Coordinatize π by the set $R \cup \{\infty\}$ where ∞ is a symbol not in R, with respect to a given quadrangle. (If π is infinite, then $R \cup \{\infty\}$ must have the same cardinality as any line.)

Let P, X, Y, I be a quadrangle in π, and consider the lines $\ell_\infty = XY$, $\ell_1 = PY$, $\ell_2 = PX$ and the points $A = XI \cap \ell_1$, $B = YI \cap \ell_2$, $J = AB \cap \ell_\infty$. Assign the elements of R to the points on $\ell_1 \setminus Y$ in an arbitrary manner except that 0 is assigned to P and 1 to A. If $c \in R$ is assigned to the point $C \in \ell_1$ write $(0, c)$ for C. Next, if $D \in \ell_2 \setminus X$, consider $D' = JD \cap \ell_1$; if D' is $(0, d)$, then D is $(d, 0)$. For any point H not on ℓ_∞, if $XH \cap \ell_1$ is $(0, g)$ and $YH \cap \ell_2$ is $(h, 0)$, then H has the coordinates (h, g). For a point $M \in \ell_\infty \setminus Y$, take the line joining M to $(1, 0) = B$, and let $(0, m)$ be the point in which it meets ℓ_1; then M is given the coordinate (m). Finally, Y has coordinate (∞).

Next, coordinatize the lines. Let ℓ be a line not on Y with (m) and $(0, k)$ the points at which it meets ℓ_∞ and ℓ_1, respectively. Then ℓ has coordinates $[m, k]$. If $\ell \neq \ell_\infty$ contains Y, then ℓ is the line $[k]$ if $l \cap \ell_2 = (k, 0)$. Finally, ℓ_∞ is the line $[\infty]$. In order to determine via the coordinates when two elements are incident, use the incidences of π to define a ternary operation T on R. Namely, if $a, b, c \in R$, then $T(a, b, c) = k$ if and only if (b, c) is on $[a, k]$. Thus $(0, k)$ is the intersection of ℓ_1 with the line joining (a) to (b, c) so that, given a, b, c, the value of k is uniquely determined.

8.17 Proposition Let π be a projective plane coordinatized by a set R. If T is defined by $T(a, b, c) = k$ if and only if (b, c) is on $[a, k]$, then the following properties hold.

1. $T(a, 0, c) = T(0, b, c) = c$ for all $a, b, c \in R$.
2. $T(a, 1, 0) = T(1, a, 0) = a$ for all $a \in R$.
3. If $a, b, c, d \in R$, $a \neq c$, then there is a unique $x \in R$ such that $T(a, b, c) = T(x, c, d)$.
4. If $a, b, c \in R$, then there is a unique $x \in R$ such that $T(a, b, x) = c$.
5. If $a, b, c \in R$, $a \neq c$, then there is a unique ordered pair $x, y \in R$ such that $T(a, x, y) = b$ and $T(c, x, y) = d$.

8.18 Remark A finite ternary ring satisfies (3) and (4) if and only if it satisfies (4) and (5).

Any ternary ring (i.e. set with a ternary operation) with two distinguished elements 0, 1 satisfying properties (1)-(5) of Proposition 8.17 is a *planar ternary ring* (PTR).

8.19 Proposition If (R, T) is a PTR then the structure π defined as follows is a projective plane. The points of π are ordered pairs (x, y), where $x, y \in R$, together with elements of the form (x), $x \in R$, and (∞) where ∞ is a symbol not in R. Lines are given by ordered pairs $[m, k]$, $m, k \in R$, together with elements of the form $[m]$, $m \in R$, and $[\infty]$. Incidence is defined as follows:

(x, y) is on $[m, k] \iff T(m, x, y) = k$,
(x, y) is on $[k] \iff x = k$,
(x) is on $[m, k] \iff x = m$,
(x) is on $[\infty]$ for all $x \in R$ and (∞) is on $[k]$ for all $k \in R$.

Finally, (∞) is on $[\infty]$.

8.20 Remark On a PTR (R, T) one can define binary operations of addition and multiplication so that $\langle R; +\rangle$ and $\langle R^*; \cdot\rangle$, $R^* = R \setminus \{0\}$, are loops. Namely, for any $a, b \in R$, define $a + b = T(1, a, b)$ and for all $a, b \in R^*$, define $a \cdot b = ab = T(a, b, 0)$. Then $\langle R; +\rangle$ and $\langle R^*; \cdot\rangle$ are loops with 0 and 1 the respective identities. If (R, T) is any PTR with $T(a, b, c) = ab + c$ for all $a, b, c \in R$, then (R, T) is *linear*, and the PTR is linear. A plane coordinatized by a skewfield may be coordinatized by a linear PTR.

The algebraic properties of PTRs and/or of the associated loops are linked to the groups of perspectivities admitted by the corresponding planes; in turn, these groups reflect the geometric structure of the planes. Linearity can be characterized by the validity of a suitable little Desargues configuration.

8.21 Proposition (R, T) is linear with associative addition if and only if π is $((\infty), [\infty])$-transitive.

A *cartesian group* is a linear PTR (R, T) with associative addition.

8.22 Proposition A cartesian group (R, T) satisfies the left distributive law $a(b + c) = ab + ac$, for all a, b, c in R, if and only if π is $((0), [\infty])$-transitive.

A *(left) quasifield* (R, T) is a Cartesian group satisfying the left distributive law.

8.23 Corollary π is coordinatized by a quasifield with the line ℓ as $[\infty]$ if and only if ℓ is a translation line.

8.24 Proposition If (R, T) is a quasifield, then $\langle R; +\rangle$ is an abelian group.

8.25 Proposition (R, T) is linear with associative multiplication if and only if π is $((0), [0])$-transitive.

8.26 Proposition If (R, T) is linear, then (R, T) has associative multiplication and satisfies the left distributive law if and only if π is $((\infty), [0, 0])$-transitive.

A *nearfield* is a quasifield with associative multiplication.

8.27 Proposition The dual plane π^* of π can be coordinatized by dualizing the quadrangle in a natural way. Associativity of addition is preserved but the left distributive law becomes the right distributive law and yields a right quasifield.

8.28 Proposition If (R, T) is linear with associative addition then (R, T) satisfies the right distributive law if and only if π is $((\infty), [0])$-transitive.

8.29 Corollary (R, T) is a right quasifield if and only if (∞) is a translation point.

A *semifield* (or *division ring*) is a (left) quasifield satisfying the right distributive law. (So, in a semifield both distributive laws hold.)

8.30 Proposition (R, T) is a semifield if and only if π is $((\infty), [\infty]))$-, $((0), [\infty])$- and $((\infty), [0])$-transitive. Alternatively, (R, T) is a semifield if and only if π is a translation plane with respect to $[\infty]$ and the dual of a translation plane with respect to (∞).

8.31 Table Classification of projective planes by PTR.

PTR	Geometric Properties of π	Name
Field	(P, ℓ)-transitivity, all P and ℓ, plus configuration of Pappus	Pappian plane
Skewfield	(P, ℓ)-transitivity for all P, ℓ	desarguesian plane
Alternative division ring	(P, ℓ)-transitivity for all incident pairs P, ℓ	Moufang plane
Semifield	(ℓ, ℓ)-transitivity and (P, P)-transitivity for one incident pair P, ℓ	Semifield plane
Nearfield	(ℓ, ℓ)-transitivity and (P, m)-transitivity $P \in \ell$, $P \notin m$	Nearfield plane
Right nearfield	(P, P)-transitivity and (Q, ℓ)-transitivity, P on ℓ, Q not on ℓ	Dual nearfield plane
Quasifield	(ℓ, ℓ)-transitivity for one ℓ	Translation plane
Right quasifield	(P, P)-transitivity for one P	Dual translation plane

8.32 Construction Given a PTR (R, T) of order n with $R = \{0, 1, \ldots, n-1\}$, for any $x \in R^*$ construct an $n \times n$ matrix $\{x\}$ as follows: the entry of $\{x\}$ in position (i, j) is $T(x, i, j)$. Then $\{x\}$ is a latin square and $\{x\}$ is orthogonal to $\{y\}$ if $x \neq y$. This yields a complete set of $n-1$ MOLS. Conversely, given such a set of MOLS in normal form, $A_1, A_2, \ldots, A_{n-1}$, such that A_1 has $(0, 1, \ldots, n-1)^t$ as its first column, label the squares $\{x\}$ so that $\{x\}$ is the square with x in the $(1, 0)$-position, $x = 1, 2, \ldots, n-1$. Defining $T(x, i, j) = (i, j)$-th entry in $\{x\}$, and $T(0, i, j) = j$, for all i, j, yields a PTR (R, T).

8.33 Proposition There exists a finite projective plane of order n if and only if there exists a complete set of $n-1$ MOLS of order n (see §II.1 and II.2).

8.3 Hall and André Planes

A *quasifield* Q is an algebra with two binary operations $+$ and $\cdot$ such that

1. $\langle Q; + \rangle$ is an abelian group;
2. $\langle Q^*; \cdot \rangle$ is a loop;
3. $x(y+z) = xy + xz$ for all x, y, z in Q;
4. $0x = 0$ for all $x \in Q$;
5. $ax = bx + c$ has a unique solution for x, given $a, b, c \in Q$, $a \neq b$.

8.34 Remark Point (4) is a consequence of the other axioms. By dropping axiom (5) one obtains a *weak quasifield*. However, a finite weak quasifield is a quasifield. The kernel K of a weak quasifield W is the set of all elements $k \in W$ such that $(x+y)k = xk + yk$ and $(xy)k = x(yk)$ for all $x, y \in W$.

8.35 Proposition The kernel K of a weak quasifield W is a skewfield and W may be regarded as a right vector space over K, in the natural way.

The *Hall quasifields*, which coordinatize the Hall planes, are defined as follows. Let F be a field and $f(s) = s^2 - as - b$ an irreducible quadratic over F. Let H be a two dimensional right vector space over F with basis elements 1 and λ so that $H = \{x + \lambda y : x, y \in F\}$. Take as addition in H the addition of the vector space and define a multiplication as follows: If $y = 0$, $x(z + \lambda t) = xz + \lambda(xt)$; if $y \neq 0$, $(x + \lambda y)(z + \lambda t) = xz - y^{-1}tf(x) + \lambda(yz - xt + at)$. Then $\langle H; +, \cdot \rangle$ is a Hall quasifield. Note that $f(\alpha) = 0$ for every element α in H not in F (so $y \neq 0$); F is in the kernel of H and every element of F commutes with all the elements of H.

8.36 Proposition A Hall quasifield H is associative (i.e. is a nearfield) if and only if $F = \mathbb{F}_2$ (in which case $H = \mathbb{F}_4$) or $F = \mathbb{F}_3$ and $f(s) = s^2 + 1$.

8.37 Remark The Hall quasifield of order 9 (in fact, a nearfield) can also be defined as follows. Let F be $\mathbb{F}_9$ and define a new multiplication $\circ$ on F by $x \circ y = xy$ if x is a square and $x \circ y = xy^3$ if x is not a square. Then $\langle F; +, \circ \rangle$ is a quasifield; moreover, multiplication is associative. If $F = \mathbb{F}_{q^2}$, q an odd prime power, and a multiplication $\circ$ is defined by $x \circ y = xy$ if x is a square, and $x \circ y = xy^q$ if x is not a square, then $\langle F; +, \circ \rangle$ is a quasifield, actually a nearfield.

8.38 Construction Assume $F = \mathbb{F}_{q^m}$, $K = \mathbb{F}_q$ and $\mathcal{G} = Aut_K(F)$, so that $\mathcal{G}$ is the cyclic group of order m and K is the subfield of F fixed by $\mathcal{G}$. The *norm function* over K of $\alpha \in F$ is

$$N_{F/K}(\alpha) = \prod_{j=0}^{m-1} \alpha^{q^j} = \alpha^{(q^m-1)/(q-1)},$$

and maps F onto K and F^* onto K^*. Let φ be any mapping from K into $\mathcal{G}$ that maps the identity 1 of K^* onto the identity of $\mathcal{G}$. A quasifield F_φ can then be defined. The elements of F_φ are the elements of F, addition is the same and multiplication $\circ$ is defined by

$$0 \circ y = 0 \text{ for all } y \in F, \quad x \circ y = xy^{\alpha_x} \text{ for all } x, y \in F,\ x \neq 0,$$

where $\alpha_x \in \mathcal{G}$ is the automorphism of F given by $N_{F/K}(x)^\varphi$ (for $x \in F^*$, $N_{F/K}(x) \in K^*$ and its image under φ is an element of $\mathcal{G}$). Then $\langle F_\varphi; +, \circ \rangle$ is an *André quasifield.* If φ is a homomorphism from K^* into $\mathcal{G}$, then the André quasifield is associative (it is an *André nearfield*).

A plane coordinatized by an André quasifield is an *André plane.*

8.4 Semifields

A *(finite) semifield* $\mathcal{S}$ is an algebra $\langle S; +, \cdot \rangle$ such that

1. $\langle S; + \rangle$ is an abelian group with identity 0;
2. $\langle S^*; \cdot \rangle$ is a loop with identity 1;
3. both left and right distributive laws hold: $a(b + c) = ab + ac$, $(a + b)c = ac + bc$ for all $a, b, c \in S$.

8.39 Proposition A proper semifield (i.e. one which is not a (skew)-field) contains three elements a, b, c such that $(ab)c \neq a(bc)$.

8.40 Theorem A finite proper semifield has order p^n, p a prime, $n \geq 3$ an integer and $p^n \geq 16$.

8.41 Remarks For a semifield $\mathcal{S}$ the following subsets are important:

left nucleus of S: $N_l = \{x \in S : x(ab) = (xa)b \text{ for all } a, b \in S\}$;
right nucleus of S: $N_r = \{x \in S : (ab)x = a(bx) \text{ for all } a, b \in S\}$;
middle nucleus of S: $N_m = \{x \in S : (ax)b = a(xb) \text{ for all } a, b \in S\}$.

The intersection of these three nuclei (sometimes called semi-nuclei) is the *nucleus* N of S and when S is finite of characteristic p, N contains $\mathbb{F}_p$ All nuclei are (skew)-fields. Obviously, S is a left (right) vector space over N_l, (N_r), N_m, and N.

8.42 Example Let $F = \mathbb{F}_4 = \{0, 1, \omega, \omega^2 = 1 + \omega\}$ be the field of four elements. The elements of the semifield V have the form $u + \lambda v$, $u, v \in F$, λ a symbol not in F. Addition is defined componentwise, whereas multiplication is defined as follows:

$$(u + \lambda v)(x + \lambda y) = (ux + v^2 y) + \lambda(vx + u^2 y + v^2 y^2).$$

The nucleus of V is $\mathbb{F}_2$. (In V, $(ab)c = a(bc)$ if any two of a, b, c are in $\mathbb{F}_4$.) V has six automorphisms σ_{ij}, $i = 0, 1, 2$, $j = 1, 2$, such that $(u + \lambda v)\sigma_{ij} = u^j + \lambda\omega^i v^j$ and an anti-automorphism τ : $(u + \lambda v)\tau = u + \lambda v^2$. The plane coordinatized by V is a translation plane (and a dual translation plane) of order 16, namely the so-called semifield plane with kern $\mathbb{F}_2$, *Semi2*.

8.43 Example The following semifield W has the same elements and the same addition as V in Example 8.42, but multiplication is defined as follows:

$$(u + \lambda v)(x + \lambda y) = (ux + \omega v^2 y) + \lambda(vx + u^2 y).$$

In W the associative law holds whenever one of a, b, c is in F. The nucleus of W is $\mathbb{F}_4$. The plane coordinatized by W is a translation plane (and a dual translation plane) and is the so-called semifield plane of order 16 with kern $\mathbb{F}_4$, *Semi4*. The semifields in this and Example 8.42 are essentially the only semifields of order 16.

8.44 Example *Dickson commutative semifields.* Assume $F = \mathbb{F}_{p^m}$, p odd, $m > 1, \sigma$ is a nontrivial automorphism of $\mathbb{F}_{p^m}$ (over $\mathbb{F}_p$) and f is a nonsquare in F. Then the Dickson semifield has elements $a + \lambda b$, $a, b \in F$, λ a symbol not in F, addition is componentwise and multiplication is defined as follows:

$$(a + \lambda b)(c + \lambda d) = (ac + b^\sigma d^\sigma f) + \lambda(ad + bc).$$

The planes coordinatized by the Dickson semifields have many interesting properties; e.g. they admit polarities, their full collineation groups are solvable.

8.45 Example *Albert's twisted fields.* Define a new multiplication on $\mathbb{F}_{p^n} = F$ by

$$x \circ y = xy^q - cx^q y$$

where $q = p^m$, $1 \le m < n$ and $c \ne a^{q-1}$ for any $a \in \mathbb{F}_{p^n}$. So $\langle F; +, \circ\rangle$ is a pre-semifield i.e. $\langle F^*; \circ\rangle$ is a quasigroup but not a loop. Take any $b \in F^*$ and define a new multiplication $*$ by

$$(x \circ b) * (b \circ y) = x \circ y;$$

then $\langle F; +, *\rangle$ is a semifield with multiplicative identity $b \circ b$.

8.46 Remark There are many other examples of semifields, both finite and infinite [2, 8].

8.5 The Hughes Planes

8.47 Remark Hughes planes are nondesarguesian planes of order q^2, q an odd prime power, which cannot be coordinatized by any linear PTR (but a coordinatizing ternary ring exists which is not linear). They can be constructed by using a nearfield, namely the nearfield $\langle N; +, \circ\rangle$, where $N = \mathbb{F}_{q^2}$, $+$ is addition in the field and the multiplication $\circ$ is defined by

$$x \circ y = \begin{cases} xy & \text{if } x \text{ is a square in } \mathbb{F}_{q^2} \\ xy^q & \text{if } x \text{ is not a square in } \mathbb{F}_{q^2}. \end{cases}$$

The kernel of this nearfield N is the field $F = \mathbb{F}_q$ and every element of F commutes with each element of N.

8.48 Construction To construct the Hughes plane of order q^2, $Hu(q^2)$, consider the set $V = \{(x, y, z) : x, y, z \in N\}$. Define addition $+$ on V componentwise ($+$ denotes addition both on N and V) and define a left scalar multiplication by the elements of N by $k \circ (x, y, z) = (k \circ x, k \circ y, k \circ z)$. Both the points and the lines of $Hu(q^2)$ are the elements of $N^3 \setminus (0, 0, 0)$, and (x, y, z) represents $\{(k \circ x, k \circ y, k \circ z) : k \in N^*\}$. The points of $Hu(q^2)$ can be identified with the points of $\mathrm{PG}(2, q^2)$. To define the incidence, fix a basis $\{1,t\}$ for N as a vector space over F. To determine whether (x,y,z) is incident with (u,v,w) express $u=a+ta_1$, $v=b+tb_1$, $w=c+tc_1$. Then (u,v,w) and (x,y,z) are incident if $xa+yb+zc+(xa_1+yb_1+zc_1)\circ t=0$. The set π_0 of all points (x,y,z) such that $x,y,z\in F$, together with the lines joining them, forms a desarguesian Baer subplane of $Hu(q^2)$ which is left invariant by the full collineation group of $Hu(q^2)$ and is also called the real Baer subplane. This full collineation group depends on q^2; namely,

If $q^2=3^2$, then $Aut\ Hu(9)=PGL(3,3)\times Sym_3$.
If $q^2=p^2$, $(p>3)$, then $Aut\ Hu(p^2)=PGL(3,p)\times C_2$.
If $q^2=p^{2m}$, then $Aut\ Hu(q^2)=PGL(3,p^m)\rtimes C_{2m}$.

8.49 Proposition $Hu(q^2)$ is not (P,ℓ)-transitive for any choice of P and ℓ and hence is neither a translation plane nor a dual translation plane.

8.50 Remarks $Hu(q^2)$ is a *semitranslation plane*, which means the following. Let ℓ_0 be any line of π_0 (i.e. ℓ_0 is a line of $Hu(q^2)$ such that $\pi_0\cap\ell_0$ is a line of π_0), then $Hu(q^2)$ admits all elations with axis ℓ_0 and center a point on $\pi_0\cap\ell_0$ which fix π_0. (For another approach to Hughes planes see [4, 5, 6, 12, 13].)

8.6 The Figueroa Planes

8.51 Remark Figueroa planes form an infinite class of nondesarguesian planes of order q^3, q a prime power, which are neither translation nor semitranslation planes. They can be constructed as follows.

8.52 Construction Take $\mathrm{PG}(2,q^3)$, the desarguesian plane over $\mathbb{F}_{q^3}$; it admits a planar collineation α of order 3 ($\langle\alpha\rangle=Aut_{\mathbb{F}_q}(\mathbb{F}_{q^3})$). Then the points and lines of $\mathrm{PG}(2,q^3)$ fall into three disjoint classes according to the structure of their orbit under α. For a point P,

$$P \text{ is } \begin{cases} \text{type I} & \text{if } P^\alpha = P \\ \text{type II} & \text{if } P, P^\alpha, P^{\alpha^2} \text{ are collinear and distinct} \\ \text{type III} & \text{if } P, P^\alpha, P^{\alpha^2} \text{ are not collinear.} \end{cases}$$

Line types are defined dually. Define an involutory bijection μ between the points of type III and the lines of type III as follows: For a point P and a line ℓ both of type III, $P^\mu=P^\alpha P^{\alpha^2}$, and $l^\mu=l^\alpha\cap l^{\alpha^2}$. Let $\mathcal{P}$ be the set of points and $\mathcal{L}$ the set of lines of $\mathrm{PG}(2,q^3)$. Define a new incidence structure $Fig(q^3)=(\mathcal{P},\mathcal{L},\mathrm{I}_F)$ in terms of the incidence I of $\mathrm{PG}(2,q^3)$:

$$P\,\mathrm{I}_F\,l \iff \begin{cases} P^\mu\,\mathrm{I}\,l^\mu & \text{if } P \text{ and } l \text{ are both of type III} \\ P\,\mathrm{I}\,l & \text{otherwise;} \end{cases}$$

then $Fig(q^3)$ is the *Figueroa plane* of order q^3.

8.53 Proposition For $q>2$ the Figueroa plane is not desarguesian. The points and lines of type I form a desarguesian subplane of order q of both $\mathrm{PG}(2,q^3)$ and $Fig(q^3)$ and if $\mathcal{A}$ denotes its collineation group, then the collineation group of $Fig(q^3)$ is $\langle\alpha\rangle\rtimes\mathcal{A}$.

8.7 Derivation

8.54 Remark *Derivation* is a construction due to Ostrom which for $q>2$ produces a new (nonisomorphic) plane of order q^2 from a given one of the same order.

8.55 Construction Assume π is a plane of square order q^2, and ℓ_∞ is a line of π containing a set Δ of $q+1$ points with the following property: there exist $q^2(q+1)$ Baer subplanes of π all of which contain Δ such that any two of them either share the points in Δ only or share also one point in π^{l_∞}. Then a new affine plane, $\pi^{\mathcal{D}} \setminus \ell_\infty$, the *derived* plane of π^{l_∞} with respect to Δ, can be constructed:

1. the points of $\pi^{\mathcal{D}} \setminus \ell_\infty$ are the points of π^{l_∞};
2. all lines of π^{l_∞} which, in π, meet $\ell_\infty \setminus \Delta$ are lines of $\pi^{\mathcal{D}} \setminus l_\infty$;
3. each line of π^{l_∞} which, in π, meets ℓ at a point of Δ is substituted by one of the Baer subplanes described above;
4. incidence is set theoretic inclusion.

8.56 Remark Construction 8.55 yields a new affine plane which uniquely completes to a projective plane, the derived plane of π with respect to Δ. Indeed, any two points of $\pi^{\mathcal{D}} \setminus l_\infty$ are on a unique line which is either one line of π^{l_∞} or a Baer subplane of the given set. Any two lines which are also lines of π^{l_∞} either meet in a point or are parallel; two lines which are affine Baer subplanes of the given set in π^{l_∞} either meet in a point or are parallel (when the Baer subplanes share only the points in Δ). A line of $\pi^{\mathcal{D}} \setminus l_\infty$ which is a line of π^{l_∞} meets an affine Baer subplane (on Δ) in π^{l_∞} in exactly one point.

8.57 Remarks Derivation can also be discussed from an algebraic point of view (see [8]). By deriving the derived plane of π with respect to the same Δ, π is recovered. Not all planes are derivable. The desarguesian planes are derivable with respect to any line and any Δ. By deriving PG(2, 9) one obtains the Hall plane of order q^2. Both the semifield planes of order 16 are derivable. All Hughes planes are derivable when ℓ_∞ is a line of the real Baer subplane π_0 and $\Delta = \pi_0 \cap l_\infty$.

The four existing planes of order 9 form a chain in the following sense:

$$\mathrm{PG}(2,q^2) \overset{\text{derive}}{\leadsto} \mathrm{Hall}(9) \overset{\text{dualize}}{\leadsto} \mathrm{Hall}(9)^* \overset{\text{derive}}{\leadsto} Hu(9),$$

and the arrows can be reversed. The derived plane of $Hu(9)$ is also known as the *Ostrom-Rosati* plane (it is a dual translation plane) [13].

8.58 Remarks Twenty-two planes of order 16 are known (see, e.g., [11]). Up to duality there are eight translation planes (one is PG(2, 16)), four semitranslation planes and a plane recently found by Mathon which is neither a translation plane nor a semi-translation plane. Also, all translation planes of order 25, 27, and 49 are classified [1, 3, 10].

8.8 See Also

§I.5	Projective planes form an important class of symmetric designs.
§II.2	MOLS and transversal designs, with connections to projective planes.
§IV.25	Hall triple systems.
§VI.7	Desarguesian planes and other geometries.
[9]	A comprehensive reference on translation planes.
[7]	The standard reference for Figueroa planes.

References

[1] T. Czerwinski and D. Oakden, The translation planes of order 25, *J. Combin. Theory A* 52 (1992), 193-217.

[2] P. Dembowski, *Finite Geometries*, Springer-Verlag, Berlin, 1968.
[3] U. Dempwolff, Translation planes of order 27, *Des. Codes Crypt.* 4 (1994), 105-201.
[4] R.H.F. Denniston, Subplanes of the Hughes plane of order 9, *Proc. Cambridge Phil. Soc.* 64 (1968), 589-598.
[5] M.J. de Resmini and L. Puccio, Subplanes of the Hughes plane of order 25, *Arch. Math.* 49 (1987), 151-165.
[6] M.J. de Resmini and A.O. Leone, Subplanes of the derived Hughes plane of order 25, *Simon Stevin* 67 (1993), 289-322.
[7] T. Grundhöfer, A synthetic construction of the Figueroa planes, *J. Geom.* 26 (1986), 191-201.
[8] D.R. Hughes and F.C. Piper, *Projective Planes*, Springer-Verlag, New York, 1973.
[9] H. Lüneburg, *Translation Planes*, Springer-Verlag, Berlin, 1980.
[10] R. Mathon and G.F. Royle, The translation planes of order 49, *Des. Codes Crypt.* 5 (1995), 57-72.
[11] T. Penttila, G.F. Royle, M.K. Simpson, Hyperovals in the known projective planes of order 16, *J. Combin. Des.*, to appear.
[12] T.G. Room, The combinatorial structure of the Hughes plane, *Proc. Cambridge Phil. Soc.* 68 (1970), 291-301.
[13] T.G. Room and P.B. Kirkpatrick, *Miniquaternion Geometry*, Cambridge University Press, London, 1971.

9 Computational Methods in Design Theory

Peter B. Gibbons

9.1 Introduction

9.1 Remarks Over the past 25 years the computer has become an indispensable ally in the search for combinatorial designs of many types. The use of clever computational techniques has not only enabled many existence and enumeration questions to be settled, but has also allowed larger classes of designs to be analyzed, often leading to the formulation of conjectures which have been proved for infinite families of designs. This interaction between the fields of design theory and computer science is genuinely two-way, with designs having a variety of applications in computer science (see [13]).

This section outlines some of the main computational methods that have been successful in solving existence and enumeration problems in combinatorial design theory. Closely related are analysis techniques, which are examined briefly.

9.2 Remarks The construction of a combinatorial design can be formulated as a special type of combinatorial optimization problem. With such problems there is a set Σ of *feasible solutions* (or states) together with an objective function (*cost*) $c(S)$ associated with each feasible solution S. The task is to find an optimal solution, corresponding to a feasible solution with overall (i.e., global) minimum cost. In some cases the objective function can be viewed as being a *profit*, in which case the task is to find a feasible solution with maximum profit. One feature of design problems is that an optimal solution is easily certified as being optimal, unlike the case with many other optimization problems. On the other hand approximations to an optimal solution are generally of little use in design construction.

9.3 **Table** Topics covered in this section.

Topic	Section
NP-completeness	9.1
Backtracking	9.2
Isomorph rejection	9.2
Lookahead	9.2
Branch and bound	9.2
A_{tk}-matrices and basis reduction	9.2
Hill-climbing	9.3
Simulated annealing	9.3
Tabu search	9.4
Group structures and tactical decompositions	9.4
Isomorphism	9.5
Invariants	9.5

9.4 **Remarks** Optimization problems, even with similar descriptions, can vary greatly in difficulty. For example, a minimum spanning tree of a weighted graph $G = (V, E)$ can be found by a worst-case $O(|E|+|V|log|V|)$ time greedy algorithm, whereas the current best algorithm for obtaining a minimum cost hamiltonian circuit of G runs in worst-case exponential time. The current belief is that the second problem (the well-known traveling salesman problem, TSP) cannot be solved by a worst-case polynomial time algorithm. This belief is based on knowledge of the theory of NP-completeness (see Garey and Johnson [32]), which establishes the existence of a set of NP-hard problems with the property that if a worst-case polynomial time algorithm exists for any one of the problems in the set, then a worst-case polynomial time algorithm exists for all problems in the set. The enormous amount of time spent by prominent researchers in attempting, without success, to find polynomial time algorithms for a large variety of NP-hard problems (of which TSP is one) have led many people to conjecture that such algorithms do not exist.

To date relatively few design construction and analysis problems have been shown to be NP-hard (i.e., probably do not have polynomial time algorithms). Such problems include the following:

Problems	References
Completion of partial latin squares	Colbourn [11]
Embedding of partial triple systems	Colbourn [9]
Is a partial STS t-colorable for any fixed $t \geq 3$?	Colbourn et al. [16]
Is a block design t-colorable for all $t \geq 9$?	Colbourn et al. [16]
Is an STS 14-colorable?	Phelps et al. [73]
Is an STS k-chromatic?	Phelps et al. [73]
Is a partial SQS 2-colorable?	Colbourn et al. [15]
Deciding the decomposability of Kirkman systems	Colbourn et al. [12]
Orienting triple systems	Colbourn [10]

Most design construction problems appear to be intractable, i.e., do not have polynomial time algorithms. It is therefore acceptable to use methods that run in worst-case exponential time. The goal of the methods described is to minimize the average time for a single design construction or overall time for a complete enumeration of a family of designs.

9.2 Exhaustive Search

▷ Backtracking

9.5 **Remarks** Backtracking is one of the oldest methods for design construction. It works by building up feasible solutions one step at a time, exhaustively covering all possibilities in a systematic fashion. It is guaranteed to find an optimal solution if allowed to run long enough (which, of course, is the catch). However, by using clever pruning techniques, such as rejecting partial feasible solutions which either cannot lead to a complete solution or are equivalent to partial solutions already generated earlier in the search, substantial improvements in execution time can be made.

Backtracking may be formulated as a search of the product space $X^n = X_1 \times X_2 \times \cdots \times X_n$, where the component selection spaces $X_1, X_2, \ldots, X_n$ may or may not be the same. For any $k \in \{1, 2, \ldots, n\}$, define a Boolean-valued *feasibility property* $\Pi_k : X_1 \times X_2 \times \cdots \times X_k \to \{\texttt{true, false}\}$ such that for any $(x_1, x_2, .., x_k) \in X_1 \times X_2 \times \cdots \times X_k$, $\Pi_k(x_1, x_2, .., x_k) = \texttt{true} \Longrightarrow (\forall\ 1 \leq j < k)\ \ \Pi_j(x_1, x_2, \ldots, x_j) = \texttt{true}$. With an enumeration problem, the task is to find all $(x_1, x_2, \ldots, x_n)$ such that $\Pi_n(x_1, x_2, \ldots, x_n) = \texttt{true}$. With an existence problem, the task is to find just one such $(x_1, x_2, \ldots, x_n)$.

The key to the efficiency of a backtrack algorithm lies in the choice of an appropriate feasibility property. Suppose one has constructed a partial solution $(x_1, x_2, .., x_k)$ $(1 \leq k < n)$, and finds that $\Pi_k(x_1, x_2, .., x_k) = \texttt{false}$. Since $\Pi_h(x_1, x_2, .., x_h) = \texttt{false}$ for all $h > k$, there is no point in extending this partial solution, and so in one stroke one eliminates consideration of $| X_{k+1} | \times | X_{k+2} | \times \cdots \times | X_n |$ infeasible solutions.

In general, only n, X_1, X_2, $\ldots$, X_n, and Π_n are given, and it is our task to find appropriate feasibility properties $\Pi_1, \Pi_2, \ldots, \Pi_{n-1}$. Good feasibility properties should often have $\Pi_k(x_1, x_2, \ldots, x_k) = \texttt{false}$, particularly for small values of k.

9.6 **Algorithm** The general backtrack algorithm for enumeration is:

$\{$Find all $(x_1, x_2, \ldots, x_n) \in X_1 \times X_2 \times \cdots \times X_n$ such that $\Pi_n(x_1, x_2, \ldots, x_n) = \texttt{true}\}$

```
procedure Search((x_1, x_2, ..., x_k), k);
begin
    if k = n
    then record (x_1, x_2, ..., x_n) as a solution
    else
        for each x_{k+1} ∈ X_{k+1} do
            if Π_{k+1}(x_1, x_2, ..., x_{k+1})
            then Search((x_1, x_2, ..., x_{k+1}), k + 1)
end

Search((), 0)
```

9.7 **Example** Dinitz, Garnick and McKay [23] recently devised a method to enumerate the 526,915,620 nonisomorphic 1-factorizations of K_{12}; it is outlined here (see §VI.4 for definitions).

1. Number the vertices of K_{12} $0, 1, \ldots, 11$. Write an edge e as an ordered pair (x, x') with $0 \leq x < x' \leq 11$. For any two edges $e_1 = (x_1, x_1')$ and $e_2 = (x_2, x_2')$, define $e_1 < e_2$ if either $x_1 < x_2$ or $x_1 = x_2$ and $x_1' < x_2'$. A 1-factor f is written as a list of ordered edges, i.e., $f = (e_1, e_2, e_3, e_4, e_5, e_6)$ where $e_i < e_j$ whenever $i < j$. For two 1-factors $f_i = (e_{i1}, e_{i2}, \ldots, e_{i6})$ and $f_j = (e_{j1}, e_{j2}, \ldots, e_{j6})$, define

$f_i < f_j$ if there exists a k $(1 \leq k \leq 6)$ such that $e_{il} = e_{jl}$ for all $l < k$, and $e_{ik} < e_{jk}$.

2. A 1-factorization F of K_{12} is written as an ordered list of 11 1-factors, i.e., $F = (f_1, f_2, \ldots, f_{11})$, where $f_i < f_j$ whenever $i < j$.
3. F and H denote 1-factorizations, and f_i and h_i denote 1-factors contained in F and H, respectively.
4. For two 1-factorizations F and H, say that $F < H$ if there exists some i, $1 \leq i \leq 11$, such that $f_i < h_i$, and $f_j = h_j$ for all $j < i$.

For $1 \leq i \leq 11$, an ordered set of i edge-disjoint 1-factors, $F_i = (f_1, f_2, \ldots, f_i)$, is a *partial* OF of rank i. Partial OFs of rank i can be ordered in the same way as OFs.

The general backtrack algorithm can be used to construct a 1-factorization of K_{12}. A solution is a 1-factorization $F = (f_1, f_2, \ldots, f_{11})$, where $f_i, 1 \leq i \leq 11$ are 1-factors, and $f_i < f_j$ whenever $i < j$. The feasibility properties are defined as follows:

$\Pi_1(f_1) = \texttt{true}$
$\Pi_k(f_1, \ldots, f_k) = \Pi_{k-1}(f_1, f_2, \ldots, f_{k-1})$ **and** $f_k > f_{k-1}$ **and** $f_k \cap f_i = \emptyset, 1 \leq i < k$.

In extending the partial OF F_i to F_{i+1} one can reduce the number of 1-factors f_{i+1} that need to be considered at this level by backtracking to construct a suitable f_{i+1}. This is then a two-level backtrack algorithm.

9.8 Remarks The algorithm described is intended to construct just one OF, if it exists. Suppose, instead, that one wishes to enumerate all OFs. Here the task can be made considerably easier by enumerating all nonisomorphic OFs (from which can be obtained all distinct OFs, if necessary). To do this, *isomorph rejection* is introduced by strengthening the feasibility properties so as to reject partial OFs that are isomorphic to partial OFs that have been generated earlier in the search. That is, try to extend a partial OF only if it is the lexicographically lowest (or *canonical*) representative of its isomorphism class. Some details of how this is done are now provided.

A partial OF $F_i = (f_1, f_2, \ldots, f_i)$ of rank i is *proper* if f_j contains edge $(0, j)$ for $1 \leq j \leq i$. If F_i is not proper, then it is *improper*. A complete 1-factorization is necessarily proper.

The automorphism group of the complete graph K_{12} is S_{12}, the symmetric group on 12 elements. Thus given a proper partial OF F_i (of rank i), one can rename the 12 points using a permutation $\alpha \in S_{12}$, and obtain another partial OF (not necessarily proper) of the same graph, denoted F_i^α. Then F_i is *canonical* if $F_i \leq F_i^\alpha$ for all permutations $\alpha \in S_{12}$. A canonical partial OF must necessarily be proper.

The following results on canonicity are from Seah and Stinson [81]. If two proper partial OFs of rank i, F_i and H_i, are distinct and are both canonical, then F_i and H_i are nonisomorphic. If a partial proper 1-factorization $F_i = (f_1, f_2, \ldots, f_i)$ is canonical, and $1 \leq j \leq i$, then $F_j = (f_1, f_2, \ldots, f_j)$ is also canonical. If a partial proper 1-factorization $F_i = (f_1, f_2, \ldots, f_i)$ is not canonical, then any complete OF extended from F_i is also not canonical. The backtrack search can now be altered so that only canonical partial OFs are considered at each level. That is, strengthen the feasibility properties to:

$\Pi_1(f_1) = (f_1 = \{(0,1), (2,3), (4,5), (6,7), (8,9), (10,11)\})$.
$\Pi_k(f_1, \ldots, f_k) = \Pi_{k-1}(f_1, f_2, \ldots, f_{k-1})$ **and** $f_k > f_{k-1}$ **and** $f_k \cap f_i = \emptyset, 1 \leq i < k$
and $(f_1, \ldots, f_k)^\alpha \geq (f_1, \ldots, f_k)$ $\forall$ $\alpha \in S_{12}$.

Much of the art of successful backtracking is in implementing heuristics to reduce the amount of work done at each level. A good example in this case is provided by the use of cycle structures of pairs of disjoint 1-factors to reduce the number of mappings in S_{12} to be considered at each level.

Consider a canonical $F_i = \{f_1, f_2, \ldots, f_i\}$, $i \geq 1$. Since the union of two disjoint 1-factors is a union of disjoint cycles of even length, for any 1-factor f which is edge disjoint from f_1, $\{f\} \bigcup \{f_1\}$ forms a graph isomorphic to either three disjoint 4-cycles; a 4-cycle and an 8-cycle; two 6-cycles; or a single 12-cycle. Thus, each F_2 is in one of four isomorphism classes. The following are the four 1-factors which, when unioned with f_1, yield in turn each of the four canonical rank 2 1-factorizations of K_{12}.

1. $\{(0,2),(1,3),(4,6),(5,7),(8,10),(9,11)\} \bigcup \{f_1\}$ forms three disjoint 4-cycles.
2. $\{(0,2),(1,3),(4,6),(5,8),(7,10),(9,11)\} \bigcup \{f_1\}$ forms a 4-cycle and an 8-cycle.
3. $\{(0,2),(1,4),(3,5),(6,8),(7,10),(9,11)\} \bigcup \{f_1\}$ forms two disjoint 6-cycles.
4. $\{(0,2),(1,4),(3,6),(5,8),(7,10),(9,11)\} \bigcup \{f_1\}$ forms a 12-cycle.

The isomorphism class of a pair of 1-factors is their *cycle structure*, and these classes are *type 1*, *type 2*, *type 3*, and *type 4* respectively. This ordering of the types is the same as the lexicographic ordering of the canonical representatives of the types. Extend the definition of *type* to apply to all canonical partial OFs. For all canonical $F_i = \{f_1, f_2, \ldots, f_i\}$, $i \geq 2$, $\{f_1\} \bigcup \{f_2\}$ is one of the four canonical rank 2 1-factorizations of K_{12}; the *type* of F_i is the type of $\{f_1\} \bigcup \{f_2\}$. All canonical rank i 1-factorizations of K_{12} which have type s lexicographically precede all canonical rank i 1-factorizations of K_{12} which have type t, for $s < t$.

Suppose some proper canonical F_i, $2 \leq i \leq 10$, is to be extended by adding a 1-factor f. Further, assume that F_i has type t, $2 \leq t \leq 4$. Let g be a 1-factor in F_i such that the type s of $\{f\} \bigcup \{g\}$ is minimal. If $s < t$, then $F_i \bigcup \{f\}$ is not canonical because there exists a permutation α that maps $F_i \bigcup \{f\}$ to a canonical rank i OF of type s. This leads to strengthened feasibility properties:

$$\begin{aligned}
&\Pi_1(f_1) = (f_1 = \{(0,1),(2,3),(4,5),(6,7),(8,9),(10,11)\}).\\
&\Pi_k(f_1, \ldots, f_k) = \Pi_{k-1}(f_1, f_2, \ldots, f_{k-1}) \textbf{ and } f_k > f_{k-1} \textbf{ and } f_k \cap f_i = \emptyset, 1 \leq i < k\\
&\qquad \textbf{and } (f_1, \ldots, f_k)^\alpha \geq (f_1, \ldots, f_k)\ \forall\ \alpha \in S_{12}\\
&\qquad \textbf{and } \text{type of } (f_1, \ldots, f_k) \leq \text{type of } f_k \cup f_i\ \forall i \in \{1, \ldots, k-1\}.
\end{aligned}$$

Thus, if F_{k-1} is of type t, then the only permutations which need to be considered are those which map some pair of 1-factors in F_k onto the canonical rank 2 factorization of type t. This enables the use of *dynamic programming* techniques, saving information from the generation of permutations at rank i to speed up the generation of the permutations at rank $i+1$. In particular, the algorithm uses a stack of the $\binom{i}{2}$ pairs of factors in the current F_i, where each pair is stored as the set of cycles formed by the union. When factor f is added to F_i, the algorithm pushes onto the stack the i unions of f and f_j where $1 \leq j \leq i$. The desired permutations are generated by traversing the cycles.

9.9 **Remarks** Example 9.7 illustrates many of the important factors of backtracking. The implementation of strong feasibility conditions can considerably reduce the size of the search tree traversed by the algorithm. Of particular significance is the use of isomorph rejection techniques to avoid generating isomorphic solutions. The described algorithm generates the canonical representatives of each isomorphism class. It therefore employs a *total isomorphism rejection* scheme and is an *orderly algorithm* (see Read [75] and McKay [64]). Sometimes it is not possible to use the full automorphism group at each level (due to its size or difficulty of calculation). Using subgroups at each level results in a *partial* isomorph rejection scheme, allowing the possibility of generating solutions which are isomorphic. These isomorphs must be removed by an explicit isomorph check at the end of the search. To obtain an effective backtrack algorithm one needs to be careful in choosing a subset of the group to provide a significant amount of isomorph rejection along the way. The technique of isomorph rejection is also useful in an existence search, since equivalent non-solutions are culled out in the search process.

Other work on orderly enumeration of incidence structures by Ivanov [47] and Pietsch [74] has resulted in the enumeration of many classes of block designs, PBDs, linear spaces, and GDDs.

9.10 Remarks One technique not illustrated by this example is that of "lookahead". Suppose partial solution $P_k = (x_1, x_2, .., x_k)$ $(1 \le k < n)$ has been constructed. A lookahead check involves extending P_k only if an x_j can be constructed separately, with respect to P_k for each $k < j \le n$. This essentially corresponds to a strengthening of the feasibility properties, but using information that cannot be obtained locally from the current partial solution.

In the case of the 1-factorization example, before extending a proper canonical $F_i, i < 11$, one can lookahead to check whether it is possible to construct a 1-factor f_k for each $i < k \le 11$ which is disjoint from F_i and which contains the edge $(0, k)$. This heuristic was not tested by Dinitz, Garnick and McKay [23]. However, Gibbons and Mathon [35] have found that in signing GDDs, for example, lookahead is useful, particularly if the check is "turned on" for values of k in the middle of the range $1 \dots n$.

Another type of lookahead is achieved by noting that, on completion of a partial solution $P_k = (x_1, x_2, .., x_k)$ $(1 \le k < n)$, one has the freedom to choose from any of the unchosen $X_{k+1}, X_{k+2}, .., X_n$ at the next level. A promising heuristic is to choose from the smallest of these sets. This *minimum degree heuristic* is used to great effect in the construction of cyclic Steiner quadruple systems (see Colbourn et al. [14]).

9.11 Remarks The technique of *branch and bound* is a variation of backtracking which has been successful in solving optimization problems. With this method, given a partial solution $(x_1, x_2, \dots, x_k)$ $(k < n)$, one calculates a *bound* $B(x_1, x_2, \dots, x_{k+1})$ for each $x_{k+1} \in X_{k+1}$. $B(x_1, x_2, \dots, x_{k+1})$ is an upper bound on the profit of any solution obtained by extending $(x_1, x_2, \dots, x_{k+1})$. Partial solutions for which $B(x_1, x_2, \dots, x_{k+1})$ is no greater than the profit of the best solution found so far are immediately rejected. The remaining partial solutions are explored in order of decreasing bound value.

Heuristics such as "lookahead" mentioned above are examples of *bounding* if one views the construction problem as one of maximizing a $0-1$ function, where 1 implies existence, and 0 implies nonexistence. An infeasible partial solution is one with an upper bound of 0, implying that it should not be further extended.

9.12 Remarks Arguably the most famous backtracking search conducted to date was that by Lam, Thiel and Swiercz [58] in 1989 to show that a projective plane of order 10 (see §VI.8) does not exist. Their search took a total of 800 days of CPU time on a VAX-11/780 plus about 3,000 hours on a supercomputer, the CRAY-1S.

The method relies on results obtained from studying the binary error-correcting code (see §V.1) associated with a putative projective plane of order 10. Let A be the line-point incidence matrix of such a plane, and S be the vector space generated by the rows of A over F_2. A vector in S is a *codeword*. The number of ones is its *weight*. The *weight enumerator* of S is defined to be $\sum_{i=0}^{111} w_i x^i$ where w_i is the number of codewords of weight i. Each codeword of weight i corresponds to a configuration with i points.

The weight enumerator of S is uniquely determined by the values of w_{12}, w_{15}, and w_{16}. By 1985, Lam and others had shown that these three values were all zero, which allowed them to compute the weight enumerator, and in particular show that if a projective plane of order 10 exists, then it must contain 24,675 codewords of weight 19. 19-point starting configurations were therefore used as the basis for the search.

Let V be a set of 19 points forming a codeword of weight 19. There are 6 *heavy* lines each containing 5 points of V, 37 *triple* lines each containing 3 points of V, and 68 *single* lines each containing 1 point of V. There are 66 different incidence structures of the 6 heavy lines with the 19 points of V, which can be reduced to 45 by theoretical arguments.

Each of these is represented by a 6×19 submatrix of the incidence matrix A, and is extended in all possible ways to a 43×19 submatrix A_2 by considering the possible incidence of V on the triple lines. Isomorph rejection is performed at this stage so that only nonisomorphic A_2s are retained.

Next, each A_2 is completed to a 111×19 submatrix by considering the incidence of V on the 68 single lines. By using row permutations there is only one possible extension for each A_2. Finally the remaining portion of A, involving the *outside* points (viz. those other than V), is constructed. Partial isomorph rejection is used during this stage. For further details, see Lam, Thiel and Swiercz [58].

A feature of the search is the careful attention devoted to correctness, essential with an effort of this magnitude and result. To help detect programming errors, they used, wherever possible, two independently constructed programs, as well as applying internal consistency checking when generating the A_2s. The other type of error which needs to be considered is a hardware error, which apparently occurs on the CRAY-1S at the rate of about one per thousand hours of computing. In fact, Lam et al. discovered one such error, and reran the 1,000 A_2s just before the failure. However, there remains the possibility of other undetected errors. The authors argue that, even with such errors, it is extremely unlikely that a projective plane of order 10 has been missed. For, if an undiscovered plane exists, then it can be constructed as the extension of each of 24,675 A_2s. If, in the unlikely worst-case, all the 24,675 A_2s are isomorphic and a hardware error happens to affect this A_2, then a plane could be missed. Since there are about half a million A_2s, the probability of a hardware error affecting this special A_2 is very small.

Despite the care taken in this search, the time is approaching when an independent verification of this result is warranted. Given the improvements in computing power over the past 5 years this should be a less daunting task. For example, whereas the recent enumeration of 1-factors of K_{12} (Dinitz, Garnick and McKay [23]) required over 160 MIPS-years of computation time, the search was completed in less than 8 months by distributing parts of the problem to workstations running at rates of 12 to 50 MIPS.

9.13 Remarks Other examples of the successful application of backtracking:

Configurations	References
Steiner triple systems	Cole, Cummings and White [21]
Orthogonal Steiner triple systems	Gibbons [33]
Subsquare free latin squares	Gibbons and Mendelsohn [39]
GDDs, projective planes	Gibbons and Mathon [35], Pietsch [74]
Bhaskar Rao designs	Gibbons and Mathon [36]
Strong starters	Kocay, Stinson and Vanstone [49]
Howell designs	Seah and Stinson [81]
Block designs	Gibbons, Mathon and Corneil [38]
	McKay and Radziszowski [66], Pietsch [74]
Incidence systems	Ivanov [47]
Latin squares (order 8)	Brown [6], Kolesova, Lam and Thiel [51]

▷ A_{tk}-Matrices

9.14 Remarks The study of A_{tk}-matrices has resulted in the discovery of many new t-designs.

Given a $t-(v,k,\lambda)$ design (X,B), let X_h $(0<h<v)$ be the set of all h-subsets of X. Let $X_t=\{x_1,x_2,\ldots,x_m\}$ and $X_k=\{y_1,y_2,\ldots,y_n\}$, where $m=\binom{v}{t}$ and $n=\binom{v}{k}$. The A_{tk}-matrix $A_{tk}=(a_{ij})$ is the $m\times n$ 0-1 matrix where $a_{ij}=1$ if and only if $x_i\in y_j$. It is clear that a $t-(v,k,\lambda)$ design exists if and only if there is a solution z to the matrix equation $A_{tk}z=\lambda J$, where z is an n-dimensional vector with nonnegative integer entries, and J is the m-dimensional all ones vector. A solution, in theory, can be obtained using integer programming techniques. For most designs of interest, however, A_{tk} is prohibitively large. To reduce the size one assumes the action of a group G on the set X. Let ρ_t and ρ_k be the number of orbits under the induced action of G on X_t and X_k, respectively. Denote by $A_{tk}(G)=(a_{ij})$ a $\rho_t\times\rho_k$ matrix where a_{ij} is the number of k-sets in the j-th orbit of X_k containing a representative t-set in the i-th orbit of X_t. $A_{tk}(G)$ can be obtained from A_{tk} by adding the columns in each orbit of X_k and keeping one representative row from each orbit of X_t. A t-design with G as a subgroup of its automorphism group exists if and only if there is a solution z to the matrix equation $A_{tk}(G)z=\lambda J$, where z is a ρ_k-dimensional vector with nonnegative integer entries, and J is the ρ_t-dimensional all ones vector.

9.15 Example (Kreher and Radziszowski [55]) Suppose that $(X=\{1,2,3,4,5,6,7\},B)$, a 2-(7,3,1) design, is to be constructed, with a subgroup $G=\langle(145)(276),(26)(45)\rangle$ $\simeq S_3$ of its automorphism group. $A_{23}(G)$ is the following matrix:

	123 347 136 356 357 234	124 457 156	125 147 146 456 157 245	126 247 567	127 467 167 256 257 246	134 345 135	137 346 235	145	236 237 367	267
{12 47 16 56 57 24}	1	1	1	1	1	0	0	0	0	0
{13 34 35}	2	0	0	0	0	2	1	0	0	0
{14 45 15}	0	1	2	0	0	1	0	1	0	0
{17 46 25}	0	0	2	0	2	0	1	0	0	0
{23 37 36}	2	0	0	0	0	0	1	0	2	0
{26 27 67}	0	0	0	1	2	0	0	0	1	1

$z=\{0,1,0,0,0,0,1,0,0,1\}^T$ gives a solution to the equation $A_{tk}(G)z=\lambda J$, and thus yields a 2-(7,3,1) design with $B=\{124,457,156,137,346,235,267\}$ and G a subgroup of its automorphism group.

9.16 Remarks Kramer and Mesner [53] first suggested using the A_{tk}-matrix equation to find t-designs. Since then the method has led directly to the discovery of many new t-designs (see Remarks 9.21).

To use this equation to find a t-design effectively, three problems must be solved:

1. Find a group G that is likely to be a subgroup of the automorphism group of the constructed design.
2. Find an efficient algorithm for constructing the $A_{tk}(G)$ matrix.
3. Find an effective procedure for solving the equation $A_{tk}z=\lambda J$.

Problem 1 is often solved by using one of the projective special linear (PSL) groups, as well as some other groups (see Kramer, Leavitt and Magliveras [52]). There is a trade-off relating to the size of the group used. The larger the group, the greater the achieved "fusion" of the t- and k-sets into a smaller number of (larger) orbits. This reduces the size of the matrix A_{tk} and hence the size of the equation $A_{tk}z = \lambda J$ to be solved. On the other hand, existence is less likely because larger segments are being fused that must fit precisely to form the design. Put another way, smaller groups are more likely to be subgroups of the automorphism group of the putative design.

The efficient computation of A_{tk} (problem 2) is achieved by solving two subproblems:

2.1. Design an efficient data structure to represent G-orbits of X_h.
2.2. Develop an efficient method to decide which G-orbit contains a given h-subset.

Problem 2.1 can be solved by using a hierarchical structure in which a node N at level h is a representative h-subset S_h from a G-orbit of X_h, and the descendants of N at level $h+1$ are representatives of the G-orbits of X_{h+1} which contain S_h. If one labels the edges with the additional elements that need to be added to S_h to produce a representative from the corresponding orbit of X_{h+1}, then this (directed) graph can be searched to obtain a solution to problem 2.2. Additional information on this method is provided in Kramer, Leavitt and Magliveras [52]

An alternative method is described by Chouinard [private communication]. Consider the action of G on the set $X = \{1, 2, \ldots, n\}$, with the usual lexicographical ordering on X. Define the orbit representative for any subset S of X as the lexicographically first subset in the orbit of S, and construct the following data structure inductively. At level 1 take each singleton subset of X, each paired with an element of G that takes that subset to its orbit representative together with a pointer to that representative. At level $m+1$, beneath any m-subset of X that is an orbit representative of m-sets, take all $(m+1)$-subsets of X that contain the m-set, each paired with an element of G taking that $(m+1)$-set to its orbit representative, and a pointer to that representative. (The set of the first m elements of the orbit representative of an $(m+1)$-set is lexicographically first in its own orbit and thus is present.) If one has constructed this structure out through level k, then it can be used to quickly find the orbit representative for any k-subset of S, and an element of G which takes it there, by inductively going through this procedure with the first t elements in the k-set, for t from 1 to k. To construct it at level $k+1$ requires finding the elements of G to be paired with each set at that level. In many cases, after the first few levels, this can be easily deduced from the links generated by taking each k-subset of each $(k+1)$-set present, using the structure present to find the element of G taking that k-set to its orbit representative and applying that element to the full $(k+1)$-set. (By using the character table one may count the orbits of k-sets for any k, thereby deducing the number of links needed.) If this does not provide enough links, additional links can be generated by applying randomly chosen elements of G to the $(k+1)$-sets which are not linked to a lexicographically smaller representative. This method has a high probability of producing a missing link on each pass. The described structure contains all the information to enable one to easily calculate the A_{tk} matrix.

Problem 3 can often be solved by using a *basis reduction* method first introduced by Lenstra, Lenstra and Lovasz [59], and first applied to designs by Kreher and Radziszowski [55]. They noted that if z is a (0,1)-vector solution to $A_{tk}(G)z = \lambda J$, then $[z, E]^T$ is a vector in the integer lattice L spanned by the columns of the matrix

$$\mathbf{B} = \begin{bmatrix} I_n & 0 \\ A_{tk}(G) & -\lambda J \end{bmatrix},$$

where E is the m-dimensional all zero vector. That is, $[z, E]^T$ is a linear combination of the columns of $\mathbf{B}$, satisfying

$$\begin{bmatrix} \mathbf{z} \\ E \end{bmatrix} = \mathbf{B} \begin{bmatrix} a_1 \\ \vdots \\ a_{n+1} \end{bmatrix},$$

where $a_1, a_2, \ldots, a_{n+1}$ are integers.

Kreher and Radziszowski noticed that integer solutions z to $A_{tk}(G)z = \lambda J$ are often short vectors in L. Their algorithm therefore searches for a reduced basis for L all of whose vectors are as short as possible. The three techniques used are

1. An $O(n^4)$ algorithm, $\mathbf{L}^3$, due to Lenstra, Lenstra and Lovász [59], for producing a reduced basis with short vectors from a given ordered basis. When the number of rows of A_{tk} is $m = 1$, solving the equation $A_{tk}(G)z = \lambda J$ reduces to solving the subset sum problem. The application of $\mathbf{L}^3$ to this problem is described in Kreher and Radziszowski [54]. If $m > 1$, further reduction techniques are necessary.
2. An $O(n^2)$ *weight reduction* algorithm is used to produce a basis with shorter vectors.
3. Finally, a *size reduction* algorithm is used to produce a basis with a smaller number of vectors.

9.17 Algorithm Basis reduction algorithm of Kreher and Radziszowski [55].

> Input basis $\mathbf{B}$ in the form specified in Remark 9.16;
> $\mathbf{B} := \mathbf{L}^3(\mathbf{B})$;
> $\mathbf{B} :=$ Size-Reduction($\mathbf{B}$);
> `currentWeight` $:= \sum_{\mathbf{b} \in \mathbf{B}} \| \mathbf{b} \|^2$, where $\| . \|$ denotes ordinary euclidean length;
> **repeat**
>   `oldWeight` := `currentWeight`;
> $\mathbf{B} :=$ Weight-Reduction($\mathbf{B}$);
>   Sort the vectors in $\mathbf{B}$ into order of nondecreasing euclidean norm;
> $\mathbf{B} := \mathbf{L}^3(\mathbf{B})$;
>   `currentWeight` $:= \sum_{\mathbf{b} \in \mathbf{B}} \| \mathbf{b} \|^2$;
> **until** `currentWeight` = `oldWeight` **or** `solution found`;

The basis reduction technique has been successful in finding some "small" t-designs (such as the 6-(14,7,4) of Kreher and Radziszowski [56]). Further improvements to the basis reduction algorithm have been suggested by Schnorr et al. [22, 77, 78, 79] and Betten et al. [3].

9.18 Remarks For other designs a more recent algorithm, SYNTH, used by Magliveras and Mathon has been successful.

Suppose one needs to solve a linear problem of the form $Az = \lambda J$, where A is a A_{tk} matrix, and z and J are as described in Remarks 9.14 above. Begin by assuming that $\lambda = 1$ and that n is large, for instance in the range $1000 \leq n \leq 10000$, with m being several hundred. Proceed using Algorithm 9.20.

9.19 Remarks The SYNTH algorithm is easily parallelizable by, for example, having individual processors carry out the computations specified for each of the ones in row 1 of A. Also, in practice it is very costly in terms of storage to pass the entire matrix A' on each recursive call. Instead vectors of flags can be maintained to record the rows and columns that have been deleted.

In general one must solve the matrix equation $Az = \lambda J$ for nonnegative integral A. Here the algorithm is modified to delete all columns v' for which $v + v'$ has entries exceeding λ. Other heuristics can be used to reduce the computation. For example, if G is the group from which A is constructed, the normalizer $N(G)$ moves the orbits of k-sets (resp. t-sets) among themselves. Consequently, there is an induced group action of $N_{S_n}(G)$ on a solution z. By selecting one representative from each orbit of k-sets for which $A[1,i] = 1$, the search can be cut down by a factor of approximately $|N(G)|$ since most $N(G)$ orbits on k-sets are regular.

9.20 Algorithm SYNTH algorithm of Mathon and Magliveras [private communication].

```
{ Find all solutions to Az = J}
    s := number of 1's in row 1 of A;
    for i := 1 to s do
        begin
            A' := A;   z := all zeroes;
            h := index of column of A' containing ith 1 in row 1;
            v := column h of A';   z_h := 1;
            Delete from A' all columns not orthogonal to v;
            Delete from A' all rows j for which v_j = 1;
            Delete column h from A';
            Recursively find all solutions to A'z' = J';
            For each solution z' combine with z to get solution to Az = J
        end
```

9.21 Remarks Some examples of t-designs found by the use of A_{tk}-matrices:

t-(v,k,λ)	References
5-$(17,8,80)$	Kramer [50]
5-$(33,6,42)$, 5-(33,7,126)	Magliveras and Leavitt [60]
5-$(13,6,4)$	Kreher and Radziszowski [56]
5-$(108,6,1)$	Grannell and Griggs [43]
5-$(132,6,1)$	Grannell, Griggs and Mathon [44]
5-$(168,6,1)$	Mathon (private communication)
6-$(14,7,4)$	Kreher and Radziszowski [56]
6-$(33,8,36)$	Magliveras and Leavitt [60]
6-$(20,9,112)$	Kramer, Leavitt and Magliveras [52]
6-$(28,8,\lambda)$, $\lambda = 42, 63, 84, 105$	Schmalz [76]
7-$(33,8,10)$	Betten et al. [3]

9.3 Non-Exhaustive Search

These techniques are very useful in situations where one expects that solutions exist. Of course they cannot be used to establish nonexistence results.

▷ Hill-Climbing (Local Search)

9.22 Remarks Define, for each feasible solution $S \in \Sigma$, a set T_s of transformations (or *moves*), each of which can be used to change S into another feasible solution S'. The set of solutions that can be reached from S by applying a transformation from T_s is the *neighborhood* $N(S)$ of S. S is a *local minimum* if $c(S) \leq c(S')$ for all $S' \in N(S)$.

9.23 Algorithm General hill-climbing algorithm.

> Generate an initial feasible solution S;
> **while** S is not a local minimum **do**
> **begin**
>   Choose any $S' \in N(S)$ such that $c(S') \leq c(S)$;
> $S := S'$
> **end**

The hope is that the final value of S is optimal, or close to optimal.

9.24 Remarks Although contrary to the notion of "climbing", hill-climbing is presented in the context of a minimization problem to be consistent with the presentation of other probabilistic methods, such as simulated annealing.

With designs a feasible solution is often defined to be a partially completed design D, and the associated cost $c(D)$ to be the number of blocks needed to complete the partial design. The neighborhood $N(D)$ of D is the set of designs obtainable from D by altering and/or adding a small number of blocks. Unlike general optimization problems, an approximation to an optimal solution does not suffice — only a complete design, which corresponds to a feasible solution with cost 0, is useful.

9.25 Algorithm Hill-climbing algorithm for designs.

> Generate an initial partial design D;
> $tries := 0$;
> **while** $(tries < maxTries)$ **and** $(c(D) > 0)$ **do**
> **begin**
> $tries := tries + 1$;
>   Select a $D' \in N(D)$ such that $c(D') \leq c(D)$;
>   **if** $c(D') < c(D)$
>   **then** $tries := 0$;
> $D := D'$
> **end**

`maxTries` is a "threshold" value specifying how many consecutive "sideways" moves are allowed without increasing the number of blocks before abandoning the search.

9.26 Remark It may be difficult to select a $D' \in N(D)$ such that $c(D') \leq c(D)$. One approach may be to first choose a $D' \in N(D)$ randomly and then check if $c(D') \leq c(D)$. If this approach is taken, a threshold value must be chosen, which is the number of attempts at finding a D' such that $c(D') \leq c(D)$ before abandoning the construction. "Abandoning the construction" really means starting again with a new initial partial design. In cases where finding an optimal solution is not guaranteed, a third threshold value is needed to specify the number of restarts to tolerate before completely abandoning the search. The choice of suitable threshold values and transformation functions are critical in obtaining good performance from the algorithm.

9.27 Remark None of the three threshold parameters are needed in Stinson's algorithm for finding Steiner triple systems (Algorithm 9.28. Although one can identify theoretical situations where Stinson's algorithm can get stuck, the algorithm seldom encounters these situations in practice and, from empirical evidence, constructs a Steiner triple system on the average in $\Theta(v^2 \log v)$ time. For example an STS(901) can be constructed in about 10 seconds on a DEC ALPHA 3000/600 workstation. Stinson [85] provides evidence that the systems generated in this way are genuinely randomly distributed. The algorithm can be adapted to construct an STS with a specified automorphism group structure or one which satisfies other restrictions.

9.28 Algorithm Stinson's algorithm [85] for constructing an STS(v).

```
blocksRemaining := v(v − 1)/6;
D := ∅;
while blocksRemaining > 0 do
begin
    Pick a random element x appearing < (v − 1)/2 times in D;
    Pick random pairs {x, y}, {x, z} not covered by blocks in D;
    if pair {y, z} appears in D
    then begin
            D := D - block containing pair {y, z};
            blocksRemaining := blocksRemaining + 1
        end;
    D := D ∪ block {x, y, z};
    blocksRemaining := blocksRemaining − 1
end
```

Since $c(D') \leq c(D)$, there is no difficulty in finding a move to make.

9.29 Remarks Other examples of the successful application of hill-climbing:

Configurations	References
Strong starters	Dinitz and Stinson [25]
One-factorizations, Room squares	Dinitz and Stinson [28], Seah and Stinson [80]
Howell Designs	Dinitz and Lamken [24], Dinitz and Stinson [29]
Leaves of partial triple systems	Colbourn and Mathon [18]
Quadrilateral-free STS	Chee and Lim [8], Griggs and Murphy [45]
Enclosings of triple systems	Bigelow and Colbourn [5]
PBDs with block sizes 3 and 4	Colbourn, Rosa and Stinson [19]
Room frames	Dinitz and Stinson [27]
STS	Stinson and Ferch [86]
Orthogonal STS	Gibbons and Mathon [37]
Sequencing groups	Anderson [2]

Dinitz and Stinson [25] described the first successful use of hill-climbing in design theory; a simplified presentation appears in [26].

▷ Simulated Annealing

9.30 Remarks *Simulated annealing* is an extension of hill-climbing in which uphill moves are allowed in a controlled fashion. The aim is to reduce the chance of the search becoming trapped in a local minimum.

Annealing, in condensed matter physics, refers to a thermal process for obtaining low energy states of a solid in a heat bath. In this process, the solid is first heated to melting temperature and then slowly cooled until the low-energy ground state is reached. If the initial temperature is not sufficiently high or the cooling is carried out too quickly, the solid freezes into a meta-stable state rather than the ground state. The converse of annealing is quenching, where the temperature of the heat bath is instantaneously lowered, resulting in a meta-stable state.

The physical annealing process can be modeled successfully by using computer simulation methods. In one of these approaches, Metropolis et al. [67] introduced a simple Monte Carlo algorithm which generates a sequence of states in the solid in the following way. Given a current state i of the solid, characterized by the positions of its particles, with energy E_i, a subsequent state j, with energy E_j, is generated by applying a small distortion to the current state, for example by the displacement of a particle. If the energy difference, $E_j - E_i$ is ≤ 0, then the state j is accepted as the new current state. If the difference is greater than 0, then the state j is accepted with probability $e^{(E_i - E_j)/k_B T}$ where T denotes the temperature of the heat bath and k_B is a physical constant known as the Boltzmann constant. This acceptance rule is known as the Metropolis criterion and the associated algorithm is known as the Metropolis algorithm.

If the lowering of the temperature is done sufficiently slowly, the solid can reach thermal equilibrium at each temperature. In the Metropolis algorithm this is achieved by generating a large number of transitions at a given temperature value. In thermal equilibrium the probability of occurrence of a solid being in a state i with energy E_i at temperature T is given by the Boltzmann distribution $P_T\{X = i\} = e^{(-E_i/k_B T)}/Z(T)$, where X is a stochastic variable denoting the current state of the solid. $Z(T)$ is the partition function which is defined as $Z(T) = \Sigma\ e^{-E_j/k_B T}$, where the summation extends over all possible states.

An analogy between the physical annealing process and the task of solving combinatorial optimization problems was first explored by Kirkpatrick, Gelatt and Vecchi [48] and Cerny [7]. They made the following correspondences:

Physical Process	Optimization Problem
States of solid	Feasible solutions
Energy of state	Cost of feasible solution
Melting point state	Randomly generated feasible solution
Slight distortion of state	Elementary change to feasible solution
Temperature of heat bath	Control parameter T

The simulated annealing algorithm is an iteration of Metropolis algorithms, evaluated at decreasing values of the control parameter. Initially the control parameter is given a large value, and, starting with an initial randomly chosen feasible solution, a sequence (or Markov chain) of trials is generated. In each trial, a configuration j is generated randomly from the neighborhood of the current configuration i. If $\Delta_{ij} = f(j) - f(i)$, then the probability of configuration j being the next configuration in the sequence is 1 if $\Delta_{ij} \leq 0$ (for a minimization problem) and $e^{-\Delta_{ij}/T}$ if $\Delta_{ij} > 0$. The control parameter is lowered in steps with the system being allowed to approach equilibrium through the sequence of trials at each step. After an appropriate stopping condition is met, the final configuration is taken as the solution of the problem at hand. There are a variety of possible stopping conditions. A common one is to terminate the algorithm when the observed improvement in cost over a number of consecutive Markov chains is small.

In design construction one can recognize when an optimal solution is found, so the stopping condition is modified to detect this. Also, if the algorithm terminates without finding an optimal solution, then the annealing process is repeated, starting with a fresh random initial feasible solution.

9.31 Remark In the general simulated annealing algorithm (Algorithm 9.32) T_k is the value of the control parameter and L_k the length of the Markov chain generated at the kth iteration of the Metropolis algorithm. `random`$[0, 1)$ returns a random number generated from a uniform distribution on the interval $[0, 1)$. Often, the algorithm is altered to save the best solution found so far and an approximation to an optimal solution can be analyzed if the algorithm is terminated before an optimal solution is found. This is not always the final value of S.

9.32 Algorithm The general simulated annealing algorithm.

```
repeat
    Set values of temperature T_0 and Markov chain length L_0;
    Generate an initial feasible solution S;
    k := 0;
    repeat
        repeat L_k times
        begin
            Randomly generate S' from N(S);
            Δ := c(S') − c(S);
            if Δ ≤ 0 or random[0, 1) < e^{−Δ/T}
            then S := S';
            if c(S) = 0 then exit loop;
        end
        k := k + 1;
        Set L_k;
        Set T_k to some value < T_{k−1};
    until (stop condition is met) or c(S) = 0
until c(S) = 0
```

9.33 Remarks The performance of this algorithm is influenced by the following factors:

1. The value of T_0.
2. The method of obtaining T_{k+1} from T_k.
3. The length of the Markov chain L_k.
4. The choice of stopping condition.

The specification of these parameters constitutes a *cooling schedule.*

9.34 Example An N_2 latin square contains no 2×2 latin subsquares (see §II.1.5) and often contains no other latin subsquares (see Elliott and Gibbons [31]). A simulated annealing algorithm for constructing latin squares which are N_2 (and therefore usually latin subsquare free) is described.

A feasible solution is a latin square, and its cost is the number of 2×2 subsquares contained in it. The transition function is described as follows.

Suppose x and y are elements, and r a row, of a latin square L of side n. Define the (r, x, y)-cycle of L as the sequence $((r_1, c_1), (r_2, c_2), ..., (r_m, c_m))$, where $r_1 = r, r_2, ..., r_m$ are rows of L; $c_1, c_2, \ldots, c_m$ are columns of L; and

1. If i is even, $L(r_i, c_i) = y$ and $c_i = c_{i-1}$.
2. If i is odd, $L(r_i, c_i) = x$ and $r_i = r_{i-1}$ (for $i > 1$), $r_1 = r_m$.

8	2	3	4	5	6	7	**1**
2	3	1	5	6	7	8	4
3	5	4	**8**	7	1	6	2
4	6	8	3	1	5	2	7
5	**8**	7	**1**	3	2	4	6
6	7	5	2	8	4	1	3
7	**1**	2	6	4	3	5	**8**
1	4	6	7	2	8	3	5

For example, suppose L is the latin square of side 8 given on the left. In this square, the elements of the (1, 1, 8)-cycle, ((1,8), (7,8), (7,2), (5, 2), (5, 4), (3, 4), (3, 6), (8, 6), (8, 1), (1, 1)), have been shown in bold type.

If the elements making up a cycle in a latin square are rotated, then a new (in general nonisomorphic) latin square is obtained. Such a rotation can therefore be used as a transition operation in a simulated annealing algorithm.

Select a random latin subsquare $S = ((r_1, r_2), (c_1, c_2), (e_1, e_2))$ of side 2, a random element $x \notin \{e_1, e_2\}$, and a random element $y \in \{e_1, e_2\}$. Performing a rotation on the (r_1, x, y)-cycle destroys the latin subsquare, but may create others.

Decrease the temperature by setting $T_{k+1} = \alpha T_k$, where $\alpha < 1$ is a constant, the *control decrement*. (While this is a popular way of decreasing the control temperature, there are many other methods, including adaptive ones.) The length of the Markov chain is held constant at all temperatures. For the stopping condition, terminate the algorithm once the cost function of the best solution obtained in the last Markov chain remains unchanged for a number of consecutive chains. The number of such chains is the *freezing factor*.

As with hill-climbing, repeat this process until an N_2 latin square is generated. Using the cooling schedule on the right it is possible to generate N_2 latin squares of orders $n \leq 18$.

Initial temperature:	1.0
Control decrement:	0.9995
Markov chain length:	350
Freezing factor:	12

9.35 Remarks Other examples of the successful application of annealing:

Configurations	References
Covering designs	Nurmela and Östergård [71]
Packing designs	Honkala and Östergård [46]
Block designs	Mathon [62]
Antipodal triple systems	Gibbons, Mendelsohn and Shen [40]

9.4 Miscellaneous Construction Techniques

9.36 Remarks A variation of hill-climbing is *tabu search* (see Glover, Taillard and de Werra [42], and Glover [41]), which has been successfully applied to many optimization problems. The aim of tabu search is to use a history of previously visited states to help select the next state to visit. In one form this can operate by maintaining a list of states or state attributes from which movement is discouraged or prohibited. This can avoid retracking along fruitless paths, particularly later in the search. The full theory of tabu search is extensive. However, simple adaptations, in conjunction with hill-climbing and simulated annealing, have been useful is solving some design problems (for example, Mathon [private communication], Östergård [72], and Honkala and Östergård [46]).

9.37 Remarks The use of a prescribed automorphism subgroup to cut down the search space was described in the context of A_{tk}-matrices in §9.2. This method can be applied widely in design construction and used in conjunction with both exhaustive and nonexhaustive techniques.

Let G be a subgroup of automorphisms of a $t-(v,k,\lambda)$ design $D = (X, B)$. Let $X_1, X_2, \ldots, X_m$ and $B_1, B_2, \ldots, B_n$ be the point and block orbits respectively of G. A *tactical decomposition* of D with respect to G is an $m \times n$ matrix $T(G) = (t_{ij})$ where t_{ij} is the number of elements from X_i contained in a block from B_j.

A search for a design with automorphism subgroup G proceeds in two stages. In the first stage, possible tactical decompositions of D are generated. In some cases, for large designs, a group action may be imposed on T itself, thus inserting another stage. In the next stage, a set of *base blocks* is constructed for the design conforming to the prescribed tactical decomposition. Rather than checking the occurrence of pairs within blocks, one checks that all *pure* differences (corresponding to points from the same point orbit) and all *mixed* differences (corresponding to points from different point orbits) are covered by base blocks the correct number of times. If backtracking during this stage, the group of T can be used for partial isomorph rejection. The full design can be obtained by applying the group G to the set of base blocks.

The value of mathematics in saving computing time cannot be overstated, and there are many examples of this. A good one is Dinitz and Stinson's search for perfect 1-factorizations of K_{n+1} (Dinitz and Stinson [30]). They constructed starters by the use of cosets in the multiplicative group of $\mathbb{F}_n$ (n a prime power). Search feasibility was achieved by a theorem which allowed them to check only a few (fewer than eight) pairs for hamiltonicity to deduce that all pairs formed Hamilton circuits (see §VI.4.6).

9.38 Examples Here are examples of designs constructed by using "favorable" groups G in conjunction with some of the search techniques described above (both exhaustive and nonexhaustive):

Configurations	References
Block designs	Mathon [61], Mills [68, 69, 70], Spence [82, 84]
GDDs with block size 4	Kreher and Stinson [57]
Orthogonal Steiner triple systems	Gibbons and Mathon [37]
Hadamard matrices	Spence [83]

9.39 Remark Methods for tabulating existence results are also important in design theory. An example is the algorithm used to construct Table II.2.72 of values of $N(n)$. It constructs many new values of $N(n)$ without ever constructing any explicit latin squares. The algorithm is supplied, as input, every known general theorem dealing with the construction of MOLS. It then generates new orders recursively, while maintaining a database of values of $N(n)$. For a detailed discussion, see [17].

9.5 Analysis Techniques

9.40 Remarks Closely tied to the problem of generating sets of block designs is the problem of analyzing the results and, in particular, determining the isomorphism classes of the produced designs. Block design isomorphism is polynomial time equivalent to graph isomorphism, or *isomorphism complete* (see Booth and Colbourn [4]). Due to the considerable effort spent attempting (unsuccessfully) to develop a polynomial algorithm for general graph isomorphism, it is unlikely that a polynomial algorithm exists for block design isomorphism. Indeed many of the graphs which have caused difficulty for isomorphism-testing algorithms are based on block designs or related configurations (see Mathon [63]). Research effort in this area has therefore been directed at developing algorithms which are fast on the average or with special classes of block designs.

9.41 Remark A block design can be modeled as a colored graph. Given a design $D = (X, B)$ where $X = \{x_1, x_2, \ldots, x_v\}$ and $B = \{B_1, B_2, \ldots, B_b\}$, define $G(D)$ to be a graph with vertex set $\{x_1, x_2, \ldots, x_v, B_1, B_2, \ldots, B_b\}$, with the element vertices having one color, and the block vertices having a second color, and edge set $\{\{x_i, B_j\} | x_i \in B_j\}$. The following fundamental results are not too difficult to prove.

9.42 Proposition

1. Designs D_1 and D_2 are isomorphic if and only if graphs $G(D_1)$ and $G(D_2)$ are isomorphic.
2. The automorphism group of a design D is isomorphic to the automorphism group of the graph $G(D)$.

9.43 Example Using this representation one can use any graph isomorphism package to determine block design isomorphism and to calculate automorphism groups. Probably the most practical graph isomorphism tool currently available is the `nauty` software developed by McKay [65]. `nauty` is a set of procedures for determining the automorphism group of a vertex-colored graph. This information is provided in the form of a set of generators, the size of the group, and the orbits of the group. It is also able to produce a canonically-labeled isomorph of the graph which can be used for isomorphism testing. This program is available via anonymous ftp at the address `ftp.cs.uwa.edu.au` in the subdirectory `pub/graphs/nauty`. Instructions are in the `read.me` file.

`nauty` is written in a portable subset of the language C and runs with a variety of C compilers on a variety of machines. The user can call `nauty` from within a C program specifying a large variety of options. Alternatively one can use `dreadnaut`, a simple program that can read graphs and execute `nauty`. Take the design $D = (X = \{1,2,3,4\}, B = \{\{1,2,4\},\{1,3\},\{2,3,4\}\})$, and label $G(D)$ so that the elements of D correspond to vertices $1-4$, while the blocks correspond to vertices $5-7$. The following example of a `dreadnaut` session is taken from McKay [65] and computes the group of D and displays its canonical labeling.

```
> $=1                                          label vertices starting at 1
> n=7 g
1: 5:                                          go to vertex 5 (block 1)
5: 1 2 4;
6: 1 3;
7: 2 3 4;
> f=[1:4]                                      fix the varieties setwise
> cx                                           run nauty
[fixing partition]
(2 4)                                          group generators
level 2: 6 orbits; 2 fixed; index 2
(1 3)(5 7)
level 1: 4 orbits; 1 fixed; index 2
4 orbits; grpsize=4; 2 gens; 6 nodes; maxlev=3
tctotal=6; canupdates=1; cpu time = 0.02 seconds
> o                                            display the orbits
1 3; 2 4; 5 7; 6;
> b                                            display the canonical labelling
1 3 2 4 6 5 7                                  the vertices in canonical order
1 : 5 6;                                       the relabelled graph
2 : 5 7;
3 : 6 7;
4 : 6 7;
5 : 1 2;
6 : 1 3 4;
7 : 2 3 4;
> q                                            quit
```

9.44 Remarks Often, using the point-block incidence graph is not the most efficient way of using `nauty`. For many families of block designs, it can be proved that the isomorphism class of each design is uniquely determined by the isomorphism class of its block-intersection graph, which is a graph with the blocks as vertices and pairs of intersecting blocks as edges. For example, this is the case with $(v, k, 1)$-designs with $v > k(k^2 - 2k + 2)$. When $v << b$ and v is small, McKay uses a faster version of `nauty` that works with partitions and permutations of the point set alone, essentially viewing the design as a hypergraph.

Crucial to the efficiency of using an isomorphism tester is the availability of a method for partitioning a list of designs into classes such that two isomorphic designs are in the same class. A property which partitions a list in such a way is an *isomorphism invariant.*

An *isomorphism invariant* is a function I for which $I(D_1) = I(D_2)$ if D_1 and D_2 are isomorphic. When $I(D_1) = I(D_2)$ if and only if D_1 and D_2 are isomorphic, the invariant is *complete.*

9.45 Remark As there is currently no known efficiently computable complete invariant for designs in general, one must at present resort to incomplete invariants. In the resulting partitioned list isomorphism of pairs of designs within the same class must still be tested. With this in mind, a good measure of the effectiveness of an invariant is its *sensitivity*, viz. the ratio of the number of classes it distinguishes to the number of nonisomorphic designs under consideration.

9.46 Proposition A complete invariant has sensitivity one.

9.47 Remarks Usually isomorphism invariants can be used to partition the vertices and blocks of a given design into classes such that two points or blocks in the same orbit are in the same class. The process can be extended to partitions of pairs, triples, etc. so as to provide an automorphism group invariant partitioning of points and blocks within various centralizer subgroups of the automorphism group. This information is critically important to an algorithm such as `nauty` which is working to generate a canonical labelling of the design and to generate its automorphism group.

A great variety of design invariants are available, many being based on graph invariants such as those described in McKay [65]. Two examples follow.

9.48 Example A general invariant which is effective for many classes of (nonsymmetric) designs is a clique analysis of the block intersection graph of a design. Given a design D, a series of *block intersection graphs* $G_i, i = 0, \ldots, k$ is defined in which the vertices are the blocks of D, with two vertices adjacent if and only if the corresponding blocks intersect in exactly i points. An isomorphism invariant of D is the total number of cliques of various sizes in G_i, as well as the spectrum of numbers of vertices, vertex pairs, vertex triples, etc. belonging to cliques of various sizes. For many classes of designs, such as the $(9, 3, 2)$-design family, the $(9, 4, 3)$-design family, and cyclic STS for $v \leq 27$, this invariant is complete. Generally the invariant is effective, although the computation of it, while polynomial-time, can be expensive due to the often large number of cliques for relatively small values of v. One of the fastest clique-finding programs has been developed by Mathon [private communication].

9.49 Example An example of a specialized invariant used by Gibbons [34] is the *fragment* structure of an STS. A *fragment* is a set of four blocks of the form $\{\{u, v, w\}, \{u, x, y\}, \{z, v, x\}, \{z, w, y\}\}$. Computing the number of fragments in an STS, together with a count of the fragments containing each point, provides a part of the cycle-structure of an STS (M.J. Colbourn [20]). Although computing the fragment structure has the same time-complexity ($O(v^3)$) as computing the complete cycle structure, it is considerably quicker and requires less space ($O(v)$). The invariant is complete for $v \leq 15$ and very effective for larger v (Stinson [85]).

9.6 See Also

[20]	A comprehensive survey of block design construction and analysis techniques. An excellent introduction to the area.
[55]	More details on the basis reduction method.
[75]	Classic paper on orderly enumeration.
[58]	Description of the famous search for a projective plane of order 10.
[85]	Foundation paper describing construction of STS by hill-climbing.
[1]	One of the first detailed descriptions of simulated annealing.
[61]	Practitioner's survey of computational methods.
[65]	Description of state-of-the-art software package for graph enumeration and isomorphism.
[23]	Recent example of orderly enumeration in action.

References

[1] E. Aarts and J. Korst, *Simulated Annealing and Boltzmann Machines — A Stochastic Approach to Combinatorial Optimization and Neural Computing*, John Wiley & Sons, New York, 1989.

[2] B.A. Anderson, A fast method for sequencing low order nonabelian groups, *Ann. Discrete Math.* 34 (1987), 27-42.

[3] A. Betten, A. Kerber, A. Kohnert, R. Laue and A. Wassermann, The discovery of simple 7-designs with automorphism group $P\Gamma L(2,32)$, preprint.

[4] K.S. Booth and C.J. Colbourn, Problems polynomially equivalent to graph isomorphism, Technical Report CS-77/04, Dept. of Computer Science, University of Waterloo, 1979.

[5] D.C. Bigelow and C.J. Colbourn, Faithful enclosings of triple systems: Doubling the index, *Acta Math. Univ. Comen* 60 (1991), 133-151.

[6] J.W. Brown, Enumeration of latin squares with application to order 8, *J. Combin. Theory* 5 (1968), 177-184.

[7] V. Cerny, Thermodynamical approach to the traveling salesman problem: an efficient simulation algorithm, *J. Optimization Theory Applic.* 45 (1985), 41-51.

[8] Y.M. Chee and A. Lim, Quadrilateral-free Steiner triple systems of orders 31 and 37, *J. Combin. Inf. Sys. Sc.* 14 (1989), 244-248.

[9] C.J. Colbourn, Embedding partial Steiner triple systems is NP-Complete, *J. Combin. Theory A* 35 (1983), 100-105.

[10] C.J. Colbourn, Orienting triple systems is NP-complete, *Ars Combin.* 22 (1986), 155-163.

[11] C.J. Colbourn, The complexity of completing partial latin squares, *Discrete Appl. Math.* 8 (1984), 25-30.

[12] C.J. Colbourn and M.J. Colbourn, The computational complexity of decomposing block designs, *Ann. Discrete Math.* 27 (1985), 345-350..

[13] C.J. Colbourn and P.C. van Oorschot, Applications of combinatorial designs in computer science, *ACM Computing Surv.* 21 (1989) 223-250.

[14] C.J. Colbourn, M.J. Colbourn and K.T. Phelps, Combinatorial algorithms for generating cyclic Steiner quadruple systems, in: *Proc. Conf. on Discrete Math. Analysis and Comb. Computation*, Fredericton, 1980, 25-39.

[15] C.J. Colbourn, M.J. Colbourn, K.T. Phelps and V. Rödl, Coloring Steiner quadruple systems, *Discrete Appl. Math.* 4 (1982), 103-111.

[16] C.J. Colbourn, M.J. Colbourn, K.T. Phelps and V. Rödl, Coloring block designs is NP-complete, *SIAM J. Alg. Discr. Meth.* 3 (1982), 305-307.

[17] C.J. Colbourn and J.H. Dinitz, Making the MOLS table, in: *Computational and Constructive Design Theory* (W.D. Wallis; ed.), Kluwer Academic Publishers, Norwell MA, to appear.
[18] C.J. Colbourn and R.A. Mathon, Leave graphs of small maximal partial triple systems, *J. Combin. Math. Combin. Comput.* 2 (1987), 13-28.
[19] C.J. Colbourn, A. Rosa and D.R. Stinson, Pairwise balanced designs with block sizes three and four, *Canad. J. Math.* 43 (1991), 673-704.
[20] M.J. Colbourn, Algorithmic aspects of combinatorial designs: A survey, *Ann. Discrete Math.* 26 (1985), 67-136.
[21] F.N. Cole, L.D. Cummings and H.S. White, Complete classification of the triad systems on fifteen elements, *Memoirs of the National Academy of Sciences U.S.A.* 14, 2nd memoir (1919), 1-89.
[22] M.J. Coster, B.A. LaMacchia, A.M. Odlyzko and C.P. Schnorr, An improved low-density subset sum algorithm, *Proc. EUROCRYPT '91, Brighton*, in: *Springer Lecture Notes in Computer Science* 547 (1991), 54-67.
[23] J.H. Dinitz, D.K. Garnick and B.D. McKay, There are 526,915,620 nonisomorphic 1-factorizations of K_{12}, *J. Combin. Des.* 2 (1994), 273-285.
[24] J.H. Dinitz and E.R. Lamken, Howell designs with sub-designs, *J. Combin. Theory A*, to appear.
[25] J.H. Dinitz and D.R. Stinson, A fast algorithm for finding strong starters, *SIAM J. Alg. Disc. Meth.* 2 (1981), 50-56.
[26] J.H. Dinitz and D.R. Stinson, Room sqaures and related designs, in: *Contemporary Design Theory: A Collection of Surveys* (J.H. Dinitz, D.R. Stinson; eds.) John Wiley & Sons, New York, 1992, 137-204.
[27] J.H. Dinitz and D.R. Stinson, A few more frames, in: *Graphs, Matrices and Designs* (R. Rees; ed.), Marcel Dekker, 1993, 133-146.
[28] J.H. Dinitz and D.R. Stinson, A hill-climbing algorithm for the construction of 1-factorizations and Room squares, *SIAM J. Alg. Disc. Meth.* 8 (1987), 430-438.
[29] J.H. Dinitz and D.R. Stinson, A note on Howell designs of odd side, *Utilitas Math.* 18 (1980), 207-216.
[30] J.H. Dinitz and D.R. Stinson, Some new perfect 1-factorizations from starters in finite fields, *J. Graph Theory* 13 (1989), 405-415.
[31] J.R. Elliott and P.B. Gibbons, The construction of subsquare free latin squares by simulated annealing, *Australas. J. Combin.* 5 (1992), 209-228.
[32] M.R. Garey and D.S. Johnson, *Computers and Intractability: A Guide to the Theory of NP-Completeness*, W.H. Freeman, San Francisco, 1979.
[33] P.B. Gibbons, A census of orthogonal Steiner triple systems of order 15, *Ann. Discrete Math.* 26 (1985), 165-182.
[34] P.B. Gibbons, Computing Techniques for the Construction and Analysis of Block Designs, Ph.D. Thesis; also Technical Report 92, Computer Science, University of Toronto, 1976.
[35] P.B. Gibbons and R. Mathon, Signings of group divisible designs and projective planes, *Australas. J. Combin.*, to appear.
[36] P.B. Gibbons and R. Mathon, Construction methods for Bhaskar Rao and related designs, *J. Austral. Math. Soc. (A)*, 42 (1987), 5-30.
[37] P.B. Gibbons and R. Mathon, The use of hill-climbing to construct orthogonal Steiner triple systems, *J. Combin. Des.* 1 (1993), 27-50.
[38] P.B. Gibbons, R. Mathon and D. Corneil, Computing techniques for the construction and analysis of block designs, *Utilitas Math.* 11 (1977), 161-192.
[39] P.B. Gibbons and E. Mendelsohn, The existence of a subsquare free latin square of side 12, *SIAM J. Alg. Disc. Meth.* 8 (1987), 93-99.

[40] P.B. Gibbons, E. Mendelsohn and H. Shen, Antipodal triple systems, *Australas. J. Combin.* 9 (1994), 137-151.

[41] F. Glover, Tabu search: A tutorial, *Interfaces* 20 (1990), 74-94.

[42] F. Glover, E. Taillard and D. de Werra, A user's guide to tabu search, *Annals Oper. Res.* 41 (1993), 3-28.

[43] M.J. Grannell and T.S. Griggs, A Steiner system S(5,6,108), *Discrete Math.*, to appear.

[44] M.J. Grannell, T.S. Griggs and R.A. Mathon, Some Steiner 5-designs with 108 and 132 points, *J. Combin. Des.* 1 (1993), 213-238.

[45] T.S. Griggs and J.P. Murphy, 101 anti-Pasch Steiner triple systems or order 19 (at least), *J. Combin. Math. Combin. Comput.* 13 (1993), 129-141.

[46] I.S. Honkala and P.R.J. Östergård, Applications in code design, in: *Local Search in Combinatorial Optimization* (E.H.L. Aarts, J.K. Lenstra; eds.), John Wiley & Sons, New York, to appear.

[47] A.V. Ivanov, Constructive enumeration of incidence systems, *Ann. Discrete Math.* 26 (1985), 227-246.

[48] S. Kirkpatrick, C.D. Gelatt Jr. and M.P. Vecchi, Optimization by simulated annealing, *Science* 220 (1983), 671-680.

[49] W.L. Kocay, D.R. Stinson and S.A. Vanstone, On strong starters in cyclic groups, *Discrete Math.* 56 (1985), 45-60.

[50] E.S. Kramer, Some t-designs for $t \geq 4$ and $t = 17, 18$, *Congr. Numer.* 14 (1975), 443-460.

[51] G. Kolesova, C.W.H. Lam and L.Thiel, On the number of 8×8 latin squares, *J. Combin. Theory A* 54 (1990), 143-148.

[52] E.S. Kramer, D.W. Leavitt and S.S. Magliveras, Construction procedures for t-designs and the existence of new simple 6-designs, *Ann. Discrete Math.* 26 (1985), 247-274.

[53] E.S. Kramer and D.M. Mesner, t-designs on hypergraphs, *Discrete Math.* 15 (1976), 263-296.

[54] D.L. Kreher and S.P. Radziszowski, Solving subset-sum problems with the L^3 algorithm, *J. Combin. Math. Combin. Comput.* 3 (1988), 49-63.

[55] D.L. Kreher and S.P. Radziszowski, Finding simple t-designs by using basis reduction, *Congr. Numer.* 55 (1986), 235-244.

[56] D.L. Kreher and S.P. Radziszowski, The existence of simple 6-$(14, 7, 4)$ designs, *J. Combin. Theory A* 43 (1986), 237-243.

[57] D.L. Kreher and D.R. Stinson, Small group divisible designs with block size 4, *J. Stat. Plann. Infer.*, to appear.

[58] C.W.H. Lam, L. Thiel and S. Swiercz, The non-existence of finite projective planes of order 10, *Canad. J. Math.* 41 (1989), 1117-1123.

[59] A.K. Lenstra, H.W. Lenstra and L. Lovász, Factoring polynomials with rational coefficients, *Math. Annalen* 261 (1982), 515-534.

[60] S.S. Magliveras and D.W. Leavitt, Simple 6-$(33, 8, 36)$ designs from $P\Gamma L_2(32)$, in: *Computational Group Theory* Academic Press, New York, 1984, 337-352.

[61] R. Mathon, Computational methods in design theory, *Proc. 13th British Combinatorial Conference, London Math. Soc., Lecture Note 166*, Cambridge University Press, 1991.

[62] R. Mathon, Nonisomorphic designs by simulated annealing, preprint.

[63] R. Mathon, Sample graphs for isomorphism testing, *Congr. Numer.* 21 (1978), 499-517.

[64] B.D. McKay, Isomorph-free exhaustive generation, to appear.

[65] B.D. McKay, `nauty` user's guide (version 1.5), Technical Report TR-CS-90-02, Computer Science Department, Australian National University, 1990. Available by anonymous ftp from `dcssoft.anu.edu.au:pub/nauty19/nug.alw`.

[66] B.D. McKay and S.P. Radziszowski, The nonexistence of 4-(12,6,6) designs, in: *Computational and Constructive Design Theory* (W.D. Wallis; ed.), Kluwer Academic Publishers, Norwell MA, to appear.
[67] N. Metropolis, A. Rosenbluth, M. Rosenbluth, A. Teller, E. Teller, Equation of state calculations by fast computing machines, *J. Chem. Physics* 21 (1953), 1087-1092.
[68] W.H. Mills, A new block design, *Congr. Numer.* 14 (1975), 461-465.
[69] W.H. Mills, The construction of balanced incomplete block designs with $\lambda = 1$, *Congr. Numer.* 20 (1977), 131-148.
[70] W.H. Mills, The construction of BIBDs using non-abelian groups, *Congr. Numer.* 21 (1978), 519-526.
[71] K.J. Nurmela and P.R.J. Östergård, Upper bounds for covering designs by simulated annealing, *Congr. Numer.* 96 (1993), 93-111.
[72] P.R.J. Östergård, New multiple covering codes by tabu search, preprint.
[73] K.T. Phelps and V. Rödl, On the algorithmic complexity of coloring simple hypergraphs and Steiner triple systems, *Combinatorica* 4 (1984), 79-88.
[74] C. Pietsch, Über die Enumeration von Inzidenzstrukturen, Dissertation, Universität Rostock, 1993.
[75] R.C. Read, Every one a winner, or, how to avoid isomorphism search when cataloguing combinatorial configurations, *Ann. Discrete Math.* 2 (1978), 107-120.
[76] B. Schmalz, t-designs with prescribed automorphism group, new simple 6-designs, *J. Combin. Des.* 1 (1993), 125-170.
[77] C.P. Schnorr, A hierachy of polynomial time lattice basis reduction algorithms, *Theor. Comput. Sci.* 53 (1987), 201-224.
[78] C.P. Schnorr, A more efficient algorithm for lattice basis reduction, *J. Algorithms* 9 (1988), 47-62.
[79] C.P. Schnorr and M. Euchner, Lattice basis reduction: Improved practical algorithms and solving subset sum problems, *Proc. Fundamentals of Computation Theory 91*, in: *Lecture Notes in Computer Science* 529 (1991), 68-85.
[80] E. Seah and D.R. Stinson, A perfect 1-factorization of K_{40}, *Congr. Numer.* 68 (1989), 211-214.
[81] E. Seah and D.R. Stinson, An enumeration of nonisomorphic 1-factorizations and Howell designs for the graph K_{10} minus a 1-factor, *Ars Combin.* 21 (1986), 145-161.
[82] E. Spence, A complete classification of symmetric (31,10,3) designs, *Des. Codes Crypt.* 2 (1992), 127-136.
[83] E. Spence, Classification of Hadamard matrices of order 24 and 28, *Discrete Math.*, to appear.
[84] E. Spence, (40,13,4)-designs derived from strongly regular graphs, in: *Finite Geometries and Designs* (J.W.P. Hirschfeld, D.R. Hughes and J.A. Thas; eds.), Oxford (1991), 359-368.
[85] D.R. Stinson, Hill-climbing algorithms for the construction of combinatorial designs, *Ann. Discrete Math.* 26 (1985), 321-334.
[86] D.R. Stinson and H. Ferch, 2000000 Steiner triple systems of order 19, *Math. Comput.* 44 (1985), 533-535.
[87] R.J.M. Vaessens, E.H.L. Aarts and J.H. Van Lint, Genetic algorithms in coding theory — a table for $A_3(n, d)$, *Discrete Appl. Math.*, to appear.

Index